# VIABILITY, COMPLEXITY AND US

## The Human-Environment System Constraining Our Future

JOHN KNIGHT

# ACKNOWLEDGMENTS

I would not have completed this book without the help and support of family, friends, generous strangers and professionals. The encouragement and support of Kathy and Alice made this book possible. Special thanks to those who have read and commented on drafts and parts of drafts: Morgan Hite, Rosanna Nicol, Becky Choinard, Nick Brown, Jim Pojar, Sibille Haeussler, Steven Barber, Jen Barber, Charles Preston, Daniel Helm, Jay Baker, Lynn Shervill and Nikki Skuce. Their feedback resulted in significant improvements. Thanks to those who have engaged in discussions, and offered encouragement, criticisms and suggestions: Morgan Hite, Berkley Adams, Scott Green, Katie McLean, Jackie King, Ken Frew, Gail Hochacka, Chris and Gail Kibble, Adrian de Groot and Duncan McCrindell. Thanks are also due to the librarians at Smithers Public Library, Coast Mountain College (formaly NorthWest Community College), the University of British Columbia, Simon Fraser University and St Mark's College. Critical to the production process were editing, Emma Woodley, securing permissions, Mary Rose MacLachlan and layout, Meghan Behse. Thanks to you all.

# TABLE OF CONTENTS

# INTRODUCTION

A rather sorry looking birch tree is visible from the front window of my family's home. It leans to one side, has a large scar and only a partial head of leaves, even though spring is fast turning into summer. Each year, parts of a few branches die to leave the birch looking ever more like it belongs in a haunted forest. This year brought a change. Early in the spring, a short horizontal row of small holes appeared in the tree trunk's bark at around eye level. I knew from the characteristic of the holes that they were made by the beak of a woodpecker called a sapsucker (Udvardy 1977).

Usually the sapsucker makes holes to dig out its principal food, insects, from the bark and rotting parts of trees. However, the row of holes in the birch served a different purpose. They were initially dry, but as spring advanced and the sap rose, a sticky, sugar-rich liquid oozed into a few of the holes that then became a favourite haunt for the sapsuckers. They enlarged some of the weeping holes and hammered at many more, most of which remained dry.

Soon, the steady flow of sap had stained the smooth grey-brown bark a light-brown to pink colour. There were now many more creatures visiting the weeping birch to partake of the sap, including flies of various sizes, a bumble bee and even a hummingbird. They preferentially congregated around a fold in the bark where some of the sap had collected and, from the smell, was in the process of fermenting. The flies found this brew particularly attractive. (Maybe this is where the term *barfly* originated.) I also noticed that one of the enlarged holes now contained a young fungus or lichen. How did its spores get there?

I wondered if this was the only tree that the sapsuckers visited. Near the oozing birch were three other clusters of birch. I checked them out but found only the occasional shallow, dry hammer hole. Despite other scars, these trees were all in fine health.

I then realized that the sapsuckers spent quite a lot of time flying to the back of the house. By following them, I quickly located a birch with a light-brown stain at eye level. Sure enough, above it were a few oozing sapsucker holes. Interestingly, this deeply scarred tree had a zinc-plated washing line attachment half buried into it. Otherwise, the tree seemed healthy. Perhaps the two sap-producing birches were stressed in some way that enabled the sapsuckers to tap into the sap.

A sound like loud crickets led me to the base of an aspen. It contained the sapsucker nest. I could see a small hole near the top of the tree from which the noise of their hungry young was being broadcast. Aspen wood is normally quite hard, so somehow the sapsucker knew that this particular tree was softer at that spot. Soft enough for it to hammer its way through the bark and excavate a hole in the tree's core.

I later noticed that these sapsuckers, and a pair of flickers nesting nearby, had both chosen nest entrances near the wound caused by an old branch breaking off. Could a fungus have entered through this scar to create wood that sounded different to the sapsuckers' hammering? Perhaps the sapsucker had introduced the fungus into the tree as it pecked at the bark to clean it of insects (Jackson and Jackson 2004; Kerry et al. 2004)?

It was quite a sight to see the parent sapsuckers, with their bright red head and breast, leave the nests and swoop down at full speed to the dying birch, landing perfectly next to the oozing holes. Sometimes they would swoop past the front windows of the house to first visit the washing line birch. Although they are amazing flyers, they can make miscalculations, as I discovered one day when investigating a dull thud from something hitting the front window. Outside, on the deck below the window, lay an unconscious sapsucker. When I picked it up to lay it on a clean rag, I had a close-up look at the bird's long needle-like beak and the fine colouring of its plumage.

The stunned sapsucker eventually regained consciousness, spent an hour or so clinging to a nearby tree and then flew away. What if, I imagined, this unexpected event had proved fatal. The history of that sapsucker family would have been quite different.

In recent years, there seem to have been many more sapsuckers, flickers and other woodpeckers in our forest. I hear more of their raucous calls and spring drummings on dead wood. I think there are more because of the balsam bark beetle which, as its name suggests, eats the living cells of the balsam tree's bark. If the beetle eats enough cells, then the flow of sap to the leaves stops. One year later, the tree turns red as the needles die. By the next year, the tree is leafless, grey and dead.

The outcome from a previous balsam bark beetle outbreak in our forest can be seen in the patch of large balsam conifers on our property. About half of them stand grey and lifeless, with pieces of bark hanging off their trunks. Perhaps there are more birds because they can now feed on the many species of insects that are living on and in the older, dry dead trees. The bird population may also have increased because the balsam bark beetle is back. Perhaps the increased population of birds will ensure that most of the remaining balsams survive.

The lives of the sapsuckers are so intimately tied to our forest that their past must be closely linked to the forest's history. I was able to appreciate that history when a live balsam in the patch fell over because its roots and butt had rotted out after becoming infected with a fungus. I cut a round out of the butt and, by counting the rings, could tell that the balsam had started growing in 1884, the year the exotic, blue-flowered water hyacinth was introduced into the United States at the Cotton States Exposition. By looking at the changes in the widths of the balsam's rings, I could tell that in the first 18 years of its life the balsam had struggled, perhaps because it was surrounded by a forest of older balsam, cottonwood and spruce trees. However, after 1902, it grew rapidly up to 1966, after which its growth declined until the day it finally fell over.

Combining this information with other snippets of the forest dwellers' personal diaries revealed that the 1884 balsam and its patch are a remnant of an old forest that was partly cleared in the early 1940s. Over the succeeding decades, the open spaces were filled by the present forest, which is dominated by the faster-growing birch, aspen and cottonwood trees. This suits the sapsuckers fine because they can more easily use these trees to make their nests, compared to the slower-growing balsam.

Today the remains of the 1884 forest are slowly being replaced by younger trees as the few remaining veterans succumb to the bark beetle outbreaks or root rot. The newer 1940s part of the forest is also slowly changing as the birch and aspen die from old age to be replaced by a new generation of slower-growing balsams. The most recent event to affect the sapsuckers and their forest was the small clearing made to build our house, with its glass windows and washing line.

The full complexity of the relationship between the forest and the sapsuckers only becomes clear when I ask myself what happened before 1884. Despite the apparent timelessness of my surroundings, by looking at landforms and by digging into the soil, I find that the forest community is growing on glacial till left behind from the time when an ice sheet up to 1.5 km thick covered this region. By around 10,000 years before present (yBP), the ice had melted, which allowed the ancestors of today's forest to move onto the bare land. When the sapsuckers arrived in that first forest, they brought with them an even earlier history.

There are three types of sapsuckers in North America, each with a distinctive geographical range: the yellow-bellied (east coast and prairies); the red-naped (Rocky Mountains and Great Basin), and our forest's red-breasted sapsucker (northwest coast and Sierra Nevada). They are sometimes considered to be three separate species and sometimes three varieties of a single species. Perhaps they arose because a single sapsucker species became split into three populations during the 20 ice ages that occurred over the last 2 million years (My). During these icy periods, which lasted up to 100,000 years, almost all of Canada, including this valley, was covered by a vast 2 to 3 km thick ice sheet and bounded by related cold-climate biomes, such as tundra (Pielou 1991; Johnson and Cicero 2004).

As I look up from the house through the trees to the nearby mountains, I can see a relative of that ice sheet: a glacier, nestled between the peaks. It supplies water to the valley. By comparing a photograph of the glacier taken in 1915 to one taken in 2000, I can see that the glacier has receded significantly (www.glacierchange.org). This tells me that the climate has warmed since at least 1915 and perhaps since the 1884 balsam was a sapling, a date well after the climate warming that ended the Little Ice.

The idea of local climate warming is no surprise to the old-timers living here. they say that winters are now much milder than in the 1940s, when temperatures colder than −25°C were more common. Surely the forest must also be recording this change? Unlike the ease with which I relate to the impact on the forest from the human land clearing of the 1940s, I find it hard to recognize, let alone relate to, the forest's slower changes in response to that longer warming.

The neighbours think that our forest's underbrush is now thicker. I hear from the old-timers that devil's club, a plant that likes the wet, is now more common in this area. Could the bark beetle outbreak be related to this climate change? Certainly, there are now hemlock saplings appearing among the balsams in the 1940s forest. But there are no mature hemlock trees here, so the hemlock is new to the neighbourhood. Did it move in because the climate is now warmer and wetter or because of the human land disturbance?

These changes prompt me to think of the future. Could it be that the remnants of the 1884 balsam forest and the younger forest of aspen, cottonwood and birch trees around it will both be replaced by a forest dominated by hemlock? Whatever the outcome, the sapsuckers will have to adapt to fewer birch and aspen, and to more conifers.

With this future in mind, I reconsider our plans to use the dead, dry balsam trees for firewood to keep the −15°C winters at bay. Perhaps they should just be left. The bugs they host will keep the sapsuckers fed until the trees are rotten enough to fall over. Once on the ground, they will slowly continue their disintegration to become part of the soil in which the young living trees can grow and feed. I am faced with the choice between firewood for now and soil for the future.

Even as all of this knowledge provides me with a strong sense of connection to our forest, it also leads to confusion. To understand the sapsucker's present and past lives, I must try to simultaneously hold in mind so many aspects of our forest. Events that occurred over a wide range in time and space, from long-lived ice sheets covering continents thousands of years ago to the short-lived sapsuckers of today hammering holes in the dying birch in our yard. At the same time, I have to make all the connections between sapsuckers, balsam, birch, flies, ancestral sapsuckers and glaciers.

Trying to include myself in the forest adds to the sense of confusion. What on earth is going on in my head when I can, in one minute, look at the forest as a living system outside of myself and then, in the next minute, see it as a living system in which I participate as it provides me with berries, mushrooms and flowers. Finally, it appears to be just another place to cut firewood, regardless of the sapsuckers' or the forest's needs. I feel overwhelmed by these tasks, so I retreat to the familiar for relief.

Sitting in front of the TV, the forest disappears from my consciousness and the images of my culture come flooding in, brought to me by the marvels of human technology. Streaming into my head is an extra-cheap, extra-large hamburger; a large crowd crying with laughter as a stand-up comedian goes through her routine, "Time flies like an arrow; barflies like birch

sap" (groan); another hamburger, this time with double bacon; and a mother and child smiling into the camera as they hold up a carrot pulled from the garden behind them. Then there is the *Giganotosaurus*-sized digging wheel filling a mind-bogglingly large truck with tarry sand. The truck grinds its way through the resulting moonscape to a steaming oil-sucking plant that produces the gasoline I could use to fuel the car that the reassuring voiceover in the next ad says will make me happy.

This world is nothing like the forest dwellers' world, but there must be some connection between them. After all, aren't we all living on the same planet? The images on the TV provide hints of a very fuzzy connection that only appears if I think really hard. The carrot in the woman's hand did come out of the ground, didn't it? The tar-rich sand sure did. I wonder how other people connect these two worlds. Does their mind become confused and overwhelmed when they try?

When I tire of the TV, I turn to a newspaper or magazine to seek answers to the question of connections. As I part the pages, my hair is swept back, like that of a cartoon character being yelled at, by the blast of stories about environmental problems and our relationship to them: pollution, extinction and declining resource availability. I am told that these issues could shortly be cured if only we implemented, paid for or believed in the promise of some invention, discovery, financial miracle, behavioural change or a current expert's solution. What led to this state of affairs? Unfortunately, these stories pour forth in the same way that media reports about health care stream out of the television: like beads on a string, isolated, with no apparent coherence or connection between them other than the string of time linking their telling, or is that their yelling.

The pervasive normality of this fragmented view of our world was personally brought home to me at a bluegrass festival. There I overheard a long-in-the-tooth, ex-commercial fisherman talking to a newly rediscovered, long-ago friend about the intervening years. In part of the discussion, the ex-fisherman ventured that there were still more than enough fish in the ocean off the local coast, it's just that no one was allowed to catch them and the licensing system let just a few get rich without fishing. His friend remained quiet as a singer in the background sang a lament for the loss of clean river water and the disappearing fish. The men made no mention of the well-publicized global decline in ocean fish stocks due to overfishing. It seems that my struggle to deal with my perceptions of our forest and my decision making about it is a microcosm of our human efforts to deal with the state of our world and its future. We flit from idea to idea, crisis to crisis, because we can't quite grasp how they all fit together.

It seems to me that I, the sapsuckers and the forest hosting us are all part of a single, functional whole called the earth. The ice age, the tar sands truck and the fisherman cum bluegrass aficionado are also parts of this whole. My feelings of confusion and of being overwhelmed arise because I am frustrated in my efforts to mentally imagine this unity and make decisions accordingly. I realize that to relieve myself of this unsatisfactory mental state, I must find a way to more clearly and deeply appreciate how we humans and our environment are interconnected. A way that would also help me to make my earth-affecting decisions with greater awareness, clarity and understanding.

Initially, I thought my goal could be reached by writing a short book. One that used our general knowledge of how our world works to describe it, our place in it and how the connections between things have resulted in our current situation. One that I could share with others in the hope that they may avoid my frustration.

However, my research quickly showed that, although useful, these big picture explanations and their straightforward solutions are significantly incomplete and can be misleading (Diamond 2005; Lynas 2011; Klein 2014; Harari 2015; Eisenstein 2018). Their satisfying order and firmly made claims, tend to mask the world's complex, dynamic nature. To varying degrees they help to maintain the impression that, at some level, humans can live more remotely from our world than we actually can. This is a view that can be seen in many fields of study. For example, in science, we treat ourselves as disinterested observers. In economics, we equate paying attention to cash with living. In religion, we relieve ourselves of some of our worldly responsibilities by focusing on an unworldly reality such as the afterlife. And, in technology, we conceive of ourselves as distant managers or entitled manipulators.

Although general theories and principles help us grasp the basic nature of our world, they do so by leaving out the details that describe our day-to-day thinking and living. As the sapsucker story illustrates, it is through those everyday, real-time experiences, actions, observations, thoughts and memories that we appreciate and change our world. It is only when general descriptions are tightly tied to the personal details of living and thinking that they can help us to fully appreciate how interconnected the functioning of the world is, and the extent of our participation in it.

If my writing neglected the complex, dynamic nature of our world or left out the personalizing details and the hard-to-grasp ambiguities, I would not achieve my objective. Instead, I too would be adding to the growing pile of well-written books, informative conference proceedings and expert panel discussions. I would be contributing to the time and effort we spend arguing about largely self-created differences among the confusion of ideas about our issues and the ways to implement often unrealistic solutions to our circumstance.

What kind of description of our circumstance would be useful and not misleading? One that would discuss how we function as a social animal, including how and why we think about ourselves and our world in the way we do. It would discuss how our world works, its current state and our relationship to it. It would also discuss how we and our environment change.

The writing would thus be based on life sciences, human studies and earth sciences. It would have to meaningfully link these different views of our circumstance into a coherent framework. One that would provide a comprehensive, personalized overview of our current circumstance. It would both guide and constrain our evaluation and decision making about the issues we face.

In essence, it would describe our circumstance as the product of our participation in a dynamic, complex human-environment system. And it would do so in a way that we can relate to at a personal level. The description should make us feel that it is relevant to our day-to-day lives, especially our decision making.

I realize that writing such a useful description is a formidable challenge. Returning to TV reality seems like an easier option, but I desperately want to resolve my sense of frustration and my feeling of being overwhelmed. Hearing the sapsuckers' calls reminds me of my responsibilities.

This book is the result of my efforts to satisfy the above conditions by describing, in an attention-holding style, the human-environment system and the limits, boundaries and links that constrain how it functions. In doing so, I have also chosen not to follow a single track through the system or focus on a narrow selection of issues, as is usual.

There is a disadvantage to this choice: it enhances the writer's dilemma of too much versus too little detail. Keeping too much detail buries the reader. Throwing away too much detail can create satisfying order and simplicity where none exists (Norretranders 1998). In the first case, the reader is distracted from the flow of the story or struggles to appreciate the larger order. In the second, we end up with abstractions so disconnected from their source details that they can take on other unintended meanings: false order, myths and self-evident truths. This outcome, as mentioned above, can be seen in our beliefs about ourselves, our environment, and our decisions about the causes and cures for our circumstance.

The dilemma of detail is addressed by using the goal of adequately describing the human-environment system as the primary detail-reducing filter. This choice results in a longer book. But that cost is more than offset by the reduction in the chances of it misleading the reader and by the significant gains from discussing, at a personal level, the constraints on the system's functioning. It also prompts us to pay attention to the constraints on our ability to conceptualize and personalize our complex world, make decisions and change our views. This encourages us to consider these factors when forming our views and making our decisions (Fischhoff 1982).

Consider that, in this information age, we face a flood of "facts" about our human-environment system and about what we think of as issues. We are overwhelmed by this information, in part because of our inherently constrained ability to both evaluate its "truth" and incorporate it into a coherent view of our personal lives and cultural circumstances. We are also overwhelmed because we have to make sense of information covering scales of time and space that are well beyond our experience, such as continental ice sheets appearing and disappearing over tens of thousands of years. The limitations to our human abilities are also a reason why we struggle to deal with contradictory information and paradoxical circumstances.

By paying attention to these constraints we can come to terms with them. In the process we can come to think of ourselves as being more of a participant in a complex human-environment system than a distant observer or manager in charge of it. For example, we may come to realize that we are the decision makers whose decisions are constrained and influenced by our culture at the same time as our decisions help to form that culture.

Taking multiple paths while focusing on constraints allows the writing to provide another benefit. The human-environment system is described in this work by a series of easily digested, near stand-alone essay-like chapters. Each of which is focused on a different key aspect of the system.

But, to ensure that the reader fully appreciates the dynamic complexity of the system as a whole, the writing also links the chapters together in such a manner that they, collectively, coherently and consistently describe a single system. These connections are made in two ways. Each chapter refers to two key aspects of the system: our surrounding (our environment) and our relationship to it. Each chapter contains expressed and implied links to the contents of the other chapters. These links work both forward and backward from each chapter.

By providing these themes and links another benefit is gained. Both the structure and style of the writing mimics the nature of a complex system. Thus the human-environment system is represented to the reader as a single hologram-like whole, and our circumstance and our relationship to it as a product of its functioning.

However, when reading the description, you might be tempted to ignore what appears to be unneeded information, repetition or useless loose ends, such as the reference to the water hyacinth above. Hopefully the above explanation will prompt you to instead wonder if such items are the links to other aspects of the puzzle discussed in earlier chapters or to be contemplated in later chapters. The repetitions may not be poor style but the inescapable consequence of describing the same complex system from different perspectives. Think of a series of essays. The structure and style thus serves another purpose: by recognizing and considering these links, we broaden our thinking.

There is a disadvantage to this approach. It will take longer to read this book than the many others addressing our current circumstance. However, you will end up with a distinct advantage: a coherent framework that will enable you to more easily and quickly evaluate and assimilate the analysis and suggestions of others, as well as develop and evaluate your own.

This raises the question of how the writing meets the other needs of the reader. One of the primary difficulties we face in dealing with our complex circumstance is that we are each more familiar which some aspects of our world than others. Are you more of a humanities or a science person? Regardless, this book should result in all readers coming to a similar basic understanding of the human-environment system and their relationship to it; otherwise the writing's purpose is defeated. However, at the same time, those who are more knowledgeable about an aspect being discussed should be able to limit what they have to read without losing the thread of the overall argument.

The difference between people's knowledge is partly recognized by grouping the chapters into three parts, and by the topics selected to be chapters. It is further acknowledged by providing an introduction to a chapter and summaries of its main sections. The reader should thus be able to satisfy their level of interest and their desire for more or less detail by picking the chapters or even sections of chapters to read. This, again, marks the reader as a participant rather than a passive receiver of complete knowledge from an "expert."

To facilitate your task of assimilating the information and conclusions, the writing is personalized with many examples, which are, as much as possible, connected to daily life. Help in dealing with the uncertainty that inescapably accompanies information and conclusions is provided in the form of statements of uncertainty and the style of the writing. To ease the process of treating the information and conclusions as a coherent whole, the writing style is more constructionist (combining) rather than reductionist (breaking into parts) in its approach. To this end, the key aspects of the human-environment system and the topics in each chapter are looked at from more than one perspective. Each perspective on its own is expected to be incomplete and uncertain. However, by combining and rationalizing the many perspectives, the reader is provided with a more representative, complete and multi-faceted description/conclusion, and a feel for the key uncertainties.

Overall, these approaches should help the reader more easily appreciate the constraints on the system's functioning and on our ability to relate to our circumstance. In particular, it should help you deal with the apparent conflicts and contradictions that plague our efforts to do so. And it should help you to meaningfully accommodate the ever-present uncertainties.

To be clear, the intended outcome from reading this work is not a simplified explanation of our circumstance and a list of specific simplistic solutions that, if followed by rote, would solve our issues. The goal is to provide a coherent, consistent framework within which the reader can more easily understand how our world functions and evaluate how we participate in it. It provides a common framework within which individuals of all stripes can more easily conceptualize the dynamics, boundaries, limits, uncertainties and relationships that characterize and constrain us and our world. This includes recognizing the constraints on how we personally think and make decisions. The hope is that, with this greater appreciation, we can more easily adjust our beliefs and thinking so that we can more wisely evaluate the complex issues we face. Our decisions are then more likely to select the most appropriate objectives and solutions to deal with our current circumstance.

If this description of our circumstance is to provide you with the opportunities I claim, then your co-operation is required. As a participant, you are asked to accept the challenge of completing tasks that the writing can't. Consider that the writing alone can't personalize this story and initiate changes in your feelings, awareness and decision making. I can only provide the framework, boundaries, guidance and relevant details. This writing is just a tool, hopefully a useful one. What it provides to you depends on you, the tool user.

It is more likely to provide the claimed benefits if you take the time to think about the information from a personal perspective. For example, the writing describes the constraints on the natural resources available to us and how we gain a sense of personal well-being. But only you can fully answer questions such as, How much stuff is enough for me to be happy? and its complement, Is my lifestyle appropriate for our circumstance?

I also ask the reader to be patient. You may feel that there are more examples and more detail in the presentation of the basics than seem necessary. However, in making these choices I have taken into account how we learn and change, and how we gain understanding rather than knowledge. For example, the details and examples provide the depth of appreciation needed to make the connections in and between chapters, and to flesh out the developing wider perspective of our circumstance. They also provide the personalization and context that facilitate the unexpected "Aha!" moments of wider understanding.

I recognize that devoting time to detail and directed introspection is a great challenge for humans (Gelter 2003). This is especially hard for those of us from the information age. We believe that greater understanding can be gained by simply increasing the speed with which we process and share the disordered deluge of facts we face. However, how much you know is less important than considering how you think about and use what little you know. To apply this maxim involves time and introspection.

The following tips are provided to assist the reader to make their reading choices. The first few chapters provide the basic knowledge, definitions and concepts needed to appreciate the later chapters. Reading is helped by consulting the detailed table of contents at the beginning of each chapter. It acts as a reading guide, a chapter overview and a filter. In addition, the beginning of a chapter's titled sections provides an introduction, while the end provides an overview or conclusion. This helps to consolidate the information just read. The glossary provides a quick reminder of the meaning of key words and phrases.

Because the chapters are sufficiently near to stand-alone essays, the reader can choose to read them in a different sequence than presented. This works best if the reader pays attention to both the linking prompts and the structure. The glossary also helps.

The main table of contents, chapter tables of contents, chapter introductions and summaries, and glossary serve another purpose. They enable busy readers and those knowledgeable about a particular topic to more easily identify the chapters and sections of chapters they want to pay close attention to and those they wish to just scan. This also enables the reader to ensure that skipping something doesn't result in them missing some important detail or the thread of the argument.

Here are some specific reading suggestions. By reading just each chapter's table of contents, introduction, overviews, summaries and conclusions, a skeleton-like view of our circumstance is obtained. By reading chapters 1, 2, 11, 17 and 18, a more complete overview is obtained. For those well versed in a particular sphere of knowledge, there is another option. The writing is divided into three near stand-alone parts: Us, Our Present, and Our Past and Future. If the reader is well versed in the life and earth sciences, then you can focus more on parts one and three as well as reading chapter 11. If you are well versed in human studies, you can focus more on parts two and three. If you are mostly interested in the future, then chapter 11 and part 3 might suffice.

Ultimately, a comprehensive reading will enable the reader to more easily change their perception of themselves and our circumstance in a manner that will more fully reflect both how they function and our current circumstance. The reader is then more likely to make decisions that will appropriately address the issues we face. The following chapter-focused description of the book will help with that reading.

## Part 1: Us

Our brain's functioning occurs largely in our unconscious, yet it constructs the internal reality we consciously experience. This is a different reality from the external world. Our perception of our world is thus limited and biased (chapter 1). These characteristics constrain our decision making (chapter 2). The ubiquitous presence of a culture both influences and further limits our perception and decision making, often in ways outside of our awareness (chapter 3). The functioning of our brain and our culture limit, bias and influence the primary formal methods (science and religion) we apply to gather and sift the information we use to make our decisions (chapter 4) and to predict the future (chapter 5). Our lives do not stand still. Despite inertia, we and our cultures do change, but in a highly constrained manner (chapter 6).

## Part 2: Our Present

The complex relationships among the earth's organisms and their surroundings enable them to function as an ecosystem (chapters 7). Ecosystems are a part of the earth's dynamic environment. We too are part of the earth's environment (chapter 8). This understanding allows us to more clearly appreciate the current disrupted state of our environment and our contribution to that state (chapter 9). We are stretching the ability of our environment to supply the resources we want. We have taken on the responsibility for maintaining our supporting ecosystem's functioning to ensure that it can continue supplying these resources (chapter 10.) Our understanding of the environment's functioning is combined with our knowledge about our human characteristics from part 1 to create a model of the human-environment system that helps us to appreciate and explain its current state and our circumstance. This chapter is a review, compilation and condensation of previous chapters (chapter 11). An evaluation of the advice our primary cultural guide, neoliberal economics, gives us about our circumstance shows it to be misleading and incomplete (chapter 12).

## Part 3: Our Future and Past

The knowledge and trends gained from part 2 are used to outline the likely future state of our environment (chapter 13). They are also used to outline the future availability of the resources we want (chapters 14, 15). After combining these potential futures with our human characteristics from part 1, it is possible to select, from global-model scenarios, those features most likely to apply to the future of complex cultures. These features describe a period of resimplification for complex cultures (chapter 15). A reality check for that conclusion is found in the history of past complex cultures. The view of them provided by the broader time-scale of history confirms that our human characteristics and the nature of complex cultures favours us making choices that lead to resimplification and explains why we find avoiding that path so intractable. The human-environment model is updated to take resimplification into account (chapter 16). After personalizing these conclusions, Tikopia is used as an example of why our probable future is not inevitable. But avoiding it requires more than placing our hope in technology (chapter 17). The final chapter summarizes our current relationship with the environmental and human constraints within which we must live for our culture to be viable and why our current decisions fall short. It then discusses how we can go about changing our thinking and decision making so that we live more viable lives and our culture can become more viable (chapter 18).

A note on categories and numbers used in the discussions. The categories reported in the literature and used here, such as the percentage of the earth's population under water stress, often have unknown uncertainties and likely have a built-in bias. Many of the numbers used are averages or represent aggregates. The pitfalls of using averages to describe features that may not be normally distributed or that have large variability is recognized. They may, for example, bias our opinion of the most common value and mask the significance of variability. The numbers are therefore used as broad-brush indications of relative importance, not as exact numbers. They are most useful when considered in conjunction with the other evidence provided. Within this context, the numbers used provide both coherent general conclusions and a feel for their limits.

Predictions are made based on information at hand and on general principles. Predictions too are used only to estimate broad-brush features of a future. The predictions used are only useful within the limits and assumptions associated with them. Thus the focus is more on their bounding conditions and their usefulness rather than on their accuracy or precision.

Metric system units and prefixes are used: nano (n) = 1 x $10^{-9}$, micro ($\mu$) = 1 x $10^{-6}$, milli (m) = 1 x $10^{-3}$, centi (c) = 1 x $10^{-2}$, kilo (k) = 1 x $10^3$, mega (M) = 1 x $10^6$ (million), giga (G) = 1 x $10^9$ (billion), tera (T) = 1 x $10^{12}$. For example, 3 x $10^6$ tonnes of coal is 3 megatonnes or 3 Mt. The exceptions are popular or common usages, such as *mb* for million barrels and *My*

for million years. There are also limitations to using one set of units or expression of values. For example, if the literature reports values in tons but doesn't specify whether this means long or short tons, or even metric tonnes, how should it be converted to metric tonnes? In those instances, the reported values are used.

Dates greater than 20,000 years before present are presented as calibrated years before present (yBP). The dates have been adjusted (calibrated) to match everyday calendar years using 1950 as the zero reference. Dates younger than 20,000 years are presented as yBP when the context is geological or ecological, and BC/AD when the context is human. The majority of the C14 dates in the literature used were reported as calibrated dates. The few dates that weren't were calibrated using tables. The decadal accuracy for the Abu Hureyra dates in chapter 3 illustrates the hollow precision of these conversions.

Discussions about the future state of dynamic, complex systems are associated with significant uncertainty. This is accommodated both by the general nature of the discussions and the accompanying estimate of their applicable time frame. *Short* or *immediate term* refer to a period up to 15 years from 2015, *medium term* up to 60 years from 2015, and *long term* beyond 60 years from 2015.

The cultural focus is biased towards Western European complex industrialized cultures. However, the discussion is still relevant to the members of other complex cultures. The word *culture* is used in two senses distinguishable by context. In the context of a specific culture that is not specified, it refers to Western culture. In a non-specific context, it is a generic reference to human culture. All cultures are complex systems, so the word *complex* preceding *culture* refers to the complexity of the culture's organizational structure (see chapter 3).

Hopefully this work will enable the reader to more easily appreciate our world, our relationship to our environment, resource issues, and the decisions we will be faced with as we move into the future. Writing this description will have been a success if the next time environmental issues and our future comes up in a discussion your position will not be, "If I really wanted to do something about the environment, I would go and live in a cave. But I'm not willing to do that, so I am going to keep on living the same way." Hopefully your response will be along the lines of, "Life is too short to be taken seriously. But if I don't take it seriously, it's hardly worth living. So let's see what I can do to make a difference and enjoy doing it."

# PART 1

# US

# CHAPTER 1

# OUR BRAIN, OUR WORLD

My hand was still on the light switch when it dawned on me that I had forgotten why I had walked into the room. It felt as if there was blank space in my mind. Fortunately, this feeling passed and my memory returned. I gathered up the sanding block and left. It could have been quite a troubling experience, except that I know I am not alone. Consider the people we see on TV and the internet doing stupid things to themselves or failing to prevent something nasty happening to someone else. Surely their brains must have stopped working too. It seems to me that our faith in our brain's amazing abilities is a little overstated.

This is a bit of a worry because, as mentioned in the book's introduction, our present circumstance is a product of the complex human-environment system's functioning, and the primary tool we use to make sense of it is our brain. Our past understanding and decisions contributed to our present circumstance, while our present understanding and decisions will contribute to our future circumstance. So, although we need to know something about that system if we are to have a meaningful discussion about our present and future circumstance, we first need to know something about the tool we will use to understand the system and make our decisions about it.

With these musings in mind, a logical place to start the discussion about our present and future circumstance is our brain. It will be introduced by describing it from three different perspectives. The results will then be used to outline how the brain functions and the limits to that functioning. The focus will be on our conscious and unconscious states of awareness, and the relationship between the two. Once this context or framework is established, it will be easier to discuss the limitations to our conscious perceptions and experiences of our world, and thus our decision making about how to live in it.

## Appreciating Your Brain

When removed from the skull, the top surface of the spherically shaped brain has a convoluted, wrinkled pattern, while the underside has a number of bulges. These lower protuberances, such as the cerebellum and the brain stem, stick out incongruously, as if they were afterthoughts. This suggests that, unlike other organs, the brain is divided into a variety of functional units.

Their collective functioning, which is our brain's functioning, is constrained in a number of ways. Consider that both the skull and the brain it encases must pass through the birth canal. In humans this is a tight fit.

To accommodate this squeeze, humans are born prematurely, relative to other species, and our brains only become fully functional much later in life, when we become adults. There are constraints that are more difficult to appreciate, such as those related to how the brain's functioning provides us with our sense of awareness of ourselves and of the world around us, called consciousness.

How can we increase our appreciation of our brain's functioning and its limitations? Westerners tend to think that the computer is an appropriate analogy for its functioning. Although useful, this analogy is significantly flawed and can be highly misleading, especially when thinking about the limits to our brain's functioning. Consider that the various parts of our brain are much more highly interconnected than a computer's parts. In addition, our brain is neither run by linear programs nor does it, like a computer, use a single master clock to synchronize its functioning.

Most importantly, when our brain manipulates symbols, both their objective meaning and their emotional significance affect how it completes a task. In contrast, the manipulation of (data) symbols by a computer doesn't change how it completes its tasks (Searle 1995). Unlike a computer, our brain's basic functioning is influenced by the information it receives. Knowing how a computer works is an imperfect way to appreciate our brain's functioning and its limitations.

Perhaps our efforts to appreciate how the brain functions will be helped by applying the same method we use to conceptualize how our other organs function. For example, our stomach's functioning can be relatively easily understood by combining visual images and descriptions of how it works with our experience of its workings (even if we are embarrassed at times). Applying this method is less effective when thinking about our brain because the connection between the brain's form and functioning are not as easily visualized or described. Our brain's "feeling hungry" and "make sandwich" are not so easily connected together in our imagination as our stomach's "have food" and "digesting."

There is another difficulty. Unlike other organs, it is hard for us to imagine our brain as being separate from our self or vice versa. For example, Westerners, at least, consider the brain to be the seat of our sense of self (the I) and the source of our awareness of the world. Collectively, these features make for some special difficulties when using our brain to appreciate its functioning and limitations. In particular, we are faced with a self-referencing issue.

We are more likely to appreciate our brain by applying the same techniques we use to develop an appreciation for a well-crafted story or song, a soul mate, or even machines. You know you like them, but before you can fully appreciate their depth and nuance, you have to spend time with them. The time you spend on the next few sections should increase your appreciation for our brain's complexity, functioning and limitations. That increased appreciation will help you more easily relate to subsequent chapters and grasp their full significance.

Brain Divisions. The human brain can be described in many ways. This subsection will use three overlapping divisions or perspectives: its structure, its functions and its halves. Individually, they help us become familiar with the details of our brain. Collectively, they introduce us to the complexity of its organization and functioning, and the influences on that functioning. Most importantly, we gain a greater appreciation for the processes, and their limitations, (the framework) by which we become consciously aware of our world and ourselves, and how we think about both. Unfortunately, we are also introduced to the mind-numbing names we use for the brain's parts. The following discussion is easier to follow by paying attention to figure 1.

*Structure.* The brain can be divided into three main structures: the brainstem, the cerebellum and the forebrain. The brain stem, which forms the top of the spinal cord, connects the spinal cord to the base of the forebrain. The cerebellum lies behind the brainstem and below the forebrain, at the back, and is connected to both. The forebrain, which makes up the majority of the brain's volume, caps the brainstem and cerebellum (figure 1a). Each of these three structures perform a distinctive set of functions.

The brain stem primarily deals with basic functions, which include maintaining the vital signs of heartbeat and breathing. It also integrates or distributes the signals received through the nervous system from and to all parts of the body. The cerebellum coordinates body movements, posture and balance.

The forebrain consists of three distinctive substructures: the diencephalon, the cerebrum and the cerebral cortex. The relative location of these substructures and their distinctive group of functions are as follows. At the base of the forebrain, just above where it attaches to the brain stem, lies the diencephalon. It performs functions such as integrating the signals to be distributed to the cerebral cortex (completed by the

thalamus) and coordinating the generation of the chemical signals that are sent through the blood to control and stabilize both brain and body functions (completed by the hypothalamus).

Surrounding the diencephalon is the cerebrum. The outer region of the cerebrum is covered by the thin cerebral cortex whose convoluted outer surface lies nearest to the skull. This is the wrinkled surface we see when we look at an intact brain. The primary functions performed by the cerebrum and the cerebral cortex are the so-called higher level or more complex functions that we more readily associate with our brain, such as sight.

The distribution of these structures and substructures within the brain is not random. Those structures primarily involved in the more basic functions (e.g., maintaining the heartbeat and controlling movement) are generally located nearer to the brain stem. Structures dealing with the more complex functions (e.g., perceiving objects and events) are mostly located further from the brain stem and nearer to or within the cerebral cortex. However, in order for the brain to complete all aspects of a particular task, a variety of brain structures are linked together into different functional groups or systems, such as sight and the limbic system. For example, the limbic system, which includes parts of the diencephalon and those parts of the cerebrum nearest to it (e.g., the hippocampus), plays a significant role in learning and behaviour and in our emotional experiences.

The nature of the linkages between the structures forming a functional group are systematically organized. For example, the strength of a link is determined by how closely the connected structures must co-operate with one another to complete a particular brain function. The more they must co-operate, the greater must be the coordination of actions (synchronization) between them and thus the stronger the links.

For example, few linkages and limited coordination is required to maintain your heartbeat. In contrast, to engage in a higher brain function such as sight requires numerous linkages and considerable co-operation between a number of structures. These include the eyes, parts of the diencephalon and the areas of the cerebral cortex involved in the interpretation and management of sight data. The links between these structures are strong.

Although co-operation between structures enables the functional group they form to perform a particular function, it also limits the tasks each of its member structures can perform: the limitations of specializing. This means that, for us to be fully functioning humans, a number of functional groups must be complexly interconnected with one another and coordinated in their operation. Consider that the complex process of visual perception requires a combination of sight, the interpretation of both emotional and other sensory signals, a mental body map and abstract conscious thought.

Our brain's structures and their organization are not unique to humans. The brains of all vertebrates, from fish to mammals, contain structures performing equivalent functions that are located in the same relative positions and are similarly interconnected. However, the relative size of each structure varies between the classes of the animals.

If the brains of the animals in these classes are laid out in the order in which the class made its first appearance in the fossil record, then the relative size of the structures within each brain are seen to systematically vary through time. Reptiles first appeared some 400 million years ago. Their brains are dominated by the equivalent of the human brain stem. The reptile's equivalents of our forebrain structures, such as the cerebrum and its functional groups such as the limbic system, are much smaller than their brain stem.

All of the structures in the reptilian brain are found in the brains of mammals, an animal class that appeared some 200 million years ago. But, unlike the reptilian brain, a mammal's forebrain structures, in particular the cerebrum and the cerebral cortex, are much larger than its brain stem. Humans are among the most recent mammals to appear. Our brain contains all of the mammalian brain structures, but our cortex is much larger than average.

In 1969 it was noted that this systematic brain variation through time was matched by a parallel variation in the animals' observed behaviour. This led to the idea of three distinct functional brain types: the reptilian, the old mammalian and the new mammalian or primate brain (which includes humans) (figure 1a) (Wilson 1999a; Newman and Harris 2009). The systematic differences between them were thought to reflect not just how the animals' brain structure and functioning had changed over time, but also how their behaviour and experience had changed over time.

This biological insight had an interesting cultural side effect for Westerners. Among at least the public, the order of the changing brains became more than a sequence through time. It became a hierarchy. Reptiles, with the simplest brains, were at the bottom; and primates, with the most complex, were at the top. This view, which mimics how our culture is structured, came

to symbolize/justify the cultural belief that we are more than just different from other animals. We are superior.

We occupy this position in the animal kingdom, so the logic went, because we have either progressed from or transcended the basic, more primitive reptilian urges and emotions to become conscious, more advanced, logical thinkers. For a similar reason, those emotions and behaviours that are present lower in the brain-animal hierarchy, such as anger and guile, came to be labelled as our base characteristics (McCallum 2005). These base emotions and behaviours were thus viewed as less than human. The more primitive animals we associated with these lower characteristics, the reptiles, were seen as less desirable or less worthy of respect. This way of thinking about animals still affects our decisions today.

*Functions.* By noting the effects that damage to specific parts of the brain has on human performance, and by using tools such as functional magnetic resonance imaging (MRI) to study the brain, it is possible to locate and map the areas of the brain associated with a specific function, such as sight. This work reveals that a specific aspect of a function, like sight, takes place in a highly localized area. For example, during sight, the detection of an object's movement in a specific direction is localized to a few neurons (Kast 2001). However, in order for us to see, these specialized cells must be linked to nearby specialized cells, and their common host structure must be linked to the other structures helping to perform the sight function. Among those structures are the retina in the eye, the geniculate nuclei (which lies in the diencephalons) and the visual cortex at the back of the brain (figure 1b) (Damasio 2000).

Unfortunately, describing the brain as consisting of specialized structures linked together to complete a specific function reinforces the view that our brain is made up of isolated modules like a computer. We are not prompted to think of our brain as the highly complex, interconnected organ it is. Consider that a structure we associate with a specific function is commonly linked to more structures than the ones needed for it to complete that function. Those links enable the other structures to perform their functions. For example, in the case of sight, the visual signals from the eyes also stimulate those brain structures (the X in figure 1b) that synchronize or generate the daily (circadian) rhythmic cycles so critical to our lives, such as the sleep-wake cycle.

The strength and importance of these cross-functional ties are more fully appreciated when we realize that, for the brain to provide us with a useful experience, it must coordinate or synchronize a number of distantly related functions. For example, the visual signals from our eyes and the sound signals from our ears are connected to the structures in the brain that direct the movement of the head and eyes. This connection helps the brain guide our mental and physical attention toward the source of these stimuli. Similarly, the interpretation aspects of the sight function include a connection to our emotional memory. This ensures that our visual perception of an object includes our feelings about it. Clearly, the view that our brain is comprised of a number of largely independent functions is incomplete. Our brain consists of specialized functions linked together by a diversity of connections into a single, complex, functional whole.

*Halves.* If you look at a brain from the front or the top, you can't help but notice the striking vertical cleft that divides it almost completely into right and left, roughly hemispherical halves. These halves are joined together on their flat faces, mostly by the corpus callosum (figure 1b). If that link between them is somehow severed, then, as tests have indicated, much of human performance remains completely to nearly completely unaffected. This occurs because the brain structures providing those unaffected functions are symmetrically divided between the brain's halves. This division replicates the similar symmetrical division (i.e., left and right) of the activities they control, such as sight and hand movement.

However, there is an apparently unnecessary twist to this symmetry. The signals from one side of the body are processed by the half of the brain on the opposite side of the body to the signal's source. For example, if your right hand reaches out to touch this page, then it is the left half of your brain that is using the signals from your eyes to instruct your right hand to move.

Surprisingly, severing the corpus callosum reveals that a few brain functions are found mostly, but not exclusively, in just one of the brain's halves. The half hosting a lateralized function completes its work largely independently of the other half. This enables each half of the brain to make a different contribution to our experience of the same event (Gazzaniga 1998). This uniquely human characteristic provides us with experiences different from those of other animals.

Consider that, in our lateralized brain, the right half is primarily involved in visual motor (movement) tasks, but it also controls true memory and the distinction between the outside world and our imagination. In essence, the right half provides an interpretation of the information it receives from the body and its external sensors. In contrast, the left half is primarily involved with language and speech, although it also provides us with an overview interpretation of events. In essence, the left half tries to provide order and coherence to our experience of events while at the same time supplying us with a sense of meaning.

The right half thus supplies us with a consistent version of an event, while the left half places the event into a larger context and allows us to talk about it. These two interpretations of a single event are combined by the brain to help provide the single conscious experience that we treat as reality (Hock 1999a). We are provided with an enhanced experience of the world around us. However, there are circumstances under which this internally generated reality is less reliable than we might expect.

If, for some reason, the information needed to make sense of an event is uncertain, unavailable or ambiguous, then our brain provides either an incomplete experience or, more usually, a complete but biased one. This occurs because the left brain uses the presence of uncertainty and ambiguity as a cue to fill in the missing bits by using our imagination. In essence, the left brain simply fabricates a reality that it thinks is the most plausible explanation or scenario for the available data (Gazzaniga 1998; Damasio 2000). We are not informed of this filling-in, so we are none the wiser that the reality we experience is partly a self-creation.

The extent of these fabrications can be substantial if the connections or processes used by our brain have been disrupted. This can occur due to electrical stimulation of electrically sensitive parts of the brain or due to stress, drugs, damage or oxygen deprivation. Under these circumstances, we can have quite unusual experiences, such as being aware of someone's presence when no one is there or engaging in an event that can't be seen outside your head (a vision) (Sacks 1986; Regush 1995; Persinger et al. 1994). These types of experiences can have profound impacts on our beliefs about ourselves, how our world works and our relationship to it. Think of Joan of Arc and her nation-changing quest.

The discovery of lateralized brain functions in the 1960s had a strong influence on Western culture's views about human nature and how we think about ourselves. It is still possible to hear people say that the primary characteristics of their personalities (i.e., their sense of self) arise from the side of their brain they consider to be dominant. Saying that you are "left-brained" implies that you see yourself as logical and have the characteristics of the associated stereotype. The "right-brained" stereotype implies that you consider yourself to be impulsive and artistic. In reality, our brain develops and operates as a single whole, not as two competing, independent brains or even two halves competing for dominance (Hock 1999a).

<u>One Brain.</u> The three descriptions above do not refer to three separate brains. They are three different perspectives of the same brain. For example, the function of sight involves both halves of the brain as well as specific structures in the forebrain (figure 1). Thus, although each perspective is useful, on their own they provide an incomplete and biased view of our brain. Only when we consider them together can we develop a comprehensive appreciation for its complex organization and functioning.

The few minutes you might have spent struggling to combine the above three descriptions into a single, coherent picture provides important insight into the constrained nature of our brain's overall functioning. Consider the difficulty we experience making and then holding in our conscious awareness the many connections between the above three descriptions of the brain. We also struggle to make sense of this information because much of it appears to be contradictory. After all, the brain's features are, at the same time, both localized and distributed, and discreet and interconnected. The brain's functions are both independent of one another and yet highly interdependent, and they appear to form a hierarchy that is both randomly organized and flexible.

Our difficulties are partly the result of the uncertainty and gaps in the knowledge used to provide the three descriptions of the brain. They are also partly the result of our brain's normal functioning, as illustrated by its tendency to try and smooth over any ambiguities by using imagination and avoidance. These factors are not unique to us thinking about the brain: they are a part of our everyday thinking. But they become especially important and noticeable when we think about a complex topic/system (complex system), such as our brain, our environment or our culture.

Fortunately, there are aids that can help us deal with these and the related challenges of trying to appreciate the organization and functioning of a complex system. These aids therefore deserve more of our attention before continuing the discussion about the brain. The idea of spending time with the system in order to become familiar with it has already been mentioned, but our busy, fast-paced lives limit the time most of us can devote to the task. We can be helped by how we direct our thinking, which will be discussed later.

The aid discussed here is how we can go about gathering and compiling information about a complex system of interest in ways that help us appreciate it as a whole. A commonly used method is to systematically break the system into smaller pieces: a reductionist approach. The pieces are studied in isolation in order to discover the system's essential parts and the essence of how they fit together. The essential parts and functions are then used to describe or model the system as a whole. Simplifying the system in this way makes it easier to understand, but it also tends to mask its complexity and creates a false sense of order. On its own, it provides a rather simplistic connection to the real world.

These downsides of reductionism can be lessened by applying a multiperspective constructivist (putting pieces together) approach to information. It looks at a complex system from a number of perspectives. For each perspective, it uses the available information to create an example-enriched description of the system. This helps us gain familiarity with the system's real-world characteristics. By also outlining how those overlapping or complementary descriptions are connected, we can link them together to more fully appreciate the system's overall structure and functioning. The result is a personable real-world descriptive framework or model of the complex system's structure and functioning as a whole that includes reminders of both the constraints on the system and the model's limitations. The use of this method to discuss the brain's structure and functioning in this chapter will serve as an illustration of its use in the rest of the book. The downside is there is more to pay attention to, but the rewards are worth it.

In summary, our brain has a complex structure and functioning: it is a complex system. We experience difficulty thinking about complex systems, like our brain, as a functioning whole. To alleviate this difficulty, a multiperspective constructivist approach to thinking about complex systems in general and the brain in particular was introduced.

The nature and limitations of our brain's functioning as a whole remain to be more fully discussed. This includes paying attention to the constraints that our brain's characteristics place on our experience of the world and our decision making. A suitable vehicle to reveal these characteristics is a familiar product of a functioning human brain: consciousness. We could approach this topic by asking a question such as, what is consciousness? But, considering the cues offered by the above discussion about the brain, we can instead pose a more personally relevant version of that question: What is the nature of and limits to the "who I am" reality and the "world I live in" reality that my brain allows me to experience?

## Unconscious Awareness

The words *unconscious* and *conscious* are descriptions of the possible state of the brain or our relationship to it. These states are not associated with a specific brain structure or location but are provided by the brain functioning as a whole. The question about the nature of and limits to our "who I am" reality can be answered by looking at the nature and limits of these two states/relationships. Particular attention is paid to the difference and links between our unconscious and our conscious awareness of the world.

In the context of our daily conscious experiences, it is easy for us to believe that we are in full control of our brain. In doing so, we underestimate the role and significance of our unconscious. Consider that we go to sleep and do not die. Thus, when prodded, we have to admit that at least some of our many internal body functions are unconsciously controlled or at least outside of our conscious awareness.

A quick inventory shows that, in fact, most of our body's functions are controlled by our unconscious. We are, for example, neither aware of nor able to consciously control the majority of the very many functions needed to keep us alive, such as those that maintain blood acidity (pH), electrolyte balance and sugar levels. There are other functions we are partly aware of that we can influence but not control, such as sleeping and breathing (Damasio 2000). Even some of the functions we feel that we can fully control, such as standing or riding a bicycle, are significantly outside our awareness. Are you really conscious of how you maintain balance when you stand up or ride? Consciousness can direct our activities and some body functions, but it is in full control of very few.

The majority of our brain's activity occurs in the unconscious. With this knowledge, we should find it easier to accept that an appreciation for our consciousness and its limits requires us to first appreciate the unconscious, its limitations and the influence it has on our conscious experiences. We are helped in this task by considering the things the unconscious is aware of and some of the functions it is engaged in.

Control, Stability and Memory. A primary function of the brain is to help ensure that conditions inside your body are stable. To accomplish this task, the brain must have sufficient information about the state of your body, such as your blood pressure. It acquires the needed information from an extensive network of sensors that routinely monitor the condition of many of your body's parts and processes. These sensors are linked to the brain through connections that are electrical (the nerve system) or chemical (the blood and lymph systems) in nature. Because many of these connections are bi-directional, the brain both senses (more correctly, the brain is influenced by) the state of your body and it influences that state. The nature of the brain's influence is determined by it comparing the sensed information to reference states stored in some sort of memory that is usually, but not always, in the brain. The brain, the body sensors, memory and the connections linking them together collectively form a control system.

The brain-centred control system works by the ceaseless flood of information arriving in the brain from our body parts being compared to their set point reference states stored in memory. When a body part's state deviates from the set point reference, the brain sends out the signals needed to return it to the desired state. By this process, your body parts, and thus your body, are kept in a state of dynamic equilibrium within limits around the reference state (Damasio 2000). You are mostly unaware of the control system's functioning because those activities all take place in your unconscious. You may be aware of the results.

The significance of the brain's unconscious control system can be seen in a person whose brain injuries render them brain dead (the person is alive but permanently unconscious). Although devoid of consciousness, the person can still maintain most of their basic body processes in a stable state and, with assistance, can live for many years. Thus, although your conscious decisions can influence some of the control functions, such as regulating breathing, and can help you participate in others, such as supplying food, our conscious ability to actively control most body processes is both limited and unnecessary.

A popular analogy for understanding the body's control function is your fridge: how its temperature sensor, temperature setting, compressor, and cooling coils work together to keep the fridge's temperature roughly constant. However, there is an important difference between your fridge and your brain–body-state control system. (Aside from the fact that your control system can't keep beer cool.) The fridge's control system acts independently of the things you put into the fridge. This is not the case with your brain. The signals received by your brain-centred control system about the state of your body and your surrounding environment are used to do more than maintain your body in equilibrium. They are also used to modify the functioning of your brain, the control system itself and even its reference states.

The presence and significance of the modifications triggered by the sensings the brain receives is illustrated by the role that a child's interactions with the world around them have on their development. The child's experiences provide their brain with the critical signals (cues) needed to stimulate the normal development of both their brain and body. For example, if the speech centres in a child's brain are to develop fully, then the growing youngster must hear speech. More generally, the flood of unconscious sensory information that both children and adults receive from their surroundings helps to maintain their sense of mental well-being. As a result, sensory deprivation is considered to be psychological torture (Cole and Cole 2001).

As long as you are alive, your brain is engaged in a pervasive, unceasing sensing of the state of your body and the immediate external environment around it. If it stops this activity, then you are dead. Don't try to disprove this. The gathered information is unceasingly used to help regulate your body's state, control your actions and make you consciously aware of your body, its state and surroundings. This continual interaction is significant because it means that, at the unconscious level, your brain, your body and your immediate environment are irrevocably intertwined into one. For example, although you routinely consciously imagine yourself as being separate or isolated from your immediate surrounding (a unitary organism), at a very basic level, you are always engaged with and embedded in it.

**Sensing.** For our unconscious to be aware of situations such as the internal state of our body (e.g., blood pressure), the relationship of our body to the external world (e.g., knowing where "up" is) and our world's state (e.g., it's raining), it must be provided with the relevant information. By appreciating the limits to the

body's ability to sense, we will be in a better position to appreciate the limitations to the information that can enter our unconscious and conscious awareness. Some of these limitations are quite surprising.

(1) Our body has a limited array of sensors. We are all aware, at least to some degree, of the five conscious senses we use to interact with the external world. There are three physical senses (sight, hearing and touch), which collectively detect parts of the electromagnetic spectrum and force, and two chemical senses (taste and smell). We are also partly aware of our other senses, such as those that provide us with a relationship to space (e.g., gravity, to help us stand upright) and time (supplied by visible, light-cued chemical clocks, as mentioned above). However, we are completely unaware of the sensors for the many internal variables, such as blood acidity and blood pressure (Sacks 1986).

There are attributes of the world around us for which we have no sensor at all. Our bodies have no X- or gamma-ray sensors. This explains why the workers who stood behind the door blocking access to the source of the radiation from the exploded Chernobyl reactor did not intuitively realize that they were being exposed to a fatal dose of radiation. For the same tragic reason, many inhabitants of Hiroshima who survived the atomic blast were surprised to find, a few days after the event, that they were severely burned.

(2) Our sensors can only detect signals within a limited range. For example, our sight can only respond to light if it falls within the visible spectrum. Compare this to bees who, among many insects, can see a flower's ultraviolet "colours," which we can't.

(3) Our sensors have other functional limits. For a signal to be detected, its strength must cross a sensing threshold and stay within limits. If a signal is too weak or too fast, then we do not sense its presence. For example, our eyes are limited to seeing light that is quite bright. Contrast this to the incredibly light-sensitive eyes of nocturnal animals such as owls. However, the light can't be too bright, otherwise even our contracted pupils and squinting might not prevent damage to the eyes, which is the reason why sensible humans do not look at the sun.

(4) It takes time to sense something and process its signal. Therefore, if two events occur too close together in time, our sensors will be unable to tell them apart (Damasio 2000). In this case, the signals just blur the events into one another within our consciousness in a manner analogous to our experiencing the discrete frames of a film as a moving picture.

(5) The response of our unconscious to a signal from a sensor varies with the context. For example, if we are standing, it tries to keep us upright and conscious. However, if our blood pressure drops, we faint and fall to the ground. Under other circumstances, our unconscious may simply ignore what is being sensed. For example, when we are stressed or drunk, we may fail to respond to a red stoplight.

(6) There are delays in the sensing system. It takes only a few milliseconds for a single neuron to fire. However, the sum of the firing times for all the neurons involved in the sensing, transmission and processing of a signal results in a significant delay between the time something is sensed and the time the brain sends out a response to it. For example, imagine that you accidentally prick one of your fingers on a pin. There is a delay between your finger being pricked and your reaction of pulling it away from the pin. You usually remove your finger about 0.25 seconds (250 milliseconds) after the prick has occurred because you react using your unconscious reflex action or startle response. In contrast, the removal of your finger by a fully conscious action would take longer than 0.25 seconds. Your conscious response is much slower than your unconscious reaction because more neurons are involved in both making the signal representing the pinprick available to your consciousness and you deciding to remove your finger.

There are some surprising implications from this difference in response times. By reacting unconsciously to the prick, the pricking event is over before you are consciously aware of it. You only become consciously aware of the event in retrospect. Your conscious experience of the event is thus a post-event reconstruction (Libert 1999; Damasio 2000).

The presence of processing delays and, more importantly, delays of different durations should result in us routinely having confusing experiences. But we don't. We generally sense and experience synchronous events as synchronous. To us, the prick on the finger and the pain occurred at the same time. The mechanism that makes this possible is still being debated. However, it seems that your unconscious brain deals with these delays by adjusting the timing of an event when it forms your memory of them (Libert 1999; Klein 2002). We consciously experience an adjusted reconstruction of the pricking event.

There are other ways the brain makes adjustments for delays. Imagine a need to react in real time to a very fast event in space. For example, a baseball batter faced with hitting a ball has to deal with the inescapable delays in their reaction. Their delay is a combination of the time it takes their brain to process the signal and the time for their muscles to move the bat. Thus, to hit a very fast ball, the batter must start to swing in advance of knowing exactly where the ball will be. This is accomplished by the batter swinging toward the place where they unconsciously anticipate the ball will be (Klein 2002). In making that decision (prediction), they likely rely on subtle cues from the pitcher's body that they have learned through experience.

These limits to our sensing abilities force us to recognize that the knowledge the brain has about the world around us is limited. And what it does sense requires adjustment and interpretation. In other words, the unconscious is aware of a representation of the world, not the actual world.

The conscious and the unconscious use the same sensors, thus the conscious reality we experience is also only a representation of the world and not as real as we would like to believe. (We will return later to the question of the relationship between the unconscious and conscious.) We are therefore burdened with the task of recognizing which of our conscious experiences are imaginings, such as hallucinations or false memories, and which are not.

**Unconscious Processing.** We can intuitively describe the general features of the conscious and the unconscious states. However, we still struggle to uniquely define and separate them (Erdelyi 1992; Merikle 1992). Part of this difficulty arises because self-reflection cannot provide you with reliable information about your unconscious, so you are left to imagine it.

Scientific study has provided us with some insight into the unconscious (Lewicki et al. 1992). But it too struggles to provide a simple yet comprehensive definition of the unconscious. So rather than search for a definition as a way to appreciate the unconscious, we will look at unconscious processes, their limits and their links to the conscious.

Our experience of waking and sleeping indicate that there is a boundary between conscious and unconscious processing. Consider too the investigation of people who have blindsight: Those who lack the ability to be consciously aware of having seen an object, so they are technically blind, yet monitoring of their brain indicates that they have "seen" the object. Their inability to consciously see suggests that a visual stimulus must cross a strength threshold before it can enter consciousness (Kolb and Beaun 1995).

However, research also indicates that the divide between the conscious and unconscious is far less sharp than implied by a threshold. The boundary is more of a zone within which conscious and unconscious processes are mixed (Loftus and Klinger 1992). An illustration of this blurring of boundaries is provided by our ability to influence but not to maintain control over some basic functions, such as breathing and heartbeat, and our limited conscious involvement in others, such as intuition and perception.

One way to appreciate where the boundary between the unconscious and conscious lies is to ask what proportion of our brain's activity takes place in the unconscious. In an eye-popping contrast to our busy, conscious experience of thinking, as little as one millionth of the sensory information received by the body-brain team actually reaches consciousness (Norretranders 1998). The great majority of your brain's activity occurs outside of your awareness, within your unconscious (Logothetis 1999; Kihlstrom 1992; Taylor 1989; Bechara et al. 1997).

Much of this unconscious activity is directed at processing the ongoing flood of sensory data about the condition of your body and maintaining it in a stable state. This involves processing tasks that are relatively simple. Some of the unconscious activity is directed at preparing the information your brain will use to provide you with your conscious experiences of yourself and your surroundings. This involves processing tasks such as compiling information and making simple choices, some of which are moderately complex (Damasio 2000). For example, our unconscious uses the information about our body parts and their relative location to one another that it maintains in memory to form, in essence, an "image" of our body, a virtual image. The unconscious attaches to that virtual body image information, such as the state of our body parts and our surroundings, which our brain then uses to provide us with our conscious sense of a physical and mental self in our surroundings.

Our unconscious is thus in a strong position to influence our conscious self and our conscious experiences. The power that it has over the conscious is illustrated by the observation that humans tend to act intuitively first, then follow up with excuses (Jacoby et al. 1992). The realization that the unconscious dominates the brain's activity and has such a pervasive influence over our conscious is a considerable blow to our faith in our conscious awareness and our control over our brain.

You could argue, in defence of the conscious, that the unconscious is simply a filter that removes "useless" information before making a summary of the rest for our conscious awareness. But, even in this case, the unconscious must still decide which information is apparently useless and then create and pass along a coherent summary. If it fails, then the brain is unable to provide us with a meaningful conscious experience.

In addition, while the unconscious is processing information into the summary our brain needs to provide us with our conscious self, the unconscious must still continue to work. It must continue processing the body's sensings, regulating body processes, creating and recalling memories, and making fast decisions. Think pinprick. The unconscious must continue to do so until the slower conscious has at least had the opportunity to absorb and make a decision about the information made available to it in the unconscious summary. The unconscious carries a large processing load indeed.

You might imagine that the unconscious conducts its processing activities in a manner similar to our experience of conscious processing, such as focusing on just a few things at a time in a linear manner. This is not the case. Unconscious processing is unfocused, wide ranging and, because the procedures are automated, fast (Kihlstrom 1992; Bruner 1992). All levels of unconscious processing are conducted in this way, from the most simple to the most complex. This includes even its more complex manipulations that require a true, but limited, decision-making capacity.

The unconscious is able to complete its processing tasks fast because, in part, they are scripted (automated). This enables it to deal with a larger amount of more widely dispersed information than is possible through conscious processing (Lewicki et al. 1992). For example, automation gives the unconscious the ability to quickly process the flood of information about a novel situation and make fast decisions about it, for example, when suddenly facing the possibility of a car accident, or experiencing a pinprick. If the unconscious waited for the slower conscious processing to advise our body how to deal with these situations, the results might not be to our liking. A more benign example of the power of the unconscious is our ability to identify and to categorize a familiar piece of music based solely on its first two or three notes. These types of snap or intuitive decisions are made by the unconscious and then passed along, fait accompli, to the conscious.

But these advantages are accompanied by limitations. The automated processing procedures are based on prior knowledge stored within various types of memory in the brain. This information includes previously established body-function set points and habits as well as previous experiences and decisions, both conscious and unconscious; (Kast 2001). If we lack the required knowledge (memories) about a situation, then our unconscious decisions might not be appropriate. Think of its use of imagination. This reliance on memories also means that the past strongly influences our present unconscious processing decisions by guiding and limiting it. And through this influence, the past affects our current conscious experiences and decisions.

This power of the past is illustrated by our preference for the familiar, even at an unconscious level. For example, in an experiment, people were shown a geometric shape at a speed which was slow enough for the shape to enter the unconscious but too fast for it to be recognized consciously. The observers were thus familiar with the flashed shape but had no conscious awareness of it. When the observers were later shown a collection of various shapes and asked to pick out the one flashed for them, they could be somewhat successful (not perfect but much better than chance) if they simply selected the shape that they intuitively preferred or liked (Kunst-Wilson and Zojonc 1980). We are predisposed to favour the familiar: what we have been previously exposed to.

But our unconscious processing is not entirely isolated from our conscious self. Our conscious activities and decisions contribute information to our memories. These contributions are then available for our unconscious to use. This is significant because, through these memories, the conscious can influence unconscious processing, to some degree. Our unconscious will, in turn, affect our conscious experiences and decisions. The conscious and the unconscious are thus joined together by a feedback loop, a link that helps intertwine them intimately and create the grey zone between them (Kast 2001). And we can consciously influence our unconscious.

Intuition. Most information received by the brain will remain in the unconscious. Only a small selection will automatically contribute to our usual conscious experiences. Intuition refers to the unconscious information or knowledge that enters consciousness as unexpected glimpses or spontaneous insights. Many cultures, including Western ones, believe strongly that the information received in this way is more valuable or truer than information received in a manner they consider to be more usual. Members of these cultures can believe in this sufficiently strongly that they actively try to tap into this hidden store of information.

By definition, intuition and intuitive knowledge lies in the ephemeral boundary zone between the conscious and the unconscious. This information could indeed be useful to conscious decision making. Consider that the unconscious is aware of events that occur too quickly or are too weak to easily enter the conscious (Kunst-Wilson and Zojonc 1980; Norretranders 1998; Kolb and Beaun 1995). The unconscious can also absorb quite complex information that you may not be in a position to consciously understand. For example, even children who are four to five years old and far too young to intellectually (consciously) understand conditional (if-then) relationships can still learn them (Lewicki et al. 1992).

However, the simple presence of intuitive knowledge does not make it useful. In order to judge how valuable intuitive knowledge might be, we need to know something more about the limits to the unconscious decision-making processes providing it. For example, does the unconscious represent sensed information truthfully? Are the decisions made by the unconscious smart? And, in particular, are its decisions more reliable than conscious decisions?

It turns out that our unconscious is not a sophisticated decision maker or classifier. The unconscious is capable of registering the presence of things (unconscious awareness). It can process information and make rapid decisions based on the few available cues and a well-established history stored in memory (Greenwald 1992). But our unconscious awareness and decisions are influenced by our current feelings and circumstances, and by the biases in the memories being used. Consider too that the influence the left brain's imaginative powers can have on our conscious decisions, as mentioned above (e.g., filling in gaps) also affects unconscious decisions. Unconscious decisions are therefore inherently no more reliable than conscious ones (Lewicki et al. 1992; Loftus and Klinger 1992).

The unconscious is also limited in the decisions that it can make. It can't fully evaluate the larger significance of the information it receives or the decisions it makes (Germano 1999). That critical task is completed by the conscious brain because it has, among other attributes, the ability to develop the more complex and successful predictive strategies needed to make those judgements (Merikle 1992). We can therefore say, with reasonable certainty, that overall the unconscious is not inherently smart, in the sense that it provides us with correct answers to arbitrary questions posed to our intuition (Greenwald 1992).

Fortunately we can help our unconscious to provide better answers. We can provide our memory with more accurate information and more complete experiences that are relevant to the questions we plan to ask of our intuition. We can accomplish this task by routinely directing our conscious experiences to that end, by choosing appropriate habits and by committing relevant information to memory through repetition (Kihlstrom 1992). These methods work because the conscious act of training our body-brain team to be proficient in something accomplishes something else.

Once our conscious effort has resulted in the information being sufficiently well-developed, it is absorbed into our unconscious. The unconscious can then routinely use it even though we have little, if any, conscious awareness of it doing so (Bruner 1992). Remember the world-class baseball batter who can hit balls travelling too fast for them to make a timely, conscious decision about the ball's future location. Other examples are walking, knowing a subject well, riding a bike and touch typing. Think of habits, regardless of whether they are beneficial or self-destructive, such as some addictions. It also applies to our thinking, regardless of whether the results come to us through intuition or some other means.

In overview, accessing the unconscious through intuition will always provide us with an answer. But if the unconscious is untrained, it cannot provide us with the reliable answers we seek. However, once our unconscious is well-informed and experienced, then it can provide more reliable intuitive guidance. There are limits, though. While a trained unconscious can allow us to use skill more effectively and can provide us with information worthy of conscious follow-up, it still cannot provide answers to complex decisions that belong firmly in the realm of consciousness.

In summary, our thinking about the unconscious, its processes, limitations and influence on the conscious is helped if we remember a few basic features. The unconscious is aware, within limits, of our body, our surroundings and their state. It uses this information to take care of the basic, mostly routine functions needed to keep our body and brain stable and functioning. The unconscious, from which intuition arises, and the conscious, which becomes aware of that intuition, fulfill different roles but they are intimately interconnected into one brain. The unconscious processes and their limitations affect our conscious experiences and decisions, and vice versa, in a feedback loop.

## Conscious Awareness

We all have some notion of what consciousness is, yet it is still a difficult topic to discuss. Not that the subject is risqué, we simply lack a suitable framework and vocabulary to discuss it.

As mentioned, if we wish to consider digestion in our stomach, we can easily mentally isolate the process of digestion from our stomach. In contrast, we are unsure how to separate consciousness from the brain. More importantly, when thinking about consciousness, there is no easy way for us to set ourselves aside from ourselves in order to observe our self and evaluate our conscious awareness. This difficulty is the biological equivalent of a philosophical, self-referencing mind knot. For example, can you believe someone who says, "I am a liar"? (If you do believe them, then you know they have told you a lie, and if you don't, then you perceive them as a liar).

Then there is the problem of vocabulary. There is a paucity of words to describe our conscious experience: consciousness, awareness and self-awareness, but not intelligence. Although human intelligence requires consciousness, it is not the same as consciousness or wisdom (Damasio 2000). Consciousness is a type of awareness. Intelligence is the ability to use the information of which you are aware. Wisdom is the ability to know when you don't have adequate information, when you must collect it, and how to appropriately use what you do know.

Overall, we know consciousness exists but, like seeing a hologram, we are not intuitively sure what it is or how it can exist. And, like a hologram, our conception of consciousness is rather ephemeral: it disappears as you look closer at it. Ironically, this means that a key to gaining a better understanding of consciousness requires more than defining a suitable framework and vocabulary to describe and think about it. It also requires finding an appropriate analogy to represent it and having a unifying narrative to coherently link these tools together. Without them, it is no wonder that our discussions about consciousness are filled with contradictions and circular logic.

Fortunately, we can increase our general appreciation for the conscious brain by simply listing some of the brain's features involved in providing it. Consciousness is built on the brain's ability to store in memory a large amount of detail about the form and state of our body, its surroundings and the connections among them. The brain has sufficient capacity and interconnections among its parts that its manipulation of the information in our memories can somehow be synchronized and coordinated to provide us with a self-referencing representation of our body, its state and its surroundings. This is the hologram-like experience we call consciousness (Damasio 2000; Searle 1995).

But not all organisms display consciousness. It seems that there is a threshold in a brain's complexity which, once stepped across, allows its functioning to more easily provide new features and abilities. For us, one of those complexity-related features is consciousness (Greenwald 1992).

Consciousness provides you with many wonders. It allows you to realize that you "are" and that you are the organism that is responding to the environment you live in. It also allows you to relate to yourself and your surroundings from a perspective other than your own.

But there are limits. Consciousness is built on and interconnected with the unconscious in ways that include feedback. And they both use our virtual body image and our body states stored in memory as references. Thus consciousness is, as discussed above, neither independent of the unconscious nor a replacement for it. Neither can you or your consciousness (if they are different) be separated from the unconscious. Additionally, the feedback-enhanced self-referencing links between them contributes to the self-centred bias in our conscious experiences (Damasio 2000).

We can further increase our appreciation for the nature of consciousness and its limitations by looking more closely at some of the common features of the conscious experiences our brain provides for us (Kihlstrom 1987; Greenwald 1992). Two of the most important are feelings and emotions, and memories.

**Feelings and Emotions.** We are all consciously aware of feelings and emotions, but not how they are tied to our brain's functioning. Our unconscious uses the ongoing flood of information received by the brain about our body to create an instantaneous "index" of our body's physical state. Our view or experience of that index is best described as a feeling. This feeling does not occur in isolation because it is attached by the unconscious to the virtual image of our body, which is also created for us by the unconscious. We thus consciously experience the index as a feeling about the state of our body, of ourselves. We may even associate our feeling with a particular body part.

From a different perspective, the range in our possible feelings and states is made available by both our genetic inheritance and our life experiences. These possible feelings are stored in memory. When the unconscious becomes aware that the body has entered a particular state, it automatically attaches to the virtual image of that body state the feeling(s) in memory associated with that state. For example, when our body becomes cold it shivers first, only then we do we feel uncomfortably cold (not the other way around). Think too of going into shock.

We are usually either unaware or dimly conscious of our feelings. When we do become consciously aware of them, we experience fatigue, energy, excitement, wellness, sickness, tension, relaxation or pain, to name a few. As the names suggest, one of the purposes of these feelings is to prompt us to pay attention to the body's current state and the body function(s) generating it (Damasio 2000). For example, the feeling of fatigue is a cue to pay attention to food and sleep. The details of our response are, influenced by our circumstance, and our personal and our culture's world view. Thus, as we all know, our conscious can ignore our feelings, such as fatigue, to a point. We can also hide our feelings from others.

If feelings are strong enough, we call them emotions. Although we experience emotions more intensely than feelings, they serve many of the same purposes. For example, they direct our brain's conscious attention toward the body state and the cause(s) of the emotions. Thus both emotions and feelings guide our future decisions.

We are aware of our emotions. Other people are aware of our emotions too because we unconsciously display them outwardly through universal facial expressions and body language that register, for example, our fear, anger, sadness, disgust, surprise, happiness. Thus, someone looking at your face can relate to your emotions and know something about your unconscious and your body state. But, surprisingly, the observer can often only infer why you have those emotions because the facial signals are insufficient on their own to inform them of the cause that resides in your body and brain.

Another surprising limitation to our feelings and emotions is illustrated by the simple prick on a finger discussed above. The difference between the newly sensed state of the finger and its reference state stored in memory is used by our unconscious to select the relevant feeling(s), stored in memory. It then associates those feeling(s) with the virtual middle finger of our right hand, stored in memory. The result becomes our conscious, unpleasant experience of a pin pricking the middle finger of our right hand, its current state.

We are left with a puzzle. This description implies that both the feelings we experience when our finger is pricked and the finger on which we experience the prick are separated from the real pinprick and the real finger both in time and space. Evidence for the separation in time is the delay in our reacting to a pricked finger, as discussed above. The evidence for the equivalent separation in space is provided by amputees who experience pain in a leg or finger that they no longer have. The body part registering the pain is located in their head. Our awareness (feeling) of pain in our middle finger is a virtual experience, not a direct experience of the process generating the sensation in our real finger (Searle 1995). More generally, our experience of a feeling or an emotion about the state of a body part occurs entirely in our brain: it is a virtual experience. Think placebo effect.

The unconscious appears to resolve both of these separation issues for us by including a corrective adjustment in its reconstruction of the experience, as discussed for the time delay. Thus, despite the separation in time and space, our brain provides us with a conscious experience of feeling a sharp pain and mild anxiety from a real pinprick occurring on the real finger in real time. So we pay attention to the real pinpricked finger, as intended. But, as the amputee example indicates, the unconscious manipulations are not always successful.

In overview, we are consciously aware of the image of our body created for us by our brain. Inescapably attached to it is the sensed state of our body, which we experience as a feeling and/or an emotion. To create this reality, our brain relies on the relevant record of our body parts and our feeling/emotions stored in memory. The fact that our conscious awareness of our body, its state and the association between them is a virtual experience in our brain both enriches and constrains the nature of our conscious awareness. For example, it leads to all of our conscious experiences, including perceiving the outside world and thinking about it, being associated, to some degree, with a feeling or an emotion (Damasio 2000).

**Memory.** We are all consciously aware that such a thing as memory exists. We can also accept that if it wasn't for memory, consciousness would be a never-ending stream of new, disconnected, isolated experiences. However, despite the importance of our memories, we are consciously unaware of how they are stored, organized, interrelated, selected and retrieved. As far as we are concerned, the existence of a memory and our awareness of it, even if we consciously decide to create and recall it, just happens (Kihlstrom 1987).

Today it is common to imagine that human memory functions like a computer: storing accurate, complete and permanent data on a digital memory device. This wasn't the analogy used in the mid- to late 1800s. At that time, memory was thought to work like the newfangled device called a camera: storing complete, accurate and unchanging information in a photograph (Daston and Galison 1992). Both of these ideas reveal more about our thinking and our cultural interests than they do about the actual functioning of memory. Think of the 1600s clockwork view of the universe.

We don't know the details of how the brain makes and stores a memory. It seems that memory making is a sensing-like event in which the sensed aspects of an item become part of our brain by changing it in specific ways: memories exist by the information affecting the brain's functioning. We also know, contrary to the computer memory device and photograph analogies, that these changes are not located in a single, isolated memory-storing structure. A memory consists of many pieces distributed among different parts of the brain (Damasio 2000). These memory pieces are connected or linked to one another in a manner that allows the memory item to be recreated when needed. Thus, when reading the following discussion, it will be useful to remember that neither a memory, nor the memory making and recalling processes are isolated modules attached to the brain. They are but an integral part of the brain and its functioning.

All memories include a few basic components. They include an item, such as an ice cream cone; an inescapable feeling or emotion associated with it, such as unhappiness; and an immediate context, such as dropped ice cream. In addition, and not suggested by the analogies mentioned above, a stored memory includes links to a variety of other memories that have some sort of relationship to the item being stored. These additional links enable our brain to go beyond simply recalling a specific memory (feeling unhappy about the loss of dropped ice cream…), they provide us with a larger context into which the recalled memory can be fitted (…with a delicious orange flavour similar to the one you ate when on holiday at the sea five years ago).

When you consciously try to commit an item to memory, you initiate an unconscious memory-making process. It is an extension of your brain's routine process of perception in which the sensed item's characteristics are separated from one another and distributed around your brain for analysis. Memory making is the process that embeds the item's separated characteristics, and the links between them, into your brain. Also embedded are the connections between the item's characteristics and both the associated feeling/emotion and the context.

In order to consciously commit an item to memory, you must pay sufficient attention to it so that all of the memory creation processes can be completed (Miller 1956a). Storing a lasting memory can take time and, unlike creating a digital or photographic memory, can require effort, as illustrated by the successive attempts that may be needed.

You may be aware from personal experience that both the success and the reliability of the memorizing act is also influenced by factors that appear to be peripheral to the process. For example, memory creation is influenced by our general feelings at the time of trying (Vander et al. 1998). Imagine attempting to memorize something when you feel bored. Compare that to your performance when you are excited about the item and have a strong desire to memorize it. Creating a memory is also influenced by the coherence of the information. For example, trying to remember conflicting information is difficult.

These types of issues mean that we tend to preferentially memorize views and experiences that are simple, coherent and important to us, regardless of whether they are true or representative of what actually happened. Remember our difficulty dealing with uncertainty or missing information and our response to it, mentioned above? These features help to explain part of our conscious experience of memory.

On the surface, the process of recalling a memory seems to be the reverse of making one. Recall is initiated when your brain receives a cue, trigger or prime that starts the unconscious process of finding the distributed parts and recreating the memory. As the primed memory grows, each newly recovered part and the developing memory itself act as cues for the recall of more pieces. The process continues until the memory is complete (Myers 1999; Damasio 2000).

Priming and memory growth are illustrated by a common trick used to find misplaced objects. You are more likely to prime your memory to remember where you missing car keys are if you physically retrace your movements over the period since you last saw them. If you mime your actions as you retrace your steps, then both your actions and surroundings will prime you to recall your thoughts and actions at that and an earlier stage in the past. You are usually rewarded by the past becoming clearer and you finding your keys.

But the unexpected can happen. You might find yourself suddenly contemplating a memory that appears to be unrelated to the task at hand. These memories appear because your efforts to recall a particular memory have provided cues to recall other memories that are only distantly related to the desired ones.

This hints at another complication for memory recall. The externally generated primes that successfully prompt us to recall a particular personal memory are influenced by our makeup, mood, our surroundings and our culture. For this reason, the prime for a specific personal memory is unique to you. It could be a smell for you but a visual stimulation for someone else (Hall 1981).

In contrast, the successful prime for the memory of an experience shared by many people is something common to that group. Think of Michel Jackson's glitter glove for those who attended his concerts. A special class of primes for the members of a group are their culture's symbols. These symbols can act as memory primes because a culture trains its members from a very early age to remember the common behaviours, beliefs and experiences to associate with a particular cultural symbol. This enables a display of the symbol, such as a Christmas tree or a flag, at a later time to prime the recall of the appropriate response (Cole and Cole 2001). The priming of group memories by symbols is thus one of the foundations of a functioning culture.

But, as hinted at, the process of recalling a memory is not the exact reverse of making the memory, which is why recall is deliberately described as a re-creation process rather than a re-assembly. Consider that, during recall, the required memory pieces are retrieved in the form they exist in when the recall takes place. This may not be their original form, the one when the memory was created. A difference can arise because stored memories display less integrity and permanence than we imagine. They are dynamic, both fading and changing with the passage of time. An original, older memory drifts as bits of it are adjusted to match new realities and linked to pieces of new memories. Similarly, the context tied to a particular memory can change as the memories providing it are altered (Gazzaniga 1998).

These changes can occur deliberately or inadvertently, and unconsciously or with conscious direction. For example, a memory can be (consciously or unconsciously) adjusted in order to fill in gaps, create gaps or meet the need for coherence. The reasons to do so can be emotional, such as the need to quieten a painful memory or to match an old memory to a new or changed feeling. These adjustments become an integral part of the memories we will recall in the future.

Even if we assume that the original memory parts are still intact at recall, the process itself could alter the memory. The context, knowledge, beliefs, feelings or expectations at the time of priming can all influence which pieces we will recall and how they will be used to recreate the memory (Myers 1999; Hock 1999b). Thus, even if a memory is retrieved accurately according to the right half of the brain, the left half may need to change or adjust its meaning. It may be desirable, for example, for the recalled memory to match the emotional and objective context that exists in our brain today (Gazzaniga 1998). These aspects of and limits to memory recall are exemplified by the fallibility of eye witness accounts in court cases and the ability of lawyers to influence the memories of a witness by asking suggestive questions (Myers 1999; Hock 1999b).

Despite these limitations, we have considerable, although selective, faith in our recalled memories. When we use a computer hard drive or USB stick as analogies for memory, we express our belief that our memories themselves are generally accurate and stable with few mistakes. This allows us to treat failures in storage and retrieval as aberrations, not the norm. As a result of our faith, the inevitable imperfections and biases in our memories that arise from our normal memory creation, storage and recall (re-creation) processes can have a greater effect on our lives than we acknowledge. They can affect our perception of self, our conscious experiences and our world view: they can significantly affect our conscious awareness of ourselves and our world.

Consider the downside of the innate ease with which we recall positive rather than negative memories (Taylor 1989). For example, the ease with which we recall positive features about ourselves joins with us being self-interested beings to form the root of aggrandizing self-deception (Myers 1999). Similarly, being interested in a particular subject, such as money, helps us to both create and recall memories related to that subject. But, in doing so, it also biases our memory and perception of the world toward that subject (Schneider and Bjorkland 1998).

There are also basic constraints on memory storage and recall that arise from our brain's finite size and its functioning. They affect our conscious awareness, so they deserve special attention. These constraints are illustrated by the limits to how much we can easily remember. For example, imagine being briefly shown a collection of different items from the same category, such as 15 different types of candy. A short time later we are asked to recall as many of the candy types as we can. We will, on average, remember no more than about 7 types. The "rule of $7 \pm 2$" expresses the limits to the item-capacity of our short-term memory, with the exact number depending on the category of the items. This rule applies to a wide variety of categories including dots, tones, numbers, loudness, words and colours (Miller 1956a; Miller 1956b). This is a significant limitation because short-term memory is one of the steps towards longer-term memory.

Humans have developed some intriguing techniques to deal with these types of memory capacity and processing limits. They include memory aids such as symbols, associations (context) or codes (e.g., words); dividing or aggregating data (chunking); keeping the context constant; comparing objects; arranging tasks sequentially; and expanding the number of variables used to classify each item. By applying these techniques, we can increase the amount of information and the ease with which we can store it in both short- and long-term memory.

Consider chunking, in which we break a complex memory problem into pieces that meet our memorizing limitations (Miller 1956a). For example, North American telephone numbers are seven to ten digits long, at the limit of easy memory. By printing the number in chunks of three or four digits and assigning each chunk specific tasks, such as an area code, the complete number becomes easier to remember.

There is a price to pay for using chunking techniques. At some point, their increased used results in an inevitable increase in errors during memory storage and recall. You can remember more information, but your recall tends to be more inaccurate and more incomplete (Miller 1956b). For example, although we might remember the chunked pieces of a memory, we may have had to omit some important details or, on recall, we may not remember the exact sequence of the chunks.

A similar problem arises with the use of symbols such as words and gestures. Symbols are summaries or representations of an understanding. They can help us to remember and recall the understanding they represent. No wonder they are one of our most important aids for both short- and long-term memory.

But symbols can be used to represent more than one understanding, so they can have more than one meaning, as illustrated by looking in a dictionary. Consider too the hash mark (#) representing both pounds weight, a number, and a tag on social media (such as Twitter's hashtag). We distinguish between these different meanings by paying attention to the symbol's details and the context in which it is displayed. This illustrates that when we use symbols, extra care has to be taken to ensure the needed detail and contextual information is both provided and clear. Otherwise the recipient will simply miss or misinterpret our intended meaning. A glitter glove may just be that, unless it is displayed in Michael Jackson's manner.

It should come as no surprise that we often miss or misinterpret the meaning of a symbol. Our computer-age world could provide many books full of examples. A classic example is the context-poor, short text message or email you send that is taken the wrong way by the receiver. For example, does "LOL!" mean "lots of luck!" (ironic or not?), "lots of love!" or "laugh out loud!"? Think too of the one word that primed an unintended reaction because the "I'm kidding!" context was lost. Then there are the opaque O and I symbols on the switches of electronic devices representing on and off (or is that off and on?). Push to find out.

A common thread running through all of these memory aids is simplification. This is the general method we apply to help us make sense of, remember and recall both simple and more complex or overwhelming subjects. However, there is a limit to the degree of simplification that can occur without a loss of meaning or understanding.

The downsides of simplification are illustrated by the unintended consequences of reductionist thinking. Reductionism generates largely isolated, easily remembered chunks of factual information about our world. But, if care is not exercised, the critical context that binds them together in a meaningful manner can be distorted or lost. For example, recalling the information related to your focus on increasing crop yields in a field without remembering that you are also creating heaven for the crop's pests could spell disaster. Think too of the now indecipherably brief notes you took at an important meeting, or your inability to implement a simple computer program task that was demonstrated to you as "8 easy-to-follow steps."

In overview, having a memory enhances our appreciation of the world around us. But the processes of memory creation, storage and recall, and their limitations, ensures that our memory is less reliable than we imagine. It is affected in many ways. This has a significant effect on our conscious awareness of ourselves and the world around us. This is especially true when we live in a complex culture and routinely use our memory when making decisions that significantly affect complex systems.

**Constraints on Conscious Experiences.** Our brain provides each of us with our personal sense of being conscious. But the nature of consciousness means that we are largely unaware of the mental processes providing it, as discussed above (Logothetis 1999; Lewicki et al. 1992). In particular, our conscious experiences only hint at the existence of our unconscious.

We are therefore largely unaware that our conscious experiences are the product of both our brain's unconscious abilities and their complementing extensions, which we know as our conscious abilities (such as our conscious decision-making tools, including logic). Consider that our unconscious can rapidly deal with unambiguous choices, while the flexibility of consciousness allows us to deal more efficiently with ambiguous choices, such as choosing between two different previously untasted cookies (Norretranders 1998). We are largely unaware that our conscious experiences also reflect the limitations to our brain's (unconscious and conscious) abilities and the influences on them. These limitations, as discussed above, are the result of the constraints inherent in the brain's functioning (its processing methods). This includes the feedback loop between the conscious and the unconscious. Although this feedback contributes to consciousness, it can also help to mask the limitations of consciousness from our conscious awareness.

To make the significance of these relationships clear to the brain's user, this subsection revisits some of the basic features of our brain's functioning. But it now looks at them from the perspective of our personal conscious awareness of our self and our world, that is, from the perspective of our personal conscious experiences. This should change how we view and experience consciousness.

Consider that your consciousness is focused and that this focus is largely on you, the present and the immediate task at hand. Being focused has the advantage that it allows us to bring more of our limited conscious resources to bear on a specific task. But, as a consequence, our consciousness receives only limited information about the context outside that task (Damasio 2000; Myers 2002a). We try to compensate for this loss by moving the focus of our conscious awareness and attention around in the same manner we use a flashlight.

We also rely heavily on the information and understanding we have accumulated and stored in memory over time (our accumulated experience) to provide us with a more complete and coherent conscious experience. But, as discussed, the information that enters our consciousness from memory has been affected by many types of (conscious and unconscious) influences and limits during its creation, storage and recall.

Our conscious experience is also constrained by the nature of memory itself. Consider that all memories are unavoidably accompanied by feelings and emotions. They influence our acceptance of information, the events we choose to focus on and our decision making. This too affects our conscious experiences and our reaction to them.

It is possible to argue that reason can control the influence of our feelings and emotions. But, in reality, it can do so only to a modest degree. Feelings and emotions are so strongly tied to our conscious decisions that, although we can perhaps be logical without feeling, we cannot be rational without feeling (Damasio 2000). The Western cultural ideal of making emotion-free decisions is unrealistic. The idea of being completely objective is unobtainable.

Our conscious experiences are further constrained by how they are provided to us. We must first sense, then our unconscious must manipulate the sensed information, before we can have a conscious experience. This takes time. Thus, in addition to consciousness being focused and tied to memories and emotions, it responds more slowly than the unconscious. In particular, it takes between 0.25 and 0.5 seconds for a simple event to be sensed and processed into consciousness. This time delay imposes limits on our conscious experiences.

Firstly, if an event repeats itself faster than around every 0.25 to 0.5 seconds, then it becomes distorted or ignored by our conscious self (Norretranders 1998).

This type of speed constraint has implications for those of us in our fast-paced world who think they can improve efficiency and save time by multi-tasking. The brain takes an irreducible amount of time to change tasks. If we switch too rapidly or too frequently between tasks, then we will suffer a drop in performance and an increase in errors (American Psychological Association 2006). Aircraft maintenance personnel, those who use cell phones while driving (or walking), and office workers who feel obliged to instantly answer incoming texts and emails while completing other tasks, take special note.

Secondly, by the time we are consciously aware of an experience it has already passed. Although strictly true for all events, this effect only becomes noticeable and important for fast events. In these cases, you are likely to be faced with making a conscious decision based on information that may no longer be relevant. You can experience this phenomenon by using a computer game that flips electronic playing cards on the stroke of a specific key. If you hit the key at a constant flipping rate for a while, that task will quickly be performed unconsciously. Now add a new rule that says, stop flipping if a colour card appears. Once you have mastered this at a slow speed, slowly increase the card-flipping rate. At some point you will find that you stop flipping cards one card too late. An alternative to this experiment is to try playing baseball with professionals.

The limit to us experiencing fast events points to a more general constraint on our conscious experiences. A conscious event is constructed in the brain by the unconscious from signals that originate in different parts of the body and memory. Each signal has its own transmission and processing delay. To avoid the impending chaotic conscious experience that this mix would provide, your unconscious synchronizes the signals and adjusts your memory so that you experience a single event (Klein 2002; Libert 1999). The unconscious also makes the event intelligible by completing a basic interpretation of it using the relevant cues in the signals and the information in your memory. In effect, the unconscious ensures that the event makes basic sense to you before it enters your consciousness, before you consciously experience it.

The need for these routine, apparently innocuous "adjustments," "normalizations" and analysis by the unconscious can have profound effects on our conscious experiences. If the unconscious faces difficulties completing these tasks, the brain will simply provide its best guess. Examples of the left brain failing to provide us with a coherent, intelligible conscious experience are those which we easily recognize as illusions or hallucinations (see below). But these simple examples are only the tip of an under-recognized iceberg.

Consider our comments concerning an event about which we have limited knowledge and no personal experience, such as a news report about a horrendous accident in which a single car drove off the road or a news report of a drop in the stock market. Regardless of the complexity of the event or our lack of knowledge about it, our unconscious will use its models and biases to rapidly construct an explanation that, to our conscious, appears largely coherent and logical. We then tend to believe that this explanation is basically true. For example, the driver was probably drunk or fell asleep, and the drop in the stock market was due to profit-taking by the traders. Once we have formed these views, they are difficult to change.

Other basic constraints on our conscious experiences arise from the limits to the amount of information that the brain can process and how it is processed. The greater processing speed of the unconscious allows it to receive up to about 11 million bits of information per second (bits/sec). It uses this data for simple decision making such as control and simulation. On the other hand, consciousness, which makes more complex decisions, can process a maximum of about 40 bits/sec. In reality, it only deals with about 1 to 16 bits/sec. Therefore, for information to move from the unconscious to the conscious in a timely manner, it must be reduced by about one millionth (Norretranders 1998).

The information reduction takes place in the unconscious through selection and summarizing, that is, through simplification. But these processes are biased rather than even-handed and random. The bias is, naturally, toward ourselves and our personal preferences, and toward our human need for coherence. As a result, our conscious experiences are likely to be based on a coherent representation of the world that is biased to our liking rather than one that is messier, less comfortable but more reliable.

Even if the reduction processes were more even-handed, our conscious experience of the world would still be subject to limitations. This occurs because, for example, removing information from a complex scene automatically results in the appearance of order and organization in the remaining data (Norretranders 1998). For this reason alone, we may perceive order and

organization around us whether it exists in the original information or not. This outcome is enhanced by our innate desire to see pleasing patterns in everything, even in random data. Think of cloud animals.

Overall, the amount of information received by the conscious from the unconscious has been reduced to a manageable level. But in the process, its original meaning and context becomes, to some degree, altered, biased and lost. This limits and biases our conscious experience of our self and our world. Because of these many constraints, our conscious experience of the world is not quite what we think it is.

In overview, our conscious experiences are affected by many factors other than what we sense about our current state and our surroundings. They include the limitations to both the brain's memory capacity and its processing ability, in particular its memory creation, storage and recall processes. Our brain must substantially reduce the information made available to our conscious and synchronize sensings separated in time and space from their source. Our conscious experiences are also affected by our innate tendency to strongly focus our attention and favour a perspective that concerns ourselves. There is also the inescapable presence of feelings, emotions and the (left brain's) drive for coherence.

Thus, despite our feeling (belief perhaps) that our conscious experiences are unbiased and coherent, these factors ensure that this is significantly not the case. We can't remove these constraints on our experiences, but we can improve the reliability of our experiences by taking them into account. In general, we could treat our conscious experiences as more biased and error prone than we expect, and our faith in them should be conditional because they are only true within limits.

To summarize, our personal conscious awareness of ourselves and the world around us is based on our brain's memories and its ongoing sensing of our body state and surroundings. The brain's unconscious processes compile, manipulate and interpret our sensings with the help of information found in memory. The nature of these processes means that there are limitations (including deficiencies) to the products, such as memories and decisions, that our unconscious generates.

Our brain's functioning uses the products of its unconscious processes to provide us with our conscious awareness (experiences). In doing so, it propagates their limitations, but it doesn't automatically make us aware of them. Our brain's functioning also augments its unconscious processes to provide us with our conscious abilities.

It is easy to conclude that our conscious experiences are not a highly accurate or complete representation of ourselves and the world around us. We are consciously aware of a simplified (selected), biased, rationalized, "coherentized" representation of them (Norretranders 1998). This does not mean that our conscious experience is a fraud. It is just not as reliable as we think: it suffers from limitations.

Consciousness gives us the ability to focus, to be aware of conflicting options, to choose between them and to imagine. It allows us to enjoy a sunset, companionship and the satisfaction of solving a puzzle. It enables us to create and comprehend a more complex model of the world; (Norretranders 1998) and so become aware of and inhabit environments we are biologically unsuited to (Damasio 2000). But these benefits of consciousness come with limitations to our resulting conscious awareness and experiences. We thus inescapably bear the burden of deciding which aspects of consciousness reliably represent us and our world, which don't and how significant are the differences. This has significant implications for our tendency to believe that our conscious decisions are appropriate, rational, coherent and unbiased.

## Perceiving the World around Us

The preceding description of consciousness, its limitations and its feedback linkage with the unconscious helps us to understand how our awareness of our self, ourselves and the world around us appears. However, if we are to apply this knowledge to our day-to-day decisions, it must be made relevant to our everyday lives: it must be personalized. This can be done by considering the simplest of our conscious experiences, those provided by our senses. After all, it is these sensory experiences that form the real-life foundation of our conscious awareness of the external world and the decisions we make to live in it and affect it. The power and limits of our sensory perceptions will be illustrated by sight.

**Sensing and Perception.** A popular model for both our visual sensing (the detection) and our perception (the conscious awareness) of an object is an instantaneous camera. In this model, the light from the object passes through the lens of the eye to generate an exact but inverted image of it on the retina at the back of the eye.

The retina is composed of rods and cones that detect the image and its colour. An exact replica of the image is then passed to the brain along the optical nerve. In the brain, the received image is compared to our memories of other images in a manner analogous to the classic method of fingerprint identification. When a match is made, we perceive the object for the thing it is. This model is somewhat correct, but in detail it poorly represents the process.

The process of perceiving an object using sight is complex. It starts when the inverted image sensed by the retina is partly analysed in the eye. For example, colour is distinguished from black and white. The information sensed by each eye is then split up before it travels along different paths to the geniculate nuclei in the relevant half of the brain. From here, the signals move to the primary and then the secondary visual cortex at the back of the brain for further analysis (figure 1). During the signal's travels, it passes through a number of secondary structures, each of which analyses it for different characteristics.

The process of sensing and analysis conducted by these connected structures is dynamic. As long as your eyes are open, the brain coordinates their continual information gathering by using its ongoing analysis of the eyes' sensings. Consider that when an object of interest is sensed, the brain-eye combination first attempts to locate the features that define the object, such as its edges and whether it is moving in a specific direction (Kast 2001). Then the centre of each eye is directed to continue scanning the object, primarily along those edges identified by the initial analysis. At the same time, the periphery of the eyes is sending to the brain contextual information, such as the object's surroundings and its relationship to those surroundings. The cues from the periphery help prime the brain to recognize features, such as the depth of field, shape and orientation of the object, that are needed to identify and to interpret it within its context. For example, the top of the object will be identified as that part nearest to the light source because our brains assume that the light falling on an object always shines from the top (Norretranders 1998).

The actual process of sensing and perception is far more complex than even this description suggests. Consider that the process also includes the following body and brain involvement. We perceive the image in front of our eyes as a continuous whole even though this is not how our eyes sense it. For example, the image each eye produces is slightly different. Each has a hole in it that corresponds to the place on the back of the eye where the retina connects to the optic nerve (Nesse and Williams 1998). (There are no light sensitive cells on this connection area, so the eye cannot sense the part of the image that falls there.) How can it be that there is no hole in the image we perceive? The brain compensates in two ways: it reduces the response of the brain cells that would detect the blind spot, and it fills in the holes by merging the separate but overlapping images from our two eyes (Haseltine 2000).

Even more intriguing is how the brain ensures that we remain largely unaware of the routine complete disruption of our seeing process that occurs when we blink. In this case, the muscle action of blinking triggers the brain to shut down both the process that recognizes changes in an image and the process of perception. The blinking event is thus completely masked from our awareness (Bristow et al. 2005).

This dynamic sensing and analysis process continues until the unconscious has gathered enough features and cues that it can use the information in memory to provide us with an "image" of the object and an identification (Cole and Cole 2001). The compiled image is then placed within its context and joined with any additional information, such as feelings, to become our interpreted visual and emotional experience of the object. It is only at this last stage in this time-consuming process that we become consciously aware of (i.e., we perceive) the identified object in its surroundings.

There is one other aspect of our sensing and perception to consider. All of these processes are subject to the influences from the wider context in which our seeing takes place. They include our emotions, memories, cultural norms and even the complexity of the process itself. These influences affect many features of what we perceive, from the things we choose to look at to how our brain interprets our sensing of them. For example, our perception of a standing person with a raised arm meant to represent the social symbol for stop will provoke different emotional and action responses depending on whether the other hand is holding a gun or a beer. These influences can also result in us not perceiving things or perceiving things that are purely imaginary, such as cloud animals. Even more worrying are those distorted perceptions that we accept as real and use to make life- or environment-changing decisions.

We have confidence in a sense-based perception of the world because of our extensive personal experiences using it, the supporting evidence provided by our other senses and the reinforcing input from our companions.

In the case of sight-based perception, I have confidence that that keys I look at and touch as I type these words are indeed real. However, we should temper our confidence in our perceptions with the recognition that most of the processes of sensing, analysis and interpretation required to provide it occur in our unconscious. We are therefore consciously unaware of the processes, their limits, the influences on them and the resulting limitations to our perceptions.

**Illusions.** The fallibility of our confidence in our ability to accurately perceive our world is most dramatically illustrated by optical illusions. In order to provide the reality we perceive, the brain manipulates the data it receives from the world and compares it with our mental models (experiences) of the world in memory. For example, our childhood experience teaches us that even though the view of an object from the side might look different than the view from the front, only one, not two, independent objects is being looked at (Cole and Cole 2001). As adults, we perceive these multiple views as one object because the information coming from the object and its surroundings primes memories of what we learned about the world from our previous experiences.

If we receive cues from a scene that are contradictory or ambiguous, or if we do not have sufficient experience to accurately interpret them, then we could experience a visual illusion (Rock 1974). For example, if the cues suggest two equally likely but different interpretations for an image, as occurs for a Necker cube, then we can perceive first one image, then the second, as the brain flip flops between the two while searching for an unambiguous interpretation (figure 2a) (Gibbs 2001a; Wilson et al. 2001). If, on the other hand, the cues are all provided but the object doesn't exist, then we will still see the object, as illustrated by us seeing the undrawn triangle in figure 2b.

Alternatively, if a primary or reference cue from an object is either missing or not in the location expected by our brain, then the object will be perceived or interpreted incorrectly. For example, we expect the illumination for a landscape to come from the top. Thus, when an aerial photograph of a landscape (taken looking straight down) is turned so that the illumination comes from the bottom, then valleys will be seen as hills.

The basic process of sight perception and the types of influences on it are common to all forms of perception. As a result, there is ample room for us to perceive a reality in our day-to-day lives that is different from that which actually exists. Some of these experiences occur inadvertently, such as misinterpreting a cultural symbol. Others can be deliberate, as occurs when magicians, advertisers, politicians and others with an agenda take advantage of our brain's processing foibles to influence our perception of specific things, including not perceiving them. Our confidence in our conscious perceptions of visual and other sense experiences should be qualified and tempered with skepticism. We should pay particular attention to deducing what we likely don't know or can't verify: what our brain fills in.

Colour. The degree to which we should qualify and restrain our confidence in our conscious awareness of ourselves and the world around us is illustrated by the limits to and influences on our perception of colour. Our eyes sense the light reflected from an object. The colour of an object varies with the wavelengths of light it reflects.

The retina of the human eye contains about 120 million rods. They reveal brightness and, when the light is dim, distinguish black from white. The retina also contains about five million cones, which reveal colour and, in brighter light, sharpness. There are three types of cones. Each is sensitive to a different, but overlapping, wavelength range of visible light that we associate with either yellow, green or blue-purple. But, despite our experience, the cones don't detect colour because it is not a property of light (Smith 2005). We are left with the questions, what is colour? and how do we come to perceive it?

Surprisingly, colour, like feelings, originates in memory. Colour and its brightness is our brain's method of informing us about the wavelengths and the intensity of the light that our eyes have sensed. An early explanation for how we saw colour suggested that when light reached an eye, each of its three cone types responded by cueing us to recall from memory a different amount of the three primary colours: red, green and blue. The mix of the three was then thought to give us the appropriate colour of the object. But this naming, and the implied explanation for colour perception, is misleading.

The process of perceiving colour, as it is now understood, is as bizarre to us as it is difficult to accept. Imagine looking at a red tomato in a sunlit garden. The tomato's image falls on a small area of our eye. Within that area, cones detect its presence. The signal from each of the three cone types are combined in a specific

manner to form the spectral signature of the tomato: its colour signature. In addition, sensings from all of the eye's cones are combined to form a spectral signature for the light illuminating the garden: the illuminating signature. Our brain uses both the tomato's colour signature and the illuminating signature to assign a colour from memory to the tomato (Fairchild 2005).

Simplistically, the colour assignment process, and thus our perception of the tomato's colour, can be thought of as our brain using the two signatures to select a colour tone (hue) from the rainbow of colours and decide on how much white to mix with it (colour saturation). A tomato is deep red because only a small amount of white has been "added" to its red-indicating signature, while the sky is light blue because more white has been added to its blue-indicating signature.

The actual assignment process is more complex than this and imperfectly known. How complex is illustrated by considering a few of the other constraints and influences on our brain's assignment of colour to objects. In order for us to perceive objects within a scene in full colour, the illuminating signature must include the wavelengths that are detected by each of the three cone types. This condition is satisfied if the scene is illuminated by white light, such as sunlight and white LED lights. But the illuminating signatures of these two sources are different, and the nature of sunlight's illuminating signature changes with time between, say, a sunny midday and a cloudy afternoon. As the illuminating signature changes, so does the object's colour signature. Yet, remarkably, despite these changes, our perception of an object's colour remains somewhat constant. Our unconscious is making quite complex compensation "decisions" when assigning colours.

How it makes them is not fully understood. Of importance here is that the brain's use of both an object's colour signature and the illuminating signature to assign the object a colour from memory is influenced by a number of other factors. They include the memory of the colour assignment made in the past. The strength of that memory's influence depends on our familiarity with the object we are looking at. It seems that the more familiar the object, the more likely the brain will be cued to recall and use the colour assigned to it in the past. These influences can result in a biased perception of colour. For example, we tend to see lawn grass, which we are very familiar with, as being greener than its reflected light suggests (Fairchild 2005; Lotto and Purves 2004).

The brain's assignment of colour is also influenced by the colour context in which an object is viewed (Land 1959; Gage 1999). As a result, paint chips and computer colour-matching techniques are limited in their ability to help us, for example, paint a room a particular colour. Consider that the size of the paint chips and any colours adjacent to them have a strong influence on your perception of their colour and thus your choice of a particular chip. After the room is painted the colour of the selected paint chip, these influences will help to determine how closely you think the room's colour matches either the colour you wanted or your chosen paint chip (Fairchild 2005; Lotto and Purves 2004). These influences mean that our selection and matching of colours has as much to do with eye-brain functioning and the influences on it as it has to do with the precision of colour computers and pigment mixing.

The question arises: what is the real colour of the object we are looking at? To answer it, we need to remember that our ability to detect light's wavelengths is limited. For example, our blue-purple cones poorly detect light wavelengths in the purple wavelength range. Thus, although the visible spectrum we measure for the sky suggests we would perceive it as having a purplish blue colour, we perceive the sky as blue (Smith 2005). Which of the two colours for the sky is the real one: the sky's wavelength spectrum or our colour experience of it? The external reality or our perceived internal reality? This conundrum is a prime to the more general issue of deciding on truth, which will be addressed in a later chapter.

It comes as a surprise that our perception of an object's colour neither accurately represents the wavelengths of light our eyes receive from it nor solely depends on the object's colour signature. The colour we perceive for the object is also determined by the signature of the light illuminating the whole scene. It is also influenced by the context in which the object's colour signature is generated and by our memory of the past colour assigned to its colour signature or to the object itself.

When we use the words *choice* and *like* to describe our use of colour, we should remember that there is more to our perception of colour than the brain's assignment of a colour. Consider that our perception of a coloured object always includes the feelings we have learned to associate with that colour and the object. For example, compare your immediate feeling about red blood to that of a red tomato.

Our culture's world view also exerts a strong influence on our perception of colour. Consider the number of colours we see in a rainbow and our relationship to them. (Excluding the diffuse banding caused by the wavelength detection limits of our three cone types (Simpson 1999).

For example, Western Europeans between 1200 and 1600 perceived between 4 and 11 colours in the rainbow (Gage 1999). The current Western cultural idea that there are seven colours in the rainbow had its immediate origin with Isaac Newton. In 1675, he demonstrated that a prism can split white light into a rainbow, and then, by passing that rainbow through a second prism, he recombined it into white light. After this success, he was still left with the problem of deciding how many colour categories he should assign to the rainbow.

In his search for an answer, Newton was influenced by his culture's religion, its notions of order and harmony and by its (and his) preoccupation with the search for practical ways to generate coloured paints using the limited number of pigments available at the time. His Christian yet occult beliefs resulted in him choosing the seven colour categories of red, orange, yellow, green, blue, indigo and violet. His choice matched European music's five whole tones and two semitones of A, b, C, D, E, f and G, with red corresponding to A. By making this decision, he satisfied his cultural and personal concept of the ultimate heavenly harmony in God's universe. An order that was represented by the seven musical notes. Through this cultural legacy, Westerners still see seven colours when we look at a rainbow, and we still talk about a colour's tone.

The impact of cultural influences on our personal perceptions and preconceptions about colour extend well beyond Westerners unquestioningly seeing seven colours in the rainbow. For example, the association of pink with female and blue with male characteristics by Westerners has its roots in the German Romantic period (late 1700s to early 1800s). The Romantic view of colours was dominated by abstract and symbolic ideas rather than by concerns about the mechanics of colour perception. For the Romantics, colour symbolism covered a wide variety of features including human gender and national characteristics (Gage 1999).

Colour symbolism is not restricted to Westerners. It is found in all cultures. For example, the colour green is often used in Muslim countries to represent Islam and is associated with their paradise (Surahs 55:76 and 76:21). Thus, when the Americans naturally labelled their safe zone in Baghdad the Green Zone, they helped to create an Islamist suicide bomber's dream target.

In summary, our ability to visually sense and perceive provides us with a wonderful experience of the world around us. However, the presence of illusions and our perception of colour illustrate that there are many influences, from processing limitations to our culture's world view, which limit and bias what we see. This ensures that our experiences and our awareness of the world around us is an inexact representation of it.

These kinds of limitations and biases apply to all of our methods of sensing and perceiving the world around us, which has significant consequences for our relationship with it. After all, our perceptions play a key role in the formation of our internal reality and our making of decisions. These consequences are made clear in the next section.

## Consciousness and Reality

Our brain's functioning constructs our internal reality from the following: our sensing of both the state of our body and the world around us; our virtual body image; the associated feelings; and the other relevant information in memory. That internal reality exists largely in our unconscious. We consciously experience only a small part of our internal reality, but it allows us to knowingly live a life at the boundary between our internal and the external reality.

However, living in the world this way comes with limitations. They arise from both the characteristics of the brain's processes that provide the internal reality and the influences, such as cultural norms and past experiences, on those processes. These limitations were illustrated by considering our perception of objects and their colour. Thus, although the conscious experience our brain constructs for us is wonderfully real and coherent, it is still an imperfect and incomplete representation of ourselves and our surroundings.

Despite these limitations, our internal reality and our conscious experience of it serves us well under many circumstances. It hardly seems to matter that in our internal reality the sky is blue, while in the greater external reality the sky has properties of (or should be seen as) purplish blue. However, these limitations can become significant when we use our brain's processes and our internal reality to make decisions.

When our internal reality is seen in this context, it adds a burden to our consciousness: we are responsible for deciding which parts of our conscious awareness and experiences are true outside of our brain, which are not and when the deviation/difference matters. We are obliged to pay attention to what we likely don't know or can't verify: what we imagine or guess. This is particularly important when we are making decisions that will affect how we live and change our world.

However, because our internal reality is our reality and we are conscious of only a small part of it, we tend to simply believe that our decisions are based on a normal accurate representation of the external reality and that we made them in a logical and rational manner. We treat our decisions as being by and large satisfactory. Without due care, we simply won't know that our decisions are inappropriate and our internal reality flawed until after we suffer the rude awakening that comes when the events we expected to happen didn't, or when the reality we thought was normal or real wasn't.

Consider how your view of the world is disrupted when your idea of normal sight is challenged by the experience of colour-blind people. We think that they have a sight experience the same as ours, minus the colour. This is not the case. A colour-blind person experiences the world differently. They have a different internal reality. Aside from being colour-vision "impaired," they have a heightened sensitivity to brightness and a loss of visual resolution. As a result, they have a greater appreciation of grey tones and a more acute night vision. To the colour-blind, their world is normal until they are made aware of the more common ability of humans to experience colour (Sacks 1996). If more people were colour-blind, would their view of the world be correct (true) or would it be just their accepted (perhaps defective to some) norm?

It is difficult for us to know the truth about the external reality, but this doesn't mean that we can't make appropriate decisions. It just means that, to do so, we have to devote more care and attention to our decision making. Within this context, the idea of finding "The Truth" by routinely disrupting the brain's functioning (whether through chemical or other means) in an effort to alter or expand our consciousness is misleading. Our effort is better directed at continuing to broaden our appreciation for the effects that our brain's normal functioning, its limitations and the influences on it (such as culture) have on our internal reality, our conscious awareness and our decision making. Our effort could also take these features into account in our thinking and decision making. We will therefore leave the question of what is the "real normal" and the argument about the true colour of the sky to prophets, ideologues and philosophers and move on.

This chapter used the multiperspective constructivist method to help us gain a more complete appreciation of the brain's complex structure and functioning as a whole. This method will be used to contemplate the complex topics discussed in the following chapters because it provides the clarity of simplicity while adequately representing complexity and the limitations of our understanding. It will be joined by other methods that will be introduced as needed. If we use these methods, pay the required attention and take the needed care, then they will help us increase our appreciation of the topic and retain in memory the knowledge we gain. We will be helped if we remember the biases and limitations to our (unconscious and conscious) awareness outlined in this chapter. The next chapter discusses what we inescapably do with our awareness of ourselves and the world around us: make decisions.

# CHAPTER 2

# DECISION MAKING

Our decision making is the process by which our brain selects, manipulates and interprets information about an issue, as discussed in chapter 1. The information used is our sensings of our self and our surroundings (including our feelings and emotions). Our decision making is influenced by how our brain reduces the information it must deal with and how it forms, retrieves and maintains our memories of it (Kahneman 2003). From this perspective, our decision making is constrained by our brain's characteristics.

Our decision making is also influenced/constrained by other factors. The most basic is that humans are a unitary organism whose decisions are innately focused on our individual human concerns, such as the need to acquire the resources to satisfy our needs and wants. Humans are also social animals, so our decision making is also influenced by our dynamic interactions within our social subgroup, such as our standing in the community and mate selection.

Overall, our personal decision making is influenced by our particular characteristics as an individual and by our parents, our physical environment and our culture's world view. These affect our decision making in wide-ranging, profound and long-lasting ways. Consider that, as growing children, these influences affected how our brain developed and the formation of our unconscious habits, scripts for behaviour and patterns of thoughts (Cole and Cole 2001; Gopnik et al. 2001; Gerhardt 2004).

It is relatively easy for us to accept that the world around us affected our decision making as we grew up. However, as adults, we find it less easy to accept that our surrounding can still actively influence and constrain our current day-to-day decision making. We feel this way because we are largely unaware that the world around us provides us with more than information (*sensu stricto*). It also provides cues to our unconscious that selectively activate thoughts, behaviours and processes that were formed or learned when we were children.

Regardless of how we personally view our decision making, experiments indicate that our decisions are inherently biased (including the errors of process and omission) in some way. Fortunately, many of the biases are of minor significance (Myers 1999). Is the sky blue or blue-purple? However, there are enough occasions where these biases can result in a significant impact on our lives, so their existence should be taken seriously. Consider that we are predisposed to decide in favour of eating tasty, energy dense foods. In an age where these foods are super

abundant, that bias contributes to the global obesity epidemic. Think too of the other lifestyle diseases. But recognizing and appropriately dealing with our decision-making biases is more difficult than we expect, as the following illustrates.

Our brain creates for us our marvellous, conscious experience of the world. However, the gift of a personalized, human-focused reality comes with a cost: we lose the innocence of just doing (Damasio 2000). We are unavoidably faced with choosing between following either our unconscious urges or our conscious will. We want to lose weight, but we feel drawn to the nicely decorated cake. Making these types of decisions is difficult because it requires us to face and question our personal emotions and drives. No wonder we often feel better when we just act (Norretranders 1998).

The conundrum we face when trying to make a choice between urges and will is not unique. Consider that we live at the boundary between the internal reality created by our brain and the external reality. Both are real in the sense that they both exist for us and are intertwined in our day-to-day lives. However, our decisions about how we should act in the external reality are made in our internal reality. If we are to make appropriate decisions, we are obliged to face the conundrum of which reality to pay attention to. We are also required to question how reliably our internal reality's features and decision making represent the external reality (Searle 1995). Think of the number of colours in the rainbow.

We think that we can resolve conundrums like these by diligently applying our conscious decision making. However, we face significant constraints in doing so. This chapter discusses those constraints by expanding on the discussion about the brain in chapter 1.

Our brain's functioning can be divided, for convenience, into its unconscious roots from which our conscious arises. The decision-making processes that occur as part of our brain's unconscious functioning are automatic, fast and inescapable. These unconscious processes provide both the foundation on which our conscious (cognitive) decision-making processes are built and the decisions/information on which they work.

In contrast, our conscious decision-making processes, which we associate with logical and rational decision making, occur slowly and are largely used at our discretion. If our unconscious decisions are firm, then our conscious decision making tends to accept and use them as is. Our conscious decision making comes into its own when an unconscious decision is uncertain or ambiguous, which often means complex. Our conscious decision making usually produces more considered decisions.

This discussion indicates that, like our perceptions discussed in chapter 1, our unconscious and conscious decision-making processes are interlinked parts of a single complex decision-making process. We are reminded of this when we experience the swirl of feelings that comes with trying to resolve the urges-will conundrum. Scientific research informs us that our decision-making is an inherent product of the innate functioning of our brain, and subject to its constraints and biases. Think of viewing the Necker cube and the missing triangle (figure 2). It also informs us that most of our decision making occurs in our unconscious.

These characteristics have some significant consequences. It is difficult for us to use conscious self-reflection to identify either our decision-making processes or their constraints and biases. It also means that we can't reduce those constraints and biases by simply altering the responsible processes or getting rid of them. They are part of who we are. If we want to make more appropriate decisions, then we are obliged to become more familiar with what scientific research has discovered about our decision-making processes, their limitations and biases. We can then concentrate on forming techniques to help us take them into account and recognize when it is critical to do so.

With that in mind, the focus of this chapter is the characteristics and limitations of our decision making. Unfortunately, the complexity of our unconscious decision making and our incomplete understanding of it means that we are faced with a fuzzy, somewhat incomprehensible picture. To make the discussion easier to follow and more personally relevant to the user, only a selection of its many processes are presented, and they will be discussed within a familiar framework. It consists of the three broad steps we commonly use to describe conscious thinking and decision making about information: references, reasoning and the drawing of conclusions.

The holism of our brain's functioning and decision making discussed above ensures that this division into three steps is a convenience with limitations. For example, establishing/choosing a reference is treated here as a step all on its own, but it can be intimately intertwined with the other parts of thinking and decision making. Similarly, although the aspects of anchoring and adjustment are only mentioned in the reasoning step, and preconception is only mentioned in the reference step, both are relevant to all of the decision-making steps.

This discussion and its layout should therefore be treated as a series of example snapshots illustrating aspects of a more complex, interconnected decision-making process. It is aimed more at helping us to appreciate the complexity, constraints and biases of our conscious decision making rather than describing exactly how we make a particular decision. The holistic nature of our decision making will be made more clear in the second-to-last section of this chapter by discussing our decision making as consumers. An introduction to how we might direct our conscious decision making to better reflect the external reality is provided at the end.

## References

Because our (conscious and unconscious) decision making is based on comparisons (of items and patterns), a key part of making a decision is establishing a reference (Tversky and Kahneman 1981; Gopnik et al. 2001; Kasser 2002). We increase our appreciation for our decision making by knowing something about how we establish these references, their nature and their influence on our decision making.

In overview, our references can be divided into various types. We are born with the references that our brain uses to interpret our basic sensings of the world around us. Many of these references, such as those used for determining the depth of field during sight, are universal (the same for everyone) and fixed. We acquire additional references from our culture, our colleagues and our environment through learning experiences or absorption. The essence of the core references we acquire in this way, such as displaying respect for elders, is universal and enduring. But, in detail, they are customized for each culture, and those details can change. Our unconscious also generates references on the fly from our memories and our other references. They are the more specific personal references we use to make many of our day-to-day decisions, such as buying a new dress or choosing what to eat. These references are usually specific and short-lived or relatively easily changed.

When all of our personal, longer-lasting decision-making references (set points, views, visions, norms and beliefs) are considered collectively, they can be treated as our personal reference normal. Our shorter-term references may be included in or associated with this reference normal as context requires. We treat our reference normal as reliable and accurate.

This section illustrates our reference normal's component references, their nature, and the limitations and biases (including errors) that can result from their use. The discussion is divided into four subsections: our sensing of the world, our culture's world view, our membership in a subgroup and the biases in our personal perspective of the world. This division too is only a convenience because, in reality, when making decisions, we use a mix of references of various types that modify each other's influence on our decision making.

**Senses.** Our senses provide us with basic information about the world around us. However, the limitations to our sensing methods ensures that the information they provide to us is less complete and accurate than we imagine, as discussed in chapter 1. Additional biases (including errors) can occur when we use our sensing references to interpret our sensings. Think of the Necker cube.

Because the process of perception takes place in our unconscious, we are often unaware of these biases in our conscious perception of the world. It is less complete, certain, accurate and consistent than we experience or believe. For example, when visibility is poor or when the sun is in front of us, we overestimate the distances to objects. But when visibility is good or the sun is behind us, we underestimate the distance (Tversky and Kahneman 1982c). Similarly, distinctive and vivid or spectacular events tend to capture our attention more than subtle and common features, even though the subtle events may be just as important for decision making (Myers 1999).

Consider too that the relative importance we attach to our senses is itself a reference for the value our unconscious and conscious give to our sensings. In general, we tend to give more weight to the things we see than the sounds we hear. We give less weight to smell and the least weight to taste (Gardner and Stern 1996). In fact, sight dominates our perception to such an extent that vivid and concrete images and reports of visual experiences are often more influential in swaying our opinion than abstract and logically compelling ideas or arguments (Myers 1999). Sight is so important that we can be said to live largely through our eyes rather than through reasoning. The phrase "seeing is believing" captures its significance.

**Culture.** We live our lives embedded in a culture, so it should be unsurprising that many aspects of our culture's world view (such as its beliefs, values and explanations) become part of our lives as we grow up. Because of our close familiarity with our culture's world view, the aspects we absorb become buried in our unconscious as part of our reference normal. Through this connection, our culture's world view has a strong, ongoing influence on our personal decision making. However, because its aspects mostly reside in our unconscious, we largely underestimate and under-appreciate that influence.

The strength of its influence is revealed by how strongly the decisions of a culture's members reflect their culture's world view, as illustrated by comparing farming to hunter-gatherer cultures. The world view of hunter-gatherer cultures guides their members to preferentially focus their attention on the landscape around them in its unaltered state. The members' decisions reflect the strong influence of this guidance by the high value they place on having close ties to the landscape around them and their limited effort to transform it. In contrast, the world views of farming cultures guide their members to place considerable value on transforming their natural environment. The members reflect the strong influence of this guidance on their decisions by their limited regard for the unaltered landscape and the high value they place on their human-built environment: their fields, buildings and domesticated animals (Wilson 1988).

A more personal example of our culture's influences on our reference normal is our tendency to value things as "good" if our culture recognizes them as beautiful (Myers 1999). Similarly, when Westerners evaluate a dog, one of our focuses is on the dog's intelligence. We do so because our culture's world view guides us to both value intelligence and use it as a reference when evaluating both ourselves and other living things (Donaldson 1996).

A culture's influence on its members' reference normal is not fixed, it is dynamic and ongoing. For example, in the early to mid-1900s, Western culture guided its members to see gorillas as metaphors for the dark side of human behaviours: the epitome of the aggressive alpha male in a competitive society. By the late 1900s, the Western world view had changed. Its members were now guided to see gorillas as a shy, gentle and friendly animal (Browne 2006).

Consider too how Westerners' personal beliefs about the relationship between our minds and our health, such as the placebo effect, hypnosis and positivism, changed as our culture's prevailing views on the subject changed. For example, mesmerism appeared in the 1600s as one of the Enlightenment's forms of healing. By the early 1800s, it had fallen out of favour as a medical treatment but remained in favour as a form of entertainment. There it remained until it came back into cultural favour as a treatment called hypnosis. Today in the United States, for example, mind-body healing practices shift in and out of favour with Americans as their cultural context, the American world view, changes (Herrington 2008).

### Subgroup.

As individuals, we spend much of our lives comparing ourselves to others through the process of social referencing. This term is often used to mean the simple act of comparing ourselves and our stuff with other people and their stuff. A favourable comparison makes us feel good, but an unfavourable comparison can give rise to social jealousy. However, as with so many human behaviours and activities, social referencing is not as simple and as direct a process as we like to think. For example, when using other people's view of us as a social reference, it usually means how we think other people see us, the looking-glass self, rather than how people actually see us (Myers 1999).

There is another kind of social reference, our cultural subgroups. We are social animals who belong to a cultural group (the greater cultural group). It is made up of a number of smaller subgroupings of individuals (subgroups). A subgroup, in this context, ranges in size from a family, to the small, intimate gatherings of friends with whom we regularly meet (social subgroup), to a collection of people with similar interests (e.g., a club), to larger groups or organizations (e.g., employees in a company, or males and females). We are all members of a number of the subgroups that make up our greater cultural group. The subgroups we belong to influence both the references we include in our reference normal and how it changes. They particularly influence the references we use when dealing with our perceptions of ourselves and others (as illustrated above), and when interpreting our experiences.

For example, when we consider joining an informal or semi-formal subgroup, we tend to choose one we think we will feel comfortable in. This invariably means one whose members are like us or, more accurately, like how we think of ourselves. By joining a subgroup, we thus tend to reaffirm the beliefs and values in our current reference normal.

In addition, as a member of a subgroup, we come, over time, to preferentially compare ourselves to its members and define ourselves, in part, by the subgroup's characteristics, in particular its world view. Through this process, our reference normal for acceptable thinking and behaviour, and thus our decision making, becomes further biased toward the subgroup's characteristics. One inadvertent consequence of this selection/biasing process is that our experience of social norms that are different from ours is quite limited. We are also prompted to believe that many more people think and act like us than actually do (Myers 2002a).

A similar process occurs when we join a larger formal subgroup, such as a government, a corporation or the military. When a person becomes a member, they accept that their personal reference normal must be adjusted so that it includes, to at least some degree, the formal subgroup's structure, values and goals. After all, the member must accept the subgroup's hierarchical structure and that they are obliged to make decisions in a manner which helps the organization reach its goals (a reference). The power of this influence on the member's decision making is illustrated when a junior person biases a report in an effort to satisfy their boss and the company's stated objectives.

There are, as always, complications. Although a subgroup's characteristics shape its members' reference normal, its members' decisions collectively shape the subgroup's characteristics (Myers 1999). Over time, this feedback loop significantly affects a subgroup's characteristics and thus how it influences its members' reference normal and their decision making. This holds for subgroups of all sizes, even the greater cultural group.

Consider too that we innately desire to feel a sense of belonging to the subgroups we are a member of. We thus tend to engage in behaviours that favour us being accepted by the other members of the subgroup, such as displaying pride in the subgroup and striving for consensus when making decisions. In doing so, the members collectively favour group cohesion, which feeds back to engender a stronger sense of belonging in each member, as expressed by their increased loyalty. As a result, their reference normal is more likely to include the subgroup's characteristics and their loyalty to it, which gives the group stability.

However, in some circumstances, especially in small groups, these tendencies and the feedback can result in a subgroup becoming dysfunctional. A number of factors can contribute to this outcome. For example, the members can be isolated from important dissenting opinions or information that conflicts with their subgroup's world view. A strong leader or authority figure can bias decisions by signalling their preferred outcome to the members' decisions. There can be an overly strong desire for unity.

Under these types of conditions, the members' innate desire (a reference) to belong and maintain group cohesion and loyalty can, with the help of the feedback, result in their subgroup displaying dysfunctional characteristics. They include the members experiencing excessive (expressed or unconscious) pressure to conform or placing too high a value on consensus. The members tend to neither realistically appraise information, especially if it is disconfirming, nor consider alternative ways of thinking and acting. They tend to be closed-minded and feel a heightened sense of being right. Their personal and their subgroup's decision making can then be described as groupthink. Its essence is captured in the currently popular phrase "making decisions in a bubble."

Groupthink can be seen in all types of subgroups from gangs to religions and internet chatrooms. Its presence is most serious in subgroups comprised of the elites in political parties, governments, the military and corporations. One classic example is the United States' Kennedy administration's planning for the 1961 Bay of Pigs invasion of Cuba. It was expected to depose Fidel Castro, who had recently taken control of the country, but the attempt was a disaster (Myers 1999). A more recent US example is the decision by the Bush administration's top decision-making subgroup to invade Iraq in 2003. Intelligence information released from the period revealed that the subgroup had prior knowledge that Iraq lacked functional weapons of mass destruction and that the war would likely take the course that it did.

Maintaining appropriate and effective decision making both as a member of a subgroup and as a subgroup is an ongoing task.

**Personal Biases.** The above subsections looked at our reference normal mostly from the perspective of influences on it from the "outside." When looked at from our personal perspective, as the user, our reference normal looks a little different. This discussion considers our reference normal from the following overlapping and interconnected perspectives of our self-interest, beliefs and values, preconceptions and ability to adapt.

*Self-Interest.* A basic reference in our reference normal is our self. We use it to gauge whether something is in our self-interest. As a unitary species, the self-interest reference is critical for our survival. However, when we take a wider or longer-term view of self-interest, it is clear that there are limits to the benefits from a self-serving bias in our decision making. For example, our short-term self-interest is limited by our long-term need to remain on good terms with the social subgroups we rely on. The nature of our self-interest reference and the constraints on our self-serving bias when using it are discussed here.

When mentally healthy, we have a strong tendency to judge ourselves to be better than average. For example, are you among the great majority of us who think we are better-than-average drivers or that we are more likely than average to go to heaven? (Of course, a majority of us can't be above average.) Similarly, we tend to accept more responsibility for our good deeds than our bad ones and to accept credit for successes while attributing failures to outside sources (Taylor 1989; Myers 2002a).

These examples suggest that the presence of self-interest in our reference normal contributes is a kind of narcissism. However, within limits, the associated self-serving bias provides us with distinct functional benefits. For example, it helps protect us from depression because it enables us to still find good things to cheer us up when times look bleak (Myers 1999).

A different side of self-interest is seen in a social setting. For example, a study of household garbage revealed a substantial difference between the food individuals reported they had consumed or provided for the family, and the food their garbage showed they had actually consumed or provided. The difference likely arose because the individuals used a self-interested image of themselves as a reference when describing those past actions.

For example, those who reported that they had supplied their household with more food or less pre-prepared food than was actually the case were described as using a "good provider" self-image. In contrast, those who had under-reported their consumption of pop (sugary carbonated drinks) and fats while over-reporting their consumption of fruit and diet soda were described as using a "health conscious consumer" self-image. Interestingly, although the study revealed that we tend to retroactively report our personal food provision and consumption inaccurately, we accurately report the consumption of a family member, especially if a negative image of them is involved (Rathje and Murphy 1993a).

These examples illustrate that the self-serving bias accompanying the use of the self-interest reference can be largely harmless. However, this is not always the case. Consider that it is certainly in our self-interest to remain in favour with the subgroups that support us in some way. However, we can bias our decisions (including our opinions) toward the subgroup's world view to a degree that is, when seen in the wider or longer term, not in our best self-interest. An example is biasing our decisions toward a subgroup's world view that is at odds with the external reality. This situation is likely the reason why some Americans display skepticism toward climate warming, even though this view is not in their long-term best self-interest (Kahan et al 2012). In the extreme cases, we can engage in groupthink.

Consider too a self-interest bias that adversely affects the stability of the person's subgroup. Think excessive self-interest. Examples are theft and greed (the striving to gain control of far more money than one could possibly need or use, which usually occurs at a cost to someone else). Overall, the self-serving bias that arises from using the self-interest reference in our reference normal can range from beneficial to toxic, especially if we take a wider view of it (Singer 2001).

Fortunately, we are capable of keeping the self-serving bias in our decision making within benign limits by constraining the influence of our self-interest references on our decision making. This task is made easier if our culture's world view provides us with guidance about the acceptable level of self-interest to incorporate into our reference normal (Myers 1999). There are strong incentives for a culture to provide this guidance. For example, a culture must ensure that the members maintain the supply of the community's shared resources (the commons), such as public land and the air we breathe (Herlocker et al. 1997; Hardin 1998). Part of achieving success is for the members to adopt and use references that limit their efforts to satisfy their self-interested demand for resources. Think of limiting water use during a water shortage. Consideration of how this is achieved introduces another personal aspect of our reference normal: our reference beliefs and values.

*Beliefs and Values.* Our personal assumptions (beliefs) and standards (values) are an important part of our reference normal. They simplify and guide our decisions. They also help us make sense of both our decisions and experiences. However, there is no innate reason why our reference beliefs and values should always be the most appropriate or realistic, or accurately represent the external reality.

As a result, our beliefs and values can, and do, introduce significant biases into our decisions. Consider

that when faced with a situation, our decisions in response are often based more on our beliefs and values than the situation's external reality characteristics (Myers 1999). For example, we have a limited appreciation for how our environment works, so we propose fixes to environmental issues based on our beliefs and values (Cohen 1995). A belief in technology, for instance, prompts a person to favour technical solutions to the pollution of smokestack emissions, such as adding filters. In contrast, a belief in capitalism favours economic solutions, such as cap and trade. A belief in ecology would favour making fewer emissions.

Surprisingly perhaps, the beliefs and values in our reference normal can endure. One reason is that a belief or value being used as a reference can have attached to or incorporated into it the justification for it being a reference (Damon 1999). We also tend to accept information that supports rather than contradicts a belief or value we favour (Swann 1984). In either case, our faith in those references is maintained if not enhanced, and our decision making that uses them continues to be biased.

Consider too the difficulty we face changing our beliefs and values when trying to counteract a decision-making bias we have become aware of, such as our instinctive belief that a person's behaviour reveals their true, innate character. We continue to hold this belief despite the overwhelming evidence from social psychology indicating that our decisions, and thus our behaviour, are strongly influenced by our social history and our current circumstance (Myers 1999). For example, the phrase "the banality of evil acts" reminds us that an evil deed is generally not prompted by the evilness of the perpetrator but by the situation in which they had to make their decision (the context). Despite this reminder, we tend to believe that only inherently bad people do bad things. We would gain more understanding if we considered their situation and questioned their decision making.

The mention of context serves as a reminder that the power of our beliefs and values (or any reference for that matter) to influence our decisions is not sourced solely from within us, as we tend to accept. Their power is also heavily dependent on the context we find ourselves in when we apply them. For example, if a group of patients is given a harmless and ineffective substance, some 30% to 40% of them will find relief from their symptoms, especially pain. But the placebo's healing power will only be experienced under specific conditions: the context in which it is administered must support the patient's belief in its effectiveness. For example, the placebo must be administered with a supportive bedside manner. This includes the caregiver demonstrating their acceptance of both the patient's belief in the placebo and their expectation that they will be cured by it (Brown 1998; Gibbs 2001b).

We are a social animal, thus our personal beliefs and values are most powerful as decision-making references when they are used in a context where others have the same beliefs or values. This could occur when the context is a subgroup of like-minded people, or if the beliefs and values are part of the greater culture's world view. For example, Western culture's world view holds that there is a fundamental, knowable order to the universe. With its deep religious and mystical roots, this is a powerful cultural vision that invariably becomes part of a Westerner's reference normal. It allows both secular and religious Westerners to hold beliefs about the order of things so strongly that they can become self-evident truths. These beliefs then become references that influence our perceptions of the world around us (Myers 1999). For example, in the 1520s, Copernicus defended his heliocentric (sun-centred), concentric model of the universe, in part, by using his culturally supported, strongly held belief in the heavenly (aesthetic) properties of the circle (Gingerich 1973). Similarly, as we saw in chapter 1, in the late 1600s, Newton's choice of seven colours for his rainbow was influenced by his strongly held beliefs about nature's order. Westerners today still favour a highly ordered view of the world around us.

This does not mean that those of our personal beliefs and values being used as references are always clones of our culture's world view. This is illustrated by Robert Oppenheimer, a physicist who was the leader of the Manhattan Project, which built the first atomic bomb. He faced considerable personal disquiet about the impending consequences of the project and the subsequent development of a nuclear arsenal. Although a Westerner, he gained some solace from the *Bhagavad Gita*. He thus held his misgivings in check by a mix of Western and Eastern cultural-religious ideas: his strongly held belief in the values of duty, faith and fate (Hijiya 2000).

The power of beliefs and values when they are acting as a reference is perhaps most clearly indicated by you thinking of a belief you currently hold strongly. Now try to provide evidence and an explanation that supports the opposite of this belief (Myers 2002a). A serious effort takes considerable energy, and it can be uncomfortable.

If nothing else is achieved, you will increase your appreciation for the power of the beliefs and values in your reference normal, and the biases that accompany them.

*Preconception.* Preconceptions are our assumptions or expectations about ourselves and the world around us. They are unavoidable in the sense that we routinely (consciously or unconsciously) form them from information in memory using our reference normal. We often use our preconceptions as a reference when dealing with the unknown. A classic example is the stereotype (a generalized preconception) we use to characterize someone before we actually interact with them. Think of a stranger walking toward you. In their role as references, our preconceptions influence the way we perceive, pay attention and thus make decisions (Myers 1999; Kahneman and Tversky 1982b). The clearer our preconceptions are to us, the stronger we believe in them and the greater the impact they have on our decisions (Taylor 1989).

There are significant advantages to forming preconceptions. They ease our decision making in many ways, for example, by reducing the brainpower needed to gauge a new situation. Preconceptions also focus our attention, which makes it easier for us to think about ways to deal with the future and to achieve our goals. They also give us the comforting sense that we are in control.

Preconceptions have a downside. The information and interpretation used to form them are less reliable than we imagine. Our preconceptions can therefore be biased (including errors) in many ways: in particular, inaccurately representing the external reality. When biased perceptions are used as a reference, our decisions will reflect these biases.

The biases in our preconceptions can endure, in part, because the process of priming our brain to recover or form a preconception also guides us to preferentially search for and pay attention to information that supports the preconception (confirmation bias) (Myers 1999). At the same time, we are prompted to ignore data that doesn't support the preconception (Gardner and Stern 1996). As a result, we tend to simply believe that our preconceptions are reality rather than try to discover the limits to how reliably they represent it (Myers 1999; Cole and Cole 2001).

Both the shortcomings and the benefits of using preconceptions come to the fore when we are faced with making decisions about a complex situation that exceeds our mental capacity to understand. Preconceptions can both help us make sense of the situation and mislead us. For example, preconceptions can help clinical psychologists in their difficult task of diagnosing an illness. But they can also lead the psychologist to incorrectly correlate a symptom with an illness, see symptoms in random data, select data that matches a preconceived diagnosis and ignore data that does not match that diagnosis. The result could be a wrong diagnosis or overconfidence in a diagnosis (Chapman and Chapman 1967; Arkes and Harkness 1980).

The role and impact of preconceptions on our day-to-day decision making is illustrated by the story of Olive Fredrickson who, in 1949, moved from an urban environment to the wilds of British Columbia. Although she had never seen a wild wolf before, her culturally influenced preconception allowed her to believe that they killed for fun, played with their victims and ate older wolves. She thought that this was morally wrong, so she hated wolves. She never took the time to study the wolves that lived around her to see if this idea was correct or not, so she was happy when a wolf died (Fredrickson 2000).

We struggle to acknowledge our personal preconceptions because they are often a part of our reference normal and stuck in a self-referencing loop. Despite this limitation, with effort we can recognize their existence and thus deal with the inherent biases they bring. Consider that when we make a prediction, we bias it toward the future we want to see (our preconception of the future) not the future that is objectively the most likely to occur (Taylor 1989). If we write down our prediction and then compare it with the actual outcome, then we can search for the reasons why our prediction was wrong. This simple experiment provides us with an opportunity to see for ourselves the preconceptions and biases in our reference normal (Tversky and Kahneman 1982c). Are you brave enough to test yourself?

*Adjustments.* On casual reflection, we imagine that if we were thrust into an unpleasant situation, we would either remain miserable or actively work at changing our circumstance. On deeper reflection, we realize that we usually just adapt to the new circumstance. A key aspect of doing so is the making of mostly unconscious adjustments to our references so that our reference normal accommodates the reality of our new circumstance.

This flexibility in our reference normal allows us to adapt to the many uncorrectable changes we experience in our lives. Consider that we all have to deal with personal aging and the loss of a loved one. Some of us will have to deal with a disability. Think of theoretical physicist Dr. Stephen Hawking who had to adjust his reference normal in response to ALS, a crippling motor neurone disease (Hawking 2009).

How quickly and how well we adapt by adjusting our reference normal depends on the circumstances (Diener 2000). In the case of events that affect our core sense of self, such as becoming disabled, we adapt slowly and usually incompletely (Easterlin 2003). On the other hand, we rapidly adapt to the presence of newly purchased possessions (Myers 1999).

The adjustments we make to our reference normal can be proactive as well as reactive, as noted above. Consider that proactively (consciously or unconsciously) adjusting our reference normal can help us undergo longer-term changes, such as working towards a challenging goal. For example, when learning a difficult skill, we can consciously divide the learning into subgoals. We can then focus our efforts on achieving an easier-to-reach subgoal. As we reach that subgoal, we adapt to it by unconsciously adjusting our reference normal to match our new circumstance. This switch, together with the pleasant feeling arising from the success of reaching the subgoal, makes it easier for us to both switch our attention to the next subgoal and maintain our motivation to reach it.

There are also unwanted and inappropriate outcomes from adjusting our reference normal. Consider the moving target of how much stuff we need to feel satisfied (Fast Company 1999). For example, as we become cash richer, we adapt to our new wealth. We unconsciously adjust (perhaps switch) the reference we use to evaluate our financial well-being from our poorer past to our richer present. Thus, our judgement of "enough" wealth tends to be in proportion to our present rather than our past wealth. This makes it easier for us to decide (feel?) that we are dissatisfied with our current financial circumstance. It also means that to increase our wealth-driven satisfaction with our financial circumstance by a fixed amount takes a greater increase in our wealth when we are richer than when we are poorer (Diener et al. 2010). More generally, our decisions have become driven by a reference that is being adjusted by our rising expectations. This makes it extremely difficult to decide how much wealth is enough.

Consider too that reference adjustments can help us adapt to an unsafe work environment, an unhealthy social circumstance or a degraded natural environment.

This can happen if we adjust our reference for acceptable so that it more closely matches our current conditions rather than maintains our expectation/goal of more favourable conditions. Our adjustment allows us to more easily say to ourselves, "It's not quite bad enough for me to act." Thus we accept rather than leave or correct our situation. Think too of drifting into groupthink.

The examples suggest that these adjustments occur in isolation. This is usually not the case, as illustrated by the shopper's cycle. After we make a desired purchase, we feel a burst of happiness. But, as mentioned above, we rapidly adapt to the presence of the purchase: we adjust (perhaps switch) our reference from the state of pre-ownership to ownership. As we adapt, the burst of happiness the purchase initially brought rapidly fades.

There is another factor at play in this cycle. Our ability to predict the nature and intensity of our emotional responses to future events is quite poor (Myers 2002a). This includes the degree of happiness we can realistically expect from the act of purchasing or owning a product. If we decide that either "purchasing" or "owning" a product can provide us with a long-lasting improvement in our sense of happiness, then we are likely to choose to deal with our post-purchase let down by making another purchase (Schmookler 1993; Kasser 2002; Myers 1999). The combination of using a reference that is inappropriate (products are poorly related to lasting happiness) and making an adjustment to a reference (pre-ownership to ownership) ensures that we will be let down by a purchase, again. The personal cost of the shopper's cycle is an underlying frustration with life, while the social and environmental cost is the over-consumption of resources (Sheldon and Mc Gregor 2000). But it is also a marketer's dream come true.

Overall, our innate ability to adjust aspects of our reference normal allows us to adapt to our changing circumstance. In combination with an ever-changing world, it ensures that we change throughout our lives. But the ability to adapt doesn't guarantee we will do so in the most appropriate way or quickly enough, as discussed in chapter 6.

In summary, the references within our reference normal are diverse. We are all born with the same fixed (universal) references used by our brain to interpret our basic sensings of our body and the world around us. We all learn or absorb additional references from our culture, colleagues and the environment. The characteristics of these references are, in essence, universal and enduring, but in detail they vary with the culture and (consciously or unconsciously) change over time. We also generate references on the fly from our memories and our other references, as needed. We use these more specific personal short-term references to make many of our day-to-day decisions. The aggregate of our references is our reference normal. In this case, it includes the made-on-the-fly references.

Some references in our reference normal, such as our sensing references and the basic norms of our culture's world view, are held in common by many people. Others are more personal, such as our made-on-the-fly references or those we have adopted from our culture or subgroup but adjusted for our personal circumstance. Our reference normal and its use thus has, in general, much in common with other people, but in the specifics it is unique to us personally.

Our reference normal can change. Overall, we have most control over the changes in those references in our reference normal that are directly accessible to our conscious selves, such as our goals and some aspects of our values and beliefs. We can consciously adjust them, add new ones and delete existing ones. We can even change their relative importance to match the context in which they are being used. However, the references we have control over, and many of those we don't, can also be changed in a similar way by our unconscious, but without our full awareness. Overall, the changes to our reference normal, especially its core aspects, usually occur slowly.

We place considerable faith in our reference normal as a decision-making aid. It serves us well most of the time. The diversity of the references in our reference normal and the flexibility, although limited, in how they can be used or adjusted allows us to adapt to changing circumstances. In contrast, our reference normal's human-generated, self-referencing characteristics, and the fact that its component references mostly lie in our unconscious, ensure that we experience difficulty consciously recognizing the biases and constraints that accompany its use. This lack of awareness also contributes to our considerable inertia to consciously question and adjust the references in our reference normal and their use. This oversight is reflected in the enduring biases (including errors) in our decision making. If we wish to make more appropriate decisions, we need to devote time and effort to becoming more aware of the nature and role of our reference normal in our decision making.

# Reasoning

We are constantly faced with making decisions. Some decisions are trivial, others life altering. Regardless, making them requires our brain's processes to select, manipulate and evaluate information, and compare it to a reference. These actions, collectively called reasoning, are our effort to make sense of the information before drawing a conclusion.

We are familiar with our conscious (cognitive) reasoning. When pressed, we would agree that it is a mix of rational, emotional and intuitive processes. After reading chapter 1, we might also agree that, despite our lack of familiarity with our unconscious, it is where most of our reasoning occurs. Indeed, its faster reasoning is the foundation of our slower conscious reasoning. This section describes a few of the fast reasoning processes, called heuristics. It also discusses their limits and biases, the influences on them and their effects on our decisions. This section also discusses the influence that both context and our view of cause and effect has on our reasoning.

**Heuristics.** We don't fully understand how the brain functions. But, by studying the products of our decision making under controlled conditions, we can discover the general methods our brain uses for reasoning. One of these methods is heuristics, which is a fancy word for *rules of thumb*. In essence, heuristics are simplifications and generalizations about the world, such as stereotypes. This subsection discusses the heuristics we unconsciously use for reasoning and drawing conclusions. We also routinely use rules of thumb when making conscious decisions.

There are advantages to using heuristics. As shortcuts, they speed up the process of reasoning, especially when they are applied automatically by our unconscious (Kahneman 2003). Heuristics also act as guides that allow us to reason about complex situations that would otherwise present us with decision-making difficulties. The results from these faster reasoning processes are passed to our conscious, where they can be accepted as decisions or used as the basis for conscious (cognitive) reasoning or decision making.

There are also disadvantages to using heuristics. We can't directly question the heuristics that are applied unconsciously or dimly within our conscious awareness. We can only question the results they provide to our conscious self. However, we tend to treat the result of our heuristic reasoning as final rather than as provisional. These results can thus easily become our

faulty decisions or contribute to us making faulty cognitive decisions. This outcome is more likely to happen in complex situations, such as those we face in our technological society.

Four types of heuristics are discussed here: representative, availability, affect, and anchoring and adjustment. By appreciating the presence, characteristics and biases associated with applying these heuristics, we are more likely to closely evaluate our conscious thoughts about an issue before we accept them as our final decision about it. The same is true for the conclusions that arise from consciously applying the rules of thumb provided in this book.

*Representative Heuristic.* The representative heuristic describes how, in order to make sense of a new circumstance, our unconscious compares the information about it that we recently acquired through our senses to a prototype (model) of the circumstance constructed from our older memories.

For example, you learn that the single-car accident (mentioned in chapter 1) in which a car drove off the road had occurred late on a moonlit night and that all of the people in the vehicle were teenagers. By thinking about this information, we cue our brain to recall memories that it thinks are relevant to understanding the accident. These memories are combined into a prototype of the accident, which our unconscious then uses as a temporary reference to interpret, by comparison, what we know about the actual accident. The interpretation appears in our conscious awareness, accompanied by our brain's unconscious evaluation of how closely the features of the prototype match the description we were given of the actual accident. The closer they are, the more likely we are to consciously believe that the interpretation itself is correct, and the more strongly we will likely hold that belief. We are also more likely to assign a higher probability to the described accident happening again (Kahneman and Tversky 1982c). All of this reasoning happens in a split second. You likely formed an opinion about the accident. Did you think the teenagers were probably drunk and speeding, so they drove off the road?

The representative heuristic is a useful reasoning tool. However, it is easy to imagine how it can lead us astray: the car accident actually happened because a deer suddenly ran in front of the car, and the sober young driver swerved to avoid it. We were unaware of this information and, unless we were paying close conscious attention to our reasoning, it is unlikely we would have asked if there was more to know. To avoid falling into these traps requires us to consciously remind ourselves to closely evaluate the initial conclusions of our reasoning and our faith in it.

The following experiment illustrates another, more widely applicable way that we can be led astray by the representative heuristic. Researchers provided an occupation-neutral description of a person's (the target's) character to a random selection of people (the audience). Although the description itself contained no details about the target's occupation, the audience were told that the target was from a group of 100 people: 70 lawyers and 30 engineers. The audience were then asked to decide if the description of the target applied to an engineer or a lawyer. Most chose lawyer.

In a subsequent test, a different audience were given the same occupation-neutral description but were told that the group the target belonged to consisted of 30 lawyers and 70 engineers. Despite the change in the proportion of lawyers in the group, most of the audience still thought the target was a lawyer (Myers 1999).

One could argue that those favouring a lawyer had based their conclusion on already existing, strongly held stereotypes about lawyers and engineers rather than on building a prototype from the provided information, that is, by reasoning using the representative heuristic. The more significant conclusion is that neither method considered the description's statistical information: the size of the occupation groups into which the 100 people had been divided into (the base rate). If the audience had considered that division, then they would have deduced that in the first case there was a 70% chance that the described person was a lawyer, while in the second only a 30% chance. We routinely ignore statistical effects in our reasoning. The representative heuristic is no exception.

There are other limitations to our use of the representative heuristic that we should be aware of. When we contemplate the past, our unconscious constructs a prototype view of the past from our memories. When we consciously contemplate that reconstruction, we intuitively think it is a fairly reliable representation of the past. Unless we consciously devote the needed effort, we tend not to take into account that the knowledge of the past in our memories is limited and that its use to form the prototype was influenced by our current lives and our feelings about the past.

We can also use the representative heuristic to contemplate the future, such as the future level of air or ocean pollution. In doing so, our unconscious creates a prototype of the future by combining our knowledge of today's pollution levels and trends with related information, such as our feelings about pollution and the future in general. Our brain then projects the prototype forward to that future time of interest. That projection becomes the estimate of future pollution provided to our conscious for our contemplation.

Because our knowledge of current pollution levels and trends is often limited, we use our intuitive perception of them, or our biased/incomplete view of their measured values, to make our unconscious projection. As a result, we tend to hold an imperfect or biased view of future pollution levels. In addition, no matter how far into the future our unconscious projects the prototype, we tend to treat each view of a different time in the future as equally representative of the future. In truth, because we live in a complex system, the further into the future the projection is made, the less representative (the more uncertain) it becomes. For our view of the future to be useful, either we must increase the uncertainty with which we express our opinion, or our description of the level of future pollution must become more vague, for example, by specifying a range of possible pollution levels. (The availability heuristic will illustrate the innate difficulty we face trying to reliably express probability.)

We can't get rid of the representative heuristic and its biases. It is a part of our unconscious reasoning. And if we use the representative heuristic without considering its (our) limitations, then we can expect some significant biases and errors in our reasoning and the resulting decisions (Kahneman and Tversky 1982c; Tversky and Kahneman 1982c). Fortunately, we can lessen its downside by accepting that we intuitively use the representative heuristic and that, under some circumstances, it can lead to inappropriate reasoning and decisions. We can take more care with our conscious use of the conclusions we receive. For example, we can learn when to consider statistics, we can consciously check the validity of our first conclusion, we can check our projections, and we can consciously search for additional information. Think cognitive reasoning. If we do, then we can avoid some of the representative heuristic's more significant biases (Tversky and Kahneman 1982a). This conclusion applies to our conscious use of all rules of thumb.

*Availability Heuristic.* The availability rule of thumb is used by the brain to reason about the frequency or probability, and the significance of, a present, past or future event (Tversky and Kahneman 1982c; Ross and Sicoly 1979).

When using the availability heuristic to evaluate a target event, our unconscious recalls from memory the instances of similar events. It then combines the total number of these instances with the ease (speed) with which they were recalled. This combination is used to evaluate the frequency of the target event's occurrence.

If our unconscious can recall only a few similar events and if the recall is difficult, then our conscious will be informed that the target event is uncommon or unlikely to happen. If the reverse is the case, then our conscious will be informed that the target event is common or likely to occur. In addition, the more likely or common the occurrence of the target event appears to the unconscious, the greater the significance it (and thus we) will tend to attach to it (i.e., common events are more important than rare events.)

The availability heuristic works reasonably well in simple circumstances, as long as we remember its biases. Consider that the making and recalling (reconstruction) of the memories relevant to the target event is subject to many influences, as discussed in chapter 1. For example, we more easily recall, and thus give more weight to, the following memories: events that are famous, familiar, more recent and spectacular; anecdotal information about events (e.g., urban myths); and events about which there has been speculation. We are less likely to remember subtle or mundane events, especially if they are old, or aspects of those events, even though recalling that information may help us to better represent the frequency of the target event (Gardner and Stern 1996). We also have difficulty remembering events we are trying to suppress, and we can't recall events we are not aware of.

The availability heuristic is particularly unreliable at estimating the frequency of events that are unfamiliar to us, such as those that are rare or slow. This is the result of more than having few, if any, relevant memories to recall. Consider that we find it hard to reliably imagine things about which we have limited experience or knowledge. Thus, when trying to recall memories similar to an unfamiliar event, our unconscious may extrapolate from our past experiences, even if they aren't really relevant. Or it might simply make up an apparently relevant memory. Regardless, in both cases, the ease with which we can construct the "facsimile" event still helps to determine the frequency of occurrence we will assign to the unfamiliar event (Tversky and Kahneman 1982c).

Our resulting biased estimates of the frequency of events are used to generate our stereotypes and our assessments of risk. They therefore reflect the availability heuristic's limitations (Taylor 1982). For example, our unconscious can underestimate the chance of us dying from common causes of death and overestimate the chances from rare causes (figure 3) (Slovic et al. 1982). Whether we will over- or underestimate the risk of such an event occurring is influenced by a number of factors. They include the factors affecting our making and recalling of memories discussed above and in chapter 1, and whether the information came from personal experience or descriptions. Our estimate is also influenced by how recently we acquired the information we use, how fresh it is in our memory.

For example, if we have no personal experience of a rare and sudden natural event, such as a flood or earthquake, in our local surroundings, then we can't recall any memories. If there are no easily accessible records, or if we choose not to search for them, then our unconscious will provide us with a conclusion based on our related knowledge and experience extracted from memory, such as anecdotes from the time before we lived there. Our unconscious will likely tend to underestimate the event's frequency and thus the risk.

In contrast, if we seek out historical media records, then there could be a larger amount of information about floods or earthquakes in our area. That information is now fresh in our mind, so it predisposes our unconscious to overestimate the frequency of such events and thus the risk (Hertwig et al. 2004; Slovic et al. 1982). These biases are not restricted to sudden events. We can similarly mis-estimate the frequency and risk from slowly (relative to a lifetime) developing events, such as climate change and soil erosion.

In overview, if our initial unconscious estimate of the frequency of an event is unquestioned by our conscious (cognitive) reasoning, then it becomes our biased, even incorrect conscious prediction of the event's frequency and thus our estimate of risk. Consider too that by tending to accept an experience-based assessment of the frequency and significance of events, we are more likely to assume that the future will largely be like the immediate past. This predisposes us to believe that the world is less variable than it is (Hertwig et al. 2004; Slovic et al. 1982). However, because news preferentially reports rare and spectacular events, a news-based assessment predisposes us to believe that the world is more chaotic than it is.

We are stuck with the availability heuristic. However, by being aware of its limitations and biases, we provide ourselves with an opportunity to deal with them. We can, for example, consider the limitations and then consciously decide on a level of uncertainty to attach to our conclusions about frequency and significance. Or we can decide to apply statistics.

*Affect Heuristic.* As discussed in chapter 1, our brain automatically and rapidly associates feelings (and emotions) with our sensings, perceptions, memories and imaginings (information). In essence, the feelings become part of that information or, if you prefer, linked to it. The affect heuristic describes how our unconscious uses the feelings associated with information to reason about it. In outline, the feeling (or emotion) is assessed by placing it on a scale between positive (good) and negative (bad). Where the feeling falls determines how our reasoning will use the information.

A general sense of contentment is considered good, so the reasoning favours that information. In contrast, the general feeling of being stressed is considered bad, so the reasoning rejects/discounts that information. If more than one piece of information is being used or if conflicting feelings are associated with the information, then the affect heuristic accommodates the differing assessments by trade-offs. For example, when reasoning about whether to undertake a task, the assessments that indicate future costs/risks (or their increase) are traded-off against the indicated future benefits (or their increase). Whatever the result of our reasoning using the affect heuristic, it enters our conscious where it is either accepted as our conscious decision as is or modified by our slower conscious reasoning.

This outline is too simple. The affect heuristic relies on more than the feelings we associate with information. It also uses both the ease with which our brain associates a feeling with the information and the strength of that feeling. The easier it is to make the association and the stronger the feeling, the greater the information's role in the affect heuristic reasoning. From a different perspective, the stronger the feelings (or emotions) associated with the information, the more likely we will use the affect heuristic to reason about it.

It is thus easy to imagine that reasoning using the affect heuristic is subject to constraints and limits. Consider that the feelings associated with information about an issue can poorly represent the information's actual importance to the issue. For example, we reason that a smiling person (positive feelings) is more trustworthy than one who isn't. We can also find it hard to determine the boundary between good and bad

feelings, especially when our feelings are weak. The assignments of feelings and the affect heuristic assessment of them are also influenced by the context in which they take place and by our past experiences. This can either help or hinder the application of the affect heuristic.

These limits and constraints are illustrated by our non-numerical relationship with information when it is expressed in numbers. We tend to find it difficult to assign feelings to numbers or quantities of things when the numbers are raw: just isolated numbers. In contrast, it is relatively easy to assign feelings to numbers when they are presented already compared to other numbers or quantities (i.e., pre-referenced), or presented in ways that allow us to easily do so. For example, it is easier to assign feelings to quantities when they are expressed as probabilities and percentages or in graphs rather than as raw numbers.

This introduces a bias into our affect heuristic reasoning. For example, surprisingly, we can more easily support a safety measure if it is presented as being able to save 98% of 150 lives (i.e., a percentage) rather than 150 lives (i.e., a number) even if the number is reporting that more people would be saved. The strength of this bias is illustrated by our continued preference for the percentage-based option even if only 85% of 150 are saved.

Context too can affect both our assignment of feelings to information and our affect heuristic reasoning about it. For example, repeated exposure to specific information can create in us a positive feeling for it. Putting pressure on us to make a decision reduces our opportunity to think analytically, thus enhancing the role of the affect heuristic in our reasoning (Slovic et al. 2002). Think of the time-dependent sales offer.

These limitations are important because the affect heuristic plays a significant role in our lives. Consider that a culture's world view exerts much of its influence on its members through emotion-filled culturally relevant stories and myths: it relies to a significant degree on the members' affect heuristic reasoning. Thus all cultures are at risk from stories and myths that are emotionally powerful but inappropriate or misleading. All cultures face the dilemma of ensuring that their stories and myths are both sufficiently true in the external reality and emotionally powerful to influence the members. This issue will appear again in chapters 5, 16 and 18.

Certainly, the affect heuristic is a speedy, subtle and sophisticated aid to reasoning in our complex world. But its limitations ensure that its usefulness is constrained.

The conscious decisions we make based on it can easily be biased and spectacularly unreliable. "What a thrill this will be," he said, before attempting to run across the freeway at rush hour. "No, I won't get pregnant," she thought.

The affect heuristic provides us with its most useful and reliable results when we have sufficient experience to do two things: assign a feeling to information that reliably represents its importance; and reliably anticipate how we will like the consequences from using it. In addition, we must also take the time to evaluate the products from using the affect heuristic and do so in a way that takes its limitations into account. These conditions are difficult to satisfy.

*Anchoring and Adjustment.* Anchoring and adjustment describes a heuristic reasoning method in which we make an initial or preliminary decision (the anchor) and then (consciously or unconsciously) adjust it as we reconsider our initial choice or deal with new information (Tversky and Kahneman 1982c). Certainly this is a useful technique, but commonly both our choice of anchor and our adjustments don't meet our expectations or our faith in them.

Our selection of the anchor can be biased in many ways. We tend to unconsciously base our choice of anchor on the information that is initially presented or familiar to us. We tend not to evaluate all of the available anchor options because we inherently favour those options that support our point of view or our preferred outcome (the good outcome). As a result, our choice of an anchor is more like a biased best guess, which we then give a privileged place in our reasoning (Kahneman and Tversky 1982b; Slovic et al. 1982).

One might think that these errors and biases in the anchor would be automatically corrected during the process of adjusting the anchor. This is not the case. When considering adjustments, we act as though we are limited in the questions we can ask ourselves about the anchor.

We can ask, How right is my choice of anchor? but not, How wrong is my choice of anchor? or even the more neutral, Is my anchor appropriate? (Kahneman and Tversky 1982b). This bias directs our attention away from evidence that contradicts our initial anchor choice and ensures that the adjustments we can easily make to it are small (Germano 1999). It is like wanting to change a flower arrangement but only being willing to consider small changes to the flower arrangement and rejecting the option of changing the vase.

This same kind of bias is seen in our response to new information about our choice of an anchor. We interpret it with a bias toward our current point of view, such as preferentially selecting from it just those bits that support our beliefs and goals. As a result, we both confirm our current anchor and discourage the consideration of disconfirming information that would prompt us to shift it (Ross and Anderson 1982). This is one reason why the presence and significance of anomalies in current scientific models can take a while to be recognized and resolved. Recognition and change might only occur after a compelling new model appears to point out the anomalies (Lightman and Gingerich 1992).

There is an emotional (affect) component in our inertia to adjusting our initial anchor. Similar to our tendency to emotionally feel a loss more than a gain, we tend to emotionally favour our initial anchor over adjusting it. The greater our emotional investment (the strength of our belief) in the initial anchor, the less likely it is that new information will spur us to adjust it. The emotional bias in favour of the original anchor can be so strong that we can fail to adjust our choice even if the evidence supporting a change is compelling. An analogy is people who stand by their belief in a prophecy long after it has failed to come true (Festinger et al. 1956).

In overview, the combination of our tendency to bias our choice of an anchor and to make adjustments to it that are too small results in the anchor-and-adjust method of reasoning being sub-optimal (Tversky and Kahneman 1982c). By using our knowledge of these limitations to consciously take them into account, we can increase this heuristic's usefulness and more thoroughly assess its results.

*Context.* We like to think that facts are the most powerful influence on our decision making. In reality, the greater context within which the facts are presented and decisions made is just as important (Myers 1999). So are the feelings and emotions associated with both the facts, as discussed above, and the context.

Context constrains all stages of our conscious and unconscious decision making: from selecting the references and information, to our reasoning and drawing of conclusions (Tversky and Kahneman 1981; Slovic 1990). The greater context has this power because it gives us more than a simple, passive reference framework for decision making. It provides us with (conscious and unconscious) cues that influence our reasoning and help us give meaning to our conclusions.

To fully appreciate our reasoning process requires us to acknowledge the importance of the context in which it takes place.

The influence of context on our reasoning is illustrated by the impact that a change in context has on our conclusion about risk (Gardner and Stern 1996). For example, your perception of the risk of harm to yourself during an armed bank robbery will be different depending on whether you are a customer in the bank or a spectator outside, or whether you are watching the event unfold on TV. The influence of context on our reasoning extends beyond its physical aspects to the feelings associated with it. For example, information presented in what we feel is a positive context tends to result in a different perception of risk than when that same information is presented in a context we feel is negative. For example, an event in which 10 lives out of 20 are saved (positive context) is seen as less risky than one in which 10 lives out of 20 are lost (negative context) (Myers 2002a; Slovic et al. 1982).

In the same vein, our perception of risk is also affected by the certainty with which the information about the event is presented. (Not to be mistaken for the certainty of the information). The greater the certainty expressed, the more weight we give to it. Thus, the positively interpreted statement "10 lives out of 20 *will* be saved" tends to produce a lower perception of the risk of death than "10 lives out of 20 *might* be saved." But both will result in a lower risk assessment than "10 lives out of 20 will be lost" (Gardner and Stern 1996). Context thus also affects the degree to which our reasoning will overestimate or underestimate the risk.

The last example is a reminder that our reasoning can be influenced by (us or outsiders) deliberately manipulating the context in which we receive information and/or reason about it. Consider that the human-environment system we live in is highly complex. This means that most of the information we receive about it is both uncertain, to some degree, and difficult to interpret unambiguously. Under these circumstances, small changes in the context in which that information is presented and is reasoned about can significantly affect our final decision. If we wish to make appropriate decisions, then we are obliged to pay close attention to the context. In particular, we can ask, is it being deliberately manipulated? If so, for what purpose?

A common method of influencing the public's reasoning is to add cultural symbols to the context in which the information is presented. However, our interpretation of a symbol's meaning is itself strongly influenced by the context in which it is displayed. Thus, if our reasoning is to be successfully influenced, then the added symbols, the presented information and the greater context must all complement one another (Polanyi 1968; Norretranders 1998).

Have you noticed that when a politician makes a significant announcement or gives a significant speech during an election, there is usually a match between the main points they express to the target audience, the context in which they have chosen to deliver that information to them and the supporting symbols? For example, politicians customarily give speeches about jobs or infrastructure projects in a work context, such as a factory or construction site. While speaking, they wear or hold the symbols representing a worker, such as a shovel, hard hat or safety jacket. They often stand in front of a podium to represent power. However, politicians are not the only people who manipulate context to influence our reasoning. Advertisers are masters at this craft, as will be discussed below.

We are reminded that the ability to manipulate context in order to influence our reasoning presents us with a dilemma. Adjusting context can be judged as responsible if it is directed at helping us produce valid and useful harm-reducing decisions. However, it is irresponsible to manipulate context in order to produce decisions that mislead, obscure what is known, avoid responsibility or cause harm. Similarly, providing no context can make information puzzling, reasoning difficult and its results misleading (Ziman 1978).

We are left with the problem of choosing an appropriate context in which to truthfully display information and the dilemma of deciding whether a manipulation of that context will enhance the user's understanding or mislead them. For example, in what context should information about climate warming and the development of antibiotic resistance be presented so that it informs and stresses the need to pay attention but does not mislead. Under what conditions is the manipulation of context by advertisers and politicians responsible? What timeframe should we use for our judgements about the consequences from context manipulation?

This discussion may have given the impression that it is relatively easy to influence our reasoning by manipulating context. However, the context laid out to influence us is less important than how we personally react to it. In particular, the cues we personally receive from the context will determine the facts and emotions we will recall from memory and use for our reasoning.

Who we each are and our personal circumstance matters. Even if the physical aspects of a manipulated context are the same for everyone, it can still cue different memories in different people. Each person's reasoning will be affected at least slightly differently. Hence our reasoning can be significantly influenced but not really controlled by manipulating context.

The above features are illustrated by the difference between the way a layperson and an expert evaluate the risks from a project billed as economy boosting. The public tend to evaluate risk in a wide context that includes the things they value and the fairness of the risk's distribution. In contrast, experts evaluate risk in a much narrower context, one that is commonly restricted to patron mandates and statistical measures, such as the number of likely deaths and significant injuries (Slovic 1990; Morgan 1993).

For example, when evaluating the risk from a tar sands oil pipeline spill, experts are concerned about the probability of a spill and the probable significance of specific impacts. The public are more concerned about the impact on both their material and non-material quality of life. From the public's perspective, all lotteries are eventually won, so probabilities are only part of the context for their evaluation.

Facts are important and they affect our reasoning. However, changing the context in which the facts are presented changes our acceptance of them, their meaning and our reasoning about them. Thus our efforts to deliberately influence decisions by changing the context in which information is considered and decisions made raises significant ethical questions.

## Cause and Effect.

We intuitively assume that events have a single cause. We also preferentially believe that there is a simple, close, direct relationship, both in time and space, between that cause and its effect (result). I was holding a knife, it slipped, and now my finger is sliced and bleeding. It is reasonable and useful to make those assumptions when reasoning about simple systems.

However, much of our reasoning deals with events that occur in more complex systems. In these cases, the relationship between cause and effect is often just as complex: causes rather than a cause, effects rather than an effect. A particular effect (event/outcome) is more likely to be the product of many direct and indirect contributions of differing significances made at different scales of time, space and strength. Applying our intuitive rule of thumb (that there is a simple

relationship between cause and effect) to events in complex systems can be misleading. For example, it leads to the incorrect assumption that we can reliably deduce the cause(s) of an effect from a small amount of information.

It also leads to the assumption that many more of the correlations in the changes between a complex system's variables are simple cause-and-effect relationships, when most aren't. Hence the admonition that correlation does not prove causation. Unfortunately, the innate strength of our belief in a simple cause-and-effect relationship provides us with sufficient reason to ignore this admonition.

In addition, the effects we see in complex systems are often separated from the causes by a large amount of time and space and many steps. This usually means that we fail to see the connections between them. And when we do see a series of connected events, we have an irresistible urge to interpret them as a sequence of simple cause-and-effect events (Tversky and Kahneman 1982b).

We also tend to assume that the significance of each cause increases nearer to the present, with the most important being the most recent. Think of the availability heuristic. The limitations to this thinking are illustrated by a close soccer game. We act as if the last goal, the "winning" goal, was the most important, when from the outcome's perspective, any one of the goals could be seen to have changed the final result. The soccer example also illustrates that we give little consideration to the importance of statistical properties in our reasoning about cause and effect. Statistically speaking, each goal is equal in its importance, even though the emotional anticipation of whether one will occur or not is very real (Myers 2002a).

There is a rule of thumb that can help us to remember the limitations to our cause-and-effect assumptions: The more complex a system, the less likely it is that any of our preferred varieties of simple cause-and-effect assumptions will fully explain the connection between an effect we see and its causes. This is true for systems of even low complexity.

The wisdom of these words is illustrated by a simple example from our environment, a complex system. It is easy for us to recognize that a landslide occurred during heavy rains. It is harder for us to accept that historical deforestation of the area, changes to drainage during housing development, road building or all three may have contributed significantly to the landslide. We prefer to think that the recent heavy rain caused the landslide rather than it was only one contributing factor. It is even harder for us to accept that our routine, daily activities can make a significant contribution to such an event (Gardner and Stern 1996). But consider the following: Who wanted most of a house lot covered with a home and parking, thus increasing runoff? Who is responsible for the discarded garden and other waste that blocks storm drains or diversions into which rainwater flows?

This example leads to the broader realization that our cause-and-effect assumptions influence our reasoning about the relationships between historical events and our present circumstance, and between our present actions and our future circumstance. We tend to reason that a particular historical event, such as the invention of the steam engine, is the dominant cause of our current circumstance. Similarly, we assume that a particular significant present action, such as our $CO_2$ emissions, will be the dominant cause of our future circumstance. The human-environment system is too complex for such simple connections to be the norm. The many historical factors contributing to our current state occurred at different times, speeds and places, and had different levels of significance. The same is true for the influence of today's actions on our future.

The human tendency to inappropriately link effect with cause(s) in complex systems does not exist in isolation. It exists alongside a cultural world view that, in essence, guides the culture's members towards a coherent and usually simple view of the world's order and their relationship to it. By accepting their culture's sense of order, a member biases their personal world view about the world's order and thus their view of cause and effect. This has a significant influence on their reasoning and decisions about how to deal with issues. Think of applying prayer, sacrifice, technology and science.

Western culture's preferred sense of the world's order is epitomized by the solar system. It supports Westerners' innate human desire to reason about the world using simple, ordered models and simple theories of cause and effect. After all, we innately find these simple, ordered explanations more satisfying than complex ones.

The consequences of these influences on Westerners' reasoning about cause and effect is illustrated by a commonly held view of how the stock market functions. Experts, commentators and the public alike tend to say that a small rise in a market index, such as the S&P500, is the direct result of the market's (more accurately, its participants') reaction to a particular piece of good financial news. If the market falls or does not react to this good news, then it is said that the news had already been discounted by the market (Fischhoff 1982). This view of the market is widely applied even though there is evidence to the contrary. Consider that at least some changes in the stock market, such as the daily spread in the prices of a stock, are the result of stock trading rules rather than any coherent and strategic decisions by the market participants (Farmer et al. 2005; Ho 2009).

Indeed, the global financial crisis of 2008 was, in part, the result of a related misapprehension: the failure to recognize that there are many non-simple connections between cause and effect in the complex global financial system. This lack of recognition meant that, among other things, the flaky financial instruments that contributed to it were seen by the participants to be, in isolation, safe investments.

Overall, Westerners' world view and our innate cause-and-effect preferences make it easier for us to believe that the world is inherently controllable. We can thus more easily believe that the simple application of personal effort in combination with our ability to create technology can, alone, solve our human and environmental problems. This assumption hasn't resulted in the expected effects because the world is less ordered and controllable than we think it is (Taylor 1989). Cause and effect are usually not so simply related.

It is logical to assume that, over time, the disappointing consequences from believing in a faulty cause-and-effect relationship or model would prompt us to adjust our thinking. Unfortunately, "updating" a faulty assumption or mental model is difficult. We face considerable barriers to both recognizing its shortcomings and revising it (Tversky and Kahneman 1982b).

Consider that we would have to first recognize that there is an incongruity between our model and the actual outcome, and then conclude that the fault lies with our personal model. We would have to decide to change the model, and then actually change it. Considering the above discussion, such as the wide separation between many causes and the effect we see, and our mental investment in simple models, we should accept that these are all, practically and psychologically, difficult steps to make, as discussed in chapter 6. They take time and effort to achieve. Our intuitive preference is to keep our faulty model and find ways to assimilate (perhaps even force) a new fact into it. Long-lasting, faulty cause-

and-effect models are therefore a part of our normal reasoning processes.

In summary, our reasoning uses our brain's innate reasoning processes, such as heuristics. Our unconscious reasoning is the foundation of and starting point for our conscious (cognitive) reasoning. Thus the limitations, such as biases and constraints, associated with our unconscious reasoning processes are reflected in our cognitive reasoning.

Despite these limitations our reasoning provides us with satisfactory decisions under many circumstances. However, there are enough circumstances, especially those that are complex, where these limitations can result in decisions with unpleasant or unwanted outcomes, particularly when seen in the long term. We have difficulty recognizing when this might be the case because, being largely unaware of our reasoning's limitations, we can believe that our decisions are more reliable than they are.

The consequences of these reasoning limitations have effects beyond their direct impact on our decision making. For example, we can reason that what is typical of the world around us is normal, which allows it to become an accepted part of our reference normal. We can thus bias our decisions and predictions in favour of seeing the typical as good and form a feedback that gives this status quo inertia to being changed (Myers 1999). It favours the biases in our decisions and predictions enduring.

## Drawing Conclusions

By reasoning about information using our reference normal, our unconscious provides our conscious with a rapidly made interpretation/decision. It is up to us to use that interpretation/decision to make a final decision we feel we can act upon consciously: we must draw a conclusion. The process of drawing conclusions can be as simple as accepting the decision from the unconscious, but it doesn't have to be.

The difficulty of finalizing a decision, and the nature and significance of the biases in that process, is illustrated by a study of people who fled Germany just before the outbreak of World War II (WWII) and ended up in the United States. It reported on the emigrants' decisions for leaving and their justifications for making them. The study found that those who chose to flee did so because they had experienced significant negative life events in Germany, including torture for some.

However, despite recognizing the potential danger they faced if they stayed in Germany, they waited an average of three years after their negative experience before they finalized their decision to leave. During that delay period, they tried as best as they could to maintain the lifestyle they lived prior to their negative experience. Before they could make the final decision to leave they often required an additional, substantial shock.

Just how difficult it was for many of the emigrants to finalize their decision to leave Germany is poignantly illustrated by a man who had fled from Germany to Switzerland at great personal cost. There, as reported by a colleague of his who continued on to America, he received a job offer for a position in Germany he had always wanted. He accepted, returned to Germany and probably died there during the war.

When questioned about their decision to delay their departure and to instead try to continue their lifestyle, the emigrants gave many different reasons. For instance, some saw emigration as a big unknown. However, when their answers were weighed against their experiences and the danger they said they knew they faced, their justifications seemed quite inadequate.

This suggests that there were other unexpressed reasons why they had delayed. Perhaps, in Germany, they had suppressed full recognition of the dangers their experiences were warning them to expect if they stayed. Perhaps they had been mentally unable to accept the lifestyle adjustments needed for them to realistically adapt to their new circumstances. Among those possible adjustments was either to live a substantially different lifestyle in Germany or to emigrate (Allport et al. 1941).

Regardless, drawing a conclusion was difficult and slow because of their discomfort reconciling their bias in favour of what they liked or wanted with their conscious awareness of the unpleasant aspects of their real-world circumstance. More generally, we all experience difficulty drawing conclusions when we are faced with the discomfort of a conflict between our will and our urges or between our internal reality and the external reality. This difficulty can exist even when our discomfort is mild.

As expected, this difficulty and the nature of our decision-making process can result in us drawing conclusions that contain significant biases and errors. Most of the time they are harmless or even beneficial. For example, being able to draw an optimistic conclusion by avoiding or sidestepping some aspects of the external reality is an essential part of maintaining our mental health (Taylor 1989; Cohen 2001). However, as the German example illustrates, at some point, failing to accept the external reality for what it is can result in significantly unpleasant outcomes, including death (Gardner and Stern 1996).

We can find it easier to avoid those unpleasant outcomes if we have a greater appreciation for the processes that allow us to make and hold conclusions that are at odds with the external reality or cultural/social imperatives. To that end, this section discusses denial, illusion, overconfidence and hindsight bias. They all in some way distort our internal reality relative to the external reality.

Although denial, illusion and repression each has a different meaning, the differences between them are not well defined. For example, repression and denial sometimes mean our efforts to alter reality, while illusion simply means that we interpret the world in the best possible light (Taylor 1989). Alternatively, repression and denial can mean you imagine that you see or experience nothing, while illusion and hallucination mean you imagine that you see or experience something other than that which exists (Cohen 2001). Regardless, our use of denial, illusion and repression biases our conclusions so that they are more to our liking but, to some degree, at odds with the external reality. In this section, *denial* will be used to discuss our exclusion of an aspect of the external reality from our conclusions, while *illusion* will illustrate the inclusion of something that doesn't exist in the external reality. Denial and illusion will be discussed separately even though they can't be meaningfully separated in real life.

Denial. Denial is, metaphorically, a way to make life simpler and more ordered by selectively getting rid of information. There is usually a pattern to our denials. We tend to deny after an experience pushes us across a mental threshold that is set by a strong belief or principle we hold. Our personality helps to determine the nature of that threshold and our reaction to it. So does the guidance from our subgroups or our greater culture about behaviours and thoughts that are either forbidden or exalted, such as aspects of sexuality and how to conduct personal relationships. The characteristics of both our denial threshold and our denials are flexible, not fixed. They change over time as we, our circumstance and our culture change (Cohen 2001).

There are benefits to denial. For example, it helps us to stay mentally healthy. Imagine how denial would help you deal with an emotionally disturbing experience over which you have no direct control, or being in possession of information that you cannot reconcile with a belief you hold strongly (Gardner and Stern 1996; Cohen 2001). Denial can screen such painful events from our conscious awareness thus providing us with the benefit of reduced stress, at least in the short term (Hare 1993). No wonder that we all deny at some time in our lives. We also deny as a group when we accept the denials embedded in our culture's world view (Cohen 2001).

But denial can also come with severe costs that are paid by us personally, socially and culturally, and by our environment. For example, consider the costs to the citizens of New Orleans that resulted from their politicians denying for years that a hurricane like Katrina (2005) was probable and would have devastating consequences. Their denial resulted in the routine sidelining of preparations for such an event, with disastrous results (Elliott 2005; Fischetti 2001).

With the cost-benefit dilemma of denial in mind, how should we deal with denial? It is unrealistic to simply ignore (deny?) that we deny, but so is the blanket cure that we should immediately "face reality." A more useful approach is to acknowledge that denial is a part of our lives, our decisions and the political process. This means accepting that we will routinely face the choice to deny, that the choice can arrive suddenly and that there is always some tension between the benefits and costs of denial (Cohen 2001). It takes work to avoid inappropriately denying, and it takes work to deal with the consequences of taking advantage of the ease with which one can deny.

For example, in the aftermath of the global financial crisis of 2008, Dick Fuld and Clifford Asness, both Wall Street hedge-fund bosses, denied that their (culturally supported) beliefs and actions had contributed to it. They did so even though the firms under their direction had promoted the sale of flaky financial instruments. The decisions of Allan Greenspan, the chair of the United States Federal Reserve System at the time, helped to create the regulatory climate that made the crisis possible. He visibly aged between the crisis and his later acknowledgement that he had found a "flaw" in his strongly held view of how the economy worked. He used the flaw to rationalize (perhaps deny) his contribution to the crisis (Patterson 2010).

As these, the German and the New Orleans examples suggest, possessing facts is insufficient on its own for us to avoid denying. And being given irrefutable facts that show we are denying is insufficient to shatter even our harmful denials. One reason is that facts often lack the emotional component that helps us relate to them.

Without this emotional connection, the raw facts may simply prompt us to reinforce our denial. And, as Greenspan illustrated, even when we make an emotional connection, we can easily be tempted to hedge our bets.

More generally, once we deny, our ability to recognize or accept that we are denying is limited by the power of the psychological needs that sustain it. For example, Albert Speer was Hitler's architect and, subsequently, the minister for armaments and war production. He struggled until his death to come to terms with his wartime knowledge of Germany's WWII concentration and extermination camps and his contribution to their existence (Sereny 1995). Similarly, Robert Oppenheimer never fully faced his rationalization, discussed above, of his contribution to the development of the nuclear bomb and the destruction from dropping them on Japan (Hijiya 2000).

Our efforts to face our denials can be hindered, and the making of them can be helped, by the support we find for them in our culturally derived fantasies, myths, ideologies and rationalizations (Cohen 2001). Consider how both George W. Bush's status as president of the United States and the American exceptionalist world view helped him to deny, in spite of the facts, that the last Iraq war was going very badly. Saddam Hussein and his son Uday could find similar culturally derived support for their view that the Iraqi army had the ability to withstand the American invasion. That the strength of their conviction also involved denial can be seen in the degree of astonishment on their faces when the Iraq army was so easily defeated (McAllester 2003).

Our efforts to respond more appropriately to our denials in general, and avoid our more serious denials in particular, are helped by us having had previous experience appropriately dealing with denial. We can gain that experience by routinely facing our small, even trivial denials rather than sloughing them off. For example, although we denied to others that we were told about the meeting we missed, we can at least admit to ourselves that we were told but didn't want to go. (Cohen 2001). Facing these smaller denials is easier than we expect because they are likely not as ironclad or threatening as serious denials. Think of the urge-will tussle. This practice makes it easier for us to acknowledge, even when denying the existence of some reality, that we are actually aware, to some degree, of its existence.

Denial is part of how we simplify decision making. By accepting that we deny, by routinely facing trivial denials and by finding alternate responses, we can help ourselves reduce unnecessary denials and avoid more serious denials. At the same time, we gain the skills to deal with the denials we do make. We also become better able to deal with similar difficult-to-change aspects of our lives. This includes being willing to accept the external reality for what it is, when necessary.

**Illusions.** Our process of drawing conclusions is influenced by our illusions. Most of our decision-affecting illusions are distortions of reality rather than fantasies. These more common distortions will be the main focus of this discussion.

The illusions we can hold about ourselves, the external reality and the future take many forms. Among our more common illusions is our belief that we have more understanding/control over the world around us than we actually do. For example, we tend to underestimate the possibility that an event will arise to delay our completion of a complex task, so we set an unrealistic completion date. Consider too our overly optimistic outlook when we make up a list of the tasks we expect to complete today. We also tend to believe that we can have a larger and more positive direct impact on the world around us than we actually can and that we are more competent than we are. For example, we tend to believe that we are better-than-average drivers (Taylor 1989; White and Plous 1995).

When we identify these kinds of illusions in others, we tend to label them as character defects. However, we all hold them. Fortunately, many of our illusions have only annoying consequences. Many others are mild, benign, positive biases that can significantly benefit our mental health. Consider that by these illusions allowing us to avoid the vagaries of living (at least for a while), they provide us with the time we need to sort, face and eventually deal with unpleasant truths (Seligman and Csikszentmihalyi 2000). Similarly, mild, positive illusions, such as unrealistic optimism and believing ourselves to be above average, may help us to deal with the ever-present uncertainty in our lives (White and Plous 1995). A mildly optimistic outlook also seems, unlike repression, to enhance our social and intellectual awareness (Taylor 1989). It appears to help smooth our relationships with those we meet and thus the functioning of society (Myers 1999).

Consider too that illusions play a critical role in the impressive learning ability of children. As adults, our mildly exaggerated positive beliefs about ourselves typically increase our motivation, persistence at tasks and effective performance. This, in turn, can lead to greater success in our endeavours. In addition, even though most of our individual successes may seem trivial, they do matter to our subgroups. Those few of our successes that are gauged by the group to be valuable can, collectively, have a significant impact on society (Taylor 1989).

But it is also true that illusions, even mild positive ones, can have a serious downside. For example, if I consider myself to be an above-average driver when I am not, then there is a greater possibility that I will be involved in an accident (Slovic et al. 1982). Similarly, although I may believe differently, I may actually have a less-than-average level of compassion for those who suffer an adverse event (Perloff 1987).

When the examples given at the start of this sub-section are seen in this context, they hint that the mildness of an illusion is less important than whether it causes or leads us to harm ourselves, others or our environment. If we fail to make this distinction, then illusions and their kin, such as denials, can result in us ending up in situations that we would, in hindsight, never have chosen. Traudl Junge was an ordinary person who could hardly be described as evil, yet she drifted into working for Hitler as one of his secretaries. After the war, much of her life was spent struggling to come to terms with the path she had travelled (Junge 2004). Her story, and Speer's given above, illustrates that failing to recognize and question our illusions and denials can result in any one of us "sleepwalking slowly into sin" (Cooke 2003).

Much like the double-edged sword of denial, we are caught between the value of illusions and their potentially negative consequences. This means that the option of simply accepting or ignoring all of our illusions is unrealistic. Unfortunately, the nature of our illusions also means that it is easier to say that we should focus on dealing with our harmful illusions than it is to do so. Our brain constructs our illusions so well that it is often difficult for us to separate them from the rest of our internal reality.

This task is made more difficult if our illusions originate with our culture's world view or are supported by it. Consider Westerners' tendency to believe in our culturally supported illusion that we have more than enough skill and understanding to manage the functioning of the environment and our manipulation of it. This belief allows us to underestimate the risk and under-value the impacts from our efforts to manipulate and mange it (Gardner and Stern 1996). For example, an unrealistically optimistic belief in our ability to mitigate or restore any damage to an ecosystem caused by resource extraction can result in us under-appreciating the actually possible consequences. Think of BP's 2010 oil well blowout in the Gulf of Mexico. Think too of ideas to geo-engineer the earth's atmosphere or oceans in order to deal with climate warming.

Fortunately, as with denial, we are at some level and to some degree aware of our illusions. Thus, there is a time and a place where a dose of realism might help us temper our over-optimistic illusions (Taylor 1989; Myers 1999). There are other ways we can reach a point where the disconnect between illusion and reality becomes too strong for us to ignore. A cognitive dissonance experience is then forced upon us. We are prompted to deal with the difference.

It is indeed possible for us to recognize and discount or even dismantle our illusions, but the process can be slow and distressing. Consider that the successful removal of an illusion will likely reveal a world that can appear to be, at least initially, bleaker than we would prefer (Taylor 1989). However, balanced against those difficulties is the realization that if our faith in the illusion is maintained, we might experience an even bleaker reality. Think of the alternate outcomes that could have been experienced by those pre-WWII Germans who struggled to successfully conclude that they must leave Germany for the United States. Fortunately, it is also reasonable to think that we can be both happy and more realistic at the same time, if we make the effort (Seligman and Csikszentmihalyi 2000).

**Overconfidence.** By now you should be unsurprised to learn that, like our faith in our beliefs and assumptions, we tend to have greater confidence in the result of our reasoning and conclusions than is warranted (Myers 1999). The factors that contribute to and support our overconfidence are many. They include our powerful, innate desire for certainty and coherence, our optimistic outlook and our limited awareness of the constraints on using our references.

Even our process of perception contributes to overconfidence. It fails to provide us with a feeling for or measure of uncertainty in our perceptions. Instead, we generate that measure through our reasoning processes (Kahneman and Tversky 1996; Fischhoff 1982). But, as discussed above, the innate thought processes that provide that assessment are limited and biased, a fact we routinely fail to take into account. Remember our tendency to consider that our first guess at an anchor is largely error free: our best guess (Kahneman and Tversky 1982b).

Our overconfidence in our conclusions is most forcefully expressed by the failures that occur in the complex systems we build and run, and/or manage. One reason for this overconfidence is our tendency to under-appreciate or ignore the limitations, such as reliability and completeness, of the information that we use to draw our conclusions about complex systems (Slovic et al. 1982). Another reason is our poor innate grasp of the nature of complex systems, as discussed in chapter 5. For example, we tend to forget that the more complex a system, the greater the level of uncertainty we must attach to our conclusions about it. Even if we accept this relationship, we greatly underestimate how quickly the level of uncertainty rises with an increase in the system's complexity.

You might be surprised to learn that this overconfidence is not limited to lay people. Despite their training, experts display overconfidence about a wide variety of matters: their beliefs about the validity of currently available scientific data, their knowledge of how technical systems behave as a whole, their understanding of how human errors affect technical systems, and their ability to predict the future of nations (Tetlock 2006). For example, experts in risk evaluation for nuclear reactor safety and environmental impact assessments display overconfidence in their conclusions about risk. They significantly under-appreciate the role of factors such as cumulative effects, context, system complexity and alternative values in contributing to both actual risk and their evaluation of risk (Gardner and Stern 1996; Slovic 1990; Slovic et al. 1982).

With this background, it is unsurprising that we all experience difficulty recognizing and dealing with our overconfidence. We could use record keeping and statistics to reveal its presence to us, but our innate relationship to statistics is poor, so this is not our usual approach (Fischhoff et al. 1977). We could conduct a realistic assessment of our personal circumstance. It would reveal to us, for example, that our past is usually a better predictor of our future behaviour than our faith in our stated intentions. It would also show that we routinely have more faith in our estimate of how long a task will take than is warranted. However, we struggle to conduct such an assessment, even though it is the key to successful personal change (Myers 2002a).

There is a bizarre twist to our relationship with overconfidence. We do recognize that some of our past conclusions led to undesirable outcomes. However, our conclusions are poorly linked in our memory to the sense of confidence we assigned to them when they were made.

Thus, the knowledge of an undesirable outcome can result in us changing our mind about the past conclusion but not our past confidence in it. Thus we aren't prompted to be more skeptical of our current level of confidence in our current conclusions. As a result, we can be, unjustifiably, just as confident in a new conclusion as we were, in the past, to a now discarded conclusion.

In overview, if we wish to deal with our overconfidence bias, we need to exert considerable mental effort to both recognize its presence and respond effectively (Kahneman and Tversky 1982b). It helps if we remember that doubt and uncertainty are poorly represented in our unconscious decision making. Fortunately, they can be represented on paper and in focused thought. Thus honest records are a useful tool for bringing overconfidence to light. So is our effort to practise skeptical thinking when consciously evaluating the decisions provided by our unconscious. Some aspects of achieving this goal are presented below in the discussion of hindsight bias.

**Hindsight.** As we experience an event, we integrate our biased view of it into our memory. In the process, we adjust our existing memories. The event and its impact on us rapidly become a part of our current thinking, including our current perspective about the past: our hindsight view of the world.

When contemplating the immediate or distant past, we usually fail to recognize or compensate for the effects that these memory changes have had on our views. This is illustrated by you imagining that you are back in the past, a teenager thinking about your unknown future. Your recollections today about those long-ago teenage speculations do not match your actual imaginings at that time. Think of your past angst. This is the case because today one of those futures exists. The events along the way to the present have been incorporated into your current memories and references, the ones you now use for considering the present and imagining the past. Our present conclusions about the past (including the circumstances surrounding our past decisions and our past view of the then unknown future) and its influence on the present are hindsight biased (Myers 2002a; Fischhoff 1982).

The hindsight bias is common. Its presence is suggested, for example, when, in the present, we exaggerate our ability to have foreseen the now known outcome of a past event. "I knew it all along," we might say, or "that should have been obvious" or "that was inevitable." (Myers 1999).

There are ways to counter the influence of the hindsight bias when drawing conclusions about either the past or its influence on the present. A key step is to accept that we have this bias. We are then provided with a reason to consciously consider how we, personally, are reaching our conclusion about the past. For example, how much reliable data from the past, such as a diary, are we using, or are we simply imagining the past? We are helped by considering the past, and our researched information or memories of it, from a number of different perspectives. In doing so, we expand the context of our decision making beyond the narrow frame of our current point of view (Fischhoff 1982). If we put in this effort, we are more likely to recognize that an unconsidered conclusion about the past or present need not be as reliable, as inevitable or as obvious as we thought. Other outcomes were also likely (Slovic and Fischhoff 1977).

An ability to appreciate and deal with the hindsight bias is important because we use our impressions of historical cultures as well as their decisions to provide us with guidance when making today's decisions. For this hindsight view to be useful, we need to take into account that past peoples and cultures may have had similar concerns to ours, but their circumstances and world views were not the same as ours. If we don't acknowledge these differences, we could be misled by our view of the past and the information from it when we use them to make decisions about the present and future.

In summary, we find it easier to draw conclusions that fit with our internal reality and match our positive feelings. It is always more difficult to draw a conclusion when we try to recognize the biases and limitations in the data and our reasoning, and then accommodate them in our drawing of conclusions. Examples are the uncertainty in the available information and references, and the presence of denial and illusions in our thinking. Either way, we tend to be overconfident in our conclusions.

## Marketing, Advertising and Consumer Decisions

Our decision making was discussed by breaking it into the three themes of references, reasoning and the drawing of conclusions. This was convenient for discussing the topic, but it makes real-life decision making seem more ordered than the dynamic, complex (conscious and unconscious) process that it is. We gain an appreciation for decision making as a complex whole by discussing marketers' (including advertisers') efforts to influence our consumer decisions and our reaction to their efforts.

Marketers apply their scientific knowledge of our decision-making characteristics, discussed above, in order to influence our purchasing choices. The results of their efforts indicate that marketers do have a valid, though imperfect, understanding of our decision making. In contrast, our reaction to the marketers' efforts indicate that we, the public, are largely unaware of the nature of our decision making, its limits or biases. We are also largely unaware, or more correctly unconcerned, about the influence marketers have on it.

**Objective of Advertising.** Corporations use advertising to increase their profits. The path to this goal is less direct than we imagine. The first objective is to influence your purchasing decisions in favour of their products. The second objective is to ensure that you purchase their products again (Martineau 1957). And the third is to lower their fiscal risk from producing products by ensuring that there is a market for them. Then the profits come.

It is well known that corporations try to keep expenses to a minimum. Yet the amount they spend on advertising is staggering. In the United States during the 1990s, advertising spending was in the range of 2%–2.5% of GDP (more than the value of farming at about 1.5% of GDP). By 2009, a recession year, the measured US advertising expenditure for just the first six months was US$60.87 billion, which was larger than Canada's budget deficit for 2010 (Daddi 2009). In 2013, total media advertising expenditure in the United States reached US$171 billion and was growing faster than GDP (emarketer 2014). Worldwide, advertising expenditure in 2000 reached between US$331 billion and US$465 billion (depending on the expenses included) (Elliott 2000). And in 2015, total media advertising expenses were reported as US$513 billion and rising at roughly 5% per year (emarketer 2016).

Does the benefit of advertising exceed this cost? It is reasonable to argue that at least the companies who use advertising believe that it pays. And they do so even though they recognize that there is no surefire recipe for creating an ad that will increase sales. Just how exquisitely fine the line is between a viral success and a dismal flop is laid out in the Canadian Broadcasting Corporation radio series *The Age of Persuasion* and *Under the Influence* by Terry O'Reilley.

This fine line is unsurprising if we remember that merchandisers are dealing with the complexity of our decision making. Amongst its generic features that they face is the ubiquitous potential for unintended consequences. This is illustrated by the movie *Finding Nemo*. Nemo, a coral-reef-dwelling clownfish, is caught and kept captive in a fish tank along with other captive wild fish. The movie records the adventures of Nemo, who is trying to escape from the fish tank, and his father, Marlin, who is trying to find Nemo. The moviemaker hoped to make money from both ticket sales and from the associated merchandise the movie would advertise. However, counter-intuitively, *Finding Nemo* created a boom in the aquarium fish trade, including an increase in the removal of fish from coral reefs. This raised the ire of many, which was bad for the moviemaker's sales (Fickling 2003).

### Marketing, Advertising and You.
Advertising is part of the larger field of marketing. At its most basic, the objectives of advertising are the same as marketing: to encourage us to buy and to influence our purchasing choices.

Here are some examples of marketers' big picture views of what they must do to reach that goal. In the early 1900s, John Kennedy realized that "Advertising is salesmanship in print" (Clark 1988, 39). In the 1950s, Martineau said that "Perhaps advertising's most important social function is to integrate the individual into our present-day American high-speed-consumption economy," and that "Just as the training of its formative years socializes the child, so advertising continuously socializes the adult" (Martineau 1957, 190 and 192). In the 1990s, a text dealing with marketing to children quoted a marketing consultant as saying, "There are only two ways to increase customers. Either you switch them to your brand or you grow them from birth" (McNeal in Seipp 2001, 150). And, in a similar vein, a corporate executive is reported to have said that, "When it comes to targeting kid consumers, we at General Mills follow the Proctor & Gamble model of 'cradle to grave.' We believe in getting them early and having them for life" (Youth Markets Alert 1998).

Over the years, marketers have developed methods to reach for their goals. By comparing the marketers' methods (and their reasons for applying them) with our reaction to these methods (and our beliefs about their influence on us), we gain an appreciation, at a personal level, for our decision making in the real world. In the following discussion watch for the references to our decision-making foibles discussed above.

As consumers, we experience marketing efforts to influence our purchasing decisions in many forms. Examples are the cheap-to-produce freebies (swag) used to build a corporation-consumer relationship; items sold at a loss to bring you into the store (loss leaders) (Anderson 2001); advertising that prompts the child to influence their parents' purchasing decisions (Ruskin 1999; Seipp 2001); and an avalanche of advertising in the form of flyers and TV ads. We also experience a flood of ads on the internet/social media to such an extent that these media are now shaped around the demands of advertising. Indeed, the foibles in how we direct our attention has made the attention economy possible (Lewis 2017).

Our beliefs as consumers about the influence these marketing efforts have on our personal purchasing decisions differ from those of the marketers. Westerners, for example, do recognize that others are affected by advertising. However, we each believe that it either does not affect our own decisions or that we can control the limited influence it might have on us (Myers 2001). We believe that we can casually, but rationally and logically, select out the fact from the fiction in advertising and thus make sound, unbiased, informed, rational purchasing decisions. In general, we consider advertising to be an annoyance, maybe an inconvenience or even a service. We treat marketing as an activity that has limited influence on us as individuals and that certainly does not affect the core of our culture.

However, it seems that the marketers know something about us that we don't acknowledge. Marketers believe that marketing has a considerable effect on our decisions to the point that they depend on its power to influence us. They believe in their dictum that to change the consumer you must know and understand them (Myers 1986). To achieve this aim, marketers spend a significant part of their budgets gathering information and researching their target customers. To help advertisers interpret this data and better target their customers, advertising agencies have, since the early 1900s, employed psychologists to delve into our nature and advise advertisers on ways to turn citizens into consumers (Staudenmaier 1996).

Important sources of this information are academic studies into how we make decisions (as described above) and the seemly innocuous customer surveys with which we are all familiar. Other more recent sources of data include the information gathered from rewards-based and frequent-user membership cards handed out by a wide variety of companies from supermarkets to airlines.

There is also a flood of information being gathered about you from your internet/social media or even email activity. Think of Google's Gmail, its internet search engine and Facebook. Computer experts use algorithms to sift through both big and little data to discover what to target, to whom and how (Pasquale 2015).

Today, marketing strategies are guided by models. They are based on a combination of the analysis of consumer information and an understanding of the techniques that influence our decisions. Examples of the actions that marketers claim they must engage in to be successful are described below as themes or rules of thumb. They are taken from a range of marketing literature from popular to academic. As you read, watch for the themes that underlie their models. These rules and themes reveal who marketers think we are, how they think we and the external world works, and how they assume that they can influence our consuming decisions.

Harry Beckwith's *Selling the Invisible: A Field Guide to Modern Marketing* was a popular book in its day (Beckwith 1997). It is filled with actions or techniques that he bills as key to successful marketing. Here is a rapid-fire, two paragraph summary of his advice. He admonishes the marketer not to spend too much time on reason, details or appeals to common sense. They should instead inspire potential customers. To successfully inspire means remembering that customers try to reduce the perceived risk to themselves by biasing their choices away from their fears rather than toward their desires. Customers, he says, believe that they are above average, so marketers must reinforce this belief by making them feel important. But remember that the more said, the less people hear. So information must be separated out and simplified for the customer. A simple message is all that the prospective customer can handle.

He claims that if customers are unable to sort out facts, they become befuddled. A confused customer then reverts to making their purchasing decisions based on the familiar, which is often the product they hear about or see the most, not the product that is the most commonly available. So, advertise often. To ensure the company's products are the ones that the consumer reverts to, the seller should stand for one distinctive thing that will give their products a competitive advantage. Branding is ideal. To choose the right branding, marketers must know and then target how people make their purchasing choices. The consumer makes their decisions based on their mental image of their desire. McDonald's sells an experience not

hamburgers; people buy oranges based on the orangeness of the fruit, even though there is no relation between orangeness and flavour. Marketers should "manipulate the colour of their oranges." Finally, marketers must always, always, advertise. Regularly. The roots of this thinking can be seen in the earlier section on reasoning, especially the discussion of heuristics.

If all of this sounds too incredible to be true, recall that Perrier in the United States was prepared to go to some length to ensure that the bottled water they sold came from places they could call "springs." They made this effort because displaying the word *springs* when marketing their water, was a very important part of increasing their bottled water's desirability in the minds of their customers (Glennon 2002). Similarly, in the cut flower production business, the demand for particular types, colours, shapes and sizes of flowers (the market) is driven by consumer demands, which are based on vague and sentimental "needs." These needs are influenced, if not manipulated, by those in the flower-selling industry. Their marketing efforts include working with the growers and their geneticists to produce the desired types and colours of flowers being promoted through marketing campaigns (Maharaj and Dorren 1995). In this context, advertising is seen to play on our ill-defined sense of discontentment and our critical self-image.

Compare Beckwith to a small selection from the more academic literature that addresses marketing. There are six basic human tendencies that can be used to influence someone to respond to a request you make of them. They are presented here with examples of how they can be used for marketing: reciprocity (when mailing sales material include a gift such as swag); consistency (get people to make a public commitment to a product); social validation (imply that many people are buying a product); liking (the Tupperware company whose salespeople sell to friends); authority (an authority figure who can provide the consumer with permission or instruction to buy a product, such as Tiger Woods in his heyday); and scarcity (limited time and limited supply offers).

Although these tendencies are universal, the importance of each varies with culture and person. Therefore, the marketing strategies that take advantage of them must be adjusted as needed (Cialdini 2001). For example, authority, such as a salesman's touch on the arm, can help sell pizza, but only if the touch is perceived by the consumer as culturally legitimate. It must be perceived as friendly rather than an invasion of privacy.

These tendencies are just as important to longer-term marketing strategies. For example, inducing the customer to make a small act of commitment, such as accepting freebies, is only the first step to creating a momentum of compliance. The hope being that future steps will lead to a future sale.

Always remember that whatever consumer tendencies are being targeted, add repetition, but not too much. Repetition keeps the item at top of mind and helps build acceptance through familiarity (think of the availability heuristic) (Myers 1999). There is a reason you see that beer or truck commercial on TV and the internet umpteen times.

Finally, from the literature discussing marketing: "Beneath the mask of rationality that our society teaches us to wear, the consumer is a living, breathing, feeling individual. He is not a technical expert. He wants far more from life than bargains. And his behaviour stems more often from emotional and non-rational causes than from logic" (Martineau 1957, 201). The essence of this view is illustrated by the factors current marketers believe influence our impulse buying.

If you are to successfully influence impulse buying, then your product must apparently satisfy eight hidden human needs. The (physical or virtual) store, product and service must all sell to the consumer their ability to provide ego gratification, emotion, security and reassurance of worth. The product itself must be up-to-date and make consumers feel powerful and immortal. And it must do so with a sense of authenticity and creativity (Mazoyer 2000). In essence, the marketer must recognize that the emotions the product represents to the customer are just as important as the product's functionality (Myers 1986). For that reason, marketing to promote impulse buying of a product tries to create in the customer's mind the notion that the product will satisfy these eight needs.

Symbols are perfectly suited to this task, and to marketing in general, because they simplify and compress the world for us. Advertisers can use them to present to the consumer an easily absorbed stereotype of the product. One they hope will "inform" us, at an ephemeral, emotional level, how our owning of the product will satisfy our desires (Ramonet 2001). A classic example of a stereotype is a brand. Its symbolic representation is the brand's mark, its symbol. It can be used to market both the company and its products.

There are other advantages to marketing an emotion-laced stereotype of a product rather than the actual product. The actual product can be displaced in the market, while its stereotype will endure in the consumer's mind (Clark 1988). Think of a new model of the same cell phone or car. Stereotypes also leave mental and physical room for the advertisement to provide the customer with the quick-access, low-effort information that they often want (Wright et al. 1984).

The principle of selling a stereotype is particularly relevant when a number of similar products are competing for the same consumers. Under these circumstances, the consumer's purchasing decision is largely based on their emotion-biased perception of the products, not on the products themselves (the technical basics). Hence the marketers focus on features, like colour and accessories. Of course, the consumer's perception of the products and their features is absorbed from the symbols and stereotypes in ads. For example, the collection of symbols (including that of the brand) used in soft drink advertisements implies that the brand's product will provide you with, for instance, the satisfaction of a pleasant, cool taste and the feelings you associate with being part of a fun group (the stereotype). Think Coke or Pepsi. Think too of the features plugged in car or smart phone marketing.

The use of stereotypes in advertising is successful when the symbols representing them are instantly and unconsciously absorbed by the viewer-consumer into their memory. What they symbolize, such as the brand, can then become a significant part of our perception of the product and thus our purchasing decisions. From this point of view, advertising is, at its heart, a complex, symbol-dominated communication system (Martineau 1957). However, as discussed in chapter 1, there is one critical disadvantage to the use of symbols: they can easily be missed or misconstrued. Think context.

To help design an advertising campaign that reduces these risks, ad companies compile their knowledge of our decision making into guidelines for classes of products. This is evident when comparing the ads for different types of vehicles, such as trucks for young males and SUVs for families. However, the routine use of these guidelines does not ensure success. Advertising's success at delivering the touted financial rewards is illusive because human decision making is more complex than can be routinely captured in an advertisement's simple directive.

Consider that advertisers have to take into account that consumers and cultures change. Our preferences for goods and services, the meaning evoked in us by the symbols that advertisers use to promote them and the social context within which a product is advertised all change over time. Today a gorilla symbolizes one thing, tomorrow something else. Think too of the switch from newspaper and TV to internet and social media. Thus, although the method of selling, say, washing soap remains basically the same through time, the details of the advertisements plugging it must change with the times.

Advertisers are up to this task, as illustrated by the changes over time in the hints advertisements provide to us about how we might justify a purchase. For example, in Canada during WWII, advertisements often suggested that buying the target product would help the war effort. However, immediately after the war, advertising changed to suggest that a person had a right to buy the product. After all, it was hinted, making the purchase was the earned reward, or one of the reasons, for having fought and won the war (Broad 2005). Today, the equivalent reasons provided by advertising to justify our purchase of a product include claims such as "environmentally friendly" or "energy efficient."

Despite these efforts, most advertising is ignored by most people most of the time: the target audience doesn't immediately buy the product. Yet, it still pays to advertise. All that an advertising campaign has to accomplish to be successful is to influence you to remain a potential consumer and thus to increase corporate profits by buying their product in the not-too-distant future.

Marketing exerts one other significant influence on our purchasing decisions. It contributes significantly to more basic personal and cultural change. By influencing our purchasing decisions, it changes our personal view of the world. This affects our culture's world view. As our culture's world view changes, so does the guidance it provides to us about our beliefs and values. This affects our purchasing decisions. Through this feedback, the long-term, collective impact of successful advertising campaigns, and the detritus from failed ones, has helped to create the Western world's consumer culture of today, as discussed in chapter 11.

**Advertising Example.** Advertising does not create demand for a product directly. It influences our decision making in a manner that prompts us to personally create the desire for, or direct an existing desire toward, a particular product. Consider the following two examples.

In 1938, Americans did not associate diamonds with engagement to be married, marriage or even romantic love. In 1938, there were also too many diamonds on the market. In order to solve their oversupply problem, the leaders of De Beers, a diamond mining company, used advertising to change our personal and cultural view of diamonds. Success came when De Beers' advertisers created ads that, when read by women, connected diamonds with romantic love and marriage. But, when read by men, they provided practical information about buying a diamond, such as its size in carats and the need to spend about two months' salary.

By 1946, diamonds had become Western culture's symbol for romantic love and marriage. The diamond ring had become part of the marriage ceremony. A new cultural tradition had been born through advertising.

In 1947, the slogan "diamonds are forever" was coined. By merging love with permanence, the slogan strengthened the connection of romantic love with diamonds. It also removed purchased diamonds from the market indefinitely because they had now become heirlooms. To sell an heirloom diamond would be unromantic and show a lack of respect (Twitchell 2000).

This embedding process continues to this day. All diamond sellers, including De Beers, still spend a large amount of money on advertising in order to maintain retail sales for traditional diamond jewellery. They also try to stimulate an increase in sales by suggesting other occasions for which diamonds should be bought, such as 10-, 25- and 50-year wedding anniversaries (De Beers 1994; Moore 2005). The branded diamond was introduced in the late 1990s in part to avoid the association of romantic love with blood or conflict diamonds and in part to create a lasting, favourable identity for the branded suppliers.

The second example illustrates the degree to which we, as individuals, can be influenced by advertising. The Philip Morris Company attached the smoking of their cigarettes to Marlboro Man, their branded symbol of a cowboy, the stereotype of American culture's value of rugged independence. Smokers then lost their independence through addiction to those cigarettes. In the process, Philip Morris made a handsome profit, as intended (Twitchell 2000).

You and I might instinctively think that the advertising campaigns conducted by De Beers and Philip Morris were successful because they preyed on our greed, vanity and stupidity. Not so. Their campaigns were a success because they applied knowledge of our human decision-making foibles in a manner that influenced our purchasing decisions (Hart 2001). In applying this knowledge, they used our personal and cultural values, beliefs and symbols. By manipulating them, they also helped to change us and our culture.

It is tempting to think that in this internet age, with its bandwagon of relationship marketing and big data, that the essence of marketing and advertising has changed. This is not the case. All that has changed is the messenger's clothes.

In summary, the essence of all marketing techniques is to repetitively supply us with incomplete, selected information about a product (which includes a thing, service, event, belief or action) in an effort to influence our consumer decision making in favour of that particular product. To be successful, marketers rely on their understanding of our human decision-making foibles discussed in this chapter. We, on the other hand, are largely unaware of those foibles. We are also largely unaware that consumer advertising is influencing our specific consumer decision and shaping both our personal and our culture's world view.

This summary of chapter 2 notes that Westerners, at least, tend to believe that normal decision making is conscious, logical and rational. We are also primed to think of deviations from this as at least undesirable if not abnormal. However, when our decisions are considered from the perspective of the external reality, then much of our decision making is unlike we imagine and falls far short of this ideal.

The information we use to make decisions is a mix of hard facts, emotions, assumptions and imaginings we acquire from our direct sensings of the world and our memory. We reason about this information by comparing it to the references in our reference normal. Those references are innate, learned or created on the fly. The process of reasoning uses the information itself, the amount of information recalled from memory, and the ease and speed of recalling it. This information includes our feelings/emotions, their strengths and the ease of locating/assigning them. Our decision-making process is also broadly influenced by our personal experiences, our (conscious and innate) desires, our membership in subgroups and our culture's world view.

Our decision making occurs mostly in our unconscious. It is automatic, fast and inescapable. Only a small part of our decision making is consciously controlled. This conscious (cognitive) decision making (e.g., being logical) occurs more slowly and provides us with an opportunity to make a more considered decision about the products of our unconscious decision making. If our unconscious decisions are firm, they are usually accepted as our conscious decisions, or, if not, they are used as a starting point for our conscious decision making. Thus our conscious decision making and decisions are an extension of, not independent from, our unconscious decision making and decisions. Many of our decision-making foibles are common to both.

The characteristics of our decision making (its foibles) and the influences on it constrain our (unconscious and conscious) decision making and ensure that our decisions are accompanied by significant biases and shortcomings (including errors). However, we are neither the emotional idiots targeted by marketers and advertisers, nor the rational, informed consumers so beloved by economic theorists (van den Bergh et al. 2000; Hernstein 1990). Our decision making can be better appreciated if it is viewed in the context in which it developed.

Our current mental abilities arose over many tens of thousands of years. During this period, our ancestors lived close to the land in small groups within which each person had close ties to all of the other members of the group. Our ancestors had limited power to manipulate their supporting environment. Thus, the topics about which they needed to make decisions and the contexts in which they made them were limited. Only in the last few thousand years have the topics of our decisions and the context in which they are made included large populations in urban centres, complex cultures and large amounts of information. Only recently has complex technology that can alter our lives and extensively modify the environment become a ubiquitous presence in our lives. Similarly, we are now faced with making our decisions while under the pressure of sophisticated, faceless human influences. Think advertising. Today we are applying our decision-making abilities in unfamiliar territory (Myers 2002a).

How then are those of us living in complex cultures to ensure that we make appropriate, valid and useful decisions? The most realistic solution starts with us accepting our decision-making foibles for what they are and that we can't remove them or their limitations and biases. To that end, when words such as *decision*, *choice*, *bias*, *constraints*, *thinking*, *limits* and *prediction* and the phrase *decision-making foibles* are encountered in the following chapters, they should be treated as prompts to accommodate our decision-making foibles in ways that are appropriate to the context and consistent with the external reality.

We can accomplish this task by applying our conscious decision making to more closely check and question our unconscious decisions (our initial decisions) and adjust them as needed. Help in the form of information that increases our familiarity with ourselves and the issues we are addressing was started in chapter 1 and provided in this chapter, and it will continue in the remaining chapters. Assimilating this information is made easier by consolidating it into rule-of-thumb guidance, summaries and models. Techniques to help with appropriate decision making are also introduced as needed. They are summarized in chapter 18.

Hopefully the combination of the reminder words, the increased familiarity with our decision making and the aids for conscious decision making will help you to more easily find and assimilate reliable information and make appropriate decisions about the issues discussed in the later chapters. Time to move on to the next topic needed to increase our understanding and acceptance of our human characteristics, and thus make more appropriate decisions: human culture.

# CHAPTER 3

# CULTURE

Culture is a mechanism used by a social species to guide the efforts of its individual members to form and maintain a unified, stable and adaptable group. Humans are a social animal so we form groups and we each inextricably participate in our group's culture. That culture is meant to guide the interactions among its members, and between them and their environment, to stay within limits. In particular, it should prompt the members to make decisions that consider both their individual and collective interests, as well as the constraints on their lives arising from the nature of their supporting environment (their culture's environmental endowment) and how humans function (Wilson 1988). We gain a better appreciation for how a human culture is formed and accomplishes its task by increasing our understanding of its structure and functioning.

This chapter starts with a generic discussion of culture's roots and roles, and how it functions. During this discussion, it will become apparent that it is useful to divide human cultures into different types based on their characteristics, in particular their level of structural complexity. By discussing these types, the similarities, differences and connections between them will be revealed. The discussion will also reveal the dynamic nature of a culture's functioning and the forces driving it to change. This leads into a section dealing with the question of a culture becoming and staying functional.

A human group's culture has an influence on its members' personal view of the world, their decision making and their interactions with our environment. It thus affects their present and future. The strength and importance of this cultural influence is illustrated in the last section by discussing the origin of Western culture's notion of progress and the nature of its influence on its members' decisions. This will be in keeping with the rest of the chapter, where Western culture will be used to illustrate the features of a culture with a complex structure.

## Roots and Roles

We treat a group's culture as an independent object, but it isn't. It has biological roots, but it resides in the members' heads. It exerts significant power over its members' lives, but it is the product of their collective decisions and actions. These are the features of a culture discussed in this section.

**Biological Roots.** Humans are social animals whose inherited biology both facilitates and constrains how we form groups and how they function (van Schaik 2006; Dunbar 2003a). A basic method social animals use to maintain cohesion is to socialize their group's members to both display standardized cues about their intentions and to respond in an acceptable manner to other members' cues. This is a biological root of human culture. In human cultures, these cues can be vocal, but they are mostly symbols and symbolic gestures because, as chapter 1 discussed, human biology ensures that we preferentially use our visual sense to assess our surroundings (Rescher 1998). The system works because we have the mental capacity to learn and remember both the cues and the expected responses (Gopnik et al. 2001).

**Social Roots.** The nature of culture is easy to understand in the general form outlined above. However, when inspected at the personal level, culture, like consciousness, seems to vanish. The general nature of human culture is best appreciated by looking at it from multiple perspectives while at the same time thinking about it as an undivided whole.

(1) Describing a human culture can be thought of as detailing the way of life of a group of people rather than the life of an individual within the group (Boyd and Richerson 2000). But it is equally true that describing a culture lays out how an individual must behave and believe in order to be accepted into and to feel they belong to the group. Common to both perspectives are a culture's basic features, such as its laws, beliefs, customs, values, symbols, organization and myths. Collectively these features are called the culture's world view: its way of looking at the world and existing in it.

When looked at from the group's perspective, a culture's world view is their method for satisfying the generic conditions required for a human group to function as a cohesive unit. It guides them to feel and display group-favouring characteristics, such as caring, honesty and respect for each other. It also guides them to satisfy their material needs in a group-favouring manner, such as by sharing (Wilson 1988; Myers 2001).

In contrast, from the individual members' perspective, their culture's world view is the collective expression of their individual views and behaviours. It includes their views about how the world around them works, why it works that way and the nature of the future (Plotkin 2000). It also includes their views on how they should conduct themselves when they are in the presence of other members, dealing with their environment or interacting with the members of other cultures. By living in this manner, they collectively express their culture's world view.

These two perspectives are complementary. A group's culture is the product of how its members choose to live and believe. Those details are, in aggregate, the characteristics of the group's world view. It guides the members how to live, and it helps to give meaning and purpose to their lives.

(2) Although our biology facilitates the formation of a culture and determines its basic features, it does not provide much of the details (Kelly 1995). They originate with the day-to-day decisions the individual members make as they try to complete their lifeway. Guiding each member's decisions is an abstraction of how to live that they carry in their brain: a personal world view. From this perspective, a group's culture is the common features of the members' collective (aggregated) abstractions about how to live: their culture's world view. Thus, even though a culture is described as a characteristic of the group, its details are actually carried in the brain of each member. A culture's world view is ephemeral.

Even as each member's decisions contribute to their group's ephemeral, cultural world view, it is influencing their personal world view. A feedback is involved. Their personal world view is also influenced by their interactions with their environment and by their interactions with the other members. Both the members' and their culture's world view change, and they do so dynamically.

(3) As each generation grows up, they learn their culture's world view from their interactions with their elders, their peers and their environment. What they learn influences their lifestyle and beliefs. From this perspective, a culture can be described as the group's common, socially learned behaviour; (Cole and Cole 2001; Kelly 1995; Plotkin 2000; Weingart et al. 1997).

(4) A culture's current world view is the aggregate expression of today's behaviours and beliefs. As each member lives their life, these characteristics are passed on from one generation to the next. Culture is therefore strongly influenced by its past (Ames and Maschner 2000). From this perspective, even "modern cultures" are profoundly historical constructions that can only be understood in a historical context: they are conservative. But cultures are not fixed. They change dynamically.

(5) Although we can each identify the culture we belong to and can distinguish it from other cultures, we are largely unaware of both our culture's influence on us and our contribution to it. Consider that we learn our culture's norms so well while we are growing up that they reside in our unconscious where they continually influence our decisions (Myers 1999). As a result, we each have limited awareness of our own culture's influences on our lives. "Riding along with one's culture is like riding a bike with the wind: As it carries us along, we hardly notice it's there" (Myers 2001, 164). From this perspective, culture is something that we are embedded or participate in rather than belong to or share in (Thomas 1991).

(6) In Western culture at least, our conscious personal experience of our culture inclines us to think of the material trappings around us, such as our buildings and dress, as the essence of our culture. But they are, in fact, only tools and symbols. Tools help us to fulfill our role as members. Symbols serve to unify our individual, unconscious cultural abstractions by acting as cultural cues. They remind us of how we should behave and what we should believe (Kelly 1995). From this perspective, by treating the material trappings as the essence of a culture, we demonstrate how well much of the functioning of our culture is buried in our unconscious.

**Power of Culture.** The culture we are born into has a profound influence on our lives. Consider that, as we grow up, it influences our physical and mental development. For example, culture strongly influences when and how we express our emotions. It partly exerts this control by (directly and indirectly) influencing the formation of the biological processes and their set points that produce and control the hormones which regulate our emotions, such as anger. It also guides us to recognize the social context in which we can feel comfortable expressing a particular emotion and to choose the appropriate method to express it (Gerhardt 2004). This becomes part of our personal reference normal.

As adults, the cultural world view we learned while growing up, as modified and added to by our personal experiences, continues to guide our decisions (Cole and Cole 2001). Consider that culture plays a significant role in determining such features as the relationship between our bodily needs and how we satisfy them (Cohen 1995). For example, although our biology limits the organisms we can eat, it is our culture that tells us the foods we should eat, why, how and when (Mintz 1989).

The powerful role culture plays in our lives can be illustrated by the following observation. It is common to think that the only way we can leave a lasting mark on the future is through a biological legacy, a child carrying our genes. However, as far as the group is concerned, it is just as important to pass on the group's consensus: its culture's world view. Then consider that cultural ideas outlive children (Angela and Angela 1993). Do you know the name of Newton's son or if he even had one? In contrast, most members of Western culture can identify the number and names of the colours in a rainbow.

In summary, from the above descriptions, culture is seen to have a dual personality. It is both ephemeral and real. It is both independent of each of us and within each of us. Our culture displays characteristics that we, as individuals, don't have, even though culture is contained within our collective, individual memories. In one sense, culture is imposed upon us. In another sense, it is formed by us. We both share in a culture, implying that we subscribe to its common features, and we participate in it, implying that we can change it as we live our lives.

These characteristics of culture prompt the individuals of a human group to work with one another to meet common goals, which binds them into a unit: a stable cultural group. But, those characteristics also allow the members to change their culture: it is dynamic and adaptable. Over time, each member contributes to the cultural world in which the "we" and the "I" live.

Using the word *culture* as if it describes a stand-alone, independent entity is only a convenience. This limitation should not be forgotten. Neither should another limitation: a large group may be described as having a culture, but it can contain many subgroups, each with their own subcultures. Think of any nation state today. In this work, when the term *culture* is applied to these large groups, it will refer to the dominant cultural characteristics, the world view, of the group's power structure and decision making. When the term *culture* refers to a specific but unspecified culture, its default meaning is Western culture and its world view.

## Functioning

It is a simplification to describe a culture's functioning as the members' efforts to both follow the guidance from their culture's world view and adjust its world view to suit their needs. We gain a more complete appreciation for how a culture functions by discussing three overlapping features found in all cultures: its individual members, its use of symbols and its organizational structure.

**Individual.** Each of us appreciates that there is a connection between us and the culture we identify with. However, our conscious awareness of the nature of that relationship and the key role we play in our culture's functioning tends to be quite limited. In particular, we don't fully appreciate how we, as individuals, are influenced by our culture and how we influence it.

We are influenced by our culture's world view throughout our lives. It starts with our parents' reproductive values, such as the ideal number of children in a family, and it ends with how we are disposed of when we die. It also affects much of what occurs in between. Consider how the world view of our parents' culture influenced their decisions about how and when the basic aspects of child rearing should occur. The examples that come to mind are the preferred sleeping arrangements for their babies; the optimal timing and method for toilet training; and the manner, beyond biology, in which boy and girl babies should be raised (Cole and Cole 2001).

As children, we learn our culture's characteristics from our parents, through our experiences, or in some other culturally approved way, such as schooling. Regardless of how we learn them, doing so is part of becoming a full member of our culture (Kelly 1995). We must learn our culture's more simple features, such as its symbols and values, and the number and names of its colours (Simpson 1999). We must also learn its more complex characteristics. Among them are how to address critical life events, such as marriage; the appropriate behavioural characteristics to display in human interactions, such as the preferred mix of co-operation versus competition; how and how much to plan for the future; and the features of our culture's perceived ideal future, its utopia (Cole and Cole 2001). In essence, as we grow up, the many details of our culture's current world view become embedded into our unconscious.

This process of culturalization is so successful that, as adults, we are largely unaware that much of our culture's world view has become a part of our personal identity and our personal world view, including its reference normal (Myers 1999). We are thus largely unaware of how wide-ranging and strong our culture's influence is on our lives. For example, Western culture's world view constrains a wide diversity of Westerners' beliefs and behaviours. Among them are the number of colours we see in a rainbow, how we deal with our garbage, and the theories that archaeologists and ecologists consider reasonable explanations for their observations (Trigger 1989). It also influences what we see as beautiful in the natural world (Dubos 1971) and the manner in which we treat our environment (White 1967).

Consider too the strength of its influence on our individual thinking and decision making. For example, central to many Westerners' lives is our notion of progress, our system of land tenure and our belief that increasing material wealth automatically increases our personal happiness. These Western cultural characteristics are so much a part of most of us that we rarely consider their validity, significance or influence on our decision making (Csikszentmihalyi 2000).

In general, the influence that a culture's world view over its members is sufficiently significant that they can, by default, ride along with their culture's wind even when doing so contradicts their personal moral sense (Myers 1999) or their self-interest (Simon 1990). Consider that a culture's world view affects the ease with which its members can think about making war against their current culturally identified enemies. Then consider the cost to those who, out of duty, fought a war they did not believe in and returned home shattered (Buss 2000).

Although being influenced by a culture's world view is one of the prices paid by human social animals to ensure their group remains functional, we are not the unconscious pawns of our culture's world view. As the word *default* in the above paragraph hints, we, as individuals, have considerable latent influence over our culture, and we can ignore its guidance. After all, we are our culture's decision makers and the carriers of its world view. There any many ways in which we exercise our influence.

We directly influence it, for example, through our efforts to resolve personal dilemmas, especially those that arise from contradictory cultural guidance or differences between cultural and personal situational influences (Myers 1999). When we face the multiple competing interests that make up these dilemmas, we, as individuals, try to balance our interpretation of our culture's guidance about the greater good with our personal interest (Reader 1990). If our resolution to the dilemma is to choose a way of living that is different from our culture's norms, then we influence those around us accordingly (Singer 2001).

We, personally, influence our culture indirectly when we share with others our mundane daily experiences and decisions, our opinions on current events or issues, and newly acquired information.

Examples are the products we do and don't buy, and why; for whom we do or don't vote; and whether we employ our acute ability to detect and report cheating, corruption and fake news or not (Wilson 1999a). These exchanges can influence the decisions of others.

Whether directly or indirectly, our individual decisions and choices, and our justification for them, can affect the social interaction among the members of the subgroups in which we participate. Our decisions and choices can thus contribute to adjustments in these subgroup's structure, goals, values and beliefs. This type of change routinely occurs in all of a culture's subgroups, which results in them routinely adjusting how they interact with one another. It is through these connections that the members' personal decisions about a wide range of issues (including social matters, the choice and use of technology, and ideological beliefs) contribute to reinforcing or changing their greater culture's world view (Thomas 1991).

Western culture provides some easy-to-see examples of member-driven cultural changes to their culture's world view. Over the last 50 years, its world view's guidance has changed in ways that make drinking and driving less legal, permit a diversity of sexual preference, make smoking less acceptable and weaken the influence of formal religion. Factors driving these changes include the personal dilemmas about lifestyle and religion that came to the fore during the 1960s to the early 2000s and the increase in deaths from lung cancer above levels that were considered usual. Currently, Westerners are personally faced with the dilemma of deciding how to balance our culturally supported need to protect our environment with our culturally supported personal desire for more goods and wealth.

**Symbol.** If an observer extracts more meaning from either objects or human actions than can be inferred from their innate characteristics, then these objects and behaviours are said to have symbolic value. For example, a white towel on a stick can be more than that, it is used to indicate surrender. All cultures use cultural symbols in their effort to become and remain functional.

Cultural symbols can accomplish this task because the members are conditioned from birth to recognize them and then respond in a specific way. Thus, when a cultural symbol is displayed to the members within an appropriate social context, it cues them to each recall from their unconscious a similar response and a similar explanation to justify it. Think of the ubiquitous presence of some kind of greeting when people meet.

The shared familiarity arising from a culture's many symbols helps to generate in each member a personal sense of belonging to the group. It also helps to provide them with a unified sense of direction and purpose to their activities (Crumley 2000). This sense of a group identity helps to keep the culture functional.

Cultural symbols can take many forms. Landforms, such as a sacred hill or lake, can represent ancestry and belonging. Monumental architecture, such as a temple, can represent power, stability and security (Wilson 1988). Symbols also include images, such as insignia and icons; rituals, such as a festival; and literary or oral works, such as myths and legends. They each contribute to the members' sense of a group identity (Togola 2000).

Cultural symbols can be actions. For example, an individual publicly demonstrates their acceptance of an authority structure and respect for it by symbolic actions such as bowing to royalty or saluting in the military. Symbolic actions are thus also part of a culture's functioning.

Cultural symbols and symbolic actions can have deep roots, such as decorative gardens. From Babylon to the British gardens at Kew, decorative gardens have been used by the elite to represent their and thus their culture's world view, power and status. This symbolic use of gardens has its ancient roots in the hunter-gatherer's symbolic use of natural features in their landscape to represent their world view (Crumley 2000). Cultural symbols and symbolic actions can also be recent inventions, such as the peace sign.

All of today's cultures use cultural symbols, and their members react to them through the same basic mechanism. Thus, despite Westerners' claim (belief perhaps) that our culture is science-based and rational, we too are surrounded by and react to cultural symbols. They range from victory monuments symbolizing respect for the war dead and what they died for, to the business suit representing status and authority. When Westerners react to such symbols, we too grant those displaying it some power over our lives. Even accepting a prescription from a medical doctor is a symbolic action recognizing the doctor's authority (Butler et al. 1998). Consider too that the order we, as Westerners, choose to impose on the world around us satisfies more than our earth-management needs, it is also a symbolic expression of the order we associate with a civilized society. Think of the order in a suburban decorative garden.

Cultural symbols, like all symbols, have a downside. Their relative simplicity means that they have a limited range of meanings that are non-exclusive. For example,

the eagle can represent power or security while both the dove and the pigeon can stand for peace. Cultural symbols will thus only unambiguously cue the viewer to remember or infer the intended meaning/response under restricted conditions. The members must have learned both the symbol's intended meaning and the context in which it has that meaning. And the symbol must be displayed in that context (Weingart et al. 1997).

The significance of context is illustrated by comparing the message conveyed by a salute from a friend to you with a salute from a soldier to a military general. The message conveyed by a salute from a civilian stranger to you is ambiguous. Consider too the meaning you ascribe to an arm raised skyward by a pope, or a policeman. How do you know the difference in meaning?

There is, therefore, always the possibility, especially in cultures with a complex structure, that a cultural symbol will be missed, ignored, misinterpreted or manipulated. Consider that cultural symbols are deliberately used in advertisements in an effort to influence your consuming preference. For example, in Western culture, symbols such as eagles, crowns, gold and castles can cue in us the feelings associated with power, wealth and luxury. Advertisers hope that by using these symbols in an advert, the feelings they generate will be transferred to advertised products such as alcohol, diamonds and cosmetics, thus unconsciously prompting us to buy them. Similarly, symbols of scientific authority, such as the white coat and technical equipment, are used to advertise detergent and toothpaste.

Because so much of a cultural symbol's meaning is specific to a culture and the cultural context in which it is displayed, it cannot be read as if it were part of a language. Neither can a symbol's current meaning in one culture be used to reliably interpret its historical or current meaning in a different culture. For example, it is hard to imagine how Westerners could use our material-based cultural world view to help us interpret the symbolic meaning of a seal carving created by the high-Arctic Dorset hunter-gatherer culture. For us to even guess at the significance of the carving, we would have to first accept that, to the Dorset, transformations and communications between human and animal were likely a part of their culture's world view. They likely believed in a spirit world. They also likely believed that some animals represented spirits that were helpful, even related, to humans (McGhee 2001).

Similarly, it is true that the Bible is a window into the cultures that existed when it was written thousands of years ago. But it is also true that to fully interpret the symbolic aspects of the Bible's books would require the reader to have a broad knowledge of the cultural context in which they were written (Lang 2002). If only Newton had known that when he interpreted the biblical description of the interior of king Solomon's temple as describing the then-known universe.

### Organizational Structure.

A greater group, and its greater culture, is comprised of a number of flexible subgroups, each of which has its own culture and structure. The subgroups include political parties, religions, work specialties, interest groups, kin groups (e.g., families, households), males, females and sports clubs. The members of the greater group/culture are also members of a selection of these subgroups, so they participate in the greater group/culture both as individuals and through their membership in its subgroups.

The interactions among the members of a subgroup form its structure and its functioning. Similarly, the (formal or informal) interactions among the members of a collection of subgroups establish the relationships and boundaries among the subgroups. Collectively, these interactions are the basis of the greater culture's organizational structure and functioning, and the features of its world view. The dynamics of the members' interactions results in changes to the relationships and boundaries among the subgroups, which gives the greater culture's structure and functioning its dynamic nature.

There are a number of ways a human group can structure itself. However, if it is to be functional, then the possible choices are, as hinted above, constrained, and those constraints must be collectively satisfied. The presence and influence of these constraints is broadly expressed in the nature of the culture's structure and functioning. Generally speaking, a culture (i.e., the group) is functional when its structure and world view enables the members to satisfy both their individual needs, such as finding a mate, and the group's needs, such as ensuring that each member respects the other members. To be successful, a culture's structure and functioning must be seen by the members to be benefiting them personally and as a group. An example is ensuring that there is a fair settlement of disputes among its members or subgroups (Eisenstadt 1988).

The smallest independently functional cultural group, a band, is found among nomadic hunter-gatherers. Their bands range in size from 25 to 50 members. These bands are thought to reflect the oldest of human cultural structures.

This small size means that each member can know everyone else and how they are related to one another (Kelly 1995). The band's structure is the kin relationships among the members. It functions through the give and take of direct personal interactions, such as sharing, that are enforced through social interactions such as social monitoring and social pressure (Wilson 1988). Although a person with special skills, such as an elder, can fill an authority role as needed, that position only lasts long enough for the group to address a specific need.

Because a band lacks the need for permanent authority positions and because all members have equivalent access to resources, the band's structure and functioning is described as egalitarian. The members routinely reaffirm their faith in their culture's egalitarian structure and world view by participating in rituals and ceremonies that are acknowledged by all and in which there is a place/role for all. In the process, they reaffirm their own and their culture's identity.

The size of a band that can have an egalitarian structure is constrained. One reason is that the workings of our brain limit the number of people with whom we can maintain a stable social relationship through direct person-to-person contact: about 150 people, on average (Dunbar 1993). Because direct relationships lie at the core of an egalitarian band's structure and functioning, this human constraint limits its functional size.

One way to circumvent this constraint is to form an aggregate of fluid, socially equal bands. The bands would disperse around their collective home range and only occasionally congregate into a larger group. The larger aggregate of bands can be a unified cultural group, or "tribe," with an egalitarian-like structure. This type of cultural structure, like the band, is common among nomadic hunter-gatherer cultures, for example, the Kung in Botswana.

If the environmental conditions can support a larger number of people, then, with adjustments to the culture's world view, the size of a group with a kin-based, egalitarian structure can reach at least 1,500 members. However by this size, the members find it increasingly difficult to keep their culture functional. For example, it becomes increasingly difficult to ensure compliance with cultural norms through social monitoring and pressure. One way those living in villages during the Neolithic likely resolved this problem was to conceive of the village as a single unifying community structure or "house" (like Çatalhöyük in Anatolia), or construct a unifying community symbol (like the long barrows in Britain). The maximum size of a group with an egalitarian structure appears to be a few thousand.

Alternatively, the group can start to change its organizational structure. They could, for example, make more permanent the greater authority that a few members would occasionally and temporarily receive. The permanence of that position and the special powers the position's occupier holds could be made visible through symbols, such as distinctive dress. The group could also divide itself more formally into subgroups by, for example, work tasks. As long as the power of these authority figures over others is limited to persuasion and the members still have equal access to most resources, then the culture has adopted a ranked organizational structure (Wilson 1988; Weingart et al. 1997). There is a wide diversity of ranked cultures.

As the size of the cultural group increases, both the number of its subgroups and the complexity of its organizational complexity irregularly increases. Around a cultural group's size of between ten thousand and a few tens of thousands members, the culture's effort to remain functional by adjusting its ranked structure starts to significantly change the culture. It starts to take on the characteristics of a state culture: it becomes a complex culture.

Although state cultures display a diversity of features and degrees of organizational complexity, they all have the same basic characteristics. The culture's authority positions are permanent, and the members are more formally divided into subgroups by occupation. The power relationships between these divisions are arranged into an upright pyramid with the most powerful positions at the top. Those higher in the pyramid have the power to order those lower in the pyramid to do some things and to enforce their compliance because the lower-downs have limited rights to refuse. The culture's structure is stratified or hierarchical. However, two things do not change with the appearance of a power hierarchy: each member's right to satisfy their personal needs and their responsibility to act in the group's best interest.

The members who fill the few positions near the top of a state culture's hierarchy, such as a monarch (e.g., a hereditary ruler, a divine ruler, a religious leader, a military leader), have the most power and influence over the culture's other members. As the examples of a monarch suggest, the culture's leaders are usually drawn from the culture's privileged subgroups (the elite), such as royal, religious, military, political, legal and wealthy subgroups. In contrast, those near the bottom of the

hierarchy (referred to by terms such as *peasants* and *public*) have little or no direct influence or power over others, but they collectively have indirect power as discussed below.

The appearance of a power hierarchy also reflects the corresponding changes in the members' specific responsibilities and access to resources. The elite who have the responsibility of making and implementing decisions for the benefit of the rest of the members have greater direct access to resources. Correspondingly, by being relieved of decision making, those lower in the hierarchy are expected to respect both those who make the decisions and their decisions. The peasants and public have less direct access to the culture's resources.

As in all cultures, the members of state cultures reaffirm their culture's world view, but the process is more complex and formal than for simple cultures. The members must now also reaffirm, and maintain, their culture's authority structure (its hierarchy). This is most easily achieved if the individual members and their respective subgroups voluntarily accept, through belief, faith or persuasion, that the status quo is acceptable. This is facilitated through practices such as elaborate ceremonies focused on the leader and by the authority figures demonstrating effective governance, including a fair distribution of resources. The alternative is for the members to be forced to accept the structure against their will.

The reaffirmation of a state culture's world view, and thus its members' cultural identity, usually occurs either with the authority structure's reaffirmation and maintenance process or in parallel with it. The two are collectively referred to simply as the culture's reaffirmation and authority maintenance efforts. State cultures are preoccupied with their reaffirmation and authority maintenance efforts and the significant cost of completing them (Tainter 1988). In particular, they devote considerable attention to securing the resources (beyond those required for basic needs) to pay for their higher reaffirmation and authority maintenance costs. One reason these costs are higher for state cultures is their use of symbols and largess to help them stay functional.

In summary, there are many aspects to a culture's functioning. A culture's world view guides its members to consider both their personal and their cultural group's best interests. For example, they are guided to balance co-operation and competition; balance privacy and surveillance (Wilson 1988); control the self-interest they express through deception, lying, alliances and greed; and maintain their supporting environment. (Note the emphasis on *guide*.)

A culture has the ability to guide its members because they embed their culture's preferred behaviours, beliefs and values into their internal reality (which includes their personal world view and reference normal). In essence, the culture's young learn the features of their culture's world view as they grow up: they are acculturated (Cole and Cole 2001). As adults, the members are largely unaware that their culture's world view has become a part of their personal identity. This is illustrated by the adult members' unconscious response to the cultural symbols they learned as children.

This mention of symbols is a reminder that a culture's visible features (such as its customs, dress and buildings) can have a practical and a symbolic function. Features that act as symbols remind the members of those aspects of their culture's world view that are hidden from view (the culture's norms), such as its beliefs and values. Features that are symbolic acts represent the members' acceptance/recognition/rejection of these norms. These symbols and symbolic acts are a part of the culture's functioning. They influence an individual member's views and decisions about themselves, their social identity, reality and normality and help them to express those views (Myers 1999). Symbols and symbolic acts both bind the members together as a culture and enable them to change their culture.

Once a member has learned their culture's world view, they become one of its carriers. A member is also one of their culture's decision maker. As a result, the members, collectively and individually, both rely on their culture's world view for guidance and adjust or maintain it as they see fit. From this perspective, the functioning of their culture is the members' decision making, which is subject to the influence of their culture's world view. The functioning of a culture therefore includes a member-decision–cultural-world-view feedback loop (member-culture feedback). This contributes to a culture's functioning being dynamic.

The current generations of a cultural group inherit their culture's world view from their ancestors. And, like their ancestors, the current generations adjust it. Thus the functioning of a culture is a dynamic intergenerational process.

A culture's functioning is related to its (organizational) structure. A greater cultural group's structure is the relationships and boundaries that were established among its subgroups by the (formal and informal) interactions among their respective members. The characteristics of that structure, and thus the (greater) group's functioning, varies with its size. Small cultural groups (e.g., hunter-gatherer bands) have egalitarian or ranked organizational structures (simple structures), while large groups (states) have hierarchical structures (complex structures). This variation in cultural structure with group size exists, in part, because our human abilities and limitations together constrain how a group of people can organize themselves and remain functional. The nature of a culture's structure and functioning is included in its world view.

Although the essence of a culture's functioning is the same for all cultures, the details vary with the complexity of the group's/culture's structure. For example, the members of a culture with a simple structure engage in a world view reaffirmation process. The members of a culture with a complex structure engage in a world view reaffirmation, an authority structure reaffirmation and an authority maintenance process (reaffirmation and authority maintenance process).

All of these perspectives of a culture's functioning are valid. Each provides us with a different insight into the nature of a culture's functioning. In combination, they reveal that its structure and functioning are the expression of the interactions among its members (as individuals), their culture's world view and their environment. Those interactions bind these factors into a mutualistic, perhaps symbiotic, cultural system. The details of these interactions vary systematically with the culture's population-dependent structure (Weingart et al. 1997; Kelly 1995). Overall, a functional culture is unified, stable and adaptable. It becomes functional when its members' decisions and its world view's guidance ensure that both the members' individual and their group's best interests are served.

## Types and Connections

The previous section revealed that the essence of how all groups/cultures function is the same, but the details vary with their organizational structure. By dividing the diversity of the earth's cultures into cultural types based on their organizational structure's level of complexity, and then studying the connections between the types, we gain a more complete appreciation for how a culture functions and changes. A three-part division is used here: simple cultures (e.g., hunter-gatherer cultures); intermediate cultures (e.g., sedentary hunter-gatherer and ranked agricultural cultures); and complex cultures (state cultures of varying complexity, such as the less complex state cultures and the more complex industrial cultures). The term *complex culture* is used with the understanding that all cultures are complex *systems*.

In this section, the three cultural types and the connections between them are discussed in the order of their increasing complexity. By focusing on this order, we are prompted to pay attention to the nature of the relationships between them and the features that facilitate and constrain an increase in their complexity. This knowledge helps us to appreciate, more generally, how a culture's functioning facilitates and constrains its characteristics (its world view's features) and how one type changes into another.

### Hunter-Gatherers.

Hunter-gathering was the dominant cultural type for most of the last 200,000 years. This type is considered to be the root from which all the other more complex cultural types, including our complex industrial cultures, arose. It is used in this book as a cultural reference (Wilson 1988).

There is a tendency to imagine that the culture of our hunter-gatherer ancestors was, in essence, the same as ours but without our material possessions and technology. We tend to think of today's hunter-gatherer cultures as relics or representations of our ancestral culture (Kelly 1995). But by paying close attention to hunter-gatherer cultures, it becomes clear that this is not the case.

A cohesive group of hunter-gatherers is called a band. Bands, both past and present, exhibit a variety of hunter-gatherer cultures. In detail, each band has a different world view, so the members from each band experience the world a little differently and have different explanations for their experiences. In detail, each band has a slightly different structure. Each band and its culture is a complete and fully functional unit in its own right.

However, all hunter-gatherer cultures exhibit common features. As the name suggests, one of these features is the way in which they secure their food. Another is an animistic "religion." It is in this collective sense that hunter-gatherer cultures are the same, represent the roots of more complex cultures and are distinct from the more complex cultural types.

Because hunter-gatherers acquire their food directly from their surroundings, they have a close relationship with their environment. For example, the size of hunter-gatherer populations roughly matches the ability of their environmental endowment to support them. This close connection to the land is also reflected in the characteristics of hunter-gatherer cultures. The connection is strong enough that, broadly speaking, the variations between hunter-gatherer cultures reflects the differences in their environmental endowment.

A generic description of the least complex hunter-gatherer cultures, usually those that are nomadic, provides details about the basic characteristics of a hunter-gatherer culture and how they are linked together. A nomadic hunter-gatherer band generally has fewer than 400 and commonly around 25 to 50 members. The small size allows each member in the band to be in direct contact with all of the other members and to know the kin connections among them. Through their hunting and gathering activities, they are also directly linked to the land that supports them. Their close ties to one another and the land are reflected in the band's world view, which is itself tied to the landscape of their supporting environment.

The structure of these nomadic hunter-gatherer cultures is kin-based (kinship provides the primary method for defining the relationship between people). It is also egalitarian, which means that individuals have an age-related decision-making status but equivalent access to resources. Although reputation and age allow for social differentiation, it does not provide grounds for status or a power hierarchy. A good hunter is primarily seen as just a good hunter rather than a better hunter than other hunters (Wilson 1988). The lack of hierarchy means that social duties too are mediated by kin and age relationships rather than power positions, the reverse of the case found in more complex cultures.

In keeping with an egalitarian structure, the world views of nomadic hunter-gatherer cultures emphasize individual autonomy while actively discouraging one person from lording it over another (Kelly 1995). The significance of autonomy is complemented by the high value placed on mobility and flexibility in decision making. These characteristics are all needed if their methods of securing food are to be successful.

Nomadic hunter-gatherers secure their diverse plant and animal foods by travelling around the area that provides them (their home range) in a pattern and at times set by the location and availability of these foods. The required flexibility in their movements and decision making is made possible, in part, by their culture favouring limited division of labour or specialization, few possessions and significant autonomy (Kelly 1995). For example, the essential possessions of the nomadic Kung are limited to no more than the 13–14 kg they routinely carry over long distances (Angela and Angela 1993).

Autonomy, flexibility in decision making and being mobile help the band to secure adequate amounts of food, however there can be considerable day-to-day variation in an individual's foraging success. To ensure that everyone is fed, nomadic hunter-gatherers, like all hunter-gatherer cultures, divide and distribute the secured foods between individuals and/or cultural subgroups in ways that strongly reflect their culture's kinship relationships and egalitarian structure. This means a pragmatic, rather than an altruistic, form of sharing: the act of sharing is accompanied by an unexpressed expectation of future reciprocity. The degree to which a culture practises sharing varies from very common to rare depending on features such as the food types and their availability (Kelly 1995; Wilson 1988).

The egalitarian nature of nomadic hunter-gatherer cultures has led to the popular Western perception that individuals in these cultures are independent and "free." This is true in the sense that they emphasize flexible decision making, favour autonomy and lack a power structure with its obligatory maintenance costs (Wilson 1988). But their "freedoms" come at a cost. This takes the form of heightened, obligatory personal awareness of their responsibilities to the band.

Each member is obliged to ensure that everyone (including themselves) conducts their interpersonal relationships within the limits needed for the culture (the group) to remain functional. These limits include maintaining the members' feeling that they belong, ensuring that sharing is conducted properly, suppressing the lording of one person over another and avoiding interpersonal conflict. These are not arbitrary requirements. They reflect requirements for their survival: their mutual dependence on one another in a cohesive band.

Each band member fulfills part of their cultural responsibilities by monitoring their personal behaviour. They are also obliged to maintain a heightened awareness of their fellow band members' behaviour and to apply social pressure when a member deviates from the cultural norm. Think self-serving bias. This applies especially to those in one's immediate surroundings. The personal responsibility of a member to monitor their own behaviour and that of the other members never stops.

Nor can individuals forget that they live in a very open society. For example, there is little if any distinction between public and private lives. Hunter-gatherer cultures therefore have to deal with the conundrum of maintaining individuality in the face of such openness and scrutiny. This requirement is partly satisfied by the way individuals conduct their interpersonal relations. For example, their criticism of others is done obliquely and quietly (Wilson 1988). Anger is restrained.

The characteristics of nomadic hunter-gatherer cultures are also linked to and reflected in their relationship with the environment that sustains them. The nature of that relationship can be seen in their religion and their resource management practices, but it is illustrated here by their land tenure system. In contrast to Western culture, there is no incentive to secure a title deed or territorial ownership of land. Their tenure systems focus on the right to use the products of a patch of land while including the requirement that their use of the land reinforces social relationships and the group's functioning (Wilson 1988). Thus, although a certain kin subgroup has rights to use a particular patch of land to collect berries, for example, those rights come with cultural responsibilities. Through these obligations, land tenure is strongly tied to sharing and exchange (Kelly 1995).

A nomadic hunter-gatherer culture's world view is made coherent and tangible to its members by tying it to the landscape (land, water and sky) where they live. They accomplish this by linking particular aspects of their world view, such as a myth or a land-use tenure allocation, to specific landscape or skyscape features, such as a hill, a star (or stars) or an animal. The landscape thus becomes more than an arrangement of physical features: it is a "map" that embodies their world view.

The map's real places can also act as symbolic cues reminding the members of an aspect of their world view, such as an explanation for how the world functions, their history, proper behaviour in interpersonal relationships and their obligations. Think of the Australian Aborigines' dreamtime journeys. In its role as a map of their world view, their landscape, along with their ceremonies, help to reaffirm their culture's world view, their sense of identity and that they belong in their home range (Wilson 1988).

Their view of the land helps bind them together. But the land also supplies them with food. Thus, from many perspectives, the land is an integral, inseparable part of their lives. For example, their multiple uses of the land and their relationships to it combine with the details of their kin-influenced, interpersonal interactions to create a complex hunter-gatherer land-resource tenure system (Kelly 1995; Dean 2000; Togola 2000; Wilson 1988).

However, as discussed above, the characteristics of a nomadic hunter-gatherer world view, like the world view of all cultures, are constrained in many ways. For example, interpersonal monitoring only works for small band sizes, probably no more than a few hundred people from a few families or households (Kelly 1995; Scott 1997). Their world view is also constrained by their environmental endowment because it limits the amounts and forms of resources available and thus the band's size and customs.

For a nomadic hunter-gatherer band and its culture to remain functional, its world view has to accommodate these types of constraints in an appropriate manner. Even for nomadic bands, this is a much more complex task than expected. Consider that their culture's world view must somehow keep the band's size within the limits set by its food supply.

To accomplish this, their world view must do more than consider mouths to feed. It must accommodate the many other factors that contribute to the dynamics of a band's size: in particular, human biology, their ecosystem endowment and individual behaviour, as well as the complex interactions among them. Examples include the degree of mobility needed to gather resources, food availability and the frequency of bad food times, the intensity of food exploitation, the help available to mothers during child rearing, breastfeeding intensity and their relationship to neighbouring bands (Kelly 1995).

There are a number of ways a culture's world view could juggle the variables they can control to satisfy the band size requirement. For example, during the good food years, the band could live in a manner that kept its size within a range that their environment could support during the bad food years (Hassan 1980). Alternatively, it could maintain the memory of how the members must live or adapt during the bad years. Another alternative is for the members to accept the possibility of forced population declines, such as starvation. Or they could use aspects of all three.

Clearly, even the least complexly organized of hunter-gatherer cultures, like all cultures, is a highly complex, integrated, flexible social system with no extra bits that can be arbitrarily removed or improved. Despite the environmental constraints on them, hunter-gatherer cultures have been present over the last 200,000 years. They were only joined by a new cultural type around 13,000 years before present (yBP) (11,000 BC) with the rise of agriculture. We should thus expect that the essence of our hunter-gatherer ancestors' world view, and the human and environmental constraints on it, still influences the complex cultures of today.

*Sedentists and Agriculture.* This subsection discusses cultural groups that are described as egalitarian farmers or as having a ranked structure. A group is considered to be sedentary when it is both associated with a specific place and lives there in usually permanent structures called houses. The time the group spends there can vary from seasonal to full-time. Thus being sedentary can be part- or full-time. Although full-time sedentary groups can be hunter-gatherers, they are more likely to practise agriculture. Sedentary farmers live, primarily, on the food they grow in nearby prepared fields and obtain from their domestic animals. Farmers who obtain most of their food from domestic animals are called herders or pastoralists. They are often nomadic (not settled), but they take their houses with them.

As this description suggests, the division between hunter-gatherer and agricultural cultures is fuzzy. Those practising farming can engage in aspects of hunting and gathering, while hunter-gatherers can engage in some aspects of farming. However, agricultural cultures generally display greater levels of sedentism and structural complexity than hunter-gatherer cultures.

One way to reveal the overlap and differences between the two ways of living is to compare the more structurally complex sedentary hunter-gatherer cultures with those agricultural cultures that have similar levels of complexity. Another is to look at how the lifestyle and world view of a hunter-gatherer culture changes to become a farming culture. This occurred in the Middle East around 11,000 BC when some of our hunter-gatherer ancestors became farmers. An understanding of the interactions involved in that change will allow us to better appreciate, in general, the diverse yet holistic nature of a culture's complex functioning and changing.

*Abu Hureyra.* The process by which a hunter-gatherer culture can become a farming culture is illustrated by the village of Abu Hureyra, Syria. Limited excavation of the site was conducted immediately prior to the site disappearing underwater behind a dam. The following provisional history of Abu Hureyra is based on the results of that excavation.

Abu Hureyra is located on a river terrace that formerly overlooked the 6 km wide Euphrates flood plain (figure 21, insert). Today, the area has a semi-desert climate. Most of the average 200 mm of precipitation it receives falls in winter. The amount is the minimum needed to practise rainfall-supported wheat farming. This rainfall also supports the two ecosystems accessible from Abu Hureyra: to the south lie the plains that host a degraded steppe ecosystem, while to the north lie the remnants of the Euphrates flood plain ecosystem that is now mostly under water. The current state of these ecosystems only poorly represents the ecosystem conditions when the Abu Hureyra site was first occupied around 11,100 BC.

In 11,100 BC, three ecosystems were accessible to the settlers. The steppe, to the south, hosted a diverse variety of plants, including trees and grasses, that supported animals such as gazelle, wild sheep and goats. The Euphrates flood plain, to the immediate north, supported a rich flora and fauna, including wild cattle and pigs. A short distance further to the north, a third ecosystem, an oak park woodland, provided a variety of plant foods. Also found within 80 to 150 km of Abu Hureyra were sources of useful minerals and rocks, such as the basalt used to make grindstones.

This bountiful environmental endowment allowed hunter-gatherers to settle at Abu Hureyra at a time when the climate warming that ended the last ice age and the resulting ecosystem changes were ongoing. Over the next 5,400 years of occupation, which covered the 1,400-year-long Younger Dryas cold period, the inhabitants experimented with growing crops, domesticating animals and adjusting their world view to match their new way of living as farmers. This is the story of their culture's long-term change from hunter-gatherer to farmer. The slowness of the process makes the changes and the contributions to them more visible. Thus, although the story could be condensed, doing so would mask the nature of the process, which is the object of its telling. The length is part of the story.

Around 11,100 BC, hunter-gatherers established a permanent settlement at Abu Hureyra. The villagers initially lived in subcircular pit dwellings that were approximately 2 m in diameter and covered by pitched roofs. They soon switched to an above ground village, where the residents lived in rectangular timber and reed huts. The village housed around 100 to 200 people, a relatively large size for a hunter-gatherer band.

A permanent settlement of hunter-gatherers implies that most of the food and resources its inhabitants required could be found within a 30 km round trip from home. Indeed, their annual round of seasonal hunting and gathering activities in the three highly productive ecosystems of their environmental endowment provided them with more than 100 species of wild seeds and fruits. These seasonal foods included wild varieties of

lentils, einkorn (a wheat), rye, barley, feather grasses, hackberries, almonds, acorns, pears, terebinth (cashew family) nuts and pistachio nuts. Despite this diversity, the bulk of their plant foods was provided by only a few species of grains and seeds. Most of them required preparation by pounding, grinding and roasting.

Gazelle were their primary source of protein. These animals were hunted each spring when their dense herds spent a short time in the area during their migration northwards. The hunt likely involved the whole community in a concerted effort to quickly trap, slaughter and process (including meat drying) a large number of animals (Bar-Oza et al. 2011). Fish were, surprisingly, generally absent from their diet.

During this and all subsequent periods of occupation, the tools they used to harvest and prepare their foods were made from stone, animal and plant materials. For example, they used a saddle quern and a grinding stone to prepare grains for eating. They already possessed whorls to make yarn (implying but not proving that they wove) and needles. A standard tool was likely a sickle, faced with chipped stone, which was used to harvest wild grains.

Around 10,900 BC (12,900 yBP), some 200 years after settling in Abu Hureyra, the area's climate began a period of rapid cooling and drying. (This date coincides with the start of the hemisphere-wide Younger Dryas climate cooling at 12,900 yBP.) This change in the local climate prompted a series of rapid, significant ecosystem adjustments. These included the rapid northward movement of the oak park woodland away from Abu Hureyra and the drying of the southern steppe, which reduced the number of terebinth trees. These changes were likely responsible for the decline in the variety and abundance of plant foods gathered by the Abu Hureyra villagers as recorded in their fireside waste dated at around 10,600 BC. The first to decline were acorns and terebinth nuts. These were shortly followed by declines in wild rye, einkorn and pulses, in an order that matched each plant's ability to resist drought.

It seems that changes to their endowment pushed the villagers to resolve a food availability issue. They may have been able to solve this problem by migrating or becoming nomadic hunter-gatherers again. However, at least some of them decided to continue living at Abu Hureyra and to address their food shortage by substituting one food for another. Perhaps they made this decision because they realized that they lived in one of the more resource-rich areas in the region.

As the climate continued to cool and dry, their opportunities for food substitutions dwindled. They responded by encouraging the growth of those plants whose seeds they were now routinely collecting for food. This augmentation took place both in the wild and through deliberate cultivation, most likely in suitably moist sites near their settlement. The first plant to be cultivated was wild rye. This was joined shortly thereafter by wild lentils. Wild wheat became a cultivated crop later, most likely because it required more effort to process into food than wild rye.

It could be concluded that the villagers of Abu Hureyra had been pushed into unusual agriculture-like activities by climate-change-induced food shortages. A second perspective suggests that their adoption of agriculture was simply an extension of their existing seasonal food gathering round. From a third perspective, their settled lifestyle had presented the villagers with the option of taking up cultivation. All three perspectives are correct in their own way. In any event, their decision to stay required them to permanently change their diet, to develop new tools and skills, and to form a different world view: they had to substantially change their culture.

Although there is no direct evidence of how they made their decision, it seems unlikely that their choice to begin cultivation was the result of a fully conscious evaluation of their circumstance. It seems more likely that the two centuries since they started living an unusual settled hunter-gatherer lifestyle in 11,100 BC had preconditioned them to drift into a lifestyle of cultivation. This was only possible because of their unusual ecosystem endowment.

It was also made possible because, during the previous centuries as hunter-gatherers, they had become accustomed to the strongly seasonal, short-lived and intense food hunting and gathering effort demanded by their endowment. Similarly, they had become accustomed to the communal activity needed to take advantage of the short opportunity to hunt gazelle. In addition, by having lived in a relatively large, permanent settlement, they had likely already solved some of the organizational and interpersonal issues needed to take up farming. Finally, although there is limited direct evidence in the archaeological record, it seems they were also likely familiar with the process of storing their foods for the leaner times between harvests. Think gazelle meat. When taken together, this history suggests that they had been preconditioned to more easily take up the practice of rudimentary agriculture as a solution to their food issue and to make the cultural changes that go along with it.

Regardless, the availability of food remained tenuous, likely as a result of the ongoing, centuries-long cool and dry period. In response, the villagers increased their cultivation efforts. This choice is recorded by the steadily increasing amounts of weed seeds found in their garbage. The plants producing these seeds represent an ecosystem's response to the kind of disruption associated with making the land available for agriculture. Despite the likely ongoing cultural adjustments prompted by their new method of securing plant foods, many other aspects of their lives remained unchanged. One of these was the spring gazelle hunt, for which they likely used their only domesticated animal: the dog.

The Younger Dryas cold and dry period lasted for about 1,500 years. By around 9460 BC it had ended. The climate became sufficiently warm and wet that the ecosystems around Abu Hureyra could support trees again.

By 8950 BC, some 500 years after the end of the Younger Dryas, the villagers' reaction to these changes was clear. They had retained their timber and reed huts but had adjusted the siting of their village. This phenomenon occurred simultaneously throughout the region, suggesting that the reasons were climate related. Although the ecosystems around Abu Hureyra had become more productive, the villagers did not return to being full-time hunter-gatherers. Instead, they continued their cultivating activities.

Perhaps this choice is unsurprising because, by this time, they had been practising some form of cultivation for at least 1,500 years. Raising plants for food was now a cultural norm and part of their world view. The limited evidence from around 8950 BC also suggests that the plants they cultivated now possessed high levels of domestication traits. For example, the wheat they grew was more easily threshed and winnowed (free threshing). These traits increased their food reward for the same farming effort, which provided them with an incentive to continue cultivating the plant.

Their increasing commitment to farming over the 850 years after 8950 BC is indicated by the steadily increasing amount of field weeds found in their crops. During this time, the gazelle hunt still continued to provide them with protein. It likely also provided them with a sense of cultural continuity and stability. However, their choice to increase cultivation also heightened the seasonal conflict between their ongoing annual round of hunter-gathering activities and the demands of their farming activities. Consider that, during spring, they were still faced with dividing their limited labour and time between the hunt, the harvest of their cultivated winter cereals and the gathering of wild foods. Part of their solution was to decrease gathering, as indicated by the substantial decrease in the indicators of gathered wild plants in their waste. These decisions were likely influenced by the, perhaps ongoing, farming-favouring adjustments to their culture's world view that had been initiated when they began to cultivate crops.

By 8090 BC, the villagers' primary sources of carbohydrate were the five fully domesticated cereals that they cultivated: rye, einkorn, emmer (a wheat), bread wheat and two- and six-row hulled barley. They also grew at least three pulses: lentils, peas and vetches. Some wild seeds, such as club rush and knotgrass, were still gathered. Their primary source of protein was still the gazelle they hunted.

Around 8090 BC, Abu Hureyra villagers took a new path. Their population started to grow quickly and the village underwent a rapid expansion. Food production from cultivation continued to increase as the villagers cleared more fields, which is indicated by the presence of yet more weeds in their garbage. At this time, the climate was reaching its warmest temperatures between the last glaciation and present times, the climate optimum, where it remained until around 3000 BC.

About a thousand years after 8090 BC, by 7100 BC, some 2,500 to 3,000 people lived in Abu Hureyra, which was now 8 hectare (ha) in size, at a population density of 312–375 people/ha. The gathering of wild carbohydrate sources ceased. The villagers' carbohydrate needs were completely satisfied by cultivating the same grain types grown in 8090 BC, with the exception of rye which was now rarely grown. The villagers still continued to cultivate the pulses they grew in 8090 BC, to which they had added chickpeas and beans. The gazelle hunt still continued to supply them with protein, although the few sheep and goats they now raised could also supply meat. However, they may have mostly supplied materials, such a wool. Regardless, the Abu Hureyrans were provided with food throughout the year.

Although their new four-season round of food-providing activities was only slightly modified from their hunter-gatherer three-season cycle, the differences were significant. The changes, such as the cessation of gathering, eased the difficulty of accommodating the schedule of cereal farming and herding, such as field preparation and lambing. However, the conflict that resulted from the overlap between the spring farming demands and the spring gazelle hunt remained. That conflict likely continued to influence their decisions about time allocation, labour divisions and communal activities.

Around 7100 BC the villagers were living in rectangular, flat-roofed mud-brick houses that were separated by narrow, filthy passages and small courts. The houses tended to face south, likely to catch the winter sun. Although small grain storage bins were found in some houses, no substantial food storage buildings were identified. It seems likely that animals were housed and foods stored on the edge of the village. Artefacts used for farming and other activities were found in the houses, but the site where they were manufactured remains unidentified. More than one generation likely occupied a house.

Although a house was only some 70 m$^2$ (roughly 8 x 8 m) in area, it was divided into a number of tiny rooms. They had hard floors, often made of coloured gypsum plaster, that were kept clean. The rooms were likely bare of furniture because the villager's skeletons indicated that they squatted for many activities.

Their skeletons also indicated that the inhabitants ate a well-balanced diet and that those who reached adulthood could live for 60 to 70 years. However, they paid a price for their diet. They had to develop strong chewing muscles: firstly, because dried meat is tough; and secondly, because they likely consumed the increased amount of grain they ate as a dry bread made from coarse-ground grain; hard, uncooked cracked wheat; or roasted whole grains.

Another consequence of chewing a much coarser and harder diet than their hunter-gatherer ancestors was the rapid wearing down and pitting of their teeth. This damage was exacerbated by the presence of grindstone rock fragments in the ground grains, which resulted in tooth chipping and breakage. The villagers therefore suffered from premature loss of teeth and most likely experienced painful tooth ache. However, caries (tooth decay) and plaque were largely absent.

Adopting grain-dominated agriculture also brought with it more sustained heavy labour than they would have experienced as hunter-gatherers. The impacts of these activities, such as field preparation, load carrying and food processing, left their mark on the villagers' skeletons. They reveal that the loads were often carried on the head which produced spinal deformities, especially when too heavy a load was carried at too young an age.

The most demanding and intensive activity in the settlement was the daily dehusking of the grains and grinding them in saddle querns. The suffering of the preparers, generally women, from the resulting repetitive strain injuries is recorded in their skeletons by the distinctive deformities to their toes, knees and back. Although the villagers' female hunter-gatherer ancestors had suffered from these injuries, their frequency and severity increased once grains became the main food source for the villagers.

The dominance of these deformities in women suggests that at least this form of labour was divided by sex. This conclusion is probably supported by the recovery of more female than male skeletons from under the floors of the houses. It seems that women spent more time at home, while the men likely tended to the fields. Although there was likely a greater division of labour than in hunter-gatherer times, there is still no evidence at this time of hierarchy in the culture.

Cultural adjustments are indicated by more than changes to skeletons and the food types consumed. Their stone and bone tools indicate both changes in skills and greater travel. For example, there was an increase in chipped stone blades, the appearance of tanged arrow heads, new types of bone tools and the use of more distant exotic material. More striking than these changes is the sense of cultural continuity and general stability. This is indicated by the ongoing gazelle hunt; the similarity of many stone artefacts to their hunter-gatherer age equivalents, such as the grinding stones; and the continued use of some specific types of bone and chipped stone tools. It is also indicated by the continued burial of the dead within the community, the same house being lived in over many generations and the use of the same sources for some exotic materials.

After 7100 BC but before 6840 BC, life in Abu Hureyra underwent a dramatic change. Hunting joined gathering as a non-essential activity. Over a period of less than 100 years, gazelle hunting was replaced as their prime source of protein by the rapid expansion and intensification of domesticated sheep and goat herding. The villagers now satisfied their sustenance food needs in an annual round of seasonal, agriculture-dominated food-securing activities that are recognizable to us today as normal (norms perhaps) for a farming culture engaged in cultivating and herding.

A number of factors probably combined to facilitate the final switch. The most likely material reasons for increasing cultivation were twofold: Firstly, the soils in both the steppe and the valley bottom could likely still support the level of cereal cultivation needed to supply the villagers' carbohydrate needs. Secondly, the climate optimum allowed the steppe ecosystem around Abu Hureyra to support intensive herding. There is also the speculative possibility that, by this time, they were maintaining their fields' fertility by spreading their domestic animals' dung on those nearest to their corrals (Bogaard et al. 2013).

In contrast, the most plausible explanation for the rapid decline in the gazelle hunt is the likely reduction in the gazelle population due to overhunting. This could have occurred because of the rapidly growing human population both near Abu Hureyra and in the rest of the gazelle's migration zone. A second possible reason for the decline is that any increase in animal husbandry would create demands, such as spring lambing, that would compete with the demands of the spring hunt and the cereal harvest. The significance of this reason would depend on their culture's guidance about the relative cultural value of the hunt, the value placed on domestic animals as protein sources and how the culture determined the ownership and division of resources. (They may have still hunted gazelle but processed the animals in the field, so there would be no record in the village. However, this possibility does not resolve the timing problem.)

Adding to these pro-cultivation/anti-hunting reasons would be their cultural relationship to their environmental endowment (Bar-Oza et al. 2011). Their chosen path suggests that their 6840 BC world view was substantially different from that of the hunter-gatherer ancestors of 3,800 years before who, around 10,640 BC, had started experimenting with plant cultivation. After all, they were now, in at least a material sense, full-fledged farmers.

After the Abu Hureyrans had completed the material switch to farming, the population continued to increase. A few hundred years later, by about 6450 BC, Abu Hureyra covered some 16 ha. The population had risen to its maximum size of between 4,000 and 6,000 people, at a population density of 250–375 people/ha.

They fed themselves from the plants they cultivated and the large flocks of domesticated sheep and goats they herded. The presence of large groups of animals is recorded at Abu Hureyra by the large number of their bones and by the changes in the steppe flora. At this time, the villagers also kept a few cows and pigs.

Their domestic animals provided non-edible resources such as hides, as well as meat and perhaps milk products. The uncertainty about milk products arises because the villagers likely could not drink raw milk due to lactose intolerance. The earliest known efforts in the region to make milk digestible by processing it, for example, into cheese and ghee, used pottery vessels. Although the making of pottery occurred before 6500 BC, by how long is unknown. The first known evidence for pottery at Abu Hureyra is younger (Evershed et al. 2013). The cattle were probably not used as draught animals.

The form and arrangement of their houses and the burial of their dead had not changed since 7100 BC, thus indicating at least some cultural continuity. However, the culture had most certainly become more agriculture-centric. Consider that to complete the tasks associated with increased animal husbandry, new domestic animals and more intensive crop production, they had to develop new skills and tools. They also had to organize themselves differently. An adjustment to their world view is strongly implied. The types of changes to be expected would include a more formal division of labour and specialization. Evidence for specialization includes the presence in only a few skeletons of grooved tooth-wear patterns associated with basket weaving or the distinctive tooth abrasion associated with thread making.

The growing importance of technology is also suggested by the new use for their old skill of weaving. They built sieves that were probably used to screen out the stone fragments and husks from the ground grain. The appearance of this invention would explain the reduction in tooth breakage seen at this time. However, the excessive tooth wear from chewing the hard grains continued.

Before 5850 BC, Abu Hureyra's population was in decline. Around that date, the villagers discovered that their skill at making baked clay ornaments could be adapted to make pots. The villagers could then turn their dietary staple of grains into a porridge by either soaking or boiling (cooking) them. The evidence that they took advantage of this opportunity is provided by the decline in the occurrence of ground down and pitted teeth and an increase in the occurrence of caries (tooth decay).

There were other, often surprising, cultural changes that correlate with the appearance of pottery. Among them was the large rise in the number of infant deaths. This likely resulted from some combination of an increase in both fertility and the death rates of the young.

The possible causal connection to pottery arises because of the large increase in the available carbohydrate from cooked grains, but not a proportional increase in nutrition. By eating cooked grains, a woman could directly raise her fertility. Or, by feeding it to her child instead of a tougher diet of uncooked grains, she could lower the time of weaning, which usually occurred at three to four years of age. That too could indirectly increase her fertility. At the same time, too much of a reliance on cooked grain, with its lack of nutrition, could result in undernourishment and the premature death of children.

Routinely eating cooked grains would also have changed the culture. One connection is through their obligation to collect more firewood in order to boil water. This would have increased the villagers' chores, altered their routines and the division of labour, and further altered their supporting ecosystems.

Shortly after 5850 BC, Abu Hureyra had shrunk to 7 ha in size. The villagers were still living in mud-brick houses, but the open spaces between them were much larger. By about 5700 BC, the village was abandoned. Why and what happened to them?

The inferred presence of large sheep and goat populations, the recorded decline in the steppe flora and the known sensitivity of the steppe biome to overexploitation all suggest it was being overgrazed. The presence of a large human population suggest that the overhunting of gazelle, excessive wood removal and too high a level of agricultural intensity may have been occurring. The short-lived cooling event of 8,200 yBP (6200 BC, just before abandonment) recorded in Greenland may have resulted in the drying of the region's climate and a decline in its productivity.

It seems that by the time the population at Abu Hureyra was in decline, the villagers were being faced with stark choices. Food, especially nutritious food, was likely scarce. Many decades would pass before the land could again support their farming lifestyle. Neither could they easily return to being hunter-gatherers. Not only were their supporting ecosystems not hunter-gatherer friendly, but the villagers had likely lost many of the needed skills. Finally, their level of cultural complexity and their world view was not of a type that could allow other options such as constructing irrigation works. Their most realistic options were to migrate or become nomadic pastoralists (Molleson 1994; Molleson et al. 1993; Hillman et al. 2001; Moore et al. 1975; Moore et al. 2000).

*Complexity's Road.* The story of Abu Hureyra provides us with a greater appreciation for a culture's functioning and changing. After 5,400 years of changes, the culture of the villagers who abandoned Abu Hureyra was significantly different from that of the hunter-gatherers who originally settled there. During this time, some aspects of their culture hardly changed, such as the use of the saddle quern, while other aspects changed steadily, such as their farming activities. Some changes happened rapidly, such as the cessation of gazelle hunting, while others occurred more slowly, such as the changes in the use of baked clay. Some changes seem to have taken place in bursts, such as the switches in housing characteristics. Overall, it took around 3,500 of those years for the Abu Hureyran villagers to change from relying exclusively on wild foods to relying almost exclusively on the farming of domesticated grains and animals.

The physical evidence in the archaeological record indicates that from 6840 BC on, Abu Hureyra's villagers could be called farmers. Yet, some 1,100 years later, when they abandoned their village around 5700 BC, the material evidence suggests that Abu Hureyra's cultural structure was still largely egalitarian. There is evidence suggesting that people were divided by their work specialization and their sex. In particular, women were buried with jewellery and their skeletons more commonly bear the marks of food preparation efforts. In contrast, men were buried with flint tools. However, there is a marked lack of evidence for a formal hierarchy in Abu Hureyra. For example, their house design remained similar throughout Abu Hureyra's history, and no substantial public buildings were discovered. It is possible that some members may have had authority over others, but its manifestation did not leave an archaeological record. Their culture's organizational complexity seems not to have exceeded the level of a ranked society.

Thus, even though Abu Hureyra's estimated population size of 4,000–6,000 around 6450 BC strongly suggests the need for a more formal cultural structure, the material evidence suggests that the community was likely egalitarian or perhaps weakly ranked. Individuals were likely still distinguished by kin, age, skill, specialization and sex rather than by position or wealth. We can think of Abu Hureyra as an unusually large Neolithic village whose residents were farmers in practice but had not yet fully adopted all of the features we associate with a farming culture. Calling Abu Hureyra's culture "late Neolithic hunter-gatherer" acknowledges that the complexity of its structure and the nature of its world view was more like a hunter-gatherer culture than that considered usual for a farming culture. The earliest known material remains in the Middle East that indicate the presence of hierarchy, and thus a more complex culture, are found in southern Mesopotamia. They are dated at around 5000–4000 BC, some 1,000 years after Abu Hureyra was abandoned.

We are obliged to conclude that the change in Abu Hureyran culture over this period was the outcome of a combination of many material and social adjustments. These changes occurred at different times and rates, and interacted with one another. Abu Hureyran culture did not undergo a switch-like change: it was transformed through a transition process.

How to describe this transition to farming? It was a product of how the Abu Hureyran culture functioned and the influences on it. That functioning is broadly described as the interactions among the individual members, their supporting environment and their culture's world view. These interactions included a feedback in which the members' decisions shaped their culture's world view, while their culture's world view influenced their decisions: the member-culture feedback.

How this description can help us to appreciate the features of Abu Hureyra's cultural transition is illustrated when we apply it to the accompanying increase in the villagers' population. Around 11,100 BC, some 100–200 hunter-gatherers lived in Abu Hureyra village. After 8090 BC, the population increased to an estimated 2,500–3,000 farmer-hunters. Around 6450 BC, the population reached its maximum of some 4,000–6,000 farmer-herders. By 5700 BC, the village was abandoned (Hillman et al. 2001).

A standard explanation for such a population change would consider a number of variables. They would include the villagers' environmental endowment, ecosystem change, the seasonality of foods, diet, the food economy, demographics, migration possibilities and human nutrition needs. A complementary view appears by looking at population change from the perspective of the individual members, their culture and the environment each making contributions that favour and/or constrain population growth

The following are examples of contributions that favour Abu Hureyra's population growth. The individuals' contribution is illustrated by the increasingly dominant role of grains in their diet. It likely facilitated an increase in a woman's fertility relative to other diets (Hassan 1980; Hillman et al. 2001). The environment's contribution is illustrated by the soil's fertility. It allowed the farming of the land to increase the availability of food and thus feed more people. The cultural contribution is illustrated by farming being more labour intensive than hunter-gathering. It provides a cultural incentive to have more children (Hassan 1980).

The following are examples of contributions that dampen population growth. By individuals relying too heavily on grains for food, they can suffer from poor overall nutrition. That, in turn, increases the possibility of infant mortality (Hassan 1980; Molleson 1994). A culture's world view can favour their traditional reliance on grains. A decline in soil fertility through overuse reduces the number of children their endowment can support.

In addition, the favouring and the dampening contributions overlap and interact with one another, and change over time. It is thus difficult, in overview, to decide which of the variables, contributions or mix of them is most responsible for the changes in Abu Hureyra's population size. For example, did pottery facilitate population growth through boosting fertility or contribute to its decline through premature death? Then remember that the population was already declining at the time pottery and cooking became available. This implies that other aspects and interactions were involved in the population changes long before the time of pottery's invention. Perhaps environmental degradation played the key role.

Likewise, remembering that the individual members are their culture's decision makers helps us to understand how the population came to grow: individuals make the decision to engage in sex (Kelly 1995). But taking our decision-making foibles into account adds another layer of complexity into our deliberations. Overall, we learn from this discussion that the process of population change was complex and dynamic. Similarly, cultural change is a complex and dynamic process.

With this in mind, Abu Hureyra's transition to farming will be treated as a dynamic, complex process that involves the interactions (including feedbacks) among the members (as individuals), their culture's world view and its environmental endowment. And the discussion of what is responsible for the transition will be adjust to focus on how Abu Hureyra's cultural change was sustained in a particular direction for so long that it went through the transition. It will take the perspective of the decision maker, the individual member.

The complexity and vagaries of the world's functioning ensures that living life is dynamic, thus the Abu Hureyrans were continually faced with making decisions. In making them, the villagers were, like us, mostly concerned with addressing short-duration (a few years at most) issues such as resolving a dispute, reacting to a short-term change in the availability of wild food, or whether to repair an old tool rather than make a new one. They likely used the same short-term focus when dealing with ongoing, long-term issues. For example, soon after the villagers began intensifying cultivation, they were faced with deciding each year how to divide their limited time and skills between the spring activities of the gazelle hunt, the farmed-cereal harvest and wild-food gathering.

Whether they were aware of it or not, implementing their short-term decisions affected more than the issues they were meant to address. The world's complexity ensured that their decisions could produce outcomes indirectly as well as directly and that those outcomes could be separated from the initiating decision both in time and space. This made it difficult for the Abu Hureyrans to see the connection between a decision and its long-term consequences (between cause and long-term effects) and then learn from it. Thus the Abu Hureyrans, like us, tended not to deliberately take the possible long-term outcomes from their decisions, such as their impact on future generations, into account when making decisions.

If the Abu Hureyrans were concerned about the future, they had a recipe for dealing with it. They "instinctively" followed the guidance from their culture's world view. However, there was a catch.

Consider that a result of their original, annual decisions to engage in farming-like activities was the slow human-decision-assisted development of self-threshing wheat varieties. Their increasing dependence on growing that wheat required them to change their habits in order to accommodate its needs. Similarly, by successive generations resolving the annual conflict between the spring demands of hunting, gathering and agriculture in favour of agriculture, the villagers drifted further into being farmers. These consequences took centuries to appear.

Critically, by routinely making their personal, short-term farming-favouring decisions, the Abu Hureyrans' collectively adjusted their culture's world view so that, over time, it became more farming friendly. Their culture's world view, in turn, increasingly guided the members to make farming-friendly decisions. The members had created a farming-favouring member-culture feedback loop. Thus, at some point, by relying on the guidance from their culture's world view when dealing with concerns about the future (or present), they biased their decisions toward farming and made it difficult to question that bias.

This farming-favouring feedback joined with the other influences on their decision making (short-term focus and complexity) to collectively affect both the possible choices that the later generations could make and which of those choices they were more likely to favour. For example, by the early villagers favouring farming, they preconditioned the later generations of villagers to make specialized farming tools and to increase their degree of social differentiation. At the same time, favouring farming slowly rendered the choice of returning to hunter-gathering impossible. Overall, it was the individual Abu Hureyran's culturally supported tendency to routinely (consciously or unconsciously) make decisions that favoured farming over other options that sustained their culture's long-term (generational) change toward farming.

But the individual Abu Hureyran's were not entirely responsible for their culture's sustained change toward farming: their environmental endowment facilitated it. Consider the influence that the environment's longer-term changes had on the members' lives. Early in Abu Hureyra's history, the climate underwent a rapid (for a climate) cooling and drying as it entered the Younger Dryas period. This had a profound effect on food availability, which the Abu Hureyrans could not ignore. It lay behind their initial farming-friendly decisions. This cold, dry period lasted some 1,500 years, so it had a significant long-term influence on their lives: it preconditioned them to continue making farming-friendly decisions.

After the climate rapidly rebounded to a more agriculture-friendly warmer, wetter period near the middle stages of Abu Hureyra's history, its influence on their culture was different. This long-term stable period facilitated their ongoing increase in farming activity and the increase in population, but it didn't remove the limits to the soil's fertility. This was expressed by the likely long-term decline in soil fertility from its overuse, especially overgrazing.

These long-term environmental changes occurred slowly, in human terms. Thus, even if the Abu Hureyrans may have been somewhat aware of them, the individual villagers were unlikely to have directly considered them during their decision making. Instead,

these changes occurred in the background, where they formed the foundation upon which the villagers made their shorter-term decisions about day-to-day issues. The longer-term environmental changes thus contributed to the Abu Hureyrans' farming future by changing and limiting their decision-making options.

There were also short-term environmental changes, such as fluctuations in the amount of rainfall and floods. The villagers were aware of these short-term fluctuations. How they decided to react to them also affected their farming future. In particular, once they had established a culturally supported farming-friendly bias in their decision making, they likely tended to respond to these environmental influences on their lives by making farming-friendly decisions.

In overview, the interactions among the three factors of individual, environment and cultural world view resulted in the Abu Hureyrans being subject to a flexible hierarchy of dynamic short- and long-term influences on their daily lives. They expressed their reaction to these influences in their short-term decisions about day-to-day issues. For a variety of reasons, their responses to the initial environmental changes triggered by climate cooling tended to favour farming, while their enduring, changed environmental conditions prompted them to continue to routinely make farming-favouring decisions. In doing so, they slowly adjusted their culture's world view in ways that resulted in its guidance coming to favour agriculture. For example, they adjusted their culture's world view to treat transforming and managing their environment as the norm. At some point, the members tended to both adjust their culture's world view to favour farming and respond to issues by following their culture's farming-favouring guidance: they had formed a self-sustaining, transition-favouring feedback loop.

The transition from a hunter-gathering to a farming was slow. Sufficiently slow that Abu Hureyrans were likely largely unaware that it was taking place or that their decisions, a feedback loop and their environmental endowment were facilitating and sustaining it. They were likely unaware that they were journeying down complexity's road.

The essence of this description about Abu Hureyra's cultural transformation can be applied to the transformation of other cultures, but the details can't because the particulars vary from culture to culture. For example, as previously mentioned, although farming and settling are linked, there is wide variation in the nature of that linkage and its significance. In general terms, hunter-gatherer cultures have different degrees of moving around, while farming cultures, even though sedentary, have different degrees and kinds of being settled and different degrees of farming (Kelly 1995).

There is also no imperative that hunter-gatherer cultures must eventually take up farming. Some hunter-gatherer cultures can settle and form quite complex organizational structures yet never take up agriculture. For example, the original hunter-gatherers who lived on the northwest coast of North America settled and eventually developed a ranked (perhaps even hierarchical) structure, even though they had minor commitments to activities we might call land farming. (They engaged in limited aquaculture.) They could form such a complex culture because it could be supported by their highly productive environmental endowment with its oceanic riches, salmon in the rivers and large trees (especially cedar) in the forests (Ames and Maschner 2000).

When we think about a culture's transformation, we tend to pay more attention to the influence that direct and short-term events, such as the *act* of settling, had on changing a culture's world view or functioning. However, it is just as important to pay attention to the indirect and long-term influences an event had on the culture: the slow, long-term changes. For example, once a group has settled, this change in context alone could, over time, prompt the members to change their perception of themselves, one another, their environment and their relationship to it. This is a key part of cultural change. Consider how a change in perception of one's fellow members is required for social differentiation to appear, which is a critical part of changing a culture's structure. Consider too the recent connection between Westerners changing our perception of the land and altering our relationship with it (Wilson 1988; Kelly 1995). Think of the environmental movement, and gorillas.

## City States to Empires.
Although Abu Hureyran culture underwent many of the adjustments considered necessary to become a more complex culture it did not complete the process. Its organization may have become ranked, but a power hierarchy did not appear. In one sense, this is unsurprising because some of the less complex farming and pastoralist cultures have shown themselves to be remarkably stable. Becoming a more complex culture is not a necessity or even inevitable.

In another sense, it is surprising that Abu Hureyran culture didn't because the flexibility in how a culture can satisfy the requirements for being functional leaves the door open for it to keep following complexity's road. A culture can form a self-sustaining driver of change that can result in it being transformed into a far more complex one. It could become a city state, a territorial state, an industrial state or an empire. This section discusses the rise of these more complex cultures.

Picture an imaginary Middle Eastern village, Newville, that was similar to Abu Hureyra at a time near its abandonment around 5700 BC. Over time, the combination of a number of factors resulted in Newville's culture becoming more complex. Each house in the village was occupied by multiple generations of a single family, so it would have been easier for the villagers to organize themselves into household groups that were identified with a specific complex task, such as sheep rearing and basket making (Ames and Maschner 2000). Alternatively, a member of a house could have been recognized by all in the village as a special person, such as a shaman. At the same time, the increased privacy that came with living in houses weakened the effectiveness of social- and self-monitoring as a means of policing behaviour. When these types of changes were combined with a population greater than about 1,500, then the members would have found it more difficult to deal with issues that affected their group's unity, stability and adaptability. Think of decision making. It was more difficult for their culture to remain functional. But these changes did allow the members to resolve their issues by forming a ranked or slightly more complex culture in which some people had greater control over decisions and some control over other people.

Around 4000 BC, a few such villages in southern Babylonia are inferred to have faced the difficulty of managing their communal food store. To address this difficulty, the villagers are thought to have handed their personal control over the storage and distribution of their communal harvest to someone they respected and trusted, such as a priest. In return for the priest taking on the responsibility of monitoring the food store and ensuring a fair distribution of its contents, the members implicitly agreed to abide by the priest's decisions. The members may also have relieved the priest of the need to grow their own food. By making this decision, they increased the complexity of their culture's structural organization.

More generally, when the members of a ranked culture gave a few in their group significant power over most of the other members and formalized their culture's pre-existing work-speciality divisions, they created an authority-based hierarchy. This is the basic distinction between simple and complex cultures. To make this change, the members also had to adjust their culture's world view to match the reality that their culture was no longer egalitarian and was more than ranked. This included elevating the decision maker's status so that they became a focal point, perhaps the embodiment, of the group's unity. By making these choices and adjustments, the members and their culture completed its transformation from a hunter-gatherer culture to an agricultural culture with a hierarchical structure of low cultural complexity.

A ranked or low complexity culture can transition to the least complex of the state cultures through a process that is, in essence, an extension of the process of becoming a farming culture. However, it is more complex and the results are significantly different. In order to simplify the discussion about this transition, it will be assumed, for the moment, that the main driver is resources.

Imagine if an agricultural culture of low cultural complexity experiences an ongoing agricultural resource shortage due to, say, population growth, a long drought or both. The members could resolve it by reducing their population (e.g., through a split) or by increasing the agricultural productivity from their lands (e.g., adopting or increasing irrigation). Or they could decide to obtain the resources by economically, politically or militarily dominating the hinterland around their lands.

If the domination option is to be successful over the long-term, then the culture attempting it must have a population that is already sufficiently large to complete the needed tasks. This is usually indicated by the presence of an urban population centre. In addition, the culture's world view and organizational structure must already have the capacity (complexity and will, among other features) to ensure that those living in the annexed hinterland will both produce the required agricultural goods and be loyal to the leader of their new "partner," who lives in the urban centre. The culture attempting domination must also have the capacity to manage the transportation of the resources back to the urban centre, as well as store and satisfactorily distribute them among the members.

If a culture is successful at dominating its hinterland, then it has become a city state. All city states display most of the following structural and functional features. They have a well-developed pyramidal power-status hierarchy. The upper echelons of the hierarchy are occupied by a handful of people (the elite) who have significant symbolic and real authority over those lower down. At the top of hierarchy is the supreme authority figure who is a unifying figure that has the prerogative of using force.

The elite live in an urban power centre among monumental buildings, such as a temple. A small portion of the rest of the population also live in the centre. The majority of the population, in the order of 80%, who occupy the lower echelons of the hierarchy, usually live rural lives on farms surrounding the centre.

A city-state culture is usually relatively homogeneous, has a formal religion and a formal, more costly method of cultural reaffirmation and authority maintenance than a simple culture. However, each city state has its own distinctive characteristics: world view, history and environmental endowment. For example, the leadership position can be filled by one of the many kinds of monarch, or by shared leadership from among them. The wide range of city-state cultures is illustrated by the variety present in Greece during the 600s BC. Think of Sparta and Athens.

The variety and range of the characteristics a would be city state could adopt produces uncertainty in defining this type of culture unambiguously and thus a transition to or from it. For example, the existence of a city state, as its name suggests, implies an association with a well-defined urban centre. Therefore, the presence of an urban centre in the archaeological record could be taken as evidence that the culture of its inhabitants had at least a city-state level of complexity. However, in the Middle East, well before the appearance of city-state cultures, such as Uruk in southern Babylonia, there existed a few widely scattered population centres of several thousand people. These centres, such as Catal Huyuk, Abu Hureyra and Jericho, were not city states, because excavations there have provided no evidence of either hinterland domination or a formal power-status hierarchy. Not all urban centres are associated with a city state or a more complex culture.

It could be argued that sufficient evidence of a culture's organizational ability to control its hinterland would be provided by the presence of monuments and administrative buildings in its urban centre (Hassan 1981; Wilson 1988). However, although the power centres of the Mayan kingdoms are well-known for their clustered temples, administrative structure and their domination of the surrounding rural population, they were not classic, well-defined urban (high population) centres. By this definition, the Mayan kingdoms are not strictly city states.

The above discussion provides important lessons for thinking about how complex cultures, including those of today, function and change. Like the division between hunter-gathering and farming cultures (and between the unconscious and conscious functioning of the brain), the strict definition of a city-state culture is irreducibly fuzzy and its defined boundaries disappear if scrutinized too closely. Thus the identification of a city state relies heavily on finding most of the suite of features associated with a complex culture mentioned above, together with indications that they did indeed dominate their hinterland.

If a culture is identified as having completed the transition to a city state, then it is implied that it must also have satisfied one other criteria. For a city state to be and remain a functional whole, its members must remain faithful to their culture at its new level of complexity. This means that they must support and maintain a city-state feature that did not concern hunter-gatherers: a social hierarchy with an elite that has both real and symbolic power over the members (Weingart et al. 1997). A city state secures the support of its members by adding the hierarchy into its culture's world view. Thus, when the members reaffirm their world view, they also reaffirm the legitimacy of their culture's hierarchy. At the same time, they become responsible for maintaining the hierarchy, which is expressed as a fixed cost that the culture must pay. Thus, in order to remain functional, a city-state culture forms a reaffirmation and authority maintenance process that matches its level of cultural complexity.

All of the more complex types of cultures form an equivalent reaffirmation and authority maintenance process for the same reason. Because it is so important to the functioning of complex cultures, a generic description of it is provided here. It looks at the process from three overlapping perspectives: the individual member, the use of symbols, and the economics of reaffirmation and authority maintenance.

The *individual members* of a complex culture have a basic faith in their culture's world view and authority structure because it becomes part of their lives as they grow up. But the members' ongoing faith in their culture's world view and structure, and their willingness to maintain that structure, depends on their ongoing evaluation of a number of factors specific to complex cultures. They include the members' perception of the leader's legitimacy, the behaviour of those in the upper echelons of the hierarchy, and the perceived effect the leader and the elite have on the members' life circumstance. The nature of the members' evaluation is illustrated by the details of what the individual members consider in their leader.

A leader's legitimacy is established in the minds of the people when he (historically they have mostly been he's) claims the right to that position because of, for example, his ancestry or his historical links to the divine (Wilson 1988). That right is confirmed if he completes the culture's required ascension ceremonies. The leader's performance is reassessed by the members throughout his reign. In general, he is judged by the degree to which he has met the culture's general expectations of a ruler, such as being wise, just, strong and benevolent. He is also judged by how his reign has affected the material aspects of the members' lives.

In particular, the members' reassessment considers whether he defeated the culture's enemies and provided peace, stability and a good harvest. It also considers whether he exercised his judicial powers in a fair manner. His stature rises if he had been successful at organizing the people and providing the resources, both of which are needed to complete culture-boosting projects such as irrigation works and monuments. If the members believe that their personal and the culture's future requires their leader to remain in favour with their culture's god(s), then they would consider whether he had fulfilled his religious duties. Did he, for example, build and maintain the religious monuments and then use them to conduct the ceremonies and rituals needed to satisfy the gods and thus keep the world in balance. If judged favourably, then the members reaffirm their leader's legitimacy and support his ongoing rule. Legitimacy, competency and concern for the members (think fairness) are key themes of their evaluation of a leader. The leader, as an individual, tries to persuade or demonstrate to the members that he meets these expectations.

The culture's *symbols* play a significant role in the reaffirmation and authority maintenance process: they help to maintain the members' loyalty to the ruler and state. Consider the symbolic function of monumental architecture, myths, ceremonies, public works and even the position of the leader in the hierarchy. The example used here is the pyramids, ziggurats, temples and palaces built in the Middle East between 3000 and 2000 BC.

The symbolic role and power of these buildings has its roots in hunter-gatherer cultures. Hunter-gatherers found their sense of meaning in the landscape, in part because they used it to represent their beliefs about how the world worked, their place in it and how to live, and the reasons why. As a hunter-gatherer culture transitioned to farming, its members increasingly found their sense of meaning and place in their transformation and management of the land, as guided by their changing world view. This included structures such as Stonehenge. As simple farming cultures became city states, or more complex cultures, the human-made environment continued to fulfill this powerful symbolic role, but now in the form of more elaborate monumental architecture (awe-inspiring buildings and infrastructure).

Monumental structures symbolized to the members key aspects of their culture's world view. In particular, the monuments made tangible to the members the mystical and aesthetic nature, power and structure of their surroundings, and their place in it, that was laid out in their culture's myths, sacred texts and folklore. For example, the alignment and details of the Egyptian pyramids represent that culture's world view of the universe and its members' place in it. By burying their monarch inside a pyramid in a specific place and orientation, the Egyptians reaffirmed their culture's world view and their personal links to the rest of the universe, as they perceived it. (Note that, in general, a complex culture's view of how its surroundings were structured mirrored the culture's own structure: hierarchical.)

Thus, the members who built the monumental structures, attended ceremonies there or just saw their grandeur could feel that they were participants in their culture's successful striving to live in the world and shape it in a way that their culture's world view said they should (Wilson 1988). In essence, the symbolic power of these structures brought to life and confirmed each member's childhood knowledge of their culture's world view. Collectively, this aroused in the members a common feeling of belonging (a sense of place), purpose and meaning. It unified them and their striving.

From the perspective of *economics*, a culture's reaffirmation and authority maintenance process is a fixed (unavoidable) cost. And it must be paid if the ruling elite and its leader are to continue to receive the ongoing support and trust of the people, if they are to remain in power. This process has a cost because it includes ensuring that food, goods and services are provided to the people; funding the hierarchy and its bureaucracy; protecting the members against enemies; dispensing justice; conducting ceremonies, such as those needed to maintain good relations with the gods;

and building infrastructure and monumental structures.

These fixed costs are directly paid for by the leader of a city state. But the leader acquires those funds by exercising the power of their position in the hierarchy to appropriate a portion (theoretically the excess) of the mostly agricultural production generated by the state's members. The leader can tax the people.

Up until about the 1800s, most of the physical work needed to generate this wealth was provided by the 80% or more of a complex culture's population who lived rural lives as farmers, along with the help from their animals (Hassan 1981). So, indirectly, the members themselves paid for the reaffirmation and authority maintenance process. However, the farmers' ability to provide for themselves and supply an "excess" depended on the elite fulfilling their obligations, such as management, protection against enemies and ensuring the harvest would be bountiful. The resulting mutual dependence between the elite and the farmers illustrates how the reaffirmation and authority maintenance process helped to bind a more complex culture's subgroups together.

It can be convenient to think of the transition to a city state in terms of a particular driver, such as resource shortages or adjustments in a world view. However, it should also be clear that the transition process is a complex one, involving a number of factors that interact at various scales. And staying functional at the new level of complexity requires the culture to complete specific cultural changes.

Theoretically, an adequate supply of resources, along with successful reaffirmation and authority maintenance efforts, would allow a city state to remain functional indefinitely. However, a city state, or any complex culture for that matter, can also be prompted or preconditioned to become more complex. Or it could even form a self-sustaining cultural drive that would guide the culture to inexorably continue down complexity's road. This is a discussion about how a drive to increase complexity beyond a city state can arise and be sustained. Drivers such as an increase in the culture's population or a shortage of resources have been mentioned above. This discussion will focus on the members' decisions.

The nature of a complex culture's real-world functioning allows the members to (inadvertently) make decisions that increase their culture's complexity. Consider that the elite (including the leader) occupy positions of trust. When a city state's elite are given control over the culture's wealth (historically, mostly agricultural production), they are expected to use their power and that wealth in both the culture's and the members' best interest. Remember how the leader and the rest of the elite are judged. They are obliged, for example, to buffer the culture against an unpredictable future and pay for the culture's fixed costs, such as its reaffirmation and authority maintenance efforts. Think bureaucracy.

Imagine that the elite feel that the amount of wealth under their control is insufficient, in their eyes, for them to fulfill their obligations. They have the authority to order an increase in the amount of wealth supplied to them. They could do so by either demanding more production from the front-line producers, decreasing the amount allocated to them, or distributing less to the "non-productive" members. The elite can even decide to act as intermediaries and use some portion of the wealth under their control to secure the extra wealth by trade. Any of these options could be necessary and implemented fairly.

Perhaps there was insufficient wealth because the culture's environmental endowment was incapable of producing it under current conditions. In this case, the elite could decide to increase the state's wealth by increasing agricultural production through, for example, building a large irrigation works. To be successful, they must both build *and* maintain these works. Building an irrigation system requires a co-ordinated workforce. Running the works requires a number of tasks to be routinely performed: repairs, desilting and the fair distribution of water to the farmers at the right time. In order to ensure that these tasks are properly completed, the members would have to agree to a more standardized lifestyle and to be more intensely managed. In other words, in order to build and run a large irrigation works, the members would have to allow the hierarchy's (the elite's) power over them to expand, accept the demand for greater uniformity in their behaviour, and agree to the loss of some personal freedoms.

By the members accepting these changes, they would be adjusting both their personal and their culture's world view to favour seeing an increase in their culture's complexity as normal. They would also be accepting the associated increased authority maintenance (fixed) costs as fair. Their culture would be becoming more complex.

But, perhaps there was insufficient wealth because of decisions that the elite made. They could have used their power and the members' common wealth for something impossible in hunter-gatherer cultures: enhancing their personal material wealth, power and status. For example, they could act as "middlemen" rather than intermediaries and thus receive illegitimate personal benefit from fulfilling their duties. They could distribute some of the members' wealth as patronage "gifts," that is, with the biased intent of influencing those receiving the gifts to enhance the distributors' personal wealth, power or status (or all three) (Kelly 1995). These activities and their consequences will raise the culture's fixed costs. Under these conditions, the elite of a city state could find the idea of increasing production by building a new irrigation system to be irresistible. Regardless of whether the intent of the elite's decisions was to keep the culture functional or to satisfy their self-interest, this decision facilitated or favoured an increase in cultural complexity.

There is another solution to a real or imagined shortage of wealth. The city state's (public and elite) members could decide to further enlarge the hinterland under their control. If successful, it would result in a burst of revenue and labour to the dominating culture. However, there is another consequence.

Along with territorial expansion comes a non-linear increase in the (dominating) culture's population, the number of subgroups (such as a greater diversity of work functions, ethnicities and religions) and its infrastructure needs (such as communication). Dealing with these changes increases the culture's complexity, which translates into an absolute increase in costs across the board. Consider that there is an increase in the culture's fixed costs, for example, its reaffirmation and authority maintenance costs (Tainter 1988). There is also an increase in the variable costs, which result from the increase in the number and variety of random problems the culture must deal with. Thus, over time, the culture may again face a wealth shortage. If the elite and the members remember just the initial burst of revenue and labour from the previous territorial expansion, they will be tempted to deal with the new wealth shortage in the same way: expanding their land base.

All three of these scenarios illustrate that the members' decisions have the potential to result in an increase in their culture's complexity. These types of scenarios are neither special nor exist in isolation. They just represent the broader potential, within a more complex culture's functioning, for the members to make complexity-favouring decisions. They can even create a self-sustaining feedback loop that favours an ongoing increase in cultural complexity. This potential is described by a rule of thumb. By routinely choosing a particular type of solution to an ongoing or reoccurring issue, the members can embed that solution into their culture's world view as a preference. If the preferred solution type both (directly or indirectly) favours an increase in the culture's complexity and results in issues that the members will preferentially resolve in ways that increase their culture's complexity, then they have formed a self-sustaining complexity-favouring feedback loop. Think of increasing agricultural production and favouring hinterland expansion.

Because an increase in a culture's land base is accompanied by an increase in its complexity, the size of the area that cultures dominate can be used to subdivide complex cultures. A city state controls just its immediate hinterland. Territorial states (states or countries) control larger areas. Empires have boundaries that include numerous country-size areas. However, beyond some size, occupying the land is not critical to increasing a culture's complexity, having stable access or control over the resources in foreign lands is.

An important question remains. Is a culture's world view really sufficiently powerful to bias its members' decision making to favour an increase in their culture's complexity? Yes, as is illustrated by contrasting two differing but equally reasonable cultural perspectives of the ancient Greek Antikythera celestial calculator. This amazing device was recovered from a 70 BC Roman shipwreck. It is thought to have been built by the Greeks around 150 to 100 BC.

The calculator was powered by a hand crank. When turned, the device's gearing mechanically completed celestial calculations and then visually displayed the results as the movement of the sun, the moon and possibly the then-known planets around the earth. It also predicted the dates of future solar and lunar eclipses. The celestial movements it displayed were based on older Babylonian observations that were augmented by more recent Greek measurements. Time was represented by an Egyptian calendar (Freeth et al. 2006).

The calculations were completed by a very sophisticated arrangement of at least 30 bronze gears. Ingenious gear shapes were used to implement the more complex calculations. The device was so technically advanced that neither the overall design nor some of the specific gearing features were seen again until the 1300s AD, when Europeans invented mechanical clocks to measure time.

Today Westerners puzzle over the fact that the Greeks could build the complicated Antikythera calculator, as well as other sophisticated technological devices powered by water and steam, yet they did not automatically turn them into labour-saving engines or money-making machines. We struggle with this because current Western world views guide us to primarily value technology for its ability to resolve the issues we face as we strive to maintain our cultural goal of progress (which favours an increase in cultural complexity) (Tainter 1988). What other reasonable reason could there be for technology?

In contrast, the ancient Greek world view likely guided its members to value technology because it allowed them to display their deeper philosophical understanding of the cosmic order, such as the laws of astronomy. From the Greeks' perspective, the main value of the Antikythera device was social, not material (Marchant 2006; Wilson 1988). They used technology to gain social prestige by, in effect, building and owning intellectual status symbols.

In summary, the diversity of the earth's cultures can be divided into cultural types based on the level of their structural complexity: simple cultures (e.g., hunter-gatherer cultures); intermediate cultures (e.g., sedentary hunter-gatherer and ranked agricultural cultures); and complex cultures (state cultures). This is possible because the other characteristics of cultures with a particular level of structural complexity are, within limits, similar. Those features include the following: the group's size, the land area it controls, its functioning, its demand for resources and its world view (e.g., the order it ascribes to the world and its perceived relationship with its environmental endowment).

For example, the world views of simple cultures provide guidance to the members that the world has a community-like structure and that their relationship with their environment is one of a participant in a community. Their resource demands are low. In contrast, the world views of the more complex cultures (state cultures) provide guidance to the members that the world has a hierarchical structure and that they are managers and transformers of the land. Their resource demands are high. The characteristics of cultures on the cusp of being complex (egalitarian farming cultures and ranked cultures) lie in between.

In this wider, longer-term view of cultures, the arrangement of their types in order of increasing complexity can be treated as a continuum (Tainter 1988). For example, work specialization ranges from none for hunter-gatherer cultures, to some for settled hunter-gatherer or ranked farming cultures, to formalized specialization for complex farming cultures (city states), to increasingly diverse formalized specialization in industrial cultures. However, when cultures are looked at more closely and over a shorter term, there is much more variety both within a cultural type and in the connections between them than the wider view suggests. In addition, some aspects of a culture's changing occur over the short term and others over the long term. Similarly, a particular culture's characteristics display more flexibility than the classification implies.

A more complete picture of a culture's functioning and changing occurs when the wide and the close-up views are combined. Indeed, they should be treated together as the products of a single dynamic, complex cultural system. In this view, a culture functions and changes because of the interactions among the factors of its group's individual members, their culture's world view and their environmental endowment (Kelly 1995). These interactions occur, in part, at a time frame relevant to its members' day-to-day living (short term) and, in part, over generations (long term). They also include feedback.

Only some of the possible combinations of these interactions result in functional cultures. The reasons for this restriction are a product of the interactions within the cultural system. Consider the cost of the reaffirmation and authority maintenance effort that is needed to ensure the group's cohesion. A simple culture (hunter-gatherer) has an undemanding reaffirmation process and lacks an authority structure, so its reaffirmation costs are low per person. In contrast, a complex culture's reaffirmation process is more costly, in part because it includes a complex authority structure, which it must also maintain. Its reaffirmation and authority maintenance costs per person are much higher. Because these fixed costs are ultimately paid for using natural resources, the more complex cultures always require more natural resources per person to function than do simple cultures. No wonder that a complex culture is preoccupied with paying for these fixed costs and that its relationship with the environment favours transforming its environmental endowment.

Environmental factors can facilitate or restrain a particular cultural change. However, the power of a culture's world view and its members' decisions to act as drivers and mediators of cultural change should not be underestimated, as will be seen throughout this book. In particular, the functioning of the members' cultural system allows them to unconsciously make decisions that (directly or indirectly) increase their culture's complexity. An increase is also more likely to occur if the members' decisions change their culture's world view so that its guidance about resolving issues favours an increase in complexity. Under these conditions, if that increase in complexity results in the members being guided to resolve future issues in a way that increases their culture's complexity, then a self-sustaining, complexity-favouring feedback loop is formed. The culture's world view guides its members to increase their cultural complexity.

## Staying Functional

A functional culture is one that meets, at least over the short term, the conditions needed for the group to be unified and adaptable: it functions in both the members' individual and their group's best interest. A culture that remains functional over the long term is said to be viable. This section starts by discussing the features of a culture's world view that favour cultural viability. It then turns to the challenges the members face when they try to recognize when their culture's world view doesn't favour viability and make adjustments so that it does. It ends by discussing the challenges of following the guidance from a culture's world view that favours viability.

**Viability-Favouring World View.** The current members of a culture live their lives (complete their lifeway, in ecological terms) within a cultural and environmental context. If their culture is to be viable, then they must complete their lifeway in a manner that will ensure their culture's functioning satisfies viability's human and environmental constraints over the long term. Because of the member-culture feedback, not only must the members strive to live within those conditions, but their culture's world view must guide them to do so. This means that their culture's world view must include viability's conditions. What are these conditions?

Consider the human conditions for viability. In broad terms, a viability-favouring world view guides the members to ensure that, for many generations (i.e., the long term), all members can live mentally and physically healthy lives. The specifics of that guidance are illustrated by considering one aspect, reproduction.

The members' urge to reproduce is easily and often unexpectedly satisfied with little cultural influence (Myers 1999). Thus to favour viability, a culture's guidance must discourage unwanted pregnancies and encourage healthy ones. This means, for example, discouraging the drinking of alcohol when pregnant. A culture's guidance must also encourage the raising of mentally and physically healthy children who will be assets to their community. This means guiding the parents to favour the appropriate child-rearing behaviours and the rest of the members to support their efforts. This support includes helping the parents gain access to the resources they need to raise a healthy child, and not hindering their efforts with either unhealthy work or home environments and overly long work hours. This translates into the need for the culture's guidance to limit the number of pregnancies. The goal is to ensure that the size of the population matches the ability of the culture's social structure and environment to support all of its members when they are living mentally and physically healthy lives. The conditions for viability are thus not isolated items, and the connections among them include feedback.

It may seem that these viability-favouring child-rearing constraints on a culture's world view are excessive, until consideration is given to the cultural fallout from inappropriate child-producing and child-rearing practices. Consider the challenges a culture faces when having to deal with adults who grew up as a lost generation of child soldiers and abused children. What values and beliefs will they pass on to their children?

The environmental conditions for viability are partly set by a culture's environmental endowment. It supplies the quantity and quality of critical resources, such as fresh water, air, food and energy, needed to support the members' lifestyle and beliefs (Trigger 1989; Kelly 1995). (Hall et al. 1986). A culture's world view is thus obliged to guide the members to respect and live within the constraints on their environmental endowment's ability to provide those resources. Consider the consequences for a culture whose population is too large to be supported by its environmental endowment, as illustrated by Tikopia in chapter 16. Think of undernourished children and social instability (Spillius 1957; Borrie et al. 1957).

It may seem that this description of the guidance needed to satisfy the environmental conditions for viability is exaggerated. After all, humans can significantly modify both their environment and their culture in order to maintain the supply of the resources we need. This has allowed us to live in a wider variety of environments and to adopt a wider variety of cultural characteristics than one would expect from the constraints our environment apparently imposes on us.

Despite our abilities, the nature of a culture's environmental endowment still constrains the characteristics of a viability-favouring world view. Consider the more complex cultures. They have a much greater capacity than a simple culture to modify their environmental endowment in order to provide the resources they need to function. However, they also have a much greater need for resources because of their greater population and the larger (resource) cost per person of their reaffirmation and authority maintenance efforts (Tainter 1988). When this elevated resource demand is combined with the nature of a complex culture's environmental endowment, such as its finiteness and its functioning, then it is unsurprising that a complex culture invariably finds itself preoccupied with securing resources. If a complex culture is to meet viability's environmental conditions, then its world view's guidance about dealing with that preoccupation must respect the long-term constraints their environmental endowment imposes on the supply of resources.

At the same time, the guidance about providing those resources must also respect viability's human constraints. For example, if providing the desired resources requires the use of slave labour, then the culture will struggle to make emancipation and equivalent access to resources for all members a part of its world view. Both human and environmental aspects are part of a viable culture's world view.

In overview, the characteristics of a viability-favouring culture's world view are inescapably limited by human and environmental conditions for viability that are interlinked in ways that include feedback. For a culture to be viable, its world view must guide its members to live within those constraints. This is true regardless of whether a culture is simple or complex.

**Adjusting a World View.** As discussed earlier, a culture's world view can be treated as the combined product of its members' efforts to make sense of the world and their decisions about how they should live in it. From this perspective it seems that the members can easily ensure that their culture becomes and remains viable by adjusting its world view so that it guides them to satisfy viability's conditions. But they would find the task of making those adjustments complex and difficult.

One reason is the nature of the human and environmental conditions for viability. The presence of feedbacks has been mentioned. Consider too that the members would have to adjust their culture's world view so that its guidance satisfies conditions that may conflict with one another. For example, it would have to, simultaneously, provide guidance that balanced the following conflicting needs: tolerance of diversity versus a strong sense of identity (Myers 2001); cultural flexibility versus faith in the culture's existing world view (Alland 1973; Norretranders 1998); group focus versus individuality (Myers 2001); living as a group versus individual privacy (Wilson 1988); and facilitating the fulfillment of personal aspirations versus protecting individuals and the environment against exploitation (Reader 1990). Add to the list the requirement that their culture's world view refrain from forcing good people to do bad things. Satisfying the conditions requires making compromises and trade-offs.

Another reason the members would find this task difficult is the way in which their human decision-making characteristics bias their decisions, as discussed in chapter 2. Think self-interest. Consider too the constraints on making a change to a culture's world view, as discussed in chapter 6. For example, the member-culture feedback ensures that there is inertia to changing a culture's world view.

There are also features of a culture's functioning that limit the adjustments the members can make to their culture's world, if it is to guide them to satisfy viability's conditions. This is illustrated by a key feature of a functional culture: maintaining effective communication among its members and their subgroups. Effective communication is more likely to occur if those communicating are all familiar with the relevant information about the subject and the context in which it is being discussed (Hall 1981; Ziman 1978). It also helps if the participants are familiar with one another.

Effective communication is relatively easy to maintain in simple cultures because the needed familiarity is inherent. However, as a culture's structural complexity increases, that familiarity is lost. One reason is that an increase in complexity is accompanied by an increase in the number of the culture's subgroups, each of which is, in some way (e.g., socially, politically and even spatially) somewhat isolated from the other subgroups. In addition, the members in each subgroup are mostly focused on their subgroup's own interests, knowledge and vocabulary. Think of the military, political parties, non-government organizations and a diplomatic corps.

If a complex culture is to remain functional over the long term, then its world view is obliged to promote and maintain effective communication between its disparate subgroups. It must, in essence, replicate/facilitate the familiarity found in simpler cultures. The preferred means to achieve this goal is to establish a communication system that can be used by the members to collect, store and exchange the information, and keep up the personal contacts, required to maintain familiarity.

The process of a culture becoming more complex therefore invariably includes changing its world view to favour the members establishing such a communication system. Or, if one exists, adjusting it to allow more messages to contain more information (i.e., increase the capacity of the system). But, for the following reasons, such a communication system faces limits in its ability to ensure effective communication.

At its most basic, as a culture's structural complexity increases, so does the number of its subgroups, while the number of possible communication links among the subgroups increases exponentially (3 subgroups have 3 links, 4 subgroups have 6 links, 5 subgroups have 10 links). This rapid increase in the number of links is accompanied by a rapid decline in their members' familiarity with one another. Thus the more complex the culture, the exponentially greater must be the increase in the capacity of its communication system, if it is going to successfully promote and maintain effective communication among its members. In effect, a culture's level of complexity limits its ability to ensure effective communication by simply increasing the capacity of its communication system.

Consider too that an increase in the capacity of a communication system adds to its absolute maintenance and usage costs. This adds to a complex culture's fixed costs, which its leaders (elite) are always wanting to reduce. Complex cultures are thus faced with two strong, but conflicting, cultural incentives: reduce absolute communication costs but increase communication effectiveness. The options available to deal with this conflict include reducing inessential communication, such as providing distribution lists for issues; sending fewer, less detailed messages; using specialist language; not sending or replying to information perceived as useless; and limiting access to the communication system by introducing fees or restricting access. However, all of these options can also interfere with the ultimate goal of promoting and maintaining effective communication.

An alternative solution is to increase the efficiency of the communication system, as measured in messages per unit resource, cash and/or time. The demand for greater efficiency can be met through technology. But the contribution that an increase in efficiency can make to effective communication is also limited. For example, at some point, increasing efficiency adds to fixed costs and is counterproductive. Consider too that, despite what we think, efficient communication is not the same as effective communication.

This is the time to remember that it is individuals who are sending, receiving and processing communications and that their ability to complete these and other communication-related tasks is constrained by their human characteristics. Think of the brain's finite signal processing times discussed in chapter 1. These human constraints limit the degree to which all efforts, such as increasing communication efficiency, can increase communication effectiveness. Consider, for example, how the time-strapped individuals in the more complex cultures of today deal with their large communication load. Think email and social media.

On the one hand, the swamped human message senders tend to resolve this problem by sending more highly condensed, simplified and standardized communications; by sending a generic reply; by providing limited or no context; or by not sending a message. On the other hand, the swamped message recipients tend to resort to discarding, ignoring, losing or skimming messages, or they place them in the dead-end "to-do" folder. As most of us have experienced, these techniques resolve the load problem but can result in a decrease in the effectiveness of communication.

One alternative is to hire more people, but that increases fixed costs. The current preferred solution is to use technology to augment or replace human decision making. Think of artificial intelligence (AI) and decision-making algorithms. In the long term, this just shifts the problem of maintaining effective communication to the designers and managers of the machines and those controlling all three. Think of Google and the G5 system. Back to first base.

Even if a complex industrial culture can pay for the energy and technology needed to implement and maintain faster, more efficient high-capacity electronic communication systems, the constraints on effective communication remain. A culture is also stuck with a dampening feedback loop in which an increase in its complexity increasingly constrains the members' efforts to ensure effective communication. In particular, an increase in complexity increasingly limits the members' ability to adjust their culture's world view so that its guidance resolves the resulting communications issues (i.e., supports the culture's complexity) in a way that

also maintains effective communication (i.e., favours viability). The members eventually have to consider adjusting their culture's world view so that it no longer favours an increase in their culture's complexity, if it is to be viable.

This discussion also serves as a reminder of a rule of thumb that is useful when considering adjustments to a culture's world view. The members can address cultural issues by adjusting their culture's world view in ways that increase their culture's organizational complexity. But, if they want the culture to be viable, then the increase can't push the culture beyond the constraints on its viability (Tainter 1988). Ignoring this limit leads to a dysfunctional culture. Think of the self-sustaining complexity-favouring feedback loop mentioned in the previous section.

In overview, the efforts of the members to adjust their culture's world view so that it guides them to make viability-favouring decisions are constrained and limited in a number of ways. It takes effort for the members to recognize and appropriately respond to them.

**Responding to a World View.** In real life, some of the guidance the members receive from their culture's world view favours viability and some doesn't. If the members want their culture to become viable, then they must follow the guidance that favours viability and recognize that which doesn't. This subsection discusses the challenges the members face when trying to comply with this requirement.

The nature of the relationship between the members and their culture's world view means that it can only provide guidance to the members, not orders. It can't force compliance. This alone ensures that, even if the guidance from their culture's world view favours viability, the members can (deliberately or inadvertently) make decisions that don't follow it. Some of the reasons why they would respond in this way are related to our human characteristics. Our self-serving bias, groupthink and denial, for example, can result in the members forming culture-disrupting, antagonistic cliques, gangs and business cartels. The elite can fall into the traps of narrow, self-serving visions and succumb to the temptations of riches and power. Examples that come to mind are Conrad Black for Hollinger, Ken Lay for Enron and the leaders of the many large financial institutions who contributed to the global financial crisis of 2008. In all of these cases, these members easily ignored guidance that favours viability and accepted that which didn't. More generally, all members of a culture (elite and public) have a preference for short-term thinking, which means that we can make decisions and react to events in ways that we think are benign but, in the long term, don't favour viability.

Our relationship with our culture can itself help to lead us astray. For example, when a culture's guidance is given in the form of a symbol, then much of the members' response takes place in their unconscious. This means that we can respond more rapidly than if we used conscious thought, but it also reduces the opportunity for us to undertake a conscious review of our response. This increases the possibility of us responding inappropriately to the display of cultural symbols. For example, a leader can use the power of a flag to help them gain the members' support to be the aggressor in an unnecessary war that is counter to viability. Think of populism.

Consider too that, under conditions of cultural distress, a culture's members can respond to issues with decisions that are at odds with viability. For example, a common response we have to the loss of a shared cultural value or belief, such as our faith in our culture's institutions, is to adopt a more mystical personal world view (Myers 2001). If enough people adopt this view it can become our culture's world view. This occurred, for example, in Europe between the late medieval era (mid-1300s) through the Renaissance to the late 1600s as a result of the profound religious, health and structural changes underway during that period.

Between the early 1300s and the 1360s, Europe experienced famine and plague (the Black Death). This was followed by the Hundred Years' War that was waged from about 1350 to 1450. From 1500 to 1650 the religious core of European culture was in flux as Catholics and Protestants battled for religious and political ascendancy, for example, the Thirty Years' War from 1618 to 1648. During the 1500s and the first half of the 1600s, the later Renaissance into the Enlightenment, the process of absorbing the flood of new knowledge (think of the discovery of the Americas) and new ideas about how the world worked added to the cultural disruption.

These upheavals and disruptions contributed to many Europeans believing that the ills in their lives were caused by witches, who made pacts with the devil to secure help. A sufficient number believed in this threat that European cultures adopted formal witch hunts and witch trials. But this response also created feelings of fear and insecurity in communities and families, thus reinforcing the other forces tearing at the fabric of their culture. Witch trials peaked in the late 1500s and then declined until they disappeared in the early 1700s. One reason for the decline was the appearance of a new source of hope in the form of the nascent notion of progress being offered by the emerging European world view (White 1968a; Roper 2004).

Ironically, we may follow our culture's guidance that favours viability but do so in a context where it doesn't. For example, until the 1960s, the women in a few tribes of Papua New Guinea ate the brains of their dead female relatives as part of their culture's grieving ritual. Grieving rituals are an important part of a functioning culture.

Unfortunately, some of these relatives had died from a BSE-like degenerative brain disease known as *kuru*. By eating their beloved's brains, the bereaved could acquire the disease. Those infected slowly lost their ability to move, until they eventually became totally incapacitated shortly before their death. Over time, the spread of the disease among the women in the tribe resulted in a lack of women and a cultural preoccupation with the search for a reason and a cure for their deaths (Burnet and White 1972). At the same time, both the deaths and their nature likely strengthened the members' dependence on their culture's grieving ritual.

Alternatively, because a culture's world view has significant influence over its members, they can follow its guidance even though there is readily available evidence that its guidance is false or misleading. An example of this response is Westerners' current belief that a political ideology based on neoliberal capitalist economic views is the sole reason why we have been able to generate economic output (i.e., create wealth). But political ideology mostly determines who gets the wealth, rather than its generation. The more important general determinants of the absolute amount of wealth generated are the availability of natural resources (in particular energy) and the technical ability to extract them cheaply (Ayres and Warr 2005; Hall et al. 1986). By ignoring this evidence, we are distracted from the more basic viability-affecting concerns, such as the limits to the disparity in wealth distribution and the importance of maintaining the functioning of our culture's resource-supplying environmental endowment.

In summary, if a culture is to remain functional over the long term (i.e., viable), then its world view must guide its members to make decisions that satisfy viability's human and environmental conditions (i.e., favour viability). The members are their culture's decision makers, so they are responsible for ensuring that their culture's world view favours viability and that they follow its guidance. We face limits and limitations when trying to discharge this responsibility. We can easily fail to both comply with our culture's viability-favouring guidance and recognize when its guidance is not viability-favouring. If we do recognize that our culture's world view needs to be adjusted to favour viability, we find it difficult to make the needed changes.

The conditions for viability and the difficulties meeting them are most clearly seen, appreciated and evaluated when they are viewed in the context that includes the members' characteristics as individuals, the features of their culture's world view, the nature of their environmental endowment and the dynamic interactions among them: that is, in the context of a dynamic, complex cultural system. When seen in this context, a culture is a work in progress whose functional status varies over time. The members' efforts to ensure their culture is viable is an ongoing task.

The consequences of the members failing to establish a viability-favouring cultural world view and to follow its guidance is illustrated by the city state of Sparta, a neighbouring rival of the ancient city state of Athens. In the Spartan world view, all males fulfilled their role in the culture by being full-time soldiers. To this end, all Spartan males were evaluated at birth to determine if, as adults, they were likely to meet the physical standards of being a soldier. Many died during the assessment process. Many more died later in battle. The impact that the reduced number of males had on the functioning of Spartan culture was amplified by its limits on citizenship. Only those who could demonstrate blood ties to the founders of the culture, and the few who successfully completed a highly restricted sponsorship program, were considered citizens. The result was a biologically predictable steady decline in the number of Spartan citizens.

To accommodate the practical repercussions from this aspect of their world view, the Spartans steadily increased the number of slaves used to complete the many non-military tasks needed to keep the culture functional. Eventually, military duties were added to that list. The slaves' ongoing dissatisfaction with their treatment and status posed a constant threat to Spartan culture. Both the Spartan world view's lack of attention to biology and the nature of their efforts to deal with the repercussions contributed significantly to Spartan culture disappearing after about 800 years (roughly 30 generation) (Wikipedia 2009).

## The Notion of Progress

A culture's world view influences how its members derive their sense of identity, meaning and belonging by providing them with guidance about how the world functions, their place in it and why. It also provides them with a conception of an ideal future and how to reach for it. A culture's world view thus influences its member's recognition of problems and their choice of solutions. It favours them choosing solutions that match their culture's world view, their utopian future and their culturally preferred manner of reaching for it.

If the discussions in subsequent chapters about humanity's present condition, our likely future and possible solutions are to be meaningful, then they must take the nature of human culture into account. One of the ways this will be facilitated is to illustrate those discussions with examples from Western culture. But, for this to be successful, it requires us to have an appreciation for the essence of Western culture's world view. This section will discuss aspects of that essence. Other aspects were mentioned above or will be introduced later, as needed.

Western culture is one of the growing number of the relatively new complex industrial and industrializing cultures. They all arose from recent historical complex agricultural cultures in the same general manner as those historical cultures arose from older hunter-gatherer and simpler farming cultures. The essence of Western culture's current world view is thus the world view of its past complex agricultural culture, adjusted to fit its industrial present. It is within this historical context that its current world view and its path into the future is easiest to grasp.

During the Middle Ages, the members of Western culture believed that the chances of the living actually reaching their utopian future depended entirely on them securing the grace (love of and favour toward humans) of their Christian god. Today Westerners believe that we can reach our utopian future through human endeavour, for example, by working hard and by competing (Marx 1996). This switch in methods accompanied the appearance of a core part of Western culture's current world view: the idea of continual improvement, the notion of progress.

Westerners can use the word *progress* in a casual, narrow, practical sense to mean moving toward the end of a specific task or project. However, when used in its wider, more general cultural sense, progress means continual change toward an ephemeral or utopian future.

This is the meaning meant by the phrase "the notion of progress."

The notion of progress is so much a part of Western culture's world view and our personal reference normal that we conduct ourselves and make our day-to-day decisions largely unaware of its pervasive influence on our lives. Nor are we aware that our belief in progress is an important source of the self-sustaining drive that directs us to favour decisions that increase our culture's complexity. Fortunately, when the notion of progress is seen within its historical context, then its presence in our current world view, our ongoing reference to it and its influence on our decisions becomes clearer.

### History.

Western culture's notion of progress started to appear in a recognizable form in the 1600s. But to place that era in context and to understand why the notion of progress was even possible requires an appreciation of Western culture's deeper roots. The starting point chosen here is the ancient Middle Eastern "Great Game" played out between about 1500 BC and 1100 BC by early territorial states that included Egypt, Babylonia, Elam and Assyria (van de Mieroop 2004). These Bronze and early Iron Age states competed and co-operated with one another through diplomatic, trade and military means. In the process, they became more complex cultures, adjusted their world views accordingly and discovered new things about the world. These ideas and practices were later adopted, adapted, augmented and spread around the Mediterranean by younger cultures such as the Phoenicians, Greeks, Romans and Arabs, as illustrated by the Greek Antikythera celestial calculator. The influence of these west-central Asian cultures on Western European cultures can still be seen today.

A particularly significant contribution began during the first few hundred years AD when the Roman's adopted Christianity as their official religion. The result was the Roman Catholic Church. During the Middle Ages, the Catholic church's formalized version of Christianity became the foundation for Western Europeans' thinking about the organization of the world, our place in it and how to live. The church's views came to dominate European culture's world view. However, despite the church's dominance, there was still room for interpretation of the scriptures and for discussions about the nature of the world.

Starting in the 1300s, near the end of the Middle Ages, Europe experienced a renewed interest in ancient Greek and Roman knowledge (the Renaissance, roughly 1350 to 1600) as well as even older knowledge, such as that from ancient Egypt. There was also an increasing awareness of more recent Arabic knowledge. European interests thus also included Middle Eastern beliefs in the supernatural and related practices, such as Kabbalah, astrology and alchemy. These practices were used in an effort to discover secrets that were believed to be hidden from all but a chosen few (esoteric knowledge). Today we disparagingly call these kinds of subjects the occult traditions. However, during and following the Renaissance period, the formal church (and thus the culture) embraced some aspects of the occult and tolerated others, with certain reservations (Katz 2005). For example, as late as the 1600s, after the Renaissance had ended, Newton could believe in numerology without fear of being declared a heretic. He used it in his lifelong search of the Bible for esoteric knowledge about God's ordering of the natural world.

During the Renaissance, this rediscovered older knowledge was joined by new information, in particular the deluge of observations and specimens arriving from the recently discovered "New World." They were accompanied by a flood of plundered gold and silver. New information was also arriving with trade goods from the Orient. There was thus plenty of fuel and funds for an intense Renaissance investigation and discussion about the relationship between the characteristics of the natural world and God's plan for humans. There was also an increasing dissatisfaction with existing European authority structures and thus a search for new ways to live. One outcome was the formation in the early 1500s of the Protestant Churches with their new religious ideas.

The printing press, invented in the mid-1400s, ensured that the new information, the debates and the changes in thinking were widely disseminated and enduring. As a result, a wider group of people beyond the elite began to participate in social decision making, while the lives of everyone were affected (Barzun 2000). By the 1600s, this history had set the stage for the appearance of the notion of progress in western European culture's world view.

During the 1600s, there were four influences on the members' decision making that enabled the notion of progress to emerge. They also facilitated its acceptance by the religious and eased its incorporation into the culture's existing world view. These four influences were the following: the ending of the Renaissance, natural philosophy, political turmoil and religious beliefs (Mazlish 1996).

Toward the end of the Renaissance (around 1600), the influence of ancient Roman, Greek and Arabic cultures on European world views was slowly displaced by contemporary ideas. The growing focus on the present started in the literary sphere and then spread to other subjects.

One of these emerging ideas/subjects was natural philosophy (nascent science). Its practitioners believed that the correct reading of God's two creations, nature and the Bible, could allow humans to discover God's order in nature. Through this knowledge, humans could find their place in God's universe and thus more easily return to Eden, the garden many Christians treat as a paradise (an expression of being one with God). This perspective allowed for the notion of progress. Examples of this influence include Descartes's 1637 *Discourse on Method*, on reasoning, and Newton's discoveries, such as his splitting of white light into a rainbow in 1675.

Consider too that the direction of political and religious change was strongly influenced by the 30 years of European war between 1618 and 1648. The religious overtones of the war about power contributed to the eventual rejection of the sect-driven religious views of the time. The political outcome culminated later in the 1700s with the rise of secular institutions such as the nation state (Goldstone 2002). Along with this emerging new order came hope for freedom and plenty, both of which can be supported by the notion of progress.

Finally, the Bible-based millennial beliefs, such as those taken from Revelations, had undergone a change since the Middle Ages. Previously, millennialists had believed that the advancement of humanity toward God, and the thousand years of heaven on earth it would bring, was only possible if very restricted faith-based conditions were met. With time, they came to believe that this advancement was inevitable and could take place in the physical as well as the spiritual realm. This allowed millennialists to believe that divine inspiration could occur through natural philosophy as well as through religious faith. Christians could now believe in the possibility of improvement that was not yet backed up by its visible signs or actions. It was thus easier for them to accept the notion of progress. One contribution to this change in direction was John Bunyan's 1678 *The Pilgrim's Progress*, on morals.

The nature and direction of these religious, political, literary and scientific changes interacted with one another to result in the nascent western European notion of progress. During the 1600s, it emphasized (with a strong religious bent) order, hope for the future and advancement through effort, faith and inspiration (Mazlish 1996). Think of the Puritans and their devotion to work and self-sufficiency. Over the next centuries, this early view of progress changed into its present form.

Throughout the 1700s, the fledgling notion of progress slowly became a part of the mainstream European world view. The Lisbon earthquake of 1755 helped it along by sparking a Europe-wide debate about man's (as they expressed it) relationship with God and the nature of natural phenomena. (The reasons why are discussed in chapter 6.) Voltaire's *Candide* (1759), on an attitude toward life, provides one example. After this event, the Western world view increasingly included a more optimistic (positivist) view of the future. This view was now based on the idea of man's gradual improvement to perfection through human effort, but subject to God's wisdom in management (God's providence) (Kendrick 1956). This was in contrast to the older view that saw the hope for salvation as arising solely through the grace of God.

After the mid-1700s, the notion of progress continued to change under the influence of ideas from the religious, moral, scientific and economic spheres of study, as well as from the larger political sphere. For example, between then and the late 1700s, the writings of Carl Linnaeus (*Species Plantarum* [1753], on the classification of plants), Adam Smith (*The Wealth of Nations* [1776], on economics) and Thomas Malthus (*An Essay on the Principle of Population* [1798], on human imperfections) all made a contribution. Like Voltaire, these authors did not write their books in a theoretical vacuum. Their writings were affected by, if not products of, the realities of the day, including their culture's current world view.

The influence of local conditions on these writings is illustrated by the impact that everyday life had on those authors living in England during the mid- to late 1700s. This was only 100 years before England could be described as having a complex industrial culture. In the 1700s, England was a complex farming culture with rising political power and a growing empire. About 80% of its members lived on farms, where they, and their animals, produced the food needed to feed themselves, their domestic animals, the elite and the small number of factory labourers. In effect, the rural farmers supported the culture and its small industrial economy by producing the country's main fuel source, food, as well as much of its agricultural raw materials.

During the mid- to late 1700s, England suffered a lengthy food shortage (to which the 1783 eruption of a volcano in Iceland contributed). About one third of the population was chronically malnourished. Contemporary writings describe a large number of beggars, some with barely enough energy to move around. These stunted or wasted people suffered chronic medical conditions such as heart and lung ailments, and hernia. Although the food bill for feeding this stunted population was smaller, the people were lethargic, their productivity on the farm and in the factory was low, and they died young (Fogel 1994).

The works of authors, such as Malthus and Smith were influenced by their and others' views about these conditions, their possible causes and potential cures. Among the topics discussed were food production, too large a population, and imperfect moral or religious behaviour. For example, Smith was a proponent of individuals striving to acquire wealth because he believed that this effort would improve the material, social and, in particular, moral-religious lives of all (Friedman 2005). He thought this increased wealth could mostly be acquired through increasing trade in their dominantly agricultural economy. His solution, as with those expressed by others, such as the shipping of the disadvantaged to newly claimed Australia in the 1780s, meshed well with England's existing progress-favouring world view. That view also favoured the ongoing rise of the nascent industrialization, which can be thought of as a material expression of the notion of progress.

In the 1800s, the notion of progress became central to positivist thinking and a standard part of Western culture's world view (Mazlish 1996). In England, Protestantism helped the idea of progress to blossom because it more readily accepted those believers who had taken to heart the idea that both hard work and inquiry into the physical world were part of fulfilling God's wishes. The influence of the notion of progress on at least the English in the early 1800s is illustrated by Darwin's turn of phrase when describing the lands and people he saw during his 1831 to 1836 round-the-world voyage on the *Beagle*. In 1839 he wrote, "From seeing the present state, it is impossible not to look forward with high expectations to the future progress of nearly an entire hemisphere. The march of improvement, consequent on the introduction of Christianity throughout the South Sea, stands by itself in the records of history" (Darwin 1972, 438). In the early 1800s, the notion of progress still included the implication of moral progress.

During the 1800s, the secular view of progress slowly appeared. In this view, man's effort alone, free of God's providence (intervention), could lead to salvation. The rise of the secular view is illustrated by the coining of the term *scientist* in the early 1830s and its subsequent displacement of *natural philosopher* as the word to describe those interested in exploring the workings of the world in a scientific manner. Later in the century, as the practice of science gained in popularity, it became more specialized and secular. It also split into disciplines.

The increasing acceptance of the notion of progress by Europeans during this period is illustrated by some of the views about the world that became acceptable to hold. For example, a person could believe that cultures evolved economically from hunter-gatherer to agriculture to commerce (Friedman 2005). Later, in the 1800s, they could also believe that cultures went through an intellectual evolution from believing in myths, to religion, to understanding through science.

The increasing acceptance of the notion of progress was sustained, in part, by the "fact of progress." For example, starting in the mid-1800s, the increasing number of accepted scientific discoveries and the products from the expanding industrial transformation of Europe made the notion of progress "visible." In the late 1800s, the departmentalization of science was accompanied by an increase in formal research focused on specific questions. This made it easier to evaluate the outcome of scientific endeavours in a specific field. The record of the successes provided to both scientists and the public further proof of the notion of progress (Harpham 2006).

However, European faith in progress was regularly shaken when the promised non-material "goods" remained undelivered or the opposite occurred. The early years of the French Revolution (1789 to 1799) initially provided Europeans with an example of social progress, but the violence of its later years shook that belief. Later contributions to skepticism about progress in all of its forms included the mechanized slaughter of World War I (1914 to 1918) and the fallout from scientific eugenics at the end of World War II (1939 to 1945).

Faith in progress was also shaken by some of its more ephemeral consequences. For example, the rise of the belief in secular progress during the 1800s led to a waning in the importance of the Christian notion of hell (a place of everlasting punishment) as a personal life concern. But, for many, the emotional support and sense of purpose previously supplied by religion and the avoidance of hell could not be replaced by the emerging scientific conclusion that humans were inconsequential bit players in an effectively timeless universe. Left without a comforting cultural world view or guidance (religious or otherwise) on how to deal with the vagaries of day-to-day life, angst and dissent appeared. One resolution was a return, by many, to the more accommodating religions and to mysticism (Mazlish 1996). This trend is illustrated by the appearance of sects such as Mormonism and a revival in apocalyptic millenarianism (Katz 2005).

**Progress Today.** The European notion of progress changed over time. From the 1600s until the mid- to late 1800s, Europeans believed that, by discovering and applying the laws of God (which govern the world he created), they could progress toward living a better life, closer to God. These laws were to be found by studying both of God's creations: the Bible and nature. (Note that God's world was believed to have the same basic structure as Europe's complex cultures.) After the mid- to late 1800s, the Western notion of progress still expressed the idea of improvement toward an imagined better future through human effort, but it did so in an increasingly secular and material manner. It also became more nuanced.

This is the notion of progress that dominates Western culture's world view today. It expresses a much weaker belief in the inevitability of progress. For example, we no longer tie history and progress in a culture's characteristics so tightly together (Marx 1996; Mazlish and Marx 1996). Instead of believing in a simple progression from error to perfection, we now accept that progress can occur though the erratic piecemeal accumulation of improvements, and that there can be setbacks (Mazlish 1996). As expected, the events Westerners now take as indicators of the reality of progress are mostly material rather than moral or religious in nature (Taylor 1989). A deeper appreciation for the current notion of progress will become clearer after discussing some specifics of its expression and influence on our lives.

A critical if not defining aspect of current Western culture is its vision of "economic progress." It describes our belief that an increase in financial wealth (economic growth) will automatically bring numerous benefits to the other spheres of life. This view includes the condition that we will only receive these benefits if we strive hard to facilitate economic progress: we must maintain economic growth. Two interesting side effects of believing in economic progress are that, firstly, modernity (often equated with being up to date) and economic prosperity are often falsely considered to be the same (Goldstone 2002). Secondly, the demonstration of economic growth has become an essential part of our culture's reaffirmation and authority maintenance process (Marx 1996). In turn, the need to provide proof of this progress strongly affects Westerners' demands for resources.

"Scientific progress" is still seen by some as the irreversible and inevitable accumulation of knowledge that moves us toward truer theories that more accurately represent the world (Kuhn 1996). However, more generally, the claims of scientific progress have become more circumspect with the recognition that "errors" in the scientific process can occur. We tend to accept that the path to scientific truth can be indirect (Mazlish 1996; Kuhn 1996).

Despite this admission, the results of scientific research still reassure us that scientific progress is real and that applying science will successfully resolve our issues and concerns. You can verify this belief for yourself by noting how often a public announcement about a scientific discovery includes the hope of a possible/imminent cure for a problem important to the culture. You can hear the mantra of progress when scientists expound on their belief in the value of accumulating knowledge as they describe or justify their creative work to the public.

We more fervently believe in progress when it is associated with science's applied relative: technology. We hold our belief in "technological progress" more easily and strongly because the products of technology make the reality of progress tangible, just as monumental buildings do for our culture's world view. Technological progress is seen to be the creator of progressive improvements in many aspects of our material, social and mental lives. It is also seen to be the physical engine of economic progress.

We believe that technological progress is good because we feel it both satisfies our material wants and reassures us that our culture is on the right path. We are comforted by our belief that technology's ability to give us physical mastery over our world will ultimately solve our intractable (social and material) problems, keep us free and make us secure (White 1968b). Our belief in these features, and their connection to economic progress, ensures that the notion of technological progress exercises significant influence over the issues

Westerners recognize and how we think they should be solved (Oreskes and Conway 2010). However, despite these perceived positive characteristics, there is a recognition that some technology is bad or can cause bad things to happen.

The phrase "the progress of nature" expresses the positive perception we hold of certain changes to living things over time. Historically, this meant that evolution and our interaction with nature resulted in more than just change. It was change *toward* the goal of nature's general "improvement." Remember Westerners' view that the changes in the reptilian brain that gave rise to the old mammalian and then the primate brain were actually following a path of improvement, as mentioned in chapter 1.

Today the progress of nature is rarely used to describe changes occurring through natural processes such as evolution. Instead, until recently, the use of "progress" and "nature" together referred exclusively to the efforts of those trying to improve, manipulate or conquer nature for the benefit of humans (White 1968b). This is still the case when referring to the hoped for steady improvement in our efforts to stimulate, manage or adjust organisms so that we can more easily or quickly satisfy our needs or wants. Examples are the creation of more beautiful flowers, faster growing trees and the other specific improvements through the genetic modification of organisms. Think of the excitement around CRISPR-Cas9, the recently discovered gene-editing tool (Travis 2015).

Today a link between "progress" and "nature" can be held by environmentalists. Think of progress on the road to sustainable resource management, mitigating our environmental impacts, and restoring disrupted or destroyed ecosystems. This usage illustrates the power of the notion of progress. However, the idea that the notion of progress applies to our interactions with nature has a number of critics who question it on the grounds of limits to resource consumption and the constraints on environmental processes. Some even raise the spectre of nature's death at the hands of progress. Often, it is unclear if the critics' objective is to redefine progress when applied to nature or simply discredit it (White 1996).

That question is difficult to answer. Consider how the notion of progress is invoked when we propose to manipulate nature to either mitigate our planned future impacts on it or restore those features disrupted or destroyed by our past impacts. The belief in progress toward that goal can be held regardless of whether the damage was caused by approved and acceptable efforts to maintain economic, technological and agricultural progress, by accident or by wanton behaviour. When seen in this context, it is unclear if the term *progress* refers primarily to a hoped for improvement in our material interactions (e.g., the application of technology) with nature (we are still the boss) or a change in our basic personal or cultural relationship with nature in order to impact it less.

The notion of "religious progress" is used to describe us coming to more closely following the word of God, usually in order to achieve the goal of later salvation (Marx 1996). Strong notions of religious progress are still found in some Christian sects. Consider the Rapture, which is a Revelation-based, apocalyptic millennialist belief among some American evangelical Christian sects (Rev. 20:1-5). More commonly, the secular notion of material progress is accepted by the religious, but with a religious bent. Similarly, the secular expression of progress is not entirely free of its religious roots. For example, when secular scientists discuss the importance of their work and the motivation that drives them, they routinely express a religious-like sense of being part of a greater striving toward an improved future, as hinted at above (Noble 1998). This universal appeal of the general notion of progress is one reason why it still plays a central role in Western culture.

With the ongoing declines in the power of formal religions, the related notion of "moral progress" is now ill-defined, except when tied to religion or economics. Consider that the connection between being a moral person and a hard worker is now quite weak. The exception is seen in the context of economics, where hard work, which leads to wealth, is still seen as part of being a better person. The connection between moral progress and some aspects of political and religious life still remains, especially in the United States. For example, there is a moral overtone to a common American belief that with hard work you can become rich or even the president of the United States.

The pervasive unconscious influence that the non-specific notion of progress has on Westerners' everyday lives is illustrated by the wide use of its many synonyms in everyday speech. These include *growth, development, advancement* and *improvement*. The normality of this usage is illustrated by two western Canadian examples. A 2006 brochure published by British Columbia's electricity supplier reads in part, "Having enough made-in-B.C. electricity will help ensure we continue to enjoy the benefits of a strong and growing economy as well as maintain our quality of life" (BC Hydro 2006). Consider too the Statoil advertisement at the Calgary, Alberta, airport promoting their development of the Athabasca tar sands. The ad's tag lines are, "Always moving, never satisfied" and "Progress is in our genes" (Statoil 2012).

Westerners' belief in the various features of our cultural world view is, like in all cultures, boosted by their symbolic representation. Progress is not an exception. Consider that our notion of progress becomes material and thus visible to us when we watch our science, labour and technology transform the land into, for example, a farm, a subdivision, a skyscraper or a rapid transit system. The reality of progress is further internalized when we buy and personalize the latest technological gadgets, or when we spend time transforming and managing the landscape around our houses and then contemplate the results. These manifestations of progress are so important to us that we, like the members of ancient complex cultures, tend to view these items as indicators of being civilized: they are symbols. As a result, statements that are considered to be against these material representations are labelled as anti-progress or anti-civilization.

However, the post-1950s industrial disasters such as Love Canal, Bhopal, Chernobyl, Fukushima Daiichi and Deepwater Horizon have softened our belief in material progress. Our reaction to these and equivalent historical events indicate that we have gained some awareness that the notion of progress is a cultural construct and that, in many ways, it is at odds with the state and functioning of our external reality. Yet, we still retain the hope that, with time and effort, these flaws will be rectified. The notion of progress still plays a central role in Western culture's world view and our personal decision making. Perhaps it endures because it merges so well with our human bias of perceiving an optimistic future, our inclination to simplify the world by focusing on the specific and our characteristic of raising our expectations when we reach a goal. Think rising expectations.

In summary, this section should allow us to more easily recognize the ubiquitous but unconscious presence of the notion of progress in Western culture's world view. We should more easily accept its inescapable influence on our personal decision-making and appreciate its contribution to our self-sustaining cultural drive to increase our culture's complexity.

This chapter, in summary, discussed the structure and functioning of a group of people: their culture. The members of the group/culture hold a personal view of the world and how to live in it. The aggregate of their personal world views is their culture's world view. For this reason, the members' material expressions of their culture, such as their clothes and buildings, reflect their culture's world view.

A culture functions and changes through the dynamic interactions that occur among the group's individual members, their culture's world view and their environmental endowment (i.e., as part of a dynamic, complex cultural system). In particular, a culture functions and changes as a result of the members' decisions. Those decisions are the product of their decision-making foibles while under the influence of their culture's world view and in the context of its environmental endowment. Their culture's world view provides them with guidance about how the world functions, their place in it, how to live in it and an explanation for why. But the members are, individually, their culture's decision makers and, collectively, the carriers of their culture. Thus their decisions affect their culture's world view: the member-culture feedback loop is involved.

A culture has structure. The complexity of that structure (cultural complexity) can be simple, complex or something in between. The characteristics of a culture are consistent with its level of cultural complexity. For example, the more complex cultures have a larger population and a larger area of land/resources under their control. The world view of the more complex cultures includes a hierarchical view of the world's structure and functioning. It also includes an expensive reaffirmation and authority maintenance process, high per capita resource demands and a relationship with their environment that enables them to meet these resource demands.

A culture's world view is historical, dynamic and biased toward human concerns. Its current version was provided by the members' immediate ancestors and, like those ancestors, the current members are collectively continuing to change it. Some of their adjustments can result in an increase in their culture's complexity. For example, their decisions can (usually inadvertently) result in their culture's world view changing so that its guidance favours them resolving issues in ways that increase their culture's complexity.

A functional culture is one that can, over the short term, meet the conditions needed for it to be unified and adaptable: it functions in both the members' individual and their group's best interests. For a culture to be functional over the long term (i.e., viable), it must meet viability's (human and environmental) conditions. It is the culture's decision makers, the members, whose decisions can (deliberately or inadvertently) adjust their culture's world view so that it provides viability-favouring guidance. However, because of our decision-making foibles, our decisions can result in our culture's world view incompletely and/or inaccurately reflecting these conditions. Its guidance to the members about how to live will then be significantly at odds with viability's constraints. Our decision-making foibles also allow us to personally ignore viability-favouring guidance. Either way, the culture can become dysfunctional.

If we want our culture to become and remain viable, then we must face both our decision-making foibles and the influences on our decision making, discussed in chapters 1, 2 and here. We are helped in this task if we appreciate how we decide on truth and how we predict the future. These are the subjects of the next two chapters.

# CHAPTER 4

# DECIDING ON "TRUTH"

We are all faced with making decisions about the truth of something: our memories, experiences, information and conclusions. When making them, we rely heavily on our feelings and experiences. When in doubt, we turn to the guidance provided by our culture's world view. Western culture provides its members with two options: religion and science. Although the arts are important, they are not discussed here because, in Western culture today, they are generally considered to be either the messenger, the mediator of behaviour or a method of expression, not a guide to deciding on truth.

The practitioners of both religion and science rely on our brain's general (conscious and unconscious) abilities when making their decisions. This common dependency ties science and religion closely to one another. The main differences between them are their centres of attention. Religions are primarily focused on providing guidance about our relationship with a deity or deities and how we can use that knowledge to satisfy our internal and external reality needs. Science is focused primarily on satisfying our curiosity about our external reality and how we create our internal reality. We can then use that information to help satisfy our needs. Before discussing the overlapping roles that science and religion play in our decision making about truth, we need to broaden our appreciation for the nature of our internal and external realities and our conscious relationship to them.

## Two Realities

We live in two overlapping realities: an internal one and an external one. Our internal reality is the one built by our brain's functioning. The external reality is everything outside of that internal reality, including the processes by which our internal reality is created. Think of a movie we see in a theatre as our internal reality. Everything else, including the projector, film and screen, is part of the external reality. A convenient boundary between the two realities is drawn by the disappearance of one reality when we die or, by analogy, when the projector stops.

The part of our brain-created internal reality we know best is the part we consciously experience: our conscious reality. We have faith in this reality and assume that it is an accurate reflection of the external world and our place in it. But, as discussed in chapter 1, even the simple act of seeing a rainbow does not accurately reproduce the external world. Our internal reality is a highly simplified, human-biased, incomplete, ever-changing mixture of fact and feeling that includes only selected aspects of the external world.

Despite these limitations, our internal reality enables us to participate in the external reality. It achieves this by providing us with a reference framework that helps us apply our mental tools when dealing with the external reality. It gives us the opportunity to complete our lifeway in our supporting environment as a member of a group of humans.

For us to succeed at that task, our internal reality does not need to faithfully reproduce our external reality. However, it must reflect the world around us to the degree that we can appropriately deal with the issues we face. After all, the cold hard facts of the external reality create, among other things, today's food and tomorrow's nuclear accident.

Our internal reality comes to reflect many of the external reality's properties because, as we grow up, we absorb our experience of them into our internal reality. For example, we learn that we can't walk through walls. However, that is insufficient for us to survive, especially if we participate in a complex culture and use environment-altering technology. We are therefore faced with the task of using a mix of our conscious experiences and our abilities, such as intuition, reason, imagination, common sense, experience, experiment and belief, to more fully incorporate the external reality into our internal reality and thus into our decision making.

Our casual use of these mental tools to achieve this goal serves us well, but only within limits. At some point, the constraints on our abilities to relate to the world around us, such as our faith in our rules of thumb and our reference normal, bump into the external reality's messy, highly complex characteristics. Our effort to resolve the resulting conflicts, conundrums and dilemmas includes denials and illusions and a tendency to follow our culturally influenced inner preferences. As a result, it is normal (usual) for us to address an issue by making decisions that are mentally quite satisfying but significantly disconnected from the external reality aspects of the issue. The ease with which this can happen and its significance is illustrated by the story behind the dummy still being used to train Europeans and Canadians in mouth-to-mouth resuscitation.

When the Norwegian toy maker Asmund Laerdal developed a resuscitation dummy in the 1950s, the face he chose was represented by a plaster mask image of an unnamed but hauntingly beautiful young woman. That he chose this face is not surprising as she was well-known in

Europe as "the drowned Mona Lisa," or France's L'Inconnue de la Seine. The young woman, so the tragic story went, had drowned in the Seine during the 1800s. Her body had remained unclaimed and unnamed, but her beauty and look of innocence had been preserved by a plaster cast of her face made in death. Poignancy was added to the story by its connection to the Seine. The river had for years held a morbid fascination for Parisians because of the numerous bodies found in it from suicides, drownings and murders.

The beautiful woman's death naturally inspired speculation about her name and life. Famous literary figures combined her beauty and anonymity into poems, novels and short stories about her imagined life and death. As a result, her face, and the tragedy of her death, became famous throughout Europe. Between the First and Second World Wars, she came to epitomize the ideal female beauty of that era and, as a result, the cult of the drowned Mona Lisa appeared.

Her fame recently spurred investigators to find out more about her tragic history. In their search, the investigators discovered a professional mask maker who had ties to the likely maker of the drowned Mona Lisa mask. By looking at the detail of the skin's texture preserved in the mask, the professional concluded that it could not have been made on a corpse. Based on his experience, he suggested that the mask was made on a living woman around 16 years of age. He also suggested that she was possibly a mask maker's daughter or a person working as a model (Chrisafis 2007; Zeidler 2009).

These observations and deductions suggest that the accepted history of her drowning and the making of her mask were the products of our imagination. This conclusion doesn't change the significant role that her facial image, the myths and the stories about her life played in the internal reality of thousands. She was a cultural icon of her time who, for decades, touched hearts and influenced decisions, including those of Asmund Laerdal.

In many instances, as happened with the drowned Mona Lisa, the influence of our imaginative creations on our decision making is harmless, if not beneficial. However, there are also instances where our failure to include the features of the external reality in our decisions has resulted in serious, long-lasting, undesirable consequences. In order to make appropriate decisions, our decision making needs to recognize that our internal reality, and our conscious experience of part of it, consists of an intimate mixture of features representing the external reality and the power of our imagination. We need to know when to accept the external reality for what it is and when to distinguish it from our imaginative creations, and vice versa. We could use our two cultural guides of religion and science to help us with those decisions. But, if we do, then we should understand their respective power and limits, and know when to use which guide.

## Religion and Belief

The religious manner of thinking and believing, and its limits, is the subject of this section.

**Religious Way of Thinking.** Religion is a formalized way of thinking and living that relies on accepting the existence and superiority of a supernatural being, beings or force (being). We assume that a person who uses a religion as a guide to decide on truth does so not only by accepting the religion's beliefs but also by applying them as convictions rather than assumptions. But the beliefs of a religion are intangible, often difficult to define and can change through time. This is illustrated by a religious person who may struggle to answer the question, what do I believe in?, but who can easily answer the question, what do I live by? (Barrett 1978). The core of a religion has just as much to do with its sanctioned practices as with its beliefs. Thus, the relationships among the members, and between them and their supreme being, are just as important as beliefs. This means that when religious people use their beliefs to decide on truth, they do so as members of a culture.

We can come to fully accept a religion's world view (its beliefs, practices, symbols and other basic features) through many paths, including upbringing and revelation. In the process, we come to accept, perhaps expect, that the supreme being and the religious organization representing it will provide us with a sense of community and unity, direction about how to live, a personal sense of spirituality, and confirming rituals. If it meets these obligations, then a religion gives us a sense of emotional security and belonging as well as meaning and purpose. It also provides us with an understanding about the nature of our world and our place in it (Barrett 1978; Raymo 1998). In other words, it guides us in our efforts to decide on truth.

**Religion's Limits.** In trying to determine the limits to religion's guidance, we can be dragged into debates about whether a religion's founding beliefs are true or false. But, practically speaking, there is little incentive or need for a religion to confirm that its founding beliefs, such as its creation myths, are in fact true in the external reality. They are taken on faith.

When looked at from the believers' perspective, what matters to them is whether their religion's guidance satisfies their personal and their culture's needs. In this context, the limits of a religion's guidance can be determined by asking two questions. Firstly, does it help its members to develop a sense of personal, physical and social-psychological well-being? And, secondly, does it help its members (and their culture) develop an appropriate relationship with one another and with their environment? In summary form, does a religion's guidance help its members to simultaneously satisfy their (material and mental) needs while living within their human and environmental constraints (the objective)?

A religion's power to guide the members is based on its formal statements of belief, or doctrine. The religion gains this power because its members accept the doctrine's tenants. They partly demonstrate their acceptance through their ritualized reverence for specific sacred symbols, such as a book, place or object, that represent their doctrine and its foundation. This spiritual connection empowers the religion to guide its members about how to live.

But that link also constrains the ability of its guidance to meet the objective of living within their human and environmental constraints. Consider that humans can believe whatever they want, but belief alone is no guarantee that the external reality won't intrude into our lives. Thus, if a religion is to meet the objective, then there are limits to the guidance it can provide about the external reality, our relationship to it and how to live. Consider a doctrine that claims, for example, physical immortality for those who believe. In general, if religions are to meet the objective, even in a spiritual sense, then they have to provide guidance that is in keeping with the constraints imposed on us by the external reality.

However, deciding on how closely its guidance must reflect the external reality is less straightforward than one might expect. Consider that religious doctrine provides guidance about influencing the external reality through practices like prayer, ritual and sacrifice. Unfortunately, it has never been reproducibly demonstrated that these practices have exerted their claimed influence on the external reality. This indicates that there are significant limits to the material benefit that these methods and the assumptions (beliefs) behind them can provide. Despite this evidence, these practices

continue. They do so because they do provide the practitioners with intangible benefits, such as improving their sense of social-psychological well-being. They can help us maintain our sense of self (Garre 1976). Thus these practices could enable a religion to help its members' satisfy their needs while encouraging them to live within their human and environmental constraints.

In contrast, if we assume that religion's guidance about the nature of the external reality and our relationship to it is accurate, then different types of constraints on its ability to meet the objective are seen. Consider first the limitations that arise from our human foibles. For example, the clergy are human, so they suffer from the same decision-making limitations as the rest of us. Evidence to support this view is provided when a clergy member's personal behaviour and their religious organization's subsequent decisions about the consequences of that behaviour unambiguously detract from the well-being of the religion's members. Examples that illustrate this limit are religious leaders' muted response to the sexual abuse of children by some members of the clergy; the acceptance of a large disparity in physical or social-psychological well-being between the clergy and the lay members; and the manipulation of the members, especially when they are vulnerable, by the clergy for their personal or their religious institution's gain.

Next consider the consequences when our assumption that a religion's guidance accurately reflects the workings of the external reality is proved false. Perhaps the most significant, dramatic and heart-wrenching examples are the catastrophic natural events in which the pious lose their lives. Consider the following examples.

The Christian church explained the visitation of the Black Death upon Europe in the mid-1300s as the result of God either providing humanity an opportunity for salvation or as his punishment for their wicked behaviour. This is illustrated by the flagellants when, at their meetings, they read aloud a popular tract called the heavenly letter, which began "O ye of little faith" (Kelly 2005; Nohl 1971). Similarly, human sin and our turning away from God were given as explanations for the earthquake, tsunami and fire that struck Lisbon in 1755 killing thousands. This included the many pious who died in church because the devastation occurred on all Saints' Day, as discussed in more detail in chapter 6 (Kendrick 1956).

Compare this to the various religious explanations reported for the devastating 2004 tsunami and loss of life along Indonesia's coastline. Hindu followers were advised to reconcile themselves to fate. Christians noted that God promises to redeem those suffering but not to prevent it. The imam of Meulaboh, Indonesia, interpreted the disaster as divine displeasure with wicked humanity (Bates 2005). More generally, Muslims are advised that nothing happens unless God wills it.

These religious explanations may or may not detract from a particular religion's efforts to administer material and psychological care and comfort to those suffering as a result of these terrible events. But they do indicate that religious explanations of natural phenomena rely on beliefs that can be disconnected from observations about the external world. This limits the ability of a revered religious authority's doctrine-guided interpretations of sacred texts and/or natural signs to explain the workings of the external world.

More generally it constrains religion's ability to help us reach for the objective by developing an appropriate relationship with our environment. Consider the diversity of advice Christian religious authorities provide to their members from their readings of Genesis about the relationship God intends us to have with our environment. The interpretations vary so widely that the resulting advice is often contradictory, as discussed in chapter 11. One reason for this outcome is the ambiguity in the accepted versions of these passages. Another is that different passages in Genesis are seen as relevant. A third reason is that differing interpretations are given for the same passages by different Christian sects and religious authorities.

A more basic reason for the diversity is that the cultural context at the time the passages were written is not taken into account when they are interpreted. For example, it appears that the writers of the Old Testament believed that God had bestowed on his believers three gifts: victory in war, life (which encompasses animals, personal well-being and the harvest) and wisdom. Of these three, perhaps the notion of biblical wisdom could be taken to imply that the Bible was intended to accurately describe the external reality and our relationship to it. However, biblical wisdom, in its historical cultural context, likely meant the practical, political wisdom of the administrators and rulers and the hidden wisdom of the scholars. In this context, *scholars* meant those who interpreted the religious books (Lang 2002).

The writers of Genesis appear to be primarily concerned with describing our ties to and relationship with God, which is what we expect from a religion.

Certainly, Genesis is not the product of detailed observations about the actual workings of our environment. This is the context within which the references to our relationship to the environment in Genesis should be seen. Thus, there are significant limits to the guidance Genesis can provide to us about the nature of our environment, such as how it functions and our material relationship to it.

The limits to religious pronouncements about our environment can be made visible and evaluated by asking variations of the questions used to evaluate the ability of religion's guidance to help an individual improve their sense of well-being. Consider first that we are dependent on our environment and its functioning for our long-term survival and well-being. Then ask, do the religious explanations and guidance promote the ongoing protection of and respect for our supporting environment, both for present and future generations? Do they help us to live within its material and functional constraints? Overall, a religion's guidance is limited in its ability to help its members to meet the objective of satisfying their needs while living within their human and environmental constraints. Unless, that is, it more accurately reflects the nature of the environment in its guidance.

In summary, religions invoke a divine, provide descriptions of our and our world's nature, and give us guidance about how we should live. However, the information religions provide tells us more about our relationship with the divine and about religiously acceptable ways to live than it does about how the external reality functions or its constraints on how we can live (Tattersall 1998). Despite these shortcomings, religions are not the last refuge of the deranged, as some would claim. Religions do recognize that we struggle to deal with our internal reality and the vagaries of life, and they do attempt to help us deal with them.

The limits to religious guidance concerning our decision making about truth should thus be drawn outside its considerable ability to attend to our mental well-being and some aspects of our physical well-being. Within this boundary, religions help to provide their members with a long-term sense of security, meaning, belonging and unity. They accomplish this in the face of a seemingly heartless environment and despite us being part of an often chaotic and apparently uncaring complex culture. Within their limitations, religions are capable of protecting their members and sustaining our environment.

A more specific exploration of the relative value, or correctness, of specific religious beliefs and practices is beyond the scope of this work. Further discussion about religious limits and its contributions to our efforts to deal with our environmental reality will appear later, after discussing the state and functioning of our environment. The discussion now turns to science, on which we will dwell for a little longer because of its greater attention to our external reality.

## Science and Experimentation

In the West, we imagine that religious thinking is as old as humans, while scientific thinking is an invention of the Greeks, the Renaissance or the Victorians. But the practice of science has accompanied the development of complex cultures from at least 2000 BC. This is not surprising since, from the day we are born, we can conduct experiments and think in the same manner as scientists (and then believe in our conclusions with religious conviction) (Gopnik et al. 2001; Cole and Cole 2001). The scientific manner of thinking and believing as a guide to truth, and its limits, is the subject of this section.

<u>Scientific Way of Thinking and Acting.</u> As practised today, Western science is a formal manner of gathering, interpreting and sharing information about the nature of the external reality (including us) or verifying a supposition about it. The practice includes a particular way of believing, thinking and experimenting (Gibbons 1999). A scientist adheres to these practices when trying to address a specific topic.

Science, it has been suggested, does or should do more. It should help us reach a goal related to the previously mentioned idea of scientific progress: diagnose and correct the differences between our beliefs (our internal reality perhaps) and the external reality (Wilson 1999a). This is not the view taken here. The focus of this subsection is to illustrate scientific thinking by discussing the idealized process of arriving at a theory and applying it.

The scientific process can be thought of as ending when the available, acceptable knowledge (the data, conclusions and beliefs) about the topic has been encapsulated into a coherent summary or theory (Ziman 1978). However, a theory can also serve as a guide for both the use of existing knowledge and the search for new knowledge (Myers 1999). It is thus better to think of a theory as being both the beginning and the end of the scientific process. It is the beginning because a theory can be used to direct initial thinking about a topic, frame questions and plan data collection. It is the end because the thinking about the new data results in an old theory being adjusted or, more rarely, a new one being created.

The thinking used in this process is scientific in nature. It is expected to winnow out a grain of knowledge from the chaff of misapprehension. This is accomplished through the discarding, discounting, sifting, sorting, selecting and ordering (evaluating and organizing) of information (Kuhn 1996; Norretranders 1998). These processes of evaluating and organizing information are accomplished through our conscious and unconscious perceptions and decision-making processes. But, as we already know, there are biases and limits to these processes. To reduce them, scientific thinking must be as free of human bias as possible. Scientists must thus strive to be objective and skeptical. These two goals are core aspects of scientific thinking (Ziman 1978). The need to specify them already hints at limits to the scientific process, which will be discussed later.

In the scientific context, being objective has a few different but related meanings (Daston and Galison 1992). Today, the essence of being objective means trying to remove (or, more realistically, to recognize and take into account) the existence of personal influence and bias in our thinking, such as unsubstantiated personal beliefs. It also requires the practitioner to observe (or, more realistically, try to observe) the external reality as it is, not as he or she wants it to be (Raymo 1998). The idea of being objective is illustrated by the question, do you love me? Both a subjective (personal opinion) and an objective (outsider's perspective) answer exists. One of the ways of answering this question requires that the respondent attempt to reduce the influence of money and sex on their answer.

Skepticism is the ability to remain open to questioning the validity of a theory, method or piece of information while at the same time remaining open to accepting it. To be skeptical, in the science context, also requires questioning one's own decisions and beliefs as well as those of others. It thus implies being able to admit that one might be wrong, or at least identify how one's thinking might be proved wrong (Ziman 1978). The importance of skepticism is illustrated by remembering that, in the real world, a complete escape from belief is not possible. However, by displaying skepticism toward our beliefs, we can question them. The requirement to be skeptical ensures that science's relationship to belief and truth is fundamentally different from religion's.

As hinted at above, the scientific thinking used to process information and adjust or create a theory must be more than objective and skeptical. It must use the conscious thinking processes of reason and logic. In this context, reason means basing one's thinking about new information on existing, accepted knowledge while restricting the use of imagination (Moshman 1998). In contrast, logic means limiting the links made between bits of information to those connections that have been previously accepted as true and those being apparently revealed by the new information. Logic is powerful because, among other things, it helps us determine, for a specific question, which relationships likely exist, which should be questioned, which are important or which can be ignored (Peters 1983). By default, logic constrains the use of imagination and emotion, as stereotyped by *Star Trek*'s Dr. Spock.

There is nothing in the above description that limits scientific thinking to professional scientists. Anyone can practise the scientific way of thinking. In fact, all of us apply it to at least some degree (Gopnik et al. 2001).

However, it must be admitted that it is easier to read the above list of the requirements to practise scientific thinking than it is to routinely and consistently satisfy them. We often find it difficult to decide when we are being objective and using reason and when we are holding to a personal belief generated by our internal reality. It is thus easy to drift away from fulfilling the requirements to think scientifically. Time, effort and guidance are needed if we are to develop the ability to skeptically and objectively assimilate new information while both holding to our current views and being open to changing them (Gibbons 1999; Ziman 1978).

It is no wonder that the practice of professional science, like religion, takes place in a science community with a culture that has a science-favouring world view. As expected, the members have faith in that world view. For example, they believe that an external or objective reality exists outside of our heads and that this reality can be discovered and described through the scientific process (Raymo 1998; Ziman 1978). Scientific thinking is believed to be a key part of that discovery process.

Because it is a community, science culture's world view guides its members to adopt and practise more than skepticism, objectivity, reason and logic. They must satisfy science culture's social practices. Among the most critical are co-operation and trust.

Trust allows scientists to accept that the information and theories being presented by other scientists are true to the best of the presenter's ability and knowledge. In this context, the act of being skeptical is meant to be directed at the presenter's data and arguments, not at the scientist providing them. Without trust, skepticism could be aimed at the practitioner, which would disrupt the co-operation within the community. Skepticism would thus become adversarial.

A sense of co-operation is also critical for the practice of science because it sustains the scientists' desire to freely exchange data and the details of methods. It also maintains their expectation of support for one another's endeavours. This includes their peers voluntarily checking the validity of a scientist's work prior to its publication, called peer review. If trust and co-operation were to disappear, then the pool of reliable information on which to build theories would be smaller, the formal peer-review process would falter and the scientific process would, at least, slow down (Shapin 1994).

Science culture's world view is distinctive. In particular, its favoured personal and social behaviours, such as co-operation, helps to distinguish Western science from Western culture's adversarial legal and political traditions. They also help distinguish it from religions.

But when the process of science is viewed within the context of our human characteristics, in particular the features of our innate decision making, then the scientist's favoured attributes are seen to be ideals. This conclusion is supported by the failures of the scientific process, such as publications withdrawn because of faked, incompletely reported or misrepresented results. Think of the paper that falsely claimed autism can be caused by vaccinations, which the Lancet retracted in 2010 (Wakefield et al 1998). Therefore, to fully appreciate science requires understanding its limits.

**Science's Limits.** The scientific process as a guide to truth will play the dominant role in reaching the conclusions in this book. To appreciate the constraints on those conclusions, a more thorough understanding of the limits to both the scientific process and science's role as a guide is needed. These limits can be appreciated and illustrated by dividing them into four broad, overlapping and interconnected groups: science's practices, its processing and presentation, its data and conclusions, and the culture in which the scientific process takes place.

*Science's Practices.* Scientists are human and cannot escape their human foibles. These include illusory thinking (Myers 1999), the biases that arise from using rules of thumb (Moshman 1998) and the important role of belief (Rescher 1998). Nor do scientists intuitively adhere to the tenants of science, for example, the renowned scientists whose ideology and blind faith made it difficult for them to scientifically evaluate the health damage from tobacco or the environmental consequences of ozone layer thinning (Oreskes and Conway 2010).

Our practice of science is therefore limited by the need to pay attention to ourselves: how we think and the influences on our thoughts and beliefs. Not an easy task. Consider our human predisposition to imagine the presence of order or to impose it beyond where it exists. This desire to find order is so strong that we can see meaningful patterns in nature when there are none (Rescher 1998). Examples are cloud animals, scientists seeing order in random data and psychologists deducing their patients' inner selves from Rorschach plots (Lilienfeld et al. 2001).

By paying attention to ourselves, we can make ourselves aware of both this innate tendency to find order in the world and the difficulty we experience restraining ourselves from doing so, when needed. We can also realise that when we do find a pattern, it is difficult to engage in the personal task of being appropriately skeptical of it. This difficulty is made worse when we recognize that our pattern recognition ability is a key component of scientific thinking (Ziman 1978).

We are similarly limited in our ability to be objective. For example, we can talk about being objective in our thinking about the patterns we find, but practically achieving a fully objective assessment of them is rarely, if ever, possible. After all, we are participants in the universe not observers, disinterested or otherwise (Norretranders 1998). Nor, as we explored in chapter 1, can we escape into an emotion- or value-free state of mind.

It is thus easier for us to accept the need to display skepticism and objectivity than it is for us to practise them (Ziman 1978). It is especially difficult to be both skeptical and open to new ideas (Myers 2002a). It is easier to slip into unconditional belief and to uncritically apply scientific theories beyond the data they were designed to encompass (Ziman 1978).

Even the application of the sacred cow of logic has its limits. Consider that in the late 1800s and early 1900s the rationalist school of thinkers, centred in mathematics, tried to sideline the need for assumptions. The proponents hoped to achieve their goal by building up a general framework of truth. It would be formed by applying the rules of logic to a set of inescapable truths, or axioms. The endeavour hit a snag in 1931 when Kurt Godel's incompleteness theorem showed, mathematically, that there are limits to the conclusions you can draw from a set of axioms by applying logic. For example, to prove the statement "these axioms do not contradict one another" requires a broader principle.

By extension, this limit applies to at least those statements that could be accurately represented by natural numbers (Dawson 2006). Because most scientific observations use numerical representation, it is reasonable to assume that there are limits to the use of logic in the broader field of science (Chaitin 1988). An illustration that this limit to logic is more widely applicable is provided by more general self-referencing statements, such as "this statement is unprovable." Their validation too requires a greater authority (Dawson 2006; Norretranders 1998).

Our use of logic is constrained by more than the self-referencing problem. It is also constrained by our brain's ability to apply it. Our effort to strictly apply logic to science issues can result in us experiencing apparent contradictions and dichotomies. For example, for strict logic to apply to a sequence, each step must have a definitive outcome (or set of outcomes) and each outcome must be the result of a fixed relationship between the two steps. An analogy would be a simple computer program.

We can think of the world around us as a complex meshwork of interconnected sequences. Our human ability to track just one is limited to a sequence that has a limited number of steps with just a few outcomes at each step. These limitations ensure that, even when thinking about a relatively simple topic, we maintain only a summary image of the part we are currently focused on resolving. As a result, when we consider a different aspect of the topic, a second, potentially contradictory (by our application of logic) summary image may appear. In fact, the two summaries may not be contradictory but complementary. They may simply be different summaries of the same complex topic.

In these cases, the apparent contradiction can be resolved by holding both summaries in mind at the same time or by mentally merging the two. Think of a young child's sudden realization that the two views they just saw, such as two doll houses, are in fact just different views of the same house (Cole and Cole 2001). However, for more complex topics, we find this merging task to be sufficiently difficult and time consuming that it can often be beyond our ability. Examples are my multiperspective perception of our forest posed in the introduction to this book; or the merging of the various descriptions of the brain discussed in chapter 1; or thinking about the related mind-brain conundrum (Polanyi 1968). Consider too the difficulty of treating subatomic entities as both particles and waves.

The practice of science is also limited by the dynamics of individuals in a group of scientists. Consider that the wider use of new facts and theories, or the adjustments to existing ones, depends on the relevant specialist subgroup of scientists coming to a significant degree of agreement or consensus to make and accept these changes. Satisfying this requirement is subject to the same limitations that apply to any human group. These group limitations join with our limitations as individuals and the complexity of our external reality to ensure that there are inevitably undetected biases, errors or assumptions in accepted data and theories.

Despite acknowledging this possibility, scientists, both as a group and as individuals, can still experience great difficulty recognizing and correcting anomalies in data or accepted theories (Lightman and Gingerich 1992). This is the case even in times of a scientific crisis (Ziman 1978; Kuhn 1996). One would expect that scientific training should allow us to compensate for these human characteristics. However, a trained brain, even one trained in the practice of science, still focuses its attention and is hard to change (Hall 1981). As a result, considerable inertia can be displayed to changing even error-riddled theories. Consider the longevity of the economist's theory that we are rational consumers (i.e., we make our purchasing decisions in an entirely rational manner) even while marketers rely on the emotional aspects of our decision making to sell consumer products (Hernstein 1990).

In overview, Western culture's world view implies that we can believe in, practise and apply science by rote. But there is more to being a science practitioner, either professional or lay. A person must understand the process and its limits and become familiar with them. Like religion, science only becomes real and a part of our lives with thought-filled training and rehearsal, and a supportive culture. Think of being scientifically skeptical (Ziman 1978; Coppola and Daniels 1999).

These requirements impose a limit on the scientific process. More broadly, our self-referencing experiences and thinking (self-referencing feedback) limit our ability to understand ourselves and our world.

*Processing Information and Presenting Findings.* The products from practising formal science are data, conclusions and theories (collectively referred to as findings). Findings are generated by gathering, processing (such as manipulating, organizing, evaluating, etc.) and interpreting information in order to draw conclusions that can be used to build or adjust theories. The essence of findings is made available for verification and assimilation by others in a presentation. That essence is made intelligible to others and its relationship to current theories made clear through the presentation's judicious use of, among other things, words, symbols, images and diagrams.

The need to process information and to present findings creates a dilemma. Consider that each of these steps involves discarding, selecting, manipulating, interpreting and summarizing the information. These simplifying processes can accomplish more than revealing the essence of the external reality. They can create the appearance of order and reveal relationships when none exist. In contrast, by not simplifying information, its essence may be misunderstood or masked. The presence of simplification's dilemma constrains how truthfully findings will represent the external reality. This, in turn, limits their usefulness.

The constraints imposed by this dilemma are illustrated by an analogy. Although we can't appreciate the forest if the trees are unacknowledged, it is equally true that the forest will remain unseen if we focus solely on the trees (Coppola and Daniels 1999). By looking at the forest at only one scale, we will limit the species we see, the context in which we see them and the complexity of the interconnections between them. Our perception of the forest will be, to some degree, biased, unclear or distorted (Norretranders 1998).

This suggests that an analogy may be a useful method to resolve the dilemma of portraying complex ideas and systems during processing and presentation. However, analogies can be misleading. For example, the nature of ancient hunter-gatherer cultures can be illustrated by an analogy with the common features of current hunter-gatherer cultures. Unfortunately, that analogy can guide us to preferentially use an "average culture" when thinking about past hunter-gatherers. We would then neglect that past hunter-gatherer cultures, of which there were many, each had their own distinctive characteristics, and that there were likely significant differences between them and present hunter-gatherer cultures (Kelly 1995).

Similarly, visual displays, such as images and diagrams, can make it easier to generate, present, grasp and assimilate findings (Ziman 1978). But they can just as easily mislead. Consider our tendency to unconsciously equate an image with the object it depicts when the image is actually a depiction of selected features of the object for a particular purpose. This effect is succinctly demonstrated by Magritte's painting of a Sherlock Holmes-style smoking pipe with the caption "this is not a pipe" (Coppola and Daniels 1999). It can also be demonstrated by scientific illustrations (figure 4). As a result, we can mislead ourselves into thinking that an image represents the truth when it is just one representation of the truth.

Misperceptions of visual displays can also occur because of our bias toward the sight sense and pattern recognition. We can easily focus on things in a display that are of minor importance or see things that are not there (figure 2). This is particularly the case for complex images. For example, outside the hands of an expert, the significance of a very complex image, such as an X-ray photograph, can easily be missed or misconstrued (Daston and Galison 1992).

Simplifying a complex image or diagram can help, but visual displays also suffer from the dilemma of simplification. Consider that if an image or diagram is to clearly convey its message, it has to be presented in a well-constrained and explicit context, often with its salient features highlighted through selection, abstraction or interpretation. But the resulting highlighting or diagrammatic summarizing can also misrepresent reality. For example, simplification can reduce an image to a cartoon with unconstrained properties, as seen in the representation of bacteria (germs) in TV ads. Similarly, compartmentalizing information with firm borders can mask the complexity of the topic's connections to a greater functional whole. The power of symbols means that their inclusion in a display can alter our (factual and emotional) perception of the topic.

These types of limitations are particularly relevant when using visual displays to present a critical part of findings: the deduced relationships among things (including ideas and theories). When the representation of these relationships are displayed within an appropriate context, they can help us understand the subject. This is especially true when the topic is complex (figure 9).

However, the presentation can unexpectedly create faulty understandings about the nature of that relationship. Consider the diagrams that use points to represent each of a topic's many states of existence and lines to represent the relationship links between those states. For example, Y diagrams are used to display the relationships between our various hominid ancestors and ourselves through time (Shipman 2002), and tree diagrams are used to display the time relationships between various languages (Gray and Atkinson 2003). Similarly, arrows are used to join the reservoirs outlined in a carbon cycle diagram such as figure 7a.

In all of these diagrams, the false impression can be created that ancestors, languages and reservoirs are static, uniform entities isolated in time and space. The diagrams also imply that the relationships between these features are simple, step-like and perhaps predetermined, whereas the actual relationships are often graded, distributed, very complex and dynamic. The difficulty of avoiding this misperception and the resulting consequences is illustrated in the previous chapter by the convenience of dividing cultures of differing organizational complexities into precisely labelled groups linked by increasing complexity. As discovered, this neatness masks the considerable diversity in each group and the continuous (not step-like), multi-faceted, complex relationship that actually links them together.

In overview, the ability of science to generate and present knowledge that meaningfully represents the external reality is partly limited by the need to process information and to present the resulting findings. The limits arise because of the conflicts among the processing and presentation imperatives such as detail, overview, simplicity, complexity, clarity, uncertainty, connections and context. This is especially significant for complex topics. We can't remove this limitation. If we wish to draw our internal reality closer to our external reality, rather than to push them further apart, then our search for truth should include being aware of these limitation and taking them into account. This is especially true in an age where the use of complex algorithms and AI (artificial "intelligence") are believed to circumvent this constraint.

*Data and Conclusions.* The nature of data and conclusions constrains, in a number of ways, the scientific process and its use as a guide to truth. Those constraints are illustrated here by the limitations we face in our efforts to collect data, to check conclusions and to meaningfully assimilate valid conclusions.

One of the most basic limits to data acquisition is the problem of diminishing returns. As science accumulates knowledge about a subject, the effort and cost of accumulating more knowledge about it rises exponentially. This occurs because of the increase in resources and time needed to fill in details, to make more precise measurements and to answer more difficult questions (Tainter 1988). Compare the cost of the physics experiment used to detect the electron (trivial) with the cost to search for the Higgs boson (billions). Think of the gains for the effort and cost of Galileo's telescope versus the Hubble Space Telescope.

There are other limits to us collecting data and reducing their uncertainty. It is unusual for us to maintain the funding needed to monitor variables for more than a few years. This makes it difficult to study something that requires either a few reliable measurements that each takes a very long time to complete, or a long record of short-duration measurements. For example, the nature of an ecosystem's functioning and changing only becomes apparent after decades' worth of data have been collected for many variables. This time requirement is one of the limits that cultures face when trying to develop science-based methods for managing their commons (Ostrom et al. 1999). These constraints on our ability to collect relevant data with a sufficiently low uncertainty means that data can represent reality but not mirror it (Ziman 1978).

It is easy to draw a conclusion from data. However, for it to appropriately affect our decisions, that conclusion must be valid (credible). The validity of a conclusion, or theory for that matter, should not be confused with its correctness. Correct means error free. Neither scientific conclusions nor theories can be judged to be absolutely correct because all, at some level, are estimates and therefore provisional.

Thus, a conclusion does not have to be correct to be valid. However, the validity of a conclusion is limited to a range of conditions and contexts. Confidence in a conclusion is boosted when both a conclusion's validity and the conditions under which it is thought to be valid are subject to a routine validity check: a complex, interactive evaluation of quality, explanation and prediction. It is a powerful tool that also helps us to identify the degree to which inappropriate belief was used to draw a conclusion.

The validity check of a conclusion (of all types, including theories) includes the adequacy of the data used, how well the data supports the conclusion, and the conclusion's internal consistency. The check also evaluates the conclusion's ability to retrospectively predict relevant, accepted (known) features of the external reality. In simple terms, the less the conclusion is able to predict the known, the less likely it is to be considered valid (assuming the data is error free) (Rescher 1998). Consider astrology. Its ability to predict a person's known nature is extremely limited. If astrology is useful, then its value is not in predicting a person's character traits from natal charts (Carlson 1985).

It is convenient to think of validity checking as occurring in two stages. The preliminary validity check, the peer review, occurs quickly. The ongoing validity checking continues as part of the scientific process of re-evaluating conclusions when testing old theories and making new ones. This stage takes considerable time to complete, which is unsurprising considering the complexity of our world and how easy it is for scientists to strongly hold incomplete ideas and use erroneous data (Kuhn 1996; Ziman 1978). It thus takes time for the limits, misconceptions or errors in a conclusion to be fully discovered and, potentially, to be corrected.

This test of time holds a special place in Western science's world view. It is believed that, with the passage of time, more precise conclusions and theories will eventually displace less precise ones (Peters 1983). The foundation of this belief is the notion of scientific progress. In its most fundamental versions, it holds that the passage of time will inevitably result in correct conclusions and theories. Time will reveal the truth of these beliefs.

It would intuitively seem that once a valid conclusion is available, it would be used by scientists and the public. This is often not the case. It takes time and effort for scientists and the public to meaningfully include the conclusion, the constraints on its validity and the significance of its simplification-related biases into our thinking and decision making: it takes time and effort to assimilate a conclusion. The need for time and effort points to another constraint on the assimilation process. The number of conclusions currently available to be assimilated is large, so large that even organizations dedicated to just collecting and distributing scientific results can experience inadvertent errors, omissions and delays.

Consider this problem at a more personal level. As early as the late 1970s, individual scientists seemed to be approaching the limits to their human ability to assimilate the conclusions needed to stay current in their field (Thomas 1978). A now commonly applied solution employed by conclusion-overloaded researchers is to split their scientific discipline into subgroups, that is, to specialize. To compensate, they routinely form teams to work on a question requiring their combined knowledge. But neither of these solutions fully addresses the human limits to a scientist's, let alone the public's, ability to meaningfully assimilate a scientific conclusion. For example, specialization can easily divert attention from the presence and nature of the wider context within which the conclusion should be assimilated if its full significance is to be appreciated. Think of this book. Hence the idea that science is leading us to knowing more about less.

This, and related types of assimilation limits, can be pushed back somewhat for scientists and the public alike by summarizing and generalizing conclusions and by applying technology, such as computers. But simplifying processes can also lead to distortion and bias, especially if the topic is complex. Think of the reptilian brain. And technological solutions can only partly address our human limitations. Think of algorithms and AI.

There are other intractable constraints that limit meaningful assimilation. They are illustrated by the following examples. Consider that the statistical representation of information is an important part of presenting conclusions and the constraints on their validity. But our ability to meaningfully assimilate statistical information is limited by our human characteristics, as mentioned in chapter 2 and discussed further in chapter 5. It is also limited by the nature of statistics.

Consider that a statistical representation of a population's state can't be directly applied to each individual in the population. For example, imagine that a scientific medical study of a large population to which I belong comes to the conclusion that 1 in 10 of the people in that population will likely develop a particular cancer. This statistic tells me that there is a risk that I might develop that cancer. But it doesn't provide me with information about my personal risk. How can I meaningfully assimilate this population-level probability into my personal lifestyle-affecting decisions?

The meaningful assimilation of conclusions is also constrained by the complexity of the system to which they apply. Consider that the conclusions from a single study of a simple system, such as the relationship between a metal bar's change in length with temperature, is far more likely to reliably represent that system than a single study of a complex system, such as an ecosystem. The single study of a complex system is more likely to be a snapshot of the system's present state.

One way to increase the reliability of the knowledge one has about a complex system is to assimilate the conclusions from many different studies conducted from different perspectives over a variety of scales in time and space. However, our human characteristics limit our ability to assimilate these studies in a manner that reliably reflects the whole. The usual solution to this difficulty is to compile the conclusions into a model or theory. This changes the nature but not the substance of the constraints on the meaningful assimilation of conclusions discussed above. We are back at first base.

In overview, the ability of science to usefully represent the external reality is partly limited by our ability to collect the needed data. It is also partly limited by the validity of the conclusions we draw and our ability to meaningfully assimilate them into a coherent representation of that reality. This constrains science's ability to reliably guide our decisions. Although these constraints can't be entirely avoided, they can be taken into account during decision making. We are left to discuss the limits imposed on the scientific process by the cultural context in which it takes place.

*Culture's Contribution.* Scientists are guided by the cultural world view of their subgroup. But this science subgroup exists within a greater culture with its own world view. These two world views and the relationship between them constrain both the practice of science and the influence its conclusions have on the members of both cultures: scientists and the public (Shapin 1994). The nature and significance of those cultural constraints will be considered here.

The more recent roots of today's Western science culture and its relationship to the wider Western culture can be traced back to Europe during the early 1600s. Even though the terms *science* and *scientist* had not yet been coined, investigations of a scientific nature were being conducted by upper-class gentleman scholars, especially in England. These English natural philosophers belonged to small informal or semi-formal social groups that held meetings where the members could share their findings.

The manner in which the gentlemen discussed their findings reflected the social culture of their upper-class background. In particular, they placed great emphasis on the notions of trust and truthfulness. Their belief in these qualities was so strong that a gentleman's words were sufficient to ensure the truth of his scientific findings.

These face-to-face meetings also served a more basic cultural function. They provided an opportunity for the gentleman scholars to reaffirm their faith in their culture's ethic of trust and truthfulness. The meetings were thus part of the individual-member–cultural-world-view feedback loop discussed in chapter 3.

Today, the practice of science in the Western world still relies on the trust ethic. Scientists still reaffirm it with one another through face-to-face meetings such as conferences. Similarly, the current formal peer-review process of verifying a scientist's report mimics the presentation and discussion of results at the gentleman scholars' meetings. However, today the guarantee to the public that science is being practised truthfully rests with the greater culture's institutions, such as the universities and government regulators who collectively facilitate the scientists' research and the peer-review process (Shapin 1994). Lay people trust that these cultural institutions will ensure both that the scientific reports will be subject to a validity check and that the process will be policed.

The public's culturally influenced faith in these institutions and in the scientists conducting the research within them is one of the reasons why the public generally supports science research. It is also a source of their expectation that science will resolve their personal and the greater culture's problems. However, the ability of the scientists and the scientific process to meet the public's expectations faces culture-related limits.

The most obvious constraint is that scientists and their culture are imperfect. Scientists don't always meet the ideals of their culture's world view. Consider that despite the cultural values of trust, truthfulness and co-operation, the members experience the same ubiquitous, but often subtle, human influences on their decisions found inside any social group. These influences include the pressure to conform, the politics of power and the desire to please (Kuhn 1996; Pearce 2010). Their presence can interfere with, even bias, a scientist's efforts to undertake research and have their results accepted by the scientific community during peer review.

The ability of scientists and their practice of science to meet the public's expectations is also limited by the nature of the relationship between science culture and the greater culture. Consider how the practice of science is influenced by the guidance the greater culture's world view provides about desirable knowledge (Holton 1996). This guidance constrains what can or should be studied. For example, in the United States, their cultural world view lay behind a temporary restriction being placed on stem cell research. And, its military is well-known for funding research that has even a hint of a military application.

The greater culture constrains not just which topics can be investigated, but how. For example, in Western culture, researchers must acquire full, free and informed consent from the human participants in their studies. They must also follow rules about the treatment of laboratory animals.

The topics that scientists study and the acceptability of their findings are influenced more forcefully by the views of those within the greater culture who provide or manage political and economic support for scientific research (Wilson 1999a). The influence of this patronage is present in all types of scientific research, but it is most easily seen in research undertaken by, or for, culturally sanctioned, private vested interests. This is called applied science or techno-science to distinguish it from public science (science).

The practitioners of techno-science are well aware of the preferred outcome of their research. They are also more directly connected to those higher up in the organization's hierarchy who approve the topics of research, favour a particular outcome and supply the funding to attain it. The higher-ups having this influence may not subscribe to the scientific ethic of trust or truth in knowledge, and they more often view knowledge as intellectual property rather than as a common good to be shared.

The presence of this form of patronage-related influence on the practice of techno-science and its leakage into public science is illustrated by studies that have scrutinized research papers reporting the results of drug studies and drug trials published in medical research journals. Positive outcomes for a drug's effectiveness and safety are more commonly reported when pharmaceutical companies have funded the studies and trials (Kelly et al. 2006; Lexchin 2005; Schott et al. 2010).

Similar constraints are also seen in public science. For example, its work practices, grant applications and results are more easily accepted by the greater culture if they do not clash too strongly with either its world view or the maintenance of its hierarchical organization (its power structure). Galileo found this out in the early 1600s when he was barred from disseminating his results because of his unpopular conclusions about the heavens. This reality conflicts with science culture's emphasis on being truthful and data driven.

As a result of these differences and influences, clashes between scientists and the greater culture are inevitable. Recent examples include the experiences of scientists researching tobacco, ozone depletion and climate change (Oreskes and Conway 2010). Consider too that in 2010, the Canadian government cut funding to government scientists, limited their research topics and excessively restricted their freedom to talk about their work. It resulted in these scientists creating the Evidence for Democracy movement (Evidence for Democracy 2015).

Overall however, the relationship between the greater culture, the science culture and the practice of science is more complementary than confrontational. Consider that the general (overarching) biases evident in the practice of science and its conclusions tend to reflect the preferences in the greater culture's world view. For example, the general theories that archaeologists and anthropologists develop to explain the data they obtain from a studied culture can be (consciously or unconsciously) influenced by the world view of the greater culture to which the scientists belong. The scientists' theories about a studied culture can also be influenced by their personal, greater-culture-influenced views about contemporary social issues in their own culture (Trigger 1989).

The resulting bias can even be self-reinforcing. For example, a Western researcher could (unconsciously or consciously) use their industrial culture's favoured economic model as a reference when describing the economies of hunter-gatherer or simple agricultural cultures. The researcher could thus easily conclude that these simple economies are primitive versions of the more complex Western economy, thus implying that progress separates the two. The greater culture, in turn, would be hard pressed not to accept a scientific conclusion that portrayed its economic system in a good light and supported its world view. The Western scientist and their greater culture would have created a feedback loop that self-sustains the researcher's conclusions and the culture's world view.

But these conclusions and beliefs would be more than biased, they would be false. Hunter-gatherer economies are complete in their own right for their circumstance. The differences between their economies and those of more complex cultures arise because of the differences between their circumstances (including world views). Hunter-gatherer economies can certainly be thought of as simpler and different, but not as primitive versions of Western economies. The notion of progress does not explain the differences between Western and hunter-gatherer economies (Wilson 1988; Firth 1965).

One could expect that faulty conclusions like this would rapidly disappear with further research. However, the ability of scientific conclusions to change the world view of science culture or the greater culture is constrained by features such as the feedback described above (Kuhn 1996). Cultural change is also constrained by the general inertia cultures display toward changing their core characteristics.

A telling example is the ongoing Western cultural belief that hunter-gatherer societies practise the idealized Western vision of altruistic sharing. This is believed, even though it is well documented that the sharing practices of hunter-gatherer cultures are generally pragmatic and expect reciprocity sometime in the future. Overall, the details and degree to which the members share is quite diverse and depends on many factors such as the environmental context within which the culture is found (Kelly 1995).

The nature of the relationship between the greater culture and science culture also limits the ability of science to meet the greater culture's expectations of science: that its practice will solve the greater culture's problems. Consider that the greater culture's relationship with science, unlike with religion, economics, law or politics, does not empower scientists to instruct the greater culture's members. Scientific findings tend to be treated as information not instructions. Thus scientists can't direct the greater culture's members (or its leaders) to conduct their lives in a certain way or to include (or exclude) certain beliefs from their personal or their culture's world view (Ziman 1978). Scientists can only inform, suggest and recommend.

Certainly, scientists can use their findings to persuade, but even that ability is significantly constrained, for example, by the difficulty non-scientists can experience personalizing scientific findings. Personalization is important because, in essence, it gives the findings sufficient personal meaning that we are encouraged to spend the time and energy needed to assimilate and apply them to our personal decision making.

One reason non-scientists experience this difficulty is that scientific findings often seem to us to be dry. We can also feel that the objective description of the world provided by scientists is negative and depressing. This is especially true when scientists describe the illusory nature of our self-perceptions or predict a "bleak" outlook for the future. We find it easier to just ignore that information or to hanker after an idealized past. Think of our optimistic bias.

These limitations on persuasion can be addressed by adjusting how scientific findings are presented, however this is not as easy a task as it sounds. For example, the emotive presentation of findings in a context familiar to the public can make the findings more accessible and thus more easily personalized. This is especially true if the findings are simplified. But our ability to satisfy these conditions in a way that still retains the integrity of the findings is constrained. As discussed, without due care, simplification and an adjustment of context can easily lead to the information becoming caricatured, trite or misleading. Think of the split brain discussed in chapter 1.

To avoid that outcome, scientists argue that a precise, objective description, especially a mathematical one, presented within an appropriate context, most clearly and accurately conveys their findings. But, by following this too closely, the findings may not be personalized or reliably, if at all, assimilated. Adding to the overall difficulty of choosing a presentation method that satisfies all concerns are our decision-making foibles, as discussed in chapter 2.

Consider too that scientists can also experience difficulty trying to turn their findings into persuasive, personable information for non-scientists for another, more basic reason. The scientists can be so engaged in the discovery and application of scientific knowledge that they themselves do not fully appreciate the larger consequences of their work or the influence of the larger culture's world view on their personal thinking about it. Or if they do become aware, then they too can simply rationalize away that knowledge and its associated feelings.

For example, the scientific theory of nuclear fission was the foundation of the Manhattan Project: the mission to develop and build the first atomic bomb. Robert Oppenheimer was a physicist and the lead scientist on the project. He was a moral man who managed to maintain a mental distinction between his job as a nuclear scientist and his knowledge that his greater culture intended to use the bomb he was in charge of building to kill thousands. He accomplished this distinction by separating the single cultural whole of which he, his profession and the project were a part, into two distinct aspects. Despite his training as a scientist, he failed to recognize that this distinction was a convenient fabrication (Hijiya 2000). He is not alone. Think about the scientists engaged in the direct genetic

modifications of organisms using the power of CRISPR-Cas9.

In overview, despite our beliefs to the contrary, science culture and the practice of science are neither separate from us nor dominant over us (White 1968b). Scientific inquiry largely reflects, supports and is biased by the world view of the greater culture in which it is embedded. This occurs because there are direct and indirect connections, including feedback, between the greater culture and science culture. Those connections are mediated by the scientists and the greater culture's members. The influence of science on the greater culture is both enabled and limited by the nature of that relationship.

In summary, the scientific process is well suited to provide us with reliable (reproducible) information about our external reality and the nature of our internal one. This scientific knowledge can help us to understand our world, how it functions, our place in it and the meaning of wisdom. But the practice of science that provides that information is significantly constrained.

The constraints on the practice of science can be divided into three overlapping types: individual human characteristics, the nature of information and the influence of culture. For example, our human decision-making foibles impose constraints on scientists' practice of science that can be reduced, but not eliminated, by the scientists applying skepticism and objectivity, reason and logic. Similarly, the nature of information constrains scientists' ability to gather and process it, which ensures that their findings are always subject to limits, uncertainty and bias. The constraints that arise from science culture, the greater culture in which science culture is embedded, and the relationship between the two are wide-ranging. Consider that the practice of science follows the greater culture's patronage and is guided by science culture's world view, which itself tends to reflect the world view of the greater culture.

There are other complementary constraints that further limit the degree to which scientists and their findings can influence personal and social change. Consider that the greater culture's world view provides guidance to its members about how to live. But what it provides might not guide the members to respect the external reality, for example, to live within their human and environmental constraints or create a viable culture within a functioning environmental endowment (Appell 2001; Stern 1993).

The ability of scientists and their findings to influence the greater culture to discover and correct any misdirection arising from its world view is constrained. This arises, in part, because of the relationship between science culture and the greater culture: scientists are relegated to the role of persuaders not guides. It also arises because scientific findings are often perceived by the majority of the public to be remote from their lives and their day-to-day concerns. They thus experience difficulty personalizing, assimilating and applying science's findings to their day-to-day living. When this remoteness is viewed from the perspective of the public's role as their culture's decision makers, then it is seen to constrain the ability of the findings and thus scientists to influence the greater culture. It also influences the relationship between the greater culture and science culture. A self-sustaining feedback is formed.

Overall, the ability of science to meaningfully guide us toward truth is significantly constrained. Science, like religion, doesn't fully meet the hopes that at least Westerners have in its ability to help us decide on truth.

## The Split between Science and Religion

By understanding the historical relationship between science and religion we can more fully appreciate the roles that they play in Western culture's decision making about truth. In particular, how did science and religion, with their different but equally strongly held beliefs about the nature of truth, come to diverge from their common heritage? A brief summary of the relevant history is provided here (Katz 2005).

Their more recent common roots can be traced back to the European world view of the 1600s that gave rise to the notion of progress and individualism. For example, it is easy to recognize a scientific aspect to Descartes's (1596–1650) early 1600s view that the universe worked like a clock. However, in making this observation, it is also easy to overlook the model's deeper connection to Christian thought.

Descartes thought that God had endowed the world he created with an order that allowed it to function automatically, like a clock, a view which was common enough to be called deism. (Descartes's idea and the analogy are unsurprising. Clocks were the marvels of that age and influenced his thinking, just as the marvel of our age, computers, influences our thinking today.)

A number of Europeans (mostly English) who were also thinking about the relationship between God, nature and man disagreed. They could not accept an explanation where God's only role was to build and start the clock. They knew that God was all-powerful. This meant that he could exert his influence on his world any time he chose. By the late 1600s, Descartes's model had been modified to allow for divine providence, that is, God's ability and wisdom to benevolently interfere in the clockwork workings of his universe.

However, in order for these natural philosophers to continue studying nature, they had to be sure that God had created a universe that did indeed operate according to a set of rules. They turned to God's word, the Christian Bible, for answers. Among the evidence for their case was God's use of the rainbow as a reminder of his promise to Noah that there would not be another flood to destroy all living things (Genesis 9:9–17). Natural philosophers interpreted this to mean that God had decided that his laws for nature would be unchanging, except under special circumstances.

As Christians, natural philosophers of the 1600s looked to the Bible for more assurance that the order existed. They believed that if they could only read it correctly, then the Bible would reveal to them the details of God's order in his universe. They also believed that nature, which was God's creation, could similarly be "read" to reveal God's order.

In their attempts to correctly read the Bible and nature, natural philosophers relied on more than the basics of Christian doctrine. They used the theories about the ordering of the world expressed in the older traditions (such as Greek, Arab and Roman) that had recently been revived or rediscovered during the Renaissance. The three most influential traditions were the Neoplatonists (with the Divine Mind which was the source of the world soul, or nature), the Gnostics (secret knowledge) and the Hermetics (powerful primordial wisdom associated with Hermes Trismegistus).

These traditions expressed the belief that there was more to the world than there seemed. The universe was a living whole whose components had properties and relationships that were hidden from the senses: that is, there were secret correspondences between things in nature and between nature and sacred texts. These traditions, in particular the Hermetic, held that if we could find these correspondences (the hidden order), then we could, through the use of intermediaries, influence the operation of the living universe to reach (progress toward) the divine. (Note, too, that it was usual to believe that the structure of their universe mimicked the hierarchy of complex cultures in which the adherents lived.) These beliefs, especially those of

Hermeticism, were linked to practices such as astrology, numerology and alchemy. Today we label these traditions and practices as occult.

Although natural philosophers used and held these occult views, this did not mean that they were secularists or that there was a split between them and the church. In the 1600s, the distinction between the formal church and natural philosophers was one of focus. Natural philosophers concentrated on explaining the order and workings of God's physical universe, while the church concentrated on representing the word of God. Certainly, there were aspects of the older traditions that the church considered unacceptable, even heretical, such as witchcraft and aspects of Gnosticism. But, on the whole, the practice of natural philosophy, including the use of their occult tools, was conducted within the Church and its doctrine. Think of how religious belief in prophecy, millenarianism and divine providence overlap with occult beliefs.

A good illustration that occult ideas were mainstream can be found in the work of Isaac Newton (1642–1727). Today he is considered to be one of the founders of contemporary, secular science. In his day, Newton was a Christian natural philosopher who believed that nature and the Bible were teeming with esoteric secrets and occult mysteries. He believed that if he could discover and interpret them, they would reveal to him the order of God's universe. In working to uncover those secrets, he applied the same exactness of thought to his interpretation of the Bible as he applied to reading nature and to his attempts to match the one to the other.

For example, Newton thought that the biblical description of Solomon's temple reproduced the universe. To him, the perpetual flame in the inner sanctum of the temple represented the sun, while the seven lamps around it represented the seven planets. His belief in this type of heavenly inspired worldly order influenced his division of the rainbow into seven colours. Ironically, despite Newton's prowess, many of his contemporaries thought that his views on gravitation were an irrational, mystical extravagance.

Newton was not alone in his occult-Christian beliefs. During the 1600s, there were a number of informal and formal European natural philosophy societies, such as the Royal Society of London (formally founded in 1660). The members of these societies, who were Christians, were also aware of occult ideas or studied occult texts, such as those with Hermetic views.

During the 1700s, the ongoing discoveries about the natural world and the cultural reactions to contemporary events, in particular the 1755 earthquake in Lisbon, gave credibility to the natural philosophers' work and beliefs (Kendrick 1956). As a result, by the early 1800s, natural philosophy's description of the natural world was rapidly gaining wider cultural acceptance. Yet, from the perspective of an early to mid-1800s natural philosopher, it still didn't matter if he (still mostly a he) included occult-Christian beliefs in his natural philosophy views. He was still considered to be gathering the information needed to construct explanatory theories about God's world.

The term *science* was coined in 1833 as part of a movement to provide more rigour to natural studies. However, a fully formed professional scientist culture only started to appear after about 1870. By 1874, the separations between the emerging practice of science, mainstream Christianity and the occult were becoming clearly visible. In that year, the head of the British Association for the Advancement of Science made a famous speech in which he indicated that he was firmly against "spiritualist thinking," which he considered to be sloppy in nature. He indicated that he was in favour of the new and more rigorous science. He also attacked Pope Pius IX of the Roman Catholic Church whose recent pronouncements he saw as anti-progress and anti-science. The trend to secularism was now apparent.

The split between religion, the occult and science took a long time to fully develop. Even into the late 1800s, the occult and science could be treated as equals by those studying nature. For example, in the 1890s, William Crooks, a prominent scientist whose research led to the discovery of radiation and X-rays, was convinced that part of his work proved the existence of a psychic force. He could argue for psychic research, and even be involved in it, with no negative impact on his considerable standing in the scientific community.

Finally, near the very end of the 1800s, a departmentalized science, formally free of religious or occult teachings, was firmly established. By this time, astrology had given rise to astronomy, alchemy had spawned chemistry, comparative religion had led to anthropology, and spiritualism was transforming itself into psychology (Katz 2005). Christianity still exerted considerable influence on the culture, while the occult, despite its rejection by science and Christianity, remained a subject of popular interest.

Today, despite the roughly 150-year split between Christianity and science, their common cultural heritage means that Christian views still permeate science and science still influences Christianity. Think of religious progress toward the perfection of humanity and scientific progress toward the perfection of theories. Think of Christian attempts to prove the scientific validity of all aspects of the Bible and scientists' attempts to find a place for a God. In the same vein, beliefs labelled as occult are still with us.

## Knowledge or Wisdom?

Science today is a co-operative effort. The practitioners engage in a mixture of intuition and rigorous testing, creative hunches, objectivity and skepticism. Its practitioners provide us and themselves with the most reproducible information available about our external reality, our place in it and how we create our internal reality (Myers 1999). But this scientific information is neither certain nor complete: it is only probable or provisional knowledge (Peters 1983). Neither should having this information be mistaken for the wisdom we require to use it appropriately. Accomplished scientists with considerable knowledge can be narrow, foolish people, while wise scholars can be considered weak scientists (Wilson 1999a). The equivalent is true of religious leaders.

If wisdom is more than the awareness of information (knowledge), the casual, unquestioned use of knowledge or the skilled use of knowledge (intelligence), then what is it? Wisdom is the ability to know when you don't have adequate information, to collect the needed knowledge and to know when and how to use it appropriately. Wisdom is the skill to use the knowledge you have as an aid to making judicious assessments of many factors before deciding between, or delicately balancing, the accessible options.

Gaining wisdom is a slow process that develops with experience over time. It is the outcome of continually trying to solve problems in a manner that takes into account the overarching (broad, longer-term) social, cultural, psychological and environmental context within which the problem is embedded. Wisdom can be thought of as the product of making the effort to be both rational and compassionate while looking at an issue from multiple perspectives.

By practising science, we are better able to gather the information needed to describe this context and address the problems. But doing so provides little direct help in our efforts to satisfy our associated human need for a sense of personal meaning and place and our need to experience feelings of belonging, contentment, compassion and fairness. These are the very attributes that religions attempt to satisfy. Thus, the wisdom needed to appreciate and deal with the larger cultural and environmental challenges we face today should include both the scientific and the religious guides about truth. To use both implies that we need to reconcile their differences.

Reconciliation means more than looking for the benefits from our practice of religion and science. It means ensuring that we recognize and face their respective failings and limitations. Consider our use of belief. It can satisfy our desires. But it also has the power to make each of us blind to cultural, environmental, personal, political, scientific or religious limits.

For example, on the one hand, unchecked religious belief, such as a strong belief in revelation as the source of truth, can lead to religious fervour, inappropriate cults and the persecution of the innocent. An unquestioned religious faith allows the believer to be more easily co-opted into supporting unfair economic and political ends that drive their culture outside its viability boundary. Unquestioned faith also facilitates our failure to recognize the price of ignoring our external reality and the impact of our decisions about the environment that supports us.

On the other hand, unchecked faith in science can lead to the belief that science alone can create the order Westerners equate with civilization (Norretranders 1998; White 1968b). Through unquestioned belief, science practitioners can forget that they are subject to human limits and cultural biases in their quest for knowledge. They, like all humans, can display self-interest and self-deception. They too are capable of conveniently ignoring the psychological, biological and cultural contexts of knowledge and the limits of rationality (White 1968c).

A too strongly held belief in the goodness of scientific knowledge can allow scientists (and the public) to reject any notions of their responsibility for their methods of acquiring that knowledge or for the uses to which it will be put. They (and the public) can forget that knowledge is not neutral and that we can use it to both create and destroy, including the destruction of a culture by undermining its belief system (Rees 1993). Just as religion can be co-opted, so too can science end up serving the self-interests of economists, corporations, politicians or the religious elite at the price of ignoring our humanity and the fate of our environment.

This does not mean that reconciling science and religion requires us to turn the clock back to a mythical golden age of unity, nor does it require the destruction of one or the other. Our use of science and religion does require that we acknowledge and try to avoid their mutually destructive characteristics while enhancing those characteristics that mutually foster wisdom. We could strive for a science that accepts the significance of spirituality and a religion that finds a place and use for material facts. We are also obliged to ensure that our beliefs in science and religion are collectively guiding us to a realistic relationship with the external reality, providing us with well-grounded emotional comfort and helping us to achieve wisdom (Raymo 1998). They should guide us toward, not away from viable living: that is, toward accepting our responsibilities as well as wanting our rights

It seems that the road to successful reconciliation would see the believers in science take from religion the reminder that humans require comfort from too much raw reality. Science should acknowledge that we can become distressed when contemplating the overwhelming immensity and interconnectedness of our complex world and the vastness of our universe (Raymo 1998). Religion reminds us that, in addition to our search for the immediate, factual and material, we seek comfort in hope, compassion and community (Myers 2000). A sense of mystery and a desire to belong are essential parts of being human (Raymo 1998).

The religious would take from science the reminder that in order to live the truth of God's external reality (of God's will or universe if you like), we have to observe the world as it is, not how we believe or want it to be (Raymo 1998). This includes remembering our role in creating our vision of our world. Science provides the basis for answering paradoxical questions such as whether a viable polio virus made from off-the-shelf chemicals is living or dead (Cello et al. 2002). In providing this service, science shows us that our religion-based perspectives of the external reality, such as our belief that we are separate from the external world or special, are largely arbitrary. Science prompts religions to remember that we are good at fooling ourselves. Religions would accept that the scientific process provides us with the observations needed to form a reference that anchors our imagination, beliefs and our intuition closer to God's external reality (Hassan 2000; Myers 2002a).

The conclusion drawn here is that scientific knowledge is the most reliable guide for understanding the two (internal and external) realities. However, there are significant limits to this knowledge and its application. It is important that any interpretation and use to which scientific knowledge is put should include an acknowledgement of the limits to that knowledge and the reminders from religion about what it can't provide. This is also true for the use of scientific knowledge in an activity in which we all engage when considering our circumstance: making predictions and speculating on how the world ought to be. This will be discussed next.

# CHAPTER 5

# MAKING PREDICTIONS

We all have to deal with the future. We could just wait for it to arrive, but we prefer to make predictions and then use them to help us deal with it. Not all of our predictions are about the future though. More generally, predictions are decisions about what we think of as the unknown.

The value we give to a prediction depends on more than the guidance it provides. Our value assignment is affected by our feelings about what is predicted and our general attitude toward prediction making (Rescher 1998). For example, at a personal level, if you don't believe in statistical predictions but you have faith in astrology, then you might be quite happy to spend a fortune on lottery tickets because today is your lucky day. At a cultural level, in the 1800s many Westerners believed in predictions of a bright future because the discoveries of the emerging sciences and the inventions of emerging technology indicated to them that progress to that end was real.

This makes a discussion about predictions both complex and broad. It requires that we think about more than the predictions themselves. We have to think about a diversity of related aspects, including how they are made, their limits, our relationship to them and their influence on our decision making. Fortunately, we can discuss the essence of predictions by dividing the subject into three overlapping and connected topics: their purpose, their evaluation and their types (the different ways of making them). In the process, we learn that most predictions concern complex systems. Our relationship to these systems and the predictions about them is strained, so a section is devoted to discussing that issue. The chapter ends with a discussion about personalizing scientific predictions about complex systems so that we can more easily assimilate them and then more meaningfully use them in our decision making.

## Purpose

Predictions have a purpose, to help us resolve a particular issue. Three kinds of purposes are discussed here: planning for the future, making sense of the world around us and setting limits to our activities. The three are interconnected and overlap.

**Planning.** An agricultural culture depends on its farmers to routinely provide food. They produce it by completing an annual sequence of seasonal, agricultural tasks: an agricultural round. Farmers must ensure that they undertake each task in the round at its appropriate time. Even today, many do so by watching for features in their non-agricultural surroundings that appear (or disappear) each year at the time when a particular task should be started or stopped. Examples of these cues are the arrival of a migrating bird, the location of the sun on the horizon, soil moisture and the arrival of a constellation at a particular place in the sky.

To be successful, a farmer has to do more than predict the time to begin a particular task in the round. The farmer has to prepare for the task so that it can be started and finished at the predicted times. Predictions used for planning thus include estimating when preparatory actions must be undertaken and deciding on the order in which they are to be done (Rescher 1998).

All planning predictions are made by looking forward to, and backward from, the present. Making them thus requires a refined sense of time. Indeed, planning predictions are sufficiently important to complex cultures that their method of measuring time, their sense of time and how they make sense of the world are all consistent with their planning needs.

Consider that the Mayan, Egyptian and Babylonian complex agricultural cultures all needed to know when to prepare for and start the tasks in their agricultural round. But they also needed to know on which day to conduct specific rituals, such as those needed to ensure agricultural success or to maintain the world's order. They also needed to plan for the building or repairing of temples and the acquiring of the special accoutrements used to conduct these rituals. Their making of the relevant planning predictions was eased by each culture including in its world view a cyclical, astronomy-based calendar and an associated cyclical sense of time that matched the seasonal nature of agriculture.

In contrast, a complex industrial culture includes in its world view both an agriculture-supporting cyclical sense of time and a linear sense of time. For example, in Western culture, our linear sense of time is consistent with our notion of progress, our long-term view of past and future, and the finality of physical death. It also caters to the needs of a factory boss who, unlike a farmer, tends to focus more on the number of days until a future industrial activity will begin than the season in which it will occur.

The ultimate illustration of the connection between an industrial culture's planning predictions and its linear view of time is the "just in time" delivery system. For example, car assembly plants maintain only a small inventory of the many parts needed to build a car.

Instead, the assemblers rely on both distant factories to manufacture the parts and transportation companies to deliver them, just in time, to meet the assembly plant's needs. For this system to work, a highly accurate timing method, reliable communication, predictable manufacturing-delivery systems and predictable workers are far more important than knowing the time in the seasons' cycle.

We are reminded that, regardless of its sense of time or the accuracy of its clocks, if a complex culture is to have faith in its planning predictions, then it must be able to reliably predict the behaviour of most of its members most of the time (Rescher 1998). For example, in complex agricultural cultures, including ancient ones such as Babylonia, it is imperative that management be able to routinely provide sufficient labour during the periods of planting, harvesting and irrigation maintenance. This means that the members must display a degree of behavioural predictability beyond their innate drive toward social conformity. The characteristics of a complex culture's world view and power structure are consistent with satisfying this planning-prediction need: they both promote and induce the required conformity and predictability in their members.

In general terms, those higher in the hierarchy are given the authority to ensure conformity. If the leaders present and explain a decision to the members in an appealing manner that is consistent with the culture's world view, then the members can more easily justify (personalize) and accept that decision. At the same time, the members are provided with the sense of purpose needed to implement it. This arrangement helps to both ensure the predictability and conformity of the members' behaviour, and to unify them. Think of recruitment during WWI. If this doesn't work, the leaders can impose conformity, but this is a double-edged sword.

In contrast to a complex culture's administrative planning predictions with their focus on behaviour and resources, the planning predictions of individuals and households focus more on future emotional consequences. We tend to ask ourselves, what planning do I have to do to ensure that I have the life experience I predict will make me happy? For example, choosing a wedding date is often seen as less of a planning deadline to complete wedding preparations and more of an auspicious day that also has symbolic or emotional value, such as love.

Our personal planning predictions therefore tend to be focused more on satisfying our perception of what a desirable future state of affairs should look like rather than on the complex of issues that must be addressed to reach that state. These characteristics collectively suggest one reason why our personal planning predictions are often significantly flawed, at least from an administrative point of view (Gilbert et al. 1998). Our plans often just don't quite go as we predicted. For example, couples commonly choose their wedding day from among the same suite of auspicious or available days on which to be married, thus creating unpleasant organizational difficulties for their weddings. Think too of our relationship to our daily to-do list: we treat it as more of a chore/goal list, and a source of good feelings when we get to cross items off, rather than a reminder to ensure that all of the steps and details needed to ensure timely completion of the task are accounted for.

These examples illustrate a broader conclusion applicable to all planning predictions. If we fail to root them in the external reality and to consider the uncertainty in our predictions, then we may still be subject to the roller-coaster ride and disappointments that planning is supposed to avoid (Myers 1999). This outcome is illustrated by an observation confided to me by someone who provides flowers to weddings. Brides often request a particular kind of flower with a particular shade of colour. In making such a specific demand, the brides act as though flowers come from a factory where they, like widgets, can be manufactured and coloured on demand. Although this may be partly true today, disappointment is still common, which creates uncertainty about their planning prediction of a perfect wedding.

**Making Sense of the World.** Our curiosity about the world's workings and the niggling unsettled feeling that can accompany our contemplation of the unknown drives us to make sense of the world. We are helped by finding patterns in the world around us. These become rules of thumb and sophisticated scientific theories that are, in essence, predictions about the unknown.

If the members of a culture place a high value on the predictive ability of a pattern, rule of thumb or theory, then it can become incorporated into their culture's world view. It thus becomes part of how the culture guides its members to make sense of their world. The prediction becomes an explanation. Consider that if we use the location of the stars, the planets, the sun and the moon in the sky as predictors of when farming activities should occur, it makes it easier for us to accept that they can actually influence aspects of farming and our lives.

With that understanding about the world, we are reassured that we can deal with the future because we can make predictions about it, and even, perhaps, control it. This is a powerful reassurance. Consider the commonplace, "Historically, the stock market always recovers from a crash." Feeling better about your investments?

Unfortunately, our prediction-assisted, personal and cultural sense of how our world works often includes unsubstantiated observations or assumptions about the external reality. Examples we saw in chapter 2 include treating correlation relationships as causal when they are not and assuming simplicity when complexity reigns. As a result, humans are routinely faced with outcomes that are at odds with their sense of how the world works and thus with their predictions. This translates into unsettling uncertainty.

Both individuals and cultures have to deal with this uncertainty. Agricultural cultures, for instance, must inevitably face it when they attempt to control the weather by using rituals based on prediction-related understandings of how the world works. This uncertainty is commonly dealt with by the members (consciously or unconsciously) hedging their explanations and predictions. For example, an agricultural culture, such as Egypt, can include in its world view both a spiritual and a naturalistic aspect to their explanations for how the world works. Any deviations from their prediction can then be dealt with by either practical actions, such as adjusting how they distribute irrigation water, spiritual actions, such as increased ritual activity or tweaking their world view. Think too of the cryptic advice from the oracle at Delphi to the rulers of Greece.

This approach takes the edge off the uncertainty in an agricultural culture's predictions and guidance about how to make sense of the world without significantly changing their world view. The members' confidence in their culture's world view is maintained which, in turn, helps keep the culture stable (Freidel and Shaw 2000). When the members of complex industrial cultures use phrases such as *acts of God, unforeseeable, avoidable error* and *unexpected*, they hint that they too use an equivalent method to reassure themselves of the truth of their prediction-based efforts to make sense of the world.

### Defining Limits.
A special aspect of making sense of the world is to define limits to our activities. Predictions help us to set those limits. For example, how strong should we build a bridge in order to avoid it collapsing,

what is the earliest date for harvesting a crop, and how much alcohol can we can drink yet drive home safely?

The successful use of a prediction to set these limits depends on a number of factors. They include how ordered are the characteristics of the system being predicted, the prediction method, our expectations and our willingness to accept the limit set by the prediction. For example, when setting the limit to the amount of alcohol we can drink and then drive safely, we are obliged to do more than predict the average amount that impairs our driving. We need to take into account the variation in impairment among people when given the same amount of alcohol. This is accomplished by choosing an acceptable safety margin. We also need to consider people's willingness to limit their drinking.

As the examples above suggest, predictions used to set limits become more valid, useful and less uncertain when they more accurately take into account a wide range of the external reality's properties and our human limitations. For example, after being shown a number of similar but distinct objects, we are capable, in the short term, of remembering on average only $7 \pm 2$ (Miller 1956a). Thus, although a prediction may correctly suggest that internet security issues could be solved by limiting the use of passwords to a random mix of 12 or more characters, this solution is unrealistic without a password management system or a chunking method to help us get them into our long-term memory.

The more complex the system, the harder it is to set limits to activities based on predictions. Consider setting limits to the degree to which an ecosystem can be disrupted and remain functional. Think too of setting $CO_2$ emission limits to keep global warming below 2°C. Complexity itself may make it too difficult or impossible to reliably predict a specific limit or even a range of limits. In these cases, the limits may have to be set and judged by predictions that use general principles or rules of thumb, such as "do no harm" or "no regrets" (Alley et al. 2003).

In summary, predictions have a purpose: to help us resolve a particular material or emotional issue. We are more likely to choose an appropriate prediction method for a particular purpose if we consider more than the issue itself. We need to appreciate the system it is a part of, our human foibles, our culture's world view and our relationship to both the issue and to our predictions. If we do so we are also more likely to make a more realistic evaluation of the prediction.

## Evaluation

After a prediction is made we evaluate it. Our routine evaluation of a prediction takes place rapidly in our unconscious and focuses on its correctness (lack of error). This is unfortunate because a prediction's correctness can only be truly determined in hindsight. A more complete and pragmatic evaluation of a prediction would give conscious attention to the prediction's other characteristics. Such an evaluation is the topic of this section. By understanding it, we can better appreciate the role and significance, and thus the choice, of the different prediction methods discussed in the next section.

In making an evaluation, it helps to remember that a prediction is made for a purpose, as discussed above. It is also a decision about the unknown, which means that it is an inherently uncertain estimate based on a foundation of limited information of various kinds and qualities. When seen in that context, the two key pragmatic aspects of a prediction are its validity and its usefulness. After discussing these aspects, the discussion will then turn to the more general process of evaluating a prediction.

**Validity.** The validity of a prediction is a measure of its credibility (excluding correctness), but only within the context it was made in. To be considered valid, a prediction has to satisfy a number of general conditions. It has to be more than an unsupported speculation, a guess or a statement of the obvious. For example, the statement that the sun will rise tomorrow is not a valid prediction. But saying that the sun will rise in San Francisco at 3.30 p.m. GMT tomorrow is a valid prediction, if it also meets the following conditions.

For a prediction to be considered valid, it must be made in a logical and coherent manner. In addition, the information it is based on must be sufficiently relevant and accurate to both support the reasoning used to make it and to fulfill its purpose. The information must also be consistent with the prediction's type and purpose. Thus, an astronomical prediction using astronomical measurements and a religious prediction based on scripture can both be valid. However, an astronomical prediction made using scripture alone is unlikely to be valid (Rescher 1998; Tainter 2000).

However, a valid prediction doesn't have to be based on complete knowledge. In fact, missing information and assumptions are an inescapable part of predictions. After all, we make predictions because the state of our knowledge and understanding, especially about the future, is always incomplete. For the same reason, a prediction doesn't have to correctly predict the future in order to be valid: the future is rarely perfectly knowable. Thus the concerns of correctness and validity are related but different.

Similarly, the discovery of errors and biases in a prediction may or may not affect its validity or usefulness. Consider first how they can occur. Our contribution to errors can be divided into two groups: errors of omission, such as leaving out critical information or failing to consider all of the relevant influences; and errors of commission, errors related to processing the information, such as making a mistake in calculation or logic. Our contribution to biases and errors can also be divided into those arising from our innate human foibles and those arising through a deliberate effort to mislead. The details, and source, of the errors and biases in a prediction will determine the degree to which its validity, usefulness and correctness is reduced by their presence.

As the possibility of errors suggests, a valid prediction should be somewhat flexible or general in nature so that it can be adjusted to accommodate new and changing information. One way to achieve this flexibility is to make the prediction conditional (Rescher 1998; Tetlock 2006). Despite this inescapable need for flexibility, it is still generally true that a prediction about a particular topic is more likely to be valid if it is based on detailed information about that topic rather than on general information and principles.

This concern for specificity arises, in part, because the process of generating general principles may require the discarding of information that is actually relevant to the concern being addressed by the prediction. For example, your current prediction about your future enjoyment of this and the following chapters was likely based on more than the general principles of human decision making (discussed in chapter 2). It probably included the more specific influences of how tired you are and the presence of phrases you find annoying. In general, a prediction is most likely to be valid if it is based on detailed, factual knowledge relevant to the concern being predicted and if it is complete and coherent within that context (Rescher 1998).

**Usefulness.** Predictions are an important part of our decision making. They can be extremely useful if they guide us toward more appropriate decisions, that is, if they inform rather than mislead. For a prediction to be useful, it must meet certain conditions, such as not being so general that, in hindsight, it would always be seen as correct.

For an estimate of a prediction's usefulness to be meaningful it must be completed in the context within which the prediction was made. Estimating usefulness is thus an extension of checking the prediction's validity (credibility). It focuses more on the prediction's suitability and relevance; how difficult it is to apply; the degree of self-criticisms (e.g., conditions and caveats); and the seriousness of errors and biases (Rescher 1998). Thus, to estimate a prediction's usefulness, it must have been described in sufficient detail that all of these features can be identified and checked.

The inevitable presence of some type of error or bias in a prediction could be used either to excuse a prediction's shortcomings or to declare it to be of limited usefulness. A more helpful response is to make an assessment of the impacts that the known and possible errors could have on the prediction's usefulness. An example is the likelihood and degree to which a particular error will result in the prediction misleading rather than informing us (Nisbett et al. 1982; Tetlock 2006). The error's likely impact can then be balanced against the consequences of other choices, such as not making a prediction at all (Rescher 1998). The importance of these types of considerations is illustrated by the prediction that Saddam Hussein had weapons of mass destruction. An error-impact study may have prevented a protracted war and the subsequent instability in Iraq.

A prediction does not have to correctly predict the future to be useful. But its usefulness is affected by the likelihood of it being correct (its reliability) and the conditions under which that likelihood will arise. Unfortunately, possible correctness has such a strong influence on our thinking about predictions that it biases our efforts to estimate a prediction's usefulness. To minimize this problem, a prediction's reliability should only be assessed late in the estimation process, when it is more likely to be placed in its appropriate wider context and less likely to become a fixation. Factors to consider when considering reliability are the type of prediction, its purpose, the assumptions and uncertainties, its robustness, how far the prediction looks into the future and "if-then" type statements.

The degree of influence that reliability should have on an estimate of a prediction's usefulness is strongly determined by the prediction's type, purpose and the context within which it is being used. For example, when estimating the usefulness of a prediction about an aspect of a poorly understood complex system, such as the future supply of fresh water on a global scale, the focus would be on the prediction's validity, limitations and relevance. In this context, a large uncertainty in the prediction's correctness (low reliability) may have only a limited effect on the prediction's usefulness.

In contrast, when estimating the usefulness of a specific, detailed prediction much greater attention would be paid to the prediction's reliability, such as its accuracy, precision and safety margins. For example, what is the maximum load that a specific beam in a bridge could bear. In this context, information about correctness would strongly affect estimates of the prediction's usefulness.

**Making an Evaluation.** The above discussion indicates that making a proper evaluation of a prediction can be a laborious, specialized, and even unrealistic task. In contrast, our routine, everyday (unconscious) evaluation of a prediction is rapid. Unfortunately, as our unconsidered tendency to focus on a prediction's correctness suggests, our everyday evaluations are often not as reliable as we imagine.

Fortunately, the above discussion also provides us with a few pragmatic general principles and rules of thumb that we can easily consciously apply to our everyday evaluation of a prediction's validity, usefulness and reliability. (1) The more relevant, accurate and issue-specific the information used to make a prediction is, the greater the likelihood that the prediction will be valid. (2) In contrast, the more detailed the prediction, the less likely that it will be correct. These two conditions should reinforce our understanding of the difference between a prediction's validity (its credibility) and its reliability (its likely correctness). (3) It is easier to predict general trends than specific outcomes. (4) It is easier to predict the results for aggregates, such as group averages, than results for individuals. (5) It is easier to make predictions by including qualifying statements and limitations (conditions), such as "if-then" statements. Used judiciously, these can help in the evaluation of the prediction. (6) Vagueness and ambiguity ease the making and accepting of predictions. For example, a "may" prediction is likely more valid and useful than a "will" prediction. But beyond some point, vagueness is counterproductive. (7) The greater the system's complexity or instability, the harder it is to make predictions about it that have a high degree of validity, usefulness and reliability. (8) A reliable prediction is more easily made about the partly known than the completely unknown. The further into the unknown a prediction is applied, the more uncertain, the less reliable, it becomes. If the level of reliability is to remain fixed, then the further into the unknown it is applied, the more conditions should be attached to that level of reliability. (In terms of time, the partly known becomes the near-term future or past, and the unknown becomes the long-term future or past.) (9) Our human decision-making characteristics contribute to our prediction evaluation difficulties.

The use of a particular rule is illustrated by our need to make decisions about complex systems. In making them, we rely heavily on predictions about a system's future state. To avoid overvaluing the prediction's validity and usefulness or undervaluing the associated uncertainty, we could apply the seventh rule of thumb, above. For example, it is difficult to make useful predictions about the impacts of climate change on a complex biological system such as a biome. It is much more difficult to make a useful prediction when humans are actively destabilizing its component ecosystems by disrupting and destroying them.

The general value of the rules of thumb is highlighted by discussing the evaluation of predictions with significantly different characteristics and contexts. The first discussion considers the evaluation of predictions about complex systems by elaborating on the above example. The second considers the evaluation of skill-based predictions.

The presence in complex systems of multiple variables, interactions and feedbacks occurring over a wide range of time and space ensures that both making valid and useful predictions about these systems and evaluating them is difficult. The rules of thumb provide a reminder of why this is the case. These difficulties can sometimes be reduced by using models to generate more than one prediction of the system's state. A suite of such predictions can be produced from a single model by using a different set of input values or scenarios to represent a number of different possible starting conditions. Alternatively, different models (e.g., constructed from the top down or from the bottom up) can be used to generate a suite of different predictions about the system's state.

Once a number of predictions are available, they can be systematically evaluated. For example, if they are displayed in some sort of a logical order, the behaviour of the system may be better understood, the characteristics of the most common (probable) outcomes may be outlined, and the uncertainties better defined (Rescher 1998). This evaluation process is helped by paying attention to a broad sweep of contextual information that is related to but not part of the prediction. This information provides semi-independent checks on aspects of the prediction, such as its coherence and consistency.

These techniques are commonly seen in the data referenced in this book. For example, the Intergovernmental Panel on Climate Change used them formally in their effort to evaluate predictions about the consequences of the human contribution to changes in the earth's complex climate system (IPCC 2013a; Stainforth et al. 2007). They are also qualitatively applied in this book as part of the multiperspective method.

The value of the rules of thumb is illustrated next by considering predictions about complex system's whose making relies heavily on the skill, experience and expertise of the predictor, and whose evaluation must rely heavily on minimal information and difficult-to-verify explanations. Good examples are the predictions made by expert economic and political forecasters. It is often difficult to decide if these types of predictions are useful or if they are unsupported speculation, and thus not predictions. The making and evaluating difficulties increases if the prediction's topic or context is associated with anxiety and pressure. Examples are the political and economic predictions made at a time when a country may go to war or a corporation may lose a large sum of money.

Our evaluation of this type of prediction, by default, depends largely on our perception of the predictor's credibility and our relationship to the prediction (Gardner 2011). The word *perception* should remind us that we need to consider the influence that the innate biases in human decision making have on our evaluations. Here are some prediction-related examples of these biases.

We have a preference for prediction makers and predictions that foretell what we believe or want to believe will happen (or has happened). We similarly tend to have a strong preference for predictions that are made with confidence, such as "will" statements, regardless of the appropriateness of that confidence. We also have a tendency to pay attention to predictions and aspects of predictions that reflect current cultural concerns, such as finances during an economic crisis. When we make and evaluate these predictions, we tend to favour current trends, that is, the short term. For example, when times are good, we tend to think the future will be good and vice versa.

There is an even more insidious, evaluation-biasing factor of which we should be aware. Assume that our perception-based evaluation of a specialist's prediction is favourable. After the actual outcome is known to us, then our prior evaluation of the prediction can influence our making and evaluation of that predictor's related, future predictions. For example, if we perceive, in hindsight, that the prediction we believed in was actually incorrect, then we are more likely to discount (excuse) rather than change our view of the predictor, the prediction method and our evaluation of the prediction. In contrast, if we perceive, in hindsight, that the prediction was correct, then our confidence in the prediction method, the person making it and our evaluation of it will receive a boost.

This, again, brings up the role of correctness in biasing our evaluation of predictions. We should always consider the possibility that our evaluation of correctness may be faulty. For example, our hindsight bias may have allowed us to believe that a prediction was correct when it wasn't, or that the prediction may have been so general that it can always be seen as correct.

Consider, too, that we tend to misinterpret the reasons for a correct prediction. A classic prediction-evaluation mistake is to assume that a correct prediction means that the theory and assumptions on which it was based must have all been correct. That conclusion allows us to confidently apply them more generally (Rescher 1998). This may be true if the system is simple, but it is far less likely to be true for complex systems, such as the environment and personal relationships. For example, King Hubbert's model correctly predicted that the United States' production of conventional oil would peak around the mid-1970s. But the model's accuracy was, in part, fortuitous. A fact only discovered in 2001 (Kaufmann and Cleveland 2001).

A correct prediction about a complex system may be the result of a fortuitous correlation, not a cause. In this case, the theory behind the prediction does not provide an explanation for the "correct" outcome. A more bizarre twist arising from complexity is that, under certain circumstances, a correlation can indeed be reliably used for prediction. For example, predictive allometric correlations are used in biology and botany, even though the details of why they exist remain unsure or unknown. This outcome can be thought of as the result of a proxy relationship: one that falls between a meaningless correlation and a cause. The above rules of thumb prompt us to take the care needed to consider these kinds of personal and system limitations when evaluating predictions.

In summary, a meaningful evaluation of a prediction considers its validity (credibility) and usefulness within the context the prediction was made in. The evaluation process should consider many factors in a complex balancing act that is limited by incomplete information and bias. This process involves weighing the prediction's compromises and uncertainties when their influences on the final outcome are themselves uncertain.

The likelihood of a prediction being correct (its reliability) is part of its evaluation. However, due to our obsession with correctness, reliability is most appropriately considered at the end of the evaluation process. The degree of reliability should also be judged within an appropriate context.

We are helped in our evaluations by knowledge about our human decision-making biases, especially those that directly affect prediction making and evaluation. This includes the influence of our culture's world view. We are also helped by applying rules of thumb that represent the general constraints on the viability, usefulness and reliability of predictions. For example, conditional statements make it easier for us to express the many shades of grey that inherently accompany the making and evaluating of a prediction. But they can also render the prediction less useful or not useful.

Although the essence of the evaluation process and the rules of thumb don't change with the method used to make a prediction, the details of the evaluation do. Nevertheless, having knowledge of the evaluation process helps us to appreciate the characteristics of the different types of prediction making. The prediction types are discussed in the next section.

## Classification

There are a number of methods and combinations of methods that can be used to make a prediction. For example, economists resolve the problem of predicting future food supplies differently from agronomists. Economists tend to use past production to predict future production, while agronomists tend to use their knowledge of plant growth and present conditions to make a prediction about future yields (Socolow 1999). Each method has its benefits and limitations, which should be considered when making and evaluating a prediction. To that end, this section discusses the methods of prediction by dividing them into three, non-exclusive types, each with their own distinctive characteristics: prophecy and divination, common sense, and scientific.

## Prophecy and Divination.

Two of the more commonly used and believed in methods of prediction are prophecy and divination. Usually, the individuals making these types of predictions have been socially identified and accepted as being gifted with predictive powers or the ability to interpret predictive signs, symbols and texts (Reader 1990). The predictions made by the gifted are often used by individuals, groups and even cultures to guide their decisions. Historical examples are the oracle at Delphi in Greece and the shamans of extinct aboriginal American cultures. Today, in Western culture, tarot card readers, psychics, astrologers and channellers, among others, provide these types of predictions for individuals, while economic forecasters provide them for the culture.

The predictions of the gifted can be more than cynical imaginings, as some claim. They can be credible attempts to link the internal with the external reality in a way that will satisfy human needs. For example, the gifted can provide guidance that relieves the anxiety we feel when the uncertainty or irregularities in the world impinge upon our lives. In this circumstance, a prophecy or divining may be valid and useful, even though it is unrepresentative of the external reality.

Similarly, to the recipient, a prophecy or divining is meaningful and accepted because it reflects or can be easily fitted into their personal beliefs about how the world works, rather than because it matches how the world actually works. Their beliefs are, to at least some degree, influenced by their culture's world view. For example, the ancestors of the Naskapi Indians of northeastern America lived as hunter-gatherers. They believed that humans and animals were only different on the outside. Spiritually they were the same because they shared the same spiritual existence after death. For the Naskapi, the presence of this after-death bond between animals and humans allowed the living to use divination in order to obtain from the spiritual world important information about a future real-life hunt (Reader 1990). For these reasons, predictions of the prophecy or divination type should primarily be evaluated within the context of our internal rather than the external reality.

In Western cultures, prophecies are considered to be religious predictions. For example, some Christians believe that a time will come when the existing world order of humans will pass to be replaced by a thousand years of a Christian paradise inhabited by believers. This millenarian prediction is seen as a religious prophecy.

But consider that the essence of the millennial message is one of a better, or even utopian, future. By changing the context and wording, this prophetic, good-news message can serve or represent political or even secular purposes, not just religious ones (Daniels 1999). Consider that if the context is religious, then it could refer to the return of the messiah; if environmental, to the return of a natural paradise; if political, to the spread of democracy; or if economic, to continually growing wealth. Limiting prophecy and divination to religious predictions is arbitrary.

## Common-Sense Predictions.

Decisions we call common sense are those ubiquitous, rapid choices we make based on information recalled from memory, our assumptions about the world and the guidance from our culture's world view. They are usually unconscious decisions but they can be intuitive decisions that involve our conscious input to some degree. Your initial opinion of this book was most likely a common-sense decision.

Common-sense predictions are, in essence, common-sense decisions that deal with the unknown. These predictions are largely created by our unconscious. Consider our common-sense predictions about the past or the future. They are made, in essence, by our unconscious using patterns, rules of thumb and our culture's guidance to select from our memories what it decides or we feel is a relevant snapshot of our current (material, psychological and social) circumstance and its context. Our unconscious then projects that snapshot (with or without our conscious help) into the past or future where it becomes our common-sense, scenario-like prediction for that time.

As expected, we evaluate our common-sense predictions using the same suite of common-sense decision-making skills we used to create it: pattern recognition, rules of thumb, preconceptions and cause-and-effect models, among others. Surprisingly though, both the prediction's creation and evaluation happen at the same time. As a result, a common-sense prediction rapidly appears into consciousness accompanied by our feelings of confidence in it (Kahneman and Tversky

1982c). Common-sense predictions therefore provide us with the advantages of speed and ease of acceptance.

Because common-sense decisions and predictions have much in common, we should expect that they would suffer from similar limits and biases. This is indeed the case (Gilbert et al. 1998). Consider that common-sense decisions are constrained by the validity of our often untested assumptions about the world. The same is true of our common-sense predictions, for example, our prediction that a steel ball will fall faster than a feather in a vacuum (Rohrer 2002). (They fall at the same speed.) In our common-sense decision making, we tend to focus on information that supports our personal view of a correct decision, and we tend to neglect the information that contradicts it (Dawes et al. 1989). Similarly, in common-sense predictions, we tend to predict a future much like the one we want to experience, not the future that is most realistically likely to happen (Taylor 1989). In our common-sense decisions and predictions about a complex system, we tend to take minimal account of its functional characteristics, such as its dynamic nature, feedbacks, the lags between cause and effect and the contribution of chance (Gardner and Stern 1996).

These limits and biases can have a significant impact on the validity and usefulness of our common-sense predictions. They can help to explain the limited success of scientists' mid- to long-term forecasts of future discoveries in their field of expertise (Rescher 1998). They can also help to explain why we visit astrologers rather than prophets of doom. Astrologers are more likely to offer us a reassuring prediction (Ellis 1973).

The practical outcome from relying on common-sense predictions that are often both flawed and poorly evaluated is illustrated by the popular alternative solutions to the dumping and forgetting of waste: disposing of it through burning or flushing. Our common-sense prediction is that fire and water will solve our waste problem because they appear to both quickly and safely remove it. They also remove any associated unsettling feelings that arise from the presence of the waste. Our faith in this prediction is helped by our apparently innate association of fire and water with cleaning and purifying. In reality, both of these disposal methods can have serious consequences for the complex system into which the garbage is burned or flushed. For example, the open backyard burning of plastic-containing household waste in rural parts of the United States is a major contributor to the increased concentration of dioxins (a carcinogen) found in humans

living in Nunavut, thousands of kilometres to the north (Commoner et al. 2000).

In those cases where we both make a common-sense prediction and experience the actual outcome, it is easy to imagine that our predictions would improve over time. Unfortunately, like our decisions, unless the prediction and the outcome occur very close together in time and space, we experience each aspect as a largely separate, isolated event. We are therefore unaware of any connection between the two and thus lack a reason to change our prediction method.

Even if our prediction and the outcome occurred close together, we may not perceive that they are different. There are many possible reasons why. For example, we may alter our memory of our original prediction so that it matches the present situation (adaptation bias). This effect, along with our hindsight bias, makes it easy for us to say of an event, "Oh, yes. I thought that was going to happen," or "Just as I expected," regardless of what we originally thought or predicted. If we do manage to remember our common-sense predictions correctly, then we will often be surprised that the outcomes are different than we forecast.

It is realistic to assume that once we have got over the surprise and accepted a deficiency in our prediction method, we would make the effort to correct our thinking or our predictive model. But making these changes takes considerable effort. Perversely, it is easier for us to find some way to squeeze a new fact into an old model or a habitual way of thinking than it is to change either. Thus it is more realistic to expect that the predictive cause-effect models we personally use today are often much the same as the inaccurate models we used to make faulty predictions in the past (Tversky and Kahneman 1982b). It seems that our ability to learn from the errors in our common-sense predictions is limited by how much conscious effort we are prepared to devote to remembering the original prediction, accepting the implications of a different outcome and changing our thinking.

Overall, common-sense predictions are often neither as valid nor as useful as we like to imagine, especially when we apply them to more complex situations. For this reason, it can be said with a high degree of certainty that the validity and usefulness of most of our common-sense predictions about the future state of a complex social system will be significantly incomplete or invalid, if not plain wrong. An example is our common-sense predictions of a culture's future poverty levels and resource availability. The same limitations apply to our predictions about the outcome of our proposed solutions for dealing with these types of social issues (Gardner and Stern 1996). We should expect this conclusion to also apply to common-sense predictions about the future state of our environment.

Despite this glum assessment, all is not lost. Common-sense predictions do serve us well in most, especially simple, circumstances. Similarly, just because common-sense predictions about complex systems are only somewhat valid, useful and reliable, it does not mean that they are useless (Myers 2002a).

When humans make common-sense predictions, we are very good at considering those human concerns that are important to our internal reality's sense of well-being. For example, when individuals predict and evaluate the risk from a proposed development project, they innately consider the fairness of the risk distribution and their personal control over that risk. These features are largely ignored by regulators because the science- and statistics-based approaches to risk assessment struggle to recognize that concerns like fairness exist, let alone that they are important (Slovic 1990; Morgan 1993). There is, therefore, merit to common-sense predictions. We just need to evaluate them more carefully. We can treat them as windows into our biases. And we can recognize and accommodate the constraints imposed on our common sense predictions by our human limits and biases, while still recognizing our strengths.

The word *accommodate* was used because many of our decision- and prediction-making biases and limits are an innate part of our unconscious. They can't be dealt with by getting rid of them. But we can acknowledge their existence and learn to recognize when ignoring them is likely to have important impacts on the validity and usefulness of our predictions. We can then accommodate our decision/prediction-making foibles. We can, for example, train ourselves to know when to be appropriately skeptical of our conscious thoughts and conclusions and broaden the scope of the information we consider.

We can similarly constrain our imagination's influence on our predictions by consciously guiding it rather than giving it free rein. For example, rather than using first impressions to provide the information about a social circumstance whose future we wish to predict, we can use reasoned compassion to place ourselves in the participants' shoes. We can also consider the circumstance from perspectives that are different from our preferred view and our favoured outcome (Myers 2002a).

Through these types of efforts, we are more likely to prompt ourselves to consider how our decision/prediction-making foibles could influence our making and evaluation of a common-sense prediction. The act of making these conscious efforts has an added benefit. It encourages us to decide whether, in a particular circumstance, it is more appropriate for us to use common-sense or scientific prediction methods and when to use a mix (Myers 2002a). Our choice of method should be made easier by knowing that common-sense predictions based on numerical facts are less useful than predictions made by using linear, scientific models, even when the models are improperly applied (Dawes 1982).

There are other methods to ensure that common-sense predictions are more valid and useful. Firstly (and a little facetiously), we could ask mildly depressed people to make the predictions for us because they have a tendency to make accurate rather than self-serving judgements, predictions and attributions (Myers 1999). Secondly, we could incorporate statistics into our making and evaluation of common-sense predictions. It is a powerful tool for finding patterns in the world around us and for describing the uncertainty associated with their description and use.

**Statistics.** Uncertainty is a characteristic of our world and a hallmark of making predictions. Statistics is a way of describing and dealing with this uncertainty. Given the availability of this tool, it would seem natural for humans to routinely use statistics as part of both our intuitive thinking and our making and evaluation of common-sense predictions. But, other than using simple measures such as frequency and averages, we don't. And when we do, our usage is usually inappropriate or wrong.

This is unsurprising, once we realise that our intuitive grasp of most statistical concepts is exceedingly poor. Consider too that we can complete only the simplest of statistical calculations in our head. A simple linear regression is beyond our brain's computational ability (Fischhoff 1982) (Germano 1999). No wonder we find it disconcertingly difficult to include even the most fundamental aspects of statistics into our everyday reasoning processes, such as our rules of thumb and intuition (Nisbett et al. 1982; Myers 2002a). Finally, whatever limited use we do make of statistical concepts, such as probability, their potential contribution to our decision making is easily displaced by other more personal considerations, concerns and explanations (Myers 2002a). For example, our attempts to assess and use the known probability of our winning a lottery is easily displaced by us imagining ourselves to be a winner.

The depth of our strained relationship with statistics is illustrated by the difference between skill and luck. The outcome of many personal, day-to-day activities can be attributed to a mixture of our personal control over the situation (skill) and statistical chance (luck). If we are to apply statistics appropriately to our personal lives, we have to be able to distinguish between the influence of skill and luck (Langer 1982). We struggle to make this distinction.

Consider our reaction to the "near miss." When a near miss occurs while playing a game of skill, such as archery, it can be associated with an improvement in skill. In contrast, a near miss in a game of chance, such as playing dice, is a product of randomness and has no predictive meaning at all. This is true even though we are in control of the game's play, such as rolling the dice. Despite this difference, we still treat a near miss in games of chance as if it were somehow related to our skill at initiating the play, such as blowing on the dice before rolling (Clark et al. 2009).

The difficulty we experience separating skill from luck has broad practical consequences for our lives. We tend to believe that we have more control than we do over events whose outcome is largely determined by chance (Langer 1982). In contrast, we tend to believe that luck controls some events that are actually influenced by our skill, especially our skill in decision making. For example, all else being equal, the enviable lives of "lucky" people often have just as much to do with them making better thought-out choices than to chance events in their lives (Wiseman 2004). The lucky exceptions are the few chance winners of a lottery and, for most, being born to their parents.

The presence and significance of our limited ability to intuitively grasp and apply statistics is illustrated by our common-sense estimate of the probability of a person dying from a particular cause. Consider that our common-sense estimate of a person's chances of dying from a cause that is responsible for many deaths (high risk), such as a stroke, tends to underestimate the risk (figure 3). With that underestimate in mind, it suggests that we are likely to pay less attention to the possibility of us personally having a stroke than we should. In contrast, our common-sense estimate of a person dying from a cause that is responsible for relatively few deaths (low risk), such as a tornado, tends to overestimate the risk. We are then more likely to be more fearful than needed (Slovic et al. 1982).

Our predictions of probabilities are generally biased in this manner because of our unconscious method of estimating them. Consider the availability and representative rules of thumb (heuristics) discussed in chapter 2. In a nutshell, when the availability rule of thumb is used to predict the relative frequency with which an event occurs (its probability), we base our estimate on the ease and the speed with which we can recall similar events. But our memory and recall of past events is biased. For example, we preferentially remember those events that are dramatic and recent. Consider too that our estimates based on descriptions of events are different from those based on our personal experiences.

When the representative rule of thumb is used to predict the relative frequency with which an event occurs, we first unconsciously create a reference (the prototype). Its characteristics are based on our memories of events that are in some way related to the event of interest. Our unconscious judgement about how similar the prototype is to the event of interest becomes our prediction about the event's probability of occurring. Because of the variation in people's exposure to events and our biases in remembering past events, our prediction of an event's probability of occurring can vary widely.

Our estimates of the probability of events are also biased by factors other than our judgement about the event's material features. For example, the bias in our estimation of the risk of an event is enhanced by the fear we experience when we perceive that the outcome is beyond our control (Myers 2002a). When estimating the probability of a sequence of events occurring, we intuitively add the probabilities we have estimated for each stage of the sequence. As a result, we think our chances of dying from a sequence of events is much greater than it actually is (Cohen et al. 1982; Tversky and Kahneman 1982c).

With this background, it would seem reasonable to advocate for pragmatic training in statistics that is specifically directed at reducing biases in our common-sense predictions. There are constraints to this suggestion beyond those related to the inherent limitations humans face when trying to think statistically. Firstly, statistics itself has limitations. It has little to say, for example, about either the usefulness of a complex system model being used to produce statistical results or the reasonableness of the many emotions that humans can express about those results, such as fairness.

Secondly, it is very easy to apply statistics outside the context or assumptions within which its use is valid or useful. For example, when the average income for Mexico and the United States are combined and presented as a North American average, the result would mislead rather than provide understanding about either Mexican or American incomes. Despite this limitation, the use of statistics can create the illusion of objectivity (Germano 1999).

Thirdly, statistical calculations are largely immune to subtle changes in the circumstances surrounding an event, even though those changes could influence the final outcome (Dawes et al. 1989). Think of the nearly continuous opinion polling needed to determine the influence of factors, such as advertising, on the outcome of a political election.

Fourthly, correctly applied statistical methods can provide more appropriate assessments and predictions for well-defined questions than can humans. However, many of the less well-defined issues we face can be addressed quite adequately by our preferred innate methods for prediction if the prediction is appropriately evaluated (Dawes et al. 1989).

It is thus counterproductive to think that the application of statistics can or should replace or displace all common-sense predictions. It seems better for individuals to accept that our common-sense predictive methods rarely include meaningful statistical considerations. Then, by acknowledging the power of statistics, we can concentrate on better appreciating the nature of statistics, when statistics should be used and which prediction types benefit most from the use of statistics.

For example, when deciding whether or not to make a series of bets at the casino, we could accept the statistical knowledge that we will lose in the long run. We could then focus less on our chances of winning and more on deciding if we can emotionally and financially afford the inevitable longer-term loss. In contrast, for those issues where statistics is an essential part of making a useful prediction, such as the strength of a beam in a bridge, we could ensure that the relevant data is generated and used by those skilled in statistical methods.

We are helped in this task by recognizing that a combination of our common-sense predictions, a trained intuition and skeptical thinking can allow us to both humanize statistically supported conclusions and better evaluate them. Consider that we are more likely to identify the factors important to human well-being and our environment's functioning that are being missed by a statistical analysis. For example, the scientific predictions of risk to humans from proposed economics-driven activities focus on the statistical expression of potential deaths, significant injury or significant adverse environmental effects. In contrast, common-sense predictions focus on other features, such as the distribution and fairness of risk, and overviews, such as who benefits and who pays (Slovic 1990; Morgan 1993). A combination of the two enables a better prediction of the consequences of a proposed activity, which then contributes to more realistic decisions.

**Scientific Predictions.** Scientific predictions are, in essence, based on pattern recognition (Rescher 1998; Ziman 1978). But this use of patterns is more formalized than their use for making common-sense predictions. Scientists use a host of methods and tools in their attempt to identify patterns and to lessen the influence of our human biases and errors on their recognition. Among these methods and tools are hypotheses, formal models, theories, mathematics, logic and statistics. The evaluation of the resulting scientific predictions is similarly formalized. For example, to be considered valid, a prediction must be scientifically testable (Peters 1983).

The results of this formal, somewhat arms-length process of making and evaluating scientific predictions remind us that there are practical, real-world limits to fantasy (Kahneman and Tversky 1982a). Yet, despite all of the scientists' efforts to reduce human influences, scientific predictions are still affected by errors, uncertainty, biases and limits. Some of these features reflect our inescapable human attributes. Some arise because the models and theories used to make the predictions are invariably imperfect and incomplete. These features are always present, in part, because of the fundamental characteristics of the world, such as complexity, chaos and choice. Thus, whatever their source, the ubiquitous and continual presence of these characteristics ensures that, at some scale, the world is only partially and imperfectly predictable, even by scientific methods (Rescher 1998).

Scientific predictions are likely to be more valid and useful, and more easily evaluated, if the characteristics of the world are accommodated during prediction making and evaluation. One way this is accomplished is to divide the world into three basic kinds of systems: deterministic, chaotic and random, and complex. Each of these systems has its own distinctive characteristics and is associated with a preferred scientific prediction method. Thus, the recognition of the kind of system being investigated allows the appropriate prediction method to be selected and relevant evaluation concerns to be raised.

However, a real-world system can rarely be perfectly matched to just one of the ideal kinds, and the dominant kind seen can be different depending on the scale used to view the system. The three kinds are really three perspectives of the world. This division and their descriptive neatness is thus only a useful convenience. This should be borne in mind when using the following descriptions to think about scientific predictions.

### The Perspective of Determinism.
All systems exhibit characteristics that fall somewhere between stable and unstable, disordered and ordered, regular and irregular, predictable and unpredictable (Rescher 1998). Those systems that we feel are dominated at most scales by the features of regularity, order and stability can be called deterministic. These systems display well-defined relationships between their variables to the point that a relationship can be called a natural law. The presence of these well-defined relationships means that deterministic systems are characterized by their predictability and their direct connections between cause and effect.

Once the relevant natural law(s) and the initial conditions (prior state) of a deterministic system are known, its future state can be predicted with a high degree of certainty (Ford 1983). Examples include the motion of the planets in the solar system, as illustrated by our ability to land spacecraft on the moon, Mars and a comet. Think too of an expert pool player watching the key balls rolling to the exact places they had predicted before making their chosen shot.

Despite their ordered and certain characteristics, our predictions about deterministic systems are still subject to limits. For instance, our predictions are only as accurate as the knowledge we used in the predictions. Because scientific knowledge is never perfect, it is unrealistic to expect error-free prediction and flawless control of even deterministic systems (Rescher 1998). This does not mean that predictions about deterministic systems are always wrong, just that they are limited, at some level, in their ability to be correct.

Consider too that real-life deterministic systems are never completely free of the characteristics associated with the other system kinds/perspectives, such as randomness (Ford 1983). The presence of randomness in a mostly deterministic system means that, under some conditions, more than one outcome is possible, with each outcome having its own less-than-100% probability of occurring (Rescher 1998). Think of pool table vibrations caused by large vehicles passing by the pool hall. Similarly, the more complicated a deterministic system, the more like a complex system it becomes, and the more likely that difficulties or errors in prediction will arise. Think of trees along a power line which could result in one falling on the line and disrupting the electrical power distribution grid.

It would be useful to clearly define and separate out the deterministic from the non-deterministic parts or scales of a system. To a degree, this is possible. However, in the end, defining a system as deterministic is one of convenience. At some point, we have to come to terms with the fact that the world is less deterministic than we imagine or would prefer. We would do better to think of our place in the universe as existing in a boundary zone displaying a mix of system types.

### The Perspective of Chaos and Randomness.
The terms *random* and *chaotic* are commonly used interchangeably to refer to systems that display considerable disorder, instability and a high degree of significant, irreducible uncertainty in prediction. This difficulty in prediction exists because critical information about these systems is permanently lost to us (Ford 1989). Consider that, in deterministic systems, any effort to increase our knowledge about them is rewarded by more certain predictions. But, in chaotic and random systems, we are simply unable to collect a sufficiently large number of sufficiently accurate measurements to fully represent the system. For example, we can't simultaneously measure all of the influences on each molecule in a gas.

Despite their common characteristics, when random and chaotic systems are looked at more closely, they are not the same. Random and chaotic are not really synonyms. In random systems, such as a game of dice, we lack sufficient information to fully describe it, but it is sufficiently ordered that we can still make useful predictions by applying the statistics of random events (stochastic calculations). To make a prediction about a random system is to list each possible outcome and its probability of occurring.

In chaotic systems, we may be able to reasonably describe the system but not in sufficient detail to accurately represent its functioning. You can appreciate why by thinking of a chaotic system as a sensitive ordered system. One that is sufficiently sensitive, even unstable if you like, that a very small changes in its variables result in extremely large changes to the system's overall state (Carlson 1999). An example of such a system is one in which its output is both fed back into the system as an input and amplified by it: that is, the system contains an amplifying feedback.

The sensitivity of chaotic systems means that even very small errors or deficiencies in our data (or model) are rapidly amplified by the model representing the system. This creates large errors in the output. Our inability to describe the sensitivity and amplification with sufficient accuracy means that our predictions about chaotic systems are always significantly uncertain.

The usefulness of predictions about real-life chaotic systems can be increased through the judicious use of statistics and other mathematical tools. For example, when the state of the earth's atmosphere is studied at the scale relevant to our experience of weather, it is a chaotic system. Yet, within the limits of a few days to a week, we can still make reasonable weather forecasts. Think of the evaluation rules of thumb.

The presence of randomness within our day-to-day life beyond the gambling hall presents us with similar challenges. Consider, as discussed above, our routine need to decide which aspects of our experiences are truly random (luck) and which we can control by skill. In essence, we are left to decide if an unexpected experience is the result of randomness, chaos or ignorance (including a lack of skill). A task made more difficult than we expect because features of all three are commonly present in real-life systems, such as those discussed in this book. A rule of thumb to deal with these tricky decisions is that the presence of any one of them ensures that our predictions about a system are subject to some level of enduring uncertainty.

But wait! Surprisingly, under special circumstances, real-life systems with a dominant random or chaotic aspect can give rise to a certain, predictable result. For example, the random movement of water molecules in our atmosphere can, under the right freezing conditions, produce the ordered snowflake (Rescher 1998). Similarly, the freezing and thawing of unsorted (randomly arranged) glacial till can result in the beautifully ordered patterned ground and sorted circles seen in the Arctic (figure 5) (Kessler and Werner 2003).

In both of these examples, the phenomenon of a deterministic-like self-ordering is only recognized if the appropriate scale is used to look at the system.

This provides a timely reminder that we should consider more than a system's kind when considering the making and evaluation of a prediction about it. We should also consider the boundaries of the system and the scale used to view and describe it. The importance of this suggestion will become more apparent in the following discussion about complex systems.

*The Perspective of Complexity.* Complex systems, such as our atmosphere, ecosystems, an individual human, human cultures and their economies are distinguished by their many variables and the large number of interactions among them. These interactions, which take place over a wide range of time and space, exhibit a wide variety of linear to non-linear relationships between the variables, including feedback. Collectively, these interactions link the variables together into a single complex system and determine the overall mix of the deterministic, random and chaotic characteristics which the system displays. They therefore also determine the degree of difficulty and uncertainty that we will experience when predicting the system's future.

The manner in which a complex system displays its mix of basic characteristics is illustrated by our atmosphere. The presence of deterministic characteristics is evident in the global-scale, long-term order and symmetry of the earth's wind regimes and climate types around the equator. Mathematical models can reliably describe this order. They can also predict the general manner in which it will change in response to variations in both the amount of sunlight reaching the earth and the composition of the earth's atmosphere.

The ubiquitous presence of significant random and chaotic characteristics in the earth's atmosphere is evident in the vagaries of its local-scale, short-term changes, that is, the weather. Where will the next gust of wind come from? The presence of random and chaotic features ensures that all weather predictions are accompanied by a statement of uncertainty, or probability, such as a 60% chance of rain on Thursday. These features also help to limit the certainty with which climate models can predict the details of how the earth's climate and wind regimes will change over time (Mitchell and Hulme 1999).

We can more fully appreciate how the interactions among a complex system's variables give rise to the system's mix of characteristics and how it changes over time (i.e., the nature of the system's functioning and changing) by considering just one of its variables: variable X. Some of the system's variables interact directly with variable X, thus causing it to change. The sum of the direct influences each variable has on variable X will partly determine the amount, direction and speed by which it will change.

Yet variable X is also being indirectly influenced to change. Of particular importance are the interactions where a change in variable X induces other system variables to change, and then those changes directly or indirectly impact variable X. That is, variable X is linked to itself by a feedback loop through some of the variables it interacts with. Thus the total influence driving variable X to change is more complex than expected: it includes the sum of the direct and indirect interconnected influences on it. In particular, the presence of feedback can, over time, either reduce (dampen or buffer), reinforce (or strengthen), amplify or maintain variable X's current amount, direction and speed of change. As a result, variable X will vary over time.

In reality, a complex system's many interconnected variables don't change one at a time. They are all changing together and, as expected, by different amounts and at different speeds. This diversity ensures that a complex system's variables are not only themselves constantly changing but they are prompting one another to change. Thus, at any given time, each variable is at different stages in its response to the factors driving it to change and driving others to change. It is the aggregate of the ongoing responses of the system's variables to the continually changing influences prompting them to each change that determines the nature of the complex system's functioning and changing. This aggregate determines both the system's overall characteristics (the mix of its deterministic, random and chaotic aspects) and the specific manner in which it changes.

When viewed from the perspective of its overall functioning, a complex system is dynamic. It is always changing, to some degree, at some scale, in a manner that is rarely smooth. In addition, the presence of interaction-caused delays means that a change to the variables in one part of the system takes time to propagate through the system to the other parts. Different parts can be changing in different ways at the same time.

In light of these characteristics, as reflected in its dynamics, it is no wonder that what we see in a complex system depends on where in time and space we choose to look at it. It also depends on the variables and the scale we choose to focus on. By limiting our focus and attention, we bias our perception and appreciation of the system's functioning and its current state.

Consider a complex system's equilibrium, balance and stability. If it is to continue to exist, the system's dynamics must direct it toward an equilibrium condition. But the dynamics ensure that the system rarely, if ever, reaches that equilibrium at all scales. As a result, the system's current state should be described either as partly out of equilibrium, as constantly moving toward yesterday's theoretical equilibrium state or as constantly catching up to ongoing changes. The extent to which we recognize this state, and the degree of balance and stability we assign to the system, depends on the variables and the scale we choose when looking at the system.

These aspects of a complex system's dynamics are illustrated by a stable lake ecosystem. It consists of the lake's water and a mix of aquatic organisms in a near equilibrium state. That state is maintained by a feedback loop. For example, if the amount of nutrients entering the lake ecosystem is increased slightly, then its water will change in composition. The lake ecosystem's aquatic organisms will respond to the nutrients by growing more, which will decrease the nutrients in the lake's water. The lake ecosystem's dynamics will maintain it in its current near-equilibrium state.

However, if the amount of nutrients entering the lake were to significantly increase, then the extra growth of the organisms would deplete the lake water of its oxygen (the lake's water undergoes eutrophication). In turn, the oxygen-depleted water would force the mix of organisms living in the lake to change in favour of those species that require little oxygen. The lake ecosystem (organisms and water) would have switched states. In the long view, the lake system can exist in one of two possible dynamic, near-equilibrium states. Despite our perception that the lake system's current state is stable, it is only quasi-stable.

In general, a complex system can exist in one of a number of possible dynamic, near-equilibrium, quasi-stable states. The overall speed, ease and strength with which a system will switch from its current to an alternate quasi-stable state is determined by the cumulative impacts, at all scales, on the system. These impacts can be sourced from internal (its variables' interactions) and external sources.

There are rules of thumb that helps us think about a complex system's switch in state. A complex system will remain in its current quasi-stable state as long as its feedbacks maintain the near-equilibrium conditions supporting that state. If the cumulative impacts on the system sufficiently disrupt its functioning, then its feedbacks will favour an alternate near-equilibrium condition. The system's characteristics will then change toward the alternate quasi-stable state that the new conditions support. If the disruption is sustained, then the system will eventually switch to that alternative state. If the disruption results in an amplifying feedback appearing in the system, then the system will become highly unstable, even chaotic and unpredictable.

Overall, the complexity and dynamics of a complex system's functioning ensure that efforts to predict its future state are both difficult and frustrating. This is illustrated by ranking the possible influence on the complex system's current state into a hierarchy of importance. In the case of the lake ecosystem, assume that the influence of the historical nutrient-poor inflow into the lake ranked below commonly occurring severe droughts. The hierarchy can then be used to make predictions about the system's future: it will be affected by drought.

But the dynamics of the system ensure that the hierarchy's ranking is not fixed. It is flexible and changes with time (Costanza et al. 1993). For example, if a pig farm is built next to the lake, then nutrient-rich water could be ranked above drought in the hierarchy. The ranking also depends on our perspective, such as the scales of time and space we use. When should we include the area of the pig farm within the lake ecosystem's boundary? We are left with the realization that making viable, useful predictions about a complex system and evaluating them requires a substantial amount of knowledge of its nature, boundaries, dynamics and context and ongoing, detailed monitoring.

There are other features of complex systems and their dynamics that add to the difficulty of understanding and making predictions about them. For example, the numerous linked interactions in a complex system result in it being able to display more features than the simple sum of its parts. Those features are expressed in a manner similar to self-organization (figure 5) (Norretranders 1998). Our consciousness is likely a by-product of our brain's complexity.

If, or perhaps when, such a new feature appears in the system, it could feed back to facilitate further changes to the system. This is illustrated by the appearance and increasing sophistication of writing in an early, more simply organized Babylonian culture.

Writing eventually allowed the Babylonians to more easily practise accounting and to comprehensively record Babylonian speech. Writing thus facilitated an increase in Babylonian culture's organizational complexity: the culture became more complex (Nissen et al. 1993).

In some cases, the new features may appear to us to be more than just the product of the system's usual response to the forces driving it to change. The features can be thought of as the product of the system adapting to those forces. These complex systems, such as an ecosystem that undergoes change as its surrounding host environment changes, are sometimes called adaptive complex systems. Note that adaptive complex systems have been considered by some to indicate the possibility of the system having consciousness. Think of Gaia.

Another prediction-affecting feature that results from a complex system's dynamics is unexpected events. They occur due to the system's chaotic and random aspects. Think of earthquakes or abrupt climate change events (Alley et al. 2003). It must be said that our labelling of events as *unexpected* more often represents our faulty, biased thinking and perception about the system or the event rather than the event actually being the product of chaos or randomness.

Which brings us to another key consideration. When making scientific predictions about complex systems, all of the above characteristics must, to some degree, be taken into account. But we find many of them hard to grasp, baffling, surprising, intractable and even frightening, as hinted at by the question of the lake ecosystem's boundary. Overall, we find it difficult to think about a complex system's dynamics in a coherent, inclusive manner. Both a complex system's features and our limited abilities constrain our efforts to make valid and useful predictions about them and to evaluate those predictions.

In overview, scientific predictions about a part of the world can be made by treating that part as a deterministic, chaotic-random or complex system, as applicable. This is a convenience. Much of the world is complex in nature, which includes deterministic and chaotic-random aspects as well as the distinctive features of complexity. If we wish to ease the difficulties we face when making predictions about the world, we could do well to broaden our appreciation of complex systems. In particular, we need to better appreciate our relationship to complex systems, especially when we participate in them. That is the subject of the next section.

In summary, we are more likely to appropriately make, evaluate and use a prediction about an aspect of the world if we choose the most appropriate prediction type (prophecy, common sense or scientific) and acknowledge its constraints. When making scientific predictions, it helps if we choose the most appropriate issue-relevant perspective (deterministic, chaotic-random or complex) to use for the system while at the same time acknowledging the constraints that choice imposes on the prediction and its use. It is particularly difficult to make valid and useful predictions about complex systems and to evaluate them. This is partly the result of the system's complexity. But it is also the product of the human-sourced constraints on our ability to relate to complex systems.

## Relating to Complex Systems

We, like all living things, live out our lives as participants in complex systems. We are participants in our culture, our supporting ecosystems and our greater environment. Most of the more serious issues we now face, from environmental change to resource availability and cultural instability, are formed and exist as part of these complex systems. Our efforts to deal with these issues rely heavily on not only the predictions we make about the issues, but the predictions we make about their complex system host.

But it is difficult for us to make viable and useful predictions about complex systems, and to effectively evaluate and apply them. One reason is the nature of complex systems, as described in the previous section. But those characteristics are insufficient on their own to account for all of our difficulties. Another key reason is the human-sourced constraints on our ability to relate to complex systems, as introduced during the discussion about the multiperspective method of describing and relating to complex systems in chapter 1. Of particular interest is our limited and biased ability to mentally perceive and conceive of complex systems, their functioning and our participation in them.

This section discusses the human sources of our strained relationship with complex systems from three overlapping perspectives: our efforts to describe complex systems, deal with them and think about complexity itself. The aim is to increase our awareness of those constraints. It also aims to provide a broad, coherent overview of the connections among them. Collectively this should help us to more easily relate to complex systems and thus make, evaluate and apply predictions about them that are more viable and useful.

**Describing Complex Systems.** One aspect of our strained relationship with complex systems becomes apparent when we try to describe one. The practical, human-sourced constraints on doing so are illustrated by the most basic of that descriptive process: choosing the categories to divide and classify a complex system's variables into. They are also illustrated by the most complicated of those steps: generating sophisticated models to represent our descriptions of the system and our understanding of its functioning.

*Divisions, Categories and Classification.* In our effort to understand a complex system, we look for patterns in its features by dividing, categorizing and classifying them. We find this task straightforward and pleasing if the features of interest are discrete, such as cows versus goats. However, frustration can set in when a feature is part of a continuum or when a feature falls into a clearly discrete natural category at one scale but has diffuse boundaries at a different scale. For example, when seen from space, the dividing line between the North American boreal forest and the Arctic tundra is easy to locate as a tree line. However, locating the tree line on the ground is a lot more difficult because the boundary is in fact a wide transition zone made up of scattered patches of trees. When dealing with these types of features, we are stuck with the difficult task of deciding on the number of categories to use and where to draw the boundary between them (Jones 1999).

An arbitrary selection of divisions and categories would seem to suffice, but that method neglects the presence of practical constraints. One example is our limited short-term ability to remember more than a few categories of things, as described by the $7 \pm 2$ rule of thumb discussed in chapter 1. If our gathering of data obliges us to remember more than this range of categories and this limitation is not accommodated, then we can be sure that the resulting data will be unrepresentative.

The number of recorded categories could be extended through the use of category lists kept on paper or in electronic form. But as anyone who has used this procedure for non-discrete data knows, there are limits to both matching observations with the categories to be recorded and doing so consistently. An alternative is to simply record a wide range of attributes and then use a computer to categorize the data. But there is still the question of how closely the computer's categories will represent meaningful, real-world divisions. Overall, efforts to address classification issues adds to the complexity and cost of data collection.

The waters of classification are further muddied by the covert influence culture has on our dividing and categorizing efforts. For example, the Western world view encourages its members from an early age to make sense of the world by treating it as divisible into well-defined (i.e., discrete) components that have a largely fixed (i.e., well-ordered) relationship to one another. Westerners also have a cultural preference for dividing the world into opposites or extremes: by analogy, into black and white. Our thinking also has a bias toward the world's material aspects. Newton's rainbow illustrates that these types of cultural influences impose significant, unconscious constraints on our possible choices of divisions and categories.

The full significance of both our innate and our culture's influences on our categorizing decisions is brought home by the controversy surrounding Caster Semenya: the South African runner who is hyperandrogenic. She looks too manly to be a successful woman athlete, so she has been subjected to gender verification tests (McRae 2016; Fahrenthold 2009). If it is technically confirmed that she has sexual traits of both male and female, then she will have no place in the two-sex division system used in sport.

A simple, two-division sex classification is useful (it is a key part of reproduction for most species), but it ignores significant aspects of our body's functioning. The process that creates a person's sex is part of a complex system of which the X and Y chromosomes are only a part. The system includes the reading of DNA, the manufacturing of hormones and the recognition of those hormones by the relevant cells. To be born as fully one or the other sex, our body must successfully produce all of the many relevant chemicals and body bits that create and maintain a sex type. After successfully completing all of these many interconnected steps, the biological part of sex creation is completed. Our culture's world view then guides parents to instil in their child the accepted gender behaviour associated with the child's culturally accepted sex.

Considering the complexity of the sex creation process, it should be unsurprising that there are many physical, chemical and psychological indicators of sex type. We should also easily accept that neither sex nor gender are quite as absolute or as easily divisible into sport's preferred two groups. Yet, even if we did, our

eyebrows might still rise when told that perhaps as many as 1 in 100 of us has at least one sex indicator suggesting that our sex is something different than the one we accept. For example, somewhere between 1 in 500 and 1 in 1,000 males born on any given day have one or more extra X chromosomes, which gives their genetic makeup some female potential.

In many cases, these types of sex-determining variations are sufficiently small that they pose no problems in dividing us into one of two sexes. However, for some people, the mismatch in their sex indicators is sufficiently large that they can fit between the sex boxes as intersex. This can arise, for example, if your genes say you should be male, but during your time as a foetus, an insufficient amount of the hormone androgen was produced to turn you fully into a male. Without an appropriate classification box for intersex people, cultures and individuals can find it difficult to categorize and to deal with the sexual reality of intersex people (Callahan 2009).

As the sex example implies, our choices about dividing and categorizing information can influence more than the description of a system. They can affect our understanding of the system, our predictions about it and our relationship to it. For example, how you decide to distinguish between domesticated and non-domesticated plants and animals (i.e., how you categorize them) influences your thoughts about the domestication process and its timing (Mannion 1999).

This also implies that we can partly alleviate the difficulties associated with the division and categorization of a complex system's features by changing our general perception of it. By accepting that a complex system's features are more continuous than discrete and that the outcome of its interactions is more grey than black or white, then our record of its characteristics is likely to be more representative.

We are also helped by expanding our usual focus on the system's specifics, such as a set of variables and the precision of their measurements, to include their context. For example, when trying to evaluate the risks to the functioning of an environmental or social system we could use more than one classification scheme to record the variables needed to make the evaluation (Morgan 1993). By contrasting and comparing the resulting variety of descriptions or interpretations of the system, we are more likely to understand the context in which the variables play their role in the system. That is, we are more likely to understand how the system functions as a whole. The resulting change in how we relate to the variables and their host system's functioning should assist us in our evaluation of risk. This is a variation of the multiperspective method of looking at a complex system used to discuss the brain.

However, there is a limit to these techniques. Consider the commitment required from the predictor to engage in the sometimes sweat-producing mental gymnastics needed to link the myriad of recorded pieces of a complex system into a coherent picture. Some would suggest that this mental effort and our difficulty relating to a complex system can be reduced, if not eliminated, by constructing a model of it.

*Models.* A model of a complex system can be descriptive, mathematical (computer), common sense or even just a reference to an analogy, it is our rationalized, simplified representation of the system. It is based on our reasoned, intuitive, culturally-influenced understanding of how the system functions (Begon et al. 1996; Ziman 1978). Think of Descartes's clockwork view of the universe mentioned in chapter 4.

These models are central to our thinking, decision making and predictions about complex systems. After all, we hope, or perhaps believe, that our model of a system will reliably represent how its functioning produced the order we discovered in the categorized data and our record of its interactions. And we expect that it will generate useful conclusions and reliable predictions about the system. But, practically speaking, the number of variables, interactions and scales involved in a complex system's functioning, as described above, is sufficiently large that a comprehensive model is usually not possible. This is especially true for our mental models, but it is also true of our descriptive and computer models (Gardner and Stern 1996). This section will focus on the most complex of our models.

Many complex systems are too large or complex to be modelled in their entirety, even when using a computer. We can partly address this issue by choosing boundary conditions that we think isolate the portion of the system that represents the issue the model is expected to address. For example, when modelling the atmosphere for weather forecasting, the boundaries are chosen to exclude those of the atmosphere's ongoing interactions that occur at a time scale shorter than minutes and longer than a few decades. However, it must be admitted that the choice of a complex system's boundary is somewhat arbitrary, as illustrated by the choice of the boundary between the boreal forest and discussed in the lake ecosystem example.

We can also increase the validity and usefulness of a model by increasing our knowledge about the system and by improving our modelling methods. But, as the discussion of data categorization illustrated, our description of the system is always, to some degree, incomplete or inaccurate. We are thus constrained in our ability to deduce and thus represent a complex system's functioning in a model.

Consider too that we are participants in many of the complex systems we are interested in. Think of the dynamic, complex, human-environment system. Our participation adds another layer of modelling difficulty. After all, we find it difficult enough to come to terms with our decision-making characteristics and to accommodate them in our thinking, let alone include them in a model of a complex system in which we are a key participant.

By default, the primary method we use to address the difficulties of modelling a complex system is to decrease the details we must deal with. We accomplish this through aggregation, assumption and selection, as illustrated by the first step in the process: boundary selection. But our use of simplification is limited by the need to ensure that there is sufficient detail in the model for it to be valid and that its output is meaningful and useful at a human scale. The making and implementing of a model thus brings us face to face with another perspective of the previously discussed simplification dilemma: the process of creating a model is also one of making choices, including compromises.

Certainly, appropriate simplifications can facilitate the making of a model and increase the value of the output. However, too much or inappropriate simplification will result in a model that fails to meaningfully represent the system's functioning. Its output is more likely to mislead than to help us address our issues. Consider that economists commonly rely on tenuous, broad economic assumptions, such as the market's supremacy and us being rational consumers, when building their highly sophisticated mathematical models. This is one reason that economists both contributed to the global financial crisis of 2008 and failed to predict it (Patterson 2010).

In generally, the simplifications made during the modelling of a complex system result in a model which only represents specific aspects of the system's functioning under restricted conditions. This means that even mathematical models of complex systems cannot simultaneously maximize the three goals of modelling: realism, precision and generality (Costanza et al. 1993).

It also means that the model's results, such as a prediction about the earth's climate system and our economy, are invariably accompanied by uncertainty (Mitchell and Hulme 1999; Stainforth et al. 2007). This reality limits the usefulness of models.

Their usefulness is further limited by how we relate to models. We do so in a manner that has much in common with our previously discussed relationship with visual displays of information discussed in chapter 4. We forget that visual displays are partial representations of reality, not mirror images of it (Ziman 1978; Mitchell and Hulme 1999). Similarly, we tend to treat a model of a system as if it were the actual system. It is thus easier for us to believe that the model is more valid and useful than it is and therefore to place greater hope than is warranted in its predictions (Cohen 1995).

Our decision-making foibles also mean that we are more likely to have faith in a model if it has an elegance that meets our cognitive and cultural biases, such as our sense of order. For example, we prefer a model whose results are presented logically, quickly and in an appealing manner, for example, dynamically on a computer or TV screen, in colour. Think weather forecasts. The resulting sense of satisfaction means that we are more likely to unconsciously (or even consciously) accept that the model's assumptions, predictions and conclusions are correct, or at least more correct than they actually are.

An overly optimistic evaluation of a model and its results can have a surprising consequence. We precondition ourselves to become disillusioned with the model and its predictions, and to do so regardless of its actual validity and usefulness. This can occur, for example, when our personal experience of reality doesn't match our high expectations of the model. It can then appear to us to be excessively imperfect: "The weather forecast is hardly ever right."

We might just as easily under- as over-evaluate the usefulness of a model and its predictions. We could be prompted to draw such a false conclusion by relying on our first impression of the model. Think of our reaction if the model's results were displayed opposite to those described above (unappealing, black and white, etc.) and then add in "too much" statistical information.

However, on its own, the elegance of a model or its presentation doesn't help us to factually appreciate the nature of the system being predicted or the limitations of the model. Neither does it help us to grasp the uncertainties that come with the model's output. Think again of our relationship with weather forecasting models. Appropriately relating to a model and its prediction is, like riding a bicycle, an acquired skill.

The importance of being aware of a model's limitations and our model-related biases is illustrated by two examples. Consider first our common use of scenarios to either constrain the model's input conditions or to study alternative possible outcomes. Certainly, the use of scenarios can help to constrain the influence of our human foibles, such as the hindsight bias, on our choice of starting conditions. They can thus help us clarify our choices or options. However, scenarios can also give our imagination free rein. Thus, unless good judgement, perhaps wisdom, is exercised when creating the scenarios and interpreting the model's results, it is easy for illogical and imaginary outcomes to be taken seriously (Tetlock 2006). Good judgement requires being aware of both our and our model's limitations.

The second example illustrates the consequences of not paying attention to a model's limitations. The Limits to Growth computer model assumes that resources on the earth are finite. The model's results describe how natural resources decline and population increases until they reach a natural limit that is determined by the earth's finite resources (Meadows et al. 1972; Meadows 2007). In contrast, neoliberal economic models, such as the Dynamic Integrated Climate-Economy model (DICE), assume we are rational consumers and that natural resources are infinite, or that there is always something to act as a substitute if a particular resource runs out. For these economic models, a limit to growth arising from resource shortages is not possible (Costanza et al. 2007).

The failure to acknowledge these types of basic assumptions has contributed to many an unnecessary misinterpretation and debate about the results from models. In the Limits to Growth example, the question to ask is to what degree and under what conditions are the earth's resources, in a practical sense, finite. A question to be addressed later in chapter 12.

This discussion could lead to the conclusion that our issues with modelling are largely a product of this computer age rather than our human characteristics. This is not the case. The process of making models and then believing in their results, while ignoring their assumptions and limitations, is thousands of years old.

For example, a Babylonian text from 2100–2000 BC describes a model of the changes a four milk-cow herd will undergo over 10 successive years. Its results predict the annual growth in the number of cattle in a herd and the associated increase in the production of milk products (cheese and fat). The limits to the model can easily be seen in its implicit assumptions: there are no cow deaths; milk yields and calving rates remain constant even as cows age; and the number, sex and spacing of calves is fixed. In the real world, the model's predicted increase in cows and milk products would not be met (Nissen et al. 1993).

In overview, the human-sourced constraints on our ability to realistically describe a complex system contribute to our strained relationship with the system. For example, our efforts to gather the data we feel reliably describes the system are constrained by the difficulties we experience accommodating features that display a mix of continuous, overlapping and discrete relationships with one another. Our efforts to describe the system by making a representative model of it are constrained by both the nature of the data we collect and the need to make simplifications and compromises when building the model.

The human constraints on our efforts to describe a complex system arise from both our innate individual human characteristics and the influence of our culture's world view on our perceptions and decisions. These constraints limit and bias the representativeness of our descriptions and our view of our descriptive efforts. This alone ensures that uncertainty is always present in the conclusions we draw and the predictions we make based on those descriptions.

If we manage to interpret a valid model's predictions within the context of the model's and our limitations, and the external world's constraints, then it can provide us with useful conclusions and predictions. For example, the Babylonian model does inform us of the likely maximum limit for milk-product yields. To achieve the goal of useful and meaningful predictions requires us to take into account other aspects of our strained relationship with complex systems.

**Dealing with Complex Systems.** A different perspective of the human-sourced constraints on our relationship with complex systems is illustrated by our experience dealing with human-designed and human-built complex systems. We express considerable pride in the (human, cultural and personal) management and technological skills we use to design, build and control these systems. We are responsible for these systems. However, being human, it is no wonder that we tend to blame outside forces or unforeseeable events for their failure. In reality, when we create a complex system, we focus intensely on getting it to work in one way for one purpose. We don't completely accept that the complexity we create allows the system to fail in many more ways.

Many of the human-built complex system failures can thus be attributed to the constraints on how we think when we conceive of a system, build and run it. Consider that our human decision-making foibles and our culture's world view ensure that we are unlikely to anticipate most of its failure paths. Thus, despite hopes to the contrary, they can't be fully taken into account during the system's design, construction or operation (Perrow 2007; Gardner and Stern 1996). This contention and its relevance to our relationship with complex systems in general are elegantly illustrated by the 1990s attempt to replicate (model) the earth's biosphere in a human-designed biosystem called Biosphere 2.

The Biosphere 2 experiment was isolated from the earth's biosphere by placing it inside a futuristic, dome-shaped building. The interior of the building was divided into a number of sub units. Each contained the plants, animals, earth and/or water needed to create a replica of an ecosystem chosen for the experiment. It was expected that the sub-units representing the earth's ocean and land ecosystems and Biosphere 2's atmosphere would interact with one another in a manner that would result in the experimental biosphere, like the earth's biosphere, balancing itself and then self-regulating that balance.

Humans were included in the experiment and provided with the tools to control the system if needed. The Biospherians lived in Biosphere 2 for the duration of two closed-system experiments. The first lasted two years, the second, six months.

Biosphere 2 failed to reach balance, even though the human participants played an active role in trying to achieve it. The lack of success was demonstrated at the end of the experiment by the degraded and stressed state of at least some of Biosphere 2's ecosystem sub-units.

There were many reasons for the failure. Among them was Biosphere 2's inadequate (too simplistic?) representation of the complexity of the earth's biosphere, in particular its stabilizing feedbacks. The balancing problems were exacerbated by technical problems, such as the unanticipated reaction of the biosphere's atmosphere with the structure's concrete parts. The proper functioning of Biosphere 2 was also hindered by an unconsidered, non-technical design factor: the vagaries of the normal interactions among humans. The relationships among the Biospherians complicated decision making about how to deal with the system's operational problems (Marino and Odum 1999; Nelson et al. 1993).

The failure of the Biosphere 2 experiment illustrates that, even at a scientific level, there are biases and limits in our thinking about human-designed complex systems and in our dealing with them. It also illustrates that these constraints apply to our relationship with natural complex systems. These constraints thus contribute to the current, seemingly intractable issues we face as participants in our complex human-environment system. More on this in later chapters. In this chapter our dealings with human-designed systems can teach us more about our constrained thinking about complex systems in general.

The generic features of the Biosphere 2 failure are widely applicable to the failures of other human-designed complex systems. Spectacular examples are the Chernobyl and Fukushima Daiichi nuclear explosions, the Deepwater Horizon oil well blow out in the Gulf of Mexico, and the northeastern USA power blackout of 2003 (Cauley 2004; USCPSOTF 2004). Along with Biosphere 2, these examples illustrate our ability to design complex systems that provide the output we desire; our involvement in their functioning; the limits to our scientific predictions about their futures; and the limits to our abilities to manage them. They also illustrate the common features of the failures of our complex systems, our reaction to those failures and our resulting explanations for them.

We consider that human-designed complex systems have failed when they enter a state that doesn't meet our expectations and predictions. Our descriptions and explanations for how these systems entered that failure state reveal the generic differences between our thoughts about human-made (and natural) complex systems and their actual characteristics. For example, by describing the above failures as explosions, a blow out and a crash, we illustrate our preference for thinking about a failure as if it were a single, well-defined cause-and-effect event. Think, too, of our search for a threshold, trigger or tipping point which, when crossed, can be used to explain the failure and perhaps be its cause.

But these spectacular, rapid failures are, like most failures in human-designed systems, generally not the simple, direct, single cause-and-effect events the above descriptions imply. Most are the result of a combination of many events spread over time and space. More usually, a mix of technical situations and human decisions interact in ways that precondition the system

to fail. Preconditioning can be subtle, reversible and slow. When the system is in a preconditioned state, the convergence of many, often small, scattered incidents can result in an event we call the failure (Perrow 2007; Gardner and Stern 1996).

The final failure event may occur quickly, as an explosion, but it doesn't have to. It can be a long-drawn-out affair, such as the faster than expected filling of a dam with silt or the unexpected slow salinization of irrigated land. In these cases, a focus on tipping points and simple causes can be particularly hard to define and can be misleading.

In keeping with our faith in our ability to design failure-free systems, once a failure mechanism is established, we often label it as a design "flaw" and operational "error" rather than as a result of our prior, imperfect thinking about complex systems (Gordon 1978). For example, the space shuttle disasters were described as the result of an "itinerant" foam block and a "faulty" O-ring, rather than the result of a failure or limit in our ability to anticipate these types of possible events in a complex system. We could just as easily have labelled these failures as "acts of God."

For example, after hurricane Katrina had passed, some 1,800 homes were contaminated by crude oil from Murphy Oil's failed oil storage tanks. Despite the relative simplicity of the situation and common knowledge about hurricanes in the area, the company described the event as an unforeseeable "act of God." They preferred not to call it the product of an error in the company's design and siting of the tank farm (Perrow 2007).

Careful listening to the focus, content and logic of the explanations given for failures reveals that our contribution to them occurs as a result of more than our unsubstantiated faith in our abilities. For example, by talking about an itinerant foam block rather than an unconsidered foam block, we express our human preference for treating ourselves as somewhat distant from the systems we design and operate. This distancing can contribute to the system's failure.

Consider that we think of ourselves as participating in a human-designed system only as, for instance, a designer, manager, director, maintenance person, operator or user. Each of these personnel have a limited role and limited responsibility for the system's functioning. In these roles, we can each admit that we are an integral part of the system but, at the same time, we can think of the whole system as being separate from us. This is analogous to the way the law allows a corporation's directors, managers and investors to see themselves as separate from the corporation and the responsibilities for it but still capable of running and benefiting from it.

This thinking is a convenient fiction. Each of us who participates in human-built complex systems, from designer to operator to owner, and even users, are tied to it like we are to our family and our human-environment system. Through these wide-ranging, inescapable ties we contribute to and are, in this context, more responsible for the human-designed system's functioning and failures than we admit.

We might question this conclusion, until we consider the failure of systems such as the Three Mile Island nuclear plant and the northeastern USA power grid. Of particular interest are the types of reasons why the operators running these plants had not responded differently to the incidents which led up to these systems failing. They included the operator having incomplete scientific understanding about how the system functioned; the system's design supplying them with inadequate information about the state of the system; and the system providing limited capacity to isolate failed components from the functioning part (Gardner and Stern 1996; USCPSOTF 2004).

These reasons have a common feature: neither designers nor operators are solely responsible for the oversight(s) leading to failure. This is quite a reasonable conclusion from a human perspective. But, from the system's perspective, a significantly different conclusion is drawn: both are contributors and both are responsible because they are participants in the system.

The conclusion that our level of participation and responsibility for the functioning and failures of human-designed complex systems is greater than we accept is personalized by an analogy. The responsibility we consciously accept for our personal health and safety is demonstrated by our conscious health-affecting decisions and behaviours. The level we accept is commonly much less than our actual responsibilities, as shown by some of the health issues we may have, such as dental fillings, unwanted pregnancies, being unfit and carrying extra weight. However, these health issues are also the outcome of us fully participating in a culture that provides us with an abundance of incentives and resources that facilitate the appearance of these health issues.

In overview, generally, the biases and limits to our dealings with human-created complex systems during the design, building and operational phases contribute to their failures. This conclusion also applies to our manipulation of natural complex systems. But our strained relationship with complex systems is the result of more than how we think about and deal with their functioning. It includes how we think about the notion of complexity itself.

**Thinking about Complexity.** The nature of the human-sourced constraints we experience when relating to complex systems is also illustrated by our thinking about complexity itself. Consider our need for food energy to survive. It is reasonable to think that we could build a mathematical model that would explain and predict a hunter-gatherer band's foraging patterns. The model would do so by considering a variety of foraging options within the band's home range. For each option, the model would balance the energy the members would expend when foraging, with the energy the foraged food item would provide. The option with the greatest energy return would be their preferred foraging pattern.

The model would be more realistic if we added in nutritional needs and variations in rainfall (food availability) over time. But even this more complex model cannot realistically explain the members' foraging behaviour. After all, hunter-gatherer foraging patterns are also influenced by the cultural context in which they occur, for example, the presence of cultural taboos. Thus, to be realistic, the model should include cultural and other non-food influences on their foraging activities. We prefer not to add them in because it is difficult to meaningfully implement human behaviour in a mathematical model (Kelly 1995).

More generally, we have to be pushed to recognize that our thinking about a complex system involving humans, and our interpreting of the output from models of it, should, at some point, fully consider the complexity of the system (i.e., with us as a part of its functioning). The same logic applies to our thinking about the resource issues that today's industrial cultures face (Ferrer-i-Carbonell and van den Bergh 2004). Instead, the degree to which we consider the complexity of the system varies with the aspect (e.g., material vs social) of the system we focus on.

In the end, at the heart of our strained relations with complex systems lies our relationship with complexity itself. This sub-section discusses the constraints on our thinking about and dealing with complexity that arise from our innate thinking and decision-making foibles, and the influence of our culture's world view on them. The emphasis will be on Western culture.

A culture's world view exerts a significant influence on its members' (conscious and unconscious) thinking about the world around them. Consider that a more complex culture's world view strongly guides its members to believe that their culture's stability, their personal well-being and even civilization itself depend heavily on their culture's hierarchical structure and its functioning. The members are also guided to find a similar structure and functioning in both their spiritual and the natural worlds, for example, the hierarchical order in their culture's heavenly host. Indeed, as a culture's organizational structure becomes more complex the nature of the relationship among its gods is adjusted accordingly, as the history of Babylonia illustrates. The strength of this unconscious connection to that structure helps to constrain/bias the members' relationship with complexity and their thinking about it.

Western culture is no exception. The more recent roots of our present world view's guidance about the nature of the world around us can be traced to the AD 1500s. At that time, Western culture absorbed mathematically supported (e.g., geometry) deterministic descriptions of the world's functioning and related ideas, such as progress. These views changed over time to become our current world view.

We are left with a question. Considering that, in many instances, the relationships of importance in the world are not deterministic, why would determinism still dominate Western thinking? Why didn't the thinking associated with the calculus of chance and many-bodied problems come to dominate Western culture's world view? After all, the basic mathematics had been recognized and later developed by the Bernoullis between 1650 and 1750 (Ford 1983). There are at least three reasons why.

Firstly, Newton's and Laplace's deterministic method of describing the world was available. The elegance of its mathematical implementation and the certainty of its descriptions could easily mesh with the idea of hierarchical order and the emerging notion of progress. Certainly it could mesh more easily with them than could the characteristics of the probabilistic view of the world.

Secondly, until recently, most of the calculations that could be easily performed were deterministic in nature. Those studying the natural world were thus largely unaware of the dominant role that its non-deterministic characteristics played in its functioning (Norretranders 1998). Thirdly, at the level of our day-to-day activities, our common-sense experiences allow us to believe that the world is largely deterministic (simple cause and effect). This personal influence will be discussed in more detail below.

It was thus easy for natural philosophers in the 1600s and 1700s to embrace determinism rather than chance. The idea of randomness as a general way of thinking languished, while the deterministic perspective became slowly incorporated into all branches of investigation from life science to philosophy. Determinism became identified with the practice of science and, along with progress, became a core part of Western culture's world view (Ford 1983). Think of economics at the transition to the 1900s striving to be physics-like in its descriptions.

In the early 1900s, Westerners' belief in determinism was shaken by the theory of relativity, the recognition of quantum effects and Godel's discovery of the constraints on logic. It was also affected by the chaos of WWI. Since then, the invention of the computer has allowed the calculations that represent fractals and aspects of a chaotic system to be routinely completed and studied. Among them is the Lorentz representation of the weather with its butterfly effect. As a result, Westerners have become more intellectually aware that much of the world's functioning displays non-deterministic characteristics and that those characteristics influence our lives (Norretranders 1998).

Today, Westerners might consciously acknowledge that human decisions are made within a complex cultural, ecological and historical framework or system (Kelly 1995). However, by default, we often still think and act as if the world is more like a deterministic system, with a hierarchical structure, than it is. We are still inclined to believe that the world can be described, and its future predicted and managed, by some sophisticated variant of the early 1600s clockwork view of the universe.

This considerable cultural and personal inertia to embracing a more complex view of the world might seem surprising until we recognize how deeply rooted and how much a part of Western culture's world view hierarchy and determinism are. The ongoing strength of that influence is illustrated during a period of cultural upheaval. Both historic and present Western political leaders quickly refer to the importance of maintaining law and order or of avoiding chaos and subversion. Westerners' usual response is, as cued, to call for simple, deterministic solutions to their ills.

The strength of this cultural influence on a Westerner's personal thinking about complexity is illustrated when we are asked to think of something positive. Perhaps an enjoyable movie or a raise in pay? If you are a Westerner, you would also likely add to a list of positive things an increase in the availability of the resources we want, such as trees for lumber. Similarly, you would likely describe a feedback loop that increases the availability of resources, maintains economic growth and raises your pay as positive: the feedback loop economists describe as the virtuous cycle.

Perhaps you didn't transfer the positive *value* you associated with a pay raise and an increase in the supply of resources to the "objective" *description* of a complex system's feedback as positive. If you are unsure then ask yourself the following question: Could I benefit from a decline in the availability of the natural resources I personally consume, a decline that was prompted by a failure of the virtuous cycle's feedback? Take into consideration that Western culture's world view guides us to always *value* the virtuous cycle with its provision of increasing resource supplies as positive. And remember that excessive positive feedback within a system leads to its instability. (Because of these types of biases in meaning, feedbacks are described in this book as dampening rather than negative and reinforcing rather than positive. They are described as self-sustaining if they largely maintain the current state or as amplifying if they substantially heighten the current trend.)

Westerners' deterministic thinking persists for more than cultural reasons. It also persists because our innate characteristics allow us to believe that our world is largely deterministic. For example, our hindsight and adaptation biases combine with the pleasure we get from finding simple patterns in the world to reinforce our preference for finding direct connections between cause and effect. We feel comfortable with determinism. We are ambivalent about the random. And we fear chaos.

We feel uncomfortable with complexity and its associated uncertainty. It takes only a low level of complexity for us to feel intellectually overwhelmed and emotionally uncomfortable. By way of an analogy, consider the unpleasant feelings we can experience when faced with too many consumer choices, such as flavours of ice cream or types of breakfast cereal on the supermarket shelf. To avoid this discomfort, we commonly react by reverting to products we are familiar with (Schwartz 2000).

We react in an equivalent manner when faced with the feeling of "too much" system complexity and uncertainty. For example, when faced with our dawning awareness that the world's characteristics don't fit our deterministic perspective of it, we prefer not to make a substantial change to our views and embrace complexity. We prefer to instead make accommodations to maintain the status quo. For example, we downplay,

ignore or sidestep the issue with statements such as "I just don't know," "Too much for me," "I don't really care," and "Things will turn out fine." A reaction you might experience while reading this book.

Alternatively, as suggested by the supermarket example, we can simplify the issue until we feel comfortable. This can be achieved by focusing on only those aspects that are of interest to us or that we can understand, while largely ignoring the rest of the system (Popper et al. 2005). A specialized version of this response has become a scientific research method called reductionism. Think too of the externalities of economics.

Even if we become personally aware of the innate tendencies in our personal thinking about complexity and the cultural influences on it, their impacts on our thinking is still strong. Sufficiently strong that we experience difficulty compensating for them. As a result, we tend not to include complexity's basic characteristics into our everyday thinking and decisions. Among those characteristics are the inescapable presence and significance in our lives of random change, conditional predictability, fuzzy boundaries, dynamic stability, quasi-stable states, feedback and the wide range of time and space scales involved in change. The notion of complexity remains remote from us.

An overview of our thinking about complexity notes that it is constrained by both our personal characteristics and our culture's world view. The members of complex cultures, in particular, prefer to think about the world as if it were largely deterministic and hierarchical rather than complex. As a result, humans routinely face inconsistencies between their perceptions and predictions of a (human or natural) complex system's functioning and future, and their experiences from being involved in it. No wonder we, as participants in a complex human-environment system, struggle to meaningfully deal with the issues we face.

In summary, the human-sourced constraints on our ability to relate to complex systems contributes to the difficulty we experience making viable and useful predictions about them. They also make it more difficult for us to appropriately evaluate and use our predictions. The human sourced constraints were illustrated in this section by our efforts to describe, deal with and think about complex systems and complexity. An overview of these constraints in action (their interactions) is provided by the way the members of Western culture tend to relate to complexity and complex systems.

Westerners may intellectually see ourselves as participants in the dynamic, complex human-environment system, but in our day-to-day decision-making lives we see ourselves as somewhat remote from the system and its dynamic complexity. Consider our tendency to prefer a simple deterministic description of both the world and our personal experiences of it. We prefer a perspective which focuses on order and stability rather than dynamics, a view that also sees us in control.

The human source of these personal difficulties arises in part from our innate human characteristics and in part from the influence that our culture's world view has on our lives. Consider the considerable support for the above preferences provided by Western culture's world view. In particular, it guides us to see the ideal cultural and environmental structure as a power hierarchy and to seek simple technological cures for complex problems. Western culture has an overall narrow focus/emphasis on the notion of progress and economic growth.

In keeping with our cultural characteristics, Westerners tend to treat the world as inherently describable, predictable and controllable. We thus tend to believe that our predictions about the future of both human-made and natural complex systems are more viable, useful and reliable than they are. We thus tend to imagine that we have a much greater ability to manage and control the functioning of a complex human-environment system than we do.

These beliefs have contributed to Westerners participating in the functioning of natural complex dynamic systems, such as ecosystems, as manipulators and harvesters at a level that is well beyond our prediction and management abilities. The result has been the failure of many ecosystems and the loss of the resources they provide to all of us (Germano 1999). For example, overfishing the cod found on the Grand Banks off Newfoundland's coast not only resulted in a crash in the cod population but also prompted the ecosystem to switch to a low-cod, quasi-stable state (Bundy 2005; Bavington 2005).

The same is true for our creation and participation in human-built complex systems. This comes to light when our complex systems fail. We find it personally and culturally convenient to attribute the failures to an uncontrollable, unexpected, external cause or an act of the incompetent. When Westerners do acknowledge our contribution to a failure, we prefer to treat it as a prompt to improve our techniques rather than to change our views and beliefs. We prefer not to meaningfully acknowledge that behind a material contribution to the failures lies the difficulty we experience appropriately accommodating both complexity's presence and its characteristics into our lives and decision making (Rescher 1998). For example, we prefer not to accept that our personal, culturally supported, progress- and deterministic-favouring view of the dynamic, complex human-environment system contributed to the failure.

This leads to a conclusion about our relating to complex systems that is applicable to more than Westerners and our culture. We are more likely to appropriately deal with the many complex-system-related issues we face, such as making more valid and viable predictions about them, if we change how we relate to these systems in general and the human-environment system in particular. This is easier said than done because of the changes we are required to make.

We need to do more than just acknowledge that dynamic complex systems display a mix of deterministic, random and chaotic characteristics. We need to think of our experiences and the issues we face in terms of its functioning and changing. That functioning and changing is the expression of the system's many variables interacting with one another over a wide range of time and space in a variety of ways that include feedback. The system changes in response to a flexible hierarchy of ongoing, system-wide (internally generated and external) influences on its functioning. In this context, changing is a different view of functioning and the age-old question of whether the chicken or the egg came first is meaningless. They were formed together, over time.

This acts as a reminder that we are also obliged to rise to the challenge of personalizing and assimilating the nature and functioning of complex systems. We are obliged to personally accept that, at some level, we are active participants embedded in the system, not disinterested observers or managers. Our thinking is required to include that there are limits to the order and predictability the system displays and that our predictions about it contain inescapable uncertainty. We must come to terms with the fact that the constraints on our thinking about complexity and complex systems are a reflection on our human perceptual and decision-making foibles, the influence of our culture's world view on our lives and our choices.

Many of the human constraints on us doing so are not correctible (removable) imperfections but part of who we are. However, they can be accommodated in our decision making, as discussed in chapter 18. In essence, to more appropriately deal with our issues through predictions, we need to come to terms with ourselves and personalize the nature of complex systems. We can start by personalizing scientific predictions.

## Personalizing and Assimilating Scientific Predictions

We make three types of predictions: prophecy and divination, common sense, and scientific. We use them for a variety of purposes, including planning for the future, making sense of the external reality and setting limits to our activities. If we wish our predictions to exert the most appropriate influence on our decision making, then we should focus on those predictions that are more valid and useful.

Assume that we have already taken these lessons to heart. There is one more aspect to consider. For a prediction to influence our decisions it must be, or be made to be, personally relevant at some emotional and logical level: it must be personalized. We can then more easily assimilate it into our lives and our decision-making.

Prophecy and common-sense predictions are easy to personalize because, by their nature, they are inherently personalized. Their characteristics enable us to easily assimilate them into our lives. We tend to use them for our day-to-day decision making and do so with limited regard for their validity or usefulness.

In contrast, scientific predictions tend to be difficult to personalize because, by their nature, they tend to be depersonalized. After all, when making and presenting these types of predictions, scientists try to put aside personalizing factors, such as emotions, and personally relevant details. In addition, they include an expression of uncertainty, which goes against our preference for predictions said with certainty and confidence. Furthermore, both the prediction and its uncertainty are often presented in a statistical form, which we find difficult to personalize.

A scientific prediction is also more difficult to personalize and assimilate if it is complex, bland or contains features that contradict our current wants or expectations. Think of the ozone layer's thinning and global warming. We would expect that if the details of a scientific prediction lay beyond our personal experience it would always be difficult to personalize. But this is not the case. If the forecast seems familiar to us or it matches our expectations to some degree (i.e., if

we can easily imagine it), then it can be relatively easy for us to personalize (Norretranders 1998).

Overall, although scientific predictions are an important part of our efforts to deal with our issues, their role is often constrained by the difficulty we can experience personalizing and assimilating them. If we wish to more fully use scientific prediction in our decision making, then we have to devote the time and effort to overcome this difficulty (Ziman 1978). That task is the subject of this section.

We will find it easier to personalize scientific predictions in general if they are demystified. We are helped by considering an earlier hint that the three types of prediction are not as separate from one another as common sense would have us imagine. For example, all predictions are products of human thought, so each prediction can contain elements of all three types. For this reason, scientific predictions can have characteristics in common with prophecy and common-sense predictions.

Imagine a reputable scientist making a prediction about the future benefits and breakthroughs likely to occur in their field of research. If their expert forecast is difficult to ignore because of the scientist's culturally supported reputation, then their prediction can be argued to contain an element of prophecy (Ellis 1973). The case for a scientific prophecy would be enhanced if the prediction was based largely on the scientist's unsubstantiated beliefs or the unconditional extension and heightening of present trends (bleak if the present is seen as bleak, or rosy if the present is seen as rosy) (Heilbron and Bynum 1999). The ubiquitous presence of these types of human judgements and cultural influences blurs the distinction between scientific and prophetic predictions (Ellis 1973).

Similarly, the fuzzy boundary between common sense and scientific predictions is well illustrated by research on predictions made in the complex fields of politics and history. In these fields there is a strong culturally supported demand for highly reliable predictions about the future (i.e., predictions with a high likelihood of being correct). But making them is difficult because of the chaotic and random elements present in politics and history. We therefore rely on experts to use their skill and experience to provide us with the required predictions by applying scientific methodology to the most reliable information available.

Research indicates that experts' predictions about political and historical events are less reliable than either we, or the experts, imagine. The head-shaking conclusion is that the experts' predictions in these fields are often less reliable than crude extrapolations from the present and not much better than the common-sense predictions made by laypersons who are well-informed in the relevant field. A basic reason for the experts' poorer-than-expected showing lies in their use of intuition and common sense to make up for the many deficits in both the available information and their understanding about these complex subjects.

It would be reasonable to expect that in trying to overcome these deficits, the experts' knowledge and experience would at least give them a substantial advantage over lay people. Surprisingly, any benefits that experts derive from their expertise soon reaches the point of rapidly diminishing returns. Their predictions quickly come to reflect the characteristics of common-sense decision and prediction making, such as preconceptions, hindsight bias, anchoring and treating complex systems as deterministic (Tetlock 2006).

Similarly, when experts in history or politics evaluate predictions in these fields, they exhibit the same features that lay people display when they evaluate their common-sense predictions. The influence of features such as hindsight bias, spin and overconfidence are easily seen. For example, when the experts are confronted with a failure in their predictions, they too, like non-experts, try to preserve their existing beliefs and defend their predictions rather than adjust them. We can say, with confidence and high certainty, that the making and evaluation of scientific predictions can display a significant degree of overlap with the features of common-sense predictions.

Now that they are demystified, we can concentrate on how to both find and personalize the more valid and useful scientific predictions. The above research on political and history experts can help. This research discovered that the usefulness and reliability of the predictions made by experts were not uniformly poor. Some experts, which the researcher called foxes, provide relatively more reliable predictions than others, called hedgehogs. The difference between the two groups is attributable to their different ways of thinking. Foxes are more willing to take into account the complexity of politics and history, the limits to their own thinking and the need to adjust their prediction in the light of new information. In general terms, a fox's thinking is more likely to be characterized as broad, flexible and tolerant of conflicting information. In contrast, a hedgehog's thinking is more likely to be characterized as focused, decisive, deterministic and

confident (Tetlock 2006). We are more likely to find the more valid and useful predictions made by experts by paying attention to the manner in which they think. That is, favour the predictions made by foxes.

The fact that an expert in history and politics can improve the reliability and usefulness of their predictions by simply changing how they direct their thinking suggests something else. If we adopt a more fox-like way of directing our thinking, broad and flexible, then we will likely reduce/accommodate our own biases when relating to an expert's predictions. And we will be more likely to find, personalize and assimilate those expert predictions that are more valid and useful.

For example, when evaluating a prediction we are more likely to acknowledge that we intuitively favour a prediction made by an expert who both firmly expresses confidence in their prediction and holds an ideology with which we can agree. We can more easily accept that ideology, especially when strongly held, can limit the alternate futures that a predictor will consider, thus biasing their predictions. We are thus more likely to avoid perceiving the predictions of these experts as being more valid and useful than is warranted. And, when personalizing a prediction, we can more easily come to terms with its difficult-to-personalize aspects, such as those that contradict our personal and our culture's world view.

Broad and flexible thinking (nuanced thinking) can also help us relate to complex systems and complexity. But, in order to practise it, we first have to recognize our personal hedgehog characteristics. We can then strive to develop skeptical, objective and pragmatic thinking, and train our intuition accordingly. We are helped by striving to increase our knowing and adjust our personal world view. These techniques have been mentioned in passing and will be discussed in later chapters, especially in chapter 18.

There are some aspects of personalizing the most valid and useful scientific predictions that the discussion so far hasn't addressed. One of them is the timeframe for our thinking about predictions that deal with the future or past. How far into the past or future can we or should we think about an issue, or a prediction about it (Cohen 1995)?

"How far" is not an arbitrary choice. It is limited by how natural systems and humans function. Biologists and ecologists constantly remind us, and geologists take it as self-evident, that weak but persistent (long-term) forces are just as important in shaping the future as are strong, short-term forces (Parmesan and Yohe 2003). If we want to find and personalize the most valid and useful scientific predictions about the future, then we have to be able to evaluate and relate to prediction timeframes that are at least as far as 30–120 years into the future.

But the timeframe with which we, politicians and neoliberal economists are most comfortable is usually only a few months to a maximum of about five years (Gardner and Stern 1996). We will consider and can relate to longer timeframes, but only if we work at it. One method to become familiar with long timeframes is to routinely work with or read about past events that fall within longer time spans. This is one reason why history is used extensively in this book.

Our efforts to familiarize ourselves with longer time spans is also helped by us recognizing that 20–30 years separates a generation (the time between the parent's birth and the birth of their first child), 70 years is roughly a human lifespan and 100 years is roughly the time that passes between your parent's birth and their child's (your) death (at least in the developed world). This creates a basic personalized 100-year mental framework of time into which we can add marker dates from personally important historical events. They include the multi-generational births, deaths, marriages and emigrations in our family, and the worldly events of personal interest such as humans landing on the moon, Mount St. Helens exploding and Michael Jackson or Prince dying. By using this framework, we can help ourselves to personalize scientific predictions about the future or past.

Another undiscussed aspect of personalizing and assimilating the more valid and useful scientific predictions is human cultures' time-honoured reliance on dramatic stories and myths. The personalization and assimilation of the selected predictions could be helped if they are expressed as, or accompanied by a story (Myers 1999). Consider the 1968 movie *2001: A Space Odyssey*. Its futuristic journey into space and the ordeals the astronauts went through as they dealt with HAL, a human-like computer, neatly expressed the period's hopes and fears raised by predictions of imminent space travel and the power of computers.

But there is a downside. Stories and their telling can result in the trivialization, loss of meaning and distortion of the prediction or its significance. Unfortunately, for a number of reasons, we find it difficult to devote the care needed to take this into account. For example, we are more interested in hearing stories about a future that turns out the way we want or expect, or that entertains us, rather than stories about the way the future might actually unfold. Think too of the dilemma of simplification.

The use of stories is also limited by their generally focused, linear and sequential nature. This makes them easy to personalize but distances them from the nature of complex systems. After all a complex system's "characters" are linked together in multiple ways by many events happening at the same time but at different rates, with a dash of chaos and randomness thrown in.

Standard storytelling can be adapted to tell such complex stories by frequent referrals to the greater context and by providing links to contrasting perspectives within the story. Those techniques are found in this book. For example, when you see apparently loose ends and seemingly extraneous examples, treat them as walk-on reminders of the links between topics. Think acts as a cue: as in, "think of the blue-flowered water hyacinth."

Perhaps we need a new way to tell stories. Perhaps we could learn some lessons from the Ancient Greeks and Hindus. Think of their interrelated myths describing the messy lives of their many gods and the convoluted connections among them.

In summary, this section notes that if we want scientific predictions about a complex system to exert the most appropriate influence on our decision making, then we should focus on finding, personalizing and assimilating the more valid and useful predictions. We are helped in this task by remembering that although scientific predictions have distinctive features, they are not divorced from our other prediction making methods.

Our task of evaluating and personalizing scientific predictions is eased by attempting to think in a more broad and flexible manner that includes considering our human characteristics and using a relevant time scale. This way of thinking provides us with a greater awareness of the interconnectedness and dynamics of a complex system over a meaningful time span. It helps us take into account our decision-making foibles and the influence of culture. It should help us, for example, to more easily accept that our complex, external reality is, especially over the long term, less stable and ordered than we casually think but, especially over the short term, more ordered and stable than we often fear.

Scientific predictions will neither displace common-sense predictions and prophecies nor remove their powerful influence on our decision making. However, the skills learned from dealing with scientific predictions should help us to find, personalize and assimilate the more (context dependent) valid and useful predictions of these other types. And we are more likely to show skepticism toward utopian or ideological predictions and futures, and those that are at odds with the nature of our external reality.

The summary of this chapter starts by noting that if we personalize and assimilate its conclusions, then we should be able to radically change how we make, evaluate and relate to predictions. They will then become the hoped for reliable guides to the future and trustworthy aids to our decision making. However, if we put our faith in this prediction, and our abilities to usher it in, we would have forgotten a key message in this chapter about the nature of complex systems and how we relate to them. We would be ignoring the presence of both uncertainty and our human characteristics. We would thus risk making yet another suite of grand decision-making errors and experiencing the consequences. Think of the late 1990s dot-com bubble (Doward 2001), the Iraq war, the global financial crisis of 2008 and the predictions of who would win the 2016 US election. The results of such grand predictions are helping to darken the growing cloud hanging over our environmental future.

This is the time to remember that being conscious, more intelligent and more aware does not automatically make us masters of our world (Norretranders 1998). Nor does it make us conquerors or designated managers of our complex environment. We are and will remain participants. And despite appearances to the contrary, in the long run, our control over the world will still be limited. Nevertheless, our impacts on it will remain significant (Rescher 1998).

This summary should have started by saying that by personalizing and assimilating this chapter's information, we should more easily make and find more valid and reliable predictions. It should also have said that the resolution to our issues requires us to do more than improve our making and evaluating of predictions. It requires changing our views of ourselves and our relationship with the external reality (Taylor 1989). With this in mind, we should place less emphasis on using predictions as truth-sayers about an imminent utopian future to strive toward, or a hell to avoid. We should use prediction making as another tool to help us and our cultures recognize, accept and then change our lives and world views so that we live more within the constraints of the dynamic, complex human-environment system in which we live. It is we, not better forecasts of rain, that make the decision to wear a raincoat.

All this talk of personal and cultural change raises a critical question: Is it realistic to expect us to change in the suggested ways? In light of this and previous chapters, an unconditional "yes" seems to ooze utopian optimism. Perhaps we should choose another option to deal with the future? We could, for example, focus on finding a fox expert whom we could all trust to lead us to a better future by telling us what to do. Alternatively, we could decide not to bother changing because it has been predicted that we will muddle through okay. But, before you decide which option to choose, you may want to read about how we and our cultures change. That is the subject of the next chapter.

# CHAPTER 6

# PERSONAL AND CULTURAL CHANGE

The dynamic nature of the world we live in ensures that personal and cultural changes are inevitable. However, we have the capacity to decide how we want to respond to or direct at least some of those changes. We are thus left to ponder how much influence we have over personal and cultural change, and when we should exert it.

We could believe that we are innately capable of appropriately directing ourselves to make the needed changes in our lives. That belief works until we experience a disappointing outcome. We might rationalize our disappointments as unfortunate mistakes or someone else's fault, or that we should just try harder.

Alternatively, we might feel that we can't make sense of the world because it is too complex, or we might realize that the external reality can override our conscious inner attitudes or beliefs and force us to change (Myers 1999). We might therefore believe that there is little we can do to directly influence change. This view works until we realize, in retrospect, that we could have done something. We could then blame the outcome on chance (luck) or a supernatural power.

When should our day-to-day decisions direct us to passively drift along with the winds of change, and when should we actively attempt to influence the direction, pace and outcome of change? Making this choice can be difficult. We are stuck with a conundrum.

We are more likely to make an appropriate decision if we take into account our human characteristics and limitations, in particular, our decision-making characteristics (chapter 2) and the nature of our culture (chapter 3). This chapter builds on those chapters to provide a coherent overview of another characteristic we should take into account when making decisions: how we, both as individuals and as a culture, change. The first section discusses personal change and the second cultural change. Each is subdivided into two subsections: the processes of change and the influences on change.

It is convenient to discuss the features of human change by dividing it into these four subdivisions, however we should not forget that these features are all interconnected. They overlap and interact with one another. After all, they are part of a complex system. Unfortunately, we find it difficult to meaningfully think about and apply the descriptions of change in that context. For these reasons, a later section discusses personal and cultural change as a normal aspect of a complex human-environment system's functioning. Indeed, this is the context within which we should view

and make decisions about past, present and future personal and cultural changes (both ours and those of others). The reason to think about change in this way and the challenges of doing so are personalized by a section discussing the appearance of writing in the Middle East.

## Personal Change

We all accept, at least in the abstract, that our behaviours and beliefs do change. However, when we examine our lives in the decision-making instant of the present, we appear to ourselves to be mostly unchanging. Thus, the first step to appreciating personal change is to accept that, at some scale, it is ongoing and inevitable.

Personal change can be voluntary, as guided by our cognitive decision making. It can also occur without our active participation, in our unconscious, as guided by its decision-making rules. It can even be imposed on us. Think of aging. Personal change is a mix of conscious and unconscious contributions, which makes it difficult to discuss.

This difficulty is resolved in a manner similar to how the difficulties of discussing decision making were resolved in chapter 2. The outcome of personal change is accessible to us, so it can be discussed using terms that we can relate to. A selection of these generic or specific descriptors are divided, for convenience, into *processes* of change, *persuasion*, and *influences* on change.

In keeping with the introduction, treating a descriptor as a process does not mean that it is exclusively a process. For example, persuasion can be treated as either a process of change or an influence, depending on the context. Descriptors can also overlap with one another, like persuasion and directing. Despite what the discussion or examples might imply, no judgement of good or bad is intended. Nor is such a judgement intended by terms such as *enable* or *facilitate* and *limit* or *constrain*.

<u>Processes of Personal Change.</u> The processes described here can enable personal change under one circumstance and constrain it under another: context is important. Invariably, more than one process is needed to explain a particular change. They can build on or hinder one another as they interact, so the overall process of personal change is the complex (interacting) sum of all the processes involved, as discussed in a later section.

*Drift.* Personal drift is change that occurs over the medium- to long-term through the slow accumulation of adjustments to our actions, beliefs and behaviours. Drift is an inevitable by-product of aging. For example, the slow changes in our conception of reality as our brain gradually matures from childhood to adulthood and the changes in our preferred leisure activities as our adult years increase (Cole and Cole 2001). Drift also takes place through the inescapable, subtle changes to the material and emotional parts of our memories, such as our tendency to forget the bad. It also occurs through the innumerable, small conscious and unconscious day-to-day adjustments we make to our beliefs about ourselves, about what is right and about how the world works. Over time, the cumulative impact of these adjustments can result in large changes to our behaviour, our view of ourselves and our personal view of the world (Taylor 1989).

Drift may result in harmful as well as helpful changes. Consider the downside of drifting into obesity (Popkin 2002) or sleepwalking into sin, as the life of Hitler's minister for armaments and war production, Albert Speer, illustrates (Sereny 1995). The upside of drifting is that we may end up in a place we call "lucky."

Although drift is an unconscious process that mainly takes place in the background, we can (usually unknowingly) influence the process by our conscious actions. Consider that consciously acting out a behaviour can initiate unconscious changes to our memories so that they better match the behaviour we are acting out. These consciously initiated unconscious changes can help us drift into a new way of thinking or doing (Myers 2001). For example, the act of using the fingers and mouth as if smoking a cigarette can, over time, help a person decide that they would like to smoke. Hence the connection to change by preconditioning and adjustments.

*Insight.* Our personal world view can undergo a sudden and profound change as a result of a revelation or an epiphany. This overwhelmingly powerful life-changing experience can suddenly appear in consciousness from the unconscious. The process is not well understood; however it seems to arises under unusual circumstances.

In some cases, it seems to result from the effort of dealing with a life under stress, such as a deep personal discrepancy in one's life (Miller and C'de Baca 2001). It can also occur during a highly unusual physical circumstance. For example, when astronauts look down at Earth from space, they can have what is described as an "overview experience" (Yaden et al. 2016). Think too of near-death experiences.

The powerful and emotive terms used by a person to describe their insightful experience clearly express how it has transformed their view of themselves, their relationship to others or their relationship to the world. But the details of their transformation are determined by more than the experience itself. Both the person's description of their experience and the practical changes they subsequently make to their lives show that their transformation was also influenced by their prior beliefs and cultural background. Their transformation was influenced by how they interpreted their experience. For example, religious people tend to explain and apply their experience in a religious manner while secular people do not.

Less profound but analogous experiences of sudden insight can also occur. They include the "Oh yeah!" you express when you wake up in the morning with the solution to an annoying problem that had bugged you to distraction the previous day. These types of experiences can also result in lasting personal change, with the nature of the changes, again, depending on how the experiences are interpreted.

*Cognitive Dissonance.* We can also change as a result of the unpleasant feelings that arise with our conscious realization of a contradiction/conflict between our conscious awareness of something and our conscious thoughts about it, called cognitive dissonance. Think of urges versus will. It is one of the two most common methods by which we change. (The other is persuasion, which will be discussed later).

The classic example of change prompted by cognitive dissonance is the breaking of an addiction and subsequent recovery. The addiction can be to anything: drugs, work, material goods, making money, sex, gambling or power. The cognitive dissonance arises because of the difference between what we expect our addiction to deliver to our sense of being and what it actually delivers.

The key to lasting personal relief from an addiction that has been brought to our attention by cognitive dissonance is a self-assessment. In conducting it, we must satisfy four conditions: We must become consciously aware of the negative consequence arising from our behaviour; feel bad about that consequence; be willing to take responsibility for the behaviour causing it; and accept that the unpleasant feelings are the result of our personal behavioural choices (Hock 1999c). By coming to these conclusions, we can successively become aware of our addiction and prompt ourselves to decide to change. A "should change" can lead to a "want to change." We are then more likely to actively seek and implement a direction and a method to actually change.

However, our initial incentive to change is dominated more by the desire to reduce our unpleasant cognitive-dissonance-related feelings as quickly and as easily as possible rather than by the desire to deal with their ultimate cause. Thus our initial efforts to change will most likely result in our making adjustments that allow us to feel better but still leave the essence of our circumstance intact. Our initial efforts to change just mask the real problem.

Fortunately, by persevering and facing the contradiction/conflict, we can eventually make the lifestyle adjustments needed to effectively deal with the roots of our cognitive dissonance (Cheer 1999). Those efforts to change can be helped or hindered by the context within which we face the dissonance. Consider the difficulties faced by alcoholics, moneyaholics or shopaholics who want to change but live in a culture or social group that accepts as normal the routine excessive consumption of alcohol, a disproportionate desire for money or the heightened drive to purchase material goods.

Addiction is not the only source of cognitive dissonance. We can experience and react to it in an equivalent manner as a result of mundane to serious contradictions/conflicts arising from a wide range of experiences, behaviours and beliefs. A good example is the case of Traudl Junge, first mentioned in chapter 2. She was an average person who, although disinterested in Nazi politics, still drifted into Hitler's orbit to become one of his secretaries. She was in Hitler's bunker when he shot himself.

After the war, she became exposed to the wider reality of events in Germany during her childhood under Hitler's rule. In trying to deal with the unambiguous aspects of that reality, she was initially content to accept the idea that she had been too young to have appreciated the implications of the signs and changes taking place around her. But, in subsequent years, she grew more aware of how Nazi policy had affected day-to-day living in Germany. As a result, she came to see some of her earlier personal life experiences in a different light. She came to question the quality of her decision making.

In her later life, she suffered through years of mental and physical distress as she struggled to come to terms with her inability, as a young woman, to appreciate the significance of the signs around her. She questioned how she could have become a willing, though passive, participant in the events during the war. Toward the end of her life, she came to recognize that she had displayed a lack of critical thinking. She finally interpreted and then accepted this lack as an indication of her immaturity and her personality (Junge 2004).

Our reaction to cognitive dissonance when addicted provides insight into why it is hard for us to readily change our attitudes and behaviours under equivalent non-addictive circumstances. Consider that we might recognize that our environment is undergoing logically identifiable, slow, incremental and growing changes that are negatively impacting our lives or will do so in the future. However, we still seem unable to deal with our contribution to these changes. It seems that we find it too difficult to make all of the connections needed to satisfy the equivalent of the four preconditions for change prompted by addiction-related cognitive dissonance.

One reason might be that the separation between our contribution to the cause and the resulting environmental effects on us is just too great. For example, the mental distance between us driving a car and the health consequences of the tailpipe emissions is close enough for us to accept that we should change but not close enough for us to want to change or to actually change our driving habits in a meaningful way. The same could be said for issues like our contribution to drug resistance among disease-causing agents or our pollution of the oceans.

*Directed Change.* Although illusions can help us deal with the vagaries of life, they can't stop the impact that the external reality has on our lives (Taylor 1989). For example, your belief in your invincibility is only realistic until you have a severe accident. Fortunately, it is not necessary to have an accident before you can change your illusionary belief in your invincibility. By deliberately applying our thinking and decision-making skills, we can become consciously aware of our circumstances and then initiate the personal changes needed to appropriately deal with them: we can direct personal change.

For example, when faced with cognitive dissonance, we can decide to exert our conscious effort toward recognizing and dealing with the roots of the underlying contradictions in our lives or beliefs rather than simply masking or sidestepping them. Similarly, we can decide to admit to ourselves that an earlier decision was inappropriate. Both of these choices are powerful learning experiences, ones which direct us to change ourselves in ways that can help us to avoid unnecessary or dangerous illusions in the future (Myers 1999).

This potential to consciously direct personal change is available to humans of all ages and skills. Consider that both scientists and babies use similar methods to change their views about the structure of the world and how it works. They both formulate theories, make and test predictions, seek explanations and revise their knowledge and views in the light of new evidence. Note, too, that they both prefer to revise their theories rather than create new ones, and that the process of change takes time (Gopnik et al. 2001). All that holds babies back from exercising those skills in full consciousness, as scientists do, is babies' limited degree of conscious awareness. Growing up resolves that limitation for many.

It is easy to say that techniques such as systematic, conscious, reflective thinking about the world around us and our place in it can help us gain control over our personal life and direct how it changes. However, the process of reflective thinking has to be learned, and it takes effort to meaningfully apply (Gelter 2003). Consider too that the appropriateness of a conscious decision that arises from our reflective or other thinking also depends on how willing we are to meet the considerable challenges of dealing with our decision-making foibles, as discussed in chapters 2 and 4.

Here is a reminder of some of those challenges. How broadly are we prepared to cast our information-gathering net? Are we prepared to take into account the influence of our culture's world view on our gathering and use of information? To what degree are we willing to recognize, reduce or accommodate the human biases in our selection and use of information? Are we willing to devote the effort needed to pay close attention to the logic of our thoughts, to be open to alternatives and to think skeptically and objectively? Have we come to terms with complexity? Meeting these types of challenges is harder than we expect. Consider too that we struggle to incorporate all of the many connections and influences that are relevant to personal change into our thinking. As a result, the outcome of a conscious change-directing decision may not be what we expected.

In reality, we are less likely to change by engaging in systematic, conscious, reflective thinking than we imagine. We are more likely to change (unconsciously or consciously) in the same manner as babies and young children do exclusively: through personal experiences. This is in keeping with our stronger preference to learn through personal experience rather than through borrowing other people's self-discovered knowledge or through formal study. Yes, we all engage, to some degree, in a conscious effort to change. But we neither use conscious systematic thinking as routinely as we believe, nor do we use it in as systematic a manner as practised by scientists at work.

***Degree of Knowing.*** Information is a key part of our decision making, as discussed in chapter 2. Its use is a process of personal change (as discussed here), while its presence in our lives is an influence on personal change (as discussed below).

Having higher quality (valid, reliable) information about the external reality is an important part of gaining understanding and making appropriate decisions, as discussed in chapters 4 and 5. We assume that simply having that information is (or should be) sufficient to guide personal change in the most appropriate manner. However, there is commonly a marked contradiction between what we know and the way we live. Examples include smoking cigarettes while being aware of (knowing) its connection to lung cancer, and gaining excess weight while knowing its adverse health consequences, such as diabetes (Finger 1994). (When *know* and *knowing* are used in this way they mean that we are sufficiently aware of information to reliably recall it.) This disconnect exists because knowing even higher quality information is insufficient on its own to give it the power to guide our decision making. It hasn't affected our degree of knowing.

What is our *degree of knowing*? It represents more than how well we know something (our ease of recalling it), how much we know (being well informed) or whether we use that knowledge routinely in our lives. Think of a birdwatcher knowing the names of many birds. The term also represents more than whether we have mastered a difficult technique, such as counting or the ability to build a space ship (being smart). Our degree of knowing reflects our use of knowledge to make decisions and live in ways that better match the external reality when it is treated as a functional whole. In this sense, our degree of knowing represents our skill at deciding (if, when and how) to use what little we know to reach for that goal. Or, if you like, the probability of us using what we know in this way. Our degree of knowing is thus a measure of wisdom and wise decision making, a topic mentioned in chapter 4 that will be discussed more fully in chapter 18.

Our efforts to increase our degree of knowing (increase our knowing) is a process of change. It starts when we become aware of a piece of higher quality, useful information about the external reality. The first step is to find in that knowledge, or assign to it, personal meaning or relevance: we need to personalize it. If we accept that personalized information we, in effect, take ownership of it. We merge it with our personal world view and our sense of self to become part of who we are: we have assimilated it.

Personalized and assimilated information increases our awareness of ourselves and our connection to the external reality, thus giving it the potential to guide our decision making and how to live. We have given ourselves the opportunity to increase our knowing. When we use that knowledge to make decisions and live in ways that are more in tune with the external reality as a whole, then we can be sure we have increased our knowing. We will, for example, try to stop smoking.

The process of increasing our knowing is illustrated by the more general version of change through cognitive dissonance mentioned above. A cognitive dissonance experience means that information we are aware of has resulted in an unpleasant personal mental experience: in this way it is personalized. Such an experience does not mean we will automatically decide to resolve the conflict by making the needed analysis or any related changes. We are just as likely to try and ignore it or rationalize it away (Myers 2002a).

For that personalized knowledge to have any effect on our lives, we must assimilate it. For example, we must accept that our routine late-night gaming is causing our exhaustion, or that despite the displeasing nature of a piece of high quality information about the external reality, it requires our further attention. By assimilating this personalized higher quality knowledge, it becomes part of our sense of self, our internal reality.

We have given that information the opportunity to influence our decisions. We are thus more likely to use it to direct ourselves to change in a way that will directly address the source of our cognitive dissonance. For example, we are more likely to game less or live in ways that are more in tune with the external reality. If we do make those changes, then we can be sure we have increased our knowing about ourselves and our circumstance.

Assimilating and acting on knowledge under conditions of cognitive dissonance is understandably difficult. After all, our exposure to uncomfortable truths about ourselves, our beliefs and our denials invariably impinges upon our sense of well-being. Fortunately, not all of our efforts to increase our knowing under the conditions of cognitive dissonance are like the taxing

and convoluted but successful efforts of Traudl Junge. Nor are they like the unsuccessful attempts by Albert Speer, Hitler's architect, and Robert Oppenheimer, the leader of the Manhattan atomic bomb project (Sereny 1995; Hart 2008; Hijiya 2000). More generally, our sense of discomfort as a result of cognitive dissonance is quite mild, more worthy of a deleted expletive.

Neither is a cognitive dissonance experience of any kind a prerequisite to increasing our knowing. However, we should always be prepared to face the potential discomfort of personalizing, assimilating and acting on information. After all, like completing a personal assessment, these processes all require that we attempt to link information, including about ourselves and our circumstance, into a coherent whole. We thus invariably end up facing contradictions in and among parts of our world view, the strength of our faith in some information and aspects of our lifestyle. This discomfort can't be entirely avoided. Think of the $CO_2$ coming out of the tailpipe of your car.

We can also experience discomfort for a reason other than cognitive dissonance. For example, we can find the process of compiling, consolidating and then linking together our disjointed knowledge of various qualities about a topic of interest to be frustrating or even onerous. Fortunately, our discomfort can be eased by practising a more ordered manner of thinking.

For example, we can start with the knowledge about a subject already considered acceptable and then systematically build a wider view outward from it, one piece of checked, higher quality information at a time. Most of us prefer this sequential expansion approach, especially for unfamiliar complex issues, because we can easily stop and restart the process. This makes it easier for us to explore options and deal with discomfort. But there can be many strands to track, which makes it easy to miss things.

An alternative method is to simultaneously hold in our minds a variety of perspectives or attributes of the subject and then combine them into a coherent whole. This method has the advantage of being simple to describe. But, beyond some low level of complexity, it requires considerable practise to execute; an innate ability not all of us have. And the lack of checks on our imagination can result in us having faith in what may be our mis-compiled knowledge. Being smart is no guarantee of making smart or even wise decisions.

In overview, whichever way you look at it, increasing our knowing takes time and effort. Both are required because the personalizing and assimilating of information unavoidably require us to test the quality of our current knowledge and, perhaps, deal with uncomfortable contradictions and inconsistencies among our beliefs, behaviours and circumstance. They are also needed if we are to adjust our perceptions of ourselves and widen our relationship to our world.

The tasks associated with increasing our knowing were treated above as being largely discrete but joined in a linear fashion: personalization, assimilation, then change and application. However, there is considerable overlap and interaction between them. Accepting that increasing our knowing is actually a complex process makes it easier to fulfill the tasks.

Success at increasing our knowing results in both significant personal change and insight into our circumstance. It broadens our conscious awareness of ourselves and the external reality. This is illustrated by a big-game hunter and trapper who lived in the wilderness of British Columbia during the 1940s. Theodora, his wife, shared his wilderness experiences. She described how her husband's years of paying close attention to the lives and habits of animals in the wild allowed him to develop a deep appreciation for their behaviour as individuals (Stanwell-Fletcher 1978). Because he took his increased understanding and personal connections to his surroundings to heart, it led him to change his world view and behaviour. He came to see the needless, perhaps mindless, taking of life as distasteful.

*Preconditioning.* Overall, personal change is a complex, ongoing process. Preconditioning is a useful concept to describe how events unevenly distributed in the past can, collectively, make a particular future change more probable, but not certain. That is, they can facilitate specific future changes but not actually cause them. For example, compare, "I only realized later that my experience on the farm as a child made it easier for me to become a biologist," (preconditioning) with, "It is because of my experience on the farm as a child that I became a farmer" (cause and effect). As the example illustrates, preconditioning events are distant (in time, space or both) from the changes they facilitate, and they are usually only identified in retrospect. Although preconditioning events are distinct, they have characteristics that can overlap with other contributors to change, such as drift, and cause and effect.

Preconditioning events that can facilitate personal change encompass everything from the incidental acquisition of background knowledge, to major alterations to one's life circumstances. We can even precondition ourselves to later accept, build on and expand an idea about ourselves by mentally or physically acting out the idea. An example is miming or imagining (thinking about) how we would like to be in the future. If we become that person, then we can say, in retrospect, that our acting or imagining preconditioned us to become our dreamed of selves, or that we made a self-fulfilling prediction about our future (Myers 1999).

We can also precondition ourselves for change in more subtle ways. For example, the scientists who engage in extensive and sometimes intense discussion about the merits of a new theory while still using the old one are, in a sense, preconditioning themselves to more easily accept the new theory (Kuhn 1996). In this sense, self-preconditioning is part of adapting to an anticipated change.

The presence and importance of preconditioning in our everyday lives is well illustrated by the "foot in the door" phenomenon mentioned in chapter 2. By handing out small, free items (swag), such as pens and bags, in sales rooms and trade shows, the swag donor hopes to precondition you. They hope your small step of accepting the swag will create in you a "momentum of compliance" that will predispose you to take the big step of purchasing their products in the future (Myers 1999). Of course, there is no guarantee you will. Significantly, this example also illustrates that change does not always come voluntarily from the inside. We can be persuaded, even manipulated, from the outside.

## External Inducements: Persuasion. Personal change can be influenced by or result from external inducements, with the outcome often depending on the context. These inducements can come from the environment, other people and our culture's world view. Being persuaded to change by other people is one of the most common ways we change, so this will be used here to illustrate personal change induced from the outside (Hock 1999c). The person persuading is called "the persuader." The audience are those, whether an individual or a group, being persuaded.

The goals of the persuader vary widely. They include changing the audience's world view, biasing their purchasing decisions, securing their co-operation, gaining control, changing their behaviour and influencing their vote. The variety in these goals is matched by the variety in the nature of the persuader's efforts to persuade: marketing, propaganda, bullying, friendly advice and manipulation, among others.

The simplicity of this description might create the impression that the art of persuasion is simple or that a persuader can be successful by simply following a persuading recipe that relies on rules of thumb. This would be misleading. The diversity of methods and goals reflects the reality that the process of persuasion is complex, to the point that the outcome of applying it is difficult to predict, as mentioned in chapter 2.

Consider that, to be successful, a persuader has to pay attention to more than the content of their message. They must consider the subtleties in the audience's reaction to the message. They must also pay attention to how the audience perceives both their delivery of the message and their personal image as they deliver it. However, the audience's reaction can be difficult for the persuader to interpret. For example, the audience's reaction depends, in part, on their interest in the issue, but they may not display it to the persuader. Thus the audience does not always provide the persuader with the reliable persuasion-improving guidance they need. Consider too that the persuader cannot know or control all of the features that influence the impact of their message on the audience. Further complicating the persuader's attention to these details is their own personality type, such as their use of analysis, their tolerance for uncertainty and their need to be true to themselves (Myers 1999).

With these connections between the persuader, the message, its presentation and the audience in mind, each of these factors will be discussed in the following sub-sections. The perspective of the audience will be taken in the discussion, which complements the discussion in chapter 2 about the persuasion of an audience through advertising. Think of yourself as the audience.

**The Persuader.** Persuasion is a skill. Some people are more persuasive than others. In general terms, a persuasive person is more likely to be well-known, attractive and seen as credible to us, the audience. But even skilled persuaders face difficulties in their efforts to persuade us.

Consider their need to establish credibility with us. Not a straightforward task because we have a variety of often contradictory ways of evaluating a persuader's credibility. We tend, for example, to accept a persuader as credible if we perceive the person to be an expert and trustworthy. But are you surprised to learn that we also tend to perceive a person to be an expert if we agree with them? Our evaluation of a persuader's credibility is also influenced by our perception of why we think the person holds their views. For example, are they biased by self-interest or are they reflecting evidence that is largely free of their vested interests? Is the persuader being a cynical manipulator or true to themselves?

The difficulty a persuader has establishing credibility is illustrated by imagining your reaction to two persuaders making pronouncements about human carbon dioxide emissions contributing to global warming. In the first instance, the persuader is a meteorologist studying climate predictions. In the second, the persuader is a business person running an oil or coal company. Then consider how your reaction might change if you lived in a region whose income comes from coal production, or an area that historically did not experience droughts but is now drought-stricken.

To gain a persuasive advantage, a persuader may manipulate their personal attributes, such as their credibility, attractiveness and notoriety. In the case of their credibility, a persuader can enhance our view of them by speaking confidently or by wearing a symbol of authority. For example, the authority embedded in the cultural symbol of a white lab coat or a uniform is used by advertisers to increase the credibility of their sales message and to enhance our reaction to it (Myers 1999). However, success is not guaranteed.

**The Message.** We consider the central part of the persuader's message as its factual content. But the message we, the audience, receive from those facts, and their persuasive power, is strongly affected by the message's other features, such as its context, symbols and presentation. They help to establish a relationship between the factual content of the message and our present beliefs and personality. They also trigger our emotional response to the message. Collectively they help us to personalize and assimilate the message as a whole.

Consider that the balance between reason and emotion in a message helps to determine who in the audience will most likely be persuaded. For example, a more reasoned message is more readily accepted by an educated or analytical audience. On the other hand, reasoned arguments with an emotional appeal are more likely to persuade thoughtful, involved people. In contrast, uninvolved or image-conscious people prefer a message with peripheral cues that can be easily evaluated by rules of thumb. For this audience, it is less important whether the presentation is right or wrong than whether it makes the right impression on us. For example, the phrases "lower taxes," "more freedom" and "greater choice" can be immediately interpreted to mean "good" regardless of the truth of the statements or the likelihood of them being true. Where we each fit in this mix changes with time and circumstance.

A persuader holding a minority position in an audience is more likely to be successful at persuading us if a reasoned, thought-provoking message is used rather than an emotional one (Myers 1999). Surprisingly, the message's reasoned arguments do not need to have much content to be persuasive. An excellent example of the power of minimal content persuasion is the successful attempt by climate change deniers and associated lobby groups to counter arguments in favour of climate warming. They simply raise doubt in the public's mind about the validity of climate change science and global warming. They achieve their success by focusing attention on the "uncertainty" in the science without even clearly defining "scientific uncertainty" and its acceptable levels (Begley 2007; Oreskes and Conway 2010).

Unsurprisingly, the emotions in a persuasive message play a major but more complex role than the facts themselves. In general terms, if the audience associates a positive emotion with a message, then it is more persuasive. Fear can bring about change too, but it can also result in denial. Hence the difficulty of trying to persuade people to change their current choice by directly pointing out its negative consequences. Think of medical practitioners trying to change health-reducing lifestyles or environmentalists pointing out the economic and health consequences of environmentally disruptive behaviours and actions.

The collective experience of those using emotion-grabbing messages to persuade results is a useful rule of thumb. The persuader should express their message's negative content in positive terms by supplying the audience with hope or a means for a solution. This is a lesson well learned by all proselytizing Christian churches with millennialist beliefs: the end is near, but so is salvation (if you take the steps the church provides). Watch for this method in advertising: the yuck, then the solution.

### The Presentation.
The way in which a message is delivered also affects how it will influence the audience. For example, in all presentations, including advertising, repetition is a powerful indirect persuader. Up to a point, the more the message is repeated, the more likely it will become embedded into our unconscious, where it more easily avoids our critical, conscious evaluation processes.

Thus, through repetition, we can come to accept and believe a message, even if it contains false information or gross simplifications of complex issues. Indeed, we can be preconditioned by repetition.

Repetition works especially well for uncritical or passive audiences. In contrast, it is less successful for more thoughtful audiences. In addition, the more we become aware of an issue or the more personal its significance (as our degree of knowing increases), the less likely that we will be passively persuaded to change our opinion (Myers 1999). Ironically, the repetition of information is also a tool to help us increase our knowing. Consider that the repetition in this book has that objective.

It is important to note that the persuader's message and its presentation are not as separate from one another as this discussion suggests. They blur into one another because both the message and the presentation use and manipulate reason and emotion. However, emotion is much easier to manipulate than reason. And, as we have all experienced, in the appropriate context, an emotional message given in an emotional presentation can be especially persuasive.

### The Audience.
Most of us spend our time focused on our day-to-day activities and concerns. If a persuader's attempts at persuading us fall outside of this focus, their message will remain in our background and will barely be noticed. If their efforts do manage to reach into our focus area, the persuader will find that some of us are more easily persuaded than others. A feel for the relationship between our general personality traits, the type of message, and its presentation, that we find persuasive was provided above. Those observations are complemented here by examples of the relationship between specific personality traits and the messages we find the most persuasive.

In general, happy people see the world through rose-tinted glasses. They tend to be more impulsive in their decision making and are more easily persuaded by small gifts. In contrast, unhappy people tend to think more before they act and require stronger arguments to be persuaded. There is also a tendency for people with moderate self-esteem or who are younger in age to be more easily persuaded.

These observations, as with most generalizations, tendencies and isolated facts, should be treated as useful simplifications or rules of thumb rather than natural laws. Consider that analytical people prefer direct forms of persuasion. However, surprisingly, their response to such a message has less to do with their intellectual reaction to its content than it has to do with their emotional reaction to the (analytical) way it is being presented (Myers 1999). No wonder that the audience-reaction rules of thumb are just one part of a persuader's wide-ranging attempts to tailor their message to fit their estimate of the audience's likely response.

For example, a dermatologist trying to preserve the future state of his patient's skin by having them avoid the sun observed that, "It seems as if very few people listen if I say, 'You're going to get skin cancer if you don't change your sunbathing habits.' But if I tell them, 'Sunbathing will make you look like a wrinkled old prune,' they'll listen to me almost every time" (Bark 1995, 178-9). Think, too, of consumer profiling.

## Influences on Personal Change.

There are many features in our lives that can influence how and why we personally change. These influences can be personal, cultural or environmental in origin. Influences of a personal origin include our conscious human perceptions and our decision-making foibles. Influences can be simple or complex, a directing driver or a passive presence. They can take the form of limitations, limits and boundaries, as well as inducements and encouragements.

Just as more than one process of change is usually responsible for personal change, more than one influence is invariably active during change. In addition, they can facilitate personal change under one circumstance and constrain it under another, so the context in which they act is important. They also build on or interact with one another. Thus the overall impact that these influences have on change can be thought of as their complex sum.

We are often unaware of these influences. When we are aware, our relationship with them is complicated. Consider that our difficulty recognizing and accepting the role an influence plays in our changing also affects our changing. For example, we tend to think that we are personally exempt from many of the limitations to personal change, until we find ourselves bumping into them. Think of a failed New Year's resolution, or discovering we are not immune to advertising's influence.

This variety makes it difficult to discuss the overall role that these influences play in personal change. Fortunately, many kinds of influences have common features, so they can be grouped. A selection of these groupings are discussed here to illustrate, in broad terms, the nature and role of the influences on personal change. Note that some of the influences discussed can also be processes of change, such as emotion and persuasion.

*References and Context.* Our conscious and unconscious decision-making processes make use of information by comparing it to references and paying attention to the context in which it is provided. This subsection discusses how references and context influence our decision making and thus personal change. It builds on the discussion about our personal suite of references (our reference normal) in chapter 2.

We rely on our reference normal (including its references made on the fly) and our culture's world view to help us recognize the need to make a personal change, contemplate a proposed change and evaluate an ongoing change. The context in which we use these references, for example in a social group versus alone, helps to determine the influence they will have, such as facilitating or constraining a particular personal change.

For example, a young child learns about themselves and our world by exploring their surroundings. The feelings and things they discover prompt (cue and guide) the child to add their discoveries to their mental representation of the world or work at adjusting it so that it can accommodate them. However, both a youngster's experiences and the changes they will make to their world view are strongly influenced by the context in which they grow up. This is usually a secure environment provided by their adult caregivers and dominated by adult references. Our process of reference- and context-constrained learning by experience continues when we are an older child and young adult, but in a different form. In essence, our preference is to alter the world immediately around us and then learn from the impact that its altered state has on us (Gopnik et al. 2001).

By the time we are adults, the references imposed on us by our caregivers and their culture's world view have become part of our personal adult world view and our adult reference normal. However, we have adjusted them. After all, as we grew up we used what we experienced, learned, rejected and created to modify and diversify our references as we saw fit.

As adults we apply these references in a different and much wider social and environmental context than we did as children. Our social context is now dominated by the people we are in immediate contact with: those in our family, our social subgroups and the more formal subgroups we belong to. This is the adult context in which our references influence personal change.

Consider that our desire to remain part of a social group still ensures that we display both a reluctance to change in ways that are not supported by the group's world view and a preference for changing in ways favoured by it (Myers 1999). For example, we find it much harder to change our views about a person or group of people once they have been culturally labelled or stereotyped by our group (Hock 1999d). Consider too how inconsistencies among our references or between a reference and the context within which it is being used can affect our decisions about change. For example, Westerners who are contemplating a reduction in their material consumption to meet the cultural goal (a reference) of being green also have to deal with their culture's guidance (a reference) that supports or promotes the gratuitous consumption and discarding of material goods. We are then stuck with the task of scrutinizing context or selecting between these conflicting references.

*Information.* Our conscious efforts to contemplate and guide personal change are influenced by many factors. One of them is information. Its influence on our decision making, beyond the details of its "content," depends on our skill at dealing with its vagaries. Consider its availability. At one extreme, if we are unaware of information, we can't use it in our decisions. At the other extreme, we can be overwhelmed with information. In between lies the information that we use.

Consider too that a piece of information we use is neither neutral nor isolated. It is generated and transmitted in an interconnected, value-laden personal and cultural context, as discussed in chapters 1, 2, 4 and 5. That context, and the feelings and emotions we automatically attach to the information, help to determine our relationship with it and thus its influence on our decision making.

We use our skill at dealing with information's vagaries to judge its quality and usefulness, and to personalize and assimilate it (i.e., our relationship with it). Information's quality is the product of combining estimates of its validity (credibility) and reliability (likelihood of being correct), while usefulness is a measure of its suitability and relevance to the decision, as discussed in chapter 5. Our skill at dealing with the vagaries of information merge into the skill we display in using its content.

Previous examples have illustrated that, when making decisions, we tend to give only cursory conscious consideration to the vagaries of information and the methods, such as quality and usefulness, of dealing with them. Consider that we neglect that our memories of information can be less reliable than we imagine. After all, we (unconsciously) routinely both change existing memories and create false ones (Loftus 1997). We also tend to forget that feelings and emotions are either information or attached to information. Consider our bias in favour of information we consider to be positive, and that we are strongly drawn to titillating rumours and urban myths (Myers 1999).

As a result, we can easily place too much trust in information that is conditional, incomplete, false or misleading. It can become part of our decision making and thus influence personal change. The issues of false and misleading information (misinformation) were discussed in chapters 4 and 5. That discussion will continue here. Consider the following two examples.

When using Google to search for information, we tend to look at only the first few search results. The lack of a disclaimer on the search page allows us to assume, by default, that the order of the search results is a reasonable representation of their quality and usefulness. However, the order is determined solely by an algorithm. When it evaluates a website for inclusion and a rank in the list of search results, it considers many (originally 200) of the site's features. They include the number and structure of links to the website, how close the search terms are together on the website, how recent the information is, the number of words on the site and text size (Google 2014; Brin and Page 1998). Quality and usefulness are not or only weakly represented in the algorithm.

The nature of this algorithm evaluation method also provides many opportunities for errors, omissions and biases to appear in the search results. For example, the importance of links can mean that opinion-giving websites (think blogs and analysts) discussing a topic of interest may all appear on the search list, even near the top, while the website hosting the primary source of the data being used may not even be on the list. Or if it is, the original data may be realistically inaccessible to the public. Think of peer-reviewed scientific publications. Consider too that the features of a website used by the algorithm to evaluate it can be, and are, manipulated by those wanting that website to be present in and near the top of the list of search results (Lazuly 2003). The algorithm can also be adjusted to favour some websites over others. Think profiling and favouring Google.

Thus, although Google's sophisticated searching can provide us with a listing of websites that meet our search criteria, it is much less likely that their presence or rank in the listing will reflect either their quality and usefulness, or even their accessibility. How does one choose which website to believe and how much faith to put in it? By us neglecting these limitations when using search results, we can give misinformation credibility and value. Think fake news.

The second example deals with the Nobel prizes. There are only five Nobel prizes given out by the executors of Mr. Nobel's will (the Nobel Foundation) in his name: chemistry, physics, medicine, literature and peace (Nobel 1895). Economists, among others, also refer to a Nobel prize for economics. However, this prize was endowed by the Bank of Sweden to celebrate its 300th anniversary. They call it the Riksbank Prize in Economic Sciences. In contrast, the Nobel Foundation refer to it as the Sveriges Riksbank Prize in Economic Sciences in Memory of Alfred Nobel (Sveriges Riksbank 2014; Nobel Foundation 2014). We are misled by the phrase "Nobel prize in economics" to give the economists receiving the bank's award the same prestige and humanitarian intent we symbolically associate with Mr. Nobel's five prizes. This association is hotly contested by a number of scientists (MacLean 2017; Henderson 2005).

But misinformation is not only imposed on us from the outside, we create it ourselves. Of interest here is how information we receive from the outside can undergo a strange form of manipulation in our heads to become misinformation. For example, we have a tendency to uncritically select and remember information that appears to us to be positive or that sustains a belief we hold (the confirmation bias) (Myers 2002a). When we use this biased information to guide how we change, then we are relying on self-generated misinformation.

We spread the misinformation we have received or created when we casually pass it along as truth. Think of rumour, urban myth, propaganda and fake news. In doing so, we give misinformation power. We also give misinformation credibility and power when we use it to support or justify our decisions to others, in particular the members of our social groups.

Western culture's faith in progress leads its members to expect that misinformation currently circulating would quickly and inevitably disappear. This is not necessarily the case. It can enter a self-sustaining misinformation loop. For example, some types of popular writing, such as self-help and personal improvement tomes, websites and blogs, are easily accessible, have authority and can, like pronouncements by those labelled as experts, have a strong influence on their readers. The popularity of these works can be increased by the author telling the reader the things they want to hear and by presenting the supporting "information" in a form they can easily and uncritically accept. The stage is then set for transforming a piece of misinformation into a widely accepted "truth" (Francis 1992). For example, the ideas that extra doses of vitamin C can relieve or cure the common cold and that autism can be caused by vaccinations still circulate. The false option of quick weight loss is an industry on its own. This practice of creating misinformation is so common today that it has been said that social media has brought us into a post-truth age.

Many of the limitations and undesirable influences associated with using information as an agent for personal change could be reduced by the user spending more time checking its quality and usefulness, and by using its content more carefully. There are techniques that can help us do so, as discussed above and in earlier chapters. For example, we are helped by thinking about information's content in a skeptical, objective and nuanced (flexible and considered) manner. We are also helped when we accept the characteristics of our innate decision-making processes and take them into account. These types of verification and care-exercising techniques can help us find and use more appropriate information and sources of information. Unfortunately, we can find them difficult to implement, so we tend to think of them as onerous tasks rather than aids.

Fortunately, there are some useful rules of thumb that can help us apply them. We can carefully check the quality of those pieces of information about an issue that we already know. Skepticism and objectivity are keys to evaluating that information, as discussed in chapter 4. We can then focus on checking the quality and usefulness of new information that is acquired from a variety of sources, as well as its coherence and consistency with what we already know. We are helped in this task by remembering that our familiarity with what we know predisposes (preconditions?) us to be aware of and accept new information that is an extension of what we already know (Myers 2002a). Skepticism and checking contrarian sources is in order. We are also predisposed to favour "cure all" information. By routinely applying these rules, we are more likely to rely on more reliable sources. We will likely increase the quality and usefulness of a growing pool of coherent information influencing personal change.

When considering information in the context of personal change, the surprising last word is that it is possible to place too much emphasis on its factual quality, usefulness and quantity. This constraint arises because, as mentioned above, factual information on its own has limited power to prompt us to change (Forrester 1989). It seems that information that simply supplies us with objective, verifiable facts about an experience exerts far less influence on personal change than when it helps us give meaning to an experience. For example, when new information places us or what we already know in a larger context to which we can relate, then it can help us to discover meaning and resolve contradictions in our experiences. Think of epiphanies as an extreme example.

If we are to take this limitation into account in our decision making, then we are obliged to pay attention to our relationship with information. In particular, we must acknowledge the important role that feelings and emotions play in our personalization, assimilation and use of information. Consider how our strong desire to avoid the feeling of being either wrong, fooled by misinformation, manipulated by those who bring it to our attention, or seen as a sucker can all bias our relationship to information. These desires can contribute to us unnecessarily displaying avoidance tendencies, such as distrusting unfamiliar but useful information. Similarly, we can believe in or hang on to information long after it has been proved to be false (Slovic 1993). These emotional connections to information can have a particularly strong influence on our unconscious, rapid, intuitive evaluation and use of information, as discussed in chapter 2. Think too of how we deal with cognitive dissonance.

Overall, information and its vagaries has a significant influence on personal change. Information that is higher quality, useful, and easily personalized and assimilated helps us to make meaningful sense of our experience. This is especially the case when we use the information in a nuanced manner to increase our degree of knowing. Thus, how much you know is often less important than how you relate to and use what little you know (especially if it is high quality).

**Feelings and Emotions.** Feelings, such as desire and tiredness, and emotions, such as joy and fear, are inevitably attached to all aspects of our lives. They include our memories, beliefs, values, perceptions and, as mentioned above, what we treat as information and our relationship to it. Whether positive or negative,

these feelings and emotions have a pervasive and profound influence on our decisions and thus on both (conscious and unconscious) personal change. Their hidden power is illustrated by the greater likelihood of us adopting responsible environmental behaviour if we have had a positive emotional experience of nature than if we are in possession of isolated factual knowledge about nature (Finger 1994).

But it is not just the presence of a feeling or an emotion that influences personal change, it is also their strength. Emotions are, by definition, stronger than feelings, so they have a greater influence. For example, emotions have a more powerful influence on how tightly we link together our standard of living, the material features of our lifestyle reflecting it, and our culturally influenced personal values and beliefs supporting that lifestyle. The type and strength of our emotions thus influence our ability to change the material and mental features of our lifestyle. Are you emotionally satisfied by having a lot or by wanting a little? How strong are those emotions?

We usually display a strong positive emotional attachment to our core beliefs and values, and their material or symbolic representations. Thus, when an event occurs or information becomes available that appears to require us to change those aspects, we tend to display a very strong negative emotional response. This reaction both limits our ability to effectively evaluate the information and reduces the chances that we will respond by deciding to personally change.

This can be seen in our strong emotional responses to scientific hypotheses that contradict our personal values or cultural norms. Examples are the debates generated by Darwin's *The Origin of Species* (1859) and the Club of Rome's 1972 report, *The Limits to Growth* (concerning a natural-resource-limited future) (Meadows 2007). More recently, there are the ongoing strong emotional reactions by many to the evidence of human-caused climate warming (Oreskes and Conway 2010).

Overall, our feeling/emotional self exerts a wide-ranging influence on our desire for personal change and our efforts to implement it, especially if what is to change is central to our sense of self. The influence that our emotional self exerts on personal change complements the powerful influence our psychological needs, such as the need to belong, exert on change. Their combined influence can be sufficiently strong that we can, for example, deny the existence of information and events that we perceive as unpleasant or undesired, or we can believe too strongly in misinformation that we perceive as positive or desirable. However, in a different context, their influence can also act as a spur to undertake a change.

*Inertia.* Inertia is our general resistance to change, regardless of the source. The presence of inertia in our personal lives is illustrated by our preference for routines, a familiar environment, the status quo and sticking with an initial decision (Myers 1999). Indeed, we can be described as creatures of habit who, as adults, tend toward the lifestyle, beliefs and values we experienced when we were being socialized as children. Thus, although we can say that we want the world to be dynamic and new, in reality we don't really want many parts of our personal world to change that much. Similarly, we may say that we want to change our ways, but once we have developed a rationale to explain a behaviour or belief, we will, by default, maintain the status quo. The presence of inertia is one reason why our past behaviour is the better predictor of our future behaviour than our stated intent to change (Myers 2002a).

Our inertia to undertaking personal change is surprisingly powerful. Its power is illustrated by the experiences of those Germans who left Germany in the late 1930s, as described in chapter 2. Although they recognized that their livelihood, perhaps even their lives, were under threat, they still experienced considerable mental anguish at the thought of leaving (changing the status quo). So they delayed leaving for as long as possible (Allport et al. 1941).

We can become personally aware of the presence and power of our inertia by paying attention to our reactions when we become aware of factual evidence that contradicts a belief we hold strongly (i.e., held with strong emotions). We commonly respond with disbelief, even denial. Thus, even though the disconfirming information may be powerful enough to logically and irrefutably demolish our belief, our inertia to change provided by our belief system's defences can result in the belief persisting for some time.

Under certain circumstances, the thought of belief demolition may even cause us to strengthen our faith in our outdated conviction. An example is the reaction of cult members, and some experts in the political or economic spheres, who strongly believe in a prediction that includes a date or criteria for fulfillment. When sufficient time has passed and the prediction remains unfulfilled, the believer's faith in the prediction should come to an end. This may happen, but often it is easier for them to find excuses and adjustments that accommodate the failure and thus sustain their belief in the prediction (Festinger et al. 1956; Tetlock 2006). The fallout from the global financial crisis of 2008 provides some economic examples.

In the light of our inertia to change, it is unsurprising that neither our intelligence, knowledge, morals (Ophuls and Boyan 1992; Damon 1999; Finger 1994) nor even a threat to our personal safety (Allport et al. 1941) can be sufficient to trigger an appropriate change in our beliefs or behaviours. If you think you wouldn't make this type of inertia-related mistake, consider the following example. In the early 2000s, smart, well-off young women who knew that South Africa had a very high incidence of AIDS and who were knowledgeable about HIV/AIDS still repeatedly had unprotected, casual sex. How could this happen?

There were many reasons, all leading to the same decision. Some believed that the risk did not apply to them, even though it was well known that HIV in South Africa was increasing at the same rate among both the educated and uneducated. Some of the young women were fatalistic. Others had more immediate concerns such as self-esteem issues and the need to express their freedom. Most of them saw their decisions and behaviour as rational, so they had no reason to change their lifestyle. The inertia these young women displayed to change was more powerful than their ability to recognize and accept the reality of their situation (Twiggs 2001; Taitz 2002; Eaton et al 2003).

The essence of our inertia to change lies in our unconscious and is innate. We are stuck with it. Fortunately, it provides us with the benefit of stability in our lives and, with effort, we can accommodate its downside. It is indeed possible for us to accept uncomfortable-making information and make appropriate changes to our beliefs and behaviours (Myers 1999; Myers 2002a). We must just devote the time and effort to make sure that, step by step, the processes of change, such as preconditioning, directed changes, drift and increasing our knowing, take us there.

*Time.* Conscious personal change requires that you recognize the need for change, decide to change and then implement it. Completing these steps takes time. Imagine receiving a new piece of information or advice suggesting that personal change is in order. It takes repeated views or repeated consideration of that piece of information for us to build the mental trust or confidence needed to accept it (i.e., to personalize and assimilate it) (Bella et al. 1988; Edwards 1982).

Similarly, deciding to change and then implementing change both take time. After all, each step requires the brain to complete a complex task that may also involve making adjustments to the distributed but interconnected mental web that is our internal reality. Consider the number of adjustments needed to change just a single memory. Not only must the target memory itself be changed but so must all of the links to the peripheral memories. Even the peripheral memories themselves might have to be changed (Kahneman and Tversky 1982b).

Unsurprisingly, the amount of time it takes to (consciously or unconsciously) adjust our views, beliefs and behaviours varies with the proposed changes. Some aspects of ourselves can change quickly, such as our attitudes to people and things (Myers 1999). Think of our trust in a person and our choice of a favourite ice cream. Complicated changes and changes to the behaviours and beliefs we hold dear occur more slowly. Particularly slow to change, if they change at all, are our beliefs and activities that have been relegated to routines deeply buried in our unconscious, such as our core beliefs, our personality traits and habits.

As the effort to change something as simple as a habit illustrates, the decision to change these buried routines is often the easiest step. We quickly discover that successfully breaking a habit requires dedicated effort for a longer period of time than we imagined. For example, correcting a swimming stroke requires between one and three months of concerted effort and focused attention, with no guarantee that we won't backslide.

There is another time-related influence on personal change. The amount of personal change we can undergo and the speed with which it can take place is limited, if we are to remain functional. In general, substantial personal change must occur slowly, through an adjustment or a compensation process.

However, there is some flexibility in these limits to personal change. For example, preconditioning and being young makes it easier for us to undergo some changes rather than others. Similarly, how we change affects both the degree and the speed by which we can change. For example, undertaking rapid change though frank and open discussion of any kind can result in significant distress because the amount of raw reality, especially personal reality, that we can handle at one time is limited. In contrast, substantial change through an epiphany that we see as positive can occur relatively quickly.

The considerable amount of time and effort it takes to make some personal changes could be taken as a cue

to search for shortcuts. This is not the intended conclusion. Consider that for an addict to make the personal changes need to effectively deal with their addiction they must face the causes of their cognitive dissonance. In this context, their quick-fix solutions, such as promising to quit next month, can be seen as unnecessarily delaying the needed personal change. However, to make the needed lasting changes requires that they increase their knowing, which is a slow and taxing, if not upsetting, process. In this context, the unfulfilled promise to quit is a necessary step on the journey of personal change. Thus, beyond some point, trying to speed up or skip steps in that process is counterproductive.

However, it is possible to facilitate change, and make it less stressful and frustrating. We can accept that change will take much more time and effort than we intuitively expect. By making a realistic assessment of the time needed, we will further improve the chances of successfully changing. The estimate could, for example, prompt us to divide the task into more easily achieved goals.

There is a cultural connection to the influence that time has on personal change. A culture whose world view is devoted to constant change, requires too rapid a change or restricts change too strongly can stress its members. This affects both their ability to change and how they change. That, in turn, impacts the culture's functioning and how it changes. A full appreciation of personal change requires acknowledging cultural change, which is the subject of the next section.

*Functional Boundary.* Our human characteristics, including being social animals who rely on a supporting environment, constrain our functioning. These constraints can be thought of as collectively forming a flexible multi-dimensional boundary around us: a functional boundary. If a person wishes to remain functional, then they can only change in ways that stay within that boundary. If we are pushed outside that functional boundary for long enough, or make changes that place ourselves there, then we become dysfunctional. Think of the consequences from adverse childhood experiences.

The discussions above and in earlier chapters have mentioned specific contributions to the boundary. In overview, the physical contributions include the powerful and relatively inflexible biological limits, such as the need to eat and sleep. Other contributions are more flexible, such as the degree of crowding people can experience before feeling psychologically distressed. Some are dominantly psychological, such as the need to feel wanted and have a sense of belonging in a group. Think too of the amount and speed of personal change we can sustain and remain functional.

The presence of this boundary exerts its influence on personal change in many conscious and unconscious ways. The loss of functionality from crossing it can vary from minor to severe and short term to chronic. It can be expressed as a feeling of excessive stress or by displaying any number of psychological disorders, such as attachment disorders. Both the presence of the boundary and the reaction to crossing it influence and limit personal change.

In summary, personal change was discussed here by dividing it into different processes and influences. They can each facilitate and/or constrain change, depending on the context. More realistically, the overall process of personal change occurs by all of the relevant processes and influences interacting with one another over time in a manner that includes feedback. Personal change is thus a complex event that occurs within the context of a social group, its world view and the environment.

Lasting personal change comes slowly, within limits and not necessarily in the rational, conscious, planned way we imagine. This book *may* persuade you to adjust the manner in which you think about yourself, the world around you and your relationship to it. Whether it does has just as much to do with the time and effort you devote to changing as it has to do with the quality of the writing and arguments.

## Cultural Change

A culture is represented by its world view, which is the complex sum of the personal world views of its members. Over time, as the members' personal world views change, so does their culture's world view. Cultural change was broadly discussed in chapter 3. However, discussions in later chapters about possible future changes to the cultures of today will be better served by having an understanding of cultural change as an event in its own right, like personal change. That is the purpose of this section, which builds on chapter 3.

It is easier to gain an appreciation for cultural change if the dynamic complex process is discussed by dividing it into two broad categories: processes of cultural change and influences on change. As with the section on personal change, despite what some of the examples in the discussion might imply, no judgement of good or bad is intended. Nor do such terms as *enable* or *facilitate* and *limit* or *constrain* imply a judgement of appropriateness.

**Processes of Cultural Change.** The complexity of the overall process of cultural change makes it difficult to discuss. Following the method used for personal change, the overall process for cultural change is illustrated here by a selection of the descriptors/ "processes" (processes) of cultural change to which we can more easily relate. These processes are not exclusive. Some can be treated as either a process or an influence, or both, and the processes can overlap, for example adjustment and inertia.

Like personal change, each process can enable cultural change under one circumstance and constrain it under another. Thus the expression of a particular process and its importance to change is affected by the context in which it occurs. In addition, more than one of the selected processes is invariably present during cultural change and they build on or interact with one another. The overall process of cultural change is the complex sum of these types of interacting processes, as discussed in a later section.

*Drift.* Cultural drift, like personal drift, refers to the slow, long-term undirected (purpose-free) process of cultural change. It occurs as a result of an inevitable, slow, ongoing change or the accumulation of a diversity of small changes in either the culture's functioning or its supporting environment, or both. For example, a culture's older generation is always being slowly replaced by a younger one.

Consider too the slow "distortion" of values that inevitably occurs when one generation transmits its culture's world view to the next generation (Barzun 2000). For example, parents exert a strong influence on the beliefs and behaviours their children will adopt as they grow up (Kelly 1995). But, as those children become adults, they will adjust their beliefs and behaviours to suit themselves and their circumstance. This will affect the influence that they, as future parents, will have on their children.

Drift is thus driven in part by the members' more mundane day-to-day decisions, the ones they make in response to the myriad of small influences upon their daily lives (Ames and Maschner 2000). Should I buy a new cell phone or not? Should I watch a movie online or in the theatre? Even small changes in material circumstances can, collectively, eventually lead to significant changes in a culture's world view (Inglehart 1990).

The presence and significance of cultural drift is illustrated by a comparison. During the late Middle Ages (ca. 1200–1400), Western Europeans saw their place and role in their culture as largely fixed and preordained from birth. This contrasts with the view held by their descendants, the members of current Western European cultures. We, as a matter of course, participate in our culture as autonomous or independent individuals.

In the long view, individualism is a radical change in the European culture when compared to its world view in the late Middle Ages. It appeared slowly over a few hundred years during the gradual transformation of European culture that nominally began in the early 1600s, called the Enlightenment (Eisenstadt 2000). Because the change took place through a long, slow process of drift, each generation was largely unaware that it was taking place.

Cultural drift is still occurring, as illustrated by the following two examples. As a result of mid- to late 1900s changes to European living conditions, there has been a steady, one-generation drift away from the older European cultural view that money enables individuals to achieve happiness. The drift has been toward the view that social and leisure activities fulfill that role (Inglehart 1990).

Secondly, in urban Uganda, a slow change away from a large, expensive family funeral to a less expensive, faster, Western-style funeral in a funeral parlour is underway. This change seems to be driven, in part, by the steady rise in urbanization and, in part, by the financial fallout from the ongoing AIDS epidemic (Wax 2006).

*Preconditioned Change.* Like personal preconditioning, cultural preconditioning describes how distantly related events distributed unevenly through time and space can collectively increase the probability of a specific cultural change occurring in the future. It can facilitate, but not ensure, the occurrence of that cultural change. The connection between a specific cultural change and the preconditioning events that facilitated it is usually only identified in retrospect.

This Trojan Horse manner in which cultural preconditioning and its non-causal outcome occurs is illustrated by a connection between hunter-gatherer cultures and our more complex urban cultures. Hunter-gatherers acquired the basic skills for building a "house"

at least as far back as 40,000 years before present (yBP). However, it took until around 20,000–15,000 yBP before these skills were sufficiently refined for a few Middle Eastern hunter-gatherer cultures to construct a house that we can readily identify with. After about 10,000 yBP (8000 BC), a version of that house type was used in the first permanent village settlements, such as Abu Hureyra.

Knowing how to build houses did not cause people to live in urban environments. However, it did provide them with an opportunity to form villages. There they could adjust their social skills, and thus their culture's world view, in ways that allowed a relatively large number of people to successfully live close together in harmony over the long term. The ability to build a house and the acquisition of specific social skills combined to precondition, but not cause, the members of a few post-8,000 yBP (6000 BC) cultures to create and more easily live in urban environments. The opportunity or potential to do so was present but the outcome was not preordained. Preconditioning was part of the slow process of village-supporting cultures becoming state-supporting ones (Wilson 1988).

As the example suggests, preconditioning affects the ephemeral as well as the material aspects of a culture's world view. For example, in Europe during the Black Death of the mid-1300s, some 20 million people died (roughly 30% of Europe's population). Many more died from subsequent plagues and unrelated famines. These events helped to heighten the members' religious feelings and deepen their dissatisfaction with the established Catholic Church. Those trends were expressed simultaneously through a popular desire for a more personal relationship with God and a greater interest in mysticism. Together, these ephemeral changes preconditioned European culture to more easily experience, from the 1500s on, the Reformation and the rise of Protestantism (Kelly 2005).

Examples of cultural preconditioning, like those of personal preconditioning, can be used to suggest that an element of predictability, cause and perhaps even a sense of purpose exists in past events. But these conclusions can be misleading or exaggerated because our retrospective efforts to identify the role and significance of historical events suffers from the hindsight bias. We therefore find it hard to reliably distinguish between events that precondition and those that are direct causes. For example, European infectious diseases were introduced into the Americas soon after ships started arriving there. The role that these diseases played in the success of the later 1500s and 1600s

European invasions and changes to indigenous American cultures was only recognized through the eyes of history (Diamond 1997). How much of the significance ascribed to these diseases by historians is a result of hindsight bias? How much was cause, and how much was preconditioning? Perhaps all we can say for sure is that diseases likely contributed to or facilitated change in the Americas.

One of the advantages of considering preconditioning as a contribution to cultural change is that it reminds us to keep our thoughts about the origins of cultural change both skeptical and broad. The need to be cautious about drawing simple conclusions about the significance of past events on subsequent cultural changes is reinforced by an experiment. Try to definitively identify which of today's events will be seen as either preconditioning or causal events in the future. (You may wish to read chapter 5 on predictions before answering.)

*Adjustments.* A culture's world view provides guidance to its members about how the world functions, how to live in it and an explanation for why. By the members collectively identifying with their culture's world view and following its guidance, they stabilize the functioning of their group. This stability makes it easier for them to live their day-to-day lives.

However, as they make their day-to-day decisions to deal with everyday affairs and a changing world, they, like their ancestors before them, inevitably end up changing their culture's world view. This leads to a conflict. The members change their culture's world view as they respond to varying circumstances and desires, but they also resist cultural change in order to reap the rewards of the stability provided by an unchanging culture.

This duality is illustrated by Westerners. One Westerner can embrace their culture's flexibility. They can believe that change (progress) will maintain, even improve, our culture's ability to provide us with the stability and security we expect. At the same time, they can find it difficult to accept that an alternative world view may provide us with more appropriate guidance about a way to live. They tend to ignore the evidence from other (past and present) cultures that our world view's current guidance to favour progress and progress-enhancing changes is neither the only nor necessarily the best option to ensure our long-term stability and security.

Another Westerner can dislike cultural change because they believe that moving away from our existing norms, often expressed as "Western values," will disrupt our stability and security. However, at the same time they can be proud of a history in which progress has given us better lives. They may even accept or be proud that the most sacred of our past cultural traditions, such as being ruled by royalty, are no longer essential to our culture's functioning because we have progressed to better traditions, such as individualism and democracy (Inglehart 1990; Reader 1990).

This conundrum of cultural change is generally resolved, in aggregate, by the members accepting cultural change but preferring that it only occurs in restricted ways, and that it does so slowly. Cultural change is acceptable if it is made through adjustments that remain close to the core of their culture's current world view. This leaves the culture with an ever-present limitation. Although it has the capacity to change in order to appropriately address their issues and their future, there is no guarantee that the members will ensure that it does (Eisenstadt 2000). The members of the world's cultures are more likely to successfully meet this challenge if they more fully appreciate cultural and personal change.

*Directed Change.* All the members of a culture (consciously or unconsciously) influence the manner, degree and direction in which their culture's world view will change. However, the organizational structure of the more complex cultures also gives those individuals and groups occupying its top positions (the elite) greater authority, power and opportunity to direct cultural change. China's one-child policy springs to mind. But the exercising of that extra power by those at the top is more difficult than it sounds, as the following discussion illustrates.

The Portuguese dictator Pombal wanted to respond in a pragmatic manner to the aftermath of the 1755 Lisbon earthquake that had destroyed much of Lisbon. However, the Catholic Church's representatives in his government believed that God had created the earthquake to remind Christians of the central role that redemption played in their lives. In the Church's view, the appropriate reaction to the earthquake was to encourage the people to confess their sins and ask for God's forgiveness: "Except ye repent, ye shall all likewise perish" (Kendrick 1956).

To have his way, Pombal reduced the Church's substantial presence and influence in the government. In essence, he directed the Portuguese culture's world view to become more secular. His action illustrates that those higher in a culture's hierarchy do have the power to direct a complex culture's world view to change. It also illustrates that their wielding of that power is constantly being influenced by the elite near them, the near elite. They include dissenters, potential usurpers of power, alternate power groups and acutely self-interested individuals. A similar but weaker influence arises from the elite further from power, the peripheral elite.

Different views among the elite about how to direct cultural change can itself lead to cultural change (Eisenstadt 1988). For example, Han dynasty emperors ruled China from 206 BC. During the reign of the emperor who ruled in the AD 180s, a squabble arose between him and some of the elite over their demand for changes to the who and how of exercising power. At the same time, the emperor faced other problems. The interactions between his various efforts to deal with these issues facilitated the AD 184 Yellow Turban revolt by dissatisfied peasants. Although the revolt was finally put down in AD 205, the Han dynasty was on its last legs (Hsu 1988). It ended in AD 220: a drastic cultural change.

The example of the Han dynasty also illustrates that the efforts of the elite to direct cultural change are affected by the views of the members of the culture who are the most numerous but the most distant from the powerful: the public. In complex agricultural cultures, they are the roughly 80% of the population who are almost exclusively rural farmers. The public have this influence because the elites' power to direct significant cultural change ultimately rests on the ties between the elite and the public. For example, the public and the elite are mutually dependent on the successful ongoing reaffirmation of their collective faith in their culture's world view and authority structure, and on the ongoing functioning of their culture's authority structure. Thus the elite's culture-changing efforts rely, at least indirectly, on the public's support.

This serves as a reminder that a cultural group consists of different subgroups that are each a part of its structure. In complex cultures, the presence of these subgroups ensures that cultural change directed from within the culture is less the product of a specific cause-and-effect influence, such as an elite decree. Rather, it is more the product of a mix of the dynamic influences exerted on the culture's functioning by its various subgroups and the people representing them. These subgroups can be formed by the elite, such as business groups and the military, and by those near the bottom of the power hierarchy, such as peasants and grassroots organizations (e.g., Greenpeace and Mothers Against Drunk Driving [MADD]).

At any given time, the relative significance of these individual influences on whether or how a culture will change can be ranked into a flexible hierarchy of influences. The hierarchy is flexible because the interactions among the players (such as occurs during politics, everyday affairs, lobbying, consultation, force and protest) changes the players' relative ranking in the hierarchy. As the Han dynasty example illustrated, the hierarchy of influence can at times be significantly different from the formal hierarchy of power as set out by a culture's world view. Thus, it is the aggregate of these individual influences, weighted by their real-world ranking, which can be thought of as the overall internal force directing a culture to change. (Any external influences on cultural change, such as the environment or foreign cultures, can be added to the hierarchy of influences, as discussed below.)

This perspective of directing cultural change is more fully illustrated by considering that, to remain functional, a culture needs to use natural resources while simultaneously protecting their quality and quantity, and the environmental functions providing them. When the members attempt to satisfy all of these conditions, they can face the conundrum of an intractable balancing act, as illustrated by their efforts to manage their environmental commons (i.e., the resources owned and used by all of its members, such as air and water) (Ostrom et al. 1999). The conundrum tends to be resolved by the members following the guidance from their culture's world view. If some of the culture's members, as individuals or as subgroups, are dissatisfied with this status quo resolution, then they can try to direct their culture's world view to change.

Some Canadians, for example, are trying to change their culture's world view in ways that its guidance will better protect their commons from the impacts of resource extraction activities such as tar sands oil production. Others are trying to ensure that their culture's guidance favours the increased production and use of this and other resources. Those involved in promoting these two views expend much effort to bend the public's and the decision makers' opinions to favour their view, which, if successful, will change the culture's world view in that direction.

These efforts to direct cultural change are affected by the participants' standing in their culture's hierarchy of power. This can be seen in how they assess and respond to a particular resource extraction project's possible environmental impacts (i.e., impacts on the commons). Consider that those who are higher in the hierarchy of power make and use their assessments differently than lay people, who tend to be lower in the hierarchy.

Elite participants usually hire professional specialists to, on their behalf, assess the proposed activity's possible impacts on the commons. These assessments are standardized and coordinated. They are numerical and focused on regulation and permit requirements. They are statistical, built around a formal evaluation (probability and severity) of risk, with a focus on short-term and simple impacts, such as catastrophic failures or pollution, deaths and exceptional benefits. In making their assessments, the professional specialists are influenced by the needs of the elite who hired them.

In contrast, the assessments by lay participants are informal, less coordinated and more context dependent. Their assessments tend to focus more on the personal, their community and its surroundings. Lay participants tend to consider both the emotional as well as the material impacts. This includes a wide range of quality-of-life values and features that could be affected by the activity. For example, the level of fairness in the distribution of the risks, rewards and impacts. The lay participants' evaluation of risk tends to be more common sense: they tend to rely more on the decision-making methods discussed in chapter 2 (Slovic 1990; Slovic 1993; Morgan 1993).

The differences, perhaps conflicts, that arise between those trying to direct cultural change in different directions can be resolved in a number of ways from negotiation to imposition. A common method is for those lower in the hierarchy of power to direct part of their efforts to increasing their standing in the hierarchy of influence. In Canada, it is currently popular to accomplish this by, for example, asking whether the members of the public have informally signalled their general acceptance or rejection of a project and its risks: whether they have provided or withheld a "social licence." Note that, regardless of how these differences are resolved and whether the efforts to make a specific directed change were successful or not, just trying contributes to changing the culture.

In overview, the act of directing cultural change from inside the culture may appear to be an exclusive right/responsibility of the culture's elite. In reality, it is a dynamic complex process, with the manner, direction and strength of change being the complex aggregate of the participants' efforts. Those participants include the culture's members as individuals and those representing subgroups. The aggregate impact is, simplistically, a mix of the number of participants along with their standing in the power hierarchy and their efforts to increase their influence. Those efforts range from their casual everyday decisions to dedicatedly following a detailed plan.

*Inadvertent/Unexpected Change.* The functioning of a complex system, such as a culture, is not exclusively deterministic, it is also random and chaotic, as discussed in chapter 5. A culture's world view thus inevitably faces the possibility of undergoing inadvertent and unexpected changes.

The nature and significance of those changes are illustrated first by a longer-term and then by a shorter-term example. Prior to the 1700s, Europeans mostly thought of marriage as an economic-political union that involved the members of the extended family or kin. After the mid-1700s, along with the rise of Romanticism, this view underwent a transformation. The ideal or model marriage came to be seen as more of a romantic union between two individuals.

The romantic view of marriage was a well-established ideal in early 1900s US culture. Although Americans may have strived toward that goal, few could hope to achieve it. Most of them still relied too heavily on their kin to help them meet life's challenges.

But, by the mid-1900s, Americans were experiencing increasing economic prosperity. Accompanying this change was their ability to satisfy the aspirations of materialism, consumerism and individualism. For example, the rise of manufacturing provided increased employment opportunities for women. With a job came the ability to satisfy their desire for financial independence, and the interdependence among kin was reduced. In this context, the aspiration to marry based on companionship rather than for economic or political purposes became a realistic goal.

So many Americans implemented the romantic view of marriage that it became the cultural norm, thus changing American culture. However, an unintended and unexpected outcome of these economic and cultural changes was a rapid increase in America's divorce rate. "Without anyone intending it, economic changes weakened restraints on divorce" (Myers 2001, 52).

The second example concerns the invention of ultrasound scanners, the use of which held the promise of improved prenatal care. This benefit justified, even required, the increasing use of the scanners. But an ultrasound test can also easily and cheaply determine the sex of a foetus (i.e., before birth). "Unexpectedly," the use of ultrasound tests for prenatal care in cultures that favour one sex over the other was also accompanied by a boom in sex-selective abortions. For example, only 20 years after its introduction into northern India, ultrasound machines had combined with pre-existing cultural values to cause a growing regional reduction in the number of girls being born. As northern Indians respond to the resulting social fallout, such as a shortage of brides (an excess of grooms, perhaps), they are inadvertently slowly changing their culture's world view in unexpected ways (Larsen and Kaur 2013; Singh 2008).

As the example of the ultrasound scanner implies, inadvertent and unexpected cultural changes are not always chaotic and random events over which we have no control. We contribute to these events happening though our decision making. Our contributing decisions can be made as individuals or as a group (the greater cultural group).

A major source of our individual contributions is our decision-making foibles. Because this was discussed in chapter 2, only a reminder is provided here. Designers and planners are unlikely to identify all of the conditions that will allow their technological or management plans to fail. One reason is that the complexity of the project and the world in which it will operate makes those failure conditions difficult to identify, as discussed in chapter 5. Another is their strong personal preference to discover those conditions that are most likely to hinder their project's success. This decision leaves the less likely but knowable failure conditions to be inadvertently and unexpectedly experienced, perhaps "discovered," in the future (Gardner and Stern 1996).

A major source of the contributions from a complex culture's group is its world view. This is illustrated by a complex culture's planning for the future. The beliefs and values in the complex culture's world view will influence the nature of the plan the members will make: for example, whether it will include the functioning of their supporting environment or the needs of future generations. Think too of Western culture's imperative to satisfy its notion of progress.

The culture's world view will also influence the group's decision-making process, such as the evaluation of the plan: for example, whether the concerns of the plan's detractors as well as its supporters will be heard. The culture's guidance could mean that, despite the plan's significance to the group, there may not be an open, transparent debate among the members. This increases the likelihood of the planners not identifying all of the features needed for success and the members not complying with the plan's requirements when it is completed. If an open debate is held, it tends to be adversarial in nature. This arises, in part, because different people/subgroups focus on different issues and draw different conclusions about the plan's outcome: they tend to take different, somewhat limited and perhaps inflexible perspectives. Then there is the issue of the complex culture's hierarchy of power.

Both the presence of an adversarial debate and the influence of the hierarchy of power on decision making are illustrated by the acceptance phase of the draft 2007 report from the Intergovernmental Panel on Climate Change (IPCC). That draft was part of the global effort to plan for future climate warming. Some of the panel members charged with accepting the draft objected to its representation of our climate-affecting actions and the expected consequences. The scientists who had authored the report vigorously defended its conclusions. Nevertheless, before accepting the report, the panel members altered it to be more in keeping with the views on climate warming held by the governments who had appointed them to the panel (The Associated Press 2007).

In light of these features, it should be unsurprising that when a complex culture's group decisions, such as planning for the future, are evaluated within a longer-term, wider context, or in retrospect, they are rarely seen to be as coherent, well-thought-out or comprehensive as many might have believed (Dietz et al. 2003). The decisions often reflect the culture's imperatives too strongly, with a correspondingly too narrow or too rosy a view of the possible outcomes. In particular, they reflect an incomplete consideration and accommodation of our human limitations, limits and biases; the influence of cultural world views; and the insufficient attention paid to the functioning of our wider environment, especially its complexity.

As a result, the members of a complex culture routinely implement decisions whose short-term outcome may achieve the expected result. However, the longer-term, more enduring outcome can be significantly different to, even the opposite of, what the members had expected. Our (individual and group) decision making can thus contribute to the presence of inadvertent or unexpected changes in the culture's world view, or the preconditioning for such a change in the future. Think ultrasound.

Certainly, the possibility that inadvertent and unexpected cultural changes will occur cannot be entirely eliminated. However, we can accept that they do occur and that we have a role in facilitating them. This acknowledgement alone can help complex cultures avoid the more severe and embarrassing consequences of improperly thought-out decisions and the resulting cultural changes they bring. Imagine how different history would have been, for example, if Canada had not decided to move First Nations' children into residential schools; if Australia had more thoroughly debated the importing of rabbits; or if America had not decided to reduce bank regulation and mortgage lending requirements in the 1990s and 2000s.

### External/Imposed Change.

A culture can be changed by aspects of the cultural system that are not part of the culture itself, such as its interactions with another culture and its environmental endowment. These external/imposed interactions can be treated as a process of cultural change. For example, a change in climate was a key part of the Abu Hureyran's switch from hunter-gathering to farming. Similarly, a decline in the availability of resources can prompt or force a complex culture to change in ways that address the shortage and its consequences. For example, in Egypt, around 2150 BC, a reduction in the Nile floods caused by drought was the main contributor to the disintegration and transformation of its culture (Hassan 1996).

Consider too that one culture can impose itself on another. An example is the conquest of the Aztec and the Inca empires by the Spanish through a combination of superior weapons, horses and, perhaps, smallpox (Diamond 1997). The culture least able to deal with the new circumstance has to undergo a painful imposed cultural change.

But the meeting between two cultures is not a one-way street. The intricacies of the many interactions between the two mean that both cultures must deal with contradictions, benefits and downsides. They both undergo, to different degrees, an externally imposed process of cultural change.

For example, between the 1600s and the late 1800s, the British exploring the Arctic struggled merely to survive for the few years of their expeditions. During this period, they had routine contact with the Inuit, who not only survived in the Arctic but maintained a viable population and culture. The British were faced with reconciling the contradiction between their culture's view of themselves as superior to the Inuit and the reality that the Inuit were thriving in a harsh land that was killing the British.

From the Inuit's perspective, it is only in the last 100 or so years that their culture has been fully exposed to Western culture. They are experiencing the inevitable contradictions, benefits and downsides as they continue to adapt to it. For instance, on the one hand, there are Inuit who value their traditional relationship with the seal, in which the Inuit, the seals and the practical aspects of the seal hunt are fully integrated into their spiritual world view. These Inuit feel and express great dissatisfaction with the Western world view generally and the Canadian government specifically because they treat the seal solely as a financial resource (Mallon 2002). On the other hand, the Inuit who began to exploit the estimated 34,000 tons of clams on the ocean floor near Broughton Island could only do so with the help of high-tech equipment, such as dry suits and compressors. That clam field was described as an underutilized natural resource potentially worth C$900 million (Hill 2002).

We all accept that the functioning of the earth's environment can directly force a culture to change, as mentioned above. But we are less aware that aspects of the earth's environment can also indirectly become a key player in cultural change. For example, on All Saints' Day in 1755, the wealthy city of Lisbon was destroyed by an earthquake, a subsequent tsunami and fires. Because the earthquake struck on a religious holiday, most of the faithful were attending church. Given that 35 of Lisbon's 40 churches were destroyed, it is unsurprising that a disproportionate number of the estimated 10,000–15,000 who died (from an estimated population of 275,000) were devout Christians.

This shocked the pious throughout Europe. The dead were the very people whom they had reasonably expected God to look upon with favour and therefore spare. Why, the question was asked, would God allow the pious to die? The emotion behind the question was heightened because they had been killed by an earthquake, fire and a flood (caused by the tsunami): events that, for Christians, have potent symbolic power. They are the forces God is expected to use when he destroys the sinners during the apocalypse at the end of the world (Mathew 24; especially verses 7, 39, 42).

The Lisbon earthquake and its aftermath thus sparked widespread, heated discussions at all levels of the Christian-based European society, including philosophers and church leaders. The more famous participants included Immanuel Kant, Jean-Jacques Rousseau, François-Marie Arouet (or Voltaire, who wrote *Candide* about the Lisbon earthquake) and John Wesley. However, the debate was not whether nature (science) or religion was supreme. At this point in time, both philosophers and clergy held to a Christian-dominated cultural world view. The focus of the discussion was mankind's (as they expressed it) relationship with God. The topics included the nature of evil, our place in the world and the reasons for earthquakes (Kendrick 1956). These discussions could include the emerging natural philosopher's view that an earthquake was a natural phenomenon and thus only indirectly from the hand of God.

The earthquake, in conjunction with the subsequent debate about it, contributed significantly to both the long- and short-term changes in the European culture's world view. For example, the optimistic view of the future that prevailed at the time of the earthquake was based on the religious view of achieving salvation through God's grace. God's inexplicable killing of the pious by earthquake, fire and flood, and the resulting confusion over how to achieve God's grace, helped to weaken that belief and engender pessimism in the Europeans' outlook.

In contrast, the natural philosophers' emerging optimistic view of the future seemed more appealing. In their view, mankind could, through effort and with God's providence, improve both humans and nature. On this path, humanity could gradually progress until they achieved the full happiness and perfection God intended. The earthquake and subsequent discussion also contributed to other more pragmatic cultural changes. For example, when Pombal removed the Church members from his government, he introduced Europeans to the idea of a more secular government.

In overview, the overall process of cultural change can be described by a complex mix of the processes described above. The nature of that complex mix is illustrated by the fallout from the externally imposed Black Death of the mid-1300s that killed some 20 million people over just a few years. That unexpected experience helped to change European culture's view of death. The event also significantly reduced Europe's population size and altered its demographics. At the same time, the resulting labour shortage unexpectedly improved the cultural standing of the peasants and labourers in the society to such a degree that it helped to precondition the demise of European feudal culture (Kelly 2005). The event helped to change the direction in which the culture was drifting.

## Influences on Cultural Change.

Human cultures are flexible and they do change, as discussed above. This subsection, like that for personal change, discusses a selection of the broad groupings of often invisible influences that can either facilitate or constrain change, depending on the context in which they are applied. Similar to personal change, the overall impact of these influences is the complex aggregate of their interactions with one another.

*Cultural Inertia.* Cultures have an inherent resistance to significantly changing their characteristics, in particular their world view, and to making those changes quickly. The presence and strength of cultural inertia is illustrated by the outcome of the almost complete separation of East from West Germany during the Cold War. During the decades-long partition, the East German government expended considerable effort to reshape East German culture to match Soviet ideals. Despite their valiant attempt, the fall of the Berlin Wall and the subsequent reunification of Germany revealed that the core differences between the two cultures were small (Inglehart et al. 1998).

Consider too, the Polynesian ancestors of the Maoris who, around AD 1200, settled in what is now called New Zealand. (The Norse settlements in Greenland reached their maximum size.) A few centuries had to pass before the founding culture had differentiated into the variety of Maori cultures now found on the island (Diamond 1997). Despite this differentiation, the core characteristics of each present day Maori culture still retains the basic Polynesian characteristics that were, even in 1200, many hundreds of, if not a few thousand, years old. This example illustrates both a culture's inertia to changing its core and the inevitable time lag/delay between the arrival of a circumstance prompting/permitting cultural change and a culture changing to accommodate it (Inglehart 1990).

The root of cultural inertia is the members' innate personal inertia to change. This is illustrated by the behaviour of the late 1800s and early 1900s Mediterranean and Asian immigrants to California who settled along its coast. Despite being a minority living in a new country with a different culture, they recreated, in detail, their homeland's fishing communities and cultures (McEvoy 1986). The same can be said of the British settling in their colonies. Think acclimatization societies. Even today, immigrants form immigrant enclaves in their new home country.

But the members' inertia to change varies, on average, with their age. Compared to the younger members, the adult members are the most familiar with, and thus the most invested in, their culture's current world view. The older members tend to feel the greatest discomfort and are the slowest to adjust to changes in their culture's world view and their surroundings (Inglehart 1990). For example, the full metrification of Canada failed, in part, because the older generation did not want to learn a new way of weighing vegetables at the supermarket. To do so required them to accept not just a new cultural reference for weights and measures but to discard the cultural attachment to the old reference. Overall, the spread in the ages of a culture's members (the culture's demographics) influence its inertia to changing its world view.

Another aspect of a culture's inertia to change is its history, as discussed in chapter 3. What matters in this case is not the culture's actual history but the version embedded in the culture's world view: the one the members are emotionally connected to. When the members of a European culture, for example, draw attention to their culture's founding story or their monumental buildings, they are doing more than pointing out evidence of the actual historical roots of their culture's beliefs, values and power. They are indicating their cultural identity: their emotional attachment to the version of their historical roots provided by their culture's current world view. Hunter-gatherers do the same when pointing out a culturally important landscape feature on their home range.

In this context, a culture's version of its history helps the members to remember, relate to and remain faithful to their culture's world view and its guidance about how to live. At the same time, the members' personal connection to that history helps to ensure that they will usually display considerable resistance to any efforts to directly make significant changes to the core of their culture. In general, cultures display inertia to change, unless the adjustments fit within their world view's guidance.

The presence of cultural inertia also influences the manner in which cultures prefer to change. It is easy to imagine that when a culture does change its cultural myths, values and symbols, then the old ones are, like cell phones, displaced or discarded and replaced by the new. This is not the case. Instead, as indicated by the ongoing presence of religious symbols and myths in today's largely secular Western culture, significant cultural change most often occurs by adapting and adjusting the form and meaning of the current values, symbols and myths (Inglehart 1990).

The use of cultural adaptation and adjustments as a way to balance cultural inertia with cultural change is illustrated by the way in which science culture's world

view usually changes. Rather than scientists resolving a current scientific problem by accepting the solution that fits the data best, they tend to accept the resolution that best addresses the problem within its historical context (Kuhn 1996). To make that fit often requires the scientists to make "compromises" or assumptions. A now familiar example is Newton choosing seven colours for the rainbow in order to fit with his culture's long-standing and prevailing views about the order in the universe. Consider too Einstein adding the cosmic constant to his equation to ensure that it fitted his and his peers' view of the universe.

Change by adaptation and adjustments also means that the older, core features of a culture's world may appear, to a casual observer, to have disappeared from its more recent world view. But, on closer inspection, they are still present in an adjusted form. Consider the ancient Egyptians' long-term focus on the afterlife but the different ways in which it was expressed. Consider too Western culture's notion of progress. It was most clearly (overtly) expressed by the Victorians in the mid- to late 1800s. Today Westerners think that the notion is largely a historical relic. But the result of it being questioned wasn't its abandonment but its adjustment, a change in its details and how it is expressed. The notion of progress is still a critical part of Western culture's current world view and its guidance to Westerners about the preferred ways to live and resolve our issues, as discussed in chapter 3. Our efforts to satisfy our notion of progress feed back to maintain its presence in our culture and thus our inertia to changing it.

Looked at from a different perspective, inertia affects how all aspects of a culture change, but it doesn't do so with equal strength. It is relatively weak when a cultural change is simple, desired or lies on the culture's periphery (i.e., the aspects of the culture that are seemingly weakly connected or unconnected to the core). An inertia-related influence on change in these situations is discussed under Time.

In contrast, as the above examples illustrate, inertia's influence is strongest when the core aspects of a culture's world view are involved. They usually change slowly through limited adjustments that are spread over time. However, these ongoing changes can occasionally be interrupted by rapid, significant and thus traumatic changes. These types of changes to the core will most commonly occur only in response to the presence of accumulated issues or very large, rapid shifts in its socio-political, economic, technological and environmental conditions (Inglehart 1990). They are more likely to occur in cultures with a high proportion of young members, especially if they are dissatisfied.

In overview, despite the facetious remark about older Canadians, inertia is more than an impediment holding back much needed cultural changes. It helps to provide a culture with the continuity and stability that are critical for the well-being of both the members and their culture, especially during times of uncertainty. In this context, cultural inertia represents more than the members' resistance to a change. It reflects the compromise between the ever-present influences prompting change and the members' need for stability and continuity. It also reflects the limit to how fast a culture's world view can make a change if it is to avoid disrupting its functioning or deeply disturbing its members, and how that change can most easily take place. Think of cultural change through a process of adaptations and adjustments.

*Time.* The resistance to cultural change described under Cultural Inertia ensures that the core aspects of a culture's world view will usually change slowly, through processes such as drift and adjustment. Many decades if not centuries can pass before those changes to the core are evident (Inglehart 1990). This was illustrated above by the rise of individualism in Western culture and the differentiation of the Maori cultures.

However, there is much less cultural inertia to change on the periphery of a culture's world view, so change there can take place much faster, especially when compared to the time it takes for human biology to change (Barash 1986). A simple example is the relative ease with which cultures all over the world have adjusted to cell phones. But, even if a cultural change is desired and unhindered by inertia, it still take time to complete: it takes time just to complete the process of change.

The total amount of time it takes to complete a change is more easily appreciated by imagining that a complex industrial culture has an increasing population of elders. If the culture is to provide long-term care centres for them, then the members must first recognize that demographic trend. If the cultural guidance to the members doesn't already give elder care a high priority, then the members must change their culture's world view. Think inertia. Once motivated to provide the care centres, the members must take into account the lead times and cut-off dates needed to plan any needed infrastructure and secure the funds. Then there is the time it takes to build and staff the infrastructure. The overall time between being prompted to recognize the need to make a change and completing the change is the complex sum of the time to complete all of the practical and mental changes in between.

But the process of cultural change is influenced by more than the time it takes to complete a change. It is also influenced by failing to realistically consider the constraints on the time available to complete the change. This oversight can result in the members trying to deal with today's issues too late, using a response more suited to a historical issue or playing catch-up. It could also result in missed opportunities, which could influence future cultural changes by limiting the members' options. This likely applies to our efforts to deal with climate warming. In extreme cases, this oversight could force the members and their culture to make substantial, unexpected and unplanned changes.

*Natural Resources.* The making of some cultural changes requires an increase in the amounts or a change in the types of natural resources that the culture uses. Thus either the lack of those resources in the culture's endowment or the members' inability to provide them influences how their culture can change. For example, a culture can only industrialize if it has access to large amounts of inanimate energy. Similarly, the reduction in the availability of resources currently supporting an aspect of a culture's world view can force it to change.

Surprisingly, a resource's influence on cultural change does not depend exclusively on its scarcity. Consider that the abundance of a resource can result in a culture's world view treating its use as a cultural norm: the members can come to culturally/emotionally depend on it. Once that bias is in place and the members are intensely exploiting that resource, they can find it difficult to change their culture's world view away from that focus. For example, the Maya and corn and, as illustrated below, a complex culture switching from a long-standing energy source. Consider too that there can be a cultural conflict over the relative value of resources, for example, Western culture's dilemmas of food versus biofuel and animal protein versus vegetable protein. These types of conflicts also have an influence on the members' efforts to change their culture.

A resource can gain an even greater influence over cultural change when it is given a special place in a culture's world view, such as a critical symbolic or spiritual role. For example, the exalted position given to cattle by the ancestral Xhosa culture and the spiritual meaning given to corn and jade by the historical Maya. Consider too that many cultures value food for more than its ability to sustain life. It is often part of the social grease that keeps the culture's wheels turning. In the more complex cultures, some foods are considered critical to satisfy non-food resource wants.

Collectively, the availability and use of natural resources is a powerful, multi-faceted influence on cultural change. The members' efforts to consider and address this influence are limited by their culture's world view and their decision-making foibles. Natural resources represent one aspect of the environment's significant influence on cultural change.

### Organizational Complexity.

A culture's changing is indirectly influenced by its level of organizational complexity. Consider that the more complex a culture's structure, the greater the costs, responsibilities and issues the members must address if it is to stay functional. In particular, a complex culture's expanded reaffirmation and authority maintenance processes come with a greater cost per capita compared to a simple culture, which is ultimately paid for using natural resources. A culture's focus on supplying them can prompt it to change or contribute to its inertia to change.

A complex culture maintains the desired supply of natural resources by using its complex structure to manage the members' efforts to produce them. Successful management requires that the members display significantly more uniformity and predictability in their behaviour than those of simpler cultures. The members are thus obliged to accept more restraints on their personal freedoms for the common good. This type of cultural demand on the individual members can eventually come into conflict with their innate desires, such as wanting to behave as they see fit. Thus, if a culture wants to become more complex, it must formally deal with these issues by persuading the members to accept both these constraints on their behaviour and the need to enforce them (Weingart et al. 1997). The need to satisfy this requirement constrains a culture's efforts to become more complex.

And once a culture is complex, its complexity influences how it changes. In particular, its hierarchical power structure outlines its decision-making organization. The people higher in the structure (the elite) are endowed with more decision-making power and are expected to use it on behalf of and for the good of the members. However, despite statements or beliefs to the contrary, those who become the elite do not lose any of their humanity when they rise to their illustrious positions. The elites' human foibles can combine with their new-found power to result in them redirecting their efforts at governance away from meeting their obligations to the people. Today's saviour can become tomorrow's thief, which is a problem that influences how a culture can or will change.

In addition, the decision making of the most powerful in a complex culture's hierarchy can be affected by the individuals and groups both near and peripheral to power. This influences cultural change. For example, as a result of being lobbied, the most powerful can make decisions that inappropriately favour those near and peripheral to power at the expense of the culture's members as a whole (Yoffee 1988b; Goldstone 2002). Their culture will be influenced to change accordingly.

An illustration of the influence a culture's level of complexity has on its changing is provided, in overview, by a complex culture contemplating changing its dominant energy source. Because of its structure, those of the elite who control the pre-existing energy sources are inclined to work against the change: there is significant inertia to change. Unless, that is, they are culturally restrained, unusually enlightened, provided with incentives or can co-opt the changes to satisfy their self-interest (Hall et al. 1986). Current examples of this would surely include ExxonMobil's funding of lobby organizations that cast doubt on the science of climate change (Begley 2007; Oreskes and Conway 2010). After all, forced reductions in carbon emissions would herald significant changes for the oil industry.

If a culture did indeed adopt a new energy source, then its structure would influence the details of how the members must adapt: how they would have to change their lifestyle to accommodate both the source and its wider impacts on their lives. Consider the many changes to domestic and working life in England that accompanied the switch in the energy source from wood to coal during the middle of the 1800s. Think of steam-powered machines in factories, increased laundry, mobility and standardization. In the process of adapting, the Victorians changed their culture's world view and increased their culture's complexity.

### Individuals.

Each member of a more complex culture hopes (perhaps believes) that their culture will change in a well-reasoned, timely, painless and appropriate manner. When this doesn't happen, the members tend to blame opponents, or the elite. In one sense, this is reasonable because, being in charge, the elite have greater power and access to resources. This gives them greater influence over how their culture changes and greater responsibility for that change.

In the end, though, each of the members are individually responsible to a significant degree. After all, they are each the carriers of their culture's world view, and their individual decisions will collectively change it. Thus, the limitations and biases in our personal perceptions, predictions and attempts at deciding on truth can all contribute to our culture failing to change in the manner we desire, or changing as we desired but with undesirable consequences. Our individual contribution can thus be thought of as our best intentions that are biased and limited by our decision-making foibles and the influences on them. After all, we experience difficulty, or perhaps just reluctance, recognizing and meaningfully accommodating those foibles and influences when we make our decisions. Dealing with this oversight is, of course, our personal responsibility.

This subsection illustrates some of the ways in which our individual decisions influence cultural change. Consider that we, regardless of whether we are a member of a complex culture's elite (media, politicians, etc.) or its public, tend to rely on the statements and predictions of experts to guide our decision making. In choosing an expert, we prefer those who make simple, decisive statements and confident predictions rather than those who make conditional predictions with admissions of uncertainty. As a result, the statements and predictions that we prefer to rely on when making our culture-affecting decisions are likely to be less reliable and less useful than we assume (Tetlock 2006). Thus, perversely, the less reliable experts can have more influence on our decisions than they should. This is especially true in the spheres of politics and economics (Rescher 1998). More generally, this type of bias in our selection, evaluation and use of information affects how our decisions will influence our culture's world view to change.

Consider too our relationship to the past. As discussed, our culture's version of history has a significant influence on our culture-affecting decisions. It is easy to imagine that if we could forget the past when making decisions, we would reduce its contribution to both our biased decisions and our culture's inertia to change. Perhaps this is why we can find it easier to evaluate and accept new-to-us political, economic and technological ideas and their products: being new they are largely free of the burden of a history.

Surprisingly, we tend to accept new ideas and products for a different reason. We use the (real or imagined) lack of history attached to them as an opportunity for us to (consciously and unconsciously) realign our views of both our relationship to the new and its possible effects on our present and future lives (Barzun 2000). If we can realign our views so that the new appears to satisfy our current personal, culturally-influenced (i.e., history-biased) needs, concerns or beliefs, then we can accept the new. Think of the internet. This response to the new reveals the inescapable presence of insidious, persistent biases in our influence on cultural change.

This discussion indicates that, rather than trying to escape the inescapable past, we are better off spending time and effort putting the present and the new into its proper (appropriate) historical perspective. In the process of doing so, we are more likely to restrain our imagination and consider context. We are more likely to remind ourselves that there is more to the past than we imagine, and that all we dream about the future can't become real. We are then more likely to use the past to help us personally face the present and the new for what they are, not what we want, hope or imagine them to be. We are thus more likely to make appropriate culture-affecting decisions.

Our individual decisions do indeed influence cultural change. However, it could be argued that this discussion is largely irrelevant because most of our day-to-day decisions are trivial. Our decision about which bread to buy, whether to drive to the store to pick it up and how many kids to raise just doesn't have the same influence on cultural change as a leader deciding whether to sign a trade deal or not. This is true, when seen in isolation, but collectively, and over the long term, our mundane decisions do have a significant influence on cultural change. They contribute to cultural drift, cultural inertia and resource usage, among others. In this context, even our tendency to dismiss the significance of our everyday decisions is an influence on, or even a process by which, a culture changes. After all we are the carriers of our culture, and its decision makers.

In overview, our individual decisions do significantly influence how our culture changes. However, our decision-making foibles mean that the changes to our culture resulting from our decisions might not be what we expect or want. We can't change our human characteristics by much. But, as individual decision makers, we can recognize that our decision-making foibles exist and accept how cultures change. And we can include this knowledge in our thinking and decision making. If we do, then we are more likely to make appropriate culture-influencing decisions. The remaining question of what "appropriate" means is answered in the next subsection.

*Functional Boundary.* There are a wide variety of ways that a group of people can organize themselves so that their manner of living and believing (as represented by their culture's world view) binds them into a functioning complex cultural system. However, each cultural mix comes with conditions that must be met if the culture is to remain functional over the long term (viable), as discussed in chapter 3,. The human and environmental conditions for viability can be thought of as collectively forming a multi-dimensional boundary around the culture: a functional boundary. If the members fail to meet these conditions or change their culture so that it is pushed too far outside its functional boundary for too long, then it will become dysfunctional. The presence of the functional boundary thus limits cultural change, if the culture is to remain functional.

The presence and significance of this somewhat pliable boundary is illustrated by nomadic hunter-gatherer cultures. Two core aspects common to these cultures' world views are key to them being functional: flexibility in decision making and mobility. These characteristics help to ensure that their members can both acquire the variable seasonal resources they need to survive and maintain the human contacts they need to complete their lifeway.

Imagine if a nomadic hunter-gatherer culture's capacity to be mobile and flexible was steadily being restrained by a slowly changing environment. It is difficult to see how the members could make the changes needed to adapt without pushing their nomadic hunter-gatherer culture beyond its functional boundary. The changes to its core characteristics would likely be so substantial that the members would lose their nomadic hunter-gatherer cultural identity (Kelly 1995). Think of the Abu Hureyrans.

The situation would be worse if the restrictions on both flexibility and mobility were applied suddenly by, for example, a neighbouring culture occupying part of their home range or forcing them to live in a permanent village. The size and speed of the needed cultural changes would likely be so great that, in attempting to complete them, the culture could become dysfunctional. Examples of nomadic cultures experiencing these

changes are the Kung living in the Kalahari Desert and the Moken living in coastal Thailand.

Here is another example of the general conditions that make up a culture's functional boundary. In order to function, all cultures require natural resources, in particular fresh water, food and some non-food form of energy. The availability of these resources to a culture is limited by both the characteristics of its environmental endowment, such as a river to provide water, and by the members' ability to grow, capture or extract them, such as building dams and wells. Consider too that the natural resources a culture needs to function are provided by environmental processes, such as water purification and soil creation. Thus, if a culture wishes to be viable, then it can only change in ways that both stay within the (amount and type) of resources its environmental endowment can provide and protect the environmental functions that provide them. More details and other examples of the functional boundary conditions can be found above and in chapter 3.

The essence of the functional boundary around both complex and simple cultures is the same. But the details differ in a way that matches the culture's complexity. For example, a complex culture is divided into many more subgroups than a simpler one. If a complex culture is to stay within its somewhat pliable functional boundary, then its world view must unify the members. It must also ensure that they are aware of and respect their culture's functional boundary rather than squabble over whether it exists and what it consists of. To achieve these goals, a culture's world view must promote effective communication among the members and their subgroups. However, as concluded in chapter 3, maintaining the required communication in a complex culture is difficult and faces inescapable limits. These complexity-related constraint are all part of a complex culture's functional boundary.

Consider too that those higher in the culture's organizational hierarchy (the elite) are granted both power over those lower down and preferential access to resources. These privileges come with the responsibility of acting in the best interests of the members, not themselves. For example, the elite must ensure that the members, especially those lower in the hierarchy, have fair and adequate access to resources. In post 1600s Europe, this relationship between leaders and people came to be called a "social contract," an idea expressed by Thomas Hobbes (1588–1679) in *Leviathan* (1651) and by John Locke (1632–1704) in *Two Treatises of*

*Government* (1690). Think too of Canada's idea of social licence mentioned above. Regardless of how it is expressed, this contract is a part of a complex culture's functional boundary. Enforcing this contract and preventing the abuse of privilege is difficult, as discussed in chapter 3.

The overall difficulty complex cultures face when trying to stay within their functional boundary is illustrated by the feedback that exists between the practice of science and Western culture's world view. In the early 1900s, scientists described existing hunter-gatherer cultures, in particular those in North America, as claiming exclusive use over an area of land: they were described as strongly territorial. Because, at that time, hunter-gatherer cultures were thought to represent our most basic selves, Western Europeans interpreted this territoriality to mean that humans in general were inherently strongly territorial and that aggression between groups was therefore unavoidable. This belief appeared at a time when national territorial aggression was raising its head in Europe. That aggression produced World War I and World War II.

The early 1900s European view that human cultures are innately strongly territorial remained popular, even with some anthropologists, until the 1960s. Only then did it became more widely accepted that North American hunter-gatherers were likely driven to strong territoriality as they adapted to the fur trade: their need to secure trapping rights to an area (Kelly 1995). Considering the millions of deaths and wasted resources that resulted from World War I and II, can it be said that the relationship between science and Western culture contributed to pushing European cultures across their functional boundaries? Or did it just mirror the members' unrealistic ideas of the time?

Overall, for a culture to remain functional, it is constrained to change within its somewhat pliable functional boundary. The culture cannot decide to live beyond this boundary for too long without risking inappropriate change and instability. Think of the European witch trials. It takes considerable, ongoing effort for a culture to stay within its functional boundary. A culture that does stay within them over the long term is said to be viable.

In summary, it is easy to think that cultural change is the simple sum of the various processes of cultural change and the influences on them described above. But, as the discussion indicates, these contributions interact with one another, so their summing or balancing is not simple but dynamic and complex. This is illustrated by Christianity in Europe.

Christian culture (Christianity), like most religions, displays considerable inertia to changing its world view. But it is hosted by European cultures that are continually changing. Christianity has responded to these conflicting influences on it by making both internal adjustments to its world view and by continuing to split into sects, each of which claims that their interpretation of Christianity's central tenants is, biblically, the correct one. This dynamic balancing of change and continuity through adjustments has allowed Christianity to remain functional and a part of European culture.

Influences prompting a culture to change can be generated internally, or imposed from the outside, for example, by a hostile culture or an environmental change. Regardless, because the individual members are their culture's decision makers, their responses to these influences will, collectively, make a significant contribution to, or even determine how their culture's world view will change over time. For example, if the members' responses result in their culture's world view changing in ways that guide them to stay within their culture's functional boundary, then, over the long term, their culture is more likely to become viable.

When responding to these influences, the members should remember that cultural change is a dynamic, complex process that includes feedback. Their responses could place their culture into a feedback loop in which its changing can take on a life of its own. The culture and its members can then end up in a place the members neither intended nor wanted their culture to go, such as outside the culture's functional boundary, and be held there.

Short-term examples include the French Revolution, when the proponents of liberty, equality and fraternity started claiming one another's lives. The spiral into World War I should also be considered. A more contemporary example is the post-1970s neoliberal economic revolution in the United States. The cry for deregulation to keep economic growth going gathered so much momentum that it resulted in the global financial crisis of 2008, which nearly bankrupted the country and its citizens.

For a culture to avoid that type of outcome requires its members as individuals (us) to be aware of ourselves, the influence we have on our culture and how it changes. And we must apply that knowledge: a theme found in many of the following chapters.

## Change within the Human-Environment System

The many processes and influences contributing to personal and cultural change are linked together and interact with one another in ways that include feedback. We might recognize this connectivity, but the difficulty of thinking about change in this way means that we, instead, tend to focus on a few select processes and influences. This focus is itself an influence on how we and our cultures change. It makes a significant contribution to the misunderstandings and arguments about all aspects of personal and cultural change, from the need to change to how to change. It therefore lies behind much of our current state of affairs and our preferred response to it. And it affects our future.

It would be easier for us to routinely think and make decisions about personal and cultural change as an interconnected whole if we could relate it to the dynamic, complex human-environment system in which it occurs. We are helped in this task by remembering the characteristics of a generic complex system described in chapter 5. Those generic characteristics, are reviewed here before using them to connect personal and cultural change to the human-environment system.

All complex systems consist of many variables that interact with one another in a variety of linear to non-linear ways, over a wide range of time, space and strengths. They also form reinforcing, self-sustaining and dampening feedback loops. Collectively, these interactions link the system's variables together into a single, dynamic complex system. The interactions are how the system functions and thus, from a different perspective, how it changes.

When seen in overview, a complex system's functioning and changing displays a mix of deterministic, random and chaotic characteristics. It also displays a wider range of features than the simple sum of the interactions would suggest. These characteristics are expressed by different parts of the system changing in different ways at different places, at the same time. Exactly how a particular part of the system will change and how fast that change will spread through the system is determined by the system's details.

On closer inspection, a specific aspect of the system's functioning and changing is usually the result of many variables and their interconnections rather than just a single, simple cause-and-effect relationship. What we see in this closer view can be described using generic descriptors/terms for the system's functioning and

changing. They include *drift, adjustment, preconditioned, unexpected, directed, inertia, imposed, complex/simple* and *external/internal*.

The details of a particular complex system's functioning, the influences on that functioning and the resulting changes will determine which of its features will be easy for us to see. What we do see and the significance we attach to it also depends on our choice of the variables, the time scale and the space scale we use to look at the system (our perspective). The system's complexity means that we can choose to view it from any one of a number of equally valid but different perspectives. It also means that our resulting different views of the system can appear to be contradictory. However, if we treat a complex system as a functional whole, then we realize that these differing perspectives can be collectively used to form a coherent view of the system's current state. In this view, its state is a product of the system's functioning and the (internal and external) influences on it. Those influences can be thought of as forming a flexible (dynamic) hierarchy.

The above description indicates that it is possible to meaningfully think of the changes occurring within a complex system as being either a part of its functioning or a particular view of that functioning. It is also possible to predict a complex system's future, but only with difficulty and uncertainty, as discussed in chapter 5. The degree of difficulty and uncertainty varies with the system and the scale of the prediction. In addition, the presence of random and chaotic features ensures that the unexpected should be expected.

These generic characteristics of a complex system apply to the complex human-environment system, and its subsystems. The details are determined by the human-environment system's specific variables and interactions. We are a key variable, so we are an important contributor to those details.

This knowledge can be applied to the human-environment system by dividing its variables into three overlapping, familiar, personally relevant factors: individual, culture and environment. These three factors can then be treated both as basic components of the human-environment system and as standard perspectives to contemplate it from. They each provide us with a different but related insight into the culture's functioning and changing and thus insight into cultural and personal change.

By taking the *individual's* perspective, we can focus on our internal reality, our human concerns and our human foibles. This will enable us to more fully grasp the limits and biases to our personal understanding of the human-environment system, the external reality and our decision making. We can then more easily appreciate the significance of our influence on the system's functioning and changing. And we can more fully appreciate our personal changing and the influences on it, especially those arising from our culture's world view. Information on this perspective is found elsewhere in this chapter and in chapters 1, 2, 4 and 5.

By taking the *culture's* perspective, we are better able to grasp that our culture's world view acts as an intermediary among the individual members, and between them and their culture's environmental endowment. At the same time, we can more easily understand that a culture's world view is a creation of its members. We can then more fully appreciate the influence that our culture's world view has on how we conceive of, think about and relate to the external reality, ourselves and our culture's world view. And we can also more easily appreciate the nature of our culture's changing and the influences on it, especially our decisions. Details of this perspective can be found elsewhere in this chapter and in chapter 3, with relevant aspects in chapters 2, 4 and 5.

By taking the *environment's* perspective, we can focus on the foundation of the human-environment system. This perspective helps us to understand how our environment functions and changes, as will be discussed in more detail in chapters 7 and 8. That knowledge helps us to appreciate its influence on our lives, in particular, how the environment's functioning can both facilitate and constrain how we and our culture change. We can also better appreciate how we change the environment, and the significance of the influence a changed environment can have on our and our culture's changing.

All three perspectives are taken in this book. They are used both to break the discussion into chapters and to organize the discussion within each, as seen in part 1. In addition, the human-environment system is used as the unifying common reference that is found, to some degree, in all chapters.

This broad component-perspective view of the human-environment system makes it much easier for us to appreciate it as a complex system. We can also more easily appreciate that the processes of (personal and cultural) change and the influences on them, as described above, are part of the system's functioning and changing. Each description is a different but overlapping and interconnected perspective of personal and cultural change that takes place at various scales and includes feedback.

This knowledge should result in us significantly adjusting our usual view of personal and cultural change. Our tendency to think of the process as relatively simple and deterministic should be replaced by the view that it is complex and dynamic. For example, as I change, I influence my culture, its other members and the environment around me. I am, in turn, influenced by changed members, a changed culture, a changed environment and so on in an endless feedback loop operating at all scales of time, space and strengths. In addition, what I think about change and the contributions to it depends on the perspective I choose to look at this system and my role in it.

However, there is a limitation to this three-part reductionism. To gain the full benefit of this understanding, we still have to deal with our innate tendency to look at the system intently from one preferred perspective and with a bias. This tendency means that some aspects of the human-environment system or its changing will be favoured, some excluded and others distorted. To compensate for this limitation, we need to continually remind ourselves to take it into account. This can partly be accomplished by using the environment and the unity of the human-environment system as reminders and references. We can also refer to comprehensive models, as discussed more fully in other chapters. We are further helped by practising a more nuanced, multiperspective manner of describing and thinking (nuanced thinking and decision making) about a topic, as illustrated by the following example.

In the 1600s, a clockwork model was used to express a European view of how the world worked. Here a simple windup wristwatch illustrates, by analogy, how the technique of nuanced thinking and decision making can help us to appreciate the human-environment system's workings as a whole and routinely treat (personal and cultural) change as a part of it. Look for the component-perspective and descriptors, detailed above.

We tend to think of a wristwatch as a simple, deterministic cause-and-effect system that has one purpose: to keep track of time. We believe that we can remove or disrupt some of the watch's parts and yet still retain its essence: the ability to tell time. For example, removing the strap has no impact on the watch's ability to tell time. However, if a gear is removed, the watch no longer functions and is considered useless.

But looking at the watch from just the perspective of a working timepiece neglects that there are other equally valid but less obvious perspectives to take. The watch could be seen as a historical artefact, in which case its value is in its mere existence. Consider too that the watch has a symbolic function. From this perspective, a strapless watch is different from a strapped watch. Imagine that the strapless watch was Mr. Big's flashy Rolex. To him, its ability to tell time is of secondary importance to its value as a status symbol. Without a strap, he can't appropriately display that symbol on his wrist. The watch is essentially useless to him, even if it can still tell time (Garfield 2016).

Now replace the watch with an ecosystem. When this single functional whole is looked at from a short-term perspective, it seems that its functioning will only be significantly disrupted if its keystone species is removed. However, from the long-term perspective, the removal of a few, or even one, seemingly irrelevant species might change the ecosystem's future or options for the future in important ways.

This nuanced, multiperspective manner of thinking can help us appreciate each of the component-perspective views of the human-environment system and their interconnectedness into a single, dynamic complex system. It can help us to understand and appreciate that human and cultural change and the influences on it is part of that system's functioning.

In summary, the human-environment system can be divided into three component-perspectives (factors): individual, culture and environment. The interactions among the many variables these factors represent occur over a wide range of scales in time, space and intensity, and include feedback. These interactions are the system's functioning and, from a different perspective, its changing. A broad, generic view of the system's functioning and changing describes it as a mix of deterministic, random and chaotic characteristics, as discussed in chapter 5. The closer view is described using terms that include *drift, adjustment, preconditioned, unexpected, directed, inertia, imposed, complex/simple* and *external/internal.*

These terms were used above to help describe the nature of personal and cultural change. A particular change is described by an interconnected mix of the processes of change and the influences on them. By acknowledging both these features of change and our participation in the dynamic, complex human-environment system, we can more easily treat personal and cultural change as a normal part of this system's functioning and accept that the environment is also involved.

In this context, we can appreciate that personal and cultural change are both contributions to and products of the dynamic, complex human-environment system's functioning. We should also recognize that there are human and environmental constraints within which we must live and change, if we and our culture are to remain viable. With this understanding, we should find it easier to adjust our personal and our culture's world view so that they guide us to change toward that end and adopt a viability-favouring lifestyle.

In more practical terms, before a person can easily undergo a specific change they must be preconditioned/ primed/willing to make it, and the circumstances/ context must favour the change occurring. In addition, if the resulting change is to favour viability, then the information used, and how it is used, must also favour viability. And, if a persuader/leader is involved, they must not only inspire viability but follow their own advice promoting it. The equivalent applies to cultural change.

## Writing, Reading and Accounting

After reading the descriptions and analysis above, one is still left with the sense that something is missing. The nature and significance of personal and cultural change within our human-environment system still seems to be distant from our day-to-day lives and the issues we face. We need a real-life example that is more personalized than the broad overview provided by the story of Abu Hureyra's change through time in chapter 3. We need an example that helps us appreciate the challenges and limitations we personally face when trying to realistically imagine, think about, or engage in personal and cultural change.

The example used here is the appearance of written accounting in the Middle East. It was chosen for more than its description of an important cultural change that we can relate to. It was also chosen because the world view and circumstance (context) of the culture going through those changes was different from an industrial culture's world view. Similarly, the views of the long-dead individuals who lived through those changes are unfamiliar to the members of an industrial culture. We will therefore struggle to adopt a perspective that allows us to both recognize the context in which those long-ago changes took place and relate to the thoughts of those who participated in them. But, through our focused effort, we will come to more personally appreciate the less visible influences on our personal thinking about our current circumstance and those changes we think would appropriately address it.

Each member of a culture that writes routinely uses groups of standardized written numerical signs (written numbers) to record the amount of something. They also use groups of standardized written sound signs (written words) to record speech and thoughts. The written signs can be used for this purpose because the literate members all perceive the same connection between the arrangement of a group of these signs and the name/number they wish to record. Using this system helps us to communicate and remember.

Our distant ancestors were just as human as we are, but they were illiterate. To become literate, they and their cultures had to attach a specific, commonly held meaning to a sign. This was a difficult change to make because, unlike the connection between clay and pottery, most of the connection between signs and what they represent occurs in our heads (Fischer 2001). Try conceiving of the reverse: becoming illiterate. In the Middle East, the process of grasping the idea of symbolic abstractions and using signs for writing was gradual. There are thus no specific invention dates, just convenient reference periods.

Prehistory's preconditioning of the process is recorded in the widely distributed but uncommon use of pictograms, tokens and enigmatic signs (Rudgley 1999). The early part of the transition from prehistory's abstractions to written records likely started around 8000 BC, within the context of the domestication and settlement of humans in farming villages like that of Abu Hureyra. In this context, familiar objects came to be more routinely represented by stone and clay substitutes, or tokens. After about 4000 BC, the transition continued in a different context. Around this time, larger settlements with more complex cultures became more common in the region. In these settlements, tokens were increasingly used to record items (mostly food) in a manner considered to be the beginnings of administrative accounting (Fischer 2001).

Specifically, starting about 4000 BC, the sedentary agricultural people of the southern Mesopotamian region started to congregate in increasingly dense urban centres, such as Uruk in Babylonia and, to a lesser degree, Susa in nearby Elam (figure 21). The formation of these proto-cities was accompanied by urban issues, such as maintaining its food supply. The citizens' resolution, such as managing the growing of crops in the fields surrounding the city and distributing the harvest within the cities, included the beginnings of a central administration. This is suggested by the introduction of the cylinder seal to denote the authority of those using

it. Evidence that these tools included more formal accounting is the appearance of a greater variety of specifically shaped clay tokens, each of which represented a specific item that was likely being accounted for.

One interesting accounting variant was the enclosing of the tokens in a spherical clay envelope or bullae, probably to secure the record they represented. In some cases, impressions of the tokens were made on the outer surface of the bullae. This meant that if you could relate the abstract, token-shaped impression to the object that it stood for, then you could know the content of the bullae without opening it. You could "read" the "written" impression.

In Babylonia, after around 3100 BC, accounting using writing appeared. Although it was increasingly used, it did not displace tokens (Nissen et al. 1993). The suitability of tokens for Babylonian accounting ensured that they would stay in use alongside writing until around 2000 BC. It took a while for the level of abstraction in written accounting to become the norm.

Early Babylonian written accounting records were kept on flat clay tablets using proto-cuneiform. In this script, objects were represented by a somewhat abstract pictorial symbol, such as a cartoon drawing of a fish to represent fish. Over time, proto-cuneiform changed to become a more complex form of writing called cuneiform. For example, a later cuneiform writing system used in Uruk consisted of about 1,200 symbols. This system recorded objects or actions using pictorial symbols. Some symbols reflected parts or all of a physical object, others were enigmatic (apparently arbitrary), while many were symbol combinations. About 60 of these symbols were used to record the count of an item (count symbols). The remainder of those symbols that were used for accounting represented items and words such as *deliver* and *receipt*.

Some of the count symbols have been directly related to real-life measures of quantity. For example, a representation of a ration bowl was one of the symbols used to record small amounts of barley when it was treated as a cereal. A representation of a beer storage jar was one of the symbols used to record amounts of beer.

The rules governing the use of the Babylonian accounting symbols are radically different from those governing the use of the numerical signs of today. In order to deal with the diversity of items to be accounted for, Babylonian count symbols seem to have been arranged into more than 14 different counting systems, of which five were most commonly used (figure 6a). In order to create an accounting record for a particular item, a Babylonian scribe had to choose the relevant system.

Within each counting system, the count symbols were arranged in a specific order. However, the numerical spacing between the symbols in a system varied. For example, the spaces in the counting system used to record germinated barley for beer making changed, from the smaller to the larger amounts, by respectively 2 to 5, 6, 8 and 10 (figure 6a). (Babylonian's preferred count groupings of 6 and 10, their base.) In addition, the spacing sequence for each counting system was different and unique. This is totally unlike the Western culture's single number system, which has a single sequence of number signs with a fixed space of one between them: 0 to 9, a base of 10.

One explanation for the variation in the spacings is that, in many of the counting systems, the relationships between successive count symbols matched the relationships between the amounts that the real-life measuring devices or storage containers held of that item. It seems highly likely that the written accounting system and its rules mimicked, perhaps even duplicated, the structures of the Babylonian's physical measuring and storage systems, in use long before the time of cuneiform. But it is also unlikely that the variation exclusively reflected the range in the measuring/storage item's capacity because the amounts represented by some count symbols is too large to be a container.

When a scribe wanted to record the count of an item, they would first select the written counting system appropriate for that item. From that system, the scribe would select all of the appropriate count symbols needed to represent the item's count. These symbols were then written down together in a manner that, collectively, represented not just the unit count for the item but the item's type and the system used to record it.

To picture how a count system was used, imagine today recording a volume of beer using a combination of symbols, each of which represented a beer container type: kegs, flats, six-packs and cans. You would then write the volume down using a collection of these symbols. For example, for a volume equal to seven cans, you would write one six-pack symbol and one can symbol. When asked you would say you had one six-pack one can of beer.

After looking at the accounting record in figure 6b, you would be excused for wondering how anyone could read its apparently disorganized symbols. Reading the record must have become even more difficult over time because the pictorial symbols became abstract symbols. There were other complications to reading an accounting record. Not all count symbols were unique. Some were used in more than one system. Because the count systems were independent of one another, the same count symbol could be used to record a different count of a different item. To read an accounting record, the reader had to recognize the system being used.

If that wasn't confusing enough, there was sometimes a choice of which system could be used to record the amount of an item. Consider that if item B was made from item A, then the scribe could choose to record the count for item B by using item A's counting system. This was done by using the amount of item A needed to make B. For example, a count of barley beer could be recorded as the amount of barley needed to make that count of beer, which would be written down using the barley counting system (figure 6a).

It seems that for a Babylonian scribe to read and write using this accounting system, they needed to know more than the counting systems and their symbols. The scribe needed to know the culture's measuring conventions and storage details. They also needed to know how the particular item being recorded would be used later and how the resulting product was measured. In other words, the scribe had to be fully familiar with the culture's relationship with the items being accounted for. It is only with this degree of familiarity that the scribe could accurately write down the count for an item and then later use the symbols as cues to read that account (figure 6b) (Damerow 1998).

The significance of this conclusion can be more fully appreciated if we contrast the Babylonian's method of counting with ours. For us, numbers as represented by our number signs are part of a universally applicable, abstract number system that is independent of the physical world around us. After you have learned that single system's highly abstracted signs and symbols and its standard rules, you can apply it to all things. As the Count on *Sesame Street* demonstrates, if you can count, record and read numbers of sheep, you can count, record and read numbers of trees. At the same time, a universal number system helps to create the sense of a unifying order in a diverse world. This is, at least, a Westerner's view of numbers and counting.

A Babylonian scribe thought of numbers and used them differently than we do. They likely had no notion of an *abstract* single numerical entity, such as 1 or 5 (not to be confused with recognizing one or five of something).

They certainly had no notion of zero. The scribe had no concept of a universal, abstract, integrated number system. They did not use a number system. They used a counting system that mimicked the world they were familiar with.

The evidence for this includes the following knowledge: there was more than one Babylonian counting system; a count symbol's value could depend on the item being recorded (it could depend on context); the spacing between the count symbols within a particular system was not uniform; and arithmetic was probably accomplished by the equivalent of finger counting (Damerow 1998; Nissen et al. 1993). In essence, cuneiform accounting was more a system to record their activities than it was an arithmetic tool. It was a memory aid. A Babylonian scribe recorded and read the amount of grain in figure 6b by, in essence, imagining that number of containers.

For Westerners, our highly abstract way of thinking about numbers is such a powerful reference normal that it is tempting to question whether the above interpretation is correct. Or at least, it is tempting to ask how it is that Babylonians could practise accounting at all. You can be sure that the above conclusion is not some unrealistic interpretation of Babylonian history. How can we be sure? Because, to our surprise, we have examples to show that the ability to count and do arithmetic does not require knowledge about the abstract concept of number or a universal number system (Damerow 1998).

For example, today the Munduruku of the Amazon determine the number of objects they see by an approximation or estimation process. They use the same process to do arithmetic. In their culture, they have exact number words for only one to five. Above five they talk of things in various versions of many. As a result, below five their estimation and arithmetic is exact, but above five, both their estimate of the number of things and their arithmetic is inexact. This is strange, even bizarre to us, but it is normal (standard) for them (Pica et al. 2004).

Even more surprising to us is that the Piraha of the Amazon have no exact number words at all. They have three words to describe few, fewer and many. However, they do understand the concept of an exact number match and the concept of one.

The Piraha and the Munduruku indicate that the core of our numerical capacity is innate but a universal number system is not. Number words are an invention, a learned skill, which helps us to remember amounts and thus, most importantly, to see and work with the world

differently (Frank et al. 2008). Further steps are needed before we can create, and thus learn, a universal number system.

When we look at the details of the Babylonian accounting system through the scribes' eyes, and within the context of their long-ago culture, then surprising conclusions can be drawn about the general process of personal and cultural change. To scribes, their method of written accounting was an integral part of their personal world view. The counting system they used was just as much an integral part of their fully functional Babylonian culture and its world view as our universal number system is for us today. Thus Babylonian scribes, like each of us, had no reason to question their "experience" of numbers and counting or to wonder if it was a product of learned techniques or beliefs. Certainly the system could change but only at their convenience.

The same is true for their, and our, other basic perspectives, views and experiences of the world. Certainly there are significant differences, even contradictions, between their and, say, Western culture's view of the world. Yet, each culture's respective members considered their personal experiences to be largely true and normal (standard). The members of both cultures treated their respective personal and their culture's world view as a largely accurate representations of a greater, immutable external reality. There was little incentive to change their personal or their culture's world view, and they experienced considerable inertia when they considered doing so, just like us. Think of Western culture's notion of progress.

It is common to suggest that the Babylonian's *technical* invention of written accounting revolutionized their culture. But it seems more likely that its technical aspects mostly addressed practical aspects of their accounting. After all it changed how they documented their accounting but not how they generated what they wrote down or their numerical thinking. The core of their accounting method remained largely unchanged. And they did not adopt a universal number system, which came much later.

The most profound cultural changes that resulted from adopting written accounting were likely more indirect and subtle. For example, there was likely an increase in the members' familiarity with abstraction. This is suggested by the slow increase (drift) in the Babylonian's use of abstraction in their written accounting. This increased familiarity with and use of abstraction preconditioned them to make other changes to their personal world view, which would have contributed to changes in their culture's world view and their relationship with their world.

The Babylonians were preconditioned by their greater appreciation for abstraction, symbols and writing to undergo a different critical conceptual (mental) change. The scribes, at least, came to mentally associate an object's accounting symbol with the sound of the object's name rather than with the object itself. Once that connection was made, writing could be used to record sounds and thus speech.

The monosyllabic nature of many Sumerian words facilitated this mental transition. The written symbol for an object with a monosyllabic name could, by default, simultaneously represent both the object and the sound of its name. By focusing more on the representation of the sound rather than the object, an object's accounting symbol could become a sound sign. These object-sound signs were adopted and adapted to represent the sound names of other objects and non-concrete items, such as actions. They could also create new object-sound signs. This could be accomplished by adjusting existing object-sound signs, for example, if another object had a similar sounding name, or by mixing object-sound signs, for example, if an object's name was multi-syllabic (i.e., created a complex word). In the process, the application of writing expanded from accounting to speech (van de Mieroop 2004).

It took until about 2400 BC or shortly thereafter (around the time that the great pyramid of Khufu was built in Egypt) before cuneiform was fully capable of being used to record and resound Babylonian speech. This applied to both Babylonian languages, Sumerian and Akkadian (Fischer 2001). By this time some 700 years had passed since the first proto-cuneiform characters were in use.

An important question to ask is, what mix of technical, cultural and personal factors maintained, for so long, the drive to expand the use of writing well beyond its application for accounting? Was that change dominated by drift, was it directed, or was it a mix of factors? Whatever the combination, it likely involved many variables and interactions between them.

Consider that during the period of the writing's expansion to record speech, the main language of Babylonia changed. The simpler Sumerian was slowly displaced by the more complex Akkadian. Perhaps that switch generated the need to record speech phonetically rather than symbolically (Nissen et al. 1993).

The process of change was certainly slow. Was this the result of a mix of inherent cultural inertia to change and heightened cultural conservatism, or lack of need (Fischer 2001)? Perhaps the writing of speech grew at a rate directed by the bureaucrat's need for increased communication, copying and precision. Or should we blame personal inertia or lack of mental capacity, such as the difficulty coming to terms with the abstraction?

In the end, the writing down of speech appeared, and the associated personal and cultural changes rippled through the culture. Individuals would have been prompted to expand their abstract thinking. This change likely had some impact on their understanding of their physical world and their relationship to it. Perhaps it contributed to the ongoing severing of their direct connection to their environment started by farming and continued by accounting and a more complex culture. It likely changed how people saw one another. Consider that the division between literate and illiterate could appear.

Their culture's world view would have been adjusted to accommodate the new dimensions of writing and reading speech, and the associated peripheral changes, such as a scribe and new views of the world. In particular, the scribe could be seen as being in control of something special, even magical. Think of numerology. Because the written word could symbolically carry authority, writing down speech changed the exercising of power. For example, writing could represent the authority of the ruler's spoken words. The ruler could thus exert their control over the people for much larger distances and, potentially, with greater reliability. The culture was also preconditioned to possibly later accept different potentials for management, such as the standardization of production, greater management at a distance and, importantly, different views of the world's order.

The ability to write down important utterances that were previously committed to memory changed how they could be used. In particular, it changed how a culture's history was managed and adjusted, and it changed who controlled it: the literate and their bosses. Similarly, religion and the authority of the gods were changed by the appearance of sacred texts.

Some changes did not happen for the Babylonians. Just as accounting was practised without knowledge of a universal number system, so the writing of Babylonian speech using cuneiform did not fully represent the phonetics of a language. It is also likely that the early scribes who wrote down speech failed to recognize that writing was a universal, autonomous method capable of recording thoughts, as we use it today. Consider that in the Western world prior to about AD 500, reading was practised out loud or as a chant. When it was recognized that reading could be conducted silently, that ability was considered extraordinary. This discovery contributed to the later recognition by Westerners that written words can be thoughts and thoughts can be directly written down (Fischer 2001).

In summary, this chapter discussed the nature of personal and cultural change. It did so by dividing it into processes and influences that we can relate to. However, we find it difficult to conceive of the overall process of personal and cultural change as the complex sum of these processes and the influences on them.

We can more easily relate to personal and cultural change if we treat it is an aspect of the system in which we, along with our biases and limitations, are a key player: the dynamic, complex human-environment system. We are helped to do so by dividing the system into three factors (individual, culture and environment) that we can also use as points of view: the three factors are component-perspectives. The variables that each of the three factors represent interact with another one over a wide range of time, space and intensity in ways that include feedback. These interactions are the system's functioning and changing, its dynamics. In overview, that dynamics is described as a mix of deterministic, random and chaotic characteristics. In the closer view, the generic terms used to describe a particular aspect of its functioning/changing include *drift, adjustments, preconditioned, unexpected, directed, inertia, imposed, complexity* and *external/internal*. We are also helped to relate to personal and cultural change by applying nuanced thinking and decision making.

With this help, it is easier for us to think of personal and cultural change as, at some level, an interconnected whole that is part of the dynamic, complex human-environment system's functioning. We can also more easily and realistically answer the questions posed at the beginning of this chapter: when not to influence change, when to and in what manner. A meaningful answer notes the importance of functional boundaries and viability. When our present and future are contemplated in this context, we can conclude that it is appropriate to guide personal and cultural change so that we and our culture stay within our respective functional boundaries. We can favour changes that help us adopt a more viable lifestyle and help our culture to become more viable.

The Babylonian accounting example illustrates that we face challenges in our effort to do so. Success requires that, when making decisions, we take into account our decision-making foibles, the constraints on our imagining of our circumstance, the influences on both and how we change. The lessons of this chapter will be applied and built on in later chapters, in particular when discussing a model of the human-environment system in chapter 11 and directing change toward viability in chapter 18.

This ends part 1. Most of the discussion focused on how we and our culture function and change, and the related constraints and limitations. Of special interest was the role of our decision making and its foibles. So was the presence of an internal and an external reality, and the contribution that our culturally mediated interactions with one another and our environment made to our internal reality. Our internal reality includes our reference normal and our personal world view. The collective internal realities of a culture's members are reflected in their culture's world view. Also discussed in part 1 was some of the abstract features of how the external reality functions and how we and our culture are part of a dynamic, complex human-environment system. This knowledge forms the background for parts 2 and 3.

However, in part 1 our environment was mostly mentioned in generalized or abstract terms. So part 2 will start with a discussion of its real-world functioning and changing. Part 2 will continue with a discussion of our real-world relationship to the environment and its role in our lives. It will then provide a model of the complete human-environment system that will see us through to the last chapter in part 3.

# PART 2

# THE PRESENT

# CHAPTER 7

# THE STRUCTURE AND FUNCTIONING OF AN ECOSYSTEM

We have all used the words *nature, ecosystem* and *environment* either alone or as part of a phrase. We all sort of know what the terms mean, yet we still can't seem to agree on the impact we are having on them or the significance of the resulting changes. These issues will be addressed in the next few chapters.

One of the main difficulties we face in relating to the world around us is one of scale. We can look at the world on a global scale, a neighbourhood scale or at any scale in between. When talking about nature (the non-human part of our surroundings), the following terms are used to distinguish between these scales. The *biosphere* is that thin band from just below to a short distance just above the surface of the earth that supports life. The word *environment* collectively refers to a system of life plus the geophysical systems, such as climate, on which that life depends. Earth's environment can be used as a synonym for the biosphere. An *ecosystem* is any selected part of our environment that contains the living and the non-living aspects that are immediately relevant to its functioning. Think of a lake ecosystem. A *community* is an ecosystem seen at a scale relevant to the individual community members, including their interactions with one another. A *biome* is an ecosystem on a regional scale. The terms *ecosystem* and *environment* are not mutually exclusive, and the location of the boundary between them can be argued. The same is true for *ecosystem* and *community*. The context in which the terms are used should provide sufficient guidance to the intended meaning.

This chapter, discusses what an ecosystem is, how it functions and our relationship with it. The discussion will be simplified by focusing on terrestrial ecosystems. One tool to increase our degree of knowing about ecosystems is supplied by ecology. Unfortunately, a beginner's experience of opening an ecology textbook for the first time is like entering a hall of distorting mirrors at the fun fair. You recognize features and examples, but you just can't quite figure out what the picture is supposed to look like. Part of the reason for feeling overwhelmed is that our environment is a very complex system (Levin 2005). Another reason is that ecology texts try to describe how an ecosystem functions in a logical manner. However, because of its complexity, there is no single structure to serve as a framework for putting it all together. There are only preferences for a particular view of an ecosystem's structure and functioning (Begon et al. 1996).

If we are to appreciate an ecosystem, it needs to be described in a manner we can relate to. The perspective taken here is that an ecosystem is most easily appreciated when described by themes we can readily personalize: who lives in the ecosystem (species and individuals); what they use and generate to live (resources); where they live (spatial distribution); how they get along with others (interactions) and how many there are (number). In addition, to appreciate an ecosystem we ought to know how all of the pieces are linked together into a functional whole, as well as the constraints on that structure and functioning (constraints). Once we have a framework into which these overlapping subjects fit, we can look, in the next chapter, at how ecosystems, as functional wholes, change over time.

## Living Aspects: Species and Individuals

Carl Linnaeus (1707–1778) was born into a Christian culture that believed God created and ordered all living things. Linnaeus was not a fervent Christian, but in keeping with the times, he was a creationist. He believed that the number and nature of organisms on the earth had not changed since God had put them here. Thus, by discovering the orderliness of those organisms, he could reveal the glory of God's creative power.

The outcome of his culturally supported work was *Species Plantarum* (1753), which detailed his systematic classification of all the then-known living things. Unlike previous classifications, he did not order organisms by their usefulness to humans, such as grouping together sugar cane, sugar beet and bees because they all produced sugar. He instead took the unusual perspective, for that time, of using the organisms' innate characteristics, such as their shape or, for plants, their sexual organs, to try and represent the divine order.

His classification system allowed Linnaeus to discern a remarkable pattern of similarities and differences between organisms that was not apparent to the casual observer. Consider that all flowering plants produce seeds enclosed in a fleshy fruit rather than open to the environment. Compare, for example, an apple from an apple tree, which does display flowers, with a cone that comes from a pine tree, which does not.

We still use a Linnaean-type system to classify life. In this method, the classification box into which an individual is dropped is called "species." The number of species on the earth is large. The estimated total number lies between 3 and 30 million (Begon et al. 1996). The best guess is 10 million, of which some 1.9 million have been described. The majority of the unnamed species are insects, bacteria and viruses, which is not surprising since the smaller the size of an organism, the greater the number of possible species.

The order in the Linnaean classification system facilitates our thinking about how a group of individual organisms with similar characteristics (i.e., a species) relates to individuals of a different species, and about how species change through time. Linnaeus's work also identifies two rules of thumb that help us think about ecosystems and our environment: how nature functions makes better logical sense when it is looked at from *its* perspective rather than when we impose our values upon it; and to appreciate nature we must consider both a wide and a close-up view of our world.

The Linnaean system has at least one significant disadvantage. When we use the system, we tend to focus on finding an organism's unique species box, just like birders checking off a species box in their life list of observed birds. However, by focusing on the box, we ignore that, from an ecosystem's perspective, each individual within a species is unique, and thus there is always variation within a species. For example, individuals of the black bear species can be big or small in size, and black, brown or even white in colour.

The presence of these types of variations can make it difficult for us to decide which species box an individual belongs in. Is a white bear really a black bear rather than a different species? A classifier at this point of indecision has to decide if the individual being classified represents the natural variation of a species or a different species altogether. The previously mentioned conundrum of categorization raises its head again.

This is the time to realize that species classification is a tool or model, not a replica of life. To paraphrase Jones, we will fail to understand life if we treat it like a stamp collection (Jones 1999, 54). Like most of the sharp boundaries we give to our categories of natural things, at some level, the boundaries of a species exist because we have chosen them, not because they must exist. Thus, although birders, pedigree animal breeders and GMO designers might want to minimize variation and gradation, it is an unavoidable, essential characteristic of life's complexity.

Our focus on the species box also means that we tend to forget about the way in which an individual of a species lives from birth to death and experiences that life, collectively called its lifeway. A lifeway includes an organism's life cycle (e.g., growth, reproduction and death), the details of nurturing and training its young,

how it perceives the world around it, and its interactions with the other members of its community and environment. Our oversight is unfortunate because life is expressed through an individual's lifeway, not by it species classification. A stuffed moose can be classified to a species, but it is no more a living moose than a recording of a birdsong is a bird.

One way to personalize the importance of individuals is through an appreciation of the influences that determine their personal characteristic shapes and lifeways, and their variation. We should immediately think of DNA, because both an individual's species and much of its lifeway is stored there. But those genetic controls are not the sole determiners of an individual's characteristics. They are influenced and complemented by near and non-genetic factors.

To make a rat, for example, the genes in a female rat's egg must be activated by fusing it with a male rat's sperm. From this point on, a series of automatic processes reads the rat's gene-stored DNA blueprint and then uses its instructions to build a new rat. However, the processes of reading and building are neither fixed nor self-contained, and the blueprint is somewhat flexible.

For example, the fluid in a cell is in contact with the outer part (the epigene) of the genes hosted by the cell. Because the epigene mediates the expression of the genes, any change in the characteristics of the cell's fluid that alters the epigene can affect the expression of the genes. (Lister et al. 2009).

There is a wide range of factors that can (directly or indirectly) influence whether a particular gene will be read (activated) and how strongly it will be expressed. Consider the so-called environmental factors that have this potential. They can be inanimate or animate, so they include the nurturing of parents; the physical, chemical and social aspects of the individual's host community; and interactions with other species.

The influence that an inanimate factor can have on how an organism is built is illustrated by comparing *Salmonella* grown in space with the same strain grown on Earth. *Salmonella* grown in space is more virulent and more resistant to changes in its environment, such as acidity (pH) (Wilson et al. 2007). An animate factor's influence is illustrated by a mother rat's nurturing activities. Part of her maternal care is to groom (lick) and nurse her young. The level of her licking efforts influences the formation of those aspects of her offspring's endocrine system that will control the production of stress-response hormones. Her licking will therefore determine the young rat's response to stress as an adult. Surprisingly, as an adult, the stress-response levels to which a young rat has become predisposed or preconditioned by its mother's licking can be passed along to its young (Fish et al. 2004). That is, the environmentally induced changes to gene expression in a growing individual can become heritable.

The changes in a growing individual's characteristics are thus not solely determined by reading its genes like a book. They are also determined, within limits, by factors that influence the process of reading the book or change the story. These factors can affect how the individual functions from conception to death. These types of influences make a significant contribution to the variation between individuals within a species.

It might seem that the presence of flexibility in the characteristics of individuals and thus variations in their respective species is either unnecessary or represents defects. This is not generally the case. Individuals of all ages benefit from some flexibility, or plasticity, in the expression of their genes. Consider that it gives them some freedom to adapt physically and behaviourally to their changing circumstances. For example, flexibility allows an individual organism to take some advantage of changes in its usual food resources and to somewhat adapt its basic life processes to environmental changes (Jones 1999).

Similarly, variability within a species gives it a greater ability to accommodate or adapt to longer-term (greater than a generation) environmental or ecosystem changes. Variability thus provides a species (or, more precisely, the group of individuals it represents) with a better chance of surviving from generation to generation. Think of individuals and their species having to adapt to an expanding, sprawling urban environment.

It is for these types of reasons that some flexibility and variability is an essential characteristic of individuals and species. There are limits, though. For the individual, one of the most important limits imposed on them is their ancestry, in particular their genetic inheritance. You can only be your species, and you need to associate with members of the same (or very closely related) species if you are to fulfill your lifeway. Think of reproduction. This limits the degree to which an individual can differ from the other members of its species.

When a young individual's inherited characteristics and personal peculiarities are considered within the context of its life-supporting host environment, then another critical aspect of their life becomes apparent. An individual's characteristics can be thought of as the attributes or tools it will use in its efforts to complete its lifeway, such as eating, growing, reproducing and training its future young. In using these tools, a youngster

will interact with its environment and nearby individuals, both of its own and other species. By interacting with them the youngster becomes a member of a community.

The word *community* is deliberately used to convey the notion of a whole. It expresses an individual's dependence on others, both as individuals and as a group. It also communicates that an individual participates in the formation and ongoing functioning of the group that supports them: their community.

Membership in a community thus both enables and limits an individual's efforts to complete their lifeway. Consider that as a youngster begins to participate in their community, they discover that its functioning both facilitates and constrains their efforts to secure adequate amounts of food: it provides that food but only makes a limited amount accessible. All individuals thus face choices and must make decisions. For instance, there are compromises involved in exploiting resources, and there are trade-offs when dealing with reproduction costs. In addition, the result of their decisions can affect the functioning of their supporting community, and thus their efforts to complete their lifeway. A feedback loop exists between individuals and their community.

In summary, the characteristics of organisms that look the same are used by the Linnaean classification system to assign a species name to the individuals and their populations. However, there is inescapable variation among the members of a species, so the rigidity of the classification box is more apparent than real: it is a convenience. Variation is an unavoidable, essential characteristic of life's complexity. Associated with the species name is a description of how an individual member lives its life, its lifeway.

By ordering the earth's organisms on their terms, the Linnaean classification system helps us to think about how species change though time. However, it is the individuals of a species that complete the changes. It is they who undertake the process of living, interact with the individuals of the community's other species and thus form the community in which they all try to complete their lifeway. Referring to species as engaging in these activities is a convenience.

## Resources

Living is a process by which an organism tries to complete its lifeway while participating in the functioning of its community. During this process, it uses elemental resources, occupies space, consumes energy, modifies its environment and emits waste. The organism's relationship to resources can thus be described from a number of perspectives, such as energy flows, material flows, the individual or its community. In this section, the focus will be on the flow of elemental resources, first from the point of view of the resource user and then from the point of view of the resource being used. During the discussion of the user's view, some of the element cycle that links these two views together is noted within square brackets [ ].

**The View of the User.** The plant-soil community in the forest near our home provides a good example of the user's view of resources. All plants use resources in order to construct themselves, function and interact. In the process they all generate waste.

For a plant, the resource of carbon dioxide comes from the atmosphere. It is "breathed" in through the plant's leaves. The green parts of its leaves then use the energy provided by sunlight to convert carbon dioxide, plus other resources, into the carbon compounds, such as cellulose [carbon cycle], that the plant needs to complete its lifeway.

Some of the other resources a plant needs are nutrients and minerals. The nutrient elements include nitrogen [nitrogen cycle] and phosphorous. The mineral elements include calcium and iron. These resources are found within the soil, but they must be dissolved in water before they can be drawn into the plant by the plant's roots. The water itself [water cycle] is also a resource used by a plant. It plays many roles in the process of living, from facilitating photosynthesis to becoming part of the plant.

The unwanted products from the plant's use of resources are discarded as waste. For example, during photosynthesis, oxygen (along with water [water cycle]) is respired out through its leaves to the atmosphere. Unwanted leaves and twigs are discarded to the soil as solid organic matter [carbon cycle].

It would seem that all that is left to know about a plant's use of resources is its remaining resource needs. Plants require soil to stand upright, space to grow and air for chemical communication and reproduction. But that is insufficient knowledge if you wish to fully appreciate how a plant's use of resources makes it a member of a plant-soil community.

In our forest, a willow tree had fallen over, pulling its radiating disk of roots out of the 5 cm thick, dark, damp, organic-rich layer (organic soil) that forms the forest floor. If we pay close attention to the tree's exposed root wad, we can see, nearer the base of the tree, a few broken roots sticking out. When the tree was upright, these roots extended vertically down through the organic soil into the reddish mineral-rich, organic-poor compacted dirt below it. This mineral-rich soil is the ultimate source for many of the minerals and nutrients that plants require. But the processes that release them from the mineral soil operate very slowly. So slowly that the short-term availability of many minerals and nutrients, such as nitrogen and magnesium, in a form useful to plants is one of the ultimate constraints on the number of plants that the earth can support.

The willows, like most members of the plant-soil community in our forest, satisfy most, if not all of their mineral and nutrient needs by acquiring them from the organic soil layer. They can do this because, as its name suggests, the organic soil consists of the discarded, dead plant material that is in the process of decomposing. During decay, the minerals and nutrients in the rotting plant matter are made available in a water-soluble form that the willows' organic-soil-hosted radiating roots can easily absorb. The organic soil's recycled minerals and nutrients meet most of the willows' immediate needs. Only the amounts lost during the recycling process need to be supplied by the mineral soil.

A puzzle arises. Plant matter consists mostly of cellulose, which breaks down very slowly if only exposed to air. You can confirm this by looking at a piece of wood inside your house. How is it possible for the dead plant material in the soil to break down and release the minerals and nutrients in it?

We are not buried in dead plant matter because there are organisms in the moist organic soil that specialize in breaking down plant matter. This largely invisible and unknown decomposer (detrivore) community consists of millions of individuals from many species. We know that the species range in size from 0.001 mm to around 30 mm, with the majority of species falling in the smaller size range, such as fungi and bacteria. Only a minority are larger than 2 mm, such as insects and worms. This explains why, when we look at organic soil, it seems like mostly lifeless dirt (Begon et al. 1996).

We don't know exactly how many species there are in the organic soil community. For example, we know of 15,000 nematode worm species but speculate there could be as many as 100,000. Some 3,600 earthworm species have been described, but that number may eventually double (UNEP 2002).

Because this underground community lives in the dark, they have no direct access to sunlight's energy. Almost all of the energy available to its members ultimately comes from the soil's dead plant matter. Its compounds, in particular cellulose, store some of the sun's energy that was used by the living plant to make them. The soil community's members gain access to this energy by breaking the cellulose down into sugars, which they can then use as an energy source [carbon cycle]. At the same time, they help to release the nutrients [nitrogen cycle] and minerals in the plant matter.

Despite the significance of the soil organisms and the decomposition process to plant growth, the details of how the organic soil community functions is poorly understood. One reason is that the many members making up the organic soil community are tightly bound together through multiple and often strong interconnections. For example, only about 1% of soil microbes can be cultured independently of their soil community (Wolfe 2001). It is a complex system.

Overall, the soil community functions by its many member species co-operating to complete the many decomposition steps needed to turn dead plant matter into food. The members participate in this process through their efforts to complete their lifeways. In doing so some organisms manage to directly extract some of the energy, minerals and nutrients stored in the dead plant material. The waste and (living or dead) bodies of these decomposer organisms, and the remaining partly decomposed plant matter, is then available for other organisms to use. This process is repeated many times before the plant matter is completely broken down (Begon et al. 1996).

During this complex process, the minerals and nutrients released into the organic soil are used many times (recycled) before they finally leave the soil. Most nitrogen [nitrogen cycle] departs through the roots of the plants growing in the soil to become a part of a living plant, that will be returned to the soil on its death. A feedback is created. Some depart from the soil dissolved in the water that runs off the ground after a rain [nitrogen cycle]. A little departs as a gas. In contrast, the carbon dioxide [carbon cycle], which is generated when soil organisms extract and use the energy captured in the dead plant matter, returns to the atmosphere from where it came. The energy departs as heat, which is why compost is hot.

The recycling of the minerals and nutrients in the organic soil's dead plant matter by the plant-soil community is increased and the losses reduced by a number of the soil's inherent, interconnected features. Consider porosity. As the decomposer organisms feed on the organic material, they move through the soil. Earthworms, for example, can move some 8 to 12 tonnes of soil per hectare each year (Wolfe 2001). Both the passage of decomposers through the soil and the decomposition process itself make the soil porous while the plants roots and the fungi associated with them bind the soil together limiting that porosity. The result is a soil that is porous enough to act like a water storage sponge [water cycle] but not so porous that the water can freely flow through it. The organic soil has a built-in, slow-release water management function.

Both the plants and the decomposers benefit from the presence of soil porosity. Its water storage function enables the plants to survive the dryer times between rainfalls, which are their ultimate source of water [water cycle] (Begon et al. 1996). The presence of pores makes it easier for the plant's roots to penetrate the soil and thus recover the nutrients and water the plants need. Porosity facilitates the circulation of both the water and the oxygen the decomposers use to extract the energy from the sugars they recover from cellulose. The constraints on the soil's porosity, such as its binding by roots and fungi, also provide benefits to both plants and decomposers. The slowed water runoff reduces the loss of nutrients and prevents soil loss by erosion.

Although the plant-soil community is a single, largely self-contained resource-using/recycling system for nutrients and minerals, we tend to think of it as two worlds. An above ground plant community and, isolated from both it and the greater above ground world, a below ground soil community. However, plants inhabit both worlds: below ground (roots) and above ground (stems and leaves) so they can mediate changes in the above ground world to the soil community and vice versa.

When plants are fed on by above ground organisms, such as caterpillars and deer [nitrogen cycle], they can respond by changing the compounds in their buds and leaves, and the amount of biomass they produce. There can also be a change in the mix of plant species. All of these changes affect the quality and quantity of the dead organic matter supplied to the soil. The plants can also respond to being fed upon by changing the compounds both within and released from their roots. Through these changes, the plants prompt the below ground community to adjust to a changing above ground world (Wardle et al. 2004).

Consider too that a change in the below ground soil community, such as a reduction in nitrogen [nitrogen cycle], affects the plants growing in it. Any change in the quality (e.g., nitrogen content) and quantity of their above ground biomass will influence the lives of the above ground organisms that eat it. For example, herbivores preferentially eat nitrogen-rich plants [nitrogen cycle]. There is a feedback linking the above and below ground worlds.

The above descriptions illustrate some general rules of thumb about resources and the process of living. The individual members of a community can complete their lifeways because the overall functioning of their host community makes available the resources they need (Wolfe 2001). Thus, at some scale, an organism's ability to complete its lifeway relies on it being a participant in a functional community.

But, because a community's functioning is the aggregate of all of the members' efforts to fulfill their lifeways, they are dependent on one another. The strength of this interdependence is illustrated by the plant-soil community: its above and below ground members rely on each other to complete their roles in the recycling of minerals and nutrients. For example, if the organic soil is removed from an area, few plants can grow. If the plants stop contributing material to the soil community, it eventually ceases to function.

The willow tree's root wad mentioned at the beginning of this section illustrates additional rules of thumb about the use of resources and the process of living. The wad was described as dominated by horizontal roots with only a few roots travelling vertically into the mineral soil. Although only few in number, these vertical roots serve a vital purpose. During the occasional drought, which dries out the organic soil, these roots supply the willow with water found deeper in the mineral soil. By using and transpiring it, the willow reduces the impact of the drought on itself and its community. The amount of something often has little to do with its biological importance to the user.

The willow's ability to reduce the impact of droughts is more important to it than is readily apparent. Although rare events such as a drought are minimally reflected in climate averages, they are critical factors in limiting the survival of organisms. The willow's ability to alleviate the impacts of drought allows it to increase its distribution (range) across the land. Averages do not necessarily represent what is important for an individual to complete its lifeway.

With the recognition that the plant-soil community is a complex system of life, the boundary between it and the mineral soil below takes on a special significance. It separates the upper, life-dominated aspect of our planet (the biosphere) from the life-poor earth below. When walking on the land, it is useful to remember that the biosphere ends, in a practical sense, rarely more than a few meters below the soles of our feet. That boundary reminds us that there are resource limits to the process of living. It also symbolizes how special life on Earth is.

**The View of the Resources.** All living things rely on resources, such as carbon, water, nutrients and the mineral elements, to complete their lifeway. They take them in, use them and then generate waste. An organism's use of resources can be described from the perspective of these resources: where they come from (the source reservoirs); where they go as waste (the sink reservoirs) and how they move between them (the pathways). Because the earth is a closed system, the resources' reservoirs and the pathways linking them form a closed cycle. See, for example, the carbon and nitrogen cycles enclosed within the square brackets in the section above and illustrated in figure 7 (Socolow 1999). Once a resource's cycle is fully described, its critical pathways can be flagged and the limits to its availability to living things can be detailed. A few features of the carbon and nitrogen cycles that are important to this and future discussions are presented below.

An important part of the carbon cycle is plant activity (figure 7a). With the help of sunlight, plants use carbon dioxide from the atmosphere's carbon reservoir to create the complex, organic carbon molecules plants use, such as cellulose. When plants die, the dead plant material is added to the organic soil. These soils do not normally accumulate carbon because, as discussed above, the decomposer community utilizes the dead plant material to complete their lifecycle. In the process, much of the carbon is returned to the atmosphere as carbon dioxide. This is one pathway loop in the carbon cycle.

However, in some environments, such as the sites of rapid burial, and the cold Arctic and warm tropical acidic, oxygen-free bogs, the decomposers cannot complete their work. In these land environments, the organic matter accumulates and begins the process of moving into an underground carbon reservoir in the form of peat, coal or natural gas (methane). If the plant material is marine, then it forms oil (Begon et al. 1996). For example, on average, the transformation through burial of roughly 24 metric tonnes (roughly two dump trucks) of ancient oceanic plant matter results in an amount of crude oil sufficient to produce one litre of gasoline (Dukes 2003).

If, in the future, the peat, coal, oil or methane is exposed to oxygen, then the combustion process releases the carbon back to the atmosphere as carbon dioxide, as happens in the soils. In the process, the sunlight captured by the original plants is released as heat. This completes a different pathway loop in the global carbon cycle.

Plants make and host small amounts of biologically important nitrogen-containing carbon compounds, such as protein. The plant matter hosting these nitrogen compounds can be eaten by animals or, when dead, become a part of the soil. The soil decomposers make the nitrogen in the dead plant matter available and use it. All the while, a small amount is either released to the atmosphere as a gas or washed into a river and, finally, the sea. The successive usage of nitrogen on land is repeated in the ocean until all of the nitrogen is released back to the atmosphere (figure 7b).

Animals can only acquire nitrogen in a usable form from what they eat. The ultimate source of that nitrogen is plants. Some of the nitrogen in the plants that the herbivores eat is passed up an ecosystem's trophic web by the feeding interactions of its other species. All of these animals also excrete some of their nitrogen as a waste product and release the rest when they die. Some of this excreted and released nitrogen is used by plants, some is returned to the atmosphere as a gas and some is transported to the oceans to be used there before being released as a gas.

This description of the plant and animal (biological) loops in the nitrogen cycle is missing the most difficult step: making nitrogen available to living organisms. The atmosphere contains the vast majority of the world's nitrogen, but it is in the form of nitrogen gas, which is not directly usable by plants or animals (Faure 1998a). Plants can only use nitrogen when it is in its water-soluble forms of nitrate or ammonia, which are rare and unstable. At any one time, only a very small percentage of the total nitrogen on the earth is available to plants, and thus to animals, in a form they can use. Most of it is provided by the organic soil's recycling process.

A few plants can augment that supply. They have entered into a symbiotic relationship that can completes the extremely difficult and energy intensive reactions needed to turn nitrogen gas into a plant usable form: the fixing of nitrogen. The classic example of this

relationship occurs between bacteria that can fix nitrogen (e.g. rhizobium) and plants such as alders and legumes that can host them. Rhizobia live within nodules located on these plant's root tips (Begon et al. 1996). In that environment, rhizobia can complete the difficult nitrogen-fixing steps by using the energy supplied by the plant. The energy needed is considerable, amounting to about 20% of the carbohydrate the plant manufactures (Wolfe 2001).

No wonder that only a few organisms are the ultimate source of the majority of the usable nitrogen available to plants and animals. In this context it is unsurprising that the soil community's nitrogen-recycling function plays such a critical role in both the biological part of the nitrogen cycle and an ecosystem's functioning. Without the community's involvement, most of the fixed nitrogen would be used only once. After which, it would leave the soil as a gas or dissolved in water to end up in the oceans.

Despite these fixing and recycling efforts, the scarcity of plant-usable nitrogen still constrains the organization and functioning of ecosystems. Consider that it influences both plant distribution and animal behaviour. For example, living organisms preferentially concentrate and interact with one another around nitrogen-rich locations: urination and defecation sites, nitrogen-rich dead things and places where plant-usable nitrogen is fixed (Begon et al. 1996). This reality of nitrogen is important when considering the relationship between humans and nitrogen in later chapters.

The element cycle view of minerals and nutrients reveals other influences on resource availability and ecosystem functioning. Consider that the presence of protein in plants links the carbon cycle to the nitrogen cycle: the cycles exert their influence collectively. Consider too that the dynamics of an element cycle determine an element's availability. At its most simple, there are delays between an element being released in one place and its later appearance at an ecosystem in a usable form. The time scales involved can be months to centuries and even millions of years. The space scale can be local, but it can also be global, in which case it ignores political and even some geophysical and biological boundaries.

The significance of the element cycle linkages, dynamics and the scale over which they operate is illustrated by the Amazon forest. The primary reservoir for the minerals and nutrients in the Amazon basin is the standing vegetation of the forest itself. They are made available to the growing plants by the recycling of dead vegetation. In the tropics, that process is hyper-rapid, so tropical organic soils are extremely thin. The recycling process is also imperfect, with the result that a small amount of nutrients, such as nitrogen, and minerals, such as magnesium, are routinely lost from the Amazon forest soil system into rivers. These losses can't be fully replaced from the Amazon basin's mineral soils. The high rainfall in the tropics ensures that rocks rapidly become deeply weathered, so the resulting mineral soils are typically nutrient and mineral poor.

A puzzle arises. Despite the inability of the Amazon's leached mineral soils to replace this loss, there are still sufficient minerals and nutrients to sustain the bounty of the Amazon forest. Considering that Amazon biogeochemistry plays such a significant role in the carbon cycle and global climate regulation, it is important to identify the mystery source of the replacement nutrients and minerals.

The answer lies in Africa. The deficiency is made up by nutrient- and mineral-rich dust blown 5,000 km from the Sahara Desert. An outline description of this process starts with the satellite images of a windy winter over northern Africa. The majority of the dust is probably sourced from a paleo-lake in the Bodélé Depression of Chad. Large clouds of this dust blow westward from the depression toward the Atlantic. On their journey, the dust clouds rise in elevation and are joined by the products of biomass (forests, grass, etc.) burning in the Sahel and by smog from Europe. This mixed air mass then travels across the Atlantic Ocean at an elevation of up to 3 km. As it moves, the air mass loses soil dust to the ocean but gains potassium, sodium and chlorine from it.

The dusty air mass's journey ends over the Amazon Basin, where the particles in it help to seed the rain clouds needed to complete the local water cycle. The resulting rain transports the minerals and nutrients from the air onto the forest floor, where they replenish the soil with minerals and nutrients (Koren et al. 2010; Ben-Ami et al. 2010; Remer 2006). In the summer, the winds shift and the soil from the desertification of the Sahel now fertilizes the Caribbean islands and Florida (Griffin et al. 2001). An important rule of thumb drawn from this is that all individuals participate in the element cycles, some of them planetary in scale, on which they and their decedents depend to complete their lifeway.

In summary, an ongoing supply of elemental resources is needed to sustain life. From the viewpoint of an individual in a terrestrial ecosystem, most of the resources it needs to complete its lifeway are ultimately made available by its community's recycling and conversion processes, and by the local features of the earth's geological and atmospheric processes. From the viewpoint of the elements, a different picture emerges. The resources in any community are available because they are part of the interlocking element cycles that operate on a local to global scale over a wide range of timeframes.

The members' efforts to complete their lifeway clearly affects the distribution of resources at a local scale, but they seem not to affect it at a global scale. But the members are part of the resource cycles, so they do, collectively, affect the global-scale distribution of resources.

These two views complement one another. They, together, illustrate that the process of living is part of the earth's integrated, complex system of life, which functions at scales from local to global. We refer to the different scales used to look at this system as the individual, the community, the ecosystem, the environment and the biosphere. Even if the individual community member is unaware of it, their process of living is both constrained and facilitated by the characteristics and functioning of the earth's system of life at all scales.

The following sections will discuss other aspects of the process of living that should increase our appreciation for the nature and functioning of the earth's complex system of life.

## Spatial Distribution

We tend to think about the distribution of a particular species as the area in which we can expect to find individuals of that species. This view is represented by species range maps, such as those found in field guides for identifying birds and plants. By implication, a species range map also represents the area within which all of the conditions required for an individual of that species to complete its lifeway, such as temperature, resources and space, can be found.

However, if you tried to find an individual of that species by going to a random place within its range, you would likely be disappointed. Instead of the monoculture of species implied by the map, you would find a community consisting of a variety of species. To find an individual of a particular species in this community, you might turn for help to the concept of a *species niche*. This abstract term represents a species' place and role in its community. It also represents one view of the conditions needed for an individual of that species to complete its lifeway. To help you use this information, you could refer to a generalized description of the physical relationships between the species forming the host community. The community structure it reveals would include the niche for the species of interest. All very useful, but your search could still be disappointing.

The abstractions of species range, niche and community structure are most useful in our search for an individual of the desired species if we appreciate how the species making up its community are actually distributed. The essence of that distribution is most easily illustrated by an example. A walk in our forest to find individuals of a particular willow tree species reveals that the forest consists of an apparently chaotic mix of individuals from many different species. It doesn't matter whether I look at our forest community from afar, or if I look at one tree from close-up, there is always a diversity of species present. Neither is the distribution of the individuals of the same species simple, as I discover after I have found a few willows of the right species.

In some places I can see substantially more willows than other species of tree, while in different places I see only a few scattered willows. When I mentally outline the willows' distribution, I can't see any order. The area occupied by the willows of interest is a collection of erratically distributed patches of various sizes, irregular shapes and ill-defined boundaries (figure 8).

The patchy distribution of the individual willows illustrates a fundamental mental barrier to at least Westerners' appreciation of communities and their functioning. How do we describe and think about a community if a basic ecological property of its member organism, their spatial distribution, appears to be, at best, only somewhat organized and somewhat definable? By accepting the reality of patchiness, we are forced to put aside our desire to find simple, regular order in the distribution of individuals and species. We are also denied the simple solution of ignoring patchiness, because the distribution of individuals and species has a significant impact on an ecosystem's functioning and how an individual fulfills its lifeway.

There is order in the distribution of species. It just doesn't easily match our sensibilities. The order becomes more clear if we look at the distribution of species over a larger area.

If land species are mapped on a global scale, then the surface of the earth can be divided into at least eight large, mixed-species communities called *biomes*. (It could be divided into many more, but remember from chapter 1 the rule of $7\pm2$, which limits the number of items we can easily remember C01*). These biomes are somewhat systematically distributed between the equator and the North or South Poles. This pattern is matched by a corresponding, systematic change in the biome's descriptions. For example, each contains key plant species and as you approach either pole from the equator, the number of plant species in a biome declines.

The poleward distribution of biomes and the decline in the number of their species is the product of the somewhat systematic poleward reduction in the maximum amount of available sunlight and water. This effect is recorded by a general decline in the amount of potential (or actual) evaporated plus transpired (evapotranspired) water. A decline in this measure means that plants find it increasingly difficult to fulfill their lifeway (Gaston 2000; Currie and Paquin 1987). The result is a systematic poleward change in the numbers and characteristics of the plant species used as biome indicators.

However, the actual amount of evapotranspired water at a location is determined by the climate there. Thus the details of evapotranspired water's surface distribution is, like climate, controlled by more than distance from the equator. It is also controlled by altitude and continental or regional landscape features, such as mountains and valleys. For example, the higher you go up a mountain, the more the climate and thus the amount of evapotranspired water become like that nearer to the poles. The detailed pattern in the distribution of plant species and biomes more closely matches the distribution of climate zones than latitude alone.

Plants form the base of an ecosystem's food web (trophic web). Thus, any changes in a biome's plant diversity and distribution strongly influence the diversity and distribution of the biome's non-plant species. Unsurprisingly, the climate-related decline in the number of plant species in a biome is matched by a decline in the number of its animal species.

Our sense of this global-scale order in, or control over, the distribution of organisms changes when we look at a biome's boundary over a range of scales. Consider the tree or shrub lines used to separate the boreal forest biome from the tundra biome (Sturm et al. 2001). At a continental scale, the boundary appears to be sharp, line-like and well ordered. On a regional scale, it is diffuse and gradual. On a local scale, the boundary is indistinct and experienced as a change in the patchiness of the key species (figure 8) (Stevens and Fox 1991). This is in contrast to our preference, which is for the boundary to be sharp at all scales, in keeping with our culture's sense of order.

The features of a biome's boundary change with the scale we use to view it because at each scale a different set of controls over it is being expressed. In this context, at a global scale the location and nature of the biome's boundary is determined by the gradient in climate-related variables, such as evapotranspired water and light. At a continental scale, the sharp changes in the mix of species represent the point on that gradient (the biome's boundary) where that species is unable to survive (Begon et al. 1996).

At a regional scale, the more gradual change in the species mix across the more fuzzy boundary reflects the impact that short-lived, regional-scale extreme weather events have on the biome's species. A good example of such an event is the extreme droughts that commonly occur in continental climates. In particular, the boundary reflects the place at which these events are strong enough and frequent enough to limit a species' ability to survive and expand their range. But the events are also too short-lived and weak to eliminate a species from an area (Begon et al. 1996).

At a local scale the biome's boundary can be difficult to recognize because it is represented by a change in species patchiness (figure 8). To fully appreciate the local controls on the biome's boundary at this scale, we need to remember that the global-, continental- and regional-scale controls on biome boundaries still apply. They still form the basic framework that controls the spatial distribution of species but their visual presence is now part of the inconspicuous background.

The changes in a species' patchiness across the biome boundary at the local scale reflect the influence on its living conditions that we are more familiar with, such as microclimate, landforms and their orientation. Examples include a swamp, which signifies greater water availability; a hollow that limits exposure to sunlight and a hillside facing the sun that increases it; a bare, soil-free rock outcrop and a boulder. These influences are unevenly distributed over the land, which results in the uneven distribution of the conditions needed for an individual of the species to complete its lifeway (its niche). I see the overall result, at a local scale, when I walk in our forest: a seemingly random distribution of the individuals of a particular species and the patchy distribution of that species

within its host community. As a result, the biome boundary is all but invisible at the local scale. (In case you are curious, willow patches are larger and more continuous in and around open, wetter areas.)

From this discussion, it might seem that the patchy distribution of organisms in a community is solely the result of their passive response to conditions that have been imposed on them by global- to micron-scale land and atmospheric features. Reality is more complex because, at some scale, the organisms themselves make a significant contribution to the influences on one another. At some scale, the efforts of one community member to complete their lifeway affects where the others can live.

Consider that the mere presence of an organism changes the local climate and resources around it, and thus who can live near it. For example, in our forest, the presence of nitrogen-fixing alders alters the distribution of available nitrogen around it. A large cottonwood tree has a dark and a light side, while a spruce tree creates a shaded, snow-poor zone around the base of its trunk. Each of these spaces is preferred or avoided by different organisms. The distribution of species is thus also influenced by the species themselves as they participate in their community.

This allows the local distribution of species and individuals to change at a biologically controlled speed. Consider that in all communities the old individuals of a species complete their lifeway and die, while young individuals of the same or different species can take their place. In the process, a species' distribution can change. This is one reason why the distribution of individuals and patchiness is dynamic: it is always changing at some scale, as will be discussed in chapter 8.

Patchiness is an inescapable feature of the distribution of individuals and species in a community. It provides the benefit of more individuals having more opportunities to interact with individuals of other species. It thus facilitates the provision of resources and the availability of niches to the community's members. Patchiness, like the variation in the characteristics of individuals and species, makes a critical contribution to the longer-term functioning and survival of both individuals and their supporting community.

But there is a downside to patchiness that all species must deal with. It can limit an individual member's ability to complete some of their lifeway functions, such as reproduction. Consider that patchiness can result in small groups of a species' individuals (patches) being sufficiently isolated from one another that they can't interact to breed. The consequence could be inbreeding, population decline, extinction or forced migration.

Certainly, it is possible to suppress patchiness, as can be seen in agricultural monocultures, but, like trying to suppress variation in species, suppressing patchiness can't remove the forces driving its formation, or its importance to both an individual's lifeway and its community's functioning. Instead of suppressing it, species treat the benefit-disadvantage conundrum as a trade-off. They resolve the patchiness problem by developing characteristics to accommodate the downside. For example, plants use airborne fertilization to resolve their reproduction issue. Animals simply walk around.

This discussion about spatial distribution has focused on land species. The same general conclusions apply to ocean species. They too display an ordered but patchy distribution that reflects the global- to local-scale influences on the distribution of their living conditions and the gradients among them. Like the distribution of land species, the distribution of most non-plant ocean species is ultimately linked to and dependent on marine plants. They too require sunlight, for photosynthesis, plus mineral and nutrient resources if they are to thrive. However, the details controlling the distribution of sunlight and resources in the oceans is different from those on land, so ocean species are distributed differently.

The intensity of sunlight rapidly decreases with the depth below the ocean's surface. As a result, ocean plants are concentrated within 200 m of the surface of the ocean and decline rapidly with depth. Below about 200 m, the ocean's detrivores eventually come to dominate.

As expected, the distribution of the ocean's nutrient and mineral resource is uneven. It is partly controlled by how, at a global scale, the nutrients and minerals arrive at the ocean. Some come from the air, but most are delivered by rivers and the base of the Antarctic ice sheet. These additions are redistributed by the ocean currents that flow both from the ocean deep to the surface and from the poles to the equator, and back.

The combination of limited light penetration and the uneven distribution of nutrients and minerals restricts the locations in the ocean that can support a great diversity of species and large population sizes. They include the continental shelf offshore from rivers and off the coast of Antarctica. They also occur along coastlines with upwelling nutrient-rich currents, such as off of Namibia and Peru and adjacent to some islands, ocean ridges and seamounts. Overall, ocean species diversity is greater in warmer waters but productivity is greater in circumpolar regions and cold upwelling waters (O'Dor 2003).

In summary, a mix of the global- to local-scale influences, from climate to nitrogen distribution, determine the distribution of living conditions and the gradients between them. This determines the distribution of species and individuals. However, the distribution of organisms affects the interactions among them, which is also an influence on their distribution. So a feedback is involved.

There is order in the spatial distribution of species and individuals and the boundaries between them. Distributions can vary from patchy to uniform, and the boundaries from fuzzy to sharp. Overall, the spatial distribution and order we see, and the reasons we give for it, depend on both the perspective we take (community to biome) and the scale (local to global) we use to look at it.

## Interactions

As individuals attempt to complete their lifeway activities they interact with individuals of the same and different species, and with their environment. This section focuses primarily on interactions between individuals of different species. Because interactions only occur between individuals, not species, the term *species interaction* refers to the common or "average" features of the interactions between individuals of different species.

Knowledge about the interactions between species will help us to appreciate how an individual of a species completes its lifeway. These interactions are also an aspect of their host community, so this knowledge will also help us appreciate how it forms and functions. The types of species interaction discussed here are predator-prey, co-operation (mutualism-symbiosis), parasitism/disease, competition, communication and interactions with the surrounding world.

**Predator-Prey.** The words *predator* and *prey* conjure up images of lions chasing and killing antelope. However, in a wider sense (which is the one used here), a predator-prey relationship exists whenever one organism eats another: for example, when antelope "dismember" and eat grass.

Humans easily relate to the predator-prey interaction in the first example but not the second. After all, antelope don't have to chase and kill grass. Perhaps we should think of different types or degrees of predator-prey interactions, or label the category as a one-sided feeding interaction.

In the same vein, we think of the classic predator-prey relationship as a simple dynamic balance between eating and being eaten, trying to catch more and trying to be caught less. This may be true from the perspective of the individual predator or prey, but from the perspective of their respective species, the relationship is far more complex.

Consider that a predator species is faced with limits. These include the type of prey it can eat, the number of prey an individual predator can eat at one sitting and the need to develop a long-term strategy that successfully balances the energy (and nutrition) expended to find and catch its prey with the benefit acquired from eating it. The level of impact a predator can have on its prey's population is also limited, if it wishes to eat in the future. Collectively, these types of interaction-related limits constrain the predator species' characteristics, such as the requirement to learn and the maximum size of its population.

From the prey species' perspective, a predator's efforts to eat it have impacts on it beyond the eating of an individual. The effects depend on the prey's as well as the predator's species. For unitary (singular, non-modular) prey species, such as antelope, the classic effect of successful predation is usually death. To ensure its survival as a species, the prey species' lifeway must include an appropriate (effective and energy efficient) response to predation attempts and predation deaths. Possible reactions include an increase in birth rate, a decline in density or an improvement in predator avoidance tactics. The last option adds a psychological twist to this relationship. The predator doesn't actually have to be present for the prey to act as if it is: the prey just need to fear that the predator is or might be present. Within limits this sensitivity is an advantage, however being too sensitive can disrupt the prey's efforts to complete their lifeway.

Modular prey species, such as grass, must also respond appropriately to predation. And they do, in a manner that is equivalent to unitary organisms but less straightforward. Consider that, for plants, the overall result of predation by herbivores is usually not death but something between death, dysfunction and a mutually beneficial relationship. Plants that are chewed on partly compensate for their loss in a variety of sometimes unexpected ways. For example, they can regenerate their missing bits, make their usually desirable bits

distasteful and warn their neighbours to do the same. As a result of these responses, determining the effect of herbivores on plants requires studying their relationship over an adequately long time (usually many years) and over a large enough space (Begon et al. 1996).

If we take into account that both predator and prey are part of a community, then both the direct and indirect effects of the predator-prey relationships need to be considered. For example, the snowshoe hare is a herbivore in the boreal forest biome. As the hare population increases, so does the population of its main predator, the lynx. However, there is a delay of one to two years between the increase in the snowshoe hare population and an increase in the lynx population. During this delay, the growing snowshoe hare population has a significant impact on the vegetation it eats. This includes the shrubs that squirrels must consume in order to maintain their fertility (Krebs et al. 2001). Thus the degree to which the lynx preys on the snowshoe hare indirectly affects the squirrel population. A general rule of thumb arises: the results of direct interactions between species, such as predator-prey, cascade through the community as indirect interactions to affect many species.

It is convenient to think of predator-prey interactions as eventually reaching a static equilibrium between predator and prey populations. But there are other influences on this interaction that ensure the equilibrium is dynamic. The repercussions from the slower reproductive rate of the lynx relative to the hare provides one example. Consider too that the degree to which insect predators affect the plants they prey on can change with the level of pollution. The presence of these types of complications should remind us that predator-prey interactions, like all interactions, are actually part of a larger complex, dynamic community. From this perspective, the dynamics of the predator-prey interactions can even result in the predator-prey system switching to an alternate stable state (Begon et al. 1996).

## Co-operation (Symbiosis and Mutualism).
There are many ways in which organisms can co-operate with one another. The wider context in which their co-operation takes place, and the details of the benefit each receives, confounds efforts to precisely define and classify an interaction as dominantly co-operative. With that in mind, this section outlines two specific forms of co-operative interactions: symbiosis and mutualism.

A mutualistic relationship occurs between two species when the direct interaction between them provides each with a benefit. This is not a trivial benefit, but one they each depend on, to some degree, to complete their respective lifeways. Think of the relationship between pollinator and pollinated when the pollinator receives food for the service.

A symbiotic interaction is one in which two species are dependent on one another to complete their lifeways to such a degree that their lifeways are intertwined: for example, the interaction between a herbivore and the bacteria living in its gut. The bacteria overcome the problem of cellulose breakdown for the herbivore, while the herbivore provides the environment necessary for the bacteria to complete their lifeway (Costerton and Stewart 2001).

There are two particularly important symbiotic relationships found in the soil. One occurs between nitrogen-fixing *Rhizobium* bacteria and plant species such as the alders mentioned above. The other occurs between specific fungi species and more than 90% of higher plant species. These fungi and the tips of the plant's roots form intergrowths called mycorrhizae. The plants benefit because the fungus enhances the supply of minerals and nutrients to it. And the fungus benefits by receiving from the plant a steady supply of energy in a form that it can use, such as a sugar (Wolfe 2001).

We think of co-operative interactions as occurring between two species, but there are some suggestions that multispecies mutualism may exist. For example, at high altitudes the living conditions for plants are harsh. Under these conditions, the mere presence of different species in a plant cluster may be sufficient to alleviate this harshness for the mutual benefit of all species (Callaway et al. 2002). Think of the boreal forest's shrubline.

This example also illustrates the factors complicating our definition and identification of co-operative interactions. Does the benefit to the plants result from benign indifference or to active co-operation? Is shade, for example, a product of interaction or is this outcome the unintended consequence of complexity? Regardless of our views, these benefits do exist and should be recognized when considering how a community functions. Think of the relationships that exists between the various decomposer species who successively break down different fractions of the soil's plant debris.

**Parasitism and Disease.** Parasitic and disease-causing interactions occur when one species (the parasite/disease) completes part of its lifeway by using the body of another (the host), for example, for reproduction or shelter. For example, malaria plasmodia, flu viruses and tapeworms all use the human body to complete their lifecycle. Although many parasites and diseases are host specific, there is some flexibility. Under the appropriate conditions, they can, like predators switching their prey, change or increase the types of hosts they infect. For example, the HIV-1 group M virus, which causes the current AIDS pandemic, is hosted in chimpanzees from Cameroon. The virus was probably transferred to humans in the early 1900s (Keele et al. 2006). Parasitism is common, with perhaps more than 50% of species displaying some degree of parasitism during their lifeway (Begon et al. 1996).

For a parasite/disease to noticeably infect either an individual or the population of a host species, a few conditions must be met. There must be a sufficient number of the parasitic/disease-causing agents. There must be a sufficient number of the vector that transfers the parasitic/disease-causing agent to the host. The size of the host species' population must be big enough and the individuals must live close enough together (high population density) to sustain the lifecycle of the parasite/disease. For example, a cycle of measles epidemics can only occur in a human population that is both larger than about 300,000 people and living in close contact with one another, as in a city. Interestingly, the large size of this number indicates that measles was likely not a common human disease until larger cities started to appear a few thousand years ago (Burnet and White 1972; Begon et al. 1996).

The host is often seen as a passive victim of its interaction with a parasite/disease, but that is not necessarily the case. An individual host often tries to rid itself of the parasite/disease and then retrospectively protect itself with features such as immunity. The host species can reduce the chances of a serious parasite/disease infection if its lifeway keeps its population size and density beneath a disease-favouring threshold.

The lifecycle of a parasite/disease combines with a host's response to being infected to result in distinctive, dynamic infection-response patterns. For example, because both the number of parasitic/disease-causing agents in a host species' population and the number of infected individuals in that population is usually very low, a host's level of infection can be low but dynamically stable (i.e., chronic). A host species' population can also be subject to cyclic infections, which may last from months to many years. The cycles can be both regular and irregular in form. The pattern of a host's infection can also display apparently random, irregular epidemic outbreaks.

The dynamic nature of the interaction between the parasite/disease and its host species means that although a substantial number of individual hosts can die, it is only under special circumstances that parasites and diseases are exclusively responsible for extinctions. One possible example may be occurring today. The large changes in the climate and environmental conditions in the neotropics of Central America are thought to have resulted in an increase in the virulence of the chytrid fungus to a degree that many species of frog are being driven to near local extinction (extirpation) (Lips et al. 2006).

The above discussion alluded to the idea that the interaction between the parasite/disease and an individual host invariably results in significant harm or loss to the individual. In reality, lasting debilitating harm is often difficult to demonstrate (Begon et al. 1996). The resulting difficulty of unambiguously classing an interaction into the parasitic/disease box is clearly illustrated at the microbe level.

The "parasites" are the DNA fragments or whole genomes that are found in bacteria, viruses and a non-living, self-replicating ring of DNA or plasmid. (Plasmids are formed during cell decomposition under certain conditions.) The "hosts" are single cells, which can be independently living or part of a larger organism. Infection occurs when the DNA parasite is transferred to the host cell by the parasite's carrier. The transfer process occurs in a number of ways. For example, viruses simply inject their DNA into a cell. This is the method used by the human flu virus. A plasmid completes the transfer of its DNA when it is touched by a suitable host micro-organism, such as a bacterium (Cohen and Shapiro 1980). These types of DNA transfer interactions are especially common in soils because the density of bacteria, viruses and plasmids is high in this environment.

Once transferred, the parasitic DNA either takes over the cell or becomes part of the host's DNA. In either case, it modifies the host's behaviour. The flu in humans has been mentioned. In the plant-soil community, plasmids transfer the nodule-making gene from the nitrogen-fixing bacteria into a plant's root tip cells, where it is used to make the plant's *Rhizobium*-supporting nodule (Wolfe 2001). The formation and spread of antimicrobial drug resistance among disease-causing bacteria has been attributed, in part, to DNA transfers by plasmids occurring in the soil and our guts. In particular, these transfers spread the "parasitic" genes that confer drug resistance to disease-causing microbes (Cohen and Shapiro 1980). Which of the above DNA transfers are parasite/disease, symbiotic, co-operative, a mixture of types or in a class of their own? Perspective and context plays a significant role in the answer.

**Competition.** The presence of competitive interactions between individuals in a community needs no introduction. The details depend on whether the competing individuals are from the same species (intraspecific) or different species (interspecific). For example, competition for mates is an intraspecific affair.

The nature of intraspecific competition is wide ranging. In the case of finding a mate, it could be as simple as who can get to a potential mate first. Or it could depend on a selection process: which individual can acquire and display those attributes considered desirable to the potential mate making the selection. The outcome of a mate competition also depends on more than the competition itself (Begon et al. 1996). For example, the salmon's mate selection process is conducted in shallow streams where they congregate to spawn. Predation from the shore plays an important role in determining the outcome of the competition.

The degree and outcome of intraspecific competition also depends on the density of the species' population. Consider that a greater population density results in an increase in the number of individuals competing for the same resources, among other things. Competition under these conditions usually leads to some individual not breeding or dying. This is one way in which a species' population and its density can be stabilized.

For example, the successful germination of many pine tree seeds in a clearing near our forest resulted in a very high density of young pine trees. Over time, as the seedlings compete for space and sunlight, many will die. The self-thinned forest will then mature at a tree density more in balance with the resources available to sustain it.

Among social animals, there are more opportunities for intraspecific competition than there are among solitary ones because potential competitors are always present. For example, jays are social birds who cache nuts. The competition for nuts is expressed not only while collecting them but also through cache pilfering. The jays have to expend energy trying to prevent or avoid pilfering (Emery and Clayton 2001). The cost to individuals from expending this effort is offset by the benefits to the social group, such as cooperating in the defence of juveniles against predators.

In the case of interspecific (between species) competition for resources, the idealized model predicts that the outcome will be the reduction or elimination of one species. This does happen, especially if one of the competitors, such as an introduced exotic species, arrives on the scene fully formed with some distinct advantage over its adversary species. But interspecific competition is usually influenced by many more factors than those included in the idealized model, so other outcomes are possible. For example, a low-intensity competition between two species that lasts for a few generations can induce them both to change their behaviour. They can both adjust their lifeway and niche requirements in a manner that reduces the competition between them. Competition leading to niche differentiation results in coexistence, not the extinction of one (Begon et al. 1996). Interestingly, niche differentiation is another mechanism that contributes to the creation of patchiness.

No matter how interspecific competition is resolved, the result is never really permanent. After all, the community in which two competitors live is always changing in some way at some scale. As the context of the competition changes so do the influences on the competitors and the competitive balance.

There is a cultural aspect to our views of competitive interactions between species. Westerners, at least, tend to think, perhaps demand, that competition should produce only a winner and a loser. Think of the penalty shootout in sports competitions. Westerners also tend to judge both between- and within-species competition using a restricted time scale and a narrow focus. We tend to bias our view of who is or should be a winner in favour of the species competitor with the characteristics that we favour or consider most deserving.

The consequences of these culturally supported tendencies are illustrated by a description of a "recovery" succession that follows an ecosystem's disruption. Note the adjectives used. The plants that appear immediately after a disturbance such as a fire are the fast growers and fast reproducers: the "opportunistic" or "weedy" species. In the short term, they can be judged, reluctantly, to have the competitive advantage and to be winners. However, as these initial winners complete their shorter lifeways, they are slowly replaced by the slower growing but more efficient resource users, the late succession species (Begon et al. 1996). These late succession species often rely on the environment created by their predecessors to survive. If we take the perspective of the community, who is the winner and who the loser? Are these relationships competitive, or are they a time-separated, co-operative interaction?

**Communication.** It is common for humans to think that communication between organisms, of either the same or of different species, is generally poorly developed. Of course, ourselves excepted. We tend to think of their methods of communication as simple and mostly passive, like one-way responses to a stimulus. For example, we think that crickets sing and a mate is attracted. I think too of the racket made by the sapsuckers and the crows in our forest as they announce their presence, stake out territory and seek mates.

Perhaps we have these views because we have such pride in our own ability to communicate through speech and writing. Perhaps we are unable to appreciate other organisms' communication because we are unaware of their efforts or its subtleties. After all, as we saw in chapter 1, we do have to be prompted to acknowledge that we are only poorly equipped to detect the constant swirl of chemical messages around us.

An appreciation for the sophistication of other species' communication systems starts by looking at one of the simplest (dumbest?) of organisms, bacteria. As many a cartoon ad plugging a disinfectant has illustrated, we think of bacteria as individuals who live free and independent lives. That view was even the basis for scientific attempts to control them. However, we have recently come to realise that bacteria have the ability to chemically detect and respond to the presence of other bacteria of the same or a different species. Through this communication system, they can mutually change one another's behaviour and virulence. Consider environmental conditions that prompt a number of free living bacteria of the same species to generate the same chemical signal. If the strength of their collective signals is great enough, then it will cause the behaviour of all of the individuals to change in the same manner.

For example, they could all be prompted to express the genes that cause them to congregate and then organize themselves into a complexly structured colony or biofilm. Once formed into a film, the bacteria in different parts will behave differently. The members then rely on chemical communication between the parts to coordinate their activities. Examples of this outcome include the plaque on your teeth and the digestion films on the food in a cow's intestinal tract. It is these biofilms that defy most conventional efforts to eradicate bacteria (Federle and Bassler 2003; Costerton and Stewart 2001).

Communication systems can be more complicated than those used by bacteria. Consider that social animal species function as a group, so they are often required to act collectively. This occurs when the members are signalled by selected individuals, or by a majority, to act in a certain manner. The individual members then all exhibit the required behaviour. For this to happen, the group must have a structure, a fairly sophisticated communication system and a common understanding of it (Conradt and Roper 2003).

This ability to generate a collective response allows social animals to make group decisions through a process called "quorum sensing." The process is similar to the way soccer or opera fans collectively judge a player's performance. For example, whooper swans will only form a flock and fly when the neck and head movements of those signalling that desire to the other members crosses a threshold of intensity. That threshold is affected by the context, such as the status of those signalling and the swans' anticipation of finding food at the destination (Black 1988). Perhaps bacteria can be said to apply quorum sensing when they make biofilms?

**The Inanimate World.** We know that all organisms use senses like smell, taste, touch, gravity, magnetic fields and light to sense the world around them. And we know that they use the information they gain to direct their interactions with it. But we don't necessarily appreciate the sophistication of these sensings or how they influence the organisms' efforts to complete their lifeway.

For example, when sandpipers migrate between South America and the Arctic, they use a variety of environmental cues. Near the pole, they use a sun compass to determine the most efficient route to travel. If they decide to take a rest on their journey, they switch to a magnetic or a geographic navigation method because it allows them to more easily navigate within their resting areas. When they resume their journey, they switch back to their sun compass. However, for their sun compass to work properly, they have to reset their internal clocks to the local solar time.

Thus, for sandpipers to complete their migration journey, they require, at the minimum, an easily adjustable internal clock, the ability to measure the sun compass variables, the ability to complete the required calculations, and a good memory for either geography or magnetic fields (or both). To make this navigation feat even more remarkable, their northward journey follows a different route than their southward journey (Alerstam et al. 2001). Incredibly, a bird with a small brain can accomplish this feat, yet most of us, even with our big brains, don't even know what a sun compass is.

An organism's environmental sensing can have an even more direct and profound influence on its response to the environment than that illustrated by migration. Perhaps the most remarkable example of this is the reaction of *Pfiesteria* (a dinoflagellate) to changes in the water chemistry of the estuarine environment where it lives. *Pfiesteria* will react to both normal changes, such as those caused by the presence of fish, and pollution caused changes, such as abnormal increases in the concentration of nutrients like nitrogen. Its response is to adopt one of 24 unique shapes, each with significantly different characteristics. These shapes vary in size from as small as 5 microns to as big as 750 microns. A remarkable variation in size of over 100 times. For example, in the presence of fish, this ultimate of shape-shifting organisms adopts a specific shape that excretes a fish-killing poison. The *Pfiesteria* then eats the dead fish (Burkholder 1999).

**Trophic Web.** Within an area, the individuals of a group of species are routinely engaged in some kind of (direct or indirect) interaction with one another and their environment. For example, on the sub-Arctic's flat, treeless barren lands (moist tundra) of northern Canada, a large lichen-covered rock in a prominent, elevated location is prime real estate. It is the focus for interactions between many species. Birds use these rocks to proclaim their existence and intentions, to find and eat a meal, and to rest. They also defecate there, which adorns the rock with a white streak that kills some of the lichen species and sustains others. What reaches the base of the rock fertilizes mosses, grasses and shrubs. This is an ideal site for hares to find shelter, food and to survey the land. The hares attract the animals that prey on them, and so on.

Collectively, it is this multiplicity of interactions, acting over a wide range of scales, that binds the individuals of a group of species together into a complex, four-dimensional community or ecosystem. This description of a barren lands community suggests that it has a structure. What is that structure and how does it function?

One view of an ecosystem's structure and functioning is seen by focusing on a basic activity common to all of its member species: the interactions that provide them with the energy they need to complete their lifeway. When these feeding interactions are noted, then the member species can be divided into groups with similar feeding interactions. These groups have a hierarchical relationship with one another so they become levels which, collectively, outline the flow of "food" energy through the ecosystem. This is the structure of the ecosystem's trophic web (informally called its food web) (figure 9).

The plants, which rely on ("eat") sunlight for their energy, are the base of their host ecosystem's trophic web, at trophic level 1. The rest of an ecosystem's species obtain their energy from food, by eating other species. Thus, above level 1 the trophic level of the eater is determined by the level of the species it eats. The species, such as antelope, that eat plants for food (the herbivores) are at level 2. The organisms in each succeeding higher level obtain their food by eating the organisms in the level(s) below. For example, lions (which eat antelope) are at level 3 or above.

An ecosystem's trophic web reveals other features of the ecosystem's structure and functioning. Consider that all of the species at a particular trophic level have more lifeway characteristics in common than their feeding interactions. For example, compare the lifeway features, such as reproduction, of the herbivores, which occupy level 2, to that of carnivores at a higher level. Consider too that an ecosystem's species at level 2 and above rely, directly or indirectly, on the plants for their food. Thus, all of an ecosystem's species ultimately rely on sunlight for their energy.

The amount of plant-produced biomass at level 1 and the proportion distributed up the web as food limit the maximum number of trophic levels into which an ecosystem can be divided. The proportion distributed depends, in part, on how much of the food available at one level is eaten by the species in the levels above. And it also depends, in part, on how much of the food a species eats is turned into potential food for the species in the levels above.

The efficiency with which food energy is transferred between trophic levels is low. It varies from 2% to 24% with an average of about 10%. This low level of transfer efficiency means that, regardless of the amount of food in level 1, only a few trophic levels can be supported before the available food (energy) runs out. By the mid-trophic levels, the available food can support only a few individuals of a limited number of species. The higher trophic levels are occupied by even fewer species with very small populations.

The highest trophic level in an ecosystem therefore represents the limit on a species' ability to secure the food needed to sustain a viable population that is imposed on it by the distribution of energy up the web. As a result, a higher trophic level can't exist. The absolute maximum number of land-based trophic levels an ecosystem can sustain is five (Begon et al. 1996). Unsurprisingly, the land-based ecosystems with the greatest number of trophic levels are those in climate zones with the greatest plant productivity and density, for example, the tropical forests.

The number of trophic levels in an ecosystem's trophic web can tell us something else about the ecosystem's overall functioning. For example, the temperate grassland biome has three trophic levels: plant, herbivore and carnivore. If nutrient availability is sufficiently low, then the productivity of the plants at level 1 limits the herbivore and thus the carnivore population. It also limits the degree to which the herbivores can impact the plants. The ecosystem is thus said to be (self) regulated from the bottom up. However, if nutrients are plentiful and carnivores limit the herbivore population, and thus their impact on plants, then the ecosystem is said to be regulated from the top down (Sinclair and Krebs 2001). In both cases a feedback is involved.

This kind of self-regulation mechanism helps to explain how an ecosystem and its trophic web remain stable through time. It has been suggested that an ecosystem is more likely to be stable if its self-regulation feedback is formed from many weak interactions among its members (i.e., unlike the lynx-hare relationship). In other words, its stability depends on the complexity of its structure and functioning. But, for that complexity to exist requires a diversity of species. From this perspective, stability is said to depend on the ecosystem's biodiversity. The larger the number of member species (the greater the biodiversity), the greater the ecosystem's stability will be (Cardinale et al. 2012). However, it has also been suggested that an ecosystem's stability depends on specific keystone species and their interactions.

There is another basic feature to consider. Whether a change in diversity, complexity or keystone species will add to or detract from an ecosystem's stability depends significantly on both the details of an ecosystem's characteristics and the greater context in which it exists. For example, an ecosystem's complexity and surroundings ensure that there are always many, internal and external, stability-affecting influences operating on the ecosystem over a wide range of scales. How it responds to them matters. In this context, our preference for single or simple explanations for an ecosystem's stability and change, such as either biodiversity, keystone species or complexity, is unrealistic (Finke and Denno 2004; Begon et al. 1996).

Our consideration of an ecosystem and its trophic web's stability should pay attention to the features that collectively increase its ability to withstand disturbance (its robustness), its ability to recover from disturbance (its resilience), and its ability to adapt to the drivers of change (adaptability) (Begon et al. 1996). Those features include trophic redundancy (alternative pathways through the web) (Sinclair and Krebs 2001), ongoing low-level disturbances, heterogeneity of the ecosystem in time and space, and the presence of variations in the characteristics of individuals. It is in this sense that biodiversity, complexity with its feedbacks, and keystone species are important for an ecosystem's functioning and changing. They are all aspects of an ecosystem's stability and changing, as expanded on in chapter 8.

In summary, an individual's effort to complete its lifeway includes it deliberately interacting with a limited number of other species in the manner preferred by its species. But there is often more to a species' preferred interactions than there seems. For example, a herbivore's preferred interaction is preying on specific plants. But, if the herbivore's feeding on a plant benefits the plant, for example, by spreading its seed, then there is a mutualistic component to the herbivore's interaction with the plant.

Consider too that a species can engage in interactions with other species both directly and indirectly. For example, a crocodile has a direct predator-prey interaction with a herbivore who drinks at a waterhole. However, the crocodile also helps to maintain the waterhole on which the herbivore relies for its survival. The crocodile thus also has an indirect, co-operative interaction with the herbivore. Together, the inevitability of indirect interactions and the diversity of species in a community ensure that an individual invariably engages in more types of interactions with a wider variety of species than implied by its species' preferred interactions.

The complex sum of all the direct and indirect interactions among the individuals of a group of species, and between them and their environment, binds them together into an ecosystem. The sum is the ecosystem's structure and functioning. It is responsible for the ecosystem's features, such as patchiness and the niches in which the members complete their lifeway.

Those of the interactions among an ecosystem's species that are directed at securing the energy their individual members need to complete their lifeway form another feature of an ecosystem: its trophic web. The base of its trophic web's structure is the plants, who acquire the energy they need by "consuming" sunlight. The rest of the species acquire their energy by consuming food (other species). Who gets eaten by whom groups the ecosystem's consuming species into a hierarchy of feeding (trophic) levels that are increasingly distant from the plants, which are the web's primary food source. This is the trophic web's perspective of an ecosystem's structure. From this perspective, the ecosystem's functioning is the consuming by its species, and the consequences.

The trophic web reveals that the structure and functioning of an ecosystem ultimately depends on the sunlight used by the plants. It also reveals other energy-related constraints on its structure and functioning. For example, that the number of land-based trophic levels in its structure is limited to five, and its functioning only remains stable and adaptable with the help of feedbacks among its trophic levels from the top down or the bottom up. The trophic web view of an ecosystem's structure and functioning complements the descriptions of these features from other perspectives, such as the nutrient- and mineral-cycle view, with its soil community's recycling process, discussed above. Think too of the co-operative and competitive interactions. They are all different views of the same system.

Knowing more about the interactions among species increases our appreciation for an individual's process of living, and the structure and functioning of their host ecosystem. It also helps us to counter the cultural influences on our thinking about these interactions and an ecosystem's functioning. Consider the biases in Western culture's thinking about species, communities and ecosystems that result from our cultural stereotypes of species and their interactions. Think of vicious wolves in fairy stories. Of particular interest is the widespread usage of social metaphors and analogies that promote the assumed superiority of predator-prey and competitive interactions. For example, the use of the phrase "the survival of the fittest" when it is taken to mean the survival of the strongest or the best competitor, rather than its older, original meaning of the best suited. There is also the exalted place given to competition by economists.

With this in mind, it is no wonder that, when Westerners casually think of the hierarchical structure and functioning of an ecosystem's trophic web, we can imagine ecosystems are dominated by predator-prey and competitive interactions. It is thus easy for us to conclude that the larger, more visible species found in an ecosystem's higher trophic levels are its most important. For example, lions are the "kings" of the jungle. We consequently give these species more attention and value. Think too of our other animal symbols of power, such as tigers, eagles, sharks and polar bears. Think of the names for sports groups.

In holding this superior opinion of predation and competition, Westerners neglect the critical role that the other interactions, such as co-operation, play in the long-term functioning and survival of individuals and species, and in the stability and adaptability of their host ecosystem. We would gain a more realistic view of how ecosystems function if we remember that the perspective, context and scale we choose to look at a community strongly influences what we see and how we interpret it. Our effort would be helped if we took to heart the biblical admonition to pay closer attention to the foundation of our house, the lower trophic level species, the plants, and the smaller species in the soil that support land-based plants.

## Constraints

The individuals of different species interact with one another and their environment as they try to complete their lifeway. In the process, they form an ecosystem. As discussed, the structure and functioning of that ecosystem is facilitated and constrained by the types of species, the resources available to them and geographical factors, such as climate and topography. This section discusses additional constraints on species, and the structure and functioning of the ecosystem they form. The constraints discussed are allometric relationships, food and space. The overall significance of these constraints is provided by the example of the limits to the number of larger species in an ecosystem.

**Allometric Relationships.** An allometric relationship exists when the average value of two lifeway or body variables for each of many species collectively display a strong correlation. For example, it might seem strange, but among unitary species, there is a correlation between their average body mass (weight, size) and their average population density (figure 10) (Peters 1983). If you

found a new unitary species and measured its average body mass, then you could use the interspecies body-mass-to-population-density correlation curve to predict the average population density of the new species. If you knew the size of the area with habitat suitable for your new species, then you could easily estimate its average population size.

The mathematical expression of allometric relationships is usually exponential (logarithmic), which means that a small change in one variable is related to a much larger change in the other. But, despite the mathematical precision, the correlation being expressed is not perfect. The deviations from the ideal represent both the uncertainty in our measurement and each species' inherent, but limited, flexibility in forming its unique features.

A wide variety of a species' variables display allometric relationships, especially when they are paired with body mass. For example, there are allometric relationships between the body mass of unitary species and their other characteristics, such as organ mass, feeding rates, defecation rates, nutritional requirements, time to traverse their home range, population density, critical body temperature and life span. The variables displaying allometric relationships extend to aspects of reproduction, such as litter mass, the manner in which the species moves, and how much they sleep (Swihart ct al. 1988; Peters 1983; Bejan and Marden 2006; Siegel 2005).

Allometric relationships form a flexible hierarchy with one another that is based on how many categories of species (e.g., family or genera) a relationship simultaneously applies to. Because variables paired with body mass often apply to many categories, as illustrated in figure 10, they form primary allometric relationships. If the primary relationships are mathematically removed from the description of a particular category of species, then secondary allometric relationships can be seen. For example, among mammals, including humans, the age at which the young of a species matures increases as the gestation period increases. Similarly, the number of young produced decreases as the time the parents spend raising their young increases (Harvey and Zammuto 1985).

The suite of allometric relationships relevant to a particular species collectively outlines its basic characteristics. Now, imagine what would happen to the species if one of these characteristics were changed significantly. Because the allometric relationships that apply to the species are connected to one another (e.g., through body mass), all of the species' other allometrically connected characteristics would also have to change. Allometric relationships can be thought of as allometric constraints, and a species can be thought of as representing a unique set of allometrically constrained characteristics. In this view, a species represents one of life's many possible quasi-stable states.

From a different perspective, a species' unique set of basic characteristics are not the product of an arbitrary mix of features. All of a species' form and lifeway features simultaneously satisfy all of the relevant pairs of allometric constraints on it. A species is the product of the compromises, trade-offs and co-operation among its possible features that allow it to be functional over the long term (viable) (Read and Harvey 1989; Begon et al. 1996).

This points to a puzzle. How can it be that allometric relationships are not causative, in the strict sense, but they are still predictive constraints. It is thought that both their wide-ranging, ubiquitous presence and their predictive ability exist because the individuals of a species are axiomatically constrained by chemistry and physics to function as a single whole. Think of a wrist watch. The allometric relationships illustrate this constraint quantitatively.

However, despite their mathematical description, allometric constraints do not deterministically (rigidly) set the details of a species' characteristics. As hinted at, complexity ensures that, in real life, there is some flexibility in satisfying the constraints. This is expressed by the variation in the characteristics of the individuals making up a species' population and by the average characteristics of the species falling around, not on, the ideal allometric curves.

There are a few other features of allometric relationships that should be mentioned here. Most of the known allometric relationships refer to unitary species. However, when looked for, they are also found among plants, at least when they are treated as unitary organisms. For example, the annual growth rates of plant species, from algae to conifers, are related to their "body" mass. Similarly, the daily amount of water used by land-based plant species is related to their above ground dry body mass. Finally, the maximum flow of sap up the stem of land plant species is constrained by their stem diameter.

Thus, although I might see the variety of trees in our forest as the outcome of a random combination of features, this is not the case. The characteristics of each plant species is constrained by its need to simultaneously satisfy all of the relevant allometric relationships. As soon as one basic feature of a tree is selected, then the choices of the other features that can join with it to form a functional tree are allometrically constrained. For example, a tree with dense wood, such as a birch, will have a smaller water storage capacity (used to balance daytime water demand) and will be slower growing than a tree with a light wood, such as cottonwood (Meinzer 2003).

Unsurprisingly, allometric relationships also constrain the ability of the members of a species to fulfill their lifeway. At its most basic, if an individual's mix of characteristics fall too far outside its species' allometrically constrained functional zone, then the individual will likely experience difficulties completing its lifeway. For example, humans that are much taller than two metres tend to die younger because they are prone to exceeding their heart's sustainable pumping ability.

Consider too the allometric relationships between the body mass of a mobile unitary species and both the amount of energy (food) it must secure and the distance it can cover per unit of time. The members of these species are thus allometrically constrained to find that amount of food within their home range (Peters 1983). Among North American mammals, the larger an animal's body size, the larger its home range (Brown and Maurer 1989).

More broadly, allometric relationships constrain the interactions an ecosystem's member species can engage in to complete their lifeway. This affects the characteristics of the host ecosystem they form. Consider, if a species is to participate in an ecosystem, then there must be a niche for it and sufficient resources, such as energy (food), for the members of the species to maintain a viable population. (For a mobile unitary species, the niche and resources must both be located within an area no larger than its home range.) The niches and resources a host ecosystem provides are determined by the ecosystem's structure and functioning, but that structure and functioning is a product of its member species interacting with one another and their environment. Thus an allometrically influenced feedback loop exists between an ecosystem and its member species. It facilitates and constrains both an individual organism's ability to fulfill their lifeway and their host ecosystem's ability to support them. The ecosystem's structure and functioning is allometrically influenced if not constrained.

This way of looking at the structure and functioning of a host ecosystem complements the trophic web view.

This view can also be expanded to include the details of the non-biological constraints on the ecosystem's structure and functioning. For example, the climate zone, topography and soil type at the general location of the ecosystem all affect the availability of resources, such as sunlight, water, minerals and nutrients, to the ecosystem's members. This affects their ability to complete their lifeway, which affects the functioning of their host ecosystem.

In overview, many of a species' characteristics are allometrically constrained. A species' form and lifeway, including its preferred interactions, can be thought of as its expression of having satisfied all of the allometric constraints on it. This is true for all of the species that form an ecosystem.

The allometrically influenced interactions among a group of species form the dynamically balanced complex system that is their host ecosystem. Thus there is an allometric influence on its structure and functioning. But, the host ecosystem provides the niches and resources its member species require if they are to interact, which completes a member-ecosystem feedback. Allometric relationships, feedback and the availability of non-biological resources all contribute to an ecosystem's structure.

**Food Availability.** As discussed, the members' relationship with energy is the structure and functioning of their host ecosystem's trophic web. The plants, which form the base of the trophic web, consume sunlight, while all of the other species consume food, starting with the plants. Like nutrient cycles, there is a global dimension to the availability of the food energy in an ecosystem, which is highlighted by discussing how food production is facilitated and limited. There are three aspects to consider: the ingredients needed to make food, the machines that make the food, and the efficiency of making the food and distributing its energy.

Images of Earth from space show that, from the perspective of the minerals, nutrients, water and carbon dioxide (ingredients) needed to make food, Earth is effectively a closed and finite system. In addition, the amount of the minerals and nutrients available to the food-producing machines is limited and depends heavily on the soil-plant recycling processes. This was discussed above, so all that is needed here is a reminder of one key aspect. The recycling processes must continue to function if the supply of the minerals and nutrients needed to make food is to be maintained.

The primary food-making machines are the plants, both in the ocean (e.g., light-using plankton or phytoplankton) and on land. Plants transform the recycled food ingredients into the compounds needed for the plants to grow and complete their lifeway. That growth, which is the biomass they produce, is also the foundational food for all of the ecosystem's other species.

The basic process by which the plants transform the ingredients into food is photosynthesis, which occurs in their leaves but is powered by sunlight. From the perspective of sunlight, the earth is an open system: it is provided by the sun in a fixed amount from its, effectively, infinite supply. However, the earth's spin, tilt and orbit around the sun, in combination with the earth's weather, result in an uneven distribution of sunlight on the earth's surface. The amount of food the plants can produce is limited by the amount of sunlight available where they live.

If plants are grown under ideal conditions (with optimal amounts of sunlight and elemental resources), then they all produce roughly the same amount of new biomass per unit of the plant's mass. This means that the process of photosynthesis does not change all that much between plant species. But, even under optimal conditions, the efficiency with which photosynthesis uses sunlight is low, <10%. And the supply of resources at a location is usually less than optimal. As a result, the real-world ecosystem-scale efficiency of photosynthesis falls between 0.01% and 3%. For example, the efficiency of photosynthesis for a tropical forest lies between 1% and 3%, while a temperate forest lies between 0.6% and 1.2%. By way of comparison, the average efficiency of cultured crops lies around 0.6%, with short-term efficiencies reaching between 3% and 10%. Overall, the highest efficiencies represent the maximum for photosynthesis under optimal conditions, while the lower numbers reflect the constraints imposed on photosynthesis by the limited availability of critical ingredients, such as water and nutrients.

Plants can partly compensate for some of these resource limitations by adapting to the local environmental conditions through metabolic and morphological specializations, such as having waxy leaves to reduce water loss (Begon et al. 1996). But they cannot overcome the fundamental limit imposed by the nature of photosynthesis. One way or another, the ability of plants to produce new biomass is therefore irreducibly constrained.

The amount of new plant biomass an ecosystem produces each year is called its net primary productivity

(NPP). Unsurprisingly, the global distribution of the NPP from indigenous ecosystems, and its variation through time, correlates closely with the distribution of climates, biome types and evapotranspiration, and their variation through time (Tivy 1999). As suggested by the photosynthesis efficiencies given above, biomes at the equator produce and recycle a large NPP quickly. In contrast, the biomes near the Arctic produce and recycle a small amount slowly.

An ecosystem's NPP is the ultimate source of most of the food (energy and resources) used by its non-plant members to complete their lifeways. The distribution of this plant-derived energy to the members starts when the species in the ecosystem's trophic level 2 eat plants as food. Some of the energy in this food is used to maintain their body functions, to provide heat and to power growth. Some is excreted in the by-products from these uses and as undigested plant material.

The transfer of food energy further up the trophic web occurs by individuals of the species at higher trophic levels eating individuals in a lower level, down to and including level 2. The overall efficiency of energy transfer between trophic levels lies between 2% and 24%, with an average of only about 10%. As mentioned above, this low level of energy-transfer efficiency limits the number of trophic levels an ecosystem can support.

In the Western world at least, the low efficiency of food energy transfer between trophic levels can be treated as an ecosystem/species shortcoming, an inefficiency. Think waste. This neglects that the transfer process doesn't occur in isolation. It is one of the features that arise to ensure that the feedback-influenced balancing of the many interconnected constraints on species and ecosystems results in a functional host ecosystem. Think allometric constraints.

In the zone of balance, the efforts of all of the ecosystem's member species to complete their lifeway will also keep their host ecosystem stable and functional: robust, resilient, and adaptable. And their host ecosystem will provide sufficient support (e.g., food, space) for all of its member species to maintain a viable population over the long term. This means that, routinely, a sufficient number of individuals from each member species will successfully complete their lifeway, while a sufficient number will become the food the other members need to complete their lifeway. It is the overall balance between these interests that ensures the ecosystem is functional, not just the efficiency of energy transfer between species/trophic

levels. When seen in this context, both high and low efficiency have their place.

We are reminded, again, that to appreciate an ecosystem's structure and functioning requires us to treat it as a whole and look at it from many, sometimes conflicting, perspectives and scales. The pitfalls of focusing on one or a few issues/constraints of interest to us, in isolation, will be repeatedly seen throughout the manuscript. Think of the unused land in the Sahara.

**Space.** Space (an area) is a resource. It is just as important a constraint on ecosystems and its individual members as nutrition and energy. This section discusses its importance by addressing two questions: how much space is there?, and what are the constraints that space imposes on organisms completing their lifeways and thus on the functioning of their host ecosystem?

The amount of space available to a species should be determined at a scale appropriate for the size of the species being considered. When standing some distance from a cottonwood tree trunk in our forest, it seems to be a smooth cylinder. You can imagine calculating its surface area and, if you know the area of the home range for a small bug, estimating how many bugs could live on the tree trunk. It turns out that the result would be wrong. From the view point of the bug, the cottonwood trunk is extremely rough with a larger amount of available living space (Morse et al. 1985). The observed relationship between the scale used to look at a rough surface and its surface area is described by a fractal relationship. It applies to the space available for all the sizes of organisms living on our rough planet.

The importance of space is easier to appreciate if we recognize that an individual requires space to complete their lifeway. For example, when a plant's seeds have germinated too closely together for the available resources to support them all, then they will be thinned through death. Alternatively, disease can more easily spread among them to achieve the same result. In contrast, when individuals are spread too far apart, they fail to breed. There is an optimal amount of space for an individual to complete their lifeway.

This prompts the critical but difficult-to-answer question, what is the minimum area required for a particular ecosystem to remain functional? It could be answered by adding together the minimum area needed for each of its member species to maintain a viable population, that is, to satisfy their resource and reproduction needs. The total area on the ground (disregarding overlaps) needed to keep all of the ecosystem's member species viable is then the minimum area required by the ecosystem to remain functional (Begon et al. 1996). That area can't be smaller than the largest area required by a member species to be viable.

The question can be answered from a different perspective. The size of an area helps to set the maximum number of species that it can sustain over the long term. This is called the island effect because it is most obvious when comparing the species diversity of an ecosystem hosted by islands of different sizes.

Either way, an individual's need for space has important implications for ecosystems. If an ecosystem is divided into fragments below its minimum functional size, the result is not an increase in the number of similar ecosystems. It usually results in the decline in ecosystem biodiversity through the migration and extinction of species that require an area that is larger than the size of the fragments. A corresponding decline in ecosystem functioning soon follows. The result is the appearance of a number of ecosystems with characteristics that are different from the unfragmented ecosystem. This outcome is illustrated by the consequences of us fragmenting an ecosystem during resource extraction, as discussed in more detail below (Brooks et al. 1999).

It could be suggested that knowledge of this species-area relationship has solved a significant hurdle for environmental management and conservation: how much area to set aside. However, this is only partly the case. Our (complex cultures') strained relationship with what can seem like the arbitrary nature of natural boundaries and categories raises its head once again, accompanied by the consequences of the simplifying we engage in so that we can make our decisions. Consider the difficulty we face reaching agreement on an ecosystem's often ephemeral and dynamic boundaries to a level that is amenable to human management (i.e., a sharp line at a human scale). Finding a resolution to this issue is not helped by our struggle to relate to a functional ecosystem from multiple perspectives, as illustrated by my relationship to our forest described in the introduction. As a result, it is quite easy for complex cultures to allocate an area to an ecosystem that has little relevance to the lifeway of many of its members and is incapable of supporting their host ecosystem as a functional whole.

**Number of Species.** The discussion about trophic webs, food, allometric relationships and nutrient cycles implied a food- and efficiency-imposed limit on the number of species that could be supported by an ecosystem. The discussion about climate noted an evapotranspiration-imposed limit on the number of species that could be supported by a biome. There is also a space constraint on that number. The following is a continental-scale example of the presence and significance of such community-imposed limits on the number and kinds of species, and thus trophic levels, in a biome.

Before 5 million years (My) ago, North and South America were isolated continents. On each continent, the number of families and genera of mammals were in a dynamic equilibrium with their environment. By 2.5 to 3 My ago, the continents were joined together by the Panamanian land bridge. The resulting movement of species between the two continents resulted in an increase in the number of families and genera of mammals on both continents. For example, armadillos and porcupines moved to the north, while skunks and camels moved to the south. However, by the present day, the number of families and genera on both continents has declined through extinctions back to their pre-collision numbers (Marshall et al. 1982).

This example illustrates the rule of thumb that, in the long (human) term, a functional ecosystem can be treated as saturated with species: there is likely a long-term upper limit to biodiversity. If additional species (think exotics) are introduced into the ecosystem, then its structure and functioning will be adjusted to maintain a dynamic quasi-equilibrium state. Think orcas and otters. The adjustments could include selectively reducing the number of species.

For an illustration of why this limit exists, consider what would happen if a species much larger in size than any of an ecosystem's current members was added to an island's ecosystem. The allometric relationship between a species' body mass and feeding area, or home range, mean that the new arrival can only be supported if the ecosystem's area is increased in size, exponentially. If that is not possible, then the new arrival must go extinct (Begon et al. 1996). The example of North and South America also reinforces, again, the idea that an ecosystem functions as a dynamic whole: it is a dynamic, complex system.

In summary, by the individuals of a group of species interacting with one another, they form their host ecosystem and its features, such as its species mix, trophic web and nutrient recycling. The structure and functioning of that ecosystem are ultimately the product of the compromises and trade-offs (balancing) that are needed to simultaneously accommodate a wide range of factors that facilitate and constrain the individual members' characteristics and their efforts to complete their lifeway. Those factors include allometric relationships, climate, topography, soils and the associated availability of resources, such as space, minerals, nutrients, water and energy/food.

An ecosystem's structure and functioning constrains the types of niches and the amount of resources it can make available for its member species. This constrains their contribution to their host ecosystem's structure and functioning, which completes a species-ecosystem feedback loop. This loop is part of the process that forms their dynamic, complex host ecosystem's characteristics and determining its resilience, robustness and adaptability.

Because an ecosystem is a dynamic complex system, it can display characteristics that are more than the sum of its members' interactions (Begon et al. 1996). One of these features is its apparent ability to "adapt" to new circumstances. The appearance of this ability is analogous to the appearance of consciousness in humans. Also associated with the complexity of ecosystems, as already hinted at, is our difficulty relating to them. The degree of that difficulty will become clearer in the next section.

## Numbers

To reliably describe the functioning of a community, we need to measure its components/features. For example, we need to know the number of species and the size of their populations. Changes in these numbers help us to determine the importance of the interactions they engage in and to predict the future of the community, such as the chances of a species' extinction. This section discusses the constraints on acquiring reliable measurements about an ecosystem.

__Counting.__ One measure of something's significance is how many of them there are. Counting cows to determine their number is straightforward because they are individuals: they are unitary organisms. However, as the Babylonian model of establishing a cow herd discussed in chapter 5 showed, we might also have to record its life stage. Is it a calf, a yearling or a mature cow?

But what to do in the following case? A large aspen in our forest sends a sucker up from one of its roots into the clearing around our house. It appears some 10 m away from the trunk. Should I count this young aspen as an individual or as part of the parent aspen?

This counting problem arises, at least for Westerners, because many organisms, in particular plants, are not independently mobile or born with their lifetime supply of parts, like cows. They are modular, meaning that, throughout their lives, they can grow functional units, such as leaves, as needed. There is also some flexibility in which functional unit they can grow. For example, a bud on a tree can turn into either a leaf, a flower or a branch depending on the circumstances. Consider too that a large part of a modular organism can be removed or die, yet what is left can still be fully functional. Consider a tree with dead limbs.

Before we can count modular organisms, we have to come to terms with their, to us, strange characteristics. We are stuck with the difficulty of defining their identity and boundaries. For example, at which point is a modular organism more dead than alive (Begon et al. 1996).

This counting problem can be partly resolved by determining which aspects of an organism are most ecologically significant and then naming and measuring those features. For example, biomass (the total biological weight of an individual, community or population) is a useful measure of biological functioning and its significance for some species or groups of species. Biological density (the amount of a biological attribute per unit area) is useful because density affects many relationships, such as breeding rate and population growth. However, it can be difficult to relate aggregate measures to the details of a community's functioning or an individual's lifeway. Think of aggregate measures in economics.

These measures also fail to resolve the boundary problem: which pieces of a modular organism to include in the measure and which not. Consider the dead parts of a modular organism. In our forest, these would include the rotting parts of the aspen and cottonwood tree trunks and their dead branches. Is this dead material a part of the tree's biomass, or is it a part of the rotting organic soil that has not yet fallen to the ground?

To change the question slightly. It would be difficult to argue that an aspen tree whose trunk is rotten at its core is a completely dead tree and should therefore not be counted as a tree. It could be argued that the tree is partly dead or dying. So does the tree count as only a portion of a fully living tree? What about the dead branches? Are they dead like fingernails or dead like an amputated limb? Can I chop them off because they are useless (except as firewood for me), or should I leave them because they are "functional" parts of the tree?

Finally, to broaden the scope of the question a bit, the fungi attached to and surrounding the roots of the aspen are inseparably tied to the tree in a symbiotic relationship. Are the fungi a part of the tree? If the tree can't live outside the soil, does this mean that we should treat the soil as an organ of the tree?

These might seem like the types of questions that have entertained many philosophers and the religious for hundreds of years, while boring the masses for just as long. However, if we consider why these questions are so hard for us to answer, we can both more easily resolve the counting problem and broaden our appreciation for ecosystems and their functioning. Our difficulty is summarized in the ultimate "Is this alive or dead" question. Are viruses dead or alive? In particular, is a polio virus dead or alive? The scientists who manufactured a virulent polio virus from inert, off-the-shelf chemicals answered the question this way: "If the ability to replicate is an attribute of life, then the polio virus is a chemical [$C_{332,685}H_{492,388}N_{98,245}O_{131,196}-P_{7501}S_{2340}$] with a life cycle" (Cello et al. 2002, 1018). From the human perspective, viruses appear to be both dead and alive at the same time. For Westerners, at least, this observation just does not compute.

One of the reasons is that we are familiar with ourselves and treat our human experience as normal: we are a unitary organism and our death is well defined. Another is the world view of our cultures. Western culture's world view, for example, favours us treating the world as being highly (deterministically) ordered, and describes it using a vocabulary of absolutes and opposites: the black and white view of a grey world. Certainly, our concepts of discrete, unitary individuals and a sharp divide between life and death are useful, but they are accompanied by significant limitations. Like intersex, we have no counting box in which to put one-and-a-half modular organisms. Or was that, perhaps, one long-lived, somewhat dead organism? We have neither a coherent concept of "one-and-a-half organism(s)" or a "long-lived, half-dead organism" nor the language needed to meaningfully describe and thus relate to them. With our culture's prompting, we can easily choose simple absolutes to describe the fuzzy features of our environment.

If we wish to more appropriately think about and relate to ecosystems, and better represent them in our decisions, then we will have to accept the significant degree to which our perceptions and thinking about them are influenced by our innate human characteristics and the guidance from our culture's world view. However, these cultural and personal human constraints are not restricted to our views about ecosystems. They apply to our views about the external reality in general. We need to find a way to more widely come to terms with characteristics such as duality and fuzzy boundaries for what they are, not what we want them to be. We need to think about the world around us in a more flexible and nuanced (more realistic perhaps?) manner.

We are helped by trying to imagine a modular organism's view of its lifeway (if it could have a view) and the role it plays in its host community. By putting ourselves in a plant's roots, so to speak, we might more easily accommodate their modular growth, their slow, piecemeal experience of death and their fuzzy unitary-ness into our thinking. We could then, perhaps, more fully accept that a plant's ties to its immediate environment are sufficiently strong that our efforts to separate it from its soil are either a theoretical convenience or a practical transfer of responsibilities to us.

Alternatively, we can continue to use our preferred unitary view and counting system when looking at modular organisms and their functioning, but adjust it. We would have to accept that whatever unitary box we put something into is a convenience. Thus we should interpret the significance of the something in that box with flexibility and caution. To do so, we could look at the organism and its lifeway relationships from a variety of unitary perspectives. And in each case we should acknowledge that we are choosing between different perspectives rather than between right and wrong options. Then, to compensate for each perspective's imperfect representation of the organism's real life, we could mentally acknowledge each of their limitations. We could also make an effort to merge the multiple perspectives and their limitations into one view.

In this more nuanced way of thinking the fungi and perhaps the soil attached to a tree's roots could be seen, under some circumstances, as having some properties of organs. Under those circumstances, they could be called pseudo-organs of the tree. Similarly, under some circumstances, the air around us can be thought of as a part of us. Imagine what would happen if we shook it all off of us, like soil from a plant's roots.

In considering these methods, we come to a singular realization: if a modular organism, such as a tree with dead branches, and its host community are left to themselves, then the tree and its community are quite capable of looking after themselves. Somehow, we need to acknowledge, accept and then incorporate that reality into our representation, counting and thinking about modular organisms and their host ecosystems. There is a coupled rule of thumb to help us address this task. Focus less on deciding which is the "correct" method to represent an aspect of an ecosystem and more on how to accommodate the biases and limits in how we have chosen to represent and think about it.

How can we go about implementing this change? Consider the wristwatch example from chapter 6. When viewed by Mr. Big from his narrow perspective, his status-boosting watch is fully functional only if it has the bits he values: the strap is important, but its ability to tell time, not so much. In contrast, if we take the wristwatch's perspective, it is only fully functional if all of its bits are present.

Similarly, our representation and thinking about a modular organism or its host community would be more complete if we take a perspective, or amalgam of perspectives, that considers more of the bits that the organism or its community would consider necessary for it to function fully. If we take a contrary perspective, we can happily remove bits even if doing so disrupts the organism's or community's overall functioning. That representation and thinking is not wrong, just incomplete.

In overview, practically speaking, it is easier for us to take the perspective of a unitary organism than a modular one. This indicates that the counting problem arises for more reasons than an ecosystem's diversity and complexity. It is also the result of human perceptions, conceptions and understandings. In this context, the resolution to the counting problem includes us taking the time to find and acknowledge the biases and limits that accompany our preferred perspectives of organisms and ecosystems, and the method(s) of representing them, including numbers. This approach may not fully resolve the counting or the related boundary problems (fuzziness and duality), however it will prompt us to think in a more nuanced manner about individuals and their host community. That would prompt us to consider the human impediments to us treating or thinking about individuals and communities as functional wholes, which would help us to interpret what we do count in the context of the ecosystem's structure and functioning.

## Population.

A number of individuals of the same species forming a group are a population. Although we preferentially pay attention to a population's current size, we should also pay attention to the manner in which it grew or shrank to that size, its dynamics. A species' population dynamics is captured in measures such as changes in its birth and death rate over time. When these measures are used in conjunction with general information about the species and its community, then it is easier to identify the influences driving its population to change, such as a significant decline in successful feeding interactions or an outbreak of disease.

The nature of a population's dynamics is illustrated by considering a species that reproduces sexually. A critical part of its population's dynamics is its members' ability to find a mate. If we assume that all of the species' other lifeway functions are satisfied, then the density of those capable of mating can be used as a measure of the ease with which mates can be found and offspring produced. For example, there is a minimum density below which it is too difficult for potential mates to find one another, and thus the species becomes dormant or extinct. Conversely, if the density rises above some density threshold, then it is possible for births to exceed deaths, and the population will grow. The population will keep growing until it reaches an ultimate restraining factor, such as disease or a lack of resources (Berryman 2003). The dynamic size of a population is thus bounded by at least two long-term limits: (1) the minimum number and density needed for a population to both reproduce and pass on the skills needed to raise functional adults (its minimum viable population); and (2) the maximum number and density that the community's resources can support (its carrying capacity). In between lies the range in the size and density of a potentially viable population.

We tend to think of a stable population as one that has a fixed size, but in an ecosystem it means a varying size that remains between those two limits. That variation can be erratic, slow or even cyclical. For example, the population size of the measles pathogen (as measured by infected people) in a large and dense enough unimmunized human population varies in a roughly two-year cycle as infection plays against immunity and death (Burnet and White 1972; Begon et al. 1996). Similarly, the population sizes of both the lynx and its snowshoe hare prey are tightly bound together in a roughly 10-year cycle of rise and decline in their respective numbers (Krebs et al. 2001).

A greater appreciation for how the population of a species can be both dynamic and stay within its stability range starts by considering population density in more detail. An increase in a species' population density increases the success of activities that both increase its population, such as mate selection, and reduce the population, such as predation, starvation and disease. Models of simple systems indicate that as population density increases, the success of the population-promoting effects is eventually overtaken by the population-reduction effects. Thus, over time, the population's growth slows and stabilizes at a fixed number (Begon et al. 1996).

But a real-life population does not reach or stay fixed at some equilibrium point, as the simple model suggests. There are many influences on population size, so the stabilization process is complex. Consider that, in real life, a species' population is not monolithic but fragmented. It consists of individuals dispersed in patches that can be divided into somewhat distinctive groupings or sub-populations. The groupings are connected to one another in some manner, such as migrants. However, the presence of this patchiness ensures that the ability of an individual to find a mate is more difficult than inferred by an abstract, average density of undispersed mates (Begon et al. 1996). This serves as a reminder that although changes in a population record the outcome of a species' overall reproductive efforts, it is the individuals who engage in these efforts. It is their individual choices and activities that affect their population's dynamics.

It is within this context that the complex dynamics of population stabilization and the influences on (contributions to) it become easier to appreciate. One influence is the amount of time an individual takes to complete a population-related activity: for example, finding and securing a mate, and bearing and raising young. Delays are an inescapable aspect of population stabilization.

So is the involvement of individuals from other species. Consider the trophic and allometric relationship between unitary predators and prey. Predators usually reproduce more slowly than their prey. In addition, the survival rate of a predator's young will only increase after the population of the prey they feed on has increased sufficiently to provide the needed extra food. These delays contribute to the population dynamics of both predator and prey. This is clearly illustrated by the predator-prey relationship between the lynx and snowshoe hares because lynx, which feed almost exclusively on hares, are also the hare's primary population control agent.

A rise in the snowshoe hare population occurs one or two years before a rise in the lynx population. The result is an overshoot in the size of the hare population followed by an overshoot in the lynx population. The increase in hare predation results first in the hare, then the lynx population crashing to a low before the hare population starts to recover. The overall result is a dynamically stable population for both animals that has a roughly 10-year cycle. The lynx population cycle matches that of the hare, but delayed by a few years. That oscillation is weakly modified by the hare's interaction with squirrels, who prey on newborn hare.

Each species in an ecosystem is subject to a suite of influences on its reproductive success and death rate. The full explanation for its population's dynamics, as recorded by changes in its population, is the complex sum of all of these influences on the population. That sum also determines whether the species' population will stay within its current dynamically stable range or not.

The need to consider a wide variety of sometimes surprising influences on population, like squirrels have on hares, becomes particularly important when trying to predict the minimum size of a viable population. It is often thought of as the minimum number of individuals capable of breeding, but it is strongly influenced by other factors. Consider the need to raise the young and teach them their life skills. Success at either can depend significantly on receiving help from non-breeding adults. If this help is needed, then their presence is critical to the population being viable. For example, part of being a whooper swan is the ability to understand their communication signals. Then there are those ultra-high-density flocks of birds and schools of fish we see on our TV screens executing unimaginably fast twists and turns in unison. If their considerable abilities are not innate, then the young have to learn them by participating in a group of skilled practitioners. Too few experienced adults to execute the required manoeuvres properly and the young will be unable to learn the needed survival skills.

The large number and wide range of the influences on a species' efforts to complete its lifeway constrain the ability of models to reliably predict the size and limits of its dynamically stable population. This includes the minimum size of the species' viable population and its host ecosystem's carrying capacity (the maximum size of its viable population). Fortunately, there are practical rules of thumb that can help us to think about real-life population dynamics. For example, a cyclically varying population is closest to reaching its minimum viable size when it is at the low in its cycle. At this low level, a small disturbance could push the population below its minimum viable size and lead to its disruption or extinction. Think drought or overharvesting.

The inescapable presence of unknown, poorly understood or rarely occurring population-affecting factors gives rise to another rule of thumb: the actual minimum viable population size of a species should be expected to lie well above the theoretical minimum viable size. This indicates that a large safety margin is needed when managing or exploiting a species. In the same vein, an ecosystem's real-life carrying capacity for a particular species is difficult to determine. One rule of thumb suggests that the actual carrying capacity should be considered to lie well below the theoretical carrying capacity, especially for managed or exploited species. Another useful rule of thumb suggests a reason why.

As a population grows above its long-term stable population-size range, it should become increasingly difficult for more individuals of the species to complete their life cycle successfully. Under some circumstances a factor, such as a delay in the feedback response suppressing growth, could allow the population to grow substantially above its host ecosystem's carrying capacity. This is unsustainable. At some point the population will be reduced, usually catastrophically. The phenomenon of population overshoot and collapse is not unusual and it can push the population outside the envelope for a viable size (Hardin 1993). If it stays there too long, the crash can result in severe population disruption, even extinction.

In overview, the details of a species' population stability and dynamics are determined by more than its members' density and resource availability. They are also influenced by the collective characteristics of its host community, its resources and the species' lifeway.

**Extinction.** Species don't exist forever. The fossil record for the last 500 My indicates that 99% of the species known for that period are extinct. It also indicates that species go through a cycle. After a species comes into existence, its population increases slowly. With time, the increase becomes steadily more rapid until it slows just before peak population, after which the species begins a slow decline to extinction (Boulter 2002). On average, this cycle takes between 1 and 10 My (Begon et al. 1996). Using that as the range for a species' existence, and assuming a total number of species on the earth today of about 10 million, then a crude calculation predicts that, on average, between 100 and 1,000 of today's species would go extinct each 100 years (0.001% to 0.01% of total species/100 years). This is referred to as the global background extinction rate.

For a number of reasons, there is significant uncertainty in this background rate (Barnosky et al. 2011). For example, the number of species on the earth today is poorly known, with estimates falling between 5 and 30 million species. The number of 10 million is only a best estimate (Ehrlich 1986; Pimm et al. 1995). Consider too the bias in the fossil record.

The significance of an average extinction rate is also subject to debate. Consider that, at any given time, the extinction rate varies as a result of both random events, such as the Cretaceous asteroid impact 65 My ago, and cyclical events, such as those related to the earth's orbital imperfections. For example, mammals exhibit pulses of extinction every 2.4 to 2.5 My. Changes in the rates of rodent species' appearance and extinction indicate that these extinction pulses correlate with the earth's drier and colder periods. The regular occurrence of these colder, drier periods and the related pulses of rodent extinctions are caused by interactions between the earth's 1.2 My and 2.37 My orbital cycles around the sun (van Dam et al. 2006). (These cycles are much longer than the better known Milankovitch orbital cycles.) The presence of these types of random and cyclical features means that the average extinction rate serves more as a general reference than a true representation of background extinction rates at a particular time.

There is also the question of how we determine more recent extinctions. There can be significant uncertainty in declaring a species extinct or not based solely on a counting of individuals. Consider that the organism may simply be hard to find. Then, if individuals are found, do they adequately represent the viability of the remaining population? This is a critical question because the individuals found may simply be completing their lives as the "living extinct" of a doomed population. Remember George, the tortoise from the Galapagos islands?

Rather than trying to determine the point of extinction for an individual species in isolation, we can try to predict the average number of species extinctions to expect after an ecosystem has been disrupted. This more useful prediction is possible because of relationships such as those between body mass and home range size, and the limits to the number of species that can be supported by an ecosystem (discussed above). If an ecosystem is physically divided into smaller communities, then some species (particularly those that are endemic) can face extinction. The average number and average rate of their extinction depends on the number and size of the remaining communities. For example, if a tropical forest is broken into 1,000 ha (10 km$^2$ or 3.2 x 3.2 km) fragments, then about half of the species predicted to become extinct will do so within 50 years (Brooks et al. 1999). The historical fragmentation of North America's east coast temperate forests during settlement and logging displayed an equivalent area-disruption-to-species -extinction relationship (Pimm and Askins 1995). The rule of thumb for these relationships is that the combination of increased ecosystem fragmentation, smaller fragments, frequently repeated disruptions and disrupted surroundings results in an increasing number of expected extinctions.

Measures of a species' population size and dynamics do help us describe the state of species, especially when the trends in these measures are strong. However, this information, even if routinely presented in news bites seen on the internet or in wildlife management brochures, is insufficient on its own for us to fully appreciate the current and future state of a species, such as its nearness to extinction. Population numbers and their changes are most useful when they are presented and interpreted within the context of a species' lifeway and the state of the community within which they complete it. In addition, the information is most reliable and trustworthy if the species, its population and its community have been studied together over several generations, as exemplified by the 10-year study of the Kluane boreal forest (Begon et al. 1996; Krebs et al. 2001). Even with this information, it is still difficult to reliably predict a species' nearness to extinction or to determine the carrying capacity of its host environment. A large buffer is in order.

In summary, the numerical measures of an ecosystem's features, such as the mass of its species and their population sizes, are useful for assessing their role and significance in their host ecosystem's structure and functioning. But an ecosystem's dynamic complexity can, alone, make it difficult for us to decide which variables are important and how to meaningfully define a feature's or subsystem's functional boundary.

It can also be difficult because of our culturally guided or supported perceptions and views of an ecosystem's or a species' features. Westerners, for example, can feel that these features are contradictory (express a duality) or "incomprehensible" especially if we neglect the importance of scale and context. For example, organisms can be modular as well as unitary; modular organisms can be living and dead at the same time, and have ill-defined (fuzzy) boundaries; an "organism" can be something that is both animate and inanimate but not really either. Consider too that a stable population is not static but dynamic within a range; and although individuals of a species are still alive, it can be, in effect, extinct. There is also the issue of low and high efficiency both contributing to viability. These philosophical and practical difficulties are analogous to those we face when dealing with the particle-wave duality in physics.

Our human and culturally supported biases and limitations can make it difficult for us to appreciate an ecosystem's structure and functioning, and evaluate the appropriateness of our decision that will have an impact on it. The resolution to these human difficulties includes us thinking in a more nuanced manner about individuals, species and their host community. Think scale and context. This would help us, and our culture, treat an ecosystem as a dynamic complex system, even if its boundary is ill defined.

## Overview

I now recognize that an ecosystem consists of a large number of species engaged in a diverse range of interactions with one another and their environment. These interactions, which include feedback, operate over a wide range of time and space. They collectively link the individuals together to form the structure and be the functioning of the ordered yet chaotic whole that is their host ecosystem. The ecosystem, in turn, supports the members who form it, which completes another feedback. An ecosystem is a single, coherent, dynamic complex system. I now understand why it is so difficult for us to appreciate an ecosystem and to personalize the bits that we do understand (Berryman 2003).

With this new appreciation, a walk in our forest becomes more interesting. My view of our forest's kaleidoscope of trees, underbrush, bugs and animals has changed. I no longer see a random mixture of species or a chaotic arrangement of imperfect individuals constantly engaged in a fierce competition while surrounded by wasted resources.

I now see a space that is fully occupied by a dynamically stable, functioning community of species living and interacting in many ways over a wide range of scales. I appreciate that the individual organisms are neither cast in an inflexible mould nor free to do as they please. There is some flexibility in their behaviour and lifeway, and they can adapt to change, but this flexibility is constrained by who they are and by the functioning of their supporting community. I'm reminded of the importance of feedbacks. Individual organisms are participants in their community: they are an active part of its functioning. Thus they, like the other present and the past members of that community, both form their community and are constrained by it.

For example, in our forest community the availability of nitrogen is limited. Successive generations of plants migrate through the community in a pattern that tends to favour those locations with higher nitrogen concentrations in the soil. Similarly, when the deer and moose wander through the community at their lifeway speed they display a preference for those areas hosting the more nitrogen-rich individuals of the plants they eat. But these animals will excrete their waste nitrogen someplace else in the forest. When the availability of nitrogen is viewed in this context at a longer time scale, it is seen to both constrain the members' distribution and facilitate that distribution being both patchy and dynamic (Begon et al. 1996). The limit to available nitrogen is thus not an imperfection to be fixed, nor is it a barrier to be overcome in order to reach a better or more perfect ecosystem. It is a description of one aspect of the complex ecosystems' current state of dynamic balance.

Ecosystems, such as our forest, are indeed a single functional whole. From this perspective, it is clear that we cannot permanently alter one part of a community (either by addition, subtraction or disruption) without in some manner changing many other parts of the community. Neither is it realistic to find causal (deterministic) laws that define how the system functions. It is more important that we understand and appreciate the relationships that describe how the system functions, and the constraints within which it does so (O'Hara 2005).

The rules of thumb identified in this and the following chapter are the simplified, qualitative versions of some of those relationships. They may appear to be deterministic laws, but in the spirit of Linnaeus's method of classification, they are more accurately thought of as convenient boxes that *collectively* outline a flexible boundary around an ecosystem. Individuals must remain within this boundary, if they are to more easily complete their lifeway and ensure that their host community remains functional for future generations.

This chapter treated ecosystems as if they are largely static and isolated. They aren't. They change and are part of the earth's environment. What happens if we,

like a clock maker, introduce time and change to this picture, include more of the environment and then set the system in motion? How will species and ecosystems adapt? That is discussed in the next chapter.

# CHAPTER 8

# THE EARTH'S ENVIRONMENTAL SYSTEM

The previous chapter outlined the structure of an ecosystem and the essence of its functioning by treating it as largely static. But ecosystems are constantly subject to forces driving them to change. If we are to meaningfully contemplate an ecosystem's future, then we should consider that change. We should treat an ecosystem as dynamic. The first section in this chapter discusses the basic features of an ecosystem's changing and illustrates it using a real-world example. It then considers the limits to predicting ecosystem change.

Two of the basic constraints on a land-based ecosystem's species are the supply of water they can use (available water) and the temperature of the atmosphere around them. The values of these two variables are significantly determined by the climate at the ecosystem's location. A section therefore discusses the earth's climate system. After outlining the basics of the climate system, the details of its functioning and changing are discussed using past climate changes.

A change in the world's climate system prompts the earth's ecosystems to change and vice versa. The connections between the two are key parts of the earth's complex environmental system, which is discussed in a section. The chapter ends by discussing where we fit into the earth's environmental system.

## Ecosystem Change

We talk about an ecosystem changing as if it is the ecosystem that responds to the influences driving it to change. But life is lived by individuals, not by ecosystems. It is the individual members of an ecosystem, like the members of a human culture, that respond to the influences that are prompting them, either as individuals or as a group, to change. Their responses, in turn, change their host ecosystem.

Although an individual member of an ecosystem responds most actively to those influences that directly impact its day-to-day efforts to complete its lifeway, such as changes in its food supply, it also responds to a host of other influences. Its responses too can take many forms. The individual can adjust its behaviour, migrate (move), die, function imperfectly, do nothing or even try to alter the influence. Which mix of options the individual chooses and how it will express them is both facilitated and constrained by its species, its lifeway characteristics, the nature (type, speed and size) of the influence and the context in which the influence occurs.

The individual member's response, whatever it is, will affect its community in some way. The aggregate of all the members' individual responses changes their host ecosystem. But a changing ecosystem influences its members to change. A feedback loop exists between the individual members and their host community (ecosystem, biome).

As a result, the initial source of the influence prompting the members to respond can originate both internally as a result of the ecosystem's functioning, such as death and new life, and externally, such as a change in climate. Regardless, the members' response to the prompt is mediated by their host ecosystem's functioning as a whole. Indeed, an ecosystem's functioning and changing are different views of the ecosystem's overall dynamics.

If we are to appreciate ecosystem change as an event in its own right, we need to have a greater appreciation for its dynamics. It might seem from the above description that this is not possible, until we remember the ubiquitous presence of distinct relationships and limitations that constrain an ecosystem's functioning, as discussed in chapter 7. Consider that the members are constrained in their choice of how they can respond to an influence by allometric relationships, resource limits, past choices and their inherited circumstance. Similarly, their host ecosystem is constrained by the presence of a trophic web structure, its dependence on local resource recycling, and its reliance on global nutrient and mineral cycles. It is also constrained by its inherited circumstance, its past (Begon et al. 1996). Collectively, these features also ensure that there is understandable order to the details of an ecosystem's functioning and changing.

We can also more easily appreciate ecosystem change if we recognize that an ecosystem is a dynamic complex system whose functioning and changing displays the generic features of all complex systems. In a broad view, an ecosystem exists in a dynamic, quasi-stable state that can be adjusted to a degree or, if changed substantially, the ecosystem can switch to a different quasi-stable state (Scheffer et al. 2001). Think of a disrupted lake ecosystem. Like all complex systems, an ecosystem's functioning displays, in overview, a mix of deterministic, random and chaotic features, and the connection between cause and effect is complex. This allows us to predict future change, but only with a significant degree of uncertainty.

With this dynamic complexity in mind, this section first illustrates how the generic description of a complex system applies to the details of an ecosystem's functioning and changing. It then personalizes that dynamics through an example. The section ends by discussing the limits to us predicting ecosystem changes.

## Features of Ecosystem Change.

In light of an ecosystem's complexity and our human characteristics, it is no wonder that we experience difficulty appreciating ecosystem change. This difficulty is alleviated by demonstrating that the generic terms used in previous chapters to discuss personal and cultural change can be applied to ecosystem change. The overlapping terms discussed here are *drift*, *preconditioning*, *inertia/delay* and *scale*. Although randomness will not be discussed, it should be remembered.

### Drift.

Drift refers to slow change. An ecosystem can change through drift as a result of changes to its species' characteristics (including their interactions) or their location. There are a number of ways this can happen.

A species' characteristics can drift because of random genetic changes. Genetic diversity is also inherent in a species' population: each of the members invariably displays slightly different genetic characteristics. Regardless of the source of these slight genetic differences, if the members are exposed to an influence that affects their ability to complete their lifeway, then their genetic differences can enable each to respond to the influence in a slightly different manner. As a result, some individuals will be better able to complete their lifeways than others.

If the influence persists, even if weak, for generations, then a selection process is created. It favours those individuals whose characteristics are more likely to result in them both surviving, and raising and training young who will successfully complete their lifeway under the new living conditions. Over generations, successive births, selection and deaths result in the average characteristics of the species' population undergoing a drift toward those favoured features.

But both an individual's and a species' adjustments through drift are constrained. As they change, their resulting characteristics must, as always, meet all of the relevant allometric relationships and other constraints on it completing its lifeway as a functional whole: change must be coherent. Thus change through drift eventually affects more than one of a species' characteristics. The significance of the constraint of coherence is illustrated by the consequences of drift imposed on the bighorn sheep herds that live on Ram Mountain, Alberta, Canada.

Over a 30-year period, trophy hunting for the bigger horned rams has reduced the possibility of them passing on their genes for a big body and large horns to their young. We see the result as a reduction in both the body and horn size of succeeding generations of rams. However, there are additional consequences for the sheep. The size of the rams' horns and body is a critical part of the bighorn sheep's mate selection process. Our imposed changes on their horn and body size opens the door to changes or failures in their mate selection ritual (Coltman et al. 2003; Festa-Bianchet et al. 2014).

Even without a change in genetics, the succession of living and dying can result in drift. For example, successive generations tend to raise their young a little differently, thus behaviours can drift with the passing generations. Consider too that when a community's old individuals die, they leave behind a vacant space with particular characteristics (a niche-space). These are filled by the young, but not necessarily of the same species. For example, in one part of our forest, balsam trees are dying or being blown over, while the openings their passing has created are slowly being filled by both balsam and hemlock trees. A small slow change is taking place. The distribution of their host ecosystem's species is changing.

Similarly, an individual's living can also slowly change the characteristics of the space they inhabit by, for example, altering the availability of nutrients, the microclimate or the resident fungi. The changes (by drift or preconditioning) in the space's characteristics can allow individuals of other species to survive in that location. By moving in, these individuals change the shape and location of their species' patches, as discussed in chapter 7.

This acts as a reminder that a mobile organism, such as a fox, can drift across the landscape by exerting effort to stay close to its changing food resources. An immobile plant can't change its location in response to, say, changes in nitrogen availability. Plants drift across the landscape, as a species, by dispersing their seed. If a seed lands on available, suitable ground, even if it is quite distant from their parents, then an individual of that species can establish itself (Begon et al. 1996). Thus, over generations, the patches and ranges of a plant species can be seen to expand, shrink and move (figure 8).

Because plants are the base of an ecosystem's trophic web, as they move, so do the species that depend on them. As a result, their host ecosystem can slowly expand, shrink and move. Overall, as the characteristics and distribution of an ecosystem's member species drift, so does the ecosystem's functioning, size, location and/or its mix of species.

Although ecosystem drift is ubiquitous, we only fully appreciate its presence and significance when we have a longer-term record of its species and their distribution in an area. This record might reveal that a drift in the location of the ecosystem's member species, their patches or their range boundaries is random. However, the record might also show that there was some directionality to their drift (figure 8). For example, its species, and thus the ecosystem, may have migrated. Over the long term, the drift of hemlock into this part of our forest may not be random but a migration, perhaps driven by climate warming.

In overview, as species slowly change their characteristics or location, they slowly change their host ecosystem's characteristics, such as its functioning, size, mix of species, and/or location: the ecosystem changes through drift. Over time, that overall change can be substantial. For it to take place in a particular direction, rather than being random, the influence driving it has to satisfy four conditions. It must be widely distributed, affect each species in the community differently, be sufficiently long lived, and have directionality. History indicates that these conditions are easily met.

*Preconditioning.* The present state of an ecosystem is the outcome of its past responses to historical influences on it. If those influences facilitated but did not cause an ecosystem's present state, then the past can be said to have preconditioned the present state.

Consider that some ecosystems are routinely and significantly disturbed by natural events, such as fire. After each event, they recover through a sequence of distinctive and somewhat repeatable ecosystem changes collectively called a succession. During the succession, the early stages don't force or cause the ecosystem to go through the later stages. They just precondition it to do so. An alternate outcome is possible.

For example, after our sub-boreal forest has gone up in flames, it responds through a succession. Recovery starts with the ash-covered ground being occupied by weedy flowers, such as fireweed, and fast-growing deciduous trees, such as willow, aspen, alder and cottonwood. These plants are eventually replaced by coniferous trees, such as pine, spruce and balsam, which are slower growing but more efficient users of resources. Today, our local forest is a mixture of deciduous and coniferous trees, suggesting that it is only partway through the ideal, decades-long succession.

At each stage in the sub-boreal succession, plants play a dominant role in preconditioning the appearance of the next stage. For example, some of the early plants in the sub-boreal succession, such as alder and the legumes, fix nitrogen, which preconditions the soil for the plants of the later successions to be successful. Other plants provide the food and/or the habitat for higher trophic species, such as deer and birds. The appropriate food, for example, must be available before these animals can migrate into the area to become a part of the ecosystem's later succession stages. The food-providing plants facilitate these animals coming but don't guarantee their arrival: they precondition it. In turn, some of the later arrivals, such as birds, help to disperse the seeds of plants, thus preconditioning but not guaranteeing these plants will complete their lifeway (Begon et al. 1996).

Preconditioning, like feedback, is a ubiquitous feature of ecosystem change. It, too, complicates our efforts to identify and separate a cause of ecosystem changes from the effects of an ecosystem change. We are helped in our efforts to make an appropriate distinction between the two if we remember a rule of thumb. Our perceptions and interpretations of the nature and outcomes of ecosystem change depend heavily on the perspective and scale we use to study the ecosystem. Whether an event is a cause or an effect can depend on our perspective.

*Inertia.* An ecosystem does not respond immediately to a prompt favouring change: it displays inertia to change. Inertia is present because it takes time for its members to react to an influence and for the interactions between the individual members responding to it to be completed. And it takes time for their response to be transmitted throughout the ecosystem. An ecosystem's inertia to change can be thought of as the aggregate of these delays.

The functioning of an ecosystem provides many sources for these delays. They can be inherent in an individual's life cycle. An example is the time it takes for a species' young to mature. This results in a delay between an individual's birth and when it can reproduce. Similarly, the seeds of some boreal forest plants, such as the weedy species, germinate under similar specific soil conditions. The seeds lie dormant for many years until these auspicious soil conditions arise.

Delays can have an allometric aspect. For example, larger animals have longer lifespans (generation times) than smaller species. Thus, larger species take longer than smaller species to adjust their populations or behaviours in response to ongoing ecosystem changes, such as a sustained change in food availability.

Westerners, with our focus on efficiency and productivity, tend to think of these delays as a shortcoming. But consider the ubiquitous delay seen between an increase in a predator's population and its prey's population. For example, within the boreal forest an increase in the lynx population occurs later than an increase in the population of its prey, the snowshoe hare. This happens, in part, because lynx kits will only survive if there are sufficient hares to feed them. The delay in the increase of the lynx population is its way of accommodating that biological constraint, rather than a shortcoming.

We also tend to think that, once the time needed for an ecosystem to change has passed, it will be in equilibrium with the driver of that change. This is usually not the case. Consider that an ecosystem's greater host environment is sufficiently complex that, at some scale, it is continually varying and thus continually prompting the ecosystem to change. The state of an ecosystem is thus never fully in equilibrium with its greater environment. In general, ecosystems are more in tune with conditions from their immediate past than their present. Thus, although the concept of equilibrium is useful, such as the so-called climax (last stage) in an ecosystem succession, the expected ideal mix of species forming that equilibrium state is likely rarely reached, even in the long term (Begon et al. 1996). Our forest may remain a mixed one. It may never complete the ideal succession.

The inescapable presence of inertia means that an ecosystem can be better thought of as existing in a transient state of catch-up rather than equilibrium. Its members are not only reacting to continually changing influences on them, they are also catching up to past influences. For example, the present global distribution of biomes is the result of species redistributing themselves in response to climate warming after the last ice age. This redistribution process lagged behind the conditions driving it by decades to at least hundreds of years (Hughen et al. 2004). This may sound like an excessive delay, but consider that trees can live for hundreds of years, and soils take just as long to form.

*Scale: Time and Size.* Generally, when we search for the (internal or external) drivers of ecosystem change, we tend to be drawn toward those drivers that are more obvious to us. They are the localized, intense short-duration ones, such as a hurricane or pest outbreak. In contrast, we tend to pay little if any attention to the less visible, widely dispersed, weak but persistent drivers, such as an evolving species and a decades-long, regional-scale drying trend that slowly reduces the water available to all species. Our preference is unfortunate because, over the long term, weak persistent drivers of change can be just as significant as strong, short-lived ones. It is also unfortunate because ecosystem change is usually the product of a number of drivers operating at a variety of scales, at the same time.

There is a similar bias in the scale of our attention to the indicators of an ecosystem's response to a driver (i.e., to how the ecosystem is changing). For example, we preferentially focus on the responses of the larger species rather than the smaller ones. Our focus can result in a significant bias in our perception of ecosystem change. This is illustrated by the allometric relationships which indicate that smaller organisms, such as bacteria and small mammals, invariably live faster lives and have shorter lifespans than larger organisms (Blueweiss et al. 1979). This means that, after a given amount of time has passed since an ecosystem has been prompted to change, its smaller species will have lived through many more generations than its larger species. The smaller species will have had many more opportunities for selection to adjust their basic characteristics to better fit their changing host ecosystem than the larger, longer-lived species. Thus, by the time we see changes to the characteristics of an ecosystem's larger species, their host ecosystem will have been undergoing change for a while.

This is illustrated by the response of species to the relatively slow, global-scale 10,000 or so years of warming that ended the last ice age. The characteristics of the smaller mammal species, such as the rodents, changed the most. In contrast, the characteristics of the surviving larger mammal species remained relatively unchanged (Barnosky et al. 2004a). The larger organisms managed to adapt through migration.

If a driver is considered to be simultaneously prompting the whole ecosystem to change (an ecosystem-scale driver), then the ecosystem's response looks different. Each species responds to that driver at different speeds and in different ways (e.g., migration versus changing characteristics). Their response also depends on the scale of the driver: fast or slow, strong or weak, steady or variable, local or regional, internal or external. Under some conditions, a driver can require a species to respond too fast or make too large a change.

Most of the individual members of these species will be unable to appropriately adapt, so they will function imperfectly or even die prematurely. A species' population can decline or become extinct. If there are too many deaths and species extinctions in an ecosystem, or if they occur too rapidly, then the ecosystem's functioning is disrupted: it becomes less diverse and less stable.

In addition, as each of an ecosystem's species responds in its particular way to an ecosystem-scale driver of change, they influence one another to different degrees, and they do so over a range of space and time. Consider that the organisms who live faster and respond more quickly to a driver tend to dominate the lower levels in the trophic web. As these smaller, more numerous organisms respond, they can change the food supply for the species at higher trophic levels. Thus, changes in the lower trophic levels can influence how the rest of the ecosystem's species respond to the driver. The smaller species can thus affect how the ecosystem as a whole changes.

For example, the first fish to respond to the increasing temperature of the North Sea were the smaller, shorter-lifespan fish in the lower trophic levels (Perry et al. 2005). They responded by moving to colder water. The larger fish in the higher trophic levels who fed on these smaller fish were thus forced to follow them into colder water. The larger fish moved, even though they might have been able to tolerate the increasing temperature where they used to live.

Overall, an ecosystem changes because there are many (internal and external) influences acting on it and its species at different scales of time, space and intensity. The ecosystem's species respond to these influences in keeping with their characteristics (e.g., size), which means in different ways, and each at different scales of time, space and intensity. In addition, as a result of the interactions between the species, those undergoing long-term weak changes are affecting and being affected by those undergoing shorter-term strong and short-term weak changes. No wonder that, at first glance, ecosystem change seems incomprehensible. The process of ecosystem change is dynamic and complex, with feedback.

This provides us with a rule of thumb about contemplating ecosystem change. Our perception of it depends on the scale we choose to look at it. Imagine casually looking at a local ecosystem from your car as you drive by. You would likely see only the larger, more rapid, direct ecosystem changes, such as destruction through fire. In contrast, by walking though that ecosystem each day, you could develop a more detailed, personal familiarity with it. You would more easily identify the slower, ongoing changes resulting from weaker but persistent forces, such as the slow changes accompanying the aging of its immobile members or the consequences of overharvesting a species. More effort is required if we are to become aware of changes involving its smaller species. To appreciate the indirect and ongoing influences of slower, regional- to global-scale (background) changes being experienced by an ecosystem, we have to gather and analyze sufficiently detailed multi-scale information to support a big-picture, long-term view of the ecosystem.

An overview of this Features of Change subsection notes that we can more easily appreciate ecosystem change if we remember that it is individual organisms that react to drivers of change, not ecosystems. It is through the aggregate response of its individual members to the influences on them that their species and thus their host ecosystem changes. Only by remembering this foundation can talk of species or ecosystem change be meaningful.

An ecosystem is a dynamic complex system that displays the features of a generic complex system. Its functioning is the product of the individuals of its member species interacting with one another in a variety of ways, including feedback, at different scales of time, space and intensity. A particular aspect of that functioning can be described by generic features such as drift, preconditioning and inertia.

An ecosystem's changing is just a different view of its functioning, so the description of an ecosystem's functioning applies to its changing. The influences prompting an ecosystem to change (the drivers of change) can originate internally, from the system's functioning, or externally. Its members' (individuals and species) responses to the drivers interact with one another to spread the changes through the ecosystem. They respond by changing their characteristics or migrating, or both.

Being aware of these features and their description helps us to acknowledge the biases in the perspectives we use to look at and think about ecosystem change and its drivers. This makes it easier for us to successively view the ecosystem from different perspectives while consciously being aware of the benefits and biases of each perspective. We are then more likely to adopt a more realistic and nuanced overview of ecosystem change. For example, in the case of the North Sea fish, you could say that the marine ecosystem is moving with the fish. Alternatively, you could focus on the location and say that the existing ecosystem is changing as the fish leave. Both are correct but have different implications.

**Example of the Sea Otter.** The Aleutian Islands archipelago is some 800 km long. From an island's shoreline to a few kilometres out to sea, the ocean and its floor is occupied by a nearshore ocean ecosystem. One of its member species is the iconic sea otter. The following simplified description of this ecosystem's feeding structure (trophic web) is used to both personalize ecosystem change and illustrate its generic features. This example also illustrates the influence that the perspective we take of an ecosystem has on our appreciation of its changing.

The Aleutian nearshore ecosystem can exist in two quasi-stable states: sea otter dominated or sea urchin dominated. Both are described by a trophic web with four levels. The discussion starts with the otter-dominated state.

The ocean plants (primary producers) at level 1 of the web are macroalgae (kelp) and phytoplankton. These plants directly sustain the organisms in level 2. In particular, mussels consume phytoplankton, sea urchins eat kelp, and crustaceans feed on kelp detritus. Level 3 is dominated by the nearshore fish (e.g., greenling, great sculpin and Pacific cod) that eat the crustaceans. At trophic level 4 are the harbour seals, which eat fish, and the sea otters, which eat sea urchins, starfish, mussels, bivalves (e.g., clams), crabs and slow-moving fish (figure 9).

In this otter-dominated ecosystem, the otters eat enough urchins that the kelp can flourish. The resulting kelp forests provide the habitat needed by many species of nearshore fish and some offshore fish to complete their lifecycle. For example, the kelp fields provide the young fish with shelter from predators and protection from waves. The fields also facilitate the settling of silt, while kelp fronds provide a substrate to which herring attach their eggs. In turn, some of these fish support the harbour seals and, peripheral to the ecosystem, eagles.

The otter-urchin nearshore ecosystem is not restricted to the Aleutian Islands. It is found all along the northwest Pacific coastline from Alaska to California. Starting in the early 1700s, hunting steadily reduced the sea otter population along this coast. As a result, fewer sea urchins were being eaten, so their population grew. The abundant sea urchins consumed so much kelp that its fields declined. In response, the

ecosystem switched from being an otter-dominated to a sea-urchin-dominated quasi-stable state. This included a change in the ecosystem's mix of fish species and a reduction in fish populations. At the same time, the number of mussels increased. Along with these primary changes came a host of other changes, such as an increase in octopuses and chitons, which need not be detailed here (Estes and Palmisano 1974; Simenstad et al. 1978).

This sea-urchin-dominated nearshore ecosystem was found along most of the northwest Pacific coastline until the overhunting of sea otters ceased around 1911. At that time, the otters were extinct in some areas, but the ecosystem as a whole was still robust, resilient and adaptable. Thus, in the decades after 1911, the sea otter numbers slowly increased all along the Pacific coast. The sea urchin population declined and the urchin-dominated ecosystem slowly changed back to an otter-dominated one. By 1965, the sea otter population around the Aleutian Islands was estimated to be at the ecosystem's carrying capacity for sea otters.

Then, in the late 1980s, the Aleutian sea otter population started to decline. This change was not noticed until the early 1990s, when scientists detected an unexpected slight decline in the otter's population along the western Alaskan coast. By the mid-1990s, the otter population had rapidly declined to only 25% of its 1970s size (Estes et al. 1998). The otter decline continued into the 2000s until, by 2003, their population off the Aleutian archipelago was estimated to be only about 3% of its 1965 carrying capacity. Clearly the decline in otters was dramatic and significant (Estes et al. 2005).

The reasons for the decline were still unknown in the late 1990s. Nevertheless, the ecosystem was sufficiently well understood that a useful prediction was made: the ongoing sea otter decline would result in a return of the sea-urchin-dominated ecosystem. As the years rolled by, the urchin numbers did increase, and the ecosystem started to change.

The lack of a direct "cause" for the otter decline was only resolved in 1998 with the surprising discovery that they were being eaten by orcas (also commonly referred to as killer whales). As incredible as it may sound, the disruption of the Aleutian nearshore ecosystem could have been caused by as few as four otter-eating orcas (Estes et al. 1998). But orcas are not part of the nearshore ecosystem. They are a deepwater species from the offshore ecosystem. The nearshore ecosystem was not as closed as the above description of its four trophic levels had implied.

By eating otters, the orcas became a significant influence on, even a part of, the nearshore ecosystem. But it is unclear what was driving them to do so. There is evidence that, by the 1990s, the orcas' usual food, sea lions and harbour seals, had substantially declined. But that just leads to another question, why did those populations decline?

One suggestion, based on anecdotal evidence, is that the fish species preferred by sea lions (schooling fish such as cod, mackerel and herring) and by harbour seals (small schooling fish such as herring, hake and whiting) have declined. In support of this argument is the decline in the populations of birds who also eat these fish. There is also evidence that the declines in the preferred fish are associated with an increase in the less desirable pollock (Kaiser 1998).

One possible contribution to these changes is overfishing. Perhaps it reduced the preferred fish populations in a manner similar to that seen in California between 1850 and the present (McEvoy 1986). Perhaps the west coast pollock stocks are increasing for the same reason they are shifting northward in the Bering Sea: they are responding to ongoing ocean warming (Grebmeier et al. 2006). Another suggestion is that the local extinction of baleen whales from human hunting may have somehow changed the offshore ecosystem. Perhaps the young of these whales were a significant source of food for the orcas.

Regardless of the reasons for the orcas' changed behaviour, by inserting themselves into the nearshore ecosystem at trophic level 5, they initiated a top-down restructuring of the nearshore ecosystem. It changed from an otter- to a sea-urchin-dominated state (Estes et al. 1998). The orcas are said to have triggered an externally sourced trophic cascade in the nearshore ecosystem.

We can also conclude that the nearshore ecosystem and the orcas' offshore ecosystems are currently interacting. What will happen to the nearshore ecosystem in the future will be partly determined by what happens offshore. The drivers behind those future changes are unknown but will be multi-faceted.

We can expect that the state of the complex, dynamic nearshore ecosystem will continue to change over a wide range of time and space in a manner that involves feedback. Think climate warming. In changing, the ecosystem will continue to display a complex system's features, such as drift, preconditioning, inertia and randomness. The explanations we choose for the change we see in the nearshore ecosystem will thus reflect the scale and perspective we take to study and think about it.

This is an opportune time to point out that humans too have triggered a trophic cascade switch from an otter- to an urchin-dominated nearshore ecosystem. As mentioned, we inserted ourselves into the nearshore ecosystem as a trophic level 5 predator during the Russian sea otter fur trade between the early 1700s and the early 1900s, as mentioned above. The cascade was the result of overhunting the otter.

There is also evidence that we triggered a similar cascade much earlier in our history. Archaeological excavations in Aleut settlements on Amchitka Island in the Aleutians suggest that between about 700 BC and AD 1300 the local ecosystem was sea urchin dominated. This implies that the Aleut overharvested otters, thus initiating a local trophic cascade. The likely consequence for the Aleut is recorded by the fish remains recovered from their settlements. A low proportion came from the preferred fish that they depended on. The Aleut likely experienced a reduction in a valuable, high-quality food source (figure 9) (Simenstad et al. 1978). How should we think about the reasons for our overharvesting of otters?

**Limits to Predicting Change.** After contemplating the features of ecosystem change and the example of the sea otters, one might wonder about our ability to predict the future state of other ecosystems. Consider the meteorite impact that largely destroyed the dinosaur-dominated ecosystems at the end of the Cretaceous period around 65 million years (My) ago. Would we have predicted that it would have led, many millions of years later, to mammal-dominant ecosystems?

The general answer is that the structure and order of an ecosystem, such as the hierarchy of the influences on it, means that its future state is predictable. However, being a complex system, those predictions are subject to the limits imposed on predictions about any complex system, as discussed in chapter 5. Consider that the functioning of ecosystems contains a chaotic component. This means that its response to a driver of change is highly sensitive to the ecosystem's initial conditions. It is difficult for us to know these conditions with sufficient accuracy to reliably predict the details of an ecosystem's future. For example, after an event such as a fire has severely disrupted our forest, our knowledge of its succession allows us to predict the general features of the ecosystem that will likely recolonize the unoccupied space in the long term, but we can't predict the details (Begon et al. 1996). Similarly, we know that an ecosystem will change if a species is added (think exotic species) or removed from it (think over exploitation). We may even be able to predict the general form of the resulting changes. But we are unable to make reliable, detailed predictions about those changes, especially over the long term.

We might think that we could at least make reliable, detailed predictions about those aspects of an ecosystem that clearly have deterministic characteristics, such as cyclical population changes. This is not the case because these types of changes are never truly deterministic. Cyclical changes are never truly cyclical. Similarly, although the hierarchy of influences driving an ecosystem to change may allow for predictions, the flexibility in the hierarchy limits the prediction's certainty. The most useful predictions about ecosystems tend to be general in nature and conditional.

Fortunately, there are rules of thumb to guide our making and use of general predictions about ecosystem change. In order to make valid, useful predictions about an ecosystem's future, the nature of its functioning and how it changes needs to be known in sufficient detail. To gather the basic information about an ecosystem's functioning requires studying it continuously for many years, certainly for decades (Begon et al. 1996; Russell et al. 1995). During this time, a wide range of variables, not just those that appear to be most significant, have to be monitored. That monitoring must also be sufficiently detailed that it produces more than isolated snapshots of the ecosystem. Such a long, continuous and detailed study is needed in order to capture both the wide range of scales over which an ecosystem changes and the complexity of the ecosystem's dynamic interactions: its functioning.

The reason for these rules is illustrated by the conclusions from a continuous 10-year study of the boreal forest at Kluane, Canada. It is one of the longest continuous studies of an ecosystem in which many aspects of the system were recorded. (Most ecosystem studies are less than around three years.) Despite this effort and the relative simplicity of the ecosystem, the team of researchers could only outline the basic framework, not the details, of how the boreal ecosystem functions and changes.

The study confirmed that the boreal forest ecosystem is a three trophic level, predator-controlled (top-down) ecosystem in which hares comprise the largest amount of non-plant biomass. But, surprisingly, despite this dominance and the fact that the cyclical lynx-hare, predator-prey interaction is the more obvious one in the

ecosystem, the functioning of the boreal forest is not dominated by their interlocked cycles. There were other unexpected significant contributions to the ecosystem's functioning. For example, squirrels have a large impact on the hare population because they prey on young hares that are less than a day old. Of great surprise was the discovery that the ecosystem is characterized by considerable redundancy, especially among the birds. The importance of berries and mushrooms and the ecosystem's sensitivity to climate change was recognized, but the details remain unknown. As a result, even after 10 years of intense study, only uncertain, general predictions about the ecosystem can be made (Krebs et al. 2001).

Rules of thumb also remind us that having the required knowledge is no guarantee of a successful prediction. For that to happen, we must take the knowledge into account. Even then, as the degree and speed of an ecosystem's disruption increases, so does the difficulty of making even general predictions about its future state (Wilson 1999b).

In summary, our intuitive or common-sense appreciation of ecosystem change as an event is limited and biased by more than the complexity of ecosystems. It is affected by our innate and culturally constrained perception of the world around us. In particular, we tend to focus on those parts of a changing ecosystem we can understand and relate to. Unsurprisingly, these are its features that are familiar, easily visible, important to us and/or occur on the time scale of days to a few years. Examples are floods, large animals and edible fish.

Overall, our intuitive appreciation of an ecosystem's functioning and changing is biased and incomplete. We don't completely grasp the complexity of its dynamics. In particular, we fail to grasp that an ecosystem is simultaneously changing in a variety of ways over a wide range of scales and intensities, and that these changes involve interactions and feedbacks. It is dynamically stable within limits. We tend to prefer and accept simple cause-and-effect descriptions and explanations for why and how an ecosystem is changing.

Our intuitive views of an ecosystem's functioning and changing are complemented by the guidance we receive from our culture's world view. For example, Westerners are guided to view an ecosystem with a bias toward those species our culture favours, such as trees that produce timber, and toward the ecosystem changes that promote their presence. In contrast, we are biased against or indifferent to those species that our culture doesn't favour or is indifferent to, such as the non-timber-producing alder and willow trees. We are, respectively, against or indifferent to the ecosystem changes that promote their presence.

Consider too that Western culture's world view provides guidance to the members about how ecosystems function and change. For example, we are guided to think that they normally exist in a well-balanced, static equilibrium state. We are also guided to treat ecosystems as being either infinitely robust or infinitely fragile, until proved otherwise. This cultural guidance influences our goals for ecosystem management, protection and rehabilitation, and our related practices. Through these sometimes conflicting connections, our cultural guidance influences the changes we feel are acceptable to impose on ecosystems. Consider the confusion of guidance about how to ensure that an ecosystem's robustness, resilience and adaptability are restored or maintained.

Regardless of our views, the arrival of the future will force us to treat the ecosystem changes now underway for what they actually are: dynamic complex events. But will we grasp their significance before the consequences are unavoidable? Is it possible for us to develop a more realistic appreciation for ecosystem dynamics in advance?

We could place our faith in detailed ecosystem models and their predictions. But, even if these models were built, the burden of appreciating their predicted outcome would still be subject to our human decision-making foibles and the influence of our culture's world view. We are more likely to fully appreciate ecosystem change if we focus on treating it as an inherent aspect of a complex ecosystem's normal dynamics. We are helped in this task by remembering the basics of an ecosystem's functioning and changing. Here is a reminder of those features mentioned so far. They are presented in the form of ecosystem rules of thumb that are, for convenience, divided into four overlapping groups (O'Hara 2005).

(1) Rules of thumb concerned with the basics: An ecosystem is a dynamic, complex system whose functioning and changing (functioning) is the interactions between its many (individual and species) members and the influences on them. Those interactions take many forms, including feedback, and occur over a wide range of time and space scales, simultaneously. The resulting generic features of ecosystem functioning are the same as for all complex systems: they include drift, preconditioning, inertia/delays and randomness. At any given time or place in an ecosystem, all of these features of change are present. What you see in an ecosystem depends on the perspective and scale you choose to view it. An ecosystem's functioning and ability to change is constrained by its history.

(2) Rules of thumb used at a species scale: Change is experienced and expressed by the individuals of a species. As an ecosystem's member species change in response to the influence on them, their host ecosystem changes. The small, short-life-cycle organisms, such as insects, will react to an influence first. Rapid, strong changes favour weedy species. As the amounts of sunlight, nutrients and water change, so must the plants. Plants are the ultimate source of food for almost all of an ecosystem's higher trophic level species, so as the plants change so must those species.

(3) Rules of thumb used at an ecosystem scale: Both external forces and internal member-community interactions, such as feedback loops, influence an ecosystem to change. Ecosystem change is constrained in fundamental ways, such as by allometric relationships, the trophic web and resource availability. Weak, long-term influences on an ecosystem are just as important as strong, short-term ones.

(4) General rules of thumb used for change: An ecosystem exists in a quasi-stable state; it can change within that state or switch to a different quasi-stable state. Key measures of an ecosystem's ability to cope with change are its resilience, robustness and adaptability. The less disrupted an ecosystem's functioning is, the more resilience, robustness and adaptability it will display as it responds to a driver of change. The functioning of an ecosystem includes deterministic, random and chaotic aspects. To make a valid and useful prediction about an ecosystem's changes requires long, continuous, comprehensive data collection periods in the order of decades. Useful predictions are conditional, "if-then". Climate change is a fundamental influence on individuals and thus on ecosystems.

## Climate Change

Two of the primary influences on land-based ecosystems are sunlight and water, which are also the core aspects of a climate. A primary influence on ocean ecosystems is sunlight. Thus, to appreciate changes to ecosystems and biomes, and their distribution, we require some knowledge about the earth's climates, their distribution and how they change. An overview of the climate system and past climate changes provides us with the needed understanding.

<u>Climate System.</u> We naturally focus on the weather around our homes because its current state and how it changes have an immediate effect on our daily living. We are also aware that the weather displays a yearly cycle of seasons and that the average seasonal weather where you live is different from where I live. We thus understand the idea of climate.

However, our memory of the climate around our home is affected by our attention to memorable weather events. In addition, the changes in our home area's climate are masked from our experience by the weather's day-to-day and seasonal variations. Thus, the members of at least complex cultures don't have a reliable personal reference/memory on which to base their intuitive opinions of climate change. Our appreciation for climate and climate change is helped if we have an understanding of the climate system and how it functions, and a record of historical climate conditions to refer to.

*Basics.* The weather and climate at a particular location, such as your hometown, are summary descriptions of the state of the atmosphere there: its direction of movement, its temperature, its water content and the water's form (e.g., snow or rain). *Weather* refers to the short-term values of these variables, while *climate* refers to their decadal (medium-term) averages. The difference between the medium-term average weather for each of the climate's seasons is the climate's seasonal variation.

It is possible to represent the state of the earth's atmosphere through averages, such as its average temperature. A more detailed representation of its state is the somewhat systematic distribution of the earth's different climates. A basic description of how that climate distribution arises and changes is provided by combining four atmospheric features. They are the total amount of heat in the atmosphere (its average temperature); the uneven global distribution of the heat in the atmosphere (the uneven distribution in its temperatures); the drive to balance the uneven distribution of heat; and the temperature sensitive variation in both the amount (actual and the theoretical maximum) of water and its form (e.g., snow, rain) in the atmosphere.

An overview of how they combine notes that the source of the atmosphere's heat is sunlight. Because the earth is a sphere, the sunlight-induced temperature of the atmosphere at the poles is colder than at the equator. This temperature difference is reduced by a global-scale movement of hotter air at the equator toward the poles and colder air at the poles toward the equator. The distribution of the earth's climates can be thought of as the product of this heat redistribution.

However, the above description is incomplete. The heat redistribution process also involves the spin of the earth, land masses with topography, and oceans with heat-distributing ocean currents. When they are taken into account, the result is a climate system that includes a convoluted pattern of global-scale, heat-redistributing air currents or winds, and ocean currents.

We are made aware of these air currents when we see the continental- and regional-scale swirls of clouds moving across the satellite images during a TV weather presentation for our home area. The direction, temperature and water content of the particular air current that moves over our house is our weather. Its daily changes at our homes becomes our experience of weather. They include the appearance of the towering clouds that produce stunning thunder and lightning storms with their roof-rattling gusts of wind and rain downpours. The medium to long-term average direction, temperature and water content of the regional-scale air flow over our home region is our area's climate. The steady changes in those averages over many years become our experience of climate change. Yet, despite this exposure to these features of the climate system, it is still difficult for us to appreciate how the system's functioning results in our home region's climate and weather, and the global distribution of the earth's climates.

Our efforts to personalize the basics of this process are helped by considering a small part of the heat redistribution pattern. The Gulf Stream brings warm surface water from near the equator up the east coast of North America and then across the North Atlantic to Europe. It is widely believed that the Gulf Stream is the sole reason why Europeans experience a warmer climate for their latitude than they should. This is not the case.

About 50% of Europe's warmth is associated with the warm, moist air mass streaming over California from the Pacific Ocean. As this easterly moving air mass is forced to rise over California's Sierra Nevada and the southern Rocky Mountains, it cools. This forces its water content to fall as rain or snow (hence the white mountains, Sierra Nevada). As the dry air descends to the deserts on the east side of the mountains, it is warmed by compression. The warm, dry air then gains more heat as it flows toward the east-southeast across the southern United States. When the warmed flow of air is near northern Florida, it turns toward the northeast and then flows across the Atlantic Ocean to Europe. During this last stage of its journey, the warm air crosses the warm Gulf Stream, where it picks up moisture. It is this warm, moist air current that delivers the water and half of the heat responsible for Europe's warm, moist climate and its grey damp weather (Seager 2006).

This example illustrates that the influences on the weather and climate at our hometowns operate at a global as well as a local scale. It illustrates that the climate system providing that weather and climate is dynamic. It includes the global-scale, characteristically convoluted heat-redistributing air and ocean currents as well as the exchanges between them. It also includes the interaction of these currents with the land, such as its topography. The example also reminds us that the climate system is neither simple nor deterministic. It is a dynamic, complex system.

In order to appreciate the global-scale diversity of the earth's climates, their distribution and how they change, we need to expand on this description of the climate system's basics. In particular, we need to appreciate the system's sunlight-trapping process and some details about the system's dynamics.

*Sunlight Trapping.* The proportion of the sunlight reaching the earth that will be trapped as heat in the atmosphere is largely determined by a two-step process. The sunlight reaching the earth has a wavelength/energy range that allows most of it to travel unimpeded through the atmosphere to strike the water, ice, land and plants on the earth's surface. Some of that sunlight will be reflected back into space. How much is determined by the ability of each object's surface to act like a mirror, called its albedo. A useful rule of thumb is that lighter-coloured surfaces, such as snow and ice, reflect more sunlight (high albedo) than darker surfaces, such as liquid water and most land plants (low albedo).

The sunlight not reflected by an object is absorbed by it, which is expressed as an increase in the object's temperature. We experience this when the rocks in the sun are hotter to our touch than those in the shade. Overall, the less direct sunlight the earth's surface reflects, unchanged, back into space, the more is available to heat its surface.

In the second step, the warmed land, water and plants act like room heaters. They radiate much of their heat back toward space (earth's radiation). But, unlike sunlight, the earth's radiation occurs over a wavelength range that allows most of it to be absorbed by the

atmosphere, thus warming it. By trapping the earth's radiation, the atmosphere acts like a blanket.

The exact amount of the earth's radiation trapped by the atmosphere is determined by its composition. Water vapour is a significant component of our atmosphere, with the amount varying widely over the earth. If the water is ignored, then the atmosphere's composition is approximately 99% nitrogen plus oxygen. The remaining 1% is a mix of the minor gases, such as carbon dioxide, nitrogen oxides, methane and methyl chloride (the only common natural chlorine-containing organic compound) (Faure 1998b). Included in the 1% are the human-made gases, many of which contain chlorine: for example, chloroform, chlorofluorocarbons (CFCs), hydrofluorocarbons (HFCs) and the gas phases of pesticides, such as DDT.

Each of the atmosphere's gases has a different ability to trap the earth's radiation. Nitrogen and oxygen have a low ability. Carbon dioxide, methane, nitrous oxide ($N_2O$) and some HFCs are very efficient trappers. These so-called greenhouse gases are responsible for trapping most of the earth's radiation.

The contribution a particular greenhouse gas makes to the atmosphere's ability to trap the earth's radiation is determined by a combination of the amount of the gas in the atmosphere and its ability to trap the earth's radiation. For example, methane is present in the atmosphere in much smaller quantities than carbon dioxide, but its ability to trap the earth's radiation is at least 20 times greater than carbon dioxide. Thus, despite methane's low concentration, it makes a significant contribution to the atmosphere's ability to trap the earth's radiation. The total amount of the earth's radiation trapped by the atmosphere as heat is determined by summing the amounts trapped by each of its greenhouse gases.

This simple picture is complicated by the presence of solids and liquids in the atmosphere. The liquids are dominated by water, as seen in clouds. The solids include ice, soil dust, combustion products such as ash from forest fires, and tiny solid particles called aerosols, which are mixtures of both natural compounds and pollutants. These solid and liquid particles affect the amount and distribution of sunlight reaching the earth and the amount of the earth's radiation trapped by the atmosphere.

For example, all particles in the atmosphere block direct sunlight from reaching the lower atmosphere and the earth's surface. Think smog. Some of these particles have high albedo, such as water clouds and white sulphate particles, so they simply reflect sunlight back into space. Others have low albedo, such as soot, so they absorb sunlight, warm up and thus add a little heat directly to the atmosphere. Consider too that cloudy nights are warmer than clear nights because the clouds help the atmosphere to trap the earth's radiation.

Overall, the amount of heat in the atmosphere (its temperature) at a particular location is determined by the sum of the heat from two sources: the heat contributed by the sunlight at that location; and the amount of heat brought in or removed from that location by air and ocean currents. Those currents are trying to balance out the heat differences between pole and equator (the heat redistribution processes).

*Dynamics.* As the ongoing variation in the weather and the passing of the seasons at your hometown shows, the earth's atmosphere is constantly changing in some way. Much of the nature and scale of this change is internally generated by the functioning of the earth's climate system. Simplistically, its functioning results in changes to the earth's albedo and the composition of its atmosphere, which then changes the amount of the earth's radiation its atmosphere traps. This affects the earth's weather and climates. The climate system exists in a quasi-stable dynamic state. The dynamics of this state (the system's functioning and changing) are discussed here with a focus on carbon dioxide ($CO_2$) and the carbon cycle.

The boreal forests are dominated by coniferous trees such as pine, spruce and balsam. During the northern winter, these forests lie dormant. During the northern summer growing season, they remove such a large amount of $CO_2$ from the atmosphere that they cause the annual dips in the record of the atmosphere's $CO_2$ content being collected at Mauna Loa, Hawaii. During the growing season, the boreal forest also emits compounds and aerosol particles into the atmosphere. Some of them, such as terpenes, help clouds to form, which changes the amount of sunlight reflected back into space (Tunved et al. 2006). The functioning of the boreal forest has a cyclical influence on the atmosphere's composition and temperature.

The boreal forest also has a random influence on the atmosphere's composition and temperature. For example, randomly occurring forest fires are a part of the boreal forest's functioning. When a fire burns off a particular area's trees, understory and woody debris, it immediately releases $CO_2$ and black soot into the atmosphere. Over the medium term, the burnt area's annual withdrawal of $CO_2$ from the atmosphere drops to

near zero and its albedo is dominated by land that is bare (in summer) or snow-covered (in winter). This new state of affairs will exist until the new, young forest enters its rapid growth stage.

The combination of the cyclical and random aspects of the boreal forest's functioning changes the atmosphere composition and the earth's albedo, thus changing the atmosphere's temperature and the earth's climates. However, at the same time, any changes to the climate where the boreal forest is growing affects the forest's functioning. The boreal forest is one part of a global-scale biome-climate feedback loop within the earth's climate system.

The boreal forest illustrates just a few of the many interactions among the features of the earth's atmosphere, land, oceans and its living things that affect the amount of heat in the earth's atmosphere and its distribution. These interactions collectively form the earth's complex climate system. They are the system's functioning and changing (functioning), its dynamics.

A global-scale illustration of how these interactions are bound into a single, complex climate system is provided by the global carbon cycle. Consider that the boreal forest, along with all of the other land plants, are one of the carbon cycle's land-based carbon reservoirs. The atmosphere itself and aspects of the oceans also act as carbon-cycle reservoirs (figure 7a). There are many links and feedbacks between these land, atmosphere and ocean carbon reservoirs that bind them together into the carbon cycle. The functioning of the carbon cycle helps to form the climate system, its dynamic functioning and its overall features such as inertia, delays, preconditioning and unexpected events. The carbon cycle's contribution/role is discussed by focusing on $CO_2$, a greenhouse gas.

Imagine a year when mega forest fires burned up a significant proportion of the boreal, Amazon and Indonesian forests. During that time of smoky fires, a large amount of carbon in various forms was transferred from the land-based carbon reservoir to the atmospheric reservoir. Within a few days to a few years the $CO_2$ portion was mixed by air currents into the atmosphere on a global scale. This widely distributed increase in its $CO_2$ content meant that it could trap more of the earth's radiation, thus increasing the atmosphere's temperature.

Over time, a portion of the forest fires' $CO_2$ will be removed from the atmospheric reservoir by land plants. If new forests are allowed to grow on the burnt land, then they will participate in that removal. The atmosphere's $CO_2$ content will decline over a period of decades.

It will also decline because a portion of the $CO_2$ will be transferred into the ocean (figure 7a). This portion will dissolve into the oceans' surface layer in a process that will take months to a few years to complete. As the dissolved $CO_2$ content of the oceans' surface layer increases, two things happen. The surface layer becomes more acid. And a portion of the surface layer's dissolved $CO_2$ content will be mixed into the ocean layer immediately below it (the middle layer) by a process that will take decades. A similar process transfers dissolved $CO_2$ from the middle ocean to the deep ocean. By the time the deep ocean has mixed in its portion of the forest fires' $CO_2$ emissions, many centuries will have passed (Friedlingstein and Solomon 2005).

This slow process of moving $CO_2$ (and heat) into and out of the oceans endows them with inertia to change. The oceans are also large, thus, even though they contain low levels of dissolved $CO_2$ and are relatively cold, they store a large amount of the earth's $CO_2$ (and its surface heat). Because of the oceans' inertia and large storage capacity, they play a key role in keeping the earth's climate system stable. The oceans thus favour the earth's climates changing slowly. The word *favour* rather than *force* is used deliberately. Consider that if the atmosphere's $CO_2$ content changes faster than the rate of transfer into the oceans, then the earth's climates will change at a rate set by the change in the atmosphere's $CO_2$ content.

For example, if the amount of $CO_2$ emitted during the year of the mega fires imagined above was more than the oceans could remove in a year, then there would be an "excess" of $CO_2$ in the atmosphere. If this excess amount was significant, then the atmosphere's temperature would rise and the earth's climates would warm. Unless, that is, there was, fortuitously, some sort of compensating response/event. Think of an increase in the land's albedo (e.g., snow on land cleared by burning) or an increase in particles (soot or sulphate) in the upper atmosphere.

The slow rate of transferring the excess $CO_2$ into the oceans also means that the atmosphere will remain $CO_2$ rich long after the $CO_2$ contributions to the atmosphere have stopped (the fires go out). It could therefore take a while for the atmosphere to cool. Similarly, the overall shift in the distribution of carbon among the carbon cycle's reservoirs (its balance) as a result of the fires will also endure.

Such a year of mega fires would be an extremely rare event. Its value lies in it helping us to appreciate how climate-affecting interactions involving aspects of the earth's atmosphere, oceans, land and living things are bound into a dynamic complex climate system. It also helps us to appreciate the complex dynamics of its functioning and changing.

A more complete picture of the system and its complexity can be obtained by embellishing the above description with methane in the carbon cycle. The water cycle could also be added, for example a change in the area of sea ice or ice sheets would change the earth's albedo. Precipitation, of course, facilitates plant growth, thus providing a link between the water-cycle and the carbon cycle's contribution to the climate system's functioning.

However, even if such a detailed description were completed, we would still be left with an unanswered question: how large, fast and long must a (cumulative) change in the climate system, such as $CO_2$, be to result in the earth's climates and their distribution undergoing significant changes? You might have already guessed what the answer would be: it depends on our focus, the scale and reference we choose to use. A pragmatic answer to this question at a human-relevant scale is provided by considering sunlight, the source of the heat driving the climate system. If the amount of sunlight reaching the earth were to change, then the climate system would respond and the temperature of the atmosphere would change. The characteristics and distribution of the earth's climates would change.

The variation in sunlight the earth currently experiences is not dominantly caused, as one might expect, by changes in the sun's emission of sunlight. It is the changes in the distance and orientation of the earth relative to the sun that matter. These changes are caused by "irregularities" in both the earth's rotation around its axis and its orbit around the sun. There are four primary orbital irregularities that are collectively known as the Milankovitch cycles. There is the eccentricity in the shape of the earth's orbit around the sun, which results in a 400,000-year and a 100,000-year cycle. There is the change in the tilt of the earth's axis of rotation or obliquity, which results in a 41,000- to 42,000-year cycle. And there is the circular movement of the earth's axis of rotation or precession, which results in a 19,000- to 23,000-year cycle (Adams et al. 1999).

The observed variation in the amount of sunlight reaching the earth as a result of these irregularities is the simple deterministic sum of these four contributions. That variation will, for our purposes, be called the Milankovitch sunlight cycle. The sunlight cycle ensures that, over the long term, the amount of sunlight reaching the earth is always changing to some degree. There is always a force driving the earth's climate system to change. By looking at the historical changes in climate that result from the Milankovitch sunlight cycle, we are better able to answer the question about significant climate change at a human-relevant scale. This discussion will be started in the next section and continued in the next and later chapters.

An overview of the Climate System subsections notes that, at its most basic, the earth's climate system functions because sunlight heats the earth's land, oceans and, indirectly, its atmosphere. The lands, oceans and atmosphere around the equator have a higher temperature than around the poles. This heat difference is reduced by a convoluted pattern of global-scale air and ocean currents. This heat redistribution system, in combination with complicating factors (e.g., the earth's topography, ice sheets and spin) provides an explanation for the variety of the earth's climates and their global distribution.

A more complete view of the climate system's functioning and changing (its dynamics) includes the climate-affecting interactions among the features of the earth's atmosphere, oceans, land and living things, as illustrated by the carbon cycle. These interactions alter the albedo of the earth's surface and the composition of its atmosphere, which changes the amount of heat in the atmosphere and thus the nature and distribution of the earth's climates. However, the combination of the oceans' large heat capacity and dissolved $CO_2$ content, and the slow speed with which these two features change, means that interactions with the oceans help to stabilize the climate system. But, counter-intuitively, the ocean's inertia to change also limits its ability to dampen climate change under some circumstances. When the change in the $CO_2$ content and temperature of the atmosphere is larger and occurs faster than the oceans' inertia allows it to effectively respond, then the earth's climates can change faster than the presence of the oceans' moderating influence would suggest. Add to the description the ongoing variation in the amount of sunlight reaching the earth that results from the Milankovitch sunlight cycle, and we arrive at an even more comprehensive view of the climate system's dynamics.

Overall, the earth's climate system is a dynamic, complex system that can be described as a heat redistribution system that depends on sunlight to function. Its functioning is the interactions among the features of the earth's land, oceans and atmosphere that affect climate. That functioning displays the generic features of a complex system. There is a wide diversity of interaction types, including feedback, that occur, simultaneously, over a wide range of time, space and intensity. In a wide view, the system's functioning displays deterministic, chaotic and random features. A particular aspect of the system's dynamics can be described by generic descriptors such as *inertia* (think oceans), *delays* and *preconditioning*.

As expected for a complex system, whether we think that a change in the climate system will result in significant changes to the earth's climates depends on what we choose to look at in the system, and the time scale, space scale and reference we choose when looking at it. The following rule of thumb might help: The effect of a small but long-term change may have a long-term outcome that is just as significant as a short-term, very large change. But the details for each scenario will be different. The outcomes are discussed next.

**Past Climate Changes.** By studying the record of past changes in the climate system, we can learn more about its current functioning and changing as a whole. We can then better appreciate when a change to the climate system is significant for humans and how it can occur.

A record of past climates based on direct measurements of rainfall and the atmosphere's temperature is only available for the last 300 years or so. Longer records are based on proxy measures of climate variables. The proxies for the atmosphere's temperature include the temperature measurements down boreholes on land; hydrogen and oxygen isotope ratios in deep-sea sediment or ice cores; and carbon isotopes in either tree rings or carbonate minerals in fossil coral heads (Mann et al. 2008; Thompson et al. 2002; EPIC community members 2004). The process of reconstructing past climates is helped by records of the atmosphere's composition, such as the levels of atmospheric $CO_2$ trapped in the annual ice layers within glaciers, ice caps and ice sheets.

By combining these types of climate and climate-related information, it is possible to reconstruct past climates and how they changed over time (Marcott et al. 2013). Of interest here is an 800,000-year record of the atmosphere's composition and temperature obtained from ice cores drilled out of the Antarctic ice sheet (EPIC community members 2004; Petit et al. 1999). This subsection focuses on the youngest 400,000 years of that record.

*Glacial-Interglacial Record.* The ice-core record from Antarctica reveals 400,000 years of variations in the atmosphere's temperature at the core's recovery site. It also reveals the global scale variations in the atmosphere's CO$_2$, nitrous oxide (N$_2$O) and methane content. The pattern in the record for each of these variables is remarkably similar: they all display a number of synchronous, long-period oscillations, interrupted by short-period spikes and troughs, or abrupt events (figure 11). These features represent the global-scale changes in the atmosphere's composition and local-scale changes in its temperature over the past 400,000 years.

The long-period temperature record consists of four large-amplitude cold-warm cycles. These correspond to four global glacial-interglacial cycles. Each cycle consists of two parts: a cold or glacial period of 60,000 to 90,000 years' duration, during which the polar ice sheets grew; and a warm or interglacial period of 10,000 to 20,000 years duration, during which the ice sheets shrank (Siegenthaler et al. 2005). For example, during the climate cooling towards the last ice-age minimum (at 25,000 yBP), two ice sheets grew to cover almost all of North America, north of the Canada-US border, under ice up to 3 km thick. During the subsequent climate warming that ended the last ice age, these ice sheets almost completely melted. Today, we are partway through the resulting youngest interglacial period.

The four glacial-interglacial cycles are the product of cyclical change in the amount of sunlight reaching the earth, as described by the Milankovitch sunlight cycle, and the response of the earth's climate system to that change (Ganopolski et al. 2016). That response is revealed by looking at four ways in which the earth's glacial-interglacial cycle deviates from the Milankovitch sunlight cycle.

Firstly, the interglacial temperature peaks (and subsequent glacial troughs) occur later than the Milankovitch sunlight cycle peaks (and troughs). There is a delay between the change in sunlight reaching the earth and a change in the temperature of earth's atmosphere.

Secondly, unlike the more symmetrical Milankovitch sunlight cycle, the glacial-interglacial temperature peaks and troughs are significantly asymmetric. In general, the rise to an interglacial peak is consistently steeper and smoother than the slower, rougher fall from a peak (figure 11).

Thirdly, the rise to an interglacial temperature peak and the fall to a glacial trough is much larger and is reached must faster than the changes expected from the Milankovitch sunlight cycle. For example, the final burst of warming to the interglacial peak of 125,000 years before present (yBP) could have been completed in as little as 400 years (Adams et al. 1999).

Fourthly, the long-period glacial-interglacial cycle is interrupted by much shorter duration, lower-amplitude, abrupt temperature irregularities. These abrupt events are not seen in the reconstructed Milankovitch sunlight cycle (Dansgaard et al. 1993).

By considering examples of these deviations, first for the long-period cycle and then for the abrupt events, the functioning of the climate system is more easily personalized. The examples used come from the time between the peak of the previous interglacial (125,000 yBP), the minimum of the last ice age (25,000 yBP) and today's interglacial, which started at roughly 10,000 yBP.

*The Long-Period Cycle.* The Milankovitch sunlight cycle had already passed its minimum and was increasing when the atmosphere's temperature reached the minimum of the last ice age at around 25,000 yBP. The atmosphere's temperatures then started to rise toward the current interglacial (warm) period. Despite this warming, the size of the Greenland ice sheet continued to increase until it reached its maximum aerial extent around 21,500 yBP, roughly 3,000 years after the time of the earth's temperature minimum (Johnsen et al. 1992).

These relationships illustrate the presence of inertia in the climate system. One source of this inertia is the oceans: they change slowly. Another is the ice sheets. At the height of the last ice age's extreme cold, their surface was dry, white snow that reflected sunlight, like a mirror, back into space. The ice sheets' surface area was so large that enough sunlight was being reflected back into space to delay the impact of an increase in sunlight on the atmosphere's temperature.

The asymmetry of the peaks and troughs in the record of the long-period cycle arise largely because of differences within the climate system between how an ice sheet grows compared to how it melts (figure 11). During the atmosphere's cooling from an interglacial temperature high to the low of an ice age, snow accumulates to form the ice sheets. As they grow to cover an increasingly larger area of land, their dry and white surface reflects increasing amounts of sunlight back into space. This promotes atmospheric cooling. Accompanying the cooling is drying, which slows the ice sheets' expansion and thus their contribution to the atmosphere's cooling.

A Milankovitch-cycle-induced increase in sunlight presages the end of an ice age. But time must pass before there is a corresponding increase in the atmosphere's temperature and the ice sheets can begin to melt. The ice fronts closest to the equator are the first to melt and retreat toward the poles, which exposes the dark land or dark ocean water they covered. At the same time, the surface of the ice sheets nearest to the retreating ice fronts melts to become wet and thus darker with a lower albedo. Both of these changes reduce the area of the ice sheet that is covered with dry, white snow of high albedo. Both changes lower the earth's overall albedo. As a result, the earth absorbs more sunlight, which means there is more earth radiation for the atmosphere to trap as heat, so its temperature rises. As the earth's atmosphere warms, the ice sheet continues to melt and more of its surface becomes wet and darker, thus completing a feedback loop (Hansen et al. 2007). This albedo-reduction–ice-melting feedback loop is sustained until the ice sheet has largely disappeared.

The feedback loop formed during the melting of an ice sheet promotes its melting, which is unlike the feedback loop formed during the formation of an ice sheet that dampens its growth. It is this difference in the feedback that is responsible for the asymmetry of the glacial-interglacial peaks. The difference can also be thought of as the difference in the inertia experienced by the process of cooling to an ice-age trough compared to warming to an interglacial peak.

The observation that the temperature rise to an interglacial temperature peak (or fall to a trough) is larger than expected for the increase in sunlight attributed to the Milankovitch sunlight cycle is explained by the contributions to the atmosphere's warming between the last ice age and the present interglacial. The Milankovitch sunlight cycle contributed approximately 0.25 watts per square metre ($W/m^2$) to the warming. In contrast, the changes in the earth's albedo contributed about 3.5 $W/m^2$, while the combined increases in the atmosphere's $CO_2$, $N_2O$ and methane content (greenhouse gases) contributed about 3.0 $W/m^2$. The large size of the contribution from the earthbound part of the climate system indicates that those processes amplified the small increase in temperature directly attributed to the Milankovitch sunlight cycle (Hansen et al. 2007; Alley et al. 2003). The climate system includes a strongly amplifying feedback loop that explains both the larger-than-expected fall to an ice-age trough and rise to an ice-age peak.

What are the features of the climate system responsible for that amplification? The decline in the earth's albedo that contributed 3.5 $W/m^2$ to the warming was mostly related to the ice sheets: the wetting of the snow and the exposure of dark land during melting. The rest was due to the reduction of the area covered by white sea ice and the increase in greenery as the atmosphere warmed.

The increases in the atmosphere's greenhouse gas content that contributed the 3.0 $W/m^2$ to its warming since the end of the last ice age are recorded in figure 11. Those increases only began about 1,000 to 4,000 years after the initial rise in the atmosphere's temperature had occurred (Mudelsee 2001). (The warming came first and then the increase in greenhouse gases.) After that delay, the increase in the atmosphere's greenhouse gas content, in particular $CO_2$, $N_2O$ and methane, contributed to an increase in the atmosphere's warming, which then promoted the release of more greenhouse gases. This formed a self-sustaining feedback loop that amplified the sunlight's contribution to ongoing warming.

The process that released these gases from their ocean and land reservoirs is illustrated by the carbon-containing gases. As mentioned, the carbon cycle's ocean reservoir stores a large amount of dissolved $CO_2$. Counterintuitively, $CO_2$ is less soluble in warm water than cold, thus as the ocean reservoir warms, it releases $CO_2$.

One of the carbon cycle's land reservoirs contributing greenhouse gases during warming was the permanently frozen ground (permafrost) found in the unglaciated polar and sub-polar climate zones and near the ice fronts of that time. Permafrost contains frozen organic matter and the greenhouse gases of $CO_2$ and methane. During the warming at the end of the last ice age much of that permafrost melted. The trapped greenhouse gases were released and the thawed organic matter became available for the changing soil community to decompose into $CO_2$ and methane. These greenhouse gases were emitted to the atmosphere.

But a warming atmosphere and increasing $CO_2$ also promote plant growth. This moves $CO_2$ from the atmospheric carbon reservoir into both the land and ocean plant reservoirs. Despite this process, the Antarctic ice-core record shows that during the warming to the present interglacial, atmospheric $CO_2$ and methane rapidly increased (figure 11). Simplistically, $CO_2$ was released faster than the plants could remove it, while methane was released faster than it could break down in the atmosphere.

The 0.25 W/m$^2$ of atmospheric warming that the Milankovitch sunlight cycle had contributed to climate warming since the end of the last ice age 25,000 years ago was amplified by feedbacks in the earth's processes. They decreased the earth's albedo and increased the atmosphere's greenhouse gas content so that the atmosphere's temperature rose to its interglacial temperature value. In the process, the earth's ice sheets shrank to their current size or all but disappeared.

In overview, the Milankovitch sunlight cycle is the external driver responsible for the atmosphere's long-term temperature cycle. But there are significant differences between the long-term cyclical temperature record that the variation in sunlight is expected to produce in the earth's atmosphere and its actual long-term temperature record. This difference is caused by the functioning of the earthbound portion of the earth's complex climate system, in particular its amplifying feedbacks. Its functioning is also responsible for the match between the temperature record and greenhouse gas record.

*The Abrupt Events.* By focusing on the long-period parts of the earth's temperature curve, we gain a more personal understanding of how the earth's climate system functions and changes as a whole. Still left to be explained are the short-term, irregularly occurring temperature spikes and dips that are imposed on the 400,000-year Antarctic long-period glacial-interglacial temperature record (figure 11). Understanding these abrupt events helps us appreciate the climate system's functioning at a scale relevant to the more immediate past and future.

The nature of the abrupt events is most clearly recorded in the younger portion of a 250,000-year record of the atmosphere's composition and temperature proxies obtained from ice cores extracted out of the Greenland ice sheet near its centre. In that record, abrupt events are seen to last from 500 to 2,000 years (Johnsen et al. 1992; Dansgaard et al. 1993). A good example is the Younger Dryas abrupt cooling event that began around 12,900 yBP, near the end of the last ice age (Adams et al. 1999). During its cooling phase, the atmospheric temperature at the location where the core was recovered fell an extraordinary 10°C in less than 150 years. After staying at near ice-age temperatures for roughly 1,400 years, the climate warmed by 12°C to 15°C in less than 200 years (Firestone et al. 2007; Taylor et al. 1997). It rose to the level expected by the long-term, sunlight-cycle-induced warming trend. The Younger

Dryas event did not alter that trend, it disrupted the trend by being pasted onto it.

These details of the Younger Dryas event, and the characteristic steepness of other short-term abrupt events seen in the Antarctic record (figure 11), all indicate that an amplifying feedback in the climate system likely played a key role in their formation. However, the initial driver or trigger for these events is unknown. This is partly because of our lack of knowledge about the chaotic aspects of the climate system and partly because some may be the result of random events (Adams et al. 1999).

Potential drivers for abrupt events include unknown minor contributions to the Milankovitch sunlight cycle, perhaps in combination with changes in the characteristics of the sun's emissions (Rind 2002). Possible drivers also include catastrophic natural events, such as the outbursts of glacial Lake Agassiz into the Arctic Ocean (rather than outbursts into the Atlantic Ocean) (Teller and Leverington 2004; Stanford et al. 2006). They could also be triggered by random events, such as extreme volcanic eruptions, extraterrestrial impacts (as suggested below for the Younger Dryas) (Firestone et al. 2007), large vegetation changes, such as an enhanced version of the widespread death of pine trees in the interior of western North America as a result of pine beetle activity (Kurz et al. 2008), or some combinations of these events.

In summary, both the amount of sunlight reaching the earth and its variation are described by the Milankovitch sunlight cycle. The sunlight heats the earth's surface and, indirectly, its atmosphere. The earth's climate system can be thought of as a sunlight-driven heat redistribution system centred on the atmosphere.

The climate system is a dynamic, complex system consisting of aspects of the earth's oceans, lands, atmosphere and life. The interaction among those aspects are the climate system's functioning and changing. The system displays the features of a dynamic, complex system, such as a wide diversity of interaction types, including feedback, that occur over a wide range of scales in time, space and strength. In overview, the system displays deterministic, chaotic and random features. A specific change can be described by generic terms such as *inertia*, *delay* and/or *preconditioning*.

At its most simple, climate change occurs when the variation in sunlight described by the Milankovitch sunlight cycle prompts the earthbound part of the earth's climate system to respond. Its functioning does so by altering the earth's albedo and the atmosphere's composition in ways that amplify the temperature change expected from just a change in the amount of sunlight. The earth's climates and their distribution change accordingly. The inertia of the oceans and the presence of ice sheets stabilize the system.

The climate system's functioning is brought to life, in overview, by the 400,000-year climate record deduced from Antarctic ice cores. The climate system's response to the variation in sunlight is seen in the long-period glacial-interglacial temperature cycle and the matching variations in the atmosphere's $CO_2$, methane and $N_2O$ content. The aspects of the climate system's functioning responsible for the details of that 400,000-year record are illustrated by the features of climate change for the time between the previous and the present interglacial (figure 11).

There was a delay between the Milankovitch-sunlight-cycle-induced increase in sunlight reaching the earth and the warming of the atmosphere at the end of the last ice age. There was also a delay between that warming and both the ice sheets starting to melt and the release of the greenhouse gases that changed the atmosphere's composition. These delays reflect the inertia in the climate system.

There is an asymmetry in the atmosphere's cyclical cooling and warming phases that is not seen in the sunlight cycle. It is the result of differences in the climate system's response to a decline versus an increase in sunlight. Finally, the fall in temperature to the last ice age and the corresponding rise to the present interglacial is larger and more rapid than would be expected from the change in the amount of sunlight alone. These features occur because the functioning of the climate system amplifies the effects from a change in sunlight (Hansen et al. 2007).

Also present in the 400,000-year record are short-period, abrupt temperature events that interrupt the long-period temperature cycle. They are possibly triggered by chaotic or random events within the climate system, such as a volcanic eruption, or by event's outside of it, such as a strike by an extraterrestrial body. Regardless of the trigger, the features of the abrupt events are a product of the climate system's functioning, especially its amplification of change. After all, that functioning determines the system's response to any driver of change.

The climate system can be thought of as existing in a quasi-stable state, which it adjusts in response to the influences on it. Beyond some degree of adjustment, climate change takes place by a swing between quasi-stable states, such the switch from a glacial to an interglacial state or an abrupt event. In this view, the system's amplification of effects ensures that weak drivers can result in large climate changes (Broecker 1987; Taylor et al. 1993). No wonder that the details of climate change are rarely smooth and predictable.

## The Climate Change–Ecosystem Change Connection

The above two sections provide us with a greater appreciation for how the earth's ecosystems and the climate system function and change. Along with that appreciation comes a realization: to fully appreciate these climate and ecosystem changes, they should be thought of as inseparable parts of a single, dynamic, complex environmental system (Levin 2005). After considering the difficulty we experience trying to imagine the functioning of an ecosystem or the climate system on their own, there should be no surprise that we get stuck trying to mentally imagine how the earth's dynamic, complex environmental system functions and changes (functions).

Our difficulties might be relieved through techniques such as mathematical modelling. However, for a number of reasons, an environmental model that is both comprehensive and produces an easily understandable output is unlikely to be built (Costanza et al. 1993). Instead, we could take the view that the innate self-regulating ability of the earth's ecosystems and climate is reason enough for us to attribute to it an aspect of incomprehensible self-regulatory consciousness. We could then simply refer to the earth as Gaia. In this view, an appreciation of the environmental system through respect would be considered more useful than an appreciation through mathematical description.

Each of these views is valid and useful, but each is limited in its ability to provide us with both a reasoned, empathetic attachment to our environmental system and a personalized, disinterested understanding of its functioning. We need both if we are to more clearly recognize the central role that the environmental system plays in our lives and if we are to respond appropriately to how it is changing. Somehow, we must increase our personalized appreciation for our environment system as an integrated whole. Somehow, we need to increase our degree of environmental knowing and engage in more nuanced environmental thinking when making decisions about the system.

We are helped in this task by recognizing that the ongoing change in sunlight reaching the earth is a key very long-term influence on both the earth's climates (e.g., the atmosphere's temperature) and ecosystems (e.g., the life of plants at the base of the trophic web). It is thus a key influence on the earth's environmental system. By considering the environmental system's dynamics in this light (pun intended), it is easier for us to appreciate its functioning and changing.

In overview, the distribution of the earth's climates represents the global distribution of temperature and available fresh water, and thus the distribution of the earth's land biomes. As the amount of sunlight reaching the earth changes, as described by the Milankovitch sunlight cycle, so does the nature and distribution of the earth's climates. A change in the climate's variables of temperature and available water at a particular place, forces the individuals of the species living there to respond (Parmesan and Yohe 2003; Whitlock and Bartlein 1997). If these variables change sufficiently, then individuals and species will respond through a mix of migrating to follow their climate-influenced ecosystem niche and adjusting their lifeway characteristics to better match their new living conditions. But in doing so, they change their host community's (ecosystem or biome) species mix, functioning, location and size.

These changes to their host community feed back to influence the already changing climate system. Consider a tropical forest. On a local scale, the lifeways of the species affect the temperature and moisture content of the forest's atmosphere from the forest's floor to above its canopy. On a regional scale, the prolific growth of plants in a tropical forest removes $CO_2$ from the atmosphere while emitting water vapour and cloud-forming chemicals, such as methyl chloride, into it. The forest's growth also results in its low albedo. Thus, as the forest changes, it can influence the climate system to change.

On a global scale, these types of ecosystem changes collectively change the climate system's basic variables: the atmosphere's composition and the earth's albedo. This completes a sunlight-driven climate-ecosystem feedback loop. It amplifies the environmental change prompted by the Milankovitch changes in sunlight, as discussed above.

This section personalizes the general description of these connections by discussing the environmental changes that occurred during the deglaciation of northern North America. It will start at a global scale, move to a biome (regional) scale, then on to the ecosystem and individual (local) scale. It will finish by discussing the expected complicating factors.

**Global Scale.** At the height of the last ice age, around 25,000 yBP, the southern third of Alaska and all of Canada, except for its northwestern Arctic fringe, were covered by two joined North American ice sheets. The Cordilleran ice sheet covered the mountains of western Canada and parts of Alaska. The larger, 1 to 3 km thick Laurentide ice sheet covered the flatter central and eastern parts of Canada. (The Greenland ice sheet lay further to the east.) To the south of these ice sheets' southern margins, vegetation grew almost to the ice sheets' edge. In this area, bitterly cold, dry winds blew off the bright, white, compacted snow surface.

A Milankovitch cycle increase in the sunlight reaching earth ended the last ice age. If an observer in space had been watching northern North America in fast forward at this time, they would have seen the result of the ongoing warming as a dramatic sequence of changes. By around 20,000 yBP, the southern margin of the ice sheet started to slowly melt and retreat poleward. Despite the ongoing warming, the retreat of the ice front was slow, until 15,000 yBP.

Around 14,600 yBP, the atmosphere received a short-lived temperature boost. By around 14,000 yBP, the great northern ice sheets finally started to melt in earnest (Stanford et al. 2006; Overpeck et al. 2006). This would have been most easily seen in the west and south. In the west, the Cordilleran ice sheet became separated from the shrinking Laurentide ice sheet by the opening of a north-south trending, dark ice-free corridor on the prairie (eastern) side of the Rocky Mountains and parallel to them.

At the same time, the southern ice front of both the Laurentide and the Cordilleran ice sheet were rapidly retreating northward. They left behind ice-scoured land that was exposed to sunlight for the first time in tens of thousands of years. Its dark colour changed the earth's albedo, which fed back to help amplify the warming. Some of the plants whose ice-age homes lay to the immediate south of the ice front, took advantage of the warmer climate and the presence of the vegetation-free land to migrate northward. Some continued up the ice-free corridor between the ice sheets while, at the same time, others moved southward into the corridor from the ice-free parts of Alaska. Once these pioneers had established the base of a trophic web, they were soon followed by the other plants, insects and animals from both the pioneers' and the neighbouring biomes.

The ongoing warming and melting meant that the cold climate biomes to the immediate south of the ice front continued spreading northward across the landscape, as if trying to catch the northward retreating ice front but never quite managing to do so. Behind them, the biomes from further to the south were also moving northward. When seen together, they formed a northward-moving parade-like succession of biomes.

For example, at the height of the ice age, the Cordilleran ice sheet's southern edge was only a short distance north of the eastern Cascade Range, Washington state. At the height of the ice age, the Cascade area was covered by a light-coloured, cold, dry steppe (grassland) biome. To the south of it grew a dark, pine-dominated biome. If the space observer looked at the eastern Cascades while warming occurred, they would have seen the grassland biome migrate northwards to be rapidly replaced by the pine biome moving northward from further south. The eastern Cascades changed from a light-coloured grasslands into a darker-coloured pine forest (Whitlock and Bartlein 1997). As these biomes moved, they also changed.

As the melting ice sheets shrank, silty water poured off them to form large, milky rivers. Some of this water filled lakes. In the case of the Laurentide ice sheet, enormous lakes formed along its southwestern front. Lake Agassiz, for example, was 1,000 km long. Occasionally its ice dams failed, sending torrents of water flooding across the land into the Arctic or Atlantic Oceans, or even into the Gulf of Mexico.

The ice sheet melting and ecosystem migrating began to slow in 12,900 yBP when the sudden start of the Younger Dryas abrupt cooling event plunged the area back into ice-age conditions. This cold period lasted for roughly 1,400 years. Around 11,500 yBP, the Younger Dryas ended with a sudden warming, after which the ice sheet melting and ecosystem migrating resumed at a rate in keeping with the Milankovitch-driven change in sunlight. This warming and migrating continued until around 10,000 yBP, by which time the increase in sunlight was near its peak. The atmosphere's temperature plateaued and the climate entered the current interglacial period. The atmosphere's temperature remained remarkably constant from then until, around 5,000 yBP (3000 BC), it started to cool toward the Little Ice Age.

By around 6,000 yBP, no more cold winds blew off the Laurentide and Cordilleran ice sheets because they had almost completely melted (Dyke and Prest 1987). The land that the ice had buried was now covered by vegetation: tundra biomes, in the north, were successively replaced, toward the south, by boreal forests and the grasslands of the prairies. The glacially dammed lakes were all but empty. Sea level had risen about 130 m since the depths of the last ice age, enough to flood much of northern North America's coastline and to separate Alaska from Siberia (Lambeck et al. 2014).

No matter where on the globe the observer in space looked during this warming period, they would have seen similar warming-related changes taking place. For example, cloud patterns shifted and changed to reflect changes in the atmosphere's temperature and moisture. In mountainous areas, glaciers and snow lines moved closer to the summits. Biomes on the equator, such as the small ice-age tropical forest biome in South America and Africa, expanded in all directions (Adams et al. 1997; Ray and Adams 2001).

**Regional Scale.** When these end-of-the-ice-age environmental changes are viewed at a regional scale and at a slower pace, a more complex picture emerges. The environmental changes that took place around the edges of the shrinking Laurentide ice sheet are illustrated by examples from two significantly different climates and biomes. The first is central Alaska (north of the ice sheet); the second is around Appleman Lake, Indiana (south of the ice sheet).

At the height of the ice age, most of the northern two-thirds of Alaska was ice-sheet free. To its east lay the Laurentide ice sheet. To its west, sea level was 130 m lower, which meant that Alaska was linked to eastern Siberia by an ice-sheet-free land bridge. To the south, the third of Alaska adjacent to the coast was covered by the Cordilleran ice sheet. The ice-free area of Alaska, called Beringia, was extremely cold, water limited (dry) and underlain by permanently frozen ground (permafrost). Yet, at least in its central portion (central Alaska of today), it supported a dry tundra biome.

The climate-warming-related changes to the central Alaskan dry tundra biome are most easily appreciated by dividing the growth forms of the plant species that the tundra-type biome can support into four groups: graminoids, forbs, dwarf shrubs, and shrubs/trees. Graminoids include grass (true grasses) and grass-like plants, such as sedges and rushes. Forbs are the (non-woody, non-graminoid) flowering plants we associate with a colourful Arctic, such as asters and saxifrages. Dwarf shrubs are branched, woody plants that grow low to the ground, such as blueberries and dwarf birch. Shrubs/trees are woody plants that grow larger than dwarf species, such as willow, birch and poplar. Some species, like willow and birch, can switch between the classification of dwarf shrubs and shrubs/trees depending on their growing conditions.

At the height of the ice age, the central Alaskan tundra was so cold and dry that it could support only a small subset of the plant species that grew in the Arctic during warmer times. At that time the area's plant diversity was the lowest it had been in 50,000 years. For example, some species of shrubs, trees and dwarf shrubs were rare or absent (Willerslev et al. 2014). However, there were likely regional-scale variations in plant distributions. For example, the part of the central Alaskan tundra nearer to the lower and warmer southern half of Beringia likely supported some shrubs and perhaps trees.

There is still disagreement about whether the central Alaskan tundra at the height of the ice age was dominated by forbs or graminoids. The current view is that it was likely forb dominated, although short-stemmed graminoids were common. The proportion is important because it has implications for the variety of larger animals (megafauna) that the tundra could have supported. Regardless of the answer, the ice-age central Alaskan tundra biome certainly provided adequate food for horses and mammoths but only marginal forage for the few bison that lived in the area. Wapitis and moose were absent because suitable food was absent.

The story of the changes to central Alaska's ice-age cold, dry tundra biome starts around 15,700 yBP. By then, a long-term observer over Alaska would have noted that the climate was warming and the front of the Laurentide ice sheet to the east was retreating. After this date, they would have seen more dramatic changes.

The surface layer of the permafrost, on which the ice-age tundra grew, now melted in summer. As warming continued, more of the area's organic-rich permafrost melted to a greater depth. The thawed permafrost provided more soil for any new arrivals to grow in. But it also released its trapped $CO_2$ and methane into the atmosphere. At the same time, the soil community ate the permafrost's thawed plant material, which generated more $CO_2$ and methane. The thawing permafrost contributed to the ongoing climate warming.

This unfrozen soil, and the increase in the area's temperature and amount of available water, allowed some plant species, such as willows, from southern Beringia nearer to the coast, to increase their range northward into central Alaska. Wapitis and moose are thought to have migrated into the area by following the willows, their ideal forage. Around the same time, humans arrived,

perhaps along with the wapitis and moose. The new plant arrivals mixed with the adapting plants already there to form the foundation of a transitional tundra biome. It could support bison, wapitis and moose, but horses and mammoths struggled to survive (Hoffecker et al. 2014; Guthrie 2006).

After around 13,500 yBP, the permafrost had melted to a sufficient depth that the cold, dry central Alaskan tundra had completed its transition into a warmer, moist tundra. The landscape was now seasonally frozen with lakes and peat bogs surrounded by irregularly frozen, nutrient-poor, acidic soils. The thawed soils supported a greater diversity of plants than grew there at the height of the ice age. The tundra biome was likely still dominated by forbs, but now dwarf shrubs were as, or more, important than graminoids. Shrubs and trees were present. Moose and wapitis were present, but horses and mammoths were now extinct in Alaska (Willerslev et al. 2014; Guthrie 2006). These changes in ecosystem and climate continued until 10,000 yBP, when the warming reached the present interglacial period's temperature maximum.

The 3,000-year history covering the central Alaskan tundra's transition between types illustrates how, from a specific area's perspective, its biome and landscape responded to a warming climate by gradually transitioning through different biome types. The living adapted both as individuals and as species by migrating. Others went at least locally extinct.

Why did mammoths and horses become extinct? Climate change was involved, but was it responsible? Another of the many suggestions is that the cold tundra ecosystems in which they lived was previously self-supporting. In particular, the productivity of the tundra's nutrient-poor soils was increased and maintained by the dung of the megafauna who fed on the plants. Warming may have disrupted this nutrient feedback and extinction resulted. We will return to the question of megafauna extinction after the next example of environmental change.

The Appleman Lake example provides a perspective of climate-ecosystem change and its complexity in a much different climate-biome. At the height of the last ice age, the Laurentide ice sheet's southern boundary stretched across North America, roughly following the present day Canada-US border. In the east, the nearest ice-free land south of the ice sheet, which still covered Lakes Michigan and Eire, was in the area around Appleman Lake, Indiana.

During the depths of the ice age, the biome around Appleman Lake was dominated by more northern, cold-adapted (boreal) plants, such as spruce and larch trees. This boreal forest biome supported megafauna species, such as mammoths. The climate warming that induced the central Alaskan cold, dry tundra biome to change also induced equivalent, but more complex, changes in the boreal forest biome around Appleman Lake. Those changes are recorded by the types of fossil pollen and microbe species trapped within the region's lake sediments.

An astute, long-term observer in the Appleman Lake area would have noticed that the warming climate was accompanied by changes in the mix of boreal forest plants and declines in the mammoth numbers. By 14,000 yBP, mammoths were locally extinct and the Appleman Lake area was covered by a biome that is totally unfamiliar to us. It consisted of cold-climate-adapted spruce trees growing alongside more temperate-climate-adapted broadleaf deciduous trees, such as ironwood and ash trees. The factors that may have helped to form this unusual transition biome include forest fires as an ongoing, significant event; an unusual climate; abrupt climate change (see the Younger Dryas below) and the decline, then absence, of megafauna.

As climate warming continued, the unusual biome continued to change. By about 12,000 yBP, the cold-adapted spruce were absent from the Appleman Lake area. The area's biome now consisted of pine (a species found in sub-boreal forest biomes), ash and ironwood trees.

By about 10,000 yBP, at the beginning of the interglacial period, the pine trees were absent. The Appleman Lake biome had completed its transition into a true temperate broadleaf deciduous forest that hosted oak, ash and ironwood trees. Forest fires played a significant role in this biome's functioning too. This temperate forest biome is the immediate ancestor of the Appleman Lake area's current temperate deciduous forest biome.

The species that successively disappeared from the Appleman Lake area as the climate warmed did not necessarily go extinct. They may have just been displaced. For example, a number of the boreal species, such as spruce, energetically migrated. They each followed their climate niche as it, and the retreating ice front, moved northwards. Today, these boreal species are a part of North America's current boreal forest, which is now first encountered some 1,000 km to the north of Appleman Lake, in Canada (Gill et al. 2009). Further to the north, in Canada, the current boreal forest grades into the warm, moist tundra biome that stretches westward to central Alaska on ground previously covered by the Laurentide ice sheets.

**Local Scale.** It is individuals who responded to the climate warming that melted North America's ice sheets. This is the scale where individuals make the "decisions" that influenced their own, their species and their host ecosystem's future. The context for this discussion of change at a local scale is the time between 10,000 yBP and 6,000 yBP, when the Laurentide ice sheet was rapidly disappearing.

In order for individuals and species to occupy the deglaciating land, they had to migrate. This is a technique well used by individuals and species when responding and adapting at a local scale to a change in their surroundings, such as a changing climate. The process of migration is illustrated by the story of the shrubline in northern Alaska: the present climate-warming-induced northward movement of the boundary between the shrub tundra and the shrub-free tundra (Sturm et al. 2001).

If you were standing near that shrubline today, you would not see a line or even a distinct zone. Instead, you would see an unremarkable collection of irregular shrub patches. Each patch is a mix of species: a dynamic community in its own right (figure 8). As the local climate warms, the temperature and soil moisture content around a shrub patch increases, thus allowing a patch to add new members and new species, to expand in size and to merge with other patches.

At the same time, in some locations to the immediate north of the shrubline, the climate will have warmed sufficiently and the ground thawed enough to support some shrub species. In a few of these locations, all of the other requirements (such as nutrient availability and adequate space) for those shrub species to thrive will be present. If those places are within the seed dispersal distance of reproducing shrubs, then individuals of those shrub species could establish themselves there. If they do, then those shrub species would have migrated northwards.

If an isolated shrub growing to the north of its species' historic range not only survives but thrives, then a new shrub-patch community can form. Whether that will happen depends on more than the continuation of global climate warming. It also depends on the microenvironment immediately around the initial shrub changing into one that can support the other plant and animal species that make up a shrub-patch community.

These changes to the microenvironment can occur through a number of mechanisms. For instance, shrubs are dark and trap snow, so the founder shrub creates a warmer and wetter microclimate around itself. A shrub also alters the local nitrogen and carbon cycles, thus changing the availability of nutrients immediately around it.

When the microenvironment around the isolated founder shrub has changed sufficiently, then it can be joined by more of a shrub patch's species. Their arrival further changes the existing patch's microenvironment in favour of other shrub-patch species. By this process, a new shrub-patch community is established and grows (Myers-Smith 2007).

If warming continues and outlier shrub-patch communities continue to be established further to the north, then the shrubline will continue to move northward. At the same time, the shrub patches to the south of the shrubline will continue to expand and merge to eventually form a more continuous shrub biome. As the shrub biome enlarges and migrates, its darker colour and microclimate adjustments contribute to local climate warming. This both helps the shrubline to migrate and contributes to global climate warming.

Our appreciation for migration as a response to climate change is further enhanced by considering the limits faced by migrating individuals and species. There are physical barriers limiting migration, such as mountain ranges. There are also lifeway barriers. Consider the speed of migration. During deglaciation, the lodgepole pine in western North America had to migrate more than 2,000 km to the north at rates of 150–350 m/year (Pielou 1991). Similarly, on the eastern side of North America, the boreal spruce of Appleman Lake migrated northwards at the astonishing maximum rate of around 1,000 m/year, with an average for the leading edge of around 500 m/year (Pitelka and PMWG 1997).

Not all species can migrate that fast. A species' speed of migration is limited by, among other things, how its young disperse and the presence of a community with a suitable niche into which the migrant can move. If a less-than-perfect niche is all that is available to a species, it could, for example, compensate by adjusting its lifeway accordingly. But there are limits to the speed and degree to which a species can adjust its lifeway, as discussed in chapter 7.

The mention of a niche and community acts as a reminder that all individuals are part of and rely upon a host community for their survival. They rely on others. This reliance also constrains the efforts of individuals and species to migrate. In this sense, the migration (and adjusting) of individual and species is a community affair.

It is within this context of a community that the concept of an ecosystem migrating (and adjusting) becomes meaningful. An ecosystem's migration is made possible by its members' ability to migrate and adjust their lifeways, but an ecosystem's migration is also constrained by its members' differing abilities to do so. For example, during the period of rapid warming during deglaciation, some of a community's species would have been unable to migrate fast enough. Others would have been unable to adjust their lifeways to fit a niche that was changing as their host community migrated. The populations of the species constrained in this manner would have declined and faced at least local extinction (Guthrie 2003; Pielou 1991).

Consider too that, at some point, a host ecosystem and its members might have to deal with strangers. In the case of climate-induced migration, these strangers may be co-travellers from adjacent ecosystem. Or they may appear because the ice-sheet barrier separating two migrating ecosystems/species had finally melted.

At the local scale, the outcome of a meeting between two species that were strangers is determined by the interactions between them. Think exotics. That outcome is influenced by their respective characteristics, the ecosystem context where they met, and the differences and similarities between their preferred niches. It is constrained by how and how fast species and ecosystems can change, which are set by allometric constraints and trophic relationships. Think of balance.

One example of the outcome is illustrated by the meeting between two plants, skunkweed (*Polemonium viscosum*) migrating northwards and cotton grass moving southwards, as the Laurentide ice sheet that separated them melted. Their niches were sufficiently different that, when they met, they could coexist in areas where both of their niche requirements were satisfied. In contrast, although the bison from Alaska had been separated by the ice sheets from the bison in the US, south of Canada, for some 100,000 years, when they met they were sufficiently similar that they could successfully mate (Pielou 1991).

The resolution of meetings between other species was more complex and/or conflicted. The outcome of these interactions could lead to extinctions, the establishing of exclusive range boundaries, or a change in their lifeway traits. One example, described in the introduction chapter, is the relationship between my local forest's sapsucker species and its North American relatives. Another is the meeting between thirteen-lined ground squirrels moving northwards and heather voles moving southwards. Whatever the outcome of the meetings, their mutual ecosystem host was changed.

In overview, an ecosystem can be thought of as moving and changing because its member species can migrate and adapt. However, because of the differences in the abilities of its members to migrate and adapt, ecosystem movement and changing is also a process of being torn apart. The community's slower migrators and adapters follow a different path into the future than the faster migrators and adapters. Under some conditions, the diverging ecosystem pieces can become new ecosystems. Think of the unusual Appleman Lake biome and the transitional central Alaskan biome. The process of forming these new or adjusted ecosystems can require the potential members coming to terms with individuals of other species, which they do by interacting with them.

Our human characteristics make it hard for us to appreciate the relationship between climate warming and ecosystem change at a human-relevant, local scale. As discussed, it is hard for us to tell if the climate is changing. Similarly, it is hard for us to tell by casual inspection if the changes we see in an ecosystem are directed or random. Consider focusing on the increasing presence of hemlock trees within the forest near my house. Is this change a part of our forest's ongoing, dynamic adjustments to random changes in local conditions: its normal variability? Or is it a directed change, such as migration, being driven by a longer-term directional driver, such as a changing climate? We might recognize that a species is undergoing a climate-directed migration if it is moving as fast as the spruce and pine did during deglaciation. If it is moving slower, though, probably not.

If we are unable to recognize species migration, we are unlikely to recognize ecosystem migration. We can also fail to recognize ecosystem migration because our casual inspection pays little attention to the majority of an ecosystem's members and their migration peculiarities. In particular, we tend to ignore the smaller, but faster responding, unobtrusive species. Our appreciation for an ecosystem's local-scale response to climate change would improve if we paid closer and extended attention to an ecosystem as a whole and to more of its member species, and kept ongoing long-term records.

**Complications.** The above discussion may have created the impression that the driver-response relationship between ecosystems and the climate system occurs in a smooth steady manner. Because the environmental system is complex, this is usually not the case. Invariably, there are some factors operating at a local to global scale that can complicate the expected simple relationship.

An example is the Younger Dryas abrupt cold event that interrupted the long-term climate-warming trend driven by the Milankovitch sunlight cycle. Around 12,900 yBP, the Younger Dryas rapidly returned the climate in northern North America to ice-age temperatures for 1,400 years. After that time, the temperature rose just as rapidly to re-establish the warming trend driven by the Milankovitch sunlight cycle (Firestone et al. 2007; Taylor et al. 1997). The Younger Dryas thus interrupted North America's sunlight-cycle-driven deglaciation and disrupted the efforts of individuals and ecosystems to adapt to it.

It is unclear if the Younger Dryas was a global-scale event. It was at least a significant event in the western hemisphere. For example, in northwestern Europe, the Younger Dryas cooling occurred so fast and temperatures dropped so low that the Milankovitch-driven migration of species toward the north dramatically reversed. Pine forests abruptly died and were rapidly replaced by southward-moving tundra plants such as dwarf birch and the dwarf shrub *Dryas octopetula*. The surprising presence of this dwarf shrub among the pine gave the Younger Dryas event its name (Weart 2003).

The Younger Dryas also affected ecosystems in the elevated hinterlands of northern South America. The Amazon River carries eroded sediment and plant debris from its watershed into the Cariaco Basin of the Atlantic Ocean. Ocean sediment cores recovered from this basin indicate that, over the Younger Dryas period, the hinterlands' climate and vegetation rapidly switched from a humid tropical forest biome to a grassland biome and back (Hughen et al. 2004). As figure 8 suggests, the Younger Dryas may also have affected the climate in Antarctica.

The Younger Dryas illustrates that significant climate change can occur very quickly. It can disrupt an ecosystem's efforts to adapt to one climate change driver by forcing it to also respond to another. The question arises whether the shock of an abrupt event can be serious enough to change ecosystems through species extinctions.

The Younger Dryas provides a possible answer. South of the Laurentide ice sheet, details of the ecosystem changes that occurred over the period of the Younger Dryas are preserved in North American cave sediments. The record shows that, before the start of cooling, ice-age megafauna, such as mammoths, were present in the south. In contrast, the sediments younger than 12,900 yBP (the start of cooling) do not contain megafaunal bones (Firestone et al. 2007; Kennett et al. 2009). Based on this and other evidence, it is generally accepted that the megafauna were essentially extinct in North America by around 11,900 yBP, which is close to the end of the Younger Dryas. The Younger Dryas's rapid climate cooling and warming may indeed have played a role in North America's megafaunal extinctions (Gill et al. 2009; Barnosky et al. 2004b).

However, to complicate matters, the extinction of mammoths and horses in mainland Alaska occurred just before the Younger Dryas event. This uneven occurrence of mammoth extinctions suggests two possible ways in which the Milankovitch-driven climate change may have also played a role in their extinctions. Locally, it may have reduced their food supply, as discussed for Alaska. More broadly, the stress of the Milankovitch-driven climate warming may have preconditioned the megafauna to experience extinction if a significant event, such as the Younger Dryas, occurred (Gill et al. 2009).

But maybe it wasn't just the rapid climate change associated with the Younger Dryas that finished the job. The cave sediments also record that the start of the Younger Dryas is marked by an extremely thin, black, carbon-rich layer that contains unusual chemicals and minerals, including microdiamond. The data collectively suggests that the start of the Younger Dryas cooling may have been triggered by a comet airburst or landfall and subsequent continental-scale vegetation fires. Perhaps the blast and fires were the sole or the dominant cause of the megafauna extinctions. In this explanation, the rest of the 1,400-year Younger Dryas abrupt cooling event was the amplified response of the earth's climate system to the random comet strike. It simply finished the comet's extinction job.

Nice story, but there are other events to consider. Around the time of the Younger Dryas, there were at least two other possible contributors to megafaunal extinctions. The first were the humans who migrated from Siberia into Beringia. Sometime after 15,000 yBP, they migrated into the warming ice-free areas of Alaska. There, these new arrivals coexisted with and hunted the native mammoths for at least 1,000 years before the mammoth went extinct (Guthrie 2006). Our Alaskan ancestors also saw the last of the North American horses, which they apparently did not hunt.

Before the start of the Younger Dryas, humans were living in the United States' lower 48 states. There, as the Clovis people, they hunted mammoths. Thus, humans may have played a role in the extinction of the environmentally stressed mammoths (but maybe not the horses) both in Alaska and in the rest of the United States.

There is other evidence to support the idea of human involvement in the mammoths' extinction. North America is not the only continent where post-10,000 yBP megafauna-poor ecosystems are found. There is evidence that humans may have contributed to these and other even earlier megafauna extinctions, for example, in Australia (Barnosky et al. 2004b). Perhaps we can think of the human contributions to megafauna extinctions in Australia and the Americas as the outcome from our ancestors being an exotic species (invasive perhaps) in these lands.

Another event to consider is the large amount of meltwater trapped in the ice-dammed lakes that formed along the southwestern edge of the Laurentide ice sheet during deglaciation. One of the largest of these lakes was Lake Agassiz. At the time of the Younger Dryas, it stretched for 1,000 km along the ice front, where Lake Winnipeg is today. Occasionally, Lake Agassiz and the other ice-dammed lakes would burst, sending catastrophic floods across the landscape into the Arctic Ocean, the Atlantic Ocean or the Gulf of Mexico.

Lake Agassiz generated at least 18 floods. The second largest occurred around 12,900 yBP, at the start of the Younger Dryas. During this flood, some 9,500 km$^3$ of water were released. To put this into perspective, the water took a century to drain out and dropped the level of Lake Agassiz by 110 m (Teller and Leverington 2004). The sudden transfer of such a large amount of fresh water into an ocean likely played some role in climate change, ecosystem change and ice-sheet dynamics. Collectively, the bursting of ice-dammed lakes may have indirectly contributed to megafauna extinctions.

These types of environmental events can, either on their own or more likely in combination, complicate a particular process of ecosystem or climate change at anywhere from a local to a global scale. They can accomplish this by influencing the process, by taking over as the primary driver of change (e.g., the Younger Dryas event), or by a mix of both. This leads to the following conclusion. It was noted earlier that the Milankovitch sunlight cycle drives long-term, global-scale climate change and prompts changes in individuals, species and ecosystems. However, the sunlight cycle is not directly responsible for the observed global-, regional- and local-scale climate and ecosystem (environmental) changes. They are the product of the interconnected climate and ecosystem systems: the functioning of the earth's environmental system and the influences on it.

Relevance. There is one last question to answer: Is the functioning and changing of the environmental system described at the beginning of this section and illustrated by the deglaciation of North America more broadly applicable? In particular, did these types of changes continue after the end of the Younger Dryas and the end of deglaciation? Are they present and relevant to us today?

Consider the following examples. In the Middle East, the Younger Dryas cooling and the associated ecosystem changes likely helped reduce the availability of traditional wild-food sources. In the case of at least Abu Hureyra, that change seems to have provided part of the incentive for humans to start cultivating cereals around 13,000 yBP (Hillman et al. 2001).

In North America, it seems that the Younger Dryas may have combined with the long decline and final loss of megafauna prey animals, in particular mammoths, to prompt the Clovis people to undergo a cultural adjustment. The change was needed for them to survive in the emerging climate and associated ecosystems of an ice-sheet-free North America (Firestone et al. 2007).

In Western Canada, the cedar tree migrated northwards up the west coast of British Columbia over the period 7,500 yBP to 3,000 yBP. Cedar became one of the pillars of the northwest Pacific coast's indigenous cultures (Turunen and Turunen 2003). In more recent times, the Little Ice Age period of climate cooling, between roughly 1300 and 1800, lay behind at least the following events of that time: in the coastal mountains of British Columbia, the glaciers and vegetation marched down the mountains, the Norse Greenlanders went extinct and the people living in Europe adjusted their living and farming to colder temperatures. The functioning and changing of the environmental system is relevant today: a subject that will be discussed in more detail in later chapters.

In summary, if we wish to more fully appreciate both climate and ecosystem change, then we need to think of them as intimately interconnected parts of the earth's dynamic, complex environmental system. One reminder of this interconnection is the close relationship between the distribution of the earth's climates and its land biomes. Another is that both the earth's climate system and its ecosystems depend on sunlight to function.

The environment system's components are the earth's land, oceans, atmosphere and life. The features of these components interact with one another over a wide range of scales in time (from the ultra-fast to millions of years), space (from local to global) and strengths (from weak to strong). Like all complex systems, these interactions, which include feedback, are the functioning and changing (functioning) of the environmental system. In overview, that functioning displays deterministic, random and chaotic features while particular aspects of the functioning can be described using generic terms, such as *inertia*, *delays* and *preconditioning*.

The Milankovitch sunlight cycle describes the amount of sunlight reaching the earth and how it varies over time. This variation, together with the functioning of the earth's environmental system ensures that there are always a number of influences prompting the environment to change. A particular change in the system can be thought of as the product of a flexible hierarchy of interacting (internal and external) influences on the system. The system's complexity means that the change we see in the system and what we think of as driving it depend on what we choose to look at and the scale of time and space we use when doing so.

Hopefully, the details provided in the above discussion have personalized the environmental system's functioning. A greater appreciation should temper our tendency to fixate on simple, direct cause-and-effect explanations for climate and ecosystem changes. It should help us to, instead, contemplate environmental change in a more nuanced manner. We are then more likely to treat it as the product of the dynamic balancing of many variables as they interact with one another at many scales, under the influence of a flexible hierarchy of influences.

We are also more likely to view the consequences of an environmental change as the long-term outcome from many (covert and overt) trade-offs, not just the immediate, short-term outcome from the change itself. We can then more easily accept that the success of a species at adapting to environmental change depends less on it winning a particular battle and more on it simultaneously addressing the many influences driving it to change, while still satisfying the basic environmental constraints (biological, ecosystem and climate, among others) on its lifeway.

There have been enough hints that humans are personally involved in the environmental system. The question is, where do we fit in?

## How We Fit In

We live on the earth. Somehow we must fit into its dynamic, complex environmental system. Where we belong, and the associated environmental constraints on how we can complete our lifeway, are the subject of this section.

**Classification.** One way to determine our place is to find our box in the Linnaean classification system. To achieve this we, in effect, compare ourselves to other species. In the process of making this comparison, we also determine in which ways we are similar and different from them, and in which ways we are truly unique. During our search for the right box, it is instructive to remember Linnaeus's creationist beliefs and his views about nature's order.

We are most similar to the mammals. We, like the other mammals, give birth to live young and have four limbs. Similarly, our skeleton's structure, the organization of our primary organs within it, and our sensing functions, such as eyesight and hearing, can easily be related to those of the other mammals. Our body functions are also similar to those of mammals. They are so similar to those of rats, pigs and monkeys, for example, that these animals are used as medical models for human diseases. Even human behaviours, such as aggression and social hierarchies, are similar to those exhibited by some mammals (Cole and Cole 2001).

Within the mammals we fall into the primates (monkeys, apes and related species). Our similarities to primates are sufficiently close that we can personally relate to the relationship between their body functions and behaviour. For example, the males of our primate relatives engage in dominance contests. After the contest, the testosterone level in the winner goes up and in the loser goes down. The same testosterone changes occur in human males when they participate in one-on-one competitions in which there is a winner and loser, such as a chess game or a wrestling match (Ellison 2001).

Within the primates, we are classified as an ape. They include chimpanzees and gorillas. We are more similar to these animals than to the non-ape primates. For example, although we can understand the emotions expressed by many primate species, it is much easier for us to relate to the emotions expressed by the apes. We find it remarkably easy to conclude that apes experience the same range of basic emotions as humans do, such as psychological pain, anxiety and happiness, although with nuances befitting their specific ape species (Tattersall 1998).

Of the apes, we are, genetically, most closely related to the chimpanzees. Our genes have a greater than 98% match with theirs (De Waal 1995; Jones 1999). This prompts us to conclude that humans and chimpanzees likely have much in common. You can confirm this by watching movies of chimpanzees going about their day-to-day business either in the wild, for example, in Senegal making and using spear-like tools, or in a lab, for example, Kanzi at work on a human-made tool, a computer keyboard (Roach 2008). Our human classification box is next to the chimpanzee box.

**Allometry.** If we belong among the many unitary species on the earth, then our basic characteristics should be subject to the same allometric constraints as them. For each of the relevant allometric curves, our characteristics should fall within the accepted variation. This is the case.

Consider an allometric curve of data from a wide variety of mammal species in which the mass of an organ of interest is plotted against the body mass of the species from which it came. The measurements for most human organs fall on the relevant mammal-organ curve. Alternatively, the mass of the organs for a species can be normalized to that species' average body mass. When the data for human organs is compared to the data from the other mammals, then, for most of our organs, the two are numerically indistinguishable. However, both methods reveal that the human brain deviates the most from this trend/relationship (Stahl 1965). Human brains are twice their expected size for our body mass (Leonard and Robertson 1994).

The size of our brain is one of our unique features. But does it exclude us from the mammals? If all of the allometric constraints that apply to the other mammals apply to us, then, to maintain coherence, our larger brain must be balanced by changes to our other body and lifeway features. Is there evidence for the required adjustments?

Consider the allometric relationship between a primate's body mass and the energy use while at rest. It shows that the heavier a species, the more energy its uses at rest. Humans do not deviate from this trend. However, there is a catch. Our large brains consume 20% to 25% of our resting energy, which is substantially more than the

8% to 9% recorded for the brains of the other primates. In order for our brain to use such a high percentage of our at-rest energy and still fall on the at-rest energy curve, we must use less of our resting energy for other purposes.

It turns out that we use less energy for digestion. We make these digestive energy savings by adopting a diet that is significantly different from the other primates. We focus on eating foods that are more easily digested, have a much higher proportion of extractable energy and have a greater nutritional quality. Fruits and fatty foods are an obvious choice. When our distant ancestors cooked some of their food, such as grains and meat, they, in essence, pre-digested it. This is why we have a shorter gut relative to the other primates (Leonard 2002; Leonard and Robertson 1994). Both our brain size and our gut length are uniquely human. But, despite these differences, when we are treated as a functional whole, we still fall within the allometrically constrained functional limits for primates in general and apes in particular.

If we are truly one of the primates, then our basic lifeway characteristics should also satisfy the same allometric relationships that constrain the lifeways of other primates. For example, primates find their food within an area called their home range. There is an allometric relationship between primate body mass and home range size. The average home range size of today's human hunter-gatherers is larger than expected from this relationship (Anton et al. 2002). Our home range is larger, in part, because the high-quality foods the members of the band depend on are generally less common and often more widely distributed than poorer quality foods.

To access that food, the hunter-gatherers must also be inherently capable of foraging over a larger area than other primates (Milton and May 1976). Hunter-gatherers can do this and stay within the other allometric constraints on their lives because, when foraging, they move more efficiently than the non-human primates. Hunter-gatherers, like all humans, achieve this efficiency by walking and running on two rather than four legs (Leonard and Robertson 1997).

But, if we are subject to the same allometric constraints as other primates, then this solution must have been balanced by trade-offs in other aspects of our lifeway characteristics. This is the case. For example, although our manner of walking and running is relatively efficient, it is also relatively slow. Slower walking limits the return distance that a hunter-gatherer can cover during a one-day foraging trip to no more than 20 to 30 km (Kelly 1995). Being slow and relatively

poor climbers also made our ancestors more vulnerable to predation. In addition, the changes to the pelvic structure that make bipedalism possible reduces the size of the birth canal. Because our heads are large, we require a large birth canal. As a result, human females and their babies experience difficulties during birthing. Females thus require more support during birthing and when raising our more helpless young.

Our distant ancestors had to compensate for, or accommodate, these trade-offs if they were to remain viable as a species. They did so, in part, by behaving in a more co-operative (social) manner and by using our larger brains to strategize. The connection between allometric constraints and culture starts to appear. The links between our greater brain power and our efforts to satisfy our lifeway requirements (such as higher energy foods, predator evasions and relying on others) provided the feedback loop(s) that could sustain our distant ancestors' efforts to make the necessary types of accommodations and compensations to remain a viable species.

In overview, our body and lifeway characteristics are constrained by the same allometric relationships that apply to species in general and to primates in particular. Like them, in order to function as a species, our characteristics must satisfy all of the allometric constraints on our lives, simultaneously. However, the way we meet these constraints is, like all species, unique. Our characteristics, such as our bigger brains, a diet of higher-quality foods, walking upright, and our ability to co-operate closely and strategize make us different from other species but not separate from the earth's allometric constraints on living.

**Population.** A species' population is subject to the constraints imposed on it by its host community, ecosystem and environment. For example, the food available to a species is limited by its host ecosystem's productivity. That productivity is, in turn, constrained by the resources, from sunlight to nutrients, available to the ecosystem producing the food. A species' population is also allometrically constrained by the average mass of its individual members and by the trophic level at which the species feeds.

Are human populations subject to these types of constraints? If true, then we would expect them to be most clearly displayed by hunter-gatherer populations because they live closest to the land. In particular, the population densities (people/km$^2$) of hunter-gatherer groups should be related to the ecosystem type they occupy, and that density should be predictable by allometric constraints (Livi-Bacci 1997).

In 10,000 yBP, all humans were members of hunter-gatherer cultures. They lived in a wide variety of biome types, but their densities in these biomes are unknown. Some density estimates for existing and historical hunter-gatherer populations are known but uncertain. They range from 0.01 to 9 people/km$^2$, with the most common densities falling between about 0.1 to 1 people/km$^2$. The land biomes that could possibly host hunter-gatherer cultures are listed here in order of their NPP (annual plant-biomass productivity) and accompanied by the average of the estimated hunter-gatherer population densities that could be assigned to a biome. From lowest to highest productivity, they are the extreme deserts (e.g., Antarctica), 0; the Arctic, 0.1 (including those living around polynyas); boreal forests, 0.04; temperate deserts, 0.08; tropical and subtropical deserts, 0.03; grasslands, 0.12; temperate forests, 0.14; tropical forests, 0.27 people/km$^2$. (Polynyas are coastal marine-resource-rich areas in the Arctic that are ice free year-round.)

The (unweighted) average of these biome-related population densities is 0.1 people/km$^2$, while the average of all known hunter-gatherer densities is 0.54 people/km$^2$ (Kelly 1995; Hassan 1981). The biome-related average is low because the listing (except for the Arctic) generally excludes cultures whose local, highly productive home range are not readily related to a land biome. For example, the hunter-gatherers living along the Pacific coast of North America live in such an environment. Their high average density of 0.79 people/km$^2$ is possible because their home range contains rich marine resources and migrating salmon. The overall average is also likely low because, by the time population density measurements were made, many hunter-gatherers had likely been displaced from prime living sites. So they are under-represented.

The average hunter-gatherer population density can also be predicted by applying the estimated average hunter-gatherer weight of 59.5 kg to the allometric population-density-to-body-weight relationship for all species. By this method, the average ecosystem-constrained hunter-gatherer population density should be around 0.58 people/km$^2$ (figure 10) (Peters 1983). However, considering that our food needs are quite specialized, our actual average allometrically constrained hunter-gatherer population density likely falls below the ideal allometric curve, that is, below 0.58 people/km$^2$.

In overview, hunter-gatherer population densities are generally below 1, and high population densities are located in rare sites with high human food productivity. When the cultures that could not be related to a land biome are excluded, population densities vary with the productivity of their land-biome host type. Even after considering the limitations of the data, there is a general correspondence between the average of known population densities and the average predicted using allometric constraints. It is reasonable to suggest that early human population densities were constrained by their host biome's/location's productivity. Hunter-gatherer populations were, and still are, constrained by a combination of their supporting ecosystems' productivity and its size (area).

**Complex Behaviour.** The uniqueness of our brain extends beyond its large size and energy use. Along with our bigger brains comes a much larger memory, a more diverse information processing ability and an extensive vocal repertoire. These abilities enable us to exhibit unique characteristics such as extensive tool use, well-developed consciousness, intelligence and complex cultures. There is a tendency among Westerners, at least, to point to these characteristics as justifying the view that we are fundamentally different from other organisms. This is not in the sense of "different but one of them," as concluded above, but "different and apart or separate from them." Here we will look at a few of these characteristics to see if our unique features do indeed make us as separate as we like to think.

*Tools.* When considering our apparent unique use of tools, we immediately run into the problem of defining the boundary of tool use. For example, how are we supposed to view organisms that can manipulate the environment without any tools? Are they more or less unique or advanced? For instance, it seems strange that the abilities of bees to build hives, termites to build mounds, or birds to weave complex nests are not considered special enough to class them as outside the norm. It seems that the idea of tool use as unique to humans is focused specifically on the human ability to pick up an object, perhaps modify it in some predetermined manner, and then use it for a specific purpose. The boundaries we use for defining tool use are thus biased toward human practices.

However, even if our preferred human-biased definition of tool use is adopted, it does not exclude all organisms. Chimpanzees, for example, make and use a variety of tools to achieve a number of objectives. Their tool use includes stones to break open nuts, sticks to fish for termites and spears to kill animals (Roach 2008). But tool use is not found in all chimpanzee groups. It seems that those groups that use tools do so because they face difficulties acquiring some critical food resource in their home range. Those groups that don't use tools do so because they have no need to, rather than because they lack the required mental ability (Whiten and Boesch 2001).

Perhaps we can say that human tool use is unique because only we can evolve new technologies from old. In order to display this distinguishing feature, called cumulative technological evolution, humans must satisfy three important tool-making characteristics: the diversification of tool designs, cumulative change in design through time and accurate transfer of the tool-making skill to the younger generation.

But these criteria are also satisfied, at least in a rudimentary way, by the tool-using crows of New Caledonia, a large Pacific island near Tikopia. These crows use a multistep process to manufacture an insect-catching tool from the leaf of a screw pine tree. This tool appears to have evolved over time, and the process is learned by the young from the old (Hunt and Gray 2003).

Tool use, even as we define it in our biased manner, is not unique to humans. It is true that our tool-making technology and tool use is far more extensive and complex than that of any other species. But this is an indicator of our greater memory and learning capacity, our bigger brains, rather than an indicator of our separateness from other species.

*Consciousness.* Consciousness can mean either awareness of one's surroundings or the capacity for self-reflection. When we consider humans to be uniquely conscious, we are usually referring to the narrower, self-reflective meaning (Taubes 1998). Other primates also appear to be conscious in the self-reflective sense (Logothetis 1999). However, because we have no accurate method of determining what a chimpanzee, for example, is thinking, we can't yet answer the question of whether consciousness, as we choose to define it, is uniquely human (Pearson 2002).

*Intelligence.* Perhaps intelligence, meaning our mental ability to learn, reason and solve problems, separates us from all other species. But by using our mental abilities as the intelligence standard, we bias the comparison in

our favour. Firstly, our standard ignores the relevance of the organism's lifeway (i.e., context) when measuring intelligence-related accomplishments. Consider that all organisms that have brains use them to successfully solve problems within the context and constraints of their lifeway and community. We are barely aware of the problems other species face, let alone have the capacity to solve them.

Secondly, our standard does not take into account the variation in our intellectual abilities. For example, humans are skilled at manipulating information (our imagination) but not at manipulating numerical data (which is why we use computers). Similarly, although we do engage in rational and reflective thought, our ability to complete this task is significantly limited and biased. Thirdly, by including consciousness as a prerequisite for intelligence, our assessment that our intelligence separates us from other organisms is self-fulfilling.

Despite the bias in the standards we set, there are many animals up to the intelligence challenge. Consider the sandpipers, mentioned earlier, who migrate using a sun compass (Alerstam et al. 2001). Most of us don't even know what a sun compass is, let alone how to do the calculations needed to use it.

Then there are the members of the crow family (corvids). Corvids, in general, have such exceptionally sophisticated social behaviours that they have been described as "the second-best liars, cheats and thieves in the animal kingdom" (Gee 2002). Included in this group are the jays, who have a particularly sophisticated power of deduction. Scrub-jays often store the nuts they eat in food caches. They remember the what, where and when of their nut caching. They also remember the experience of having their cached nuts pilfered by other jays and then use this knowledge to change their nut-storing process in order to avoid nut pilfering. In particular, if they know that other jays are watching them when they cache nuts, they will return later when no other jays are around and move the nuts to a new cache (Emery and Clayton 2001). This ability indicates that jays are capable of mental time travel and planning for the future, a trait normally considered to be exclusively human.

Among the other contenders for the intelligence prize are the chimpanzees. One chimp in a Swedish zoo clearly displayed the ability to plan for the future. He gathered and cached rocks in anticipation of later using them to throw at zoo visitors as part of his dominance displays. Evidence for the extent of his forethought is found in his rock cache. Some rocks were pieces of concrete that he had broken from larger blocks. He had determined where on the blocks the most easily recovered loose pieces were located by the hollow sound they made when he tapped them (Osvath 2009).

In overview, on balance, we do have and use a much wider variety of mental abilities than other organisms, but these abilities are present in other animals to differing degrees for different purposes. We are thus certainly unique in the sense that humans have one-of-a-kind mental abilities. But we are not unique in the sense that these abilities completely separate us from other species. The perception of separateness has more to do with the biases in our sense of self and our cultural world view about our relationship to our environment than with the reality of the world around us. If we bear this in mind, then the pedestal on which we put ourselves should not be too high. Neither should we put too much faith in our "unique" abilities to appropriately solve the problems we face.

### Culture.
The lifeway characteristics of non-human organisms are limited by both environmental and species-specific constraints, as discussed in chapter 7. This includes the behaviours that underpin cultures, such as group decision making. The biological characteristics of a human lifeway are, as discussed above, constrained in the same basic way as other species. If these constraints extend to the social aspects of our lifeway characteristics, then the complexity of our culture can't be used to argue that we are separate from other organisms or that culture frees us from environmental constraints. This subsection builds on the connection between allometric constraints and our culture established above.

The influence of both the environmental and human-species constraints on human culture will be more easily seen in hunter-gatherer cultures than in the more complex agricultural or industrial cultures. After all, the daily activities of hunter-gatherers are more directly connected to their supporting ecosystems and their cultural groups are more closely knit together by their smaller sizes.

Most hunter-gatherer bands fall in the range of 25 to 1,500 people. Genetic considerations set the minimum, genetically viable band size to between about 25 and 50 people. The members of hunter-gatherer bands in this size range must find suitable mates outside of their band (Kelly 1995). To be successful, their culture's world view must guide the band's members to live in ways that will secure these mates. It can't, for example, direct the members to conceive of themselves as genetically pure or to shun outsiders.

The smallest hunter-gatherer bands are usually nomadic, while the larger bands can live more settled lives. Regardless, the characteristics of the ecosystems making up their home range provide food but constrain its nature and availability, which affects the features of their cultures. The general form of this influence is illustrated by the Mande culture, from its hunter-gatherer roots into its early Iron Age culture.

The hunter-gatherer Mande lived in an area centred on the Niger River, near the Senegal River, West Africa. The climate and biomes of this region vary from the harsh semi-desert in the north to the more productive savannah in the south. These areas are all characterized by highly variable amounts of annual rainfall. The extended periods of reduced precipitation, such as a drought, determined the minimum amount of food available to support the Mande. Rainfall thus limited the maximum number of Mande who could live in a particular area over the medium to long term. Over a much longer period, a particular area's human carrying capacity was also affected by changes in the climate. When the climate became wetter, the region's biomes moved northward, and when dryer, southward.

For the Mande to have remained a functional culture in this dynamic, variably productive host environment, their culture's world view had to guide them to accept their hosts characteristics as reality (i.e., to live within its constraints). This guidance was needed if the members were to make decisions that would enable them to easily survive during their host ecosystem's periods of lowest productivity. Similarly, during the good years, their culture's world view had to guide them to retain, at least in their culture's memory, the lifestyle choices needed to survive during the bad years, after all, they would inevitably come again (Togola 2000; McIntosh 2000). The environment thus constrained the Mande culture's world view.

When looked at in more detail, a hunter-gatherer band's supporting environment imposes a wide range of subtle but powerful constraints on their culture's world view. Consider that the types, amounts and locations of the food provided by the band's home range ecosystems vary through time. In addition, these variations are only somewhat regular because, in detail, food availability is affected by many factors. Thus, in order for the members to secure the allometrically determined amount of food they require to survive, their day-to-day foraging decisions must consider the environment-affected variations in the availability of their food. For example, they must take into account where on their home range the plant foods are in season, the appropriate ways to forage for them and where the food animals are likely moving to. In essence, their decisions about foraging must routinely consider the state and functioning of their home range ecosystems, which means that their decision making must be based on extensive knowledge and be flexible.

It also means that if the band is to be functional, then their culture's world view must routinely guide the members to collect the required knowledge, help them to remember it and encourage them to exhibit flexible decision making. Their world view can't, for example, guide the members to adopt a highly regimented way of living and working, like that found, even demanded, of the members in more complex cultures. The band's world view is environmentally constrained.

Consider too that foraging decisions constrain the members' non-food activities, such as their social activities. After all, if foraging is to be successful, then their social activities must be compatible with the demands of foraging, for example, the length of time it takes. The members receive their guidance about social activities from their culture's world view (Smith et al. 2017). Think of myths and parables. Thus, the guidance a hunter-gatherer's world view provides about social activities is, through their need to forage, subject to environmental constraints.

But, social activities impose constraints on a hunter-gatherer's lifeway and thus on their foraging decisions. Among the social features they need to consider when making those decisions are the following: are women pregnant or breastfeeding, who will look after the young if women with children go foraging, the type and intensity of labour performed by women with children, how the secured food will be prepared and shared, and the lessons from history. As a result, a hunter-gatherer's foraging decisions are invariably made within the constraints of both an environmental and a social context (Kelly 1995). This is unsurprising because, as mentioned above, a viable human lifeway is formed when the required balancing of the allometric constraints on our lives also includes our social activities and needs, such as co-operating and strategizing.

Evidence that a functional culture's world view must and does provide guidance about these constraints and their balancing is illustrated by the BaMbuti hunter-gatherers, or pygmies, who live in the tropical forests of the Ituri province, Democratic Republic of the Congo. When they decide to hunt, the BaMbuti have to choose between using nets or bows and arrows. From the

outsider's perspective, their decision should logically be based on the efficiency of their efforts (the size of the catch), so there appears to be no advantage to net hunting. However, from the BaMbuti's perspective, their decision is based on wider considerations, in particular, whether the women are able to join the hunt or not. A woman's decision is based, in part, on the amount of food she thinks she can procure by other means (Kelly 1995).

In overview, if a hunter-gatherer band is to be viable, then their culture's world view must contain and represent the constraints on it being functional. It must guide the members to make foraging and social decisions that are compatible with both the environmental and human (social and physical) constraints on how they can live. And the members must follow that guidance.

When non-sedentary hunter-gatherer cultures are seen within this context, their defining emphasis on suppressing elitism, being flexible in decision making and practising some form of sharing are not disconnected, abstract social ideals or ideological imperatives. They are functional necessities. They are embedded in their culture's world view in an attempt to bind the members into a long-lasting, cohesive, functional whole that stays within the environmental and human-species constraints on their lives (Kelly 1995).

A hunter-gatherer band's world view can come, over time, to reflect these environmental and human-species constraints on their lives because that world view is formed by the interactions among the members and between them and their home range ecosystem. In essence, the particulars of their culture's world view is the product of a dynamic relationship between the individual members, their host environment and their inherited cultural world view. Over time, that world view is adjusted as it passes from generation to generation, as discussed in chapter 3 and 6.

A hunter-gatherer culture's world view must provide appropriate guidance to its members about the environmental and human constraints on their lives, if it is to be viable. But does this constraint apply to more complex cultures, especially industrial ones? An answer starts to appear by considering the connections between hunter-gatherer (simple) and more complex cultures. The primary differences between them are the sizes of their groups and their organizational structures, as discussed in chapter 3. Hunter-gatherer groups are relatively small in size, and the majority have an egalitarian structure. In contrast, only large groups have complex cultures with their hierarchical structures.

None of the changes needed to turn a simple culture into a complex one remove the requirement for a significant degree of coherence and consistency among a culture's decision-making, its structure and its world view. Neither do these changes terminate the culture's participation in the human-environment system, nor enable it to escape the environmental or the human-species constraints on the members' and their culture's functioning. It just changes their expression. A complex culture's world view must also guide its members to consider the human and environmental constraints on their and their culture's functioning.

Is there evidence for this? Consider that, in complex cultures, the members can more easily become distant from their supporting environment. As a result, their culture's world view can be coherent and consistent, but significantly disconnected from their supporting environment and its constraints on their lives (Weingart et al. 1997). The consequences provide evidence that a complex culture's world view must guide its members to consider the human and environmental constraints on their and their culture's functioning, if it is to viable. An easy-to-see example is provided when its guidance is at odds with the biological and social reality of reproduction.

The natural birth ratio for the human sexes is around 952 girls for every 1,000 boys. The consequences of significantly altering this ratio can be found in contemporary India. As was touched on briefly in chapter 6, the cultural world view of some northern Indian provinces prefers boys to girls. The efforts of the members to satisfy their culture's preference was eased by the development of ultrasound technology because it facilitated the process of preferentially aborting girl foetuses. The result was so significant that it led to a rapid decline in the average number of girls being born in India as a whole. This can be seen in the Indian census data for children under six years old. The 1981 census reported an average of 962 girls being born for every 1,000 boys (i.e., normal). In the 2001 census the average had declined to 927, while in the 2011 census it was 919.

The consequences of male selection are particularly marked in the data from the adjacent north-central provinces. For example, the regional data from the 2001 census revealed that, in Delhi province, 868 girls were born for every 1,000 boys; in Haryana, 819; and in Punjab, 798 (Singh 2008; Census of India 2001). The regional data from the 2011 census

revealed that the preference for male children is enduring: in Delhi province 871 girls were born for every 1,000 boys; in Haryana, 834; and in the Punjab, 846 (Census of India 2011).

If the cultural world view of these northern provinces continues to favour male children, and if the imbalance in births endures, then their members can expect to experience some form of significant cultural adjustment, even destabilization. Consider that by the late 1990s, the excess of young males meant that many of them would never marry. By 2012, the ongoing complementary deficit of young women was contributing to changes in marriage practices, such as dowry payments and the loosening of cultural restrictions on the clans and districts within which a marriage partner can be sought.

The severe bride-bridegroom imbalance has not yet led to general changes in the cultural preferences for sons or changes in sex selection practices. However, in keeping with the discussion of cultural change in chapter 6, it is at least prompting important changes to specific aspects of the patriarchal culture's beliefs that peripherally support the current sex preference and sex selection practices. Those aspects include some changes to inheritance rules, the granting of limited autonomy to women and a change in how aging parents are cared for. For example, traditionally those sons who receive an inheritance from their parents are expected to care for the parents in their old age. Part of this care is provided by the son directly and part though the son's kin that result from his marriage. No marriage means that unmarried sons, and now daughters, bear more direct responsibility for caring for their aging parents, which also complicates the granting of inheritances (Larsen and Kaur 2013).

Consider too the 2012 and 2013 media reports of rapes of women by younger men in northern India, the attitudes of law enforcement to these events and the unexpected public outrage. It can be speculated that they are all an expression of the culture's current view of women, the consequences of supporting male sex selection, and the changes in these views now underway (Burke 2013). Even the functioning of complex cultures is subject to environmental and human-species constraints.

In overview, the characteristics of the human species make us unique, but not in a way that separates us from the earth's environmental system or its other species. Our characteristics are constrained in the same manner as all other organisms: by the need to be a functional whole while also being a member of both a functional social group and a host ecosystem. Think of allometric constraints. This holism is expressed by the interconnected and complementary nature of our physical and mental characteristics, our lifeway and our culture, and our dependence on our physical surroundings.

We are an integral part of the earth's environmental system. We are embedded in that system. Thus, if we are to endure, our culture's world view must guide us to stay within the human and environmental constraints that accompany us being a social species sustained by an environmental system. And we must follow that guidance.

In summary, this chapter notes that a primary driver of climate change is the ongoing variation in the amount of sunlight reaching the earth. Changes in sunlight and climate affect the individual members of the earth's ecosystems. Their responses change their host ecosystem, which then contributes to climate change. As a result of these interconnections and feedbacks, the climate system and ecosystems are inseparably bound together to form the earth's dynamic, complex environmental system. It displays the generic features of a dynamic complex system.

Its functioning, which includes features such as allometric constraints, trophic levels, nutrient recycling, radiation capturing and heat redistribution, helps to maintain the environment system in a quasi-stable state. But it also ensures that the system is dynamic. Changes to one part of the environmental system, regardless of scale, will eventually spread to the other parts and can be amplified along the way. And, beyond some threshold of change, the system will switch into a different quasi-stable state.

The human species is an integral part of this environmental system. We, like all species, are participants embedded in the system. We contribute to its functioning and changing and, like all species, our characteristics and how we can live are constrained by how we function as a species and how our host environment system functions and changes.

The complex cultures that arose from our hunter-gatherer roots have enabled us to transform our lives and our environment in ways undreamt of by our ancestors. Our power has given us abundant food to support an extremely large population, lengthened our average life span and provided more material goods than we can possibly use. No wonder we tend to believe that this world of plenty is possible only because we are smart enough, our cultures complex enough and our world views correct enough to have overcome the human and the environmental constraints on our lives. It is easy for the members of complex culture's to believe that we have become largely separate from the earth's processes and its other organisms. We can believe that if there are any undesirable consequences from transforming our environment, then we can use our technologies to mitigate or correct them.

However, as discussed in this and the previous chapter, we can neither change the basic rules of how we and our environment function nor remove the constraints that this functioning places on how we can live. Thus, any changes we make to our environment and how we live our lives are not costless improvements or escapes, but trade-offs. Regardless of our awareness of these trade-offs, they adjust the dynamics of the human-environment system and change how the constraints on our lives are expressed (Pecl et al. 2017). This affects our future.

Beyond some degree of adjusting the human-environment system and circumventing its constraints on our lives, we burden ourselves with responsibilities. We become responsible for maintaining or replacing the environmental system's functions that make available the resources (including services) we require to complete our lifeway. In essence, we become environmental managers. But in doing so, we also have to consider the constraints on how humans, as a social species with a culture, can function. Complex cultures also become human managers.

Have we been successful in our role of environmental and human managers? The next two chapters will provide information relevant to answering that question at a human scale.

# CHAPTER 9

# THE STATUS OF THE BIOSPHERE:
# ITS PERSPECTIVE

You are lying in bed, wide awake. As you try to shake off the thoughts about the last chapters that are hindering you from drifting into dreamland, you feel a presence, an all-powerful, not-to-be-ignored presence. It informs you of concerns that the earth, under the current management, is going through profound changes. Because you are a good and worthy person, it continues, you have been chosen to complete a special assignment: investigate and report on the status of the earth's biosphere. But not too long a report, mind, because the presence is busy.

In the interests of brevity, the effect that this presence had on your sleep that night will not be described. Let's just say that, by sunrise the next morning, you had an outline of a report in mind. Entitled "The Status of Our Biosphere," it would apply what you learned from the previous chapters. Because you know that in most creation stories humans and non-humans are linked together but treated as distinct, you decide that your report will be similarly divided into two parts. The first part, presented here as this chapter, takes the non-human perspective of the biosphere's status. The second part, the next chapter, takes the human perspective. Here begins part one.

A reminder that *short* (or *immediate*) *term* refers to a period up to 15 years from 2015; *medium term*, up to 60 years; and *long term*, beyond 60 years.

## The Biosphere

Our home, the earth's biosphere, is a complex system. As discussed in chapters 7 and 8, it is constantly changing as a result of the Milankovitch sunlight cycle, geological processes and the functioning of earthbound environmental system. Since the rise of agriculture, we have added our increasingly significant contribution to those ongoing processes. To describe the biosphere's current status from a non-human perspective is, in essence, to describe our contribution to its changing. The reasonableness of this approach becomes clear after considering an overview of our past and ongoing impacts on the earth's biosphere.

The human species is only one of some 1.9 million described species and an estimated 10 million species in total. Yet by 2000, we had appropriated for ourselves an estimated 23.8% of the net global land-based annual production of new plant biomass, or net primary productivity (NPP). Our appropriation is the result of our biomass harvesting of food and fibre crops (49.8%); land-use changes to ecosystem productivity, such as forestry and pasture (39.1%); human induced fires (7.3%); and the space occupied by our infrastructure (3.7%) (Haberl et al. 2007; Imhoff et al. 2004; Vitousek et al. 1986).

The significance of our NPP appropriation becomes clear when we realize that the NPP is the annual new production of biomass at trophic level 1 in all terrestrial ecosystems. Our appropriation comes from the foundation of these ecosystems. Because they don't harbour wasted or unused resources waiting to be put to better use by us, our level of appropriation involves a trade-off. Our appropriation is balanced by our disruption or destruction of the terrestrial ecosystems that previously covered between 30% and 50% of the earth's land surface (Vitousek 1994). For example, in order to conduct just our agricultural activities (food, fibre and pasture, not wood production, etc.) we had, by the year 2000, disrupted or destroyed the ecosystems on an estimated 34% of the earth's ice-free land area (Ramankutty et al. 2008).

In making our appropriation of terrestrial NPP, we annually use roughly 23% of the world's renewable fresh water (Postel et al. 1996). We also affect the atmosphere's functioning (Imhoff et al. 2004). Remember that some 25% of our $CO_2$ emissions come from biomass burning, much of it from land clearing.

Our impacts on the biosphere extend to the oceans. In 2008, 17 databases were synthesized to create a global-scale view of the cumulative impacts that human activities, such as fishing and pollution, are having on the near-surface parts of the oceans. The data suggests that some 41% of the world's oceans have been adversely affected to a medium-high to very high degree as a result of a combination of two or more of our activities (Halpern et al. 2008). This includes impacts on the oceans' ecosystems (Myers and Worm 2003).

Over the last 6,000 years of the present interglacial period, human activities have increasingly affected the state of the earth's biosphere. The question is how significant are these impacts? This chapter provides an answer by dividing the biosphere into three interconnected environments: atmosphere, ocean and land-based. Each of the three sections describes both the impacts of our everyday activities on an environment and its current overall state. The following section consolidates our impacts on the three environments into our basic contributions to the biosphere's current state and uses them to describe that state.

The sections will be illustrated, where possible, by global-scale numbers and averages as well as examples.

They are used even though it is recognized that global-scale numbers and averages have limitations and can be misleading. In particular, they fail to represent that we participate in ecosystem change at a local and individual level, and that variations around the average or median are important. Hopefully these limitations are overcome by timely reminders and the numerous examples.

## The Atmosphere

The atmosphere is a thin film of constantly moving gas surrounding the earth. The word *film* is used because the lower atmosphere that can support life (troposphere) is only 10 km thick at the poles and 20 km at the equator. This distance is less than you could comfortably walk in a day, smaller than the hunter-gatherers' maximum foraging range and well within the distance many people routinely commute to work. It is so thin that when you look up at the anvil-shaped top of a thunderstorm cloud or see a transcontinental jet winging overhead, you are looking at the top of the atmosphere's film. What is the current status of this film?

Pollution. The word *pollution* is used here to describe the additions of human-generated gasses, liquids and solids to the atmosphere, ocean and land, irrespective of the relative or absolute quantities or impacts involved. Pollution is divided into two types: human additions of the environment's normal constituents and human additions that are novel. A novel addition is a human-made compound that is not normally present. For example, if our pollution into the atmosphere contains carbon dioxide ($CO_2$), then it is increasing a normal constituent. If it contains DDT, it is adding a novel compound.

*Increasing the Normal.* This subsection discusses the degree to which our pollution has changed the level of the atmosphere's normal constituents. It also discusses some of the consequences.

As outlined in the previous chapter, ice cores drilled from the Antarctica ice sheet provide us with a record of the variations in the atmosphere's temperature and composition over the last 400,000 years. The temperature record reveals a series of Milankovitch sunlight-cycle-induced, cold-warm (glacial-interglacial) periods (figure 11) (Petit et al. 1999). We live in the youngest interglacial period.

The composition record reveals that the atmosphere's $CO_2$, methane and nitrous oxide ($N_2O$) content changed in concert with the changes in temperature. This record can be used as a reference to decide whether our pollution has resulted in anomalous changes in the composition of the current interglacial atmosphere. In the process we gain an appreciation for the significance of those changes.

First to be discussed is *carbon dioxide*. During past interglacial periods, the concentration of $CO_2$ in our atmosphere reached between 260 and 290 parts per million (ppm) (figure 11). This was true of the present interglacial until 1750: its $CO_2$ level hovered about 280 ppm. After that date, the concentration of $CO_2$ began to rise increasingly rapidly. It reached 379 ppm by 2005; 390.9 ppm by 2011; and 396.0 ppm by 2013 (IPCC 2007a; WMO 2012; WMO 2014). Today's $CO_2$ concentration of roughly 400 ppm is 110–140 ppm above the usual interglacial $CO_2$ concentration of 260–290 ppm (figure 11). The current level of atmospheric $CO_2$ is anomalous.

How anomalous can be appreciated by noting that the current level of $CO_2$ is the highest known for the last 650,000 years (Siegenthaler et al. 2005). It might be the highest in the last 20 million years (IPCC 2001). The atmosphere's current 110–140 ppm of "excess" $CO_2$ over the usual interglacial levels is greater than the difference in its $CO_2$ content between the interglacial periods (the maximum $CO_2$) and the glacial periods (the minimum $CO_2$) of 80–110 ppm during the last 400,000 years (Petit et al. 1999). The current level of atmospheric $CO_2$ is highly anomalous.

The rate at which $CO_2$ is increasing is also noteworthy. The increase from the pre-industrial (pre-1750) level of about 280 ppm $CO_2$ to the 2011 level of 390.9 ppm took 261 years, at an average rate of around 0.425 ppm/year (IPCC 2007a). Compare that rate to one of the fastest increases in $CO_2$ recorded for a glacial to interglacial transition. During part of the 630,000 years before present (yBP) transition, $CO_2$ rose from 200 to 235 ppm in less than 2,000 years, for an average increase of around 0.018 ppm/year (Siegenthaler et al. 2005).

Just as significant as the rapid average post-1750 rate of increase in $CO_2$ are its much faster, recent short-term average increases (IPCC 2007d). Between 1960 and 2005, the average rise in $CO_2$ was 1.4 ppm/year; between 1995 and 2006, it increased to 1.9 ppm/year; and between 2005 and 2011, it reached 2 ppm/year (roughly 0.5%/year at current levels) (IPCC 2007a; WMO 2012). The current rates of increase are much faster than in the early 1800s. The primary, direct cause of the post-1950s rapid rise in $CO_2$ is the large increase in the combustion of fossil fuels to power our industrial cultures. Significant secondary contributions come from forest clearing and related activities such as farming (IPCC 2007a).

The amount of $CO_2$ we have released into the atmosphere over the last 250 years is so large and has occurred so quickly that the normal carbon cycle balance has been significantly disrupted and is undergoing a major rebalancing. For example, over the past few decades, about 50% of the $CO_2$ we have emitted into the carbon cycle's atmosphere reservoir remains there; 30% has entered the ocean reservoir; and 20% has been removed by terrestrial vegetation (figure 7a) (Feely et al. 2004). The carbon cycle is adjusting to the increase in $CO_2$ in its atmospheric reservoir by favouring processes that move carbon into the ocean and terrestrial reservoirs. In this context, the current rapid $CO_2$ rise can be thought of as our post-1750 emissions occurring faster than those redistribution processes can remove it. Think of the oceans' inertia to change.

The rebalancing of the carbon cycle is also being driven by pollution-related increases in the atmosphere's *methane* content. Methane concentrations during the six interglacial periods of the last 650,000 years lay between about 700 and 790 parts per billion (ppb) (figure 11) (IPCC 2007a; Spahni et al. 2005). Methane's concentration in the present interglacial stood at around 715 ppb just before 1750, when industrialization began (Evans 2007). Its concentration rose rapidly to reach 1774 ppb by 2005; 1812 ppb by 2011; and 1824 ppb by 2013 (IPCC 2007a; WMO 2012; WMO 2014).

The current concentration of atmospheric methane is the highest in at least the last 650,000 years (Spahni et al. 2005). Between the 1800s and early 1980s, methane rose at an average rate of 17 ppb/year, the fastest rise in the methane record (IPCC 2001). The current level of methane and its historical rate of rise is highly anomalous. About 60% of the current increase can be directly attributed to human activity. The most important sources are cattle breeding, rice cultivation, landfills, fossil fuel exploitation (fugitive emissions in particular) and biomass burning (IPCC 2007a; WMO 2012).

The concentration of methane in the atmosphere is low compared to $CO_2$. However, its increase is still significant because methane plays an important role in the carbon cycle and the climate system. For example, it is at least 20 times more effective as a greenhouse gas than $CO_2$.

Our pollution is also contributing *nitrogen oxides* to the atmosphere. There are a variety of gaseous nitrogen compounds in the atmosphere, including nitrous oxide ($N_2O$), ammonia and nitrogen dioxide ($NO_2$). The history of $N_2O$ is representative of the increases above normal concentrations being recorded for all of the nitrogen oxides.

During the last 650,000 years, atmospheric $N_2O$ remained below about 270 ppb until the early 1880s (IPCC 2007a). After that time its concentration started to increase. It reached 310 ppb by 1999; 319 ppb by 2005; 324.2 ppb by 2011; and 325.9 ppb by 2013. The average rate of increase up to 1999 was about 0.8 ppb/year. Between 1999 and 2005, the rate increased to 1.3 ppb/year; and between 2005 and 2011, it had slowed to 0.74 ppb/year (IPCC 2007a; WMO 2012; WMO 2014).

Prior to the 1880s, about two-thirds of $N_2O$ emissions came from soils and one-third from the oceans (Spahni et al. 2005). The rise in the atmosphere's $N_2O$ was initiated by an increase in fossil fuel combustion. The post-1880s industrial-scale production of nitrogen compounds, such as dynamite, added a significant amount. After 1913, the production of human-made nitrogen fertilizers in large quantities became possible, so it too made a contribution. However, it wasn't until the post-1940s green revolution push to intensify terrestrial food production that the manufacture of fertilizers for agriculture became, by far, the largest single source of human-related $N_2O$ emissions. Currently, the two main sources of human-related $N_2O$ emissions are human-made fertilizers and the combustion of fossil fuels (McNeill and Winiwarter 2004; WMO 2012).

Today about 40% of all $N_2O$ emissions are contributed by the above human activities and 60% by normal emissions (Spahni et al. 2005). The increase in our $N_2O$ emissions, together with our other nitrogen emissions, have severely disrupted the nitrogen cycle (figure 7b). This is seen by their increased contributions to acid rain and the ozone layer's depletion (Faure 1998a). We experience nitrogen oxide emissions as a light brown smog in urban centres.

Our $N_2O$ emissions have also affected the climate system. Although the absolute amount of $N_2O$ in the atmosphere is small, its increase is significant because $N_2O$ is about 290 times more effective as a greenhouse gas than $CO_2$.

We are also increasing the *sulphur* content of the atmosphere. In the pre-industrial sulphur cycle, the majority of sulphur additions to the atmosphere came from ongoing, low-level volcanic emissions. These steady background emissions were randomly increased by much larger contributions from major volcanic eruptions.

Our current sulphur contribution is sufficiently large that it has changed the atmosphere's sulphur content and disrupted the sulphur cycle. For example, in 1995 our sulphur emissions were roughly equal to the pre-human background emissions. Our largest emissions come from activities related to fossil fuel extraction, such as sour gas (hydrogen sulphide) venting and fossil fuel combustion. For example, the sulphur content of coal is commonly 1% to 5% and oil, 2% to 3%. Another significant source of emissions is the roasting of sulphur-containing metal ores (Faure 1998a).

Human sulphur emissions are mostly found near ground level in urban and industrial areas, where they contribute to smog. Downwind from a source area, the sulphur rich air forms a plume. In these plumes, the sulphur forms droplets of sulphuric acid smaller than one micron in size. During rainfall, these acid particles are flushed from the atmosphere to create an acid rain that disrupts and destroys forest and river ecosystems (Begon et al. 1996).

For example, acid rain percolating through forest soil changes the soil's biochemistry. It speeds up both the leaching of minerals and nutrients, such as calcium, magnesium and nitrate, from the soils and their transport into rivers. At the same time, acidic soils make elements such as aluminium available in a form toxic to plants. In a perverse twist, acid rain from fossil fuel combustion also contains nitrogen combustion products that act as a fertilizer. They stimulate plants to take up nutrients and minerals from the soil at a time when they are being lost through acid rain leaching (Wolfe 2001). No wonder many plants struggle to grow when exposed to acid rain.

Regulation has successfully spurred efforts to reduce acid rain and smog by, for example, requiring that only low-sulphur fuels be burned and by ensuring that sulphur in smokestacks is captured rather than emitted. As a result, between 1985 and 1994, the sulphur dioxide emissions from Europe dropped by half, with most of the reductions being achieved by factories (Clarke 1999). By 2009, the level of sulphur dioxide emissions in the United States had dropped to 76% of their 1980 value (EPA 2011).

There is an ironic catch to this sulphur cleanup. Winds move some of our sulphur emissions higher into the atmosphere, where they form tiny white solid sulphate particles that reflect sunlight and contribute to atmospheric cooling. This is the same process that contributes to the multi-year climate cooling after a large volcanic eruption. By reducing our sulphur emissions, we have decreased the occurrence of ecosystem-damaging acid rain but, ironically, facilitated the process by which our $CO_2$ emissions warm the atmosphere.

Any tiny solid particle in the atmosphere is called an aerosol. They, of which the sulphate type is one, have a range of compositions, sizes and colours. Our pollution contributes to an increase in the amounts of normal and novel aerosols in the atmosphere. For example, our burning of biomass, such as wood and grass, increases the atmosphere's content of normal combustion-related aerosols and larger particles. And our land disturbances increase its normal content of the larger dust particles.

In overview, the atmosphere's current $CO_2$, methane, $N_2O$, sulphur dioxide and particulate content is significantly elevated compared to both pre-industrial interglacial and past interglacial normals. We directly experience these changes as an increase in acid rain, smog and exposure to ultraviolet light. The rapid and substantial post-1750 changes in the atmosphere's composition have also initiated a dynamic, global-scale rebalancing of the element cycles, such as carbon, nitrogen and sulphur, that link the atmosphere to the oceans and land (figure 7a). As a result, our atmospheric emissions are affecting both the terrestrial and ocean environments. Our emissions also affect the climate system. Those impacts will be discussed after describing our novel contribution to the atmosphere.

*Adding the Novel.* We like to think that our atmospheric pollution is a recent phenomenon that it is localized to industrial sites or cities. Neither is true. Consider that our novel additions began with our earliest industrial activities and were, then as now, distributed around the globe by air currents. For example, the lead released into the atmosphere by Roman and Greek silver and lead smelting between 500 BC and AD 300 is present in Greenland ice cores at four times its late 1900s value. After the passing of the Roman Empire, European emissions of lead decreased and remained low throughout the medieval period. Shortly after AD 1000, atmospheric lead started its steady increase to its late 1900s level (Hong et al. 1994).

Even during that historical period our novel atmospheric pollution was a mix of compounds from different sources. For example, the Roman and Greek additions of lead were accompanied by copper from Roman, Greek and Chinese smelters (Hong et al. 1996). However, today our mix of novel pollutants includes a much wider variety of human-made compounds. The increase in that variety began in the mid-1700s, the early days of the Industrial Revolution. Ever since, there has been an ongoing increase in the number and amounts of the novel pollutants we have emitted into the air. The increase became more rapid with the post-1850s chemical revolution.

The novel pollutants of today are produced during the manufacture, use and disposal (the lifecycle) of a wide range of products, such as plastics, pesticides, propellants, cleaners and paints. They include the by-products created at each stage in their lifecycle, such as the products from its decay. Collectively, novel pollutants are the human-made compounds we dispose of directly or indirectly, deliberately or inadvertently.

Not all of our novel atmospheric pollutants are gasses. They can be particles, many of which can't be seen because they are only a few micrometres in size: the aerosols. When present in sufficient amounts, these particulates contribute to a hazy sky and smoke. The smoke emissions from chimneys and tail pipes identify some of the more significant source of novel particulates: the combustion of fossil fuels, particularly coal and diesel. Other sources of novel particulates include road, tire and brake dust. They also include the pesticide- and fertilizer-rich dust blown off the dry bed of the Aral Sea and dispersed as far as 1,000 km downwind from that ex-lake (Stone 1999; Micklin 1988).

The generation, presence and dispersion of novel atmospheric gaseous and particulate pollution is most easily illustrated by the brown cloud over eastern Asia. The cloud is a mix of the region's gas and particulate emissions (novel and normal) from activities in urban, industrial and rural areas. This might prompt us to think of the news reports about thick smog over one of the region's cities, but there is much more to the cloud. During the dry season, this brown cloud can cover an area of around 10 million km$^2$ (3,000 x 3,000 km), up to a height of 3 km above the earth's surface. Each year during the monsoon, part of this cloud becomes a convection plume that can reach to a height of 8–12 km, which is just above the tops of the tallest mountains in the region.

The Asian brown cloud is basically a gigantic pollution cloud. The gaseous part includes contributions from natural sources, such as fires started by lightning, but the majority is a mix of human-generated normal and novel gasses. The visible part of this staggeringly large brown cloud consists mostly of particulates such as fly ash, organic carbon, black carbon, dust and sulphates. Up to 75% of the particulates classed as aerosol are the products of human activities: the novel portion is mostly the result of fossil fuel (coal and diesel) combustion, the normal portion is the result of biomass burning.

Similar large brown clouds form over Europe and the Americas where their presence is indicated by a regional haze hanging over densely populated regions. All of these brown clouds have regional effects. Consider that they reduce the amount of sunlight that reaches the earth and the ocean beneath them. Under the Asian brown cloud, this causes local ocean cooling and a reduction in rainfall.

Brown clouds are not stationary. They move with the prevailing winds. The lower-level of the Asian brown cloud is moved by Westerly winds to North America in about six to seven days (Ramanathan and Ramana 2003; Lawrence 2011).

As a brown cloud moves with its host air current, it sheds some of its pollutants. For example, as the Asian brown cloud moves over the Pacific Ocean, some of its novel pollutants are washed into the ocean. This was one source of the polychlorinated biphenyls (PCBs) found in the north central Pacific, where salmon mature. (Think too of the ancient Greek's lead and Chinese copper pollution found in Greenland).

The basics of this global-scale atmospheric redistribution process are the same for all novel airborne pollutants regardless of source, but the details differ. These details are significantly influenced by a pollutant's chemical and physical properties, such as its stability and reactivity. Of particular importance is how these properties influence the partitioning (distribution) of novel pollutants between its atmospheric and its other (ocean- and land-based) reservoirs. The practical details of this partitioning are summarized in the description of the pollutant's lifecycle.

The novel pollutants in the Arctic helps to illustrate these lifecycle-related redistribution features. The wide variety of pollutants found there include the insecticides, herbicides and pesticides used in agriculture and gardening, such as dichlorodiphenyltrichloroethane (DDT), lindane, toxaphene and endosulfan (Li and Macdonald 2005). They also include chemicals used in industrial processes, such as PCBs and polybrominated diphenyl ethers (PBDEs). PBDEs act as a flame retardant when added to the plastics used to make objects from furniture to computers. Also present are the by-products from the manufacturing, use and disposal of these and other chemical products, such as dioxins from incinerating waste and polycyclic aromatic hydrocarbons (PAHs) from the combustion of coal.

Few of these novel pollutants are generated or released in the Arctic. Most are released into the atmosphere (or water) in the warmer, more populated areas of the world, where they are manufactured and used. For example, in 2000, about 80% of the carcinogenic dioxins found in Nunavut, a territory within the high Arctic of Canada, were emitted from smokestacks in the United States, where chlorine-containing compounds, such as plastics, are burned. The dominant combustion sources were municipal waste incinerators, backyard trash burning, cement kilns that burn hazardous waste and the secondary smelting of recycled copper (Commoner et al. 2000). Similarly, the most significant sources of organochlorine pesticides and brominated compounds manufactured and used in Europe that might find their way to the Arctic are easily identified by air sampling over Europe (Jaward et al. 2004).

Once airborne, these novel pollutants, and the other pollutants in their host brown cloud, are transported toward the Arctic by long-term, global-scale air currents that corkscrew from equator to pole. It is the details of how pollutants travel from their sources in the south to the Arctic that varies. Some pollutants, such as sulphur dioxide, stay in the air and complete the journey to the Arctic in months. Others, such as the pesticides lindane and DDT, take years (Semeena and Lammel 2005).

The slower moving airborne pollutants complete the journey through a combination of wind transport and a cyclical process called cryopumping or global distillation. In this process, pollution is drawn toward the cold of the poles in stages. During the summer, these atmospheric pollutants move. During the cold winters, they condense onto airborne ice crystals, some of which fall to the ground as snow. During the spring snowmelt, the pollutants vaporize to continue their grasshopper-like migration further poleward, toward the cold. Visitors arriving in the Arctic during late winter and early spring can see the outcome: vaporizing pollutants from the warming ice crystals are mixed with the pollutants arriving directly to create a haze similar to the smog over a big city and in a regional-scale brown cloud (Bidleman et al. 1989; Shaw 1995).

The air dispersal of pollution to the Arctic is further complicated because its constituents decay at different rates. Those that decay very slowly are called persistent. Persistent organic pollutants (POPs), such as DDT, are found widely distributed around the world, but they are most prevalent in the Arctic and, to a lesser degree, in mountain ice caps (Villa et al. 2003). There are three reasons why. Firstly, persistent pollutants are more likely to complete the journey to the Arctic than pollutants that decay quickly. Secondly, because the Arctic is a cold trap, the airborne pollutants that arrive there tend to stay. Thirdly, the cold slows down their decay.

The pollutants that arrive in the Arctic by air become distributed among their Arctic reservoirs. A pollutant's possible reservoirs include ice, snow, soil, liquid water, plants and animals (including humans). Think of lead in the Greenland ice cap. Because pollutants tend to remain in the Arctic, they (especially POPs) can become concentrated in their preferred reservoirs. There they can reach levels that are much higher than those found in the pollutant's reservoirs in the warmer regions, near where the pollutants originated.

The impact a particular pollutant has on an Arctic species or ecosystem depends on the pollutant's properties, its concentration and the specific reservoirs into which it is preferentially partitioned. A pollutant can be benign, or affect a few to many species. It can kill some directly, while disrupting the functioning of others. It can act quickly or slowly, at high or low concentrations. The effects of many pollutants, either individually or in combination, are poorly known or unknown. Think of cumulative adverse effects.

A useful feature to remember when evaluating an Arctic pollutant's possible impacts on species and ecosystems is whether it is more soluble in fat or water. The preferred reservoirs for fat-soluble pollutants are the body fat or milk fat of living organisms (Bidleman et al. 1989). This preference is more important in the cold parts of the oceans and in the Arctic than in equatorial regions. After all, if an Arctic organism is to survive the cold, either on land or in the water, it needs to produce a large amount of fat for insulation and energy use (Kidd et al. 1995). Examples of novel airborne pollutants that both concentrate in the Arctic and are fat-soluble include PCBs, PBDEs and dioxins (CBC News 2003). Many of the fat-soluble pollutants are thought to have significant biological effects, as will be discussed a little later. The details of a pollutant's properties and life cycle do matter.

The novel pollutants in the Arctic illustrate a basic aspect of our relationship with atmospheric pollution: our struggle to appreciate it and its significance.

Consider the separation in time and space between a specific novel pollutant being released into the atmosphere well south of the Arctic, and its subsequent impact on organisms in the Arctic (Li and Macdonald 2005). Regardless of the pollutant, the greater the separation in time and space between our creation or release of a pollutant and its impact on species and ecosystems, the more we struggle to acknowledge our connection to or responsibility for either the pollutant or its impacts. Overall, our ability to create and release novel pollutants has outstripped our ability to recognize and accept the consequences of our creations.

This reminds us that there is an inherent unfairness associated with pollution of all types. The people and ecosystems who are affected most by the pollution are often not those who receive most of the benefits that arise from the related product's creation and use. This is certainly true for the impacts in the Arctic. Perhaps by visualizing the formation and migration of brown clouds and the concentration of some of its pollutants into the fat of Arctic animals, and people, we will more easily recognize this physical and mental gap. Think of an Arctic resident breastfeeding her child with tainted milk. Certainly a wider vision of pollution that incorporates longer-term, continental-scale thinking is needed if we, the creators and releasers of pollution, are to accept our responsibilities.

## The Overall State of the Atmosphere.
After 1750, human activities resulted in significant increases in the amount of $CO_2$, then methane and finally $N_2O$ (the main greenhouse gases) being emitted to the atmosphere. This prompted the earth's climate system to adjust its quasi-stable state, which led to a rise in the temperature of the earth's near-surface atmosphere (troposphere). This temperature increase is put into perspective by describing a reconstruction of the troposphere's global average temperature over the last 11,300 years. The reference is the global temperature of the earth's troposphere for 1750–1800. It is chosen because the troposphere was weakly polluted up to that time.

On a millennium scale, the warming that led up to the current interglacial (warm) period ended around 10,000 yBP. From then until about 5,000 yBP, the temperature remained roughly constant at around 0.8°C above the 1750–1800 reference. After 5,000 yBP, the temperature steadily fell, to reach about 0.35°C above the reference, around 2,000 yBP (or around AD 1).

On a century scale, from AD 1 the temperature rose slightly to 0.4°C above the reference during the medieval warming interval, which lasted from about AD 500 to AD 1000. By around 1300, the cooling that became the Little Ice Age began. Cooling continued until the temperature reference (0°C) was reached in 1750–1800. After that date, the temperature rose increasingly rapidly to reach >0.9°C above the reference around 2015. Consider that the average temperature for the period 2000–2009 was already at least as warm as all but 30% of the last 11,300 years. The atmosphere is certainly anomalously warm (Marcott et al. 2013; Mann et al. 2008).

The increasing speed of the atmosphere's recent rise in temperature is described by the following averages. The global average temperature rise between 1900 and 2000 was 0.06°C/decade (IPCC 2001). A steeper increase occurred after the mid-1970s. For example, the rate reached 0.2°C/decade between 1990 and 2005. As expected, there were variations in this rate. For example, between 1998 and 2012, the increase was only 0.05°C/decade (IPCC 2007a; IPCC 2013a).

It is difficult for us to appreciate the significance of this extraordinary global-scale, post-1750–1800 average temperature increase (climate warming) because we intuitively relate to temperature changes on a local or regional scale. The following discussion personalizes global climate warming by both connecting it to regional and local temperature changes, and by discussing its physical manifestations.

When the average global-scale atmospheric temperature increase is broken into regional increases, the polar latitudes are seen to be experiencing a much faster temperature rise than the tropics. For example, between 1906 and 2006, the Arctic warmed at 0.12°C/decade. This is double the average global warming rate of 0.06°C/decade for that period. This doubling is a rule of thumb for the difference between the global rate and the Arctic region's rate of warming (IPCC 2007a). During the period between 1972 and 2002, the Arctic warmed at an astounding 0.5°C/decade (Otto-Bliesner et al. 2006). Between 1998 and 2012, it warmed at around 0.1°C/decade.

The significance of these rates can be gained by comparing them to the warming rates during the 10,000 years from the depths of the last ice age to the start of the present interglacial, 10,000 yBP. By doubling the global average warming rate for that period, the Arctic's Milankovitch-induced warming rate is estimated to be 0.03°C to 0.05°C/decade. This warming was interrupted by the Younger Dryas abrupt cooling event of 12,900 yBP to 11,460 yBP (likely a non-Milankovitch event).

The Arctic's rates of temperature change during this event are recorded in the previously mentioned ice core taken from the centre of the Greenland ice sheet. The cooling period at the start of the Younger Dryas included an astounding 10°C fall in less than 150 years, a fall of more than 0.66°C/decade (Firestone et al. 2007). At the end of the Younger Dryas, the temperature warming included an equally astounding 12°C to 15°C rise in less than 200 years, a rise of more than 0.6°C/decade (Taylor et al. 1997).

By comparing these rates, the following broad conclusions can be drawn. The recent regional warming rates in the Arctic of between 0.12°C and 0.5°C/decade are much faster than the Arctic's long-term, Milankovitch-induced regional warming rate of 0.03°C to 0.05°C/decade. But they are less than the local Arctic rates recorded for Greenland during the cooling and warming stages of the Younger Dryas event at around 0.6°C/decade. Even so, the current regional rates are highly anomalous. The significance of the current global warming has as much, if not more, to do with the rate at which it is occurring than the absolute increase.

Personalizing global climate warming continues with examples of the resulting regional and local physical changes. The most visible local reminder of climate warming is the ongoing retreat of most mountain glaciers. For example, Glacier National Park, USA, boasted 150 glaciers in 1850. But, by 2009, only 26 remained (National Park Service 2010). Similarly, a study of the terminuses of 244 glaciers on the flanks of the north-south trending Antarctic Peninsula showed that, after the 1950s, the majority have been in retreat and their rate of retreat is increasing. Consider too an east-west line drawn across the peninsula to separate those glaciers that are retreating (north of the line) from those that are not (south of the line). Over time, that separating line has steadily moved further southward as more glaciers began to retreat (Cook et al. 2005).

The retreat of a glacier's terminus is only one indicator of atmospheric warming. A more significant measure is the loss of a glacier's total mass, called wasting. The masses of 300 glaciers and ice caps (excluding the Greenland and Antarctic ice sheets) from different mountain ranges around the world revealed that not only were they wasting but, since the 1970s, their rate of wasting has increased (Kaser et al. 2006).

The continental ice sheets of Greenland and Antarctica are also wasting. Between 1992 and 2011, they each lost a significant amount of ice from their near-coastal regions due to melting: Greenland lost

2,700 gigatonnes (Gt); Antarctica, 1,350 Gt. Although there is considerable short-term variation in this loss, the longer-term trend indicates that this rate of loss is increasing, especially for Greenland (Shepherd et al. 2012). In fact, the weight lost from Greenland's coast as a result of the post-1990s ice melt is sufficiently large and sustained that the coastal land is rising out of the sea at an accelerating rate (Jiang et al. 2010).

The direct melting of ice in the near-coastal regions of these ice sheets is not the only mechanism by which they are wasting. A large number of the glaciers along the coastlines of Greenland and Antarctica flow out from the interior of the ice sheets into the ocean. Because of ocean warming, their terminus is disintegrating and melting faster than in the past. Both this increased melting and warmer ice have resulted in these glaciers flowing faster. They are thus delivering larger than expected volumes of ice from the interior of the ice sheet to the warmer oceans. Even though the ice sheet's interior is well outside the zone of direct melting and still permanently frozen, it is thinning because more of it is flowing more quickly into the ocean. The combination of this dynamic thinning and the direct melting has resulted in the Greenland and Antarctic ice sheets wasting faster than expected (Prichard et al. 2009). And the rate of wasting is increasing.

As land-based ice melts, it contributes to a number of environmental effects. The most obvious global-scale effect is a rise in sea level. For example, between 1992 and 2011, the wasting of Greenland and Antarctica contributed, on average, 0.59 mm/year to global sea level rise (Shepherd et al. 2012). This is a substantial contribution, considering that the total global average rate of sea level rise between 1993 and 2010 was 3.2 mm/year.

The wasting of glaciers and ice caps also has local ecosystem effects. These ice sources act as a regional freshwater reservoir, which by melting in the summer and storing snow in winter, ensures a more reliable annual flow in the rivers they feed. This is especially important if the rivers flow across an area with a dry climate, such as Alberta, Canada, and coastal Peru.

In some places, so much of the local ice caps and glaciers have melted (think Glacier National Park) that they can no longer act as reservoirs for the rivers they feed. In response, the amount and timing of the flows of water in those rivers changes, so does the water's temperature and composition. The combination of these changes disrupts local freshwater aquatic communities and ecosystems down the length of these rivers. For example, lower flow means less water to dilute and flush dissolved compounds and pollutants. It also means less water to support aquatic and human communities (Schindler and Donahue 2006; Baraer et al. 2012).

There are other changes to the climate system that accompany an increase in the global average temperature of the troposphere. On a regional scale, both temperature and rainfall change. As a result, the distribution of the earth's climates is changing: they are moving toward the poles (Zhang et al. 2007). As chapter 8 discussed, a change in climate forces species and ecosystems to adapt. A less visible consequence of a warming lower atmosphere (troposphere) is a detectible cooling of the outer atmosphere (stratosphere).

Despite this and the other changes to the climate system, the atmosphere's circulation patterns that redistribute the earth's heat were, at least in 2002, basically the same as they were during the Little Ice Age some 400 years ago (Mayewski and White 2002). Hints that this might not last are the speed of temperature change and the changes in the northern jet stream. It is displaying wider variations in its track to the north and south and longer periods at its extremes. The result is longer hotter periods in the Arctic and longer colder periods further south. However, there is no firm evidence that the climate system is about to enter an abrupt climate change event.

In summary, our pollution has and is changing the composition of the atmosphere. The increase in its normal components is affecting the earth's climate system. The average temperature of the earth's atmosphere is warming (climate warming). The amount of warming is more than 0.9°C above the 1750–1800 reference, which is warmer than all but 30% of the last 11,000 years. The speed of the warming is just as important as its amount. Current warming rates are not quite as fast as the rise and fall in temperature during the Younger Dryas event, but they are much faster than the average warming that ended the last ice age. The effects are particularly noticeable in the Arctic, which is warming at roughly twice the speed of the global average. The earth's climates are moving towards the poles.

Our atmospheric pollution (normal and novel) is distributed, in part, by air currents around the world, as illustrated by brown clouds. It is affecting ocean and terrestrial ecosystems globally, as will be discussed in more detail below. The Arctic is a special case because it is a sink for atmospheric pollution.

# The Ocean Environment

The ocean environment is the earth's largest reservoir of water, heat and soluble chemicals. It also hosts a variety of ecosystems. The oceans moderate (buffer) the changes to the atmosphere's temperature and composition, and supply, through evaporation, the fresh water essential to land-based ecosystems. The earth's oceans are thus an integral part of the earth's biosphere. The oceans and their ecosystems are showing the effects of our influences on them. This is discussed in subsections dealing with climate warming, pollution and the overall state of ocean ecosystems.

<u>Climate Warming.</u> The most visible indication of climate warming's impact on the oceans is the accelerating melting of Arctic sea ice. This is the product of the Arctic's longer and warmer summers.

The temperature of the lower atmosphere (troposphere) in the Arctic warmed at 0.5°C/decade between 1972 and 2002. People and animals experience this change as a spring melt that, in 1996, was occurring, on average, 3.7 days/decade earlier, and a freeze up that was occurring, on average, 4.5 days/decade later. As a result, the number of days that sea ice can melt increased from around 76 days in 1979 to around 85 days in 1996, a rise of 5.3 days/decade (Smith 1998). This increased opportunity to melt translates into a loss of Arctic sea ice.

The aerial extent of the Arctic Ocean's summer sea ice varies significantly from year to year. However, over the 1900s, the general trend is a decline in area. In addition, since 1979, the rate of decline has steadily increased. The average rate between 1979 and 2001 is estimated at about 6.5%/decade, by 2005 it had risen to 8%/decade, and in 2012 to 13%/decade. The sea ice area in 2012 was a record low. The summer ice cover for that year was reduced to 50% of its extent in 1979 (NSIDC 2005; NSIDC 2010; NSIDC 2012). A reconstructed Arctic sea ice record for the past 1,450 years indicates that over the last 40–50 years both the ongoing decline in the area of summer sea ice and the magnitude of that decline are unprecedented (Kinnard et al. 2011). The declining trend in the area of Arctic summer sea ice is matched by a declining trend in the remaining ice's thickness and mass (Tilling et al. 2015).

The amount and rate of this ongoing reduction in the Arctic's sea ice cover matters because its presence plays a critical role in the regulation of the earth's climates. Consider that white sea ice reflects sunlight more effectively than dark sea water. As the area of the Arctic's summer sea ice declines, less sunlight is reflected back to space and more is absorbed by the increasing area of darker ocean water. The Arctic Ocean is warming. This increases the earth's radiation, which facilitates atmospheric warming. A warmer atmosphere favours the melting of Arctic sea ice, thus creating a climate-warming–sea-ice-melting amplifying feedback. The feedback contributes to changing the nature and distribution of the earth's climates. A warming Arctic also alters the global-scale process by which heat is redistributed from equator to pole. Think of the consequences from changes to the northern jet stream's wiggle and changes to the Northern Atlantic overturning circulation.

Ocean warming is not restricted to the Arctic Ocean. All of the earth's oceans are warming. So far, most of this warming has occurred near their surface. For example, between 1955 and 2003, the top 300 m warmed an average of 0.171°C; while between 1955 and 1998, the top 3,000 m warmed an average of 0.037°C (Leviticus et al. 2005). These temperature increases appear insignificant compared to the rise in the atmosphere's temperature, until we remember the oceans' much larger heat capacity and slower mixing times, its inertia.

The warming of the oceans has contributed to changes in both their physical characteristics and their ecosystems. The increase in ocean volume resulting from both warmer sea water (i.e., its thermal expansion) and the addition of water from the melting of ice on land (i.e., not floating ice) has caused sea level to rise. Between 1901 and 2010, sea level rose 190 mm, which is the largest rise over a 100-year period for at least the last 2,000 years. That rise occurred at an average rate of 1.74 mm/year. In contrast, from 1971 to 2010, the average rate was 2.0 mm/year, while from 1993 to 2010, it was 3.2 mm/year (IPCC 2013a). The rate of sea level rise is speeding up.

The significance of these rates becomes clearer when we compare them to the rate of sea level rise that occurred during the last glacial-interglacial transition to the present, when the great ice sheets melted. Between about 13,800 yBP and 7,000 yBP, the average rate of sea level rise was around 11 mm/year. After 7,000 yBP, sea level rise slowed. In the recent past, it averaged a steady background rate of 1 mm/year (Miller et al. 2005a). The current average rate of sea level rise is more than double that background rate but not as fast as the rate during deglaciation. Along with the increase in the rate at which sea level is rising has come the increased

inundation and erosion of coastal land and more opportunities for sea water to enter freshwater aquifers (IPCC 2014a).

The warming of the oceans affects more than their physical properties. It also affects their composition. For example, a combination of ocean and atmosphere warming has meant that, between the 1950s and the 1990s, the salinity of the Arctic Ocean decreased (increased melting of ice provides more fresh water). At the same time, equatorial oceans are becoming saltier (increased evaporation) (Curry et al. 2003). A change in temperature also changes the maximum amount of a compound that can dissolve in it.

Pollution. The composition of the world's oceans is changing for another reason. We are changing the amounts of its normal constituents and adding novel ones. Some are carried to the oceans in the air, as discussed above; others are carried to the oceans by rivers.

Rivers polluted by runoff from the land and the discharge from our waste-water systems flow into coastal waters. As a result, the areas of the oceans offshore from all the major rivers of the world contain the many pollutants from activities as diverse as washing clothes and cars to industrial cleaning and growing food. On reaching the sea, some of these pollutants immediately breakdown, react quickly to form harmless products or enter a different reservoir. Those that don't are persistent. They are then spread by ocean currents around the globe, including into the Arctic and Antarctic Oceans (Li and Macdonald 2005).

For convenience, the impact of pollution on our oceans will be discussed by dividing the pollution into two types: dissolved and solid. The effect this pollution has on our oceans is illustrated by its impact on ocean organisms and ecosystems. These impacts affect large and small species over a wide range of concentrations and time spans.

Dissolved. The oceans are the earth's largest reservoir of water-soluble pollutants. The vastness of the oceans and the low concentrations of the pollutants in them make it easy for us to assume that they are harmless. For a number of reasons, this is not always the case. After all, the oceans circulate a larger amount of pollutants for a much longer time than the atmosphere, thus making them more available to the large amount of life the oceans host. There is, therefore, more opportunity for organisms to be affected by the normal and novel pollutants dissolved in the oceans.

Consider, first, that as a result of our $CO_2$ and methane emissions to the atmosphere, we have forced the carbon cycle to rebalance. Although the transfer of $CO_2$ from the atmosphere to the ocean has slowed climate warming, it has made the ocean's surface layer more acidic (a decrease in pH) (Feely et al. 2004). This is another illustration, like sulphur, that it is not possible to escape the impacts of our actions by shuffling the causes or consequences between atmosphere, ocean and land. There are only trade-offs.

How significant is this increase in the oceans' acidity? In the recent past, the pH of the oceans' surface waters lay between 8.0 and 8.3 (neutral is pH 7, highly acidic is 1). Today the oceans are more acidic by about −0.1 units. This change seems inconsequential, but it has the potential to affect the ability of some marine organisms to complete their lifeways.

Foraminifera are a small marine species that eats dissolved free amino acids, bacteria and unicellular algae. A 0.2 to 0.4 mm sized species of planktonic foraminifera called *Globigerina bulloides* lives in the near-surface waters of the Southern Ocean. It makes a shell out of calcium carbonate, a material that dissolves in acidic water.

When *Globigerina bulloides* die, their shells sink to become part of the sediments on the ocean floor. Cores extracted from the Southern Ocean's sediments record how the weight of a specific size range of *Globigerina bulloides* shells has changed over the last 50,000 years. The variations in its shell weights were compared to the variations in the atmospheric $CO_2$ content, recorded in an Antarctic ice core (figure 11). The comparison reveals that the shells reached their greatest average weight, 30.3 micrograms (μg), during the depths of the ice age, when the atmosphere's $CO_2$ content was at its lowest. In contrast, the average shell weight was significantly less, 24.4 μg, during the early part of the current interglacial when the atmosphere's $CO_2$ content was much higher at typical interglacial values. This correlation makes sense if one remembers that higher atmospheric $CO_2$ content means more acidic ocean water, which makes it more difficult for *Globigerina bulloides* to form and maintain its protective shell.

However, between 1997 and 2004, the average shell weight was only 17.5 μg, which is substantially lower than the 24.4 μg expected for our interglacial period. Their anomalously low weight correlates with the anomalously high $CO_2$ content of the atmosphere, which is the result of our $CO_2$ pollution, and the associated increase in the acidity of the oceans. *Globigerina*

*bulloides* illustrates that as the oceans become more acid, those marine organisms that use calcium carbonate shells, such as corals and other foraminifera species, experience greater difficulty forming and maintaining them (Doney 2006; Feely et al. 2004).

This has global-scale implications for marine ecosystems, which are illustrated by continuing the foraminifera example. There are many foraminifera species that make carbonate shells. They occupy the lower trophic levels of ecosystems in various parts of the oceans. They are a significant source of food for organisms such as snails and small fish. Those organisms are, in turn, food for the organisms at higher trophic levels. Thus the disruption of these foraminifera lifeways will affect organisms throughout their host ecosystem's trophic web. This type of relationship, and the existence of equivalent non-foraminifera species that make carbonate shells, ensures that ocean acidification is having a significant impact on ocean ecosystems globally.

There is another consideration. Foraminifera, like all marine carbonate-shell-forming species, use the $CO_2$ dissolved in sea water to make their shells. When those shells sink to the ocean floor, they transport carbonate into the deep sea ocean sediments, which is a carbon cycle reservoir (figure 7a, bottom left corner). This process removes dissolved $CO_2$ from sea water. But, as our $CO_2$ emissions force the oceans to become more acidic, the foraminifera shells become lighter, so they transport less dissolved $CO_2$ into the open ocean's sediments. On a global scale, because all of the species whose carbonate shells end up in the open ocean's sediment are removing less $CO_2$ from sea water, the rate of ocean acidification can increase, which further disrupts ocean ecosystems. This feedback loop can also affect the rate of transfer of $CO_2$ from atmosphere to ocean and thus atmospheric warming and land-based ecosystems (Moy et al. 2009).

Next to be considered are those novel compounds in our pollution that are found in the oceans. They are usually present at low concentrations. Novel pollutants affect marine organisms and ecosystems either at those low levels, as discussed later, or after being concentrated up to doses that can affect them, as discussed here.

Compounds become concentrated in their reservoirs because they prefer those hosts over others. For example, hexachlorocyclohexanes (HCHs) prefer cold water. Thus, HCHs transported from their source areas to the Arctic by air and ocean circulation tend to accumulate in the Arctic ocean's cold surface waters. Their concentration in this reservoir exceeds the levels found in the warmer waters off the HCHs' main source areas, India and Asia (Li and Macdonald 2005).

Similarly, a concentration process called bioaccumulation starts in the ocean environment when a novel marine pollutant consumed by an organism low in its host ecosystem's trophic web is preferentially stored in the organism's body. Think especially of those persistent pollutants that are preferentially soluble in fat. When this organism is eaten, the pollutant in its fat passes into the eater's gut. The amount not excreted is stored in the eater's fat. As this pollutant is passed up the trophic web from eaten to eater, its concentration in the fat of its successive hosts increases. This process of bioaccumulation up the trophic web can result in the higher trophic level species, typically the large carnivores such as predatory fish, polar bears, orcas and beluga whales, containing pollutant levels that may not kill them but can interfere with their body functions, such as reproduction.

Because compounds tend to show a preference for a particular type of reservoir, for example, fat rather than water, a number of novel compounds can reach elevated concentrations in the same reservoir. Thus the biological effects of these compounds on an organism are only partly determined by each pollutant's properties and their elevated concentrations in the reservoir. They are also determined by the interactions among the pollutants (e.g., creating new pollutants) and by the interactions among the impacts each pollutant has on the organism (e.g., one affecting reproduction, another the ability to smell). The impacts of pollutants on an organism are determined by the cumulative impacts on it over time from the total chemical load it is exposed to and the related interactions. That cumulative effect is reflected in changes to the overall state of its species and its host ecosystem.

The process of bioaccumulation and the resulting impacts are exemplified by the Pacific sockeye salmon that spawn in the gravels of specific rivers flowing into particular Alaskan lakes. The process starts far from land in the salmon feeding grounds located toward the centre of the northern Pacific Ocean. Despite its isolation, the water there contains traces of PCBs. Think brown cloud. (PCB's are long-lived, so although their use is restricted or banned, they are still wide spread in significant amounts.) The lower trophic level species that live in the area ingest and absorb these PCBs into their bodies. Some of these species are the prey on which the salmon will feed during their years of maturing in these distant waters. The PCBs the salmon eat are stored in their fat. As adults, they will carry this PCB load with them when they undertake their epic migration back to the Alaskan lake where they grew up.

Once they reach their lake, the salmon will find their birth stream, spawn and die. Some of the PCBs in their bodies will end up in the bodies of the salmon's terrestrial predators, such as eagles and bears. A portion of the PCBs will be stored in the lake's sediments. The rest will end up in the lake's water and be consumed by its ecosystem's lower trophic level organisms. Many of these organisms are the prey on which the next hatching of salmon will feed, until they are old enough to migrate back to the ocean.

Each year the returning adult salmon bring back a little more PCBs from the Pacific to their spawning-lake's ecosystem. Consequently, each year the amount of PCBs eaten by the young salmon increases. Currently, their PCB load is sufficiently high that it is likely now, or will soon be, affecting their body's functioning (Krummel et al. 2003).

Another aspect of bioaccumulation and its consequences is illustrated by orcas. This is the same species that is eating otters in the Aleutian archipelago off Alaska, but these orcas are from further south. They live in the Pacific coastal waters on either side of the political boundary between Washington state, US, and British Columbia, Canada.

This resident orca population eats a diet that is 95% salmon. Three quarters of these salmon are chinook. They, like sockeye salmon, concentrate PCBs in their fat. The chinook thus provide the orcas with both food and unwanted dietary PCBs. The PCBs accumulate in the orcas' bodies to levels that exceed the threshold above which the animals are expected to experience adverse health effects (Cullon et al. 2009). These possible impacts include the disruption of their endocrine system, in particular its control over reproduction (Hickie et al. 2007).

It is reasonable to expect that the presence and impacts of a high PCB concentration (load) would be restricted to the adult orcas. After all, only adults would have had their lifetime to bioaccumulate the PCBs from the salmon. This is not the case. High loads are evenly distributed among all the orcas in a population, including the young. This is the case because the fat-soluble PCBs are concentrated in orca milk. As young orcas breastfeed they receive a high starter dose of PCBs from their mothers. Thus, the mothers unknowingly help ensure that the health impacts of PCBs not only affect the most vulnerable orcas, but become a part of them for life.

Unfortunately, the orcas are likely less healthy than expected for the amounts of PCBs in their bodies. It is the cumulative effects from all of the pollutants they host that influence their health. This includes the interactions among them and among their impacts.

*Solid.* Not all of the pollution in the ocean is dissolved in sea water or ends up in the fat of marine organisms. Anyone who has walked along an ocean coastline will be aware of the solid garbage washed ashore from distant disposal points. The sources include waste-water systems discharging directly into the oceans or into rivers that then transport the waste to the sea. They include deliberate and accidental discharges from ships, as well as materials discarded on land that are blown or washed into the oceans (Moore 2008). Some 60% to 80% of this solid garbage is some type of plastic. It will be the focus here.

Sampling of the surface layer of the world's oceans shows that these plastic reminders of our globalized economy and daily living are ubiquitous (Day et al. 1990). A close inspection shows that those reminders are not restricted to the larger, more visible pieces of plastic. For example, surface samples from a part of the north Pacific subtropical gyre (sea water flowing in a circle several hundred kilometres across) over the period 1999–2010 revealed that plastic particles <5 mm in diameter (microplastic) were present at an extraordinary median concentration of 0.116 particles/m$^3$ (Goldstein et al. 2012). This high concentration is the reason that the gyre is also known as the Great Pacific Garbage Patch.

Are ocean species and ecosystems being affected in some way by this plastic pollution? The larger chunks can maim and strangle the larger marine animals, such as whales and dolphins (Moore et al. 2001). Lost nets continue to catch fish (ghost fishing) as they float around the ocean. Meanwhile, some 45% of all seabird species are known to treat some floating plastic objects as prey and eat them (Moore 2008). By consuming plastic, an organism can both reduce its food intake and block its gastrointestinal tract.

Plastic pollution has other, less visible, but potentially more serious environmental consequences. To appreciate these effects, we have to answer the question, where do all the large pieces of plastic garbage end up? The general answer is that plastic persists in the ocean for a long time because, like dry wood, it is not readily broken down into its elemental parts. Instead of decaying, the larger plastic bits are mostly reduced by abrasion and by the slow UV-stimulated chemical breakdown of its framework into ever-smaller plastic fragments.

Fragmentation continues until the particles are a few to tens of microns in size (1 micron [μm] is 1/1,000 of a millimetre). These smallest of plastic particles consist of fibre fragments from plastic rope and cloth, and plastic fragments small enough to approach the size of plastic's basic constituents. The smallest of these particles can be ingested by the smaller animals that are found at all but the lowest levels of a marine ecosystem's trophic web, for example, juvenile fish and filter feeders such as barnacles (Thompson et al. 2004).

Some plastic particles sink to the ocean floor because of their innate high density, for example PVC. Others float and accumulate on the ocean's surface. Some can marginally float, so they can sink with a change in conditions, for example, when they become encrusted with sea life

Our consideration of the impacts that plastic particles have on species and ecosystems starts with the portion floating in the oceans. Their physical presence alone affects ocean organisms and ecosystems. For example, a marine insect called a sea skater lives in the Great Pacific Garbage Patch. It lays its eggs on hard floating objects that are smaller than about 0.2 mm in size. Prior to 40 years ago, there were few objects on which a sea skater could lay its eggs. This helped to constrain the sea skater's population. However, since then, the number of plastic particles in the gyre has increased roughly one hundredfold. The presence of many more particles on which the skater can lay eggs has likely resulted in changes to the trophic web of the Garbage Patch's ecosystem. After all, more particles precondition the sea skater's population to increase. There would thus be an increase in the skaters' demand for food and in their predators' population (Goldstein et al. 2012).

Another portion of the floating plastic particles ends up in what may turn out to be an Arctic plastic reservoir. As the Arctic's sea ice forms each winter, both the smaller and the larger particles floating on the Arctic Ocean's surface become trapped and concentrated into the ice. Preliminary samples reveal that the concentration of plastic particles <5 mm in size in Arctic sea ice ranges from 38 to 234 particles/m$^3$. This is around 100 times their concentration in the Great Pacific Garbage Patch of 0.116 particles/m$^3$. These particles are released at these high concentrations when the sea ice melts (Obbard et al. 2014).

A third portion of the floating particles (mostly small but some large) eventually sink to the bottom of the oceans, regardless of depth, as photos of the North Atlantic sea floor indicate (Pham et al. 2014). Of greatest concern is the accumulation of the smaller particles in the bottom sediments of the nearshore or intertidal ecosystems because they host an abundance of bottom-dwelling marine organisms. Ocean-bottom particles can disrupt an ocean floor ecosystem by physically smothering it or by their physical impacts on the species who ingest them. The physical impacts on the smaller organisms who ingest plastic particles are largely unknown.

This line of thinking leads to a surprising realization. We know little about the chemical impacts that all of these plastic particles have on the functioning of the large and small organisms who are exposed to them or eat them (Thompson et al. 2004). An indication of what to expect is provided by our general knowledge about the chemical nature of plastics and plastic waste.

Fifty percent or more of a plastic particle is its polymer frame. It encapsulates the additives used to give the plastic its desirable physical and chemical properties. A wide range of binders, fillers, colourants and plasticizers, such as phthalates, PBDE and bisphenol A, are used for that purpose. Although some of the chemicals making up the frames and additives are known to have biological effects, the impacts of most are largely unknown (Thompson et al. 2009).

These additives are not the only chemicals of concern. Plastic particles act as a local reservoir for some of the novel pollutants found in the oceans, in particular the water-avoiding (hydrophobic) toxic pollutants, such as PCBs, DDT, nonylphenol and PAHs. These types of chemicals can become concentrated onto a plastic particle from the surrounding ocean by the processes of adsorption and absorption (Teuten et al. 2009; Rios et al. 2007). When these "contaminated" plastic particles come in contact with or are eaten by organisms, such as barnacles, salmon, birds and orcas, the chemicals they carry are available to enter the organism's body and disrupt its functioning. The oceans' plastic particles can act as both the carrier and delivery system for a biologically effective pollutant.

What little research has been conducted to date indicates that these chemicals do have a biological impact on marine organisms. For example, lugworms are a small sized but important member of estuarine ecosystems. They live in the estuary's sediment, which they eat. Experiments indicate that if the worms ingest contaminated PVC plastic particles from the sediment, then the plastic, as a solid, interferes with their digestion. In addition, both the plastic's additives and the pollutants it has sorbed can enter the worms' gut tissues. The cumulative impact of these additives and pollutants disrupts a suite of the lugworms' lifeway functions, such as feeding and immunity (Browne et al. 2014). Disrupting the functioning of enough lugworms likely affects its host ecosystem.

There are exponentially more smaller plastic particles than larger ones in the ocean. It thus seems likely that both plastic's constituent chemicals and sorbed pollutants are affecting a much wider size range of marine organisms and thus their host ecosystems than we currently imagine (Moore 2008). We also know that the amount of plastic in the ocean is increasing. So, whatever these solid and chemical impacts are, they are also likely increasing (Thompson et al. 2009). We can conclude that plastic waste is likely exerting a slow but persistent influence on marine organisms and ecosystems at all scales and intensities, without us noticing.

***Short and Long Term.*** The example of plastics illustrates that both dissolved and solid pollution is likely affecting marine organisms of all sizes. But it doesn't clearly illustrate that organisms can be affected by ocean pollution, whether solid or dissolved, over the short and long term, and at low and high concentrations. Also still to be discussed is how the species-specific effects spread through their host ecosystem to result in the overall effect of ocean pollution on marine ecosystems. An examination of the impacts from the black, smelly, sticky liquid that is crude oil will accomplish this.

Consider the 42 million litres of crude oil spilt by the Exxon Valdez oil tanker into Prince William Sound, Alaska, in 1989. It polluted hundreds of kilometres of the Alaskan marine nearshore and intertidal ecosystems. As with many forms of pollution, the most obvious short-term effects of this crude oil on marine organisms, such as drowning in black goo, are not necessarily the most significant long-term impacts. A study of the spill's longer-term effects discovered that species were being affected 14 years after the spill and they are likely still being affected today. The affected animals include salmon, sea otters, harlequin ducks and black oyster catchers. The consequences include premature death and reproductive impairment.

This disruption continues because oil spill cleanups remove only a small amount of the oil spilt and what remains behind contains many persistent chemicals. As a result, black, semi-solid-to-thick liquid residues from the Exxon Valdez's oil spill are still present in the sediments forming the physical foundations of the spill area's life-rich, intertidal marine ecosystem. During the daily tidal fluctuations, the flow of water over the oil-polluted sediments disperses the chemicals leaking from them.

This movement ensures that intertidal zone organisms, such as mussels growing on rocks above the sediment and fish eggs buried in nearby spawning gravels, are chronically exposed to the oil residue's chemicals. Even though the concentrations may be well below the accepted lethal levels for adults, the chemicals are still affecting these organisms. For example, even at concentrations in the 1 ppb range, the oil's PAHs are lethal to the developing eggs of pink salmon.

The individuals of the intertidal and nearshore ecosystems who were exposed to or ingested these chemicals are eaten by species higher in their ecosystem's trophic web, such as birds. Thus the impacts that the oil has on the lower trophic levels of an ecosystem cascade up its trophic web to affect the higher trophic level species (Peterson et al. 2003). Because the oil residues are widespread, a significant proportion of an ecosystem's species and a particular species' population can suffer these impacts. This too contributes to the overall result: the ecosystems' recovery from the spill has been delayed well beyond our expectations, and promises.

### The Overall State of Ocean Ecosystems. Ocean
ecosystems are difficult to access. Fortunately, a reasonable global-scale estimate of their general state can be made by considering how our activities are driving ocean ecosystems to change from the top down or bottom up of their trophic web, or both.

*Top-Down Changes.* When we affect the fish at the top of an ecosystem's trophic web, we affect their host ecosystem from the top down. This is illustrated by considering the impact our fishing activities are having on the fished species. But for the discussion to have meaning, those activities must be placed in context.

Our historical efforts to catch fish were mostly restricted to coastal ecosystems. For example, European marine fishing efforts only began to have a substantial effect on coastal fisheries and marine mammal populations, such as whales, after about AD 1000 (Barrett et al. 2004). From then on, Europeans steadily increased their efforts to catch fish in ever more distant coastal waters, such as the cod off Newfoundland.

The Industrial Revolution of the 1800s was accompanied by a substantial increase in fishing efforts, but it was still largely focused on coastal waters. Beginning in the late 1940s, a multinational, global-scale exponential increase in both coastal and deep (open) ocean fishing efforts began, powered by oil. This subsection discusses the impacts that this ongoing increase in our fishing efforts had, and is continues to have, on ocean fish stocks and their host ecosystems from the top down.

The discussion relies on reports that interpret global-, regional- and species-scale fish-catch records. The nature and quality of the records mean that they are difficult to meaningfully interpret. However, with sufficient care, researchers have been able to provide a valid, retrospective, global-scale view of changes and trends in the state of the fisheries hosting the oceans' fish stocks (Froese et al. 2012; Kleisner et al. 2013; Pauly et al. 2013; Watson and Pauly 2001).

Based on their findings, the total global ocean-fish catch is made up of industrial (more than half of the total), artisanal, subsistence, recreational and bycatch (unwanted, discarded) catches. A reconstruction of the total annual global fish catch from 1950 to 2010 indicates that it peaked at 130 million tonnes (Mt) in 1996 and then declined steadily to around 110 Mt in 2010. The decline was accounted for by a steady reduction in the industrial catch, despite an exponential increase in the industrial fishing effort (a measured quantity) (Pauly and Zeller 2016; Watson et al. 2012).

During this period, the status of the ocean fisheries providing the industrially caught fish changed dramatically. In the 1950s, the fisheries classed as undeveloped and developing collectively accounted for 95% of fisheries, while 5% were classed as fully exploited. By 2008, the undeveloped, developing or rebuilding fisheries collectively accounted for only 16% of fisheries, while those classed as fully exploited, overexploited or collapsed collectively accounted for 84%. The steady increase in the percentage of the fully exploited, overexploited and collapsed group of fisheries for the period between 1950 and 2008 indicates that the fish catch was, and is, being sustained by overfishing the fish species' stocks. This is a view supported by a recent analysis which concluded that many more fisheries belong in the collapsed class than expected (Froese et al. 2012; FAO 2010).

What does a collapsed fishery mean, and what does it tell us about the state of a fishery's host ecosystem? Ocean fisheries can be broadly divided into two types:

coastal regions (from coastline to the continental shelf margin) and open ocean. The fish species most sought after by humans in both types of fishery occupy the higher trophic levels of their host ecosystems. Examples of these preferred fish are cod and skates from the coastal region, and tuna and billfish from both the coastal and open ocean.

The post-1950 catch data for a particular preferred predatory fish species, reveals the trend in its catch size over time. A predatory fish fishery newly opened to industrial-scale fishing produces its maximum catch size near the beginning of exploitation. During the first 15 years of fishing, the catch size for that species rapidly declines. After 15 years, the catch size is as much as 80% smaller than the maximum catch. Despite the decline, fishing may not stop. It can continue even when the catch size has declined to less than 10% of the maximum. At that low catch level, the species' population (the stock) is considered to have collapsed: it can no longer easily recover. This sequence of events was and is being repeated for predatory fish in fishery after fishery. As a result, the current global biomass of the desirable large predatory (upper trophic level) fish is only around 10% of what it was prior to industrial-scale fishing (Myers and Worm 2003).

A popular misconception is that overfishing isn't a problem because, once fishing ceases, the fish population will rapidly recover. In reality, most fish populations take at least 15 years to recover from a collapsed state. For some species, the population decline continues long after fishing has stopped (Hutchings 2000; McEvoy 1986).

An illustration of the impacts that our (measured) intense industrial fishing effort is having on the preferred fish species is provided by sharks. There are some 350 species of shark, a number of which are top trophic level predators. The adults of shark species vary in size from a few centimetres to 12 m in length. Sharks are slow growing, slow to reproduce and produce few young (Manire and Gruber 1990). This means that they are particularly sensitive to population disturbances.

For example, after the 1960s, eight shark species in the northwest Atlantic were recorded as being commercially targeted. Between the 1980s and 2000, the catch size for seven of the eight had declined by more than 50% from the time when fishing for each species had started (Baum et al. 2003). But only a minority of shark species are caught by this type of targeted fishing. The majority are caught as bycatch. This has had a significant impact on sharks because they are most commonly found in the most heavily fished parts of oceans (<3 km deep, mostly coastal) (Priede et al. 2006). Together, the targeted and unwanted shark catch has significantly disturbed their populations. As a result, an estimated 33% of shark species are endangered (Baillie et al. 2004).

The substantial declines in the populations of the large predatory (upper trophic level) fish are likely driving important changes in their host ecosystems from the top down. This conclusion can be drawn because, firstly, there are fewer species in the higher levels of an ecosystem's trophic web, and their populations are small in size. Thus any significant disruption to their numbers and diversity is likely to be ecologically significant. Secondly, a drastic change to the highest trophic level of a trophic web is known to trigger a reorganization of their host ecosystem's web through a top-down trophic cascade. (Illustrated later by the cod-dominated ecosystem off Newfoundland's coast. Think too of the sea otter–sea urchin ecosystem off the northwest coast of North America.)

The top-down changes to marine ecosystems are also being driven by another aspect of our fishing efforts: there is no respite from it. As the catch of a preferred predatory fish declines, our increasing (measured) industrial fishing effort switches to another fishery. This includes exploiting desirable fish at lower trophic levels in the ecosystem hosting a collapsed predatory fish fishery and exploiting new predatory fish fisheries. Thus, on the one hand, by fishing down the trophic web of the ecosystem hosting an overexploited predatory fish, we increase our disruption of that ecosystem. On the other hand, by opening new fisheries, we push more predatory fish fisheries into the overexploited class, thus disrupting their host ecosystems (Pauly and Watson 2003; McEvoy 1986).

Our top-down impacts on ocean ecosystems from our fishing activities include more than overfishing the predatory fish. Consider a global compilation of the 1950 to 2003 data for all of the ocean species (vertebrates and invertebrates) being fished (i.e., extracted) within 64 large coastal marine fisheries. These fisheries provided 83% of the total global catch of marine fish plus invertebrates. The peak catch for the species in this data set occurred in 1994. Despite an increase in (measured) fishing effort, the total size (tonnes) of the coastal fish catch had declined 13% by 2003.

Unsurprisingly, in 2003 some 65% of the fish species recorded as being fished at some time between 1950 and 2003 were considered to be collapsed or recovering from collapse. Surprisingly, of the desirable species whose populations had collapsed, the percentage still being fished had steadily increased from the 0% reference in 1950 to 29% in 2003. In concert with these observations, extinction rates have been slowly increasing since the 1800s (Worm et al. 2006).

This state of affairs indicates that, on average, the functional state of the ecosystems supporting the 64 large coastal marine fisheries has declined. It is reasonable to assume that the ecosystems experiencing the largest biodiversity loss and the largest increase in extinctions would be those hosting the fisheries suffering the greatest impacts from our fishing effort. They are the fisheries reporting either the greatest declines (collapses) in the desirable fish species or the ongoing fishing of a collapsed fishery, or both.

Consider too that these top-down changes to marine ecosystems are likely being enhanced by the collateral damage that results from our fishing effort catching untargeted and unwanted organisms. This bycatch consists mostly of unsaleable and undersize fish, but it also includes a wide variety of mammals, reptiles and birds, for example, whales, turtles and albatross. The collateral damage to a fishery's host ecosystem from the bycatch occurs both when the organisms are removed from it and when they are thrown back into the sea, where many become a food bonanza. The scale and likely significance of the disruption caused by the bycatch is brought into focus by its size: roughly 10.1 Mt of bycatch was discarded during industrial fishing in 2010 that retained 99.9 Mt of fish from the total 110 Mt catch (Pauly and Zeller 2016).

In overview, our fishing activities are disrupting, on a global scale, the ecosystems that support (vertebrate and invertebrate) fisheries. The substantial decline in the populations of large predator fish numbers has likely initiated a top-down trophic cascade in their host ecosystems. Adding to this disruption are the more general impacts on their host ecosystems from the bycatch and, particularly in coastal fisheries, fishing down the trophic web. These changes ripple through each fishery's ecosystem, disrupting its functioning. It is easy to conclude that our activities have significantly disrupted the functioning of the ecosystems that host at least those 84% of marine fisheries classed in 2008 as exploited, overexploited and collapsed.

*Bottom-Up Changes.* By significantly affecting the base of the marine trophic webs (i.e., the ocean's plants) we are changing ocean ecosystems from the bottom up.

The most common ocean plants are the widespread, free-floating phytoplankton species. They account for about half of the earth's total (land plus ocean) net primary productivity (NPP). Because they dominate the lowest level of a large majority of marine ecosystems, they help to constrain the distribution of ecosystems, the number of species each hosts and their population sizes (Myers and Worm 2003).

Phytoplankton, like all plants, require sunlight to live. Sufficient sunlight is available at the oceans' surface but rapidly decreases to a negligible level by 300 m depth. Phytoplankton productivity is thus represented by their abundance in this surface layer. That abundance can be measured by sampling and satellite imaging.

A global assessment of phytoplankton concentrations in the world's oceans revealed that their productivity in 8 of 10 ocean regions has erratically declined since 1900. These trends are tentatively thought to be at least partly caused by global-warming-induced changes to those characteristics of the oceans' surface layer that affect photosynthesis. For example, the warmer the surface layer hosting the phytoplankton, the harder it is for the nutrient- and mineral-rich colder, deeper ocean water to mix upward into it, thus depriving the phytoplankton of those critical resources. In addition, the size of the fish catch has likely disrupted the nutrient cycles in shallow oceans. Whatever the causes, the decline in phytoplankton abundance is likely changing ocean ecosystems from the bottom up (Boyce et al. 2010).

The study also located localized anomalous phytoplankton increases in the coastal marine environment. The affected ecosystems are found worldwide in the estuaries and deltas of many of the world's major rivers and in the neighbouring offshore coastal marine areas. These ecosystems are normally some of the most biologically productive areas of the oceans. Why are these areas experiencing increased ocean plant abundance and what is its significance?

The answer is connected to the constraints imposed on photosynthesis by nutrient and mineral availability. Consider nutrients. As mentioned above, the amount of nutrients that ocean currents can supply to the surface water of the oceans is limited. However, unlike the surface layer of the open oceans, the surface layer of coastal zones receives additional nutrients from rivers.

Prior to the late 1800s, the dominant source of the nutrients, in particular plant usable nitrogen (available nitrogen) and phosphorous, transported by the rivers was geological formations and the leakage from the soil recycling processes. This limited the amount of nutrients the rivers could deliver to the coastal zone. Today a significant source of the river-supplied nutrients is our pollution. Consider its contribution of available nitrogen.

A large proportion of that nitrogen comes from the human-made nitrogen fertilizers farmers apply to their fields in order to intensify agricultural production. On average, about 50% of the fertilizer farmers add can be accommodated by soil processes, while about 25% ends up in the food we eat (Socolow 1999). Most of the rest, the excess nitrogen, is washed from the fields into the rivers. The remaining portion of our pollution-supplied available nitrogen comes from our industrial and domestic waste systems when their waste water is discharged into rivers.

The total amount of nutrients that rivers now carry into deltas, estuaries and the marine surface waters around the mouths of rivers is well above their pre-industrial levels (Diaz 2001). The nutrients have a number of effects on ecosystems. They allow *Pfiesteria*, the shape-shifting dinoflagellate that favours polluted estuaries, to proliferate. They also change the nutrient constraints on phytoplankton and algae growth in the adjacent coastal marine environment. The phytoplankton and algae can proliferate to cause the anomalous productivity increase in the nearshore ocean areas off river mouths, as reported in the global study mentioned above.

The bottom-up consequences for ecosystems are illustrated by the impacts that the Mississippi River's nutrients are having on the coastal marine ecosystems in the Gulf of Mexico, off the river's mouth. The initial impact is seen as a greenish algal bloom in the Gulf's surface water. Under normal circumstances, these blooms provide a boost to the productivity of an area's ecosystems, including those supporting fisheries. But if too much nutrient is added, then the blooms become excessive. Some blooms become so large and dense that they can be easily seen from the air (UNEP 2007). This level of algal growth changes the Gulf's coastal ecosystems from the bottom up.

When the algae in a bloom die, they fall through the water column toward the ocean floor. This rain of food is a feast for the bacteria that live deeper in the water column. Their population explodes. But in order to digest this food, the bacteria need oxygen, which is only available in limited amounts at that depth. Thus, if the bacteria's population boom is too large, they eventually deplete the available oxygen, creating a zone of very low oxygen or oxygen-depleted water. All creatures requiring oxygen, including fish and bottom-living

(benthic) organisms such as clams, are forced to leave the zone or die. The resulting dead or hypoxic zone is the typical end stage of this nutrient-driven process called ocean eutrophication (Diaz 2001).

The dead zone need not stay at depth. It can be brought up to the surface of the Gulf by winds. We experience a surface dead zone as a huge area devoid of life. "Huge," in the case of the Gulf dead zone, means a surface area of about 10,000 km$^2$ (100 x 100 km) in 1999 and 4,400 km$^2$ (66 x 66 km) in 2000 (Simpson 2001b). It is safe to say that the ocean ecosystems in the dead zone area have been severely disrupted, if not destroyed.

The formation of a eutrophied zone is subject to the usual delay between driver and response. It can take a number of years after the start of pollution-related nutrient additions to an area before the affected marine ecosystem is forced into the most severe degree of eutrophication's oxygen depletion. Thus, although the Mississippi River's additions of nutrient pollution to the Gulf started to increase after the mid-1940s, the size and severity of the resulting disruption of the Gulf's ecosystems only became obvious after the 1950s, when the dead zone phenomenon started to appear (Diaz 2001). Since then, the size of the Gulf's dead zone area has doubled each decade, indicating that excessive nutrient additions are at least continuing. This increase can be reversed, but only after an ongoing long-term reduction in our addition of nutrients, particularly nitrogen and phosphorous, to the Mississippi River.

Algal blooms, eutrophication and the resulting dead zones do occur naturally. However, our post-1940s production of fertilizer from our industrial-scale fixing of nitrogen and mining of phosphorous has ensured that the algal blooms and thus ocean eutrophication has become a far more common, larger and longer-lasting global phenomenon. Today, there are at least 400 eutrophication sites around the world, with different degrees of oxygen depletion. The global trend suggests that many more are nearing the severest dead zone stage.

The 400 sites, collectively, cover some 245,000 km$^2$ (500 x 500 km). The area affected seems small, but eutrophication's impacts on fish rank with those of overfishing and climate warming. This is the case because the affected ecosystems in estuaries and deltas are nurseries for many fish species, while the affected coastal marine ecosystems support some of the world's most productive fisheries (Diaz 2001; Diaz and Rosenberg 2008).

In overview, in the coastal marine environment, those ecosystems off the mouth of major rivers are being driven to change from the bottom of their trophic web up by nutrient pollution supplied by the rivers. This change is expressed by increased phytoplankton abundance and excessive algal blooms that result in dead zones. In contrast, ecosystems in the open ocean are being driven to change from the bottom up by an ongoing decline in its phytoplankton production. This seems to be the result of climate-warming-related changes to the characteristics of the ocean's surface zone, where the phytoplankton live, that are inhibiting the supply of nutrients to them. Overall, ocean ecosystems worldwide are being induced to undergo significant changes from the bottom up.

These bottom-up impacts are joined by other ecosystem activities that affect the lower trophic levels of marine ecosystems. They include direct habitat disruption, such as the discarded bycatch and, in the coastal zone, the destruction of the ocean floor by bottom trawling. Other related impacts include the filling in of saline wetlands and the impacts of pollution, such as plastic. There is enough evidence to conclude that marine ecosystems are being significantly affected from the bottom up by human activities.

In summary, climate warming, pollution and our direct actions, such as our fishing efforts, have resulted in the oceans and the ecosystems they host undergoing significant changes. The oceans' acidity, temperature and sea level are rising. Ocean organisms are being affected. At the same time, their host ecosystems are being driven to change both from the top of their trophic webs down and from the bottom up. The analysis of long-term records indicates that the integrity and functioning of many near-surface ocean ecosystems in both the open ocean and coastal-zone fisheries are likely disrupted to a significant degree.

On a global scale, oceans and their ecosystems have been significantly changed over at least the past 50 years. These changes are beyond that to be expected from normal variations, such as multi-year, pre-industrialization climate fluctuations. Trends suggest that these changes are ongoing and increasing.

The disruptions are also likely much greater than suggested here. A review of many ocean variables suggests that the main reason this state of affairs exists is the limited attention we give to the wide range of our ocean-disrupting activities and the interactions among them. We fail to consider our cumulative impact on the oceans (Rogers and Laffoley 2011).

## The Land-Based Environment

The land-based environment includes both land and fresh water, and the terrestrial and freshwater ecosystems they host. The land-based environment provides nutrients to the oceans and is part of the climate system. It is an integral part of the earth's biosphere. This environment too is showing the effects of our influences on it. Its state is discussed in four subsections dealing with climate warming, freshwater availability, pollution and the overall state of terrestrial ecosystems.

<u>Climate Warming.</u> The world's climates are warming, which drives species to adapt. Consider that species are being prompted to begin their spring activities, such as breeding and flowering, ever earlier, and delay their autumn activities, such as their fall migration and dormancy, until later (IPCC 2007b). For example, between the 1950s and early in the first decade of 2000, the initiation of spring activities for 694 globally distributed species started on average 5 days/decade earlier (with a range of 24 days/decade earlier and 6.3 days/decade later) (Root et al. 2003; Walther et al. 2002).

Today's species are adapting to the current climate warming in the same basic manner as their ancestors adapted to the climate warming that ended the last ice age: through migration and lifeway changes, as discussed in chapter 8. In the case of migration, as warming moves a climate zone poleward and to higher elevations, the species suited to that zone migrate in order to follow it. This is most easily seen today where a species is nearest its biological survival limits, that is, at the edges of its range (Davis and Shaw 2001). For example, European alpine plants are moving up the sides of their mountains at rates of <1 to 4 m (elevation)/decade (Grabherr et al. 1994); butterflies in Europe and North America are moving northwards at up to 7.4 km/year (Walther et al. 2002); and fish in the North Sea are moving northward up to 16.8 km/year (Perry et al. 2005).

A species also adapts by adjusting its lifeway. Consider that, as an ecosystem's species migrate to accommodate a change in climate, so does their host ecosystem. But each species migrates at a different pace. Thus, as the host ecosystem moves, its mix of species, and thus the niches it provides, changes. This places pressure on the individuals of the member species to adjust their lifeways accordingly (Davis and Shaw 2001).

Examples of climate-related pressure on a species to change its lifeway are provided by the loss of synchronization between it and its prey. For example, the earlier spring in Europe has disrupted the synchrony between the timing of the bursting of an oak tree's leaf buds and the hatching of the winter moth whose young (caterpillars) feed on the leaves. A similar timing mismatch has developed between the great tit, a bird, laying its eggs and the availability of the food it feeds to its nestlings (Walther et al. 2002; Visser and Holleman 2001). If the young of the moth and the great tit are to survive, then these species must adjust their lifeway.

Climate warming is most marked in the Arctic, so the overall dynamics and outcome from the efforts of an ecosystem's species to adapt to the current warming are most easily seen there. For example, around 1850, shortly after the end of the Little Ice Age, the (short lifespan) algal and invertebrate community in the majority of 55 circumpolar Arctic lakes started to change in response to climate warming. As a lake warmed, its community's primary productivity increased, its species mix changed and its organization became more complex. Because the lakes' species are small in size, it is unlikely that any humans living in the area at that time would have recognized these changes or what they meant (Smol et al. 2005).

Ongoing warming prompts these circumpolar lake ecosystems to continue changing. At the same time, the previously discussed northward movement of the shrub line in the Alaskan Arctic continues (Sturm et al. 2001). The southern edge of the Arctic's permafrost belt continues to melt, creating more opportunities for more southern species and ecosystems to migrate northward.

The equivalent process is occurring in the Arctic's marine environment. For example, the presence of sea ice in the northern reaches of the Bering Sea, and an associated ocean temperature gradient to warmer seas in the south, defines an ice-front-demarcated boundary between two ecosystems. To the north of the east-west-trending sea ice front lies an Arctic marine ecosystem. It is characterized by birds and large ocean mammals (e.g., whales and walrus) that are connected to bottom-dwelling organisms (e.g., clams). To the south lies an open water, Subarctic fish-dominated (e.g., pollock) marine ecosystem.

As a result of the current climate warming in the Arctic, the temperature of the Bering Sea has increased and the sea-ice front has moved northward. Subarctic fish are appearing where whales used to live. The Arctic marine ecosystem is moving northwards and

being replaced by a Subarctic marine ecosystem (Grebmeier et al. 2006). Ongoing climate warming ensures a continuing loss of Arctic sea ice and thus the ongoing northward shift of the Bering Sea's Arctic marine ecosystem.

There is a reason for mentioning these Arctic land, permafrost, lake and marine events here: their synchronous occurrence complement one another. Simply put, the reduction in the area of sea ice and the shrub line's movement northwards reduces the Arctic's albedo, while the melting permafrost increases the atmosphere's greenhouse gas content. These changes increase the proportion of sunlight trapped in the lower atmosphere, thus amplifying climate warming. Increased warming feeds back to continue forcing the members of the Arctic's marine and land-based ecosystems to adapt by migrating northward and changing their lifeway (Stokstad 2004). This changes their host ecosystem.

This complementary nature of climate-induced environmental changes and their resulting amplification of climate warming is not restricted to the Arctic and its land-based ecosystems. Just like the warming at the end of the last ice age, the process is global in scale and affects all land-based ecosystems and their species. But this time, the changes are primarily being driven by our greenhouse gas emissions, not by the Milankovitch sunlight cycle.

**Freshwater Availability.** Fresh water is critical for the survival of land-based organisms. Of all the water on the earth, only about 0.77% (about 10 million km$^3$) is fresh and liquid. It is found in the natural reservoirs of underground aquifers, soil pores, lakes, swamps and rivers.

This water would soon be used up if it were not for the rest of the water cycle. Simply put, evaporation from the sea results in precipitation on land. The rain falling on land can evaporate, be stored in the soil, or be taken up by plants and mostly transpired back into the atmosphere. The water that runs off the land fills its reservoirs. The water collected in lakes, swamps and rivers sustains their aqueous ecosystems. Once its reservoirs are full, the excess water becomes rivers. Their water hosts freshwater ecosystems, before it is discharged into the sea.

Humans have manipulated this flow of water since at least the dawn of agriculture. However, in the last 100 years, our appropriation and redistribution of water to satisfy our wants has increased to the point that we are significantly disrupting the details of the water cycle. Consider that, as mentioned earlier, we appropriate some 23% of rainfall on land (renewable water). We do so, directly, by using the rain to grow rain-watered crops and, indirectly, by withdrawing water from its reservoirs, particularly aquifers, lakes and rivers, for domestic and industrial purposes (Postel et al. 1996). Consider too that, since 1900, our freshwater withdrawals from rivers and lakes have risen fivefold (Gleick 2001).

Our appropriation, redistribution and use of fresh water has two broad impacts on indigenous land-based ecosystems. Firstly, it means that less water is available to those ecosystems that previously used the water, thus disrupting or destroying them. For example, our appropriation and redistribution of the water in the Colorado River, US, has reduced the flow at its mouth to a trickle. This has, in essence, dehydrated the Colorado River's delta and its aquatic ecosystems.

Secondly, although our use of water does not destroy it, we commonly return it to the environment, and thus to the organisms that could use it, in an unusable or polluted state. It is in this state because it has, for example, been heated by our efforts to cool fossil fuel fired electricity generating plants or had waste from farms, homes, industrial plants and city streets added to it.

The amount and distribution of available fresh water is also being changed by climate warming. The more obvious changes are to rainfall and the melting of glaciers (Zhang et al. 2007). The more subtle effects include changing the evaporation rate from surface water and soils, and the possible reduction in the ability of plants to generate water vapour (transpire water) as they respond to increased atmospheric $CO_2$ (Gedney et al. 2006).

Land-based organisms are forced to adapt to these human-related changes to fresh water's availability. In the process, they and their communities change. Over the last 100 years, this has affected the nature and distribution of all of the earth's land-based ecosystems.

## Polluted Land and Water.
The effect of our pollution's normal constituents on land-based ecosystems was discussed by focusing on climate warming and acid rain. This subsection discusses the effect of our pollution's novel constituents on land-based ecosystems by focusing on our product testing and polluted freshwater.

The discussion begins with the reminder that some of our pollution's human-made constituents are called novel because the earth's environment and its organisms, including our bodies, have had no previous experience dealing with them. For example, we manufacture a large number of organic chemicals that contain the halogens chlorine and bromine. These halogenated chemicals include PCBs, chlorofluorocarbons (CFCs), most plastics and most pesticides. All but a handful are novel. Some halogenated chemicals are produced in large quantities. For example, of the top five organic compounds manufactured in the United States during 1997, two contained chlorine (ethylene dichloride and vinyl chloride) (EIA 2000). We could therefore expect that, before we use and then release them into the environment (i.e., as pollution), they would have been subjected to long-term testing over a wide range of environmental conditions and concentrations to ensure they have no significant short- or long-term environmental impacts, including on ourselves.

These observations apply to all of the many novel chemicals in our pollution. How we actually address their presence is discussed in three subsections. Testing and Evaluation discusses how we decide which of our novel chemicals can be safely used and released to become pollutants, and under what conditions. Usage and Release discusses the amount of the compounds we use and how they are released into the environment to become pollutants. Impacts illustrates the kinds of effects our novel chemicals are having on land-based species and their host ecosystems. These discussions provide another benefit: they are relevant to our thinking about the normal constituents in our pollution and their environmental impacts.

*Testing and Evaluation.* The European Union (EU) is one of the largest producers of chemicals in the world. It will be used as the best-case example of the recent state of chemical testing. In order for the EU to evaluate a human-made chemical today, its testing must generate a specific set of data. However, that requirement only came into effect in 1981. Chemicals registered before that date are recorded on what the EU calls the pre-1981 list. In 1998, some 100,195 chemicals were on the pre-1981 list. This represents the majority of the chemicals in use within the EU today.

Note that the pre-1981 list deliberately excludes the by-products created when a chemical is manufactured or breaks down. This exclusion is an important oversight because these by-products can be harmful. Consider that the combustion of products such as coal and garbage generate a wide variety of compounds with a wide range of toxicities. One example is the carcinogenic dioxins generated when chlorine-containing plastics are incompletely burned (Commoner et al. 2000). Another is the up to 100 varieties of PAHs generated during the incomplete combustion of oil, coal, gas, garbage, tobacco and when charbroiling meat (Rios et al. 2007). Think too of 4-NP.

The current degree to which the EU has evaluated the impacts of manufactured chemicals on biological systems can be judged by how fully the top 2,465 chemicals (by the amounts produced) have been tested using the EU's current protocols. By 1999, only 3% of them had been tested sufficiently to generate the EU's full data set, 14% had the basic set, while 21% had no data (Allanou et al. 1999).

The limited testing of the high-volume chemicals is mirrored by the limited, retroactive testing of the lower-volume chemicals on the EUs pre-1981 list. Very few have been evaluated to current EU standards. One is left to wonder how many of the low-volume, new and pre-1981 chemicals may be causing significant, adverse biological effects. For example, how many of these incompletely tested or untested chemicals act as endocrine system disruptors or neurotoxins at very small doses (CNMAR 1992)?

To their credit, the EU has plans to fully test all chemicals with a sales volume in Europe that exceeds one tonne per year (called the REACH program). However, this commitment has created its own environmental problem. If the EU were to use the current testing methods to provide the required data set, then the number of test animals that would be killed lies somewhere between 2.5 million (the original EU estimate), 9 million (the revised EU estimate) and 54 million (based on an independent assessment) (Morin and Ricard 2009; Hartung and Rovida 2009).

But even the EU's proposed updated testing is unlikely to fully identify a chemical's possible significant impacts on species, including humans, or ecosystems. Like the current tests, the updated tests would not fully take into account interactions between chemicals, bioaccumulation, chemical cycles in the environment, breakdown products, manufacturing by-products and long-term effects. The previously mentioned example of significant impacts on salmon eggs from exposure to just 1 ppb of PAHs reminds us of the need for long-term (multi-year), low-dose, ecosystem-relevant tests. A more personal example of the need for time would be the decades-long delay between someone inhaling the chemical soup in cigarette smoke and the soup's generation of the approximately 23,000 DNA mutations that eventually lead to the appearance in the smoker of small-cell lung cancer (Pleasance et al. 2010).

The testing of a chemical for undesirable effects is followed by an evaluation of the results. This process is not as simple as it sounds. There is inevitably uncertainty in the results from short-term tests. There is also the difficulty of deciding how to apply the results to organisms on which a chemical is usually not tested and what constraints to place on its use or disposal. This uncertainty allows cultural factors and "arguments" from parties with vested interests in an approval to strongly influence the regulator's decisions. The presence of this culturally supported influence on the EU's testing and licensing process is illustrated by the fact that neither less harmful alternatives nor the need for the product are considered during the evaluation. The net result is that, for all but those chemicals with clear-cut impacts, there is an overall bias toward increasing the number of available chemicals, some of which may have significant, long-term human or environmental effects (Lyons et al. 2000).

*Usage and Release.* Perhaps this concern about testing is unwarranted. Maybe the amount of novel pollutants we release into the environment is trivial. We gain an appreciation for the current chemical load on our environment by considering the trends in the numbers and amounts of chemicals we manufacture, use and discard.

In 1930, the global, annual production of chemicals was around 1 Mt. After the 1940s, chemical production increased rapidly. By 2003, about 400 Mt/year were being produced globally (Jansen 2003). Interestingly, the United States had earlier reported that the weight of the top 50 chemicals they alone produced in 1997 was already 330 Mt (EIA 2000).

The rise in the production and use of individual chemicals follows this general trend. For example, the production of phthalate esters has increased from very low levels in the 1940s to 3.5 Mt/year by the early 2000s (Bornehag et al. 2004). The production of endosulfan, of which about 0.338 Mt has been used since the mid-1950s, is still rising. About 10 Mt of HCHs were used between 1948 and 1997. For at least some chemical products, the amounts we use are substantial.

At some point in the lifecycle of a human-made compound, we transition from treating it as a good of value to treating it as valueless waste: it becomes pollution. Similarly, at some point we treat the by-products generated during a compound's manufacture, use and discarding as waste: they too become pollution. This transition includes both a social or internal-reality aspect and a physical or external-reality aspect.

The social aspect occurs when we decide (feel, believe, assume) that the compound no longer needs to be under our control. The term *social release* emphasizes this aspect. The physical aspect occurs when we physically release the compound into the environment. That act physically and symbolically finalizes our decision that we are no longer responsible for the compound. The distinction between social and physical release is illustrated this way: social release allows us to feel comfortable about the compound's possible impacts on the environment (including us) that might occur after we physically release it from our custody. Note that there is no inherent requirement for us to actually know what those impacts will or won't be, or to take them into account if we do know. This applies equally to individuals making this decision either as a member of the public or as a member of a corporation or government.

Our preferred methods for removing waste from our surroundings and releasing it uses water: think of sewage and sewage treatment plants. For example, in the early 2000s an estimated global total of 2 million tons (2 million tonnes, if long tons) of waste per day were disposed of into water (United Nations 2003). Between 1996 and 2005, we generated polluted water at an average global rate of 1,363 $km^3$/year (Hoekstra et al. 2011). In the late 1980s, over 90% of US measured hazardous waste was classed as waste water (Ausubel 1996).

We personally experience waste removal by water each time we use some of our domestic water appropriation to flush the toilet, wash clothes or pull the plug after washing dirty dishes. The water that disappears into the pipes in an urban home has many components. It contains the waste from our digestion of food. This includes the residues of medications, such as birth control chemicals and antibiotics, that are excreted in our urine and faeces because they are never completely used by our bodies. Also entering the pipes is a wide variety of novel chemicals: the constituents of cleaning products such as soap powders, washing liquids and detergents; toothpaste; and beauty products. Then there is the range of materials we simply discard into the toilet, such as expired medications, unwanted liquid and "flushable" products.

This domestic process of removing waste using water is repeated in customized form in factories and offices. The resulting wide range of discarded compounds are mixed together in the waste water system. Although this waste water soup may go through some purification process, such as a sewage treatment plant, generally only a limited amount of a few compounds are removed, and more by-products are created (Soares et al. 2008).

Our (hopefully partly treated) waste water soup is released into the environment at the end of a sewage disposal pipe. It usually enters a water body, such as a river, where it is joined by runoff from our gardens, farms, parking lots and streets. Included in this runoff is a variety of compounds that include spilled oil, fertilizer, farm animal waste, waste dump leachate, pesticides, herbicides, plastic fragments of various types and even chrome from automobile finishes. This moving, water-borne pollution soup is free to disrupt freshwater aquatic and terrestrial ecosystems at any stage on its journey to the sea.

In general, as discussed above, the spread of our pollution's novel compounds into an environment, its species and ecosystems is controlled by the pollution's physical and chemical properties; the nature of its reservoirs, such as fat and snow; and the earth's air and ocean currents. For example, approximately 4.5 Mt of DDT were produced between 1947 and 2000. Of the 2.6 Mt used on land in agriculture, 1.03 Mt entered the atmosphere. A portion of the rest entered organisms, the rivers and oceans (Li and Macdonald 2005).

*Impacts.* This subsection illustrates the possible effects pollution can have on land-based ecosystems by discussing the impacts that the above freshwater pollution soup can have on freshwater ecosystems. Unlike the section on ocean pollution, which emphasized the impacts resulting from the concentration and accumulation of a pollutant, this section emphasizes the impacts at low doses.

Consider that freshwater aquatic organisms, especially those living in a vision-restricted environment, inform themselves about the state of their world and one another by sensing and emitting specific chemicals. For example, salmon smell their way up rivers to their spawning grounds, and killifish generate an odour trail that their companions follow to form a shoal. Any pollutant that could affect these abilities would seriously disrupt the organism's efforts to complete its lifeway.

Surfactants are used in cleaners to help keep dirt away from the item you are cleaning. A commonly used surfactant is nonylphenol ethoxylate. One of the more common of its breakdown products is 4-nonylphenol (4-NP) (Soares et al. 2008). Experiments indicate that even at concentrations as low as 0.5 micrograms per litre (μg/litre), 4-NP interferes with the killifish's ability to create the odour trail they use to maintain a shoal. The disruption of this ability is a significant loss because shoaling is a part of its survival skillset, for example, to avoid predators (Ward et al. 2007). 4-NP is commonly present at 1 to 2 μg/litre, well above the level that disrupts the killifish's ability to shoal, in the water discharged from sewage plants, even if the water is treated (Soares et al. 2008). If cadmium were present in the pollution, then even if an odour trail was laid down, the killifish may not be able to detect it because even at very low doses cadmium disrupts the ability of fish to smell (Ward et al. 2007).

Consider too, that many of a complex organism's internal body functions, such as a rat's response to stress and a primate's reaction to competition, are controlled by function-specific signalling agents that operate in their bodies at very low concentrations. The production and distribution of these chemicals, called hormones, is managed by the individual's endocrine system. Some of the pollutants that are absorbed into the body are hormones, others act as (mimic) hormones or block a hormone's action. These pollutants are thus capable of disrupting an organism's body functions. And they can do so at concentrations of the chemical that are well below levels considered to be toxic (Vandenberg et al. 2011). Of interest here is the endocrine system's use of the hormone estrogen to help control the development of reproductive organs and their functioning.

First to be discussed is the extensive, worldwide use of the herbicide atrazine. It is one of the most commonly applied herbicides in the United States. There it is most often used to control weeds in corn and sorghum fields, but it is also used in residential areas and on golf courses (Hayes et al. 2002a). The United States considers atrazine to be non-toxic to organisms in a single dose, except at very high levels (Barr et al. 2007). This is very reassuring because atrazine is also the most common pollutant found in the United States' surface and ground water. For example, in the United States' corn belt where it is commonly used, atrazine reaches levels of 2 to 5 μg/litre in surface water (IARC 1999).

However, laboratory experiments on African clawed frogs have shown that atrazine at concentrations perhaps as low as 0.1 ppb (0.1 μg/litre) but certainly as low as 2.5 ppb (2.5 μg/litre) disrupt their sexual and reproductive development. These disruptions are expressed by the feminization of male frogs, which is indicated by the presence of oocytes (eggs) in their testes, changes in the male larynx (used for male mating calls) and coincidental chemical castration. These laboratory results have been confirmed in the field by the positive correlation between the presence of feminized wild leopard frogs, the amount of atrazine present in their habitat and the level of atrazine use in the immediate region (Hayes et al. 2002a; Hayes et al. 2002b; Hayes 2004; Hayes et al. 2010).

The full significance of atrazine's impact on leopard frogs becomes clear with the realization that the basics of the endocrine system's control of reproduction is common to many animals. Atrazine's ability to disrupt the sexual and reproductive development of frogs at low doses by acting as or mimicking a hormone likely extends to a wider variety of vertebrates, including fish and humans. Unsurprisingly, atrazine is not the only pollutant to feminize organisms at very low concentrations. Nor is the reproductive system the only system affected by endocrine disrupting pollutants (Vandenberg et al. 2011).

To continue the example of estrogen and estrogenic pollutants, consider sewage discharge. It can contain 17a-ethynylestradiol (EE2), an estrogen sourced from human birth control pills through urine. The discharge from sewage treatment plants also contains numerous other compounds that mimic estrogens to varying degrees. Examples are some plastic additives, such as bisphenol A; some pesticides, including atrazine; and some cleaning-related chemicals, such as 4-NP in some of its isomeric forms. At very low concentrations, all of these estrogen and estrogenic compounds contribute, on their own and collectively, to the feminizing of male organisms, for example, promoting the formation of eggs in the testes of male fish and frogs (Blazer et al. 2007).

These pollutant-related disruptions of an organism's senses and its bodily functions affects its ability to fulfill its lifeway. This, in turn, can change the species' host community. The process is illustrated by an experiment conducted in a Canadian test lake using EE2. It was deliberately added to a lake until its concentration reached levels similar to those found in the discharge from municipal sewage treatment plants (i.e., <5 ng/litre). For comparison, this is 1,000 times lower that the concentration of atrazine in the waters of the corn belt.

The males of the fathead minnows who inhabited the lake were feminized to such a degree that their ability to reproduce was impaired. The minnow population eventually collapsed (Kidd et al. 2007). Minnows are an important food source for the top predator fish, the lake trout. As a result of a trophic cascade-like effect, the loss of the minnows also affected the lake ecosystem's lower trophic levels, for example the rotifers (Kidd et al. 2014).

These examples illustrate that some pollutants can, on their own, impact freshwater organisms and ecosystems at extremely low levels. They can also combine to create the same outcome or enhance it. The examples also illustrate that the responsible pollutants can be sourced at harmful levels from treatment plants designed to reduce pollution. Testing for these types of ecosystem impacts, especially at such low levels, is usually not part of the chemical (or waste plant) permitting process.

In overview, there is sufficient evidence to conclude that the impacts on the functioning of freshwater organisms and their host ecosystems from pollution can be significant. These impacts are likely more widespread than we currently imagine. After all, despite the potential significance of pollution's cumulative impacts on individuals, species and communities, many of these possible outcomes are largely under-anticipated and under-recognized by our current product evaluations and cultural guidance.

Perhaps this is unsurprising considering the large variety and amounts of the chemicals we are producing and the constraints on our relationships with them. Consider too that it is easy to argue, for example, that the outfall from a sewage treatment plant will be diluted by the river it flows into. This neglects the processes that can concentrate them, such as bioaccumulation, evaporation or trapping in reservoirs, as discussed for ocean pollution. Think PCB.

Although the focus here has been on novel pollutants, especially at low concentrations, in freshwater aquatic communities, the evidence indicates that these conclusions can be generalized to terrestrial and marine ecosystems.

## The Overall State of Land-Based Ecosystems.
An ecosystem is formed and functions by the interactions among the individuals of its many species and between them and their environment. A measure of the overall impact that our actions are having on an ecosystem is the change in the number and rate of its species going extinct, or threatened with extinction. Data about extinct and threatened species is most complete for terrestrial species, so these measures are used here to gain an overview appreciation of our impacts on land-based ecosystems.

*Species Extinction.* We gain a feel for the significance of our current impacts on ecosystems by comparing today's average extinction rate with an average rate for an earlier geological time period (the background rate). The word *feel* is used deliberately because there are significant limitations to using this method (Barnosky et al. 2011). They include the difficulty of precisely determining present extinction rates (Pimm and Askins 1995); the constraints on comparing current extinction rates, based on a 400-year record, with a past rate based on a fossil record millions of years long (Regan et al. 2001); and the limitations to applying average measures of extinction to the extinction of a particular species.

The most reliable data about recent extinctions concerns mammals, most of whom are terrestrial. Based on this information, one method estimated that their current average extinction rate lies between 17 and 377 times (best estimate of 36 to 78 times) greater than the average background extinction rate for mammals (Regan et al. 2001). A different method estimates that the maximum mammal extinction rate over the last 1,000 years lies between 24 and 693 extinctions/million species years, with the fastest rate occurring nearest to the present. This translates into an increase over the background rate of between less than 13 and 385 times.

For comparison, the maximum mammal extinction rate during the megafaunal die-off at the end of the last ice age was less than 9 extinctions/million species years, which translates into 5 times the background. Thus, even the lowest of the estimates for current mammal extinctions (13 or 17 times background) is greater than the maximum of 5 times the background during the megafaunal die-off (Barnosky et al. 2011). Whatever the actual number, the current extinction rate for mammals is likely to be unusually high.

Support for this conclusion is found in the known causes of current extinctions for a variety of vertebrates (birds, reptiles, etc.). The three most common current causes are human related: the ecosystem hosting a species is either physically destroyed, disrupted through fragmentation or disrupted through the introduction of foreign species (Pimm et al. 1995). Examples of destructive activities are large-area clear-cut logging and land clearing for agriculture (Ramankutty and Foley 1999). Examples of activities that lead to fragmentation are patchy but extensive clear-cut logging and road networks.

The contribution of exotic species to extinction is illustrated by Lake Victoria, central Africa. Since the early 1900s, it has experienced the "successful" introduction of two exotic fish (Nile perch, Nile tilapia) and one exotic plant (water hyacinth). In combination with eutrophication and overfishing, these introductions have been largely responsible for the extinction of some 300 of Lake Victoria's original fish species and the significant disruption of the ecosystem's functioning (Schwartz et al. 2006; Kaufman 1992).

The Lake Victoria example reminds us that extinctions need not result from a single cause. They more likely result from the cumulative impact from a number of ecosystem-disrupting factors. Thus the secondary factors of over-utilization, such as hunting, as well as pollution should be added to the three current primary extinction-causing factors listed above (Thomas et al. 2004).

Our actions are also more likely to cause extinctions if they occur on isolated islands and in restricted, sensitive, species-rich areas called biodiversity hotspots (Pimm et al. 1995). Knowing this helps us personalize our contribution to extinctions. Consider our impacts on bird species living on those Pacific islands (including Easter Island, Hawaii and a part of New Zealand) that humans colonized between about 2000 BC and AD 1000. The archaeological and paleontological records of extinctions on these islands indicate that, after humans arrived, there was a decline in the size of bird populations and an increase in their extinctions. Overall, a maximum of 2,000 bird species were driven to extinction (an average of 66 species/100 years). If such an event occurred today, it would represent a roughly 20% drop in bird species. (A more recent analysis suggests that roughly 1,300 species, or a 13% drop is more likely (Duncan et al. 2013).

The likely change in the extinction rates on Pacific islands due to our presence is illustrated by the dramatic increase in the vertebrate extinction rates on the Galapagos Islands after we settled there. During the 4,000 to 8,000 years before the arrival of humans, an estimated 0 to 3 populations became extinct (maximum rate of 0.0375 species/100 years). In contrast, after humans arrived on the islands in the 1530s, the following 460 years saw a total of 21 to 24 populations become extinct (4.6 to 5.2 species/100 years). The recent yearly vertebrate extinction rate on the Galapagos Islands is thus roughly 2 orders of magnitude (100 times) greater than the islands' pre-human rate (Steadman 1995).

When thinking about extinctions, it is useful to remember that the inexorable decline in a species'

population size leading up to its extinction can take some time, up to many hundreds of years in the case of some Pacific island extinctions. From this perspective, today's extinction numbers and rates represent the final outcome of past, not current, activities on that species' future. Today's declining populations of mammals, reptiles and birds are reminders of the impacts that our past behaviour had and is having on them. To prevent their extinction in the future, we have to reduce our impacts on those species today: in particular, the degree to which we engage in the destruction of a species' host community, as well as its disruption through fragmentation and the introduction of exotic species. We truly are the guardians of our children's inheritance.

***Threatened Species.*** With the above in mind, a more complete appreciation for the present health of land-based species and their host ecosystems can be gained by looking at those species whose population size is near their viability limit (i.e., classed as vulnerable, endangered and critically endangered). These are the species threatened with extinction. When you read these numbers, imagine that this is the interest rate that you are paying on your credit card debt.

A 2004 assessment of 3% of the world's 1.9 million described species concluded that some 15,589 species are threatened with extinction. When the better-described classes of species are considered, then 32% of amphibians (the most threatened species), 23% of mammals, 12% of birds and 18% of chimaeras (fish with cartilaginous skeletons) are threatened. When looked at by genus, 42% of turtles and tortoises, about 33% of sharks, 15% of conifers and 52% of cycads (related to conifers) are threatened (Baillie et al. 2004).

A reassessment of vertebrates in 2010 reported an increase in the number of threatened species to 41% of amphibians, 25% of mammals and 13% birds. These reported increases are supported by the observation that the population size of all monitored vertebrate species, except reptiles, have been in decline since at least 1980 (Baillie et al. 2010). Over time, more of the animals to which we most easily relate are becoming threatened with extinction.

Because only about 1.9 million of the earth's estimated 10 million species have been described, it could be argued that there is a large pool of undiscovered species from which any extinctions could be easily replaced. However, allometric relationships suggest that most of the undiscovered and undescribed species will likely be very small organisms, such as bacteria, rather than the larger organisms listed above. In addition, the undiscovered larger species are likely to be naturally rare and thus more easily driven to a threatened status. This contention is illustrated by the estimated number of undiscovered flowering plants.

There are 350,000 to 400,000 known flowering plants (multiple names for one species excluded) with an estimated additional 10% to 20% (between 35,000 and 80,000 species) still to be discovered. The majority of the undiscovered flowering plants are thought to exist within biodiversity hotspots as small populations of highly specialized plants found in isolated places (endemic species). They will likely be easily driven to a threatened status. At present, 22% of the known flowering plant species are considered to be threatened. If the estimated total (known plus still-to-be-discovered) number of flowering plant species and their estimated status is used, then between 27% and 33% of all flowering plants are thought to be currently threatened.

The general conclusions about flowering plants can be applied to other species. It is highly unlikely that the loss of larger species due to either recent extinctions or possible future extinction of those currently threatened will be compensated for by new discoveries of equivalent-sized species with large populations (Jowit 2010; Joppa et al. 2010). It is also more likely that the currently estimated number of threatened species within a class of organisms, such as amphibians, is an underestimate.

The question remains, what exactly is causing species to be threatened with extinction? The primary causes are the same that lead to extinctions: host ecosystem destruction and its disruption through fragmentation and the introduction of exotic species. For example, an assessment of all threatened species in the United States concluded that the destruction and disruption of their host community had contributed to the current status of 85% of them, while exotic introductions had affected 49% (Wilcove et al. 1998). As these percentages illustrate, a threatened status often arises from a combination of factors.

The numbers of threatened species, the increasing trend and the connection to our activities is personalized here by a few details about threatened birds, amphibians and primates. The state of the 9,914 known bird species was reviewed in a 2004 world survey. This study revealed that 1,211 species (12.4%) were threatened with extinction. These threatened species included 95% of all albatrosses, 60% of cranes, 29% of parrots, 26% of pheasants and 23% of pigeons. The most significant

single cause was habitat destruction (in this case including disruption) which affected 1,045 (86%) of threatened species. The two next most common causes were overexploitation and exotics. Each of these have, on their own, affected 30% of threatened bird species. About 40% of birds were affected by a combination of two of the top three threats (BirdLife International 2004).

The remaining contributions to bird declines are more specific, such as the impact of the pet trade. For example, 46 of 145 parrot species in the neotropics were reported in 1998 to be endangered. Trapping and nest poaching of the young as well as habitat loss were listed as the two most significant causes of their declines (Wright et al. 2001).

The 2004 bird survey also reported that, on average, the status of those birds previously listed as threatened was worsening, and more species were drifting nearer to the threatened status. For example, the three vulture species found in India and Pakistan have become endangered in less than 10 years. Their decline is the result of using Diclofenac as a general anti-inflammatory medicine for cattle. Its use was originally thought to be environmentally benign. In reality, as little as one-tenth of the normal dose of the drug in a dead cow can kill a vulture that feeds on the carcass. Only a few carcasses with this dose would have been sufficient to devastate the three vulture populations (BirdLife International 2005; Green et al. 2006).

On a global scale, the birds' story is being repeated by frogs and primates. There are 5,743 known amphibian species. A world survey of these amphibians published in 2004 found that since the previous survey in 1980, the populations of 435 species (25%) had declined. The number of threatened species had increased from 1,772 (31%) in 1980 to 1,856 (32.5%) in 2004. The primary causes for the rise in the threat level for amphibians were host community destruction and disruption (41.5%), over-utilization (11.5%) and unexplained (47%). The nearer a species is to extinction, the more likely the reason for its population decline was listed as unexplained. This lack of an explanation exists despite the fact that amphibian population declines were first noted in the 1970s and have since been shown not to be the result of random population fluctuations (Stuart et al. 2004).

Possible and partly studied causes for the unexplained portion of the amphibians' rise in threat level include the following. An amphibian skin disease called chytridiomycosis (chytrid) is caused by the fungus *Batrachochytrium dendrobatidis* to which most amphibians have no resistance. Chytrid was, and still is, being rapidly spread around the world by humans introducing infected exotic frogs into uninfected areas (James et al. 2009). For example, the fungus likely spread from South Africa in the 1930s on infected clawed frogs that were being used for human pregnancy tests. The fungus likely arrived in Britain in 1999 when humans introduced an infected, exotic North American bullfrog, which is resistant to the fungus, into a trout rearing pond (Cunningham et al. 2005).

There are other factors, on their own or in combination, that may adding to chytrid's significant impacts. For example, in Costa Rica about 120 species of amphibians and reptiles have become extinct since 1980. The main cause is thought to be a climate-warming-induced reduction in the forest floor leaf litter required by many Costa Rican frogs to complete their lifeways (Whitfield et al. 2007). In nearby Panama, frog extinctions are related to an outbreak of chytrid, whose spread may be facilitated by climate warming (Lips et al. 2006).

In the United States, there has been a sharp rise in frog deformities since 1995. This increase appears to correlate with frog population reductions. The common theme linking the many possible causes of these deformities and the population reductions is our alteration of frog habitat. For example, the pollution of artificial wetlands with fertilizer and manure, which is likely the cause for the spread of the frog-deforming parasitic worm *Ribeiroia ondatrae* (Blaustein and Johnson 2003). Similarly, our pollution of amphibian habitat with the herbicide atrazine is resulting in frog feminization and their chemical castration (Hayes et al. 2002a; Hayes et al. 2002b; Hayes 2004).

The 2008 International Union for Conservation of Nature (IUCN) assessment of our nearest relatives produces a more certain evaluation of the 634 primate species. They found that 303 species (47.8%) were threatened with extinction. The primary threats were listed as hunting, habitat destruction and habitat disruption by fragmentation (Mittermeier et al. 2009).

It is easy to take the view that the threatened status of a species is inconsequential, at least to humans. However, at least some of the consequences to humans are direct, easily seen and significant. For example, the loss of the vulture on the Indian subcontinent means that the Parsee culture is losing its preferred way of disposing of its dead. At the same time, the greater Indian community is losing the preferred method to dispose of animal carcasses. One likely outcome is that feral dogs will fulfill the function provided by the vultures. They carry rabies, so there

could be associated human health consequences (BirdLife International 2005).

In overview, the ongoing increase in the global number and rate of land-based species being classed as threatened and extinct largely reflects our impacts on their host ecosystems, particularly their destruction, their disruption and the introduction of exotics. Also important are our direct impacts on the species themselves, such as hunting them or extracting them from their host ecosystem for the pet trade. Our pollution plays a role, but its clandestine contribution to extinctions in land ecosystems is currently not fully accounted for, as is also the case for marine ecosystems.

In summary, climate warming is affecting the earth's land-based environment. It is simultaneously affecting the earth's land-based (terrestrial and freshwater) species and their host ecosystems, especially those in high northern latitudes. Our appropriation of fresh water is leaving less water for land-based ecosystems, especially freshwater ecosystems. Pollution is affecting the functioning of land-based species and their host ecosystems in fundamental ways, even at low concentrations. The increase in the number of land-based species that are threatened with extinction and the increase in their rates of extinction reflect the overall increase in our impacts on land-based ecosystems, such as their destruction and disruption. The cumulative effects of all of our impacts on the land-based environment and its ecosystems has changed and is continuing to change it in significant ways.

## Our Contributions

The current state of the biosphere was outlined above by describing the state of the earth's atmosphere, ocean and land-based environments. Although it is clear that the state of those environments is, collectively, the biosphere's current state, it is difficult for us to relate to that state and our contribution to it, especially at a personal level. That difficulty is dealt with in this section by providing an overview of the biosphere's state from the perspective of our contributions to it.

This section starts by dividing the descriptions of our impacts on the three environments given above into our three basic global-scale contributions to the state of the biosphere. Attention will be paid to the relationships between our contributions. Their collective contribution to the biosphere's state can then be discussed in generic terms.

The discussion will be illustrated by the impact our contributions have on the current state of the biosphere's species and ecosystems because we can more easily relate to them. After all, their functioning and changing has previously been discussed in chapters 7 and 8, so we are already familiar with the rules of thumb, principles and themes indicating how an ecosystem's members, and thus their host ecosystem, would respond to our activities. The relationship between ecosystems and the climate system has been previously discussed, so we can make the link between them and the biosphere's functioning. By focusing on ecosystems, we can more easily appreciate the nature of our contributions and the dynamics of how they combine to result in the biosphere's current state.

## Our Basic Contributions.

Our impacts on the three environments that led to their current state, as discussed above, can be consolidated into three basic, linked and overlapping contributions: our pollution, our direct actions and climate warming. They, collectively, help to describe our contribution to the state of the biosphere as a whole.

### Pollution.

Our pollution consists of a wide variety of solid, liquid and gaseous compounds. Some are novel; others are excess amounts of the compounds normally found in the environment. After we release these novel and normal compounds into the environment, they are dispersed around the globe by wind, water (rivers and ocean currents) and organisms. The chemical and physical properties of a pollutant determines its preferred method of dispersal, its preferred host reservoir, and the types of its physical and biological impacts. This combines with the amount of a pollutant and its rate of release to determine the severity of its impact. The collective, cumulative impact of all our pollutants on the biosphere is our pollution's contribution to its state.

Our contribution arises because of our relationship with pollution and the environment: our awareness of both and/or our attention to them. We can experience difficulties acknowledging our pollution's contribution to the biosphere's state because, unless blatantly obvious, polluted land, water and air may not seem to us to be polluted, and we can miss the resulting, often inconspicuous, impacts. For example, both the polluted water and the feminized fish or frogs living in it can seem to us, on casual inspection, to be normal. So can the oceans, despite an increase in their acidity and the amount of plastics, and the resulting impacts on marine species. Think orcas and salmon. To appreciate the full impacts of pollution requires us to pay close attention to our environment.

We face other difficulties in achieving the needed awareness. Consider that just the separation in time and space between a pollutant's release and the response of an organism or ecosystem to it means that we often lack a simple, direct, immediate reminder of the link between the two. We must work to acknowledge the connection between our pollution (cause) and its impacts (effects) on the organism/ecosystem (Carson 1962).

Similarly, it is relatively easy for us to imagine that a pollutant can impact an organism at a high enough concentration. We may even accept that, given enough time, natural processes can concentrate a pollutant to a level that results in chronic or toxic effects. But we struggle to reconcile that reality with the observation that a pollutant at concentrations well below a toxic or chronic concentration can have a significant impact on an organism. These impacts are possible because, at very low concentrations, a pollutant may, for example, disrupt an organism's endocrine system. Consider too, if the pollution in our greater environment changes the chemistry in an organism's cells, then it can influence the epigene of the DNA in the cell and thus how its genes are expressed. No wonder we tend to under-appreciate the range of behavioural and physical impacts that our pollution can have on an organism and the pathways along which this can take place. And when we do appreciate the impacts, we struggle to conceptualize what a safe dose is, let alone set one.

Laboratory studies help us overcome these difficulties and set a safe dose by providing information about the impacts that a particular pollutant can have on a particular species. However, as discussed, this information is incomplete, and the resulting standards and their application rarely reflect a polluted ecosystem in the real world. That ecosystem can contain many novel and normal pollutants that may each affect the ecosystem's species, and may each do so differently. Because the pollutants break down and can react with one another, there can be more pollutants to affect individuals than expected. Because individuals interact with one another and their environment, there are more pathways for pollutants to affect them than expected. Think eat and being eaten.

Thus, in the real world, the individuals of a polluted ecosystem's many species are being simultaneously affected, directly and indirectly, by many pollutants, in different ways, to different degrees, over both the short and long term. The cumulative effect from the total pollutant load on an individual changes them and contributes to changing their species and their host ecosystem. It is difficult for us to fully appreciate this reality, for which our formal tests and evaluations offer only limited help. Try imagining the possible simultaneous impacts that PCBs, PAHs and estrogenic as well as smell-affecting chemicals have on a salmon's lifeway. Add to your cumulative picture how these effects on salmon change their host ecosystem, including those species that eat the salmon, such as orcas.

The personal difficulties we experience relating to the impacts of our pollution on species and ecosystems are the result of more than the complexity of the system and our individual human foibles. They also arise because our culture's world view exerts a strong influence on our awareness and evaluation of its pollution's impacts. This is illustrated for Western culture by the earlier discussion about the EU's chemical testing and evaluation process. In particular, consider Western culture's influence on Westerners decisions and justifications for our social and physical release of compounds to become pollution. For example, does it favour us releasing a compound as waste even when we are not really sure of its possible direct impacts, let alone its possible contributions to our cumulative environmental effects? Does it favour us ignoring or downplaying what we do know? At the same time, does its guidance helps us to believe that we are, in fact, being good proactive environmental managers?

Overall, humans tend to think that the impact of our pollutants on an ecosystem are usually the result of a quick-acting toxic dose of a single pollutant in isolation. We tend to think of those impacts as a regrettable but acceptable cost for the perceived benefits, such as convenience and economic growth, that we receive from the actions generating the pollution. We are encouraged to think that any impacts can be mitigated and the status quo restored.

As a result the impacts of our pollution on individuals, species and their host ecosystems can be significant and occur over a wide range of time frames, intensities and mechanisms. Many of those impacts are more obvious and occur faster than the impacts from climate warming. Most appear more slowly than the fastest of our direct actions.

*Direct Action.* Our decisions can result in actions that directly affect ecosystems. They include those actions that physically alter an ecosystem, such as land clearing, which is one of our fastest-acting direct impacts on

them. The results are easy to see, so we can easily be aware of them. In contrast, we tend to be unaware of our direct actions whose impacts are less visible, more subtle or slower to appear, such as our harvesting of species and the introduction of exotic species. The global-scale impacts from the harvesting of species has been illustrated by the high percentage of collapsed ocean fisheries. The global-scale impact from the introduction of exotics is discussed here.

An exotic species is any organism, regardless of size, introduced into a community that has not previously adapted to its presence. Today, human-introduced exotics are found globally. They neatly illustrate many features of our direct, ecosystem-affecting actions. They also provide an unambiguous symbol of human attitudes and thinking behind those actions.

A species must pass through at least one of three stages to be called an exotic. First, the species is *introduced* into an indigenous ecosystem. If it survives and reproduces, then it is *established.* If it manages to *spread* without our assistance, then the exotic is invasive.

The significance of this progression is illustrated by a study of known (i.e., obvious) vertebrate species that were introduced, in both directions, between Europe and North America over the period 1400 to 2000. They are divided into fish, bird and mammal groups. A feel for the success of vertebrate introductions is provided by the following percentages. Between 40% and 80% of the species in a group became established. Of the species in each of these groups that became established, some 60% to 80% spread. When taken together, some 20% of all introduced fish, birds and mammals eventually spread: they became invasive.

A feel for the time between a vertebrate's introduction and becoming invasive is provided by averages for the groups. Fish and mammals took, on average, about 10 years to become firmly established, while birds took about 50 years. Once established, the average time it took fish and birds to become invasive was negligible, while mammals took 50 years. Simplistically, fish most quickly become invasive, birds less so and mammals most slowly (Jeschke and Strayer 2005).

Although it is easier for introduced plants to establish themselves than introduced vertebrates, it seems that a lower proportion of introduced plants eventually spread. For example, by the mid-1990s, some 2,000 exotic woody plants had been introduced into various parts of the world. Of these, at least 135 (6.7%) have become invasive (Bright 1998).

An appreciation for the global scale of exotic introductions can be gained by noting the proportion of a country's plants that are established exotics. Australia, for example, has 15,000 to 20,000 native vascular plant species and an additional 1,500 to 2,000 exotic plants (about 10% of all its plants are exotic). Canada, on the other hand, has about 9,000 indigenous and about 2,800 exotics (24% are exotic).

Islands are likely to have a higher proportion of exotics than continents. For example, New Zealand has about 1,800 indigenous and about 1,600 exotic plants (47% are exotics). Bermuda has 165 indigenous and 303 exotic plants (64% are exotics).

It could be argued that exotics have been naturally introduced into indigenous ecosystems throughout geological time. However, it is not exaggerating to say that in the last 100 years, at least, the rate of human-facilitated introductions has far exceeded natural rates and has resulted in an equivalent increase in their impact. For example, the background (pre-human) rate of insect introductions into Hawaii has been estimated at about 1 every 50,000 to 100,000 years. In the early 1970s, the human-facilitated rate was about 1 every 19 to 26 days (0.05 to 0.07 years). The 1970s introduction rate for Hawaii is between 0.75 to 2 million times faster than the pre-human rate (Vitousek et al. 1996).

There is also a tendency to think that an indigenous ecosystem will only be affected by an exotic species if it becomes invasive. This is not the case. As soon as an exotic is introduced, both it and its new host community start adapting to one another's presence. Consider that, in trying to establish itself, an exotic will, at least, interact with those indigenous species occupying the niche favoured by the exotic. Consider too that there is a limited amount of space available in an ecosystem, and it can only support a limited number of species. Thus, success at becoming established invariably means that the indigenous host ecosystem has been disrupted to some degree (Chapman and Reiss 1992).

It is reasonable to expect that if a species is labelled *invasive*, then it will be severely disrupting indigenous ecosystems. However, not all species labelled *invasive* by Westerners are exotics. The term can refer to any widely dispersed species that we personally or as a society don't favour. Exotics can also be lumped in with noxious weeds, which includes undesirable indigenous species.

From an ecosystem's perspective, what matters is that the species is exotic. Overall, exotic species disrupt their new host ecosystem to a degree that varies from minimal to severe and can contribute to the extinction of some of its indigenous species (Wilcove et al. 1998). For example, in New Zealand, exotics are the number one threat to native biodiversity (Clout 2001). Think too of the exotic-prompted disruption of Lake Victoria's functioning and the extinction of many of its fish species.

The following example highlights our role and reasoning during an exotic introduction, and the potential scale and severity of the consequences from its subsequent establishment and spread. The tropical marine blue-green algae *Caulerpa taxifolia* is used by marine aquarium managers around the world as a decorative plant. It is favoured for this purpose because it is fast growing and secretes a highly toxic substance that discourages other organisms from eating it. The selective breeding of *C. taxifolia* produced a variety with another useful trait: extraordinarily rapid growth even in cold waters. This variant (novel species) became the favourite for many aquarium keepers.

Shortly before 1984, the cleaning process at the Oceanographic Aquarium at Monaco released this variety of *C. taxifolia* into the temperate ecosystems of the Mediterranean Sea. In this new environment, it could survive, grow rapidly and remain unpalatable. This variety quickly became invasive. It can smother most Mediterranean ecosystems found between the surface and the greatest depth at which marine plants can grow. These ecosystems include the *Posidonia* (seagrass) meadows, which provide important fish habitat.

By 2004, this variety of *C. taxifolia* had established itself in 300 km$^2$ (17 x 17 km) of the Mediterranean basin. It is currently being spread around the Mediterranean by ocean currents and human activities, such as the aquarium trade and ballast dumping, and by hitching a ride on recreational boats. It has also been introduced into temperate marine ecosystems all around the globe, where it is in the process of smothering those ecosystems (Kluser 2004; Yip 2001).

The *C. taxifolia* example illustrates how our decisions and activities play a significant role in the recent explosion of introductions. We are both deliberately and inadvertently helping organisms to overcome their natural barriers to movement. This is being accomplished through the increase in our modes, speeds and frequency of transporting goods, and ourselves, all over the world; our disposal of waste; and, surprisingly, by an increase in human migration (Bright 1998). Nostalgia prompts immigrants to want species that remind them of their ancestral home (Jeschke and Strayer 2005).

The last point reminds us that it is not just our human foibles that facilitate the introduction and spread of exotics. Our culturally influenced values and beliefs play a significant role. Consider that, at one level, we are largely indifferent to the value of our local ecosystems and to the impact of exotics on them. At another level, we import, distribute and encourage exotics to grow because our personal efforts are culturally supported or encouraged. Consider that Westerners' desire for the exotic is usually more powerful than our concern for the consequences of exotic introductions. When taken together, these culturally supported beliefs and values ensure that our decisions favour the introduction, establishment and dispersion of exotics, and the disruption of indigenous ecosystems.

There is another aspect to exotic introductions that highlights the central role we play in the process. Assume that we would like to predict which species will become invasive so that we can stop them being introduced. However, our ability to make that prediction is fundamentally constrained by the previously discussed complexity-related limits to predicting change in any ecosystem. This difficulty is made worse by the exotic nature of introduced species. After all, once the species to be introduced has been removed from the constraints of its original host community, such as it predators and competitors, its future behaviour becomes more unpredictable (Mitchell and Power 2003).

This reality severely limits the usefulness of the general knowledge acquired from past introductions when trying to predict the outcome of future introductions. For example, knowing that some species had displayed weedy characteristics after introduction but before they become invasive has limited value in predicting which species will become invasive if they are introduced (Kennedy et al. 2002; Tilman 1999).

It is easy to conclude that the most certain way to reduce the disruptions caused by exotics is to avoid introducing them into an ecosystem. To be successful, our response to exotics has to be proactive rather than reactive. This changes the focus of our attention away from the exotic as the cause of the ecosystem disruptions and toward the difficulty we face accepting the responsibilities that go along with making an introduction. The central issue is our beliefs, actions and values that result in their introduction, establishment and spread. This conclusion applies to our other direct actions.

*Climate Warming.* Our pollution has changed the composition of the atmosphere. Among the compounds in that pollution are $CO_2$, methane, nitrous oxides, sulphur and particulates, which change the atmosphere's ability to trap heat. At the same time, our direct actions, such as deforestation, have changed the earth's albedo. By making these changes we have altered the earth's climate system in ways that affect the rest of the biosphere.

The earth's atmosphere is warming. Its climate zones are moving toward the poles. At the same time, a climate zone's average temperature and precipitation, and its weather, are changing. The oceans are becoming warmer and more acidic. When the members of a species experience these changes, they and their descendants respond by a mix of migrating and adjusting their lifeway. The collective response of a species' members changes their host ecosystem, as discussed in chapter 8 and above.

In this way, the changes we make to the climate system indirectly affect individuals, species and their ecosystem hosts in all environments. But those changes affect the climate system. The resulting feedback amplifies our changes to the climate system. Our impact on the climate system results in our most significant indirect contribution to the current state of the biosphere: climate warming.

From our perspective, both climate warming and the resulting changes to ecosystems occur slowly. Currently, these changes are certainly less obvious to us than the consequences of our direct actions, except perhaps in the Arctic. Thus we tend to be personally unaware of and pay less attention to the significance of climate warming and climate-warming-driven ecosystem changes.

In overview, it is easier to consider the current state of the biosphere when our impacts on its environments are consolidated into three contributions: pollution, direct action and our most significant indirect contribution, climate warming. By discussing their general effect on species and ecosystems, we more easily appreciate how significant our contribution to the biosphere's current state is. We should not forget that behind these three contributions lies our thinking, values and beliefs. Our personal and cultural relationship to our contributions and ecosystems play a critical role in our effects on them and thus the biosphere.

<u>Our Cumulative Effect.</u> Our overall contribution to the biosphere's state is more than the simple sum of the above three (direct and indirect) contributions. It includes

the results of their feedback-influenced interactions and their accumulation over time. Our overall contribution is our complex, cumulative effect on the biosphere.

This subsection discusses our contributions' cumulative effect on the biosphere's ecosystems. This will help us to more fully appreciate their current state, and thus the current state of the biosphere. It will also help us to evaluate and discuss, in later chapters, both their and the biosphere's possible future states.

Our individual day-to-day activities are, or become, our contribution (direct actions, pollution and climate warming) to an ecosystem's present and future state. *Are* and *become* are both used because our current activities affect ecosystems immediately and over time: an ecosystem usually arrives in its current state through a process rather than a single event. That process is often a long-term affair in which both the direct and indirect impacts of our current activities spread through the ecosystem at different speeds to affect its functioning in a diversity of ways. The cumulative effect of our contributions result in us progressively changing the ecosystem's state.

An ecosystem's current state lies along a continuum of possible states from functional to non-existent. That continuum is divided into three states: *functional*, *disrupted* and *destroyed*. These terms were used casually above but will now be more precisely defined.

The current state of an ecosystem is the present condition of its structure and functioning. That condition is evaluated by comparing it to a reference state in which both the ecosystem's structure and its ability to respond to a driver of change through self-transformation is intact: it is (fully) functional. In this state, the ecosystem is resilient, robust and adaptable. If an ecosystem is to be called functional, it can, at most, be weakly disrupted. An ecosystem that remains functional over the long term is viable.

A *disrupted ecosystem* is one whose historical functional state has been significantly affected, usually by external drivers, but only to a limited degree. It is still capable of recovering relatively easily to the essence of its recent functional state.

In contrast, a *destroyed ecosystem* is one whose state has been affected, usually by an external influence, to the degree that its ability to recover to its original functional state has been largely lost. A destroyed ecosystem may recover, but it is likely to follow a different path into the future than if it had only been disrupted. In the extreme, a destroyed ecosystem is one that has been obliterated.

The boundaries between the three different states is difficult to define. They should be treated as fuzzy, gradational or overlapping. In addition, an ecosystem's state is dynamic and changes over time, so a time frame for some form of recovery from a disturbance is implicit in the terms. Time, the possibility of change, and the fuzziness of the boundaries will be acknowledged later by using the generic term *transformed* to describe an ecosystem's generic, significantly disturbed state.

The classification of an ecosystem's state does not rely on knowing the factors causing it. However, as discussed in this chapter's introduction, human-related drivers of change play a significant global-scale role in the current state of the earth's ecosystems. The discussion of the classifications will therefore be illustrated using examples of our impacts on ecosystems.

Note too that the reference we intuitively use to gauge changes in an ecosystem is our oldest personal memory of the ecosystem, which is not necessarily its functional state prior to its disruption by humans. An example is a contemporary European living in the Argolid, Greece, remembering from their childhood the heath and heather covered hillsides with scattered clumps of trees that was, before human's took up farming there, an open mixed forest, as discussed below. It is easy for us to underestimate the changes an ecosystem has experienced at our hands.

*Disrupted.* The progressive disruption of an ecosystem that results from our cumulative effect on it is easily illustrated by a coral reef ecosystem because of its reliance on an easily understood self-sustaining feedback. In that feedback, the coral polyps build their skeletons, which collectively become the ecosystem's physical structure: the coral reef. The physical features of the reef are the foundation for the habitat and niches required by the reef ecosystem's other members to prosper. They, in turn, maintain the conditions needed for the polyps to flourish and continue building their skeletons, which maintains the reef. The cumulative effects of our impacts on the coral reef are progressively disrupting this self-sustaining feedback: they are affecting the ecosystem's foundation and functioning.

For example, the overharvesting of reef sharks appears to initiate a top-down, reef-affecting trophic cascade. Simplistically, our removal of sharks contributes to a decline in the populations of algae-eating herbivorous fish. As a result, more of the reef area becomes covered by algae and less by coral polyps, thus limiting the polyps' reef-building activities. The ecosystem's self-supporting feedback is disrupted.

Our activities also disrupt the coral reef ecosystem's functioning from the bottom up. For example, periods of excessive ocean warming result in bleaching events in which a large proportion of a coral reef's polyps die (Ruppert et al. 2013; Walther et al. 2002). At the same time, ocean acidification is making it harder for corals to grow and easier for them to dissolve.

Our other ongoing activities that are contributing to the disruption of a coral reef's functioning include the removal of small reef-fish species to stock home aquariums (Simpson 2001a). We also overfish the larger fish species and other reef organisms for both food and resources (Newton et al. 2005). In addition, solid pollution (think plastic) and dissolved pollution (think sewage) are reaching some coral reefs in significant amounts, particularly those reefs near larger human communities. Finally, invasive exotic species, such as lion fish, have been introduced onto reefs.

The degree to which the functioning of coral reefs has been disrupted by our activities is illustrated in a study of 19 Caribbean coral reefs. For a coral reef to be considered healthy, the proportion of its surface that is covered by live corals should exceed 30%. Below about 10%, the coral polyps can no longer produce enough of their limestone skeletons to maintain the reef's structure. In essence, the corals can't stop the reef from eroding away. The loss of this supporting architecture places the reef ecosystem's survival in jeopardy. The 19 Caribbean reefs displayed a wide variation in coral cover with an average of 10%. Thus some of the 19 reefs are healthy, but many are severely disrupted and face destruction or are being destroyed (Perry et al. 2013). Similarly, our global-scale cumulative impacts on the world's coral reefs are largely responsible for the majority being described as between 30% and 80% degraded (Pandolfi et al. 2003).

Up to a point, even a significantly disrupted reef can recover. For example, coral reefs disrupted by a single cyclone event can take between one and a few decades to recover. For comparison, if humans disrupt a terrestrial ecosystem on a scale of around 30 km$^2$ (5.5 x 5.5 km), then its recovery will take roughly 50 years. If the disruption is at a scale of 100 km$^2$ (10 x 10 km), recovery will take roughly 100 years (Dobson et al. 1997).

By definition, a disrupted ecosystem has the potential to recover to near its original state, regardless of the degree to which its functioning has been disrupted. But, the more the ecosystem is disrupted, the less likely it will recover. Its recovery also depends on whether it is given both the conditions and enough time to recover. For example, if disruptions are repeated too frequently or are ongoing (chronic), then the ecosystem may never recover, or only do so slowly. Alternatively, it may turn into an ecosystem related to, but not the same as, the original. It could be argued that if this new ecosystem's functioning is sufficiently different from the original ecosystem, then the original should be labelled as *destroyed* rather than *disrupted*. The following example suggests that when the disruption is significant and future recovery is possible but uncertain then the term *transformed ecosystem* is a useful interim descriptor.

The ocean-shelf ecosystem off the coast of Newfoundland was a bottom up, four-trophic-level cod-dominated ecosystem. It supported a cod fishery for Europeans fishers for about 500 years. In 1992, as a result of post-1950s overfishing, the cod fishery collapsed and a moratorium was put in place to allow the stocks to recover. But the ecosystem had been significantly changed. By removing too many cod, in particular breeding adults, a trophic cascade was triggered. During it, the numbers of the fish that preyed on young cod increased to the point that their feeding prevented the recovery of the cod population. In the resulting prawn- (shrimp) and crab-dominated ecosystem, the cod had been replaced as the top predators by hake and seals (Bundy 2005; Frank et al. 2005).

The cod-dominated ecosystem had been severely disrupted, but had it been destroyed? It depends on the perspective you take (what you think is important in the ecosystem), which geographic part of the fishery you look at and how long you are prepared to wait for a promised recovery. In the meantime, our disruption of the ecosystem can be said to have at least transformed it.

***Destroyed.*** An ecosystem experiences rapid, complete physical destruction when we replace it with infrastructure, such as roads, subdivisions, dams and mines. But an ecosystem's destruction need not be rapid, and it can experience degrees of destruction. It can be a drawn-out affair during which the degrees of destruction erratically increase, as illustrated by the conversion of indigenous ecosystems into farmland.

In order to create our pastures and crop fields, we appropriate a functional indigenous ecosystem. To create pasture, we alter the mix of the ecosystem's plants. To create a crop field, we destroy the above ground portion of the ecosystem's plants and animals. All that is left is the underlying organic soil community, which is then used to grow agricultural crop plants. The scale of ecosystem destruction that results from our success at transforming ecosystems into farmland is significant. For example, 50% of the earth's temperate grassland biome has been dedicated to pasture, while a significant proportion of the remainder has been replaced by cropland (McNeill and Winiwarter 2004; Ramankutty et al. 2008).

After an ecosystem has been converted into farmland, another level of ecosystem destruction is possible. The ongoing farming of a field tends to slowly degrade its soil through a loss of the soil or its fertility, or both. For example, with the expansion of farming over the last 200 years, the global decline of organic carbon from soils has increased rapidly, with the rate of loss from crop fields being twice that from pasture fields (Sanderman et a. 2017; Begon et al. 1996). Soil degradation is sped up when farming practices limit soil-rebuilding processes or promote events such as salinization or erosion. Once the organic soil is removed or infertile, the destruction of the original ecosystem is completed. A well-known historical example is the 1930s Texas dust bowl (Trimble and Crosson 2000). A contemporary example can be seen in Haiti.

It is convenient to think of ecosystem destruction as occurring in ideal stages, as illustrated by farming: cover changes, destruction, soil utilization, depletion of soil nutrients and organic matter and, finally, erosion. But, in reality, ecosystem destruction does not follow such neat stages. More generally, it is the cumulative outcome from a variety of disruptive activities, such as pollution, water appropriation and exotic introductions, carried out too intensely, or too frequently, over a longer period (chronic disruption). These activities are augmented by the occasional appearance of more familiar destructive activities, such as land clearing and infrastructure construction. These activities too are usually scattered through time and space and are followed by periods in which the ecosystem can partially recover.

This more pragmatic view of ecosystem destruction is illustrated by the long history of the Argolid, a northeastern region of the Peloponnisos peninsula, southern Greece. (The general region is the cradle of the Ancient Greek culture). The regional landscape is characterized by rugged hills with geologically unstable slopes and few flat-bottomed valleys. Before agriculture was practised, the Argolid was covered by an oak-dominated, parkland forest with hornbeam, beech and holly trees.

Although the removal of the Argolid's forest began with the arrival of agriculture sometime before 6000 BC, it was only after about 4000 to 3000 BC that the practice of farming became widespread. From then till the present, the area experienced periods of land clearing, agricultural expansion and agricultural intensification broken by periods of land abandonment and some degree of ecosystem recovery. Over this period, the climate slowly dried. The Argolid's ecosystems were thus faced with both recovering from ongoing but varying human impacts and adapting to climate drying. They, and their supporting landscape, changed into what we see today.

The oak forests were replaced by cleared land, with a surrounding ecosystem dominated by hornbeam, pine, scrub oak and heather. Eventually, erosion of the slopes left behind depleted and rocky soils that, in the current dry Mediterranean climate, can only support low bushes, such as heather and herbaceous plants, and scattered clumps of pine trees. The eroded soils ended up in the valley bottoms. Outside of these valleys, the Argolid is now only marginally suitable for farming (Bintliff 2002; Runnels 1995; Van Andel et al. 1986).

What were the human contributions to the eventual destruction of the original ecosystem? Their direct contributions included deforestation, ecosystem conversion to agricultural fields and inappropriate farming activities. The indirect human contributions included the economic and population-driven expansion of farming onto ever-steeper slopes and more marginal lands; the choice of crops, such as cash crops like olive oil; the grazing of animals; and political decisions about land management and land usage.

These contributions interacted with one another. Consider that removing the above ground ecosystem in order to plant crops facilitated the erosion of soil, especially on the slopes. Land conservation regulations, such as those requiring erosion control structures to be built on the slopes, did exist during Ancient Greek times, but they were erratically enforced. Land clearing in conjunction with neglecting the need to build and maintain erosion control structures contributed to a substantial increase in the rate of organic and mineral soil erosion. This state of affairs was noted by both Aristotle and Plato in the 300s BC, near the end of Athens' experimentation with democracy. The outcome for the Argolid was the loss of the indigenous ecosystem's ability to either adequately adapt to the changing climate or to recover from human disturbance. The original ecosystem was effectively destroyed (Bintliff 2002; Runnels 1995; Van Andel et al. 1986).

The human activities on the slopes impacted more than those ecosystems. The natural and human generated products of erosion filled not only the valley bottoms but the swamps and coastal bays. This successfully destroyed or transformed the terrestrial and marine ecosystems that once existed there. The size, significance and impact of this erosion is exemplified by the fate of the ancient port city of Troy, a short boat ride across the Aegean Sea from the Argolid. Troy was rediscovered and excavated in 1871 by Heinrich Schliemann, but only after he realized that the bay on which the city was built had become filled with eroded soil. At the time of its discovery, the city lay some 5 to 6 km inland (Kraft et al. 2003; Kraft et al. 1980). (Land uplift likely also played a role).

A major contribution to the destruction of the Argolid ecosystem and its change into the current one was the long-term farming-related activities that were unevenly distributed over time and space. The farmers added to, redirected and accelerated the natural ecosystem-transforming influences of an unstable landscape, a drying climate and uplifting land. This description of an ecosystem's destruction hints at a feedback between ecosystem changes and human decisions about activities that influence it to change. This will be discussed in later chapters.

In summary, our contribution to the biosphere's current state were divided into three basic contributions: pollution, direct action and climate warming. The effects of our direct actions on the biosphere's ecosystems can be fast and obvious, such as ecosystem destruction. The effects of pollution range widely in speed and significance but are less obvious. Climate warming's effects (our most important indirect effect) occur more slowly relative to our direct actions. In the short term, climate warming's effects are largely imperceptible. But, over the long term, they inexorably affect all of the biosphere's components simultaneously. The climate warming contribution can be thought of as changing the foundation on which our pollution and direct actions act.

By looking at the general effects we have on species and ecosystems from the perspective of our three contributions, we realize that they are related to and interact with one another. Those interactions include feedback. Our contributions have a cumulative effect on species and ecosystems, and thus on their biosphere host.

Our cumulative effects on an ecosystem place it on a continuum between functional (resilient, robust and adaptable) and destroyed. The continuum can be divided into three useful rule-of-thumb types: *functional*, *disrupted* and *destroyed*. A functional ecosystem is only slightly disrupted. A disrupted ecosystem can return to its original functional state. The boundary between disrupted and destroyed is the point where a disrupted ecosystem won't return to the essence of its original functional state. A destroyed ecosystem can still be or become an ecosystem, but not its original one. In the extreme, a destroyed ecosystem is no more. An ecosystem that has been substantially disrupted but it is uncertain whether it has been destroyed can be thought of as *transformed*. The complex, dynamic process by which our contributions change an ecosystem's state illustrates the process by which we change the biosphere's state.

There is more to our contributions to the state of the biosphere's ecosystems than meets the eye. Behind them lies our personal and cultural relationship to the biosphere, our contributions and the resulting changes. Both our personal and our culture's world views are thus an important part of our contribution to the current state of the biosphere and its ongoing changes.

Our overall contribution to the current state of the biosphere has been increasing over the last 4,000 years. We have become the dominant driver of the biosphere's changing for at least the last 200 years. The overall state of the biosphere's ecosystems indicate that the biosphere's current state is significantly disrupted.

## Conclusion

Our actions are, directly and indirectly, largely responsible for the current state of the earth's atmosphere, ocean and land-based environments. Those actions are also largely responsible for the current state of the biosphere. They can be divided into three basic contributions to its state: our pollution, our direct actions and climate warming. Some of our pollution is changing the composition of the earth's atmosphere, and some of our direct actions are changing the earth's albedo. These changes affect the earth's climate system. The resulting warming of the lower atmosphere (troposphere), climate warming, is our most significant indirect contribution to the state of the biosphere. It is globally affecting the ocean and land-based environments and the ecosystem's they host. For example, the earth's climate zones are shifting poleward. Individuals and species are responding by migrating and changing their lifeway characteristics.

The consequences of our contributions are collectively and cumulatively changing the biosphere's state. For example, warmer oceans from climate warming are impacting marine ecosystems at the same time as our marine pollution is affecting them from the bottom up of their trophic webs and our direct actions of fishing are affecting them from the top down. The overall result is a global-scale disruption and destruction of the biosphere's ecosystems. These ecosystem changes affect the climate system, so a feedback is part of our cumulative effect on the biosphere.

For at least the last 100 years, we have been a dominant driver of changes in the biosphere. Regardless of the perspective we take, the earth's biosphere has been significantly disrupted. The degree to which the biosphere has been affected is personalized by revisiting the astounding percentages of the earth's surface on which its ecosystems have been directly impacted by humans.

By 2000, we had appropriated about 34% of the earth's ice-free land surface for agriculture (excluding tree plantations). One-third (12%) was used for crop land, and two-thirds (22%) provided pasture for our domesticated animals. In creating our crop land, the original ecosystems were largely destroyed. They include 30% of the pre-farming temperate forest biome in Europe and the United States; significant portions of tropical deciduous forests in South Asia; and a significant proportion of global grasslands. In creating our pastures, some 50% of the pre-farming grassland biome and substantial portions of the savannah and shrubland biomes have been significantly disrupted (Ramankutty et al. 2008). Today, the last remnants of the ecosystems hosting 44% of the world's terrestrial plant species and 35% of its terrestrial vertebrate species are found in 25 biodiversity hotspots. These areas make up only 1.4% of the earth's surface (Myers 2002b).

The area of the earth whose ecosystems have been impacted by humans also extends to the oceans. There is every indication that marine ecosystems have been and are being significantly disrupted and destroyed on a global scale. For example, by 2008, at least 84% of industrial fisheries (coastal and open ocean) were fully exploited, overexploited or collapsed, implying that their host ecosystems were significantly disrupted.

The degree to which we have disrupted and destroyed the biosphere's ecosystems is greater than suggested by the percentages referred to above because they only consider the most obvious of our direct contributions. The percentages do not include our pollution and our climate warming contributions. Nor should we forget that these changes take place in an effectively finite system (i.e., the earth, except for sunlight). They are trade-offs, not isolated impacts.

The above changes to the biosphere could be seen as a normal part of the process of environmental change. However, the intensity, scale and speed at which they took place over the last 200 years (and continue to take place) make these changes highly anomalous. So does the fact that a single species, us, is largely responsible for the biosphere's significantly disrupted state.

The difficulty of grappling with this reality is illustrated by one effort to comprehensively consider its current state as a whole. In this study, the biosphere was viewed from its perspective. To do so, its functioning was represented by nine factors: biodiversity loss, climate warming, the nitrogen cycle, stratospheric ozone depletion, ocean acidification, global freshwater use, change in land use, chemical pollution and atmospheric loading with aerosols. Efforts were made to set boundaries for each factor which, if crossed, would indicate that the biosphere's functioning is being critically compromised (disrupted). The authors suggested that our activities have already driven the factors of biodiversity loss, climate warming and the nitrogen cycle across their boundaries. The status of the remaining factors is either "yet to be determined", "satisfactory" or rising toward their boundary (Rockström et al. 2009).

This idea of boundaries is worthy of our attention because it accepts that there are environmental constraints on how we can live and that the changes to the biosphere now underway are significant. It also reflects our lack of awareness of these environmental constraints on our lives and their importance. However, this analysis and its conclusions suffers from significant limitations.

One limitation is the confidence with which it is thought that the dynamics of the complex biosphere's functioning is satisfactorily represented by nine numerical factors with somewhat fixed boundaries. Think of the GDP and environmental footprints. Another limitation is that these factors are largely considered in isolation.

Perhaps the most significant limitation is that the boundaries were chosen with the objective of ensuring that the biosphere will remain in a state capable of satisfying human goals. This was done without considering the role a culture's world view plays in setting those goals. The study could thus conclude that, "The evidence so far suggests that, as long as the thresholds are not crossed, humanity has the freedom to pursue long-term social and economic development" (Rockström et al. 2009, 475). The merging in their discussion of social development and economic development into development as the overriding human objective strongly implies that Western culture's notion of progress lies behind the authors' thinking. It implies that our current situation is a problem of actions and processes, not beliefs and goals. The selection of development as the goal implies that we should strive to stay within our human and environmental constraints by becoming better ecosystem and biosphere managers, rather than by changing the essence of our world view. There is no intimation that we need to change the essence of what we believe in or our lifestyle.

Are we successfully fulfilling the role of ecosystem managers? If not, can we do so? Our discussion is brought back to the start of this chapter. From the biosphere's perspective our efforts at management have resulted in its state becoming highly anomalous compared to just 200 years ago. The earth's biosphere is significantly disrupted and human dominated. There is no longer a distinction between pristine wilderness and human-affected areas (Vitousek et al. 1997; Barnosky et al. 2004b).

So ends the first part of your report on the status of our biosphere. You hope your description of our biosphere, from its perspective, is what your boss wants. But you know that before you can find out you will have to complete part two of your report. Maybe, you surmise, after you have looked at the biosphere from our perspective you can report that we have been successful managers. Our view of the state of the earth's biosphere is the one taken in the next chapter.

# CHAPTER 10

# THE BIOSPHERE:
# SUPPLYING WHAT WE WANT

This chapter is the second part of your report to the all-powerful presence. It discusses the earth's biosphere from our perspective. The focus is on its ability to satisfy our current demands for resources (which includes services).

A casual review of the demand-supply balance leads you to the conclusion that our ability to innovate and use technology has enabled us to extract or produce all we need. So it seems that all is well and there is not much to report. If there is an issue, it will likely revolve around the unequal distribution of the products of our labours.

You decide to divide the second part into sections that deal with the demand-supply balance for farmed food, wild food, water, energy and health. You also include a section on the metals and minerals that come from the solid earth. But, as you put fingers to keyboard, you experience a nagging feeling. You remember that the supply of the things that we eat and drink, and much of those we use, depends on more than the simple presence of our biosphere and our ability to extract them. It ultimately depends on how the biosphere functions as a system. You realize that your uncomfortable feeling arises from the conclusions to the first part of your report: our activities have and are changing the structure and functioning of the biosphere.

To regain your peace of mind, you decide to include in this report a section on the costs of supplying each resource at levels adequate to satisfy our demand. You decide that the word *costs* should cover the broad range of non-economic (e.g., environmental and social) costs rather than the usual narrow focus on economic (cash) costs. You also realize that the level of our demands and the acceptable costs of satisfying them are influenced by our culture's world view, so you add a section on culture.

## Food from Agriculture

Our ancestors were participants in an ecosystem within which they fed themselves by hunting and gathering with tools made from materials such as stone, wood and bone. Their host ecosystem's functioning constrained their lifestyle and population size. Around 8000 BC, human cultures in the Middle East started to secure some of their food through agriculture. Since that time, agriculture has played an increasingly important role in supplying resources, especially food, to an expanding human population.

Today, the limited number of plants we cultivate supply most of our food and a significant proportion of our wants for non-food biological resources. We also prefer that a restricted selection of plants form the pasture on which our domestic animals feed. The process and limitations of supplying the agricultural resources, especially food, we want from these favoured plants and animals is the subject of this section.

Production. Once humans had adopted agriculture, an increase in population required an increase in agricultural production. This was historically accomplished through three methods: increasing the area of land under cultivation either locally or in distant lands; increasing plant and animal yields through breeding and selection; and adjusting farming practices (McNeill and Winiwarter 2004). These methods are still being used today.

However, as the area of ecosystems that could be easily converted into medium- or high-productivity agricultural lands decreased, our efforts to increase production turned toward increasing productivity on existing farmland (agricultural intensification) (Ramankutty and Foley 1999). This meant focusing more on making changes to farming practices and selecting higher yield plants and animals.

All of these efforts to increase agricultural intensity were historically constrained by the energy available to practise farming, by the need to maintain soil fertility and, of course, by the amount and timing of rain (which is itself constrained by the weather). Energy use was limited to the amount provided by manual labour and domestic animals. Maintaining soil fertility was constrained by the speed of its recovery during a fallow period and by the availability of, and effort to apply, the fertilizer provided by humans and animals (dung). Despite these constraints, both changes in farming practices, such as inventing more efficient animal-drawn machinery and the selection of more productive plants and animals increased agricultural intensity.

Starting in the industrial revolution, the energy limit was circumvented by the increasing use of fossil fuels (in particular oil) to power farming machinery. By the late 1940s, a significant proportion of the Western world's human and animal farm labour had been replaced with machines. For example, in the United States, the number of agricultural workers dropped from 10% of the workforce in the 1920s to 2% in 1975 (May 1975). Similarly, after 1913, the soil fertility constraint was being increasingly circumvented by the use of manufactured fertilizers. By this time, the Victorians' experimentation with plant and animal breeding had provided many of today's higher yield heritage plants and animals.

After World War II, with the remainder of the lands suitable for farming now mostly under the plough, a new concerted effort to increase plant and animal yields began. Between the 1960s and the 1990s, a combination of plant selection and related adjustments to mechanized farming practices brought about a dramatic global increase in agricultural intensity, called the green revolution. For example, by planting strains of wheat selected for their high yields, wheat production in the United States rose spectacularly from roughly 1.5 tons/ha in 1960 to about 2.5 tons/ha in 1980 (Ausubel 1996). Today, as a result of the green revolution's success, inexpensive food is plentiful and cheap in most countries.

However, there is more to the green revolution's success than plant selection and machines. Consider that agriculture is practised on a landscape of fields that were formed by largely destroying the original ecosystem and its functioning. In its stead, we created a farmed ecosystem consisting of only a few crop species that are grown as close together as possible. We thus became responsible for the functions normally fulfilled by the original ecosystem: supplying nutrients, regulating water, dispersing seed, thinning and pollinating plants, and controlling pests. For example, land clearing creates an ideal habitat for weedy species, both local and exotic. At the same time, the low diversity and high density of our crop species creates the ideal circumstances for the generation and spread of crop diseases, parasites and pests (Begon et al. 1996). To be successful, farmers are obliged to accept these responsibilities.

As any farmer will tell you, their obligations to our agricultural ecosystems are never-ending and ever-changing: our responsibilities are both enduring and dynamic. Both of these characteristics are illustrated by their experience with black stem rust, a severe fungal disease that can decimate wheat crops. In the late 1950s, scientists thought they had beaten the disease because they had successfully selected and bred varieties of wheat that were resistant to it. However, in 1999, a new variety of black stem rust called Ug99 appeared in Uganda. At least 80% of all currently farmed wheat varieties are susceptible to Ug99. An international search for wheat varieties resistant to Ug99 began (Stokstad 2007a). By 2011, significant progress had been made, but much work remained to be completed (Singh et al. 2011).

In essence, by practising agriculture we enter into a highly interdependent or mutualistic relationship with our crop plants (and animals). We become a critical part of a highly simplified, unstable, human-created ecosystem (Begon et al. 1996). When we intensify agriculture, we strengthen our mutualistic relationship with our farmed ecosystems, increase its instability, heighten our existing obligations (responsibilities) and add new ones (Tilman et al. 2002).

The green revolution's agricultural intensification included a significant increase in our obligations. Consider that the plant selection process focused on increasing the yield of the plant parts we eat. But this reduced the plant's ability to deal with its other lifeway needs, such as changes in soil moisture and drought. We thus became more responsible for maintaining a water supply. Then, at harvest time, we remove from the field the bumper crop that our selected, thirsty high-yield plants provided for us and transport it to a distant location. In doing so, we remove the nutrients in the crop from the field. We thus heighten both the disruption of the soil's nutrient recycling system and the depletion of its nutrients (McNeill and Winiwarter 2004). We become more responsible for the soil's fertility and functioning.

The green revolution's intensification was successful because we fulfilled all of our mutualistic obligations to our farmed ecosystems. This was only possible because of an abundant supply of cheap fossil fuel energy. This energy made it possible, then and now, to more easily complete many farming activities: operate farm equipment; manufacture increased amounts of fertilizers and pesticides; and construct and operate the irrigation systems needed to supply the increased amount of water (Khush 1999; Tilman et al. 2002). Consider that the global land area under irrigation increased from an estimated 0.08 million km$^2$ in the 1880s to around 0.95 million km$^2$ in the 1940s. By the early 1990s, the area under irrigation had more than doubled to between 2.5 and 2.8 million km$^2$ (Thenkabail et al. 2006). By 2009, some 3.01 million km$^2$ were under irrigation and increasing by about 0.6%/year (FAO 2011).

Fossil fuels made it possible for us to practise our extremely high-intensity, green revolution agriculture, often called industrial agriculture, in which fewer farmers can feed more people. Unsurprisingly, the amount of fossil fuel energy needed to maintain industrial agriculture's level of intensity is large. For instance, in 1972 it took approximately 52 US gallons (197 litres, about 1.3 barrels) of refined oil just to grow the plant and animal foods that the average North American ate (Heichel 1976). Some express this by referring to "the oil we eat" (Manning 2004).

But there is more to agriculture's use of fossil fuel energy than meets the eye. All species must act to secure the energy they need to complete their lifeway. Think trophic web. For animal species, the ratio of the usable energy in the food they secure to the energy expended in securing it is the energy return for their food-securing effort, which is a measure of their efficiency of energy use. There is also an allometric relationship indicating that the greater a species' body mass, the larger the amount of food energy it must secure in order to live. Despite this relationship, the energy returns for all species range between 2:1 and 20:1, with an average of 10:1. (The bigger the left hand number the greater the energy return.) Numbers less than 1 indicate a net energy loss that, in a biological system, translates into starvation.

Now consider the difference in the energy return between humans providing food using fossil fuels and providing it though hunting and gathering or subsistence farming (those who grow only enough to support themselves). The energy return for food gathered by hunter-gatherer cultures lies between about 5:1 and 10:1. This is in the low half of the all-species range (2:1 to 20:1). The energy return for hunter-gatherers is low because we eat relatively rare, high-energy foods that require substantial amounts of energy to hunt and gather. If subsistence agriculturists are included in the comparison, then the human range of values shifts to between 5:1 and 50:1, with an average value of around 10:1. This increase in our food energy return is one reason why some of our ancestors could afford to take up and continue to practise agriculture (May 1975).

To evaluate the energy returns for industrial agriculture's crops, our energy expenditure must include the fossil fuel energy we use to produce them. When this is done, the energy returns for individual foods are seen to vary from near zero to 5.5:1. (figure 12). For example, growing corn results in a return of 4.5 units of food energy for each unit of fuel energy. In contrast the energy return for growing fruit and vegetables lies below 1:1, near 0.5:1, a net energy loss.

The energy return for beef and pork sourced from industrially raised animals (those eating feeds whose provision relies heavily on fossil fuels) lies between 0.2:1 and 0.6:1 (Heichel 1976). The returns are so low because only about 10% of the feed that cattle and pigs eat adds to their weight, to our food. Thus little of the fossil-fuel

energy used to provide the feed becomes part of the animals we eat. Energy returns will be higher for the meat from animals that eat non-fossil-fuel grown feeds (i.e., free range animals.)

If humans ate a diet based solely on high energy return crops such as corn, sugar cane and wheat, then the energy return of their diet would be around 4:1 to 5:1. However, energy is not the only requirement for a healthy diet. We require proteins, fats, minerals, nutrients and vitamins that are most easily obtained from fruits, vegetables, oil plant crops and meat. When these fruits, vegetables and protein are grown using (crude) oil, they have an energy return <1. If this is taken into account then the energy return of a balanced, fossil-fuel-supported diet will lie below 5:1, the low end of hunter-gatherers' energy return range.

However, the energy returns quoted for foods grown using fossil fuels do not include the energy needed to move the crop from the field to the distant consumer market in the cities, and into our homes. If this energy and food waste is considered, then the energy return for a balanced fossil-fuel-supported diet (especially a meat-rich version) is unlikely to be much above the minimum energy return of 2:1 quoted for all species. Thus, although industrial agriculture can supply us with more food than hunting and gathering or subsistence farming, it comes with the (economic and environmental) cost of supplying and using fossil fuel energy.

From the 1970s to the present, the green revolution made it possible for us to feed the rapidly increasing human population. Today the human population is still increasing, but the production of basic foods per capita is increasing more slowly or declining. This suggests that the great gains in agricultural intensification attributed to the green revolution are likely over.

Consider that, between 1977 and 1986, there was an increase in the global cereal yield per sown area of about 2.5%, even though the sown area shrank by about −0.1%. In contrast, between 1987 and 2001, the global cereal yield increased by only about 1.5%. During this time, the sown area shrank by about −0.4% (Huang et al. 2006). Similarly, the rice yields in China, South Korea and Japan, all early adopters of green revolution techniques, have stagnated at about 80% of the maximum possible green revolution yields (Tilman et al. 2002).

Some have suggested that the current stagnation or decline in agricultural output is mostly the farmers' response to a slowing in the increase of the human population: a slowing in the demand for food (WHO/FAO 2002). However, since the mid-1980s, world grain production per person has been flat or slightly declining, except for a slight increase between 2005 and 2008. Since 2000, world grain reserve stocks have declined from their peaks in the 1990s. Unfortunately, since about 2002, the interpretation of these numbers as an indicator of food supply is complicated by the increasing use of corn for biofuels (Worldwatch Institute 2012; Worldwatch Institute 2010). Despite the uncertainty, scientists and farmers have recognized the possibility of a future food shortage and have been investigating ways to revitalize the green revolution through techniques such as the genetic modification of plant foods (Huang et al. 2006; Khush 1999).

But not all the news is gloomy. Between 1990–1992 and 2010–2012 the number of undernourished people in the world dropped from 980 million to 852 million. When expressed as a percentage of the global population at each date, then the undernourished declined from 18.2% to 12.2% (FAO et al. 2012). The results of the green revolution linger on.

<u>Costs.</u> There are non-economic costs associated with the practice of agriculture. A feel for the scale of those costs is gained by considering the amount of land and water we use for agriculture.

The earth's ice-free land surface (excluding Greenland and Antarctica) is around 126 million $km^2$. In 2000, some 15 million $km^2$ (12%) of that land was used to grow our crop plants (food and fibre, excluding wood). A further 28 million $km^2$ (22%) was used for pastures to feed our domesticated animals. In total, agriculture appropriated 34% of the earth's ice-free land surface (Ramankutty et al. 2008).

Although only about 2.7 million $km^2$ (18%) of crop land was irrigated in the late 1990s, it produced some 40% of our crops. That irrigated land used about two-thirds of all the water we removed (withdrew) from rivers, dams and groundwater (Postel 2001; Postel et al. 1996). Of the irrigated land producing food crops, three-quarters is located in Asia, in particular India (the world's largest user of groundwater for irrigation) and China (Thenkabail et al. 2006).

The substantial non-economic costs of using this amount of land and water for farming are borne by both the biosphere and us. The overall costs borne by the biosphere are represented by our resulting appropriation of the biosphere's annual net primary production of new biomass (NPP) on land. In 2000, our food crops, fibre

crops (excluding wood) and pastures together appropriated some 18.7% of the NPP. If our non-farming appropriations resulting from forestry (2.5%), infrastructure (0.9%) and human-started fires (1.7%) are included, then our total NPP appropriation rises to 23.8% (Haberl et al. 2007; Vitousek et al. 1986; Imhoff et al. 2004). The specific cost to the biosphere from us physically disrupting and destroying ecosystems in order to create fields and pasture was discussed in chapter 9. Think of the environmental cost of clearing a tropical forest.

The non-economic costs borne by us are the long list of responsibilities associated with practising farming and the related impacts on ourselves. Of course, all of the costs from practising agriculture are linked to one another, and to each mouthful of food we eat. This is the perspective used to illustrate the costs borne by us in more detail.

The costs to us from land clearing is our need to deal with its effects, which range from an increase in soil erosion to its contribution to climate warming (McNeill and Winiwarter 2004). The costs from supplying water and energy for agriculture will be addressed later. The costs from our use of human-made fertilizers and pesticides are discussed here.

There are a wide range of non-economic costs associated with the manufacture and use of nitrogen fertilizers. They include contributions to ozone layer depletion and climate warming (Socolow 1999). Consider too that the amount of human-made fertilizer added to a field usually exceeds the needs of the crop plants. The excess can change the functioning of the soil community by, for example, raising the concentration of ammonia in the soil to a high enough level that the nitrogen-fixing soil bacterium *Rhizobium* stops fixing nitrogen (Wolfe 2001). In addition, when the excess nitrogen is washed off the fields, it can contribute to drinking-water pollution and aquatic ecosystem disruptions from riverbanks to ocean dead zones, which disrupt river and ocean fisheries.

The routine use of chemical control agents for plants (herbicides) and insects (insecticides), collectively referred to as pesticides, also comes with a cost. As previously discussed, the residues from their use can disrupt both a wild species' efforts to fulfill its lifeways and the functioning of the wild ecosystems hosting them. These ecosystem costs occur worldwide because air and water has dispersed the pesticides as pollution well beyond the farm or garden where they are applied. The global distribution of DDT is well known. So are its long-lasting impacts on wild organisms and their host ecosystems. Additional examples of these chemicals and their costs are given in chapters 9 and 14.

There are even farming costs from using a pesticide to control pests. Consider the collateral damage. For example, the inadvertent killing of organisms that are important to agriculture, such as insects that are pollinators of crops or predators of crop pests.

The appearance of pest resistance is another farming cost. Insect resistance to insecticides can appear in less than 10 years, while plant resistance to herbicides can appear in about 10 to 20 years (Tilman et al. 2002). Today, a minimum of 900 targeted pest insects and plants (weeds) have at least some degree of resistance to the control agents used against them (Bright 1998). Dealing with this collateral damage and resistance adds to our farming responsibilities.

Resistance can also arise in non-agricultural pests, disease vectors or disease agents if they are exposed to agricultural control agents, either deliberately to control them or inadvertently. For example, DDT was first manufactured in the early 1940s. It was quickly adopted in Sweden to control the housefly. But by 1946, within one to two years of DDT's first use, the Swedish housefly was resistant to it (Brown and Pal 1971).

Consider an ironic cost of resistance that can result from the use of pesticides when growing the food needed to keep us healthy. The industrial farm-animal-rearing business routinely uses medications in an indiscriminate, non-targeted manner in order to "prevent" illness and promote growth in their food animals. In essence, medications are used as pesticides against microbes that have not yet resulted in disease. Farmers thus facilitate microbes developing resistance to these drugs, which can include medications similar to or the same as those used by humans. This resistance can appear in as short as one to three years (Johnston 1998). (The process is discussed in more detail below.)

Drug resistance in industrial farm-animal-rearing facilities can develop in the animals. It can also develop elsewhere in the facility. For example, the animals' faeces contain both the drug residues from their "treatment" and those organisms in their gut that had previously developed some level of drug resistance. The disposal of the faeces creates a microbe- and faeces-rich soil in which drug-resistance-endowing genes are developed, selected for and/or transmitted between micro-organisms, including those that cause animal and human diseases. These drug-resistant, disease-causing microbes can then be passed on to humans and animals

(D'Costa et al. 2006). Acquiring a drug-resistant disease is a cost because it makes human and animal diseases difficult to treat.

A question arises: how significant are pesticide-related costs? A 1992 study attempted to answer this question in economic terms. It estimated that for every US\$1 of direct costs from pesticide use spent in the United States, US\$4 in food was returned. However, when quantifiable indirect costs were considered, such as bee deaths and the loss of pesticide-contaminated crops, then the value of the food produced for each dollar spent on costs dropped to US\$2 (Pimentel et al. 1992).

But, this type of economic evaluation of costs is subject to considerable uncertainty (discussed in more detail in chapter 12). For example, it is difficult to decide on a value to assign to some costs and a time scale over which it is relevant to assess them (i.e., the boundaries of the system). Consider too that the culturally influenced world view of the assessors is just as important as monetary accounting: it strongly influences their perception of a cost and a benefit, usually with a bias in favour of human concerns. As a result, an economic estimate of costs should be considered a minimum.

**Culture.** Our biology determines the nutrition we require and the organisms we can eat. In contrast, our culture's world view strongly influences our relationship with food. It provides guidance about which organisms we call food and explanations for why we prefer to eat them and not others (Mintz 1989). It influences how we acquire or produce our preferred foods and our acceptance of the costs of doing so. Its guidance includes the symbolic and alternate non-food values we should attach to particular foods and to the land or ecosystems providing them. Thus, although our biological ties to food are unbreakable and universal, our relationship with food is flexible and varies from culture to culture.

Some of the connections between a culture's world view, its members' farmed-food preferences, and their acceptable food production methods and costs is illustrated by the United States. A glance at an American menu would suggest that they grow and eat a lot of wheat and meat. Surprisingly, the three crops that occupy about 82% of US croplands are corn, hay and soy beans.

Corn is the foundation of the American diet. It is largely absent from the menu because it is mostly consumed after it has been processed into a different form. Their food industry turns corn into ingredients such as corn starch, corn sugar, corn syrup and animal feeds. Soy beans too are turned into food ingredients and animal feeds.

Hay is grown to feed cows during the winter so that they can produce milk, or to feed cattle until they are sent to feed lots. It is there that the cattle are given corn-based feeds, which they transform into meat and fat before being slaughtered. Pigs and chickens are similarly fattened in barns. The milk, fat, eggs and meat are shipped to factories where, with the help of corn-based and other ingredients, they are turned into the prepared food items found on the American menu (Wolf et al. 2006; Manning 2004).

This US food production system and its emphasis on corn is culturally supported in a number of ways. The US federal government strongly promotes the production of corn through a substantial income subsidy to corn farmers. It also subsidizes those converting corn into ethanol. The ethanol is then added to the fuel used to power the farm equipment needed to grow the corn and transport it to the feed lots and factories (Pimentel and Patzek 2005).

Additional support for the American food production system is provided by their culture's dominant perception of agriculture. It is largely seen as a physical process in which intensification and efficiency are prized. Farming issues are treated as problems having technological solutions, such as pesticides, computer-controlled equipment and genetically modified organisms. All of these features illustrate that, in general, the American culture's world view sees agriculture as an industry.

In this context, it is easy to understand why the costs they assign to the production of food are usually just the cash costs, plus profit. The non-economic costs are downplayed or ignored. For example, the herbicide of choice when growing corn is atrazine, the herbicide that feminizes frogs (Barr et al. 2007). Despite this cost, atrazine is accepted for use in the United States because it is effective, it increases farm yields and it passes their culture's chemical evaluation process.

The American culture's view of food production is complemented by their culturally favoured relationship to their food. Consider the predominance of fast food and sugary drinks on the standard American menu. The main ingredients of those menu items are highly processed corn and soy, along with the meat, fat and sweetener derived from them. The non-economic costs of routinely consuming these foods are widely ignored. Overall, the Americans' production and consumption of these foods, along with their views of the costs, form a culturally supported and sustained food system.

Although American culture is uniquely their own, their menu illustrates a worldwide, cultural trend in food preferences. Historically, a typical agriculturist diet consisted of a starch (carbohydrate) core, protein from a pulse (an edible legume such as beans and peas) and flavouring. Protein in the form of meat may have been added, but in small quantities. Fat was usually derived from dairy products. This basic diet is changing worldwide.

As people move into cities, there is a rapid increase in their routine consumption of "luxury" high-energy, low-nutrition foods. This means foods that contain a high percentage of the luxury ingredients sugar, meat and fat (Mintz 1989). The vanguard of this trend is the Western energy-dense, low-nutrient diet: the standard American menu. (Think of figure 12).

These worldwide, culturally supported changes in urban diets affect a farmer's decisions about which foods to produce, how much to produce and how to produce them. The farmer's choices, in turn, affect their agricultural obligations. Their production choices also feed back to affect our diet choices as well as the human and environmental costs associated with our agriculturally produced foods. Consider the increased health service costs associated with a culturally supported diet that favours the development of obesity, such as a flavourful, nutritionally poor diet that exceeds our average energy requirement of roughly 2000–2500 kilocalories per day (kcal/day) and our protein requirement of 5%–8% of our caloric intake (Socolow 1999).

In summary, the answer to the question of whether the earth is able to supply our current food demands is determined by how we interpret it. If we ignore the regional differences in supply and short-term variation, then global averages, such as those listed above for grains, indicate that, with the help of fossil fuel energy and technology, we have fulfilled our immediate responsibilities to our agricultural ecosystems. Most of us are now provided with an abundance of cheap food. Some may by hungry, but there are few large starvation events. The conclusion could be drawn that our culturally supported food production system has successfully broken the bond between ecosystem functioning and our food supply.

However, the conclusion could also be drawn that the earth can only supply the food we need because we have simplified our food-supplying ecosystem and formed a strong mutualistic relationship with our crops. Our answer would then have to take into account the non-economic costs related to maintaining a highly intensive, unstable agricultural food production system. These costs include the loss of services from a disrupted ecosystem, the added responsibilities of having to provide or complete functions normally fulfilled by a functional indigenous ecosystem, and the collateral effects from fulfilling those responsibilities. The conclusion could be drawn that the earth can supply the food we want, but at a considerable non-economic cost, including an increased burden on ourselves.

We could decide to consider the impacts that producing today's food supply will have on the ability of future generations to provide tomorrow's food supply. If we do, then the focus of our attention shifts to other influences, such as our culturally influenced views about food, our ecosystem functioning and population growth. This perspective of the question will be addressed later in chapters 14 and 15.

## Wild Food

Unlike farmed foods, wild foods grow in a self-sustaining, human-independent ecosystem. The ability of a wild food and its host ecosystem to satisfy our demand depends on our ability to balance our demand to the supply, and the costs to the benefits. The nature of this task and our success at it is illustrated in this section by ocean fisheries, the dominant source of our wild foods. (A fishery is a place, an ecosystem, where the population of a fish species can be fished.) This discussion takes a different perspective of ocean fisheries than in chapter 9, but the descriptions overlap.

**Harvesting.** The greatest difficulty humans face when using wild species for food is managing the harvest. This involves much more than securing the food. We are faced with the daunting task of setting and enforcing a harvest (catch) size that would maintain the food species' population over the long term. Imagine the difficulty of deciding whether a large catch of ocean fish represents overfishing of a marginal population or a sustainable extraction from a highly productive population. In this view the managers have to estimate the maximum catch that a fish's dynamic population can sustain (the maximum sustainable yield).

Managers can estimate the maximum sustainable yield by conducting a statistical analysis of historical fish catch sizes. But, for the estimate to be reliable, it has to be based on a sufficiently accurate and precise multi-year record. Often the available information is incomplete and biased, as illustrated by the uncertainty in the only data set available for global marine fisheries. An estimate of the maximum sustainable yield is always accompanied by significant uncertainty (Watson and Pauly 2001; Pauly et al. 2013; Froese et al. 2012).

The difficulty managers face is not entirely alleviated by a reliable data set and forecasts. After all, this knowledge about a fishery's status is often only available long after the decisions about today's effort to catch fish is well underway. Consider too that acting on an estimate of the maximum sustainable yield that is based on a fish species' population dynamics is insufficient on its own to ensure the fish will maintain a viable population and provide an excess for harvest. Fish are not created by these dynamics or by managers. They are created by individual fish living in their host ecosystem as part of their species' population. Managers thus also have to ensure that the host ecosystem and its functioning is kept sufficiently intact so that enough fish can complete their lifeway. But, it is difficult to reliably assess the state of a fish species' host ecosystem or its population, so this is generally not meaningfully taken into account.

Collectively, these aspects of management should significantly constrain a manager's decisions. For example, if they want to ensure that the catch is sustainable over the long term, then they must set the catch quota well below the estimate of a maximum sustainable yield. This requirement helps to explain why setting a sustainable catch quota is an intractable problem that is tied up in the foibles of human decision making. For example, the presence of significant uncertainty in both the data and the estimated sustainable yield is easily trumped by the greater certainty in the expected demand for fish and the desire to catch and sell them at a profit. Thus, those making the ultimate decision about wild food management can be more influenced by immediate and short-term economic and cultural concerns than by longer-term ecological realties and the significance of the uncertainties in the data (McEvoy 1986). A reactive rather than a proactive management of the catch is to be expected.

The global-scale state of the ocean-fish stocks (the supply) that results from decisions made in this context are provided by a reconstruction of the global ocean-fish catch between 1950 and 2010. It indicates that the catch size peaked at 130 million tonnes (Mt) in 1996 and then declined steadily to around 110 Mt in 2010. As detailed in chapter 9, the 2010 catch consisted of the following catch types: industrial (66.4%), artisanal (20%), subsistence (3.5%) and recreational (0.9%). A further 10.1 Mt (9.2%) of bycatch from industrial fishing was discarded. The decline in the catch size since 1996 was the result of a steady reduction in the industrial catch type, which occurred despite an exponential increase in our industrial fishing effort (Pauly and Zeller 2016; Watson et al. 2012).

Unsurprisingly, a conclusion drawn in chapter 9 applies here: the ability of the industrially fished ocean-fish stocks (the fisheries) to supply fish has changed dramatically. In the 1950s, the fisheries assigned the status of undeveloped and developing collectively accounted for 95% of fisheries. The remaining 5% were classed as fully exploited. By 2008 the fisheries considered to be undeveloped, developing or rebuilding collectively accounted for only 16% of fisheries, while fully exploited, overexploited or collapsed fisheries accounted for 84%. The steady increase, between 1950 and 2008, in the fishing effort, and the percentage of the fully exploited, overexploited and collapsed group of fisheries, indicates that the fish catch was and is being maintained by overfishing. This view is strengthened by a recent analysis which suggests that many more fisheries belong in the collapsed class than expected (Froese et al. 2012; FAO 2010).

Worldwide, efforts to manage specific ocean fisheries have had limited success at preventing their collapse or stopping ongoing overfishing. A typical outcome is illustrated by Canada's Newfoundland cod fishery. For 500 years this fishery was one of the world's richest. Post-1950s, the cod were overfished to the point that the breeding stock had been reduced to about 1% of their historical numbers. In 1992, the fishery was closed because the cod population had collapsed (Bavington 2005; Bundy 2005). The essence of this story is the same for a succession of historical fishery collapses, including those of the Californian sardine and the Chilean anchovy (Ludwig et al. 1993; McEvoy 1986).

Costs. There are non-economic costs associated with supplying wild foods. In general, the more the short-term drive to produce a wild food takes precedence over the long-term need to maintain the food-providing population and the functioning of its host ecosystem, the greater the costs. Those costs are illustrated by ocean fisheries.

The costs of supplying ocean fish arise from the act of fishing. The removal of too many of the targeted fish disrupts its population and affects its host ecosystem from the top down. The result is less fish to catch, which impacts those dependent on the fishery for food and work. In the extreme case overfishing results in the collapse of a fishery. The costs for fishing are thus paid for not only by the fish and its host ecosystem, but by humans.

The costs of fishing can arise less directly. Consider that the act of industrial fishing catches and lands a large number of untargeted and unwanted organisms. This bycatch occurs, for example, when some of longline fishing's thousands of hooks inadvertently catch sea birds, especially albatrosses (BirdLife Global Seabird Programme 2008). Deep sea ocean trawling kills birds when they hit or become snagged on the trawl line (Sullivan et al. 2006). Surface trawling removes more than the desired fish. Bottom trawling scrapes the ocean floor, which scoops up the targeted species plus a host of unwanted living things, while severely disrupting or destroying the foundation of their host ecosystem. Globally, the bycatch consists of a wide variety of mammals, reptiles and birds, plus non-target, unsaleable and undersize fish. In 2010, some 10.1 Mt of bycatch were discarded. The deaths of so many organisms disrupt the fisheries host ecosystems, and thus the fishery itself.

The environmental costs of fishing also include the impacts of the emissions from the oil burned to reach fisheries thousands of kilometres from a ship's home port. There is also the plastic pollution of all sizes associated with fishing gear and the people that fish (Pham et al. 2014). For example, abandoned nets can ghost fish for years.

The sum of these environmental and social impacts is the cost of supplying and eating wild fish. To reduce these costs, we have to accept responsibility for maintaining the fish population. This requires us to recognize and accept both the limits to fish availability and the constraints on how they can be caught.

Culture. A culture's world view exerts a strong influence on its members' demand for wild foods and their perception of the "acceptable" costs of supplying them. This is illustrated by the following two examples for ocean fisheries, one from the past and one from the present. During the Middle Ages, a change in the Christian fasting regulations resulted in an increased European demand for fish (Barrett et al. 2004).

Similarly, the current practice of cutting the cartilaginous fins off a shark before the animal is thrown, often alive, back into the sea (shark finning) is culturally driven. The fins are needed to prepare shark fin soup, a low-nutrition dish some east Asian cultures consider a delicacy. A culture's values and beliefs are just as strong an influence as nutritional needs on the members' demand for wild foods and the acceptability of the non-economic costs of providing them.

A culture's world view also contributes to the difficulty of developing and implementing an environmentally realistic fisheries management strategy, as is painfully documented by the history of the Californian fisheries from 1850 to 1980. The Californian world view at the time was characterized by the primacy of maintaining the economic benefits of a fishery and the denial that fisheries are subject to ecosystem constraints. It thus hamstrung attempts at fisheries management. The outcome was the rise and subsequent collapse of one Californian fishery after another, for 130 years (McEvoy 1986).

We imagine that a more complete scientific understanding about the state and functioning of a fishery would ensure that the Californian outcome is being and will be avoided. This is not necessarily the case. After all, a culture's world view still influences the gathering and interpretation of data and the manner in which the uncertainty in the sustainable yield estimates is accommodated by managers and fishers.

Consider the EU's experience from 2000 to at least 2011 as they tried to deal with the ongoing collapse of its North Sea fisheries. As happened in California, many of those participating in the ongoing debate focused on protecting fishing jobs, a way of life, rights, economic imperatives, political expediency and sovereignty (de Bresson and Augereau 2002; Brown 2004; Osborn 2003; Brown and Osborn 2001; Osborn 2000; Brown 2000; Osborn 2002; Froese 2011). Missing from their list were the realities experienced by the fish and the resulting human responsibilities.

The significant presence of these attitudes is illustrated by the need, in 2005, to fine almost all of Britain's Whitby-based fishing fleet for overfishing. It is also illustrated by the fishers' response: those fined chose to blame the EU regulations for forcing them to overfish. They saw the EU regulations as unnecessarily stopping them from fishing rather than as part of its effort to protect the stocks from being overfished (Brown 2005).

Consider too the three Scottish and Shetland Island fish-processing plants (among Europe's largest) and the 27 skippers of the fishing boats supplying it that were all fined for deliberately overfishing 170,000 tonnes of herring and mackerel between 2002 and 2005 (The Scotsman 2005; Carrell 2012). From where did they think the fish for the future would come? Today, the North Sea fishery is in the end phase of collapsing (Froese 2011).

When the current status of global fisheries, mentioned above, is considered within the context of culture, then the general conclusion is that, despite some management successes, most fisheries management debates only end when the fishery becomes an ecological basket case, unable to supply the fish we want (Ludwig et al. 1993). Behind this outcome lies a lack of cultural appreciation (acceptance perhaps) for the connection between ecosystems' functional constraints and the limits to the sustainable supply of wild foods. This lack of ecologically sound, culturally supported beliefs about the harvesting of wild stocks is further illustrated by the solutions being implemented to deal with the ongoing global decline in the availability of the preferred wild ocean fish. They include moving to a new fishery, fishing down the host ecosystem's trophic web and ocean-based fish farming (Christensen et al. 2011).

The consequences of the first two were discussed in chapter 9. Ocean aquaculture is discussed here. In the West, it is culturally acceptable to make the claim that open-ocean aquaculture can both easily substitute for wild fish and protect the remaining wild stocks, such as salmon. We forget that open-ocean fish farming can impact ocean ecosystems in the equivalent manner that land farming can impact land-based ecosystems (Bavington 2005). We are obliged, for example, to supply fish food; use chemical controls to deal with disease and pest outbreaks in the dense concentrations of penned fish; deal with waste; and control large predators, such as seals. Meeting these aquaculture obligations comes with costs to wild fish, their host ecosystems and humans.

For example, in 2002 some 45% of the fish meal used by the fish-farming industry in general (fresh and ocean water) was produced from wild fish stocks (UNEP 2007). In 2002, 3 kg of fish were required to make 1 kg of farmed salmon (Pauly and Watson 2003). On the coast of British Columbia, salmon in fish farms has increased the number of sea lice parasites affecting wild salmon stocks (Krkosek et al. 2007). In 2007, the highly infectious salmon anemia virus spread through the Chilean salmon farms despite the use of 385 tonnes of antibiotics that year (Legrand 2009).

This discussion about wild fish resources applies to other wild resources, such as forests.

In summary, the answer to the question of whether ocean fisheries and, by extension, other wild foods can satisfy our current wild-food demands depends on how the question is interpreted. If only availability is considered, then some species are in short supply or unavailable. But many are still available, for a price. From this perspective, there are some local limitations to supply and choice, but there is no global shortage.

If consumers expect that wild foods are locally sourced from populations and ecosystems that remain fully functional, then many will be disappointed. The non-economic costs for maintaining the supply of wild foods are often high, especially if the economic cost of the food must remain low. The question becomes more difficult to answer if we apply the caveat "as long as supplying wild food today does not negatively impact the future supply." This perspective will be addressed in chapter 14.

## Water

Humans use water to quench their thirst, clean items, dispose of waste, manufacture goods, and grow both food and fibre. For all but waste removal, these uses require uncontaminated fresh water. This section discusses our current freshwater supply-demand balance.

**Availability and Usage.** Only about 0.77%, or 10 million $km^3$ of the earth's water is liquid, fresh and thus suitable for use by humans. This water is a part of the earth's water cycle. The aspects of that cycle that are of most interest to us start with the evaporation of water from the land and sea. It is returned to the earth as precipitation (rain and snow).

The annual precipitation on land (about 110,000 $km^3$/year) is renewable water. It is either taken up by plants, evaporates or runs off the land. Of the precipitation taken up by plants, a small amount (<0.2%) is used by them to create their biomass. While creating it, they transpire the rest through their leaves into the atmosphere. The sum of the water transpired and evaporated back into the atmosphere is called evapotranspired water (about 70,000 $km^3$/year). The part of the annual precipitation that runs off the land (about 40,000 $km^3$/year) maintains the water levels in lakes, rivers, aquifers and dams as it flows to the sea to be re-evaporated. During the runoff's journey, some of it is available for us to appropriate and use (Postel et al. 1996).

*Accounting.* There are different ways to estimate the amount of water we demand. A top-down accounting method focuses more on dividing the total amount of the earth's fresh water into broad categories of our water usage. For example, the annual runoff into rivers, lakes and groundwater is about 40,000 $km^3$/year. In the early 1990s, we withdrew from it a global total of 4,430 $km^3$/year (11%). Of that 4,430 $km^3$/year, agriculture used roughly 2,880 $km^3$/year (65%), mostly for irrigation. Industry used about 975 $km^3$/year (22%) for activities such as power generation, material processing and goods production, and waste removal. Municipalities used about 300 $km^3$/year (7%) for direct human use, mostly for waste removal. Some 275 $km^3$/year (6%) was lost from reservoirs by evaporation. (Our "in stream" use, such as recreation, of 2,350 $km^3$/year can be added to our 4,430 $km^3$/year to result in a maximum estimate of our total usage.) We thus use 11% or 17% of runoff. In contrast, the total annual amount of evapotranspired water is about 70,000 $km^3$/year from which, in the early 1990s, we appropriated 18,200 $km^3$/year (26%) during the growing of our rain-fed crops, pasture and the trees we harvested.

In the early 1990s, the sum of our evapotranspired and our runoff withdrawal water uses was 22,650 $km^3$/year (excluding our in-stream use) or 24,980 $km^3$/year (including our in-stream use). Our maximum total water appropriation was thus around 23% of the roughly 110,000 $km^3$/year of rainfall that fell on land (renewable water). If we just consider the part of the renewable water that is available to us (i.e., from which we could reasonably expect to make withdrawals), then our maximum appropriation increases to 30% (Postel et al. 1996).

A bottom-up accounting method focuses more on summing the amounts of water required by each of our many activities that use it. It provides a different insight into our demand for water. However, in order to gain that insight, an explanation for a few useful concepts and terms needs to be provided before describing the method.

Every day, we will each, on average, drink a minimum of about 2–3 litres of water. We will also use water for activities such as washing clothes, cleaning dishes, bathing and flushing toilets. These are our direct uses of water. In satisfying our other wants, such as food, electricity and manufactured goods, we will also use water, but indirectly. For the bottom-up method to work, we have to know the total amount of water we use both directly and indirectly to satisfy each want.

The amount of our direct water use can be measured. How is the amount of our indirect water use determined? The preferred method is to estimate, for each want-satisfying item, the amount of water consumed and the amount used for waste removal during its production.

Water that is "consumed" becomes unavailable for further use. We tend to assume that an item's consumed water refers to the amount that becomes part of it. This is only partly true. By definition, the water consumed to provide our plant foods or plant resources is the sum of the water that the plants transpired while growing and the water that evaporated from the fields in which they grew (evapotranspired water). The water consumed to manufacture an item is the amount transformed to become part of the product plus the amount evaporated (Hoekstra et al. 2011). Consumed water should not be confused with the withdrawals of water from surface or ground water. For plants it should not be confused with just the amount incorporated into the plants' biomass.

The amount of water used to produce an item is the sum of the water consumed and used for waste removal during its production. This is called the item's embedded or virtual water "content." The average virtual water content of item types, such as rice and wheat, cover a wide range. Examples of these averages for everyday items are provided in the next three paragraphs.

Examples from plant-based foods are the following: 1 kg of dry rice (in the state found in the store) contains about 3,420 litres of virtual water; 1 kg of wheat, about 1,330 litres; and 1 kg of corn, about 910 litres. For comparison, 2 g of the earth's average biomass is associated with about 1,000 litres (1 kg) of evapotranspired water. In other words, 1 kg of the earth's biomass has an average virtual water content of about 500 litres of water. In comparison, growing our food is water intense.

Examples from our animal-based foods are the following: 1 glass of milk (200 ml) contains about 200 litres of virtual water and 1 kg of boneless beef about 15,497 litres (a standard 113 g hamburger patty contains 1,751 litres). Beef, like all meat, has a high virtual water content because, during a cow's life (three years for an industrial cow), it requires water to turn grains and roughage (pasture, hay, corn-based feed, etc.) into meat. The process of converting the plant material into protein is inefficient. It is so inefficient that only 155 litres of the virtual water in 1 kg of boneless beef comes from the cow's direct drinking of water. The rest comes from the virtual water in the human-provided grains and roughage. As a result, beef contains some 17 times more virtual water than a kilogram of corn.

Examples from our more processed foods and manufactured products are the following: 1 slice of bread with cheese contains about 90 litres; a 200 g bag of potato chips, about 185 litres; a cotton T-shirt, about 2,000 litres; a cup (125 ml) of coffee, 140 litres; and a 150 g hamburger, about 2,400 litres (Hoekstra and Chapagain 2006). A 32 MB memory chip 1.2 cm$^2$ in size manufactured in 2004 contains between 5 and 60 litres/cm$^2$ (Williams et al. 2002). (To compare these values to the amount we drink each day, divide them by 2–3 litres.)

I can calculate my personal water demand from the bottom up by adding the virtual water content of all of the foods I eat and the products I acquire (my indirect water use) to the water I directly use in all of my domestic activities, such as removing waste. Similarly, a country's water demand can be calculated by adding the water directly used by its citizens for domestic purposes to the virtual water in all of the agricultural and manufactured products the country produces. Of particular importance is a country's net (internally used) water demand: the sum of its domestically used water and its net trade (import minus export) in water. (The traded water is almost all virtual water.) The sum of the net (internally used) water demand for all countries is a measure of our (bottom-up) total global water demand.

A bottom-up estimate of the average annual global water demand for the period 1996 to 2005 was reported as 9,087 Gm$^3$/year. (Note that Gm$^3$/year is the same as km$^3$/year. The term km$^3$/year will be used here. The difference between the top-down and bottom-up water-use estimates will be discussed in the next subsection). Of that water agricultural production accounted for 92% (crop products 81%, animal products 11%); industrial production, 4.4%; and domestic use, 3.6%.

Based on the average annual global population of 6.13 billion for that period, the average individual's total (direct and indirect) water demand was 1.48 million litres/person/year. This is 4,063 litres/person/day or, at 164 litres/barrel, about 25 barrels/person/day. If only the 3.6% used for domestic purposes is considered, then each person on earth used an average of 146 litres/person/day (0.9 barrels), of which we drink about 2–3 litres/person/day. Imagine hauling that 146 litres to your house every day.

The bottom-up estimate of our total global water demand at 9,087 km$^3$/year can be divided up in other interesting ways. Some 1,762 km$^3$/year (19%) was tied up in international trade items, of which some 91% were agricultural products and 9% industrial products. Some 43% of this traded water is accounted for by the oil crops, such as cotton (>50%), soybean (20%), palm and rapeseed (canola) oil (including its products).

The remaining 7,325 km$^3$/year (81%) of our global water demand was in non-traded items (i.e., the water used internally, within the average country's borders). Of this water, 92% was in agricultural products, 3% in industrial products and 5% was used for domestic purposes. The dominant agricultural items were cereals, meat and milk. Overall, regardless of whether our global water use is associated with trade or internal use, it is overwhelmingly tied up in agricultural products.

Our 9,087 km$^3$/year global water demand can also be divided into three water types: green, blue and grey. Green water is the rainfall consumed to grow crops and pasture. Blue water is groundwater plus surface water consumed to grow crops and manufacture industrial goods. Grey water is the water used to remove the waste from homes, the manufacture of goods, and the production and use of crops (i.e., polluted water). Grey water accounts for our waste removal usage; green and blue water together account for our production and direct usage.

Our 9,087 km$^3$/year demand is divided into 74% green, 11% blue and 15% grey. In particular, of the 1,762 km$^3$/year of our water demand tied up in traded items, about 68% was green, 13% blue and 19% grey. This illustrates that rainfall is clearly critical for both trade and non-trade purposes, and a significant amount of polluted water accompanies traded goods.

By dividing the bottom-up estimate of our total global water demand according to each countries demand, the variation in their contribution to our total water demand is revealed. China, India and the United States had the three largest demands (1,207, 1,182 and 1,053 km$^3$/year respectively). India had the largest blue (ground plus surface water demand, which was mostly used to irrigate wheat, rice and sugar cane fields. China had the largest grey water demand.

On a per capita basis, the United States had one of the highest water demands at 2,842 m$^3$/year/person. Bangladesh had one of the lowest at around 750 m$^3$/year/person. In between lies China at 1,071 m$^3$/year/person and India at 1,089 m$^3$/year/person.

Of the global gross blue water in exports, 49% was provided, collectively, by the United States, Pakistan, India, Australia, Uzbekistan, China and Turkey. Of these countries, all except Turkey were net exporters of

(essentially virtual) water and all are, to some degree, water-stressed countries. This pits the importance of their internal water security against their desire to trade (Hoekstra and Mekonnen 2012; Hoekstra et al. 2011).

*Discussion.* There are significant differences between the bottom-up and top-down approaches to our global average annual water demand. For example, the bottom-up estimate of our water demand at 9,087 km$^3$/year (8% of renewable water) for 1996–2005 is much smaller than the top-down estimate of 24,980 km$^3$/year (23%) for the early 1990s. The bottom-up method assigns an average of 400 km$^3$/year to industrial production for 1996 to 2006, whereas the top-down method assigns 953 km$^3$/year for the early 1990s. The bottom-up method assigns 92% of our water demand to agriculture, 4.4% to industry and 3.6% to domestic use, whereas the top-down method assigns 65% to agriculture, 22% to industry, 7% to domestic and 6% to evaporation from dams. The following are some possible reasons for these differences.

The bottom-up method equates the demand for water with the sum of the amounts consumed and used to remove waste. Most significantly, it excludes the water demand of forests to grow timber. It also excludes the rainfall demand needed to sustain fallow crop land and pasture that does not produce a product. The bottom-up method also excludes collateral water usage, such as that lost in leaking pipes and evaporation from dams. In contrast, because the top-down method is focused on total appropriation, it includes these demands.

Some of the disparities are likely due to differences in the source data and definitions. The two methods also likely differ in the decisions about what data is missing and how to estimate it. Overall, the bottom-up study should be thought of as providing us with an uncertain estimate of the balance between the wide range of our demands for water to consume and remove waste and the various sources of that water. In contrast, the top-down study gives us an uncertain estimate of our complete water demand or appropriation (Hoekstra and Mekonnen 2012; Mekonnen and Hoekstra 2011; Mekonnen and Hoekstra 2010).

Despite these differences, the following conclusions can be drawn. Our use of water is dominated collectively by rainfall-watered timber, crops and pastures. If the water used for timber and non-productive pastures are excluded, then our agricultural animals and crops (food and fibre, but not trees) still use and consume much more water than industry. Direct rainfall is still the most important source of water for growing crops, even though irrigating lands increases our agricultural intensity. Changes to rainfall can have a significant impact on agriculture. Think climate warming.

Other conclusions are that a significant proportion of our water demand is used to remove waste (grey water). Internationally traded water is dominated by the virtual water in agricultural products. However, a significant proportion of traded water is virtual waste water.

The estimates of our current water demand given above should be placed in a historical context. They represent the period after the slowing of the rapid, post-1900s increase in all uses of water. During that period, our total freshwater withdrawals from all sources rose fivefold (Gleick 2001). This rate was faster than the rise in global population over the same period (Vörösmarty and Sahagian 2000). This increase in our per capita demand for water was driven by a combination of the greater demands from the green revolution crops, our personal activities and expanding industrial production.

The significance of that increase and the scale of our current, global water demand is illustrated and symbolized by the drying up of the Aral Sea. So much water is being withdrawn from the lake's watershed to supply agricultural needs, especially cotton farming, that there is insufficient flow in the Syr Darya and Amu Darya to fill the sea (Micklin 1988). The proportion of water withdrawn from the Aral Sea's watershed is not unique. The flows at the mouths of many major rivers, including the Nile, Rio Grande, Colorado and Yellow, are either significantly reduced or intermittent.

These reduced flows are a reminder that the earth's fresh water resource is unevenly distributed across the globe. It is also an indication that, at least on a local scale, our demand for water is not always balanced by the supply. This can be expressed by the proportion of the human population suffering from limited or no access to water.

Unfortunately, the estimates vary widely, in part because the assessments are made using a variety of terms. For example, compare being water stressed (lacking adequate water) to suffering a water shortage. A water shortage can mean either a physical shortage (lacking available water) or an economic shortage (available water exists but it isn't being made available to people).

The following are numerical examples of these assessments. Between 1995 and 1996, an estimated 1.9 billion people (33% of that period's 5.7 billion) lived under medium-to-high water stress. By 2000, between 0.5 and 1.5 billion people (8% to 25% of that period's 6.1 billion) lived under severe water stress. Most of these people lived in dry regions (Vörösmarty et al. 2000).

An estimate of water stress arising from the uneven distribution of water due to both climate and the limitations to human water systems was conducted in 2001. It concluded that about 1.1 billion people (about 18% of that period's 6.2 billion) lived without safe drinking water and 2.4 billion (about 39%) without adequate sanitation services (United Nations 2003). In contrast, a 2007 report estimated that 1.2 billion (18% of that period's 6.6 billion) suffered from a physical water shortage and 1.6 billion (24%) from an economic water shortage (Molden 2007). It seems that 18% to 39% of the world's population routinely experiences some form of water supply issues.

A country has three commonly used methods to deal with its lack of water: importing of virtual water in the form of traded products (discussed above); pumping of local groundwater (discussed here); and redistributing surface and ground water from where it is available to where it is needed (discussed below). The alternate solutions of conservation, reuse and increased efficiency are slowly gaining popularity.

The significance of pumped groundwater as a source of water worldwide is illustrated by its use in the United States. A 1995 estimate suggested that more than 50% of the US population relied on groundwater for their drinking water. Overall, withdrawals from groundwater accounted for some 22% of the United States' total water usage, including agriculture (Glennon 2002).

There is a constraint on using groundwater to meet human wants. If it is removed from an aquifer faster than it is replaced by surface water, then the aquifer is not being recharged: its water is being mined like a metal. The groundwater is being depleted. A global assessment of groundwater withdrawals for all but the earth's most humid regions estimated that the average amount of unreplenished groundwater used in 1960 was 126 km³/year. By 2000, the level had increased to 283 km³/year. Globally, on average, groundwater is being depleted, and the rate of depletion is accelerating.

The significance of this groundwater depletion is highlighted by the limitation it imposes on agriculture. Globally, between 16% and 33% of the groundwater withdrawn for agricultural purposes is not being replenished. Those agricultural regions where this depletion is concentrated face a decline in their current level of agricultural production as the aquifers supporting them run dry. Of particular concern is northern India because its food production relies heavily on aquifer-fed irrigation. The area supported by the Ogallala aquifer in the United States also faces groundwater depletion (Wada et al. 2010). Groundwater depletion is a serious, under-appreciated issue.

### Redistribution.

Fresh water can be present in a region, but, from the human perspective, it is in the wrong place at the wrong time. To resolve this problem, we build water redistribution systems consisting of dams, pipelines, pumps and canals. Substantial redistribution works were first built more than 4,000 years ago. In the recent past, the number and size of these works has increased at a staggering rate. The scale and increase is illustrated here by dams.

Today, globally, there are over 300 dams that can each potentially hold back more than 25 $km^3$ of water. Almost all of these giant dams were built in the 1900s, most of them after 1950. Today there are more than 45,000 dams greater than 15 m tall. They can, collectively, hold back more than 6,500 $km^3$ of water (average of 0.14 $km^3$/dam) (Nilsson et al. 2005). If dams of all sizes are considered, then in 2000, our dams held back about 8,400 $km^3$ of water. This is about 21% of the estimated total, global, annual rainfall runoff of 40,000 $km^3$/year (Vörösmarty and Sahagian 2000). If these statistics don't leap off the page to give you a sense of the scale of our dam-related water redistribution efforts, then look at a picture of the Hoover Dam in the United States or the Three Gorges Dam in China. Their size can only be appreciated if you find the dots that are vehicles and the specks that are people. Think too of the symbolic nature of a culture's monumental structures.

The redistribution of the water stored behind these dams produces hydroelectricity, irrigates crops, supplies our domestic and industrial needs, and helps to remove our waste. Clearly, redistribution is an important part of meeting our water wants. But not all the water in the redistribution system reaches its destination. Water is lost through evaporation and leaks from dams, canals and pipes. For example, evaporation and seepage results in losses from the Aswan High Dam on the Nile, Egypt, that amount to some 13% of the inflow into the dam. For the Kariba Dam, which occupies the border between Zambia and Zimbabwe, about 20% is evaporated (Vörösmarty and Sahagian 2000).

Further losses take place through leaks in the rest of the redistribution systems, from canals to city piping. Consider that up to 30% of water in municipal water supply systems can be lost through leaking pipes. This leakage can occur regardless of how well or badly the city's redistribution system is developed as illustrated by Mexico City and New York (Burkeman 2002; Gleick 2001).

### Costs.

There are a number of non-economic costs that accompany our appropriation and redistribution of so much fresh water. This section will illustrate these non-economic costs by focusing on the dam-related costs borne by ecosystems (ecosystem costs) and humans (human costs).

The scale of the ecosystem costs is revealed when considering the world's 292 largest river systems. They carry 60% of the world's runoff from 54% of the earth's land area. Of these 292 rivers, only 35% are completely free of dams, 52% have dams in the main channel, and 65% have one or more dams in their watershed. The collective capacity of the dams on some of these rivers can hold back more than a year's worth of the river's discharge. A significant proportion of the world's water flow has been affected by dams (Nilsson et al. 2005).

Dams not only affect a river's water flow. Almost all rivers carry sediment to the sea. Most of this sediment is generated in the upper reaches of a river's watershed. Once a river is dammed, this sediment is deposited into the lake behind the dam. For dams of all sizes, an average of around 26% of the sediment inflow remains trapped behind the dam's wall. This percentage is high enough, and enough rivers are dammed, that the total amount of sediment now being carried to the ocean each year has been significantly reduced. The pre-dam global average annual silt load of 14 billion tonnes/year has been reduced to some 12.6 billion tonnes/year today. This reduction has occurred despite the increase in the global average rate of erosion caused by the dam-facilitated increase in human activity in the watersheds upstream from dams (Syvitski et al. 2005).

The dam-caused changes to the amount, timing and nature of a river's pre-dam flow and its sediment load are accompanied by changes to the river's other water variables, such as temperature and nutrients. Joining these changes are dam-related changes to the land. They include an increase in erosion, mentioned above, and the reshaping of the downstream land in order to redistribute and use the water in the dam. The overall result is significant impacts on aquatic and land ecosystems in both the river's above- and below-dam portions of its watershed. Unsurprisingly, the most severely affected ecosystems are found in those watersheds whose rivers have been dammed the most.

The resulting watershed-scale ecosystem impacts are the product of many specific local- and regional-scale ecosystem disruptions and destructions. The most visible impacts occur immediately upstream, behind the dam. There, the valley bottom ecosystems are underwater: destroyed. The impacts downstream from the dam are just as significant. Some are obvious, such as the irrigated fields, others are not.

The less obvious impacts are most easily seen in a river's delta and its highly productive low-lying flood plain. The reduction of a river's flow by a dam results in its delta wetlands being reduced in area. This occurs partly through an absolute reduction in water volumes. It also occurs because there is a reduction in the frequency, timing and size of floods. This changes the amount and timing of the water and nutrient-renewing sediment delivered to the delta's flood-dependent grassland and wetland ecosystems, which rely on these resources to function. The low flow and reduced sediment load also facilitates increased salt water flooding and erosion.

As the impacts on a river's delta suggest, dam-related ecosystem costs do not exist in isolation. They interact with one another in a complex manner. They also interact with our other water-affecting activities, such as land clearing. As a result, the costs of satisfying our water demand by using dams are greater and more widely distributed in time and space than we expect from the above descriptions. The scale of the overall impact is illustrated by the Danube River with its many dams.

There are so many dams on the Danube that for its first 1,000 km there is, on average, one dam every 17 km. They help provide food, power and water to the large population in the Danube's watershed. At the same time, the runoff from the watershed's farms and the disposal of the population's waste water end up in the Danube. The dams also trap much of the Danube's silicon-bearing sediment. As a result, the water that leaves the Danube's last dam to flow the final 1,000 km into the Black Sea contains up to 80% less silicon, but more nitrogen and phosphorous, than it did in pre-dam times.

After the Danube flows through its delta, it delivers nutrients and minerals to the western Black Sea's nearshore and offshore marine ecosystems off the river's mouth. There, the excess nitrogen and phosphorous causes excessive phytoplankton blooms. They lead to the same eutrophication-related dead zones as those occurring in the Gulf of Mexico, off the Mississippi Delta, but with a twist.

The base of these marine ecosystems' trophic webs was historically dominated by a beautiful silica-shell-making phytoplankton (marine plant) species called a diatom. The ability of diatoms to grow and make their shells is limited by the relative proportions of silicon, nitrogen and phosphorous dissolved in the water they live in. Any one of these elements can limit the diatoms' growth, but in the Black Sea, silicon never did because it was supplied in abundance by the Danube.

Today, the decreased amount of silicon limits the growth of the diatoms. The excessive phytoplankton blooms are now dominated by species that don't construct silica shells. The change in the composition of the Danube River has caused both a larger, more intense dead zone and a drastic change in the phytoplankton species mix. The dams, along with pollution, have significantly disrupted the western Black Sea's marine ecosystems from the bottom up (Humborg et al. 1997; Teodoru et al. 2006a; Teodoru et al. 2006b).

The non-economic costs of dams are not only borne by ecosystems. Although dams can provide benefits to people, such as flood control, the redistribution of water and the provision of hydroelectricity, they also burden the people with significant costs. The most obvious are the costs borne by those whose farmlands, homes and livelihoods now lie underwater behind the dam.

Some of the less obvious human costs are illustrated by people living in the Nile Delta. They are experiencing the erosion of their farmland, loss of fertilizing silt, flooding of farmland by salt water, incursion of salt water into freshwater aquifers, loss of fish breeding areas and the loss of nearby food fisheries. They also experience a reduction in irrigation opportunities (Nilsson et al. 2005; Nilsson and Berggren 2000). These costs are being exacerbated by a rise in sea level.

Consider too the dam-related cost to human health and the delivery of health services (UNEP 2007). Schistosomiasis (bilharzia) is a parasitic fluke that severely debilitates the infected person. The irrigation canals associated with the building of dams in hotter climates facilitates the spread of this disease. For example, after the Aswan High Dam was built on the Nile, the number of schistosomiasis cases increased (Dzik 1983). Consider too that micro dams were built in an area of Ethiopia subject to famine in order to increase food production. However, the dams also resulted in a 7.3 times increase in the incidence of malaria among children living near them (Ghebreyesus et al. 1999).

Worldwide, if a large dam is built in an area where the soils that will be flooded contain inorganic mercury, then their flooding can allow bacteria to convert the mercury into the potent neurotoxin methylmercury. It can bioaccumulate up the food web into the top trophic-level predator fish thus rendering them unsafe for human consumption. Local communities who rely on these fish can be severely disrupted by either the health effects from eating the fish or the loss of the fishery (Rosenberg et al. 1997).

We would imagine that, by relying more on groundwater, we could isolate ourselves from many of the costs associated with dams. After all, the water is underground, and we gain access to it by pumping it to the surface. This is not the case because the use of groundwater comes with its own constraints and costs. Groundwater is replenished by surface water. We are thus limited to withdrawing groundwater from an aquifer at a rate that is less than its recharge rate, if we wish to avoid depleting the aquifer.

If we meet this condition but the surface water replenishing the aquifer is polluted, then the aquifer will become contaminated. For example, in England, some 81% of groundwater aquifers are threatened by surface pollution. The most common pollutants are nitrates from fertilizer and, unsurprisingly, atrazine (Environment Agency 2006).

If the withdrawal rate from an over-pumped aquifer is reduced to below its freshwater replenishment rate, then the aquifer can eventually recover. Unfortunately, not all aquifers will be recharged. For example, the very large Ogallala aquifer, United States, was last replenished from the surface some 10,000 years ago, at the beginning of our interglacial period. Its "fossil" water is essentially being mined. No wonder that the water table in those parts of the Ogallala aquifer that lie below Texas have already dropped some 45 m (Glennon 2002). As these aquifers run dry, those relying on the water pay the social cost of being forced to relocate themselves and their industry.

We use most of our withdrawals of surface water (mostly mainly from dams) and groundwater to irrigate our crops. One surprising potential cost from this use is a decline in soil fertility. This occurs because all water, but especially groundwater, contains dissolved mineral salts. When this water is used for irrigation in drier regions, its rapid evaporation deposits salts in the irrigated soil: it becomes salinized. This reduces the soil's fertility because our crop plants have a limited tolerance for mineral salts.

Although the loss of soil fertility through salinization is less common than its loss through soil erosion and nutrient depletion, salinization can present a significant long-term cost to food production in some regions (McNeill and Winiwarter 2004). Consider the extensive irrigated fields on the fertile plains of the rivers flowing through hot, dry regions of the world. For example, salinization is occurring in those irrigated areas bordering the Indus River in Pakistan, around the Aral Sea and in California's Central Valley.

The long-term significance of irrigation-related salinization is illustrated by ancient Babylonia, Iraq (Jacobsen and Adams 1958). The loss of soil fertility and the resulting drop in food production changed Babylonian history. Today, that legacy and the ongoing salinization in southeastern Iraq still limits the area's food production.

Culture. Our culture's world view guides us to use particular references and preferences when thinking and making decisions, as discussed in chapter 2. Examples are our culture's preferred views about technological, economic, political, social, spiritual and environmental matters. In this way, the guidance from our culture's world view influences our relationship with water. In particular, it influences the value we place on water, its uses and which of the non-economic costs associated with supplying, using and discarding it we find acceptable. It influences our perceptions of water issues and constrains our conceptions of possible solutions.

It also ensures that our relationship includes both a practical and a symbolic aspect. Consider the value we give to water. Its possible symbolic value is illustrated by those being immersed or touched with water as part of a religious ceremony, or venerating a river, such as the holy Ganges, a lake or a spring. Its possible practical value is illustrated when those with the rights to irrigation water carefully portion it out, and when the drivers of recreational vehicles churn up wetlands and creeks. Consider too that the water behind a dam has both a practical aspect (it is used for irrigation and recreation) and a symbolic aspect (it is part of a monumental structure).

Considering the breadth of our culture's influence, it is unsurprising that we tend to display a conflicted and disjointed, confused perhaps, relationship with water. This is especially true for complex industrial cultures. A personalized Western cultural example of this conflict is provided by the cut flower industry. The phrase *cut flower* is used for flowers with stems, such as roses, carnations, chrysanthemums and irises, that are grown for the florist trade. A significant proportion of the four trillion or so cut flowers flown into the United States in 1999 came from South America. Europe has a similar cut flower trade with Kenya. The main growing areas are around Bogotá, Colombia, and Lake Naivasha, Kenya.

In these areas, the cut flowers that satisfy the market are grown using highly intensive farming practices. Large amounts of water are used and water-quality-affecting pesticides are applied at rates among the highest in the agricultural sector. The costs of growing these high virtual water content flowers are borne by the producing area's population and ecosystems. They both suffer from a reduction in the availability of clean water and the health impacts of the pesticide-polluted water (Ogodo and Vidal 2007; Maharaj and Hohn 2001; Maharaj and Dorren 1995). Through these connections, the culturally influenced high value that Americans and Europeans place on the giving and displaying of beautiful, virtual-water-rich cut flowers from Colombia and Kenya conflicts with the resulting decreased quantity and degraded quality of the water available for the local users in the flower-growing areas.

Consider too that for many cultures the ability to use water without concern symbolizes affluence and luxury. Similarly, the ability to control it symbolizes status and power. An example of both from the Renaissance are the Italian water gardens that were built around water features such as waterfalls, ponds and fountains.

Today, the power of water's culturally supported symbolic value is most easily seen in the tourism industry catering to those who want to escape a cold, wet northern winter by relaxing at a luxury resort in a warm, often dry part of the world. Both the resorts' promotion and the tourists' expectations of such a holiday include water-expensive recreation opportunities such as swimming pools, golf courses and decorative gardens. Making this luxury available requires a large amount of water that may not be plentiful in the preferred tourist destinations. Despite this reality, it is difficult for authorities there to restrict this luxury use of water. To do so could easily appear to be an attempt to restrict the good life for the tourist and thus the development, employment and government income from providing it.

The ultimate example of this type of culturally supported luxury use of water comes from the myopic gambling heaven of Las Vegas, a desert city. Some two-thirds of the city's water demand during the 1990s was used outdoors for purposes other than growing food. At the end of the 1990s, Pat Mulroy, general manager of the Southern Nevada Water Authority, was asked to comment on the contradiction between Las Vegas's chronic water shortage, people wanting to grow grass lawns, and no attempt being made to ask patrons in the hotels to conserve water. She replied, "People don't want to live in reality" (Leslie 2000, 52).

The cultural influences on our relationship with water can be recognized not just in the drier parts of the world, but anywhere that water resources are used. Consider our decisions about water-related issues. Our perception of an unsatisfactory supply of domestic water and how we will accommodate its human, economic, technical and environmental aspects into an acceptable solution is influenced just as much by our culture's world view as by factual information (Martinez-Austria and Hofwegen 2006; World Water Forums 2002). This multi-faceted nature of our culture's influence is illustrated by a tragic example from Bangladesh.

Much of Bangladesh's surface drinking water has a long history of being contaminated by disease-causing organisms. Addressing this issue directly by making cultural changes that would reduce the surface contamination was seen as difficult, especially as the population was growing. The decision makers favoured the "easier" technical solution of supplying the region with groundwater from wells drilled into a shallow aquifer. The well-drilling program started in 1971. It was only in 1993 that water testing detected the presence of an ongoing arsenic contamination problem (Mead 2005; Frisbie et al. 2002). Today, the health of millions of Bangladeshis is affected by arsenic, with the possibility that tens to hundreds of thousands are suffering from chronic arsenic poisoning.

A description of how the arsenic enters the borehole water illustrates other cultural connections to this tragedy. Traditionally, rural Bangladeshis excavate gravel to construct the dry platforms on which they prefer to build their homes. These open gravel pits accumulate water and organic matter to become a pond. Eventually, the pond water percolates far enough into the ground that it replenishes the groundwater in the shallow aquifer tapped by the drilled wells.

Because the composition of the pond water and the groundwater are different, the mixing of the two changes the groundwater's chemistry in important ways. Its new inorganic composition makes the arsenic-bearing minerals present in the aquifer's host rock unstable. At the same time, the carbon added from the pond water allows bacterial action to release the arsenic into the groundwater.

The arsenic-rich groundwater is then pumped from the wells to the surface where it is used as irrigation water in the rice paddies. It is also used for drinking water and for the preparation of the culturally preferred, water-intense rice-based diet. These uses all expose the people to the arsenic in the groundwater. Pumping also provides the hydraulic gradient needed to move more pond water into the aquifer (Neumann et al. 2009).

This story started with a simple attempt to address a shortage of clean drinking water. It ended with the largest arsenic poisoning in history. It illustrates that a culture's world view influences water matters indirectly as well as directly. It directly influences how they view the issue and limits their possible solution. Among the generally applicable, indirect influences are population growth and food preferences. It also illustrates that the non-economic costs from satisfying a water demand, such as ecosystem disruption, destruction and pollution, can affect future supplies by disrupting the water cycle. The resolution to water issues is complex, especially if the influence of culture is considered.

In summary, can the earth supply today's water wants? As with food, the answer is determined by our perspective of the question. On a global scale, the averages suggest that it can. On a regional scale, our water capture and redistribution systems successfully supply adequate water to many. However, there is still a significant number of regions that remain undersupplied with water and face water shortages. The reasons are both human and natural in origin.

If we consider the impacts of appropriating and redistributing so much water, then we realize that the non-economic costs of alleviating water supply issues are high. These costs range from the disruption of ecosystem functioning to the additional responsibilities we foist on ourselves by altering the water cycle. These costs can significantly reduce the benefits we receive from our water appropriation and redistribution activities.

The days of easily increasing an abundant supply of water are passing. Consider the increased competing interest for water and the concerted efforts to increase the efficiency of water use. For example, the attempts to reduce the amount of water needed to grow a crop (Postel 2001).

If the question of adequate supply includes an assurance that meeting today's supply will not impact future supplies, then the focus shifts to other influences on possible solutions. These include our cultural views about an acceptable water future, its preferred use and climate change. This view of the question will be addressed in chapter 14.

# Energy

Our hunter-gatherer ancestors lived within the limits of the food and wood energy available from their host ecosystem. The discovery of other energy sources, such as animals, wind, water, coal, gas and oil allowed us to substantially increase our use of non-food energy. Today we use this energy to transport us, resources and goods all over the world. It also enables us to supply ourselves with an abundance of food, water, effort-saving devices and entertainment. How easily is our energy demand being satisfied and at what cost?

**Source and Production.** Until the Middle Ages, all work in Europe was accomplished by some combination of human labour (personal, co-operative, paid, peasant or slave), domestic animals and wood combustion. During the Middle Ages, these sources of energy were joined by machines powered with water and wind. In the late 1700s, coal was added as a fuel to power steam-driven machines. Near the end of the 1800s, coal gas was also used for lighting and heating, while oil was used to power the newly invented internal combustion engine. The three fossil fuels of coal, oil and natural gas were joined in the early 1900s by electricity. Initially generated by water, it was subsequently produced using oil and coal. Electricity generated by nuclear fuels appeared after the 1950s.

The significance of this energy transition over the last 200 years is illustrated by events in the United States. Around the 1850s, about 70% of their annual output of work was provided by animals, 20% by fuel (almost all of it wood, the rest coal), and 10% by humans. By 1980, some 130 years later, about 95% of work was provided by fossil fuels (oil, gas and coal), 2.5% by hydro and nuclear and 2.5% by wood (Hall et al. 1986). Fossil fuels had replaced animals, wood and humans as sources of non-food energy.

The more recent global-scale changes in energy sources are illustrated by comparing energy demands in 1973 with those in 2010 (More recent data to 2013 is consistent with the following discussion). The total global energy use in 1973 was 71,024 terawatt hours (TWh). Fossil fuels provided 86.7% (oil 46.1%, coal 24.6% and natural gas 16.0%). The remaining 13.3% was provided by nuclear (0.9%), hydroelectricity (1.8%), combustible renewables (10.5%, including wood, biofuels and waste) and other energy sources (0.1%, including wind, solar, geothermal and tidal) (figure 13b).

By 2010, energy use had doubled to 147,898 TWh. But the energy sources and their proportions had remained roughly the same. Fossil fuels provided 81.1% (oil, 32.4%; coal, 27.3%; and natural gas, 21.4%). The remaining 18.9% of our supply was provided by nuclear (5.7%), hydroelectricity (2.3%), combustible renewables (10.0%), and other energy sources (0.9%) (IEA 2012a).

Today, the proportion of our energy demand provided by fossil fuels hovers around 80%. The amount of fossil fuels needed to satisfy that demand is staggering, and growing. For example, in 1973 about 20.6 giga (billion) barrels (Gb) of crude oil were produced. By 2010, the amount had risen to 30.1 Gb (IEA 2012a). Because of the dominant place of fossil fuels in our energy supply, this discussion about energy will focus primarily on them.

The method of extracting (producing) fossil fuels varies. Gas is pumped. Coal is mined like a metal, such as uranium. Oil is pumped when a liquid and mined like a metal when it is near solid, as it is in the tar sands.

There are only a few unevenly distributed areas on earth that can produce fossil fuels. Currently the majority of the most desirable sources of oil and gas occur in the Middle East, central Asia and Russia. However, the largest consumers are in Europe, China and North America. In order for production to satisfy demand, fossil fuels are transported (redistributed) long distances using trains, trucks, ships and pipelines. Although the details of each fuel's redistribution system vary, they must all function smoothly because fossil fuels are not stockpiled. They are extracted from the ground in amounts that closely match the demand.

We like to think of our fossil fuel production points and their distribution systems as being static. However, as sources are depleted and demands change, then so do the points of production, the quantities produced and their distribution systems. For example, Britain produced more than half of the world's coal until the 1870s. Although global demand for coal has risen since then, Britain now produces very little coal. Similarly, the United States produced more than half of the world's oil until the 1950s (McCabe 1998). By 1991, it consumed twice the oil that it pumped. It had become a net importer of oil because its production was declining and its demand rising (Ivanhoe 1995).

Similarly, as the easily accessed, favoured sources of oil and natural gas (conventional sources) are being depleted, attention is switching to less desirable sources (non-conventional sources). For example, oil is now being produced from the tar sands of Canada and Venezuela, and from oil shale in the United States. These switches in source type and location require changes to the methods of production and the systems of distribution.

**Consumption.** Our use of new energy sources over the last 200 years was driven by an increase in both the total amount of energy we use and the size of the human population. Our energy use increased faster. This is seen in the dramatic rise in our per capita use of inanimate energy since 1850. In 1860, the global population of 1.25 billion used an estimated global annual average of 0.9 megawatt hours per person (MWh/person) (Cohen 1995). In 1900, 1.64 billion used 3.7 MWh/person. In 1973, 3.94 billion used 18.0 MWh/person. In 1990, 5.28 billion used 19.2 MWh/person. That is a fivefold increase in our per capita energy use in just 90 years; (US Census Bureau 2008). In 2005, 6.48 billion used 21.0 MWh/person (IEA 2007). In 2010, 6.83 billion used 21.6 MWh/person (IEA 2012a).

The scale and significance of our global average per capita energy consumption can be more easily appreciated at a personal level by considering the following. The 1860 energy use of 0.9 MWh/person/year would have allowed each person to have kept a single 100 W incandescent light bulb lit 24 hours a day (2.4 kWh/day) for one year (0.88 MWh/year). That energy in the form of oil would fill 0.5 barrels.

In 2010, we used 21.6 MWh/person/year. This would have allowed each of us to have kept 24.5 light bulbs lit 24 hours a day for the year. A much more brightly lit life. Alternatively, each of us could have received that energy, at home on New Year's Day, in the form of 13 barrels of oil to use as we pleased.

These averages can be personalized further by comparing them to the amount of energy your body needs to move and do work. A typical diet provides us between 2,000 and 2,500 kcal/day. If we could use all of that 2,500 kcal (2.9 kWh) for lighting, it would power a single 100 W incandescent light bulb (2.4 kWh/day) for 1.2 days.

But we can't. About 1,500 kcal (1.74 kWh) of our food energy is needed just to maintain our basic body functions when we are at rest, perhaps in front of the TV. For example, about 20% to 25% of that at-rest energy is used to keep our brain functioning so we can watch a TV drama.

Only 1,000 kcal (40%) of a 2,500 kcal/day diet is available to us for non-basic activities. Assume that an average, healthy adult can devote an optimistic 15% of that daily, discretionary energy, or 150 kcal (0.44 kWh), to producing continuous useful work. If Citizen Average had decided to devote each day's useful work to reducing their 21.6 MWh energy debt for 2010, then by the end of the year, they could have personally supplied only about 160.5 kWh (0.77%) of their debt. To fill their energy gap by using other people's useful work, then they would have had to hire 134 people for that year.

In order to apply these numbers to yourself, you have to take into account that the global averages hide substantial differences in energy use between countries. For example, in 2010, Citizen Average USA used 83.1 MWh of energy, while Citizen Average Bangladesh used only 2.44 MWh. Thus, for the average US citizen to personally fill their energy needs in 2010, they would have had to hire roughly 516 workers, while the average Bangladeshi would have only required about 15 (IEA 2012a).

Costs. There are non-economic costs associated with both providing the energy we demand and using it. The costs of dams and thus of hydroelectricity have already been discussed. Here the costs associated with the extraction, transportation, combustion and use of energy from fossil fuels (and uranium) are introduced. They and the non-economic costs associated with other sources of energy will be discussed more fully in chapter 15.

When fossil fuel extraction and upgrading facilities are built, ecosystems are disrupted and destroyed. This is especially true for the strip-mining of coal and tar sands bitumen. Similar costs occur when distribution networks are built.

More insidious but just as significant are the costs associated with the wastes, by-products and leakages from the production processes. These include the contaminated water and flared sulphur-rich sour gas from oil and natural gas extraction, and the metal-rich, acidic waste water associated with coal and uranium mining. On average, the non-economic costs (per unit of energy recovered) of producing non-conventional oil and gas are much higher than conventional oil and gas. Examples include the larger amounts of contaminated water and the leakage of natural gas from coal-bed methane and shale gas deposits. Even more incredible are the four barrels of water used by the mining method to produce one barrel of oil from the Athabasca, Canada, tar sands (a non-conventional oil source) in 2011 (CAPP

2011). I invite you to view the storage ponds for the Athabasca tar sands waste water on Google Earth.

The non-economic costs for transporting oil and gas include the commonly acknowledged leakage and spills from wells, pipelines, trains and ships. An example is BP's Deep Horizon blowout in the Gulf of Mexico. The potential for these spills to have significant long-lasting, chronic impacts on species and ecosystems was illustrated by the Exxon Valdez oil disaster considered in chapter 9 (Peterson et al. 2003). Less recognized is the cumulative impact of the numerous small spills and leaks from our domestic use of oil products.

To release the energy in fossil fuels, they are burnt. Combustion takes place in devices from the internal combustion engine of automobiles to the furnaces of power stations. The costs associated with this combustion include their now infamous emissions of carbon, sulphur, nitrogen oxides and particulates that gave us urban smog, acid rain and changed the atmosphere's composition. For example, the global average energy use of 21.6 kWh/person in 2010 resulted in energy-related $CO_2$ emissions of about 4.44 tonnes $CO_2$/person (compare this to the 4.18 tonnes $CO_2$/person in 2004). The regional differences in these emissions are significant: each US citizen contributed, on average, 17.31 tonnes $CO_2$/person; China's citizens, 5.43 tonnes $CO_2$/person; while Bangladesh's citizens, just 0.36 tonnes $CO_2$/person (IEA 2012a).

Not forgotten is the disposal of the other forms of combustion waste. These include the warm to hot water from water-cooled, fossil-fuel-powered electricity generation facilities, the slag from the combustion of coal, and the radioactive waste from nuclear power stations and reprocessing plants.

Greater access to energy also allows us to manufacture more goods, which means more environment-affecting waste from their manufacture, use and disposal. For example, more consumer goods certainly mean more consumer waste, such as packaging, and more obsolete products, such as electronic waste (e-waste). These consumer-waste costs shouldn't appear as pollution on our land and in our air and water, often far from the source of the energy use, but they do.

Overall, many of the energy-related pollution costs are unexpected, long term and follow indirect paths. For example, British records for a 160-year period show a very strong correlation between the ratio of two wheat-crop-affecting pathogen strains in British wheat fields and the level of sulphur dioxide in the atmosphere. It seems that small changes to the composition of the atmosphere from fossil fuel combustion can determine which pathogen will affect a farmer's wheat crop (Bearchell et al. 2005).

The notion is often put forward that the ecosystem costs associated with using fossil fuels (emissions, in particular) are in fact declining because the efficiency of fuel use (defined as energy use per GDP) is increasing (Ausubel 1996). However, increasing efficiency only effectively reduces emission's ecosystem costs if it reduces the absolute amount of emissions. As the numbers above indicate, the absolute amount of energy and thus the absolute $CO_2$ emissions are rising. This is the same phenomenon that is occurring with water: the withdrawals per capita are falling slightly, but the absolute amount of water being used on a global scale and its non-economic costs are rising as the global population increases (Gleick 2001).

Culture. Just as it does for water, a culture's world view influences its members' relationship with energy: their decisions about their energy wants and use, its cash and symbolic value, and the non-economic costs they consider acceptable for providing and using it. Historically, the guidance from a culture's world view reflected the reality of a restricted energy supply. For Westerners, that changed during the Industrial Revolution of the 1800s.

The increasing availability and consumption of non-human energy during that period accomplished more than allowing Europeans to do more work. It made possible or tangible what they had previously only imagined. In doing so, energy changed their view of the world and their place in it. The increasing availability of energy allowed Europeans to have greater faith in themselves, their world view and its vision for the future. It facilitated the transition of their culture into one which expressed the Victorian ideas about the nature of the world (ordered and controllable), their vision of the future (embracing the notion of progress) and their faith in their cultural identity (in control, superior).

The essence of the practical and social changes that occurred in the 1800s are still aspects of Western culture's current world view. They are reflected in the guidance it provides to us about our relationship with energy. Deeply embedded in Western culture's current world view lies the expectation, assumption perhaps, of an ongoing abundant supply of energy from one source or another. Tied to that view is Western culture's support for energy-intensive lifestyles and consumerism, discussed in more detail below. As a result, we are more than physically reliant on abundant energy. We are culturally, and perhaps psychologically, dependent on it.

The features of Westerners' current relationship with energy are made visible by our answers to the question of how much energy is enough for us to personally live a satisfying life. We find that deciding on enough is elusive, and it is unusual for us to conclude we could use much less than we do today. Our relationship with energy was also made visible during the early 2000s by our reaction to our oil-price-induced heightened awareness of peak oil and the intense debates about global warming. Westerners were presented with the possibility of energy scarcity at the same time as being made aware of the life changing non-economic costs of providing and using energy. A conundrum arose because a solution to both shortages and climate warming included reductions in energy use. Considering this solution revealed the contradiction between our culture's ongoing promotion of an energy-intensive lifestyle, even as it promoted energy conservation efforts to deal with rising energy costs as a result of an oil shortage. Westerners were faced with deciding how to rationalize our culture's conflicting guidance.

Our ongoing, conflicted, perhaps confused, thinking and decision making about a resolution to this conundrum reveals much about the key aspects of our relationship to energy. Consider our ongoing support for the exploitation of fossil fuels that have high non-economic (social and environmental) costs, even though doing so strongly conflicts with our professed concern for protecting the environment and human health. Examples are the consideration given to exploring for oil in designated wilderness areas such as the Arctic National Wildlife Refuge (ANWR) on the North Slope of Alaska (Borger 2001); efforts to open up high-risk, deepwater oil fields in the Arctic; and the strenuous efforts to continue producing highly polluting, low-energy-return oil from tar sands. Think too of biofuels being considered an acceptable energy source even when the biofuels' feedstocks are either food crops or grown on land which could produce food crops (Wald 2007).

When confronted with energy issues, we inevitably end up focusing most intently on maintaining the current energy supply. Hence the solution of green, renewable energy. This reveals that Westerners' personal and cultural identity is tied to an ongoing supply of abundant inanimate energy. It also reveals the power of Western culture's guidance about energy.

In one sense, Westerners are not alone. Other contemporary cultures may not have been through our deep-seated, energy-favouring cultural transformation during a long period of abundant energy. Yet their members have at least come to place great store in the immediate benefits that cheap abundant energy can provide.

In summary, at present the answer to the question of whether the earth can meet our current energy demands is yes, it can. Our global distribution network has even resolved regional differences in energy availability. This assessment implicitly considers the associated non-economic costs to be acceptable.

When these costs are considered, then the price of providing and using energy, especially fossil fuels, is seen to be high. So high that their use is disrupting, on a global scale, the ability of the biosphere to supply the other resources and services on which we rely. These costs are rising with our increasing use of energy.

There are also indications that our ability to supply cheap-to-extract fossil fuel oil and gas may be declining while that the non-economic costs of supplying them may be rising exponentially. This raises the question of whether our culturally driven demands for energy and our energy withdrawals to satisfy present demands are affecting the future supply of both energy and other resources. This, and the question of whether we could live satisfying lives with less energy, is addressed later in chapter 15.

## Health Services

On the one hand, Westerners attribute our good health to our health services: their ability to stave off the malign aspects of the earth's biosphere. But when pressed, we agree that good health is also the result of our biosphere supplying us with adequate amounts of clean water, clean air and healthy food, and our personal lifestyle choices. The success of our health services is dependent on, not independent of, the earth's biosphere. We can better appreciate good health if we more fully understand the relationship between our health, our health services and our biosphere. Our appreciation starts with a bit of history.

History. During their lives, our hunter-gatherer ancestors were exposed to health-impacting organisms. If our ancestors were healthy and their host ecosystem was fully functioning, then the spread of these organisms was constrained. The limiting factors included our ancestor's defence systems, their low population density and the rarity of pathogenic organisms (Thomas 1978). Within this context, it was more likely that a hunter-gatherer would suffer and die from ailments related to hunger, trauma (misadventure) and septic infections than the illnesses we think of as classic infectious diseases, such as small pox (Kelly 1995). If a hunter-gatherer reached adulthood, then they could live a long life.

The context within which the members of today's complex cultures complete their lifeways and use health services is quite different. Their current experience of ill health emerged over time as the outcome of our hunter-gatherer ancestors settling and taking up agriculture. During this switch, changes occurred to their diet, their relationship with their environment, and their experience of illness.

For example, a farmer's diet was rich in simple carbohydrates, obtained by cooking carbohydrate-rich foods. As a result, their overall nutrition could be quite poor. They also routinely suffered from rotten teeth and related infections (Molleson 1994; Angela and Angela 1993).

Consider too that the members of farming cultures lived closer together at higher densities in permanent settlements for longer periods of time than most hunter-gatherer cultures. These settlements grew in size as the farmer's food-producing success led to a population increase. Farmers also domesticated the social animals, such as sheep and goats, with which they closely worked and even lived. Farmers were also more often and more closely in touch with organic-rich soil and animal dung.

The changes in lifestyle that accompanied our ancestors' transition into farmers resulted in them being more frequently exposed to unfamiliar, animal- and soil-hosted diseases, some of which switched hosts. It also resulted in them satisfying the rules of thumb for increasing the spread of diseases: increase the density of both host and disease. Thus our farming ancestors' experience of illness was different from our hunter-gatherer ancestors' experience.

Over time, as farming cultures changed to become more complex, so did the members' experience of illness. The connection between the two is illustrated by measles. The relationship between the density of human populations and the spread of measles among us is now well known. It indicates that when our ancestors were living in small, scattered groups at low density and low numbers, the incidence of measles would likely have been quite low and random.

However, the experience for those living in the post-7000 BC village of Catal Huyuk, Turkey, was likely a little different. Its population was about 10,000 people, an exceptionally large population for that period. Imagine that none of the residents were immune to measles. When a person infected with measles walked into the village, they triggered a spike in infections. As measles swept through the population, the villagers experienced an epidemic. The infected residents either acquired long-term immunity or died. Eventually, the number and density of the villagers still susceptible to the disease was too low to sustain the epidemic, so it ended. In addition, because the population was sufficiently small, the number of susceptible people left was too few to sustain the disease over the long term. That is, there were too few people to act as hosts at a low rate of infection. Measles faded into local extinction.

Over time, as children were born and immigrants arrived, the density of the susceptible people in the population eventually increased to levels that could again sustain an epidemic. The villagers just had to wait for the arrival of the next randomly infected person. As a result, measles epidemics were likely rare events around 7000 BC.

As the size of an agricultural culture's population grew and their culture became more complex, some 20% of its population settled in urban centres. The residents of a town or city housing more than 200,000 to 500,000 people would have had a different experience of a measles outbreak than those living in a smaller urban centre. The epidemic would still have ended because immunity and death reduced the density of those susceptible to infection, but measles would no longer have become locally extinct. There would now have been a sufficiently large number of susceptible people in the town that they could sustain the disease until births and immigrants returned the susceptible population to an epidemic-sustaining density. In these larger urban centres, the residents could have experienced a measles epidemic as often as every two years. If the birth rate was high enough, then the epidemic could have, depressingly for them, occurred as often as every year (Bjørnstad et al. 2002). Their experience of illness had changed substantially from that of their ancestors.

This illustration of the relationship between "increased population in constant contact with one another at a high density" and "increased disease infection rate" applies to other infectious diseases, vector-borne diseases and parasites. Thus, those of our ancestors who opted to live in often unsanitary urban environments preconditioned themselves to more easily and more often suffer from what were to them new or previously rare diseases, particularly the infectious diseases. Some of these diseases were chronic and debilitating, such as malaria for adults. Others could form killer epidemics, such as plague, cholera and smallpox, especially for the young.

The benefits of city living were, and still are, balanced by the reality that they are an excellent environment for the spread of diseases. Consider that the death rates in European cities over the last 2,000 years were, until recently, sufficiently high that their size was often only maintained through immigration from rural areas. Prior to the mid-1800s, the most prominent limiter of at least European populations was infectious diseases, which often occurred as epidemics (Burnet and White 1972). Well-known examples are the Black Death (plague) of the mid-1300s, which killed an estimated 30% of the European population, and the 1665 plague outbreak in London (Kelly 2005). In the 1600s and 1700s, 20% to 50% of the children and 10% of the adults that were infected with smallpox are estimated to have died.

For centuries, European's struggled, both practically and emotionally, to deal with the fairly frequent, apparently random outbreaks of disease and indiscriminate killer epidemics. There was an ongoing search for relief from the death and suffering caused by illnesses, and the dread from contemplating their possible occurrence. This concern and the culture's historical experiences influenced European culture's world view of death, disease, health and their relationship to God.

This is the context in which our more recent European ancestors completed their lifeways and contemplated the provision of health and health services. The guidance from their culture's world view, with its cultural memories of plagues, contributed in many ways to the development of today's public health systems. For example, in 1720, Lady Mary Wortley Montegue excitedly returned to England from Turkey, bringing the practice of smallpox variolation (inoculation using a small amount of live smallpox-infected material). Shortly thereafter, variolation services became available in England, for a fee.

Edward Jenner practised variolation. By building on his experience he had, by 1796, developed and published details of the much safer technique of smallpox vaccination (immunization using a different agent than the live disease). He based his method on cow and possibly horse pox. He offered the service free to all (Smith 2011). Perhaps he was influenced to do so by the periods of food shortages in England, the generally poor health of much of the population, and the philosophical discussions about how to resolve them, as discussed in chapter 3.

Similarly, although the connection between sanitation and disease had been recognized in earlier centuries, it was not rigorously applied in Europe until the 1800s. In that century, the unambiguous connection between the two was made and decisions were taken to deliberately increase sanitation to improve public health (Burnet and White 1972). For example, following the 1854 cholera outbreak in London, John Snow conducted a study to determine its cause. He concluded that the source was a water well on Broad Street, so he removed the pump handle. The resulting end of the epidemic showed that indeed cholera was spread through polluted water and that intervention could effectively control its spread (Wellcome Library 2009).

In 1876, Robert Koch provided an early description of a bacillus that caused disease, in this case anthrax. By the late 1800s, the rapidly growing knowledge about disease and its cure included the lifecycle details of specific diseases, such as the plague (Kelly 2005). Along with this increase in knowledge came further targeted cures and public health services to deliver them. The late Victorian public's strong (perhaps obsessive) focus on cleanliness illustrates that these new ideas about disease had been accepted into their personal and their culture's world view. These discoveries also sustained the public's faith in Western culture's world view such as the notion of progress.

Our knowledge of disease and cures continued to grow into the 1900s. For example, the initial discovery of antibiotics in the late 1800s had led, by the early 1940s, to the mass production of pure penicillin. This drastically reduced the occurrence of septicaemia and cured those suffering from many bacterial diseases. The average lifespan increased.

By the mid-1900s, the tremendous visible successes from applying vaccination, antibiotics and basic sanitation had understandably influenced European culture's view of disease and health services. There was a popular perception among both health service workers and the public that the framework of knowledge needed to effectively control diseases had essentially been completed (Burnet and White 1972). All that remained to fulfill the cultural vision of relief from all disease was to find and implement the missing details needed to cure

each disease. Implicit in this view was the idea that diseases were largely static and could be eradicated once and for all. Think smallpox.

Almost forgotten in the euphoria was the prior recognition, in the early 1900s, that disease was dynamic. For example, a disease agent could accommodate a changing circumstance by altering its expression and developing resistance to treatment. The significance of disease dynamics only began to be fully recognized in the 1980s with the discovery that a circular DNA fragment called a plasmid can easily transfer DNA from one species of bacteria or virus (microbe) to another (Cohen and Shapiro 1980). This overturned the long-standing idea that microbe species could only change slowly, through random mutations (Williams and Ryan 1998). Since the late 1990s, the view that disease is a dynamic part of a dynamic ecosystem has slowly gained acceptance in Western culture's perception of illness and the delivery of health services.

The need to recognize the dynamic, ecosystem-based nature of disease is being reinforced by recent discoveries. Consider the view that bacteria live out their lives in solitary isolation, as shown in TV ads. This perspective had to be radically revised when it was discovered that, despite their simplicity, bacteria have elaborate chemical-sensing and signalling systems.

Bacteria have the ability to sense and respond to their environment and to other bacteria. They also have the ability to communicate. These characteristics allow at least the disease-causing bacteria *Staphylococcus aureus*, salmonella and cholera to make group "decisions." Perhaps, more correctly, it allows them to initiate group action through a process of quorum sensing in a manner analogous to whooper swans' decision making. Bacteria use their communication ability to form colonies or biofilms, such as plaque on teeth, and to control their virulence (toxicity) (Federle and Bassler 2003).

We are each a participant in an ecosystem. That ecosystem, and the diseases it hosts, are dynamic within the limits of the ecosystem's functioning. Despite the advances in the understanding and the treatment of diseases, this dynamics is still the context within which the relationship between our biosphere, our health and the provision of health services should be discussed.

**Dynamic Services.** Between the late 1700s and the 1950s, health services in Europe underwent significant changes as more effective responses to diseases and other health issues were implemented. This continued after the 1950s, but by this time the previously neglected constraints on the successful delivery of health-care services, which result from the dynamic complexity of medical issues, were coming to the fore. Those constraints are illustrated by the development of drug resistance in disease-causing microbes.

Before 1910, insects were known to become resistant to poisons (Carson 1962). By the time the first mass production of an antibiotic, penicillin, had begun in 1943, bacteria were known to develop resistance to it. Alexander Fleming, the discoverer of penicillin, forcefully brought the medical consequences of this resistance to the public's attention in his 1945 Nobel Prize address (Fleming 1945). Despite this knowledge, it wasn't until the 1990s that the full significance of microbial resistance to drugs started to become more widely recognized. By that time, an increasing number of the microbes causing human diseases, such as pneumonia, meningitis, septicaemia, tuberculosis and malaria, were displaying some degree of resistance to one or more of the drugs used to treat them (Goossens and Sprenger 1998).

From our perspective, a disease-causing microbe has developed drug resistance when the dose needed to kill it is also harmful to the patient (Hawkey 1998). The development of that resistance follows ecosystem principles. The process can be appreciated by considering the environments favoured by microbes, such as the guts of humans, wild and domesticated animals, and parts of the general environment, such as organic-rich soils. The presence in these environments of a wide variety of microbes and their diversity of genes means that they can be thought of as reservoirs for microbial genes.

If a disease-causing microbe species enters a gene reservoir that also contains the drug used to kill it, then the microbe can develop resistance to the drug. This can occur when the concentration of the drug is high enough to kill most but not all of the individuals in the microbe's population: those that survive have a greater inherent resistance to the drug. When the survivors reproduce, they pass these genes to the next generation. Repeating this selection process over generations adjusts the genetics of the microbe's surviving population in favour of those genes that confer the most resistance to the drug. It becomes drug resistant. Because microbes have relatively short lives and reproduce rapidly, the microbe's population can rapidly become drug resistant, in as little as a few years (Hawkey 1998).

There are alternative ways for the microbes within a gene reservoir to develop resistance. It can take place

slowly, through random genetic mutations, or rapidly, by the microbe acquiring the required genetic information from other microbe species or DNA fragments that already carry it (D'Costa et al. 2006; Berger-Bächi and McCallum 2006). Acquisition occurs when a virus or plasmid carrying the required genes transfers them to the microbe or when the microbe eats the virus or plasmid (Cohen and Shapiro 1980). Either way, the microbe can incorporate the resistance-conferring genes into their own DNA.

The ability of disease-causing microbes to rapidly become drug resistant through a combination of mutation, acquisition and selection highlights the need for health services to be dynamic. Consider the bacteria that caused the deadly plague epidemics of the mid-1300s and 1600s. In 2007, a plasmid containing the genes capable of conferring multiple-drug resistance on the plague bacteria were found in a wide variety of geographically dispersed environments. This preconditions the rapid spread of multiple-drug resistance among plague bacteria. The concern is that, under favourable conditions, these multiple-drug-resistant plague bacteria would be available to re-enter the human domain in large numbers (Welch et al. 2007). The reality of this potential is driven home by the 1995 discovery that some plague bacteria were already multiple-drug resistant, and by the occurrence, during 2014 to 2015, of 349 cases of plague, including 89 deaths, in Madagascar (WHO 2015).

The progression to a more widespread undesirable outcome that results from a health service being insufficiently responsive to this dynamics is illustrated by the very common, often harmless, bacteria *Staphylococcus aureus* (SA). Harmful infections of SA were routinely and successfully treated with methicillin, an important antibiotic. A methicillin-resistant SA (MRSA) did develop, in hospitals, but it was confined to them, until recently. Today MRSA infections can be acquired in the public environment, worldwide. This pathway is called community-acquired MRSA (Allen 2006). In 2003, SA was identified as a cause of flesh-eating disease. This is a potentially fatal disease, whose seriousness is increased by MRSA's drug resistance: it makes the disease far more difficult to treat (Miller et al. 2005b; Allen 2006).

These observations about resistance also apply to viruses, including those that cause AIDS, hepatitis and influenza (Pillay and Zambon 1998). If a health service is to successfully deal with microbial diseases, it has to meaningfully deal with the presence and the dynamics of drug resistance. The complexity of doing so is illustrated by our strenuous efforts to respond to the bacteria causing tuberculosis (TB).

TB, one the most deadly of human diseases, is difficult to treat. Success requires health service providers to follow the labour-intensive DOTS (directly observed treatment short course) strategy in which a health provider personally supervises each medication dose to the patient for the duration of TB's long treatment period. Failing to complete the treatment also favours the development of drug resistance, which further complicates TB's treatment.

It has been known for some time that TB bacteria could be divided into different strains, each resistant to a different drug. Thus, only a limited suite of drugs was effective in treating a particular case of TB. To be successful, DOTS had to deliver the correct drug.

In 2000, it was suspected that a strain of the TB bacteria might be resistant to many drugs. By 2006, some strains of the TB bacillus were shown to be resistant to almost all TB drugs. Extreme drug resistant (XDR) TB had appeared. The XDR strain of TB is now found worldwide, although currently it is preferentially found in patients suffering from AIDS. The presence of both XDR TB and AIDS in a patient makes it more difficult to treat both illnesses, and to limit the spread of TB, especially the drug-resistant varieties.

If a health service is to successfully deal with TB over the long term, then it must respond to its presence and dynamics in an equivalent dynamic manner. In particular, during treatment, the health service must test for drug resistance, select an effective drug and keep the required range of drugs in stock. It must be administered using DOTS. The health service must also continually work at finding new effective drugs that can cure XDR TB. Most importantly, it must proactively work at containing both the spread of TB and the development of drug resistance (Wright et al. 2006; Cohen 2006; Van Helden et al. 2006).

These observations apply to a health service's efforts to deal with any disease whose agents can become resistant to the drugs used to treat it. Critically, the efforts to be dynamic must extend beyond money and technology. Consider the attention health service providers must devote to reducing both their own and the public's contribution to the development of resistance.

For example, despite health service providers' best intentions, hospitals provide a stable, crowded (high-density) environment, rich in antimicrobial drugs. These are the ideal conditions for disease agents to acquire drug resistance, as exemplified by the super-resistant bug MRSA (Struelens 1998). Dealing with the development of drug resistance in hospitals is a never-ending task.

Health services must also deal with the wider public's contribution to the development of drug resistance, such as our indiscriminate and inappropriate use of antimicrobial drugs and suppressants on ourselves, our pets and our food animals. More personal examples, over which we have direct control, are our failure to complete a course of drugs. We also tend to discard drugs into the toilet or trash can, routinely use cleansers that contain antimicrobial drugs, and use antibacterial drugs for viral infections (Walsh 2005; Ghafur 2010).

In addition, when we personally ingest a drug, apply it to our bodies, give it to animals or discard it, a significant proportion remains unused and intact. This portion is excreted in our urine and faeces or washed off our bodies to join the drugs we have previously discarded into our waste system. These drug residues eventually enter the greater environment, where they can promote the development of drug-resistance-conferring genes, especially in the local microbe gene reservoir.

Then there is the unexpected. For example, quaternary ammonium compounds (QACs) are common ingredients in a wide variety of products, such as shampoo, disinfectants and fabric softeners, that you and I routinely use without much thought. Researchers discovered, to their great surprise, that the genes conferring resistance to QACs on bacteria also confer resistance to some antibiotics. This includes bacteria that affect human health, such as *Staphylococcus* (Gaze et al. 2005). Health service providers must deal with all of these contributions to drug resistance, if they wish to successfully limit its development.

This discussion might have created the impression that the need to deliver dynamic health services is only a local concern. However, the worldwide spread of disease-causing microbes indicates otherwise (Daszak et al. 2000). Headline-grabbing examples are AIDS from chimpanzees in Cameroon (Heeney et al. 2006; Keele et al. 2006) and avian influenza from birds via poultry (Stein 2003). In the past, psittacosis was spread with imported wild parrots (Burnet and White 1972). Today, the provision of local and regional dynamic health services has an inescapable global component.

As a result of this globalization, we are in the process of changing the context in which human's complete our lifeway, become ill and require health services. It has increased the need for dynamic services. To appreciate that statement, think of the above diseases as exotics. When seen from this perspective, a local health service must increasingly deal with a variety of health-affecting exotics. They include new disease-causing agents; old agents that have novel resistance to drugs; genes that confer novel drug-resistance; and new vectors that spread diseases, such as the Asian tiger mosquito.

The importing of these exotics and the additional responsibilities and costs they bring to dynamic health services are illustrated by the emerging spread of the exotic NDM-1 gene. Incredibly, it confers on bacteria their resistance to the complete B-lactam class of antibiotics. The gene developed in India and Pakistan largely because of chronic indiscriminate and improper use of antibiotics (Ghafur 2010).

Vacationers and medical "tourists" travelling to India and Pakistan can easily acquire bacteria or plasmids that carry an NDM-1 gene and can survive in the human gut. The visitors can also acquire a disease-causing organism that carries the NDM-1 gene. When the visitors return home, they bring the gene with them, either as a symptomatic disease or as an asymptomatic part of their gut flora (or both) (Kumarasamy et al. 2010). The NDM-1 gene is now spreading around the world. At the time of writing, this exotic gene is being rapidly spread into Europe, but it is also found in Canada, the United States, Australia and Asia.

The burden that the exotic NDM-1 gene places on health services is greater than we would intuitively expect from this description. Consider that the aquatic environment around New Delhi, India, hosting the plasmids that carry the NDM-1 gene also facilitate the gene's transfer to other bacterial species in that environment. One of these species causes cholera. Also present in this aquatic environment are the genes that confer on disease causing bacteria their resistance to other antibacterial drugs.

New Delhi's urban aquatic environment is a community-based gene reservoir from which the NDM-1 drug-resistance-conferring gene and disease-causing microbes carrying NDM-1 can spread. This reservoir also facilitates disease-causing microbes, like cholera, becoming multiple-drug resistant: they can carry the NDM-1 gene as well as genes that confer resistance to other drugs. This makes it even more difficult to treat these diseases and control their spread (Walsh et al. 2011). To be successful, a dynamic health service must deal with all of this: it must be proactive and reactive.

In overview, the last 160 years of providing health services has been incredibly successful. Over the last 60 years, this success, and globalization, have changed our awareness and experience of health, health-affecting agents and health services. We have also come to realize that the influences on our state of health are dynamic. Our health services must also be dynamic.

To be successful over the long term, a health-care service's dynamics have to match the dynamic nature of the medical issues that the services are expected to deal with. This was illustrated by drug resistance, our efforts to deal with it and the exotic nature of many disease agents. It showed that the biosphere's functioning places much greater constraints on the provision of effective health services than we generally understand. It also indicated that the delivery of appropriately dynamic services is constrained by the nature of our relationship to our health, our health services and our biosphere.

Costs. What non-economic costs could accompany the success of our health services? A casual evaluation would suggest none. However, this conclusion ignores the many surprising and unexpected costs that a closer look reveals.

Ecosystem costs arise from our efforts to control disease vectors. Consider the post 1940s malaria eradication programs. They focused on controlling malaria's vector, the *Anopheles* mosquito, through swamp draining and the application of DDT. The resulting ecosystem disruptions, such as DDT's impacts on bird reproduction, are well known (Carson 1962).

There are costs from treating illness. Consider drug residues. They contribute to what is surely one of the biggest costs to us: the development of drug-resistant disease-causing microbes, as discussed above. There is also an environmental cost, as illustrated by the previously discussed feminizing abilities of estrogen and estrogenic residues from birth control and related medicines. Drug residues are also likely affecting the functioning of microbe-rich environments, such as soils. Consider too how our use of antimicrobial drugs is changing our service-providing gut flora. Think of food digestion.

Ironically, perhaps the most significant, yet unexpected and unintended health-services-related cost to ecosystems and ourselves results from our success at extending our lifespan. Consider the discovery and development of antimicrobial drugs. Some microbiologists suggested, as early as 1937, that if the expected success of antimicrobial drugs wasn't matched by a corresponding and timely drop in the human birth rate, it would result in a human population explosion (Burnet and White 1972). The rapid population growth the earth experienced since the 1940s has contributed to our increased disruption and destruction of ecosystems. It has also led to a growing number of larger urban centres, an ideal environment for the spread of infectious diseases and our exposure to new diseases. Consider the dread among health-care officials that accompanied the spread of Ebola into West African urban centres in the 2010s.

This larger human population living at higher densities, especially in cities, joins with drug resistance and new diseases to increase the unavoidable (fixed) economic and the non-economic costs of providing a dynamic health service. For example, we have become increasingly responsible for supplying and maintaining the basic health services in the urban environments that were previously provided by our environment. Think of waste removal and water purification. Demanding tasks.

For example, many years of insufficient effort and funds weakened Zimbabwe's public health services and health-related urban infrastructure, such as waste water, solid waste and water delivery systems. Eventually, in 2008, the arrival of the rainy season overwhelmed some of the collapsing waste-water systems. By November 2008, a nationwide cholera epidemic overwhelmed the remnants of the health system (IRIN 2008). Think too of the rise in NDM-1 and the spread of chytrid by exporting the claw frog used in pregnancy tests. The presence, recognition and significance of these costs is influenced by a culture's world view.

Culture. A culture's world view has a profound influence on the provision of dynamic health services. A wider appreciation for culture's role can be gained by considering its influence on the following health-affecting factors: high-density living, drug resistance, economic costs, technology and communication about health. Although the discussion of these topics is biased toward Western culture, the essence applies to all complex cultures.

The world view of a complex culture influences its members' recognition of and response to one of the most important contributors to the spread of infectious diseases: *high-density living*. Complex cultures consider it normal to crowd ourselves together in urban centres. Indeed "densification" is a byword for dealing with housing issues in cities. Consider too the common Western practice of placing those most vulnerable to infectious diseases together in child- and elder-care centres (Goossens and Sprenger 1998). We crowd our food animals together, sometimes on an industrial scale, and we can even live with our pets and domestic food animals.

In most medical circles, the causes and prevention of *drug resistance* are recognized. However, that recognition and the corresponding response is not yet an essential part of their greater culture's world view (Goossens and Sprenger 1998; Huovinen and Cars 1998). Western culture, for example, facilitates the development of resistance in a number of ways.

A personal illustration of this occurs in a doctor's office. Consider the culturally supported expectations of a patient and their doctor. The patient expects to receive relief from illness, and the doctor expects to provide it. These expectations apply to even common ailments, such as a sore throat, flu or cold. Our expectations are often satisfied by the prescription of antimicrobial drugs, even though the prescribed treatment is often unnecessary or clinically ineffective against the disease-causing agent. Because of the culturally mediated expectations embedded in the doctor-patient relationship, it is difficult for either party to act on this reality (Butler et al. 1998).

Consider too the routine use of antimicrobial drugs in farm animals to "prevent" illness and promote farm animal growth. In the West, this inappropriate use is considered acceptable, in part because the cultural objective of providing cheap, abundant food displaces any concern about the resulting future development of drug resistance. Our culturally influenced, inappropriate use of antimicrobial drugs on food animals also applies to the treatment of our pets. This is important because, in 1997 they received 40% of the antibiotics given to animals (Johnston 1998). In summary, the biological constraints required to limit drug resistance are poorly incorporated into Western culture's world view and our personal health decisions, as discussed further in chapter 14.

There is a Western cultural imperative to provide and improve public health services. But there is also a cultural imperative to reduce the *economic cost* of providing these services. The outcome of this conflict is a preference for procedures that reduce annual budgets (Wise et al. 1998). Consider the British government's efforts to encourage patients to seek cheaper medical treatment overseas, for example, in India. Some of these patients returned home carrying the exotic NDM-1, drug-resistance-conferring gene. The British decision may have reduced the short-term economic costs of providing current medical services but, in the longer term, the resulting spread of drug resistance will likely increase those costs.

How do we decide on the level of unavoidable (fixed) economic funding needed to provide and maintain effectively dynamic health services? Consider the services needed to effectively treat and limit the spread of the world's leading cause of infectious disease deaths among adults: tuberculosis. As discussed above, treatment includes DOTS, drug-resistance testing and holding a variety of drugs in stock. Providing this treatment is expensive and time consuming.

Despite these efforts health services are only partly successful at controlling the spread of TB and limiting the development of drug resistance. The primary reason is that tuberculosis is most commonly found among the poor and malnourished living in crowded conditions. These people have limited ability to fight the disease and limited access to health services and adequate drug treatment. Greater success at treating TB, restraining its spread and limiting drug resistance would have to address both poverty and the issues restricting effective TB's treatment. Success would require a culturally backed, political initiative to promote, fund and support long-term, community-driven efforts to that end (Farmer and Kim 1998). This would require a change in both social and economic priorities.

If a culture wishes to effectively address TB, it faces a choice. It can decide to pay a lot in fixed costs now or pay much more later. The appearance of XDR TB illustrates the increasing costs of deciding to pay later. Yet surely technology will lower these costs?

Our perception of the benefits and costs (economic and non-economic) which *technology* brings to our health and health services is influenced by our culture's world view. There is a strong Western cultural belief that medical technology is largely responsible for improving our health. It is true that technology has provided us with impressive health benefits. But it is also true that the simple changes to the health basics, such as sanitation, sanitary practices, water and air quality, nutrition and lifestyle are equally responsible for large increases in the human lifespan and quality of life (Burnet and White 1972). The success of these simple changes relied at least as much on us making adjustments to cultural attitudes and beliefs as they did on technical innovation. Think of the Broad Street pump handle and the arsenic poisoning in Bangladesh.

In addition, because Western culture's world view tends to unconditionally hold most technology in high esteem, Westerners tend to underestimate the (economic and non-economic) costs it imposes on our health and health-care services. Consider that technology, in all its forms, is paid for and powered by wringing more cash value and energy out of our life-supporting environment. By disrupting and destroying our environment to make these payments, we decrease its ability to provide the basics of good health: clean water, clean air and nutritious food.

In applying technology, we also add to our environment a host of incompletely tested chemicals that may have long-term health impacts on us and add to our health services costs. Consider those chemicals that could interfere with the functioning of our endocrine system (Vandenberg et al. 2011). Think particularly of those chemicals that may disrupt the endocrine system's control over our reproductive system. One suggestion that this could be happening to a population is a significant deviation in the at-birth sex ratio from the normal 0.96 females for each male.

For example, between 1984 and 2003, the Aamjiwnaang First Nation in Ontario, Canada, recorded an increasingly abnormal birth-sex ratio. By 2003, there were 2.33 females being born for each male. Because sex at birth can be affected by many factors, such as stress and cultural sex selection, even this large a deviation does not prove that human-made chemicals are the cause. However, in this case, chemicals are suggested because the Aamjiwnaang mothers live within 25 km of 50 industrial-scale chemical plants producing a wide range of compounds, such as petrochemicals and polymers. In 2009, these factories emitted 110 million kg (0.11 Mt) of air pollution (Basu et al. 2013; Mackenzie et al. 2005).

A full change in sex is only one type of reproductive system disruption. Distortions to sex development, such as the partial feminization seen in frogs and fish, and the disruption of reproductive function, are likely more common. Consider the phthalate class of chemicals. They are found in a wide variety of household objects, from plastics to cosmetics and paints. For example, the most common type of phthalate is the main plasticizer in polyvinyl chloride (PVC), a plastic ubiquitous in our home and office environments. This phthalate can escape from the plastic as a gas. We can be exposed to phthalates in general through the air, physical contact and ingestion. As a result, phthalates are one of the more common human-made chemicals in our urine (Bornehag et al. 2005; Bornehag et al. 2004).

There is sufficient evidence to conclude that phthalates can affect our reproductive development. For example, the greater a pregnant mother's urine-recorded exposure to phthalates during pregnancy, the more likely her newborn male child will show reduced male features (i.e., be feminized) (Swan 2008; Jurewicz and Hanke 2011).

A different type of reproductive disruption in humans caused by human-made chemicals is illustrated by some of the perfluorocarbon (PFC) class of chemicals, such as perfluorooctane sulfonate (PFOS) and perfluorooctanoate (PFOA). Humans are exposed to them because they are used as surfactants in cleaners and are found in household products, including food containers, furniture and paints. PFOS and PFOA have estrogenic properties and are known to cause reproductive disruption in fish and mice. They have been shown to cause early onset of menopause in women (Knox et al. 2011).

These technology-related disruptions of the human reproductive system can have a significant impact on health service costs. The costs of dealing with infertility immediately come to mind. Consider too that a person's identity is intimately tied up with their sex and sexuality. Any deviations of these qualities from the accepted cultural norms, such as sexual ambiguity and sexual dysfunction, can have serious personal consequences. Think of the personal issues faced by those born intersex (Callahan 2009). Addressing these physiological and psychological issues is part of the health services we expect. A reliance on technology to resolve health issues is a double-edged sword.

The public communications by government, health services and corporations influence the health of a culture's members. This observation might seem superfluous to this discussion until we remember the demands placed on health services by the human population's fastest growing health issues: the lifestyle-related diseases such as lung cancer related to smoking, obesity, non-genetic diabetes and some types of heart ailments. The messages from a culture's organizations can promote either the spread, or the avoidance and cure, of these lifestyle disease. The outcome depends significantly on those aspects of the members' lifestyle that their culture's world view considers acceptable to target, and what can be said and how.

The connections between a culture's world view, its organizations' communications and lifestyle diseases is illustrated by lung cancer caused by smoking. Scientists have known, since at least the 1950s, that cigarette smoking causes cancer (Doll and Hill 1950; Doll and Hill 1954). Despite this knowledge, cigarette smoking and the boosting of cigarette sales through communications, such as advertising, remained culturally acceptable in the West

for decades. One paradoxical reason governments gave for continuing to allow the sale and advertising of cigarettes was that the revenues from cigarette taxes were needed to pay for health services.

Cigarette companies took advantage of this cultural acceptance. Remember Philip Morris, who used the Marlboro Man, a cowboy, in their successful 1950s advertising campaign promoting Marlboro cigarettes? They were able to exploit the cultural values of prestige and rugged independence associated with the cowboy image to greatly increase their sales of cigarettes not just in America but worldwide (Twitchell 2000).

It was only in the 1970s that an American cultural shift away from smoking began. Under the pressure of growing numbers of smoking-related lung cancer deaths and revelations about cover ups, the cigarette manufacturers' campaign against the science-based link between cigarette smoking and cancer slowly fell apart. As it did, the public's opinion changed, and the efforts of health services to communicate the health costs of smoking became effective. Health messages, such as one from the American Surgeon General, could be widely received, accepted and acted upon by the public (Garfield 2005; Oreskes and Conway 2010).

In general, it is difficult to produce an effective health services message because it has to consider both the receiver's state of well-being and their culture's world view. Most people who receive health-related communications are healthy. Considerable skill is therefore needed to effectively reach out to people with information about illness and disease when they feel healthy, and when the culture permits the behaviour being targeted for reduction (Belongia and Schwartz 1998).

Today a similar conundrum faces Westerners in general and Americans in particular, but this time it concerns their culturally sanctioned food and exercise preferences that are contributing to a rise in obesity. In parallel with smoking, Western culture accepts a corporation's right to profit from the promotion and sale of excessively high-energy, often low-nutrition foods that appeal to our palette. This, combined with the culturally acceptable promotion of leisure over exercise, has contributed significantly to the rise in obesity (WHO/FAO 2002). Until there is a cultural shift, health services will find it difficult to effectively communicate to the public that their culturally-supported current lifestyle choices are at odds with biological reality and have negative health consequences (Mintz 1989; Leonard 2002). The same, in essence, is true for antimicrobial drug resistance. These issues are not limited to Western culture.

In summary, has our provision of health services successfully satisfied our health demands? The answer depends greatly on how you choose to view the question. If taken at face value, with an eye on the past, then the answer is yes. We have satisfied our health service demands by inventing new treatments for diseases and by increasing the supplies of water and food.

The answer changes if we include the non-economic costs. In providing health services, we have disrupted our supporting environment in a manner that constrains its ability to provide clean water, clean air and nutritious food. We have also facilitated the extraordinary growth of our large human population. That population is moving into cities to live at high densities. This crowded urban ecosystem is a perfect place for infectious human diseases to reproduce and spread. We are responsible for providing all of the services needed to keep these crowded, disease-favouring urban environment's disease free.

More generally, by providing the health services we currently enjoy, we have, like the farmer, also burdened ourselves with responsibilities. At the same time, by globalizing trade and travel, we have globalized the spread of disease-causing agents and their vectors. Thus, ironically, our cultures' world views are contributing to both the creation/spread and the cure of illness: our world views are contributing to both the effectiveness of health services and the costs (non-economic and economic) of providing them.

Providing effective health services under these conditions is far more complex and requires a much larger cultural and personal commitment than we imagine. In particular, our health services must be sufficiently dynamic to accommodate the dynamic nature of disease-causing agents and their host ecosystems on a global scale. The rapid rise of drug resistance and obesity both signal the presence of significant unmet challenges.

The impact of all these considerations on our ability to provide future dynamic health services will be considered in more detail in chapter 14.

## Metals and Minerals

Our hunter-gatherer ancestors rarely used the few metals available to them, such as native copper and gold. In contrast, our current industrial societies rely heavily on a wide range of metals, such as copper, iron and aluminium; minerals, such as gypsum and potash; and aggregates, such as sand and gravel. The status of metals will be used to illustrate the earth's ability to supply our demand for these resources.

**Source and Production.** The primary source of most metals is minerals that contain significant amounts of metals in their structure (ore minerals). Ore minerals are found in many types of rock. However, it is only practical to mine and extract them if two criteria are met: the amount of rock containing the ore minerals is large enough; and the concentration of the ore minerals in that rock is greater than some threshold, expressed as the amount of metal in a tonne of rock (grade). The volume of rock that meets these criteria of grade and tonnage is called ore. The space occupied by ore in the ground is referred to as an ore deposit. Ore deposits, like fossil fuel deposits, are rare. On a human time scale, ore deposits are non-renewable. They take tens of thousands to millions of years to form, but are mined out, today, in less than 100 years. A secondary source of metals is the items we recycle (Evans 1993; Gordon et al. 2006).

The first step in mining an ore deposit is to remove the rock that is limiting access to the ore. This waste rock forms the large rock piles we see at a mine site. Once the deposit is exposed, the ore is mined, crushed and passed through a concentration process that separates out the ore minerals of interest. The finely crushed rock from which the ore minerals have been removed (tailings) is discarded, often behind a tailings dam or in a lake. The concentrated ore minerals complete their journey into a metal through an energy-intensive conversion or smelting process. During conversion, the metal elements end up in an ingot. Some of the unwanted elements, such as sulphur, are driven off as a gas, while the rest are collected into a glassy solid called slag (Evans 1993).

Today, most of the metals we use, such as copper and molybdenum, come from giant ore deposits that are being mined as open pits. For example, the Bingham Canyon mine, Utah, is an open pit 4 km wide and 1 km deep. These giant deposits contain huge amounts of metal but usually at low ore grades. For example, ore grade copper is typically around 0.5% and molybdenum around 0.1%. Each tonne of ore produces a small amount of metal and a large amount of waste.

Each step in the mining, concentration and conversion processes uses energy. The amount needed to mine and concentrate the ore minerals rises exponentially as the grade of the deposit decreases (Hall et al. 1986). The energy needed to convert a mineral concentrate is largely fixed for each type of mineral.

As expected, the low grades of giant deposits mean that most of the energy needed to exploit them is used to remove waste rock, move ore and crush it. As a result, most metal production is energy intensive: a relatively small amount of metal is produced per unit of energy. To reduce both the energy and time (cost) needed to handle ore and waste, most mines build their concentrating plants, waste rock piles and tailings dam as close to the ore deposit as possible.

In contrast, the amount of energy used to convert a concentrate to a metal is relatively low. It is common for mines to transport their metal-rich mineral concentrates to converters or smelters all over the world. The reasons include avoiding the cash expense and environmental constraints of building and running a converter at a mine that might only operate for a few years.

A notable exception is aluminium. To produce 1 kg of aluminium from an aluminium ore mineral concentrate requires roughly 15.5 kWh of electrical energy. This is a sufficiently large amount that the concentrate is inevitably shipped from the mines to converters located near a supply of cheap hydroelectricity.

The amount of energy needed to produce aluminium is personalized by noting that, on average, an aluminium beverage can weighs about 13 g (Aluminium Institute 2010). A bag of 13 cans contains 169 g of aluminium, which required 2.6 kWh of energy to produce. That is more than the 2.4 kWh needed to keep a single 100 W incandescent light bulb lit for 24 hours. If you added one more can, then the 2.8 kWh of energy needed to produce 14 aluminium cans is almost the same amount of energy in a 2,500 kcal (2.9 kWh) daily diet.

An aluminium beverage can's relationship to energy should provide you with a different context for thinking about the can you will empty and discard in less than 15 minutes. Fortunately, much of the energy used in the production of metal items can be saved *if* they are recycled. In the case of aluminium, up to 80% of the energy first used to generate a can's 13 g of aluminium is retained when it is recycled into another can (note that the energy cost of transportation to the smelter is excluded) (Evans 1993).

Metals can be produced in a variety of forms and qualities. In general, the purer the product, the more energy is needed to produce it. For example, producing the ultra-pure silicon needed to make electronic chips uses about 160 times the energy needed to produce standard-purity silicon (Williams et al. 2002). This extra cost of production is one of the reasons why generating electricity from silicon-based solar cells can be expensive.

**Usage.** A metal's lifecycle describes its history from extraction, to its usage in a manufactured product, to waste and back to the earth. Extraction was discussed above. Its usage is discussed here. Metals are used in a wide variety of manufactured goods from the obvious, such as copper in computers, through to the not so obvious, such as titanium oxide to make toothpaste and the white base in paint, to the unexpected, such as bismuth in Pepto-Bismol (Pepto-Bismol 2009).

During the time that a metal-containing product is in use, the metal is said to be "in service." Some metal-containing products can be in service for a long time, in particular those that form our infrastructure, such as the aluminium in high-voltage power lines and the copper in home wiring and plumbing. In contrast, the metals in many consumer products, such as cell phones, have short in-service lives.

The amount of metal required to satisfy a country's in-service needs is determined by its citizens' lifestyle. For example, around the year 2000, the developed countries (i.e., the rich complex industrial cultures) had about 200 kg of copper/person in service. On a continental scale, within North America (Canada, US and Mexico), the average was about 170 kg of copper/person (Gordon et al. 2006).

This section's title of "Usage" conjures up the image of a metal being consumed in one step, like food and fuel. Metal usage is different because, at the end of a product's in-service life, we may be able to recycle the metals in it. The amount of recoverable metal depends on the cultural incentive to recycle and how the metal is used in products. The metals in products such as cars, tin cans, plumbing and power lines are commonly recycled.

If a product is not recycled, then the metals in it end up as waste. Some metals become waste because they were used in a manner that makes them inaccessible to recycling, such as those in plastics, paints and medicines. Other metals can be recycled but become waste because the product's design makes metal recovery too difficult or because the discarder decided not to recycle the product. Think landfill.

For a country to maintain its desired lifestyle, it must mine (or trade for) new metal in an amount that both replaces the metal lost to waste (unrecycled and unrecyclable metals) and satisfies any increase in the country's in-service wants. Some, especially economists, believe that the demand for newly produced metals has dropped over time because they are being used more efficiently, and because plastic is used as a substitute.

This economic argument is based on trends in the units of metal used per dollar of economic output (economic efficiency). Others believe the economic argument that once a country has established its infrastructure, then its absolute metal use will decline (economists' Kuznets curve argument) (Ausubel 1996; Gordon et al. 2006).

The relevant data about US metal usage reveals something different. Although the economic efficiency of metal use has been rising, so has the absolute amount of metals being used. For example, the amount of in-service copper within the United States steadily rose from roughly 180 kg/person in 1950 to around 240 kg/person in 1999. That average growth rate of 1.2 kg/person/year was sustained by newly mined copper (Gordon et al. 2006). In 2000, the 281 million Americans used 3 Mt (10.7 kg/person) of newly mined copper. That amount both maintained the growth rate of in-service copper and replaced the amount lost to waste (Menzie et al. 2005). If the average copper-use growth rate of 1.2 kg/person/year applies to 2000, then in that year they discarded 9.5 kg/person of copper to waste.

Costs. Building and operating a mine has significant non-economic costs. It disrupts and destroys the ecosystems on top of and around the ore deposit. We see that legacy as a hole in the ground around which are waste rock piles and tailings dams. The amount of discarded material can be extraordinarily large. For example, on a global scale, in 1994 each tonne of copper mined produced an estimated average of 350 tonnes of waste rock and 147 tonnes of tailings (for an average ore grade of about 0.68% copper) (Menzie et al. 2005).

Another of mining's environmental costs is its impact on the quantity and quality of surface and ground water. For example, if any groundwater is present in the mine, it is pumped out and discarded. The water that passes through the mine excavations and waste-rock piles commonly becomes polluted. The pollutants include the metals and acidity acquired from the rocks and the nitrates used in blasting. After the mine's closure, the metal-polluted water can continue to form and flow from the mine for tens to many hundreds of years. If considerable care is not taken, this polluted water can enter rivers and groundwater aquifers.

It is common for mining proponents to imply that the costs from mining are restricted to the relatively small area, or footprint, of the actual mine site. This is the equivalent of claiming that the impact of a factory is limited to the activities inside the building. Although mines occupy only a small fraction of the earth's surface,

"the scars are long-lasting, and the environmental impact can extend well beyond the limits of disturbed ground through air pollution and water contamination" (Hodges 1995, 1308). The resulting off-mine-site costs include the disruption of ecosystems from increasing long-term road access, especially to remote mine sites; the dropping of the neighbouring water table by mine dewatering; the disruption and pollution of surface and ground water around the mine site; the pollution associated with conversion and smelting; and the emission of greenhouse gasses from fossil fuel use. For example, in 1996, it was estimated that the fossil fuel energy used by mining in the United States released more than 20 tonnes $CO_2$/capita (Menzie et al. 2005).

Both small and large mines can have significant non-economic costs. Consider the small Equity Silver mine in central British Columbia that operated for just 14 years before closing. During the design, permitting and operating life of the mine, inadequate attention was paid to the possibility that the mine could generate polluted water. Metal-rich, acidic waters began to spread down water courses from the mine site. All of these downstream drainages are fish bearing, and they all feed into the important salmon- and steelhead-trout-bearing Bulkley River. As a result, the current owners of the closed mine will have to treat the mine waste water indefinitely if they are to maintain the downstream water quality at levels acceptable for fish and people (ESMFSTAG 2006).

The environmental footprint of Freeport-McMoran's massive Grasberg mine in tropical Indonesia is truly unbelievable. The mine is located in the headwaters of the Otomina River, a tributary of the Ajkwa River, and is some 80 km by air from the Arafura Sea. The visible part of the mine excavation is an open pit more than a kilometre in diameter, surrounded by waste rock.

Grasberg's tailings are not discarded into any type of containment structure. They are deposited, along with the associated polluted mine waste water, directly into the Ajkwa River, where they are filling the lower 45 km to its mouth and the adjacent parts of the Arafura Sea. By 2004, about 166 $km^2$ of the Ajkwa flood plain ecosystems had been buried to a maximum width approaching 4 km. To contain the lateral spread of the tailings, levees up to 15 m in height were constructed in some places (Paull et al. 2006). The scale of this destruction is unsurprising in the light of the 225,000 tonnes/day of tailings that the company reported their ongoing operations were releasing in 2006 (Freeport-McMoRan Copper and Gold Inc 2007).

The effects of this mine waste disposal on these river and river plain ecosystems are more than burial. They are also being disrupted by the elevated amounts of copper, cadmium and mercury in the associated metal-rich mine waste water. The coastal waters and ecosystems immediately around the mouth of the Ajkwa River are similarly affected by burial and dissolved pollutants. Despite this evidence, the company claims that their method of disposal is permitted and is the best method to dispose of its mill waste (Norwegian Ministry of Finance 2006; Perlez and Bonner 2005; Freeport-McMoRan Copper and Gold Inc 2007; Neale et al. 2003).

Most mining companies do make an effort to store mine waste and tailings in a manner that will prevent ecosystem disruption and pollution. However, when viewed over the long term, they are often unsuccessful. Consider that waste dumps do fail and tailings impoundments (lakes and dams) do leak or collapse.

For example, the tailings dam at the Baia Mare mine site in Romania was commissioned in 1999. It failed in 2000, releasing some 100,000 m$^3$ of polluted water containing between 50 and 100 tonnes of cyanide and a suite of heavy metals. The water flooded into the local river before adding its pollution to the Danube. After travelling some 2,000 km from the mine site, the pollution plume dispersed into the Danube delta. During its journey to the sea, the plume disrupted aquatic ecosystems. For example, it killed an estimated 1,240 tonnes of fish in Hungary alone (Baia Mare Task Force 2000).

The record of tailings dam failures is incomplete. It currently indicates that, between 1995 and 2012, they failed at a minimum, average rate of between 1 and 2 dams/year (ICOLD 2001; WISE 2012). In recent decades, tailings dams have failed at a higher rate than water supply dams, so they appear to be among the least reliable earth structures constructed by humans. The frequency and severity of tailings dam failures, and other mine waste management incidents, indicate a key reason why they happen, "the technical and managerial challenges of responsible mine waste management is under-recognized" (Morgenstern 2001, 54).

Like tailings dam failures, mining's long-term pollution of water is occurring for an equivalent, equally mundane reason. Mining companies design and permit mines based on predictions about the volumes and compositions of the contaminated water they will generate. Although these predictions are often accompanied by significant uncertainty, they are still used as actual rather than minimum outcomes and applied with inadequate contingency plans (Kuipers et al. 2005). As a result, mines commonly end up generating contaminated water that exceeds the quality or quantity set out in their permits. Not all of these oversights can be easily rectified in hindsight.

For example, the mine permit allowing the Brenda molybdenum mine in British Columbia to start production stipulated a zero contaminated water discharge to the environment. But, after closure, the containment system filled to overflowing. The permit had to be quickly amended and a plant installed to treat the excess water to remove molybdenum. The treated water is now being discharged from the mine site into an agricultural and domestic watershed. Treatment and discharge will have to continue for an estimated 50 to 200 years (Patterson 2003). The downstream water users had no choice but to accept this "error" in prediction and its consequences.

The secondary source of metals from recycling is often viewed as benign. But, if conducted without concern for environmental or human health, it too can have a substantial cost. Examples of recycling sites where environmental and human costs are commonly high include ship breaking (Rousmaniere and Raj 2007) and e-waste deconstruction (Slade 2006; Goodman 2003). Even the smelting of recycled metals can have substantial costs. For example, in 2000, emissions to the atmosphere from the secondary smelting of copper in North America was a significant contributor to the carcinogenic dioxins that ended up in the Arctic (Commoner et al. 2000).

**Culture.** The amount of new metal we need to mine, and the degree to which we find the non-economic costs of providing them acceptable, is strongly, though often subtly, influenced by our culture's world view. Consider the expectation among economists that once a Western country's population has stabilized and become economically prosperous enough to satisfy their basic infrastructure needs, then their demand for metals, such as copper, should steadily decline. This, as illustrated above by the United States, is not the case. The growth in the demand for new metal persists, but at a slower rate (Gordon et al. 2006).

Part of the reason is that the American culture's world view supports a vision of living called consumerism. It assumes that an ever-rising material standard of living and the increasing supply of new products associated with it will automatically endow substantial benefits on individuals and their culture. Those participating in consumerism play one of two interacting roles: those seeking products (the consumers) and those manufacturing them (the producers). Consumers and producers are engaged with one another in a mutual search for more satisfying, newer and "better" products. This search takes the form of a rapid cycle of innovation, production, purchase, use and disposal of material things (Slade 2006). Hence the alternative name for this cultural vision: "producerism."

Individual consumers experience this interaction as the relentless shopper's cycle of rising expectations and the need for the new. The mirror image of their experience is the individual producer's experience. They are caught up in a relentless rush to produce new goods with features they hope the consumers will find, or be persuaded to find, desirable.

The consequences of this culturally supported producer-consumer cycle (which is part of the economists' virtuous cycle) include an increasing diversity of products and services (telephone, plus internet, plus fax, plus cellphone, plus…?) with ever-declining product lifespans (less than 18 months for the cell phone) (figure 14). This prompts an ongoing demand for more and differently configured infrastructure, and provides a greater opportunity for imperfect recycling and the loss of metals to waste. The result is the persistent growth in demand for new metal, such as copper (Gordon et al. 2006). As long as the American culture's world view supports producer-consumerism, and as long as the producers and consumers have the funds and energy to support their mutual striving, then they will sustain an increasing demand for new metals.

Producer-consumerism (consumerism) has become a global phenomenon. (As has the virtuous cycle.) It has had a significant impact on the global demand for metals and on our views of the non-economic costs of providing them. Consider how a Westerner's choice of a vehicle is influenced more by their culturally supported, consumerism-influenced self-image than by the metals the vehicle contains or the non-economic costs of producing, using and disposing of it. The greater importance of self-image, relative to the non-economic costs, can be inferred from the company ads attempting to influence our car-purchasing choice. This is most clearly seen in ads where dirt road, water, snow or sod fly in all directions.

Consider too the apparently trivial example of the thin, shiny film of disposable aluminium in the one-use packages provided for many "foods," such as bagged chips and candy bars. The higher relative value that Western culture's consumerism-favouring world view places on this disposable packaging and its contents compared to the metal it contains results in more energy-intense aluminium ending up as unrecyclable waste. It also reflects, perhaps symbolizes, our willingness to accept the non-economic costs associated with aluminium mining, extraction and conversion.

A country's world view also influences the culture of miners and their attitude to the non-economic costs of mining. Significant attempts have been made by some mining companies and mining organizations to change mining's past behaviours and to address its non-economic costs (Hodges 1995; PDAC 2010). But overall, there is little cultural incentive for mining organizations to publicly condemn those mines and mining companies or executives with a poor human rights or environmental record. Consider that in 2007, after less than two years' operation, Glencairn's Bellavista gold mine, Costa Rica, was destroyed when the ore pad, from which the gold was being leached by cyanide, collapsed onto the recovery plant. A nearby river that fed a local municipal water supply was affected (Sherwood 2008). The Canadian company no longer exists, but the mess does. Its directors and management easily moved into other mining companies and projects, some in Canada.

The corporate leaders of Glencairn's mine at Bellavista; Aurul's operation at Baia Mare; and Freeport's Grasberg mine all noted that they received the necessary permits. The companies, and the cultures issuing the permits, appear to consider that the receipt of authorization from the relevant government agent is sufficient reason to justify their actual impacts on the environment and thus exonerates the companies (Freeport-McMoRan Copper and Gold Inc 2007; Baia Mare Task Force 2000; Caldwell 2008). Think too of the Mount Polley mine in British Columbia. Can we conclude that within mining culture, consumerist- and producerist-type views still displace the drive to reduce non-economic costs? We can certainly conclude that, in general, the amount of metal we want, our views of the non-economic costs of mining and using metals, and who pays is strongly influenced by our culture's world view.

In summary, whether the earth is able to provide our current metal demands is, as usual, determined in part by how we interpret the question. Today our application of technology and energy ensures that the earth can easily supply all of our metal wants. Current metal prices do not indicate large shortages of on-the-shelf metal supplies. Regional differences in the distribution of metal deposits have not significantly influenced the global availability of metals and minerals.

However, if the question of supplying our needs includes a consideration of the non-economic costs of extracting, using and discarding them, then a different interpretation is necessary. The costs of supplying metals are substantial and can last for hundreds of years. These impacts begin during exploration, remain until long after a deposit is mined out and abandoned, and cover a much larger area than the mine site. We have yet to fully accept responsibility for the non-economic costs associated with our mining, use and discarding of metals.

If we consider the impacts that satisfying today's metal wants will have on tomorrow's supply, then the focus shifts to other issues. After all, the slowness of the earth's metal deposit renewal process constrains the amount of metal it can make available in the short term. This view of the question will be addressed later in chapter 14.

## Conclusions

Our normal, focused (flashlight), short-term view of our world can easily lead us to conclude that the earth is perfectly able to satisfy our natural resource (including services) demands. In this view, the earth is currently most able to supply us with metals and minerals, and farmed food. It is least able to supply us with wild foods. Certainly, there are some regional-scale, chronic difficulties with fulfilling our wants, particularly for fresh water and health services, but there are currently no acute, and only relatively minor chronic, global-scale resource shortages.

However, this view largely treats the supplying of a resource in isolation. This is unrealistic because, as this chapter has illustrated, our efforts to satisfy our resource demands are interconnected. For example, by increasing the supply of one resource, such as energy, by using another, such as biofuels from corn, we affect the supply of other resources, such food, agricultural land and water. Included in these interconnections are the wide range of non-economic costs associated with supplying a resource. For example, by exporting a resource, such as food, those producing it absorb the non-economic costs, such as pollution, health effects and the depletion of water supplies. Consider too that the supplying, using

and discarding of resources can impact the functioning of the biosphere, thus affecting its ability to provide the resources we want.

A more realistic view of the biosphere's ability to provide us with resources appears when we conduct a comprehensive assessment of the supply and demand for resources: our supply-demand circumstance. It would recognize the connections and interactions between our resource demands, the costs of satisfying them (maintaining the supply) and the biosphere's functioning, which provides the resources and pays much of the cost for us supplying and using them. We find it difficult to conceptualize and relate to this reality in a comprehensive manner because of its complexity.

We deal with this difficulty by making simplifying assumptions about our circumstance. An example of the benefits and pitfalls of doing so is illustrated by the footprint model. It consolidates the environment's functions that are needed to both supply the resources to a human group (town, city, country, etc.) and dispose of all of its waste into a single hypothetical area of land, called a footprint. The difference between the size of a group's footprint and the actual amount of land available to its members is an estimate of whether or not the group is satisfying their resource-supply and waste-disposal needs in a sustainable manner (Wackernagel et al. 2002; Rees 2003). For example, the model unsurprisingly indicates that the footprint of cities is so much larger than the physical area they occupy that, to function, the urbanites have to be net importers of resources and net exporters of waste (Wackernagel and Rees 1996).

There are significant limitations to the footprint model. Consider that it combines things that should be left separate, such as sustainable and unsustainable land use; it concentrates on political regions not bioregions to calculate footprints; and it is unable to measure or include planning decisions. It also masks the interconnections between factors and the importance of time frame (Jeroen et al. 1999).

Still, within these limitations, a comparison of footprint sizes (i.e., their relative sizes) does illustrate the relative changes in the sustainability of our wants. For example, the size of our global footprint has increased from 70% of the biosphere's estimated ability to supply our wants in 1961, to 120% in 1999 (Wackernagel et al. 2002). The footprint model indicates that our demand for resources and services has undergone rapid, significant growth. It also, less meaningfully, suggests that by 1999 we had exceeded the earth's carrying capacity for humans.

It is possible to argue over aspects of both the data used by the footprint model and the specifics of its conclusions. However, the general conclusion that our resource demand is increasing and that there are supply limits cannot be dismissed as a fringe fantasy from a flaky model because it is supported by independent evidence. For example, allometric relationships suggest that, from an ecosystem's perspective, there are many more of us today than our body size indicates there should be. Consider too our nearness to the nine apparent environmental boundaries discussed in chapter 9.

Unfortunately, although making simplifying assumptions can provide us with this kind of insight, it can also hinder our efforts to comprehensively assess resource supply and demand. Indeed, it can even reinforce our flashlight view of our circumstance. For example, simplification favours our tendency to focus on a limited selection or an aggregate of variables as if they fully represent a resource's supply, demand and costs, or the state of our biosphere's functioning. This focus tends to distract us from considering the complexity and dynamics of supply, demand and the biosphere's functioning. Thus, without care, making simplifying assumption can hinder our efforts to think about our circumstance in a more comprehensive and coherent manner, or do so at a personal level.

In contrast, our efforts are helped if we think about each of the above descriptions of a resource as a different perspective of our overall circumstance. Each description can be thought of as a distinct but overlapping, interacting piece of a single puzzle. We are helped to mentally tie these perspectives (pieces) together by treating the various discussions about non-economic costs and culture as all being part of a single greater framework or context within which the supply and demand for all resources takes place.

It is within this greater context that we can think in a clearer and more inclusive manner about our circumstance: our resource demands, their supply, the non-economic costs and the connections to the biosphere's functioning. We can more easily appreciate that the connections and interactions between these pieces strongly constrain the amount of resources that can be routinely made available. It becomes easier to accept that satisfying a particular resource want involves more than our ability to make it available, to extract it. The process impacts our environment, which affects its ability to provide the other resources we want: a feedback is involved.

After thinking about the above resource descriptions in this more nuanced manner, a more comprehensive overview assessment of our resource demands, supplies and costs appears. We are indeed largely securing the resources we want from the earth. But most of the earth's resources are finite (in a practical sense) and its environment's functioning is constrained. Thus we are unable to supply our wants or dispose of our waste by tapping into excess or unused environmental resources (including services). They simply don't exist. In essence, we have been successful at supplying the resources we want by engaging in poorly recognized and weakly assessed trade-offs involving us and the rest of the biosphere.

For example, from the biosphere's perspective, taking up farming trades its self-sustaining, human-supporting indigenous ecosystems for agricultural ecosystems. These agricultural ecosystems consist of us, and the plants and animals that feed us. Humans are a key species in these agricultural ecosystems because their functioning depends on us. The price the biosphere pays for the trade-off thus includes more than the disruption and destruction of the indigenous ecosystems. It includes the adjustments to its functioning that are needed to accommodate the agricultural ecosystems, our increased presence and our greater involvement.

From our perspective, by taking up farming, we relieved ourselves of the vagaries of a limited, seasonal food supply. In exchange, we accepted the burden of being responsible for the lives of the plants and animals that feed us, the fields on which they grow and our impacts on the biosphere. By adopting farming practices, we also changed how we discharged our obligation to one another: we changed our culture's world view and structure, which added to our responsibilities. We too have had to pay a price in order to satisfy our food wants through the farming trade-off.

Along with this understanding comes the more general realization that these types of culturally supported trade-offs apply, on a global scale, to our efforts to satisfy all of our resource demands. Thus, although we have maintained our desired supply of resources, our cumulative impacts from extracting, using and discarding them have, by any measure, significantly disrupted the biosphere's functioning. The biosphere is now human-dominated: it has become a human-environment system.

However, we have not altered, and cannot alter, the (external-reality) fact that any trade-offs with, or changes to, the biosphere we make are subject to the basic constraints on the biosphere's functioning. The nature of these constraints are hinted at by concepts such as allometric relationships, carrying capacity, boundaries and footprints. We are obliged to work within those constraints, if we wish the system to remain functional, stable and capable of supplying us with resources over the long term.

Overall, we tend not to respect these constraints. We demonstrate this by our decisions to predominantly focus on one side of the trade-offs/changes that we make to maintain the supply of resources: our want side, not the cost side. Supplying the resources we want thus inevitably burdens us with the often under-recognized responsibilities of maintaining the biosphere's resource-producing functions. It is within this context that we can most meaningfully think about the biosphere's ability to provide the resources we want from our perspective. Its functioning has been disrupted to a degree that interferes with its ability to provide our resource wants and dispose of our waste.

So ends the second part of your report to the boss. As you contemplate it, you have to acknowledge that we have had limited success in our role as environmental managers. We are now part of a biosphere whose functioning we have disrupted. We are living in a human-environment system on the edge of or even outside its constraints on our lives. You wonder what the boss will say after reading it.

# CHAPTER 11

# EXPLAINING THE BIOSPHERE'S CURRENT STATE

After considering your State of the World report (chapters 9 and 10), the all-powerful presence requests that you write yet another report. It must explain how in heaven the earth's biosphere arrived in the state you describe.

But you are at a bit of a loss about what to write. After all, your last report already outlined our cumulative impact on the biosphere from our past and ongoing extraction of resources, their use and the disposal of waste. The earth's environment has become a human-dominated system, a human-environment system, and we have become its environmental managers. Under our management, we have been successful at providing the resources (and services) we wanted, but our resulting disruption of the environment's functioning has strained our ability to provide the resources we now want. This has contributed to our current (material and social) circumstance. The environment's current state and our current circumstance are, together, the current state of the human-environment system, which is the biosphere's current state). What more is there to say?

Eventually you realize that the last report was incomplete. You had failed to fully consider the role that the nature of our decision making and the influences on it have played in creating the biosphere's current state. You realize that if your explanation is to be comprehensive, coherent and consistent, then it has to include the information about how we, as individuals and as a culture, function. This was presented in part 1 of the book.

You decide that your new report on the biosphere's current state will address this deficiency by looking at the overlapping contributions we have made to that state and how we came to make them. Thus the first section of this chapter discusses the most widely used explanation for the biosphere's current state: our material contribution. The following three sections then take a much broader perspective. They discuss our contribution from three perspectives: as individuals (focused on our internal reality); as a culture (focused on our cultures' world views); and as an interaction between us and our environment (focused on the feedback among individuals, their culture and the environment). These sections continue the discussion in the last section of chapter 8 by, in essence, reviewing, compiling, condensing and applying the past 10 chapters in a manner that links them together. The discussions in these three sections are illustrated and personalized by environment-focused examples from Western culture that are relevant to our day-to-day decision making. This is a complex topic, so be prepared for some unavoidable overlap/repetition in the discussions.

The final section condenses the four preceding sections into a personable, human-reality-centred descriptive model of the human-environment system. Applying it to Western culture helps us to more fully appreciate, at a personal level, how our material, individual and cultural contributions to the biosphere's current state arise and combine to result in that state. The model will be a key reference for the discussions in part 3.

## The Material Contributions

A material-focused description of our contributions to the biosphere's current state notes that the environment's functioning provides the resources (including services) we want. We rely on technology to make those resources available to us in the forms and the substantial (per capita and absolute) amounts wanted by our large and growing human population. But in satisfying our resource demands and by using them, we have significantly changed our environment and its functioning. We have thus strained the environment's ability to provide the resources. This is the reason given in the material view for the biosphere's current state (Goodland and Daly 1998; Daily and Ehrlich 1992). The following review of our material contribution is divided into three subsections: our population size, our consumption of resources and our disposal of waste.

Population. Every day our environment must, for every one of us, provide all of the food and materials we consume, bear the brunt of their use and then transform or store the waste we produce. If our supporting environment is to continually supply these resources (including services) to us over the long term, then we can't exceed or disrupt its ability to provide them. We must respect that its functioning is constrained. One way to measure whether we are living within these environmental constraints is to compare the size of the global population and its resource demands today with those that existed when our ancestors certainly lived within the constraints of our (human) environmental niche: a human population who lived without significantly disrupting its supporting environment.

The hunter-gatherers who lived around 10,000 BC, before agriculture was widely practised, were such a population. One method to calculate their population size is to assume that the estimated average weight of today's hunter-gatherer today (59.5 kg) was the same back then. This weight is applied to the allometric curve detailing the relationship between the average body mass of an individual from a species and that species' respective average population density when they are supported by an undisrupted host biome (figure 10). This relationship provides an average population density for humans of around 0.58 people/km$^2$ (it ranges between 0.06 and 6 people/km$^2$), as discussed in chapter 8 (Anton et al. 2002; Peters 1983). The highly optimistic assumption is then made that, immediately before 10,000 BC, all of the biosphere's biomes, excluding Antarctica, could have supported hunter-gatherer populations at 0.58 people/km$^2$. By this calculation, the roughly 135 million km$^2$ of the earth's ice-free land surface could have supported a maximum of around 80 million people living a hunter-gatherer lifestyle. This is an absolute maximum estimate of the biosphere's human carrying capacity at the end of the last ice age.

A more realistic estimate of the biosphere's hunter-gatherer carrying capacity is made by multiplying the known densities of hunter-gatherer bands living in each of the biosphere's current biomes by each biome's area. This method holds promise, but there is a lack of reliable data. The resulting order-of-magnitude estimate of the total possible hunter-gatherer population is 8.6 million people (global average density of 0.063 people/km$^2$) (Tivy 1999; Hassan 1981).

More direct estimates of the biosphere's hunter-gatherer carrying capacity have been made using models that include the densities of archaeological sites and contemporary ethnographic information, such as a band's resource utilization. In these models, an average hunter-gatherer density in the 0.1 people/km$^2$ range was assumed. The resulting estimates of the global, pre-farming human population lie between 1 and 15 million people (Hassan 1981; Livi-Bacci 1997).

On balance, it seems that, around 10,000 BC, the global, ecosystem-constrained hunter-gatherer population was likely no greater than a few tens of millions and certainly no greater than 80 million. Their global average density was likely in the order of 0.1 people/km$^2$ and no greater than 0.58 people/km$^2$. The productivity of the biosphere's biomes and the nature of our food requirements indicate that whatever the average human density was, there were only a few special locations that were capable of supporting hunter-gatherers at a density greater than the average, as discussed in chapter 8. The vast majority of the biosphere's inhabited biomes hosted hunter-gatherers at densities lower than the average.

Compare these hunter-gatherer estimates to the estimated 2016 global human population of 7.3 billion people, at an average density of 54.0 people/km$^2$ (US Census Bureau 2016). The current human population is between an incredible 80 and a staggering 800 times larger than the size that the biosphere could support if we were all living a hunter-gatherer lifestyle in undisrupted biomes. Even if the 800 times difference is discounted, the 80 times difference still implies that we are having a significant impact on our supporting environment.

A feel for the environmental significance of this difference in densities can be gained by comparing our current environmental impacts, as described in chapters 9 and 10, with reports of the impacts associated with a less industrialized historical population with a much lower population density. Around AD 100, the most populous area in Asia was China, with around 53 million people occupying some 5 million km$^2$, for a density of around 10.6 people/km$^2$ (Hsu 1988). Much of China lay in the temperate forest and grassland biomes.

In the AD 300s, the combined Eastern and Western Roman Empires contained an estimated 56 million people living on about 4 million km$^2$ of land, for an average density of about 14 people/km$^2$ (Harl 2009; Taagepera 1979). The Roman Empire lay mostly in the Mediterranean biome. The Romans' environmental legacy from supporting its population includes cleared forests, increased soil erosion, animal extinctions and a dramatic increase in the atmosphere's lead and copper content (helped by Chinese contributions) (Hughes 1975; Hong et al. 1994). If a historical population at an average density of around 14 people/km$^2$ significantly disrupted and destroyed its supporting ecosystems and notably affected its supporting global environment, then we could reasonably expect that the current global population density of 54.0 people/km$^2$ is causing a much more significant level of disruption and destruction on a global scale.

The degree to which our current global population is exceeding the constraints of our human niche comes starkly into personal focus if we imagine the following. Assume that we decide to obey the allometric body-weight-to-population-density relationship that constrains all organisms' population densities, but we still want to maintain a population density of 50 people/km$^2$. To be successful, we would have to change our body weight from 59.5 kg to about 0.7 kg (Peters 1983). At that size, we would fill our environmentally constrained niche as a small animal, such as a rat or a small guinea pig.

With this environmental constraint on population growth in mind, we are left to wonder, how is it that the current human population could have grown so large? After all, if a hunter-gatherer band's population density rose above the maximum that could be supported by their undisrupted host biome, then a substantial reassignment of ecosystem resources from the other species to humans would have occurred. In the process, the hunter-gatherers would have disrupted or destroyed their host biome's functioning and its long-term ability to support them. The area would have experienced an ecosystem-forced reduction in the biome's human population though some mix of migration, lower birth rates and higher death rates.

One answer suggests that settling and taking up agriculture resulted in our hunter-gatherer ancestors inadvertently circumventing the resource constraints that their supporting ecosystem imposed on their population size. By replacing some of their local ecosystems with agricultural ecosystems, and by taking on the responsibility of managing them, they increased their food supply. They overcame a basic constraint on their lives.

However, the details of why the population increased are poorly known. It likely involved a mix of a fertility-increasing change in diet, an increased value placed on children by the new labour-intensive farming lifestyle and an increased opportunity for having sex or full-term pregnancies, as discussed in chapter 3 (Moore et al. 2000). Whatever the mix of these causes was, it was not accompanied by the appearance of either cultural guidance or short-term environmental imperatives coercing the members to limit their population to a size that respected the constraints their environmental endowment and farming responsibilities imposed on them over the long term. The lack of cultural guidance reminding the members of their responsibility to limit their population has been a feature of most complex cultures ever since.

The members' ongoing responsibility to limit their population sizes arises because farming could provide more food, but it couldn't remove the many long-term external-reality constraints on how much and when. Farming could only circumvent not overcome these constraints. Thus, once our early farming ancestors neared or exceeded their agricultural ecosystem's capacity to routinely supply enough food to a growing population, these constraints reasserted themselves.

Indeed, lean times were a reoccurring theme throughout our recorded and inferred farming history. Yet, over the longer term, populations still continued to grow. How is this possible?

When farmers were still few in number, they could adapt to food shortages in the same manner as hunter-gatherers: by migrating or surviving on hard-times food from wild ecosystems. When an increased population made this unrealistic, farmers had another option. The guidance from their farming culture's world view about their relationship with their supporting environment was different from that of their hunter-gatherer ancestors. Farmers were permitted, if not encouraged, by their world view to consciously try and circumvent the current environmental constraints on an agricultural ecosystem's productivity by further altering their environment. This is what the Abu Hureyrans did when they became sheep herders, and what the Babylonians did when they built irrigation works. It is what we do today. This points to another basic reason for the overall increase in the size of the human population. We find it easier (less stressful) to face the practical difficulty of increasing resource availability from the environment than the difficulties of reducing resource demand, in particular, population reduction.

Exercising the option of increasing productivity was successful, within limits. Eventually, the human population grew large enough that diseases joined famine as the primary external-reality constraints on population increases. During the 1800s and 1900s, the constraints of famine and disease were circumvented by improved hygiene, the discovery of antimicrobials and the green revolution. This resulted in the global population increasing at an unprecedented 1.7%–2.2%/year during the third quarter of the 1900s (US Census Bureau 2016).

In overview, the resources (including services) needed to support a human population of any size are provided by the functioning of our environmental endowment. By practising agriculture and applying technology, we increased the amount of resources available to us and thus managed to support a very large population. But, in doing so, we did not remove the environment's long-term constraints on the human population that it can support, we just circumvented them. These constraints kept appearing in a different form, such as starvation and disease. We are now faced with an environment whose functioning has been significantly disrupted, and we are stuck with acting as environmental managers, a role we are unsuited to fill.

**Consumption and Disposal.** Simplistically, the larger the human population, the more resources it consumes and the more waste it discards, thus the greater the population's impacts on the environment. But population size alone can't fully account for our environmental impacts. Our lifestyle also needs to be considered.

Each citizen of a country strives to maintain a particular lifestyle. This can be represented by the resources the citizen must consume and the waste they must dispose of to maintain that lifestyle. (Because the consuming of resources is always followed by the disposing of waste, the word *consumption* will be used to represent both). Similarly, a country can be thought of as having an average lifestyle that is represented by its average, per capita level of consumption. Thus the impacts that the lifestyles of a country's citizens have on the environment are represented by the combination of its average per capita resource consumption (its average lifestyle) and its population size (Hunter 2000; Stern 1993).

For example, a country's average lifestyle can be represented by its per capita consumption of energy. In 2013, China's per capita consumption of energy was 2.21 tonnes of oil equivalent (toe/capita), while the United States' consumption was 6.92 toe/capita. But China's population was 1.36 billion, 4.3 times bigger than the United States' population of 316 million. China's absolute consumption of energy was 3,006 toe, which was bigger than that of the United States at 2,187 toe (IEA 2015). Thus China's and the United States' environmental impacts are more similar than we would expect if we had just compared either their population sizes or their average lifestyles.

Humanity's material contribution to the biosphere's current state depends on both the total number of us and our personal, culturally influenced choice of lifestyle. This begs the question, at what level of personal consumption (i.e., lifestyle) do we significantly impact our environment? An answer can be provided from two perspectives: an environmental perspective and a lifestyle perspective.

From our supporting environment's perspective, if we require it to continue providing us with the resources (including services) we want over the long term, then we can't demand more from it than its functioning can provide. We can't disrupt its functioning too much. For example, to remain functional, an ecosystem must remain sufficiently resilient, robust and adaptable so that it can respond to the ever-present forces for change

in ways that keep that essence intact. To meet this constraint, our degree of environmental disruption must be limited to at least the levels that respect the basic constraints on an ecosystem's functioning, such as allometric relationships, nutrient cycles, primary productivity, biodiversity, trophic web details and the connections among them.

As discussed in earlier chapters, from the environment's perspective and at an ecosystem to global scale, we are currently widely ignoring and exceeding these types of environmental constraints. This is occurring in part because there are at least 80 times more of us than could be supported if we were hunter-gatherers. In part it is occurring because the current global average per capita consumption of resources, such as water, oil and copper, is well above the average for our hunter-gatherer and both our ancient Roman and Chinese ancestors. From our environment's perspective, our current level of consumption continues to drive our supporting environmental system well beyond the functional limits needed to support us in the long term. We are over-consuming.

From a human perspective, the question of an appropriate amount of consuming and discarding (i.e., the question of an appropriate lifestyle) is much harder to answer. All of us desire to live a well-fed, healthy, want-free, happy existence. We can also agree that a certain amount of consumption is needed to meet these objectives. Any amount above that can be considered over-consumption. But we have great difficulty agreeing on where the boundary lies between necessary consumption and over-consumption, between need and want (World Bank 2002).

We can better understand this difficulty when we consider the ever-present fingers of our cultures' world views and our innate individual human characteristics being poked into every aspect of our consuming and discarding decisions. Consider that our personal choice of an over-consuming boundary is strongly influenced by our culture's guidance about acceptable lifestyles. For example, Western culture's world view guides Westerners to equate material wealth with our sense of social-psychological well-being, such as happiness. Compared to hunter-gatherers, we already have an abundance of material wealth well beyond that which could be considered necessary to live a well-fed, want-free lifestyle. We should be more than happy. Yet, the members of Western culture still feel driven by their internal reality and their culture's world view to acquire and consume ever more resources (Kasser 2002). Are we over-consuming or still in the process of becoming reasonably satisfied?

Consider the following answers that appear when we consider why humans consume. We consume for survival, such as clothes in cold climates, and to satisfy cultural imperatives, such as Christmas gift giving or owning a cell phone. We also consume to satisfy psychological imperatives, as illustrated by Imelda Marcos owning hundreds of pairs of shoes, or by us keeping up with the Joneses. We also consume to feel part of a community, as illustrated in the extreme by consumer fads.

Of course, we rarely think of our own purchases in these terms. For us, there is sufficient reason to make a purchase if we can justify it. Examples are a powerful enough perceived desire (need or want), the belief that making the purchase would satisfy that desire, and feeling that the funds to make the desired purchase are available. Making that purchasing decision is also made easier for us if our desire is perceived by us to be a biological or social imperative (a need), or a socially acceptable want. The required social "permission" comes in many forms such as advertising, our social group's views and a supportive cultural world view.

However, cultural or social imperatives can also strongly guide us to consider the multi-faceted external-reality aspects of a purchase. If they do, then we are more likely to include them in our purchasing decisions. Think of boycotts for goods produced by child labour and the rising concern about single-use plastic items. In the absence of such guidance, the environmental impacts from the production of the purchased goods (including services), their use and the disposal of associated wastes will generally remain in the background, unknown to us or ignored. Under these circumstances, it is easy for us to personally think that we are not over-consuming, even as observers and environmental cues indicate we are.

Our decisions about an over-consumption boundary become more informed if we also accept that both our personal and our culture's total consumption can increase even though there may be little or no active encouragement to consume. This can arise indirectly as a result of interactions between a number of pre-existing factors. For example, a couple's culturally supported preference for few or no children means that they may spend less on human-services-related consumption, such as childcare and education. Instead, they might spend their greater disposable income living a more intensive, material-resource-consumptive lifestyle, an activity

encouraged by Western culture's world view. As a result, a nation's per capita resource consumption rate can remain high or rise, even though its per capita fertility rate is falling.

Consider too that a future increase in resource consumption can be preconditioned by adjustments to a culture's world view. For example, by the culture accepting either divorce or fewer generations living together, it also accepts fewer people per household (living in a house). The culture can also accept large house sizes as a status symbol. These types of changes are currently occurring worldwide. As a result, more dwellings must be built, furnished and maintained, which translates into more resources per capita to run a household. This outcome can be seen in China, which has a nearly stable population size. By the early 2000s their average household size had decreased from 4.5 to 3.5 people, while the floor size of the average household had increased three times (Liu et al. 2003; Liu and Diamond 2005). This change in lifestyle has contributed to an increase in China's per capita consumption even as its population remains stable. Deciding on a boundary for an appropriate level of consumption, from a human perspective, is more complicated than we imagine.

In overview, our global per capita level of resource consumption and waste disposal is having an impact on our environment's functioning. Whether or not this impact is significant and a result of individuals over-consuming depends on the perspective taken. From the environment's perspective, it is and we are. From our perspective, the answer is "maybe not." "Maybe not" arises because our decisions and feelings about consumption are subject to many influences, such as our preferred lifestyle and whether our culture's world view supports our desired consumption. Similarly, whether we see our resulting environmental impacts as significant or not depends on our culturally influenced personal view of our relationship with our environment.

In summary, a focus on our material contribution to the biosphere's current state reveals the connections between resource consumption, waste disposal, population size, lifestyle and the environment's functioning. The relations between them result in an easily understood material explanation for the biosphere's current state. Consider three popular explanations for our current (resource) circumstance and suggestions for solutions based on this view.

(1) The Malthusian view is that there are finite biological limits to resource availability. This places

simple, absolute limits on their consumption and the size of the population that our environment can support. (2) Neoliberal economic thinking blames resource shortages on a constrained market. When the market is functioning properly, it should automatically induce in the population the appropriate technological and personal responses needed to resolve resource shortages and deal with environmental issues, if any. (3) The distributionist view is that there is enough of everything, it is just not distributed correctly (Homer-Dixon 1995). These views provide us with insights but don't comprehensively explain either our circumstance or the environment's state.

A more complex material explanation for the biosphere's current state is provided by the IPAT model. In it our environmental *impact* is modelled as the product of interactions between *population*, *affluence* and *technology* (Daily and Ehrlich 1992). This model improves our understanding of the interconnectedness of consumption, resources and our environment, but it too is significantly incomplete as an explanation for the biosphere's current state. In particular, it notes our impact on our environment but not the nature of its functioning. It also neglects the nature of our decision making and the influences on it.

A material-focused explanation is too simple to fully explain the biosphere's current state. A comprehensive explanation has to closely consider how we and our ancestors came to make the day-to-day decisions that materially contributed to the biosphere's current state. How is it that we chose to live outside the human and environmental constraints on our lives and then fail to respond to the consequences in a way that better respects them? And why is it so difficult for us to face and fulfill our responsibilities? Think increasing population. Addressing these questions requires taking into account our internal reality, our decision-making foibles, the influence of our culture's world view on our decision making and the environment's functioning. Doing so is the subject of the rest of this chapter.

# The Individual's Contribution: Their Internal Reality

Our brain uses our memories (including those of our feelings), our reference normal, and the sensings of both our body's state and our surroundings to provide us with a reality in our heads. Our internal reality is, in essence, the product of our thinking processes and the (past and present) influences on them.

Our internal reality resides mostly in our unconscious. We are consciously aware of only a small part of it. We are aware, as needed, of those parts that are our personal conscious understanding and explanation for our experiences and opinions: our personal world view. However, regardless of the degree of our conscious awareness of our internal reality, it is reflected in all of our decisions, beliefs, values and behaviours, including those that affect our relationship with our environment. Indeed aspects of our unconscious internal reality can be deduced from our short-term conscious experiences of ourselves and the world around us.

In order to fully appreciate our individual contribution to the biosphere's current state, we need to broaden and personalize our understanding of our internal reality. The focus here is its representation of our environment and its influence on our related decision. First to be discussed are two interconnected human characteristics that contribute to our internal reality's representation of the environment: being aware of our environment and paying attention to it. Those discussions reveal the difference between our internal reality and the external reality, which is then discussed. With that understanding of our internal reality in place, our conscious internal-reality view of our relationship with our environment can be reviewed. In the process, we gain an appreciation for our individual contribution to the state of our environment.

**Awareness.** Being consciously aware of something is a prerequisite to paying conscious attention to it. We prefer to be aware of some aspects of our lives and not others. For example, we innately want to be aware of our social surroundings and the changes in the relationships between those in our social circle. Think gossip. Our routine desire to be aware of some things and not others helps to constrain and bias our internal reality. Our biased internal reality then influences our awareness. A self-sustaining feedback loop exists between our awareness preferences and our internal-reality biases.

Ironically, we tend to be unaware of our internal reality's awareness-related biases because they largely occur in our unconscious. As a result, we tend to believe that our internal reality is a reliable and accurate representation of the external reality. It often isn't.

For instance, we can claim awareness of features that don't exist in the external reality, as illustrated by us falling for optical illusions and magic tricks ((figure 7).

In an interesting twist, we can also reject features that do exist. Consider that scientific discoveries about our world (external reality) can seem so unreal to us that they might as well be from another planet. Do you *really* believe that *Pfiesteria*, the shape-shifting organism mentioned in chapter 7, actually exists?

Fortunately, we can become more aware of our internal reality's awareness-related biases and their influence on our environment-affecting decisions by considering how these biases arise. First to be considered are the difficulties we face when considering our environment's complexity and the ranges in time and space over which it functions. Last to be considered is how we deal with unpredictable events.

*Complexity.* Our environment is a dynamic, complex system in its own right, as discussed in chapters 7 and 8. It consists of many variables linked together through various types of interactions, including feedback, that take place over a very wide range of space and time (Hunter 2000). The environment's complexity ensures that it is constantly changing. After all, it is constantly responding to the internally generated influences that its functioning (e.g., ecological and geological processes) is continually exposing it to. It is also subject to external influences, for example, changes in the amount of sunlight reaching the earth.

Our environment's dynamic complexity makes it difficult for us, as individuals, to appreciate it and relate to it. For example, although the system's dynamics is a mix of random, chaotic and deterministic relationships, we prefer to think of the system as mostly governed by simple if-then rules, that is, by deterministic relationships. Similarly, the environment's functioning displays a mix of features, such as drift, inertia and preconditioning, that are the product of multiple interconnected "causes." The environment thus rarely changes in a manner that meets our, or at least Westerners', inclination to treat it as a black and white affair. It is more grey, which is not to our liking. Finally, because the environment's dynamic complexity includes ongoing interactions occurring at different scales in time and space, it never reaches a state that corresponds to Westerners' preferred concept of simple balance or equilibrium. It is perpetually both catching up to and partly caught up with past influences, as well as starting to react to new influences.

For us to personally become more fully aware of the environment's complexity takes considerable effort, and time. Can you picture the partly ordered, partly chaotic dynamics that result from the simultaneous participation of a flowering plant and its forest host in both the nitrogen and carbon cycles (figure 7)? Consequently, our first and enduring impression of our environment is that it is a largely incomprehensible, perhaps disordered, work in progress. It is easier for us to simplify, ignore or remain mostly unaware of the environment's complexity rather than try to accept and meaningfully incorporate it, as-is, into our daily decision making.

When we do try to broaden our awareness of our complex environment, we usually use the same techniques we use when dealing with any complex system. We tend to focus our attention on those environmental features we are already aware of, or toward which our internal reality and our culture's world guides us (see next subsection). Joining these preferred features are those that readily enter our awareness or forcefully gain our attention.

To create the feeling that we have made sense of what we know, we use mental models, experience-based rules of thumb and simplifying assumptions. In doing so, we tend to avoid more nuanced and skeptical thinking. Think of our personal methods of weather prediction. Our innate way of making sense of information doesn't favour us becoming fully aware of our environment's complexity.

If you have any doubts about this assessment, ask a decision maker higher in the social hierarchy, whom you expect to know more, to describe their understanding of how an ecosystem or the climate system functions. You might be surprised to find that their view is more simple and incomplete than you imagined. Alternatively, tackle the difficult task of noting the tendency toward simplicity and incompleteness in our personal preferred solutions to environmental issues.

*Scale: Time and Space.* As previously mentioned, our environment functions at time scales from sub-microseconds to millions of years, plus some, and at spatial scales from atomic to solar system. In contrast, when going about our day-to-day lives, we personally experience events in the time range of 250 milliseconds (ms) to about 50 years, and the space range of 1 mm to about 20 km. We are thus most acutely aware of events that are recent, less than six months old (the same time span that politicians rely on for you to forget their promises), and nearby, less than a few kilometres.

Nevertheless, with sufficient effort, most of us can expand and personalize our abstract, disconnected intellectual awareness of human events so that we cover up to two or three generations. In this context, a generation is around 25 years, so our easily personalized awareness time period is likely around 60 to 100 years. We are helped by the direct memories of the elders and the extended memory provided by a culture's oral or written history (McIntosh et al. 2000). Similarly, with effort, we can expand and personalize our awareness of events or changes that occurred up to tens or even hundreds of kilometres away. The further away in time or space the event is from us, the more likely the details will be incomplete and less reliable. These characteristics apply to our awareness of our environment.

There are other constraints and biases on us becoming aware of and personalizing environmentally relevant events. The events we most easily remember and personalize are catastrophic, such as Mount St. Helens exploding and the tsunami that struck Indonesia in 2004, especially if they are local or personally relevant. Do you know someone who was affected by either of these two events? We thus tend to think of these types of events as the main causes of environmental change, even though they may be just as or less significant than slow changes that continue for longer periods (Dobson et al. 1997). For example, we tend to remain largely oblivious to ongoing, chronic changes, such as soil erosion, salinization and species extinction, especially when they are dispersed on a regional scale.

These scale-related constraints bias our overall casual awareness of our surrounding. For example, we would certainly be aware of a patch of tropical jungle being fragmented into 1,000 ha parcels (10 km$^2$ or 3.2 x 3.2 km) by logging over a period of five years. However, we would likely be only dimly aware of the resulting local species extinctions. In part because about 50 years must pass before around half of the eventual extinctions have occurred (Brooks et al. 1999). Although fast by ecosystem standards, these extinctions are too slow for us to easily grasp. Thus we easily become aware of only part of the consequences of logging.

Our limited appreciation of the scales over which our environment functions and changes combines with other human characteristics in ways that help to explain our lack of awareness of our personal contribution to the biosphere's current state. Consider that we prefer to learn through experience rather than study. In doing so, we most easily and effectively learn from our experiences when they are, or when we treat them as, simple cause-and-effect events in which our doing is soon followed by our experiencing the outcome. Any time lag or spatial distance between the two strongly affects our awareness of the connection between our actions and the consequences, and thus the lesson we may learn from them. Think of dump and forget.

Unfortunately, the functioning of our complex environmental system means that there is often a large separation in time and/or space between our environment-affecting activities and the appearance of humanly recognizable environmental changes. We can therefore easily fail to recognize our contribution to that change. Failure is made more likely by the often indirect and multi-sourced nature of cause and effect in the environmental system.

Consider too that, collectively, our individual human impacts on our environment can occur at a global scale. Not only is the scale of this impact difficult for us to grasp, but our local, small personal contribution seems at least trivial, if not benign. Think of tail-pipe emissions and the dropped Styrofoam cup. In these cases, there are few indicators prompting us to be personally aware of and assign greater significance to our personal environment-affecting decisions and actions. No wonder we, as individuals, can be unaware of, downplay or ignore our personal contribution to humanity's impact on our environment.

Even if we do make the effort, it takes time for us to accept responsibility for our contribution. For example, humans started manufacturing large quantities of organochlorine compounds (chlorinated hydrocarbons) in the late 1940s. They contribute to the smog-like haze that appears each spring over the Arctic Ocean north of Alaska. The haze was discovered in the 1950s and chlorinated hydrocarbons were detected in Arctic marine mammals in the 1960s. However, it took until 1981 to figure out that much of the Alaskan Arctic haze originated in Eurasia (think brown cloud). It was still later before we figured out that the chemicals in the haze, such as the chlorinated hydrocarbons, were being concentrated into the Arctic environment because it was acting as a cold trap (Shaw 1995).

Today, the environmental impacts from many of these older chlorinated hydrocarbons are known. By the year 2000, some of them, such as toxaphene and DDT, had been widely banned. Some of those banned, such as DDT, are still being used, although in lesser amounts. Others are in the process of being partly phased out. For example, endosulfan usage rose until 2000, where it reached the average usage recorded for the period 2000 to 2004. Production is thought to have increased until its phase-out started in 2012 (Weber et al. 2010; BiPRO 2010; Li and Macdonald 2005). Because discovering, acknowledging and acting on this knowledge took so long, even if you do not personally use any of the environment-harming organochlorine compounds, the goods and services you purchase likely did or still do. At what point in this decades-long story could we have personally become sufficiently aware of this state of affairs to have voluntarily chosen to stop (directly or indirectly) using them or buying the related products?

In overview, it is difficult for us to personally become fully aware of and appreciate the scale of time and space over which our environment functions and over which we make our personal contribution to our environment's state. Our awareness tends to be constrained and biased. By acknowledging this limitation, we can more easily accept that our personal consumption and discarding decisions (our lifestyle) do indeed make a significant contribution to disrupting our environment's functioning, locally to globally, and short to long term.

*Unpredictable Events.* The complex functioning of our environment includes a chaotic and a random component. This makes truly unpredictable events inevitable. However, our limited awareness of how our environment functions, and the spin we put on our knowledge and experiences, allows us to quite convincingly identify and rationalize more than a reasonable number of outcomes from our decisions as "unpredictable."

We tend to label these events as *unanticipated*, *unexpected*, *unforeseeable*, *inadvertent* or, most conveniently, as *an act of God* or a *natural disaster*. In the short term, these labels provide us with a number of advantages. For example, they can act as a ready-made excuse, thus helping to alleviate the unpleasantness of the experience. And they can mask its causes from our full awareness, thus relieving us from our responsibility for the event. But, in the longer term, we suffer from both the disadvantage of being unable to learn from our mistakes and having to repeatedly use these explanations for reoccurring events.

The story of the schoolboy and the atomic reactor can help us become more aware of our over- and inappropriate use of labels like *unpredictable*. He gleaned his basic knowledge about atomic reactors from an outline-level school textbook. By applying logic to his knowledge, he managed to find the materials, including the radioactive ones, needed to build the lab-scale equivalent of an atomic reactor. In the final stages of his project, it took men in protective suits to dismantle and remove his radiation-spewing reactor from a shed in the back garden of his home. During his experiments, he likely received a high enough dose of radiation to compromise his future health (Silverstein 1998).

He certainly had the considerable smarts needed to build a reactor. But, despite the available information, he only had a dim awareness of the consequences and thus what he became responsible for when he built and operated it. The dose of radiation he unexpectedly received was only unexpected from the perspective of his internal reality. Those "unexpected" consequences of his awareness-limited decision making are analogous to many of the environmental consequences we label as unexpected: they are the result of our awareness-limited environmental decision making. He is us.

When seen in this context, is it really surprising that many recent disasters turned out not to be truly unexpected events? The Chernobyl disaster occurred as a result of a safety test; the BP Gulf oil well blowout was caused by shortcuts; the Fukushima Daiichi meltdown was the result of a lack of attention to, or denial of, risk (Hixson 2015; McCurry 2015). Remember, only catastrophic events were mentioned here, not the unexpected outcomes that appeared over the long term.

In overview, our limited awareness of the world around us helps to ensure that our internal reality's representation of the external reality is biased and incomplete. In turn, our internal reality influences our perception of our environment and our relationship with it. Thus a feedback exists between the two. This influences our routine, environment-affecting decisions, which become our personal contribution to its state.

**Attention.** Awareness of something does not, on its own, guarantee we will pay it appropriate attention. Consider too that the things we do pay a lot of attention to are not necessarily the things to which we should pay that level of attention. The nature of our attention thus deserves our attention.

We tend to pay particular attention to certain aspects of our lives. For example, we innately pay close attention to those features needed to complete our lifeway. These include our physical needs, such as food and sleep, and our psychological needs, such as a sense of social and emotional well-being provided by such things as companionship and belonging to a stable community. We do the same with our surroundings.

By routinely giving preferential attention to some aspects of our lives and surroundings, we bias our internal reality with a strong dose of self-centred spin.

Our biased internal reality then influences the things we pay attention to. A self-sustaining feedback loop exists between our attention preferences and our internal reality's biases.

We are largely unaware of those attention-related biases because they mainly reside in our unconscious. Their presence and influence on our internal reality and thus on our environment-affecting decisions are illustrated by the personal attention we give to our waste and to our labelling of our environmental impacts as *incidental*. Once aware of these biases, we can pay them more attention.

**Waste.** The influence that our personal internal reality has on our environment-affecting decisions should become clearer after considering the attention we usually give to the stuff we no longer value but are forced to be aware of: our waste, our garbage, our trash, our rubbish. For our hunter-gatherer ancestors of some 10,000 years ago, the solution to waste was to discard it whenever and wherever was convenient. This was quite appropriate for our largely mobile ancestors, who were few in number and discarded mostly environmentally friendly waste in small amounts. Today there are many of us, most of whom are living sedentary lives and producing a large volume of environmentally damaging waste from our industrialized lifestyles. Despite this change in the nature of our waste, there has been limited change in the attention we give to it. Our instinctive, preferred solution to its presence is still the same as our distant ancestors: dump and forget. Think of the space junk circling the earth.

If you doubt this description applies to us today, consider the following results from some American garbage studies of the early 1990s. They found, as discussed in chapter 2, that the food and drink items the study participants reported throwing out were different from the actual waste the researchers found in the participants' garbage bins. They also reported that what the participants said they threw out cast themselves in a good light.

This type of biased awareness extended to the attention they paid to those aspects of their waste that were considered hazardous to humans or the environment. Note first that the nature of the hazardous waste in the participant family's trash bin reflected their income. The discards from the poor were mostly related to automobiles; from the middle class, mostly paints and stains; and from the affluent, mostly lawn- and garden-care products. Despite these differences, the disposers all gave the same limited and biased attention to these hazardous parts of their waste.

This was demonstrated when a garbage company announced a special "hazardous waste day" to enable the public to identify and properly dispose of their hazardous waste. Shortly after that day, researchers made inventories of the study participants' *regular* household garbage. Remarkably, their garbage contained far more hazardous waste than average. It seems that the hazardous waste day served to bring to the disposers' attention the hazardous nature of some materials they owned, to which they reacted by disposing of them. But, ironically, they did so without considering the proper method of disposal, which was the core reason for holding a hazardous waste day in the first place. It is as though once reminded of the hazardous materials they owned, they simply wanted to get rid of them: dump and forget.

Certainly, since the time of those studies, many countries have made significant improvements in their handling of waste, such as recycling. But "dump and forget" is still an innate reference normal. A current, global example of this is the plastic floating in our oceans. It is being improperly disposed of in all parts of the world. You can verify this for yourself by checking a (wet or dry) water course, near an urban centre, at a point where a road crosses it. Look for plastic in the water course, on its banks and on the adjacent roadside. When the rains come, where will the plastic go?

There is an argument to be made that Westerners', for example, lack of attention to garbage generation and disposal is excusable because of the difficulties we face in our busy lives. After all, we do have to satisfy our basic needs, such as raising children, socializing with friends, running businesses and working to pay for the things we need. We are all aware of this time crunch. However, accepting this conclusion results in a contradiction we must deal with.

Westerners are fond of pointing out that, more than ever before, our superior mental inheritance has allowed us to create technologies that enable us to more easily and completely satisfy our material wants. This is true, but in making this claim we conveniently forget that exercising our ability comes with costs. They include our responsibility for the increased amounts and types of waste we generate, and its impacts on us and our environment (Rathje and Murphy 1993a). Westerners fail to see a contradiction: we clearly have the mental capacity to generate the garbage, but it seems we have insufficient capacity to recognize and discharge our resulting responsibility for it. This includes not generating it in the first place. Think of plastic containers that are very difficult to recycle, or unrecyclable, because they contain two plastics or are lined with aluminium.

It is tempting to place the blame for most waste issues on the failure of big businesses, industry and the government to accept their responsibility for generating and managing it. But this justification can't deflect much of the blame from us as individuals. Regardless of whether the waste is mine tailings, a disposable coffee cup, from manufacturing, or unused food, the decisions about its generation and disposal are not made by corporations and governments. They are made by the people who own and run, or work and invest in, these organizations. People like you and me. It is up to us as individuals to pay attention to our waste. But we prefer not to, regardless of how smart we think we are.

### Incidental Impacts.

When our impacts on our environment unexpectedly intrude into our consciousness, we may pay attention to them. But that attention doesn't necessarily result in us adjusting our internal reality so that it is more in tune with the external reality. It doesn't automatically result in a decline in our environmental impacts. We are just as likely to rationalize our behaviour and beliefs in a way that favours our current internal-reality view or purpose and thus maintains our impacts.

A common method to rationalize our impacts is to simultaneously acknowledge and ignore them. One way to accomplish this is to turn them into "incidentals," regardless of their actual environmental significance. We do so by labelling our impacts *small, unimportant, irrelevant, inconsequential, innocuous, trivial* or *normal*. An example of these internal-reality adjustments is for us to consider our car's exhaust emissions to be trivial relative to our immediate need for our journey, or insignificant when compared to tail-pipe emissions on a global scale. Another example is thinking of the trash we throw out as innocuous compared to our need to get rid of it or to the trash dumping of others. So too is considering that the pesticides and nitrate fertilizer running off our green lawn and colourful flower beds into a storm drain are a normal part of the cost of having a beautiful garden. An action also becomes an incidental when we say, after the fact, that "it's too late now, the damage is done" and then move on.

Unfortunately, in the external reality, our incidentals are not always as benign as we imagine. Many of our personal incidentals are contributions to the significant, long-term, global-scale environmental disruptions that characterize the biosphere's current state. Consider the bioaccumulation up the trophic web of the incidental amounts of waste (pollutants) we dump into our air and water. Think of PCBs in salmon and orcas.

In overview, the limited and biased attention we pay to our surroundings constrains our internal reality's ability to represent the external reality. In turn, our internal reality influences the attention we pay to our surroundings. A feedback exists between the two. Thus, those features to which we pay attention and the level of attention we give to them may bear little or no relation to the environmental importance of those features, when they are considered from the external reality's perspective.

### The Reality Paradox.

Our brain, under the influence of our culture's world view, uses our past experiences (our memories) and sensings to create our internal reality. However, our brain's functioning suffers from limitations and biases. There are also biases in our memories, sensings and our culture's world view. Thus our internal reality imperfectly represents the external reality and how it functions. This biases our (conscious and unconscious) perception of the world and our relationship to it, as illustrated by the nature of our awareness and attention.

Our internal reality is more like a map of the external reality rather than a replica. It is a selective representation of the external reality. The limits to a map's usefulness and accuracy can be deduced by comparing it to the actual landscape. Similarly, an independent observer, such as a space alien on an Earth mission, could deduce the differences between your internal reality and the external reality.

We, however, are not an independent observer of our own internal reality. Unsurprisingly, we feel that our internal reality is real, complete and accurate, so we personally find it difficult to even admit that there could be significant differences between our internal reality and the external reality, let alone recognize them. However, we can come to appreciate their existence by considering the power behind the following observation: a disease is what we get, while illness is what we feel (Brown 1998). Which of the two is more personally real, disease or illness? Which would the space aliens report if they could medically probe us but were forbidden to communicate with us?

The common aspects of these two answers illustrate that our internal reality and the external reality can overlap. But the differences illustrate that the two need not be exactly the same. In the end, only one of these realities dominates our conscious experience. The other survives our death.

The presence of the differences between our internal reality and the external reality could be used to conclude that our internal reality is the product of a deranged mental state and thus an illusion. Certainly, parts of our lives, sometimes large and important parts, may be considered to be illusions that are contributing to the differences, but our life itself is not an illusion. It seems more truthful to say that we can recognize puzzling mysteries, inconsistencies and contradictions in our mental picture of our lives and surroundings, but we don't have quite enough ability to truly resolve them.

Perhaps we could describe the differences between the realities as a disconnect. Certainly, a space alien would report that we have strange ways of making sense of our world and relating to it. But a disconnect implies that the differences can be eliminated. Although they can certainly be reduced and accommodated, like our other human foibles, I don't think the differences can be fully eliminated. Our internal reality is the product of our human characteristics, which appeared over hundreds of thousands of years (or, if you prefer, were provided to us "as is"). Those aspects of our internal reality that are common to all humans are not the result of some defect. They and the limitations they impose on us are part of who we are.

It is more useful to think of those differences between the two realities (the divergences) that we become aware of as creating a paradox, a *reality paradox*. After all, if we do recognize differences between the two realities, we then face a contradiction or inconsistency between our perception of a circumstance and the actual circumstance that we can't, in good faith, dismiss. We experience a baffling and even disturbing reality paradox.

Despite our superior intelligence, we usually struggle to deal with this (emotional and intellectual) experience head on because doing so requires us to intently look in the mirror, an awkward and often painful task. We tend, as with other cognitive dissonance experiences, to find less traumatic ways to neutralize our reality paradox experiences. We tend to find or create personally satisfying explanations, more like side-stepping rationalizations, that mask or ignore the differences. Perhaps this is why religion, science, politicians and economic advisors exist: to provide us with either a friendly ear or an alternative, more comforting explanation. Often we resolve the paradox

by simply following our culture's guidance or our urges. But we can't always deal with the external reality by avoiding it or explaining it away. After all, at some point it may simply impose itself on us. Think of our inevitable death.

The influence that the reality paradox aspects of our internal reality exerts on our individual contribution to the biosphere's current state is revealed by our relationship to our personal domestic garbage. Our culturally influenced choice of our lifestyle, our rationalization of that choice and our resulting view of that lifestyle (such as "cool," "sophisticated" or "middle class") resides in our head as part of our internal reality. So does our perception of the garbage generated by our lifestyle (such as "an inconvenience" or "distraction" to be disposed of quickly).

In contrast, our garbage itself resides entirely in the external reality, where it is a physical manifestation of our lifestyle. On a few occasions we may become consciously aware of this external-reality aspect of our garbage. We may realize that it represents the resources our lifestyle demands and symbolizes the impact our lifestyle has on our environment. This awareness can prompt us to experience a reality paradox about our lifestyle. Our efforts to deal with it are likely to focus on rationalizing away the differences causing the paradox by, for example, denying, masking or justifying them. We can, for example, treat our garbage as trivial, assure ourselves that we dispose of it properly (according to regulation) or claim that our chosen lifestyle is essential to our well-being. Our reaction, of course, doesn't change the fact that our personal garbage results from our lifestyle, and that both of them are part of our individual contribution to the biosphere's current state.

In overview, the normal functioning of our brain creates for us an internal reality that is different from the external reality. Our internal reality combines with our decision-making foibles to enable us to make decisions and hold views that can be at odds with the external reality. When we become aware of these differences and related issues, we can experience a reality paradox. We could treat the experience as a cue to respond in a way that reduces, in the absolute sense, the features of our internal reality that are responsible for the paradox, and the related issues. But our human characteristics make it difficult for us to respond in this way. Think of our decision-making foibles, such as our inertia to change and our desire to belong. Our more usual response is to rationalize the differences in ways that mask or justify the features behind the paradox. Thus the reality paradox aspects of our internal reality endure and we continue to make decisions and hold views that are at odds with the external reality.

The reality paradox aspects of our internal reality are reflected in our relationship with our environment and our view of it. But humans have always suffered from a reality paradox, so why is this important? The significance of its influence on our environment-altering decisions increased substantially with the increase in our cultures' complexity, our population's size and our power to alter our environment. We should not underestimate the environment-affecting power of our reality paradox when we place a tool in our hands.

Environmental Views. We depend on the environment for our survival, so our relationship with it is important. There are three basic types of relationships that individuals can adopt: exploitation, conservation and preservation. As their names imply, our choices are dominated by our human-centred perspective of that relationship. The choice of the environment's perspective is limited to a sub-type of the preservation relationship (de Leeuw 1992). Our preferred personal relationship with our environment can be thought of as our culturally influenced choice of a particular, but flexible, mix of these three relationship types. This preferred mix is also our preferred personal view of our environment, which we use as a reference when we evaluate and form our view of an environmental issue.

When we are personally faced with an environmental issue, its features (consciously and unconsciously) cue us to recall memories that seem relevant to that issue. They, along with our references, are used by our (conscious and unconscious) decision-making processes to form what seems to us to be a coherent view of the issue. That view includes a cause for the issue, our feelings about it, a proposed response to deal with it, and our (material and emotional) explanation/justification for our view.

This sounds like a straightforward process, until we remember that the environment's complexity ensures that even the most simple of environmental issues has many features that can act as cues. Thus, even though an issue's features may not change, the potential for us to hold a different view of it is always present. For example, we can decide to switch our focus to a different aspect of the issue or we can revisit it at a later time in a different context. In either case, its features can provide us with a different set of cues, thus triggering different memories.

Alternatively, even though the cues we receive might be the same, our internal reality may have been adjusted, such as our preferred relationship with the environment, our memories or our mood. There is thus some flexibility in our evaluation and choice of a view to hold of the issue and/or our explanation for holding it, and they can change.

Our initial evaluation of an issue could result in us holding a more exploitative view of it, which we could explain by saying that we must use our environment to survive. Alternatively, or at a later time, our evaluation could result in us holding a more conservation or preservation view of it, which we could justify by saying that we must respect our environment because it supports us now and must do so in the future. No wonder that a deeply considered evaluation of an environmental issue can easily be filled with tension and result in considerable uncertainty in our resulting view of it.

These characteristics of how we form and hold an environmental view of an issue are illustrated in the introductory chapter by my struggle to rationalize my various views of our forest. Should I treat it as a place to be exploited for wood, conserved for food or honoured because it has its own right to exist? My different views weren't prompted by changes in the forest but by changes in my perspective. I struggled just as much when, instead, I tried to merge my differing views into a single consistent view of our forest. This task is not made any easier by the guidance I receive about environmental matters from my Western culture's world view, as discussed in the following section on culture's contribution.

It is therefore unsurprising that, within a group of people, a diversity of views about the same environmental issue can exist. Nor is it surprising that the discussions among the members to find a mutually acceptable view can become arguments about the correctness of their respective contrasting views. The group can even split up into subgroups that can each strongly defend their respective view.

Consider that, regardless of its role in an ecosystem, a specific species can be viewed as either a pest to be exterminated or just another living creature to be respected and accommodated. For example, Creative Salmon Farms raised exotic Atlantic salmon in nets off the coast near Tofino, British Columbia. In 2000, some 15 to 20 Californian sea lions, which feed on fish, were shot and killed by the farm because, in the company's view, "They [the sea lions] were attacking our fish farm." The members of environmental groups were outraged by this action as it clashed with their view of indigenous sea lions as special animals and exotic salmon as not (Keating 2000; Anderson 2000).

However, when seen from a different perspective, such diversity in strongly held opinions is surprising. We live in the same world, and we have the capacity to reduce the differences in our personal and our inter-group views of an environmental issue. We could, for example, look at the issue within its broader cultural, environmental and historical context. We could consider both an external- and an internal-reality perspective of the issue, as well as take into account the environment's complexity. We could hold conditional views of the issue.

Unfortunately, a suitable reference is not readily available and our ability to actually undertake such a nuanced, multiperspective evaluation of the issue is constrained by our human characteristics. For example, to complete such an evaluation we would have to take into account that the representation of our environment's structure and functioning provided to us by our culturally influenced internal reality is incomplete and biased. We would also have to compensate for our simple rules-of-thumb decision-making foibles and our reliance on both facts and feelings, which can suffer from varying degrees of reliability, relevance and completeness. It takes time and energy to make a meaningful evaluation of an issue, an expense we prefer to avoid.

We are instead content to form our view of an environmental issue as usual: by relying heavily on our unrestrained, casual use of our innate decision-making processes and our unquestioned internal reality. We thus tend to bias our environmental view of an issue toward our internal reality and our culture's world view, our unconsidered preferences. We can easily (perhaps usually) end up firmly holding an unrealistic, human-concern-dominated view about an environmental issue.

As a result, when the environmental views of individuals from financiers to environmentalists are looked at from the environment's (the external reality's) perspective, they are often seen to be inappropriate or incomplete. They may even be invalid. Similarly, there is a good possibility that the members of two groups may be arguing over the correctness of their conflicting environmental views when neither view accurately nor even reasonably represents the environment's perspective of the issue. This is most likely if their views are considered over the long term.

These features are illustrated by an experience of mine. One day, while walking in the forest near our house, I found seven black garbage bags, each filled with household waste. In one of the dumped bags was the flagship magazine of the British Columbia Wildlife Federation, an advocacy group for those who enjoy hunting and fishing. Still attached to the cover was the mailing label. It was addressed to a local community member who later turned out to be the dumper.

Inside the magazine, the executive director's report stated that the federation is a conservation organization directed at the protection, enhancement and promotion of the wise use of the environment. He noted that his organization (which means the members it represents) is "a ferocious opponent to those who would destroy these resources, or to those who would stifle your ability to enjoy them" (Walker 2000, 31). What was the dumper thinking when he chucked his garbage? What beliefs or statements of the federation prompted him to join? Regardless, his internal reality contained a reality paradox feature: a significant difference between his personal internal-reality view of the environment and the external-reality view.

Because we do not live in isolation, the reality paradox biases in one person's views can end up making a much larger contribution to the current state of our environment than we would initially imagine. Consider that by implementing our reality-paradox-biased, environment-affecting decisions, such as dumping garbage, we also promote or support equivalent behaviours and thus views in others. Similarly, by expressing our reality-paradox-biased view of an environmental issue in social or formal settings, we help to create or enhance similar views and thus behaviours in other people. In both cases, those people that respond favourably also strengthen our own views.

This paradox-strengthening process is enhanced in social groups. The members can come to collectively hold views, make decisions and engage in behaviours that each member can strongly feel are reasonable and culturally acceptable, even though they may be environmentally unrealistic and contradictory (Kahan et al 2012). Once we, as members, incorporate these group-supported, reality-paradox-favouring views into our internal reality, it is much harder for us to address an environmental issue from the external reality's perspective. It is thus easier for us to contribute to the biosphere's current state.

An example is provided by the need to protect an ecosystem resource from overexploitation, such as the wild-fish stocks we use for food. In the case of the fisheries off Newfoundland, New England, California,

Peru and now Europe, many round-table discussions were held to determine how the environmental issue of declining fish stocks should be addressed. Despite intense, often noisy and acrimonious arguments about who is to blame, who should pay, the need to fish, and if there is even a problem, the efforts ended because the fish ran out. The fisheries collapsed (Finer 2003; McEvoy 1986).

This point was ultimately reached because the majority of the people at the table, either as individuals or as organization representatives, were unable to accept and reliably represent in their discussions the external-reality view of how fisheries function and fish complete their lifeway. In particular, they were unable to consider the fish from the fish's perspective, the manner in which the marine ecosystem hosting them functioned and the impact of fishing on both. Despite the environmental reality of declining fish stocks staring them in the face, the participants' perceptions and decisions were dominated by their immediate, group-supported internal-reality view of their environment. Their reality paradox remained unrecognized, unquestioned and thus able to influence their decisions.

In overview, our personal relationship with our environment can be divided into exploitation, conservation and preservation types. Our somewhat flexible, culturally influenced preferred mix of these dominantly human-centred types is also influenced by the opinions of others. Consider that social activity within a group can influence its members' opinions about information, and our relationship with the environment. Our personal mix is our preferred view of our environment that also acts as a reference. It influences our awareness of, our attention to, and our decisions about the environment and environment-related issues.

Overall, our personal views of an environmental issue tend to be human centred and to favour the views of our social group. It is thus easy for us to end up holding an (environmentally) unrealistic view of both the environment and an environmental issue. And, as the examples of fish and our garbage illustrate, we personally find it difficult to recognize and accept that our views are indeed less reliable than we think. One reason is that we tend not to treat a reality paradox experience as a reminder of the limitations to our view of the environment. It is also difficult because the outcome from our decisions is usually distant in time and space from our decision making. Think too of social pressure. Our unreliable views can endure.

In summary, our brain uses our memories to create our internal reality. Although it serves us well in many circumstances, its representation of the external reality is, to varying degrees, inaccurate and incomplete: it includes divergences from the external reality. Our internal reality thus includes the potential for us to experience a reality paradox.

Some aspects of our internal reality become our conscious awareness of our environment, our attention to it and our views about it. The divergences in our internal reality combine with our decision-making foibles to allow (perhaps facilitate) us to personally make environment-affecting decisions and hold environmental views that can be (environmentally) unrealistic. This conclusion also applies to our views and decisions that directly affect humans. Think of social issues.

Because we find it difficult, even after experiencing a reality paradox, to recognize and remove the divergences in our internal reality, they are an ongoing influence on our individual decisions. We thus don't have to be bad to make biased decisions, just normal. While to make appropriate decisions requires special effort (Ophuls and Boyan 1992). Our resulting decisions, including their biases, are our individual contribution to our current circumstance and the environment's state: to the biosphere's current state. This is true regardless of the official position we hold in our culture's power hierarchy when making them. However we, as individuals, are not solely responsible for this state of affairs. As mentioned, our culture's world view also influences the formation of our internal reality and thus our decisions.

## The Culture's Contribution: Its World View

A culture's world view provides guidance to its members about the world around them and how to live in it. This guidance is a significant, pervasive influence on the members' thinking and decision making throughout their lives. It is through this influence on its members that a culture makes its contribution to the biosphere's current state. The members are largely unaware of this cultural influence because it is mostly exerted in their unconscious, as part of their internal reality. Thus, in order to more fully appreciate a culture's contribution to the biosphere's current state, we need to have a more personal understanding of the influence that its world view has on its members' thinking and decision making.

This section builds on chapter 3. It begins with the nature and limitations of a culture's influence on its members' view of the external reality. This is illustrated by the influence that Western culture's guidance to its members has on their thinking and decision making about the environment. The outcome from a culture's influence on its members' thinking and decisions becomes its contribution to the biosphere's current state, as illustrated by Western consumerism. Although this discussion focuses on environmental matters and Western culture, its essence is relevant to social matters and other cultures.

## Representing the External Reality.

A culture buffers/mediates its members' interactions with one another and with the environment, as discussed in chapter 3 (Wilson 1988). It accomplishes this by its world view providing the members with guidance about the workings of the world (the external reality) and the making of their day-to-day decisions to live in it.

The members tend to follow the guidance from their culture's world view. After all, as they each grow up, they absorb from their social surroundings the world view of the culture they were born into. They thus automatically incorporate its characteristics into their internal reality. Indeed, as the members grow up, they collectively become the carriers of their culture's world view. But they are also their culture's decision makers. A feedback loop exists between a culture's world view providing guidance to its members and their decisions shaping their culture's world view: the member-culture feedback loop.

One consequence of this feedback loop is that, despite the members' strong ties to their culture's world view, it provides them with guidance rather than rules. This serves a purpose. A culture's world view must be unchanging if it is to provide the members with a sense of stability, but it must also be flexible if it is to allow the members to accommodate or adapt to the changing complex world. The need for flexibility was illustrated in chapter 8 by the manner in which the BaMbuti, a hunter-gatherer group, make their day-to-day decisions to secure resources (Kelly 1995).

But this duality contributes to another outcome. It allows the members to make adjustments to their culture's world view that result in its representation of the external reality being sufficiently inaccurate that its guidance doesn't reliably favour cultural viability. In particular, it may imperfectly represent to the members the environmental and human constraints they must live within for their culture to be viable. As the young members grow up, the cultural world view they absorb includes these cultural divergences from the external reality.

The ease with which these divergences can appear is illustrated by assuming that a member of a hunter-gatherer band receives accurate guidance from their culture's world view about their supporting environment's constraints on their lives. This is a reasonable assumption because the members of hunter-gatherer bands are quite aware of their connection to their home range ecosystem. Think of the food-harvesting constraints on their culture and vice versa, as discussed in chapter 7 (Kelly 1995).

Now imagine that a member feels that this guidance is at odds with what they want to do now. What happens next depends on the strength of the member's self-monitoring. It also depends on how strongly the rest of the members believe in the current cultural guidance about the issue. This will determine how strongly they will exercise those social incentives, such as social monitoring and peer pressure, that will prompt the dissident member to conform to the guidance. If these aspects are all weak, then the dissident is likely to simply ignore the guidance or rationalize their preferred choice of action, or both.

For example, a few members of the Waorani hunter-gatherer culture living in the South American tropical forest use dynamite and DDT to catch fish. Even though one member recognized that this practice resulted in overfishing, he still argued that it was more important to have lots of fish now than to worry about the number of fish that will be around when his children grow up. He justified his decision by calling on his culture's world view which, he asserted, held that it was impossible for him to harm or destroy the forest that sustained him (Davis 1997).

In general, if many of a culture's members are persuaded, drawn or driven to this type of action and justification, then, as alluded to above, their collective personal choices would change their culture's world view. The old cultural guidance would be modified to be more in keeping with the members' desires rather than with their supporting environment's constraints. This guidance would then reinforce the members' new internal reality. Through this feedback-linked, internal-reality–cultural-world-view adjustment process (a reality adjustment process) the members can end up willingly subscribing to guidance from their culture's world view that is at odds with the environmental and human constraints they must live within.

The members of today's complex industrial cultures also can and do adjust their culture's world view so that it supports their wishes but less accurately represents their environment's constraints on their lives. Consider that we rely on science to provide us with the most accurately available description of the world's state and its functioning (the external reality). We then use this information in a process referred to as science-based decision making.

However, we also undermine that process, as illustrated by the previously discussed example of the United Nations Intergovernmental Panel on Climate Change (IPCC) accepting the final draft of the 2007 second scientific report. Despite the scientists' objections the (non-scientist) panel members altered the report to be more in keeping with the views on climate warming held by the countries who had appointed them to the panel (Adam 2007; The Associated Press 2007). Over time, these types of interventions can cumulatively adjust a culture's world view so that it misrepresents the external reality. The consequences have been illustrated in past chapters and will be seen in future chapters.

A more complete appreciation for these types of world view adjustments and how they come about is illustrated by looking at two conflicting chemical assessments of atrazine. This continues chapter 9's discussions about atrazine's ability to feminize male frogs. The European Union placed a limit of 0.1 parts per billion (ppb) on the presence of any pesticide residue in drinking and groundwater (the precautionary view of regulation). In October 2003, the EU banned the use of atrazine because those using it could not keep its concentration in water below this level.

In contrast, the United States re-permitted the use of atrazine in October 2003, as long as the maximum level in drinking water did not exceed 3 ppb, well above 0.1 ppb. For the US regulators to reduce their safe level, their process required conclusive scientific evidence that atrazine at levels lower than 3 ppb posed a definite risk to human health or the environment (the demonstrated-impact view of regulation). Thus, despite both the substantial evidence of frog feminization caused by atrazine at levels perhaps as low as 0.1 ppb and the US Environmental Protection Agency's recognition that atrazine is an endocrine disruptor, the US regulators decided that there was insufficient evidence to lower the acceptable level below 3 ppb or to ban its use (EPA 2007).

The difference between these two conclusions is the product of more than two different approaches to the evaluation of the environmental and health risks from human-made compounds. At its heart lies the differences between their respective cultural world views. For example, US culture places more emphasis on business innovation than on environmental functioning, which meshes with their greater faith in technology and progress. This allows forces within the US culture to limit the regulators' consideration of external-reality-based environmental concerns when, for example, evaluating and regulating the use of chemicals.

The source of these reality-adjusting forces is illustrated by atrazine. Since its initial registration within the United States in 1958, the government reviews of its use have been successfully subject to intense lobbying by those with a vested interest in atrazine's manufacture and use. In particular, the corn industry, which is the main user of atrazine, has strong ties to the US political structure. Think corn subsidies (Weiss 2004; Souder 2006; Ackerman 2007; Sass and Colangelo 2006; Hayes 2004).

In overview, a culture's world view has a distinctive and strong influence on its members, but that influence is constrained. It can, for example, guide its members to believe in a tool such as science, which can help them discover the nature of the external reality. But that doesn't ensure that the culture's world view will provide its members with appropriate guidance about the nature of the external reality or how to live within its human and environmental constraints on their lives. This is the case because the members, as the carriers of their culture and its decision makers, can and do adjust their culture's world view, and thus its guidance, to suit their immediate-term human wants. And they can do so even if they are aware of an accurate representation of the external reality and their long-term need to consider it.

**Western Culture's Environmental View.** This subsection illustrates the kind of influence a culture's world view has on its members' thinking and decision making. It discusses the guidance Western culture's world view provides to it members about their relationship to the environment, and the influence that view has on their personal views and decisions about the environment and environmental issues. For example, it influences our decisions about which features of our environment are beautiful, useful and worth paying attention to (White 1967; Dubos 1971). It also influences those decisions that will contribute to the state of the earth's environment.

Western culture's guidance to its members about their relationship to the environment can be described using the same three basic types of generic relationships (exploitation, conservation or preservation) used above to discuss our personal views of our environment and environmental issues. However, a more complete appreciation of its environmental guidance is gained by paying attention to the restricted mixes (categories) of the three basic relationship types it favours and the explanation/justification it offers for holding a particular mix. There are five philosophical-religious categories of guidance: mystical unity with nature (e.g., New Age); stewards of God's creation (e.g., the shepherd); co-operating with God's nature to improve it (e.g., development); part of nature's ecology (e.g., participants in nature's web); and dominion (masters) over God's creation (e.g., domination) (Westhoff 1983). Each category thus represents a specific environmental view and explanation.

So, which category (or mix of them) does Western culture use to guide its members? Put differently, what is the culture's preferred view of the environment and environmental issues? This question is best answered in the context of the categories' histories.

The five philosophical-religious environmental views appeared at different times in Western culture's past. As the culture changed, the older categories were adjusted and new ones appeared. The relative standing of the categories was adjusted and so was the culture's guidance to the members about which mix to adopt as the culture's preferred environmental view. An understanding of this history helps us to appreciate Western culture's current environmental guidance.

The dominion, stewardship and mystical categories of guidance have roots that go back perhaps as far as 3000 BC. This is the time when simpler farming cultures in the Middle East were well into the process of becoming more complex urban agricultural cultures (Lang 2002; Wilson 1988). Western Europeans received their more recent, post AD 350, environmental heritage primarily through Christianity and its sacred book, the Bible, but it was moderated by the Europeans' local history and the other religious-spiritual traditions in their past.

At the end of the medieval period, nominally in the 1300s, European culture began, in earnest, a transition into its current form. Many influences contributed to this post-medieval cultural transition, however it was both expressed and justified almost entirely within a Christian religious framework. The transition began when Europeans started a broader examination of their personal and cultural relationship with both their Christian God and the earthly institution representing God on the earth, the Catholic Church. *Broader* means that their discussions went beyond their previous narrow focus on the afterlife and the fate of their souls. It began to more deeply consider the nature of the physical world created by God and the role intended for them in it. Think of the notion of progress appearing.

During this cultural transition, adjustments were made to the older categories of environmental guidance (dominion, stewardship and mystical) and two newer ones (ecology and co-operation) were introduced. These changes and introductions were largely expressed, until the 1900s, in terms of defining or clarifying humanity's relationship with God (Katz 2005). The close connection between the Western world views' environmental guidance and Christianity can be seen in the long-standing use of biblical passages to explain and justify the environmental guidance categories.

For example, the idea that humans have dominion over the bounty of God's earth and the right to use it as we see fit is explained using Genesis. Particular reference is made to Genesis 1:26–28 in which God made man, and Genesis 9:2–17, in which God blessed Noah and provided the rainbow as a symbol of his promise to Noah not to send another destructive flood. Our duty to act as stewards of God's earthly creations is explained using Genesis 2:15, in which God directs us to manage and protect the Garden of Eden. These connections to the Bible remain today, despite the rise in secularism and the arguments about the proper translation and the intent of the original authors (Gardner and Stern 1996). The mystical category too has many informal connections to mainstream Christianity but it was also influenced by a much wider range of religious-spiritual beliefs.

The post-1400s idea of co-operating with nature is tied strongly to the Christian belief that God created nature as detailed in Genesis. In this view, nature's order represents God's divine will. If we could deduce that order, then humans could more readily fulfill God's will and return to Eden (perfection). We are thus more than environmental stewards or managers. We are meant to co-operate with nature in order to improve or develop it to God's end. These ideas were associated with natural philosophers and appeared along with the rise in the notion of progress after the 1600s.

The newest category, ecological, appeared as natural philosophers slowly turned away from the formal Christian churches. The rise of Western secularism (thinking in a way free of religious faith) could be said to have started in the 1600s with the appearance of deism and the idea of the clockwork universe. Certainly it was present by the 1700s, as hinted at in the description of the 1755 Lisbon earthquake provided earlier. However, secularism only really came into its own during its accelerating rise from the early 1800s to the present, which paralleled the transition from natural philosopher to scientist. In the ecological view, we are just one part of a scientifically describable complex web of nature. If we are part of nature, then the idea of co-operating with it takes on a different meaning.

As Western culture went through its post-medieval cultural transition, its guidance about the preferred view of the environment changed. For example, from the late 1700s to the mid-1900s, the favoured view was dominated by the dominion category but tempered, mostly, by the co-operation and stewardship categories. From the mid-1900s to the present, the dominion category continued to dominate, but it has increasingly been tempered by environmentalism. It is a view of our environment dominated by the ecology category but containing aspects of the stewardship and mystical categories.

One reason the dominion category has remained ascendant since at least the 1700s is that it has successfully absorbed the notion of progress and become secular. For example, since at least the early 1900s, expressing the dominion view of our environment was increasingly explained/justified by saying that holding it would achieve social progress through ongoing material betterment and economic growth (material progress). The ties to God were increasingly noticeable by their absence.

The dominion category also remained ascendant because it absorbed the key aspects of its competing categories, thus sidelining them. This process is illustrated by the fate of the co-operation category. Consider an aspect at its core: the phrase "to develop nature." The Latin root of the word *nature* means to be born, to produce, to come into being. The phrase "to develop nature" originally meant to improve or actualize nature's potential by co-operating with it. This goal was also expressed by the phrase "making progress toward God's plan for man." The meaning and goal of development was originally religious in nature.

Today, development is a Western cultural imperative. But when the phrase "to develop" is used to express an environmental view, it means to alter the environment to satisfy humans' material self-interest: it has a utilitarian meaning (Westhoff 1983). It has completely lost its esoteric dimension; it has lost its religious connection to co-operating with nature or managing it as a steward. In essence, the dominion category has co-opted the phrase "to develop," thus sidelining the co-operation category.

Since the late 1900s, the dominion category has been absorbing the environmentalist view. It is achieving this through the Western belief that humans can satisfactorily avoid and resolve their environmental impacts by applying technology (e.g., mitigation and restoration) and making science-based decisions. In this context, exerting dominion over the environment as managers can appear to satisfy the rationale behind the ecology and stewardship categories. It also appears to promise a better relationship with our environment, thus appearing to satisfy the rationale behind the mystical category (Redman 1999). Think of green technologies and the green economy. It seems that the thoroughly human-focused, progress-supporting aspect of the dominion category is what makes it so successful as an environmental view.

In overview, Western culture's world view changed over time. In keeping with these changes, new categories of philosophical-religious environmental views appeared, existing ones were changed and the guidance about the preferred mix of environmental views was adjusted. Western culture's present guidance continues to be strongly biased toward the dominion category, which is consistent with its world view's notion of progress. However, its world view also contains the stewardship, ecology, mystical and co-operating categories (roughly in that order of importance). Since the 1950s, the strongest alternative to the dominion view has been a combination of the stewardship, ecology and mystical categories, called environmentalism.

Overall, Western culture's preferred environmental view of dominion over nature is reflected in its members' general preference for a personal relationship with the environment that is closer to exploitation than conservation and furthest from preservation. It is also expressed in our tendency to use utilitarian-biased arguments when explaining our view of our relationship with the environment and our solutions to environmental issues. We tend to just gesture toward the presence of esoteric arguments.

**Making a Contribution.** An individual's contribution to the biosphere's current state is rooted in our personal internal reality and our personal consumption of resources. Our contribution can be represented by our lifestyle, when combined with our culturally influenced personal explanation for it. Similarly, a culture's contribution to the biosphere's current state can be represented by the aggregate of its members' consumption of resources, when combined with the features of its world view supporting that consumption. This subsection uses the features of Western culture's world view, such as its representation of the external reality and its preferred relationship with the environment (its preferred environmental view), to illustrate a culture's contribution to the biosphere's current state. The focus will be on Western culture's relationship with both goods (including services) and the technology that provides them.

Western culture's ideal standard of living/lifestyle can be represented by its current utopian horn of plenty. Rather than being filled with food and drink, as in the past, it is stuffed with goods: winning lottery tickets, extra-large houses, cars, bags of cash, innumerable electronic gadgets, exotic vacations and more. These consumer goods mean more to us than they seem.

Western culture's world view guides its members to believe that these goods, and the technology that provides them, will improve our general sense of well-being by reducing physical work, increasing luxury leisure time and providing effortless, pain-free improvements to our health (Strong 1995). Critically, they will also improve our social-psychological well-being by, for example, making us feel happy (Kasser 2002; Csikszentmihalyi 2000). We are guided, in an interesting case of circular logic, to believe that goods will also provide us with the means to cope with the demands, insecurity and stress placed on us by Western culture's work regime and its materialistic consumer lifestyles. Consider the recreation machines and the mind-adjusting tools (including those that chemically bend it) we think of as essential life-improving aids and coping tools. For example, many believe, at least until recently, that the smart phone (think social media and internet) can unambiguously empower us to more efficiently and effectively deal with the work and social demands of our culture, as well as become a better person.

It is of note that the cultural beliefs in the benefits, such as happiness, security and stress-reduction, claimed for consumer goods continue to exist even though those benefits have, at best, only a transitory relationship to these goods (Easterlin 2003; Kasser 2002). Remember the shopper's cycle discussed in chapter 2? In it, the first blush of happiness we experience from the purchase of a consumer good soon fades to be replaced by the need for more consumption, thus continuing our quest for happiness.

Many, but not all, Westerners follow our culture's guidance in favour of consumption and adopt a consumer lifestyle, conspicuous or otherwise. Nevertheless, we all receive the associated cultural permission to indulge in weakly restrained consumption of resources and disposal of waste. That permission is supported by our culture's dominion view of the environment: it allows us to keep our attention focused more on maintaining the supply of resources needed to provide goods and services, and less on the environmental costs of providing, using and discarding them. The broader conclusion is that the guidance from Western culture's world view contributes to the current state of the earth's environment because it tends to guide us to both see ourselves as largely separate from the environment and our culture's functioning as mostly independent from viability's constraints.

This conclusion about the Western world view's contribution to the environment's current state should be more fully appreciated after discussing the appearance of producer-consumerism (consumerism). Consumerism, with its material-focused consumers and their consumer lifestyles, is among the key aspects of Western culture's world view that favours resource consumption. The focus will be on their appearance in the United States.

Consumerism's most recent roots are found in the cultural changes that started during the industrialization of the 1800s. For example, in the mid-1800s, many US citizens felt that it was important for them to be personally capable of making many of the things they wanted. However, by the late 1800s and early 1900s, as manufacturing technology became more efficient, this view was slowly replaced by the feeling that it was more important to be able to buy manufactured things (Slade 2006; Brower and Leon 1999). By the 1950s, American culture could be described as a consumer culture.

The American's cultural and personal transition to consumerism during the 1900s was the product of many factors. Here it is viewed from the perspective of US manufacturers who facilitated the transition by their efforts to boost their sales to the emerging consumers. The manufacturers focused on three interconnected aspects of consumption: advertising, technology and an accepting cultural world view. They will be discussed. Advertising's multi-faceted role was discussed in chapter 2. It will lie in the background of this discussion.

Technology, like advertising, plays a multi-faceted role. Its material and social aspects are illustrated by a focus on obsolescence. Consider that all manufactured products have a finite lifespan, after which they are worn out and can no longer perform the task for which they have been made. They have become functionally obsolete and must be replaced.

But obsolescence can occur in other ways. Older products can become obsolete after technological progress has created a new product that users feel functions better than the older ones. Products can become obsolete because of changes in the users' perception of the old product (i.e., psychological obsolescence), for example, progress in fashion. Finally, a product can become obsolete through progress in the manufacturer's ability to plan, or control, its lifespan: planned obsolescence. This means that a product can be designed to break, wear out or been seen as unacceptable within a predetermined lifespan. Alternatively, it can be designed to be irreparable or disposable.

Planned obsolescence contributes significantly to the boosting of sales. Consider the use of the style cycle (psychological obsolescence) in the auto industry. For example, Harley Earl, the head automobile designer for General Motors during the time of the car fin is reported to have said, "Our big job is to hasten obsolescence. In 1934 the average car ownership span was 5 years: now [1955] it is 2 years. When it is 1 year, we will have a perfect score" (Slade 2006, 45).

With obsolescence in mind, consider the manufacturers' efforts to promote a cultural world view that favoured an increase in sales: a view that included consumerism. The presence of the notion of progress in Western culture's world view preconditions it to undergo a transition to consumerism. However, for the culture to actually undergo that transition, the members had to (largely unconsciously) change themselves in ways that favoured consumerism. If they did so, then they embedded consumerism into their culture's world view. The manufacturers guided them to make those changes.

This certainly happened in the United States. The efforts of the 1900s manufacturers were successful at guiding (think advertising) Americans to complete their culture's transition from a fledgling purchase-focused culture of the late 1800s to a fully fledged consumer culture of the 1950s. For example, Americans abandoned the older cultural value of thrift. They bought into the idea, literally and figuratively, that improvement in their material well-being was a worthy goal, one which would also improve their social-psychological well-being. Think sense of self and standing in the community. By the 1950s, these and other consumer-friendly characteristics of a consumer culture were firmly embedded in the American world view.

During the 1950s, the three aspects of consumption targeted by the manufacturers become harmonized in a manner favourable to consumption. That, along with favourable economic conditions, resulted in American manufacturers experiencing what they had wanted. There was a rapid rise in their sales.

In the following years, American culture's consumption-favouring world view continued to change. Its guidance came to favour Americans treating consumption as a goal in itself, thus promoting frequent, repetitive consumption and, debatably, over-consumption. Such consumption was possible because Americans became both sufficiently wealthy and (perhaps remained?) psychologically and culturally oblivious to the reality of resource extraction and the fate of their waste. This is illustrated by consumer culture being described as the throwaway culture: dump and forget.

Although the essence of consumerism is still strongly associated with an American lifestyle, it is now, to varying degrees, found worldwide. Globally, consumer goods are being manufactured just so that people (consumers) can have something to buy (Slade 2006). Consider that during Canada's 2011–2012 time of fiscal uncertainty, the Canadian government expressed concerns that its citizens were taking on too much debt in order to buy consumer goods.

Worldwide, both we and the environment are having to pay the costs of changing the resource users of an earlier period into the resource consumers and discarders of today. Consider the global-scale environmental disruption and destruction that results from providing the natural resources needed to manufacture consumer goods, and from the manufacturing, using and discarding of those goods, as discussed in chapters 9 and 10. Think of plastics and pesticides.

Not that a Western throwaway consumer culture is always destructive and messy. Consumers are becoming more aware of the costs of consumption. Today, a good consumer in the West is expected to at least dispose of their waste in a particular manner. Recycling efforts are continuing to grow.

However, there are limits to a Western consumer's awareness of and attention to the environment. When we evaluate our personal consumption or balance it against our environmental impacts, we still tend to follow the guidance from our culture's world view and our internal reality's desires. In particular, we accept our culture's emphasis on progress, maintaining economic growth, and a dominion-dominated, utilitarian relationship with the environment and environment-related issues. As a result, Westerners have a limited personalized appreciation for the full environmental costs of adopting a consumer lifestyle and supporting a consumer culture. We are still largely unaware of the environmental constraints on our lives.

We are, mentally, still largely detached from the impacts of pouring our obsolete products into environmental sinks to join the waste from their manufacture and use. It is no wonder that when the reality of these impacts impinge on our lives, Westerners experience a reality paradox. And we tend to deal with that uncomfortable reminder by either ignoring the connections or rationalizing our response in some culturally acceptable manner. Examples of these culturally favoured "explanations" include assuming that the impacts are inconsequential or unexpected, noting the need to consume to live and maintain the economy, and pointing out that no one should be expected to live a materially deprived life in poverty.

From the environment's perspective, these explanations don't change the outcome. The environment continues to be disrupted or destroyed, and Westerners' culturally supported internal reality with its consumer reality paradox is maintained (Ophuls and Boyan 1992). We should now more fully appreciate the contribution that Western culture's world view, with its dominion-favouring environmental view, made and continues to make to the environment's current state.

This discussion about consumerism may have created the impression that most, if not all, Westerners whole-heartedly support the transformation of the environment into consumer goods. A position they hold because they accept their culture's guidance in favour of a dominion, utilitarian view of the environment and a belief in progress. To think this would be to neglect the ongoing tension among Westerners about both consumerism and the most appropriate (personal and cultural) view to hold about the environment or environmental issues.

This tension is understandable, regardless of whether we look at it from the perspective of Western culture or its members. From Western culture's perspective, the presence of four other environmental categories in its world view ensures that just below the surface of the members' faith in its dominion-favouring environmental guidance lies their uncertainty about it. For example, on the one hand, the dominion view allows us to produce the goods we believe are needed to improve our quality of life. On the other hand, the remaining categories remind us that producing, using and discarding those goods impacts the environment. The rationalization used to deal with the latent conflicts and contradictions in this two-handed guidance includes appeals to jobs, economic growth and increased recycling. For example, some argue that economic growth provides the funds required to protect the environment by recycling, while others argue that it just provides the material needed to maintain recycling.

The presence of five Western categories of environmental views and the variety of cues provided by an issue means that, regardless of their culture's guidance, a Westerner can experience the often difficult and messy task of deciding which view, or mix of views, is "actually" appropriate for us to hold, as discussed above. For example, which mix will adequately satisfy both our personal utilitarian need to consume and our esoteric concerns about our impacts on the environment? What is the appropriate balance between our desire to both use the environment and protect it? Thus, although most Westerners defer to the culturally supported dominion category and favour the exploitation view when answering these questions, doubts can remain. A non-trivial proportion of Westerners hold a view in which a different category is up front.

In essence, Western culture is split into two camps. On one hand, there is the dominion-dominated, utilitarian (material), producer-consumerism view of the environment. On the other hand, there is environmentalism's more esoteric view with its mix of the stewardship, mystical [new age] and ecology categories.

The two camps are engaged in a struggle to decide which of them should be Western culture's preferred view of environmental issues and our relationship with the environment. This is a clash in which the dominion camp is trying to absorb and sideline the environmentalism camp, which pushes back. Examples of the battle grounds are "green" technologies, the "green" economy, "sustainable" development, climate warming and environmental assessment processes. The split can also be seen in Westerners' diverse and contradictory reactions to newly surfaced environmental issues, such as plastics in the environment. Outrage, concern, anxiety, ambivalence, blindness, indifference and apathy can all be seen.

Interestingly, strongly represented in both camps is faith in the Western notion of progress towards a better world, as discussed in chapter 3.

The overall practical consequence of this ongoing struggle is that Western culture tends to relate to and treat the environment in much the same way as we personally treat a nuisance relative: being a relative, they deserve our attention, but we still give it reluctantly. Westerners tend to give attention to the environment as an obligation, an imposition or afterthought, not with the enthusiasm shown toward our favourite relative, the economy.

This can be seen in the formal processes that Western culture establishes and uses to evaluate the possible environmental impacts of their pending utilitarian actions (i.e., of a development). These environmental impact assessments do consider the environment, but they tend to be biased against placing too many restrictions on implementing the desired actions. In Canada, this is expressed in the view held by some that its assessment processes should always strive to "Get to Yes." It is also illustrated by the activism often needed to have environmental concerns taken seriously. In general, environmental impact assessments tend to treat the environment as more of an inconvenience/ obligation than either the main concern or as having the same standing as the proposed activity.

When the reality of events, such as BP's Gulf oil spill and the Fukushima Daiichi nuclear plant's disintegration, impinge on Western culture, the general underlying cultural tensions about our relationship with the environment come to the fore. The culture's faith (i.e., the members' collective faith) in the dominion-favouring aspects of its world view is shattered, at least in the short term. Alternative views come to the fore and the culture suffers the equivalent of a reality paradox experience with its disruptive and painful feelings.

Western culture reacts (or, more correctly, its members collectively respond) to these events in a manner equivalent to how an individual responds to cognitive dissonance or a reality paradox experience. Western culture tends to deal with the environmental impacts by labelling them as incidentals or by believing that the damage will be repaired by technology-assisted mitigation and restoration efforts. Western culture tends to similarly deal with the causes of a disruptive event by labelling it as an accident and believing that any oversights can be dealt with by improving the responsible techniques and technologies, which will then also usher in a better future.

In contrast, if a solution that will likely avoid a future disruptive event is available but it requires us to make significant changes to our values, visions or beliefs, then it is unlikely to be implemented in a timely manner, if at all (McIntosh et al. 2000). It takes considerable time and effort for an unwanted event to result in deliberate changes being made to Western culture's preferred environmental view. Thus the underlying essence of our cultural contribution to the environment's current state tends to endure.

In overview, the current guidance that Western culture's world view provides to its members about the environment and environmental issues is still dominated by the dominion view of the environment. This is in tune with Western culture's faith in progress and its strong attention to producer-consumerism (consumerism). Indeed, its manufacture, use and discarding of goods, and the associated lifestyle, represents its contribution to the environment's current state. However, a struggle between the dominion view and the environmentalism view (a mix of the stewardship, mystical [new age] and ecology categories) is underway. There is cultural tension, perhaps confusion, over which view is to be preferred. This rounds out the discussion of the contribution that Western culture's world view has made, and continues to make, to the environment's current state.

In summary, a culture's world view provides guidance to its members about the world around them and how to live in it. This includes guidance about their relationship with the environment and their views of environmental issues. But, its guidance both imperfectly represents the external reality and is biased towards humans.

Because a culture's world view influences its members' decisions, the biases and imperfections in its world view are usually reflected in their decisions. Those decisions can affect their environment. Thus, the influence that a culture's world view exerts on its members' decisions is the culture's contribution to the current state of the earth's environment. A culture's world view displays inertia to change, so its contribution to the environment's state tends to be ongoing.

Although the discussion in this section focused on Western culture's contribution to the material aspects of its circumstance, and thus to the state of the earth's environment, the essence of the discussion is also applicable to the social aspects of its circumstance. After all, a culture's guidance influences the members' internal reality and thus both their human- and environment-affecting decisions. This discussion is also applicable to other industrial and industrializing cultures. It is through these (mental and material) connections that a culture's world view contributes to its current circumstance and to the state of the earth's environment, and thus to the state of its biosphere.

## The Contribution Through Feedbacks

So far, the discussions about how the biosphere arrived in its current state have considered the contributions to it from a material perspective; an individual's perspective, which emphasized their internal reality with its divergences; and a culture's perspective, which emphasized its world view and its divergences. But each contribution was discussed largely in isolation. This is an oversight because we (as individuals and a culture) are tightly linked to one another and to the environment, as discussed in chapters 3 and 8. Indeed, we form a dynamic, complex human-environment system, as concluded in chapter 10. By considering the contributions from the perspective of the links between them, we can more fully appreciate how they combine to collectively result in the biosphere's current state: the human-environment system's current state if you like.

This section first illustrates those links between us and our culture prompting us to routinely make environment-affecting decisions that significantly impact our environment's functioning. This is followed by an illustration of our environment's ongoing response to those human impacts, and our reaction when the environment's response is imposed on us. The discussion will use examples from Western culture and, because individuals are a culture's decision makers, it will preferentially focus on us and our internal reality, with its reality paradox.

<u>Our Ongoing Impacts on the Environment.</u> Our internal reality (with its reality-paradox-causing divergences) and our culture's world view influence our environment-affecting decisions. Despite some of our decisions having significant environment-altering consequences, we still continue to repeat them. This conundrum, and the reasons behind it, is illustrated and symbolized by our ongoing introduction of ecosystem-disrupting exotic plants and animals into indigenous ecosystems. By considering our and our ancestors' exotic-introducing activities, and the context in which they take place, we can more fully appreciate why we continue to make our many other environment-impacting decisions when the consequences suggest we wouldn't, or

shouldn't. This discussion continues one started in chapter 9. It begins by looking at our exotic introductions from three perspectives: the ecosystem, the introducer and their culture.

From the perspective of the earth's ecosystems, humans seem oblivious to the reality that, on a global scale, their activities are responsible for the large majority of today's exotic introductions. We even seem unable to make the basic connection between news stories about an invasive exotic currently impacting our local area or others' lives and our culturally influenced, personal choices that result in us introducing exotic species. Americans, for example, seem unaware that, by 1993, about 65% of the 70 established exotic fish and most of the 60 introduced, but not yet established, exotic fish in the United States (including Hawaii) arrived through the aquarium trade (Stein and Flack 1996; Courtenay 1993). Regardless of our lack of awareness, from the ecosystem's perspective we are still responsible for the resulting substantial impacts on them and the associated economic costs.

From the introducers' perspective, their decision to deliberately introduce exotic species was reasonable and justified. In some cases they were trying to solve an immediate problem or satisfy a pressing desire. For example, in the early 1970s, fish farmers raising catfish in the southern United States had a problem. Their catfish-rearing ponds were routinely choked with algae and suspended matter. The introduction of the Asian bighead and silver carp species into the ponds solved their problem. The carp ate it.

But by 1980, the large, rapidly reproducing carp, with their voracious appetites, had "unexpectedly" entered the wild. Floods aided their spread into the Mississippi and Illinois Rivers, where the carp soon dominated the rivers' ecosystems. Today, these exotic carp pose such a threat to the Great Lakes ecosystems that at least US$9.1 million has been spent trying to keep them out. Methods include building an electric fish fence across a shipping canal that joins the Mississippi River to the Great Lakes (EPA 2010; Hansen 2010).

A stranger example of the introducers' logic also illustrates the compounding aspect of our exotic introductions. The aftermath of the global financial crisis of 2008 left many Arizona homes and their ubiquitous swimming pools unattended. Maricopa County health workers expressed the concern that these pools could become breeding grounds for mosquito larvae. Think of the exotic Asian tiger mosquito of disused car tire fame. The fear was that the adult mosquitoes might spread the virus that causes West Nile fever, a recently introduced exotic. The authorities had an idea to address this mosquito issue: deliberately introduce mosquito fish into the pools to eat the mosquito larvae. Some 40,000 mosquito fish would be bred at the Phoenix Zoo for introduction into the pools (Reuters 2008).

The plan sounds like it would benignly solve an immediate problem. However, in conceiving it, the proponents neglected a few things. The mosquito fish is native to rivers in the southeastern United States that flow into the Gulf of Mexico. Worldwide experience since at least 1993 has shown that mosquito fish introductions outside this area easily become invasive. The fish can radically disrupt aquatic environments by, for example, preying on the juveniles of larger fish (Courtenay 1993). It is not known whether the proponents of the plan had considered the possibility of deliberate or inadvertent mosquito fish releases from the swimming pools, or the potential consequences. Nor is it known whether the proponents realized that, despite its name, the mosquito fish is not particularly good at controlling mosquitoes.

From a culture's perspective, the continuing, largely unchecked introduction of exotics seems surprising in light of the very long record documenting that poorly considered introductions can be highly disruptive to ecosystems and costly to cultures. This circumstance becomes less surprising when we remember that the deliberate introduction and nurturing of exotics is a very old practice that was accompanied by cultural support or indifference. The world views of at least the complex cultures of today reflect this history and still imperfectly represent to the members their host environment's functioning and their relationship with it. The young members of these cultures absorb this world view into their internal reality as they grow up, so they are disconnected to some degree from their supporting environment and its functioning. The result is their culture's continuing at least tacit support for its members' decisions to introduce and nurture exotics.

This cultural influence is illustrated by the history of Western culture's acclimatization societies. The first acclimatization societies were founded during the 1860s in Europe, a few decades after the popularization of both exotic plant collecting and decorative gardening had begun. These Victorian era societies were eventually established worldwide in European colonies. By the 1900s, most of them had ceased to operate, although some survived until as late as the 1980s.

The specific objectives of each society were different, but they all had in common the desire to improve their local, indigenous ecosystems for financial, practical or aesthetic reasons. Behind their objectives, and the members' actions, lay a mixture of mid- to late 1800s European cultural beliefs: in particular, the notion of progress, a sense of exploration and discovery, a view of the environment based on improvement and control (domination, co-operation and perhaps stewardship), and the emerging scientific discoveries and ideas about biology and zoology.

The societies were run by enthusiastic amateurs who neither followed scientific practices nor conducted rigorous scientific experiments, as evidenced by their general failure to keep adequate records. For the members, sufficient proof of a project's success was obtained when the exotic species they introduced were established in their new environment. For example, the expatriate members considered the objective of aesthetic improvement to be met if their home country's biota were successfully established in the expat's new country.

Through their efforts, the acclimatization societies were responsible for, or contributed to, a large number of successful exotic introductions all over the world. Many of these exotics subsequently become invasive pests that disrupted their new host ecosystems. Examples are the New South Wales society in Australia, which had acclimatization society objectives. In the 1870s, a Mr. E. Cox accepted custody of two brown hares on behalf of the society. He reported to the members that the hares had not been seen since their arrival because the grass was too high. A decade later hares had overrun much of the southern part of New South Wales.

The Indian mynah bird was introduced to the Melbourne area of Australia by the Acclimatisation Society of Victoria, at least as early as 1862. The objective was to reduce insect pests. That bird is now one of the more hated bird pests in Australia. Similarly, the starling was introduced into the United States by the American Acclimatization Society with a similar result (Lever 1992; O'Loughlin 2009).

Acclimatization societies mostly concentrated on introducing alien organisms that their members saw as useful. However, as mentioned above, they also maintained both a parallel scientific interest in collecting exotic plants and a popular desire to establish (propagate) them for decorative, aesthetic and social purposes. When the societies died out, their populist vision of improving and beautifying nature through exotics did not, and neither did the infrastructure they had established to accomplish their goals. The societies' lasting legacy is their contribution to today's horticulture industry and the support it offers to its clients, the members of the public engaged in decorative gardening (Bright 1998; Lever 1992).

This inheritance helps us to appreciate the practices of today's Western decorative gardeners and the pragmatic aspects of their culture's current environmental views supporting them. Consider that today, with few exceptions, even the routine, deliberate introduction and spread of exotic plants and related organisms is acceptable or tolerated. For example, American gardeners could buy, at least until 1998, more than 60% of North America's worst weeds at a garden nursery (Bright 1998). In 2010, I checked to see if any of the plants on North Carolina's list of invasive plants were commercially available in that area (Smith 2008). An internet search quickly revealed that the noxious weed oriental bittersweet was for sale there, while heavenly bamboo was available from the surrounding states.

The decorative garden itself connects the practices of today's Western decorative gardener to the philosophical roots of their culture's current environmental views supporting those practices. Those roots go back much further than the acclimatization societies. They go back to the changes in both the personal and the cultural world views of our early ancestors that accompanied their switch from hunter-gatherer to farmer. As hunter-gatherers, their understanding of the world and their sense of life's meaning was tied to the existing landscape. As farmers, those attributes became tied to the requirements of farming, in particular the need to modify and manage the land. Thus, building and working a farm accomplished more than providing food, it helped a farmer find personal meaning and cultural standing by being seen as a nurturer and a provider. It helped to provide a sense of personal and cultural identity.

Similarly, the creator of a decorative garden can find an equivalent, equally powerful sense of meaning in their effort. The gardener could, for example, display to both themselves and others their ability to manipulate and order nature's beauty, or even their power over nature itself. The Hanging Gardens of Babylon (probably located at Nineveh) illustrates that, even in the earliest of complex cultures, the elite used decorative gardens for these purposes. The equivalent gardens in more recent times are exemplified by those surrounding the post-1400s grand palaces and mansions of Europe, such as Versailles and Kew. In addition to displaying the manipulation of beauty,

these (elite) decorative gardens showcased the standing of the garden owners in the community: their achievements, wealth and power (Crumley 2000).

The domestic decorative gardens of today's non-elite express a more muted version of finding meaning through modifying the land and manipulating/managing nature. Doing so still gives the gardener a sense of pride, pleasure and satisfaction in their ability to order and nurture nature's beauty in a manner defined by their culture. The act of growing things can also engender in the gardener the feeling of a promise for the future. The domestic garden is still a display of the gardener's standing, and perhaps power, but this view is softened compared to the one held by the elite.

Overall, decorative gardening in the West is a culturally sanctioned activity that symbolically represents four of the five categories of environmental guidance provided by Western culture. Consider how gardening represents our dominance over nature, our duty of stewardship, our co-operating with nature in order to improve it, and the mystical aspect of growing things (Westhoff 1983). Conspicuous by its absence is the ecological view that our environment is a single, coherently constrained, functional whole of which we are an inseparable part.

The question arises, how can it be that, in an era bursting with ecosystem knowledge, we can still hold so strongly to a discretionary plant-manipulation-focused activity that essentially ignores the ecological view of our environment? Consider this question from the perspective of a Western gardener. If a gardener is to address it, then they must become personally aware of how their decorative gardening can philosophically and practically, directly and indirectly, contribute to the current disrupted state of our environment. It is harder for them to gain that awareness and pay it the needed attention than we imagine.

A gardener's efforts to do so are constrained by both nature's and human characteristics. Consider the presence of an inevitable delay between a decorative gardener nurturing an exotic plant in their garden and the appearance of the ecosystem consequences. The time between the two events may be too long for the gardener to independently make the connection between them. It is thus less likely that they will experience a reality paradox, let alone try to reverse the introduction and change their behaviour.

For example, the water hyacinth, mentioned in the introduction to this book, was brought into the United States from South America to provide a beautiful blue decorative flower for the 1884 Cotton States Exposition (Begon et al. 1996). Who knows when (if) the introducer came to know that, by 1900, waterways in Florida were so badly infested with the decorative plant that paddle wheelers could not pass. The exotic grew so fast that even mechanical harvesters failed to control it. In 1993, water hyacinth was still infesting and altering the aquatic ecosystems of the southern United States (Schmitz et al. 1993).

Certainly, if Western culture provided guidance to its members about the environmental risks and responsibilities from introducing exotics, it could help gardeners become aware of these costs and pay attention to them in their decisions. However, such guidance is absent or weak and seen in isolation, so there is little cultural pressure on decorative gardeners to adopt an ecosystem-focused personal world view and apply it to gardening. Quite the opposite is the case.

Western culture's world view both gives a decorative gardener permission to improve the aesthetics of their surrounding environment and provides exotic-friendly guidance about how to do so in ways that match the culture's beliefs and values. That guidance also helps a gardener mask, ignore or justify any exotic-gardening-related reality paradox experiences they might have. Thus, regardless of whether the gardener actively or consciously subscribes to the Western notion of progress or our dominion view of our environment, they can, in good conscience, introduce, nurture and spread exotic plants, along with their attached organisms. For this to happen, it is sufficient that the gardener just engages in an activity permitted by their culture's world view: the one they absorbed as they grew up.

Overall, decorative gardeners can easily be unaware of, avoid, or rationalize away the environmental responsibilities that come with being a decorative gardener. At the same time, by holding to their decorative-garden-favouring beliefs, desires and actions, they help to sustain the environment-disrupting aspects of their culture's world view. Thus the decorative gardener and Western culture are engaged in a self-sustaining feedback loop that favours the introduction of exotics, often by default.

This feedback, and the horticulture industry's role in it, is neatly summed up by an advertisement/label stuck into the pot of a plant for sale at a local garden nursery. It read, "We have searched the world over to bring you the best plants to guarantee a beautiful garden" (Proven Selections 2002). As with the acclimatization society objectives, nowhere in the ad was there mention of the ecosystem price to pay for introducing exotics, or the responsibilities the horticulturalist acquires for selling it and the gardener for nurturing it. There was no mention of indigenous ecosystem functioning. The advertisement symbolizes why our more general culturally supported environment-altering decisions continue today with few effective checks and balances: there is insufficient personal or cultural awareness, attention or will to change current beliefs and practices.

In overview, this discussion is not an attempt to exclusively finger Western decorative gardeners. Their ongoing introduction of exotic organisms into local ecosystems was used to both illustrate and symbolize how Westerners (as scientists, business people, politicians, recreationists and decorative gardeners, to name a few) not only make but continue to make decisions that disrupt the environment. In particular, Western culture's dominion-favouring view of our relationship with our environment helps to create in its members an internal reality that tends to ignore the environmental constraints within which we must live. Think of our personal bias toward exploitation. As a result, we hold beliefs and live our lives in ways that are at odds with the external-reality view of that relationship. We too easily make decisions that significantly disrupt our environment's functioning and destroy ecosystems. Our decisions also reinforce those aspects of our culture's world view that favour us making decisions that contribute to the biosphere's current state: the status quo is thus maintained by a member-culture feedback.

When we personally become aware of the full extent of our environmental impacts, then we can experience a reality paradox. But, despite this prompt, we have a culturally supported, innate tendency to rationalize the outcome as unpredictable, unintended, incidental or inconsequential. If we did decide to significantly change both our personal and our culture's world view in ways that would reduce our reality paradox, we must still deal with the considerable personal and cultural inertia inhibiting us from doing so. It takes considerable time and effort to adjust both world views so that they guide us to live more within the environmental (and human) constraints on our lives.

## The Environment's Ongoing Response and Our Ongoing Reaction.

The members of industrial cultures tend to believe that their impacts on the environment will be restricted in time and space, passively received by it, and easily mitigated or restored, piecemeal, by either humans or "nature." In contrast, from the external reality's perspective, our environment responds to our impacts on it in a dynamic, complex manner. Our impacts can thus cascade through the environmental system for years to centuries in a way that changes it. Its changing state is imposed upon our lives. If the changes are significant, then we will eventually be forced to react. This feedback-linked interaction is clearly illustrated by the long and well-documented history of our emissions of chlorofluorocarbon (CFC) compounds into the atmosphere, its ongoing response and our ongoing reaction to that response.

Almost all of the many CFCs are human made. The first of them, Freon, was created in the late 1920s. Tests at the time indicated that not only did many of the CFCs have useful industrial properties, but they appeared to be both chemically inert and directly non-toxic to humans (Midgley and Henne 1930; Sparling 2001).

By the 1930s, Du Pont was manufacturing these apparently wonder chemicals on a commercial scale. During and after World War II, the production of CFCs rose rapidly as they became increasingly used in industrial and domestic applications. They found use as coolants (e.g., in refrigerators), propellants (e.g., in aerosol cans), fire-extinguishing gases, and solvents. The enthusiasm for these chemicals is expressed in a 1959 textbook. The author, when referring to their remarkable properties, said, "There is no doubt that the future will bring many additional developments in the field of polyhalogenated chemicals" (Briscoe 1959, 133). Who would have thought that today the most well-known development would be the appearance of an ozone hole in the Antarctic?

The ongoing story of the ozone affair is astonishing. Some 20 km above the earth's surface, within the atmosphere's stratosphere, is an ozone-rich layer that is roughly 10 cm (yes, 10 cm) thin. Its ozone is formed by the interaction between the oxygen in the atmosphere and a portion of the sun's high energy ultraviolet (UVC) radiation. Ozone itself absorbs the sun's UVC radiation. This too reduces the amount of UVC reaching the earth, where it would, for example, damage the DNA of living organisms. In humans, that damage can cause skin cancer.

It is hard to imagine that the traces of CFCs we release on the surface of the earth through waste disposal and leakage could reach the ozone layer and reduce the amount of ozone in it (i.e., "thin" it). But they can and do. The inertness of CFCs means that, once released, they can survive in the atmosphere until they reach the ozone layer. There, the sun's UVC radiation

provides enough energy to decompose the CFCs in a manner that releases their chlorine atoms, each of which then catalyzes the breakdown of about 100,000 molecules of ozone (Faure 1998b). The result is a lower concentration of ozone in the layer, which can then let through more DNA-damaging UVC.

We know this now, but it wasn't known when, shortly after CFCs were discovered, they were enthusiastically put to use. In contrast, our recognition, acceptance and response to the impacts that CFCs have on the ozone layer took longer, much longer, to occur. It is within the context of the history of CFCs and the ozone layer that the lessons from the environment's response to our impacts on it are most easily appreciated.

Ozone's ability to absorb high energy radiation was known in the 1880s. But it was only in 1934 that the presence of a thin ozone layer in the stratosphere was confirmed. Starting in 1957, ground-based measurements from Halley Bay, Antarctica, increased our understanding of the layer. For example, the data showed that the layer was dynamic, continually being created and destroyed.

It took until the late 1960s to early 1970s before the scientific community came to accept that the ozone in the layer was created by the sun's radiation impinging on the outer atmosphere. By this time, it was also known that ozone's destruction could be facilitated by at least nitrogen oxides and chlorine, which are normally present in the atmosphere in low amounts. However, there were still intense debates about how the dynamic balance between ozone's creation and destruction took place, and the compounds that were actually responsible.

A 1974 laboratory experiment clearly demonstrated the connection between CFCs in the upper atmosphere and the destruction of ozone. It revealed the mechanism, described above, by which that destruction occurred (Stolarski 2001; Sparling 2001). This confirmed that CFCs played a key role in determining the balance between ozone's creation and destruction. In particular, when the concentration of CFCs in the ozone layer reaches more than around 10 to 40 ppb, the chlorine released by their decomposition breaks down the ozone faster than it can be produced, thus thinning the layer.

Outside the science community, the Western response to the external reality being revealed by this discovery included more than raising concerns about CFCs facilitating damage to DNA. After all, the discovery strongly intruded on our preferred, culturally supported internal-reality view that CFCs were safe, useful and economically valuable. The 1974 discovery and its implications were actively resisted. For example, the CFC manufacturing industry said that human emissions were too small to have the proposed impacts on the ozone layer. Starting in 1975, they suggested that volcanoes were adding much of the chlorine to the stratosphere that was destroying the ozone (Oreskes and Conway 2010).

By 1978, the major role that CFCs played in ozone depletion had been confirmed. The concerns about the long-term environmental and health consequences from increased amounts of UVC passing through a depleted ozone layer started to be taken seriously, by some. As a result, the US government banned the use of CFCs as an aerosol propellant (Sparling 2001). By this time, comprehensive satellite monitoring of the ozone layer was underway.

When 1980 arrived, Du Pont, along with the rest of the manufacturers of CFCs, were still expressing their skepticism about ozone science. They were now actively lobbying to continue manufacturing CFCs for use in all non-aerosol propellant applications (Sparling 2001; Greenpeace 1997). Their arguments against the science and for the right to produce CFCs continued even though CFCs were now no longer considered to be wonder compounds.

The remaining arguments against CFCs being responsible for ozone depletion and the consequences should have ended in 1985. This was the year when the multi-year, ground-based record of ozone concentrations in the Antarctic ozone layer collected at Halley Bay was published. The data, compiled up to 1984, showed that the seasonal variation in the layer's ozone content was marked by a steep decline each spring. The degree of these annual declines had increased over time until spring 1984 when ozone's concentrations showed a dramatic 40% decline (Farman et al. 1985).

These results, and the retroactive recognition of ozone depletion recorded in the older satellite data covering all of Antarctica, confirmed the presence of a continental-scale spring depletion of ozone in the layer. The concept of a seasonally thinned ozone disc or hole over the Antarctic was born. (In retrospect, the satellite data showed that the accelerated ozone thinning over Antarctica had started around 1976 (Stolarski 2001; Sparling 2001).

But, despite this evidence, the doubts about the seriousness of ozone depletion, its consequences and its connection to CFCs were not completely dispelled (Stolarski 2001). Du Pont continued to question the science and continued to make efforts to protect its economic investment (Greenpeace 1997). In spite of the manufacturers' pushback, the international Montreal protocol on CFCs was signed in 1987. It initiated the very slow introduction of restrictions on the release of CFCs and the phasing out of their production. For example, the protocol permitted the use of CFCs in developing countries until 2010. This placated the industry's economic concerns.

In order to accommodate the results of ongoing scientific work, the Montreal protocol also permitted its revision in later years. This was an important clause because there are a number of twists and footnotes to this story. They illustrate the complexity of the environment's response to CFCs and how that response is being imposed on us. They also further illustrate the role that our reality paradox features played in our reaction to that imposition. Of particular interest are the limitations and biases in our environmental awareness, attention and decision making.

Our focus on the Antarctic data distracted us from considering the possibility that the ozone layer's thinning might already be occurring globally. Research, which began in 1988, confirmed that the ozone layer in mid-latitudes was indeed thinning. However, that mechanism of ozone destruction was slightly different than in the Antarctic. Its mid-latitude destruction involved sulphate aerosols. These tiny airborne particles hampered the ability of nitrogen oxides to slow the destruction of ozone caused by CFCs. Even today, the impacts on the ozone layer from our ongoing emissions of sulphate, nitrogen oxide and minor gases are poorly understood (Stolarski 2001).

After the 1988 research was published, it was accepted that ozone depletion was global in scale and that the process was more complex than expected. Despite this complexity, it was expected that the decline in emissions of CFCs prompted by the Montreal protocol would result in the ozone layer repairing itself relatively quickly. This estimate took into account the long lifespan of CFCs.

This was not to be. The original prediction was too optimistic. It failed to take into account the extremely small, short-term (10–20 year) cyclical changes in the sun's radiation. A revised prediction, published in 2006, suggested that the ozone layer would thin until around 2010, after which it is hoped that the layer would be restored by 2050 to 2060, much later than expected (Dameris et al. 2006). Our environment is engaged in a long-term, complex response to the impact of our CFCs on it.

The full complexity of our environment's response to CFCs is revealed by additional information. Expeditions to Antarctica in 1987 showed that the unique weather conditions on that continent allowed the stratosphere above it to cool to unusually low temperatures. When the stratosphere is colder than $-78°C$, then crystals of nitric and sulphuric acid form, which we see as wispy, polar-stratospheric clouds. Their presence accelerates the reactions that release the chlorine from the CFCs. The cold stratospheric temperatures thus help to create the Antarctic ozone hole each spring.

The Arctic atmosphere has the same composition as the air over the Antarctic, but the cold temperatures needed to create the acid-crystal clouds are absent. More correctly, they weren't present until recently. In the spring of 2011, researchers announced that the Arctic was experiencing its greatest ozone layer loss on record. Counterintuitively, the reason was related to the global warming of the earth's lower atmosphere or troposphere. The trapping of heat in the troposphere results in the stratosphere above it cooling. At the end of the first decade of the 2000s, the temperature of the stratosphere over the Arctic was dropping below $-78°C$. The conditions suitable for the amplification of ozone's destruction through acid-crystal cloud formations that were previously only found in the Antarctic are now found in the Arctic (AWIPMR 2011).

This "unexpected" development may not alter predictions about the timing of the ozone layer's restoration, but it does remind us of another "unexpected" connection between ozone layer thinning and ongoing climate warming. The Montreal protocol's compromises allowed for the production of HFCs because they don't release chlorine on breakdown. As a result, the production and use of HFCs rapidly increased. However, HFCs are extremely potent greenhouse gases (Velders et al. 2009). They are increasingly contributing to global warming, stratosphere cooling and, through acid crystals, to Arctic ozone depletion.

The time spans of the events in the ozone layer's story provide further interesting insights into both the imposition of the environment's response to CFCs onto us and our reaction. About 30 years elapsed between the first creation of CFCs in the late 1920s and their evaluation, mass production and wide application in the late 1950s. In contrast, some 130 years is expected to elapse between the time CFCs were first made and the time the ozone layer is fully restored (roughly 60 years from creation to partial acceptance of the impacts in 1987 plus the 70 years from then until restoration).

The 100-year difference between these two periods hints at the wider role that our internal reality and our culture's world view played in the CFC story. Well before the ozone layer's thinning was recognized, we were preconditioned by a combination of our human-focused personal and cultural world views, including their divergences from the external reality, to possibly make the ozone-layer-affecting decisions we did. In particular, we were preconditioned to rapidly accept that CFCs were harmless. When the evidence of their significant environmental and human impacts came to light, we were preconditioned to react by exerting our right to continue producing CFCs on economic grounds. Think too of our (ongoing) preference to treat the atmosphere's functioning as simpler than it actually is.

With this in mind, it is also easier to understand the significant, culturally supported value we gave to CFC products compared to our perception of the possible environmental costs. It is also easier to appreciate how we came to place our misguided faith in the use of HFCs as a compromise cure for CFCs and have an overly optimistic hope for a speedy repair of the ozone layer.

In overview we, optimistically, hold up the ozone story as the prime example of our ability to appropriately deal with environmental issues. It more realistically illustrates a number of other things. We have an unrealistic faith in our knowledge about our environment's functioning and our ability to avoid disrupting it. We routinely fail to recognize and accept that the environment's response to our impacts on it are complex and ongoing. As a result, we have emotion-raising reality paradox experiences. When we respond to them, we do so in a way that ensures we will repeat rather than avoid such experiences in the future.

We can believe whatever we like and do as we please to our environment (or ourselves). But the nature of the external reality ensures that the consequences of our decisions will be imposed on us in some form at some time. We are ill-equipped to react to this imposition in a way that reduces future human-triggered impositions. This conclusion can be seen in stories from tobacco smoking to ocean fish harvesting to $CO_2$ emissions.

In summary, this section discussed the personal and cultural links between us and our environment and how that results in our material, individual and cultural contributions combining to result in the biosphere's current state. When we make decisions that impact our environment, it responds by changing. These changes are imposed upon us, and we must react. We and our environment are tied together by our impacts/influences on one another: they form a human-environment feedback loop. It is related to the feedback loop between our personal internal reality and the external reality: the internal-reality–external-reality feedback loop.

Our initial environment-affecting decisions and our later reactions to the environment's response are influenced by another feedback. We are under tremendous psychological, economic and social pressure to follow the guidance from our culture's world view. We bias our personal world view and our decision in its favour. But we are also our culture's decision makers, so our desires and decisions can (consciously and unconsciously) adjust our personal and thus (collectively) our culture's world view. Our adjusted culture's world view becomes the one that will, in the future, guide us and our children to bias our decisions in its favour. This completes a cultural-world-view–member-world-view feedback (member-culture feedback).

Neither the human-environment feedback nor the member-culture feedback contain a mechanism to force us to proactively make decisions that will fully and reliably reflect the environment's functioning and its constraints on how we can live. We can't be proactively forced to accurately reflect the external reality in our decisions. We thus can and do, as the examples illustrate, make decisions that result in both our internal reality and our culture's world view drifting away from the external-reality view of the environment. This enhances the reality paradox features of our internal reality.

As a result, when it comes time for us to deal with the environment's responses to our impacts on it, we tend to miss that its impositions on us are also prompts for us to re-evaluate our personal and our culture's world view, and our decisions based on them. We prefer to rationalize the impositions away as insignificant or unexpected, or treat them as guides to information we lack or the changes we must make to our technologies and techniques. We rarely treat these impositions as prompts to change the belief aspects of our personal or our culture's world view, such as our view of our relationship to the environment. We thus continue to make environment-disrupting decisions that pay limited attention to the environment's functioning and its constraints on our lives.

Each of the three perspectives (individual, cultural, material) of our contribution to the biosphere's current state indicate that they are products of the interactions, including feedback, among the features of the environment, us (as individuals) and our culture. Our contributions are linked together as part of a human-dominated dynamic, complex human-environment system. It is within this wider context that we begin to fully appreciate how our contributions have collectively contributed to the biosphere's current state.

## A Dynamic Model of the Human-Environment System

Despite this chapter's systematic layout, we find it difficult to grasp how its apparent kludge-like confusion of contributions, feedbacks, constraints and complex system characteristics result in the biosphere's current state. What is needed is a clear, coherent overview description of how these pieces fit together as a whole. We need a description that can be used to explain that state: the environment's current state and our current circumstance.

Filling in this gap is the job of this section. It is achieved by condensing the information in the above discussions into a simple descriptive model of the dynamic, complex human-environment system. Once the model is applied to Western culture, it should make the functioning and changing of the human-environment system more relevant to us at a personal level. It should be easier for us to apply it to human groups of any size or cultural complexity, and to human or environmental issues at local or global scales. The model will serve as a reference for the remaining chapters.

**The Model.** This model of the dynamic, complex human-environment system divides it into the three previously mentioned primary factors: us as individuals (including our internal reality, its divergences from the external reality and our personal world view); the culture of a group of individuals (its world view with its divergences); and our environment (whose functioning supports all living things). Each of these factors is a complex system in its own right. The relationships/links among them are represented by two overlapping feedback loops: one dominantly concerned with belief (the belief loop) and the other with resources (the resource loop) (figure 15).

The belief loop represents those belief-related relationships that bind the individual members into a cultural group. An example is the member-culture feedback. Basically, as we grow up, we absorb from the people around us the world view of the culture we were born into, which results in the essence of our culture's world view becoming a part of our internal reality. This is a key step in us becoming an adult member of our culture. As an adult member, we each contribute to both our culture's decision making and carrying its world view. Thus, through our individual decisions, we collectively adjust our culture's world view. That changed world view then influences us and is absorbed by our children. This completes the member-culture feedback.

The resource loop represents the interactions with our environment that bind the individual members and their cultural group to their environmental endowment and its functioning. For example, the members live a lifestyle that reflects their culturally supported relationship with their culture's environmental endowment. The members acquire the resources needed to live that lifestyle from that endowment. As they go about securing and using those resources, they make decisions. Those decisions can, collectively, change their lifestyle, and their culture's relationship with their environment, which completes the member-environment feedback.

As indicated, the two loops are not isolated from one another. They have features in common, such as our internal reality (including our personal world view), our culture's world view and our lifestyle. Through these common features, the loops interact with one another. The interactions are mediated by those aspects of our (culturally influenced) conscious and unconscious decision making that help us to make sense of what we know (i.e., our rationalization process). In particular, rationalization generates for us what we feel is, within the context of the two interacting loops, a consistent and coherent understanding (an explanation) of a particular circumstance and our proposed decisions about it (figure 15). For example, our chosen lifestyle and our explanation for it can be thought of as our rationalization of the (personal, cultural and environmental) influences on our lives.

Rationalization is thus a key part of the process that binds the complex workings of individuals, their culture and the environment together. Consider that the rationalization-mediated interactions between the two loops influence individuals, their culture and the environment to change. At the same time the feedbacks within the loops and the interactions between them ensure that the changes in one part of the system will be spread through the rest of the system. When the two interacting loops and rationalization are considered

together in this light, they are seen to represent the complex human-environment system's functioning and changing: its dynamics.

For example, any changes to the (supply or type of) resources being provided to the members of a culture by their culture's environmental endowment prompts them, as individuals, to change their resource usage (in the resource loop) (figure 15). But, because of rationalization, this change is accompanied by an equivalent adjustment in their mental relationship with their culture's endowment. If enough individuals are affected, then there are related rationalized changes to their culture's world view (in the belief loop). This changes its guidance to the members about both their mental relationship with their culture's endowment (the belief loop) and their practical use of its resources (the resource loop). The members respond by changing their personal use of resources and their impact on the environment providing them. This completes a feedback that may either strengthen or dampen the initial changes, or their effects, that began this process. These dynamic changes in the system can be described by the generic terms, such as *drift*, *inertia* and *directed*, used to describe change in all complex systems.

There is another aspect of the system's dynamics to consider. The guides we use (our internal reality and our culture's world view) when making our decisions contain many divergences from the external reality, such as those in our internal reality that favour us experiencing a reality paradox. Our decisions are thus strongly influenced by how we deal with these divergences during rationalization. Consider, for example, how our consumer decisions could change if we tried to make them in ways that recognized and resolved divergences among our lifestyle, the environment's functioning and/or our sense of self.

Indeed, one would expect that the rationalization process would automatically reduce our personal divergences so that, over time, our internal reality would come to better reflect the external reality. Think of us gaining wisdom. The member-culture feedback would also be expected to ensure that, over time, our culture's world view would undergo a matching adjustment, which would remove its divergences. As our individual and cultural divergences disappeared, our internal reality and our culture's world view would become more coherent and consistent with one another and with the environment's functioning. We would see this reflected in the changes to our lifestyle and our explanation for it.

This does happen, but not to the degree we would expect. The evidence can be seen in the ongoing (persistent) inconsistencies and disconnects between our culturally supported lifestyle and our environment's functioning, discussed above and in chapters 9 and 10. Divergences appear and persist because the functioning of the human-environment system in general, and humans in particular, allows them to. For example, in the resource loop, the environment's impositions on us cannot force us to account, during decision-making, for either its constraints on our lives or the impacts that its future responses to our decisions will have on us. In the belief loop, our culture's world view provides us, the members, with guidance about the human constraints on our lives. But, as the word *guidance* indicates, even if it were accurate, we are not bound to follow it.

As the examples suggest, we end up personally bearing much of the responsibility for ensuring that our decisions reflect the external reality. But our decision-making foibles alone ensure that our decisions can, and do, result in divergences appearing in our internal reality (i.e., its reality paradox features), as discussed in chapters 1 and 2. Because we are our culture's decision makers, the personal divergences of its members can, and do, collectively result in divergences appearing in our culture's world view.

Fine, but why do the divergences persist? Although our rationalization process can and does remove personal divergences, it is not particularly effective. There are a number of features contributing to its ineffectiveness.

One of them is the complexity of the human-environment system, as discussed in chapters 7 and 8. This makes it easy for us to miss, mask or ignore a divergence in our thinking. For example, a significant amount of time and distance can separate our decision from its outcome, which ensures that it is difficult for us to make the connection between the two. Complexity also makes it difficult for us to decide which of the many possible perspectives or contexts is the most appropriate to use when working with or evaluating a belief or decision, including the importance we have attached to it.

Consider too that when our personal internal-reality divergences become part of our culture's world view, they also become part of the member-culture feedback in the belief loop. The cultural divergences not only become part of a culture's guidance to its members, they acquire enhanced inertia to being changed. Our efforts to eliminate divergences become caught in a Catch 22 situation.

Rationalization also tends to be ineffective because of how we change, as discussed in chapter 6. As we go about our daily business, we may become aware of a divergence. We tend to display a negative reaction to it that ranges from a mildly unpleasant surprise to a more intense sense of cognitive dissonance or a reality paradox experience. Regardless, we are obliged to deal with our new awareness. To do so, we must change something.

For example, if our negative feelings about our new awareness are significantly strong, then they, rather than the facts of the divergence, tend to (consciously or unconsciously) grab our responding attention and effort, at least initially. We are prompted to focus on quickly restoring the more pleasant feelings we were experiencing prior to our new awareness. We usually achieve this by (consciously and unconsciously) selectively excusing, ignoring and explaining away the new information while justifying the old. This quickly recreates the sense of coherence, consistency, stability and order in our lives, which usually restores the pleasant feelings we desire. This tactic allows us to deal with the divergence without having to adjust our pre-existing decisions, memories, personal beliefs, behaviours, thoughts, logic and knowledge. It also allows us to imagine (emphasis on *imagine*) that we have satisfactorily dealt with the divergence. But, from the perspective of the external reality, we haven't.

Even though we may become both aware of and accept the presence of a personal divergence, we will still tend to display inertia to changing our internal reality to deal with it. This inertia is smallest for changes that are easy to make or lie on the periphery of our internal reality and largest for changes to its core. We thus prefer to deal with divergences by adjusting the periphery of our internal reality, while the core only changes slowly. Insidiously, because of our decision-making foibles, we may think that the changes we do make to our internal reality will result in it more closely reflecting the external reality, but that is not necessarily the case. Think of being persuaded to change.

Overall, our decision-making foibles, the nature of how we change and the system's complexity allow divergences to appear and persist. Thus, although rationalization can and does reduce personal divergences, it more usually just creates that impression. We tend to use it, at least initially, to simply mask, ignore or explain away our awareness of a divergence or to substitute one divergence for another. Fortunately, under many circumstances, a divergence or its presence is unimportant. However, under some circumstances, it is critical.

If we are to recognize critical divergences and remove them, then our decision making must meet certain conditions. We need to appreciate the complexity and dynamics of the process by which divergences could form and persist. When evaluating a circumstance or event, we are obliged to take into account the possibility that our divergences may have contributed to it or be influencing our thinking about it. This requires us to think about the circumstance, our decision making and the influences on it in a nuanced manner. And it requires that we personalize and assimilate what we think is reliable (fact and feeling) information. We are then more likely to recognize a divergence and remove it by making the personal changes that bring our internal reality closer to the external reality.

The conditions needed to successfully deal with divergences are much the same as those needed to gain wisdom. We find it difficult to meet these conditions because we innately prefer to treat the human-environment system as a simple deterministic system and think about it in a black and white manner. We prefer to personalize and assimilate human-centred knowledge. Basically, it is easier for us to follow our urges and our culture's guidance than to take the time and make the effort to recognize and remove even critical divergences. So they tend to persist.

If enough of a culture's members were to successfully reduce their personal divergences, then their culture's world view would change in a manner that reduces its divergences. But, in light of the conditions to be met for success, it is no wonder that divergences in a culture's world view also persist. The ongoing presence of cultural divergences is reflected in the culture's guidance to its members, for example, its guidance about preferred lifestyles when seen in the context of their real-life relationship with the environment.

This state of affairs could be taken to mean that we should focus exclusively on reducing cultural rather than personal divergences. However, as hinted at above, this is not a useful solution because of the connections between the personal and the cultural. Think of the member-culture feedback. Both have to be tackled together. The model also reminds us that critical divergences are not restricted to material (resource) aspects. They include mental (belief) aspects. Finding and implementing successful solutions involves multiperspective, nuanced thinking and personalization.

The model provides other benefits. It helps us to contemplate how we can and, critically, will likely contribute to the system's future state. For example, in the resource loop, our environment's functioning provides us with the resources (including services) that make life possible. But the environment's complex, dynamic functioning also constrains the availability of these resources. This limits a culture's choice of a world view, if the members wish their culture to remain viable over the long term. But it takes time and effort for the members to recognize and then make the (personal and cultural) changes needed for their culture's world view to guide them to live within the resource constraints on their lives. Thus, to be successful, the members must start the process of becoming more viable well in advance of needing to. However, we tend to be reactive not proactive. We are thus more likely to make decisions that will maintain the status quo into the future until we are forced to change.

This overview, in keeping with other chapters that deal with the basics, focuses on the rules of thumb implicit in this model of the human-environment system. They are divided into non-exclusive overlapping groups. The overview also acts as a reminder of the model's connection to the real world. The later summary of this chapter will provide an overview of the model as a whole.

Rules of thumb dealing with the system's dynamics: The human-environment system displays the generic features of a complex system. Its numerous variables interact with one another in a number of ways, including feedback, simultaneously over a wide range of time, space and intensity. These linked interactions are how the system functions and changes (its dynamics). The system exhibits deterministic, random and chaotic features. The specifics of its dynamics can be described using generic terms such as *drift, adjustment, preconditioned, unexpected, directed, inertia* and *imposed*. The connections between causes and effects can be both direct and indirect. Whether something is a cause or an effect can depend on the context in which it is viewed and the perspective taken.

The feedbacks in the system play a critical role in its dynamics. There is an overall human-environment feedback loop, which is linked to our personal internal-external reality feedback loop. There is a member-culture (world-view-affecting) feedback loop. These feedback loops can, collectively, facilitate change by providing internally generated drivers of change, but they can also constrain change, thus stabilizing the system.

Rules of thumb about our decisions: We are participants in the system. Individuals are the system's human decision makers. Our individual decisions, even our trivial day-to-day ones, are a key part of the system's functioning and changing. Our decisions are influenced (directed, facilitated and constrained) by our culture's world view, our internal reality, our decision-making foibles, our environmental experiences and the system's functioning. Our culture's world view and our internal reality contain aspects that are different from the external reality. These divergences, such as our reality paradox features, play a key role in our decision making and its consequences. The system's functioning spreads the effects of our day-to-day decisions throughout the system. Thus today's decisions become our (personal and cultural) contribution to the biosphere's future state, the one that we and our descendants will face well into the future.

Rules of thumb dealing with rationalization: Our decisions affect the system's functioning and changing. A key factor in our personal decision making is how we rationalize all the various aspects we are dealing with, such as our knowledge, our culture's guidance, our intended decision and our experiences. The result of our rationalization is not just masking and fabricated excuses. The result can require the members to make real changes to their personal world view and their lifestyle. Rationalizations can precondition the members to drift into a new way of living in the future, prompt them to decide to live differently now or, more usually, result in them favouring the status quo.

However, the changes that result from our efforts to create a more coherent and consistent understanding of information by rationalizing it will only reduce our personal (and thus cultural) divergences if we use the external reality, such as viability, as a reference. It also requires us to apply wisdom rather than expediency during rationalization. Think of the need for trade-offs and compromises.

Rules of thumb dealing with change: The system's changing is a different view of its functioning. The system is driven to change by internally generated and externally imposed influences. It is always changing at some scale. The system is invariably striving toward, rather than being in, equilibrium. A significant personal change (transition) requires the members to change both their perceived (mental) and their practical (material) relationship with one another and their environment. A cultural transition occurs when enough members undertake a personal transition. There are costs as well as benefits to a personal or cultural change or transition. The system provides us with many options for the future, but it constrains the choices that are viable for humans or will maintain the system's stability.

Rules of thumb dealing with constraints: We face human and environmental constraints on our lives. If we and our culture wish to remain functional and viable, then our decisions must satisfy those constraints. However, we can't be proactively forced by those constraints to do so. We can be retroactively forced to comply, but only when our transgressions become severe. Thus, it is we who are responsible for deciding if we wish to proactively include these constraints in our decision making. We are responsible for both recognizing the cues that we are not living within them and then acting to comply.

Humans are not only physically constrained in how we can live and change. We are also socially and psychologically constrained. We need to feel that there is a significant level of consistency and coherence (order and meaning) in our personal and collective lives. Our tolerance for the speed and manner of personal and cultural change is limited: if it is too fast or too dramatic, then we or our culture become unstable. For a successful cultural transition to occur, personal, cultural and even environmental changes are constrained to be somewhat consistent and occur relatively slowly in a somewhat coherent manner.

## Applying the Model.

The model of the human-environment system provides us with a means to think about our current circumstance and the current state of the environment (the biosphere's current state) in a comprehensive, coherent and consistent manner. This is demonstrated by applying the model to Western culture, an example of the world's complex industrial and industrializing cultures. This practical example also helps us personalize and assimilate the model's generic description, thus helping us to more fully appreciate and use it.

Start by considering Western culture's current circumstance from the perspective of the model's cultural factor. The core of Western culture's world view includes a belief in progress, economic growth, consumerism, democracy, individualism, innovation, technology and a dominion-biased relationship with the environment. It guides Westerners' decisions. But we are our culture's decision makers, so our collective decisions can change our culture, which completes the member-culture feedback.

Certainly, the presence of this feedback in the belief loop allows the core of the Western world view to change. However, we, like the members of all cultures, tend to follow our culture's guidance because we are indoctrinated to do so from birth. By following it during our day-to-day decisions, we reinforce the core of our culture's world view: we maintain the essence of the status quo. This favours cultural stability and the core of our culture's world view changing slowly, by drift, over generations.

The belief loop overlaps with the resource loop and interacts with it (figure 15). In the resource loop, the aspects of Western culture's world view, such as progress, economic growth (think consumerism), innovation and technology, guide us to adopt a lifestyle that demands a significant amount of resources. But the supply of the resources we want is constrained by the effort needed to provide them, which results in resource shortages, and by their practical finiteness, which results in resource scarcities, as discussed in chapters 10 and 12. Western culture's world view guides us to preferentially resolve resource supply issues in ways that maintain the supply rather than reduce our demand. Those solutions invariably involve us applying the tools of innovation and technology, which adds to our demand for resources. Think energy. When taken together, Western culture's guidance and it members following it ensure that the culture's demand for resources tends to increase over time.

However, beyond some level of demand, an increasing amount of effort per unit of resources is required to maintain the supply. The needed effort is provided by increasing the mix of workers, technology and management engaged in that task, which requires an increase in Western culture's organizational complexity (cultural complexity). This increase in cultural complexity comes with a cost, such as paying for more workers. These material costs are ultimately paid for by resources extracted from the environment. The provision of these resources completes Western culture's cultural-complexity–resource-extraction feedback (complexity-resource feedback) in the resource loop (figure 15).

But an increase in cultural complexity involves more than the production and consumption of resources. If Western culture is to remain functional, then an increase in its structural complexity has to be accompanied by corresponding rationalized adjustments to both our personal and our culture's world views. One reason they are needed is to ensure that we will remain unified: we must (individually and collectively) continue to support

our culture's now more complex structure and pay for the higher costs that come with it. For example we, as individuals, need to display an ongoing or increased faith in our leaders, in innovation and technology (think efficiency) and in the (real and perceived) social-psychological benefits that technology's material products provide. We also need to display a willingness to provide the labour, taxes and investment needed to pay for them. Basically, there has to be an increase in our culture's reaffirmation and authority maintenance efforts. These costs too are ultimately paid for by resources, which adds to the complexity-resource feedback.

This discussion reveals something else about Western culture. The resources we provide support Western culture's complex structure, its dynamic functioning and its world view. Those resources are needed, for example, to apply innovation and technology, and to pay for its reaffirmation and authority maintenance costs. However, it is the rationalization of our day-to-day decisions about resource use and personal or cultural adjustments that will significantly determine whether our culture will become or remain functional and stable. It is rationalization that gives us the opportunity to determine whether our decisions have considered both the environmental (e.g., resource) and human (e.g., mental) constraints on a culture being stable and functional. Examples are the need for humans to feel that they are being consistently treated in a fair and just manner, and to stay within the practical limits to resource supplies. Rationalization also allows us to ensure that our decisions are comprehensive, coherent and consistent. This completes the description of Western culture's current circumstance from the perspective of the model's cultural factor.

Adding the environment's perspective of this circumstance to the picture makes it more realistic. Western culture's world view favours a dominion-biased relationship with the environment. In the resource loop, our culturally supported methods of extracting and using resources, along with the discarding of the associated waste, disrupts Western culture's environmental endowment and thus its ability to provide the resources needed to support our culture. Western culture's world view guides us to preferentially deal with such environmental issues by becoming environmental managers. We are guided to preferentially discharge our managerial responsibilities by applying innovation and technology rather than by changing our beliefs and behaviours. Our dominion-biased relationship with the environment ensures that our record as environmental managers is, from the environment's perspective, poor, as discussed in chapters 9 and 10. The overall result is its contribution to the environment's current state.

Next consider Western culture's current circumstance from the perspective of the model's individual factor. At all levels of our culture's complex organizational hierarchy, individual Westerners are routinely making (day-to-day and executive) decisions that (directly and indirectly) contribute to our culture's current circumstance. Our contribution is the product of our decision making, including rationalization, and the influences on it.

The process of growing up ensures that our internal reality (including its reality paradox divergences and our personal world view) mirrors our culture's world view (figure 15). Our internal reality influences our decisions. For example, in the belief loop, we believe in progress (e.g., development and growth) and a dominion-dominated relationship with the environment. In the resource loop, we express this by our decisions favouring the extensive use of resources (i.e., consumerism) and the use of technology to acquire them. Our decisions also favour those activities that make our culture's world view real for us, such as being a consumer and transforming our environment. Think of decorative gardening.

As a result, Westerners tend to choose a resource-intensive, environment-altering lifestyle and believe that it is a key to a better social-psychological life. In doing so, we provide ourselves with the sense of cultural and personal stability, order and meaning that we desire. We also tend to sustain our culture's world view (think culture-member feedback), its increasing demand for resources and thus its complexity-resource feedback.

However, the presence of divergences in both our internal reality and our culture's world view combine with our decision-making foibles and the world's complexity to allow us to make decisions that feel appropriate to us but are at odds with the external reality. Our decisions can thus take us and our culture outside of the human and environmental constraints in which we must live, if we wish ourselves and our culture to remain functional over the long term, and hold us there. We find it difficult to recognize that we are making such decisions and to make the appropriate changes, thus our divergences persist, and we continue to make constraint-discounting decisions.

Consider what happens when we, individually, become aware that some aspect of either our personal lifestyle or our culture's average lifestyle is negatively affecting either us or our supporting environment. Think of pollution. At this time, we may experience cognitive dissonance or a reality paradox.

If we pay attention to this aspect and its impacts at a broad cultural scale, then we prefer to (consciously or unconsciously) look for explanations and solutions within our complex industrial culture's resource loop (figure 15). In particular, we look for technical, resource and managerial solutions. Think of efficiency, laws, innovation, environmental restoration and environmental transformation (e.g., dams). Our focus on these types of solutions makes sense because they will be easier to rationalize. After all, they are consistent with Western culture's guidance about the preferred ways of resolving issues: those solutions that favour progress, economic growth and a dominion view of the environment. They are also personally acceptable because they can maintain our lifestyle (business as usual) and are emotionally benign. We can easily believe that by choosing resource loop solutions we have reasonably addressed our impacts. We are helped by the ease of finding supporting "facts" and theories on the internet and social media.

If we pay attention to these aspects and impacts at a personal scale, then we are prompted to look in the mirror. However, the majority of us, whether in our role as scientist, capitalist, politician, environmentalist, farmer, priest or parent, or as a member of the public, prefer not to respond by doing so. We find it hard to look in the mirror for a solution. Instead, we (consciously or unconsciously) seek out, or are persuaded to adopt, masking and self-satisfying excuses or alternate explanations that let us personally off the hook or point the finger at others. We guide ourselves to search for solutions at a cultural scale, as given above. The internet and social media can help us here as well.

At both scales, we prefer to avoid the belief loop. We prefer to sidestep the prompts to pay close attention to our personal and cultural beliefs and values, our divergences, our degree of awareness or attention and our decision-making foibles. We thus circumvent the effort needed to implement a belief loop solution that would impinge upon our sense of self and the essence of our culture's world view.

This avoidance makes sense considering our desire to avoid the discomfort of looking in the mirror and questioning our beliefs. But, from the perspective of our impacts, such as pollution, it is a serious oversight because the belief loop is often a significant source of the influences that result in us negatively affecting ourselves and our environment. After all, it contains our personal and cultural beliefs about ourselves, our culture and the world around us that guide our decisions. There are indeed belief loop solutions that can meaningfully address issues and the personal and cultural aspects behind them.

For example, we can come to acknowledge the influences that lie behind our personal choice of a lifestyle and our culture's average lifestyle. Think advertising. We can then, for example, more easily recognize our critical divergences and make the adjustments to our personal world view that would remove them. This would help us to choose a lifestyle more, not less, in tune with the human and environmental constraints on our lives. By doing so, we would favour similar changes being made to our culture's world view, which would change its guidance about our preferred lifestyles. These changes would meaningfully address our impacts.

We are each free to choose how we want to live and solve the issues we recognize. Our decisions will contribute to Western culture's current and future circumstance and the environment's state. However, if we wish our culture to remain functional and become viable, then our choices must respect the relevant human and environmental constraints on how we can live. This completes a description of Western culture's current circumstance as seen from the perspective of the model's individual factor.

In overview, the model of the human-environment system provides us with an easily understood description of Western culture's current circumstance and its members' (material, personal and cultural) contributions to it. It does so by helping us to make sense of the complex mix of interactions among the (mental and material) features of the environment's functioning, Westerners' internal reality and their culture's world view. For example, our culturally influenced rationalizations result in us tending to focus on finding solutions to our issues within the resource loop and neglecting the belief loop. This focus contributes to us (consciously or unconsciously) sidestepping or ignoring the cues, such as those from our environment, that we are living outside of the human and environmental constraints on our lives. Our focus helps to ensure that both our internal reality and our culture's world view (including their divergences), and the feedbacks sustaining them, remain intact: our decisions tend to maintain the status quo. With the help of the model, we gain a more complete appreciation for how Western

culture's current circumstance and our contributions to the current state of the earth's environment arose and continues. We thus more fully appreciate our ongoing contribution to the earth's current state.

These features are not unique to Westerners or Western culture. To different degrees, the members of all the more complex human cultures are making decisions that contribute to each of their circumstance and the environment's state. Each culture is thus contributing to the biosphere's current state.

In summary, this chapter discussed our current circumstance and the state of the environment, which are together the biosphere's current state). It noted that the large global population of humans, their extensive ongoing demand for resources and the environmental consequences of securing and using them are our material contribution to that state. However, behind that material contribution lies our personal, culturally supported decisions. Through them, humans make a personal and a cultural contribution to the biosphere's current state. All of these contributions are formed and coexist within a dynamic, complex human-environment system that displays the generic characteristics of a complex system.

A human-centred model of this system is based on three primary factors: us as individuals, our culture and the environment (figure 15). They are represented in the model by, respectively, our internal reality (with our personal world view), our culture's world view and our environment's functioning (including its constraints on how we can live). The relationships between the three factors were divided into a belief loop and a resource loop, which are connected to one another by their common features, interactions and feedbacks.

The human-environment system is human dominated, so our decisions play a key role in the system's functioning. Of particular importance is our (unconscious and conscious) decision-making efforts to provide a consistent and coherent explanation for both a situation and our decisions about how to deal with it (our rationalization efforts). Rationalization mediates the interactions within and between the loops. It thus binds them to one another. It does so in a way that includes how we (as individuals), our culture and the environment function as complex systems. Thus, when rationalization and the two loops are considered together, they represent the complex human-environment system's functioning and changing: its dynamics.

Our decision-making foibles, the complexity of the system and the manner in which we change allow differences between our internal reality and the external reality (divergences) to appear in our internal reality (including our personal world view), and persist. They can thus also appear and persist in our culture's world view. Those divergences influence our decision making, including our rationalization, and thus our decisions that affect the human-environment system's functioning and changing.

How this results in our material, personal and cultural contributions to the biosphere's current state was personalized by applying the model to Western culture. Overall, Westerner's decisions favour business as usual. This does allow us to feel mentally comfortable. However, there are divergences in both our internal reality and our culture's world view. These divergences in their guidance join with our decision-making foibles to result in us tending to live in a manner that pushes against or crosses the boundary of the human and environmental constraints on our lives. Together, they also make it difficult for us to recognize our current circumstance, the state of the environment and the (mental and material) causes for it. We experience difficulty accepting our responsibilities for either our circumstance or the state of the environment, and to respond to them in a manner that reduces our contributing divergences and faulty decision making. We tend not to make decisions that favour viability. The results are our ongoing material, personal and cultural contributions to our culture's circumstance, the environment's state and thus the biosphere's current state.

The model of the human-environment system makes it easier for us to grasp and appreciate its complexity and dynamics. It is within the context of the human-environment system that we find it easier to understand and explain why the biosphere ended up in its current state and not a different state. It is also easier to evaluate the diversity of opinions about causes, cures and constraints contributing to this outcome. The model of this system will thus be a key reference for the discussions in part 3.

Yes, humans are smart and powerful. The biosphere's current state did not arise because we are evil or stupid, but because we are well and truly out of our intellectual and emotional depth. This is encapsulated by the presence of our reality paradox. To borrow a concern from economics, we are living beyond our mental means. This mention of economics serves as a reminder that it, as one of the most powerful influences on Western culture, assures us that all will be well. But,

economists say, this will only be the case if we direct our culture's self-sustaining feedbacks to follow their advice. Better include economics in this, your second report to the all-powerful presence.

# CHAPTER 12

# THE GUIDANCE FROM ECONOMICS

This chapter is the last part of your report to the all-knowing presence explaining how the biosphere's current state arose and why it endures. It deals with economics. This is quite an abstract subject, so you hope your boss will be patient reading it.

The discipline of Western economics and its associated culture will be the main focus of this chapter because it has had, and still continues to have, considerable influence on economies globally. Mainstream Western economics is the study of the production, distribution and consumption of goods and services, including related problems such as taxation, labour, resources and profit (Guralnik 1982). Economics has also been described as the study of choices made under conditions of scarcity (Hall and Lieberman 2008).

Economics has an influence on Western culture's world view and its members' lives that extends beyond its scope of study, an influence that is, perhaps, more powerful than either science or religion (Csikszentmihalyi 2000). Consider that economics provides or affects many of the references Westerners employ to make even our day-to-day decisions. Think of using price to determine value. Economics influences our perception of social and environmental issues, our selection of solutions and our visions of the ideal future. It influences our extraction and manipulation of natural resources, and our disposal of waste. Economics thus also affects our impacts on our environment.

This chapter will discuss economics with a focus on the influence it has on Western culture, its members and its environment. It will also address questions relevant to the state of the world as described in the previous three chapters. For example, does economics' guidance enhance or reduce the difference between our internal and external reality (our reality paradox)? Does its guidance increase or reduce our impacts on our environment?

The discussion starts with a brief history of Western economics, which will be set within the context of a changing Western culture. It reveals that the neoliberal brand of mainstream (neoclassical) economics currently dominates Western economic thought and that this brand has had a significant influence on Western culture since the 1970s. Thus, after the history section, the focus will be on the neoliberal brand, so the terms *economics* and *economist* will refer to neoliberal economics and economists, unless otherwise noted. However, parts of the succeeding chapters will still apply more generally to mainstream Western economics.

The neoliberal economists' efforts to understand the Western economy and guide it are based on a set of basic economic principles, assumption and practices. After the history section, examples of these basics are provided, before pragmatically assessing their limitations. These basics and their use are then placed in the context of neoliberal culture, because it is the neoliberal world view that is the ultimate source of neoliberalism's guidance to Western culture.

Neoliberalism's guidance strongly influences Westerners' treatment of our environment. Therefore, its ability to assign value to our environment and represent its functioning is discussed. The chapter ends with a coherent description of neoliberalism's guidance to Western culture and a critique of the importance that Western culture give to it. In the process, answers will be provided to the question of whether the nature of neoliberalism's guidance primarily reduce or enhance our reality paradox and our environmental impacts.

## A Brief History of Economics

Today's Western economies can trace their roots back to the 1500s, when it was decided that usury was no longer a sin and the poor were not necessarily nearer to godliness than the wealthy. Today's Western economists can trace aspects of their beliefs, theories and practices back to the Western philosophers of the 1600s to early 1700s. By this period, trade was already well developed.

During this period, limited companies were becoming more common and the long-standing view of economics as an aspect of household management was ending. The study of political (state) economy was appearing alongside the beginnings of the nation state. Questions about the workings of an economy, such as the source of a good's value, were beginning to be asked. The history of economics revolves around the efforts to answer these and related questions.

There were some significant changes during this period. For example, the relationship between people and their governments was changing. This is illustrated by the works of the Puritan John Locke (1632–1704; *Two Treatises of Government*, 1689). At the same time scientific discoveries were changing perceptions of the world, such as the discovery by William Harvey in 1628 that the blood circulated around the body.

In early 1700s France, some of those interested in economic matters (the physiocrats), such as François Quesnay (1694–1774), proposed a new idea: the source of value lay in the labour performed on land. It provided the agricultural products that made a country wealthy. They thus favoured farming as the primary economic activity and suggested that the income it generated circulated through the economy like the circulation of blood around the body. The physiocrats introduced the idea of not impeding the natural flow of income: laissez-faire. In particular, they were against any restrictions on trading, especially by farmers.

Western economics' classical period started in the mid-1700s. At this time a European economy was still dominated by agricultural products produced on farms by human and animal labour. Between then and the beginning of the 1800s, England experienced a food shortage, and the Americans and the French had revolutions. The increasing use of inanimate energy, especially from coal, began the Industrial Revolution. Three economists were particularly influential during this period: Adam Smith (1713–1790), David Ricardo (1772–1823) and Thomas Malthus (1766–1834).

Smith, considered to be the father of Western economics, published *The Wealth of Nations* (1776) 21 years after the Lisbon earthquake. He was a moral philosopher whose thinking was influenced by the physiocrats and the philosopher David Hume (1711–1766; *Political Discourses*, 1752). Smith was concerned mostly about the value of goods (services were excluded). He believed that prices were fixed in a market by the overall balance between supply and demand and that wealth could be increased by increasing market activity (Smith 1776).

During the 1800s, Western culture and the economies of its respective nation states were being transformed by the Industrial Revolution (industrialization), a related rural-to-urban shift in population, and the appearance of the working class as a distinct group of labourers. New scientific and technical discoveries allowed the notion of progress to more strongly influence thinking about the world, as discussed in chapter 3. The moral and social concerns of slavery and poverty were prominent.

These cultural changes affected thinking about the economy, however no unified sense of how the economy worked arose among the classical economists. Different views simply came to dominate; a feature of economics still seen today. Popular at this time was the labour theory of value, in which different goods were thought to have different prices because of the differences in their embodied labour. Also dominant was the idea of a free, self-stabilizing market. Wealth was seen as the surplus value (profit) from trading in the market. Classical economists believed that wealth could be increased by focusing on increasing production, for example, through investment.

The big battles among classical economists, aside from the debates about methodology, were over how the surplus value from trading goods was to be divided between the social classes (landlords, capitalists and workers). The work of Karl Marx (1818–1883), especially *Das Kapital* (1867), is the best known of these economists. He blamed the miserable working conditions of the new industrial-era working class on the bosses or capitalists (his term) and the economic system that supported them.

It was late in the classical period when John Mills (1806 –1873) published his popular Principles of Political Economy (1848). This work remained important until his death because it summarized, analyzed and tried to reconcile, in a literary manner, the various views of how the economy functioned. He also laid out a series of possible economic futures rather than, as is often the case, being a champion for just the "right" one (Mill 1848).

After 1850, the ongoing increase in the use of inanimate energy significantly changed the economy. It allowed human and animal labour to be augmented or replaced with machines and the industrial production of consumer goods to be increased. It was in this late Victoria period that the idea of a standard of living appeared: a measure of your well-being based on your material possessions. The labour theory of value was replaced by the idea that the value of goods and services (now included) arose from our subjective desire or want for them. The value of a good or service came to be thought of as the market price a purchaser was prepared to pay for it to satisfy their want.

Some economists thought that our purchasing decisions were based on our assessments of an aggregate desire, such as an average want, or aggregate price. Others came to think that our decisions occurred around the edges or margins of our immediate want and by bargaining. They pointed to the influence that the last sale of a product we wanted had on the price we would be prepared to pay for it. An analogy is the complexity of the decision a farmer faces when trying to decide whether to put marginally farmable land into production. The many changing influences on that decision include recent rainfall, its amount, expected rainfall, crop prices and labour availability. The appearance of this thinking about value is called the marginalist revolution.

Léon Walras (1834–1910; Elements of Pure Economics, 1874) relied on marginal theory to mathematically link together the major pieces of the economy: supply, demand and price within a market. His work and that of his contemporaries, in particular Alfred Marshall (1842–1924), resulted in the neoclassical synthesis. It provided a coherent explanation for the market's functioning and income distribution. It also allowed for substitution between production inputs, such as between labour and capital. However, this mathematical approach required the making of many assumptions about the market and its participants, such as the market being perfectly competitive, efficient and free (Rodrik 2015; Hall et al. 2001).

The marginalist revolution and the neoclassical synthesis slowly brought the classical period to an end. A milestone in the transition was Marshall's *Principles of Economics* (1890) in which he coined the term *economics* to replace the older term of *political economy*. Economics became a separate discipline and its practitioners became economists. At the same time, the discipline became increasingly isolated from the emerging social sciences, in part because of economics' increasing reliance on mathematics (Milonakis and Fine 2009).

Over time economics became dominated by neoclassical thinking. This does not mean that classical theories and thinking were destroyed because they were shown to be wrong. A new theory simply displaced the older one because it appeared to work better. Thus the classical idea that the economy largely worked on autopilot could still dominate neoclassical economic thinking.

Accompanying the ascendancy of neoclassical economics was the increasing use of mathematical (i.e., models) rather than literary descriptions of the economy: for example, the mathematical expression of the theory of diminishing marginal utility. (Often referred to as the law of diminishing returns in which, under certain conditions, a fixed effort to maintain cash returns or the supply of resources from an endeavour results in the cash returns or resource supplies experiencing ongoing declines (Smulders 2005). Over time, mathematical models of the economy became more complicated and relied more heavily on assumptions and intangible variables, such as marginal utility, rather than on real-life measures, such as wages. The study of economics became more theoretical and abstract (i.e., distant from real day-to-day living).

The optimistic early 1900s came to an end in 1914 with World War I (WWI). The European economy was disrupted by military needs. The financial, physical and social disruption of the Great War resulted in sweeping changes throughout Europe. There was a revolution in Russia, labour unions were strengthened, and the vote was granted to the public. Empires were weakened or collapsed, and the treaty of Versailles set the social and economic stage for World War II (WWII).

After WWI, neoclassical economics continued to dominate. It was confirmed as an abstract, deductive exercise in the marginalist mould. The idea of choice in the face of scarcity appeared (Milonakis and Fine 2009). The Roaring Twenties came to an abrupt end with the Wall Street crash of 1929. It ushered in the Great Depression of 1929–1939, with its high level of unemployment.

The neoclassical economic orthodoxy held that the cure for the Great Depression required government spending to be tightened until the autonomous economic system re-equilibrated and business profits returned to their "expected" levels. Think of current day austerity. However, John Keynes (1883–1946; *The General Theory of Employment, Interest and Money*, 1936) argued that the market would take too long to restore itself. He advocated for government involvement in the economy. This meant regulation and deficit spending. He also advocated for a more egalitarian distribution of wealth.

The US government's efforts to alleviate the Depression reflected this advice. Think of its New Deal spending and its strong regulation of Wall Street. The Great Depression ended with the increase in the war production needed to fight WWII.

WWII was a period of economic and social upheaval. When it ended, the Keynesian brand of neoclassical economic thinking was dominant. Government involvement in the post-WWII global economy was directed at stabilizing its economic fluctuations and providing a level playing field for trade. These efforts included the Marshall Plan to rebuild post war Germany, the Bretton Woods Conference, fixed but flexible currency exchange rates, and the creation of the International Monetary Fund and the World Bank. The West's view of economics was dominant globally.

The period immediately after WWII saw a rapid increase in the West's material standard of living and a rapid increase in the use of fossil fuels, especially oil. It also saw the start of the ideological cold war between capitalists and socialists (or between democracy and communism, if you prefer), trade globalization and decolonization. There was also an explosion (literally and figuratively) in inventions, the global population and the recognition of our environmental impacts.

After WWII, economics became even more mathematically oriented, theoretical and statistical: that is, more abstract. The increasing availability of computers allowed more complex mathematical economic models to be built, although implementing them on computers was difficult. Significant assumptions and simplifications were needed. Translating the results of the economic models into meaningful, appropriate policy was, and still is, difficult. Literary descriptions of the models' results and the economy were limited to clichés. Economics became more distant from those it affected.

Later in the post-WWII period, the Keynesian brand of neoclassical economic theories, models and thinking began to be questioned more vigorously. Particularly strong criticism came from the Chicago school of economics. Its economists tended to hold more libertarian views, which thus favoured less government intervention in the market. Prominent examples include Milton Friedman (1912–2006; *Capitalism and Freedom*, 1962) and Eugene Fama (1939–; "The Behavior of Stock Market Prices," 1965), which suggested that an unregulated market is efficient. Think of laissez-faire from the 1700s.

The Chicago school's economists produced the new classical synthesis. Its theoretical, abstract mathematical models included both microeconomic (e.g., price setting) as well as macroeconomic (aggregates) (e.g., inflation and the business cycle) aspects of the economy. This approach is epitomized by Robert Lucas (1937–; "Expectations and the Neutrality of Money," 1972) (Lucas 1972). In creating his model, he expressed the idea that rational consumers could be expected to have rational expectations about the future. This example illustrates that all of these emergent theories and models still relied heavily on significant simplifications and assumptions, such as the rational consumer and an efficient, free market. This reliance helped to complicate the translation of model results into realistic policy.

During the post-WWII period, economists also tackled other types of economic issues. For example, after introducing the concept of gross domestic product (GDP) in 1934, Simon Kuznets (1901–1985) published, in 1955, the first inverted-U-shaped curve relationship between a "tangible" variable (e.g., income inequality) on the Y axis and a measure of economic growth (the GDP) on the X axis. And, in the 1960s, the study of ecological economics became a minor discipline in its own right (Hall et al. 1986).

The 1974 oil crisis triggered a global recession coupled with inflation, called stagflation. Because Keynesian economics was unable to provide a satisfactory response, it fell out of favour. The laissez-faire, free market or monetary brands of neoclassical economic theories (i.e., those favouring a reduction of the government's involvement in the economy) filled the void (Coyle 2014). Many of these theories are the product of the Chicago school of economics. For example, despite its known flaws, Lucas's model was particularly popular, at least in academia, because it fitted the times. It had both the features expected of a modern, macroeconomic model and appeared to be in harmony with the United States' post-1980s period of economic growth (Rodrik 2015).

This free market brand of neoclassical economics is referred to as either neoliberal (the term used here), neoconservative, the Washington Consensus or libertarian economics. Its view of how the market works is described as market fundamentalism. Its distinctive features are a strong belief in free markets, free trade (globalization) and market efficiency. The neoliberal vision of the economy's workings came to dominate Western economic thinking and guidance.

The primacy of neoliberal economics took a body blow with the global financial crisis of 2008. After all, it was a direct product of neoliberal thinking, or miss-thinking. The dust from the crisis, now well into the second decade of the 2000s, is being settled by the ironic use of quantitative easing and the government bailouts for financial institutions that are now labelled as too big to fail and too complex to regulate.

In summary, the history of Western economics reveals that Western economists' thinking about value, the workings of their culture's economy and how it could be improved changed as the nature of the economy and their culture changed. This is evident in its transition from a complex agricultural economy to a complex industrial one. During this cultural transformation, there were changes to Westerners' social-physical surroundings, such as urbanization, industrialization, war and new scientific discoveries. There were also changes to Western culture's world view, such as the ascendancy of the notion of progress and secular thinking. All of these changes affected not only the manner in which the economy operated but the thinking about it. In turn, the changed economy and thinking contributed to the cultural changes underway. This feedback relationship between a culture, its members and its economy influences the economy and economists' thinking today.

The complexity of a culture's economy lends itself to a diversity of attitudes, beliefs and opinions about economic issues. Some beliefs about the economy tend to be more widely held, such as Westerners' belief in growth. Overall, at any given time there is a mainstream view of the Western economy and how to practise economics, but this view is not monolithic.

Mainstream economics can be divided into brands and schools, each characterized by their own common views and practices. At different times in history, a different brand/school has dominated mainstream economic thinking. The neoliberal brand of neoclassical economics dominates current economic thinking. It has considerable influence on Western culture and its members' decision making. It is the focus of this chapter.

## Neoliberal Economics: The Basics

The economy is a dynamic, complex, human-formed, human-dominated system that is closely tied to our environment. History indicates that, over time, the efforts of mainstream Western economists to understand and explain the workings of the Western economy have come to depend almost exclusively on the use of theoretical and abstract mathematical models. To become a mainstream economist today requires learning how to make, select and apply mathematical models to economic questions.

Unfortunately, the economy's dynamic complexity, our focus on its shorter-term state and our participation in its working ensures that it is difficult to construct an economic model that meaningfully represents the actual economy. To build or select a mathematical model under these conditions, economists apply what they feel are basic economic principles, assumptions and simplifications. A budding economist learns those basics from their elders, as well as how to apply them and to have strong faith in mathematical models (Rodrik 2015).

This training embeds the beliefs and practices (and their faith in them) into the economists' unconscious, where they influence the economists' thinking about economic issues, their efforts to discover more about how the market works and their suggestions about ways

to improve its workings. This section will detail some of the basic principles, assumptions and simplifications that lie behind the neoliberal brand of economics.

A reminder that the focus of this chapter is on neoliberal economics because its guidance has the strongest influence on Westerners today. Thus the terms *economics* and *economist* will refer to neoliberal economics and economists, unless otherwise noted. Despite this focus, much of what will be discussed in this and the next section can be applied, to varying degrees, to the other neoclassical brands of economic thinking.

**The Market and Value.** Since Smith's time, the market place (the market) has been identified both in theory and in practice as the forum in which the economy functions. We visualize it as a physical place, but today it is largely virtual, existing only on computers. Wherever it resides, neoliberal economists believe that the market conforms to natural laws (like physical laws), which they can precisely describe using mathematics. The boundaries chosen for the market define the boundaries of the economic system to be described.

In the neoliberal economists' view, the market has two basic functions: the setting of a commodity's value during the acts of buying and selling it, and the balancing (clearing, rationalizing) of successful transactions or trades. (A commodity is used here to mean anything traded in the market, for example, goods and services.) The measure used to conduct these trading activities is cash and cash equivalents. The market is said to be complete when all of the influences that affect trading in the market (effects on the market) are captured in the cash value of the goods and services being traded. Thus only those items and influences that can be translated into a cash value can be part of the market. All items and influences that can't or aren't assigned a cash value above zero are considered to lie outside the market. These are called externalities and are considered irrelevant to its functioning.

**The Rational Consumer.** The market is given life because we participate in it as sellers and buyers (traders), as producers and consumers, who make decisions on which we act. Economic theory assumes that you will reach your decisions and act on them in a rational manner. This means that you (layperson or tycoon) will decide to buy or sell goods or services because you have logically, freely and fully evaluated a complete suite of knowledge about all of the options. It also assumes that when purchasing or selling, you will act in your best self-interest, which means your best financial self-interest.

### The Self-Regulating Market.

When the market is complete and its individual participants are rational, then economists can conclude that the market is rational. The collective, self-optimizing behaviour of the participants becomes the invisible hand of the market. It ensures that the market is both efficient and tends to an equilibrium state where the benefit to all participants is optimized. But this will only occur if the process of conducting transactions in this market is unhindered by government controls or interference. Such a market is referred to as a self-regulating or a free market (Hall et al. 2001; Rodrik 2015). A free market also implies a stable and efficient market.

An example of the self-regulation principle is the economic textbook version of the supply-demand curve. If the supply of a good decreases but the demand remains constant, then the price of the good will rise. It will increase to a level that provides the market players with sufficient financial incentive to replenish the supply or find suitable substitutes.

The principle is also illustrated by the 1973 OPEC oil supply shock. The oil shortage resulted in a substantial rise in the price of oil. The self-regulating market resolved the problem by switching to materials that were less sensitive to energy costs and more energy efficient. As a result, metal producers, for example, suffered, but plastics manufacturers prospered (Cook 1987).

Neoliberal economists believe that failures in a nominally self-regulating free market are due to either unwarranted human interference in the market or lack of human effort to ensure that the conditions for self-regulation are met.

### Prosperity.

The economic prosperity of a country or an individual refers to their financial state or condition, as measured by the absolute amount of cash and cash equivalents they can muster. The economic prosperity of an individual is measured by their personal wealth, in particular their net worth (the total cash value of all of a person's savings and material possessions minus their debts) or their income. The economic prosperity of a country is usually measured by its GDP. It is the annual sum of the country's cash *flow* from income, expenditures and production: the total value of the monetary transactions within the country's markets (Schmookler 1993; Coyle 2014). Note that the GDP is not the country's net worth.

These measures of economic prosperity are used by economists to answer economic questions and evaluate economic decisions. For example, expressing a country's national debt as a percentage of its GDP indicates a country's ability to sustain that debt. Similarly, an individual's net worth or income can be used to determine how much money they can safely borrow.

Economists consider that the measures of economic prosperity represent more than the financial state of an economy or a person. They are also thought to reflect the state of our non-economic lives, our social prosperity. In particular, a country's GDP is assumed to reliably reflect its population's quality of life, its social welfare. Similarly, an individual's net worth or income is assumed to reliably reflect an individual's level of social-psychological well-being.

Economists have faith in these assumptions because they believe that the level of economic prosperity determines the degree to which humans can satisfy our quality of life and our personal well-being needs. These needs include, among other things, our desire for social standing (e.g., size of house); success (e.g., education); health (e.g., number of magnetic resonance imaging machines per capita); and our emotional contentment. They also believe that the measures of economic prosperity reflect our ability to satisfy wider social objectives, such as an end to poverty and suffering.

### Growth.

Economists are also interested in changes to our economic prosperity. An increase in its measures for a country (e.g., GDP) and an individual (e.g., net worth or income) are referred to as positive economic growth, economic growth or just growth. By maintaining economic growth, economists ensure an increase in our economic prosperity.

Economic growth is thought to be facilitated by more than guaranteeing that legal contracts are honoured. It is also facilitated by ensuring that the market is free. Meeting these types of conditions ensures that our efforts to increase growth, such as applying innovation and technology, are unhindered and optimized. A free and efficient market enables the virtuous cycle of production and consumption to maintain economic growth and thus increase our economic prosperity.

Economists favour economic growth because they believe it increases more than our (material) standard of living. An increase in GDP is assumed to reflect an increase in a country's social welfare, while an increase in personal wealth is assumed to reflect an increase in an individual's social-psychological well-being. Thus economic growth is needed to overcome the scarcity that prevents us from satisfying our quality of life needs (Hall and Lieberman 2008).

In essence, economists believe that economic growth ensures that a country and its individual members will be better off in all respects. Economists can even believe that economic growth results in both economic and social progress toward a culture's vision of the future. For example, the growth in GDP is considered to be a measure of the increase in freedom, human capability (innovation), care for one another and the support for democracy being provided by capitalist markets (Coyle 2014).

### Resources, Substitution, Innovation and Technology.
The commodities traded in the market are made available by the factors of production: natural resources, capital resources and labour resources (also referred to as natural capital, financial capital and human capital). Linking these factors together are innovation and technology (more realistically, human decisions). When production issues arise, economists expect that the workings of the market will induce the participants to resolve them, in particular, by the application of innovation and technology.

Imagine that a particular natural resource is unavailable in sufficient concentrations or quantities suitable for economically profitable extraction, trading and preparation prior to being used in factories. If the market is free to apply innovation and technology, then it is expected that this scarcity can be overcome, either by making more of the resources available or by substituting another natural resource for it. It is assumed that unlimited, (near) perfect substitution between all natural resources is possible (Smulders 2005).

The strength with which economists hold to their assumptions about resources, substitution, innovation and technology is illustrated by their preferred views of possible solutions to reducing the emissions of $CO_2$ to the atmosphere from using fossil fuel energy sources. For example, a possible technological solution is to capture the $CO_2$ and store it underground or in the deep ocean. Or we can substitute non-fossil-fuel energy sources, such as windmills and nuclear power plants, for fossil fuel energy sources. An innovative geoengineering solution is to induce atmospheric cooling by, for example, deploying mirrors in space. The required innovation and technology to make these substitutions can be encouraged by carbon trading. These proposals can all be accepted by economists because they fit in well with the neoliberal world view. They appear to deal with the issues head on while still contributing to economic growth and progress.

In summary, the basic principles, assumptions and simplifications described above illustrate key parts of the neoliberal view of our economic system. They are overtly or covertly present in the sophisticated mathematical economic models neoliberal economists use to explain the workings of Western economic systems and to guide economic decisions. They are also used when addressing economic problems and interpreting the results of models. The next section will illustrate and personalize the limitations to applying these basics and thus to the conclusions draw from the models that use or reflect them.

## Limitations to the Basics

Mainstream economists rely on mathematical models to answer economic questions. The model used for a particular question can either be built from scratch or chosen from the spectrum of available models. In either case, the dynamic complexity of the economy requires that the model satisfies at least three conditions if an economist is to have confidence in their preferred model and its results: it must include all of the variables relevant to the question, it must be suitable for the question (when the question is seen within its greater context), and it must realistically represent that context. Only models that satisfy all of these condition can be considered, with a significant level of confidence, to be representative of the problem.

These conditions are difficult to satisfy because the questions being posed often require many interacting variables to be included in the model, and there are a wide variety of complex contexts in which a question can be found. Satisfying these conditions is certainly made easier by the use of basic principles, assumptions and simplifications (the basics). But doing so unavoidably introduces uncertainty and error into the model and its results. This detracts from the precision and accuracy gained by using mathematical models.

Thus, to be realistic about their faith in their model and its results, economists must assess the level of uncertainty and errors that their choice of model and basics may introduce into the model's results and their interpretation. At a minimum, the validity and usefulness of a model must be checked by tests that use data from, hopefully, a well-understood, real-world circumstances. This validity-usefulness check can be a difficult task to complete (Rodrik 2015). If this and the other validity-usefulness checks are not completed, then the model's results can be applied to the real world with more confidence than is reasonable.

The following subsections discuss the constraints on applying the basics to economic models or interpretations and the difficulties of completing a validity-usefulness check. The presence of these constraints and difficulties is illustrated by examples. The examples also illustrate the consequences of not routinely considering these constraints and difficulties during modelling, the interpretation of results and the application of conclusions (van den Bergh et al. 2000; Colander et al. 2009; Leontief 1982). A reminder that the features of these examples were caused by more than the particular topic they illustrate. An example is just a good illustration of the topic being discussed. The full cause of an example's features lies in the collective outcome from the basic principles, assumptions and simplifications used during modelling, and the nature of the attention paid to the constraints on their use.

### The Market and Value.
The market is said to be economically complete (closed) when all of the influences on it are captured by the cash value of the commodities being traded in it. This assumption allows the market to be fiscally balanced (cleared).

However, the market price of commodities only represents their cash value to the traders (which includes us as consumers). Thus there are aspects of the wider economy's functioning that are not included in the price. Consider that the supplying, manipulating and using of a resource today can result in economic costs, as well as benefits, in the future. Those costs are not paid for in the present but deferred to the future, where they may or may not be accounted and paid for in the market. An example is the costs to labourers' (mental or physical) health from their exposure to asbestos during their working years.

Similarly, from the external reality's perspective, the short-term market price of a resource (its cash value) fails to reliably represent the resource's overall importance to the functioning of the environment (its functional value). For example, water can have a cash value in the market as a commodity. That cash value is less than the cash value of a diamond. However, their functional values are in the opposite order. Water is necessary to maintain the functioning of the environment's ecosystems. The diamond isn't; it is discretionary. This is a significant oversight because the environment enables the economy to function: it supplies many of the economy's resources and supports its consumers, labourers and capitalists.

Consider too that there are the aspects of the human-environment system that economists consider to lie outside of the market, which they call externalities. But, many of these externalities play critical, if indirect, roles in the economy, such as supporting its consumers and providing resources. Examples are the crop pollination services provided by insects, the water purification services provided by wetlands, the UV protection provided by the ozone layer and the people volunteering their services to help the poor and needy.

In overview, the cash value of something in the market can be unrepresentative of its overall and longer-term functional importance (value or cost) to the wider economy. The same is true of something the market treats as an externality. The market's functioning incompletely represents the wider economy. These constraints are often forgotten when the results of economic models are interpreted and used to guide our decisions and actions.

### The Rational Consumer.
The assumption that humans make rational economic decisions in their best (fiscal) self-interest after assimilating all of the relevant information is, optimistically, unsubstantiated. More truthfully, the idea is bunk. The assumption is incompatible with our brain's functioning and our decision making, as previously discussed in chapters 1, 2 and 6.

Our decision making is partly rational and partly emotional. The emotional aspect exists because our brain automatically and unavoidably attaches feelings or emotions to our sensings of the world and our memories. We are aware of some of these feelings and their influence on our decisions, but not others.

There are other ways in which our decision making is constrained from being fully rational (in the economic sense). Despite our experience of consciously contemplating and making a decision, most of our decision making occurs in our unconscious. There it is subject to many diverse unconscious influences, perhaps biases, which need not be rational. Among them are our habits, our past experiences, decision-making rules of thumb (heuristics) and our culture's world view (Schwartz 2000). These influences constrain both our perceptions of an issue and our decisions about it.

Even if we did have access to all of the information needed to make a fully rational consumer decision (which we don't), we would struggle to consciously assimilate it. Our brain's ability to absorb information is limited. And the processes it uses to handle the overload, such as selection and simplification, can bias the information we absorb.

Our efforts to act in our best (fiscal) self-interest are similarly affected. For example, our decision-making processes favour a short-term view of the future. This may be the most appropriate time frame to use when attempting to maximize our cash-based self-interest in the market, but the resulting decision may not be in our best long-term self-interest. Think of impulse purchases.

Overall, our human characteristics affect our decision making in ways that are incompatible with the notion of a rational consumer acting in their best self-interest after assimilating all of the relevant information. For example, when we think of selling something, our immediate emotional response tends to be more negative than when we think of buying something. It is thus harder for us to sell things than to buy them. This is one reason why we accumulate stuff and can hold stocks until they are worthless (Myers 2002a). Similarly, bidding wars in the market are not always rational, but they are normal. Neither is more consumer choice, a neoliberal mantra, always better because it can overwhelm our efforts at rational decision making.

The death blow to the idea of us being rational consumers comes from the advertising industry. Their methods of increasing consumption (that helps maintain economic growth) rely on our purchasing decisions being significantly influenced by non-rational, unconscious thought processes. If we were the rational consumers that economists assume, then advertising would not work in its present form. Westerners would not necessarily favour diamond wedding rings. Neither would Isaac Newton, a poster boy of rational thought, have fallen prey to a hyped up real estate deal (the South Sea Company's real estate bubble/scam of 1720), but he did (Kindleberger and Aliber 2005).

Collectively, these characteristics debunk any notion of us being rational consumers who can routinely maximize our financial interest through cash-focused market decisions based on full knowledge (van den Bergh et al. 2000). There have been many attempts to adjust the neoliberal rational consumer assumption so that it can accommodate the real-life features of our buying and selling decisions while still retaining the precision of the neoliberal mathematical models using it. However, the needed adjustments result in our characteristics being described by meaningless generalizations. For example, making us "rational choosers with flaws" sets no limits on human behaviour (Hernstein 1990).

There are other, more realistic descriptions of our economic decision making (van den Bergh et al. 2000). One of the more enduring is prospect theory (Tversky and Kahneman 1981). It recognizes that our decision making takes place within a frame or context and that we have only limited conscious control over the selection of that context or the decisions within it.

Consider too that our culture is always exerting some sort of unconscious influence, from practical to political to ethical, on our decisions (Hernstein 1990). For example, Western culture's emphasis on individuality permits more self-focused purchasing decisions than are accepted in a culture where individuals are more group oriented, for example, in Japanese culture (Schwartz 2000). And in the Western world, couples preferentially buy a wedding ring that contains a diamond rather than some other precious stone. This is a culturally influenced choice that is the product of advertising.

In fact, the form of an economy and thus the idea of a rational consumer, is culturally determined. This is illustrated by considering a subsistence farming culture such as Tikopia. Its far simpler cashless economy cannot be meaningfully separated from Tikopian culture and its members' obligations to it. The notion of the Tikopians being rational consumers in the Western sense is meaningless (Firth 1965). The idea of a fiscally rational consumer only has meaning (regardless of its truth) within Western and similar complex cultures.

In overview, market participants behave differently than the fiscally rational consumer of economic assumptions/models. Our decisions, both inside and outside the market, are far from rational. The assumption of rationality may be useful for investigating how the market functions under some circumstances. However, if one of a model's critical assumptions is that we are rational consumers, then the practical economic guidance derived from that model should be viewed with great suspicion, if not ignored (Hernstein 1990; Schwartz 2000).

**The Self-Regulating Market.** The rise of the Western world's current economic practices began in earnest during the 1600s. Since then, Western economies have experienced an ongoing string of market booms (bubbles) and busts (crashes). This history illustrates one reason why governments regulate markets. Think of the United States' regulatory response to the Wall Street crash of 1929.

Neoliberal economists have argued that this thinking is misguided and that the regulations are misplaced. After neoliberal economics became dominant in the late 1970s, its economists helped to orchestrate the deregulation of Western markets. In the process, the neoliberal captains of the economy were provided with an opportunity to showcase market self-regulation at work.

However, since deregulation, the rate and severity of bubbles has increased significantly (Kindleberger and Aliber 2005). Among them was the dot-com bubble of the late 1990s, with its uncontrolled, if not encouraged, excessive market exuberance. This was shortly followed by the global financial crisis of 2008, which was the result of Wall Street displaying wishful economic thinking, if not blind self-interest, fraud, deception, market manipulation and a loss of ethics (Rushe 2011; Taibbi 2011; Coyle 2014). The Wall Street bankers, with the help of the regulators, had been providing themselves with excessive financial incentives and the market with flaky financial instruments.

These crashes, but especially the one in 2008, revealed that the neoliberal captains at the helm of the economy had changed both the market into a casino and the business cycle into a bubble cycle, for which the public are still paying (Taibbi 2011; Patterson 2010). What better way is there to debunk the assumption of market self-regulation than to gently point out that the creation and bursting of the recent, spectacular bubbles occurred when neoliberal ideas dominated economic thought while its acolytes practising them dominated the market. The recent crashes occurred at a time when the neoliberal mantras of deregulation and globalization were widely implemented and the market regulators, by design, were asleep at the controls.

Either the market is not self-regulating, in the neoliberal sense, or the neoliberal economists who guided the market's deregulation where incapable of recommending reliable or realistic ways to meaningfully do the job. Either way, the idea of a self-regulating market is unachievable, in a practical sense. The neoliberal economists' vision of a self-regulating market is, at best, a convenient assumption for studying the workings of the market. In contrast, regulation is needed to maintain the real market's stability and functioning.

The need for regulation shouldn't be a surprise. All complex cultures create institutions to help with the tasks needed to keep themselves functional. They also enact and enforce rules for the institutions and their members to follow to ensure that both are working in the culture's best interest. The market is one of these institutions.

For example, in order to keep the stock market stable, governments constrain the manner in which information about a stock can be reported, for example, by limiting embellishment. For the same reason, governments enact rules that constrain how stocks can be traded. These trading rules make it possible to predict the spread in the bid and ask price for stocks without any need for a market player to be thought of as a rational strategist (Farmer et al. 2005).

What about corruption and greed in the market? It is only after a society has decided on norms and made laws that set standards for personal behaviour that corruption and greed can be formally identified and controlled. It isn't innately controlled by the market's functioning. If anything, the reverse is true.

The real market's functioning is as much the product of preferences and standards set by the wider culture as it is the product of self-interested consumers (rational or not) unknowingly balancing supply with demand. If we wish to ensure that a market is stable and working in our best interest, then our knowledge has to extend beyond believing in assumed human behaviour, unsubstantiated economic theories, flaky financial instruments and transaction statistics. It requires us to understand how individual humans function and form groups that create a culture. And it requires us to understand how the members of a culture, under the influence of their culture's world view, create and control institutions like the market.

In addition, unlike the relationship between physicists and the solar system, economists not only describe the market but they also shape it. They do so by applying their basic principles and assumptions to their models and thinking. Then they use their conclusions to influence the standards and manner in which the market is (or isn't) regulated, and the traders behave. Economists are part of a self-supporting economy-economist feedback loop. This should be taken into account when we consider their advice.

In overview, focusing on and believing in the abstract concepts of a self-regulating or efficient market hides from us the fact that an economy and its market exist within a greater culture. The market is a dynamic, complex, human-dominated institution that reflects human values, characteristics and cultural norms (Folke et al. 1994). The market displays the consequences of rational and emotional human behaviour, and cultural influences, including that of economists (Myers 2002a). The market is also tied to the environment, so it is both facilitated and constrained by the changes in the environment's state and functioning, such as drought. Recognizing these types of influences provides better insight into the functioning of the real market and the need for appropriate regulations than believing in the neoliberal assumption that the real market is self-regulating.

**Prosperity.** Neoliberal economists use the cash-based measures of GDP and personal income (or net worth) to represent, respectively, the state of a country's and our personal economic prosperity. They also assume that these measures reliably represent, respectively, the state of a country's social welfare and our personal social-psychological well-being. The growth of these measures is considered to represent both economic and social progress.

Does the GDP reliably represent a country's social welfare? Consider that the contribution of the agriculture sector (including forestry and fishing) to the United States' GDP declined slowly from around an already low 2.0% in 1987 to 1.4% in 2008 (Normile and Price 2004; Bureau of Economic Analysis 2010). From the neoliberal economists' perspective, the loss of agriculture would have a minimal impact on the American (material) standard of living or social welfare (McIntosh et al. 2000). In fact, Americans would be economically better off if they replaced agriculture with a more growth-oriented industry, say the manufacturing of electronic widgets. That would increase their GDP and thus, apparently, their social welfare. If they took this interpretation to heart, however, they would be stepping into King Midas's golden shoes and starve to death surrounded by a growing pile of stuff.

If a country's GDP imperfectly reflects the (absolute or relative) importance of something as critical as food to it members, then what does it represent? As mentioned above, the GDP is the annual sum of the transactions, expressed as cash, which make up a country's economy. Spending on the following are all recorded as positive additions to a country's GDP: constructing and running hospitals, schools and factories; waging wars; buying cancer-causing cigarettes; and fixing the outcome of bad planning decisions, as experienced by New Orleans after Hurricane Katrina.

A country's GDP doesn't distinguish between those parts of a country's economy that detract from the members' social welfare and those that contribute to it. The GDP is not a reliable measure of a country's social welfare. A country's GDP and GDP/capita represent other attributes of its economy: its production capacity; its past and current ability to spend money; its citizens' ability to act as consumers; and its ability to secure a supply of useful energy, as discussed below. The changes in a country's GDP represent a change in these abilities (Coyle 2014).

Does our net worth or income reliably represent our social-psychological well-being? One would think a safe bet is to agree. After all, economists, politicians and the public routinely use them for that purpose. But these indicators only account for those parts of a person's assets that can be given a cash value (economic assets). They leave out those parts of a person's life that are important but can't be or are not expressed as cash: the parts that lie outside the market as externalities. This deficiency limits the ability of economic measures to act as reliable indicators of our social-psychological well-being.

Consider the correlation between a measure of our social-psychological well-being, such as happiness, and economic measures of personal wealth. Early work indicated that self-reported happiness correlated poorly with personal income. As personal income level rises from destitute to poor, the level of self-reported happiness increases significantly. However, as income increases above that minimal amount, the level of happiness hardly increases at all (figure 16a) (Adams 2001; Easterlin 2003).

The result is similar if a population's average self-reported happiness over time is compared to the population's average income. For example, there has been a steady rise in the average American's inflation-adjusted, personal income since the late 1950s. Despite this increase, there has been no substantial change in the percentage of Americans reporting themselves to be very happy, up to the end of the record in 1995 (figure 16b) (Myers 1999).

However, there were arguments about the interpretation of these results. For example, the lack of a correlation between an economic measure and happiness could be the result of comparing measures of social-psychological well-being that lie on a finite scale (0–100%) with economic measures that are infinite (a dollar amount). These arguments have been addressed in more recent work by psychologists and sociologists (Easterlin 2003).

Your sense of social-psychological well-being can be divided into two largely independent components. One is your social-psychological feelings about your well-being, for example happiness. The other is your social-psychological judgement (evaluation) of your life circumstance, for example your satisfaction with your material standard of living. Once social-psychological well-being is separated into feelings and judgements, the impact of a change in income on each can be studied more thoroughly and in relative terms.

An increase in income is seen to result in a more positive judgement of your life circumstance (greater satisfaction), but there is a catch. Imagine you receive a $10,000 raise in your income. Your satisfaction with your material standard of living will increase. But the size of that increase depends on whether you originally earned $10,000 or $80,000. Surprisingly, the increase in your satisfaction from adding $10,000 to a $10,000 income is only matched by adding $80,000 to an $80,000 income, that is, by doubling your income (Diener et al. 2010).

Basically, if you want to increase your positive judgement about your life circumstance by a fixed amount, then the richer you are, the larger the absolute increase in your income will have to be. This is captured in the cliché, "the more you have, the more you want." Think of the often outrageous salaries and perks that CEOs, especially of investment banks, think they deserve. This general weakening of money's ability to satisfy is another example of the declining marginal utility of money (diminishing returns). Perhaps this is an appropriate time to remember that there *can* be too much of a good thing.

Turning to the impact that a $10,000 increase in your income has on your feeling of well-being, as measured by self-reported happiness: it has no impact. Lasting positive feelings are not generated by income, they are the result of satisfying your feeling-related social-psychological needs: being respected for who you are, belonging, learning new things and having a sense of autonomy (a characteristic that we hold in common with members of hunter-gatherer cultures, but which is not to be confused with individualism). Our efforts to satisfy these needs, and the results, are largely excluded from economic measures as externalities because they have no cash value. This is unsurprising as they can't be traded (Oswald 1997).

Overall it seems that, at a personal level, having greater personal wealth enables us to experience the satisfaction of having what we want, but it has little impact on whether we like either what we have or our circumstance (i.e., how we feel). In contrast, if we satisfy our feeling-related social-psychological needs, such as being genuinely respected, then we produce and increase the feelings we like (our positive feelings), such as feeling happy. At a national level, this distinction is reflected in the differences between a country's rank by income compared to its rank by its citizens' average score for social-psychological feelings. For example, among 89 countries, the United States is ranked first in GDP/capita but ranked 26th in positive feelings (Diener et al. 2010).

In overview, a country's GDP and personal incomes are measures of the capacity to spend. The GDP does not reliably represent the social welfare of its members, while income poorly reflects an individual's social-psychological well-being. The economists' use of GDP and personal wealth for these purposes has done our nations and ourselves a great disservice. Despite the evidence to the contrary, Westerners, encouraged by economists, devote much personal time, energy and our culture's resources to increasing our personal income and our country's GDP. We do so partly in the mistaken belief that the resulting economic growth will automatically result in an improvement in our country's social welfare and our personal social-psychological well-being (i.e., in social progress). In reality, other criteria, such as a sense of belonging, fairness in wealth distribution and having equitable personal access to resources, must be satisfied for that objective to be met. These criteria cannot be met by the practices of economics alone.

In reality, a non-economic indicator is needed to reliably measure social welfare and social-psychological well-being (Coyle 2014). Such an indicator would be a better guide for our quality-of-life decisions. The study of these two topics and the development of such an indicator is more appropriately conducted within the fields of anthropology, sociology and psychology than within economics, in general, and neoliberal economics, in particular. Anthropologists, sociologists and psychologists are better suited to advise us how we can enhance our country's social welfare and our personal sense of well-being than economists.

**Growth.** Growth in a country's economy is measured by an increase in its GDP. For an individual, growth is measured by an increase in their personal wealth. There are some strange contradictions between the neoliberal beliefs or assumptions about economic growth and the external-reality view of it. For example, an increase in a country's GDP (economic growth) is seen by economists as good because it means that the citizens' social welfare is improving, even though GDP weakly represents their social welfare. It is claimed that an increase in a country's GDP will mean an increase in the wealth of each of its citizens. However, the GDP is not a measure of how that wealth, or the costs of generating it, are distributed among the culture's members.

Neoliberal economists assume that a country's growth is best facilitated by ensuring that the market is free and efficient. How reliable is this belief? An answer is provided for a country by using its economy's components, the three factors of production (capital resources, labour resources and natural resources [including energy]), to model the long-term record of changes in its GDP.

When the United States' GDP is divided between the three factors of production, then the cash contribution from natural resources is seen to lie between just 4% and 10% of the GDP. One interpretation of this observation is that natural resources are economically unimportant. Thus, in an early attempt to model the growth in the United States' GDP, they were therefore left out, treated as externalities. When just the remaining two factors, labour and capital, were used, then the resulting modelled GDP could only imperfectly and incompletely match the measured GDP. The modellers accounted for the gap by assigning it to an abstract production factor that was assumed to be only indirectly measured by cash accounting: the "human innovation" or "technological progress" factor. Innovation has played a significant role in economic thinking ever since (Ayres and Warr 2005).

More recent efforts to model GDP growth in the United States took a different approach. These studies used a complete suite of measured data for all three factors of production, thus excluding the abstract innovation factor. These models generated an extremely good fit between the modelled and measured GDP for the period from 1900 to 1970, and a good fit between 1970 and 2000. But this was only the case when the natural resource factor included the amount of useful inanimate energy (all sources) used in production (Ayres and Warr 2005; Kummel et al. 2002). (Useful energy is energy that is capable of doing economically productive work.)

Their choice to use useful energy is in keeping with our practical understanding of how energy makes its contribution to production. For example, a unit of gasoline contains a certain amount of energy. When we burn it to power a car, a portion of that energy is lost as heat and unburned fuel. Only some of the gasoline's energy, the useful energy, moves the car.

The useful-energy models of GDP provide such a close fit to the measured GDP that no innovation factor is required. This should change our perception of how the economy functions, including the source of economic wealth and its growth. Firstly, the models fitting of the factors to the GDP suggest that the labour factor makes the smallest contribution to GDP and its growth; capital lies in the middle; and the energy part of natural resources makes the largest contribution, at around 50% of GDP. Energy's contribution to GDP is thus much larger than the 5% range estimated by using just its market cash price. That is, energy's importance to the economy is not reflected by its current market-driven cash value. It is priced far too low (Ayres and Warr 2005; Hall et al. 2001). This implies that the market is incomplete.

Secondly, economic growth in the United States depends heavily on increasing the supply of useful energy. This increase occurs by producing more inanimate energy from its various sources, such as fossil fuels and hydro. Growth is also helped by increasing the efficiency with which useful energy is made available and used. For example, between 1900 and 2000 the efficiency of electricity generation in the United States increased tenfold. This increase ensured that the price of electricity fell and remained low, despite a 1,300 times increase in its use (Ayres and Warr 2005). The broad availability of cheap electricity contributed significantly to economic growth in the United States. It allowed more people to use more electricity.

Although the abstract innovation factor does not appear explicitly in the useful-energy models of GDP, it is thought to be embedded in the energy term. It is present as an increase in the efficiency with which useful energy is provided and used. More generally, a significant part of innovation's importance to the economy lies in its ability to increase the efficiency with which natural resources and labour are used.

Thirdly, economists like to describe economic growth as the product of a self-reinforcing feedback, or virtuous cycle (figure 14). Traditionally it was thought that applying innovation and technology facilitated an increase in production and a lowering of prices. These changes would satisfy consumer demand more cheaply for a greater income and profit. Since profit pays labour, steady or increasing wages allows them to demand more goods and services, which spurs producers to invest in innovation and technology.

The useful-energy models of GDP indicate that the virtuous cycle is better described, at least until 2000, by focusing on capital in combination with low-priced energy rather than innovation in combination with technology. In particular, businesses are spurred by the hope of profit to replace less profitable capital and labour combinations with more profitable capital and lower-priced energy combinations. This switch is facilitated by innovation's and technology's ability to increase efficiency. We experience this in our daily lives as the quest for increased productivity. An imperative that translates into enhancing or replacing labour through mechanization and automation, called technological progress (Ayres and Warr 2005; Kummel et al. 2002). The rest of the virtuous cycle is believed to remain unchanged, but it has been globalized in scale.

There is another aspect of energy that needs consideration: it is consumed during production and is thus non-renewable. The functioning of the economy (including the virtuous cycle), the creation of economic wealth and its growth are thus all highly dependent on the ongoing availability of abundant supplies of cheap energy. Wealth and its growth are limited by the supplies of energy in general and cheap energy in particular. Think oil.

The useful-energy models thus give a boost to an observation made during economics' classical period: natural resources matter to an economy. This cause has been taken up since the late 1960s by ecological economists. They have been pointing out since then that the economy and its growth depends heavily on the earth's concentrations of natural resources, such as fossil fuels, metals and wood (Hall et al. 1986; Hall et al. 2001).

Economic growth does not primarily depend on us ensuring that the market is free and efficient, or on us applying innovation and technology. It depends on the combination of factors, of which the ongoing availability of cheap energy is significantly underrated.

In overview, an increase in a country's GDP is a measure of the growth in the value of the transactions within its economy. The neoliberal assumption that economic growth primarily depends on us ensuring that the market is free and efficient, and that we can freely apply innovation and technology to solve issues, is too simplistic.

The sources of a country's economic growth are its capital, labour and natural resources: the three factors of production. Using these factors to model the changes in a country's GDP indicates that the market price (cash value) of energy underestimates its importance to the economy. From an energy perspective, the market is incomplete. In addition, because energy is consumed during production, just maintaining an economy, let alone growing it, depends on an ongoing cheap supply of energy.

Neoliberal economists under-appreciate the importance of the role that energy in particular, and resources, in general, play in making economic prosperity, personal wealth and their growth possible. This is unsurprising if we remember that economists treat many aspects of the environment that supplies resources as externalities. This begs the question, how do economists propose to deal with a resource supply shortage or scarcity? That is discussed in the next subsection.

### Resources, Substitution, Innovation and Technology.

Economists assume that, when we are faced with production issues, the workings of a free market will induce the participants to efficiently resolve them with the help of innovation and technology. For example, their application will allow us to substitute more productive combinations of capital and natural resources (especially energy) for less productive combinations of capital and labour. But all production activities require the use of some labour and natural resources. Think of the labour and resources needed to build and power labour-saving devices. Thus our ability to substitute one for the other is theoretically limited (Folke et al. 1994). It also raises the question of whether the above substitution simply replaces the limits to labour resources with limits to natural resources.

Can the workings of the market really resolve production issues to the degree economists expect? Consider a possible scarcity of a natural resource. Economists experience and address this issue within the economy. By definition, this means that natural resources only enter the economy when they acquire a cash value as a result of being traded in the market. Similarly, natural resources leave the economy when they become waste, because it has no cash value (Hall et al. 2001). Thus, from the perspective of natural resources, economists treat the economy as an open system.

Economists can therefore consider resources to be infinite. They can believe that there are no limits to supplying natural resources, only difficulties. Any natural resource scarcity can thus be resolved with the help of innovation and technology.

This contradicts what we all accept: the earth is finite, so the amount of natural resources is ultimately finite. The resolution of this conundrum is to ask if this finiteness actually limits our economic system or is it inconsequential. Is the finiteness of the earth's resources practical or theoretical in nature?

The earth's supply of some resources is truly finite, such as the area of arable land. Although many other of the earth's resources are finite, such as fresh water and wood, their supply is renewable. The amount available can be infinite, but only if they are being used at a rate that is slower than their rate of being recycled or replaced. If not, they are finite in a practical sense

The renewal rates for some natural resources are exceedingly slow, making their supply finite in a practical sense. For example, the creation of metal and fossil fuel deposits takes at least hundreds of thousands of years. The long-term availability of these types of resources depends entirely on how quickly we deplete them.

Our hunter-gatherer ancestors mostly used their renewable resources at rates that were slower than their renewal rates, thus ensuring that their supply remained infinite. However, even they may have contributed to some renewable resources being finite in a practical sense. One example is the post-ice-age extinction of some megafauna species, such as the mammoth.

Complex industrial cultures use renewable resources at much faster rates than hunter-gatherers. This has lengthened the list of renewable resources whose supplies are now finite in a practical sense. Consider that the too rapid extraction of groundwater from many aquifers is leading to falling water tables and local water shortages. Too high a rate of tree extraction is leading to deforestation, as is occurring in parts of the Amazon Basin. As with all resources, a distinction should be made between local shortages and absolute scarcity.

In addition, the activities of industrial cultures are disrupting the environmental processes that maintain the supply of renewable resources. This disruption helps to push those resources toward practical finiteness, they become globally scarce. Consider activities that are disrupting the functioning of the ocean ecosystems that

supply us with resources such as fish. Consider too those activities which are altering the atmosphere's composition in ways that affect our environment's ability to renew natural resources. Think of acid rain and climate warming.

As mentioned, economists believe that, regardless of a natural resource's finiteness, when a scarcity arises, it will be overcome by the market-induced application of innovation and technology. These methods are expected to either enhance the extraction of scarce natural resources, such as those from currently unexploitable deposits, or make possible the unlimited substitution of one natural resource for another. However, there are constraints on the application of innovation and technology to this or any other end.

Consider our efforts to overcome a shortage or scarcity of a natural resource by applying innovation and technology. Those efforts are constrained because the devices that increase the supply are themselves built from, and powered by, resources such as iron and oil (Kummel et al. 2002; Ayres and Warr 2005). Thus our efforts to apply a natural-resource-using device to provide natural resources creates a feedback that eventually results in diminishing natural resource returns. The ongoing application of innovation and technology to resolve a resource supply issue can thus transform a resource supply constraint into a limit. This is most clearly illustrated by the application of a device to overcome an energy scarcity.

There is a limit to the amount of energy that an energy-producing device can use if it is to supply energy to us at an energy gain (Hall et al. 1986). Think EROI. If the device uses more energy to function than it provides for us, then it exceeds that limit and adds to rather than overcomes the scarcity.

The alternative is to substitute one energy source for another. But this too runs into limits. One of them is the economic constraint that the substitution must result in an economic gain, not a loss. Such a loss can occur with energy substitutions because, for example, the economic productivity (economic quality) of our various energy sources are not the same. On average, electricity has the highest economic quality, petroleum a lower and coal the lowest (Cleveland 2005). This means that it is usually economically easier to substitute oil for coal. In contrast, substituting coal for oil in some application is not economically viable.

There are also social factors that constrain our efforts to resolve a scarcity using innovation and technology. These include the willingness of society to devote resources to a scarcity, to be organized in the manner needed to implement a technological solution, and to accept that the distribution of the costs and benefits from implementing the solution is fair (Homer-Dixon 1995). These social limits apply independently from any biogeophysical constraints.

In the broad view, and contrary to economists' expectations, the application of innovation and technology to resolve a scarcity of natural resources through substitution and enhanced extraction is limited in a practical manner. These limits arise, in part, because the supplies of many resources are effectively non-renewable and thus finite. Renewable resources only remain so as long as they are exploited at a slower rate than their rate of renewal. Limits also arise because of both practical and economic constraints on the application of innovation and technology. These constraints can be circumvented to a degree and in the short term, but not overcome completely or indefinitely (Smulders 2005).

Despite this reality, economists still have faith that innovation and technology can routinely solve a scarcity of natural resources by increasing the efficiency of their use. There are two common forms of efficiency: economic efficiency, represented by GDP units per unit of natural resource; and material-use efficiency, represented by products produced per unit of natural resource. (They are usually reported as their inverse: economic intensity [material/GDP] and resource use intensity [material/product produced].) It is assumed that an increase in either type of efficiency (a decline in their intensity) will resolve a scarcity by reducing our absolute (total) demand for the natural resource. A process referred to as dematerialization (Ausubel 1996).

Computers are routinely put forward as the ultimate example of product dematerialization. They contain small amounts of resources, yet they substitute for large amounts of many material things, such as the paper in telephone books and regular mail. However, this claim of product dematerialization is placed in doubt as soon as the computer's complete, material-focused lifecycle is considered. For example, in 2002, a 2 g, 32 MB DRAM memory chip required about 1,700 g of natural resources (excluding water) over its lifecycle of being manufactured, used and discarded (Williams et al. 2002).

Similarly, in 2004, the energy needed to manufacture a computer was the equivalent of 260 kg of fossil fuels, 11 times the computer's weight (Williams 2004). That is large when we consider that the fossil fuel energy equivalent needed to produce many products is just 1 to 2 times their weight. Even the fuel equivalent to produce an energy-intensive aluminium beverage can is only 4 to 5 times the can's weight.

A life-cycle analysis of a particular product/service may indeed indicate that an increase in material-use efficiency has resulted in a progressive decline in the product's use of a particular resource (product dematerialization). However, whether those savings will contribute to an absolute decline in the demand for the resource depends on our decisions, which are influenced by our culture's functioning and world view.

Consider Western culture's support for economics' virtuous cycle (figure 14). The workings of the cycle provide no incentives to direct any resource savings toward reducing the total demand for resources. Indeed, it ensures that any savings are immediately used by innovation and technology to produce more goods and stimulate more consumption.

Consider too the history of copper use in the United States. The economic efficiency of copper use (GDP/unit copper) shows a steadily increasing trend with time (Ausubel 1996). During this time, great strides were also made in increasing copper's material-use efficiency (products/unit copper) for individual products. Nevertheless, contrary to the expectations of the believers in efficiency and the logic of the Kuznets curve, both the absolute and the per capita amount of copper used in the United States has been rising since the 1930s, as discussed in chapter 10 (Gordon et al. 2006). Any reduction in copper use as a result of increased economic efficiency or product material-use efficiency has been wiped out by an overall increase in copper demand: by consumer/economic growth.

Although increasing efficiency may reduce a resource demand and relieve a resource shortage in the short term, it can't, on its own, maintain that success over the long term. It can't reduce a scarcity, ensure an absolute decline in resource use or result in dematerialization. After all, increasing efficiency can't eliminate the following: the use of resources by technology, the limits set by the supply of finite resources, the constraints on substitution or the social demand for resources (think virtuous cycle). Neither can it eliminate the "law" of diminishing returns, even in its economic version. For an increase in efficiency to reduce total resource demand, ensure resource sufficiency, or maintain the environment providing them, it must be accompanied by the social will to dedicate the savings to those ends, such as permanently lowering the rate of resource extraction. Otherwise, increased efficiency can only delay the appearance of limits and scarcity.

Despite analyses to the contrary, economic arguments that innovation and technology, such as increasing efficiency, will solve a resource scarcity persist (Simpson et al. 2005). One reason is the difference in the way economists and scientists identify and predict resource shortages and scarcity. In general, economists base their thinking about the current and future state of natural resources, and the resolution of scarcity, on market variables, such as trends in price, production and demand. They focus on the market's functioning (Krautkraemer 2005). These economy-related variables are significantly limited in their ability to predict either the future availability of natural resources that currently lie outside the market (the future supplies of a commodity) or when a fundamental shortage (a scarcity) might occur.

Consider that the current (spot) price of a resource mostly reflects the short-term difference between the immediate demand for a resource and its "on the shelf" supply. Similarly, the price of a resource on the futures market reflects economic-variable-based predictions about its supply-demand balance in the market for, at most, a few years into the future. Finally, some economists claim that historical trends in the availability of resources show that innovation and technology can reduce shortages in the future. This may be valid, especially in the short-term, but it also says nothing about the fundamental constraints that might limit the validity of this method in the future when a scarcity arises, or when that future might occur.

In contrast, scientific (geology, agronomy, ecology, etc.) assessments of natural resource supplies pay limited attention to the market variables. Their evaluations are based on an understanding of both the earth's processes that provide the natural resources and the constraints on their functioning. As a result, in recent times it has usually been scientists rather than neoliberal economists who have first identified a fundamental, longer-term, earth-imposed natural resource scarcity, and the importance and difficulty of dealing with it. For example, earth scientists, not neoliberal economists, brought the idea of peak oil and its possible impacts on the economy to our attention (Cleveland and Kaufmann 1991; Hubbert 1973). The problem with the scientists' conclusions lies in their expression and their lack of attention to the power of cash.

In overview, the neoliberal economists' standard definition of the economy makes it an open system for natural resources. Thus the supply of resources can, when viewed from the market's perspective, appear to be infinite. Economists can believe that an economy is impervious to the consequences of consuming natural resources and degrading ecosystem services. They can believe that continual economic growth is a realistic, long-term objective.

However, slow natural resource renewal rates, the use of resources too rapidly and the disruption of the renewal processes all ensure that the supply of many resources are effectively non-renewable on a human-relevant time scale. Rather than being infinitely available, they are finite in a practical sense. A scarcity of a natural resource can arise because of this basic finiteness.

Economists assume that the application of innovation and technology will routinely resolve any natural resource scarcity (whatever its cause) by enhancing their extraction and facilitating the near-unlimited substitutions of one resource for another. But the ability of innovation and technology to increase resource supplies is limited by their own use of natural resources, especially energy. The law of diminishing returns applies. The application of innovation and technology is also constrained by economic considerations and by the need for social buy-in. Our ability to resolve the scarcity of a natural resource is thus constrained by practical limits: practical finiteness, diminishing returns, economic viability and social acceptance.

Applying innovation and technology does change how natural resources are used. It can increase a resource's material-use efficiency. Whether that increase will reduce our absolute demand for a particular resource depends on more than innovation and technology. It depends on whether we decide to use efficiency's savings to produce more goods or reduce our absolute demand for that resource. By reducing our absolute demand, we can delay the appearance of the resource's supply shortage or limit. However, unless our total demand for the resource per unit of time is less than the amount by which the resource is renewed in that time, an increase in efficiency cannot permanently overcome diminishing returns and practical finiteness.

As long as economists fail to comprehensively consider the earth's natural processes, their ability to realistically predict and resolve natural resource scarcity will remain significantly limited. Our environment's long-term or ongoing provision of natural resources is more appropriately measured, evaluated and appreciated within fields such as ecology and geology. Only discussions about their short-term supply belong exclusively in the field of economics.

<u>Conclusions.</u> Economists rely on basic principles, assumptions and simplifications (basics) when making and applying models to study and address real-world economic questions. By comparing the basics to real-world evidence, some of them are shown to be unrealistic or unsubstantiated. Most are subject to significant constraints, if their use and the resulting guidance is to be meaningful.

For example, economists treat the GDP as a reliable measure of social welfare, while personal wealth is believed to reliably measure social-psychological well-being. In reality, both assumptions are only weakly true. As a result, despite the economic guidance extolling the non-economic benefits of maintaining economic growth, it is limited in its ability to increase our feelings of either well-being or our satisfaction with our life circumstance. Other factors are just as important in determining social welfare.

Economists also routinely define the economic system as a closed system, but in doing so they treat the environment as an open system. They treat natural resources as infinite and the environment's functioning, which provides them, as an externality. Economists can thus assume that if a resource scarcity does arise, then innovation and technology can always overcome it. They can then conclude that there are no inherent limits to economic growth. These assumptions and beliefs may be realistic in the short term or during theoretical economic studies. However, over the long term, our ability to resolve a scarcity in the supply of natural resources is constrained by practical limits: finiteness, diminishing returns, irreducible economic costs, renewal rates and social acceptance. Over the long term, economic growth is limited by the environment's functioning in a practical manner: natural resources constrain an economy's functioning. Think of energy.

Likewise, economists assume that the functioning of a free market is self-regulating in the neoliberal sense. They also believe that the best solutions for economic and social issues will arise from a free market rather than a government. These assumption-beliefs may be realistic in modelling and in the short term, under some circumstances, but over the long term, they don't represent the functioning of the real economy. After all, the real economy is a human-designed system that is part of a complex human-environment system. Thus, to function fairly, the economy has to be subject to human controls (e.g., curbing greed, maintaining government's functions, making resources equitably available and protecting citizen's rights). The economy is also subject to natural constraints, such as our human characteristics (e.g., our human decision-making foibles and our culture's functioning), and by the environment's functioning (e.g., the practical finiteness of natural resources).

Criticisms pointing out these types of biases and flaws in economists' thinking, assumptions, models and conclusions, as well as in their application to the real world, are common (Rodrik 2015; Colander et al. 2009; van den Bergh et al. 2000; Hall et al. 2001). Neoliberal economists have made efforts to address some of these criticisms. For example, some economists have tried to broaden the definition of a model's key assumptions so that they are more realistic while still retaining mathematical simplicity. The outcomes of these efforts include ridiculous characteristics, such as no limits to human behaviour, and conundrums, such as accommodating the presence of technology but having to abandon the key assumption of perfect competition to do so (Smulders 2005).

This alone provides a judicious reminder that economic models and their assumptions are just tools to analyze an aspect of the complex economic system. They don't fully or accurately represent how the system actually works as a whole. If false certainty is to be avoided, then models and their results should be used with caution and skepticism.

For a model to provide meaningful conclusions, the model, the assumptions and the results must be evaluated and tested using empirical (measured) data and inductive reasoning within the context of the economic sub-system being investigated. This is commonly not the case (Rodrik 2015; Milonakis and Fine 2009). Support for this statement is provided by surveys of peer-reviewed papers published in academic publications directed at economists during the period of neoliberal economics' recent ascendancy.

From 1972 to 1991, between 40% and 60% of the papers (considered in five year blocks) published in two premier economics journals presented theories and models without any data. This is a much higher proportion than the 0% to 22% that occurred in equivalent political science, sociology and chemistry journals. When data was used in these economic papers, there was a strong tendency for it to be highly aggregated (Leontief 1982; Morgan 1995).

There is an attempt in some economic disciplines to include more representative data in their economic studies. However, until recently there was not much data that could be used to appropriately test mainstream theories, for example, the causes of growth (Coyle 2014). One reason is that the available data was itself usually aggregated, which significantly limits its use to represent the external reality.

Aggregating data is an issue because it is a type of simplification. As previously discussed, when judiciously used, simplifications (and assumptions) can reduce the complexity of the system being studied thus making it more amenable to mathematical manipulation, modelling and statistical bench testing. However, as with all simplifications, aggregation can also produce model results that are significantly at odds with the system's actual behaviour. For example, highly aggregated input data reduces the resolution of the model's output and increases the possibility of false order appearing. Unless taken into account, both of these features can cause economists to conclude that their model is more valid and useful than it is and thus that our economic system is simpler, more ordered and more self-controlling than it actually is.

In the end, economists have only partly and imperfectly addressed the criticisms about their neoliberal views of the economy and their related practices. Economists still fall too easily into the traps set by their reliance on abstract, theoretical, highly simplified mathematical models that have had little or no appropriate empirical testing. There are also the traps set by using those models on a selected part of the economy that is treated as separate from the dynamic, complex human-environment system in which it functions. Economists thus defend their view of their models' results and conclusions too strongly (Rodrik 2015).

This begs the question, why do economists continue to hold onto their neoliberal views of the economy so strongly? One reason is that their training embeds the basic principles, assumptions, simplifications and practices into their personal reference normal. Economists thus find it easier to believe but harder to question these basics. It is also more difficult for them to recognize the constraints on the use of the basics in models and reasoning. Finally, the basics and thus the economists' views of them are generally derived from mathematical models that include substantial simplification or are supported by models that rely of the basics.

The overall result is that their economic views (like the basics), and the model results supporting their views, have received little or no rigorous, appropriate data-based testing and investigation using inductive reasoning within a wider context. In essence, economists have set themselves up, with the greater culture's support, to sidestep the reality paradox and related cognitive dissonance experiences that are part of learning and changing. An economist can thus find it relatively easy to develop faulty views and hold onto them dogmatically. They can thus remain deeply committed to their views.

Overall, the combination of an economics-biased normal along with the complex, dynamic nature of the economic system and the difficulty of appropriately checking a model's validity and usefulness ensure that it is easy for economists to have unrealistic faith in flaky models and faulty views. For example, a model can be built on false but acceptable assumptions, or it can be simplified to the point that the order and relationships in the model are self-generated. A model can be used out of context, beyond its applicability and in a dogmatic manner. Even though the results can be valid within a model's assumptions, they may not be valid in the real world. These possibilities make the difficult job of drawing useful conclusions from a theoretical model and then translating them into the real world even more difficult. These features ensure that the faith an economist has in an economic model, its results and their views should be accompanied by many if-then conditions. As expected, these constraints are often not explicitly stated.

There is more to this story. Neoliberal economists don't practise economics in isolation. They are members of a culture. This is the context in which neoliberal economics' commitment to its basics and models is most fully appreciated.

## Neoliberal Culture

During training, a trainee economist is exposed to more than mainstream economics' standard practices (models, theories, and the basic assumptions, simplifications and rules of thumb). They are exposed to a favoured set of beliefs about the world. A trainee thus absorbs into their personal world view both an economics-biased reference normal for practising economics and the essence of economic culture's world view. As discussed in earlier chapters, under this circumstance these practices, along with the culture's beliefs and values, tend to be treated as difficult-to-question, self-evident and even universal truths.

By way of an analogy, if you grew up in the West, your reference normal and your personal world view are both influenced by your culture's Christian background. Your decision making is similarly influenced. This is the case regardless of whether you consciously subscribe to all, some or no aspects of the Christian doctrine.

It is within the context of their culture that neoliberal economists' commitment to their economic basics, models and beliefs is more fully appreciated. This is also the context within which it is easiest to understand how the neoliberal economics brand came to be accepted by Western culture and came to have such a strong influence on Western culture's world view. This section discusses neoliberal culture with those goals in mind. In doing so, it must include more than economists because, as the global financial crisis of 2008 revealed, it includes more than economists among its members. Thus to discuss its nature and role means stepping beyond its economics roots to include those other members and their roles in its formation and functioning.

### Neoliberalism.
This subsection discusses the nature and rise of neoliberal culture. It, like all cultures, has a world view that encapsulates the members' conception of how the world works, their place in it and why they should hold these views. Neoliberalism also provides its members with a vision of an ideal future.

Neoliberalism's distinctive economic features include a strong belief in free markets, free trade (globalization), economic growth, market efficiency and theoretical mathematical models. It also strongly favours market-driven solutions to economic, social and environmental issues, rather than government intervention. Thus effective solutions are expected to favour deregulation, privatization, austerity, low taxes, motivation through financial incentives, a focus on economic growth and limited control of the economy (which would only be exercised through monetary policy). Think quantitative easing. It also favours innovation, technology and efficiency because they enhance productivity and thus growth.

The bulk of these cultural features and their supporting beliefs, myths, ideology, practices and traditions are handed down by its elders. They include aspects that were inherited from an older mainstream Western economic world view, such as the belief that competition in a free market makes it efficient (which is good) and that economic growth is economic progress (which is also good). The neoliberal world view also includes foundational aspects from Western culture's world view, such as the notion of progress and the emphasis on individualism, as illustrated in chapter 3.

By default it also favours a dominion view of the environment.

Once an economist absorbs the neoliberal world view into their personal world view, its guides their beliefs and thinking about a wide range of economic and non-economic topics. They range from what is an acceptable economic assumption, to what does the ideal future should look like, to how the greater culture should make its economic decisions. It is this neoliberal cultural world view (rather than just its economic preferences) that rounds out an understanding of economists' biases and faulty conclusions, and most importantly, how they come to hold on to them so dogmatically.

Neoliberalism is the dominant economics culture. The widespread presence of neoliberalism among economists is illustrated by the significant support (78% to 93%) that polled economists gave to a collection of 14 economic propositions. For example, 93% of those polled supported the proposition, "A ceiling on rents reduces the quantity and quality of housing available," and 85% supported the proposition, "If the federal [United States] budget is to be balanced, it should be done over the business cycle rather than yearly." The economist who compiled these results suggests that the level of agreement is reason enough for the public to adopt these propositions (Mankiw 2009).

However, of the 14 propositions, only one could be thought of as supporting government intervention in the market. Thus there was a pro-neoliberal bias in both the poll and the call for support. In addition, four of the propositions were later shown to be only conditionally true (including the first example given above). Because the conditions under which each of the remaining proposals would be true were not provided, it was also considered questionable to unreservedly agree to any of them (Rodrik 2015). Considering that the results from economic models are context dependent, as discussed above, this is a reasonable conclusion.

The neoliberal culture's world view has spread beyond its roots in academic circles. Its membership is large enough that neoliberalism influences Western culture. However, Western culture also influences neoliberalism. The interactions between these two cultures creates links and feedback loops that bind them together. It is by discussing this interaction that the nature of neoliberalism's guidance to Western culture, the degree of its influence on Westerners and the limitations of that guidance can be most easily and fully appreciated.

In the wake of the 1974 oil crisis and amidst growing concerns about rising taxes and increasing government debt, Keynesian economics floundered, while neoliberal economics emerged from the background of Western economists' culture. Neoliberal-leaning economists were ready to solve the problems of the day with economic theories that they claimed better described the true nature of humans' economic interactions and thus the natural workings of the market. Among the academic economists providing the supporting mathematically derived economic theories were Milton Friedman, Eugene Fama and Robert Lucas of the Chicago school. Academic economists also contributed to neoliberal social-cultural theories. Prominent among them was James Buchanan, also a graduate of the Chicago school. His economic libertarian (far right, market knows best), usually data-free, social-cultural theories strongly influenced the social aspects of economic thinking. Both Friedman and Buchanan were awarded the "Nobel" prize in economics, giving them credibility.

In forming and promoting their theories, neoliberal economists interacted with one another and with supporters from outside of academia, mostly from the business community. These interactions resulted in the neoliberal world view: a libertarian-leaning, if not libertarian, ideology supported by mathematical and social theories based on little empirical evidence. The academic centre of neoliberalism was the Chicago and Austrian schools, while arguably the ideological centre is the Mont Pelerin Society (MacLean 2017). Among its economist founders were Friedrich von Hayek (1899–1992) and Milton Friedman (1912–2006).

Like all world views, neoliberalism included a historical justification for its views. In essence, the members created a founding myth for their neoliberal culture. They did so by adopting, and then adapting, aspects of the classical economic period, in particular the writings of Adam Smith. (Neoliberalism's historical roots are actually more closely tied to Ricardo's deductive abstractions than to Smith's more pragmatic thinking).

The neoliberal economists' successful promotion of their theories and explanations allowed them, in the late 1970s and early 1980s, to start reframing Western culture's understanding of the economy and economic progress. They began to change how Western economies functioned. These changes and their impact are illustrated in the United States by the appearance of the corporate takeover movement, junk bonds, shareholder value and deregulation.

Wall Street institutions contributed to and embraced neoliberalism's theories, myths and beliefs. They then took advantage of the changes to Western culture that they had helped to bring about, such as deregulation of the market. Wall Street's resulting spectacular financial success helped it to became the epitome of at least neoliberal economics, if not its world view, in action. It came to symbolize neoliberalism. Wall Street became the neoliberals' "spokes-street," which is how "Wall Street" will be used here.

Wall Street institutions embrace neoliberal beliefs to such a degree that they pervade the institutions' day-to-day operations: their single-minded devotion to profit, a hyper-rapid response to opportunity and the use of no strategy as a strategy. Naturally, this neoliberal focus is reflected in a Wall Street investment banker's lifestyle. Their lives include hard work (up to 100 h/week); routine corporate restructuring (downsizing and rehirings); a focus on the present (very short term); and high compensation (achieved by the number of deals signed and by changing jobs). Bankers understand these working conditions to mean that they have no job security (and thus they have a limited sense of loyalty to an institution); and that they are part of a steep pyramidal hierarchy (which provides them with both a sense of place and a sense of status, as measured by money). A banker knows that to succeed they must be aggressive (ambitious and competitive) (Smith 2012; Ho 2009).

The Wall Street bankers are up to the task. They are recruited mostly from the elite US universities, especially Princeton and Harvard. The novice bankers come to the job with prestige and the recognition of being smart. The combination of their pedigree and lifestyle allows the bankers to see themselves as occupying the high moral ground at the top of the global status heap. They imagine themselves to be participating in the financial market as the elite who are shaping it, rather than as the workers within it.

Generally, an individual banker's perspective of the world reflects Wall Street culture's neoliberal world view. This includes a belief in independence, property ownership rights, self-interest, profit, efficiency and a self-regulating market. In this world view, the greater good is served by unreservedly supporting a free (unregulated) market and those people focused on generating profit from it. Wall Street bankers can see Wall Street culture as superior to the wider culture (Ho 2009). This world view, and their faith in it, is not exclusive to Wall Street bankers. It can be found in the British equivalent of Wall Street, the City of London (Luyendijk 2015).

However, Wall Street's practices ensure that the banks forming the Street are always living on the edge of collapse. They manage to survive because Wall Street's sense of elitism allows its member institutions to project themselves as powerful, connected, rich, global and oozing opportunity: the so-called Masters of the Universe. This appearance helps to generate a faith in Wall Street's neoliberal beliefs and practices well beyond the street.

The wider faith in Wall Street is also sustained by ongoing support from prominent academic economists in America's elite universities and business schools, such as those from the Chicago school. This ongoing support takes the form of a symbiotic relationship between them. The academics in these elite institutions tend to conduct their research and teach with a neoliberal bent. This is reflected in the neoliberal-leaning economic theories, models and values they publish and their students absorb. As Wall Street generally draws its recruits from institutions teaching these views, it can use the recruits to implement the abstract theories and thinking generated by academia's elite. Through this feedback, the academics' theories gain credibility and Wall Street's practices are legitimized. Their mutual dependence furthers each other's goals, practices and dreams (Ho 2009). In the process, faith in Wall Street and neoliberal culture's world view is given credibility and acceptability so it becomes more widely accepted. This symbiosis is further supported by the wider range of lower tier economics departments and business schools that also adopt neoliberal economics curricula.

The wider American culture also came to support Wall Street's beliefs and practices and to adopt the essence of neoliberalism. This happened, in part, because of the connections Wall Street and the neoliberal academics have to the political and business elite. Think of Margaret Thatcher and Ronald Regan. As mentioned above, these connections helped to make significant changes to the American culture's view of its economy and how it functioned.

For example, the United States pre-1970s stakeholder view of corporate responsibility meant that a wide range of cultural groups (e.g., local communities, workers and shareholders) were included in decision making. It was replaced by neoliberalism's shareholder-value view of responsibility, in which corporations are only responsible to their shareholders. During the switch, stakeholders were displaced by the shareholders who, in neoliberal terms, "reclaimed their rightful place as the sole heirs to corporate wealth."

These changes helped the myths, assumptions, values and practices of neoliberalism to seep into the American public's economic beliefs and thinking, where they displaced the Keynesian influence. The American public came to equate speculation to satisfy greed with economics' idea of a self-interested economic behaviour. Greed came to be seen an essential part of maintaining progress (Kindleberger and Aliber 2005). They came to think of Wall Street bankers as the foundation of the corporate system and the prime generators of prosperity and personal wealth.

The buy-in by the public is important because, despite views to the contrary, they have considerable influence over the market's functioning and its regulation, and their greater American culture. The public have this power because they are both the carriers of Western culture, which directs governments to regulate the market (or not), and the economy's primary driver. Consider that the GDP in the United States is strongly dependent on the personal spending habits of its citizens: the members of the public. Since 1929, consumer spending in the United States has contributed between 60% and 80% of its GDP. The percentage rose steadily from around 58% in 1950 to around 68% in 2013 (Bureau of Economic Analysis 2014; Bureau of Economic Analysis 2009).

This explanation of the relationships that resulted in neoliberal thinking coming to dominate American culture's economy-related thinking and decision making leaves a question unanswered: Beyond ideology and profit, why would the public support neoliberal economic views? After all, displacing public interest with shareholder interest could be seen as against the public's interest.

Here is one pragmatic reason why they embraced many aspects of the neoliberal world view. Despite its stance against government's holding debt, neoliberalism belief in the market ironically favoured the society embracing the power of money, including leverage and debt financing. These features made it easier for the public to immediately purchase the things they wanted, even consumables. Think of credit cards. They could thus receive the benefit of living the good life now, while delay paying for it until later. They could also join the "big boys" in using the power of money and the market to become richer, quicker. These perceived benefits and the experience of carrying and repaying a debt made it easier for the public to support neoliberal views. This includes the almost fanatical effort to stimulate growth (i.e., to increase GDP) through deregulation and privatization. After all, growth makes supporting a debt easier. Once a member of the public was in debt and vested in the stock market, there was little reason for them to abandon neoliberalism's views about growth and stockholder value.

As the comments about the City of London mirroring Wall Street hinted, these changes were not restricted to the United States. They spread throughout the Western world and beyond. Neoliberalism came to dominate economic thinking in those countries subject to Western influence.

In overview, neoliberalism's belief in a free market, free trade (globalization) and market efficiency have come to dominate the thinking of Western economists as well as the political and business elite. This view strongly favours economic growth and market-oriented, rather than government-control-oriented, solutions to economic, social, and environmental issues. These economic solutions include deregulation, privatization, austerity, low taxes, motivation through financial incentives, a focus on economic growth and limited control by government, mainly through monetary policy. The essence of neoliberalism has also seeped into and slowly become embedded into the public's personal world view and thus into Western culture's world view. This has made it difficult for the members to question neoliberal views. Neoliberal guidance thus gained considerable influence over Western decisions about economic, social and environmental matters.

**Neoliberalism's Limitations.** Western culture has faith in neoliberalism's guidance, as illustrated by the considerable power they gave, and still give, to the neoliberal captains of the economy. The most recent result was an economic boom followed by the self-made global financial crisis of 2008 and its ongoing fallout. This history reveals the limitations of neoliberalism's explanations for how the economy works and its guidance to Westerners about how both the economy should be managed and we should live.

Start with its guidance about the functioning of the Western economy and its institutions standards of practice. The academics' theories, with the help of Wall Street, facilitated the deregulation of banking. One consequence was increased competition among the investment banks and an increase in shareholder demand for greater profits. This created a need, which the banks' bosses satisfied by engaging in dubious post-deregulation practices, such as creating and selling novel financial instruments that had only wispy connections to the external reality.

The creation and selling of these instruments was facilitated by a subgroup of academically trained quantitative economic modellers called "the quants." They built the models and programs that could both create unintelligibly complex, derived financial instruments, such as credit default obligations and credit default swaps, and attach an acceptable price to them. This expanded the pool of investments that Wall Street banks could offer to its clients. However, the banks, with the help of the models, unrealistically assessed the risks of buying them.

The quants also built models and computer programs that increased short-term profits by increasing the speed of trading and justifying greater leverage. The quants certainly helped to generate greater profits, but they also helped to make the market more complex and unstable. Despite, or perhaps because of, the quants' apparent brilliance, they were unrealistically certain that they could build and apply the needed models to the dynamic, complex real market. They failed to recognize and accept that their work only considered a small part of the complex economic system, and that their models were only valid within their assumptions and in isolation.

They, and neoliberal economists in general, failed to recognize that by applying these models and programs to the real market, they significantly changed the economic system. In particular, by speeding up trading and generating opaque, riskier derivatives, the quants were a key part of the self-sustaining profit-focused feedback loop that affected all aspects of the market's characteristics. This included reducing the validity of the quants models. Their ill- and unconsidered decisions contributed to the global financial crisis of 2008 (Patterson 2010; Ho 2009).

Contributions to the 2008 crash were also made by people on the fringes or, at least nominally, outside of Wall Street. Consider those members of the political, regulatory and business elite who believed too strongly in neoliberal views and theories. They placed too much faith in neoliberalism's assumptions, such as a self-regulating market, and its practices, such as those of the Wall Street bankers. Their faith allowed the business elite to support the bankers' efforts to secure policy changes in favour of neoliberal ideals, such as economic deregulation and weak enforcement. Those doing the lobbying could find a friendly ear with the politicians and regulators. Think of Allan Greenspan. The neoliberal leanings of the regulators stopped them from making a realistic assessment of the consequences of making the proposed changes, such as the power of unconstrained financial incentives to turn benign self-interest into selfishness and blind greed. Enough politicians had neoliberal leanings to pass the needed laws.

As a result, those who should have prevented both the real estate and the banking bubbles, instead enabled, even facilitated, them forming and becoming the 2008 crash. The 2008 grain futures bubble illustrates this. In the case of grain, Wall Street successfully pushed for changes to the US regulations that governed futures trading. Incredibly, these changes allowed Wall Street to expand their trading in wheat futures but never fill the contracts. Because the regulators and traders treated wheat futures contracts as if they had no connection to wheat in the ground, they allowed, perhaps encouraged, a grain bubble to form, as evidenced by soaring wheat prices. The bubble eventually burst (Kaufman 2010).

With this background, there should be no surprise that, during neoliberalism's post-1970s ascendancy, there was an increase in the frequency and severity of the economic booms and busts that have routinely occurred throughout recent economic history. Nor should it be a shock that the vast majority of economics trained analysts (both in and outside academia) failed to provide timely warnings about the potential for at least the 2008 bust. As illustrated above, their faith in neoliberalism resulted in their blindness to the cues from the external reality, their acceptance of faulty results from models and their inability to recognize their personal biases or misinterpretations.

Neoliberal economists try to avoid their responsibility for contributing to and failing to provide warnings of the 2008 crash with "explanations." For example, they claim that the relevant causal factors of the crash were political in nature and thus fell outside of the market they paid attention to. As a result, they felt, incredibly, that those factors could be considered legitimate externalities (Colander et al. 2009; Kindleberger and Aliber 2005). This begs the question, why then do we trust and under-regulate economists and Wall Street bankers? Why do we give them so much power and influence over our economy?

Perhaps it is more useful to ask, why has neoliberalism remained popular and Wall Street remained both powerful and largely unchanged *after* the 2008 crash? A window into why is provided by a global survey of financial services executives. The 2013 report revealed critical aspects of Wall Street's (as spokes-street) current world view and practices. The executives recognized that their practices should be more ethical. Yet 53% of them still said that the progression of their career would be inhibited if they were unable to be flexible on ethical standards. Similarly, 53% felt that their firm would be less competitive if its compliance with ethical standards was too rigid. Those reporting the survey's results concluded that changing financial services culture would take years to decades (Kapoor 2013).

Consider too that the lifestyle, goal and behaviour of those in the industry is, in essence, the same today as it was before the 2008 crash (Crowe 2015; Luyendijk 2015). Wall Street (in the spokes-street sense), through its members, displays considerable inertia to changing its practices and world view. Neoliberalism remains popular and influential among financial services workers.

But this is insufficient on its own to explain why these practices continue and why those practising them remain so powerful, at least in the West. A hint as to why is provided by the EU court case accusing Trafigura of direct damage to human health, as discussed in more detail below. The EU and its people cared about this damage enough to do something about it. This is much less the case with neoliberalism in general and Wall Street's practices in particular. One reason being that the essence of neoliberalism is now an integral part of Western culture's current world view and functioning: it has been largely accepted and absorbed by the public. Thus, for the status of neoliberalism and Wall Street practices to change, Westerners would have to change what we see in the mirror. In addition, we would have to find a different way to structure and run an economy. Both are hard work.

Westerners (public and elite) may be unhappy, even upset, about the 2008 crash and its aftermath, but not sufficiently to face our reality paradox experience in a way that will seriously deal with neoliberalism's reality and influence. We are still prepared to find alternative explanations and events or people to blame for the current circumstance. Think cognitive dissonance. The public may even feel powerless and thus cynical as a result of the lack of leadership from those with enough power to deal with the causes of the crash. They may feel that the powerful are just making cosmetic adjustments to the periphery. Either way, there is considerable personal and cultural inertia in at least the West to effectively deal with the influence and power that neoliberalism and Wall Street's practices have over our lives.

It is as if Westerners are content to take consolation in the fact that the use of flaky thinking and theoretical models to guide practical economic and cultural decisions is the norm. After all, it seems to have been central to Babylonian economics in the 2000s BC. For example, the foreman of a labour group grinding grain into flour was expected to ensure that his group met its per-worker flour production targets, even though they appear to have been set at an unrealistic, theoretical maximum. If the group's actual productivity did not meet expectations, then the foreman was held liable for the deficit. He could lose his possessions and be assigned to a labour group. Babylonian records also suggest that many labourers fled these labour groups (van de Mieroop 2004; Nissen et al. 1993). (See also the example of predicting milk production discussed in chapter 5.)

In Babylonian times, the foreman and labourers ended up paying for faulty economic thinking. Today the cost of having faith in neoliberal economic beliefs and models is paid for by the public, now in the form of personal debt, bank bailouts and increased national debt. Think too of quantitative easing. The environment also pays, which will be discussed next. It is not paid for by the neoliberal economists and the Wall Street bankers. It seems they are too big to be punished.

In summary, the captains of the economy, such as academic economists, Wall Street bankers and top regulators absorbed the essence of the neoliberal culture's world view into their unconscious during their training. This means that the practical aspects of neoliberal economics become more than a preferred choice for the captains. They become part of their neoliberal-culture-supported personal world view. Thus they can more easily fall into neoliberal belief-related traps and hold their conclusions too dogmatically.

Westerners have largely accepted the essence of neoliberalism into both our personal and our culture's world view. Thus neoliberalism's guidance exerts considerable influence on Westerners' decision making. The limitations of neoliberalism's guidance and the consequences for Westerners believing in it too strongly were revealed by the global financial crisis of 2008. The difficulty of reducing neoliberalism's considerable influence is illustrated by the difficulty of reforming the financial service sector after the financial crisis.

The cost of having faith in misinterpreted neoliberal economic models and unrealistic neoliberal economic and social beliefs is paid for by the public in the form of personal debt, a widening gap between rich and poor, business bailouts and increased national debt. It is also paid for by the externalities that are sacrificed in the quest for growth. One of these externalities is the environment and its functioning, which supplies the economy with natural resources and sustains us.

## The Environment's View of Neoliberal Economics

By Westerners following neoliberalism's guidance, we affect more than people. We affect our supporting environment. The nature of our impacts on our environment depend on how accurately the neoliberal world view and its guidance represents environmental realities, such as its functioning and importance to us. This section addresses the limitations to neoliberalism's guidance from the environment's perspective.

**Valuing the Environment.** A natural resource in the economic sense is a feature of the environment that has, or has the potential to, play a role in the economy. For a natural resource to play a part, it must be assigned an economically meaningful (cash) value or price: it must be traded in the market. It must become a commodity. For example, a tree or a parrot only receives a cash value and becomes a commodity after it is traded in the market: that is, after a tree becomes lumber and a parrot becomes an exotic animal, alive or stuffed (d'Arge 1994).

The process of a resource becoming a commodity requires that some amount of it must be physically removed, and philosophically disconnected, from its host environment (McEvoy 1986). As a result, the cash price assigned to a resource as a commodity suffers from a significant environmental shortcoming. The price excludes the value of the direct impacts on the host from the act of removing the resource. It also excludes the value of the removed resource to the remainder of its kind, its host and its other users. For example, the cash value of trees removed from a forest for lumber does not include compensation for removing future soil constituents, destroying habitat and food for other forest dwellers, or disrupting local water and climate management services. These non-commodity-related costs are excluded from the cash value because economics' guidance considers them to be either free, valueless or too complex to convert into cash value: they are treated by the market as externalities (Hall et al.

1986). However, to the environment they are critical aspects of its functioning.

This point is more easily personalized by considering the costs a taxidermist includes in their price of a stuffed animal commodity, such as an arctic fox. There are the costs of the materials used, the costs of acquiring the fox and the costs of the taxidermist's skill to stuff and mount it. The type of skills the price includes is illustrated by a taxidermy brochure which notes, "Our goal is to create mounts that appear lifelike and effectively capture the character and mood of the animal. Using our knowledge of arctic mammals and skills at creating our own manikins, each animal is customized so that its natural beauty is highlighted" (Robertson's Taxidermy 2004). It is difficult to imagine how the price of a stuffed arctic fox would adequately include the costs to the remaining foxes for their loss of a prospective mate or the fox's host ecosystem for the loss of the function the fox performs.

The economic thinking that occurs when an environmental resource becomes a commodity directs us to continually maximize the profits to be made from it. One method is to estimate a natural resource's future value as a commodity. In the example of a tree, this can be calculated through a process called discounting. You start by assuming that you will extract (cut down) the tree today. You can then estimate the price you are likely to obtain for selling it. Assume too, that you will put the cash profit from the sale into the bank, where it will earn interest at a rate you can also estimate.

The next step is to decide how many years into the future it might be before you actually cut the tree down. That number of years allows you to use the estimated interest rate to calculate the total amount of interest you could reasonably expect to receive for the money you could put into the bank (if you cut the tree down today). In step three, that total interest payment becomes the additional value you could want to receive from the tree when you cut it down in the future (i.e., its future value). Or, more worryingly, how much you could feel you would lose if you didn't cut and sell the tree today (Begon et al. 1996).

Discounting is a useful economic tool. It can be used to evaluate the impact on future profit if you may not have access to a commodity in the future, or if you fail to sell it in the future. Discounting is the logic behind discount sales used to sell slow moving items, especially after Christmas.

However, when it is used to maximize profit from natural resources being valued as commodities, there can be severe environmental impacts. For example, it can lead to the assessment that the best economic strategy to maximize profit from slow reproducing (higher risk) natural commodities, including whales and trees, is to harvest them all, fast, and then live off the interest. After all, harvesting today reduces the risk of losing access to the resource in the future, and the interest from banking today's sale generates income faster than the rate at which whales and trees produce more saleable commodities (Pimm 1977). By using discounting to value natural resources, an economist expresses how they value their environment as well as their misunderstanding of, or indifference to, how ecosystems function.

An environmental resource is often a physical object, but it can also be a service (an amenity), such as water purification. Some attempts have been made to assign a realistic cash value to the services provided by ecosystems, or in other words, to commodify ecosystem services. This is done by assigning to a service a value equal to the cash cost that humans would have to pay if they were obliged to provide the service themselves. For example, the value of the ecosystem services that purify water is equal to the cost humans would have to pay to build and run a water treatment plant.

A study that used this replacement method to estimate the economic value of 17 ecosystem services, such as water purification, provided in 16 biomes reported that their value was at least two times the world's GNP. By this assessment, our current spending to ensure that our ecosystems continue to provide these services bears little resemblance to the importance of these services to us (Costanza et al. 1997). Looked at from a different perspective, the market price we pay for the commodities that these services provide, such as clean water, does not cover all of the costs of providing them. Their market price, like that of useful energy discussed above, is too low because the processes providing these services and even the services themselves are treated as externalities. From this perspective too, the market is incomplete.

The replacement method can also be used to estimate the costs of the ecosystem degradation and pollution associated with production. For example, a study took into account as many as possible of the ancillary or indirect costs associated with pesticide use in agriculture. These costs include human illnesses, such as cancer, and the disruption of services provided by natural ecosystems, such a pest control. By this method, the estimated total cost of pesticide use was at least three times the amount of money spent to purchase the pesticides themselves. The cash cost of the pesticide did not reflect its total human and environmental costs (Pimentel et al. 1992).

However, there are limits to the use of the replacement method to assess the cash value of ecosystems and the services they provide. Consider that the complex nature of an ecosystem means that not all of the attributes that keep the ecosystem stable, functioning and providing those services will be identified, let alone assigned a meaningful cash value. Neither will all of the production-related impacts on those attributes be evaluated and assigned a cash cost. For example, it is difficult, if not impossible, to place an economic value on species diversity (Tilman and Polasky 2005). As a result, estimates of an ecosystem's value using economic replacement and commodification are at best a minimum (IPCC 2014a; IPCC 2007b). A more meaningful assessment would focus on the degree of disruption and the list of responsibilities we acquire.

In overview, cash is the standard used by economists to measure value. But cash is not a universal standard like volume. It is akin to human beauty: only valid within a human context. As noted above, the economic system assigns a cash value to a natural resource by physically removing and philosophically disconnecting it from its host environment and then trading it in the market as a commodity.

The cash value of a natural resource commodity does not include the costs of extracting it that are borne by the remaining resource, its host environment's functioning or the other users of that resource and environment. These costs are considered to lie outside the market and thus have no economic value. They are treated as externalities, regardless of their importance to the environment's functioning and to us. With effort, such as applying the replacement method, some externalities can be assigned a cash cost. Overall, when seen from our environment's perspective, economics poorly represents the full value of natural resources or the functioning of the environment that provides them. The market is significantly incomplete.

This is clearly illustrated in overview by imagining that we apply economic principles to evaluating the functioning of an ecosystem. We would create an ecosystem equivalent of the measure at the heart of applied economics: the GDP. To do so we would add up the weight or energy of all of the food transactions that occurred between all of the species in the ecosystem.

We could call it the ecosystem energy product (EEP). Then, if the number increased each year, we would say that the ecosystem was functioning well. If it fell, we would say that we need to somehow stimulate the ecosystem participants to create more and eat more food. What could be further from the reality of how an ecosystem functions?

## Protecting the Environment. 
Neoliberalism's guidance strongly affects not just the value we place on our supporting environment but how we manage it, protect it and respond to our impacts on it. Those impacts include our economy-related activities of extracting natural resources, producing goods, using them and discarding waste.

From an economic perspective, the variables needed to manage a natural resource commodity are collected during the process of its extraction and trading. In the case of an ocean fishery, these variables are the current yield of fish (in tonnes) and the price received for them. Although managing the fishery using these variables can optimize the short-term fiscal outcome from fishing, it can't maintain the long-term supply of fish. Why? Because economics only deals with commodified fish. Thus their price only represents the present cash balance between the current supply of fish on land and the demand for them in the market. It does not represent fish in the ocean or the state of their host ecosystem. The costs of supplying the fish only represent the cost of landing them, not the cost to the fish species or its host ecosystem from the fishing. Similarly, within the context of an economic system, the size of a fish catch will only reveal if a fishery is being overexploited well after it enters that state.

Despite these limitations, the economists' cash-focused perspective of natural resource management has a strong, ongoing influence on how Westerners manage the provision of natural resource commodities. The historical outcome was vividly illustrated in chapter 10 by the Newfoundland and Californian fisheries (McEvoy 1986). They collapsed.

There have been efforts to deal with this disconnect. Today, most Western countries require an environmental assessment to be completed before an activity that will disturb the environment, such as the proposed extraction of a commodity from its host ecosystem, begins. They may also require the ongoing monitoring of some environmental variables at the extraction site until the activities are completed. The stated objective of this assessment and monitoring is to protect the environment.

Neoliberalism's guidance provides little, if any, support for this assessment and monitoring, or encouragement to conduct either in an environmentally or scientifically meaningful manner (Folke et al. 1994). In one sense, this position is surprising because these expenditures increase the GDP (i.e., contribute to economic growth), which, they argue, is key to our economic prosperity and to our social welfare. In another sense, this is unsurprising. After all, neoliberalism's most powerful imperatives are directed at maximizing cash returns (profit). From this perspective, the activities of assessing, monitoring, protecting or restoring our environment's functioning are a cash cost: an economic burden, an obstacle or irrelevant. These activities are thus to be rendered impotent or their costs (short and long term) actively reduced.

There is thus an inconsistency in the economic arguments about reducing our environmental impacts. When economists and business people argue that, over the long term, both the number and severity of our environmental impacts will be reduced by economic growth, they support it using the GDP. However, when arguing in the near term or the specifics, they focus on profit and just its growth. They reason that profit will, for example, support innovation and technology, which will increase our efficiency of resource use and thus decrease our consumption of resources and the generation of waste. The result will be a reduction in our environmental impacts.

The generic reasons why both the short- and long-term claims are, on their own, misleading or false are the same as why an increase in efficiency can't alone solve a resource shortage. They include choosing measures and system boundaries that are unrepresentative, and using relative measures tied to economics when absolute measures tied to the external reality are needed. Specific examples are using pollution measures tied to GDP that are of little or no relevance to the environment's functioning, and neglecting the virtual environmental impacts that are inherent in imported products, discussed below.

These disconnects between economic arguments and our environmental impacts are illustrated by the following examples. An economic assessment of $CO_2$ emissions from fossil fuel use plus cement production was conducted for the period between 1990 and 2015. It showed that the global $CO_2$ emissions per GDP unit (emissions intensity) had declined, which means that there was an increase in the global economic efficiency of fossil fuel use. This is believed to indicate that there

was a decline in the environmental impacts from our $CO_2$ emissions. In contrast, the absolute amount of $CO_2$ emissions rose over this period, indicating that a real-world increase in our environmental impacts had occurred (Jackson et al. 2015). Consider too that an increase in the efficiency of copper mining or copper use will only reduce mining's absolute environmental impacts if the savings are directed to that purpose, as discussed in chapter 10. An increase in economic efficiency or GDP does not inherently (i.e., on its own) result in an absolute reduction in our environmental impacts.

Economists and business people like to claim that "economic growth" (increasing profit) will reduce our environmental impacts because it will make more money available to improve the skills and technology needed to conduct environmental mitigation and restoration work. Profits will also provide the funds needed to complete this work. Thus, so the story goes, a rise in profits will result in us being increasingly successful at both mitigating the disruption of our environment's functioning caused by our economy-related activities and restoring that functioning when it has been destroyed.

This argument is unsound for a number of reasons. The activities that will generate the funds needed to pay for the enhanced mitigation-restoration skills and technology are the same activities that contribute to the problem in the first place. (Think of selling cigarettes to pay for the health care of smokers.) The best we can achieve is a relative, not an absolute reduction in impacts. Consider too that there is a conflict of interest when generating and distributing profits. The powerful neoliberal cultural imperative to treat mitigation-restoration as an undesirable cost also treats the profits as rightfully belonging to the shareholders. An increase in profits will preferentially go to them, not environmental mitigation and restoration work, so the relative reduction in our impacts will tend to be small.

In reality, avoidance is a far more effective method of achieving a reduction in our environmental impacts and their severity than either economic growth (however defined), increased efficiency or after-the-fact mitigation-restoration efforts. It is the only way to achieve an absolute reduction. Unfortunately, avoidance is only effective if those engaged in or regulating economic activities choose to focus their efforts on that outcome. Neoliberalism, in particular, does not support those efforts.

An example from the 1980s illustrates why environmental protection efforts will only be meaningful if they are given an equal or higher standing in decision making than economic concerns. A number of companies were importing used automobile tires from Japan into the United States to make automobile parts (increased efficiency of resource use). The tires were not checked to ensure that they were free of exotic organisms, either before they were shipped or after they had arrived in the United States. This was unfortunate because travelling in the water trapped within the tires were exotic Asian tiger mosquitoes. By 1985, the mosquitoes had established breeding colonies in Houston, Texas. From there, they quickly became invasive.

The mosquitoes' spread was helped, in part, by the storage and disposal of approximately 3 billion tires in low-cost tire dumps scattered across the United States: they provided ideal breeding sites. Their spread was also helped by the many other suitable breeding places created by the debris of economic activity, such as discarded tin cans, plastic bags, candy wrappers, ornamental containers and bird baths. By 1992, the mosquitoes were found in 23 states in the continental United States.

The consequences of this invader's spread could have ended with the disruption of the local ecosystems. Unfortunately, Asian tiger mosquitoes are an excellent vector for a whole host of diseases. These include dengue fever, yellow fever and many types of encephalitis (including West Nile virus and eastern equine encephalitis). These mosquitoes are also the vector for a parasite that affects our dog companions: dog heartworm.

There is more. Because Asian tiger mosquitoes can use blood from humans, birds, reptiles and other mammals, there is also the potential for diseases to be transferred between various species. This excellent disease vector is now awaiting the arrival of these and other exotic diseases into the United States (Craig 1993). One that could soon be added to the list is the Zika virus (Vogel et al. 2016; CDC 2016).

Is it realistic to expect that economics-driven mitigation-restoration efforts will now remove Asian tiger mosquitoes? Remove the water containing tires and restore the environmental conditions in the United States to their original state? Would this restoration include the reversal of the collateral damage, such as human health effects? When answering these questions, consider who would most strenuously object to proactive tire inspections, tire-related liability insurance or bonds to cover possible impacts, and why. Neoliberalism's guidance does not favour either the theory or the practice of protecting the environment's functioning. If environmental protection efforts are to be successful, they can't be subservient to economic beliefs and imperatives.

In overview, the discussion about efficiency and mitigation-restoration efforts as ways to reduce environmental impacts resulting from economics-related activities highlighted the most important limitation to any economics-driven efforts to protect our environment. There is a stark disconnect between the activities encouraged by the economists' world view, as symbolized by Wall Street, and the actions needed to maintain our environment's functioning. This disconnect exists because the imperatives of neoliberal economics (such as growth) and its designations of much of our environment's features as externalities with a zero or negative value, are in conflict with the reality of the environment's functioning.

From the perspective of neoliberalism, it is more important that our efforts to reduce environmental impacts satisfy economic imperatives than effectively reduce those impacts. Thus, over the long term, the economy-favouring methods of protecting our environment, such as efficiency, mitigation-restoration, carbon trading and geoengineering, inevitably commodify nature and contribute to our ongoing environmental disruption. Similarly, even though avoidance is most effective at protecting the environment's functioning and minimizing environmental impacts, it is, metaphorically, at the bottom of the economists' list of environment-protecting options because it runs counter to the imperative of growth and development.

In summary, this section illustrates that, from the environment's perspective, the neoliberal economic world view is largely incapable of reliably representing or protecting the environment's functioning and its importance to us. Its efforts result in the impacts from our economic activities being mostly buried in the residuals, assumptions, externalities and immeasurable factors of economic theory and practice (Faucheux 2005). There they remain, hidden, until they are experienced and then paid for by our descendants and the environment (Hall et al. 2001). This is in keeping with the economic focus on the market, growth and the short term.

In essence, weakly or unrestrained economic activity and a focus on growth does to our environment what weakly or unrestrained inflation (of all kinds) does to an economic system. It is disrupted and destroyed. Despite these limitations, neoliberalism has a strong influence on at least Western culture's decisions about the value of our environment. It also influences our assessment and response to our economics-related impacts on our environment and its functioning. It influences what we think of and accept as environmental protection. That influence is in favour of economic imperatives.

## Neoliberalism as a Guide beyond Economics

This section discusses the strong influence that neoliberalism exerts on Western culture and its members' decisions beyond economics. It starts by revisiting neoliberalism's roots and history, but now with a focus on two critical aspects that were previously mentioned only in passing: how mainstream economists' views about social and environmental issues changed through time. It is within this context that a summary of neoliberalism's guidance can be most easily appreciated. The section then provides an answer to the questions posed at the beginning of the chapter: Does economics' guidance reduce or enhance our reality paradox and our environmental impacts? This leads into the final topic, the need to constrain neoliberalism's influence, and how that could be achieved.

**The Historical Context.** Between about 1640 and 1800, the philosophers studying the European economy came to believe that enhanced trading activity would improve more than mankind's (as they saw it) material condition. They thought that engaging in trading would also supply the context, opportunity and drive needed for individuals and societies to make what they believed were the necessary, religiously sanctioned, moral improvements to their society. In particular, they thought that a person's striving to acquire economic wealth would, invariably, contribute to social improvement rather than simply being exploitative. But success required that business be conducted in a moral manner.

Adam Smith belonged to this group. The roots of his moral views are found in his earlier work, *The Theory of Moral Sentiments* (1759). His view that society would be both morally and materially improved by an individual's effort to improve their personal circumstance (self-interest) is found in his later work, *The Wealth of Nations* (1776) (Smith 1776). In essence, he believed in the notions of material and moral progress through individual action, and that they were linked.

Early in this period, the physiocrats noted the economic importance of arable land and the labour to work it. Later in this period, Thomas Malthus (1766–1834) published *An Essay on the Principle of Population* (1798). He held that the finiteness of arable land could, and would, limit both population growth and economic growth.

In the early 1800s, philosophers interested in the economy continued to believe that striving to increase personal wealth should result in a culture's moral and material improvement. This view, along with their faith in progress, helped their study of the economy to be accepted into the evolving Western world view (Friedman 2005). Note that the notion of progress did not preclude Malthus's views being given serious consideration.

During the mid- to late 1800s, the Western world view became increasingly secular. Yet it still held strongly to the belief that the combination of scientific advancement (progress) and economic growth (progress) would result in cultural improvement (social and moral progress) for all. In the early part of this period, it was felt that continual progress in this direction was essentially inevitable. Still, mainstream economists seriously discussed whether there were natural limits to the economy. For example, John Mill in his *Principles of Political Economy* (1848) included both moral (fairness) and environmental concerns (limits to land and population growth) as well as the notion of progress in his contemplation of possible economic futures (Mill 1848).

As the late 1800s turned into the early 1900s, economics became a discipline in its own right. It split from philosophy and the other social sciences. At the same time mainstream economics became focused on the mathematical description of the workings of the market. In the process, alternate ways of describing the market, such as energy flows and its biophysical connections to the environment were sidelined (Hall et al. 2001). During this period, the application of technology continued the industrial revolution's successes at circumventing resource barriers. For example, it provided human-made nitrogen fertilizers. The concern over natural resource limits began to fade.

Neoclassical economists' view of the relationship between economic progress and social-moral progress slowly changed. Helped by the First World War, they came to feel that progress was less certain and that social-moral progress was *dependent* on economic growth rather than being a multi-faceted event associated with it, when it was conducted in a moral manner. Social progress was becoming equated with economic growth (Friedman 2005). At the same time, the intense focus within neoclassical economics on highly abstract, theoretical mathematical models ensured that the discipline continued to become more distant from the social aspects of a culture's functioning. It also became more disconnected from the environment's functioning and the natural resources its functioning provides to the economy (Milonakis and Fine 2009).

After WWII, the neoliberal brand of neoclassical economics appeared. It became dominant after the 1970s and remains so today. Neoliberal economists strongly believe that the functioning of the market is governed by "natural" economic laws, the invisible hand of the market. If the market is to be efficient and provide benefits, then the participants in the market must follow these laws.

In their mechanistic, mathematics-dominated production view of economic activity, economic prosperity becomes one with social prosperity (social welfare). Self-interest becomes divorced from the idea of social obligations: economic rights become detached from social responsibilities. As a result, economists now believe that what little, if any, responsibility they have for social-moral progress it is adequately discharged by simply respecting the laws governing the market's functioning. Neoliberals believe that, by respecting these laws, economic growth is optimized which thus optimizes economic and social progress. No additional effort is required of economics (Friedman 2005). During this period, the environment and its functioning became firmly established as an economic externality whose constraints, if any, on the economy can be resolved through the application of innovation and technology.

There have been challenges to neoliberal economics. In the 1960s, interest in the long neglected connection between the environment and the economy reappeared. It continues to be addressed by the minor discipline of ecological economics. The population issue was raised in the late 1960s but faded away by the end of the 1980s. The idea of sustainability became prominent in the late 1980s, as discussed below, while moral concerns became more prominent in the 1990s and continue today as a result of a widening gap between rich and poor. To date, these developments have had little impact on either the core of neoliberal beliefs, the guidance it provides Western culture or its influence on Westerners' decision making.

**The Guidance and its Limitations.** The guidance that neoliberalism provides to countries suggests that the most effective solutions to their economic, social and environmental issues will favour growth-focused, market-oriented (not government-oriented) actions. These solutions, as outlined earlier, feature market deregulation, privatization, austerity, low taxes, motivation through financial incentives and limited control of the economy (mostly through monetary policy). Their guidance favours innovation and technology as solutions because they enhance productivity and efficiency, and thus growth.

Neoliberal guidance also contains its justification for this advice. Maintaining market-driven economic growth (progress) will make all of a country's citizens materially richer (a rising tide lifts all boats), improve their social welfare, enhance their personal social-psychological well-being and protect their environment's functioning. Maintaining economic growth will also fund the science and techno-science needed to maintain progress towards these goals, such as increasing efficiency and solving shortages.

Westerners, as discussed above, have absorbed this guidance and the explanations for it into both our personal and our culture's world view, where they play a central role in our day-to-day decisions. Our personal opinions and decisions tend to mirror the essence of neoliberal guidance, which can be described as economic rationalism. We tend to believe that all aspects of our personal lives can be improved by freeing the market so that we can engage in the unhindered striving to increase our personal wealth and enjoy its fruits.

We express this belief by striving to exercise our freedom to buy whatever we want, whenever we want and as much of it as we would like. Taking on debt helps us to achieve this goal. In doing so, we encourage the production of the most goods and services, for the least effort, at the lowest cost (i.e., efficiently), thus helping to maintain growth and increase wealth (Wilson 1988). This was referred to above as the consumer-producer feedback loop at the heart of the economists' (now globalized) virtuous cycle (figure 14). We believe that engaging in this material quest will improve the present and ensure that not just the material but the social-psychological future we hope for will arrive.

But, as discussed above, the neoliberal world view and thus its guidance is missing or misrepresents critical elements of the economy, human behaviour and the environment's functioning (Hall et al. 2001; Hawken 1993). In particular, it doesn't direct our economic activities to satisfy the social and environmental constraints within which we must live if we and our culture are to remain functioning over the long term (i.e., viable). In short, neoliberalism's guidance and goals are largely ideological. Thus, in the long term, neoliberalism's promises are unlikely to be met because they and its guidance are unrealistic or incomplete.

Consider that neoliberalism's guidance neither makes a distinction between human needs and wants, nor indicates how to balance your wants for material goods with the social and environmental costs of producing and using them. Its guidance has little useful to say about a material standard of living that respects the constraints set by our environment's functioning. This extends to the impacts economic activity has on environmental resources held in common by all of us, such as air, water or common lands (Ostrom et al. 1999).

By accepting neoliberalism's guidance, our decision making reflects its shortcomings. In general, we treat those aspects of our environment that have low or no cash value as largely unimportant: we treat them as externalities. The significance of our acceptance is more easily illustrated and personalized by real-world examples. The example used here is our relationship to the waste generated by economic activity and the impact that waste has on both us and our environment.

Consider how readily we accept the neoliberal view that the relatively less degraded environments of Western countries are the result of our economic prosperity. Growth, so the economic argument goes, has enabled Western countries to enact stronger environmental regulations and pay to clean up their historical waste (Jones et al. 2002). This is partly true.

But it is also true that globalization has allowed businesses in the more economically prosperous Western countries to move their production facilities to countries where operating costs are lower. For example, many of the goods American companies sell to Americans cheaply, but at a profit, are being manufactured offshore in countries such as China and Bangladesh. This transfers the environmental impact of production from a Western country to the countries now producing and exporting the goods.

But there is more. One of the reasons the operating costs are lower in these exporting countries is that they have weaker environmental and labour laws. This means that a product manufactured there is likely to be associated with increased environmental and human health impacts (Liu and Diamond 2005). The costs of these impacts, for example, the premature deaths related to PM2.5 pollution

(particles <2.5 μm in diameter) generated during the production of goods, are borne by the exporting country (Zhang et al 2017). These impacts are also irrevocably embedded in the economically cheaper-to-produce products as virtual environmental (water, air, etc.) and health costs (Rees and Wackernagel 1994).

When these products are imported into Western countries from the offshore factories, these virtual costs come with them. But, in the importing country, these costs are neither physically visible nor are they fully accounted for by economic measures, such as the price of the product and the GDP. The importing businesses, their host country and its citizens treat these virtual costs as externalities. A Western country's less degraded environment and healthier population is thus partly the result of us paying a small cash price to the producing countries as compensation for them degrading their land, water and air, and taking responsibility for polluting the global commons on our behalf. For example, one reason why the EU's $CO_2$ emissions have declined is that its $CO_2$ emitting activities have been outsourced to China (Jackson et al. 2015).

The presence of weaker environmental and labour laws in offshore production centres, and the acceptability of adopting neoliberal economic attitudes and incentives toward those laws, is illustrated by ship breaking. When a Western-owned ship reaches the end of its useful life, it is broken apart for scrap. These days, ship breaking is rarely conducted in developed countries, such as Canada or the United States, but it is common in India, specifically at Alang. One reason is the lower wages. Another is occupational and environmental health regulations that are less stringent than those in Western countries. This includes constraints on the disposal of hazardous waste. The result is a lower fixed ship-breaking costs for western ship owners, but higher environmental and human health costs in India (Rousmaniere and Raj 2007).

The economic arguments in favour of taking advantage of these low cost conditions in India are many. They are all weakened once our thinking expands beyond the neoliberal economic rationalizations. For example, providing jobs in India is not the same as making the effort to provide healthy, safe and environmentally considerate jobs in India.

Moving along, the degree to which business can rely on neoliberal beliefs when making or justifying their decisions is illustrated by a case against Trafigura, the world's largest oil trading company. Try to imagine the socially or environmentally responsible reasons that

Trafigura's decision makers could give for the following aspects of the case against them. Evidence was presented that Trafigura had purchased, at low cost, the sulphur-rich gasoline in the holds of a tanker named the Probo Koala. Its holds were then used as chemical reaction containers to complete an inexpensive sulphur-removing process, called caustic washing. When Trafigura sold the resulting high-value, sulphur-poor gasoline, they would make a substantial profit. However, caustic washing also produces toxic oil tanker waste or sludge (toxic slops), which is why the process is frowned upon in Europe.

The failure of Trafigura's efforts to cover up the toxic nature of Probo Koala's slops resulted in them experiencing difficulties disposing of the slops in Holland, and in rising associated costs. The slops were then shipped to Abidjan, Ivory Coast, where evidence indicated that it was disposed of by dispersing it around the city. Without admitting wrong doing, Trafigura's decision makers agreed to compensate the families of the 15 people that had died because of the waste and the approximately 30,000 people who were claimants for health effects likely caused by the slops (Bernard et al. 2006; Leigh and Hirsch 2009; Leigh 2009a; Leigh 2009b; Evans 2010).

The normality of this type of thinking is illustrated by the global mining giant Xstrata. In 2010, its decision makers decided to terminate 670 jobs in and around the mining and smelting town of Timmins, Canada. The reasons the company gave included fierce Asian competition, the strength of the Canadian dollar, a reduced outlet for Xstrata's by-products, increased electricity and gas costs and increased environmental expenditures needed to meet new air quality regulations (Talaga 2010).

All of these waste-related corporate decisions were made by individuals. It is at this level that neoliberalism's influence on Western culture affects even our personal decisions. This will be illustrated by continuing with the theme of waste.

The economics-influenced guidance that Western culture provides to its members encourages us to increase our material aspirations. Think of the culturally-supported consumer advertising to which we are exposed. Yet that guidance provides us with few incentives to consider the waste that is generated by the manufacture, use and disposal of the extra goods needed to meet those aspirations. Neither does it inform us of the environmental or human impacts from disposing of that waste. Examples of the outcomes are the concentration of plastic waste in the ocean; the feminizing effects of

the chemical compounds we discard into the environment; and the issues associated with the disposal of e-waste. Think of cell phones with their ultra-short lifespans.

The economics-influenced guidance we personally receive about waste and its consequences enable us, by default, to treat it as an externality and distance our self from it as soon as possible. This guidance amplifies the disconnect between our internal reality and the external reality. Neoliberal economics' influence on Western culture has deepened Westerners' reality paradox and increased our impacts on our environment.

In overview, Neoliberalism's guidance exerts a strong influence on Western culture's world view. Thus Westerners are influenced by neoliberal beliefs and standards when we make our day-to-day decisions. Think of our ongoing participation in the virtuous cycle, our casual use of debt and our faith in economic growth. This has contributed to our successful liquidation and transformation of the earth's natural resources into a cornucopia of goods, services and economic wealth (Wackernagel and Rees 1996).

However, neoliberalism's guidance to Western culture often misses or misrepresents critical elements of human behaviour and the environment's functioning. By embracing neoliberal guidance, Westerners have come to accept a distorted vision of ourselves, our economy and our environment, and how they function.

We have come to confuse cash value with social, environmental and even economic importance. We equate economic growth with social progress. We tend to think that increased personal wealth will improve our personal social-psychological well-being. We tend to forget that natural resources, especially energy, are needed to generate cash value and growth (Folke et al. 1994). We are thus less likely to remember that, in the long run, it is the environment's functioning and a viable cultural world view, not economic theory, ideology, practice or growth, that will sustain us (Hall et al. 1986).

Neoliberal guidance has helped to mask our divergences and enhance our reality paradox. It has increased our impacts on our environment, and it has lessened our relative appreciation for those impacts.

**Constraining Neoliberalism's Influence.** In 1987, the Brundtland commission explicitly recognized that we are straining our environment's ability to meet our present and future human needs (United Nations 1987). To deal with this circumstance, it laid out the challenge of sustainable development: "development that meets the needs of the present without compromising the ability of future generations to meet their own needs" (United Nations 1987, 41). In this statement, *needs* refers particularly to the needs of the poor: satisfying their needs has the highest priority. Sustainable development is, in essence, an attempt to add into our thinking both the social responsibility aspect of Enlightenment economic thought (previously discarded in the late 1800s, early 1900s) and a more realistic view of our relationship with the environment.

The Brundtland commission's definition of sustainable development is politically astute because it is intuitively reasonable and broad enough to be accepted by all. It even provides the opportunity for economists to participate using the concept of economic development. As a result, the manner in which economists rose to the challenge of sustainable development joins with their contribution to the 2008 financial crash to outline why neoliberalism's influence needs to be constrained and why doing so is so difficult.

Economists of all brands have spent much time and effort trying to come up with economics-based measures to represent Bruntland's sustainable development. To date, the efforts have been largely unsuccessful. At first glance, it might seem that the difficulties are the result of the task's complexity. But, as with the choice and calculation of the GDP to measure wealth, there are more basic reasons.

In order to define and calculate a meaningful measure, the economists would have to resolve a contradiction brought to the fore by the goal of sustainable development. They would have to place a meaningful value on what economics calls valueless externalities, such as the environment's functioning and human emotional needs. Certainly, assigning cash value to the environment's functioning through methods such as replacement does provide a bridge between economics and the environment. However, as detailed above, these economic adjustments can only realistically represent those aspects of the environment that have direct material value to humans. They have limited ability to represent those aspects that are important to its functioning, such as robustness, resilience and adaptability. The contradiction of assigning economic value to externalities has not been resolved.

The broad definition of sustainable development provided economists with a way to sidestep the externality problem: focus on an economic interpretation of the definition. Consider the following economics-friendly interpretation: "Sustainability means that people in successive generations should have what they need to be at least as well off as we are" (Coyle 2014). This popular interpretation allows economists to believe that sustainable development issues can be addressed by continuing to apply standard economics-focused solutions: business as usual. Thus, despite the evidence to the contrary, they can believe that economics' current imperatives (e.g., economic growth), goals (e.g., profit) and measures (e.g., cash wealth and GDP) can represent and address the non-economic aspects of sustainable development, such as the environment's functioning, social welfare and social-psychological well-being. Here are examples of the resulting economics-friendly solutions to sustainable development issues.

Some economists suggest that sustainable development could be more effectively addressed by splitting it into private and public components. People can then act in their personal economic best interest (the private rational consumer) but vote for a government that will look after the future (the public component) (Pezzey and Toman 2005). Of course, this solution leaves the powerful and wealthy with the opportunity to exercise their influence to ensure that a government favourable to economic growth is elected.

Other economists believe that sustainable development issues can be addressed by applying innovation and technology. Doing so would result in us protecting the environment by, you guessed it, using natural resources more efficiently and deploying more effective mitigation-restoration techniques. At the same time, economic measures, such as economic efficiency, dematerialization and economic emissions intensity, could be used to evaluate how quickly these methods are moving us toward sustainability. However, as discussed above, closer scrutiny reveals that the unconstrained reliance on these methods and measures can result in conclusions that are more apparent than real.

But there is an even darker option available to economists. Existing definitions of economic variables and measures can simply be adjusted to confirm progress toward sustainable development. An illustration that this can and does indeed happen is provided by the United States. The definitions or presentations of measures for poverty, unemployment, inflation and growth have all been changed over time in ways that often mask rather than clarify the likely reality of these phenomena (Phillips 2008).

There are examples of these types of adjustments being made with less finesse. The Argentinean government allegedly calculated its inflation rate by ignoring some items and changing the numbers for others (Forero 2009). In Russia during the communist era, and in Greece during the first decade of the 2000s, the measures of the state's economy, such as GDP, were basically faked (Coyle 2014). These examples and the economics-friendly interpretation of sustainable development illustrate that those with political power or with financially backed vested interests can change the application of economic theory or its measure in ways that could magically make sustainability appear, our reality paradox invisible and the economy provide the expected numbers. Think of the environmental assessments and monitoring of projects such as the Athabasca tar sands. Think too of carbon trading and carbon credit schemes. Of course, the resulting guidance economists would provide to the public about our efforts to live environmentally sustainable lives would enter the realm of fantasy. The economic community could have chosen a different solution. They could have faced, head on, the impossible task of having to assign cash value to externalities. They could have accepted that the economy exists within a human-environment system and that their current economic thinking, methods and measures are unsuited to meaningfully represent the non-monetary aspects of how the system functions. They could have prompted themselves to accept the system's functioning for what it is and change their thinking about the role of the economy in reaching for sustainable development. There is a precedent for such a significant change in views.

In the early 1700s, Linneas, a Christian creationist, took the bold step of classifying plants without regard for their relative importance to humans. Botany went through a revolution in understanding. The consequences rippled through the life sciences. Despite the foundations for such a Linnean, reality-facing breakthrough for economists existing in the late1800s, they had not experienced it by the early 2000s. The distance from such a breakthrough is illustrated by the Western leaders of economics' culture.

The annual World Economic Forum is held in Davos, Switzerland. It is attended by the who's who of the economic world. Klaus Schwab, the founder of the forum, said in his 2001 address to the meeting that the number one challenge of globalization was to create the conditions for faster growth (Elliott 2001). His number two challenge was the need for social and environmental sustainability, whatever that meant to him. His number one wish was granted, until the reality chickens came home to roost bringing the global

financial crisis of 2008 with them.

At the 2012 Davos forum, four years after the crash, Schwab admitted that capitalism was broken. However, in keeping with others whose fervent beliefs have been shattered, he still remained, like Alan Greenspan, a firm believer in free enterprise. Schwab saw 2008 and its aftermath as a sin of capitalism for which he was struggling to find a solution. Whatever that solution is, he suggested, it would involve a new moral compass for capitalism. Just as Allan Greenspan had resolved his discomfort over the crash and his role in it by discovering a "flaw" in his economic thinking, Schwab was also looking for a patch, not fundamental changes in neoliberalism's doctrine (Charlton 2012).

The response of these two men when faced with the difference between neoliberal culture's world view and the external reality indicated that they had each experienced a reality paradox experience. Yet they both reacted to their feeling of cognitive dissonance by only partly recognizing the issue and then dealing with it through tweaking. They are not yet ready to fully deal with the external reality.

Similarly, one would think that the 2008 crash should have at least put to rest neoliberalism's simplistic economic beliefs and theories in the same way that Biosphere 2 put to rest the ecologists' idea that ecosystems can be treated as simple systems. This is not the case. There have been few substantial changes to the economists' world view since 2008, as discussed earlier in the chapter.

Neoliberal economists and their supporters are still trapped by their unshakable belief in the importance and the correctness of their discredited assumptions about the nature of humans, the environment and the economy. They seem unable to move cash value and growth, unsubstantiated models and assumptions, from the centre of their thinking. Their guidance to Western culture continues to reflect these discredited beliefs. They seem unable to accept that the market is but one aspect of a human-created cultural system that is constrained by the characteristics of its host environment and its human members.

Economists are not the only ones trapped. Consider that, ironically (or is that despairingly), even though the crash of 2008 exposed neoliberalism's limitations, its guidance still remains the most important influence on at least Western culture's world view and thus on its members' lives and thinking. Its influence remains strong in other countries as well, as illustrated by the United Nations 2015 acceptance of 169 targets to achieve sustainable development (United Nations 2015). When seen in isolation, each target seems intuitively reasonable. However, there is limited recognition that the goals interact with one another as part of a complex system. For example, the issue of population size is dealt with by the phrase "take account of its trends and projections". Consider too that, collectively, the goals retain, intact, the unresolved contradictions which arise when sustainable development treats striving for standard economic growth (development) as inherently compatible with striving toward both environmental sustainability and social well-being.

Despite the 2008 crash, neoliberal economics' world view is largely unchanged. It remains the economic reference normal for the members of many cultures. As a result, Westerners' reality paradox continues to be enhanced, and our supporting environment continues to be increasingly disrupted. The members of Western culture find ourselves stuck with the neoliberal status quo, believing in the economic equivalent of a perpetual motion machine: a self-regulating, self-optimizing market that will both provide us with continual economic growth and look after our environment and social needs. Another economic crash and continued environmental disruption is not only possible, it seems inevitable. The need to constrain neoliberalism's influence is pressing and great, but so is the challenge of doing so. What to do?

As a culture becomes more complex, it establishes institutions, such as a law court and a church, to help it function. The market is one of these institutions, the one representing a culture's economy. A culture controls its institutions (which means the members within them) to ensure that their workings are helping the culture to remain functional and, hopefully, viable.

Today, Western culture controls the market through a regulation process. But, as the global financial crisis of 2008 illustrates, if market regulation is to be effective at keeping the market and Western culture functional or viable, then those aspects of neoliberal economic views and neoliberalism used to guide the regulation must change in ways that support that objective. The idea of sustainable development was one way to make those changes, but it is built to fail.

It could be argued that lasting cultural change is most effectively driven by its members. After all, they are the carriers of their culture and its decision makers. If the needed changes to neoliberalism are to take place in this way, then individual economists would have to adjust their internal reality. They would have to deal with their reality paradox experiences and reduce the divergences in their personal world view by questioning their economic reference normal, such as its focus on growth. They would have to abandon efforts to justify neoliberalism's reality-paradox-enhancing guidance as being well intentioned. And they would have to reject efforts to ignore economics' contribution to our disruption and destruction of our environment by treating the environment and its functioning as externalities. It would mean discarding the use of cash as a proxy measure for the state of either social or environmental matters.

Evidence for success would be changes in the beliefs and behaviours of the captains of the economy. For example, they would recognize that the Probo Koala affair, the global financial crisis of 2008 and the growing gap between rich and poor are more than a "malfunction" of the market, a "bump" in the economy or a regrettable "flaw" in human thinking. They would be accepted as at least being supported by, if not the product of, the many basic divergences between neoliberalism's world view and the external reality.

Undergoing such a transition in their personal world view is a long-term process for both economists and the captains of the economy. Understandably, to date there have been few appropriate changes to neoliberalism's guidance through this means. It is also highly unlikely that the required changes to the neoliberal world view will be driven by the workings of the market alone.

This is the time to remember that neoliberalism is embedded in Western culture, as discussed above. Western culture can therefore do more than just regulate the transactions that are the market's functioning, it can constrain the corruption of the regulation processes by those with money-backed vested interests. It can also actively prompt neoliberalism to change in favour of functionality and viability.

Western culture could encourage economists to make the personal and cultural changes mentioned above by prompting and supporting their efforts to find alternative economic world views. For example, it could fund research that is focused on viability, not growth: research directed at finding ways to structure and run a country's economy so that it stays within viability's cultural, personal and environmental constraints. Because economics is incapable of meaningfully representing social or environmental matters, economists would have to defer to other professions whose measures can more reliably and accurately represent those matters. And economists would have to tie those measure tightly to their thinking and models. Western culture could thus also fund research to answer questions such as how much stuff is enough to live a good life and why. Think too of research focused on fairness in tax and fairness in income distribution from the perspective of viability.

Making the needed changes would be helped if Western culture required economists' organizations to expand their fiscal responsibility doctrine into a fiscal-environmental-social responsibility doctrine. Economists could express this in a principle that is the equivalent of the medical injunction to do no harm: an economic precautionary principle that extended beyond cash, profit and growth. The effort to do so would be facilitated if Western culture required graduating economists and anyone wanting to occupy senior financial positions (executives, directors, etc.) to demonstrate that they have a meaningful understanding of the things their economic decisions affect. In particular, they should have sufficient understanding of how our greater environment and its ecosystems function, the roots of social stability and human well-being, and the connections between these factors. These aspects were outlined in the previous chapters, summarized in chapter 11 and will be discussed in chapter 18. The tests of competence could include answering the Asian tiger mosquito questions listed above.

In essence, with prompting from Western culture in favour of viability, economists would be provided with the tools to resolve the contradictions among the sustainable development targets. They could more easily abandon the economics-friendly interpretation of sustainable development. They could more easily abandon neoliberalism.

These efforts by Western culture, along with strong market regulation in favour of viability, would result in neoliberalism's influence waning. Its broader membership could come to more meaningfully consider the environment and our social-psychological well-being in their economic decisions. They could come to accept that, in the medium to long term, our economy is dependent on our environment's functioning. It provides the resources (including services) that are needed to keep the wealth-generating consumers, labourers and innovators of the present and future alive, happy and healthy. Those resources are also needed to improve the lot of the needy. These efforts by Western culture would devalue the routine efforts of neoliberalism's members to sidestep their social and environmental responsibilities. It would make those decisions and their justification less acceptable.

However, there is a catch to this idealized plan for Western culture to constrain neoliberalism's influence and prompt changes in its and its members' world view. If Western culture were to be successful, then its efforts would have to be largely supported by its members. But, as indicated in the human-environment system model discussed in chapter 11, and in the history section of this chapter, feedbacks are involved. In particular, Western culture's guidance to its members tends to harbour, support and use the same misconceptions as its economist advisors. Western culture's guidance is strongly influenced by neoliberalism. A Catch-22 exists.

Success at prompting neoliberal economists to change their world view requires Westerners in general, as our culture's decision makers, to change our culture's world view. To accomplish that change, Westerners would have to adjust our personal view of our culture's economy, our relationship to economics and our supporting environment. Thus, ironically, although we point at economists and the market as the source of our economic and environmental problems, a meaningful response to those issues requires us all to make matching changes to ourselves and Western culture's world view. This is another difficult task to complete. One that is discussed more fully in later chapters.

A short time after handing in your report, you sense a deep hang-your-head-in-despair sigh from your boss, the all-knowing presence. This is shortly followed by the feeling that this situation really is our problem now. The boss gave us participant status in this world. Thus we are expected to get on with accepting the responsibilities that go along with this illustrious position. You now accept that we, our divergences as represented by our reality paradox experiences, and our decision-making foibles lie at the heart of the issues we face today.

It seems clear that we should accept and deal with our environmental responsibilities. It seems that we should accept and try to reduce our divergences. But you know those tasks are difficult. You also remember that the past does not guarantee the future, so you wisely decide to find out more about the nature of our future. Maybe the above criticism is misguided, passé, and we don't need to do anything. Maybe, despite neoliberal economists' guidance, things will be fine. Maybe things will just take care of themselves. Maybe our lives don't need to change all that much. Now that is an easy solution.

This ends part 2. It dealt with our external world, including how the biosphere functions, our relationship to it and its role in our lives. Part 3 deals with the future *and* the past. It looks at our environment's likely future state and that state's likely ability to supply our future needs. The validity and usefulness of the resulting predictions are then tested by using the history of past cultures as a reference. The conclusions provide us with guidance about how we can address our likely future.

# PART 3

# THE PAST AND FUTURE

# CHAPTER 13

# THE BIOSPHERE'S FUTURE

The biosphere changes over time. A complex culture's efforts to deal with those changes include predicting the biosphere's future state. But the biosphere is a dynamic, complex system, so making valid and useful detailed predictions about it is difficult at the best of times. Making them today is even more difficult, for two reasons. Firstly, we are in the process of rapidly changing a large number of the biosphere's variables, which increases the number of its possible futures. Secondly, because our influence on it is now so marked, its future state depends significantly on our decisions and actions, which are difficult to predict in detail. Both reasons add to the uncertainty in predictions about the biosphere's future. For these reasons, discussions about its future will focus more on those general features, and their possible variations, that are most likely to characterize the biosphere in the future, than on a detailed prediction about a particular future or feature.

The biosphere's future is more easily and reliably discussed at a time scale relevant to us if we remember that its current functioning and state is represented by the human-environment system. The future of its environmental aspects is discussed in this chapter by dividing it into two subsystems: the climate system and ecosystems. The discussions about them are illustrated and personalized by referring to known trends in their variables as well as emerging information about the (human and environmental) influences on those trends. Of particular interest is whether those influences are likely to maintain or adjust current trends.

But our supporting environment will change as a whole, not as two discrete subsystems. To gain an appreciation for the manner and direction in which it may change as a whole, the final section merges the conclusions about the most likley futures for the climate system and ecosystems. That merging and a discussion about it uses the knowledge gained in parts 1 and 2 (as summarised in chapter 11) about the functioning of the complex human-environment system. The next chapter will discuss the future biosphere's ability to meet our likley future demands on it.

A reminder that *short* (or *immediate*) *term* refers to a period up to 15 years from 2015; *medium term*, up to 60 years; and *long term*, beyond 60 years.

## Climate Futures

Climate is a key aspect of our environment. Any changes the earth's climate will undergo will have an important influence on the long-term future of the earth's environment. Currently, the topic of climate warming and our contribution to it is highly contentious. This extended introduction will provide an overview of the limits to climate change predictions and a discussion about the idea of human-caused climate change from both a scientific and a cultural perspective. Predictions about future climate change and the possible consequences can then be personalized.

## The Limits to Climate Change Predictions.

During the late 1990s, the population of the temperature-sensitive mountain pine beetle in central British Columbia began to increase to the level of an epidemic. The beetles' feeding activities killed large numbers of its host, the lodgepole pine tree. In 2001, the then president of the Northern Forest Products Association of British Columbia, whose members were being affected by the pine deaths, discussed the need for a cold weather event to halt the epidemic. He used a 50-year temperature record to conclude that the region's temperature changed on an eight-year cycle. This indicated to him that the next cold snap would likely occur in 2004 (Ardis 2001). His prediction, not unexpectedly, proved to be wrong.

This example illustrates our dismal record of common-sense climate predictions. It also reminds us that our human characteristics, such as our strong desire to see patterns even in random data and our focus on specifics, ensure that we have a poor intuitive grasp of how complex systems in general and the climate system in particular function. Thus, if we wish to be more considered in our opinions about today's flood of climate predictions, then we need to both appreciate the scientific limits to climate change predictions and remember our personal limitations.

The earth's climates are the product of a dynamic, complex, feedback-stabilized climate system that operates at the boundary between order and chaos. The system changes when an influence prompts some of its many variables to adjust. The adjustments are then spread through the climate system by the many interactions between the system's variables. These interactions operate over a wide range of time and space, and involve feedback. This is the manner in which the climate system changes. For this reason, it is only the most general aspects of the climate system that can be reliably predicted by even a trained intuition.

To circumvent this human limit, scientists build mathematical climate system models to predict the climate system's future state. But, because of the climate system's complexity, even these models will suffer from limitations that significantly restrict the accuracy and certainty of the predictions they generate (Mitchell and Hulme 1999; Stainforth et al. 2007; Collins 2007). The presence and significance of these limitations provides fodder for the ongoing scientific and public debates about climate change predictions (Solomon 2008). The aim here is to establish a feel for the nature of a climate model's limitations and the context in which the resulting predictions should be judged.

The limitations to climate models can be divided into two broad classes: those related to a lack of knowledge or data about the system; and those that arise from the complexity of the climate system. The lack of knowledge is illustrated by our incomplete understanding of the role that aerosols and clouds play in climate regulation (IPCC 2013a). The lack of data is exemplified by the paucity of detailed, long-term climate records. They are needed by scientists to sort out the long-term (low frequency) from the short-term (high frequency) influences on climate change (Collins 2007). These unknown or poorly documented parts of the climate system are either left out of the models or are represented by adjustment factors (essentially mathematical assumptions).

There are numerous limitations to climate models that are related to the complexity of the climate system. Some of these limits are practical in nature. For example, it is difficult to mathematically include in a single computer model the large range in time and space over which our climate system operates. There is also the difficulty of assigning coherent and consistent initial conditions to all of the system's many variables. This is important because the choice of initial conditions strongly determines the model's results.

Other complexity-related limits are more theoretical in nature. Consider that the climate system lies on the boundary between order and chaos. Randomness and chaos are thus unavoidable aspects of the system. This means that, even if we had full understanding of the causes of climate change and built a perfect model, its predictions would still be uncertain. For example, even if we repeatedly applied the same inputs to the perfect model, the successive outputs would be different (Mitchell and Hulme 1999).

These complexity-related limits ensure that although tricks, such as using the past climate to "tune" a climate model, can reduce the uncertainty in the model's predictions, they can't eliminate it (Smith et al. 2007; Collins 2007). Our predictions about climate will

always be, to some degree, probable, not certain. At some scale, our predictions about climate features, such as the timing of changes and the spatial distribution of events, will always be significantly uncertain.

These uncertainties are not evenly distributed across a climate system model. Overall, the smallest uncertainties are associated with the more easily described, large-period, slow and more stable relationships. These deterministic-like relationships are the easiest to incorporate into a model, and their uncertainties can be reduced by more careful observation. For example, it is relatively easy to describe the slow change over tens of thousands of years in the amount of the sun's radiation reaching the earth caused by the combined Milankovitch earth-sun orbital cycles. The resulting Milankovitch-sunlight-cycle-induced changes in the earth's climate, as illustrated by its glacial-interglacial cycles, are relatively easy to predict, with low uncertainty (figure 11).

In contrast, the greatest uncertainties are associated with the poorly understood processes that affect climate at a scale of 10s to a few 1,000s of years (Adams et al. 1999; Stainforth et al. 2007). There are also chaotic and random aspects to climate processes, as illustrated by the abrupt events seen in figure 11. These aspects of climate are imperfectly or not represented in the models.

One method to deal with the ubiquitous presence of this complexity and uncertainty is to run a model using a coherent suite of realistic inputs, or a scenario. Repeatedly running the model by slightly varying each input produces an ensemble of climate predictions for that scenario. This ensemble of predictions and its respective uncertainties can be used to outline the most likely range of a possible climate future. Predictions can then be expressed as the probability of a scenario occurring.

But predictions hedged with probabilities are not the type of answer the public expect from science in general and climate science in particular. We prefer simple, will-happen/won't-happen predictions. When this expectation is combined with our poor intuitive grasp of the climate systems and climate change, there should be no surprise that a common reaction to climate scientist's predictions is disbelief, skepticism, confusion or a shrug. One of the real-life limits to climate models is the public's difficulty relating to their predictions. Think of our relationship to weather forecasts as an analogy.

We can improve our appreciation for the climate system, climate system models, climate predictions and their limits if we apply a few of the science-based rules of thumb about complex systems and predictions previously discussed in chapters 5 and 8. Here are a few examples. The climate system is complex. It includes thresholds and feedbacks. There are delays between disrupting a variable and our seeing the result as a change in climate: that is, the climate system displays inertia to change. The feedbacks provide amplification in the climate system, which means that even a small disruption can have large effects, especially if the disruption continues for a long time (Dunbar 2003b). From the climate's perspective, there is no ideal (normal) stable climate state because the climate system is always changing at some time scale (Stainforth et al. 2007). The chaotic aspects of the climate system ensure that the possibility of an unexpected abrupt change in climate is real (Adams et al. 1999; Dansgaard et al. 1993).

The climate system's characteristics result in our predictions about its future having the following rule-of-thumb features. As with weather predictions, the nearer to the present, the more reliable the prediction can be, but there will always be uncertainty in climate predictions. The more detail in the prediction, the more uncertain it is. If we want our climate predictions to have a set (particular) level of uncertainty, then the further into the future the prediction looks, the more general it must become. General predictions about an average climate future at a global scale have lower uncertainties than detailed predictions about the climate future for a local area.

Unfortunately, these characteristics limit the usefulness of climate predictions to the public. Planners and politicians are interested in, and lay people most easily relate to, detailed climate predictions on a time scale of no longer than ten or so years, and a space scale of no greater than a few hundreds of kilometres from us. We are particularly interested in annual regional- and local-scale rainfall patterns no more than five or so years into the future. Unfortunately, at these scales, the uncertainties associated with climate model predictions are large, so the predictions are highly uncertain (Collins 2007). At these scales, it becomes hard for the people living there to distinguish between the variations in weather and changes in climate.

We are also interested in knowing if the current Younger Dryas-like rates of temperature increase are the harbinger of an abrupt (sometimes called a "runaway") climate warming event. Unfortunately, it is difficult to identify, in advance, the thresholds that must be crossed to initiate such a switch in the climate system's stable state or how they might be crossed. Thus, it is difficult to determine well in advance how, let alone when, either the atmosphere's current circulation pattern will change or an abrupt event will occur. We should be aware that an abrupt climate event might occur but treat predictions of exactly when it will, or that it won't, occur with skepticism (Alley et al. 2003).

When we do pay attention to broad-scale climate change predictions, we would like to know which of them deserve most of our attention. There is a simple resolution to this question. If we take the above rules of thumb into account, then we should focus on predictions that are backed up by reliable longer-term historical trends in climate variables, such as temperature. These trends are more reliable if they are supported by longer-term trends in historical observations of environmental features that are affected by climate variables, especially if the trends were based on data compiled from a number of independent studies. Examples of these environmental features are the indicators of spring, such as the first snow melt and the first leaves appearing; and the indicators of temperature change, such as the change in the length or mass of glaciers. Other features of interest, which we can also relate to, are the indicators of changes in precipitation, such as drought and river flows; and the general indicators of climate-related ecosystem change, such as changes in the seasonal behaviour of specific animals and insects.

The current trends in almost all of these types of environmental climate change indicators match the trends in the physical measures of climate, such as temperature. They indicate that the earth's atmosphere is undergoing a sustained decades-long warming. The models reflecting this warming trend predict that it will continue well into the future.

In overview, we have a limited innate ability to relate to the functioning of the climate system. Model-based climate predictions increase our ability to make climate predictions. But those predictions are accompanied by an inescapable level of uncertainty. We struggle to relate to this uncertainty. By acknowledging these limitations and focusing on indicators of climate that we can relate to, we can be more confident that the models' predictions of ongoing climate warming in the immediate to medium-term future are valid and useful but become less reliable further into the future.

**Human-Caused Climate Warming?** The earth's atmosphere appears to us to be vast and immutable, so it seems incredible that humans could complete the Herculean task of changing its composition. That is, until you realize that the earth's atmospheric blanket is only 10 to 20 km thin. It does not take much of a disruption to the earth's dynamic element cycles and atmospheric processes to change its ability to trap and retain the radiation that heats its atmosphere. We know that one large volcanic eruption will do the job.

The Intergovernmental Panel on Climate Change (IPCC) focuses on evaluating climate warming, the possible human contributions to it and the significance of any resulting impacts on us and our environment. The IPCC attempts to answer these questions by combining historic observational data with current theoretical research from a wide range of sources. Their data, models, interpretations and resulting reports lie at the centre of the current bitter climate warming debate.

The IPCC's most easily accessible public expressions of their views about climate change are found in their summary reports. The IPCC's 1995 (2nd) report recognized that there were considerable uncertainties in their conclusions: "Nevertheless, the balance of evidence suggests that there is a discernible human influence on global climate" (IPCC 1995, 22). In essence, they were indicating that human-caused climate warming was emerging from the background of natural climate change.

Their 2001 (3rd) report drew conclusions that were a little firmer: "In the light of new evidence and taking into account the remaining uncertainties, most of the observed warming over the last 50 years is likely to have been due to the increase in greenhouse gas concentrations." Here "likely" means a 66% to 90% chance of being true (IPCC 2001, 10).

By their 2007 (4th) report, the IPCC concluded that "Most of the observed increase in globally averaged temperatures since the mid-20th century is very likely due to the observed increase in anthropogenic greenhouse gas concentrations." Here "very likely" means a greater than 90% chance of being true (IPCC 2007a, p. 8).

In the 2013 (5th) report, the IPCC concluded that "It is extremely likely that human influence has been the dominant cause of the observed warming since the mid-20th century." Here "very likely" means a 95% to 100% chance of being true (IPCC 2013a, 12).

Over nearly two decades, the IPCC has consistently concluded, with increasing confidence, that the majority of the atmosphere's current warming (climate warming) is caused by human activity. The IPCC also predicts that significant, ongoing human-induced climate warming into the future is likely. They indicate that the size of these future changes depends significantly on the decisions we make today.

Some have taken the IPCC reports at face value, others as an outline of an extremely serious situation that requires immediate and drastic action. Others still remain skeptical, even hostile, to the reports. The range of skeptical views varies from "climate warming is not real" through to "if it is real, then it is not human caused" to "it is not as serious as proposed" and "we just don't know." These skeptical views are put forward by a variety of individuals and organizations. The proponents range from cautious individuals, to organization with specific political, economic or ideological viewpoints, to the mouthpieces for individuals and organizations that have a vested interest in a specific view about climate warming.

What are we to make of these claims and counter-claims about the IPCC projections? By looking at the debate from both the scientific and a public-cultural perspective, we can hopefully make some sense of statements and predictions about climate warming before deciding on their likely truth. And, more importantly, we can decide if they are true from the perspective of our environment. Although this section focuses on climate change, its essence can be applied to many other contentious environmental issues, such as supplying resources and using services.

*The Science Argument.* There is enough observational evidence, such as glaciers melting, to show that climates around the world are warming, and they are likely to continue warming in at least the short-term future. We also know that human activities, such as fossil fuel burning and deforestation, are increasing the atmospheric concentration of greenhouse gases and aerosols, and changing the ability of the earth's surface to reflect sunlight (albedo). Both of these changes affect the climate system's ability to trap sunlight and warm the atmosphere.

Are our changes to the climate system sufficient to cause the observed climate warming? How far into the future can this warming continue? We face difficulties when trying to find what we feel are reliable answers to these questions. One of the reasons is conceptual. We search for and believe in a simple, direct, immediate cause-and-effect relationship between our activities and climate warming. But change in complex systems involves the overall state and functioning of the system and its feedback response to a change or influence. There is also some flexibility in the system's response. Thus our influence on the climate system is complex, partly indirect, but also delayed, changing and distributed. If these characteristics are kept in mind, then the connections between our activities and the climate warming being observed today is much easier to make.

History helps us make the needed connections. By the late 1800s, both the presence of carbon dioxide ($CO_2$) in the atmosphere and its ability to trap radiation of certain wavelengths were known. The scientific debate of the time was whether the presence of that $CO_2$ could affect the earth's surface temperature. In 1886, Svante Arrhenius (1859–1927) concluded that the atmosphere's $CO_2$ (and water vapour) would preferentially absorb the earth's radiation (its re-radiation of absorbed sunlight, as discussed in chapter 8). Thus an increase in the atmosphere's $CO_2$ content would increase the atmosphere's temperature and the warmth we experience on the earth's surface. He even suggested that the glacial age, as the Victorians understood the last ice age, could have been caused by a drop in the atmosphere's $CO_2$ content to around half of its 1890s value (Arrhenius 1896).

The significance of his work becomes apparent by studying the record of the earth's climate, preserved in ice cores taken from the world's ice sheets. The cores reveal, as discussed in more detail in chapters 7 and 8, that during the last 400,000 years, the earth's climate went through four glacial-interglacial temperature cycles. We are living in the latest interglacial phase. The cores also reveal that the glacial-interglacial temperature cycle is matched by a corresponding cyclical change in the atmosphere's $CO_2$ content. During the glacial phases, the atmosphere's $CO_2$ concentration was around 180 parts per million (ppm), which is between 60% and 70% of its interglacial value of 280–290 ppm (figure 11). Thus Arrhenius's conclusion that halving the atmosphere's $CO_2$ content in 1900 of around 300 ppm could result in an ice age was reasonable. It is possible that changes in the atmosphere's $CO_2$ can be responsible for climate change.

Yes, but that is not the only possible cause. Studies of the glacial-interglacial temperature cycles revealed that the cyclical changes in $CO_2$ lag behind the changes in the atmosphere's temperature by between 1,000 and 4,000 years, depending on the point in the temperature cycle being studied (Mudelsee 2001). It has been accepted that the driver of the glacial-interglacial temperature cycle is the change in the Milankovitch sunlight cycle. For the last 400,000 years, the sunlight-driven changes in the atmosphere's temperature have induced the changes in its $CO_2$ content, not the other way around.

The records for the past glacial-interglacial cycles also reveal something else. During the thousands-of-years-long warm interglacial phases of the last six cycles, the atmosphere's $CO_2$ content never exceeded around 300 ppm. In contrast, over the past 200 years of the current interglacial period, the atmosphere's $CO_2$ content has increased to well above the 300 ppm interglacial maximum. In 2014, it was trending past 397.7 ppm (WMO 2015). In 2014, there was 100 ppm more $CO_2$ (30% increase) in the atmosphere than can be explained by the Milankovitch sunlight cycle. This "excess" $CO_2$ correlates with a substantial increase in the atmosphere's temperature.

How is it possible for the atmosphere's $CO_2$ content to switch its usual lagging, effect-like relationship to the atmosphere's temperature into a leading, cause-like driver of the atmosphere's warming? Arrhenius's research again helps us to answer this question. In drawing his conclusions, he pointed out that the $CO_2$ content of the atmosphere is part of the earth's carbon cycle (figure 7a). He knew, and we know in more detail today, that the carbon cycle exists in a state of dynamic equilibrium. The distribution of carbon among the carbon cycle's primary reservoirs (atmosphere, land and ocean) strives for a balance that is influenced by features such as the temperature of the atmosphere and ocean, and the fixing of carbon by plants and rocks.

During the last 400,000 years, the Milankovitch sunlight cycle resulted in variations in the sunlight reaching the earth. The earth's carbon cycle was prompted to rebalance itself by redistributing the carbon in its reservoirs, including the atmosphere's $CO_2$ content. The result was the amplification of the 400,000 years of sunlight-cycle variations into the atmosphere's 400,000-year temperature cycle of glacial-interglacial periods. It also resulted in those temperature variations leading the corresponding cyclical changes in the atmosphere's $CO_2$ content.

However, if any part of the carbon cycle is sufficiently disrupted by events originating on the earth (i.e., internally generated), then the carbon cycle will still be forced to re-equilibrate. This includes adjusting the atmosphere's composition, which affects its temperature. We arrive at the question of whether we are responsible for the atmosphere's 100 ppm of excess $CO_2$ in 2014. Are our activities disrupting the earth's interglacial-age carbon cycle's balance point by a sufficient amount and rate to explain the observed increase in the atmosphere's $CO_2$ and its temperature?

The answer is illustrated by focusing on our use of fossil fuels. Every year, (land and ocean) plants remove carbon from the atmosphere by manufacturing plant material (fixing the carbon) through the process of photosynthesis. On the plant's death, a portion of this plant material becomes buried underground, where it is slowly converted into fossil fuels. Today, the rate at which carbon is being fixed and converted into fossil fuels is much slower than the rate at which we are using fossil fuels. For example, in order to replace the fossil fuels we used in 1997, the amount of new plant material produced that year would have to have been about 422 times greater than it was (Dukes 2003). In other words, we are using fossil fuels at a rate hundreds of times faster than plants and geological processes can turn atmospheric carbon into fossil fuels.

By rapidly transferring such a large amount of carbon from the carbon cycle's underground reservoir into its atmosphere reservoir as $CO_2$, we are disrupting the carbon cycle from its interglacial balance point. Adding to its disruption is our transfer of large amounts of carbon from the earth's above ground carbon reservoir into the atmosphere as $CO_2$ by making cement from limestone and burning plant material from forests. The carbon cycle is dealing with this disruption by very slowly transferring the "excess" carbon in the atmosphere into the other carbon reservoirs, which is changing them. Think of the increasing acidity of the oceans (figure 7a). In the meantime, the earth's $CO_2$-rich atmosphere warms.

There is no doubt that we have disrupted and continue to significantly disrupt the usual interglacial-age balance of the earth's carbon cycle. The speed and size of that disruption is sufficient to both increase the atmosphere's $CO_2$ content by about 100 ppm and change its lagging relationship with temperature into a leading one. It is thus quite reasonable to conclude that human activity is contributing significantly to the atmosphere's warming, and that the warming is affecting other climate features, such as rainfall (Zhang et al. 2007).

Today, the serious scientific debate is not about whether humans are contributing to climate warming but about the details. Of particular interest is the exact proportion attributable to humans and the likely size and speed of climate warming during the next 100 years. There is also much debate about how to react to climate warming, including the most appropriate way to deal with the uncertainty in our knowledge. The IPCC reports are part of the attempts to address these questions about the human contribution to warming and its impacts on ecosystems and our lives.

The scientists who contributed their field observations and their models' results to the IPCC reports have concluded that the human changes to the earth's climate system are significant and increasing. They also concluded that these changes will continue into the future unless we reduce our disruption of the carbon cycle's interglacial balance. But, to appreciate the scientific argument about climate warming, we have to appreciate not just the predictions but the uncertainty their models' limitations introduce into those predictions: that is, the limitations to the validity and usefulness of their predictions as discussed in chapter 4. This subsection continues that theoretical discussion of those limitations but in a more pragmatic and personalized manner.

Many of the specific deficiencies in the models are discussed in the IPCC reports, such as the lack of knowledge about the role of clouds (IPCC 2013a). Other shortcomings are still being discovered. For example, ongoing work is revealing more about the relationship between the variation in the sun's ultraviolet (UV) emissions, the formation of ozone in the ozone layer, and the nature of the sun's radiation that reaches lower into the atmosphere to directly warm it. The initial results indicate that the actual relationship is somewhat different from the one assumed by the IPCC in their effort to separate the human from the natural warming factors (Haigh et al 2010).

Do these types of deficiencies significantly reduce the validity or usefulness of the IPCC's climate predictions and the resulting concerns about future climate warming? Some scientists insist, with varying degrees of emphasis, that the IPCC's shortcomings are significant enough for its conclusions to be taken with a grain of salt or even ignored (Akasofu 2009). For example, some point to the significant but poorly understood relationship between cosmic radiation, the formation of aerosols in sulphur-rich environments, and cloud formation (Enghoff and Svensmark 2008). Hopefully, further research will adequately resolve these types of deficiencies and issues.

Some of the suggested shortcomings appear to be more apparent than real or are irresolvable to the degree those raising them would like. For example, it has been proposed that an increase in the sun's radiation is equally or more responsible for the current climate warming than are greenhouse gas emissions by humans.

This seems reasonable when the change in sunspot activity is considered over an 11,400-year period. The record indicates that the sun's post-1970s level of radiation is unusually high.

However, on the scale of 10s of years, the rate at which the post-1970s solar activity changed is too small to be solely responsible for the observed post-1970s climate warming (Solanki et al. 2004). The change in the sun's radiation is sufficiently small that its contribution to warming is masked by the earth's ongoing internal amplification of the other influences (e.g., the Milankovitch sun cycle) on climate. The resulting difficulty in detecting the influence of decadal-length changes in the sun's radiation on climate warming makes it hard to definitively settle the argument about its contribution to recent climate change (Rind 2002).

A different argument about the sun's role notes that an increase in its radiation would warm both the lower atmosphere (troposphere), where most weather is created, and the upper atmosphere (stratosphere). In reality, as the recent appearance of acid clouds over the Arctic indicate, the stratosphere is cooling. Only the troposphere is warming. Collectively, these arguments indicate that an earthbound process is causing the current warming.

Some of the concerns raised about the IPCC's predictions are part of the normal, imperfect scientific process. Others are attempts to discredit the IPCC and the notion of climate warming (see below). Regardless of their purpose, the concerns about the validity of the predictions can't be used as an excuse to unilaterally disregard the substantial field observations that rapid climate warming is occurring or deny that humans have contributed to it by disrupting the carbon cycle. However, the concerns do serve as a reminder that climate models, like other global-scale models, can't give us the simple, reliable predictions about the future we expect from science. The details of the IPCC conclusions and predictions should be used as general guides, warnings if you like, not accurate statements about the details of our climate future, especially for the late medium to long term.

Consider the IPCC's past use of scenarios and current use of Representative Concentration Pathways (RCPs) to represent the influence (expressed as a forcing in watts per $m^2$) that our possible climate-affecting decisions could have on our climate's future. The IPCC currently represents the possible range of our $CO_2$-emissions-affecting decision's by four RCPs. Each

represents a different amount of our possible total $CO_2$ emissions from all sources for the period between 1870 and 2100, and the resulting change in the atmosphere's temperature (figure 17a) (IPCC 2013a). Our sources of $CO_2$ emissions include fossil fuels, cement production, and waste burning, such as land clearing and deforestation. (The other greenhouse gases, which can be factored in by an equivalent amount of $CO_2$ equivalents, are excluded from the RCPs.)

Some scientists have suggested that a few of the RCPs are unrealistic or flawed because there is likely an insufficient amount of extractable fossil fuels to generate the assumed $CO_2$ emissions (Rutledge 2011; Höök et al. 2010). Others claim there are likely more than enough fossil fuels to satisfy all scenarios. If either objection is likely true, then our use of the IPCC's predictions should be correspondingly restrained. Theoretically, this argument should be easily settled by comparing the amount of fossil-fuel-sourced $CO_2$ represented by each RCP with the amount of $CO_2$ that would be emitted from a carefully considered estimate of the total amount of fossil fuel likely to be produced.

Here is my attempt. In 2012, some 74% of our $CO_2$ emissions came from burning fossil fuels. By assuming that a similar proportion of our post-2012 $CO_2$ emissions will come from fossil fuels, it is possible to estimate how much of each RCP's $CO_2$ is expected to come from fossil fuels for the period 2012 to 2100. By this method RCP2.6 represents 731 gigatonnes of carbon dioxide ($GtCO_2$) emissions to come from fossil fuel combustion between 2012 and 2100. RCP4.5 represents 2,112 $GtCO_2$; RCP6.0, 2,870 $GtCO_2$; and RCP8.5, 4,563 $GtCO_2$ from fossil fuels. (Some 1,103 $GtCO_2$ are estimated to have already been released from fossil fuels between 1870 and 2011)

The likelihood of a particular RCP reliably representing our climate future depends on the amount of the earth's remaining fossil fuels that are likely to be both extractable (recoverable) and produced between 2012 and 2100. The uncertainties in the estimates of our post-2012 fossil fuel production are dealt with by calculating a low and a high estimate. The low, more certain estimate is around 1,828 $GtCO_2$. The high, much more uncertain estimate is around 5,278 $GtCO_2$. (For comparison, a key paper discussing the choice of a <2.0°C near-surface atmospheric temperature target used an estimate of total "proven fossil fuel reserves" for 2006 of 2,800 $GtCO_2$ (Meinshausen et al. 2009).

As discussed in chapter 15, part of the low estimate is the carefully researched, model-based predictions of the ultimate amounts of conventional oil and gas as well as the amounts of non-conventional oil and gas resources thought to be most likely recoverable (665 $GtCO_2$ from oil, 619 $GtCO_2$ from natural gas). The rest of the low estimate comes from well-researched, model-based predictions of coal likely to be recovered from historical and currently operating, well-developed coal-producing regions (543 $GtCO_2$). The high estimate is calculated by adding to the low estimate the additional amounts of non-conventional oil, gas and coal resources thought to be possibly or potentially recoverable by technological methods. There are 3,142 $GtCO_2$ from oil, 1,239 $GtCO_2$ from natural gas and 897 $GtCO_2$ from coal (Bentley 2002; IEA 2005 to 2013; BP 2014; Rutledge 2011; Höök et al. 2010; BP 2014).

If the low estimate of 1,828 $GtCO_2$ is taken at face value, it indicates that our future emissions will easily satisfy the 731 $GtCO_2$ of RCP2.6, and might barely miss satisfying the 2,112 $GtCO_2$ of RCP4.5. In contrast, if the high estimate of 5,278 $GtCO_2$ is taken at face value, it indicates that our future emissions will easily satisfy the three lower RCPs, while the 4,563 $GtCO_2$ of RCP8.5 can just be satisfied. Regardless of the estimate used, there is likely enough technically recoverable fossil fuel available to exceed the target of <2.0°C climate warming, discussed below.

However, this conclusion should be strongly tempered by the uncertainty in the estimates of future fossil fuel production. Here is the essence of a pragmatic analysis of that uncertainty, presented in chapter 15. In calculating the low total for fossil fuel production, the estimates of recoverable conventional oil and gas are the most reliable. The amounts of non-conventional oil and gas are likely too optimistic. In contrast, the estimates of recoverable coal are likely too pessimistic. Is the low total estimate for fossil fuel production too optimistic or too pessimistic?

A similar evaluation of the high total for fossil fuel production concludes that the high estimate is far too optimistic. For example, the estimates of recoverable coal and unconventional oil and gas used for the high total include deposits that are unlikely to be exploited for any number of reasons. On balance, it seems that the total amount of fossil fuels likely to be extractable in the future lies around but most possible above the low estimate and well below the high estimate, but it is difficult to predict exactly where.

We are left to make an educated guess about which RCP we are most likely to follow. But a fixation on a particular IPCC scenario is misplaced. No one scenario can correctly describe the future because both our

decisions about how much to use and the nature of fossil fuel production will change over time.

Consider that as the easier-to-exploit fossil fuel deposits approach depletion, the production from the harder-to-exploit deposits will increase, which means using more fuel and thus increased $CO_2$ emissions during fuel production. Therefore, even if the amount of fuel available for our discretionary use (i.e., not used to produce fuel) remains constant, the amount of our $CO_2$ emissions will still rise. These features alone will result in deviations from what appears to be today's relevant RCP. There may also be deviation from a favoured RCP because of the limitation in the prediction process.

The following is suggested as a useful, general working prediction about the RCP that we are likely to follow into the future. It assumes that all recoverable fossil fuels will eventually be burned. Thus, in the immediate term, as conventional oil crosses its production peak and energy demand continues to increase, we will continue to follow the RCP8.5 curve (Jackson et al. 2015; Raupach et al. 2007). During the medium term, fuel production difficulties alone will most likely prompt a switch to a warming path between RCP4.5 and RCP6.0. In the medium to long term, the unavoidable peak in production of the other fossil fuels and the subsequent decline toward their depletion will limit annual $CO_2$ emissions. It is guessed that the maximum warming reached by a fossil fuel dominated RCP will be in the 3.0°C range, relative to 1870 (figure 17). The most significant contribution to the uncertainty in this conclusion is the degree of effort we will devote to extract fossil fuels and how fast we will exert it, as discussed in chapter 15. We should know more clearly which mix of RCPs we are following by 2040–2050.

In overview, despite the controversies, deficiencies and limitations surrounding the scientific arguments about climate warming and the predictions about its future, there is enough evidence to say, with confidence, that climate warming is occurring and that it will continue into the future. The consequences will be global in scale, pervasive and significant. The scientific arguments also reveal that our warming future depends heavily on the inertia in the climate system and on our carbon-cycle-disrupting decisions.

In order to make informed and appropriate decisions about how to respond to warming, humans are stuck with evaluating and assimilating a mix of established facts and uncertain predictions about a dynamic, complex climate system. We must also face questions about our culture's world view, personal beliefs and accepted lifestyles because they too will influence our current and future climate-affecting decisions. We can gain some idea of the decisions we might make under these conditions by looking at our possible response to climate warming from a cultural perspective.

*The Cultural Response.* From one perspective, the people talking about the scientific discovery of climate warming can be divided into three groups. There are those with either strongly held *doomsayer* or *denier* views, while the *ditherers* in between find themselves struggling to decide what to believe as they listen to the two combatants vociferously defend their life-defining positions on climate warming (Kaufman 2007).

From a different perspective, the debate has two poles. At one pole are those attempting to understand global warming at multiple levels in order to find the most appropriate response to it. At the other pole are those concentrating on framing the debate and defending a position in order to satisfy some personal, political, ideological or financial objective. In between sit the many who are undecided or disinterested.

This messy cultural debate becomes easier to understand if we appreciate how we, as individuals, arrive at our views (opinions, perhaps) about new knowledge, decide on its implications for the future, and assign a place for it in our lives: that is, how we assimilate knowledge. We gain an appreciation for assimilation by remembering the discussions in part 1 about our innate methods of perceiving the world and making our day-to-day decisions. We also use them to evaluate and rationalize our decisions, divergences and any feelings of cognitive dissonance we experience. As discussed, these methods are constrained by our brain's functioning and influenced by both our culture's world view and our personal experiences. These are the methods we use to assimilate knowledge.

In a wider view, we (as individuals) use these methods when we interact with one another, our culture's world view and our supporting environment. They are part of the dynamic human-environment system and are included in the model of it as our rationalization process, as discussed in chapter 11. This provides us with the wider context in which to consider how we assimilate new knowledge into our lives (figure 15). In this context, we see that as a culture's members assimilate knowledge, it, by extension, becomes part of their culture's world view. Think of the Europeans coming to terms with the Lisbon earthquake of 1755.

The above description also describes the essence of how scientists assimilate a scientific discovery into their personal lives and into their science culture's world view. However, in detail, the scientists' assimilation process is different. Their process is fairly formalized and central to their culture, as described in chapter 4.

In contrast, the public's and thus the greater culture's assimilation of a scientific discovery is a largely unstructured public-private debate. In this type of debate, the participants are neither compelled to reflect the scientists' views of their discovery, refer to the scientific evidence, nor follow the scientists' code of conduct. The public-private debate is also wider ranging, which makes it better suited to finding and discussing the many implications of a discovery.

The greater culture's assimilation of a scientific discovery starts in the same manner as the science culture's process: a member's perception of a new scientific discovery about the external or our internal reality prompts them to bring it to the attention of a wider audience. If the discovery's apparent personal or cultural significance, or its newsworthiness, grabs enough of the public's attention, then a wider cultural assimilation process can begin. The initial stage of the resulting public-private debate is often characterized by the public holding poorly developed, widely differing opinions and emotions about the discovery.

The next stage sees the many evaluations of the discovery, such as its validity, usefulness, and predicted significance and ramifications, coalesce into a few dominant opinions. Think of change and persuasion as discussed in chapter 6. During these often unsettling debates, those who are casually involved unconsciously apply their innate human perceptions and decision-making processes while being guided by their culture's world view. Their opinions are also guided by informal discussions within their social circle, and by public comments by those promoted as experts.

The members of the public who are more deeply engaged are similarly influenced, but they are also more consciously involved. They pay closer attention to the interpretations provided by those members of our culture's scientific, religious, social, media, financial or political institutions whose authority over a new discovery or perception about it they trust. The guidance from those in authority usually takes the form of information and opinions, often presented in non-specialist public media and forums.

As a few dominant opinions about the discovery emerge, some individuals may find a ready-made place for one of these favoured opinions within their current internal reality and their culture's world view. For others, personal assimilation might prove more difficult. These individuals may customize a selected opinion, reject all opinions, become disinterested or adjust their personal views of the discovery to suit an opinion.

In all cases, each individual's objective is to recapture the settled feeling that comes with having a personal world view that *appears* to the holder to be coherent and consistent with their current selves and their social subgroups. Hence, scientific discoveries that clash with strongly held personal beliefs can be especially difficult for the individual to assimilate. The same is true for a discovery that clashes with strongly held cultural beliefs (Kahan et al 2012).

Note, there is only a weak imperative for an individual or the public to hold views of a scientific discovery that do justice to its scientific description. Neither do they have to take into account the scientists' views or how those views change with ongoing scientific investigation. Nor do the public have to consider the scientists' comments about the scientific validity of the dominant publicly held views. This freedom to diverge from the scientific view was illustrated in chapter 1 by the public's response to the discovery that some of our brain's functions were split between its two hemispheres. A significant segment of the public concluded that a person could attribute their personality traits to the dominance of one or other hemisphere.

The end stage of the greater culture's assimilation process is similar to the end stage of the science culture's process. It marks the time when the majority of the members have mostly come to terms with the discovery and only a few views dominate. The discovery and the public's interpretation of it have been assimilated into the culture's world view in some form. These interpretations are widely dispersed into the culture's various subgroups, even those whose members are uninterested. All that is left is to tidy up the loose ends.

The process of culturally assimilating a scientific discovery can be straightforward and quite fast. However, it can also be messy and may take a long time. This is especially true if the discovery is controversial or requires significant changes to our personal or our culture's world view.

Some discoveries are so controversial or have such a significant impact on a culture's world view that the greater culture's assimilation process can intrude into or even sideline the scientific assimilation of the discovery. Examples of controversial discoveries from Western history are Galileo's observations about the planets, Darwin's theory of evolution, Einstein's theory of relativity, the discovery that cigarettes cause lung cancer and the discovery that CFCs destroy the ozone layer. In all of these cases, the presence of a culture's imperatives that conflict with these discoveries (or their implications) and the differing beliefs of the culture's various subgroups play a significant role in the greater culture's assimilation process.

This general understanding of the individual-cultural assimilation process can help us appreciate how at least Western culture arrived at its present stage of assimilating the discovery of climate warming (both the event and predictions). It can also help us decide on the possible outcome of the ongoing assimilation debate. But to do so, Western culture's efforts have to be looked at from the perspectives of three cultural subgroups: the scientists, the powerful (the political and business elites, and their handmaidens) and the public.

The *scientific subgroup* consists of the scientists researching climate warming. They are members of a scientific culture that provides us with the most reproducible understanding we have about the functioning and state of the earth's climate system. But scientists are people with human foibles, just like the people in the other two subgroups. As individuals, scientists can exhibit a large ego, arrogance, a focus on prestige and narrow thinking. As a group, they can suffer from groupthink. We should therefore expect that squabbling, misrepresentation and infighting should appear to some degree during the scientific contribution to the cultural assimilation process.

However, a characteristic of science culture is that it tries to hold these disruptive tendencies in check by promoting trust, credibility and co-operation among its members, as discussed in chapter 4. One way science culture achieves this is by including in its world view the imperative that its members be skeptical of their own as well as others' conclusions. It also encourages them to direct any criticism of a theory toward the data and the process used to draw the conclusions rather than toward the scientists providing it (Shapin 1994). These practices lie at the heart of the peer-review process. In general, scientists hold to their culture's guidance about how they should behave. The success of the scientific method for dealing with controversial issues is reflected in the public's tendency to believe in statements made by scientists compared to those made by politicians and business people.

When scientists become involved in Western culture's assimilation of a controversial scientific discovery, they are forced to function outside of their science culture. Under these circumstances, they are subject to many more stressors. This is especially true when the powerful or the public have reasons to avoid accepting the scientists' discovery, regardless of its scientific validity (Oreskes and Conway 2010).

In the case of the West's assimilation of the climate warming discovery, the stressors on scientists include funding cuts to controversial research (Solomon 2008); those of the powerful trying to undermine the results of the scientific process (see below); and having to defend their work and their personal integrity in the public or political arena rather than the scientific one.

The intensity of these stressors is illustrated by both the hacking of email accounts at the Climate Research Unit in East Anglia, United Kingdom, and what the hacked emails, such as those of Dr. Phil Jones, revealed. They show that the discussions within the scientific community about climate warming had become more personal and political than usual. The peer review process had become stressed (Pearce 2010).

The political and business elite, the *powerful subgroup*, are a critical part of Western culture's climate warming assimilation process. Today, they have joined the scientists at the centre of the process, perhaps even displacing them. The powerful know that their decisions and actions will determine who will win or lose politically and financially during and after the assimilation process is completed. The powerful are thus under intense pressure to make sure that the making of decisions is favourable to them and their backers (Solomon 2008). Being powerful, they have the power to exert the influence needed to achieve their goals.

Whether driven by good intentions, ignorance or misguided self-interest, the increased efforts of the powerful to exert their influence has affected all aspects of the West's climate warming assimilation process. This can be most clearly seen in the well-documented evidence that some of the powerful are involved in hidden efforts to deliberately sideline, distort or incompletely represent the scientific evidence for climate change, the resulting predictions, the proposed solutions and the limitations to climate warming research, for their own ends. The evidence reveals connections among these efforts, the powerful's ideology and their financial interests (Montgomery 2006; Adam 2006; Begley 2007; Adam 2009a; Solomon 2008; Oreskes and Conway 2010; Goldenberg 2013a).

The most public example of the powerful's efforts can be seen in the United Nations' IPCC process. The organization nominally rests on a foundation of science. However, the decisions are ultimately made by the powerful. They sit as government representatives on the IPCC decision-makers panel (IPCC 2006). The influence of these political hands becomes most visible during the IPCC panel's efforts to finalize and then accept the draft reports prepared by the scientists. The final drafts can be (and are) changed by the government representatives to suit their own purposes.

Consider the following examples. During the acceptance process for the draft 1995 IPCC report, there were strongly voiced arguments between politicians and scientists over its wording and conclusions. Particularly contentious were those parts where the scientists stated that there was credible evidence for a human contribution to climate warming (Reiss 2001).

The 2007 IPCC group 2 summary report lists the impacts of climate warming on ecosystems. During its acceptance process, some government representatives, including those from Russia, China, Saudi Arabia and the United States, wanted specific passages to be removed or changed. Among these passages was the conclusion, on the first page, which stated with a very high confidence level that many natural systems are being affected by climate warming. The scientists objected strenuously to any lowering the level of confidence, so the panel simply removed the expression of confidence (Adam 2007; The Associated Press 2007).

The members of the *public subgroup* bring the full range of views about climate warming to the West's climate warming assimilation process. These views easily fill the categories of deniers, skeptics, the indifferent, the confused, believers, doomsayers, astute observers, the self-interested and the alarmist. The public also provide a wide array of explanations for their views. Despite appearances, their views and explanations are not a random collection of ideas. They reflect each member's personal world view, as influenced by their personal experience and their culture's world view.

This serves as a reminder that, in the ongoing debate, scant attention is paid to the influence that a culture's world view and its history has on the climate warming assimilation process and thus on its outcome. Consider that, since at least the 1700s, western Europeans have been debating the merits of ideas and predictions about climate warming and cooling. These earlier climate change debaters backed up their opinions with explanations that varied from religious to scientific in nature with varying degrees of coherence and credibility. The range of views they held reflected the variations in the western European cultures' world views over that period (Von Storch and Stehr 2002).

The same is true today. The diversity and relative popularity of Westerner's personal views about climate warming reflect the variation within Western culture's world view. An individual's view tends to reflect the views of the cultural subgroups they belong to.

In a sense, this is surprising. One would reasonably expect the public's ear to be tuned to the science subgroup's views about climate warming. After all, they are a trusted knowledge source. But the complexity of the topic, the scientists' limited skill at communicating it to the public, the promotion of misinformation, and more pressing personal issues all ensure that the public struggle to connect with the scientific knowledge or to objectively evaluate the climate warming predictions. In addition, their personal experience of weather provides little useful help. They can interpret it as they choose, which is why climate scientists routinely provide reminders that a change in the weather on its own need not indicate a change in climate.

Consider too that it can be emotionally difficult for a member of the Western public to assimilate climate warming into their personal world view and decision making. Imagine the conflicts they face trying to rationalize the scientists' climate warming predictions when it and the suggested responses indicate unpalatable changes to their personal lifestyle and culture. Especially as that lifestyle is supported by the guidance from their culture's world view: for example, consumerism-producerism and the energy-intensive lifestyle it favours. The public's difficulty is made worse by the efforts of those, found mostly within the powerful subgroup, who are trying to counter the scientific predictions of climate warming without due regard to the evidence or the scientific process. Of particular concern are those conducting the debate as if doubt was an acceptable end result rather than a useful tool to help decide on truth (Oreskes and Conway 2010).

In this context, it is unsurprising that the public find it harder to make the distinction between spin and fact, between thoughtful dissension or genuine error, and between hidden agenda or ego. For example, was the controversial hockey-stick-shaped temperature graph (flat from the 1800s until the temperature rose rapidly in the late 1900s) an unfortunate mistake that got out of hand, a genuine part of the normal scientific process, or

a deliberate attempt to manipulate the data (Solomon 2008; COSTR 2006; Appell 2006)? Does the word *consensus* represent the normal compromise found in the workings of the scientific process or a mask for a significant divergence of views?

The public have become sufficiently distracted by these human-centred issues and responses that they tend to exclude from their consideration the external-reality perspective of climate warming: the scientific observations. There is also little room for thoughtful introspection. Instead, the public tend to resolve their undecidedness and unsettled feelings about climate warming by favouring the status quo, whatever that means to them. This is slowly changing as the reality of climate warming inexorably intrudes into our lives.

In overview, Western culture's efforts to assimilate the discovery of climate warming, and the associated predictions about the future, are leaving the stage of intense debate. There is no longer the general feeling that the climate future is less knowable than it actually is, and there is no longer a sense that there is no clear or "reasonable" (business-as-usual) path forward. But the debate has not yet produced either a solid cultural consensus or an irreconcilable split in views. The assimilation process continues with the growing acceptance, driven by real-world events, that something may be afoot with the climate that deserves our attention.

Western culture's assimilation of climate warming is entering its last stage. In this stage, each subgroup will come to interpret climate warming to suit themselves. A reasonable prediction is that, in the immediate term, the science subgroup will remain sidelined or become more so.

Some in the power subgroup will treat climate warming as a serious environmental event and deal with it accordingly. In contrast, most will continue to choose actions that either ignore it or only appear to treat it as such an event. Their actions too will be promoted under the umbrella of sustainable development, the green economy and compromise, but, in reality, most of them will be treating climate warming as just another business/political opportunity. They will feel that their efforts are sanctioned by the need for economic growth. Others will continue to treat it as a threat to their values, such as power, wealth and ongoing financial growth.

Within the public subgroup, many will continue their attempts to make sense of climate warming and deal with the discomfort that arises from the confusing, conflicting or negative sounding views about it. Their efforts at a personal resolution will slowly gain steam as warming events intrude more into our lives. The individual's efforts and the increasingly obvious consequences of a warming climate will act as increasingly insistent cues for Western culture's leaders to address climate warming as a real event, and for the public to make changes to their beliefs and actions. But, until the idea of making serious changes gains traction, the public will try to maintain business as usual and thus, by default, side with the power group and Western culture's current world view.

On balance, it seems that, in the immediate term, Western culture's attempts to evaluate and assimilate climate warming will only result in small changes to the periphery of our personal and our culture's world view. It is unclear if, by the medium term, the more direct effects of climate warming on our lives will have prompted more substantial voluntary changes to the essence of our lifestyle. As a culture, we are currently following the same path as those who managed the fish in the collapsed ocean fisheries: business as usual until there are no more fish.

We can hope that Western culture's assimilation of climate warming will be facilitated by the public and the powerful turning away from both a focus on denier versus doomsayer, and excessive self-interest. Hope, unfortunately, is a poor predictor of our response to climate warming. A better predictor is our tendency to make our decisions under the influence of our personal foibles and our culture's world view. If we wish to more realistically assimilate the discovery of climate warming and its implications for our future, then we should consider it within the context of the human-environment system. This means also considering climate warming from the environment's perspective, to be discussed shortly.

*Conclusion.* The emerging picture of our material relationship to global warming (the translucent scientific argument) is that it lies partly buried under the rhetoric, confusion, uncertainty, posturing and excessive personal self-interest that are part of Western culture's efforts to assimilate global warming (the contentious issue). If we wish to make more sense of global warming (the environmental event), we should focus less on the wide selection of available views and pronouncement. We could instead focus more on the rules of thumb that describe the climate system's functioning. We could focus on the trends displayed by the past environmental observations and on how we are changing our environment. By using them as references, we should find it easier to make sense of global warming, the event. This is especially true if we remember the

nature of our decision-making foibles, acknowledge our reality paradox and recognize the influences of our culture's world view on our decisions.

With this in mind, what can we conclude about climate warming, the environmental event? On balance, climate reconstructions and modelling agree with the field observations: global-scale climate warming is happening. This evidence supports the conclusion that, for at least the last 50 years, our cumulative contribution to climate warming has been so large that the consequences can increasingly be separated out from the long-term, non-human background changes caused by factors such as the Milankovitch sunlight cycle and volcanic activity (Karoly et al. 2003; Zhang et al. 2007; Mann et al. 2008). For example, the 2014 near-record-high average annual temperature experienced in central England (an area 200 km by 350 km) can clearly be related to human-caused climate warming (King et al. 2015).

A significant human contribution to climate warming is real. Warming is expected to continue into the future. Despite claims by climate skeptics to the contrary, these conclusions are supported by some 97% of climate scientists (Anderegg et al. 2010).

Despite the ability of models to help us draw these general conclusions, their ability to predict the details of our climate future is significantly constrained because they can only imperfectly represent our dynamic, complex climate system. The models favour the smoother, slower-changing attributes of climate change. The less ordered and harder to predict attributes are less well represented. These features include the statistical probability of small random or chaotic changes to the climate system leading to large changes in climate (Adams et al. 1999).

The possibility of current climate warming becoming a sustained global-scale abrupt climate change event was not discussed in any of the 2007 (4th) IPCC summary reports for policy makers (IPCC 2007a; IPCC 2007b; IPCC 2007c). The 2013–14 (5th) IPCC summary reports for policy makers did mention the possibility of an abrupt event and its impacts, but it simply noted that the chance of an abrupt event increases as the level of warming increases. Any further evaluation beyond that general statement was limited by a lack of information (IPCC 2013a; IPCC 2014a; IPCC 2014b). As a result of these limitations, those parts of the current climate warming predictions that extend into the mid- to long-term future can appear more certain than they are.

The public are constrained in their use of climate predictions because of the difficulty they face trying to meaningfully relate to them. Consider that, in the immediate term, the presence of climate change can be masked from our experience by the weather's normal variation. Consider too that the process of global climate warming will itself vary. The lack of a reliable immediate-term personal experience of climate warming makes it harder for us to relate to medium-term climate predictions, especially as they seem so uncertain.

Ironically we are partly responsible for the uncertainty in the predictions. After all, our past actions and current reaction to climate warming has increased the number of possible climate futures. This has changed the probability of each occurring and increased the uncertainty in the climate warming predictions (Alley et al. 2003; Allen 2004).

There is one other certain aspect of climate change predictions that is worthy of mention. Our power to alter our climate system is sufficient that our decisions are, and will be responsible for, a major part of our medium- to long-term climate future. The greater the changes we make to the climate system, and the faster we make them, the greater will be our contribution to future climate warming. In addition, because of delays (inertia) in the climate system, our efforts to reduce our influence on climate warming will be more effective if implemented sooner rather than later. Beyond some point, our efforts will only affect the waves, not the tide.

It currently seems that our culturally influenced, internal-reality focused imperatives will dominate our efforts to assimilate climate science's representation of climate warming (the event). As a result, our decisions about how to respond to it are likely to remain tentative and/or ineffectual for a while yet. The historical record of Westerners' past reactions to significant scientific discoveries suggests that we will, eventually at least, accept the reality of climate warming, however our response will still be constrained by our favoured cultural and personal imperatives. Think of the economy and economic growth. Westerners' preferred choice of a response to climate warming will be implemented regardless of whether it will effectively reduce our contribution to warming over the long term or not. We will likely stay this course at least until the external-reality tide starts to drown our dreams and our climate-related reality paradox experiences start to overwhelm us.

More generally, it seems likely that the impacts of ongoing climate warming will impose themselves on humans for a while. There is a real chance that, by the time we decide to react in an appropriate manner at an appropriate level, we will still be debating whether we should adapt primarily through the application of technology or cultural transformation. By then, the opportunity to minimize climate warming and its impacts will have largely passed us by.

**Personalizing Future Climate Warming.** We find it difficult to assimilate climate warming and its implications for the future into our lives, let alone make the most appropriate decision about how to deal with it. These tasks are made easier by personalizing climate warming's possible futures in a systematic manner. This section first discusses the more certain predictions of steady warming and then the possibility of an abrupt warming. The carbon cycle acts as a useful reference (figure 7a). The likely future geophysical consequences of climate warming are illustrated before providing a summary.

*Steady Warming.* During the last 200 years, human activity has steadily, first slowly then rapidly, increased the concentrations of the most common greenhouse gases ($CO_2$, methane and nitrous oxide [$N_2O$]) in the atmosphere to well above the historical levels for either the present or the previous four interglacial periods. As a result, the atmosphere's temperature has also steadily, slowly then rapidly, increased (figure 11) (IPCC 2013a). This trend continues today. Between 2000 and 2014, the average $CO_2$ increase was around 2 ppm/year. The $N_2O$ increase was around 0.6 parts per billion per year (ppb/year) in 2000, rising to around 0.9 ppb by 2014. There was no change in the average concentration of methane between 2000 and 2007, but since then, it has been increasing at around 5 ppb/year (WMO 2015).

Just as the earth's climate during the past 200 years was significantly influenced by our past decisions, so will its climate future be largely determined by our present and future climate-affecting decisions. Of particular importance are our carbon-cycle-disrupting decisions: they include our future demands for fossil fuel energy and cement, and our future direct impacts on the earth's ecosystems, such as deforestation. Those decisions are and will be strongly influenced by our culture-guided views about the economy and our relationship to our environment. As previously discussed, we display significant inertia to changing our personal and our culture's world views, and thus our decisions.

The earth's climate future also depends on the carbon cycle's rebalancing process: how it responds to our $CO_2$

and methane emissions that have disrupted its interglacial-period balance. The new balance will be achieved by shifting carbon from the atmosphere reservoir into the ocean- and land-based reservoirs. But this redistribution process takes time. For example, at least 100 years must pass before the majority of our $CO_2$ emissions to the atmosphere from the last 25 years will have been transferred into the ocean and land reservoirs. Most of the remainder will be removed over the following centuries (Friedlingstein and Solomon 2005).

The inertia/delay that both the carbon cycle and we display when responding to increases in the $CO_2$ and methane content of the atmosphere enable us to make a quite certain general prediction about climate warming. We will continue to experience climate warming at least well into the mid-term to early long-term future. And the earth's oceans and ecosystems will continue to change for just as long.

It is much harder to make a certain prediction about the details of climate warming's future path: how large the warming might be and when it will peak. One reason, discussed above, is the limits to our understanding of the climate system and the constraints on building computer-based models of it. Another reason is that we may change our climate-affecting decisions, but the details of when, by how much and how effective the changes might be are difficult to predict.

The IPCC overcomes the difficulty of reliably predicting our future decisions by, as previously discussed, representing them as four $CO_2$ emission pathways. These range from drastically reducing our $CO_2$ emissions today (RCP2.6) to continuing to maintain or increase our emissions (RCP8.5) (figure 17a) (IPCC 2013a; IPCC 2007a). The climate predictions that result from these pathways reveal more about what we can expect for our climate future, if it arrives in a steady manner.

For all pathways, the average rate of global warming will be roughly the same until around 2025. It is only around 2050 that a divergence in the trends of RCP8.5, representing the fastest rise, RCP4.5, which will be peaking, and RCP2.6, which will be past peak, will be clear. By 2100, the average temperature of the atmosphere near the earth's surface will, at a minimum, have exceed its warmest temperature in the last 11,000 years (the Holocene). And, except for RCP2.6, it will have exceeded its warmest temperature in at least the last 120,000 years (figures 7 and 17a) (Marcott et al. 2013).

For all of these scenarios, the closer someone lives to the north or south pole, the larger the change in temperature they will experience. For example, the 100-year average warming in the Arctic region is expected to be double the global average, while at the tropics, it will be less than the average. Overall, as the atmosphere warms, the earth's climate types will migrate toward the poles.

The general changes in precipitation are more difficult to predict. In general, as they migrate, the climates that are dry today are expected to become dryer, while those that are wet are predicted to become wetter. Precipitation nearer to the poles will increase, while the proportion occurring as snow will decrease (Burke et al. 2006; IPCC 2013a; Zhang et al. 2007). The number of storms delivering this precipitation and their severity is expected to increase (Allan and Sodden 2008). Other extreme weather events, such as droughts, are also expected to become more common.

Throughout the 1900s, these changes to the earth's climates were masked from our day-to-day lives by the normal fluctuations in the weather. In 2007, it was predicted that sometime between 2012 and 2017 the public should become more confident that they are experiencing events related to global climate warming (Smith et al. 2007). Indeed, as mentioned earlier, in 2015 scientists successfully attributed a 2014 extreme warming event in England to climate warming (King et al. 2015). Similarly, the public are noticing that weather records are being routinely broken, previously unusual weather events are becoming normal, and rare events are more common. Think of hurricanes Harvey and Irma as well as the extended, intensive fire seasons in North America and Spain during 2017. It is now possible for an astute member of the public in some near-polar- and mid-latitude countries to note with confidence the ongoing multi-year changes in their local environment, such as glaciers melting, unusual species appearing and a later fall. Collectively these features are consistent with climate warming.

Today's decisions can't change these common trends of our immediate- and medium-term climate future. However our current and immediate-term decisions will collectively determine the mix of IPCC pathways that we and our descendants are most likely to follow into the long-term future. We do have a considerable influence over the details of our and our descendants' long-term climate experiences. We can choose the size of the warming peak that we and our descendants will experience, and when it will arrive.

What should we decide? We could continue the current trends in our decision making and trust that fossil fuels will run out soon enough. In this case, as discussed above, our future climate warming path could initially follow RCP8.5, then switch to a path between RCP4.5 and RCP6.0. If we follow this fossil fuel limited mix of RCPs, then it seems that the maximum warming reached will be in the 3.0°C range (1870 reference) (figure 17).

Even if this prediction is correct, it provides little consolation. Consider the argument that a warmer earth will allow us to avoid the next ice age. Even if we follow RCP2.6, the next ice age trough would likely still be relatively weak and only occur some 20,000 years from now (Ganopolski et al. 2016). The concern is less about avoiding a future colder world and more about how much warmer a future warm world would become. There is a good chance that the details of a world only slightly warmer than now would, overall, not be to our liking.

We could delay making our decisions until, say, 2030 to be more sure how the future is unfolding and which mix of RCPs we are following. But the considerable inertia in the climate system means that, by default, a decision to delay is also a decision to continue our current climate warming trend much further into the future. Doing either nothing or delaying are risky choices for such an important decision. If we remember that our decision will determine the climate burden our descendants will bear, then to be fair to them, we should decide to reduce our emissions today.

By how much should we reduce our emissions? Consider that today's complex cultures have built their infrastructure and economies around the climate system's current structure and functioning. Think of the current distribution of climate zones and the present sea level. To lessen the impacts of climate warming on our and our descendants' lives, the climate system should remain largely intact. A thoughtful answer would also note that the less severe the environmental consequences of climate warming are, the more predictable it is and the easier it will be for us to adapt to the change. This will be the case if the future average global temperature does not exceed that of the last four interglacial periods, and if the increase occurs slowly (figure 11). By meeting all of these conditions we would make it much easier for us and our descendants to adapt to the already predetermined part of future climate warming.

Climate scientists suggested in 2007 that these objectives are more likely to be met if we can limit the warming of the earth's lower atmosphere to a peak temperature of less than 2.0°C–2.4°C above the pre-industrial (1750) temperature reference. To meet this 2007 target, we would have to ensure that the atmosphere's $CO_2$ remains below 350–400 ppm. If the contribution of trace greenhouse gases, such as methane and $N_2O$, to atmospheric warming is included in the target as a $CO_2$ equivalent (the amount of $CO_2$ that is needed to trap the same amount of heat as the trace gases), then the atmosphere's greenhouse gas content can't exceed 445–490 ppm $CO_2$-equivalents (average of 467 ppm) (IPCC 2007c; Hansen et al. 2007; Rockström et al. 2009). In 2014, the atmosphere's $CO_2$ concentration was 397.7 ppm and rising at around 2 ppm/year. By the time you read this, the 350–400 ppm $CO_2$ (1750) target will have been exceeded.

In 2013, climate scientists expressed the peak temperature target differently. To ensure a >66% chance of the maximum warming being <2.0°C above an industrial revolution (1870) reference, humans can emit, between the years 1870 and 2100, a total of no more than 800 gigatonnes of carbon (GtC) (2,936 GtCO2) from all $CO_2$ sources (no other greenhouse gases included as a $CO_2$ equivalent.) Already, between 1870 and 2011, some 407 GtC (1,494 GtCO2) were emitted, thus a maximum of only 393 GtC (1,442 GtCO2) can be emitted between 2012 and 2100.

The IPCC's lowest warming path, RCP2.6, most closely meets the <2.0°C target. Along RCP2.6, the total $CO_2$ emissions between 1870 and 2100 are 2,388 GtCO2, which is less than the 2,936 GtCO2 limit to meet the <2.0°C target. Similarly, RCP2.6's future emissions between 2012 and 2100 are 994 GtCO2, which is less than the 1,442 GtCO2 permitted to meet the target. If we follow RCP2.6, then the atmosphere's $CO_2$ is expected to reach slightly less than 450 ppm $CO_2$, and the atmosphere's temperature will peak at just below 2.0°C (1870) (figure 17) (IPCC 2013a).

However, there is more to consider before deciding to use RCP2.6 as a guide to ensure that warming will be <2.0°C. Consider that, critically, the switch from the 2007 target of <2.0°C–2.4°C (1750 reference) to the 2013 target of <2.0°C (1870 reference) increases the acceptable maximum CO2 content of the atmosphere from 350–400 ppm (1750) to just above 450 ppm (1870). By simply shifting the reference for the <2.0°C target, we think that we can "safely" emit more CO2 and still meet the <2.0°C target. This, like saying that the dog ate our homework, may provide us with the feeling that we still have more time to act, but it doesn't change the external reality that the atmosphere will be warmer, in absolute terms. Thus the CO2 emissions allowed by RCP2.6 are a maximum, if the <2.0°C target is to be meaningfully met.

The emissions permitted by RCP2.6 are also a maximum for other reasons. Although they are lower than the maximum allowed to meet the <2.0°C (1870) target, they may not be low enough to fully account for the additional warming that will result from the expected emissions of the other greenhouse gas. Consider too the aspects of the earth's climate future that are absent or difficult to include in the IPCC's prediction of an emissions limit. To accommodate these unconsidered and unexpected aspects a buffer must be created by lowering the $CO_2$ emissions even further below that permitted by RCP2.6 to meet the <2.0°C (1870) target. Examples of these aspects are warming that is faster than expected, the unexpected increases in the emissions of known human-made greenhouse gas, and the unintended creation and emissions of a new greenhouse gas. Some of these possibilities will be mentioned when discussing energy in chapter 15. The example discussed here is HFCs and HCFCs, the replacements for the ozone-destroying CFCs banned by the 1987 Montreal protocol.

As CFCs were phased out, the global emissions of HFCs and HCFCs rose rapidly. Ironically, some of them are particularly potent greenhouse gases. Some are so powerful that if the recently discovered increasing trend in their emissions continues, then HFCs and HCFCs could contribute between 28% and 45% of the 467 ppm $CO_2$-equivalent maximum greenhouse gas emissions permitted by the 2007 <2.0°C target (Velders et al. 2009). Without significant, and immediate, constraints on the emissions of HCFCs and HFCs, their current 7% annual growth rate could undermine other efforts to reduce climate warming (WMO 2014).

It took until 2016 for the Montreal protocol to be amended so that it addressed HFC emissions. The amendment is intended to ensure that the global peak in HFC emissions occur by 2024: some 37 years after the original signing of the Montreal protocol and more than 7 years after the problem was detailed. Beyond this date, HFC emissions are expected to decline (Hall 2016).

When these features and possibilities are taken into account then, to be fair to our descendants, our efforts to meet either the 2007 or the 2013 <2.0°C targets should be more than just immediate. They should also be strong. So far, it took until 2016 before just an aspirational 1.5°C–2.0°C target for greenhouse gas reductions was agreed to by the world's nations (United Nations 2016).

In overview, the steady warming view of climate change indicates that if we wish to most easily adapt to ongoing climate warming, then our decision is clear cut. We must ensure that the earth's atmosphere and ocean circulation systems must stay much like they are, and the earth's climates must stay roughly like they are and where they are. This means that we must treat the 2013 <2.0°C (1870) target as a maximum. It also means that we must immediately, and strongly, reduce our greenhouse gas emissions to a level that satisfies this target. There is another reason why we should take the <2.0°C target to heart: the IPCC predictions take into account neither emerging contributions to climate warming, such as new human-made greenhouse gases, nor the less steady, chaotic and random aspects of climate change, such as an abrupt event.

***Abrupt Warming.*** The steady warming predictions assume that if our greenhouse emissions end, then after a delay, the current climate warming trend will slow, stop and reverse. There is another option to consider. Could today's climate warming become tomorrow's "runaway" warming? Could it become one of the spike-like abrupt events seen in the 400,000-year historical record of climate changes discussed in chapter 8 (figure 11)?

This is a reasonable question to ask because, as previously discussed, the current rate of the atmosphere's temperature rise is close to that of the Younger Dryas abrupt event. However, that event was likely caused by the air burst of a comet. If the current warming is to become an abrupt event, then it would have to be generated by an amplifying feedback formed within the earth's climate system (Hansen et al. 2007; Alley et al. 2003). With this in mind, two features of the current warming are discussed: the ongoing rapid decline in the Arctic Ocean's summer sea ice and the possibility of enhanced methane and $CO_2$ emissions from permafrost.

The most well-understood process that could contribute to an abrupt event involves the Arctic Ocean and its floating sea-ice cover, which, together, play a critical role in global climate regulation. Simplistically put, the Arctic Ocean's white summer sea ice reflects sunlight back into space. The current climate warming has resulted in the ongoing reduction in the area of this higher albedo sea ice and a corresponding exposure of more of the lower albedo, dark sea water below. This means that less solar radiation in the Arctic is being reflected back into space and more is being absorbed by the Arctic Ocean. The resulting increase in the Arctic Ocean's temperature feeds back to melt more of the sea ice floating on it. Eventually, when all the sea ice is melted, the Arctic Ocean can warm in earnest. A warmer Arctic ocean helps to maintain rapid global climate warming (Lenton et al. 2008).

Evidence of this feedback and the amplification of climate warming is provided by the increasingly earlier predictions of when the Arctic Ocean will become free of late-summer sea ice. In 2005, the National Snow and Ice Data Centre (NSIDC) used historical trends to suggest that an ice-free status would be achieved by the end of 2099 (NSIDC 2005). In 2007, the IPCC predicted that, for some scenarios, ice-free status would occur sometime in the latter half of this century (i.e., after 2050) (IPCC 2007a). In 2010, the NSIDC suggested that a date closer to 2080 was possible (NSIDC 2010). In 2011, a group using more complete calculations predicted that the date is closer to 2040–2050. Some scientists even suggest a date as early as 2020 (Black 2011). In 2012, the NSIDC reported that an ice-free Arctic might occur in "a few decades" (NSIDC 2012). In 2013, the IPCC predicted a summer-ice-free Arctic before the mid-2000s (i.e., before 2050) (IPCC 2013a).

A less understood feedback that could lead to an abrupt event involves enhanced emissions of $CO_2$ and methane (natural gas) from the belt of frozen ground (permafrost) around the Arctic Ocean within the Arctic and Subarctic climate zones. Permafrost occurs both on land (15 million $km^2$) and in some parts of the sea floor near the Arctic coastline where the ocean is less than about 100 m deep. Permafrost is a significant carbon reservoir because trapped within its frozen mass is enough organic matter and carbon-containing gases for its carbon content to be roughly twice the amount currently found in the atmosphere. How could this permafrost contribute to an abrupt event?

First consider the land-based permafrost. When climate warming reaches its peak then, for each degree of warming above preindustrial times, some 4 million $km^2$ of permafrost is expected to have melted (Chadburn et al. 2017). As it melts, its trapped carbon gasses are released as $CO_2$ and methane (a stronger greenhouse gas than $CO_2$) while the thawed organic matter is used by some bacteria as food. In doing so their digestion releases a mix of organic carbon compounds, including $CO_2$ and methane, to the atmosphere. The proportions of $CO_2$ and methane released from the permafrost by these two processes depend significantly on the amount of oxygen in contact with the thawing ground. Low oxygen favours methane production.

The amount of oxygen is determined by the wetness of the melting ground. Water-saturated ground is oxygen poor. Thus, the amount and rate of future $CO_2$ and methane emissions from thawing land-based permafrost will be determined by the poorly understood complex interactions between climate warming, the functioning of Arctic ecosystems (especially the bacteria), Arctic landforms and their hydrology. At a particular location, all of these factors are changing as the permafrost there melts in response to the current warming trend (Stokstad 2004; Vonk et al. 2012).

Despite the uncertainties, current knowledge about $CO_2$ and methane emissions from the melting land-based permafrost in Siberia illustrates the general trend of $CO_2$ and methane release and its possible significance. The Siberian tundra is densely patterned with shallow lakes. The bottom of these lakes is permafrost. As it melts, much of its carbon is released as methane that bubbles to the surface.

A study of a restricted area indicated that these emissions have increased 58% over a warming period from 1974 to 2000, but this estimate can't be reliably extrapolated to a larger area. The more important observation is that some of the carbon in the lake's methane bubbles is over 35,000 years old (Walter et al. 2006). It was trapped in the permafrost at the height of the last ice age, when the atmosphere's temperature was at its minimum. These two features illustrate that at least a steady melting of land-based permafrost is underway.

This might not be all that is happening. Recent observations suggest that, throughout the high Arctic, a short-lived, extremely warm summer season can trigger more ongoing melting of the permafrost than predicted by assuming a steady warming trend. This indicates that a major melting event, and thus a larger release of $CO_2$ and methane, is occurring or could occur sooner than previously thought (Liljedahl et al. 2016).

Next consider Siberia's sub-ocean permafrost. During deglaciation at the end of the last ice age, the frozen land along Siberia's ice-age coast was inundated to a depth of about 50 m by a rising Arctic Ocean. This drowned permafrost, which covers about three times the area of Siberia's terrestrial permafrost, forms part of eastern Siberia's continental shelf. This sub-ocean permafrost, along with the overlying marine sediments, contains a large amount of carbon.

The sub-ocean permafrost is much warmer than the terrestrial permafrost. It is kept near to melting by the overlying ocean, which has an average temperature near 0°C. As a result, the sub-ocean permafrost is steadily releasing methane at a rate of about 8 teragrams per year (Tg/year). (Schuur reports 17 Tg/year (Schuur et al. 2017). This is approximately 1% to 2% of the current annual global methane emissions of about 440 Tg/year.

Under current climate conditions, the portion of the sub-ocean permafrost that lies both less than 30 m below the ocean's surface and close to the coastline is expected to continue emitting methane (Shakhova et al. 2010). Simple models suggest that under the predicted steady climate warming conditions, methane will continue to be emitted slowly and steadily for decades. But these models do not take into account the possibility of accelerated warming events (Heimann 2010).

With this background in place, we can ask if future $CO_2$ and methane emissions from melting (land and subsea) permafrost could contribute to an abrupt warming event. An Arctic-wide evaluation of possible future emissions from permafrost concluded that their magnitude and timing is still uncertain, thus their impact on climate warming remains unclear. However, it does seem that a rapid (sudden) release of a large amount of $CO_2$ and methane from permafrost is unlikely. It is more certain that ongoing, probably increasing emissions from permafrost will last for at least a century, be substantial and sustain, perhaps increase, current rates of climate warming. This contribution is not included in the IPCC's models, so future climate warming will at least occur faster than those models suggest (Schuur et al. 2017).

The possibility of an abrupt climate warming event becomes more likely when the emissions of $CO_2$ and methane from permafrost melting are considered together with the loss of Arctic sea ice. After all, both already participate in a feedback loop in which they contribute to climate warming, while that warming enhances both sea ice loss and permafrost melting. Together they have the potential to not just maintain but accelerate climate warming.

Consider that after the sea ice has melted, the Arctic Ocean can warm in earnest. As it does, more of the subsea permafrost would melt. More of the land-based permafrost would also melt because the Arctic region plays an important role in both the regional climate and the global climate system. In particular, the warming of the Arctic Ocean will likely help to change the path of the jet stream in ways that result in an increase in anomalous atmospheric warming events in the Arctic. Think particularly of longer and hotter periods of summer weather melting land-based permafrost. The resulting increased emission from the melting of subsea and land-based permafrost would contribute to climate warming. Together, the enhanced warming of an ice-free Arctic Ocean and increased greenhouse gas emissions from the augmented melting of permafrost may accelerate the current, already rapid, climate warming.

But the emissions from permafrost will only contribute to such a feedback as long as there is permafrost to melt, unless another carbon reservoir starts to release $CO_2$ or methane. Frozen methane deposits are distributed worldwide in ocean sediments found at depths greater than about 500 m. These deposits are estimated to contain double the energy of the known amount of fossil fuels: they are substantial (Haq 1999). If the current rapid warming can be sustained until the mid to deeper ocean has warmed enough to melt these frozen methane deposits, then their emissions could sustain an abrupt warming event for a much longer period. Currently, this possibility is speculation.

Descriptions of a mechanisms for a possible internally driven abrupt event based on current data, like this one, are accompanied by significant uncertainties. This makes it difficult to decide on the probability of it happening and when. A longer-term view of the last four interglacial periods helps with that decision. The three oldest interglacial periods display sharp temperature and methane peaks (figure 11). Only the current interglacial period's peak is flat. Only the current interglacial began with a Younger Dryas cold event. Does this mean that there is latent potential for an abrupt event?

Consider too the relationship between changes to the Greenland and Antarctic ice cap and the sudden rapid rise in temperature and sea level at the end of the previous interglacial period (125,000 years before present) (Hansen et al. 2016). Could this be relevant to an abrupt event in the present? Overall, it seems that it is possible for an abrupt event to arise out of the current rapid warming and that this possibility should be a part of our decision making.

In overview, it is difficult to predict with reasonable certainty the conditions under which an abrupt climate warming event might occur, or when. Climate warming feedbacks that could become self-amplifying already exist in the Arctic. Climate modelling suggests that if the current rapid atmospheric warming does exceed the <2.0°C (1870) target, then runaway climate warming is more likely to occur. Such an event would greatly complicate our efforts to adapt to climate warming. The possibility of such an event is real. The <2.0°C target should be thought of as a maximum.

*Geophysical Consequences.* A warming climate will affect more than precipitation and temperature levels and their distribution. It will force a rebalancing of the relationship between the earth's climate system and its other geophysical systems. Examples of the likely geophysical consequences of this rebalancing are given here.

A local-scale example from the Arctic is the increase in land slumps and slides caused by permafrost melting. Regional-scale examples are the increasing intensity and frequency of warming related events such as thunderstorms, cyclones, tornados, windstorms and floods. This includes related events such as soil erosion. Global-scale examples include the increasing warming and acidification of the world's oceans and the ongoing melting of mountain glaciers and the ice sheets covering Greenland and Antarctica. The direct impacts from the redistribution of this previously frozen water are many. Of interest here is that the meltwater ends up in the oceans where its added volume combines with the expansion of the oceans as they warm to result in a global rise in sea level.

In 2007, the IPCC predicted that sea level will rise by an average of between 2 and 6 mm/year to reach between 0.2 and 0.6 m above their datum by 2100. But they acknowledged that their predictions were likely too low (IPCC 2007a). In 2013, they predicted that, by 2100, sea level would rise between 0.45 m, for the least-warming RCP, and 0.75 m, for the greatest-warming, RCP(8.5) (IPCC 2013a).

Part of the reason the IPCC increased their predicted rise in sea level is their recognition that the ice sheets are wasting by both a static and a dynamic process (Hansen et al. 2007). The static melting process is illustrated by two features: during the summer, dry white snow in the interior of the Greenland and Antarctica ice sheets becomes wet and darker; and, at their edges, pools of melt water form on their surface. The dynamic process is illustrated by the streams of ice that flow out from the interior of the ice sheets. Toward the margins of the ice sheets, these streams become well-defined glaciers. The glaciers terminate in the ocean, where they disintegrate and melt to form the dynamic edge of the continental ice sheets (Prichard et al. 2009). As the air temperature increases, so does the temperature of the ice and thus the speed with which ice

streams and glaciers flow. They thus deliver more ice to the oceans, where the warmer ocean melts it faster, which results in sea level rising faster.

With this in mind, the recent discovery that the melting of Greenland's ice cap probably contributed more to sea level rise in the 1900s than previously thought should be unsurprising. More importantly, since the late 1900s, Greenland's contribution to sea level rise has been accelerating. Thus sea level will likely rise even more rapidly than the IPCC currently expects (Kjeldsen et al. 2015).

The example of sea level rise during the interglacial of 125,000 years before present (yBP) supports the idea that sea level could rise further and faster than the IPCC predicts (figure 11). During that interglacial, sea level rose rapidly to between 6 and 9 m above current levels. This means that large parts of the Greenland and Antarctic ice sheets must have melted rapidly. Surprisingly, this occurred at a time when the atmospheric level of $CO_2$ was around 280 ppm and the global climate was apparently warmer than today by only <1°C. The mechanism for this rise is not fully understood. However, modelling of today's climate system with the atmosphere's $CO_2$ levels in the 400 ppm range indicate that today's continental ice sheets could experience a similar rapid melting event, which would cause a rapid rise in sea level in the range of metres. This rise could partly occur over a period of 50–150 years (Hansen et al. 2016; Overpeck et al. 2006).

There is a possibility that the future rates of sea level rise could be much closer than expected to those experienced during the deglaciation that ended the last ice age. Between 18,000 yBP and 8,000 yBP, sea level rose around 110 m, at an average rate of 11 mm/year. The nominal average rate of sea level rise between 8,000 yBP and 200 yBP (6000 BC and AD 1800) was 1 mm/year.

More recently, between 1901 and 2010, sea level rose at an average rate of about 1.7 mm/year. However, between 1993 and 2010, the average rise was 3.2 mm/year. The IPCC 2013 prediction is that sea level will rise at an average rate of about 4 mm/year for RCP2.6 and 6 mm/year for RCP8.5 until at least 2100. The maximum absolute rise by that date is expected to be well below a metre (IPCC 2013a). In light of the information about sea level rise from Greenland, the 125,000 yBP interglacial and calculations about them, the predicted rates could represent just the beginning of faster sea level rise, to a greater height, that could occur sooner than expected.

It is reasonable to expect that there will be a wide range of geophysical impacts from just the sea level rise predicted by the IPCC. They include the intrusion of saltwater into fresh water aquifers near the coast, and the flooding and erosion of low-lying land, such as river deltas with their coastal wetlands (Nicholls et al. 1999). It is clear that, even without considering climate warming's effects on ecosystems, it will result in significant changes to the earth's environment in the immediate to long-term future.

In overview, personalizing future climate warming helps us to appreciate that we are a major contributor to a future of ongoing warming and its geophysical consequences. If we want our descendants to more easily adapt to this emerging future, then we should enthusiastically aim to meet the <2.0°C (1870) climate warming target suggested by climate warming models. However, we shouldn't forget that this target represents the steady arrival of our climate future. If we take into account the aspects of the climate system that are poorly considered or unconsidered in the models and predictions, and the presence of a feedback that could become an abrupt warming event, then we are reminded that the <2.0°C (1870) target should be treated as a maximum.

In summary, despite the constraints on making reliable climate predictions, it is certain that significant and rapid climate warming is currently underway. It is also certain that the consequences of our past climate-affecting decisions are being carried forward by the climate system's inertia to become part of the earth's climate future. As a result, part of that future is already determined for decades to come. This inertia-related aspect of that future is captured in the IPCC's conservative predictions. What is not represented in the predictions is the possibility of an abrupt event.

We are better able to appreciate our possible climate future, and predictions about it, after accepting two things about ourselves: Our decisions are a major contributor to the earth's climate warming future; and we struggle to become personally aware of and pay attention to our contribution and its significance. We are helped with both by those model-based predictions of our possible climate futures that include our possible decisions. The IPCC represents them by four $CO_2$ emission pathways (RCPs). The presence of large differences between the RCPs illustrates the significant influence that our current and future decisions will have on the earth's climate future (figure 17).

Those decisions include the amount of fossil fuel we think it is fair for a person to use each year and the number of children we think it is responsible to raise to use that per person energy. From a different perspective, the earth's future climate will be affected by the fossil fuel intensity of the lifestyle we and our children will adopt, in combination with the level of environmental disruption we will find acceptable. For example, would Canadian citizens accept a level of warming that would require them to replace the maple leaf on their flag with a cactus? It is difficult to reliably know when or how we might change our climate-affecting decisions, thus the wide range in the four $CO_2$ emission paths also represents the uncertainty in the earth's climate future.

The climate models suggest that we should make decisions that limit the (lower) atmosphere's warming to <2.0°C above an industrial revolution (1870) reference temperature. If the atmosphere's average temperature rise is above 2.0°C, then the consequences of the warming will likely significantly increase the disruption to our lives. For example, the greater the warming, the more likely that a switch will occur in the atmosphere's or oceans' circulation systems from their present quasi-stable state into a different state (Rahmstorf 2000; Flückiger 2008).

It isn't just the amount of warming that matters but the speed at which it is occurring. Too little is known about abrupt events to include that potential in climate models, however, the possibility of one occurring is real. For example, the Arctic Ocean's declining area of summer sea ice and the greenhouse gas emissions from the Arctic's melting permafrost form a climate warming feedback that could contribute to human-caused climate warming becoming an environment-assisted abrupt climate warming event (Alley et al. 2003; Lenton et al. 2008). The longer the rapid warming continues, the more likely we are entering or will enter an abrupt event.

Currently we are making decisions that lead to a climate future that is likely to be warmer than the preferred maximum increase of <2.0°C. An assessment of fossil fuel availability suggests that perhaps a limit to the availability of fossil fuels might result in a warming peak in the 3.0°C range. However, leaving our future to this fate is unlikely to produce a desirable result.

The decisions we can make that will ensure we to stay below the <2°C warming target are constrained. These constraints arise in part from the climate system's characteristics, such as its inertia and complexity. Its inertia to change alone ensures that efforts to avoid exceeding the target must begin long before the desired target is approached. Consider too the difficulty of reversing the consequences of climate warming, such as sea level rise and changes in precipitation.

Meaningful decisions to reduce climate warming are also constrained in part by our human characteristics. Consider our preference to maintain business as usual. Consider too our preferred initial response to a reality paradox experience: mask, avoid and ignore. Think too of cognitive dissonance. The resulting inertia in our efforts to deal with the consequences of climate warming (only after they have unambiguously arrived) combines with the inertia in the climate system to strongly suggest that rapid significant climate warming will continue into at least the medium-term future.

If we wish our children and their descendants to avoid the consequences of longer-term and more severe climate warming, and to more easily adapt to the currently unavoidable aspects of warming, then we are obliged to act, strongly, now. We must ensure that we reduce our carbon-cycle-disrupting activities and, most importantly, constrain the human features and beliefs behind those disruptions. In the meantime, there is more to think about because climate warming is not all that is affecting the future of the earth's environment.

## Ecosystem Futures

The second aspect of the earth's environment that influences its future is its changing ecosystems. Each ecosystem is, in its own right, a dynamic, complex system. It is therefore difficult to predict the detail of how each will change over time. The easiest way to make a general prediction about the likely overall state of the earth's ecosystems is to extend the current trends in how they are changing into the future. The most significant uncertainty in the resulting prediction arises from not knowing if, when and by how much our decisions that are maintaining those trends will, either by choice or force, change.

This section discusses those trends and that uncertainty. It uses the rules of thumb that apply to an ecosystem's functioning, previously detailed in chapters 7 and 8, and the terms used in chapter 9 (O'Hara 2005). The reference for the discussion will be the manner in which species and ecosystems responded to the global-scale, climate-warming-induced deglaciation of northern North America at the end of the last ice age, as discussed in chapter 8. Up until around 10,000 BC, the human impacts on the earth's ecosystems were relatively minor.

It is thus convenient to divide the discussion into three subsections. Both the first (human-induced climate warming) and the second (our direct impacts and pollution releases) are primary influences on an ecosystem's future state. The third discusses the likely cumulative effect these influences will collectively have on the overall future state of the earth's ecosystems.

**The Influence of Climate Warming.** Human-induced climate warming is already affecting the earth's ecosystems and the species they host. The degree to which it will influence their future depends significantly on whether our decisions will enhance or reduce future climate warming. Will we choose, for example, to change our beliefs, adopt a lifestyle and reduce our population so that we meet the target of a <2.0°C (1870) maximum temperature rise?

As concluded above, the inertia in both our decision making and change in the climate system leads to the assumption that the current increasing warming trend will continue into at least the medium- and early long-term future. This discussion assumes that warming will reach a maximum average global temperature increase in the 3°C (1870) range before declining (figure 17). The possibility of self-sustaining warming also means that this discussion's conclusions about climate warming's future influence on ecosystems should be treated as a minimum impact estimate.

Species respond to a changing climate through the reaction of their individual members to it. During the global climate warming event that ended the last ice age, northern North America's species followed their climate niche across the deglaciating land toward the north pole. As they migrated, they adjusted their lifeways and physical characteristics to better suit their new circumstances. Those that couldn't went extinct. As species migrated, changed or went extinct the interactions among them changed. Thus their host ecosystems changed. These ecosystems too moved across the deglaciating land, and as they did so, their mix of species changed (Pielou 1991; Adams et al. 1997).

Prompted by the current warming, northern North America's species have already taken up their ancestors' task of migrating northwards and adjusting their characteristics. Their northern range boundaries, such as the Arctic treeline and shrubline, are moving further northward (figure 8) (Myers-Smith et al. 2015; Harsch et al. 2009). So are their host ecosystems and biomes, such as the boreal and sub-boreal forests. These biomes too are changing as they move. As long as climate warming continues, this is the basic manner in which all of the earth's species and ecosystems will respond: migrating away from the equator, up slope and changing their characteristics.

Species and their host ecosystems are responding to more than the amount of warming (and the related changes, such as precipitation). They are also responding to its speed. In the Arctic, the climate warming rate is roughly twice the average global rate. The average warming rate in the Arctic during the 10,000 years of post-ice-age warming was between 0.01°C/decade and 0.03°C/decade. On a geological time scale, this warming rate is fast.

The Arctic's more recent rates of warming have been much faster than that. Between 1972 and 2002, it warmed at an astounding average rate of 0.5°C/decade (Otto-Bliesner et al. 2006). Between 1998 and 2012, it warmed at an average rate of around 0.1°C/decade. These rates are nearer to those of the warming that ended the Younger Dryas abrupt cold event around 11,460 yBP. At that time in central Greenland, the local temperature rose between 12°C and 15°C in less than 200 years: an average rate of between 0.6°C/decade and 0.75°C/decade. (Incredibly, about half of this 12°C to 15°C warming appears to have taken place in only 15 years, which implies a questionable warming rate of 4 to 4.6°C/decade (Taylor et al. 1997).

The numbers suggest that if the current amount and speed of today's extraordinarily fast climate warming continues into the medium-term future, then it will severely strain the ability of species to adapt by migrating and adjusting their lifeways. Species and their host ecosystems can be expected to undergo much greater changes than those which a long-term view of northern North America's deglaciation would suggest. So, what can we expect?

Consider the response of trees in eastern North America to northern North America's deglaciation. The maximum shifts in the ranges of mid-latitude trees occurred at rates around 1,000 to 1,500 m/year. Only a few tree species, such as the Scots pine, registered these rates. The average migration rates for most mid-latitude trees, including hemlock and beech, were around 200 to 400 m/year. The maximum rate for the northern latitude boreal spruce was around 1,000 m/year and the average around 500 m/year.

In the future, if mid- and high-latitude trees have to keep pace with predicted climate warming, they would have to migrate at rates between 3,000 and 5,000 m/year. This is between 2 and 25 times faster than their known rates of migration (Davis and Shaw 2001; Pitelka and PMWG 1997). At these rates, it is likely that at least some tree species will be unable to adapt through migration alone (Root et al. 2003).

The trees and their companion species could certainly try to compensate for the constraints on their migration rates by altering their lifeways. But, like migration, the lifeway adjustment process takes time and is subject to biological limits (Davis and Shaw 2001). For example, allometric relationships indicate that smaller species are shorter lived and thus able to adjust their lifeways more quickly and easily than larger, long-lived species. In addition, some species, the ones we call "weedy," innately reproduce and grow more rapidly than most. If climate warming continues for long enough at the same or a faster rate than today, then these weedier plants and animals are more likely to successfully make the lifestyle adjustments needed to adapt to climate warming. In contrast, the less weedy and less flexible species are more likely to face extinction.

We gain a feel for the scale of the possible climate-warming-related species extinctions from a study of 1,103 land animals and plants living on 20% of the earth's land surface. Under the IPCC's older (pre-RCP) maximum climate warming scenario, by 2050 an average 33% of these species would be committed to extinction. In contrast, if the minimum climate warming scenario is followed, then 18% are committed to extinction. The divergence in the predicted extinction numbers between the two scenarios becomes much larger if the warming trends are projected to 2100 (Thomas et al. 2004). Whatever the actual numbers, if the current rates and amounts of climate warming continue, it will likely become a dominant, global driver of extinctions rather than a minor one as it is today (Parmesan and Yohe 2003).

To personalize this broad description of the ecosystem changes to be expected from rapid warming over the medium- to near long-term, consider the ongoing changes to the sub-boreal forest of central British Columbia. The lodgepole pine is a common tree species in this forest. The pine beetle, which feeds on the inner part of the lodgepole pine's bark, is also a normal part of the forest. It is usually found in low numbers. However, in recent decades there has been a significant weakening of the beetle's two main population controls: cold events, which a warming climate has rendered insufficiently severe; and fires, which are less common due to their suppression by humans. Consequently, in the late 1990s and the first decade of the 2000s, the pine beetle population exploded. By 2008, the beetles had killed about 46% (or 620 million m$^3$) of British Columbia's merchantable lodgepole pine timber.

When driving through a pine-beetle-affected sub-boreal forest, you can be surrounded by a sea of red and dying, or dead, grey and leafless, beetle-killed lodgepole pine trees. This forest may appear to be destroyed, but it isn't. It is now undergoing the process of regeneration. For example, the increased amount of light on the forest floor has spurred the previously tree-shaded (understory) plants to grow (Hamann and Wang 2006).

However, this sub-boreal forest is undergoing more than regeneration. It is moving and changing as its species respond to a warming climate. This transformation process is illustrated on the western boundary of the lodgepole pine's range.

Decades ago, humans attempted to extend the pine's range by planting pine trees to the west of its western boundary. Since those plantings, the summers in that area have become wetter. As a result, *Dothistroma*, a defoliating fungus, and *Inonotustomentosus*, a root rot fungus, have been able to spread into the lodgepole pine plantations that extended the western boundary of its historical range. The pines are dying, and their range boundary is shifting eastward. The area they are leaving is being reoccupied from further west by the previously displaced hemlock and cedar trees (Woods et al. 2005).

It might seem that turning a tree's response to fungus outbreaks and a forest's response to a beetle tree-pest outbreak into examples of species and forest climate warming response is stretching the data. But these types of climate-related impacts on forests are not restricted to central British Columbia. They are found in forests all along the Pacific margin of North America: from the pine-beetle-disrupted pine forests in Colorado to the drought-disrupted spruce forests in central Alaska (Glaister 2007; Barber et al. 2000). In between, a diversity of mature forest types within the western continental US states are experiencing a steady increase in tree mortality. The immediate causes of these deaths include pest outbreaks and water stress. The bigger picture indicates that all of these events likely have a common cause: regional-scale climate warming (van Mantgem et al. 2009).

The distribution of disrupted forest ecosystems along the western margin of North America suggests that continental-scale, climate-warming-driven changes to its forest ecosystems are underway. An appreciation for the possible future of these forests is provided by a model of how the climate niches occupied by British Columbian tree species will likely move if climate warming continues for this century. The model predicts that the current distribution of these niches, and thus the trees, will undergo minor to substantial adjustments. Consider the northward shift of the climate niches for those tree species whose southern (trailing) boundary lies in British Columbia. For example, the trailing boundary for the substantial presence of white spruce will shift northwards at about 4,400 m/year. The trailing boundary for the lodgepole pine niche is expected to shift at 14,000 m/year (Hamann and Wang 2006; Coops and Waring 2011).

The rate at which the trailing boundary of the lodgepole pine's climate niche is predicted to shift northwards is extraordinarily rapid. This becomes apparent when that rate is compared to the average 350 m/year at which its northern (leading) boundary actually travelled northwards, up the western margin of North America, during deglaciation (Pielou 1991). Certainly, 350 m/year doesn't represent the tree's maximum possible short-term migration rate. But the 40 times difference does indicate that the pine's ability to respond by migrating and adjusting its lifeway will be severely challenged by rapid warming that continues well into the medium-term future.

The rate of climate warming might be so rapid that the pines on the species' trailing boundary could be overrun by the members of the migrating ecosystem to the immediate south. The equivalent of that event happened during the equally rapid cooling stage of the Younger Dryas: the slower southward-moving mid-latitude pine forest was overtaken by a more rapidly moving Arctic tundra ecosystem. Dead pines in a tundra ecosystem almost defies imagination.

Tree species in the Pacific Northwest that are unable to keep up the pace of migration or appropriately adjust their lifeway will face the prospect of extinction. So will those that reach a topographic barrier to migration or are no longer provided with a niche by their host ecosystem. Their host ecosystem will be changed accordingly.

The emerging climate-warming-related changes to western North America's forest ecosystems and biomes reflect the nature of the changes that are expected to affect all of the earth's land-based ecosystems into at least the medium-term future. These types of species and ecosystem changes are not limited to the land. Marine ecosystems will be faced with equivalent challenges as the oceans become warmer and more acid (Doney 2006; IPCC 2014a; IPCC 2007b).

Consider that acidification-related effects will occur soonest and be largest in the polar regions, especially in the Arctic. For example, if the concentration of atmospheric $CO_2$ reaches around 450 ppm by 2030, then about 10% of the surface waters of the Arctic Ocean are predicted to be capable, year round, of dissolving the shells of the lower tropic level species that currently form the foundation of that ocean's marine food web (Steinacher et al. 2009). The level of $CO_2$ in 2014 was 397.7 ppm and rising at around 2 ppm/year.

In overview, despite statements and preliminary efforts to the contrary, we are likely to largely maintain our business-as-usual contribution to global-scale climate warming into the medium-term future. The inertia in the climate system currently favours ongoing climate warming into the near long-term future. These two inertias mean that the degree and rate of warming in the near medium-term will likely be most similar to that seen at the end of the Younger Dryas cold event.

Today's species and ecosystems will thus undergo significant climate-induced changes on a global scale into at least the medium-term to near long-term future. Species' efforts to undertake the rapid migration toward the poles will be accompanied by significant changes in their characteristics. The many failures will increase the species extinction rate. The weedier species will be the most successful at surviving. These changes will result in a species' host ecosystem migrating northward and changing its species mix as it moves. Novel (not seen before) ecosystems will appear.

## The Influence of Our Direct Impacts and Pollution.

The influence our direct impacts and pollution releases have on an ecosystem's future are the result of activities such as land clearing (e.g., deforestation and agricultural fields), species and ecosystem management (e.g., pesticides), resource extraction (e.g., wild animals, metals and water), exotic organism introductions and waste disposal, as discussed in chapter 9. These activities affect individual species, such as their efforts to complete their lifeways, and their host ecosystem, such as changing its structure and functioning. The more we change these features, the greater will be our influence on the ecosystem's future state.

We might voluntarily decide to reduce our influence. However, the inertia individuals and cultures experience when considering a change from business as usual means that the current trends in our direct impacts and pollution releases are more likely to be maintained.

There is a possibility that we might be forced to curtail our activities by events such as a decline in either the availability of the resources (e.g., energy) needed to conduct them, or the size of our population. However, these types of changes are most likely to occur later rather than earlier in the medium-term future, if that soon. Alternatively, cultural instability might result in a decline in the responsible activities, but it might also increase them. This discussion therefore assumes that our direct impacts and pollution releases will likely continue roughly along present trends into at least the medium-term future, and thus so will their substantial influence on the future of the earth's ecosystems.

The influence our direct impacts and pollution releases have on ecosystems occur at local to global scales, and they vary widely in their intensity. The resulting overall state of an ecosystem lies along a continuum from totally destroyed to unaffected. The continuum is divided into three broad, overlapping groups: substantial destruction (destroyed), significant disruption (disrupted) and largely unaffected (unaffected), as detailed in chapter 9. Today, few ecosystems remain unaffected by our direct impacts and pollution. This subsection discusses the influence that our past, ongoing and emerging direct impacts and pollution releases will have on the future state of the earth's ecosystems when their trends are extended into the future.

*Destroyed.* A destroyed ecosystem is one whose structure and functioning has been disrupted to the degree that its ability to recover to its original state has been largely lost. In practical terms, the population size of its various member species and their ability to complete their lifeway have been affected to a degree that the ecosystem's resilience, robustness and adaptability is severely affected. Its path into the future is significantly uncertain. In the extreme, a destroyed ecosystem is one that has been obliterated.

Our destruction of ecosystems can be obvious. Examples are converting wild ecosystems into human-created ecosystems, such as farm fields; replacing wild ecosystems with infrastructure, such as roads and houses; or digging up ecosystems, such as the Athabasca tar sands mine. The destructive nature of some of our activities is less obvious, such as our disposal of nutrients into the ocean, where they cause dead zones.

The scale of our current and possible future destruction of ecosystems can be imagined by assuming that the developing nations will increase their future consumption of plants and animals to the level of today's developed world. If they do, then the human appropriation of the earth's land-based net primary productivity could rise from 20% to 35%. This near doubling of our demand for the products of plant growth will inevitably be associated with an increase in the destruction and disruption of indigenous ecosystems (Doyle 2005).

If, instead, our future global demand for the products of plant growth is driven by the current trends in global population and economic growth, then our destruction of ecosystems will follow a related trend into the future. That trend is illustrated by the deforestation being reported for more than 80% of the earth's tropical forests, located in 34 countries. The area covered by tropical forest declined from 1,340 million ha in 1990 to 1,300 in 2000 and 1,240 in 2010. The deforestation rates increased from an average of about 4 million ha/year (0.3%/year) for the period 1990–2000 to about 6.5 million ha/year (0.5%/year) for the period 2000–2010. Although a slight decline in the deforestation rate occurred between 2005 and 2010, the overall rate of deforestation seems to be at least stable if not increasing (Kim et al. 2015).

But deforestation is a dynamic process. Once deforested, the land can either be left to grow back (i.e., the forest was disrupted) or it can be converted into agricultural land and infrastructure, such as roads (i.e., the forest was destroyed). With that in mind, the following reported deforestation levels become significant. An estimated 38% of all tropical forest land that was deforested from 1990 to 1997 was converted into agricultural lands: it was destroyed (Achard et al. 2002). The fate of the remaining 62%, the disrupted forest, depends on whether it was left to recover or was later turned into farmland.

Consider too that the technical methods used in the deforestation assessments poorly detect activities such as selective logging for wood extraction, while the longer-term impacts, such as the outcome of forest fragmentation, are unrecorded (Asner et al. 2005). It is safe to treat the estimated amounts and trends of deforestation and destroyed forests reported by deforestation studies as a likely minimum. It is also reasonable to conclude that the deforestation rate is a snapshot of the forests' health and that the process of disruption and destruction takes place over many years. Think of the Argolid.

The trend in the destruction of the earth's ecosystems, as illustrated by tropical deforestation, is expected to continue in this dynamic manner if the human population or its demand for resources continues to increase. This is the case even though the majority of that population is predicted to live in urban centres, not in rural areas nearer to the wild ecosystems. The destruction will continue because the urbanites will still need to be supplied with resources such as food, lumber, metal and hydro power. These resources are provided by agriculture and the exploitation of natural ecosystems in the rural areas.

The details of the connection between fulfilling urban needs and ecosystem destruction is more complicated than we might expect. Consider that those agriculturally land-strapped countries with growing populations or a focus on economic growth, or both, will face increasing difficulties satisfying their population's rising demand for agricultural resources. Addressing this issue by increasing the productivity of their farmlands faces limits. One solution is to turn the fragments of indigenous ecosystems within the existing agricultural lands and the still intact indigenous ecosystems on the surrounding marginal agricultural land into farmlands. This will, of course, destroy these ecosystems.

A favoured alternate solution for the resource-short country is to import agricultural resources. This may indeed reduce the importing country's need to destroy its remaining natural ecosystems, but it also exports and globalizes ecosystem destruction. To appreciate how, ask whether the imported agricultural goods (directly or indirectly) required the producers in the exporting country to convert more of their natural ecosystems into farms. Think of the soybean products imported into Asian countries and the palm oil imported into Western countries. They were produced by the destruction of forest ecosystems in Brazil to produce soybean fields and in Indonesia to produce palm oil plantations. The destruction of the earth's ecosystems is expected to continue into the medium-term future.

*Disrupted.* A disrupted ecosystem is one whose historical structure and functioning have been significantly affected, but it is still capable of recovering relatively easily to the essence of its past state. In its disrupted state, its species and their interactions have been affected to a degree that the ecosystem's resilience, robustness and adaptability is impaired but not lost.

As mentioned above, the current trends in our direct impacts and pollution releases will likely continue into at least the medium-term future. However, the extent and significance of the resulting disruption of ecosystems will likely be greater than we currently expect from those trends for two reasons. Emerging knowledge is revealing that our past activities have disrupted ecosystems to a much greater extent than we realize, while, at the same time, our newer activities are adding new forms of disruptions on top of the old. First, the gap in our awareness of past disruptions is illustrated by the increasing recognition of the global-scale disruption of marine ecosystems by our past and ongoing releases of plastic waste. This will continue a discussion started in chapter 9. Later the newer forms of actual and potential ecosystem disruption are illustrated by the emerging recognition of the likely contributions from genetic engineering and nanotechnology.

Between 1974 and 2012, the global production of plastic increased at an average rate of 6.7%/year to reach at least 288 million tonnes (Mt) in 2012 (Plastics Europe 2013). Most of this plastic is eventually discarded as waste, some of which ends up in the oceans. For example, of the estimated 275 Mt of plastic waste generated in 2010, 4.8 to 12.7 Mt (1.7% to 4.6%) was discarded into the oceans (Jambeck et al. 2015). Most of that plastic is discarded on land, where it enters a river, which transports it to the oceans, whose currents then circulate it around the globe (Schmidt et al. 2017).

Our plastic waste is now found globally in all oceanic environments, from the surface to the ocean floor, from the mid-ocean to the seashore, and from pole to pole (Pham et al. 2014). A minimum 0.27 Mt of plastic waste is estimated to still be floating on the surface as 5.25 trillion particles (Eriksen et al. 2014). It seems likely that most of the rest is being concentrated into nearshore and deep ocean sediments (Woodall et al. 2014).

The degree to which this waste is impacting marine species and ecosystems is slowly coming to light. Larger marine species, such as birds and whales, are known to become entangled in it. They also mistake some of the plastic particles for food and eat them (Wilcox et al. 2015). It is easy to imagine how the nutritionless and indigestible plastic junk "food" could affect their health. Think of gut blockages and eating until full yet still starving.

There are other health consequences from eating plastic particles. They arise because the particles host health-affecting chemicals. Some of them, such as phthalates and triclosan, are the additives within the plastic. Others, such as PCBs, are the chemicals in our marine waste that the plastic particles can adsorb and absorb directly from the ocean (Rios et al. 2007; Moore 2008). After an organism ingests the particle, the conditions in its gut can facilitate the release of at least some of the chemicals from the plastic's enhanced chemical load. If they enter the organism's body, they can affect its lifeway and health. For the organisms that eat them, plastic particles can thus act as both a nutritionless food and a toxic-chemical delivery system (Teuten et al. 2009).

Because the size distribution of plastic particles in the oceans covers almost the entire size range, from micron to multi-centimetres, of the foods that marine species eat, it is reasonable to expect that even the smallest of species in an ecosystem's lower trophic levels are consuming them. Earlier research indicated that organisms as small as lugworms are consuming plastic particles. A more recent experiment has shown that copepod zooplankton can consume 20 μm-sized plastic particles (Cole et al. 2016).

Earlier research showed that both the physical plastic particles and the chemicals they carry are having a significant impact on the health of larger marine species. The results of a recent experiment using Eurasian perch eggs indicates that the lifeway of these larger organisms can also be significantly affected by even the smallest of particles. The mere presence of 90 μm polystyrene particles lowers the eggs' hatching success. The larvae that did hatch but ate the particles displayed lower success for their smell-triggered predator avoidance reaction. They would thus suffer a lower survival rate in the wild (Lönnstedt and Eklöv 2016). What would happen if those polystyrene particles included adsorbed PAHs?

In addition, some chemicals can accumulate in the individual's body to reach more harmful doses. The bioaccumulation of the chemicals associated with plastic waste enables them to affect many more species than the initial consumer. The process of successive prey-predator interactions moving pollutants up the host ecosystem's trophic web to reach more biologically significant doses in the higher trophic level species was previously illustrated by PCBs affecting salmon and orcas. Through this process, the bouquet of chemicals that plastic particles deliver to an ecosystem can affect its individuals, species and thus the ecosystem itself.

The impacts that plastic particles have on an individual consumer and their host ecosystems is likely more severe than even the above description suggests. After all, the physical presence of plastic in an ecosystem alters the habitat available for its species to complete their lifeway. Those impacts include smothering it and changing the availability of the substrates used for reproduction, as mentioned in chapter 9, and the transmission of disease, as will be mentioned later. Comprehensive studies are awaited.

We are slowly becoming aware of the astounding degree to which our plastic waste is disrupting marine ecosystems. The discarding of that plastic into the oceans began more than 50 years ago and has increased ever since. Clearly, some action needs to be taken if we wish to proactively reduce these impacts. It could be argued that now we are aware of it, we will soon solve this problem by removing the plastic already in the oceans and reducing the plastic waste entering it.

Before placing faith in this hope, consider the dissolved nutrients we discard into the ocean, in particular the nitrogen and phosphorous from the excess human-made fertilizer on farmers' fields and in our sewage. One way they disrupt marine species and ecosystems is through eutrophication (oxygen depletion), which can also lead to hypoxia and the collapse of nearshore marine ecosystems into dead zones. Like plastics, this increasing disruption of marine ecosystems will continue its current increasing trend as long as our growing population discards its nutrient-containing (farm and personal) waste water into rivers.

This impact could be decreased significantly by reducing our current nutrient inputs into the oceans by half (Scavia and Donnelly 2007; Diaz 2001). So far, despite knowing about our contribution to eutrophication for decades, our effort to make the needed reduction is grossly inadequate. Further confounding our efforts is the recent realization that our use of human-made fertilizers has created a reservoir of excess nitrogen in farm soils. It will take decades for it to drain into the ocean (Van Meter et al. 2016). The essence of our relationship with our nutrient waste applies to our plastic waste.

The emerging knowledge about the effects that our plastic waste has had on marine ecosystems for the last 50 years indicates that, globally, many more species in many more marine ecosystems are certainly being affected by it in more ways than we expected. We could try to remove the plastics from the oceans, but the practical difficulty of cleaning up the excess nutrients in our fields and in ocean dead zones illustrates the constraints on our efforts. The alternative is to stop disposing of plastics into the oceans. Our ineffectual efforts to reduce our nutrient waste indicates that, unless there is a dramatic break from the past, the amount of plastic waste entering the oceans will likely continue along current trends into at least the medium-term

future. Now is the time to ask how much we know about the impact of plastics on terrestrial (land and freshwater) ecosystems. The story of plastics illustrates that our past disruption of ecosystems is much more severe than we currently imagine and is likely to continue into the future.

Our new activities are adding new forms of disruption to those (recognized and unrecognized) disruptions from our past and ongoing activities. The nature and potential scale of these newer forms of disruption is illustrated by genetically modified organisms (GMOs): organisms that have had their DNA directly altered by humans. This process has a short history.

The long-term impacts of GMOs on ecosystems are poorly known. However, we do know that GMOs are designed to be novel, so they can be thought of as exotics. It is anticipated that they will affect their host ecosystem in unexpected ways, as discussed for other exotic organisms in chapter 9. Their genetically modified (GM) genes can also be thought of as exotics. These genes have the potential to affect some of their host ecosystem's species at a genetic level, with unexpected consequences for both the species and their host ecosystem. GMOs and their genes have the potential to significantly disrupt their host ecosystem, especially if either they or their genes become invasive.

Corporations and governments have made promises that the spread of GMOs and their genes could and would be contained. These promises are proving to be hollow (Guardian Weekly 2001). There is already more than enough evidence that genes within a micro-organism community/reservoir are highly mobile, as illustrated by the spread of antimicrobial drug resistance among them. It is reasonable to expect that if/when genetically modified micro-organisms "escape" into these reservoirs, then its genes would be just as mobile in these environments. Think too of CRISPR-Cas9 and recently available kits for modifying organisms at home.

The GMO genes of larger organisms are also mobile. For example, golf course operators commonly cover their courses with creeping bent grass. They then face the considerable costs of trying to keep their courses weed free. To solve this problem, agribusiness giant Monsanto genetically modified bent grass to make it resistant to the broad spectrum glyphosate containing herbicide Roundup. The grass is said to have been made "Roundup ready." If a golf course were to be planted with GM bent grass, then it can be sprayed with Roundup to kill weeds without harming the grass.

As part of the testing of GM bent grass, it was grown on eight test plots within the Oregon creeping bent grass control district in 2003 and 2004, under tightly controlled experimental conditions (Zapiola et al. 2008). Despite these controlled conditions, pollen from the test plots was found up to 21 km downwind from the plots (Watrud et al. 2004). The pollen had spread the glyphosate-resistant genes to wild bent grass plants found growing up to 3.8 km downwind from the plot. In addition, GM bent grass seed blown downwind from the test plot area had germinated and grown (Reichmann et al. 2006).

A retroactive study revealed that the glyphosate-resistant gene was widespread in wild plant species related to the GM bent grass, and that the gene was persisting. The study concluded that cleaning up this gene mess was not possible. The genes that endow glyphosate resistance could be expected to spread into the area's other receptive plants to produce glyphosate-resistant hybrids (Zapiola et al. 2008).

That GM genes are indeed spreading is illustrated by a 2009 lawsuit concerning the unauthorized spread of the genes inserted into rice to make it resistant to Liberty, a herbicide produced by Bayer. The genes that confer resistance to Liberty are thought to have escaped into the environment during testing, a time when stringent protocols were apparently in place to prevent such an event. Regardless of the cause, by the time the lawsuit was brought to trial, the genes could be found in 30% of the United States' rice lands (Harris 2009).

The impacts that the presence of exotic genes in an unsuspecting plant host might have on its lifeway remain unknown. Similarly, the medium- to long-term impacts on the ecosystems hosting GMO plants or indigenous plants containing those genes remains largely unknown. We can at least expect significant changes to an indigenous ecosystem hosting indigenous plants that contain the genes conferring resistance to Liberty or Roundup, if it is routinely exposed to these products. An example is indigenous ecosystems adjoining farmed fields. Think too of a plant genetically modified to use saline water or use water more efficiently.

The close connection between our relationship to new activities and their emerging or possible future contributions to ecosystem disruptions is most clearly made by nanotechnology. Experiments with mice and algae have indicated that the products of emerging nanotechnology have the potential to significantly disrupt the lifeways of at least some organisms. For example, mice can inhale engineered multi-walled carbon nanotubes in low enough doses that their lungs

may not be physically damaged, but the exposure induces changes to their spleens. They then suffer from a suppressed immune system, which renders them susceptible to illness and disease (Mitchell et al. 2009).

Experiments using two aquatic algae species indicate that the presence of a charge on an engineered nanoparticle's surface allows it to stick (become adsorbed) onto the algae's surface. These adsorbed nanoparticles block both sunlight and air from reaching the algae, thus interfering with its process of photosynthesis. The charged nanoparticles are thus interfering with a basic life process of an organism that makes up the foundation of aquatic ecosystems' trophic webs (Bhattacharya et al. 2010).

Concerns about the potential for future environmental impacts from nanotechnology increase with the realization that the technology is now being used to produce products commercially, even though nanotoxicology test routines have not been fully developed or standardized (Navarro et al. 2008). One estimate suggests that it will take between 34 and 53 years to test existing nano materials for their toxicity alone (Nature 2009). More time would be needed to test for low dose and chronic biological or ecosystem effects. This is an astounding situation considering that carbon nanotubes were discovered in 1991, and that since then they have become commercially available, which facilitates them being widely dispersed into the environment.

The observation that inadequately tested and poorly restrained new technologies and inventions, like nanotechnology, will contribute to future ecosystem disruptions should provide the reader with a strong sense of déjà vu. Current innovations are following a path into the future that has been traced out by many innovations before them. The brightest among us still supply the innovations, put them on that path and then help them down it. Westerners know this history, yet it repeats. How is this possible?

It happens through a combination of our personal world view, our reality paradox features, our culture's world view and our decision-making foibles. They collectively tend to divert our attention away from fully considering new inventions, technologies and activities within the larger context of our dynamic human-environment system. Instead, our attention is grabbed by the marvel of our ingenuity.

In the present, as in the past, the efforts of the individuals and groups developing and promoting their inventions are driven by their culturally supported vision that their labours will usher in a better (lifestyle and cash) future. The public tend to subscribe, with few reservations, to the proponents' visions or promise of that better tomorrow. But, this belief means that the proponents and the public fail to consider, largely ignore or downplay the alternative outcomes and side effects that often accompany their inventions and activities. As in the past, our marvelling leaves the flaws and flawed thinking to be discovered and paid for by future generations and ecosystems.

We also make the argument among ourselves that our historical impacts were the result of old attitudes and outdated testing methods that have now been superseded. But testing hasn't kept up with the developments of new novel products and is still significantly limited in its duration and scope, as mentioned above and discussed in chapter 9. A reminder of this is provided by noting that, in 1997, only 10% of the 70,000 chemicals available for use in 1984 had been evaluated by the EU for their effects on our brains (Williams 1998; CNMAR 1992). That is the organ we use to make our decisions.

In overview, the influence that our direct impacts and pollution releases are having on the earth's species and their host ecosystems is disrupting or destroying ecosystems on a global scale. The level of their future disruption and destruction will depend heavily on our present and future decisions. We can reduce those levels by increasing our appreciation for the complexity of how ecosystems function, questioning our culturally supported need for an activity, and testing our innovations more thoroughly before they are applied. Until we take this to heart, we will fail to proactively recognize the full degree to which we have already disrupted ecosystems. And we will continue to routinely engage in new activities whose future disruption of ecosystems will surprise us.

## Our Total Influence: Cumulative Effects.

We tend to treat the influence that human-caused climate warming, our direct impacts and our pollution releases will have on the future of an ecosystem as separate from one another. In this view, our total influence on an ecosystem's future will be the simple sum of our impacts on it. But an ecosystem is a dynamic, complex system that functions and changes through the interactions, including feedbacks, among its members. As a result, the immediate effects of our influence on an ecosystem interact with one another to create indirect effects that can enhance or suppress our influences on it.

And, because the interactions take place over a wide range of time and space, our past influence on an ecosystem can continue to affect it today, and into the future. History counts.

Therefore, from our perspective, our total influence on an ecosystem's future state is the complex sum of our (direct and indirect) effects on it and their accumulation over time: our cumulative effect on it. From the ecosystem's perspective, our effect on its future state is the complex sum of its interaction-mediated response to our influences on it. The outcome of the ecosystem's response is its future state.

These conclusions also apply to ecosystems collectively. The nature of our future cumulative effect on the earth's ecosystems is illustrated by contemplating the current state of the earth's land ecosystems. Since 10,000 BC, we have appropriated for agriculture the ecosystems on 34% of the earth's land surface (Ramankutty et al. 2008). The proportion of our appropriation that is tilled fields was created by destroying the pre-existing indigenous ecosystems. The proportion that is pasture was created by disrupting indigenous ecosystems through, for example, partial land clearing. Many of the indigenous ecosystems on the remaining 66% of the earth's surface not used for agriculture have been disrupted or destroyed by our other activities, such as mining, logging, infrastructure building, recreation and waste disposal.

More specifically, consider a disrupted ecosystem located on the fringes of agricultural fields (destroyed ecosystem) and backing onto an exploited forest (a mix of functional and disrupted ecosystems). The fringe ecosystem's remaining individuals and species could respond to their host's disruption by trying to re-establish themselves. However, our ongoing occupation and use of the land can make this a difficult task. Indeed, a growing human population and current trends in our activities indicate that the ecosystem and its surroundings can expect to experience ongoing disruption by our direct impacts and pollution releases.

This is the context in which the fringe ecosystem's species must adapt to the ongoing influence on them from climate warming. They will try to accomplish this by undertaking a mix of migrating and adjusting their characteristics (including their lifeway), just as their ancestors did when they responded to the climate warming that led to the deglaciation of North America, as discussed in chapter 8. But species today face obstacles to migration that their ancestors did not, such as farm fields, urban areas and human-disrupted ecosystems. In some cases, these human obstacles can act like the insurmountable natural barrier an alpine plant faces as it adapts to climate warming by migrating up a mountain: when it reaches the summit, it has nowhere to go (Jump and Penuelas 2005; Root et al. 2003). In other cases, the obstacles just change the options for migration. For example, the disruption of forest adjacent to the fringe ecosystem changes the nature and availability of the niches that its individuals and species can migrate into.

The species in the fringe ecosystem will also adapt to climate warming by adjusting to their climate-affected living conditions. A species does so by successive generations of its individual members adjusting their physical characteristics and lifeway to accommodate their changing, migrating niche or by finding a suitable niche in a neighbouring ecosystems. A species today can find both of these adjustments more difficult to complete than their ancestors did during deglaciation because of the effects of our ongoing direct impacts and pollution releases on them, their host ecosystem and their surrounding ecosystems.

Consider, for example, how our direct impacts and novel pollution can affect a species' ability to adjust. Pollution can, for example, affect its individual member's lifeway functions, such as its behaviour (e.g., 4-NP), its immune system (e.g., nanotubes), and its endocrine system (e.g., feminizing atrazine). A species' genetics can be changed by either direct gene manipulation or the routine intense harvesting of individuals of a species.

In the case of harvesting, our focus on those individuals of a species with the features we prefer can bias the surviving population's genetic characteristics and limit its genetic diversity: for example, trophy hunters prefer sheep with larger horns; fishers prefer larger fish. Through this means alone, we have become a dominant force driving species change (Darimond et al. 2009; Festa-Bianchet et al. 2014). We also bias and limit the genetics of a harvested species' population when we preferentially replace what we have harvested with individuals that display the characteristics we prefer. For example, when "reforesting" a logged forest, we plant the kin of those individual trees we have selected and propagated because they display those characteristics of its species that we desire, such as trees that grow faster, straighter and with less knots.

But, by affecting all of these features, we have accomplished more than making it more difficult for species and ecosystems to respond to climate warming compared to the last deglaciation. We have also affected their ability to respond to our direct impacts and our pollution releases. There are thus many links and feedbacks between the effects that our influences have on species and ecosystems. It should now be clear how complex and wide-ranging our ongoing cumulative effect on the earth's ecosystems is.

The significance of our ongoing cumulative effect becomes clear if we look at it from a different perspective. The cumulative effect of our past and present influences on an ecosystem resulted in its present state. That state is the initial condition from which the ecosystem has to now respond to our ongoing and future influences on it. Thus, at each step in time, our cumulative effect on an ecosystem has changed more than its appearance. It has constrained how the ecosystem and its species can respond to our ongoing and future influences on them. The outcome of that constrained response will be its future state.

Worldwide, the cumulative effect of our past and present influences (climate warming plus direct impacts plus pollution releases) on ecosystems is significant. The result has not only changed them but has significantly constrained their ability to respond to our ongoing and future influences on them. There is also strong evidence that our influences will be significant and continue along present trends into at least the medium-term future. Thus the future state of the earth's ecosystems will likely be significantly different from what we would expect by looking at the changes ecosystems underwent during the climate warming that ended the last ice age. And the changes will likely be more severe than we expect from our current thinking about the influence of our ongoing and future direct impacts and pollution on them.

The future state of the earth's ecosystems as a result of our ongoing and future cumulative effect on them is personalized by considering the outlook for coral reefs. This continues the discussion about coral reefs in chapter 9. The inanimate carbonate foundation of a coral reef ecosystem is formed by its corals. They are sensitive to changes in the most basic of their habitat's parameters: the ocean's temperature and composition. Too high a temperature disrupts a coral's symbiotic relationship with the algae on which it depends. If that relationship is sufficiently disrupted, the coral dies in an event called coral bleaching. The increasing number of these currently anomalous events indicates that the ocean's temperature is beginning to exceed the upper limits for reef-forming corals to function (Adam 2009b).

At the same time, an increase in the ocean's acidity (declining pH), along with the associated change in its carbonate concentration, compromises a coral's ability to grow and maintain its skeleton. Too high an acidity is recorded as a slowing in its growth rate (its calcification rate). The records for Australia's Great Barrier Reef over the last 400 years up to 2005 indicate that its coral's calcification rates were stable until 1990. They then underwent an unprecedented 14.2% drop between 1990 and 2005 (average of around 0.9%/year) (De'ath et al. 2009).

The effects of today's warmer and more acidic oceans on corals are connected to one another and climate warming through the carbon cycle. A model of these connections indicates that, in pre-industrial times, some 16% of the oceans' total surface area had chemical characteristics considered optimal for coral growth. If current global-warming-favouring disruptions to the carbon cycle continue, then this favourable area may disappear in a few years (Steinacher et al. 2009).

Even if our influences stopped immediately, the delays in the ocean-atmosphere interaction will ensure that the changes will continue into the medium-term future. Think inertia. It thus seems likely that, in the immediate- to medium-term future, reef-forming corals will, at best, suffer more frequent and more severe coral bleaching, and face an increasing struggle to maintain their skeletons. At worst, the world's current coral reef ecosystems could, in the medium-term future, become extinct.

However, preliminary evidence also suggests that the algae in the corals may be able to compensate for the stresses of ocean warming by changing their characteristics. Perhaps the corals and their algae can adapt sufficiently to allow the corals to survive. Perhaps corals living at depth or living in unusual environments can fill the impending void, if not now, at least in the future. Should we treat these "perhaps" as the best indicator of the coral reef ecosystem's medium-term future?

The fast pace of the changes to the ocean's characteristics suggest not. So do the trends in our other ongoing influences on coral reefs, our direct impacts and pollution releases. They include those previously discussed activities that have a limited effect on the carbon cycle and climate warming but are significantly altering the reefs' functioning.

The larger herbivorous and carnivorous fish that are members of a coral reef ecosystem play key roles in its functioning. Even before the 1900s, the populations and lifeways of these fish were being significantly disrupted on a global scale, especially when our fishing for them became overfishing (Pandolfi et al. 2003). The overfishing of larger fish, which continues today facilitated by newer technology, was joined in more recent times by the overfishing of other reef species. For example, 55% of 49 island countries bordered by coral reefs are unsustainably fishing not just the reef's fish, but its molluscs and crustaceans (Newton et al. 2005).

Overfishing on its own is insufficient to explain the current declining trend in the density of fish on reefs. The other reasons are provided by the explanation for the decade-long decline of fish on Caribbean coral reefs: the general degradation of the reefs' ocean environment (Paddack et al. 2009). The responsible activities include the disposal of human and industrial waste and the harvesting of the smallest fish, such as the clown fish, to stock aquariums. There is also a special place for our plastic waste: it is emerging that corals in contact with plastic experience a significant increase in the risk of being infected with a disease (Lamb et al. 2018). These types of direct impacts and pollution releases are expected to continue and likely increase into the future.

It is the cumulative effect of climate warming, our direct impacts and our pollution releases on the world's coral reef ecosystems that are the reason why most of them were recently classed as between 30% and 80% degraded (Pandolfi et al. 2003). Based on the current global trends in our influences, we can expect that the distribution of functional reefs will continue to shrink and their health to decline. Many coral reefs will face functional and some ecological extinction. As long as our influences on them continue at current rates, it is unlikely that the "perhaps" mentioned above will be able to fully compensate for our cumulative effect on them in at least the medium term. The trend coral reefs are following into the future serves as an early, visible reminder of the path that many of the earth's ocean and land ecosystems are likely to follow into at least the medium-term future as a result of the cumulative effect of our ongoing and future influences on them.

In summary, our expanding influence on the state of the earth's ecosystems began around 10,000 BC with an increase in our direct impacts on them from us taking up farming. In the early 1800s, our pollution releases ramped up and we started to change the earth's climate system in earnest. Today, the future state of the earth's ecosystems will be significantly determined by the trends in our ongoing global-scale influences on them: climate warming, our direct impacts and our pollution releases.

Our direct impacts and pollution releases are significantly disrupting or destroying the functioning of ecosystems globally. The current high degree and rate of our direct impacts and pollution releases is expected to continue into at least the medium term, so we can expect the present global-scale trend of significant ecosystem changes and increasing species extinctions will continue. At the same time, human-related climate warming is influencing species and their host ecosystems to migrate and change. The current trend in the amount and rate of climate warming is expected to continue into at least the medium term. It is expected to drive species and their host ecosystems to change at speeds near to those that North American species and ecosystems experienced during the warming at the end of the Younger Dryas. An increase in extinctions can be expected.

But ecosystems are dynamic, complex systems that function and change accordingly. They display a complex interaction-mediated (including feedback) response, over a range of scales, to our influence on them. Thus it is the direct, indirect and accumulated (cumulative) effect of our past and present influences on them that results in their current state. But our cumulative effect has changed more than the ecosystems' visible features. It has changed how they can respond to our ongoing and future influences on them. In doing so, our past and present cumulative effect on ecosystems has constrained their options for the future.

Inertia suggests that, within the near term, our personal choices and our cultures' world views are unlikely to change sufficiently for our influences on ecosystems to be significantly reduced in the medium term. The reverse is more likely. Overall, it is highly likely that the distribution and state of the earth's ecosystems will undergo more significant changes, more quickly into at least the medium-term future than we currently expect (Rogers and Laffoley 2011).

## The Environment's Future

The current trends in our activities and the drivers behind them suggest that the earth's atmosphere will likely exceed the warming target of <2.0°C (1870). They also suggest that the earth's ecosystems will continue to be significantly disrupted and destroyed, while more of the species they host will receive endangered status and then become extinct (IPCC 2014a). These changes will continue for as long as the current trends in the growth in our population, our demand for resources, our disposal of waste and our direct impacts are supported by our personal views, our decisions, our culture's world view and resource availability.

But this is not the full story. The earth's climate and ecosystems are linked together. Any changes to the climate system will affect ecosystems, while changes to ecosystems will affect the climate system. For example, our burning of tropical forest materials during regional-scale deforestation in Indonesia and Brazil accomplishes more than disrupting and destroying those forest ecosystems. It is also the second largest single source of our climate-altering $CO_2$ and a significant source of our black particulate emissions into the atmosphere. This linkage means that we only gain a reliable appreciation for the possible future state of the earth's ecosystems and climate by considering our cumulative effect on them both, that is, when they are treated as part of a single complex, dynamic system: the environment.

Here is a reminder of how such a complex system functions and changes. Its structure and functioning is the product of its many variables engaging in a diversity of interactions, including feedback, that occur over a wide range of time and space. The system responds to both an internal and an external influence by adjusting its functioning, which is how it changes. Changes in one part of this system are spread throughout the system, so it changes as an integrated, interactive whole, rather than as isolated parts.

However, it takes time to complete the interactions and feedbacks, so there is always a delay in the making and spreading of changes: the system displays inertia to change. The effect of today's actions appears tomorrow. Thus, although the system's interactions drive the system toward an equilibrium, the inertia/delays mean that the system is always playing catch-up to past influences on it: an equilibrium state is never fully reached.

We tend to focus on the more visible stronger, short-term, local influences on the system. However, the weak influences that persist for a long period can have just as significant an influence on the system's future, especially if they are widespread. The more influences there are on the system and the faster they are applied, the more likely it will have to respond in a manner that exceeds some critical constraint on the functioning of its current quasi-stable state. And thus the less predictable is its future.

When this knowledge is applied to the separate predictions about the future of the earth's climate and the earth's ecosystems made earlier in the chapter, they take on a different meaning. Each prediction represents a different, but incomplete, view of the environment's overall possible future. In that overall future, our environment is neither infinitely robust nor is it in the process of being destroyed. It is more correct to say that we are in the process of transforming all aspects of the environment in critical ways. The system's inertia alone ensures that this transformation process will continue into at least the medium-term future: we have already predetermined some aspects and preconditioned other aspects of its future.

With this in mind, here is a reminder of specific changes our dynamic, complex environmental system will, is likely to, or could undergo if current trends continue, and a reminder of the connections among the changes. The earth's atmosphere will warm into the medium-term future to a level that will soon be warmer than any time since the 125,000 interglacial period. Climates and the biomes they support will move poleward. The dryness of dry climates and the wetness of wet climates will increase, and thus so will the difference between them. Globally, the weather events considered extreme, such as heavy precipitation, are likely to become the norm. There are also indications that the northern jet stream is changing its flow pattern. This means that events in the northern half of the Northern hemisphere that are currently considered to be anomalously warm or dry are expected to occur more frequently, further north and last longer (think of enhanced permafrost and ice sheet melting). In the southern half, equivalent extreme cold or dry periods are expected to occur (think of harsh cold snaps and drought). These types of climate warming changes will continue to arrive in an erratic manner. They do not yet suggest that larger, changes in the atmosphere's circulation pattern are imminent.

Sea level will rise. The oceans will continue to warm and become more acid which will affect the climate system and marine ecosystems. On land the larger number of increasingly severe, locally extreme weather events, such as prolonged drought will affect the distribution of species at the edge of their ranges.

In order for (exotic and indigenous) species to adapt to the cumulative effect of our climate warming, our direct impacts and our pollution releases, they will have to migrate and adjust their characteristics at exceptional speeds and to unusual degrees. It is safe to say that there will be an increase in species extinctions. Those species that are most likely to survive are the more robust, faster breeding and opportunistic species: the weedier species (Quammen 1998). The ecosystem transformation process and the results will contribute to amplifying the climate warming from our greenhouse gas emissions.

By considering the number of species expected to become extinct, we gain an overall appreciation for where their host ecosystems are currently headed. The current extinction rates are similar to those seen during the five global mass extinctions, which occurred during the last 500 million years: for example, during the Permian (at 230 My) and the Cretaceous (at 65 My). Today's rates thus suggest that a similar mass extinction is underway (Chapin et al. 2000; Ceballos et al. 2015).

But the total number of recent extinctions is nowhere near the number that occurred in the past mass extinction events. For example, only 10% of the current period's mammal, bird, amphibian and reptile species have become extinct. This is far less than the approximately 75% of species who died out in previous mass extinctions. Even if, as predicted, most of the currently threatened species in the above groups become extinct within 100 years, the total extinctions for that period will only increase to 30% (Barnosky et al. 2011).

For the total number of future extinctions to start approaching those of the last five mass extinctions, our cumulative impact on our environment will have to be much greater or last much longer than assumed here. This is possible. Think of us prompting an abrupt climate warming event.

At this time, more confidence can be expressed in a different medium-term outcome. Today's species extinction rate, with its preferential extinction of specialist species and large animals, is being accompanied by substantial population declines in species not yet considered to be threatened (Ceballos et al. 2017). The most prominent exception is humans, whose population continues to explode.

These changes are occurring because of our spread of exotic species and GMO genes; our dumping of bio-altering waste on land, in air and in water; the selective destruction of some biomes (e.g., temperate grasslands and tropical forests); and the rearrangement of water flows on the earth's surface. Rather than species and ecosystems experiencing the general destruction and disruption that accompanied past mass extinctions, our actions are subjecting them to a mix of human biased and heightened extinctions, population declines and a mass rearrangement of species. The earth's ecosystems are experiencing a human-driven and human-shaped transformation. But it is not just ecosystems that are experiencing transformative changes. As described above, the earth's climate is also being transformed.

Overall, it seems that our transformation of the earth's environment will continue into at least the medium-term future. It will result in, at a minimum, a warmer, stormier planet that will host disrupted and novel ecosystems with a changed distribution. The novel ecosystems will consist of a simplified mix of weedier, indigenous and exotic species whose characteristics will be strongly influenced by, among other things, our spread of exotic genes and species, our waste, our selection of organisms and our ubiquitous presence. Further into the future, the indigenous-exotic species will fill vacant and new niches in stressed and disrupted ecosystems. All ecosystems will be forced to undergo unexpected and exceptional changes as their member species adjust to one another and their changing, human-shaped living conditions.

A 2017, retrospective, updated version of a 1992 overview assessment of the status and trajectory of the earth's environment noted that it is following a business-as-usual trend into the future (Ripple et al. 2017). If the changes continue into the medium or long term at the same degree and speed as today, then the environmental changes will likely be more extreme than we expect. Think of the climate changing at the speed of the Younger Dryas abrupt event.

This description of the environment's possible future state serves as a reminder of another rule of thumb about complex systems. The further into the future, the less reliable and more uncertain a prediction becomes. However, the presence of inertia ensures that some of the ways in which the environment is changing to accommodate our influence on it will continue, at least to some degree, into the long-term future. For example, our past and present impacts on the carbon cycle will result in changes to the climate system continuing for centuries.

At the same time, the nature of our decision making, along with the chaotic and random aspects of the environment's responses to them, also ensures that the details of the environment's actual path into the future is likely to at least deviate, if not be substantially different, from our preferred inertia-favouring prediction of steady change along current trends. In fact, the environment may have already been preconditioned to follow a path outside our preferred range of predictions. We may have just failed to noticed. The environment's possible alternative futures should be given some consideration.

Consider the claim that a combination of present and future technologies could reduce our environmental impacts. Maybe, but whether that would be the outcome depends on how we apply the technology. Our decisions about its use could just as easily result in our environmental impacts being greater than, or sustained for longer than, even current expectations. For example, the provision of unlimited greenhouse-gas-free energy as promised by proponents of both fusion energy and renewable energy could alleviate (but not eliminate) the climate warming issue. However, the availability of that energy would also allow us to try and sustain a growing global population and its resource demands well into the long-term future. That choice, in combination with a lack of constraints on how the energy is used, could also sustain current trends in ecosystem destruction and disruption beyond the medium term. Thus, even though our contribution to climate warming might be substantially reduced, our direct impacts and non-greenhouse pollution releases could maintain current extinction rates for 250–500 years. Under these conditions, it becomes more likely that species extinctions and biome simplification will match those of the five previous mass extinctions (Barnosky et al. 2011).

A different possible future would arise if an abrupt climate change event, like those seen in the historical climate record, were to occur (figure 11). Think Younger Dryas. There are a number of possible non-catastrophic, internally generated changes that could favour the climate system undergoing such a prolonged rapid warming beyond that expected from our greenhouse gas emissions (i.e., a switch from its current quasi-stable state). Among them are an enhanced release of greenhouse gases from permafrost; methane from frozen subsea methane deposits; Arctic sea-ice melting; increased melting of the Greenland and/or Antarctic ice sheets; a weakening of the Atlantic Ocean overturning circulation and enhanced emissions of powerful, human-made greenhouse gases; or, more likely, some interlinked combination of these contributions. The possibility of a geoengineering solution to warming, such as sulphate emissions to cool the planet, that was incompletely considered or "unexpectedly went wrong" should also be added to the list.

A contrasting possible future would result if our environment-transforming activities were forced to slow. Among the reasons are a lack of resources, such as energy, and the social instability associated with that shortage or with the stress accompanying increasingly crowded, unjust or unfair living conditions. These types of events are a common occurrence in human history. However, history suggests that although these events slow our environment-transforming activities, they haven't, so far, stopped them. They simply delay our activities and change their form. If the stopping or reversing of the current trends in our activities is to be lasting, then it requires a change in our decisions and beliefs. It requires a change in how we choose to live.

In a future of alternative lifestyles and beliefs, we could deliberately choose to live differently. We would consciously make timely and appropriate decisions that reduce our currently strong cumulative effects on our environment. Any such well-thought-out reduction would have a considerable influence on our environment's future. However, for our decision to be timely and appropriate we would have to accept and thus incorporate into our thinking and our culture's world view some basic environmental and human features.

At its most basic, we would have to accept the characteristics of complex systems listed above. In particular, we would accept the presence of inertia in the environment system and that it is helping to carry us into the future. Thus, if a decision to reduce our cumulative effect is to be timely, then it must be made and implemented well before our undesirable environmental effects become blatantly obvious to us. Satisfying this condition is already becoming difficult because the speed at which some environmental changes are taking place is similar to the rate at which we can gather the information needed to make environmental management decisions (Parmesan and Yohe 2003).

Making timely decisions thus also means accepting our innate inertia to making some kinds of decisions, such as those that require us to change our lifestyle in order to reduce our influences on the environment. Think of the time it took us to respond to ozone layer thinning and global warming. Rather than acting in a timely manner in the future, we are, tragically, more likely to act too late to effectively reduce our environment-affecting activities. We are more likely to be forced to change than decide to voluntarily change.

Even if we do come up with timely solutions (voluntarily or otherwise), there is no innate guarantee that they would be appropriate and effective. If we are to successfully reduce our environmental impacts, then we must devote the care and attention needed to ensure that our solutions will indeed satisfy the required constraints and conditions to meet that goal, which we find difficult to do. Consider our faith in our ability to plan and execute grandiose solutions, but our unwillingness to face the limits to such plans and the constraints on our ability to make and implement them. For example, the once sure-fire solution to our $CO_2$ emissions was to plant trees to take up and store the excess carbon. The increased $CO_2$ would act as a "fertilizer" that, together with the warmer temperature, would allow the trees to grow faster and store more carbon. Problem solved.

This much-hyped solution neglected complexity: trees, both as a species and as a participant in a complex forest ecosystem, are constrained in their ability to take up $CO_2$. For example, a tree species' growth rate (a proxy measure of their ability to take up $CO_2$) is allometrically constrained by its total body mass (Meinzer 2003). Its ability to take up $CO_2$ is further limited by the soil fertility of its host ecosystem (Oren et al. 2001) and by extreme events such as drought (Barber et al. 2000). It is possible that beyond some temperature, or some level of its rate of increase, trees will not respond to increased $CO_2$ (climate warming) by increasing their growth but by shutting down. This has already been seen for some food crops. Or the forests could be subject to increased fires, which reverses any reductions. Thus, if environmental changes are rapid and severe enough, trees could become direct or indirect casualties or contributors, not saviours. Think of British Columbia's lodgepole pine and the mountain pine beetle, and its recent forest fires.

If we are to make more appropriate decisions, we will have to address more than the issue itself and the problem of timing. We would have to accept and incorporate into our thinking some basic features about ourselves: the influence on our perceptions and decisions arising from our innate decision-making foibles and our culture's world view. Specifically, we would have to accept the biases and limitations to our perceptions of our relationship with the environment, the complexity of how it functions, and what we think it does or should provide for us.

Overall, in order to consciously reach for the alternative future of living differently, we would have to, in essence, reduce the divergences between our internal-reality view of our world (including ourselves), and the external-reality view. And we would have to do so in ways that favour us living a more viable lifestyle, as discussed in chapters 11, 17 and 18. Completing these tasks takes time and effort.

It seems that the transformation of the environment described for the medium term is indeed the most likely for that time frame. Beyond the medium term, a greater rather than lesser level of the transformation currently expected seems more likely. The reliability of this assessment lies in our hands.

The summary of this chapter notes that, the biosphere is a dynamic, complex system whose functioning and state at a time scale relevant to us is represented by the human-environment system. We will thus play a significant role in the biosphere's (our supporting environment's) future for as long as our cumulative effect on it continues to result in significant physical, chemical, biological and ecological changes. Think of our impacts on the carbon cycle.

At present, the trend in our cumulative effect on the environment seems set to continue into at least the medium-term future. Consider that the ongoing growth in our population, our demand for resources, our disposal of waste and our environmental manipulation are all being maintained by our personal views, our decisions, our cultures' world views and resource availability. We are at least set to continue exerting our considerable influence on the environment until we can't, rather than until we decide not to.

If we do maintain this trend, then the earth's atmosphere is expected to continue its rapid warming and likely exceed our <2.0°C (1870) warming target. That exceedance will contribute significantly to further environmental changes. However, even if warming stays within this target, the cumulative impacts of our activities on species and their host ecosystems are expected to be sufficiently large in their own right that significant changes to our environment will still occur.

We can expect that the populations of most larger, specialist and less adaptable species will either be greatly reduced or the species will become extinct. Species shaped by our physical, chemical, bio-physical and bio-chemical activities will occupy the niches made available in a human-transformed environment. Novel ecosystems will appear. This future will find us slowly becoming earnest, noisy bystanders, wringing our hands as we experience our environment becoming increasingly hot, disrupted, simplified and weedy well into the medium term.

If we wish the future to be different, then we face some tough decisions. Before making them, we should remember that there are uncertainties in the above prediction. Consider that our predictions about the future favour steady, inertia-driven changes that follow current trends. The presence of chaos, randomness and our decision-making foibles ensures that deviations from those trends are inevitable and are more likely to occur the further into the future we predict.

A look at some of the possible alternative futures notes the following. There appears to be insufficient fossil fuels to support the most dire predictions about our climate future. However, we may have already initiated an abrupt warming event. In contrast, by the medium-term, we are unlikely to have voluntarily changed our world view and lifestyle in a sufficiently appropriate and timely manner that we will be living within the environment's constraints. We are therefore unlikely to significantly reduce our cumulative impacts on our environment during this period. It thus seems more likely that any deviations from the environment's expected trend into the medium-term future will favour greater environmental transformation, not less. How we might change this future is discussed in later chapters.

To some readers, this chapter might seem irrelevant because, even if our environment is radically transformed, we will do just fine. We will continue to adapt to a changing climate and ecosystems as we always have. In this context, *adapt* means successfully changing the manner in which we exploit, restore and manage our environment in order to secure the resources and services we want. It is believed that, regardless of the environment's state, by applying innovation and technology, we can maintain our current lifestyles, values, beliefs and attitudes, plus our striving toward our culture's visions and economic goals, far into the future. Is this realistic? The next two chapters attempt to answer that question.

# CHAPTER 14

# FUTURE RESOURCE WANTS:
# FOOD, WATER, HEALTH AND METALS

The resources (including services) we personally use are all ultimately supplied by the functioning of the earth's environment. That functioning constrains the ability of our environment to supply those resources today and in the future. Our decisions about the type and amounts of resources we want are guided by our culture's world view, such as our preferred lifestyle. So are our decisions about our relationship with the environment and how we acquire the resources. These decisions affect the functioning of the environment and thus the future resources it can supply. A feedback is involved. Thus, our future resource wants and our environment's ability to supply them should be considered together within the context of the dynamic, complex human-environment system, as discussed in chapter 11.

The interactions and feedbacks within that system make it difficult to predict the details of our resource future. However, it is possible to use a combination of the longer-term trends in our basic resource wants, the principles (rules of thumb) of the environment's functioning and the manner of our decision making to frame the limits within which our resource future will most likely occur. For example, the common multi-year growth-rate trends in our population and our use of water, food and fuel lie between 1% and 3% per year. If these rates are sustained, then the size of our population and resource demands will double within 70 years (at 1% growth) or 24 years (at 3% growth). In a finite world, these growth rates cannot continue for long.

Thus, the following discussion refrains from making time-focused, detailed predictions about our resource future. Instead, it focuses on the wider constraints we will face meeting the level of demand suggested by current trends. It also discusses the conditions under which our future demand could be different from that suggested by current trends.

This chapter starts by discussing our possible future population. The focus then moves on to discuss our possible food, fresh water, health services and metals futures. The discussion about metals reminds us of the benefits and constraints of making and using specific predictions about a resource's future.

The next chapter completes the discussion of individual resources by focusing on energy. That chapter, will also treat all of our resource demands and their supply limitations as part of a single dynamic, complex human-environment system. It outlines the cumulative constraints on our possible resource future. The focus will be on complex cultures with the default being Western culture.

A reminder that *short* (or *immediate*) *term* refers to a period up to 15 years from 2015; *medium term*, up to 60 years; and *long term*, beyond 60 years.

# Population

Our future demand for resources will be significantly influenced by the future size of the human population. This section discusses the possible fertility-based future trends in our population size. A more complete appreciation of our population future is then gained by considering the environmental and cultural constraints on those trends.

Population Size. In 1999, the global human population reached 6 billion (US Census Bureau 2008). By 2010, it had reached 7 billion and was still growing.

In 2001, a model-based prediction suggested that there is an 85% chance of population growth stopping by 2100, and a 60% chance that the population will not exceed 10 billion. It suggested a peak in the human population would most likely occur by 2070 at around 9 billion (Lutz et al. 2001).

In contrast, a 2014 model-based prediction suggested that there is now only a 30% chance that population growth will stop before 2100. Instead, there is an 80% chance that the population will grow to 10.9 billion (range 9.6 to 12.3 billion) in 2100. It suggested that the population would still be increasing after 2100. Most of this growth is expected to take place in Africa (Gerland et al. 2014).

Population predictions are more uncertain than they appear because those producing them must make assumptions about many factors that affect future fertility and survival. Sometimes these assumptions are implicit, for example, when the prediction simply projects current trends. In other cases they are explicit.

Population models are particularly sensitive to the estimates of factors such as our future reproductive behaviour, in particular a couple's decisions about the number of children to produce, and the children's survival rates. Modellers can't reliably know our future decisions or the future survival rates in advance, so estimated "averages" are used. However, as with all organisms, the future size of a population is significantly influenced by the year with the lowest survival rate. That rate can be strongly affected by factors peripheral to human decisions about fertility, such as food availability, disease (health services) and social functioning (e.g., war and poverty).

An emerging factor that complicates the process of estimating our future fertility is the possible increased impact of our lifestyle choices and human-made chemicals on our reproductive functioning and our children's sexual development. Examples are those chemicals that increase male sterility, decrease a female's ability to reproduce, increase the number of people born intersex, and change the sex ratio through feminization (Swan 2008). Consider too, that men in North America, Europe, Australia (plus New Zealand) have experienced an average decline of 53% in sperm count/unit volume and 59% in total count between 1973 and 2011. The possible causes include poor health and the environmental factors such as endocrine disrupting chemicals (think feminization), lifestyle factors including diet and smoking, and being overweight (Levine et al. 2017). All of these effects can lower the fertility rate.

Population predictions should thus be viewed as estimates. They are approximations, indicators, planning tools or guides, not deterministic calculations of the future (Cohen 1995). They are most useful at focusing our attention on the current population growth rate trend, the inertia inherent in that trend and the consequences if it continues.

## Population Limit.
Most living things, including humans, have the ability to produce more offspring than their immediate environment can support. Over the long term, the number of offspring that survive is limited by the interactions between the species in their host environment and the resources their environment provides. Hypothetically, human couples can produce as many as 15 children. However, this rate cannot continue indefinitely. Even if all of the other constraints on their survival are overcome, the number of people that can live on a finite earth will still be limited by the resources available to support them. The most basic of these resources are food, air, water and space.

We can predict the maximum size of the human population the earth can support by estimating the future availability of each resource. However, this provides an unreliable estimate because the factors that influence their availability interact with one another and our choices. For example, our production of food is affected by the amount of water available to grow it. This amount is affected by factors such as our use of water for other purposes, the effort we expend circumventing the environmental constraints on water availability and the availability of the resources (e.g., human labour and food) required to circumvent those constraints. Thus determining the limits to the size of the human population that the earth can support requires a complex analysis (Cohen 1995; Kelly 1995).

One such effort to estimate the earth's human carrying capacity was published in 1995, when the population had reached roughly 5.6 billion people. It concluded that the number of people the earth could support is significantly determined by both how many of us there are and by the resource demands of our chosen lifestyle. It also suggested that our ability to adapt to future resource scarcities was approaching our technological and social limits. With this in mind, the maximum size of a human population that the earth's environment could support, if they chose to live an industrial culture's typical lifestyle, was expected to be reached around 2045 (Cohen 1995). This is long before the date when human population growth is predicted to slow.

Regardless of whether this estimate is right or wrong, it is useful. It draws our attention to the importance of the interlinked constraints on human population growth. We can gain a broader appreciation for these constraints and the consequences of exceeding them by looking at the population limit issue from both our environment's and a human culture's perspective.

### The Environment's Perspective.
Over the long term, the size of the human population is limited by our environment's functioning, particularly its ability to continually renew the finite resources (including services) we need or want. Over the short term, our population's size is constrained by the immediate availability of those resources. To date, we have been spectacularly successful at applying our ingenuity and technology to circumvent these short-term availability constraints. We have made more resources, such as water, available to us and intensified the production of others, such as food and energy, than any past human population.

However, by focusing on circumventing the environmental constraints on the availability of resources in the immediate term, we have ignored the imperative to maintain our environment's functioning in a state that can support us over the long term. We have, for example, already allowed our population of 7 billion to exceed its allometrically constrained size for our ecosystem niche by at least 80 and perhaps 800 times. Our population is possibly set to reach 10.9 billion souls by around 2100, and to keep growing. As a consequence of our numbers alone, we have significantly disrupted our supporting environment's functioning, and we appear set to continue doing so for a while.

When the resource demands of such a large population exceedance combines with the current trends of our environmental disruption and destruction on a finite planet, we should expect an environment-triggered reaction. It is usual for a species that has overpopulated its host ecosystem to experience population-reduction events. These include fertility declines (social and physical), increases in disease outbreaks, starvation and enhanced predation.

Humans like to argue whether these ecosystem population controls actually apply to us. If we agree that they do, then we argue whether these controls would first appear among the materially rich few that consume in excess or among the "excessive" numbers of the materially poor. From our environment's perspective, these arguments are misleading. The current size and trends of the global population and the amount of resources we consume mean that rich, poor and those in between are all pushing us toward the longer-term environmental constraints on our population size. Each group is just doing it differently. The consequences of these population-resource exceedances will be felt by both a large population in a materially poorer country (e.g., China) and a small population in a materially richer country (e.g., the United States).

Consider too, that from our environment's perspective, anyone living in a city is demanding more support from their local ecosystem than it can provide: all resources must be imported and waste exported, while the ultra-high population density in cities is ideal for the spread of disease. Thus, regardless of whether you are rich or poor, if you live in a city, you are inherently vulnerable to our environment's population-control mechanisms. It is just the details of your experiences of them that will be different.

Consider too that the environment's reaction to our population size can be indirect. For example, at some point a combination of a cultural group's population size and resource shortages can result in social instability and thus changes to fertility and survival. This is an outcome that could be exacerbated by climate warming.

### The Culture's Perspective.
The future size of the earth's human population will be strongly influenced by our reproductive behaviour (including parenting). Although our reproductive decisions of today cannot change the current population, they do determine our lasting gift to future generations: the future number of people living on the earth (Demeny 1990). Whether we decide to voluntarily give the gift of a reduced birth rate to the future or not is influenced by a number of factors (Cohen 1995).

Most of the time, having children involves two people. Whether a couple decides to have sex or children is a decision influenced by more than logic or biology. It is also affected by a couple's psychological state, their emotional lives, their current material circumstances and, critically, forgetfulness (Lee 1980). Their decision is also significantly influenced by the guidance from their culture's world view.

If their culture provides couples with options and incentives to limit the number of children they have, and provides suitable birth control techniques, then couples will lower their fertility. The options include an appropriate mix of reproductive choices, while the incentives include the perceived social, psychological or economic advantages to limiting the number of children. For example, in the late 1800s, some European cultures thought that a small number of children per family was ideal. Successful family limitation was practised despite the lack of family planning tools (Cohen 1995).

More recently, the birth rate in many Western countries has dropped below replacement values. The reasons Westerner's give for having fewer children reflect our efforts to mentally balance the personal (social and material) costs and benefits of having, or not having, children. We make these decisions while under the influence of our culture's guidance about having children and fulfilling ourselves. Our conclusions are thus influenced by factors such as our perceptions of the value that our culture places on paid work compared to child rearing. We also evaluate the demands on our lives from raising a family while engaged in paid work and the (social and material) support for our child-rearing efforts provided by our culture. In the last 50 years, the balance in the West has tipped in favour of fewer or no children. The personal and cultural benefits of having children are seen as small and the personal (material and social) costs as high.

Voluntary fertility reduction is the ideal method to reduce population growth, but it has its limits. It takes time to implement and about a generation (25 years) to bear fruit. There is thus considerable inertia to slowing an increasing human population. As a result, voluntary human population reduction can't be used as a quick fix for resource shortfalls (Bradshaw and Brook 2014). That inertia also means that a growing population can precondition itself to crash because it creates a self-sustaining feedback that increases its size well beyond its long-term sustainable size (overshoot and collapse), as discussed in chapter 15.

Thus, to proactively reduce a human population, the appropriate culture-wide family planning incentives have to be provided to the members well in advance of the needed population decline. However, despite the importance of population size to our children's future, cultural leaders are generally reluctant to make well-thought-out population plans. And many in the public are equally disinclined to discuss them.

More precisely, when the elite of complex cultures discuss population issues, they seem more inclined to focus on maintaining a large or growing population. It suits their and the members' short-term economic or ideological ends, such as paying for its reaffirmation and authority maintenance costs, or raising an army. This inclination is illustrated in recent times by Western economists who argue, so far unsuccessfully, that Europeans need a higher birth rate in order to keep their economies growing because that growth is needed to support the elderly. To hold their views, economists must neglect the environment's constraints on a culture's functioning, such as limits to growth, and naively assume that most of the benefits of maintaining economic growth will definitively be assigned to elder care (Keyfitz 1991).

In summary, the number of people born will ultimately be determined by our personal reproductive choices as guided by our culture. However, the degree to which the earth can support the resulting population over the long term will be influenced by the lifestyle we and our descendants consider acceptable and its impact on our environment's functioning. It will thus also be influenced by how we choose to conduct ourselves toward our environment, and by how the environment responds to our demands (Cohen 1995).

The population size that our environment could continually support is dynamic. Whatever the supportable size is at a particular time, there can't be many more of us than that for very long, otherwise the environment will reduce our numbers (Hardin 1993). Hence, the 2001 and 2014 population predictions implicitly suggest that we are preconditioning ourselves to experience an environment-sourced population correction in the medium to long term. An alternative is an earlier social-disruption-driven correction. Maybe neither will happen because, in the future, there will be more than enough resources, such as food, to support a larger population living as they please. The rest of this chapter and the next chapter will help you decide if that is likely.

## Food

Food is a critical resource: our survival depends on it. The global-scale trends in our food supply suggest that we will meet our food demands into the immediate future (WHO/FAO 2002). However, it is uncertain whether we will meet the increased food demand in the medium-term future. This section discusses this uncertainty from the perspective of farmed food rather than wild food because the supplies of wild food, such as ocean fish, are already past their maximums and declining. Our farmed food production occurs in a dynamic, complex agricultural system. This is the context within which some of the constraints on future food production are discussed. Then pollination is used to illustrate the dynamics of that system and how our food future is likely to arrive.

**Production Constraints.** Estimates suggest that our global food supply would have to increase at around 1.2% to 1.5%/year (similar to energy and water) until at least 2030 in order to meet the predicted increase in food demand from just a growing population. However, there are indications that the rate of food supply increase may by slowing (Huang et al. 2006). Contributing to that slowing are the constraints on arable land, plant functioning and our mutualistic relationship with our crops. These constraints are discussed before considering the constraints that human culture imposes on how we will deal with the slowing.

*Land.* Until recently, the increasing demand for food was satisfied by expanding the amount of land being farmed. However, on a global scale, we have now appropriated most of the land suitable for farming (WHO/FAO 2002). An agricultural land shortage is appearing.

For some countries, the shortages are sufficiently large that the efforts to resolve them include purchasing or leasing large tracts of arable land from distant poorer countries. This can have a negative impact on the rights of the local people living on that land. Both this impact and the growth in the number and scale of these deals since 2005 prompted the United Nations to develop a set of principles for the parties to follow (Schutter 2009).

Still to be taken into account are the effects that climate warming will have on the availability of agricultural land. Consider that sea level rise will affect low-lying farmland. For example, if sea level rises by 1 m, then Egypt will lose 13% and Vietnam 7% of its agricultural land (Dasgupta et al. 2009). In addition, as climates move northward, the amounts and distribution of arable land will change significantly in ways that are difficult to predict.

Surprisingly, the area of agricultural land used to produce human food crops is already declining. One reason is that cities are expanding onto agricultural land. Another is that food crops are being displaced by the rapidly increasing cultivation of feedstocks for biofuels, bioproducts and food animals.

The recognized limits to land availability have also resulted in efforts to secure our food future through increasing agricultural intensity (i.e., increasing crop yields on existing farmland). Increases could come from making non-technological improvements to farming methods (Tilman et al. 2002). However, by far the more common vision for increasing agricultural intensity is its further industrialization, in particular boosting the yield of food plants and animals by applying technology (Huang et al. 2006). It is also expected that the increase in the crop area under irrigation will be ongoing. Our ability to intensify future food production in this manner is constrained, as discussed in the next two subsections and in the section discussing water.

*Plant Functioning.* The 1960s and 1970s was a period of rapid agricultural intensification through industrialization called the green revolution. One outcome is today's abundance of food produced from high crop-yield plants. The process of achieving this success is not open ended, it is subject to fundamental constraints.

One of these fundamental limits is plant biology. For example, plant growth is fundamentally constrained by their limited ability to photosynthesize plant matter from $CO_2$, water and other ingredients (Begon et al. 1996). Despite optimizing our food crop's growing conditions, their maximum photosynthesis efficiencies are still low, in the 3% to 10% range.

Another fundamental limit is allometric relationships. The green revolution's intense, selective plant breeding to boost crop yields was successful, in part, because it satisfied those allometric relationships through trade-offs. We traded those aspects of the plants' functioning that didn't contribute to the growth of the plant parts used as food or biomaterials for those parts that did. For example, green revolution plants are smaller in sizes (to support heavier crops) and/or have smaller root masses (growth directed to the crop) than their ancestors. These preferred plants produce much higher crop yields than their pre-green-revolution ancestors, but their ability to survive and flourish under a range of conditions is greatly constrained. They are, for example, less drought resistant and less able to secure nutrients from sub-optimal soils.

The green revolution thus required another trade-off. We have become responsible for providing the optimal growing conditions needed for the plants to produce high yields. We are responsible for exposing them to the maximum available sunlight; ensuring that they receive sufficient water, which often requires irrigation; and maintaining high soil fertility, which usually means applying human-made fertilizers. We are also responsible for controlling the undesirable insect and plant organisms that are attracted to the fertilizer, food or water bonanza our farming activities unintentionally provide for them. We deal with these pests through the liberal use of human-made pesticides and herbicides (Khush 1999). Providing all of these services consumes energy, so we are also burdened with the responsibility of supplying the fossil fuel energy needed to create and maintain the optimal growing conditions for our selected food crops, as discussed in chapter 10. As a result of this agricultural intensification, the farmer has become a bit player in a more complex agricultural system that is burdened with filling our many farming and farming-related responsibilities.

Today, many green revolution crop plants are grown as monocultures under close-to-ideal conditions, so the plants do indeed produce high yields. But those yields and their increase is slowing or stagnating. Some argue that the slowing may actually reflect a decline in population growth (WHO/FAO 2002). However, analysis of population growth and trends in crop yields indicate that the declines likely reflect the ending of the green revolution's great gains (Huang et al. 2006; Khush 1999). Its methods of increasing and maintaining yields are likely reaching fundamental limits.

For example, preliminary work indicates that adding human-made fertilizer to the fields growing some strains of cereal is no longer resulting in its expected benefits, for reasons not fully understood (Tilman et al. 2002).

Think nitrogen overload. Similarly, there was hope that the $CO_2$ we are adding to the atmosphere would act as extra plant food and stimulate greater productivity. For at least some grain crops however, this increase may be smaller than hoped for. It seems that their response is limited by their mix of highly selected characteristics, their specialized growing conditions and the fundamental limits to photosynthesis (Long et al. 2006). The yields of some of our most basic green revolution food crops may have indeed reached their fundamental limits.

Some have faith that, by continuing to apply green revolution techniques, yields can still be increased. Others believe in a new agricultural revolution based on new technological innovations, in particular the ability to genetically modify organisms (GMOs) (Conway and Toenniessen 1999; Huang et al. 2006). However, further industrialization of agriculture will probably be difficult to implement. After all, the green revolution has already taken advantage of most of the easy-to-manipulate characteristics of our crop plants. And we have already appropriated most of the easy-to-tap agricultural resources, such as water, needed for these plants to grow their high yields. Increasing the yields of our food crop plants by whatever means is thus constrained by a combination of the fundamentals of plant functioning, allometric relationships, the constraints on supplying critical agricultural resources and the difficulty of making the needed changes to the plants.

Although some future improvements in yields can be expected, the argument that future bioengineering, in particular GMOs, will overcome fundamental biological "barriers" to create an era of painless plenty should be taken with a grain of salt. Those holding the "success through technology" view of our food future forget that there are always costs (including responsibilities) and trade-offs for circumventing the constraints on a plant's or animal's functioning. A future technological breakthrough will come with costs/trade-offs like those that accompanied the success of the green revolution. And they will appear at a time when farmers are likely to be busy dealing with rapid climate-induced changes to farming. Efforts to increase food production through further industrialization will thus push more of the agricultural system's component parts closer to their fundamental limits. It will also push the responsibility for those components into our hands. As discussed, we are limited in our capacity to deal with those responsibilities.

*Mutualistic Responsibilities.* Within the agricultural system we and our food organisms are tightly bound into an ongoing, dynamic, co-dependant relationship. It is best described as mutualistic. Inherent in this mutualistic relationship are our unavoidable responsibilities to our food organisms, such as dispersing their seeds and watering them.

In this context, our responsibilities take on a new dimension. Rather than existing in isolation and being simple to discharge, our responsibilities are interconnected with one another and with the plants' interrelated requirements. This complexity both adds to our responsibilities and constrains how we can discharge them. Consider that a responsibility often comes with a timing constraint and the need to complete a whole host of closely related duties. For example, our responsibility to supply irrigation water includes providing it at the appropriate time in a crop's lifecycle as well as fulfilling the duties of desilting canals, maintaining pumps, providing electricity to run them, dealing with salination and managing water distribution.

We more fully appreciate the complex nature of our responsibilities by looking at our current efforts to deal with organisms, such as weeds, diseases and insects (pests) that prey on our food animals and our food plants. For the farmers practising highly intense industrial agriculture, the most common method to control pests is to apply human-made pesticides. But they kill more than pests. They can and do kill, for example, predators of pests. Thus, if the use of a pesticide is stopped or altered, then the pest population may not only return, it may explode in an unrestrained rebound effect.

If this rebound problem is avoided by the routine, ongoing use of a pesticide, then over the medium and long term, further undesired impacts can surface. The pests often develop resistance to the pesticide, rendering it useless. To complicate matters further, even if the ongoing use of pesticides manages to eliminate one pest, there are many other organisms waiting to fill the vacant niche in our agricultural ecosystem. After all, it provides easily available, abundant food in the form of the fertilized soil and our monocrops growing in it (Begon et al. 1996). As a result of these types of undesired impacts, our list of responsibilities includes the need to continually adapt, develop and apply new pest control methods.

Pesticide use can also disrupt those agricultural ecosystem functions that enable our crop plants to fulfill their lifeways. For example, neonicotinoids are currently the world's most widely used class of insecticide. Emerging information indicates that they also have significant longer-term impacts on many agriculturally important species. These include pollinators, such as bees, and organisms that maintain soil fertility, such as earthworms (van der Sluijs et al. 2014). By using neonicotinoids, or any other pesticide for that matter, we become responsible for fixing or fulfilling the agricultural system's services that their use has either directly or indirectly disrupted or destroyed.

Our mutualistic relationship with our food organisms obliges us, at its most basic level, to ensure that they thrive. But, as this discussion indicates, to reach that goal the obligations we must discharge are much wider ranging than our direct responsibilities to our crops and animals. After all, we must also maintain the ancillary/supporting services needed to fulfill our direct responsibilities, such as the pumps needed to provide water. We must also deal with the unexpected, unintended and unwanted impacts that discharging our responsibilities can have on the functioning of our agricultural system: we are also responsible for dealing with our (direct and indirect) collateral effects. One example is the development of resistance to pesticides. The wide range of our resulting obligations means that we are, in essence, responsible for keeping our dynamic, complex agricultural system functioning as a whole. To discharge that responsibility takes time, effort and resources.

An important conclusion can be draw from the discussion so far: by increasing the intensity of agricultural production, we achieve more than an increase in the food supply. We also increase the strength and complexity of our mutualistic relationship with our food organisms. We thus greatly increase both our short- and long-term agricultural responsibilities, as well as the time, effort and resources needed to discharge them. Greater complexity also makes it more difficult for us to fulfill our responsibilities: we are more likely to fail.

We could reduce the burden of our responsibilities that result from agricultural intensification. Consider our collateral impacts. For example, we could introduce regulations to limit the development and spread of pest resistance to pesticides. Although true in theory, in practice there are limits. Consider the level of compliance it is realistic to expect. This constraint is illustrated by the growing of corn into which Monsanto inserted the gene to produce the Bt insecticide.

In order to reduce the chances of pests developing resistance to the Bt insecticide, the US government introduced a regulation: no more than 80% of a farm can be planted with the genetically modified Bt-gene-containing corn. This simple rule favours the farmer's medium- and long-term interest. Nevertheless, in 2002, some 19% of all farms growing Bt corn in either Iowa, Nebraska or Minnesota failed to meet this requirement. If they didn't know about the regulation, it points to a weakness with regulation as a tool. If they did, then perhaps the farmers reasoned that the chances of developing resistance at that time were likely small because the total area planted in Bt corn was so small. Regardless, to ignore the regulation, the farmer must take a short- rather than long-term view of our food future (Clarke 2003). Under what conditions would those farming the 19% comply?

To continue, in 2014 it was discovered that the western corn rootworm had developed resistance to Bt. Surprisingly, that resistance had formed in less than seven years. A more complex farm management regime for the use of Bt corn was recommended. The farmers' and the greater culture's responsibilities had increased (Gassmann et al. 2014). To be truly effective, regulation must be accompanied by the acceptance of the regulation and its enforcement. It must be believed in. That requires a change in both the farmer's personal and their culture's world view.

The Bt corn example reminds us that our personal reality paradox and our culture's world view place limits on our personal and collective ability to recognize, accept and then discharge our agricultural ecosystem responsibilities. This is true regardless of whether regulation is involved or not. As much as our human characteristics give us the capacity to address our responsibilities, they also help us avoid and amplify them.

A popular rejoinder is that new technologies will eliminate our responsibilities and their amplification, while increasing crop yields. Certainly, technology can help us meet our agricultural ecosystem responsibilities, but it doesn't eliminate them: it transforms or replaces the old responsibilities while adding its own. After all, we are responsible for keeping technology functioning and for its undesired collateral effects.

This transformation and addition of responsibilities is illustrated by the English field trials of oilseed rape plants that had been genetically modified (GM) to make them resistant to the herbicide ammonium glufosinate (trade name Liberty). This herbicide was sprayed over the field planted with GM rape plants to see if it controlled the weeds without affecting the crop. The trials showed that the genetic modification worked. But it also showed that the genes conferring Liberty resistance had become established in weeds related to oilseed rape (species in the brassica family), such as mustards.

From the farmers' perspective, the field trials indicated that the benefits of herbicide-resistant GM crops came with responsibilities and costs. For example, after a field had been used to grow GM Liberty-resistant oilseed rape, a farmer could plant a different GM Liberty resistant crop. However, to avoid reducing the new crop's yield or quality, they had to ensure that only very low numbers of both the residual Liberty-resistant weeds *and* the leftover self-seeding GM Liberty oilseed rape plants were present in the field. Otherwise, when Liberty was used to control weeds in the new crop, both the resistant weeds and the self-seeded GM Liberty oilseed rape would survive as superweeds, which would contaminate the crop (Daniels et al. 2005). Growing a Liberty-resistant crop resolves one problem, weeds, but it raises others. It changes the nature and increases the complexity of farmers' long-term responsibilities to their food crops and to their agricultural ecosystem. Perhaps applying more technology would fix this problem?

The long-term amplification of agricultural responsibilities has a global dimension, as illustrated by corn. Despite the 1998 Mexican ban on GM corn, its genes were discovered in Mexican fields growing local varieties of non-GM Mexican corn (Pineyro-Nelson et al. 2008; Snow 2009). At least some GM corn genes can spread. Today, GM corn is shipped all over the world. It is reasonable to assume that the possible impacts that the GM corn's genes might have on distant, local-corn-based agricultural economies would have been fully assessed and addressed before the shipping started. But on this point you would be wrong. Remember the bias in environmental assessment discussed in chapter 11?

The validity of the concerns raised by corn is illustrated by the fallout from the 2013 discovery of Monsanto's GM wheat in an Oregon wheat field that was planted with non-GM wheat. GM wheat is unapproved for human consumption. Thus, after this discovery, Europe and Asian buyers of US wheat cancelled or halted their US wheat imports (Goldenberg 2013b). The source of the GM wheat and how widely the genes are distributed remains to be determined. (See also the spread of Liberty resistant rice discussed in chapter 13.)

There can be obvious short-term benefits to increasing agricultural intensity. But accompanying those gains is a strengthening of our mutualistic relationships with our food organisms. This increases our inescapable long-term responsibilities to those organisms and our agricultural ecosystem. The rule of thumb is that, beyond some financial, temporal, ecosystem and regulatory point, the benefits from increasing agricultural intensity are outweighed by our increased responsibilities. This limit may only become apparent to us over the medium to long term.

*Culture.* A culture's world view influences all aspects of its members' relationship with food, from their food preferences to production methods. This subsection illustrates how that influence affects the culture's food future.

Americans have successfully maintained their food supply by increasing crop yields. But this level of agricultural intensification, particularly the use of green revolution plants and industrial animal raising, is only possible because of a significant increase in the use of fuel energy for agriculture. It is used to run tractors, make fertilizers and pesticides, provide irrigation, transport food, and power the tools to gather knowledge. Most of this fuel energy is supplied by fossil fuels, especially oil and natural gas (Heichel 1976). This reliance on fossil fuel energy means that a decline or interruption in its availability could result in Americans facing a food shortage. Examples are the global food shortage (perhaps crisis) of 2007–2008 that was partly the result of extremely high fuel prices. Consider too the impact of the 1974 oil crisis on US food supplies. By contemplating Americans' cultural constraints on their possible responses to a future fuel-induced food shortage, we more fully appreciate the influence of their culture's world view on their food future.

The ingredients (processed corn, animal protein and fat) used to make American culture's preferred food-energy-rich menu items are provided by fuel-energy-intense farming methods, as discussed in chapter 10. Thus a change in American dietary preferences towards menu items that use less of these ingredients could help to reduce a fuel-induced food shortage. Consider what could happen if Americans reduced their consumption of food dishes containing large amounts of industrially farmed animal protein.

The pigs, chickens and cows, which are the sources of this protein, only convert between 2% and 14% of the energy in the plant food they eat into the energy stored as meat. The production and provision of the feed used to raise the animals on an industrial food-animal farm relies heavily on fossil fuel energy. Given the animals' low feed-to-meat conversion efficiency, it is no surprise that a unit of fossil fuel energy used to grow legumes can provide some 60 times more protein than if it were used to produce industrially farmed beef or pork protein, and 15 times more than chicken protein (figure 12b) (Heichel 1976). Thus, by reducing the amount of animal protein on the average American menu to levels more in keeping with our physiology, and by switching the preferred source of that protein from industrially farmed animals to plants, Americans could help to alleviate a fuel-induced food shortage. However, making this change requires adjusting their culture's guidance about food, which, as discussed, is a slow, difficult task.

Americans could consider another option to resolve a fuel-induced food shortage: they could question their non-agricultural uses of fuel energy. This too is a difficult step to take because their fuel-energy-intensive farming methods are part of a wider culturally supported preference for a fuel-energy-intense lifestyle. How this culturally supported energy preference complicates American reactions to a fuel-induced food shortage is illustrated by considering the use of biofuels to reduce the fuel and thus the food shortage.

Biofuels can be produced from agricultural feedstocks. One potential feedstock is the approximately 60% of the food crop plant left in the field as waste after the food portion has been removed. Other sources of feedstock are agricultural crops, such as corn, sugar cane and switchgrass, grown especially for that purpose.

But using these agricultural materials as biofuel feedstocks affects current and future food production in a number of ways. For example, growing a feedstock crop can displace the cultivation of food crops. Using plant waste as a feedstock precludes its use as an in situ fertilizer, thus requiring soil fertility to be maintained by applying other fertilizers. This usually means using human-made fertilizer that requires energy and natural gas to produce. Consider too that the energy gained from biofuels using farm products barely exceeds, or doesn't exceed, the energy cost of producing them, as discussed in chapter 15. Using agriculturally sourced biofuels to fill a fuel shortage and thus a food shortage is counterproductive. Despite this reality, the production of biofuels from corn feedstocks is well established in the United States. American farmers cultivate corn for that purpose because their efforts are supported both ideologically and financially by their culture's world view.

It is highly likely that if a fuel-induced food shortage arose, then Americans would, at least in the short term, continue trying to follow their culture's guidance about their preferred relationship with food and energy. The powerful influence of American culture's world view on their food-related decisions thus sets the tone for their food future.

This is more broadly illustrated by considering Westerners' efforts to plan for our food future. If the plan is to be meaningful, then it must reflect the external reality. Whether our making and implementing it meets that condition is strongly influenced by the guidance we receive from our culture's world view. This is illustrated by its influence on a Western food-focused movement that arose out of the food shortage of 2007–2008: food waste reduction.

There are a number of opportunities between farm and fridge for food to be wasted. The garbage studies discussed in chapter 11 indicate that we, as individual members of a household, are largely unaware of the amount of food we waste. In 2004, households were reported to have generated 14% of US food waste. In 2010, they generated 27% of Britain's total food waste (Jones 2004; DEFRA 2010). Bearing in mind the difficulty of measuring food waste, this information provides only a feel for the amount generated by households in Western industrial countries.

These and related observations suggest that Westerners' food waste would be significantly reduced if we became more aware of it and if the food distribution system was improved. The promoters of this view suggest that our efforts to reduce food waste could also provide other benefits, such as feeding the poor or hungry, reducing prices, reducing agriculture's ecosystem impacts and reducing agriculture's demand for fuel energy. Reducing food waste would thus help us secure our food future.

However, suggesting that these benefits will arise from reducing food waste neglects the pitfalls. Consider the culturally influenced meaning Westerners give to the word waste, "something not useful to humans," and that waste also has a moral dimension, *to waste* is considered amoral. These meanings provide us with flexibility in how we deal with waste. We can release it into the environment ("dump and forget"), or we can reduce the amount labelled waste (e.g., through efficiency or recycling), as long as, in both cases, we can justify our decision. As a result of these meanings, our decision-making foibles and the other cultural influences on our decision making, the outcome from successfully reducing our food waste could be much different than the noble goals some waste-reduction promoters expect. Consider two possible outcomes.

If shoppers and supermarkets reduced their food waste, then less food would need to be purchased and distributed. However, in keeping with the Western, market-driven world view, this doesn't mean that the farmers' resulting excess production would then be distributed to the poor and hungry, or that food prices would drop and stay low. It more likely means that the farmers' efficient use of the land would result in them either growing less food for that market, exporting the excess and/or switching to grow other more profitable non-food crops, such as feedstocks for biofuels or bioplastics. The now "surplus" land could even be sold for other purposes, such as housing developments. The poor or hungry may never see the food savings.

In a different outcome, a system could be set up to reduce food waste by distributing the edible portion to the poor and needy. The portion that is indeed non-edible (by humans) could be used to feed food animals. Most of the remainder could be composted. But this solution would have no effect on the amount of food produced, the energy used or the ecosystem impacts, and, if run by volunteers and donations, it would have no guaranteed effect on prices. It would however add to the culture's complexity.

If either of these outcomes came to pass, Westerners could argue, on both material and moral grounds, that the food waste had been successfully reduced. Every effort would have been made to use the food for a human benefit, and what had to be dumped could be justified. However, not all of the goals would have been reached. In particular, we would be no closer to securing our long-term food future. If reducing food waste is to reach that goal, for example, then it will have to be accompanied by basic changes to Western culture's world view. Our views on the value of farmable land, access to food, the compensation farmers receive and our diet would have to be changed to meet the external-reality conditions needed to achieve food security, such as limits to farmable land. There would likely also have to be changes to other aspects of our culture's world view that affect our food future. Consider its guidance about population growth. All of these changes would nip at the heels of our culture's guidance about progress, growth and our vision for the future. Reducing food waste can help to solve a food shortage and achieve other goals, such as food security, but only if our culture's world view changes in ways that guide us to meet the applicable conditions set by the external reality.

Overall, the guidance from our culture's world view has a strong influence on our relationship with food, including our preferred diet, and how that food is both secured and distributed. It thus has a significant influence on our food-related decisions. It also engenders in us an inertia to making any substantial changes to the food-related aspects of our culture's world view. As a result, a culture tends, by default, to maintain its current food trends into the future.

In order to successfully address a food issue in a way that is compatible with the external reality, we have to consider more than our immediate objective. For example, if we wish to ensure long-term food security, then we must focus on more than its material aspects. Our efforts must also facilitate the culture's members making the associated changes to their internal reality and their culture's world view. Similarly, the proponents of food waste reduction could evaluate their plans by asking themselves what will happen to the savings.

In overview, our food future is constrained by land availability, the fundamental limits to crop growth, and our mutualistic relationship with our crops and animals. It is also constrained by our culture's world view and our internal-reality views about food production and consumption. These constraints do not exist in isolation. They exist and interact with one another as part of a coherent, dynamic, complex food system, which is a part of the human-environment system.

**The Bees in the Field.** We will have a better grasp of our food future once we appreciate the complexity of a more complex culture's agricultural system and its current state. We gain that appreciation by focusing on the pollination of our food crop plants, because both the production of food and the ongoing functioning of our agricultural systems depend on it. Pollination must occur before the plants can produce our fruit and seed foods, which are also the plants' next generation. Just as a decline in the quality of soil affects our food future, so does a decline in the pollination of our food plants.

Food crops are pollinated in different ways. Our most important staple crops are the grains, such as wheat, oats and rice. They are related to grasses, so they don't form flowers and all are pollinated by wind.

Many plants form flowers. Some 80% of them are dependent on animals for their pollination (AFSSA 2008). Among these plants are our flower-forming food crops, such as apples and tomatoes. One estimate suggests that around half of the world's 89 major directly used food crops (e.g., coffee and nuts) are flower producing and pollinated by insects (Gallai et al. 2009).

The most well-known of these insect pollinators is the semi-domesticated honeybee (bee) (Dias et al. 1999). For example, in North America the exotic European or Western honeybee (*Apis mellifera*), a native of Eurasia, is responsible for pollinating about 100 commercial crops, including most fruits and vegetables (CSPNA 2007). The value we place on its role and industriousness is the reason the honeybee has come to symbolize our complex, dynamic agricultural system and its functioning.

Because bees are semi-domesticated, we have a close relationship with them. How we choose to fulfill the responsibilities that come with that relationship contributes to their current state. Consider the high-intensity, often industrialized nature of bee management. Their state is also affected by the consequences of our wider impacts on the environment that supports them. Not to be forgotten is the bees' response to these influences on their lives. We, our crops and the bees form a pollination-centred complex, dynamic relationship.

The story of the bees in the field reveals the bees' current status, the factors contributing to that status and the constraints on the possible cures. It also reveals the dynamics linking them together. This knowledge makes it easier for us to appreciate our agricultural system's dynamics and understand the trend we are following into our food future. The story is told using the loss of bee colonies in the United States as a focus point.

Honeybee keepers expect some of their bee colonies to die each year. (A hive houses one colony, so its death results in an empty hive.) In Europe, the mortality of up to 10% of colonies is considered normal. Surprisingly, in the United States, a mortality around 15% is considered normal (Cox-Foster and vanEngelsdorp 2009).

Although bee colonies are intensely managed, the annual mortality rate of colonies is higher than expected, worldwide. For example, in the United States, their average annual winter mortality rate over 10 winters, from 2006–2007 to 2015–2016, varied between 22% and 36% (Steinhauer et al. 2016; Lee et al. 2015). In Europe, 9 of 13 countries had a greater than 10% bee colony mortality in 2006/2007 (AFSSA 2008).

To date, no single reason for these increased colony deaths has been identified. Rather, the indicators point to a mixture of causes. The possible contributors include exotic pests, diseases, chemicals, overworking the bees and alterations to bee genetics.

In their attempts to identify the contributors to the abnormal number of bee colony deaths, researchers have considered bee deaths caused by bee parasites and pathogens. Many of them can be labelled as exotic. One example is the worldwide distribution of the bacteria *Paenibacillus*. One species infects bee larvae to cause American foulbrood, so named because of the bad smell emitted from infected hives (AFSSA 2008). Since 1994, American foulbrood has shown resistance to the antibiotics that were once able to manage foulbrood symptoms. As a result, foulbrood is once again contributing to bee deaths in the United States and Canada.

Another, more formidable exotic parasite is the *Varroa destructor* mite, which feeds on both larval and adult bees. Varroa mites are native to Asia, where they co-exist with the Asian or Eastern honeybee (*Apis cerana*). In the past, the European honeybee was imported into eastern Asia either to produce honey or as part of attempts to increase local honey production. As happens for so many exotic introductions, there was an unexpected (unforeseen, unplanned, unconsidered) outcome. The varroa mite was probably introduced into Europe on infected European bees or hives being moved to Europe. By the 1950s, the mite was an invasive exotic parasite pest in Europe. *Varroa destructor* is now found worldwide, except, perhaps, Australia (Rosenkranz et al. 2010).

European honeybees have limited intrinsic resistance or defence against varroa mites. Once they enter a hive, the mites rapidly destroy the colony, leaving behind dead bees. A few drugs have been developed to kill the mites, but starting around 1995, the mites began to display resistance to the drugs, so the destruction they cause continues.

There is twist to the varroa mites' impact on bees. The mites themselves don't seem to be responsible for most of the bee deaths in a mite-infected colony. That distinction might go to the viral pathogens for which the mites seem to be acting as a carrier or vector. The presence of varroa mites may also increase the virulence of these viruses.

The two most common viruses appear to be the Israeli acute paralysis virus and the Kashmir bee virus. The Israeli acute paralysis virus and its various strains are exotic to the United States. They arrived in the United States only recently but spread rapidly (Cox-Foster and vanEngelsdorp 2009).

Another significant, recently reported but poorly understood contribution to bee colony mortality is described as colony collapse disorder (CCD). The loss of a colony through CCD is distinguished from other bee colony deaths by the lack of dead bees. In CCD, the adult bees simply disappear, leaving behind an empty hive containing the queen and stored food. Each year the colonies lost to CCD have to be replaced. Until 2009, CCD had only been recognized in the United States. In that year, a case was reported in Europe (Dainat et al. 2012).

Speculations about the cause of CCD have most strongly implicated varroa mites and/or bee viruses because they can be found in CCD hives. However, placing blame on these viruses is not as easy as it seems. They are often present in hives that do not display CCD.

It appears more likely that a facilitator is increasing the susceptibility of the bees to viruses or other possible causes of CCD. Varroa mites spring to mind again. Tests have suggested that CCD can also arise through the combination of other facilitators and causes (AFSSA 2008; Ratnieks and Carreck 2010).

Among the other possible CCD facilitators are pathogens such as the fungus *Nosema ceranae*. Indigenous to Asia, it has spread, like varroa mites, worldwide as a result of moving honeybees around. The exotic fungus does directly kill bees. But, more importantly, the presence of *Nosema* seems to increase the honeybees' susceptibility to viruses and assist in their spread. There is no permitted treatment for *Nosema*.

Another class of possible CCD facilitators is related to the increasingly technical and economics-influenced nature of our bee management activities. Consider that our current demands on the bees can disrupt their body functions and subject them to increased levels of stress. For example, we overwork the bees when we stimulate them to produce more eggs and when we ship their colonies to a warmer climate in winter so they can work year round. They are also overworked when we truck them long distances during the summer so they can pollinate a wider variety of commercial crops. While trucking increases the owner's profit, the bees can experience a lack of sleep and rhythm changes, which stresses them and disrupts their lifeways. In the United States, the practice of bee trucking appears, in part, to have been a response to a steady, post-WWII decline in the number of bee colonies being kept. This decline was largely driven by economic considerations.

Bee illness is also promoted by nutrition deficiencies because improperly fed bees are less able to cope with stressors, pathogens and pollution. The possibility that diet is a facilitator of CCD is highlighted by the common practice of feeding bees nutritional supplements, such as artificial pollen. The need for these supplements can arise because bees often pollinate, and thus feed on, monocrops for extended periods (AFSSA 2008). But supplemental feeding for bees eating pollen from only a few crops is unlikely to provide them with complete nutrition. Both the limits to the nutritional value of the food crops to which they have access and the quality of the supplements provided to them could contribute to a decline in bee health (Stokstad 2007b).

Humans could also be acting as facilitators of CCD and elevated bee deaths through our use of pesticides, in general, and insecticides, in particular (CSPNA 2007). Consider that the roughly 5,000 chemical "protection" mixtures used on plants contain at least 450 active substances (AFSSA 2008). In the United States, more than 170 of these compounds have been identified in bee products, such as honey (Cox-Foster and vanEngelsdorp 2009). Some of these chemicals are known to be the cause of bee poisonings.

Bees do not seem to be routinely exposed to these chemicals at toxic doses. However, they may be experiencing chronic levels of impairment as a result of exposures to pesticides at sublethal doses or as a result of interactions among the many low-concentration chemicals in their pollution load. These possibilities have been poorly studied.

A group of chemicals whose impacts on bees at non-toxic levels has been studied is the neonicotinoid group of insecticides. This widely used pesticide is systemic, which means that it is found throughout the treated plant, including the pollen and nectar. Thus bees from larval stage to adults can be exposed to these neurotoxins over long periods. It is known that, at a minimum, neonicotinoids affect the bees' ability to navigate, learn, forage and reproduce. They also decrease the bees' lifespan and increase their susceptibility to viral diseases (Cox-Foster and vanEngelsdorp 2009; Tapparo et al. 2012; van der Sluijs et al. 2014). In a similar vein, it has been suggested that the pesticides most widely used to treat varroa mites may itself be contributing to excessive bee deaths if not CCD (AFSSA 2008).

Humans could also be indirectly facilitating both bee colony mortality and CCD through our interference with bee genes. Consider that we select bees to create strains that express traits we consider desirable. For example, bee keepers routinely select queen bees who display the favoured characteristics of non-aggression and whose offspring provide higher honey yields. In this selection process, bee keepers neglect the other traits that are more important for the colony's survival. Examples of these traits include the bees' ability to keep the hive clean and to respond appropriately to physiological stresses.

It may seem surprising, but a hive's survival depends critically on the genetic diversity of the worker bees within the colony. For example, normal genetic diversity in the hive is associated with a greater variety of microbiota within a bee's gut and in bee bread. That variety helps to ensure that the bees receive adequate nutrition. It is also associated with fewer pathogens in the hive, a critical factor since bees inherently have a weak immune system (Mattila et al. 2012). Thus, simplifying genetic diversity (e.g., through bee breeding selection) and reducing gut microbiota variety (e.g., through bee breeding selection and antibiotic use) could decrease the health of bees and thus the functioning of their hive.

Indeed, antibiotics are a part of the non-nutritional chemical load that bees carry. They are among the chemicals fed to the bees with the object of keeping them working and healthy. Think of the subclinical doses given to farm animals in general. Ironically, antibiotics may be harming bee health.

Whatever the combination of factors contributing to excessive honeybee colony deaths, including CCD, they load down the bee keepers with more responsibilities. The keeper must spend more money and time to more intensely manage their bees (CSPNA 2007). For example, some now use gamma rays to sterilize the hive boxes (Cox-Foster and vanEngelsdorp 2009).

In their efforts to meet their mutualistic responsibilities, deal with colony losses and keep operating costs low, bee keepers can come to rely on pesticides and nutrient supplements. They can also engage in practices that are inappropriate or dangerous to both bee keeper and the bees' host ecosystem. For example, in their efforts to treat foulbrood, they can use inappropriate chemicals or too large a dose.

From the consumers' perspective, the immediate result of the bee keeper trying to meet their responsibilities and cope with honeybee colony declines is more expensive honey. A less visible consequence is the increasingly rare occurrence on the store shelf of pure honey, directly from the comb. More common are honeys that contain the variety of chemicals and their residues to which the bees are now being exposed in the

field and in the hive (AFSSA 2008). Think especially of the chemicals used to control varroa mites.

One way honey producers and distributors deal with these chemical impurities is to mix honeys from different sources. This can dilute or mask the chemical impurities in the more contaminated honeys. In addition, honey can be strongly heated and filtered to further purify and homogenize it (Durham 2004). These efforts are one reason why many brands of honey are now blends, some from sources all over the world.

Unfortunately, despite the required labelling, the consumer won't know the history of the honey they are purchasing and eating. They face an unfair decision. At what point do you decide that the health risks from honey are sufficiently high or the food benefits too small that you should no longer eat it?

Of course, from our perspective, the main food-related function of bees is not producing honey but pollinating our food crops. This could be expressed in financial terms, but the calculation of the economic value of their pollination services is fraught with difficulty (Gallai et al. 2009). However, it can confidently be said that, in the long term, reduced insect pollination (honeybee, native bee, fly or other insect) would lead to less food being available and it costing more. A pollinator reduction would reduce food security.

Currently, the number of honeybee colonies being kept is growing far slower than the need for their services (Potts et al. 2010). This brings up the possibility of solving the honeybee mortality and pollination service problems by relying more on wild pollinator species, such as the wild bumblebees and butterflies, to do the job (Rader et al. 2016; CSPNA 2007). However, switching to wild pollinators would not be as easy as we think.

A moderately well-defined global decline in wild pollinators has been underway in recent decades (Potts et al. 2010). Consider a study of the eight most common wild bumblebee species in the United States. It indicated that in the last 20 years, the ranges and thus populations of four out of the eight species have declined by 23% to 87% (Cameron 2011).

The factors that contribute to these declines in alternate pollinators are mostly (at present) related to agricultural activities. Consider the destruction of their host communities through conversion to agricultural land. Of particular concern are the remnants of a wild pollinator's host ecosystem found on the edges of our fields. These remnants are being routinely disrupted or destroyed, thus imperilling the few remaining wild insects able to pollinate the crops in the adjacent fields.

The factors that disrupt the ecosystems near farm fields include the displacement of indigenous pollinators by exotic pollinators, such as the honeybee, and the impacts of dispersed herbicides on the species that supply food for wild pollinators. The herbicides also kill the weeds in the fields, thus affecting a food source for both wild pollinators and honeybees. This is especially true when the human food crops have been genetically modified to be resistant to the herbicides used to control weeds.

Consider too the impacts on wild pollinator lifeways from insecticides (Kremen et al. 2002; Dias et al. 1999). Of particular concern are neonicotinoids, discussed above (van der Sluijs et al. 2014) For example, field experiments using wild bumblebees exposed to neonicotinoid insecticides at real-life doses reveal an up to 85% reduction in their production of queens (Whitehorn et al. 2012).

Wild pollinator populations are declining for other reasons. For example, the genetics of ants, bumblebees and wasps means that small populations of these species produce a relatively large number of sterile males, thus increasing their susceptibility to extinction (Zayed and Packer 2005). Wild pollinators of all kinds are susceptible to pollution, the introduction of exotic diseases and pests (AFSSA 2008; Cox-Foster and vanEngelsdorp 2009; Graystock et al. 2013). No wonder that it is common to refer to pollinator declines, rather than just honeybee declines, as being an issue for our food future.

In overview, the conclusion to be drawn from the story of the bees in the field appears to be that a pollination crisis is imminent. But we and the honeybees are part of a single dynamic, complex system whose future status is clouded from view by a host of poorly known details and interactions. There may be no pollination crisis because the bees muddle through as they have during past declines (Ratnieks and Carreck 2010). Or we might yet claim that their future survival is a shining example of our success at managing an agricultural ecosystem or applying technology.

The first lesson from the bees in the field is that, over the years, our decisions have contributed significantly to the honeybee's current troubled status and the disrupted status of wild pollinators. This is the stage we have set for our crop pollinators' future. Those past choices were constrained by our human decision-making characteristics and influenced by our culture's world view. The same is true for our future choices and their influence on our crop pollinators' future.

Our past decisions favoured the exotic honeybees over native species. They favoured intensively managing the honeybee rather than husbanding wild pollinators and their habitat. As a consequence, we have formed a strong mutualistic relationship with honeybees and become responsible for ensuring that they can fulfill their lifeway. Their story clearly illustrates that we struggle to acknowledge and appropriately fulfill our responsibilities to them.

The consequences for wild pollinators from our past decisions are the same in essence but different in detail. As a result of our continuing disruption and destruction of their supporting ecosystems, they struggle to complete their lifeways. Thus today, wild pollinators are limited in their capacity to assist or replace honeybees as crop pollinators.

The second lesson from the bees in the field is that lasting, long-term solutions to honeybee issues involve a reduction in our mutualistic responsibilities to them. This lesson can also be applied to our efforts to deal with wild pollinator declines. Consider that one of the current solutions for the declines in wild pollinators is to actively provide them with hive material and cultivate food flowers. But our active participation is just a first step in making our relationship with wild pollinators mutualistic. A lasting, long-term solution to wild pollinator issues would focus strongly on protecting their existing habitat and providing them and their habitat with the opportunity to recover, expand and adapt, without our assistance. To do so would require a change in both our personal and our culture's world view.

The outcome from us intensely managing honeybees, not fulfilling our responsibilities to them, and reducing many wild pollinator populations and their habitat is a less stable, less adaptable crop pollination system. It is reasonable to predict that, if we continue to make decisions that increase our responsibilities for pollinators' lifeways, we are unlikely to find a long-term solution to pollination issues. After all, we are already overloaded with responsibilities to them that we can't, won't or don't fulfill. This is the framework within which the details of the pollinators' future will unfold. It is in this future that the possibility of a future pollinator "crisis" arises.

The tale of the bees in the field illustrates the complex and mutualistic nature of our agricultural system's functioning, its current state and why it is in that state. It also illustrates the path we are currently following into our food future: we are burdening ourselves with more responsibilities than we can appropriately discharge. The tale suggests the decisions we must face if we wish to change paths. The essence of this tale can also be applied to our other resources, including water, wild foods, health and energy.

In summary, our future global demand for food will change in close step, but not lockstep, with the trend in population growth. Currently, our global population and thus our food demand is increasing. Countries that are able to will likely try to meet their future food demand and deal with shortfalls by continuing to intensify agricultural production through the application of technology. Other means, such as importing food, will also be used.

In the short term, a country's efforts to maintain the food supply by these means will succeed. But the fundamental constraints on the preferred method of agricultural intensification through technology are already exerting themselves. These constraints include the inherent characteristics of our food organisms, such as plant biology and allometric relationships. They also include the limitations to the required resources, such as arable land that can be irrigated and the water to irrigate it, as discussed next. There is also the potential for longer-term declines in the amount and type of energy available for food production and transport, as discussed further in chapter 15.

Unexpected constraints arise because increased agricultural intensification strengthens our mutualistic relationship with our food organisms. This, and increased complexity, increases the number and types of responsibilities we must meet to ensure that our food organisms can successfully complete their lifeways and our agricultural system remains functional. Because it takes more time, effort and resources to appropriately discharge them, we are finding it increasingly onerous to do so.

The details of our food future will vary by region and will depend on the specifics of the interactions between a number of factors. Among the possible shorter-term factors are those that contributed to the food shortages of 2007–2008. They include increased oil prices, increased demand for high-energy foods, diversion of land from food into fuel crops (Goodspeed 2008), and market manipulation of food prices (Kaufman 2010).

Among the medium-term factors affecting our food future are pollinator declines, the declining trends in

wild-food availability (e.g., marine fish) (Worm et al. 2006), the declining availability of arable land and irrigation water (Ramankutty et al. 2008), and the depletion of soil fertility and volume (Kaiser 2004).

Medium- to long-term factors include those mentioned above that constrain intensification. They also include changes in the distribution of the conditions suitable for farming brought about by climate warming: precipitation amount and type, as well as temperature. Of key importance, as mentioned at the beginning, is the size of the future human population.

Our culture's world view, our decision-making foibles and our internal reality (with its reality paradox) will collectively influence our short-, medium- and long-term food future. These factors will determine our choice of family size, our preferences for certain foods, our relationship to our environment, and our impacts on the environment. They will also determine our decisions about how much faith and effort to place in intensifying farming through industrialization, the degree to which we will accept our responsibilities to our food organisms' lifeways, and whether we will (or won't) appropriately fulfill them. These factors will also influence our search for alternatives to intensification as a means to address a food shortage. They will thus significantly influence both our future demand for food and its supply.

Some of our culturally influenced decisions will result in voluntary changes to the current food-related trends. But these changes will generally happen slowly because changing the cultural world view guiding our decisions is subject to considerable inertia. Making significant changes will require considerable effort and can be socially disrupting.

When all of the trends, factors and influences on our food future are considered in the context of an agricultural system embedded in a dynamic, complex human-environment system, they suggest the following outcome. The present trends will continue into at least the immediate term. Some time into the medium term, the difference between our expectations about food availability and the realities of agricultural food production will likely present us with non-war-related, regional-scale food supply shortfall events of a kind rarely seen since the 1970s. Regional food scarcity will become more common.

History suggests that food scarcity events will be concentrated more in some regions than others, but no region will be fully immune. In chronically food scarce regions, there will be large swings in food prices, continuing hunger and latent malnutrition. Further into the future, the people in some regions will experience events similar to those associated with a more distant past: living through the occasional widespread famine that results in large population displacements and food-related political instability.

Westerners will start to feel the presence of this transition when we realize that we can no longer fulfill our increased responsibilities to our industrial agricultural system. Indicators will include an increased difficulty filling food export requests and declining or static yields. From our perspective, these indicators will likely appear to be technological in nature, or an unexpected event. But, as the bees in the field illustrate, the outcome we experience will have arisen over decades, largely as the outcome of the decisions we made within the dynamic, complex human-environment system.

In this system, the features of our culture's world view, our human characteristics and our environment interact to constrain both our decisions and our agricultural ecosystem's functioning. It thus constrains the outcome of our decisions, whether we like the result or not. This is the food future as seen along present trends. Different choices will bring different futures. Inertia suggests that those choice are unlikely to be voluntarily made by the near mid-term.

## Fresh Water

The decisions, and the influences on making them, that led to the current state of our dynamic, complex water system and its current trend into the future are, in essence, the same as those that led to the state of our food system, as illustrated by the bees in the field. There is therefore no need to discuss our possible water future and the influences on it to the same degree as for food. It is sufficient to point out the features specific to water: the constraints on balancing water supply with demand, and our likely response to water shortfalls.

<u>Supply Constraints.</u> The constraints on our water future are illustrated by first recapping both the limit to the available supply and the trend in our water demand. The accuracy of the numbers is less important than their relative sizes and the overall trends in our usage. Rain delivers a total of some 110,300 $km^3$/year of renewable freshwater to the land. In 1996, we used 24,980 $km^3$ (22.6%) of its annually renewable water. Roughly 18,200 $km^3$ of our 1996 usage was the rainfall onto

agricultural fields that was then either evaporated from the fields or transpired by the crops growing in them (evapotranspired usage). We withdrew the remaining 6,780 $km^3$ of our usage from rivers, lakes and dams (runoff usage). (Note that the maximum amount of runoff we could realistically hope to capture and withdraw in the future is estimated to be roughly 12,500 $km^3$/year. Thus our 1996 usage of 6,780 $km^3$ is about 54% of our maximum potential future runoff usage.)

Of our 6,780 $km^3$ of runoff usage, 4,155 $km^3$ (61.3%) was used directly. The largest share of that usage, at 2,880 $km^3$, was consumed by irrigated agriculture. Our direct industrial and municipal needs account for the rest, 1,275 $km^3$. From the remainder (2,625 $km^3$) of our runoff usage, a small but important percentage replenished groundwater aquifers. The rest kept rivers flowing for purposes such as waste removal (Postel et al. 1996). The earth's population in 1996 was around 5.8 billion, so our average total per capita usage (all purposes, direct and indirect) was 4,306 $m^3$/person.

The current, ongoing increases in human population, industrialization, irrigation and water for waste removal are all expected to continue into at least the immediate-term future. Projections of the current trends in historical total water demand suggest a possible water future. For example, the global use of (surface plus ground) water for all purposes increased from an estimated 1,824 $km^3$ in 1960 to 3,561 $km^3$ in 2000. Usage was projected to reach 3,935 $km^3$ in 2010 (Hassan et al. 2005). (A feel for the uncertainty in the numbers being used can be gained by comparing the estimated use of 3,561 $km^3$ for the year 2000 with an estimated use of 4,155 $km^3$ for the early 1990s given above. The general trends are more reliable than the absolute numbers.) How long can this increasing trend continue before there is insufficient runoff and evapotranspired water available to meet our demands?

A 2001 projection of water use trends suggested that if the human population reached 8 billion in 2025, there would be a total (irrigation plus direct use) water shortfall of some 192 cubic miles (800 $km^3$) of water (Postel 2001). However, the growth of total water use has slowed from 20%/decade between 1960 and 2000 to 10%/decade between 2000 and 2010. This suggests that a smaller water shortfall is possible.

In 2007, it was estimated that if the earth's population reached 8 billion people in 2025, approximately 1.8 billion people (22.5% of the population) would be experiencing an absolute water scarcity (i.e., lacking adequate water to live). In addition, some 5.3 billion people (66%) were predicted to be living under conditions of water stress (i.e., approaching scarcity) (UN Water 2007). By this estimate, an almost unbelievable 88.5% of people will face significant water issues by 2025.

These examples are among many that predict significant water shortfalls in the future. But beyond the obvious, that demand is expected to exceed supply, these numbers on their own don't help us appreciate why and where the shortfalls might occur, and what might change this future. The size of a region's water demand will depend on its population size, its water use per person and its cultural view of water. The size of a region's available supply will depend on its water endowment, the ability to access it and the constraints imposed on making it available. It will also depend on the actual changes in rainfall amounts and water distribution in the region resulting from climate warming. The following discussion helps us appreciate how these factors together influence a particular region's water future.

By the medium-term, climate-warming-related alterations to the hydrologic cycle will change the water endowment of many regions (IPCC 2013a). These effects will include changes to rainfall distribution patterns, how it falls, the flows in perennial drainages and the supply of water from melting glaciers (de Wit and Stankiewicz 2006). For example, on a regional scale, the climate of Alberta, Canada, is predicted to become warmer and dryer. Periods of drought are expected to increase. Currently, the flows in the rivers that cross Alberta are being maintained at a relatively steady flow by the glaciers in the Rocky Mountains. These glaciers are declining in size and expected to disappear. In the medium-term future, Albertans can thus expect to face more erratic and larger seasonal swings in river flows across their dryer and hotter province (Schindler and Donahue 2006).

Overall, the number of regions on earth experiencing drought is expected to increase. Under the worst-case climate warming scenario, the proportion of the global land area expected to experience extreme drought is predicted to rise from 1% in 2000 to 30% in 2100 (Burke et al. 2006). However, without a significant natural contribution to climate warming (i.e., an abrupt event), it seems that the largest increase is unlikely to happen during that time frame. A level in between is more likely.

Interestingly, a climate-warming-induced decline in a region's precipitation may not always be the primary cause of a future change in its water status from stress to shortage to scarcity. This change is more likely to occur in those dry countries that are already using a large proportion of their nominally renewable water supply. For example, in the 1990s, Azerbaijan was using 55%; Egypt, 110%; and Libya, 770% of their renewable water (Vörösmarty and Sahagian 2000). (A number greater than 100% means that the nominal renewable water supply is being withdrawn at a faster rate than it is being replenished, such as by pumping fossil groundwater.) This change is also more likely to occur in those dry or drying regions experiencing significant population growth, especially if the demand per person is increasing (Vörösmarty et al. 2000).

Our demand for water beyond its renewable rate can even occur in wetter regions that currently have sufficient water resources. For example, in some of the wetter parts of India, a large growing population is being fed by agricultural crops that are irrigated using groundwater. But, despite the high rainfall, the dropping regional water tables indicate that water is being extracted at a rate far faster than it is being replenished. Regardless of future climate-warming-induced changes to the annual amount of precipitation, these wetter high-demand areas of India are more susceptible to future water shortages.

This discussion focused on an overall climate-warming-induced decline in precipitation, but this is not the only way in which warming can reduce a region's water supply status. A warming-induced increase in the annual precipitation can also contribute to shortages if the precipitation's form and seasonal distribution changes: for example, if the dryer season is lengthened or becomes dryer, while the rain in the shorter wetter season comes in shorter, more intense bursts that quickly run off the land.

### Our Response to Shortfalls.

How are we likely to deal with these expected water shortfalls? In the 1900s, most water shortfalls were solved by increasing the supply: capturing more rainfall runoff (mostly through dams and groundwater pumping) and redistributing it more widely (mostly through canals and pipelines). Today, much of the runoff water that is easy and cheap to capture and redistribute is already being used (Postel et al. 1996). In effect, we are near the end of the 1900s era of easy and cheap rainfall-runoff capture and water redistribution.

Consider that there are few suitable sites left to build large dams. Once they are used, then building dams and their redistribution systems will become technically more difficult and financially more costly. The environmental costs will remain unchanged, but in a more crowded world, the human cost from these future dams, both up and downstream, will likely be much greater.

For example, Ethiopia's Blue Nile dam will affect Egypt's water supply and its political relations with Ethiopia. Dams on the lower Mekong would have a profoundly negative impact on the delta's environment and the many people who rely on it for their livelihood. These types of political and social difficulties can result in social unrest and have significant financial costs.

Other methods to increase the capture and redistribution of water include small dams, limited stream diversions, fog nets, desalination plants and iceberg towing. Some of these methods, such as desalination, require considerable amounts of energy. More importantly, none of these methods are likely to continuously supply enough cheap-to-provide water to support large irrigation projects. These methods are best suited to small-scale, local farms and low-volume essential services, such as providing drinking water to urban areas.

An alternative response to a water shortfall would be to increase the efficiency of water usage. One method is to reduce the losses that occur during the redistribution of water. For example, fixing the leaking pipes in an urban area's water distribution system would reduce water losses that can reach 30%. Similarly, using drip irrigation systems to distribute water to crops is reported to reduce water usage by 30% to 70%, as well as raise crop yields by 20% to 90% (Postel 2001). But increasing distribution efficiency comes with costs. These include the short-term costs of the resources, capital and human effort needed to build the works and the long-term costs of managing and maintaining them.

A particular crop plant's water use efficiency could, theoretically, also be increased. This could be achieved by altering their water-using processes through genetic engineering or by plant breeding. The objective would be to either enable the plant to produce a crop using less water or to grow in more saline water. But because a plant's water-use processes are fundamental aspects of its functioning, large strides in this direction are unlikely to be made quickly.

Changing a crop's water-using processes would also come with costs. Allometric relationships suggest that the yields from such a GMO crop plant would only meet our expectations if we took on increased agricultural responsibilities, such as providing specialized growing conditions or fertilizer requirements. We would also be responsible for the associated environmental costs. For example, by farming saline-water-tolerant GMO crop plants using the water from a saline estuary or from the land reclaimed from its marshes, we would disrupt the estuary's ability to function as a wild-fish nursery.

Certainly, increasing a particular plant's water-use efficiency would reduce its particular water demand, but it leaves unanswered the critical question applicable to all increases in water use efficiency. What will happen to the water savings? If they are not dedicated to reducing a region's long-term absolute water demand, then the saving will have only transformed, not resolved, the water shortfall. For example, growing plants in salinizing soils (e.g., in Iraq) would only delay declining crop yields, unless we halted the process of soil salinization.

In the same vein, a region can reduce its immediate water shortfall by importing water in the form of food or manufactured goods (i.e., as virtual water). However, there are costs. The importer's environment ultimately pays for that virtual water because of the impacts on it from the extracting and manipulating of the natural resources used to buy and transport the virtual water. The virtual water exporter also pays, but now in the form of both their inability to use their exported water for other purposes and the environmental (and/or social) impacts from growing or manufacturing the virtual-water-rich products being exported. Think of decorative flowers. When seen from the water's perspective, importing virtual water doesn't necessarily reduce the absolute demand for water, it just transforms and shifts the water supply problem. It can even add to the global, absolute water demand.

The discussion about our response to a water shortfall could continue along this circular supply-demand path. But the emerging picture moves our attention inexorably toward two often-forgotten rules of thumb about water issues and our water future. Firstly, they occur within a complex system. There are therefore no magic, one-shot or no-cost solutions to issues such as shortages or scarcity, and useful predictions about our water future are always conditional. Secondly, it is impossible to escape the significant influence a culture's world view has on discussions about water issues and thus our water future. These two rules of thumb are always relevant. They are highlighted in the following discussion.

Neatly engineered water control schemes can seem to be ideal solutions to water issues such as a shortfall. Certainly, technology is a part of addressing water issues, but it is an incomplete solution on its own. Think of the arsenic-contaminated groundwater in Bangladesh. A longer-term resolution must also address a water issue's human and environmental dimensions (Leslie 2000).

For that to happen, a resolution to a water issue must meet three conditions over the long term: it must be environmentally realistic; it must be accepted by the users as equitable (fair); and it must be designed, implemented and managed with the long term in mind. These three conditions are always present regardless of whether the resolution focuses on an engineering design, such as tube wells, or a new user-management method, such as private versus public water works, or a new political vision, such as treating water as a commodity. Consider that the potential water wars that some predict will occur are more likely to be ignited around issues of fairness and environmental reality than around disagreements over how to make water available and how to maintain the works.

A culture's world view has a strong influence on whether the three conditions will be met. It has this power because its guidance affects the members' views about the appropriate relationship to have with water, their perception of water issues and the constraints on possible solutions. Consider that the Babylonian's world view included gods associated with water, such as Abzu and Marduk. The world view of Europe's early farmers considered some watery places to be the realm of the spirits or a portal to them. Consider too that in many of the more complex cultures, past and present, the ability to use water without concern symbolizes affluence and luxury, while the ability to control it symbolizes status and power.

Western culture's world view is no exception. It guides us to preferentially view water as a physical resource. Thus even though water can have symbolic value to us, our perception of water-related issues and acceptable solutions to them is biased to treat it as a commodity.

Consider its influence on our decisions about what to do with the water savings from a proposal to increase water use efficiency. A Westerner could insist that they become part of an absolute water-demand reduction initiative by "putting the saving back into the ground." There the savings could maintain the supply of water for essential uses over the long term, ensure more equitable access to water or repay a debt to an aquatic environment. However, Western culture's world view would more likely guide us to preferentially allocate the savings to something that would satisfy the cultural imperatives of progress and economic growth, such as expanding irrigation, making it available to manufacturing. After all, the use of water by businesses provides a much greater increase in jobs and GDP per water unit than its use by the environment. The extra cash wealth, so the thinking goes, could then be used to deal with future water shortfalls. Perhaps to pay for pumps or importing virtual water in the form of lifestyle-boosting foods, such as animal protein.

It might seem impossible for a culture's guidance about something as basic as water to inaccurately reflect or even contradict the longer-term environmental reality of the water cycle. But it can, and does. Consider that the focus of the US world view on individual rights and economic interests has resulted in many states having groundwater management laws that defy environmental reality. One result is the over-pumping of groundwater.

For example, the Floridian aquifer in Florida is one of the largest in the United States. Until at least 2002, Floridian law allowed so much water to be pumped from this aquifer that the surface waters it fed declined. This draw down occurred even though Florida is a very wet state and the aquifer was being rapidly recharged by rainfall (Glennon 2002). Think too of the fossil-water-containing Ogallala aquifer in Texas.

It would be easy to propose that this type of cultural misguidance could be resolved by the members adjusting their culture's world view to better reflect environmental reality. However, the members face considerable inertia recognizing and accepting this need, and then making the required adjustments. The presence and significance of this inertia is illustrated by the option of reducing water demand by consuming fewer foods with a high virtual-water content (Hoekstra and Chapagain 2006).

As discussed earlier, a decrease in consumption of animal protein in the United States, especially agri-industry produced beef, could result in substantial water savings. But for the United States to successfully reduce its water demand by this method, its citizens would have to both change their culture's view about food and diets, and ensure that the water savings reduced their absolute water demand. This is not an easy transition because it requires the citizens to question their culture's world view, change their lifestyle and restrain their self-interest. The status quo has inertia.

If our solutions to water issues don't reflect the external reality of the water cycle, then that reality will eventually overwhelm us. This is the time to apply the broader lessons from the story of the bees in the field.

Our ability to manipulate water often exceeds our ability to fulfill, over the long term, the responsibilities that eventually come with that manipulating. And the consequences from failing to appropriately fulfill those responsibilities take time to appear.

In summary, a slight global-scale drop in our per capita water withdrawals seems to be occurring. The process of widely recognizing water limits may have already started. However, the absolute amount of water we use is still increasing as our population grows, and as agricultural and industrial demands for water increase.

In the short to medium term, we will likely continue to try and resolve most water shortfalls by increasing its supply through technological and engineering solutions. In the medium to long term, the environmental constraints on the water supply will ensure that these solutions become harder and more costly to implement. Water shortages will increasingly become a normal part of life for many. These chronic shortages will be felt most in the dryer, more populous regions (e.g., California) and those wetter regions with high demand but stressed supply (e.g., northern India). Shortages will be more acute in regions that become dryer because of climate warming.

The shortages in regions with shrinking water reserves (e.g., glaciers and groundwater), growing populations and growing industrialization will be enhanced because of the interactions among these factors. It is in these regions that water shortages will most likely drive the members to adopt human-behaviour-focused resolutions to water issues. These include conflict, migration, changes in lifestyle, and conservation.

A significant change to this water future would require a proactive change in our cultures' world views and our personal demand for water. Much of our water future depends on whether we decide to change ourselves and our cultures in ways that reduce our absolute water demand. Not to be forgotten is how we choose to use any savings.

## Health

It is well-known that our health future will depend heavily on the availability of, and our access to, life's basic material necessities, including nutritious food, clean air and potable water. As discussed above, in the future we will likely become more responsible for ensuring that our environment and its ecosystems continue to provide us with these necessities. Our health future will also be affected by other factors that are largely or entirely within our direct control. They include pollution, our lifestyle, communicable diseases, infections such as septicaemia, and our psychological and social needs. This section starts by discussing three of these factors: pollution, lifestyle and communicable diseases. The nature of a health service that could appropriately deal with the many health-influencing factors will then be illustrated by how it could successfully respond to antimicrobial drug resistance. Collectively, these discussions provide a feel for our possible health future.

**Pollution.** One determiner of our health future is the pollution that we are exposed to. We are well aware of some of the health impacts from past and present pollution, such as the illnesses arising from exposure to combustion pollution (smog), arsenic and cigarette smoke. The future will bring discoveries about the unrecognized impacts of old pollutants and the unanticipated impacts of new pollutants.

The health impacts from today's pollutants that await discovery are more likely to be chronic rather than acute. Of particular concern are the impacts on our endocrine system and our epigenes from low concentrations of a pollutant or the interactions among pollutants at low concentrations (Vandenberg et al. 2011; Lister et al. 2009). The emerging recognition of these types of impacts is illustrated by two examples from a steadily growing list.

The first example is two perfluorocarbons, called perfluorooctanoate (PFOA) and perfluorooctane sulfonate (PFOS). PFOA and PFOS are used in a wide variety of household products, including food containers, clothing, furniture, carpets, paints and firefighting foams. They are also widely dispersed throughout our environment, including in drinking water. PFOA and PFOS are likely already present in most of us. They have a well-studied ability to induce the early onset of menopause in women (Knox et al. 2011). They are likely having other impacts on our reproductive system.

The second example is the long-lived polybrominated diphenyl ethers (PBDEs), which are added to a wide variety of plastics as a fire retardant. The products containing them range from carpets to TVs. PBDEs are released from these products, thus polluting our environment and enabling them to enter our bodies.

There is a strong possibility that PBDEs might be acting as endocrine system disruptors in humans, perhaps affecting the functioning of our thyroid. The evidence comes from house cats. There is a correlation between the introduction of PBDEs into our home environment and the dramatic rise in the incidence of cat hyperthyroidism. Cats acquire PBDEs from their food and house environment in amounts that are 20 to 100 times greater than those found in their human house partners. Cats might be acting as a thyroid canary in our homes (Dye et al. 2007). If there are human thyroid effects from PBDEs, then it will likely be a while yet before conclusive proof is found (Vandenberg et al. 2011). Until we know for certain, should we allow or restrict their use?

Adding to the effects from existing pollutants will be the health impacts from the new, incompletely tested compounds that are just being introduced. Some of these will be new variants of existing chemicals. Others will be entirely new, such as nanoparticles, whose possible health impacts were illustrated in chapter 13. There are also compounds still under development that will follow the usual path to become pollutants and part of our personal pollution load further into the future.

**Lifestyle.** Another significant determiner of our health future is our choice of lifestyle. For example, choosing to smoke will significantly increase your risk of developing lung cancer and experiencing heart failure. Being overweight or obese increases your risk for a number of diseases, such as type 2 (non-inherited) diabetes, heart disease, and premature death.

Globally, body weights are increasing. In 2005, about 25% of the world's adult population (over 20 years old) were overweight and 9.8% were obese (obese plus severely obese), for a total of 34.8% carrying a health-affecting weight. It was predicted that, by 2030, some 38% of adults could be overweight and 20% obese: a total of 58% would be carrying a health-affecting weight (Kelly et al. 2008).

The proportion of adult men who were overweight or obese (obese plus severely obese) increased from 21% to 38% between 1975 and 2014. The proportion for adult women increased from 23% to 39%. It is predicted that this increasing trend will continue to at least 2025 (NCD-RisC 2016). This implies that some 48% of the population would carry a health-affecting weight by 2030.

The changes in the body weight of Americans has long been an indicator of the future state of global body weights. A 2008 study predicted that the proportion of the American adult population who are overweight or obese (all classes) would increase from 67% in 2005–2006 to 75% by 2015. This is in line with the 2016 study (above) whose data reported that about 74% of American men and 62% of American women were overweight or obese (obese and severely obese) in 2014. A similar weight gain trend is seen in American children, thus preconditioning what the future will bring (NCHS 2008; Wang and Beydoun 2007).

As Americans follow this increasing overweight-obesity trend, many more of its citizens will suffer substantial health impacts related to their body weight. Americans, on average, will increasingly suffer an overall decline in health, quality of life and life expectancy. Dealing with these issues will become a significant financial and social burden on the society. If the global, overweight-obesity trend continues to follow the United States' lead, then a similar outcome can be expected for the global population.

It is easy to say that these body-weight-related health trends can be changed by people exercising their freedom to choose a different lifestyle. But making a lifestyle change is difficult at the best of times. It is more difficult if your current body-size-increasing lifestyle choice is supported, even encouraged, by your culture's world view. It is even more difficult to change if that lifestyle was imprinted on you as you grew up. Think of the standard American menu.

Consider one way in which Western culture's world view preconditions its members to make lifestyle choices that favour us becoming overweight or obese. All cultures have utopian visions for the future. They help to provide their members with a sense of purpose and hope for the future.

Western culture's utopian vision has its members living a healthier, happier and longer life. In this vision, we will have the freedom to access an abundance of material goods, wondrous medical services and the tasty high-energy "luxury" foods that are associated with that good life. That future life will also be largely free of strenuous manual toil and enriched with enhanced, effortless, exhilarating and even luxurious (electronic or real-life) leisure experiences. This is the lifestyle to which Westerners tend to aspire or adopt by default.

However, Western culture's utopian vision fails to remind its members that the future comes with non-negotiable conditions. For example, to live the good life requires that we respect the biological constraints on our lives: eating appropriate portions of nutritious food, engaging in sufficient physical exercise and avoiding health-harming activities. Because our culture's world view fails to strongly guide its members to take these types of constraints into account, it is more difficult for us to choose a healthy lifestyle. It is harder for us to accept our personal responsibility for our health.

The equivalent shortcomings are seen in other cultures' world views. Consider the lifestyle changes that the rural members of an industrializing country's population undergo as they move into the growing cities. Although the migrants undergo a transition from a rural to an urban environment, they retain their culture's rural version of an ideal future: ample food, toil-free labour and enhanced leisure. Once in a city, it is relatively easy to adopt a lifestyle that matches some aspects of that vision. However, the reality of urban living also ensures that it is relatively difficult for their food and exercise choices to appropriately reflect the biological constraints on their lives. Thus, across the globe, the ongoing rural-urban transitions are making a significant contribution to the dramatic rise in the proportion of the world's population who are overweight or obese, as reflected in the numbers given above (Friedrich 2002; Popkin 2002).

The power of our cultures' world views to influence our lifestyle choices, and thus our health, is illustrated by considering the advice provided to us by Western health providers: in order to reduce your risk of experiencing lifestyle-related diseases, choose a healthy lifestyle. That message includes, as mentioned above, not smoking, eating a balanced diet, exercising sufficiently and getting adequate sleep. But this message is swamped by the culturally supported freedom to promote and engage in health-harming lifestyles and habits, such as smoking cigarettes; eating tasty, energy-rich junk food; and pursuing low-activity recreation late into the night.

Western culture's world view facilitates their promotion because, in part, it guides us to believe that a strong economy (as indicated by a growing GDP and high personal incomes) is largely responsible for improving and maintaining all aspects of our health and health services. Consequently, it seems that the safest way for us to avoid a decline in our personal health is to first ensure that the economy is always growing. When this kind of cultural imperative is combined with Western culture's belief in freedom of choice, then it is unsurprising that its world view can also unconditionally support the promotion of economy-boosting, but health-reducing, products and activities. Remember the

politician's argument that the sale of cigarettes shouldn't be discouraged because the taxes on cigarettes were needed to pay for health-care services?

Once health-reducing views, products and activities become a part of Western culture's world view, they can and do become normal activities and symbols of Western culture's utopian vision of the good life. They thus become part of the member-culture feedback, which then sustains their place and influence, as discussed in chapter 11. Think of Westerners' historical view of smoking.

As a result, we can display considerable reluctance or inertia to making those personal lifestyle choices or changes that Western health service providers advise would improve our health. After all, to successfully make them we must deal with the contrary guidance provided by both our internal reality and our culture's world view. This is one reason why, despite the evidence that change was needed, it took Western countries more than 40 years to significantly reduce their smoking rates. How long will it take for obesity rates to decline significantly?

The process of change can be helped by us recognizing that the connection between the economy and personal health is not as close as we believe. History indicates that relatively inexpensive factors, such as nutritious food, basic hygiene and healthy lifestyle choices, are among the most important factors for maintaining good health. And, despite the guidance to the contrary, an economic decline can even result in health improvements, if the culture and its members choose to respond to it in a manner that reduces their lifestyle-related causes of illnesses. This possibility is illustrated by the health changes experienced by Cubans during their 11-year economic crisis from 1989 to 2000.

The Cuban economic crisis led to a reduction in their food energy intake from an average of 2,899 calories/person to 1,863 calories/person: from more to less than desirable. This decline occurred through a reduction in the consumption of high energy foods. Overall, in pre-crisis 1980, the average diet and its high-caloric value consisted of 65% carbohydrates, 20% fat and 15% protein. By 1993, during the crisis, the average diet and its low-caloric value consisted of 77% carbohydrates, 13% fat and 10% protein. At the same time, physical activity approximately doubled and the number of cigarettes smoked declined.

Between 1997 and 2002, some eight years after the crisis had started, the incidences of obesity, death from type 2 diabetes, and coronary diseases showed significant declines. The details of the relationships causing these average health improvements are unsure. But it can be said that the culture's response to the economic crisis induced changes that are recorded as average reductions in lifestyle-related diseases (Franco et al. 2007).

Overall, it seems that as long as health-harming lifestyles are implicitly supported by a culture's world view, then the making of voluntary and proactive lifestyle changes will only occur slowly, over decades. It is possible for the trends in lifestyle-related diseases to change voluntarily and in the immediate term, but only if small adjustments to the culture's world view are required. Immediate-term changes will also occur more easily if the resources to practise the harmful lifestyle become scarce and the culture's world view responds appropriately to their loss. Making them is also easier if assistance is available from local health services.

## Communicable Disease.
Communicable diseases will influence our health future. The prevalence and severity of their future outbreaks will be determined by a mixture of environmental conditions and human decisions. Among the environmental conditions are ongoing global warming and the manner in which disease-causing organisms respond to it (Colwell 1996). In addition, if our current interactions with our environment persist, then it will generate, and we will be exposed to, both old diseases that are now drug resistant and new diseases.

Among the human decisions that favour communicable disease outbreaks are those that will increase the global human population and the proportion of people living in cities at higher densities. This is especially the case if those in charge of the current and emerging megacities fail to adequately maintain domestic water quality and sanitation services. Communicable disease spread is also favoured by our ongoing, unrestrained use of our global transportation network. Think of diseases as exotics. Our health future is strongly influenced by our treatment of these diseases, which today means our use of antiviral and antibacterial (antimicrobial) drugs.

The combination of current trends, such as our ongoing migration into megacities, increasing population and the increase in drug resistance, suggests that the potential for future communicable disease outbreaks is high. It seems that, over the long term, humans are preconditioning themselves to experience both more, and more widely spread, outbreaks of old and new communicable diseases.

## Health Services and Drug Resistance.
Our health service system (health system) reflects our efforts to respond to the many factors affecting our health future, including those discussed above. Its role in determining our health future is easier to appreciate if we think of the health system as a product of the interactions between the primary components of the complex human-environment system: individuals, culture and the environment. When considered from that perspective, the health system's role can be illustrated by its need to deal with two critical health-affecting factors: treating communicable diseases and infections (diseases), and preventing disease-causing microbes from developing resistance to antimicrobial drugs (drug resistance). The two factors are tightly linked. The focus here will be on the prevention of drug resistance, which continues the discussion started in chapter 10.

From the environment's perspective, drug resistance develops in an organism as part of its normal population-level response to being exposed to drugs that may harm it. Being a normal response, we are unlikely to permanently stop it occurring. We are thus left with the responsibility of limiting the conditions under which it can develop (Levy 1998).

In the 1990s, health-service professionals began to realize that drug resistance was becoming a serious health issue. The basic actions needed to limit its development were already known, so they just needed to be more broadly implemented (Huovinen and Cars 1998). In overview, these actions include reducing the unnecessary use of antimicrobial drugs (in particular, only using an antimicrobial drug to treat illnesses for which the drug is known to be effective) and avoiding the use of multi-spectrum drug mixtures. Additional actions include limiting the over-the-counter sale of antimicrobial drugs by ensuring that their purchase and use follows a prescription. This process could help to ensure that the unused drugs are appropriately disposed of. To ensure that these actions are effective and endure, health service workers and the public should be provided with the information, instruction and encouragement to personally implement them.

Despite this knowledge, the number of disease-causing microbes that display resistance is still rapidly increasing and so is the variety of drugs to which they are resistant. The seriousness of the issue was illustrated in chapter 10 by the recent generation and global dispersion of the NDM-1 gene. It is now endowing a variety of disease-causing microbes with drug resistance to a whole class of antibiotics (Kumarasamy et al. 2010).

There are many reasons why it is proving difficult to successfully implement the actions needed to slow the appearance and spread of drug resistance. These include our personal lack of appreciation for the impacts that widespread drug resistance will mean for the delivery of health services to us and our children. Neither do we recognize that we, personally, are on the drug resistance front line: we must personally take steps to limit drug resistance. Our culture's world view, by default, supports this situation because its guidance gives a relatively low priority to drug resistance. That cultural oversight is made more serious by the manner in which resistance is spread. Another difficulty is that the actions needed to successfully limit the appearance and spread of drug resistance among disease-causing microbes must be implemented together, in a coordinated way, not piecemeal.

This is the individual, cultural and environmental context within which a health system's efforts to address the development of antimicrobial drug resistance should be planned and implemented, if they are to succeed. Here are some specifics of that plan. A successful disease treatment and anti-resistance campaign in this age of crowded living and rapid, global transportation will have to be a long-term, multi-faceted, community-based effort that aims for a high degree of compliance. It also has to have full public and government support. These local efforts will have to be coordinated over a larger area, even globally (Turnidge 1998; Carbon and Bax 1998; Williams and Ryan 1998).

The nature and scale of such a campaign, and the constraints on it being successful, are illustrated by our efforts to treat tuberculosis (TB) and to prevent the TB microbe from developing further drug resistance. As chapter 10 discussed, TB was a widespread, chronic, debilitating and often fatal disease until antimicrobials became widely available in the mid 1900s. (George Orwell of *1984* fame died from it.)

Despite the availability of suitable drugs, it is still difficult to cure TB because the convalescing period is so long. To improve treatment outcomes, the directly observed treatment short course (DOTS) is used. DOTS requires a health-service worker to monitor a tuberculosis patient until cured, which includes making certain that the patient takes each dose of the antimicrobial drug (Farmer and Kim 1998). The DOTS method has another advantage: it is ideally suited to prevent the development of resistance. Despite this advantage, multiple drug resistant TB (MDR TB) and extreme drug resistant TB (XD TB) varieties recently appeared, making TB even more difficult to treat.

XD TB appeared for a number of reasons: the improper treatment of tuberculosis; its close association with AIDS, which is itself often untreated (Cohen 2006; Nunn 2007); the lack of testing equipment; and the ease with which TB can spread among people living in unhealthy conditions with limited access to health services. Think of slums and jails. These reasons ensure that the DOTS method is, on its own, unable to fully cure tuberculosis, stop its spread and limit the development of drug resistance. In order for the DOTS method to achieve these three aims, it needs to be joined by a broader community effort to deal with the contributing social issues: poverty, limited access to healthy food, crowding and lack of medical facilities, among others. Addressing these issues is a cultural responsibility that requires broad government involvement and public support.

The DOTS example illustrates the individual, cultural and environmental components of a TB treatment and anti-resistance campaign and the connections between them. It also points out the specific difficulties we face if our efforts are to be successful. These observations apply to a more general effort to limit the development of drug resistance, as illustrated next, with a Western bias.

Our personal role in this effort, and its cultural connections, becomes clearer by considering the routine act of prescribing antimicrobial drugs. In response to you (the patient) feeling unwell, a doctor might prescribe antimicrobial drugs. That process consists of more than the doctor making a medical diagnosis and writing down the prescription. They are also taking part in a decision-making process that is strongly influenced by the doctor as a person, you as a person and the doctor-patient relationship.

Your doctor wants to resolve your problem. When writing the prescription for antimicrobial drugs, the doctor might recognize that, in your case, their use is inappropriate. But, being human, the doctor might also rationalize their choice. They might do so because they know that giving you the prescription will give you peace of mind, or because they believe that they personally prescribe the same or less antibiotics than the average for all doctors. This belief is often false.

The patient also contributes to the doctor's prescribing decision. As the patient, you have the culturally supported expectation that the doctor will solve your medical problem, often with drugs. You react positively to the symbolic value of receiving a prescription for antimicrobial drugs and can feel hard done by if you don't receive a drug prescription. You may even pressure the doctor to prescribe them. You will experience these feelings regardless of whether the drugs can realistically cure your ailment or not.

These aspects of prescribing antimicrobial drugs are added to the normal complexity of the doctor-patient relationship. No wonder a major cause of drug resistance is proving so difficult to resolve: doctors' inappropriate prescribing of antimicrobials and individuals' inappropriate use of them (Butler et al. 1998). For example, despite a decline in the prescribing of antibiotics in the United States during the 1990s and the first decade of the 2000s, it is estimated that about a third of antibiotic prescriptions to outpatients are inappropriate (NCD-RisC 2016).

This example of meeting patient expectations in the treatment of ailments also hints at the critical role that a culture's world view plays in its members' efforts to slow the development of drug resistance. Culture guides how we think about illness and our use of antimicrobial drugs to treat it. If Westerners (doctors, patients, farm animal producers, politicians, drug producers, pet owners, you and me) are to better address drug resistance, then our culture's world view must be adjusted to provide us with more appropriate guidance about our relationship with antimicrobial drugs and illness.

Western culture's world view has to guide its members to do more than follow the basics needed to personally avoid creating the conditions under which drug resistance can develop. It should also prompt us to adjust some of our social customs and practices that favour its spread. Consider that special care must be taken in hospitals, because the high patient density and the high use of antimicrobial drugs facilitate the development of drug resistance (Struelens 1998). Similar efforts should be undertaken in elder care facilities (crowding and high drug use) and daycare facilities (crowding). Efforts should also be made to curb our general, non-medicinal use of antibacterial products, such as medicinal-drug-based antimicrobial hand sanitizers.

Western culture's guidance should also prompt its food-animal agri-industry to make changes. After all, it uses a larger amount of antibiotics on its animals than humans use on themselves. If these farmers were to meaningfully address their contribution to the development of antimicrobial resistance, then they would have to reconsider many of their culturally sanctioned practices. These include animal crowding, sub-therapeutic doses given in attempts to increase "food conversion efficiency" (animal growth), "preventative" herd dosing rather than individual sick animal treatment, and improper methods of animal waste disposal (McKellar 1998; Johnston 1998; Kumar et al. 2005). To support these efforts, the general population would have to re-evaluate their culturally supported expectation of abundant ultra-low-cost, animal-sourced foods (Livermore et al. 1998).

Pet owners too would have to be culturally prompted to re-evaluate the need to keep exotic pets, which can carry diseases that require antimicrobial drugs to treat, such as salmonella in iguanas and psittacosis in parrots (Burnet and White 1972). Owners of the more usual pets, such as dogs and cats, would have to question the current excessive use of antimicrobial drugs to treat their pets (Johnston 1998). At the same time, the owners would have to consider their exposure to drug resistant strains of microbes from their pets (Guardian Weekly 2003).

Unfortunately, it is hard to imagine us individually accepting all of our personal responsibilities or making all of the cultural adjustments needed to prevent antimicrobial resistance from developing. We are more likely to continue our current practices while placing our culturally supported faith in future discoveries of new antimicrobial drugs. We prefer this option, even though most antibiotics currently labelled as new are in fact modifications of existing drugs (Wise 1998).

Our complacency about antimicrobial resistance and our faith in future discoveries should be further shaken if we first remember that the NDM-1 gene confers drug resistance to a whole class of antibiotics. Furthermore, in the 50 years since the mid-1960s, there have been only two truly new functional forms of antibiotics available to doctors. A potential third form was discovered in 2006 after screening 250,000 drug-producing organisms. This potential drug must still go through the development and safety tests to qualify as a useful, safe drug (Pearson 2006; Brown 2006). The difficulty and increasing costs of developing new, safe and effective antibiotics of the current type indicates that their discovery is following the law of diminishing returns. We need to temper our belief that new forms of traditional, antimicrobial drugs will "solve" the health issues arising from an accelerating rise in drug resistance.

The use of a new method to discover new types of traditional antibiotics in a previously untapped source might prompt a change in opinion. Especially as this method discovered teixobactin, an antibiotic drug in which the development of resistance is expected to be slow: longer than 30 years (Ling et al. 2015). However, "slow" doesn't remove the efficacy-maintaining responsibilities that accompany antimicrobial use.

Perhaps we should put our faith in the ongoing research into different types of antimicrobial agents? The current antibiotic drugs work by disrupting the bacteria's cell membrane. The targets for current antiviral and antibacterial research include the microbe's growth-enabling genes, the infection process and the bacteria's quorum-sensing communication system (Geske et al. 2008). They also include vaccines (Wise 1998) and bacteria-attacking viruses (bacteriophages) (Pearson 2006).

Discoveries in these fields might, indeed, help alleviate the current difficulties our health system faces when dealing with microbial diseases, especially those that are drug resistant. Western culture's belief in that future provides us with great hope. It also provides us with a vision of our health future: technology's future antimicrobial discoveries are expected to quickly, completely and permanently relieve us of all our responsibilities to limit the development of drug resistance.

However, it is unrealistic to have blind faith in such a vision of the future. To do so would mean forgetting about the slow process of scientific discovery, disregarding the collateral effects associated with any new disease-controlling agents and neglecting the inevitable organism and ecosystem response to any new agents we might discover. The functioning of our bodies, together with the dynamic nature of species interactions in an ecosystem, mean that any new "miracle cures" will be constrained in some way. Finally, if we want to avoid losing the efficacy of these future wonder agents in a manner equivalent to the development of antimicrobial drug resistance, then we will have to accept and respect the constraints on their use. We will therefore always be burdened with responsibilities that are the equivalent of those we now face when using antimicrobial drugs.

Ironically, our hope for the future development of new types of antimicrobial agents distracts us from paying attention to the ongoing development of drug resistance today. We fail to realize that our lack of effort to minimize the development and spread of drug resistance today preconditions us to pay a much larger financial and health cost tomorrow, especially if our hoped for drug developments of the future don't materialize in time (Farmer and Kim 1998). Is this gamble worth it?

Two recent governmental, global-scale reviews of antimicrobial drug resistance urged that strong and immediate efforts be made to address this issue (O'Niell 2016; WHO 2014). This appears to be a promising development until we remember that these new calls for action reiterate the essence of similar calls made in the mid-1990s. It remains to be seen whether our response this time will result in significant changes to the current trends in drug resistance that we are following into the future.

The antimicrobial issue, and our individual and cultural response to it is a microcosm of the broader challenges our health system will face. The trend in our response to drug resistance represents the path we are most likely to follow into our health future. Unless we decide not to, which means accepting our current and future health responsibilities.

In summary, the complex, interacting mix of our individual human characteristics, our culture's world view and our environment's functioning will determine our health future. At the present time, we can expect that the medium-term trends in our environment's strained ability to provide the basic, material necessities of life will continue. We can also expect that the trends of increasing antimicrobial drug resistance, lifestyle diseases, appearance of new diseases, reappearance of old diseases, and afflictions related to pollution will continue into the medium-term future.

A change in our behaviour, our cultures' world views or unforeseen technological developments may change those trends. In particular, a different health future is possible if our vision of health and a health system is based more on preventative medicine merged with ecosystem principles, and the inclusion of our psychological and social needs. To be effective, this type of system would require us, personally and culturally, to more fully accept the burden of greater responsibility for our health-affecting decisions, from medicine use to lifestyle. This is a burden we are currently unwilling to fully accept.

At present, it seems more likely that some mix of the current trends will combine with ongoing population growth, increasing human densities and less available cheap energy (discussed in the next chapter) to reduce the effectiveness of our reactive health systems. We could thus face an imposed decline in our population's health and size. The possible outcomes include a decline in our general level of health and average lifespan, and the occurrence of events where death rates exceed birth rates. These health outcomes will likely be more unevenly distributed between countries, between cities and rural areas, and between socio-economic classes.

## Metals and Minerals

The global supply of metals and minerals exists because they have been naturally concentrated into the deposits from which they are being mined. These deposits are unevenly distributed around the world. The global demand for metals and minerals is also unevenly distributed around the world. The higher GDP (industrialized) countries have a significantly greater demand for all metals and minerals relative to the lower GDP (developing) countries. Fortunately, the way in which the global supply and demand for metals and minerals is balanced is sufficiently similar that the nature of the balance between them can be illustrated by a single metal (Graedel and Cao 2010). Copper is well studied and the basics of its production and demand was discussed in chapter 10. Therefore, its global supply-demand balance will be used to answer the question of whether there are sufficient metals and minerals in the ground to meet our future wants.

The discussion about copper's supply-demand balance complements the conclusions made for the other resources in this chapter. It also brings to light the limitations, the social dimension and the care that must be taken when making and using numerical based predictions about any resource's future, if we wish to avoid deluding ourselves with their apparent importance, accuracy and precision. This section is thus also a continuation of the discussion about predictions in chapter 5, but now in a manner that will be useful for this and the next chapter. To meet these objectives, this section has to be a little more detailed to fully expose the complexity. If you prefer to just focus on the conclusion, then skim/skip the first three subsections down to "Our Copper Future."

<u>Supply.</u> The global supply of copper ultimately comes from mining copper deposits. The decision of whether or not a deposit is economically minable is largely based on the interacting factors of the average amount of copper per tonne in a deposit (its grade), the amount of rock used to define that grade (the deposit's size), the price of copper, and the costs of building and running the mine (including the cost of the borrowed capital). Similarly, the proportion of a deposit that is economically minable is determined by complex relationships among a deposit's grade, size, the income from mining it and the (financial) costs of mining it. In general terms, the lower a deposit's grade, the larger its size and the lower the costs to mine it must be for it to be economically mined.

Experience tells us that only a few of the many known copper deposits are economically minable. Of the minable deposits, the rarest are those that are both large and have higher grades. Geological theory also suggests that copper deposits don't form below a certain grade. Together, these upper and lower limits to copper deposits suggest that the number of potentially minable deposits is limited (Skinner 1976; Menzie et al. 2005). Copper deposits are also being mined faster than they are being formed. Does this mean that the global supply of minable copper is finite, in a practical sense? If so, what is that amount and when will it run out?

The future global supply of copper is an estimate, made at one point in time, of the amount of copper we could reasonably expect to mine from the earth's copper deposits in the future. There are two types of estimates. A lower estimate, called the global copper *reserve*, considers only those parts of the discovered deposits that are currently economic to mine. In 2011, the United States Geological Survey (USGS) estimated that the global copper reserve was 680 million tonnes (Mt).

An upper estimate, called copper's *resource* has a specific meaning in metal-mining circles not to be confused with the more general use of the word *resource*. The USGS defines a global resource as a combination of reserves plus the amount of metal it is reasonable to expect would be mined in the future from known sub-economic and marginally economic to mine deposits, or parts of deposits, and their geologically inferred extensions, plus the metal in deposits expected to be discovered. In 2011, the USGS estimated that the global land-based copper resource was at least 3,000 Mt (US Geological Survey 2013). Note the large difference between copper's estimated reserves and resources.

Any estimate of the global copper reserves and resources is uncertain. Part of the uncertainty arises during the making of the reserve and resource estimates for each of the copper deposits included in the global estimates. This uncertainty arises because the data needed to calculate a deposit's reserves and resources is, for all but the most advanced exploration projects, either incomplete or not reliably known. (More uncertain than for oil and gas deposits.) Collectively, these information constraints introduce uncertainty into the calculations of how much copper a deposit contains and how much of the contained copper is likely to be mined.

The uncertainty problem is partly resolved by setting standards for the information used to calculate a deposit's reserves and resources. By definition, the information used for a reserve calculation must be more complete and certain than that used for a resource. Thus the estimate of a deposit's copper reserves is always smaller and more certain than the estimate of its copper resource.

Neither is an estimate of a deposit's copper reserves and resources final. The variables can change over time. The price of copper fluctuates with the economy, while further exploration provides better information about known part of the deposits and discovers new parts. If the deposit is being mined, then its contained copper is declining. As a result, estimates of reserves and resources have to be periodically re-calculated.

The estimates of the global copper reserves and resources are based on the known current estimates for all known deposits. Thus the global reserves and resources estimates are, like those for deposits, uncertain, and they too fluctuate over time. The global estimates also change because exploration discovers new deposits, and old ones are mined out. The global estimate of reserves is, like those for deposits, more certain than the estimate of global resources.

Critically, because the global copper resource is based on known deposits only, it is not an estimate of the total global amount of copper likely to be mined (i.e., the ultimate supply of extractable copper). After all, not all copper deposits have been discovered. If the success at exploring and mining copper remains the same, then the global copper resource is an underestimate of the earth's ultimate supply of copper. To estimate copper's ultimate supply, the yet-to-be-discovered copper must be added to the global resource estimate.

It is reasonable to think that the global amount of yet-to-be-discovered copper could, like the amount of undiscovered conventional oil, be predicted from some sort of systematic relationship between the size of a copper deposit and its date of discovery. For example, one might expect that the largest copper deposits, like large conventional oil deposits, would have mostly been discovered relatively early. Unfortunately, copper deposits don't show such a systematic relationship (Schodde 2010). One reason is that copper deposits are relatively small and irregularly shaped compared to conventional oil deposits. Even large copper deposits rarely have dimensions in the kilometre-plus size range. They are difficult to find. All that can be said with some certainty is that large copper deposits are becoming more difficult and costly to find.

It is tempting to think that the difficulty in estimating copper's ultimate supply could be solved by using the changes in copper's price to predict the ultimate date for its reduced availability. However, as with fish and other commodities, copper's price only reflects the market's view of the short-term supply-demand balance for the copper supply on the shelf in a warehouse, not its longer-term potential supply (Reynolds 1999). For example, during the global financial crisis of 2008, the large drop and recovery in the copper price barely registered as a change in copper production.

The estimates of the amount of yet-to-be-discovered, likely minable copper are so uncertain that the resulting estimates of the earth's ultimate copper supply are not a useful representation of the global copper supply for this discussion. The most useful estimates of supply are the global reserves and resources, but only if they and their trends are used in combination with our knowledge about the geology of deposits, the physical aspects of copper production, and the nature of the mining business. In this discussion, the reserve numbers used here are treated as the minimum supply. The resource numbers are used with the understanding that how much bigger or smaller they could or should be depends on the context (assumption, timeframe, etc.) in which they are used. In this context, and at least for the foreseeable future, the amount of copper in the ground is not finite in a practical sense. What matters are issues of mining it and the demand for it.

**Demand.** The historical global demand for mined copper is reasonably represented by the annual global copper production from mines (ignoring the changes to copper stored in warehouses) over the last hundred or so years. In 1900, an estimated 0.5 Mt of copper were produced from mines (0.31 kg/capita). In 1950, production was 2.4 Mt (0.92 kg/capita); in 2000, it reached 13.2 Mt (2.17 kg/capita); and in 2011, it was 16.1 Mt (2.33 kg/capita) (US Geological Survey 2012; US Census Bureau 2008).

The historical increasing trend in global copper production and its increasing per capita demand can be used to make an estimate of the future global demand for mined copper. Examples can be seen in many mining and business analysts' reports. In these cases, a selected shorter time segment of the varying annual production record is projected into the future to illustrate a possible future copper demand. The flexibility of selecting which part of the varying trend to project allows for a variety of estimates, and the ability to bias them to match a hoped for growing or declining future demand.

Even a disinterested projection of production trends is too simplistic a method to estimate the future copper demand. For example, if the best-fitted curve to the increasing production trend is projected into the future, then the estimate of the demand for mined copper in 2100 is unrealistic: 480 Mt/year. This implies a 48 kg/capita demand for a population of 10 billion, compared to 2.33 kg/capita in 2011 (Mudd et al. 2013).

An important reason why an unconditional projection of past global demand is unsuitable for estimating the long-term future demand is provided by the historical changes in a country's copper usage as it industrializes. Industrializing countries demand copper for two reasons. The first is to build both the country's increasing number of machines and consumer devices, such as metal-working lathes and phones, and the infrastructure to support their use, such as power distribution networks. The copper tied up in these products is said to be "in-service." The second reason is to replace the copper lost to waste. This occurs when the in-service copper is consumed or discarded rather than recycled.

The amount of copper an industrialized country has in-service varies, as discussed in chapter 10. It shows two successive long-term trends. The first starts from a very low pre-industrialization level. The amount of in-service copper then rapidly increases with industrialization. The second trend starts after a country's infrastructure is mostly completed. The amount of in-service copper continues to increase, but more slowly. The slower second trend reflects the country's ongoing demand for copper-containing consumer devices and infrastructure adjustments. The data from North America (Canada, United States, Mexico) suggests that the switch from the faster to the slower in-service growth trends occurs when the total amount of copper in-service reaches around 170 kg/capita.

The ongoing increase in a country's amount of in-service copper after 170 kg/capita contradicts the demand estimates that are based on economic theory. For example, the often quoted Kuznets curve model predicts that as a country's infrastructure construction is completed, its copper demand should peak and then fall. This false assumption serves as a timely warning that resource demand (and supply) estimates based on theory and models that are not supported by adequate, representative data can be invalid or attended by large uncertainties. They can result in resource predictions whose validity is questionable and whose usefulness is limited.

The current and future global demand for mined copper is the sum of the high in-service wants of industrializing countries, the lower post-170 kg/capita wants of the industrialized countries, and the replacement of the ongoing loss to waste (Gordon et al. 2006; Menzie et al. 2005). A useful estimate of the future global demand for newly mined copper can be made by using this knowledge and making assumptions about changes in the rates of recycling, economic growth and population growth.

For example, a short-term estimate, published in 2005, suggests that the annual global demand for mined copper will increase from its 13.2 Mt/year (2.17 kg/capita) in 2000 (US Geological Survey 2012) to about 24 Mt/year in 2020 (Menzie et al. 2005). For a global population of 7.6 billion people in 2020, the annual global demand for mined copper would be around 3.13 kg/capita, a substantial increase in demand (figure 18).

In contrast, a long-term estimate, published in 2006, focused on the global demand for mined copper if the global population in 2100 were to be industrialized to North America's approximately 170 kg/capita of in-service copper by 2100. This is the mining equivalent of the economists' belief that growth will raise everyone's living standard to that of the developed world. If the world's population were to reach a maximum of 10 billion people no later than 2100, then a total of some 1,700 Mt of in-service copper would have to be produced between 1900 (when copper demand became significant) and 2100 (Gordon et al. 2006).

However, this estimate neglects that the mined copper demand consists of both the increased in-service demand *plus* the waste replacement demands. The waste replacement that must be added to the 2100 estimate is calculated as follows. Of the 135 Mt of copper made available to North Americans by mining between 1900 and 1999, only some 52% (70 Mt) is still in-service, 41% (56 Mt) was lost to waste, and 7% (9 Mt) is of uncertain classification (Spatari et al. 2005). Assume that this historical average proportion of the loss to waste is globally applicable into the future. Thus to both supply the 1,700 Mt of in-service copper and replace the copper lost to waste will require the production of around 2,918 Mt of new copper from mines (14.6 Mt/year averaged over 200 years). This is an estimate of the maximum future demand for mined copper.

Because approximately 575 Mt was produced between 1900 and 2011, only some 2,343 Mt must still be produced from 2012 to 2100. But it must be produced at an average rate of 26.3 Mt/year during that period and at 37 Mt/year in 2100 (3.7 kg/capita). This is much less than the previously mentioned 480 Mt/year in 2100 from unconditionally projecting the mined copper curve, but still quite an increase to a large maximum absolute and per capita demand.

**Balancing Accounts.** Is there sufficient minable copper in the ground to meet our future wants? Even though we have global estimates of both our future demand for mined copper and the earth's supply of minable copper, providing a meaningful answer is more complex than simply balancing the two. The reasons appear by first balancing a short-term prediction (dated 2005) of the future demand for copper up to 2020 with estimates of its future supply. This is completed for each of two start years, 2000 and 2005 (i.e., for the demand periods of 2000–2020 and 2005–2020). Two supply estimates (reserve [high certainty] and resource [low certainty]) are used for each of the two start years (2000 and 2005). This is laid out in figure 18. The balancing is then repeated for the long-term prediction (dated 2006) of copper's demand up to 2100 from the same start years (i.e., the demand periods 2000–2100 and 2005–2100) using two supply estimates made for each of the two start years (2000 and 2005). The balances and their dynamics are then discussed.

If the predicted short-term (2005) increase in demand from 13.2 Mt/year in 2000 to 24 Mt/year in 2020 is uniform, then a total of 372 Mt would have to be produced from mines (the mined copper demand for the period 2000–2020). The 372 Mt demand can almost be satisfied by the reserves of 340 Mt estimated to be available in 2000 (the supply) (US Geological Survey 2001; Menzie et al. 2005). One could reasonably argue that a copper shortfall would not arise because the mined-out reserves would be replaced from the USGS's land-based resources of 1,600 Mt estimated for the year 2000.

Copper production is estimated to have risen from 13.2 Mt/year in 2000 to 15.0 Mt/year in 2005. The resulting 71 Mt of production is subtracted from the 372 Mt demand for 2000–2020 to give the 301 Mt mined copper demand for 2005–2020. The 301 Mt demand can easily be satisfied by the reserve of 480 Mt (the supply) estimated to be available for 2005 without even considering the resource of 3,000 Mt estimated for that year (US Geological Survey 2006). In the five years since 2000, the short-term supply-demand balance had changed dramatically from a "maybe" to an "easily."

Next, balance the predicted long-term (2006) demand estimate. The mined copper demand for 2000–2100 is 2,518 Mt. (It is the predicted copper demand [in-service growth plus waste replacement] for 1900 to 2100 of 2,918 Mt of mined copper minus the production from 1900 to 2000 of 410 Mt to 2,518 Mt). Clearly, that 2,518 Mt demand can't be met by either the reserve of 340 Mt or the land-based resource of 1,600 Mt (the supply) estimated to be available for 2000. This is still the case even if the USGS's 700 Mt of unconventional, ocean-based copper resources estimated to be available for 2000 are added to the land-based resource to create an estimated total resource of 2,300 Mt. Based on the reserve and resource data for 2000, it seems that, at best, there is insufficient copper to satisfy the economists' long-term vision of industrializing the planet. However, it is still possible to take the optimistic outlook that exploration will fill the deficit by discovering new minable deposits. An alternative view is that the production of roughly 37 Mt/year in 2100 is unrealistic.

The mined copper demand for 2005–2100 is 2,437 Mt. (It is the predicted copper demand [in-service growth plus waste replacement] for 1900 to 2100 of 2,918 Mt of mined copper minus the production from 1900 to 2005 of 481 Mt to 2,437 Mt.) It cannot be satisfied by the reserves of 480 Mt estimated to be available for 2005. However, it can easily be satisfied by the resources of 3,000 Mt estimated for that year. If future technology can access the USGS's 700 Mt of unconventional ocean-based copper resources estimated for 2005, then there are more than enough resources to meet demand over the long term. During the five years from 2001 to 2005, a significant potential future copper supply deficit in 2100 suddenly changed into a significant potential supply surplus, a completely different balance for our long-term copper future than was made for 2000–2100.

This effort to balance copper's supply with its demand indicates that making valid and useful predictions about our copper future is subject to significant limitations. One is illustrated by the size of the uncertainty in the resource estimates as exemplified by three estimates made for 2010. There are significant differences between the USGS's resource estimate of 3,000 Mt and the two independent resource estimates of 2,459 Mt and 1,861 Mt (Mudd et al. 2013; Schodde 2010). The one chosen to balance the estimated long-term copper demand will significantly change the conclusion drawn.

There are other limitations affecting our predictions about our copper future. They are discovered by considering the dynamics of copper's supply and demand estimates. Start with the dynamics of the demand estimates.

Past mine production is used as a proxy for demand. Mine production changes slowly. This inertia exists because of mining's nature. For example, it takes a significant amount of time and money to turn a deposit into a mine, particularly if the deposit is large and low grade. Once the mine is built and running, the presence of debt provides significant incentive to keep mining, even at low prices. Any excess production is stored in a warehouse. By using production as a proxy for demand, it appears to change relatively slowly, like production.

The estimates of copper's future supply are more dynamic. Consider the changes in the following series of USGS copper (reserve and resource) supply estimates. Between 1994 and 2000, copper's reserves increased from 310 Mt to 340 Mt. A sharp increase occurred in 2001, after which the reserve estimates varied between 470 and 490 Mt until 2006. They then rose steadily to reach 680 Mt in 2011.

Over the same period, the USGS estimate of copper's land-based resources changed radically. Between 1994 and 2004, the resource estimate remained constant at 1,600 Mt. In 2005, it rose dramatically to 3,000 Mt, where it remained, unchanged, until 2011. In contrast, the USGS estimate for unconventional, ocean-based copper resources of 700 Mt stayed constant from 1994 to 2008. It is absent from the 2009 and subsequent reports (US Geological Survey 1996–2008).

These changes, or lack of them, in the global reserve and resource estimates are partly the products of reporting decisions and partly the result of the steep increase in the price of copper that started in 2002. The price rise prompted an increase in exploration for new deposits and the re-evaluation of known deposits. It also meant that more of a known deposit could be considered economically minable so the estimates of its resources and reserves would increase. The price-stimulated interest in copper also prompted a revaluation of global resources and reserves. These factors, together, seem to be the reason for the erratic dynamics of copper's global supply (reserve and resource) estimates.

The difference between the dynamics of the estimates of copper demand (slow) and supply (erratic) helps to explain the wide variation in the supply-demand balance. But, if predictions of the future supply of copper using this method are to be useful, then this difference must somehow be taken into account. We are helped in this task by the reminder provided by mining debt and copper price: there is a close tie between mining and economic activity. The dynamic relationship between them creates the framework within which the individuals who are making mining-related decisions are constrained to think. Knowledge of this framework will help us evaluate the supply-demand balance and make valid, useful predictions about our copper future.

Both economic and mining activity go through crude multi-year cycles. The nature of the relationship between the two means that the mining cycle lags behind the business cycle by a number of years. It also results in the mining cycle's peaks and troughs being amplified into booms and busts. These booms and busts are experienced as large swings in the price of copper, the copper inventories held in warehouses, and the levels of exploration and mine-building activity.

In idealized form, a sustained upturn in economic activity signals the beginning of a new business cycle. From mining's perspective, the idealized mining cycle will start its upswing a few years later. The price of copper will rise high enough and stay there long enough to sustain expectations of increasing mining activity. This is followed by a period of frenetic exploration and mine-building activity: a mining boom.

The rise in copper's price joins the rise in other resource prices (think oil) in contributing to the subsequent economic slowdown. Following a delay after the slowdown, a copper oversupply develops, the price of copper falls and, eventually, the mines slow production or close. During this period, the funding allocated to exploration is spent but not renewed. This decline in mining and exploration is followed by a longer period of lower copper prices, little exploration and limited mine-building activity: a mining bust.

The business cycle, together with the associated mining cycle's booms and busts, directly and indirectly affect copper's global supply-demand balance. For example, booms prompt increased production from existing mines, decisions to put new mines into production, the re-evaluation of known deposits and the discovery of new deposits. These changes are eventually compiled and reported as changes in global production, reserves and resources. But the incentives to compile and report each are not the same through the cycle.

Economics requires that annual copper production (demand estimate), which changes slowly, is reported annually with a short lag. In contrast, global reserves (low uncertainty supply estimate) are reported annually, but significant adjustments are more likely to be made during booms than busts. Resources (high uncertainty supply estimate) are reported annually, but assessments are conducted far less frequently. Mining booms favour these reassessments, while busts don't. These features of the economic and mining cycles help to place the supply-demand balance and its limitations into perspective.

At different points in the mining cycle, copper's supply-demand balance prompts different predictions about copper's future. As the early stages of increasing demand turn into a mining boom, balances are more likely to suggest an imminent supply shortfall, as the above year 2000 short- and long-term balances illustrate. Nearer to the end of the boom, all is well, as illustrated by the year 2005 short- and long-term balances. Balances made during the downturn and mining bust, when prices are low and mines are closing, such as occurred in the late 1980s to late 1990s, suggest that the supply significantly exceeds demand. This is the time when experts suggest that the mining industry is too successful at finding and putting mines into production (Cook 1987).

In contrast, during the busts, miners make predictions of future supply shortfalls due to a lack of ongoing exploration. They cry out for exploration incentives and tax breaks in order to keep mines open and find more deposits. During the rise to a boom, miners predict an imminent shortage unless red tape is removed, incentives for exploration provided and more land opened to mining.

The last two paragraphs illustrate that the use of the supply-demand balance to make predictions about copper's future is context dependent and can be biased to the predictor's benefit. It is much harder to make a realistic, disinterested, useful assessment of our copper future than expected. The art of making a useful prediction about our copper future from this uncertain, context-dependent information is the subject of the next section.

## Our Copper Future.

A prediction of our copper future that is based on a copper supply-demand balance is limited in its usefulness by the lack of a reliable estimate for the ultimate supply of copper. It is also limited by the uncertainty and dynamics in the estimates of both the working supply (reserves and resources) and the demand. If we take these limitations and the reasons behind them into account when evaluating a prediction, we can improve its usefulness. The features to consider

include the decline in copper demand that occurs after industrialization is largely completed. They also include the relationship between the mining cycle, the business cycle and estimates of the supply and demand for copper, as discussed above. There is one other factor to consider: the manner in which our copper future might arrive. This will be illustrated by imagining a copper shortfall.

In Western culture, the standard responses to the possibility of a copper shortfall are dominated by the neoliberal economists' view that market mechanisms and a high enough copper price will resolve all difficulties. They believe that an increase in price will axiomatically increase the supply of copper by galvanizing successful efforts to discover new deposits and build new mines. Both economists and miners justify this view by pointing to the historical increases in the discovery of new deposits and the production of copper from deposits of increasingly lower grades.

It is true that, to date, mining companies have been very successful at finding all the types of copper deposits and maintaining the supply. There are no geological reasons to think that copper deposit discoveries will end soon. But, from the perspective of a shortfall, some deposit discoveries are better than others. The most desirable deposits are those that are most likely to reduce the shortfall (i.e., large, high grade and easily mined.)

A study that ranked 2,265 known copper deposits by size showed that the top 10% contained 76% of the world's copper resource. The majority of those deposits belong to the porphyry or sedimentary types. Porphyry deposits contain, on average, more metal than the sedimentary deposits. Porphyry deposits will be the reference for this discussion.

How much copper does a large porphyry deposit contain? The five largest porphyry copper deposits discovered to date contain an average of around 68 Mt (recently updated to 82 Mt) of copper, which is many times the annual global copper production of around 16 Mt/year. When the average copper content of the five largest porphyry deposits is compared to the median copper content for all porphyry copper deposits of around 5 Mt, it is no wonder that the five are called giants (Menzie et al. 2005; Mudd et al. 2013).

The discovery of just one giant porphyry deposit will have a significant impact on the estimate of copper's global resource. But giant copper deposits of any type are rare. Consider too that many of the easy-to-find (at or near surface and accessible) deposits have already been found. Thus discovering new deposits is becoming more difficult and costly. In particular, most of the still-to-be-discovered giant deposits are likely deeper underground or located in those under-explored parts of the world that are remote and difficult to access.

This serves as a reminder that, regardless of whether a giant deposit's discovery is new or old, it will only reduce a supply shortfall if mining it is physically and economically feasible. One of the factors determining feasibility is the deposit's geographic location: the physical accessibility, political situation and available infrastructure. Consider the giant Aynak copper deposit in Afghanistan. Although it is exposed at the surface as a copper-green field and has been known for at least two thousand years (it provided the copper for early brass Buddhist statues cast in the region) it still remains unmined by today's methods.

The feasibility of mining a deposit is also affected by its characteristics. Giant porphyry deposits are certainly large, but they are also low grade, containing only around 0.5% copper (Mudd et al. 2013). As a result, they are more likely to be mined if the economies of scale can be applied: large tonnages of a uniform grade mined and processed quickly at a low cost per tonne of copper. This usually means that the deposit must be at or very near the surface in a location where it can be easily mined as an open pit. Think of the world's largest open pit copper mines at Bingham, United States, and Chuquicamata, Chile.

If the long-term trend of mining lower-grade copper deposits continues into the future, then additional characteristics will constrain the feasibility of mining both giant and regular porphyry deposits. Consider that as the grade of a deposit drops, the energy needed to extract a fixed amount of copper from its ore rises. Below a grade of around 0.5% copper, the increase in that energy is very steep, and more so for the unconventional copper recovery methods being developed (Hall et al. 1986; Norgate and Jahanshahi 2010). Thus, the economic feasibility of mining a lower-grade porphyry deposit is particularly sensitive to its grade and the availability of cheap, abundant energy.

Overall, it seems that the rate at which new feasible-to-mine copper deposits, especially giant ones, will be found is likely to slow into the future as the time and cost of discovering them increases. Similarly, the average cost of mining deposits and extracting the copper will also likely increase. Thus, filling a copper shortfall by increasing mining will become more difficult.

The neoliberal economists' world view allows them, and us, to believe that the higher prices for copper caused by a supply shortfall will stimulate the market to find and develop the technological innovations needed to circumvent any difficulties we might face filling it. Consider the invention of technologies that will enable us to reduce the need to find and mine copper, such as those that increase the recovery of copper at existing mines, increase the efficiency of copper use in products and facilitate substitutions for copper. There is certainly room for improvement in the recoveries from existing mines. For example, a retrospective view of copper mining in North America suggests that only about 82% of the copper in the ore mined from deposits has been recovered. The remaining 18% was discarded in the metal-mineral recovery process's waste (tailings) or the smelting waste (slag).

However, to have unconditional faith that market-driven technology will solve the copper shortfall in this way means neglecting the real-world constraints on these solutions. There is the reality of diminishing returns. There is also the reliance on previously discussed dubious/simplistic economic beliefs and assumptions.

The last constraint also applies to the claim by economists that an increase in copper prices would stimulate recycling to a degree that it too would substantially reduce the demand for mined copper and thus reduce, even eliminate, a shortfall. That is not a bad idea considering that, in North America, an estimated 42% of the copper produced and used over the last 100 years was eventually lost to waste. Today it seems that the proportion lost to waste may be increasing (Gordon et al. 2006; Spatari et al. 2005).

However, a high enough copper price does more than stimulate greater recycling of currently discarded copper. It also encourages the destruction of copper-containing infrastructure: the stealing of in-service copper from power systems, plumbing and air conditioners. A country is then unsustainably feeding on its in-service copper to satisfy its more immediate demands.

This prompts the realization that a more complete and reliable view of a copper shortfall appears when we treat it and our response to it, including those promoted by neoliberal economists, as part of the larger, complex human-environment system. In this context, the shortfall and our response is seen to be affected by multi-faceted interactions and feedbacks between individuals, our culture's world view and its supporting environment.

The full range of constraints on the proposed solutions to a copper shortfall becomes apparent.

Consider that if efficiencies or substitutions are to resolve a copper shortfall, we will have to supply the resources, such as water and energy, needed for the related technologies to function. But our ability to provide those resources could be constrained in a number of ways. Most simply, they too could experience a shortfall. More specifically, we could experience an energy shortfall. Or, ironically, our effort to change our energy supply away from fossil fuels in order to address both peak oil and climate warming could itself contribute to a future copper shortfall. For example, both non-fossil-fuel energy sources (e.g., electricity-producing solar panels and wind turbines) and the new electricity-using devices (e.g., electric vehicles), which we have decided we must acquire if we are to maintain our current lifestyle, all require copper to function.

The idea that the savings of copper from increasing its recycling, efficiency of use or recovery during mining will automatically resolve a shortfall by reducing the need to mine it also looks different when seen within the context of the dynamic, complex human-environment system. There is no guarantee that the savings from these efforts will, on their own or over the long term, have that result. If copper savings are to contribute to this outcome, then the culture must decide to allocate them to that end. This decision is at odds with Western culture's imperative of growth and its faith in the virtuous cycle and progress. Instead, the guidance from Western culture favours us using the savings in a manner that would increase the demand for copper, think electric vehicles.

By considering possible solutions to a copper shortfall within the human-environment system, we are reminded of another factor that affects our efforts to address it: environmental issues (Hodges 1995). The scale of mining's future environmental impacts can be appreciated by considering the amount of rock moved during past copper mining. By 1994, the historical production of one tonne of copper is estimated to have generated 147 tonnes of tailings (waste from recovering the ore minerals) and 350 tonnes of waste rock (barren rock moved to expose the ore). The long-term, global average grade of mined ore was thus 0.68% copper which is near the ore grade in giant porphyry copper deposits. (The remaining 99.32% of the ore becomes the tailings). We can thus optimistically assume that these numbers are relevant to future copper mining.

If the short-term (2005) prediction that 24 Mt/year of copper will be produced in 2020 comes true, then producing it will generate some 3,500 Mt of tailings and 8,400 Mt of waste rock. It will be attached to the copper as virtual waste. If the long-term (2006) prediction of a copper demand of 2,500 Mt from 2005 to 2100 were to be satisfied, then an estimated 367,500 Mt of tailings and 875,000 Mt of waste rock would be generated by 2100 (Menzie et al. 2005). That is truly a mountain of waste.

Associated with generating and disposing of this staggering amount of solid waste are many other environmental issues. They include the greenhouse gas emissions from the fossil fuel burned to move and process the rock. They also include water-related issues, such as disrupting the local water table or surface water supplies around the deposit; the generation and disposal of contaminated water; and the leakage of metal rich, usually acidic, water from the site.

Many of these environmental impacts are high risk and can be very long-lasting. Consider that tailings are often stored in dams whose integrity has to be monitored in perpetuity and whose water overflow may have to be treated for just as long (i.e., for long after the mining company has left). Who should pay for that monitoring and treatment? Currently, tailings dams fail at a global rate of between one and two dams per year. Who should pay for the environmental and human consequences? There are other, safer methods to store tailings, but they are more (energy and cash) expensive. Alternatively, companies could try and persuade governments and the public to let them follow the Grasberg model of tailings disposal, dump and forget, or simply fill in a natural lake.

How should we expect those affected by this reality to respond? Is it any wonder that, in an increasingly crowded planet, those living in the vicinity of a proposed mine are becoming increasingly vociferous in their opposition to what they see as a threat to the environmental foundations of their way of life, or an unfair distribution of benefits and costs (The Economist 2016; Welker 2014)? Alternatively, we as individual copper users could insist on more responsible mining practices as well as greater regulation and monitoring. We could even help to reduce both mining's environmental impacts and a copper shortfall by changing a variable over which we have the most control: our personal demand for copper-containing products.

How we decide to address these interconnected questions, issues and suggestions is shaping, and will continue to shape, our copper future. And, because we are participants in the human-environment system, our decisions will be the product of more than our human decision-making foibles and our internal reality (with its divergences). They will be influenced by our culture's world view, as discussed in chapter 11. Our culture's influence can be remarkably persuasive.

Consider how Western culture's world view strongly guides its members to have faith in technological and economic progress. One way we express our acceptance of this guidance is to participate in the progress-supported virtuous cycle by adopting a consumer lifestyle. In this consumer-producer driven cycle, there is an inexorable drive for products, many of which contain copper, to be developed, brought to the consumer market, purchased, used and then discarded. The ability of technology to supply these marvellous consumer goods is, to us, concrete evidence for the existence of technological progress. We can thus more easily decide that technology will address a shortfall in copper by increasing its efficiency of use, recovery in the mines and recycling from waste.

Western culture's world view only weakly guides its members to maintain the functioning of our environment. It also guides us to do so in a manner that favours technological and economic progress Western culture's world view only weakly guides its members to maintain the functioning of our environment. It also guides us to do so in a manner that favours technological and economic progress. This bias makes it easier for us to decide that the technological improvements that we expect to address a shortage will do more: they will reduce the need to mine copper. We can thus also more easily decide that new technological developments will also both satisfactorily reduce copper mining's environmental impacts and either satisfactorily mitigate or restore mining's "unavoidable" impacts. There is therefore little incentive for us to make changes to our personal copper-using lifestyle and our culture's world view.

There are some ironic consequences from this bias in our decision making. Think of the lifestyle changes that will result from the effort mining companies are making to reduce the cost of mining by developing ways to employ technology's copper-using electronic products, rather than humans, to drive a mine's ore trucks (Crouch 2016). A greater irony is illustrated by the explosion of copper-using electronic consumer products. Although the copper in them is being used in a more economically and materially efficient manner, the products themselves have short lifecycles. Think of cell phones.

This is partly by design (tried to replace the lithium battery in a cell phone recently?) and partly a result of the nature of our expectation of ongoing improvement in these products. When these products become post-consumer waste, their short lifecycle and their difficult-to-recycle design results in a lower overall recovery rate of metals, including copper, from them. At the same time, the consumer's use of the expanding array of electronic products requires the copper-using infrastructure to be updated and grow.

To date, the overall result has been an increase in Westerners' per capita use of copper. This can be seen in North America where, even as recycling and efficiency is being promoted, the amount of in-service copper continues to increase at a long-term, average annual rate of 1.2 kg/person (Gordon et al. 2006). This ongoing, increasing demand for mined copper is helping to sustain our copper-mining-related environmental impacts and making a future shortfall more likely.

By treating the copper supply-demand balance as part of the larger human-environment system, we discover that its state of balance and the possibility of a shortfall will be most significantly influenced by our culturally affected and human-characteristic-affected decisions. From this perspective, those decisions, rather than the amount of minable copper in the ground or the technology to recover it, are seen to be the most important determiner of our copper future.

In summary, when all of the pieces in this discussion about copper's supply-demand balance are considered together, they suggest the following copper future. The increase in copper demand resulting from population growth and industrialization will likely continue into at least the medium-term future. There appears to be sufficient copper reserves, copper resources and likely-to-be-discovered minable copper deposits to satisfy that demand. It also seems that, as we move into that future, wide-ranging swings around the balance can be expected, including shortfalls.

It also seems that the further into the medium- to long-term future, the more difficult it will likely be to satisfy a demand that continues to follow the current, increasing trend. It will become even more difficult if most of the world's population come to use the same amount of copper as the members of the developed world do today. It will be easier if the recent steep trend in production is found to mostly represent China's initial industrialization and if fewer other nations fully industrialize.

Difficulties in supplying the needed copper could arise because of a combination of mining, economic and social factors that increasingly limit the finding of deposits and the number of known deposits it will be feasible to mine. The mining factors include technical constraints, such as the increasing difficulty and costs of mining and recovering copper from declining grades and deeper deposits. They also include the limited availability of the cheap energy needed to run the mines economically (Cleveland et al. 1984). The economic factors include the boom-bust cycle; increased financial costs and risks of mining; too high a copper price stifling demand; and the increasing costs of fulfilling environmental responsibilities, such as carbon emissions and impacts on water. The social factors include a more crowded planet in which people near a deposit protect their rights and way of life; the general public expressing concerns about mining's environmental impacts, such as tailings dam failures; and political instability that arises for other reasons.

The essence of copper's supply-demand balance applies to the other metals and to minerals. It seems that, over the immediate and into the medium term, most metals and minerals are unlikely to experience a global-scale shortfall caused by a lack of metal in the ground. However, shortfalls could be caused by the interaction of any number of other factors that affect their production, as illustrated above. The details for each metal and mineral will be different because, for example, some metals, such as molybdenum, are more abundant than copper, while others, such as platinum, are less abundant. Each metal will have its own scarcity point.

The option of dealing with a resource scarcity by changing our demand for it will always be present. This could include changes to our lifestyle and cultural world views, however, the strong pressures to satisfy our cultural imperatives means that changing either is difficult. The changes in our demand are more likely to be short term and reactive, for example, stimulated through rising prices. Instead our longer-term voluntary efforts to deal with scarcity are more likely to focus on increasing the supply. This will include increasing recycling, efficiency and substitution. We will also focus on facilitating mining by, for example, removing social and environmental "barriers" to mining. This may have the opposite effect by contributing to destabilizing some countries.

The summary of this chapter notes that the supply and demand for resources occurs within a dynamic, complex human-environment system. Within this context, the essence of the lessons from the bees in the fields, health services and copper's demand-supply balance help us appreciate our possible resource (including services) future. A significant influence on that future is our default or preferred view of how we should deal with a resource shortfall. That preference will not only influence today's predictions about our resources' futures but also our actions when the future actually arrives. In the Western world, like in most industrial cultures, our culturally preferred solution is to maintain supplies by applying technology.

But all technology, present and future, requires energy to function. The future supply of all the resources we want will change if the current availability of inanimate energy declines or its price rises. Energy is thus a critical resource to which all aspects of a complex culture, especially industrial ones, are linked. The availability of energy even limits the ability of a complex culture to maintain its world view.

Time to move on to our energy's future. This chapter's lessons will be carried forward to the next chapter. At the forefront is the lesson that, despite our expectations, numerical predictions about the future supply of metal and other resources can't, on their own, provide us with a certain, reliable view into the future. Our resource future is part of a complex human-environment system, thus predictions about the future must include all of the many influences that affect it. Care and effort is needed to make useful predictions about our resource future.

# CHAPTER 15

# FUTURE RESOURCE WANTS: ENERGY AND AN OVERVIEW

The previous chapter discussed possible futures, at a global-scale, for the size of our population, the provision of health care, and the availability of food, water and metals. It identified the factors constraining these futures and the limits within which each one will likely arrive. This chapter repeats that discussion for energy.

The discussion of our global energy future starts with an overview of our dependence on energy and the basic limits to providing it. Because fossil fuels supply such a high proportion of our energy, the following two sections discuss fossil fuel basics and their possible future supply. The next section discusses the future provision of energy from non-fossil-fuel energy sources as well as those solutions to an energy supply shortfall that also address the environmental costs of using fossil fuels. At the end of the section on energy, a summary of our likely global energy future is provided.

The possible resource futures discussed in this and the previous chapter do not exist in isolation. We and our resources are part of a dynamic, complex human-environment system in which both short- and long-term interactions and feedbacks occur among its variables. It is within the dynamics and constraints of this system's functioning that our future demands for resources will be comprehensively balanced with their supply.

This balancing process is illustrated using coherent comprehensive mathematical models of our human-environment system. Their output allows us to say something meaningful about the characteristics of its possible future when treated as a whole, how that future might arrive and how we might deal with it. However, the models are limited in their ability to represent our decision making. Thus, the drawing of meaningful conclusions must make use of the knowledge gained in earlier chapters about ourselves and our environment.

A reminder that *short* (or *immediate*) *term* refers to a period up to 15 years from 2015; *medium term*, up to 60 years; and *long term*, beyond 60 years.

## Energy Basics

The survival of all cultures depends on their members having access to energy in the form of food. The more complex the culture, the more it also depends on other sources of energy, both animate and inanimate. Current industrialized cultures rely heavily on an abundant supply of cheap inanimate energy to support their energy demanding world views and lifestyles (Ayres and Warr 2005; Kummel et al. 2002; Cleveland et al. 1984). Therefore the discussion about our global inanimate energy future focuses on complex industrial and industrializing cultures, with a bias toward Western culture.

An industrial culture's energy future can be predicted by balancing its estimated future energy supply against its estimated future demand. Predicting the future supply-demand balance is difficult because both supply and demand are products of many factors and interactions within a dynamic, complex energy system. Thus to meaningfully discuss our energy future requires having an appreciation for how the energy system functions and the constraints on its functioning.

This section serves as an introduction. In it our dependence on energy, the fundamental limits to supplying it and the energy costs of doing so are discussed. (Note that in the following sections, the year 2010 is used as primary reference date, as in chapter 10, with more recent data mentioned as needed).

**Dependence, Source and Use.** Our dependence on inanimate energy (energy) is illustrated at a personal level by the increase in the average annual global energy use per person over the last 150 years. Our energy consumption rose from 0.9 megawatt hours per person (MWh/person) in 1860 to 21.6 MWh/person in 2010. By 2013 it had reached 22.1 MWh/person (IEA 2015). This consumption was not evenly distributed. Those living in the more industrialized (higher GDP) countries require substantially more energy to support their lifestyle than those living in the less industrialized (lower GDP) countries. For example, in 1995, the high-income countries, with 20% of the world's population, accounted for 60% of the world's commercial energy use (UNEP 1999). More specifically, in 2010, the energy consumed by a citizen of Bangladesh was 2.44 MWh/person; China, 21.05 MWh/person; and the United States, 83.15 MWh/person. The average American thus used 34 times more energy than the average Bangladeshi.

To fully appreciate our dependence on energy, we need to know more than the amounts consumed. We have to consider the sources of the energy and the way we use it. The variety and importance of our energy sources is illustrated by their relative contribution to our total global primary provision of energy. In 2010, 81.1% of energy was supplied by fossil fuels. The sources were oil (32.4%), coal (27.3%) and natural gas (21.4%). The remaining 18.9% was provided by nuclear energy (5.7%), hydroelectricity (2.3%), combustible renewables (10.0% [6% wood, 4% biofuels, waste, etc.]), and alternative renewables (0.9% wind, solar, geothermal, tidal, etc.) (figure 13) (IEA 2012a).

We are highly dependent on fossil fuels. That dependence has not changed much: the proportion of our total global energy that comes from fossil fuels has only dropped from 86.7% in 1973 to 81.1% in 2010. Over this same period, the total amounts of fossil fuel we used more than doubled from 6,107 million tonnes of oil equivalent (Mtoe) to 12,717 Mtoe (71,024 TWh to 147,898 TWh) at an average growth rate of 2.6%/year. Not only are we highly dependent on fossil fuels, especially oil, we are using more of them, especially coal (figure 13).

The specifics of our dependence on fossil fuels is illustrated by how we use them. For example, in 2010 an estimated 67.4% of the world's electricity was generated by fossil fuels. The sources were coal (including peat) (40.6%), natural gas (22.2%) and oil (4.6%). Transportation consumed 61.5% of the produced oil. Global industrial activity consumed 79.5% of the coal and 35.2% of the natural gas (IEA 2012a). Note, agriculture does not appear as a major consumer of fossil fuels, despite its dependence on them. The unsurprising conclusion is that all of our activities are heavily dependent on substantial amounts of one or more of the three types of fossil fuels.

How does this dependence affect our energy future? Consider our use of oil. Between 1990 and 2008, the annual global demand for oil (all sources) per person was consistently close to the average demand for that 18-year period: 4.6 barrels/person/year. But during those years, the global population increased substantially. Thus, to maintain the supply of oil at 4.6 barrels/person/year, production had to increase at the average growth rate of the population: 1.6%/year. In absolute terms, oil production for 2010 was 28.9% above the 1990 level.

This growth trend has existed since at least the 1980s (EIA 2011). Similar growth trends exist for the other fossil fuels. To support our current lifestyle, our aspirations and a growing global population, we demand an ever-increasing total supply of all fossil fuels, or an equivalent growing supply of non-fossil-fuel energy.

## Fundamental Limits.
Providing an increasing supply of energy is a challenge. The most fundamental limit is that energy can't be recycled in the form it was used. New amounts must continually be supplied to maintain the current usage and provide for any growth. There are four basic energy sources from which this energy can come: the earth's heat (geothermal); gravity (tidal); nuclear materials and sunlight (which is the ultimate source of solar, wind, hydro, biofuel and fossil fuel energy). It would seem that maintaining a continuous supply of energy should not be an issue. However, the supply of underground fuels (fossil and nuclear) is finite, while the supply of geothermal and tidal energy is constrained by practical considerations, as discussed below.

The supply of energy from the sun is ongoing and infinite, in a practical sense, but its availability to us is constrained by the fixed amount of sunshine reaching the earth each day. The availability of sunshine-related renewable energy is also limited by other considerations. For example, the available amount of sunlight and wind varies with location, the weather and the seasons. Sunshine is also limited by the time of day. Hydroelectricity is limited by the amount of water supplied by the water cycle. It is thus affected by droughts. Bioenergy (biofuels and wood) are fundamentally limited by the rate at which plants can grow. If renewable energy sources are used faster than their renewable rates, then they are non-renewable and finite, in the practical sense.

The supply of fossil fuels, which are based on organic matter, are similarly limited by the amount and speed at which nature provides them. The average amount of organic matter (land and ocean) produced each year and the geological efficiencies for converting buried organic material to fossil fuels are known. They provide us with a feel for the limits to fossil fuel production. For example, 89 tonnes of plant matter are needed to produce 3.8 litres (1 US gallon) of gasoline. Based on this type of information, it is estimated that the fossil fuels we consumed over the 247 years between 1751 (just prior to the Industrial Revolution) and 1998 took the earth about 13,000 years to create. Thus, on average, our rate of fossil fuel consumption over that period was about 52 times faster than its rate of creation. The last time the earth's fossil fuel creation rate was in balance with our extraction rate was in the 1880s (Dukes 2003).

The supplying of energy from all of these energy sources suffers from technical limitations. They too are fundamental, but in a different sense. For example, geothermal power is limited by the heat transfer rate. The provision of wind, solar and tidal energy is limited by the uneven distribution of suitable sites and the intermittent nature of their sources. There are likely other fundamental/technical limits to capturing their energy.

The supply of energy from all sources is limited by the financial costs of providing it. It takes money to build and apply the technology needed to extract and distribute energy. The price of energy pays for at least part of that cost. When the cost of applying technology rises to a level that requires unrealistic prices for the recovered energy, then we have reached a fundamental financial limit to providing that energy.

In overview, for all practical purposes, the supply of fossil fuels and mined nuclear fuels is non-renewable and finite. Geothermal energy is non-renewable but inexhaustible, however its extraction is subject to fundamental technical limitations. Hydro, wind, solar and tidal energy, as well as wood and biofuels are renewable, but the amount of energy they can supply is fundamentally limited, in a practical sense, by renewal rates and technical considerations. There is one other fundamental limit to supplying energy to be discussed.

### Energy Return on (Energy) Invested (EROI).

The process of supplying energy uses energy. Regardless of whether the energy-supplying processes use the energy directly (e.g., powering oil-extracting pumps) or indirectly (e.g., manufacturing oil-extracting pumps), it is an energy cost that must be paid. Payment can be conveniently thought of as all coming from the energy being supplied. Once that energy cost is deducted, the remaining energy (the net energy gain) is available for our discretionary use, such as producing goods and providing services. This method of energy-cost accounting applies to the provision of energy from all energy sources.

A common measure of the amount of discretionary-use energy we gain from using energy to provide energy is the EROI ratio: the total number of energy units supplied for each energy unit consumed to provide it (Hall et al. 1986). For example, a fuel with an EROI of 5:1 means that 5 units of energy cost 1 unit of energy to provide. Thus, the equivalent of 20% of those 5 units was consumed to make the remaining 80% (4 units) available for our discretionary use. If the EROI is 3:1, then the equivalent of 33% was consumed to provide 66% for discretionary use.

Supplying energy with an EROI of 5:1 or less is energy expensive. Below an EROI of 5:1, the relative energy cost of providing energy rises quickly until, when the EROI reaches 1:1, we neither gain nor lose energy from supplying it. An EROI of <1:1 means that we are providing energy at a net energy loss.

The EROI was specifically chosen for this discussion from among other methods of measuring the energy gain from providing energy because it reflects the total energy cost rather than just a selected or partial cost. This means that the changes in energy's EROI over time more accurately reflect how the cost of providing it could affect our energy future. Unfortunately, there are difficulties using an EROI. Without some feel for the energy costs and benefits included or excluded when the EROI was calculated, it is difficult to meaningfully interpret or compare EROIs. An appreciation for the boundaries of the energy-providing system that the EROI represents is also needed (Hall et al. 2011).

The boundary issue is partly resolved by choosing system boundaries that at least match the reason for calculating the EROI. For example, in the case of oil (and gas), the costs should include the direct use of energy from an oil deposit to produce its oil, such as heating. After all, once the deposit's energy is consumed, it is no longer available for other purposes. Similarly, there is waste left behind after a farmed feedstock is turned into a biofuel. If the waste is used in the biofuel production process, then that energy must be included in the EROI calculation. Consider that the nutrients in the waste burnt to supply heat must be replaced by trucking in compost or by applying fertilizer made using natural gas.

For fuels, the system boundary issue is further resolved by calculating EROIs at three different points in the fuel-providing system: $EROI_{(extracted)}$ (e.g., oil and gas at the wellhead, coal at the mine mouth and corn at the farm gate); $EROI_{(refinery)}$ (e.g., gasoline at the refinery gate and refined ethanol); and $EROI_{(on\ site)}$, which means provided at the site of its use (e.g., gasoline at a construction site) (Murphy et al. 2011). $EROI_{(extracted)}$ is the easiest to calculate for fuels, so it is the most common.

Care should be taken when using EROIs. Meaningful comparisons between EROIs are most reliably made when they are of similar types, such as $EROI_{(extracted)}$, because they have similar boundaries. $EROI_{(electricity)}$ is used here to indicate that the EROIs for electricity produced in different ways can be meaningfully compared.

Rules of thumb also help us when evaluating EROIs. Energy provided at an EROI of <5:1 is energy expensive. EROIs are difficult to calculate and usually under rather than overestimates. Thus, a rule of thumb says that as the EROI declines below 5:1, so should the certainty of the statement that it is being provided with a net energy gain.

Although a fossil fuel with an $EROI_{(extracted)} > 1:1$ is produced with an energy gain, it need not provide us with a gain at its site of use. After all, there are the additional energy costs of refining and transporting it to the use site. A rule of thumb for oil says that its $EROI_{(extracted)}$ must be >3:1 if the fuel refined from it is most likely to provide a net energy gain at its site of use (i.e., $EROI_{(on\ site)} > 1:1$) (Hall et al. 2009). Based on the first rule, treat the value of 3 as an absolute minimum.

The EROI records a significant constraint on the provision of energy. Although it may be technically possible to provide energy from a particular source type, the energy cost of doing so may result in its $EROI_{(on\ site)}$ being <1:1 (i.e., it is being provided at an energy loss). The energy cost of providing energy severely limits the amount of technically recoverable energy that can be provided to us at an energy gain.

Counterintuitively, we may wish to produce a fuel that has an $EROI_{(on\ site)}$ <1:1. We may wish to do so because it has desirable properties. For example, oil can directly power more applications than coal. However, the energy cost of making that fuel available must still be paid. Regardless of where that energy payment comes from, it will still result in less energy being available for our discretionary use. These EROI rules of thumb and the limitations they represent apply to the energy from all types of sources, not just fuels. Note that, for the rest of the chapter, the EROI ratio will be expressed as a single number, with the :1 implied.

In summary, complex cultures, especially industrial ones, depend heavily on supplies of inanimate energy. Energy is consumed so it must be constantly supplied from a source. Collectively, the amount of energy that can be supplied from each inanimate energy source is fundamentally and practically constrained. In particular, the source can be finite, its replenishment rate is limited, and there are always technical constraints to making its potential energy available. There are also the inescapable energy costs of providing energy, as recorded in the EROI. Not to be forgotten are the financial costs.

The manner and degree to which these characteristics constrain our energy future is the subject of the next three sections. In the process of discussing them, an outline of a possible energy future will emerge. If you only want a general overview rather than a deeper appreciation of that future, then just read the introduction and summary for each of the next three sections.

## Conventional Fossil Fuels

In 2010, oil contributed 32.4%, coal 27.3% and natural gas 21.4% of our energy supply. In total, non-renewable fossil fuels supplied 81.1% of our energy (IEA 2012a). In 2013, their total contribution was 81.4% (IEA 2015). With this in mind, an appreciation of our possible energy futures requires some understanding of fossil fuels, how they are produced and the constraints on their supply. Conventional oil will be the focus of this section and will serve as the energy reference for this chapter. The application of the conclusions about conventional oil production to the other fossil fuels will be noted.

This section first discusses fossil fuel basics. It then turns to predicting conventional oil's future supply and how that future will arrive. The discussion will include the change over time in the energy return from producing conventional oil.

<u>Fossil Fuel Basics.</u> The hydrocarbons that the earth creates from marine and terrestrial plant matter can be concentrated into deposits (also referred to as fields) of oil, natural gas or coal. Most deposits are small, while only a few are very large. Regardless of their size, if these hydrocarbons are to be used as fuels, they have to be extracted from their deposits.

The process of extraction is limited by the fuel's properties and the deposit's geological characteristics. If oil and gas can be easily extracted from their host deposits by simple drilling and pumping, then those fuels and their deposits are called conventional. The cash and energy costs of producing conventional oil and gas are relatively low.

Oil and gas that must be extracted by more than simple drilling and pumping is called non-conventional. For example, tar sands deposits are mined to produce bitumen, which must then be chemically treated (upgraded) to give it properties similar to crude oil. Shale gas deposits must be hydraulically fractured and eroded with a mix of water and chemicals (fracked) before natural gas can be extracted. Non-conventional fossil fuels cost more money and energy to produce than conventional fossil fuels.

The boundary between conventional and non-conventional fuels is fuzzy. For example, sometimes natural gas liquids are included in the conventional oil category and at other times in the non-conventional oil category. For most of the following discussion, where it is placed is not critical. Consider that 53.8% of our 2010 energy supply was provided by oil and gas combined.

By far the majority was conventional. For example, between 2005 and 2010, the average daily global production of oil (total of all types) was some 80 million barrels/day (Mb/d). Of that amount, 85% (68 Mb/d) was unambiguously conventional oil; 12%, natural gas liquids; and 3%, unambiguously non-conventional oil (IEA 2010b). Where the presence/absence of natural gas liquid in the accounting is important, it will be pointed out.

The terms *conventional* and *non-conventional* are not applied to coal and coal deposits. Coal is primarily classified by the amount of energy recoverable from it and the amount of combustion waste left behind. High-energy coal is called anthracite, middle-energy coal is bituminous and low-energy coal is called lignite. Coal deposits have features in common with both metal and oil deposits. For example, coal deposits are large and continuous, like oil deposits, but the coal is extracted by mining, like metals. Thus, like metal mines, coal deposits mined from open pits are much easier and cheaper to exploit than those mined by underground methods

Regardless of how a fossil fuel is extracted, the amount produced becomes the amount we can use, its supply. Geological, technological and economic constraints limit how much of the fuel in a deposit can actually be produced. For conventional oil, only around 35% of the oil in its deposits is extractable in the conventional category. Some of the remainder is recoverable by unconventional methods.

It is therefore important to remember the difference between the following ways of reporting the amounts of fuel in a deposit or region. The total (recoverable plus unrecoverable) amount of a fossil fuel in an area is called the fuel's resource. The amount of a fuel's resource that is, or is expected to be, technologically feasible to extract is called just that, the technologically extractable/recoverable amount of fuel (the extractable resource). The amount of a fossil fuel's extractable resource that, under current and near-term conditions, we have confidence will actually be extracted and thus become the supply is called the fuel's reserve. (These terms are similar to, but not exactly the same as, those used for metals. Care should be taken when reading the literature to determine what their usage actually means.) The amount of a fuel's reserve decreases as the fuel is extracted to become part of the supply, and it increases as new deposits are put into production.

The amount of a fossil fuel produced from a deposit changes over time. Simplistically, production starts from zero, rises to a maximum, and then declines to zero as the deposit is depleted. The basic shape of this production curve depends primarily on the properties of the fuel and the geology of its deposits. The details are determined by economic, political and technical factors.

In the case of conventional oil, the ideal production curve from a deposit that is being continuously exploited traces out a slightly distorted bell-shaped curve with a flattish peak (figure 19). The distortion is seen as a slightly faster fall from the peak than the rise to it. This occurs because, during the pre-peak phase, more oil wells can be put into service thus boosting production. In contrast, during the post-peak phase when the deposit is nearing depletion, production is more strongly limited by fundamental constraints, such as the technical ability to extract the oil. The production curve for conventional natural gas is more asymmetric than the reference conventional oil curve. It has a more rapid post-peak decline.

The idealized production curves for non-conventional oil and gas, when extracted using wells, are highly distorted versions of the conventional oil curve. They display a rapid rise to a sharp peak followed by a very rapid, initial post-peak decline, which is followed by a longer period of extremely low production before depletion. Coal's idealized production curve is a flatter version of conventional oil's curve.

Regardless of the shape of a production curve, the area under it is the total amount of fuel produced from that deposit. This is the full amount the deposit added to the supply: the amount we eventually consume and try to replace.

## Making Predictions.

The essence of the following discussion about the making of valid and useful predictions about conventional oil's future can be easily applied to non-conventional oil. They can also be applied to conventional and non-conventional natural gas. It is less easily applied to coal.

Predictions about the future supply of fossil fuels are plagued by the same types of misunderstanding, misrepresentation and misuse discussed for copper in chapter 14 (Maugeri 2004). There is therefore no need to repeat them here. All that is required here is an occasional reminder of that discussion in the form of examples.

The total global amount of past, present and future conventional oil that can be extracted from oil deposits using either existing or reasonable-to-expect future extraction technology is called its "ultimate" supply. It is the total amount of conventional oil that we can expect to be produced. The complete production curve for conventional oil's ultimate can be modelled by knowing oil's production characteristics and having global estimates of past production, current reserves and oil that is yet to be found. The resulting modelled, slightly distorted bell-shaped production curve can be used to predict the size of conventional oil's likely future supply and its idealized rate of production.

The most well-known of the modelled conventional oil production curves is Hubbert's symmetrical bell-shaped curve. It was used to correctly predict that the United States' production of conventional oil would peak in the early 1970s (Hubbert 1973). His methodology was sound, but the accuracy of his prediction was fortuitous (Kaufmann and Cleveland 2001). Current models use the same basic idea as Hubbert, but they are more sophisticated.

Regardless of the model or its enhancements, the validity and usefulness of its prediction is strongly dependent on the completeness and reliability of the data used to make and apply it. The calculation of conventional oil's ultimate is most sensitive to two factors: the degree to which the reserves in known deposits will be changed by updated estimates of the oil in the ground; and the estimate of the recoverable oil in the yet-to-be-found deposits. (The estimate of a deposit's reserves changes over time because of oil extraction, the addition of new or reclassified oil, and refinements in the way the estimate is calculated.)

An appreciation for the uncertainty in conventional oil's data that a modeller has to deal with is illustrated by examples. Consider changes that result from the detection of reporting "errors." For example, Shell reported a reserve error whose correction reduced the company's projected future revenue by an estimated US$100 billion (Sims 2007).

When companies in the United States follow regulations about reporting new oil discoveries and reserves to the public, they express the information in a light most favourable to themselves. This results in the need for later adjustments that confuse the picture of how much reserve oil a deposit actually contains. Similarly, in the 1980s, it seems that some of the Middle Eastern members of OPEC either changed or inappropriately reported as fact historical speculations about oil reserves. This apparently happened because the size of the oil production quotas given to OPEC members are based on their reported reserves (Bentley 2002; Laherrere 2003).

Estimates of conventional oil's reserves could also change if future technology increases the level of oil recovery from its deposits to above the current average of about 35%. How much it could change depends on more than technology's success. It also depends on whether the oil produced is classed as "heavier" conventional oil or non-conventional enhanced-recovery oil (Bentley et al. 2009). This classification also reflects how much energy is used to recover the oil.

Intuitively, an estimate of the yet-to-be-found conventional oil would seem to be so uncertain that it would be just a guess. This is less the case than expected. Compared to metal deposits, conventional oil deposits are found in well-constrained geological locations. These deposits are larger, and their boundaries are relatively easy to define. In addition, a reasonable first estimate of the amount of extractable conventional oil likely to be contained in a newly discovered deposit is quickly provided by pumping tests (Bentley 2002; Bentley et al. 2000).

As a result, a global-scale historical record of conventional oil deposit discoveries from the late 1800s to the present that lists their sizes and production provides much useful information. It can be used to make useful estimates of the yet-to-be-found conventional oil, check estimates of reserves, and improve our understanding of oil production. It, for example, revealed that the annual global rate of conventional oil discoveries rose to a peak near 50 giga (billion) barrels per year (Gb/year) in the mid- to late 1960s, after which the rate declined to 10 Gb/year by 2002: a fivefold decline. A more sophisticated statistical analysis of the record indicated that by the 1980s, almost all of the world's larger conventional oil deposits, and the majority of the world's conventional oil (by volume), had been found (Laherrere 2003; Bentley 2002). Consider too that, in 2007, there were roughly 70,000 deposits producing conventional oil. But only 374 of them accounted for 60% of production, 10 for 20% and one deposit, Ghawar, for 7% (Sorrell et al. 2009). Most of the extractable conventional oil is hosted in a few, large deposits. These conclusions have a significant impact on predictions about conventional oil's future production. The most important is that, with only a small number of large deposits left to be found, conventional oil production must eventually decline.

Overall, there is sufficient historical data and understanding of oil production to model the complete global-scale production curve for conventional oil's ultimate. But this is only possible if the modellers have access to the data and devote sufficient time and effort to identifying and accommodating the data's errors, biases and limitations. If these criteria are met, then the models can produce valid and useful predictions about conventional oil's future supply (Bentley et al. 2009).

The essence of making conventional oil predictions can be applied to natural gas, non-conventional fuels and coal. The degree of success depends on each fuel's properties and the geology of their deposits. It also depends on the quality of the data available.

**The Peak Experience.** The modelled production curve for conventional oil's ultimate provides two dates of interest. Intuitively, the most important date would seem to be the time when all of the conventional oil deposits are depleted, its depletion or running-out date. But, from an industrial culture's perspective, the date when production reaches its peak is more important. At that production peak, roughly 50% of conventional oil's ultimate still remains to be produced, while beyond it, production begins its inexorable decline. If demand doesn't decline, then a shortfall in conventional oil's supply will occur. This section discusses predictions of conventional oil's peaking date and the shortfall.

Model-based predictions of the peaking date for the global production of conventional oil lie between 2000 and 2030. The favoured peaking dates fall in the first 15 years of 2000 (Bentley 2002; Laherrere 2003; Campbell and Laherrere 1998; Bentley et al. 2009). However, the peaking date can be delayed by declines in demand, such as the recession that occurred in 2008. An unexpected discovery of a large oil deposit can also change the peaking date, but not by much because the peak is so near and there is a multi-year delay before the discovery can be put into production. Finally, the production curve has a flat rather than a pointed peak. It is more of a peak zone. Thus, the usefulness of predictions about conventional oil's peaking-date lies less in their exactness and more in the general conclusion that crossing the peak zone has begun or will occur sooner rather than later.

When are the global peaking dates for conventional gas and for coal? Conventional natural gas (methane) is less well studied than conventional oil. The predicted peaking dates for gas fall in a range between 2010 and 2045, with those that appear to be the most well-thought-out falling between 2020 and 2030.

The global production curve for conventional natural gas has a steeper post-peak decline than conventional oil. Thus, at its production peak, only about 30% of the natural gas ultimate remains in the ground (compared to 50% for conventional oil). Both of these features mean that the conventional natural gas post-peak shortfall will become apparent more quickly than for conventional oil (Bentley 2002; Laherrere 2003; Al-Fattah and Startzman 2000; Hughes 2009).

The peaking date for global coal (all types) is highly uncertain. One reason is the uncertainty in the coal reserve data arising from political and economic biases (Heinberg and Fridley 2010). Another reason is the greater difficulty, compared to conventional oil, of estimating coal's reserves and its yet-to-be-found deposits. It is thus more difficult to reliably model its complete production curve (WEC 2013).

A reasonably well-thought-out peaking date for coal places it around 2045 (Höök et al. 2010). This is the soonest it will occur because the prediction was based on just past production from historical or active coal mining areas. Possible reserves and yet-to-be discovered coal lying outside these areas were not included.

Other predictions of coal's future focus on running-out dates. Because of the long tail to the production curve these dates are more difficult to meaningfully predict. The soonest is a date around 2070 for 90% depletion (running-out). This is a minimum because it also suffers from the same issues mentioned for coal's peaking date (Rutledge 2011).

The reliability of the following predictions of coal's global running out date is of less interest than the direction in which the predictions changed over time. In 2001, the International Energy Agency (IEA) estimated coal's global, running-out date to be around 2200 (IEA 2001). The European Council later suggested that this date was more likely to be around 2160, partly because of increasing consumption and partly because of more careful attention to the reserve figures. In 2006, BP suggested a running out date of 2150 (Strahan 2008). The biased and uncertain nature of the data suggests that the ultimate on which these predictions are based might be too large.

A conventional oil peak is predicted to be in progress or imminent while a conventional natural gas peak is expected shortly. How will we experience the crossing of these peaks? Consider conventional oil. In the recent past, the ongoing and increasing demand for oil was met by a matching 1.5%/year increase in the rate of production. That supply closely matched demand, with little held either in storage or in deposits whose production could be quickly increased and refined. Under these conditions, only a relatively short delay would occur between conventional oil production crossing its peak and the subsequent decline being imposed on us as a shortfall.

Fortunately, on the scale of a decade or two, the conventional oil peak will be more of a plateau than a point, thus the initial shortfall will be small. However, its smallness will hinder us from recognizing the indicators that conventional oil's production is in its peak zone. This identification is made more difficult by the ongoing "usual," rapid (months to a few years), politically and economically induced variations in the availability and cost of oil products (such as the 2010s glut). The presence of a supply shortfall will also be masked by our efforts to fill a growing energy demand from other sources.

Ironically, the best indicator that we are likely in the peak zone of conventional oil production is the increasing amount of expensive-to-produce non-conventional oil being used to meet our increasing oil demand. As the following suggests, at least some amount of it is being used to fill a conventional oil shortfall. Between 1990 and 2008, the annual global demand for oil (all types, conventional plus non-conventional) remained very close to the 18-year average of 4.6 barrels/person/year. But, to maintain that average, non-conventional oil's contribution (excluding natural gas liquids) to the total oil supply had to increase from around 0.05 barrels/person/year (1% of global oil production) in 1990 to 0.21 barrels/person/year (4.5%) in 2008. To maintain the average of 4.6 barrels/person/year until 2025, non-conventional oil's contribution is projected to increase to 0.44 barrels/person/year (9.5%) (EIA 2011; US Census Bureau 2008).

How we choose to respond to these predictions and indicators that conventional oil's production will soon peak will strongly influence our energy future, which, in the immediate term, could be our experience of crossing the peak. For example, if industrial cultures, who are heavily dependent on conventional oil, choose not to reduce their demand to match the decline in its supply, then they must instead find ways to fill the shortfall (Campbell and Laherrere 1998). The following precautionary tale indicates our likely response.

After the 1974 oil shock, the Organization for Economic Co-operation and Development (OECD) established the International Energy Agency (IEA) to monitor and report on global energy issues. Surprisingly,

the IEA's first detailed study of production from the world's 800 largest oil fields was only recently completed. It was published in the IEA's 2008 *World Energy Outlook*. The study acknowledged that crude oil (conventional oil, excluding natural gas liquids) was being produced faster from the global reserves than the reserves were being replaced. It also reported IEA's prediction that total (conventional plus non-conventional, including natural gas liquids) global oil production would plateau around 2030 (IEA 2008). However, in a 2008 interview, the IEA's chief economist suggested that the plateau date was more likely to start around 2020 (Monbiot 2008).

In November 2009, whistle-blowers within the IEA suggested that its oil fields study had actually shown that the total global oil supply was already in the peak zone. In their opinion, the IEA's 2008 prediction that total oil production would plateau near 105 Mb/d around 2030 was far too optimistic. Even a number of 90 to 95 Mb/d was too high. The whistle-blowers said that the IEA was aware of this actual decline but had been encouraged by the United States to downplay this conclusion on the grounds that it would cause financial panic (Macalister 2009).

By 2010, the IEA's prediction had changed, a bit. It suggested that crude (conventional) oil production had crossed its peak of 70 Mb/d in 2006 and would be at a plateau of 68–69 Mb/d by 2020. This level of crude oil extraction would be maintained by bringing undeveloped and undiscovered crude oil deposits into production.

They predicted that global oil production (all types) would still grow to around 95 Mb/d. This would be possible through increases in the production of natural gas liquids and non-conventional oil (IEA 2010a; IEA 2010b). The IEA maintained this stance in 2012 (IEA 2012b). This tale hints at an IEA prediction process that is focused more on speculating about how a conventional oil shortfall might possibly be met, rather than representing the real-world constraints on oil's future supply.

Economists too discount the possibility and impact of an imminent conventional oil peak. They emphasize the power of economics and technology to fill a shortfall. The basic idea is that price increases will result in the market stimulating the innovation of technological solutions to fossil fuel production issues (Radetzki 2010). Confidence in this type of evaluation is expressed by predictions based on simple projections of historical production trends and past technological successes far into the future.

These types of evaluations largely ignore oil's properties and the geology of its deposits. They fail to consider that these basic characteristics impose fundamental constraints on the amount and rate at which conventional oil can be supplied (Reynolds 1999; Kaufmann and Cleveland 2001; Bentley et al. 2009; Laherrere 2003). Similarly, by failing to consider the energy cost of applying technology to produce energy, their evaluations can conclude that an increase in oil's price and the application of new technology can increase the efficiency, amount and speed of conventional oil's recovery indefinitely. As the EROI reveals, this is not the case.

These types of evaluations, reactions and predictions are not a unique response to conventional oil likely crossing its production peak. They were part of the discussion about the future of other resources, in particular copper, in chapter 14. They will be seen again when our culture's contribution to our energy future is discussed in more detail.

For now, it is sufficient to recognize the two broad, contrasting approaches to making and evaluating predictions about conventional oil's future, and then try to reconcile them. On the one hand are those predictions and evaluations that focus on the geological and physical characteristics constraining conventional oil's production. They represent an outward-looking view of oil's future that is consistent with the earth's fundamental properties and its short- and long-term processes.

On the other hand are those predictions and evaluations that focus on humans' technological achievements and economic theory. They represent an inward-looking view of oil's future that is consistent with our concern about energy's short-term availability and our faith in ourselves and our culture. A good example is Westerners' now familiar belief in the ability of economics and technology to address all kinds of production issues and thus maintain our cultural imperatives of growth and progress?2*.

Reconciliation of the two approaches requires accepting that, in the short term, economically induced technological responses to conventional oil can successfully maintain supply. But, over the longer term, that ability is constrained by the geological and physical constraints on oil's occurrence and production. The longer-term reality is an unavoidable fundamental constraint on an industrial culture's future conventional oil supply and thus its longer-term energy future.

Reconciliation could also be achieved by both camps deciding to fill the conventional oil shortfall from other non-oil energy sources. However, at least in the short term, the way we use oil constrains our ability to make that switch. Consider that in 2010, the transportation sector was responsible for consuming 61.5% of the total global oil (mostly conventional) production (IEA 2012a). This had risen to 63.8% in 2013 (IEA 2015). Because of this dependence on oil, the possibility of filling the conventional fossil fuel shortfall with non-conventional oil and gas will be discussed before considering the other possible energy sources and solutions. But first the energy cost of producing conventional oil must be considered.

**Conventional Oil's Energy Returns.** Although the predicted conventional oil shortfall is described in terms of volume (barrels [b] or barrels of oil equivalent [boe]) or mass (tonnes [t] or tonnes of oil equivalent [toe]), it actually affects us in terms of energy: it is an energy shortfall. The EROI of conventional oil is an energy view of its cost of production. The energy in conventional oil is fixed, thus changes in the EROI of conventional oil will affect the size of the *volume* shortfall, reported in barrels or tonnes, needed to fill the *energy* shortfall. Similarly, the difference between the energy cost of supplying conventional compared to non-conventional oil will affect the *volume* of non-conventional oil needed to fill the conventional oil *energy* shortfall. To appreciate the significance of these statements, an understanding of the energy costs of producing conventional fuels (i.e., their EROIs) and how they change over time is needed.

Consider the combined production of oil and gas (oil plus gas) in the United States from 1919 to 2007. Before that oil plus gas (the vast majority of which was conventional) could be produced, its host deposits had to be discovered. They were found by drilling holes into the ground, which takes energy. A measure of the energy gained by the discovery of a fuel deposit compared to the energy spent discovering the deposit is the $EROI_{(discovery)}$ of a deposit's fuel (figure 19). The average $EROI_{(discovery)}$ for oil plus gas deposits in the United States was 1,200 in 1919, an energy bonanza. After 1919 the $EROI_{(discovery)}$ rapidly declined before slowing to reach 10 in 1969. It continued to decline, but more slowly, to reach 5 in 2007. The EROIs in the more recent years included non-conventional oil and gas deposits. Regardless, the decline still illustrates the increasing amount of energy being used to find new deposits, which, in the case of conventional oil and gas, are both increasingly scarce and more difficult to find in the United States.

Production of oil and gas from these previously discovered deposits rose, with some variation, from 0.42 gigabarrels of oil equivalent per year (Gboe/year) in 1919 to a peak of 3.5 Gboe/year in 1970 (near Hubbert's predicted US conventional oil peaking date), before declining to 1.8 Gboe/year in 2008. A small increase since 2005 was the result of the production of non-conventional oil and gas.

The $EROI_{(extracted)}$ of the oil plus gas produced over this period declined from around 20 in the 1910s to around 13 in the first decade of 2000. One reason for the decline was the increasing amount of drilling needed to establish production wells. This meant that the amount of energy needed to produce a unit of oil plus gas steadily increased over time (Guilford et al. 2011). This decline is a characteristic of all liquid and gas (well-based) fossil fuel production.

A discussion of California's oil production personalizes the above generalities. California started producing conventional oil in 1900. Around 1955, heavy oil (marginal conventional oil) started to be produced in earnest using enhanced oil recovery methods. As the sources of conventional oil were depleted, heavy oil slowly came to dominate California's oil production.

California's oil production curve (total, all types) indicates that production peaked in 1984 at around 0.43 Gb/year. It subsequently declined to 0.28 Gb/year in 2002. California's production is now being maintained solely by the ongoing enhanced oil recovery of heavy oil from previously discovered deposits.

The decline in Californian oil's $EROI_{(extracted)}$ started well before 1955. In that year, the conventional-oil-dominated production recorded an average $EROI_{(extracted)}$ of 64 (range 48–80). After that date, the increased energy needed to produce conventional oil and the increasing use of more energy-intensive enhanced oil recovery methods, such as steam, to produce heavy oil were responsible for the inexorable decline in Californian oil's (all types) $EROI_{(extracted)}$. This decline continued through the production peak of 1984. By 2005, the now heavy-oil-dominated production recorded an average $EROI_{(extracted)}$ of 6 (Brandt 2011). California's oil production is becoming increasingly more energy expensive to produce.

California's oil history illustrates a number of generic energy-related constraints on oil production. The fuel's highest $EROI_{(extracted)}$ occurred early in California's production. The inexorable decline in its oil's $EROI_{(extracted)}$ was sped up by the use of more energy-intense enhanced oil recovery methods to produce heavy oil. At the current EROI of 6, Californian oil deposits are nearing the point of energy depletion, which will occur when its $EROI_{(extracted)}$ falls to around 3. At that point, producing California's oil will no longer provide an energy gain at its site of use (figure 19).

The production peak for Californian oil was reached well after its EROI peak. Its post-peak production is characterized by a declining amount of fuel being produced from slowly depleting deposits at an increasing energy cost. Eventually, all of California's technically extractable fuel will have been produced and its deposits will be physically depleted. Although, in detail, this may occur either before or after the $EROI_{(extracted)}$ is <3, in general, predictions of depletion using EROI curves complement those based on production curves.

California's use of enhanced oil recovery methods illustrates that applying technology can maintain production from a region's physically depleting conventional oil deposits. But the produced fuel is recovered at a higher energy cost. Thus, technology can only partly, not fully, compensate for declining oil production. It can delay, but not stop, depletion (Hall 2011). This provides a link to non-conventional oil production, which will be discussed further in the next section.

Overall, the United States is the most depleted conventional oil and conventional gas producing region in the world. Thus the history of US oil and gas production and their EROIs provides an indicator of the future for the global supply of at least conventional oil and conventional natural gas, and how that future will arrive. In particular, it shows that conventional oil's EROI will decline into the future. As a result, as time passes, at least a 5%–20% greater *volume* of conventional fuel must be produced to satisfy even a constant conventional oil *energy* demand (Dale et al. 2011). It becomes even harder to produce that extra conventional oil once its production peak is crossed. If oil demand continues to increase, then the expected post-peak conventional oil shortfall will grow even faster and be even harder to fill than we expect.

In summary, fossil fuels (oil, natural gas and coal) provided 81.1% of our global energy supply in 2010. Oil and gas combined provided 53.8%. The vast majority of this oil and gas was conventional, meaning that it could be extracted at a low energy and low financial cost. Conventional fuels currently dominate our energy supply.

The amount of conventional oil produced each year changes over time. The complete idealized curve of its combined past, present and future global production (i.e., its ultimate production) has a slightly distorted bell shape (figure 19). The production curves for conventional natural gas, non-conventional oil and non-conventional gas are more distorted versions of conventional oil's curve. By modelling conventional oil's complete (ultimate) production curve, useful predictions can be made about oil's future production, for example, when production will peak and the size of the remaining future supply. But the predictions are only useful if care is taken with the data used in the model.

Conventional oil's production peak was expected to occur in the first 15 years of 2000. Observations suggest that we have indeed either reached the peak zone or are in the process of crossing it. Conventional natural gas production is expected to peak between 2020 and 2030. The peaking date for coal is uncertain, but at recent production rates, it is most likely to occur after 2045.

The global history of conventional oil deposit discoveries, their sizes and the production from them indicates that a significant majority of oil's ultimate supply has already been found. The data also shows that most of the deposits producing conventional oil are small. However, the largest single proportion of conventional oil's production comes from less than 400 large deposits. Almost all of these larger conventional oil deposits have been found, with most of them being discovered relatively early in conventional oil's history.

The history of the discovery and production of conventional oil plus gas in the United States indicates that the deposits discovered early were, on average, the least energy costly to find. They had near the maximum $EROI_{(discovery)}$. Over time, the $EROI_{(discovery)}$ declined because an increasing amount of energy was consumed in the form of the increased drilling effort needed to discover the fewer remaining deposits that were increasingly difficult to find.

California's oil production history personalized the conclusions drawn for the United States as a whole. The conventional oil produced from the earliest of California's deposits put into production had, in the long view, near the maximum $EROI_{(extracted)}$ for the region. Over time, the $EROI_{(extracted)}$ of the region's oil declined. More energy was used to produce oil from both the older, depleting deposits and the newer, less easily exploited deposits being put into production.

A region's oil production peaks because more oil is being produced to satisfy demand than is being replaced by discoveries. Under these conditions, once both the EROI and production peaks are crossed, a declining amount of conventional oil is produced at an increasing energy cost. The amount produced and its EROI continue to decline until all of the region's deposits are either physically depleted or depleted in terms of their energy return (i.e., $EROI_{(extracted)} < 3$ or energy depleted) (figure 19). Predictions of depletion using EROI curves complement those based on production curves.

Innovative technology can help maintain conventional oil production but applying it incurs an extra energy cost, which lowers the oil's EROI. Applying technology can delay but not stop the eventual depletion of a region's conventional oil production. Energy depletion can precede physical depletion. All of these conclusions apply to conventional gas production.

We are within the peak zone of global conventional oil production and near to the peak in global conventional natural gas production. If demand doesn't decline after these two peaks are crossed, a growing shortfall in combined conventional oil and gas production will arise. That shortfall in production is reported as a volume or mass, but, because we use the energy in oil and gas, we will experience it as an energy shortfall. This, in conjunction with the inevitable and ongoing post-peak decline in the EROI of the conventional fuels being produced, ensures that the size (volume/mass) of the fuels needed to fill the energy shortfall will grow faster than we expect.

Our energy future will be significantly determined by our response to the appearance of a conventional oil and gas energy shortfall. We could voluntarily reduce our energy demand, but our present effort is to increase the energy supply. We are doing so, in part, by turning to non-conventional fuels.

## Non-Conventional Fossil Fuels

The currently preferred solution to a conventional oil or gas shortfall is to fill it from non-conventional sources. Between 2005 and 2010, non-conventional oil (natural gas liquids excluded) provided, on average, 3% of total oil production. Can enough additional non-conventional oil and natural gas (fuel) be produced to fill the shortfall?

Presented below is a highly uncertain estimate of the amount of each of the various non-conventional fuel types that is likely recoverable using current or reasonable-to-expect future technology. However, just because a fuel is technologically recoverable doesn't mean that extracting it is sensible, either financially or energy-wise. Thus, rather than focusing on the exactness of uncertain fuel supply estimates, this section will focus on gaining a feel for the proportion of the estimated extractable fuel that is likely to actually be produced.

This can be achieved by paying attention to those features of a non-conventional fuel's currently short production history that indicate its future production. Those features were identified when considering conventional fuel's long history, as discussed above. One feature is the $EROI_{(extracted)}$ of the non-conventional fuel produced from those of its deposits put into production early in its history. Those early EROIs will reveal the approximate peak $EROI_{(extracted)}$ for that non-conventional fuel. Because the $EROI_{(extracted)}$ of a fuel will decline from there, the lower the peak, the closer the fuel is to being produced at an energy loss: the closer its source is to energy depletion.

Similarly, the characteristics of the production from those of a non-conventional fuel's wells and deposits that were put into production early will reveal the overall shape of the production curve for its fuel-source type. In particular, the early production numbers will indicate the speed with which the inexorable post-peak decline in production will occur, which constrains its ultimate production. Other features to which attention should be paid include the financial costs. When looked at together, these features suggest the proportion of a non-conventional fuel's estimated amount of technologically recoverable fuel that will actually be produced at an energy and financial gain.

## Non-Conventional Oil.
Conventional oil's ultimate has been estimated at 2,000 gigabarrels (Gb). It consists of 850 Gb of historical production over roughly 110 years, 850 Gb of reserves, and 300 Gb of yet-to-be-found oil. This ultimate is consistent with a conventional oil peaking date in the first 15 years of the 2000s.

This estimate of the conventional ultimate was accompanied by estimates of the amounts of non-conventional oil thought to be extractable from its sources using current or reasonable-to-expect future technologies. The list includes the 3,500 Gb of oil thought to be recoverable from conventional oil deposits using enhanced oil recovery technologies. It is treated here as a non-conventional oil. Also listed is the 600–1,000 Gb of likely and 400 Gb of possibly technologically recoverable non-conventional oil from heavy oil and tar sands oil deposits. An estimated 4,000

Gb of this type of oil is currently considered unrecoverable. Finally, some 3,000 Gb of marginally technologically recoverable oil is listed for oil-hosting shale (Bentley 2002; Laherrere 2003). (A separate dry- and wet-shale oil estimate was not provided. Wet-shale oil is also called tight or oil-shale oil.)

An estimate of conventional oil's total shortfall at the time of its deposits' future physical depletion can be estimated by using a model of its production curve. By assuming that current production is near its peak, and that future demand for oil will either remain constant at current levels or continue to increase at current rates until depletion, then the total conventional oil shortfall till then will lie between roughly 1,000 and 2,500 Gb. When these rough numbers are compared with the sum of the technologically extractable non-conventional oil from all sources at 7,500 Gb, it appears that there is more than enough non-conventional fuel to meet the conventional oil shortfall. However, just because a fuel is technically extractable doesn't mean that it will actually be produced. In order to gain a feel for the proportion that will actually be produced, we need to consider the basic features of current production from these non-conventional sources.

First, consider the production from the Athabasca tar sands in Alberta, Canada. Production is only possible from the more bitumen-enriched parts of this tar sands resource. A few deposits have been put into production using one of two methods. In the mining method, the tar sands are dug from open pits and passed through a bitumen-recovery plant. In the in situ method, the tar sands deposit is heated in place in order to liquefy the bitumen so that it can be pumped out. Regardless of how it is recovered, an upgrader is used to convert the recovered bitumen into a syncrude with properties similar to those of crude conventional oil.

The EROI and production history of both the older mining method, which started in 1970, and the newer in situ method indicates that they are both functioning at their maximum energy efficiency. In 2010, the syncrude produced by the mining method had an $EROI_{(extracted)}$ of around 6 (at the upgrader gate, including the processing energy acquired from the deposit). The syncrude produced by the in situ method had an $EROI_{(extracted)}$ of around 3.4. More syncrude was, and still is, being produced by the mining method than the in situ method. The resulting production weighted average $EROI_{(extracted)}$ for all syncrude production, up to 2010, was 5.2, that is much lower than conventional oil's current $EROI_{(extracted)}$ of 10–30.

To complete the picture, in 2010, the energy cost of refining mining's syncrude and delivering the resulting fuel to its point of use resulted in its $EROI_{(on\ site)}$ being 3.4. If the syncrude was produced by the in situ method, then the fuel's $EROI_{(on\ site)}$ was 1.6. This is close to 1, the point of no energy gain (Brandt et al. 2013).

These low EROIs on their own indicate that even though production of syncrude is a relatively new development, it is close to the point of not being sensible, at least energy-wise, to produce. There are two other features that will speed the arrival of depletion. They will be illustrated by considering the path that the Athabasca tar sands' production is currently following.

Although the tar sands' minable deposits were responsible for 64% of historical syncrude production, they are estimated to make up only 22% of the tar sands' "established reserves" and 7% of its resources (ERCB 2012; ERCB 2010). This indicates that, as the minable deposits are depleted, syncrude's production will increasingly come from deposits exploitable by the in situ method. Over time, the relatively high EROI syncrude recovered by mining will be replaced by the relatively low EROI syncrude from in situ extraction. During this transition, there will be an inevitable decline in syncrude's $EROI_{(extracted)}$ from its 2010 average of 5.2 toward the in situ value of 3.4.

Consider too that the first Athabasca tar sands deposits to be discovered and exploited were the easier to find and less costly to exploit of the minable deposits. Thus the $EROI_{(extracted)}$ of the syncrude currently being produced from minable deposits is expected to be near the maximum for that method. Over time, the $EROI_{(extracted)}$ of mining's syncrude will decline. The same is expected to happen to the EROI of the in situ method's syncrude. As a result, the future average $EROI_{(extracted)}$ for syncrude will drop not to 3.4, but to some lower value.

Both of these features speeds up the arrival of the day when syncrude can no longer be produced at an energy gain. Also to be considered are the relatively high financial costs of producing syncrude. Just the cost of building and running an upgrader ensures that the production cost is high. And, as the decline in tar sands oil production during the 2014–2016 period of low oil prices indicates, syncrude can only be produced at a financial profit if the oil price is high (subsidies excluded).

Collectively, the history of both the tar sands' production and its EROI all point to one conclusion: the proportion of the technologically extractable Athabasca tar sands syncrude that will be produced at an energy gain and a financial profit will be relatively small. Thus only a relatively small amount will be available to help fill the conventional oil energy shortfall (i.e., the shortfall in the energy for our discretionary use.)

This discussion about the Athabasca tar sands production illustrates how the features of current production from a non-conventional oil source can be used to gain an appreciation of its likely future production. It is not necessary to repeat this discussion in detail for each source. Only a summary for the other sources is needed, starting with shale oil.

The estimated 3,000 Gb of marginally recoverable non-conventional shale oil is made up of two types: dry-shale oil and wet-shale oil. The largest proportion is dry-shale oil. It is present in its deposits as kerogen, a solid. Dry-shale deposits can be exploited by the mining and in situ methods, both of which convert the kerogen into a dry-shale oil syncrude. Today, even the deposits that are more favourable to exploit aren't in production.

One reason is that the $EROI_{(extracted)}$ for dry-shale's syncrude is approximately 2 (after conversion and including the energy used from the deposit). When this syncrude is refined, it provides motor gasoline with an $EROI_{(on\ site)}$ of around 1.4. It barely provides a net energy gain at the point of use. For comparison, when the same calculation method is applied to conventional oil, its $EROI_{(extracted)}$ is 20, and the motor gasoline refined from it has an $EROI_{(on\ site)}$ of 4.5 (Cleveland and O'Connor 2011). If these features are considered together with the high financial costs of producing dry-shale syncrude, then it seems likely that most of the marginally technologically extractable dry-shale syncrude won't be produced at an energy gain.

The smaller proportion of shale oil is wet-shale oil. It occurs in its deposits as a conventional oil liquid. The United States is currently the world's largest producer of wet-shale oil, which it started producing in earnest in the early 2000s. This is unsurprising as its conventional oil deposits are close to depletion.

Wet-shale oil is produced by drilling a vertical hole (a well) with a horizontal extension into a deposit's oil-rich zone. The host shale is then fracked in the immediate vicinity of the drill hole. The oil from only that small fracked volume is pumped or flushed out using water. As a result of these characteristics, only a few years must pass before a well's production rate drops by more than half from its peak. Thus, if any level of production from a wet-shale oil deposit is to be maintained, then producing wells must be continually established. The drilling of these wells eventually ends because space constraints prevent an indefinite increase in well density.

Unsurprisingly, a wet-shale oil deposit's production curve is a highly distorted version of a conventional oil deposit's production curve. It displays a much steeper rise to a sharper peak followed by a steeper post-peak fall that slows as depletion is approached. For example, production from the two major US wet-shale oil deposits declines at rates of 38% and 45% per year. This is much faster than the average decline rate for large conventional oil deposits of 5% per year.

In the United States the average cost of drilling and fracking a shale oil well is US$8–9 million. It has been suggested that this high cost will be offset in the future by higher oil prices. But the building and running of a drilling rigs depends on oil products, so the benefit from high oil prices is constrained. This constraint applies to all aspects of producing non-conventional oil because building and running the production facilities also requires the consumption of energy.

Wet-shale oil's $EROI_{(extracted)}$ is not known. The preliminary values from initial production suggest that it will be near to the low end for conventional oil, roughly around 10. These values are likely near the maximum for wet-shale oil. The drop in the $EROI_{(extracted)}$ of its future production is expected to be steeper than for conventional oil.

For the above reasons, few of the deposits making up wet-shale oil's technologically extractable resource appear to meet the criteria for production. For example, in 2012 there were 19 producing deposits in the United States. As is the case for conventional oil, two wet-shale deposits account for 81% of its production, and five for 92%. The two most productive deposits also contain almost all of the proved reserves.

It is estimated that the total US production of wet-shale oil from currently exploited areas should peak before 2020 and could possibly be depleted by 2025, barring any production interruptions. It seems that wet-shale oil is unlikely to make a large, long-term contribution to US oil production (Hughes 2014a; EIA 2014a; Hughes 2013). Globally, it seems that only a small proportion of the 3,000 Gb of the marginally technologically extractable shale oil will likely be produced at an energy and a financial gain.

Next to be discussed is the 3,500 Gb of additional oil that is technologically extractable by enhanced oil recovery methods from depleted conventional oil deposits. This oil is present because only an average of roughly 35% of the oil in a conventional oil deposit is currently recovered before it is considered to be depleted of conventional oil. The Californian example suggests that most of the enhanced-recovery oil that is easier, cheaper and has a higher EROI (marginal conventional oil) is likely being or has been extracted. The peak $EROI_{(extracted)}$ for the remaining enhanced-recovery oil (non-conventional oil) will likely be low, below 10. It is also reasonable to expect that the production and EROI characteristics of future enhanced-recovery oil will fall between the conventional and non-conventional oil trends.

Over time, the energy and financial cost of producing enhanced-recovery oil will increase. As a result, increasingly fewer depleted conventional oil deposits are likely to produce enhanced-recovery oil at an energy or financial gain. Its production will decline faster and its deposits will be depleted more rapidly than for conventional oil. Thus, overall, the proportion of the technologically recoverable enhanced-recovery oil that will actually be produced at an energy gain and a financial profit will be a small proportion of the estimated technically recoverable oil.

In overview, only a small proportion of the extractable non-conventional oil resource is likely to be recovered at an energy and financial gain. Regardless of its source, whatever non-conventional oil is produced at an energy gain will help offset the conventional oil shortfall. However, even if 1,000 to 2,500 Gb of non-conventional oil could be produced, that would only fill the ultimate conventional oil's *barrel* shortfall not its *energy* shortfall. This is so because the EROIs of today's non-conventional oil are much lower than the EROIs of today's conventional oil. Thus more barrels (a few to a few decades percent more of 1,000 to 2,500 Gb) of non-conventional fuel must be produced in order to fill conventional oil's energy shortfall.

**Non-Conventional Natural Gas.** The conventional natural gas (methane) ultimate used to calculate one of its peaking dates given above was 2,000 Gboe. It consists of 450 Gboe of historical production, 950 Gboe of known reserves still to be extracted and 600 Gboe from yet-to-be-found sources. An additional 2,000 Gboe of highly speculative yet-to-be-found conventional natural gas is left out of this discussion (Bentley 2002).

The following are estimates of the amount of each type of non-conventional natural gas thought to be recoverable using current or reasonable-to-expect future technologies. From the tight gas resource (assumed to include shale gas), an estimated 180 Gboe is likely recoverable, 500 Gboe is unrecoverable, and 500 Gboe is of unknown status. From the brine gas resource, an estimated 450 Gboe is likely recoverable (more is possible).

An extensive frozen natural gas resource exists under the ocean sea floor, however the vast majority of it is currently considered to be unrecoverable because of technical difficulties (Bentley 2002). The modeling of just the gasifying of this frozen natural gas by electric heating indicates that the energy gain will be 4 to 5 times (Callarotti 2011). This represents the energy cost of melting the frozen gas but not all of the other energy costs, such as generating the electricity, needed to calculate a meaningful EROI. It thus indicates that the optimistic maximum possible $EROI_{(extracted)}$ is 5. This source is not considered further.

The features used to indicate the proportion of a non-conventional oil's technologically extractable resource that is likely to actually be produced can be applied to non-conventional gas types. The insight gained from using these features for gas will be illustrated by addressing the following question: Has the United States resolved its conventional natural gas shortfall by becoming the world's largest shale gas producer?

The production of conventional natural gas within the United States peaked in the early 1970s. Production from the deepwater gas wells in the Gulf of Mexico resulted in a second, but lower, peak in the early 1990s. More recently, production from non-conventional gas sources has become increasingly important. By the early 2010s, tight gas and shale gas were collectively responsible for nearly 40% of the United States' gas production and for the sharp increase in its total production.

Shale gas deposits are exploited in the same manner as wet-shale oil deposits: drilling and fracking its host rock. The production curve from a single shale gas well is similar to that of a shale oil well: a steep rise to its peak, followed by a decline to less than half of its peak in only a few years. Thus, to maintain high production from a shale gas deposit, production wells must be continually established. Production eventually slows then ends because space constraints prevent an indefinite increase in well density. In the United States each well costs US$3–9 million to drill.

The resulting production curve for a shale gas deposit is similar to that for a shale oil deposit: a highly distorted bell curve with a very steep rise to a peak from which a steep decline slows toward depletion. For example, the depletion rates for the seven most productive deposits in the United States lie between 23% and 49% per year. Predictions based on these characteristics suggest that the combined production from these seven deposits will peak soon, by 2020.

Unsurprisingly, the majority of shale gas's total production comes from just a few deposits. Of the 30 producing shale gas deposits in the United States, three accounted for 66% of production and six for 88%. In addition, most of this production comes from small areas within each deposit (Hughes 2014a; EIA 2014b; Hughes 2013).

The EROI of the shale gas produced in the United States has an estimated average $EROI_{(refinery)}$ of 17 (range 13–23). This is essentially the same as the value for current US conventional gas, which, when calculated using the same method, has an estimated average $EROI_{(refinery)}$ of 18 (range 14–25). Once shale gas and conventional gas are delivered to the user, their $EROI_{(on\ site)}$ is smaller than the $EROI_{(refinery)}$ by 5. So an $EROI_{(refinery)}$ of 6 is the minimum for an energy gain on site (Yaritani and Matsushima 2014).

When taken at face value, these EROI numbers suggest that there is no energy penalty for producing shale rather than conventional gas. However, the shale gas EROI represents initial production from the most favourable deposits, while the conventional gas EROI represents production from the last of its exploitable deposits. Thus, like shale oil, shale gas's current $EROI_{(refinery)}$ of 17 is near its maximum, while conventional gas's current $EROI_{(refinery)}$ of 18 is well below its maximum.

Overall, shale gas's production features ensure that the proportion of its deposits that are suitable (energy-wise) to be put into production is small and that their production will peak relatively quickly. It also ensures that, over time, the EROI of the total shale gas production will decline more sharply from its peak than conventional gas did. These features, along with the associated high financial costs of production, will severely constrain the United States' use of shale gas to fill its conventional gas shortfall in the medium to long term.

This conclusion is supported by the history of natural gas production from the Western Canada Sedimentary Basin, the largest producer of natural gas in Canada. Production of natural gas from a mix of conventional and non-conventional sources in the basin has been unable to stop the basin's total gas production from peaking in 2006 and its $EROI_{(extracted)}$ from declining. (The decline in EROI means that the fall in production can't be fully blamed on excess shale gas production in the United States.) A rough estimate of the EROI$_{(extracted)}$ for the known unexploited shale gas deposits in the basin suggests that only a few of them are likely to produce natural gas with an EROI$_{(extracted)}$ of around 15. The majority of the EROIs will be lower, around 5 (Freise 2011; Hughes 2014b). This discussion represents the general constraints that all non-conventional gas sources will face when filling the expected global natural gas shortfall.

**Technological Cures.** The argument is often made that technological developments will overcome the constraints on producing non-conventional oil and gas from their apparently depleted conventional sources. And it will also enable production from their under-exploited or currently unexploitable non-conventional sources. An example of this possibility is the invention of horizontal drilling and fracking, which enabled the production of shale gas and wet-shale oil. But the practical, energy and financial constraints of applying technology limits its ability to supply fuels at an energy gain (Anderson 1998).

The significance of these constraints is neatly illustrated by the outcome from applying the innovative technology used to extract bitumen from the Athabasca tar sands and convert it into syncrude. Technology made it possible for the mining method to produce syncrude from the tar sands at an average EROI$_{(extracted)}$ near 6. This means that the equivalent of roughly 17% of the energy the syncrude mining produces is being consumed to produce it.

Subsequent technological development made it possible to produce syncrude from previously unexploitable tar sands deposits using the in situ method. However, its syncrude has an EROI$_{(extracted)}$ of only 3.4. Roughly 30% of the energy in its syncrude is being consumed to produce it.

In the future, as the less abundant minable deposits are depleted, a large proportion of syncrude production will be provided by the in situ method. A barrel previously filled by mining's syncrude will still be filled by the syncrude produced by the in situ method, but its lower EROI means that less of the energy in the barrel will be available for our discretionary use. To maintain our lifestyle, that missing amount of energy must be provided.

It is easy to say that the 13% energy difference between the two methods can be supplied by the in situ method producing extra syncrude equal to that difference.

But the nature of tar sands technology constrains our ability to easily increase the production of in situ syncrude. Tar sands production facilities are run at capacity and have reached maximum efficiency. Therefore, to increase the in situ method's production, more production facilities must be built, which is a multibillion-dollar expense.

Some argue that an increase in the fuel price could be used to pay for additional in situ facilities. It could, but less easily than we think. As previously discussed, energy and the economy are tightly linked. Thus, as the price of fuel rises, so does the cost of most aspects of constructing and running a new production facility.

Another method to resolve the consequences of the EROI difference is to treat the energy from the tar sands deposit used by technology to produce syncrude as free: exclude it from the EROI's accounting. Regardless, in the real world we must still pay the cost for supply that energy for a technology to function. Using the tar sands deposit's energy to produce syncrude today will be paid for in the future by the deposits being exhausted more quickly. Using energy from another source simply denies future generations the use of that source's energy for other purposes. (Think of Bruntland's definition of sustainability.) Thus, regardless of whether you account for the energy used during tar sand's oil extraction now (in the EROI) or in the future, by subtracting it from the reserves, that energy is still spent to produce syncrude. The choice of when to face this energy payment is the mirror image of choosing what to do with efficiency's savings discussed in chapter 10. Both are an inescapable aspect of our resource future.

Collectively, energy and cash costs limit the ability of technology to maintain or increase the amount and rate of syncrude being made available from deposits that are increasingly difficult to exploit. More generally, in the short term, technology can overcome the constraints restricting the production of non-conventional fuels. But, over the long term, as lower EROI non-conventional fuel sources are exploited, ever-increasing amounts of fuel and cash are needed to produce the energy equivalent of the conventional fuel shortfall. Plus, at some point, technology cannot replace the energy required to run itself, thus rendering the technology, energetically speaking, useless (Hall et al. 1986).

Technology can thus only delay, not stop, the inevitable depletion of a finite fossil fuel resource. These constraints on the use of technology to supply energy are not restricted to the production of fossil fuels. They also apply, with different timelines, to other energy sources.

### Environmental Costs.

Producing non-conventional fossil fuels comes with environmental costs. They include the use of space and water, and the disposal of waste products into air and water. These costs can be evaluated by comparing them to the environmental costs of producing conventional fossil fuels.

The production of non-conventional oil and gas affects more space per unit of product than conventional oil and gas. For example, the mining of the Athabasca tar sands creates a moonscape of open pits and tailings dams that, to date, remain largely unreclaimed. Tight gas and shale gas production requires both more wells, more closely spaced than conventional gas, and a more extensive gathering system than conventional gas, thus increasing its environmental footprint.

Water usage is greater for non-conventional oil and gas production than it is for conventional oil. For example, during the production of tight gas and shale gas, a large volume of water is used to frack the gas's rock host (Sell et al. 2011). The production of dry-shale oil uses an estimated 1 to 3 barrels of water per barrel of oil produced (Cleveland and O'Connor 2011).

Between 2006 and 2011, a barrel of Athabasca bitumen produced by the mining method used a net average of around 4 barrels of fresh water, while the in situ method used around 0.5 barrels (CAPP 2011). A different report suggests that producing a barrel of Alberta bitumen in 2011 used an average of 1.7 barrels of water (ERCB 2012; AESD 2014). It is unclear if this number refers to consumed or used water. The water used to upgrade the bitumen to syncrude would increase these numbers.

Most of the water used in non-conventional oil and gas production is not consumed but polluted. During non-conventional gas production, the water polluted by fracking chemicals and the compounds in the gas's host rock is either pumped back underground, held in ponds or released to the environment. In the Athabasca tar sands, this polluted water is retained in large, sometimes leaking, tailings dams that emit noxious organic compounds to the atmosphere (Parajulee and Wania 2014; Frank et al. 2014). The soup of air pollution emissions from each of four Athabasca surface mining operations, whose activities each cover between 66 and 275 km$^2$, falls between 50 and 70 tons/day (Lia et al. 2017).

The production of all fossil fuel results in carbon emissions (mostly $CO_2$ and methane) to the atmosphere. There are two main production-related sources of these emissions: the combustion of fossil fuels to generate the energy that powers the fuel's production technology, and the fugitive emissions from the wells, pits and other extraction infrastructure. These carbon emissions are reported as the fuel's production-related carbon emissions per unit of the latent energy made available when the produced fuel is ultimately used: a fuel's production-related carbon intensity.

The production-related carbon intensities of non-conventional fuels are greater than those of conventional fuels. For example, in 2010, the syncrude produced from the Athabasca tar sands by mining had a fuel-production cycle (well to wheels) carbon intensity of 102 g of $CO_2$ equivalents per megajoule (g$CO_2$e/MJ). The in situ method's carbon intensity was 111 g$CO_2$e/MJ (367–399 g$CO_2$e/kWh). Syncrude's values are 12% to 24% higher than the average for conventional oil production (Englander et al. 2013). The average production-related carbon intensity of shale gas (produced by fracking) is greater than, in descending order, that of conventional natural gas and coal.

A fuel's production-related carbon intensity is only one part of the environmental costs from its carbon emissions. The latent energy in a fossil fuel is only made available for our discretionary use when the fuel is burned, which creates carbon emissions. Adding a fossil fuel's discretionary-use carbon intensity to its production-related carbon intensity provides a more complete measure of its emission's environmental cost. This measure also allows us to meaningfully compare the environmental costs of fossil fuels. For example, the total carbon intensity of shale gas is greater than or equal to that of coal (Howarth et al. 2011).

In overview, regardless of the relative values, the larger the amount of a fossil fuel used, the greater is its absolute environmental cost. If we, as individuals and as a culture, decide that we want to reduce the costs of producing and using fossil fuels, then we have two options. The surefire choice is to constrain the production of fossil fuels in general and non-conventional fossil fuels in particular. The alternative is to reduce these environmental costs piecemeal through the application of energy-using technology.

In summary, our efforts to develop and apply technology have allowed us to extract fossil fuels from previously unexploitable non-conventional oil and gas sources. Those fuels are then available to fill the conventional oil and the expected natural gas shortfalls. The estimated amount of non-conventional fuel that is likely recoverable using current or reasonable-to-expect future technology from these sources is considerable.

However, the energy and financial costs of just producing a non-conventional fuel from its source (i.e., excluding refining and transport to the use site) constrain the proportion of the estimated amount of technologically recoverable fuel from the source that will actually be produced. A feel for that proportion is provided by studying the characteristics of the fuel's production from those of its deposits that were discovered and exploited early. The EROI$_{(extracted)}$ of that early produced fuel will indicate the peak or maximum EROI $_{(extracted)}$ to expect for that source type. Because the EROI of that source's fuel will decline from that peak over time, the lower those initial peak EROIs are, the closer the fuel is to being produced at an energy loss: the closer the source is to energy depletion.

The characteristics of that early production will also reveal the general shape of the source type's production curve. That will indicate how fast the inexorable post-peak decline in production will occur and how quickly the source type will be depleted. This constrains its production ultimate.

When the energy and financial constraints on production are considered together, they strongly suggest that the proportion of a non-conventional source's estimated technologically recoverable fuel that will actually be produced at an energy and financial gain is likely to be much smaller than we expect. In addition, although the conventional oil and gas shortfall is discussed using volume (reported as Gb and Gboe), it affects us as an energy shortfall. This complicates our use of non-conventional fuels to fill the shortfall. Because the current average EROIs of non-conventional fuel types are lower than those of conventional fuels, a greater volume of non-conventional fuels must be produced to fill a conventional energy shortfall. But there are practical, energy and financial constraints on increasing production of non-conventional fuels to make up for this EROI difference. These constraints include slower production rates (e.g., tar sands), higher financial costs (e.g., tar sands) and faster depletion rates (e.g., shale oil and wet-shale oil) than conventional fuels.

The environmental costs of producing non-conventional fuels are also higher than conventional fuels, with the costs tending to increase with the energy cost of extracting the fuel. These environmental costs only affect our ability to fill a shortfall if we decide to address them. For example, capturing and storing the $CO_2$ released during fuel-production increases the energy cost of producing the fuel. If we decide to pay these costs, then the additional fuel to run the technology must be provided.

Overall, it seems that non-conventional fuels will be able to fill the projected conventional oil plus gas shortfall into the medium-term future. How much further into the future depends, in part, on how quickly the energy costs of producing non-conventional fuels increases (how quickly the EROI declines to the level of energy depletion) and how quickly their production declines (how quickly their production ultimate is reached). It also depends on how quickly the financial cost of producing non-conventional fuels will increase. Finally, it also depends on how quickly and by how much our demand changes. This review suggests that, at the current (roughly 1.5%) or a faster growth rate in our energy use, it is unlikely that non-conventional fuels will be able to fill the shortfall beyond the late medium-term.

This is not the conclusion we would draw if we take the first reports of production from a newly exploited non-conventional fuel source at face value. Think of the US wet-shale oil boom. Focusing on those first reports creates the impression that substantial amounts of fuel will be produced from this source type at relatively low energy and financial costs. We can form this satisfying, but overly optimistic opinion because we forget that, in the oil and gas business, the first production of fuel from a new type of fuel source comes from its rarer deposits: those that are less costly, both financially and energy-wise (higher EROI), to exploit.

These overly optimistic, even unrealistic opinions about our fossil fuel future are not restricted to the public. They can be routinely seen in the outlooks for future fossil fuel production put forward by private as well as public institutions, such as the Energy Information Agency, the International Energy Agency and the National Energy Board of Canada. This claim is easily checked by comparing their past predictions to actual outcomes (Hughes 2014a; Hughes 2013; IEA 2012b; IEA 2008).

An easier to grasp, more holistic appreciation of our fossil fuel future appears when we treat the shortfall in conventional fuel production and our efforts to fill it as part of a dynamic, complex fossil fuel energy system. Focus on the energy. It takes energy to provide fossil fuel energy. Thus, in order to satisfy a constant discretionary-energy want using fossil fuels, we must produce an amount of fuel larger than our discretionary-energy want.

That amount of extra energy is not fixed but increases over time, mainly because the average EROIs of the fossil fuels being produced inexorably decline over time. There are three reasons why: they decline as the currently producing deposits are depleted, as the declining production from higher EROI conventional fuel deposits is replaced by production from lower EROI conventional deposits, and as the emerging conventional fuel production shortfall is increasingly filled by production from lower EROI non-conventional fuels.

The amount of extra fuel also increases because producing that extra energy also uses energy. A dampening production-amount–energy-cost feedback determines the actual amount of extra energy we must produce. Thus, over time, even for a constant discretionary-fossil-fuel-energy demand, there is an inexorably accelerating increase in our demand for fossil fuels.

These dynamics of the complex fossil fuel energy system hasten the day when we can no longer produce fossil fuels at an energy gain (i.e., $EROI_{(on\ site)}$ nears 1). An energy gap appears. Its appearance is sped up by a growing population, an increasing personal demand for energy, the decline in the discovery of new deposits and the increasing financial cost of producing fossil fuels. It is further sped up if we decide to pay for the environmental costs of producing and using fossil fuels by deploying energy-using technology, such as carbon capture technology.

However, although our energy system is certainly dominated by fossil fuels, it is not exclusively dependent on them. In order to fully represent our energy future, this description of our fossil fuel energy system must be expanded to include the other sources of energy that could fill a conventional fossil fuel shortfall. They will be discussed next.

## Other Energy Sources and Solutions

When we picture ourselves as participants in a dynamic, complex energy system, we more fully realize the degree to which our production of fossil fuels is constrained. We are also reminded of the environmental costs of producing and using them. This is the context within which it is appropriate to discuss the non-fossil fuel energy sources we could use to fill the conventional fuel shortfall, avoid an energy gap and reduce the environmental costs of fossil fuels.

This section begins by considering substitutions between energy sources because they can play a key role in our efforts to address energy-related issues. It will then evaluate non-fossil-fuel energy sources. Bioenergy (biofuels and biosolids) is one such source. There are also non-fossil-fuel, non-biological energy sources: hydro, nuclear and the alternatives, such as solar, wind, geothermal, tidal, hydrogen and fusion. The option of using a variety of energy sources to fill the shortfall, avoid an energy gap and reduce the environmental costs of producing and using fossil fuels will also be considered. This section then discusses the influence of culture on our efforts to consider and implement a change in energy demand. A summary of this chapter's discussion of energy ends this section

**Substitutions.** Substitutions between energy sources are constrained by more than their degree of energy equivalence. The wide range of these interconnected constraints is illustrated by the following example. The carbon emissions released from electricity-generating plants that burn coal, the fuel most commonly used for this purpose, can be reduced by substituting either natural gas or oil for coal. Conventional natural gas is preferred. An often quoted reason for this preference is that, when combusted, it can generate the same amount of electricity as coal for only half of the $CO_2$ emissions (IEA 2001). This is true if both the conventional gas and coal deposits are near the electricity generating plant.

This assessment changes if the substitution involves shipping the gas overseas. Natural gas then incurs an energy and financial cost not faced by shipping oil or coal. Unlike oil or coal, natural gas must first be liquefied before it can be shipped in a special, more expensive tanker. Then, at the destination, the liquefied natural gas (LNG) must be heated to regasify it. During the processes of transporting, liquefying and regasifying natural gas, the energy equivalent of 15% to 25% of the energy in a tanker load of LNG is consumed, depending on the transport distances (Hughes 2009). This loss to our discretionary use of gas lowers its $EROI_{(on\ site)}$ and the EROI of the electricity generated from it.

The sum of the emission from the gas's production and discretionary use also affects the advantage of using it as a fuel substitute in the power plant. The sum of these emissions for conventional natural gas is slightly less than for coal. The sum for non-conventional gas produced by fracking is about the same as coal, which eliminates the environmental advantage of substituting it for coal (Howarth et al. 2011). If the fracked natural gas has to be shipped as LNG to the power plant, then its transportation emissions must be included, which ensures that the environmental costs of the gas, and thus the electricity generated from it, will be more than using coal. This example illustrates that substitutions between energy sources are constrained by a mix of energy, environmental and financial costs.

A different view of the constraints on substitutions among energy sources appears when the sources are looked at from the perspective of their usefulness to humans. The characteristics determining their usefulness include the following: the range of applications important to humans to which a particular source can be applied (its versatility); the efficiency of its use in an application; and its energy per unit volume (its energy density). For example, storing 0.24 m$^3$ of energy-dense oil fills one barrel, 0.85 m high by 0.56 m in diameter. An energy-equivalent amount of natural gas at room temperature and pressure occupies 170 m$^3$, or 708 barrels, which is why it is compressed or liquefied for transport (Edwards 1997).

Collectively, an energy source's characteristics, such as its versatility, efficiency and energy density, determine its cultural desirability and value, as captured by the term *quality* (Cleveland et al. 1984). In the case of fossil fuels, the highest quality fuel is oil. Its high energy density and liquid form allows it to be easily and efficiently used in many applications: electricity generation, all forms of transport, and miscellaneous devices such as kerosene lights. Natural gas is of middle quality. Coal is the lowest quality fuel.

The rule of thumb for fossil fuel substitutions is that the higher the quality/desirability of a fossil fuel, the more likely it will be able to substitute for a lower quality one than the reverse. For example, oil can often substitute for natural gas and coal-powered applications. In contrast, coal can only substitute for petroleum and natural gas in a few applications, mainly for electricity generation (Murphy and Hall 2010).

Because of energy's importance to an industrial culture, the quality rank of our energy sources has an economic definition: the relative economic gain (marginal productivity, useful work) per heat-equivalent unit among energy sources. The ranking of energy sources by their economic quality places electricity at the top, followed by petroleum and then coal. For example, in the United States, electricity is 1.6 to 2.7 times more economically productive than petroleum

(Cleveland 2005). If the EROI of the energy from an energy source is adjusted for its economic quality, then its EROI more reliably reflects the source's economic value to humans. This provides a more human-centred comparison of energy sources and the possible substitutions among them.

Similarly, an EROI is more representative of a fuel's energy gain if, during its calculation, the mathematical substitutions for the different energy sources involved take into account their different energy efficiencies. This is particularly important when comparing electricity generated by different methods. The term EROI$_{(electricity)}$, used later, indicates that this efficiency aspect of quality differences has been taken into account.

But neither a general ranking of an energy source's quality, nor its efficiency or economic-quality-adjusted EROI can represent all of the factors affecting the making of a particular substitution. A comprehensive evaluation requires considering all of those factors. Consider how a positive evaluation of the benefits from filling the conventional oil shortfall by substituting more available, cheaper low-quality coal for less available, more expensive high-quality oil changes as other factors are considered.

The point is most clearly made when the substitution is made by converting 1 ton of "undesirable" coal into 1 to 2 barrels of "desirable" synfuel. This is possible, but the energy, financial and environmental costs are considerable. The oil has an EROI at the converter's gate of only about 0.5 (Höök and Aleklett 2009; Keith 2009). The financially expensive conversion process generates significant CO$_2$ emissions and uses about 1 m$^3$ (4.2 barrels) of water per barrel of synfuel. Finally, the combustion of the synfuel itself generates higher CO$_2$ emissions than conventional oil. With this discussion about substitutions in mind, the following subsection evaluates the non-fossil-fuel sources of energy available to fill a conventional oil and natural gas shortfall.

**Bioenergy.** Bioenergy is provided by biofuels and combustible biosolids. Biofuels are created by converting organic feedstocks, which can be crops (such as corn and sugar cane) or organic waste, into biodiesel, ethanol and methane. Biosolids are mostly wood-based products, such as wood pellets. Bioenergy sources appear to be closer to carbon neutral than fossil fuels.

Substituting biofuel energy for fossil fuel energy would thus seem to be the ideal method to fill a conventional fuel shortfall and eliminate the environmental costs of using fossil fuels. In 2010, biofuels were responsible for <4% of the global energy supply. If they are to substitute for the roughly 80% of energy provided by fossil fuels, then a substantial increase in their feedstocks is needed. Securing the needed increase in farmed feedstocks is difficult. Imagine that, in the year prior to biofuels becoming popular, the United States turned its entire annual corn production into ethanol. That amount would only replace about 7% of the fossil fuels used annually by the vehicles in the United States (Wald 2007).

The basic constraints on significantly increasing the supply of farmed feedstocks are the same as those for the farming of food plants. They include the scarcity of agricultural resources, such as arable land and water, and the tasks of discharging our many associated agricultural responsibilities. The resources needed to farm biofuel feedstocks could be acquired by displacing other agricultural crops from existing farm land, but this simply transforms the issue: food versus fuel.

The use of an agriculture-based biofuel to fill the shortfall is further constrained by the fuel's EROI. For example, the $EROI_{(refinery)}$ for ethanol from corn lies between 0.82 and 1.73 (Hall et al. 2011). The $EROI_{(refinery)}$ for biodiesel from soya lies between <1 and 3 (Pimentel and Patzek 2005). Recent efficiency improvements may have increased that EROI slightly (Pradhan et al. 2011). The $EROI_{(refinery)}$ for ethanol from sugar cane lies between 0.8 and 1.7 (Cleveland et al. 1984). Note that the EROI of 10 often quoted for Brazilian ethanol from sugar cane is not an EROI. It only includes the energy from fossil fuels, not the energy from sugar cane by-products used to facilitate its farming and the extraction of ethanol (Macedo et al. 2008).

The low EROI values of agriculture-based biofuels are not surprising. A biofuel's $EROI_{(refinery)}$ can't be higher than the $EROI_{(extracted)}$ (i.e., at the farm gate) of the feedstocks used to produce it. In the United States, the $EROI_{(extracted)}$ for corn and sugar cane is around 4.5, and for soya around 2.5 (figure 12a). In general, the EROIs of agriculture-based biofuels are so low that it is questionable whether they are being used at a net energy gain. In addition, the difference between the EROIs of farm-related biofuels and fossil fuels means that a much larger volume of biofuel must be used when substituting for a volume of fossil fuel for it to be energy neutral.

Could the low EROI of biofuels made from farmed feedstocks and the constraints on supplying those feedstocks both be eliminated by providing either the feedstocks or replacement biosolids from wild ecosystems (e.g., forests)? Not if the biosolids are wood pellets. Their approximate $EROI_{(factory\ gate)}$ is around 2 and they are only carbon neutral when both the carbon from producing and burning them has been re-sequestered in a forest, which is expected to take at least a few decades (Furtula et al. 2017; Sterman et al. 2018).

There are also environmental costs from using biomaterials from wild ecosystems. For example, by 2008, the human-assisted pine beetle outbreak in British Columbia had affected some 145,000 $km^2$ of the sub-boreal forest, killing about 46% (roughly 620 million $m^3$) of the merchantable lodgepole pine timber (BC MFR 2008; Walton 2009). The British Columbian government decided that one method to extract as much economic value as possible from the dead trees was to establish a bioenergy (biosolid and biofuel) industry that would use the dead trees as a feedstock (Hamilton 2009; OBAC 2009). As one member of a community-industry partnership that was supported by the regional government put it, "Bioenergy uses fibre that would otherwise go unused" (OBAC 2008, 10).

Clearly the speaker did not consider the many contributions that dead trees make to the sub-boreal forest ecosystem. For example, they provide organic matter to the forest's nutrient-recycling, soil-regeneration system. They thus contribute to the forest's future soils that will help to sustain the young trees as they grow to form a new forest (Wong 2008).

This under-appreciation of the environmental costs from using wild-supplied organic "waste" as a feedstock or a biosolid leads to a more general observation about waste. Unless the use of biological waste from farming, forestry, industry and humans to create bioenergy is looked at in a comprehensive manner, it can appear more morally commendable, environmentally benign and constraint free than it is. Consider that waste can become a precious resource requiring us to maintain its production, or find a substitute. Two examples illustrate that this can indeed happen.

During the 1990s in New Jersey, a tariff-induced slowdown in the flow of garbage to recycling plants was met with demands for subsidies (Rathje and Murphy 1993b). And in 2012, John Ranta, then mayor of Cache Creek, British Columbia, said, on hearing that Vancouver's garbage would no longer be trucked to Cache Creek's dump, "I would prefer Metro Vancouver revisits its plans to build an incinerator in the Lower Mainland. (The dump) is our main industry. It's like a sawmill town losing its sawmill" (Sinoski 2012).

A different take on the relationship between waste and bioenergy is illustrated by NASA scientists' proposal to fill a possible shortfall in aviation fuels by cultivating salt-tolerant, oil-producing (land or marine) plants in saline wastelands. The scientists suggested that the plants could be grown on a gigantic industrial scale by flooding desert depressions or by using coastal saline environments. In their opinion, this saline resource was either unutilized or under-utilized (Hendricks and Bushnell 2009; Hendricks et al. 2009). To hold this view, they must have ignored the environmental significance of these places. Consider that saline estuaries are prime fish nurseries, and that the dust blown from the Bodele Depression in the Sahara Desert provides the soil fertility required by the Amazon forest (Koren et al. 2010).

These NASA scientists also felt that producing oil from ocean plants, such as marine algae (hopefully not *Caulerpa taxifolia*), had advantages over producing it from land plants. In particular, they suggested growing these ocean plants in the nitrogen-waste-rich waters of the ocean's dead zones, such as those in the Gulf of Mexico. The scientists don't mention that the long-term security of aviation fuel would then depend on maintaining the excess nitrogen in these dead zones. Much of this nitrogen originates on farmers' fields, where it is applied as a fertilizer that was manufactured using fossil fuels. No consideration was given to reducing the nitrogen pollution so that the marine ecosystems could again support a food fishery: food versus fuel. No EROI assessment was provided for their proposals.

New biofuel solutions to fossil fuel energy or emission issues usually lack an EROI which makes it difficult to meaningfully evaluate the proponents' claims. For example, it is currently popular to claim that chemicals and aquatic plants can be used to turn waste or "excess" $CO_2$ emissions into a biofuel. However, if energy-consuming technology, such as lights and heat are used, then the biofuel is likely to have a low EROI. A high quality fuel may be gained and atmospheric $CO_2$ reduced, but the energy shortfall may grow rather than shrink. This type of narrowly focused, simplistic and hope-filled thinking about energy issues is not restricted to bioenergy.

### Nuclear, Hydro and Alternatives.

It is expected that, sometime within the medium term, the growing production of non-conventional fossil fuels will barely be able to fill a growing conventional fuel shortfall. The result will be an energy gap, or crunch. There is also a growing recognition of the high environmental cost from using fossil fuels. Think of climate warming and ocean acidification.

Some suggest that a conventional fuel shortfall, an energy gap and the environmental costs of using fossil fuels could all be easily addressed by using other energy sources. Simply increase the supply of energy being produced by nuclear power, and ramp up energy production from the renewable or inexhaustible non-fossil-fuel, non-biological sources: hydro and the alternative sources such as solar, wind, geothermal, tidal, hydrogen and fusion. The question remains, can these sources provide sufficient energy? Can they provide it at an energy gain, a reasonable financial cost and zero environmental cost?

Focus first on the traditional non-fossil-fuel, non-biological energy sources of nuclear power and hydroelectricity. The ability of nuclear power plants to generate electricity is naturally constrained by the limited supply of nuclear fuels. This could be overcome by using breeder reactors whose waste radiation creates more fuel than is used to sustain the reaction.

But that solution joins the other reasons why nuclear power plants are a highly contentious energy source. The concerns include extremely high cash costs to build and dismantle a plant; the long time it takes to build a plant; uncertainty about long-term nuclear waste disposal; the association with weapons of mass destruction; and public safety concerns (e.g., Chernobyl, Fukushima Daiichi). The validity of the safety concerns is confirmed by the continuing limited (reduced) liability nuclear plant operators and equipment suppliers have secured for nuclear accidents (Daily Times Monitor 2009; American Nuclear Society 2005).

An overview of the impact that these and related constraints have had on the provision of nuclear-generated electricity is provided by its past contribution to global energy production: only 0.9% in 1973, after which it rose to 6.5% in 2004, before declining to 5.9% in 2007, 5.7% in 2010 and 4.8% in 2013 (figure 13) (IEA 2006; IEA 2009; IEA 2012a; IEA 2015). The future of nuclear power will likely be largely determined by government policy and public sentiment about the costs and trade-offs. There is also hope that different radioactive fuels could resolve any concerns.

The future for hydroelectricity is different. The period of rapid increases in the generation of hydroelectricity has passed. For example, there are only a few unused sites suitable for building large dams. Human activity has increased the average erosion rates within the watersheds behind the existing dams, which means that they are silting up more quickly. The expected average energy-producing lifespan of existing dams is decreasing (Syvitski et al. 2005). The changes in the distribution of rainfall resulting from climate warming will also affect the generation of hydroelectricity. There are also significant environmental and social costs associated with the dams used to generate hydroelectricity as discussed in chapters 10 and 14.

The constraints on hydroelectricity sources filling a conventional fuel shortfall and precluding the formation of an energy gap are reflected by the trend in its contribution to global energy production. It rose from 1.8% in 1973 to 2.2% in 2004, 2.2% in 2007, 2.3% in 2010 and 2.4% in 2013 (figure 13). Although the rate of increase in the global supply of hydro is roughly double the increase in global energy demand, hydro's absolute contribution to global energy production is low. This is not expected to change much in the future.

A commonly held belief, especially among Westerners, is that we stand on the threshold of a series of technological breakthroughs that will result in the alternative energy sources, such as solar, wind, geothermal, tidal, hydrogen and fusion, providing us with significant amounts of cheap, clean energy. Not to be forgotten are the expected technological innovations that will increase the efficiency of energy use and facilitate conservation. The occasional news reports of the imminent technological success in one or more of these fields fills Westerners, at least, with the hope that we can each continue to live our current energy-intensive lifestyle far into the future, worry, guilt and responsibility free.

To date, despite the recent rapid growth in the contribution of alternative energy sources to global energy production, their absolute contribution is tiny. It has only risen from 0.1% in 1973, to 0.4% in 2004, 0.7% in 2007, 0.9% in 2010, 1.2% in 2013 and 1.5% in 2015 (figure 13b) (IEA 2017). As illustrated by the steep rise in production from a new source of a fossil fuel, we can expect that, as alternative energy technologies improve and are more widely adopted, a rapid increase in the provision of alternative energy will occur. Once the optimal sites are being exploited and the technologies are beginning to mature, then the growth rate will slow. Maturity will occur when a technology's more intractable constraints come to the fore.

When the growth slows, will Westerners be living in our hoped-for energy-rich, fossil-fuel-energy-free world?

Some believe that, by 2030, fully 100% of the world's energy demand could be supplied by wind (51%), solar (40%), geothermal, tidal and hydroelectric sources (9%) (Jacobson and Delucchi 2009). Our hope for that outcome should be constrained by the following three reality checks.

Firstly, we expect the alternatives to do more than fill the conventional shortfall. They are expected to eventually replace all of the energy supplied by fossil fuels (81.1% of global energy production in 2010) and satisfy a total energy demand that is currently growing at between 2% and 3% per year (figure 13). Secondly, most of the alternatives provide electricity. Even though electricity is a high-quality energy source, there are technical issues substituting it for fossil fuels in some applications, such as transport and portable storage. Thirdly, our hope for an alternative energy technology is often based on optimistic initial reports that don't provide sufficient information for us to evaluate its viability. Among the missing information is an EROI; a realistic assessment of the constraints on the technological developments; a conditional estimate of time needed to resolve the technological difficulties (think hydrogen fuel cells); the demands on other resources; the financial costs of implementation and ongoing use; and possible environmental or social costs (Cleveland et al. 1984).

A greater appreciation for the constraints on the role that nuclear, hydro and alternative energy sources might play in our energy future is gained by assessing their possible contribution within a real-world context. Consider the goal of replacing transportation's dominant source of energy, oil, with a mix of nuclear, hydro and alternative sources. As predicted, making the switch has proved more difficult than we wanted to believe (Dresselhaus and Thomas 2001).

Imagine substituting hydrogen for the roughly 12 million barrels of oil used each day to power surface transport in the United States. Roughly 230,000 tonnes/day of liquid hydrogen will be needed. If this hydrogen is to be generated in the conventional manner, then 400 GW of continuously available electricity would be required. This is roughly twice the United States' existing grid capacity. In order to supply an additional two more grids-worth of electricity, many more electricity generation plants will be required. In addition, if hydrogen is to truly be an alternative fuel, then it should be generated using carbon-free electricity (i.e., not generated using fossil fuels) (Grant 2003).

The discussion about how the extra fossil-fuel-free electricity could be generated has a broader application. With some changes in the numbers, it applies to the extra electricity needed to power any fossil-fuel-free, electricity-using transportation system. Think of the current favourite: battery-based electric vehicles.

One method to supply the 400 GW electricity would be to build 200 more dams the same size as the Hoover Dam. Hydroelectricity currently has an $EROI_{(electricity)}$ of 6–16, so its electricity can be supplied at an energy gain. However, there are few large dam sites left in the United States, and they come with significant environmental and human costs.

Building 100 more 4 GW nuclear power plants would also supply the necessary electricity. Electricity generated in this manner has an $EROI_{(electricity)}$ of around 9 (range 1–17) or 5 (range 3.3–10), depending on data interpretation. Nuclear-generated electricity can, in most instances, be supplied at an energy gain. The full lifecycle greenhouse gas emissions associated with nuclear-generated electricity average 65 $gCO_2e/kWh$ (range 10–130). This is more than the emissions of 15–25 $gCO_2e/kWh$ associated with wind turbines and hydroelectricity, and close to the emissions of around 90 $gCO_2e/kWh$ for photovoltaic electricity (Lenzen 2008; Kubiszewski et al. 2010).

However, there are more significant constraints on the building of a nuclear power plant than its carbon footprint. Among them are the environmental and social constraints, such as the safety and waste concerns discussed above, and the cost. They ensure that finding suitable places willing to accept nuclear plants or their waste will be difficult.

Agriculture-based biofuels could be used to power electricity generators, but their $EROIs_{(refinery)}$ are so low, often <3, that the electricity will be provided for only a small energy gain, if any. In addition, to grow the biofuel feedstock needed to provide the required amount of fuel would appropriate an additional 3% of the United States' land surface, or an equivalent area of existing crop and pasture land (Grant 2003). The use of non-agriculture-based biofuels might allow for the continuing use of fuel-powered transport, but their EROIs and environmental costs have yet to be fully evaluated, as discussed above.

Perhaps we should place our hope, as some suggest, in genetically modifying an aquatic plant's genes so that it generates hydrogen directly. But who knows when that would happen, or what the EROI or the environmental costs would be. Imagine if lights were needed to grow the plant. Or imagine that its genes escaped into the wild. Would we create hydrogen-spewing wetlands?

There are two other alternative energy sources that are both available and potentially capable of generating the large amounts of electricity suitable for powering a hydrogen or equivalent electricity-based transportation system: solar (photovoltaic) and wind systems. To ensure that the EROIs for the electricity generated by wind and photovoltaic methods are representative, their EROIs must apply to systems, not just to individual wind turbines or solar panels. Efficiency-adjusted EROIs are needed to compare the electricity generated by different methods.

If these conditions are satisfied, then wind turbine systems provide electricity that has an $EROI_{(electricity)}$ of 5–32 (average around 20). Photovoltaic systems provide electricity that has an $EROI_{(electricity)}$ of 6–12 (Kubiszewski et al. 2010). In contrast, the electricity generated by conventional oil has a global $EROI_{(electricity)}$ range of 4–11. For coal, the range is 12–25.

From the perspective of the EROIs, wind is currently one of the more energy-rewarding methods of providing electricity. Currently, the energy cost of electricity generated by photovoltaic systems is competitive with electricity generated using conventional oil, but not coal. Electricity production is still dominated by coal.

These EROIs are guides, not definitive. Consider what might happen to the $EROI_{(electricity)}$ for coal-generated electricity if a carbon sequestration system is installed on the generating plant. The energy to run the sequestration system would reduce the plant's energy efficiency by around 10%.

In contrast, when photovoltaic or wind systems generate more than about 20% of the electricity in an electrical grid then, in order to maintain a constant supply, electricity backup or storage systems are required. The energy costs of these additions are not included in the above wind and photovoltaic EROIs. Also excluded are the expected changes in their EROIs resulting from future developments in wind and especially photovoltaic technology, or the likely decline in the average EROI once the less-than-optimal sites are put into production (Raugei et al. 2012).

But these are not the only constraints on using wind and photovoltaic systems as energy sources for a fossil-fuel-free transportation system. Others include the finite number of optimal sites to install these systems, and the technical and financial constraints of providing and maintaining the associated backup and storage systems. There are also environmental costs from the installations themselves.

Consider that a sufficient amount of wind-generated electricity would require about 150,000 km$^2$ (ca. 400 x 400 km) of windy land to be fragmented by wind turbines, access roads and power lines. In contrast, the more tightly spaced photovoltaic systems could supply the required electricity from only 5,000 to 20,000 km$^2$ (<140 x 140 km) of sunny land (Jacobson 2009; Grant 2003). However, its ecosystem costs would include the partial destruction or disruption of any wild ecosystems on which it was build.

There are also likely both local- and global-scale environmental impacts from these installations that we are currently unaware of. For example, the area to be covered by sunlight capturing (albedo changing) photovoltaic systems suggests that they would affect at least the local climate. Before saying that this would be negligible, remember our earlier view that $CO_2$ emissions were a harmless, naturally occurring gas and that the tiny amount of waste CFCs were inert. It also came as a surprise that our urban centres have an impact on their local climate.

Finally, a word about hydrogen fusion (the sun's energy generator) as a future source of energy. After decades of effort, the hope for fusion power rests with the US$19 billion ITER project now under construction. It is scheduled to start testing in 2025. The hope is that it will produce 10 times the energy it consumes to run. That energy will certainly be expensive (Carrington 2016).

In conclusion, nuclear, hydro and all of the currently functioning alternative energy sources are capable of providing electricity at an energy gain. However, the amount each can provide is constrained by features such as fluctuations in the provision of energy, safety issues, environmental costs, and substitution and implementation issues. In order to power the current US surface transportation system with fossil-fuel-free energy, such as hydrogen or electricity, a mix of these sources would be required. This discussion similarly indicates that efforts to fill the growing global conventional fuel shortfall, avoid an energy gap and avoid the environmental costs of using fossil fuels by replacing them would require a mix of energy sources.

## Comprehensive Cures.
We are left to wonder how the switch from a global energy system that receives roughly 80% of its energy from fossil fuels to one dominated by fossil-fuel-free electricity would occur. How would that new system function? An answer is provided by imagining that the United States did decide to change its energy system into one that supports a fossil-fuel-free hydrogen- or battery-powered transportation system.

Their new country-scale energy system would be dominated by a mix of hydro, nuclear and alternative electricity-generating sources. It would be divided into a number of interlinked regional-scale energy systems. At the heart of each would be an electricity distribution grid. Think of the thick cables strung across the landscape between huge pylons and the thinner lines on the poles outside your home.

The new regional-scale grids would be much more complex than the existing grids. Each regional super grid would not only carry more electricity but would receive it from a wider variety of sources with differing supply characteristics. The erratic supply of electricity provided by wind and photovoltaic systems would be of particular concern. To ensure an ongoing, reliable, stable supply of electricity to its domestic, industrial and transportation users, a super grid would have to include backup and storage subsystems. It would also, like the present grid, include a subsystem that would both balance the varying supply with the consumers' varying demand and keep the system stable. But the complexity of the super grid would make this task much more difficult.

One of these regional-scale, fossil-fuel-free, mixed-source energy systems and its super grid could be located in California. Whether California has sufficient non-fossil-fuel energy to build such a system can be estimated by adding up its potential supply of wind, sun and water-related energy sources, and the energy gained from its potential efficiency gains. By applying estimates of implementation timelines to this type of broad-brush accounting, researchers concluded that California's expected energy demand for 2050 could be satisfied by an energy system that was 100% free of biofuel, nuclear power and fossil fuels (Jacobson et al. 2014).

A different picture emerges when the energy sources, the constraints on their supply and the practical aspects of implementing the needed changes are treated as parts of a dynamic, complex energy system (including its super grid). This picture was brought more into focus by researchers who made a comprehensive assessment of a state-wide energy system that would satisfy California's legislated greenhouse gas (GHG) emissions target for 2020: to reduce its GHG emissions to their 1990 level. A supplementary target is to reduce its 2050 emissions to more than 80% below their 1990 level.

The assessment assumed that population and energy use would continue to grow at current rates. No significant, explicit changes in lifestyle were considered. The assessment did consider the building and running of the energy system (Williams et al. 2012).

The assessment showed that California's 80% reduction target could be achieved over a 40-year period. To be successful, California's energy system would have to be transformed into one that made minimum use of fossil fuels. Here are some examples of the primary transformations.

Overall, the demand for petroleum products from California's energy consumers (the end-users) must fall from 45% of the state's energy demand in 2010 to 15% by 2050. In contrast, the end-users' demand for electrical energy must increase from 15% to 55% by 2050 (i.e., their energy demand must become dominated by electricity). In order to achieve this switch, most current, direct uses of fossil fuel by consumers, including transportation, industrial processes and home heating, would be converted to electricity. For example, in 2050, California's transportation needs would be met by a mix of electricity (roughly 70%), biofuels (roughly 20%) and fossil fuels (roughly 10%).

At the same time as the switch to electricity occurred, the amount generated using fossil fuels would be reduced and those plants still using it would be fitted with carbon sequestration systems. The majority of the increased electricity demand would come from (single-source) solar, wind, geothermal and nuclear plants, supplemented by distributed-source rooftop solar panels. Energy conservation (reduction of unnecessary use) would play a role in reducing demand. However, for the switch to be successful, the growth in the demand for energy would have to be significantly reduced by increasing its efficiency of use. Despite this targeted reduction, the total electricity generation capacity would have to double.

This assessment of California's choices provides a reality check to the broad-brush evaluation mentioned above in which the state's *total* energy demand in 2050 could be provided by a fossil-fuel-, nuclear- and biofuel-free electricity-generating energy system: solar (55.5%), wind (35%), geothermal (5%), hydro (3.5%), and ocean (wave plus tidal) (1%) (Jacobson et al. 2014). The assessment being discussed here indicates that, in 2050, the so-called renewable energy sources (wind, solar and geothermal) could supply a maximum of 74% of just the demand for electricity. The remaining minimum of 26% of the electricity would be supplied by a mix of hydro-, nuclear- and fossil-fuel-generated electricity.

In order to assume that 74% of the electricity supply would come from wind, solar and geothermal sources, the assessors had to make some optimistic assumptions. They took full advantage of California's rich endowment of sun and wind. They also assumed that all of the emerging technological innovations to be used in the new energy system would meet their most favourable predicted outcomes, on time, including being successfully commercialized.

They also assumed that the increases in the efficiency of energy use would be substantial: a 1.3% *annual* increase in overall energy-use efficiency. This increase would have to start in 2010 and be sustained for an unprecedented 40 years. This would reduce by half the previously expected business-as-usual demand for electricity in 2050. Without these savings, there would be a heavy reliance on fossil-fuel-using electricity-generating plants fitted with carbon capture and storage systems.

The assessment also noted the practical constraints on making the transition from the current energy system to the new one. California would have to do more than simply change its energy sources and build a super grid. The commissioning of the super grid would alone require a well-coordinated multi-year transition in both the energy sources feeding into it and the demands of the consumers drawing from it.

For example, sufficient energy efficiency gains must be achieved well before the transportation sector starts to run on electricity, otherwise demand might overload the super grid during the transition. Once transportation is electrified, the electric vehicles would have to be integrated into the super grid as storage devices. In this role, they would help supply electricity during peaks in demand and compensate for the intermittent nature of the electricity generated by wind and solar systems. The level of coordination needed for this transition and integration to be successful is unprecedented. It implies a significant, regulatory role for government (Williams et al. 2012). This is not an American favourite and points to the social challenges of implementing such a super grid, even in California.

Once up and running, the super grid must be managed. The most basic management task the grid managers face is, as previously mentioned, to keep the grid stable while balancing input and output. This is a challenge because a super grid is both more dynamic and more complex than a current grid. Consider that a super grid must deal with a wider range of more variable electricity sources and its consumers are active grid participants rather than the passive users of today's grids.

Of most concern is that the super grid's increased dynamic complexity would, on its own, contribute to its instability. You may remember the electrical blackout that darkened much of the northeastern United States and southeastern Canada on 14 August, 2003. It was, in part, caused by the behaviour of a far simpler electricity grid (Perrow 2007). Because planners tend to underestimate the level and significance of a system's complexity in their planning assessments, the increased complexity of California's super grid would likely play a more significant role in determining its stability than currently expected (Gardner and Stern 1996).

There are other consequences from California's favoured future energy system that warrant attention. Consider the potential impacts on food production from farmland being used to provide biofuel feedstocks or space for solar and wind systems. Think of California's sunny, food-producing Central Valley. Wild ecosystems will be similarly affected if they are used to provide space and resources.

Consider too the energy and other resources needed to construct the electrical grid's infrastructure. For example, part of California's plan is to add transmission lines at a rate of 644 km/year. The electricity-carrying part of the lines are made from aluminium. The transport of the mined aluminium concentrate to a smelter will use fossil fuels, while the process of converting it into the aluminium metal needed to make the lines will require a large amount of electrical energy. To meet its power line expansion plans within its power consumption estimates, California would likely have to rely on importing the virtual energy and exporting the virtual environmental costs of producing the aluminium.

This is the time to place into perspective a key assumption, or hope perhaps, of California's plan: that the critical developing technologies will indeed be fully functional and cost effective within the needed time frame. That is, when required to, they will provide the required amounts of the non-fossil-fuel (solar, wind, tidal, biofuel, hydro and nuclear) energy for the chosen energy-source mix; achieve the unprecedented gains in energy efficiency; and attach carbon sequestration systems to the fossil-fuel-powered electricity generators. The real-world constraints on those hopes are highlighted by California's last nuclear plant being scheduled to close in 2025.

The constraints are also illustrated by considering the proposed carbon sequestration system. Although still in its early developmental stage, the capture and storage of the $CO_2$ emissions from fossil fuel use is touted by many as a surefire way to free us from that constraint on fossil fuel use. It is expected to make fossil fuels more environmentally friendly. However, implementing this technology comes with limitations.

The Californian assessment recognized that direct carbon sequestration can only be applied to fixed combustion sources, such as electricity generation plants, not its transportation sector. But it is unclear if they took into account that applying it will use energy. For example, if a sequestration system was attached to a newer, more efficient power-generating station, then its fuel efficiency would decline by an estimated 9% to 15% (McKee 2002). The power plant will have to supply this additional energy by burning extra fuel, which would reduce the EROI of its electricity.

Carbon sequestration faces other limitations. The captured $CO_2$ will have to be stored someplace. Currently, the preferred place is underground. One of the proposed places is in depleted oil or gas deposits, of which there are many in California. However, these deposits are pierced, like a pincushion, full of exploration and production wells, not all of which have been adequately sealed. It is reasonable to expect leaks.

Whatever the storage method, the promise that any stored $CO_2$ would remain in its leak-proof vault backed by infallible monitoring measures and contingency plans for hundreds of years should be rigorously questioned (Nelsen 2015). Questions such as, for how long?, what does "not leak" mean?, what is the contingency?, what would a failure look like? and what are the energy, financial and environmental costs? are unlikely to be satisfactorily answered to a reasonable degree of certainty after only a few years of testing. Like nuclear waste and tailings dams, the problem of storing carbon emissions for the long term in a secure, low-energy-cost, low-financial-cost manner is likely to remain unresolved for a while.

The sequestration example serves a wider purpose. It reminds us that our hopes for it are biased by our tendency to focus on what technology can do when working perfectly. We tend to neglect the conditions that a technology must meet, and stay within, if it is to function as intended. We tend to remember technology's past successes rather than the also-rans and the not-quite successful. This is a reminder that our more general hope that technology will quickly resolve our energy supply issues is often overly optimistic (Dresselhaus and Thomas 2001).

In overview, there is a marked difference between our hope for a global energy system based largely on still-developing alternative energy sources and technologies, and the reality of planning, building and running such a dynamic, complex energy system. As the Californian assessment illustrates, the reality of switching the current global fossil-fuel-dominated energy system to a fossil-fuel-free, electricity-dominated energy system based on a mix of sources would be difficult and slow. The result would be a more complex system that may not even be fossil fuel free. A successful switch would be easier with, or may even depend on, a slowing or decline in the current increasing global energy demand.

We could believe that future technological developments would resolve these shortcomings on a global scale. For example, carbon capture and storage plus geoengineering of the earth's climate will reduce the environmental costs of using fossil fuels to near zero. We could then use them to fill a shortfall with few concerns. But holding this belief won't stop fossil fuel deposits from being depleted and an energy gap from arriving, especially if demand keeps rising.

Similarly, we could believe that future technology will shortly solve the alternative energy supply issue. Think of fusion energy and greatly increased efficiency of use. Either way, as the Californian assessment indicates, we would be putting our energy future in the hands of hope. It is time to consider another solution to the shortfall.

Culture. It seems that filling the energy shortfall and gap expected to arise from declining conventional fuel production and an ongoing increase in demand will be a challenge. We could make the task easier by reducing our absolute energy demand. To evaluate this solution, the discussion has to switch from technology to culture.

The more industrialized a country is, the more energy per capita it consumes. This is reflected in the increased energy intensity of its members' average lifestyle. If an energy shortfall occurs and a gap is predicted, then the members will find it increasingly difficult to maintain their average lifestyle. Within this context, it seems reasonable for each member to proactively consider reducing the energy intensity of their chosen lifestyle.

This is not an option that the members of the more industrialized countries embrace. The reasons are hidden in plain sight. The energy available to a culture both supports and constrains its world view. The constraints arise, in part, because the guidance its world view provides to its members about achieving a sense of identity, belonging and purpose must be meaningful to them. This means that they must have access to sufficient energy to follow their culture's guidance about which activities to partake in, and which goods and services to acquire. A culture's guidance must therefore reflect the amount and form of the energy available to its members.

By following their culture's guidance, each member adopts an energy-supported lifestyle to which they are emotionally and materially attached. But, by doing so, the members also collectively reinforce their culture's world view: the energy they use thus supports their culture's world view. In reinforcing that world view, the members also constrain the guidance it can provide to them about finding their personal and group sense of identity, belonging and purpose, which completes another aspect of the member-culture feedback discussed in chapter 11. This feedback contributes to the members' inertia to changing both their personal and their culture's world view about an acceptable lifestyle choice or goal, and thus to their demand for the energy needed to adopt it. Overall, the energy available to a culture both supports and constrains its world view.

Thus, suggesting to Westerners, as an example, that we could each help to address our culture's predicted energy shortfall by voluntarily reducing our absolute personal energy demand is to ask us to do much more. In order to meaningfully reduce our demand, we have to question those aspects of our sense of self that are part of our culturally supported energy-intensive lifestyle, such as the value we place on material goods and our energy-intense activities. In particular, we are being asked to question those aspects of our culture's world view that support our lifestyle preferences: consumerism, our vision of the future, and our belief in technological innovation and economic prosperity (progress and growth) (Tainter et al. 2003; Hall et al. 1986). We are, in essence, being asked to address the shortfall by changing our personal and our culture's world view.

As discussed in other chapters, these types of adjustments are usually made slowly and with difficulty. The reasons are many and varied, but they are simply expressed by the rules of thumb about personal and cultural change discussed in chapter 6. An example is the limits to the speed at which people and cultures can change their core beliefs without disrupting their personal and their culture's functioning.

Overall, our global energy future is constrained by more than the future availability of energy, the cash and energy costs of producing it, and its quality. It is also constrained by the guidance from our cultures' world views, their members' personal world view and the process of changing those views. Of particular concern are the goals that the members are guided to strive toward and the culturally acceptable energy and environmental costs of their striving (Kerr and Service 2005; Hall et al. 1986; Goldstone 2002). The presence and power of this cultural influence on the members' energy-using decisions is symbolized by their culturally supported choice of a lifestyle, and by their inertia to changing it.

We are helped to personalize the degree to which these cultural and related personal influences will affect our energy future by imagining the following scenario. You are tasked with persuading Californians to undertake a voluntary absolute reduction in their energy demand. To achieve that goal, you decide to simultaneously encourage two interconnected personal adjustments that would change both their personal and their culture's world view.

The first adjustment is to their personal and cultural evaluation of the non-material benefits they gain from the goods, services and activities that the energy intensity of their lifestyle provides to them. There are limits to those benefits. For example, the actual improvement in their sense of happiness from undertaking discretionary air or land travel can be, like placing their hopes for happiness in owning material goods, significantly less than they expect. Improving your happiness is not inherently an energy-intensive activity. If it were so, then our hunter-gatherer ancestors must have been a miserable lot.

The second adjustment would be to their decisions about what to do with the energy savings from a combination of their efforts to increase energy-use efficiency, increase energy conservation, and adopt a less energy-intense search for happier lives. You would encourage them to put any energy savings back, metaphorically speaking, into the ground or, more realistically, leave the equivalent amount in the ground, permanently. This save-and-bury energy-reduction scheme is essential to avoid those energy savings being used by Westerners' virtuous cycle which, instead, promotes energy use.

Do you think you could successfully encourage Californians to adopt these mutually reinforcing changes that would, over time, reduce their absolute energy demand? How do you think the majority would respond to their state's choice of an energy system for 2050 if its implementation explicitly includes these two lifestyle and cultural changes? What if these changes were justified as being essential to reducing the size of California's unprecedented "increased efficiency" goal to a more realistic level? Before answering, consider the influence that Western culture's world view has on its members' energy-related perceptions and decisions.

As discussed above, there are two contrasting end-member views Westerners could take when considering our energy future: the short-term, human-centred view, and the wider fundamental-properties (the environment's) view. The members of Western culture in general, and its neoliberal economists in particular, prefer to consider energy issues from a human-centred viewpoint that is biased toward the short term. We thus tend to believe that your two proposed changes are unnecessary deprivations. When an energy issue arises, so the argument goes, the market will, through its price mechanism, stimulate innovation and new technologies that will solve the energy issues in the most appropriate and inclusive manner (Jaccard 2009).

It is true that markets can stimulate solutions to energy and environmental issues. But those solutions aren't necessarily the most appropriate. Consider this market solution to the conventional oil shortfall: the production of fuels from bottom-of-the-barrel fossil fuel sources. An example is the Athabasca tar sands' moonscape from which syncrude with a low EROI, high GHG emissions and high water usage is being produced. In the context of the wider cultural, environmental and fuel-constrained reality of the human-environment system in which we live, this is an inappropriate solution: it ignores the fundamental-properties view of our energy future.

Consider too, that although the market can guide us to reduce our personal energy use and our energy-related expectations, it cannot, on its own, ensure that the changes will be long-lasting. This is illustrated by the 1974 market-induced oil price shock in the United States. Its immediate result was a decline in oil consumption. But the medium-term outcome was a return to high oil consumption (Keith 2009). The market adjusted our energy use during the shock, but it did not induce a voluntary, lasting reduction to American's pre-oil-price-shock desire for the goods, services and activities provided by their access to cheap energy. It couldn't because American culture's world view supported the opposite. The power of the market (a culture's economy) is tied to the greater culture's world view.

In the case of Western culture, key features of its world view favour technological and economic progress. Thus its guidance ensures that, regardless of whether one takes the fundamental-properties or the human-centred view of our energy future, acceptable solutions to a shortfall invariably exclude a progress-questioning, voluntary reduction in energy demand. This is illustrated by the absence of or limited attention given to a planned/voluntary reduction in energy use within today's models, theories and discussions about our energy future, even those of a scientific nature. For example, the modelled production curve for conventional oil's total production is generally presented with the implicit assumption that, despite a decline in supply, significant future demand will continue. When, in the early 2000s, the debate about peak conventional oil and its impending shortfall turned to solutions, the focus was on developing alternative energy sources to fill it. The idea of reducing the absolute demand by changing cultural or personal goals was seen as a forced decline in living standards, an apocalyptic response, not a voluntary option.

Over time, the debate about the finite supply of fossil fuels was eclipsed by the debate about reducing the emissions of $CO_2$ from fossil fuels in order to avoid climate warming, as expressed by the <2°C warming target. But, even in this context, the idea of an absolute reduction in energy use as a contribution toward that target lies in the background. Among Westerners, at least, it has been pushed there by the unconscious influence of Western culture's belief in progress and technology on its members' decision making. Westerners, on both the left and the right, and among environmentalists and business people alike, have faith that (current and future) technology will solve both the environmental and the supply problems. It will do so by providing a mix of cheap, fossil-fuel-free alternative energy, increased efficiency of energy use and a means to drastically reduce if not eliminate the environmental cost of using fossil fuels (Klein 2014).

Today it is socially and culturally unacceptable to suggest to Westerners that we should consider voluntarily adopting a less-energy-intensive lifestyle and put the energy savings back into the ground. But it is acceptable to suggest that we can address our energy issues by satisfying our lifestyle wants in a less energy-wasteful manner (increase efficiency and conservation)

and by switching to a more benign alternative energy source. We can thus leave "dirty" fossil fuels in the ground.

Through this type of escape hatch, a discussion about voluntarily reducing our absolute energy demand has been pre-empted and transformed into a discussion about how to supply sufficient "clean" or "green" energy to maintain our current way of life (i.e., how to maintain business as usual). At the same time, by holding this view we reinforce the aspects of our culture's world view that support it. When seen in this context, it is unsurprising that California's proposed new energy system includes an unprecedented increase in the efficiency of energy use. It is unsurprising that a reduction in its energy demand through a change in lifestyle is not a key, if any, part of the plan.

Under these conditions, persuading Californians to voluntarily reduce their absolute energy demand by changing their lifestyles and putting their energy savings back into the ground would be a difficult task, even a non-starter. If it were to occur, Californian's would have to somehow be prompted to rediscover, for themselves, a low-energy-intensity way to experience material comfort as well as a non-material sense of personal meaning, purpose and belonging (i.e., a low-energy-intensity way of being happy). (This option will be discussed further in chapter 18.) By choosing to live such a low-energy-intensity lifestyle, they would be acting as symbols for a truly alternate energy future.

In overview, Westerners, regardless of our ideological capitalist or environmentalist leanings, have found little reason to switch from thinking about an expected energy shortfall as an energy supply issue, to treating it as a human demand issue. We have found little reason to switch from thinking about an actual shortage as a difficulty to be overcome, to instead treating it as a reminder of the earthly constraints within which we should strive to live. Westerners inherently prefer to address our energy issues, and our energy-related environmental issues, by applying technology in a way that will maintain business as usual. We prefer not to face and adjust those aspects of our personal and cultural world views that are behind our demand for energy and thus behind its environmental impacts.

The challenge Westerners face if we wish to reduce our demand for energy is not unique to us. It is faced by the members of all industrialized and industrializing countries. These cultures' world views and their members' response to their culture's guidance will significantly influence our global energy future. It seems that a change in our absolute global demand for energy is more likely to be imposed than voluntary.

**Energy Summary.** A summary of this and the previous two sections also serves to outline a likely energy future. There are three fossil fuels: oil, natural gas and coal. They provide some 80% of our global energy supply. Almost all of the oil and gas extracted to date has been conventional (i.e., easy to extract by drilling and pumping). These conventional fuels are financially cheap to extract and have relatively high EROIs. They are rewarding, both financially and energy-wise.

There is strong evidence that the supply of conventional oil is in its peak zone and that the supply of conventional natural gas will likely do so between 2020 and 2030. The production of coal is expected to peak after 2045. A supply shortfall is beginning for conventional oil and is expected shortly for conventional natural gas. If the growth trend in demand continues this shortfall will grow rapidly.

Non-conventional fuels, such as tar sand oil and shale gas, can be used to fill the conventional shortfall. However, providing non-conventional fuels is constrained by their properties, the geological characteristics of their host deposits and the amount of energy used to extract them. These constraints are reflected in the characteristics of non-conventional fuel production: their lower EROIs, higher financial cost, slower production rates, and faster EROI and production declines than conventional fuels. By not fully taking these constraints into account when evaluating the initial production of non-conventional fuels from a new source type, we develop an overly optimistic outlook for their future production. We tend to overestimate the proportion of the large amounts of the technologically recoverable non-conventional fossil fuels that will actually be produced at an energy gain.

When considering the use of non-conventional fuels to fill the shortfall, we forget that it is discussed in terms of volume (Gb and Gboe) but affects us as an energy shortfall. Because the EROIs of non-conventional fuels are currently lower than conventional fuels, a greater volume of non-conventional fuel must be produced to fill the conventional fuel's energy shortfall. Collectively, these production-related characteristics of non-conventional fuels ensure that, on the present trend, the conventional fuel shortfall will turn into an energy gap much sooner than our optimistic outlook suggests.

However, there is also an expectation that the shortfall will be filled, the appearance of a gap precluded and the environmental costs of using fossil fuels eliminated by replacing fossil fuels with non-fossil-fuel energy. That energy is expected to be provided from nuclear, hydro, bioenergy and alternative (solar, wind, geothermal, tidal, hydrogen, fusion, etc.) sources. The energy from all of these sources, except most biofuels, is supplied at an energy gain. However, their EROIs are usually at or below the EROIs of currently produced conventional fuels, so more of their energy is needed to fill the shortfall. In addition, nuclear fuels are finite and not naturally renewable, while geothermal sources are not renewable but essentially infinite. The other sources are renewable, so their supply is considered to be infinite, but only within the constraints on their process of renewal.

The amount of energy each non-fossil-fuel source can supply is also constrained in other ways. It is limited by features such as fluctuations in energy production, production rates, technical limitations, safety issues, environmental costs, and implementation and substitution issues. For example, the potential to generate electricity using solar and wind systems is unevenly distributed around the world, its generation from these sources is intermittent and using it as electricity to power transportation is difficult.

Our expectation is that these supply and implementation issues will be overcome by a combination of transforming our global energy system and developing new technologies. Our global energy system is expected to become a mix of non-fossil-fuel sources, most of which would produce electricity. Each country's future energy system would be built around one or more electricity-distributing super grids.

An assessment of California's options for implementing such a fossil-fuel-free, electricity-dominated, mixed-source energy system provides a window into that global energy future. The assessment concluded that, by 2050, California's current energy system could be transformed into one that satisfies their increased energy demand and reduces its carbon emissions to more than 80% below its 1990 level. But fossil fuels would still be used.

This conclusion relied on the most optimistic assumptions about the favoured system. In particular, maximum use would be made of California's abundant non-fossil-fuel energy sources, especially sun and wind. And all of the required "in-progress" technological developments would be completed on time and be commercially viable. This includes unprecedented ongoing improvements in energy-use efficiency. It is assumed that the challenges of implementing, managing and balancing the super grid would also be resolved.

The essence of the Californian assessment indicates that there are substantial energy, financial and technological difficulties hindering the transformation of the current fossil-fuel-dominated global energy system into a mixed-source, non-fossil-fuel, electricity-dominated energy system. It is not that easy to replace the 80% of the current system's energy supplied by fossil fuels with non-fossil-fuel energy. We can thus reasonably expect that our expected future energy system will, at least into the medium-term future, incompletely meet our expectations: to fill a conventional-fuel-energy shortfall, avoid an energy gap and eliminate the environmental costs from fossil fuel emissions.

We can hope that these transformation difficulties will shortly by overcome by future technological developments. They are expected to either boost alternative energy supplies by, for example, producing energy through fusion, or reduce the impacts of fossil fuel's carbon emissions by, for example, capturing and storing those emissions, geoengineering the atmosphere, or both. However, our energy future could be more effectively secured if we left less of it in the hands of hope.

The industrialized and industrializing countries have the highest per capita rates of energy use. A decline in the global energy supply would significantly affect their members' lives. Their members are also being affected by the environmental impacts from fossil fuel emissions. With the Californian assessment in mind, it would be reasonable to expect that the members of these countries would reduce their heavy reliance on hope by broadening their efforts to address energy-related issues beyond increasing the energy supply. We could expect them to consider voluntarily reducing their absolute energy demand to a more easily satisfied level.

However, the members of industrialized and industrializing countries struggle to consider and implement such a voluntary reduction. After all, they would have to question and change their present energy-use choices, as symbolized by their culturally supported energy-intense lifestyles. This would mean questioning and changing both their personal and cultural values and beliefs that support those choices.

Doing so is a difficult task because the aspects to be changed are part of both a member's personal and their culture's world view. They are thus included in the member-culture feedback, which favours avoidance

and inertia to change: it sustains business as usual. The task is also difficult because of our human decision making foibles and the characteristics of change. Consider that we give limited consideration to the earth's fundamental, medium- to long-term constraints on our energy supply because we are distracted by our preferential focus on the short-term energy events that have an immediate impact on our day-to-day material lives, and our sense of self and belonging. Compare, for example, our response to the shorter-term, human-induced fluctuations in the price and availability of fossil fuels with our reaction to a prediction of an imminent longer-term decline in conventional oil production. Overall, there is little personal or cultural incentive for the members of industrial and industrializing cultures to voluntarily reduce the energy intensity of their lifestyles. The members instead prefer to treat an energy shortfall as a *supply* issue resolvable by technology, rather than a *demand* issue alleviated by voluntarily reducing their absolute energy demand.

Our energy future will be determined by more than the potentially available energy supplies. It will also be determined by the technological constraints, the financial burden and the energy costs of providing it. Critically, our energy future will also be determined by the energy-affecting decisions we make under the guidance of our cultures' world views about how to think about energy issues: its supply, our demand for it, and the environmental costs of producing and using it.

In 1974, Rose wrote, "Thirty years from now energy will still be a serious topic; only the details will change" (Rose 1974, 21). Today, it is easy to repeat this prediction, with confidence. We can only hope that 30 years from now our energy-affecting decisions will be made with the increased awareness that our energy future will arrive within a dynamic, complex human-environment system.

## Comprehensive Review of Resources

The discussions about resources (including services) in this and the previous chapter have provided information about the possible future for individual resources. They also point to common features affecting our resource future. Critically, if our population and demand for water, food and fuel continue on their current multi-year growth-rate trends in the 1% and 3% per year range, then there would be a doubling of our population and resource demands within 70 years and 24 years respectively. In a finite world, these growth rates cannot continue for long before shortages arise.

But we and the resources we want do not exist in isolation. They are part of the dynamic, complex human-environment system. If we wish to more fully appreciate our possible resource futures, then that is the context in which we should contemplate them.

Unfortunately, our efforts to do so are significantly constrained by our limited ability to picture such a complex system. Try to imagine how, in the future, enough food will be produced to feed a growing population during a time when the amount and distribution of water is changing as a result of ongoing climate warming. After some thought, consider how industrial agriculture's efforts to produce that food would be affected by the decline in supplies of cheap conventional oil and natural gas.

Our task is made even more difficult by our limited ability to recognize the constraints on our own perceptions of our world and the biases in our thinking about it. For example, we routinely tend to focus on those aspects of a resource that we know something about and that we consider to be of immediate relevance to our personal, day-to-day lives. When we do recognize the wider connections among resources and our demand for them, we still tend to focus on the simpler relationships and their short-term changes. We often miss the more basic, often subtle, underlying interconnections and longer-term influences on our resource future, and our day-to-day contribution to it.

Overall, our "intuitive" assessments of both our personal and our global resource future reflect our fragmentary, biased and incomplete appreciation of ourselves and the dynamic functioning of the complex human-environment system we participate in. We tend to grasp, and hold onto, simplistic predictions about that future that match both the guidance from our culture's world view about a desirable future and our innate, short-term, usually optimistic outlook. We do so with little regard for whether those predictions reliably reflect the external reality view or its constraints on the supply of resources. Thus our ability to alter the earth in order to secure the resources we want is far ahead of our ability to assess the consequences that our earth-altering actions will have on the supply of those resources, and therefore us, in the future.

For example, Westerners can honestly believe that the issues of climate warming, peak conventional oil, and long-term food security can all be simultaneously resolved by developing non-fossil-fuel energy sources. We can also truly believe that these goals can be reached without significantly changing our lifestyle, our culture's world view and our environment's functioning, even as the global population grows. No wonder we are left wondering why in heaven the longer-term results of our resource-affecting decisions routinely incompletely match the desired outcome we predicted.

This section attempts to address these limitations in our wider thinking by providing a coherent overview of our resource future. It starts by discussing resource-focused global-scale models of our dynamic, complex human-environment system and their predictions of how our resource future might arrive. Their coherent, resource-focused predictions are then evaluated to see if they are consistent with our human characteristics. Of particular concern is the influence that our culture's world view, our reality paradox and our decision-making foibles have on our resource-related views and decisions. It is then possible to outline a general path to our resource future. The section ends with a discussion about the option of choosing a different future.

**Models.** We live our lives as participants in a dynamic, complex human-environment system, as discussed in chapter 11. By using first principles and real-world data, we can build simplified but coherent and integrated global-scale mathematical models of that system (often called global systems models). In order to appropriately use the results from these models, we need to appreciate their limitations.

The most basic limitation in a mathematical global systems model is the difficulty of reliably representing the human-environment system's complexity and dynamics, as discussed in chapters 4 and 5. In particular, so far we have been unable to meaningfully represent either the processes by which we make decisions, as discussed in chapter 2, or the processes by which we personally and culturally undergo change, as discussed in chapter 6. The models thus focus on the system's more tangible aspects, such as resources, while using scenarios to represent our choices and changes.

Mathematical global-systems models are also limited by the methods used to implement them. Today, that invariably means as a program on a computer. There are computer-programming-related limitations to comprehensively implementing calculations that represent the wide range of time and space over which our complex human-environment system functions (i.e., its dynamics). Compromises and simplifications are required if the program is to run and its output is to be intelligible.

We must also make simplifications in order to represent the interactions between humans and the earth's physical sub-systems. As mentioned, human decision making about the physical world is represented by dividing our possible choices/changes into simple scenarios. The influence that each scenario has on the model's results can then be evaluated and compared (de Vries 2007). This is a significant limitation because each scenario represents a fixed average choice/change, whereas we continually change, although slowly. When, how and in which direction is lost when using scenarios. Consider too the assumptions we must make if we are to represent some poorly known parameters, such as the amounts of recoverable fossil fuel resources.

Overall, these limitations ensure that the models of the human-environment system are incapable of making accurate and detailed predictions. However, they can provide us with a coherent and consistent framework within which to evaluate their conditional predictions (the model's results) about our resource future. But to benefit from that framework, we must keep the model's suite of limitations in the back of our minds when evaluating the results.

The most well-known global-scale, integrated mathematical model of our human-environment system is Limits to Growth. The primary objective of this relatively simple 1970s model was to examine the outcome from the interactions between the basic components (resources, population, etc.) of a finite world. To accommodate the range of our possible resource-affecting decisions, the Limits to Growth model used three scenarios: the stabilized world; the standard run (business as usual); and an emphasis on technology. Its results were first published in 1972 (Meadows et al. 1972).

Recently, the Limits to Growth resource trends published in 1974 were retroactively compared to the current trends calculated by the model when it was supplied with the observed resource data for the period 1970 to 2000 (Turner 2008). That comparison is used here to illustrate the value and constraints associated with using integrated, global-scale models to determine our resource future, within the context of the human-environment system. The 30-year retrospective review confirmed that the Limits to Growth model was internally consistent. It also showed that the resource trends for population, industrial output per capita and food production per capita calculated in the early 1970s for the business-as-usual scenario reasonably matched the trends based on the observed data for 1970 to 2000 (figure 20).

The review also compared the model's early 1970s depletion trends for non-renewable resources (represented by fossil fuels) to two fossil fuel trends calculated by the model using the reviewer's two more recent estimates of the total fossil fuel ultimate: a coal-biased high (75,000 Gboe); and a low (12,000 Gboe, ±30%). The fossil fuel depletion trend based on the reviewer's coal-biased high fossil fuel estimate matches the early 1970s trend based on the comprehensive technology scenario. The trend based on the reviewer's low fossil fuel estimate matches the early 1970s trend based on the business-as-usual scenario.

As discussed above, coal-biased estimates of the earth's fossil fuel ultimate are often too high. In contrast, the reviewer's low estimate (12,000 Gboe) falls between the two fossil fuel ultimates made in chapter 13: the more reliable 8,100 Gboe and the less reliable 17,700 Gboe. These two estimates were used in chapter 13 to discuss our future fossil-fuel-related $CO_2$ emissions and climate change. The Limits to Growth business-as-usual scenario estimate of our non-renewable resource future is thus compatible with that discussion.

The results from the recent review of the Limits to Growth model indicates that its business-as-usual scenario is a valid representation of our resource future. It suggests a future in which a continuing rise in population is matched by an increasing production of the food, energy and goods we want, up to a point. When this point is reached, the model indicates a rapid decline in resource supplies, and thus in the population (Turner 2008). This ending is described as "overshoot and collapse." It is the type of path followed by a finite natural system that includes an amplifying feedback, as illustrated by the change in a bacteria's population when grown in a closed container with no predators and a fixed amount of food. This future also seems to fit the human population history of some Pacific islands, as described by archaeological studies (Kirch 1997).

However, like all models, the Limits to Growth model is incomplete, and not all of its assumptions and data are correct. Some of its deficiencies are specific, such as not considering climate change in its food estimate. Other deficiencies are more generic. For example, when modelling a complex system such as the human-environment system, it is difficult to determine and therefore model the exact mix of real-world conditions that cause a turning point in current trends, especially if the change happens quickly. Thus, a model's prediction of the timing and nature of a collapse are, like that for an abrupt climate event, highly uncertain at best.

As mentioned, a model represents our decisions through fixed scenarios. But human change is an ongoing process that takes place in a number of ways, such as drift and conscious decisions. The path we will actually follow into the future can switch between scenarios, thus changing the relevant prediction about the timing of events. When we might switch is better determined by an appreciation of human nature than a resource-focused mathematical model.

Thus the value of the Limits to Growth model and its recent review lies not in its predicted timing of our resource future, which is likely incorrect. Its value lies in its insight into how the human-environment system functions when viewed from a resource perspective. That provides us with a framework, a planning tool, we can use to consider our resource future. In this context, perhaps the most useful conclusion from the Limits to Growth model and its review is that we are indeed following a business-as-usual path into the future. It also indicates that, in a finite world, slavishly following this path leads, in the resource view, to overshoot and collapse (Turner 2008). We are prompted to focus on why we are following a business-as-usual trend.

There are a number of recent, more sophisticated integrated global systems models. They include DICE, GUMBO, TARGETS and IMAGE2. They can be used to test and expand on the Limits to Growth conclusions (de Vries 2007; Costanza et al. 2007). With the exception of DICE, these models, like Limits to Growth, assume that resources are finite. Because discussions earlier in this and in previous chapters (e.g., 7, 12 and 14) show that, at the rate and in the manner we are using most natural resources, they are indeed finite in a practical sense, the DICE model will not be considered further.

The GUMBO model dynamically links all aspects of our biosphere with our other activities and includes some degree of substitution between resources. It also attempts to assess the values of ecosystem services. Its initial conditions are four end-member scenarios whose values are chosen around a neutral (business-as-usual) case. The preliminary output from GUMBO suggests that all scenarios end in overshoot and collapse. The particular variables pushing the result to follow that path are those we associate with the functioning of complex cultures, such as energy availability and population size.

A comparison of the results from GUMBO's different scenarios suggests that, if we desire maximum social welfare in our future, then we should adopt policies that are skeptical of technology. Specifically, rather than concentrating on promoting consumption and human-built capital, we should focus on knowledge, social capital and maintaining natural capital (Boumans et al. 2002).

The TARGETS model takes a different approach. It allows the user to choose the mix of variables they think represent the preferred ("utopian") choices for an uncertain future and then evaluate the results. This method is the precursor to the IPCC's early climate scenarios.

On the surface, it seems that these more sophisticated models have "resolved" some of our decision-making dilemmas about resource use. But models can only provide advice expressed as possible future outcomes. None are capable of determining how realistic are the choices of input-scenarios, the validity of the models' other assumptions, the interpretations of the output or our reaction to them (de Vries 2007; Costanza et al. 2007). Making those decisions is our responsibility.

In overview, it is up to us to decide if the models and their predictions are sufficiently useful for discussing our resource future. It is our decision whether maximum social welfare is in fact desirable. We decide if our reactions to theoretical outcomes are appropriate and if a particular future is nearer to Eden or Armageddon.

To appropriately make these decisions requires that we personally have a more practical understanding of ourselves and our human-environment system. In particular, we need to have a deeper appreciation for human behaviour, how cultures and ecosystems function and change, the scientific process, how we determine truth, our reality paradox, the role of culture in our lives, and the influence its world view has on our decisions. These topics, the subjects of previous chapters, were summarised in chapter 11 as a qualitative model of the human-environment system.

By applying that practical knowledge to the mathematical models' output, we can direct our attention away from accuracy concern and toward validity and usefulness. When viewed in this manner, the framework the models provide gives us insight into how a possible resource future might arrive, the kinds of decisions leading to that future, and where other decisions might lead. The following discussion about our predicted resource future is divided into two parts: one is business as usual, the other is overshoot and collapse.

**Business as Usual.** A retrospective review of the updated Limits to Growth study concluded that, on a global scale, we have indeed been following its business-as-usual resource-use scenario into the future. Is there a plausible, real-life explanation for why we have maintained such a track? Are we likely to continue following it?

The historical global trend in population growth is largely responsible for the increasing trend in the demand for water and food. There is considerable inertia to reducing the population growth rate. After all, raising children is one of those culturally supported milestone stages in a human lifespan. Even if the number of children being raised per family were to decline significantly today, inertia ensures that the global trend in population growth would only start to change significantly a generation (roughly 25 years) later.

The dominant reason for the historical and ongoing steady increase in the atmosphere's $CO_2$ content is our ongoing, increasing combustion of fossil fuels (figure 17) (Jackson et al. 2015; Raupach et al. 2007). This increasing use is the result of more than business-as-usual population growth. It is mostly driven by an ongoing focus on industrialization, begun in the 1800s, and a heightened focus on economic development. Industrialization also requires an increase in other resources, such as metals. Barring something unexpected, the influences of population growth and the imperative of industrialization/economic development suggest that current resource trends will continue into at least the mid-term future.

The nature of our humanity provides an explanation for why we will likely try to follow the business-as-usual resource trend even further into the future. We are creatures of habit. For example, the best predictor of your personal future behaviour is your past behaviour, not your stated intentions

Consider too the relationship between us, as individual members of a culture, and our culture's world view, as discussed in chapter 11. When making decisions, we tend to follow the guidance from our culture's world view. But we are also the carriers of our culture: it is our collective decisions that create and adjust our culture's world view. We shape the cultural world view that both provides us with our vision of a desirable future and guides our decision making. This member-culture feedback, together with our being creatures of habit, helps to stabilize our lives and thus our culture's world view (figure 15).

But it also ensures that the core of both our personal and our culture's world view display significant inertia to change. They both usually change slowly, through many small adjustments. More rapid, lasting, larger change may occur, but only when we are subject to unusual and powerful influences that induce us to question our culture's world view or expend the effort needed to confront our personal reality paradox. This can be a painful affair.

Collectively, these characteristics ensure that our culture's guidance and our decisions are more likely to favour a business-as-usual trend. But we are left with a conundrum. How can the presence of this personal and cultural inertia to change be reconciled with the considerable value we place on human consciousness and our ability to innovate and adapt? How can we be so staid and so flexible at the same time?

An answer is that the cultural world view with which we are indoctrinated as we grow up does more than guide us. It provides us with boundaries for our thinking and behaviours. We are prompted to stay within these boundaries by external social pressure to conform and by our internal desire to belong (Kahan et al 2012). Experience also teaches us that those who hold views and behave in ways that mark them as outsiders have a harder life. It is easier for us to live within the boundaries set by our culture's world view. For these reasons, we tend not to express our flexibility if it takes us too far outside of them.

The subtle but powerful nature of this influence is illustrated by Westerners' use of the word *adapt*. When Westerners describe how we should or will adapt to a changing resource future or a resource shortfall, it implies a specific kind of response. It indicates our intent to make and support those innovative changes to the way we conduct business, apply technology or live our lives that will maintain the essence of our current lifestyles and beliefs. Thus *adapt* invariably indicates our need or intent to change how we will alter the world around us in order to maintain the status quo, rather than how we will change the status quo to match a new external reality. *Adapt* rarely indicates our intention to change the essence of how we live and what we believe. That is usually the outcome, but not the intent.

For example, Westerners will discuss innovations that could reduce a resource shortage, such as copper or energy, even if implementing it will significantly affect our environment's functioning. Consider our ongoing efforts to make non-conventional fossil fuels available to fill an energy shortfall. Consider too our serious contemplation of applying climate-altering geoengineering technology, such as mirrors in space and sulphate in the sky, to mitigate the consequences of using those fuels. It is acceptable to consider these technological innovations because they are compatible with the Western world view's boundary-setting guidance about progress, growth, individualism, innovation and technology. In contrast, we avoid discussing innovative solutions to a copper or energy shortage if implementing them will require us to significantly change the essence of our culturally supported lifestyle or our culture's world view. We will avoid contemplating those types of innovations even though they may better match the functioning of the external reality.

By extolling the virtues of innovation and progress, and by using words that ooze flexibility, like *adapt*, Westerners appear to be contemplating radical and forward-thinking efforts to address the issues of future resource supplies, population growth and environmental disruption. And we appear to be addressing these issues in the most appropriate and comprehensive manner possible. But, in reality, the primary purpose of these solutions is to keep us within our personal and our culture's comfort zone.

In overview, our business-as-usual demand for resources is maintained by more than population growth, industrialization, and our inertia to change. It is also maintained by our culturally supported preference for changes that support the status quo. The presence and consequences of these influences and biases can be seen in all cultures. Thus, it is unsurprising that, on a global scale, we are following and will likely continue to follow a slowly changing business-as-usual track into the future.

## Overshoot and Collapse.
The Limits to Growth and other models suggest that maintaining the business-as-usual track into the future will eventually result in us supplying amounts of resources that are unsustainable. The supply will then be in overshoot. Eventually the amount being supplied will undergo an unavoidable rapid decline. The supply will collapse.

This outcome is a reasonable ending to expect for a system within which resource supplies are finite, in a practical sense, and where demand is increasing, as expected for a business-as-usual imperative that includes growth. However, for these modelled futures to be taken seriously (treated as likely), there must be a plausible mechanism to explain, within the real-world human-environment system, how overshoot could occur and how our efforts to pre-empt collapse would be unsuccessful.

By following business as usual in a system with finite resources, a shortage of a resource will eventually

occur. It can be dealt with by applying technology. Successive shortages can be overcome by repeatedly applying technology.

Technology can also be used to deal with our disruption of resource-renewing functions, such as the pollination of food plants, waste processing and the water cycle. But all technologies require resources, such as metals and energy, to function. Consider that as finite resources become more difficult to provide, more energy must be used to provide them.

For example, as the EROI of energy production declines and its deposits are depleted, a greater proportion of energy must be used to provide it. The price for energy rises. To pay for it and support an increasing demand from a growing population while maintaining current lifestyles, the economy must grow, which means more energy must be provided, as discussed in chapter 12. This technology–resource-demand feedback joins with a growing population to result in an increasing demand for energy and the imperative to provide it. This kind of feedback lies behind overshoot. Heightened efforts to increase the supply simply speed up the arrival of the point at which the feedback can't be sustained: the supply of energy and thus the mechanism providing it collapses.

There is one other feature to consider. We do not use resources in isolation. In particular, the technology used to provide resources requires a variety of resources to function. This ties the supply of each resource tightly to the supply of the other resources. Because of these links, it is possible for a shortage in one or a few resources to directly or indirectly disrupt the aspects of the human-environment system that provide the other resources. Think of the link between energy and all resources. Under certain conditions, the supply system could fail and a general collapse in the supply of resources could occur. The varying levels to which today's interconnected supplies of wild food, water, farmed food and energy, but not metals, are stressed, as discussed in this and the previous chapter, indicate that such synchronous shortages are possible.

But, the futures the models portray for us are not inevitable, just possible. Missing from the models and the above often-used, inevitable-sounding, resource-focused explanation for a future collapse of resource supplies is human decision making. Specifically, why wouldn't the members of at least the largest per capita users of resources, the industrial cultures, routinely recognize, on a personal level, the early signs of overshoot and react appropriately? After all, they have the required knowledge and skill, and the most to lose.

The question is answered by continuing the above discussion that outlines how Westerners come to make decisions that favour a business-as-usual future. The guidance from our world view favours more than our preferred meaning of *adapt*. It favours excluding a voluntary reduction in resource use from a list of possible acceptable solution to issues, as described in the energy section. It also favours the use of the saving from increased efficiency being used to increase resource usage, not reduce demand, as discussed in chapter 12 and 14.

In doing so, Westerners' efforts to protect the functioning of the environment are biased, often subtly, in ways that favour maintaining business as usual. Our protection efforts are formalized in what we consider to be science-based environmental protection and resource extraction evaluation processes (protection-extraction processes), such as environmental assessments and environmental management planning. Although these processes provide some protection, they only do so to a point.

These protection-extraction processes are, like Westerners' use of *adapt* and *efficiency*, subtly biased in favour of Western culture's imperatives of progress, growth, individualism and technology. The inescapable presence of this bias results in the design of the protection-extraction processes being, at least unconsciously, tilted toward the drawing of conclusions that will maintain the immediate supply of resources: they are tilted toward the default of business as usual. Their presence simultaneously tilts the processes away from recognizing and seriously considering information or circumstances that may prompt us to make decisions which don't favour resource extraction.

The tilt away is illustrated by aspects of the Canadian federal and provincial resource protection-extraction processes. Within them, decisions are made based on demonstrations of "significant" adverse effects rather than the precautionary principle. The processes thus display a willingness to accept mitigation (reduce and repair) over avoidance (retain existing environmental and social conditions) as a means to deal with environmental and social impacts. Management plans similarly favour resource extraction (Bavington 2005; Bavington and Sajay 2003). Think fish.

In addition, even the best of protection-extraction processes can be, and are, changed when they are perceived to be a barrier to progress and growth. For example, Canada's better-than-many protection-extraction processes were watered down and their tilt away increased by the former Harper government. Think of the Athabasca tar sands moonscape and the proposal for its syncrude to be shipped overseas via the proposed Northern Gateway pipeline and tankers. Think of "Get to Yes."

Westerners, as individuals, help to sustain this bias. For example, most members view the protection-extraction processes through the lens of our culture's guidance. We are thus not inclined to find reasons to question the processes. After all, providing the resources demanded by our lifestyle helps to sustain our culturally influenced sense of self and our identity as a cultural group. We also see ourselves and the cultural subgroups we belong to as competent environmental managers rather than participants in an environment whose functioning constrains how we can live.

Our opinion might change if we paid more attention to the indicators of declining resource supplies and the disruption of the environmental functions providing them. But the indicators reflecting this disruption, such as soil erosion, are often obscure and slow to change, which helps to mask them from our already culturally biased awareness and attention. The indicators are further masked by the attention-getting, erratic short-term changes in resource supplies caused by war, the economic cycle and drought, to name a few. They, along with the flood of mundane events, decisions and social concerns that fill our day-to-day lives crowd out from our awareness and attention the longer-term indicators of resource declines.

Consider too that human decision making deals with emotions and imaginings, as well as facts, about the external reality. Unsurprisingly, Westerners, like all humans, prefer to avoid unpleasant facts and feelings. We therefore tend to avoid contemplating an event or making decisions about it when doing so arouses unpleasant feelings or requires us to make undesirable changes to our routines. Think of our response to cognitive dissonance. Think too of an unanticipated disruption to our lifestyle caused by either a shortage of resources or our need to deal with a resource-related issue.

In the light of both this personal and the cultural pressure on Westerners' decision making, it is unsurprising that, when faced with a current or future resource shortage, our response is biased. We prefer to focus on those simplified solutions that will provide us with both quick relief from a shortage and reassuring explanations for it, especially if the blame is placed on external forces. Think of our response to the oil price shock of the early 1970s. In contrast, we tend not to consider explanations and solutions that require us to question our personal, culturally supported resource demands and environment-disrupting behaviours that may have contributed, or be contributing, to a shortage. Westerners, like the members of other complex cultures, tend to favour solutions to resource shortages that maintain business as usual.

In this context, it is unsurprising that our response to a resource shortage can drive its supply into overshoot and collapse. Consider the participants in the debates about declining ocean fisheries off the coasts of Newfoundland, California and Peru, and in the North Sea. They squabbled over rights, profits, jobs, management, foreigners, control, technology deployment and who the responsible party was. Their desire to catch fish remained unquestioned. All the while, the fisheries resource continued to collapse around them (Bavington 2005; McEvoy 1986).

All of these material and mental pieces are linked together by the descriptive model of the human-environment system outlined in chapter 11. It also provides a consistent, coherent overall explanation and mechanism for overshoot and collapse. The characteristics of its three principle factors (us as individuals, our culture's world view and our environment) and the feedbacks created by the interactions among them are, collectively, that mechanism. The model provides a unified explanation for the pervasive and powerful personal and cultural pressure on Westerners to favour business as usual. It also explains the complementary pressure on us to refrain from considering or undertaking personal and cultural changes that detract from business as usual. It includes the nature of the environmental constraints on resource supplies and our relationship to them. It can coherently explain, from a human perspective, the observation that humans are following a business-as-usual trend that leads to resource and population overshoot and collapse. It also allows for the following description of that future.

If global industrial development and population growth continue along the 1% to 3% business-as-usual growth trends, then the resulting doubling of resource demand in 70 to 24 years, respectively, ensures that resource shortages will eventually appear. A common response will be to increase our efforts to maintain the supply.

Our success will be accompanied by steadily increasing financial, resource and environmental costs. Our social and environmental responsibilities will also increase. We will therefore find it increasingly difficult to supply the amounts of resources that will both pay for the increasing costs of providing them and satisfy our increasing demands.

For some resources, the resulting supply issues and associated higher prices will create dissatisfaction among the public. They and their government will become stressed. We will likely blame our emerging circumstance on misdirected government spending, poor management, corruption, too many other people, chance, red tape, the lack of investment in technology, the rich and, perhaps, migrants.

These concerns will be true to varying degrees. But we will fail to see that further behind this outcome will be a combination of our human decision-making foibles, personal reality paradox, cultural world views and the member-culture feedback. These underlying factors will have collectively preconditioned us to (personally and collectively) make those past and present decisions that favour business as usual. They will also have helped to mask from our awareness the information suggesting that a reduction in demand is needed. And when we do become aware of that knowledge, these factors will distract us from paying appropriate attention to it.

As a result, we will make decisions that create and sustain too high an ongoing demand for resources even when our resource-supplying efforts indicate that we are approaching overshoot. As that state of affairs arrives, we will still continue, even heighten, our efforts to increase resource supplies, such as weaken the environmental protection part of our protection-extraction processes. These types of decisions will be justified as necessary to maintain our way of life; to adapt, in the Western sense, to the shortages; and to address the public's dissatisfaction.

The masking and distracting will be sufficiently effective that there will be a significant delay between us entering a state of overshoot and finally accepting that we must voluntarily reduce our demand. By the time this realization dawns on us, the inertia in the human-environment system will likely be carrying us inexorably forward to collapse. During this time, the sources of some resources will either be nearing depletion or the environmental functions that renew them will have limited capacity to do so.

Harbingers of an imminent, unstoppable collapse will include the culture's resource-supplying system starting to feed on itself. The amounts of resources provided to the public will begin to inexorably decline. As a result of the ongoing shortages, and our reaction to them, they will be exacerbated, thus adding to the collapse-sustaining feedback.

As the collapse gains momentum, it will widen in scope. The functioning of the globalized resource-supplying system will be disrupted. The consequences will be imposed on us. We will become helpless spectators to our unfolding future, which will result in us feeling increasingly overwhelmed and fearful. We will start to lose faith in the future and our leaders. We will question our culture's world view.

Under these trying conditions, many more of us will seek explanations and solutions for our circumstance that lie on the edges of our culture's world view, such as mysticism, unfounded conspiracy theories and "magical thinking". Throw in wars, population declines, mass migration, famine and a loss of law and order, if you like. This state of affairs is often referred to as the doomsday outlook or a cultural collapse.

In the West, these words have a powerful symbolic meaning that raises strong negative emotions. They bias Westerners' ability to systematically consider this theoretically possible outcome. This is the same reaction seen in 1972 when *Limits to Growth* was first published. Here it instead prompts the search for an alternative outcome.

## Avoiding Overshoot and Collapse.

The models of our resource future suggest that business as usual ends with an overshoot stage that is dominated by unsustainable, excessive demand. It is followed by the collapse stage that is dominated by restricted supplies of resources. The above mechanism also suggests that collapse occurs when a culture finds it too difficult to keep all of the business-as-usual balls in the air rather than when a specific event or shortage arises. Overall, it seems that avoiding collapse requires proactive action well ahead of the overshoot stage. This effort should be biased toward restraining our resource demand rather than increasing its supply. (Note that the idea of restraining demand is based on external reality constraints. It is not the same as neoliberal austerity, which is primarily based on a mostly data-free ideological financial imperative.)

In order to reduce our absolute demand for resources, the personal and cultural barriers to discussing this option must be overcome. As discussed, inertia, distraction and masking means that we are unlikely to spontaneously make the necessary cultural and personal changes in a timely manner. A conscious, planned effort might be successful. What does this involve?

One comprehensive global model suggests that a cultural world view leading to the business-as-usual outcome has the following characteristics: a commonly (but not universally) held egalitarian, environmentalist view of the finiteness and sensitivity of the environment; and the opposite view being held by those with the power over decision making (de Vries 2007). This describes the current cultural visions and power structures in many of today's post-material (post-modern) Western European cultures and their more recent changes (Inglehart et al. 1998). The implication is that if Westerners wanted to avoid a resource supply collapse, then we have two end-member options for overcoming the barrier to discussing resource demand reduction. We can either change the objectives of those who achieve positions of power; or we can merge our two disparate visions into a single, more environmentally aware and resource-aware world view. (The quick fix of unilaterally replacing those in power is tempting, but as history shows, this action need not end in the promised or expected outcome.)

Both options are difficult to implement. The core of both Westerners' personal and cultural characteristics is very stable and only changes slowly. And the members personally struggle to consciously recognize and make the needed complexly interconnected changes. It is easier for us to simply drift until, over the medium to long term, we are overtaken by an externally imposed resource future.

We are back to thinking about an unusual and powerful set of events that could both prompt and facilitate us making the appropriate cultural and personal changes in the timely manner needed for us to escape the inertia carrying us into a business-as-usual future. The nature of that special circumstance and exactly how it could arise is unclear. However, if it happens, it would have to be sufficiently powerful to push our reality paradox, our environmental constraints, and the complexity of the human-environment system into our personal and cultural awareness. It would have to prompt the public and the powerful to switch their efforts from maintaining business-as-usual resource supplies to addressing the external-reality constraints on our resource future.

Perhaps such a special circumstance will occur as a perfect storm of normal events that lead to personal cognitive dissonance, epiphanies and their cultural equivalents. An example of such a circumstance was the (lost) opportunity to change Wall Street culture during the fallout from the global financial crisis of 2008. Perhaps a special circumstance might come about through scientific knowledge mixed with personalized stories that expose us to the likely environmental and cultural futures to which our choices are trending. Maybe, but don't bank on it. It hasn't worked in the past.

Consider the decisions of the 1,000 people living on the nine tiny Pacific islands of Tuvalu, an archipelago that rises only about 3 m above sea level. They are aware that rising sea level will likely make the islands uninhabitable within 50 years. What to do?

In a stroke of luck, the internet domain name for the Tuvaluans was .tv (dot tv). After they sold the rights to it for a small fortune, they decided to use their new found wealth to develop their islands through paving roads, among other things. When the Tuvaluan Minister of Natural Resources, Sam Teo, was asked why these choices were made in the face of impending sea level rise he replied, "Just because we are sinking doesn't mean that we don't want to raise our standard of living" (Baram 2005). They are us.

In summary, this chapter provides a comprehensive, coherent overview of a possible resource future. It suggests that the supply of resources is critically limited, that overshoot in a resource supply system is possible, and that it could be followed by a sudden forced decline in the group's population and resources: a collapse. This conclusion is not intended to be a prediction of a preordained or even a certain future. After all we can voluntarily respond to the issues we face today in ways that will help us avoid such an undesirable outcome. But neither is there a guarantee that we will deal with it in this way.

However, contemplation of this conclusion raises some serious questions about the explanation and mechanism provided by the descriptive model of the human-environment system. A key variable in that model is a generic culture's world view. In addition, the explanation for the global-scale model's outcome of overshoot and collapse is illustrated by Western culture. Thus, strictly speaking, the mechanism and explanation only applies to Western culture. It suggests that Western culture is likely to experience overshoot and collapse.

To have global-scale relevance, the idea of overshoot and collapse, and the explanation and the mechanism for it, must at least recognize that the world's resources are unevenly distributed and that there are many cultures. It must also recognize that each of the earth's countries will react differently to resource shortages and experience a country-specific outcome. The conclusion and explanation must accommodate the influence of this diversity of cultural reactions and resource distribution.

Consider that the energy-intensive lifestyles of the citizens in complex, energy-importing industrial and industrializing countries are likely to be more affected by a reduction in inanimate energy supplies than less energy-dependent countries. Countries with large or rapidly growing high-density populations and small per capita resource endowments (e.g. Egypt) will likely be most affected by scarcities in food and water. This distinction can already be seen today.

A country's reaction to the harbingers of a resource supply collapse will also depend on a mix of its characteristics. The main ones are the features of its culture's world view, the proportion of young with their expectations, the level of inequality in the distribution of wealth and access to resources, the state of its environmental endowment and its preferred method of resolving social issues. Consider that the more unequal the wealth distribution, the larger the proportion of young, the greater the population density, and the higher the people's unfulfilled expectations, then the more likely it is that resource shortages could result in social instability. In particular, a country is more likely to face social and political unrest if the largest proportion of its dissatisfied population is under 30 than over 50.

However, although countries are independent, it is also true that they are no longer largely stand-alone, self-sufficient entities. Today, economic globalization and climate warming, as two examples, have tied us tightly together. As a result, some combination of events could affect so many countries that it results in a global-scale resource supply collapse. Think of the global financial crisis of 2008 and the stressed state of the resources described in this and previous chapters.

There are other objections to the models, explanation and mechanism provided above. It could be argued that the mathematical global models are incomplete, if not flawed. Consider that they don't fully take into account our ability to innovate on the fly, and the results are too black and white.

For example, the decline in resources is unlikely to be as synchronous and smooth as the mathematical model suggests. And it may not proceed to the model's inevitable conclusion. After all, as resource availability declines, people will have to live with less. As the population's size and lifestyles are forced to adjust to this new way of living, the absolute demand for resources will decline, perhaps to a maintainable level rather than a full-scale collapse. What happens to resource supplies under these conditions depends on a number of factors. If the resources are being sourced from largely depleted non-renewable resources, then the supply will remain small. If the environmental processes that provide a renewable resource are intact but overworked, then as they recover, the supply could recover fairly quickly. If the renewing processes have been significantly disrupted, then the supply could remain reduced for a while. Think of the supply of Newfoundland cod.

Most importantly, it is argued, the discussion is about the future, so the results of the mathematical models, and the explanations and mechanism provided by the descriptive model, can't be proved by measurement or observation. Therefore, any conclusions are mere speculations and any action based on them is a waste of time (Tetlock 2006). And anyway, the argument continues, we have muddled through before, so we will muddle through again, especially with our superior technology. This perspective is appealing because it fits with our innate optimism and the positive illusions we tend to hold about the future. It is particularly appealing for Westerners; it matches our culture's world view.

Others argue that our resource future could be different if we simply got real and trashed our false hopes. But, on its own, this would be counterproductive. Holding an optimistic outlook provides us with the positive boost we need if we are to deal with the vagaries of life and muddle through to the future (Wiseman 2004). Trashing false hopes needs to be balanced by other hopes. But what? More false hopes?

Perhaps we should, indeed, put our faith in our humanity. Sit back, relax and watch TV or seek solace on the internet or social media. Fortunately, there is an alternative. History provides us with an opportunity to "ground truth" these arguments and counter-arguments about the nature of our future and how to address it.

History allows us to identify the common characteristics of how past cultures changed over the longer term. By helping us to make sense of the past, history helps us make sense of what, in our short-term, close-up view of the present, appears to us to be a chaotic jumble of influences on us and our culture's present and future. It helps us to more reliably interpret the mathematical models of our resource future and adjust the descriptive models of the human-environment system to include more realistic mechanisms. With more useful predictions and explanations in hand, we have a better chance of evaluating the above arguments and appropriately addressing, today, our longer-term resource future. Time to move forward to the past.

# CHAPTER 16

# LESSONS FROM HISTORY

Predictions about our global-scale future range from the theoretical idea that the relationship between resources and population initiates overshoot and collapse, to the intuitive idea that we will successfully muddle our way through any difficulties. Which of these predictions provides the most realistic view of our possible future? Our efforts to make that decision are helped by contemplating the histories of past cultures. After all, knowing how the futures of past cultures arrived provides us with insights into how the futures of present cultures might arrive.

For history to help, we must have knowledge of a few basic characteristics of human cultures. As discussed in chapter 3, all cultures are complex systems formed and sustained by interactions among the members. Although each culture is unique, they have a suite of common characteristics. For example, cultures have a world view and a matching organizational structure. The complexity of that structure is a key feature for distinguishing cultures from one another. In general terms, hunter-gatherer cultures have simple structures (simple cultures), while agricultural cultures have complex structures of different levels of complexity (complex and more complex cultures). Industrial cultures have the most complex structures.

The level of a culture's organizational complexity (cultural complexity) is not fixed. It can change over time. A change in a culture's complexity is indicated when a number of its characteristics change together.

An increase in a culture's complexity is accompanied by a corresponding change in its world view and is associated with an increase in its population size, population density and the land area it controls or influences. The material features indicating these changes include the appearance of, or increase in, work specialization, manufacturing activity, agricultural production and its intensity, and the storage of a resource surplus. For more complex cultures, the features include the appearance of urban centres and their monumental buildings, or an increase in their numbers. We are provided with evidence of such appearances and increases by recorded history and archaeology. Having that evidence enables us to infer the degree to which the culture's organizational complexity increased and how it took place.

If the evidence indicates the presence of an elite at the top of a multi-level hierarchical power structure, then that culture is considered to be a more complex culture. In these more complexly organized cultures, the rulers capture any resource surplus and use a portion of it to help them govern the people. For example, they use it to pay for the culture's ongoing (world view and structure) reaffirmation and the authority maintenance costs, which include the cost of the bureaucracy. The culture's reaffirmed world view explains to the members the validity and importance of their culture's characteristics, including its structure (level of complexity). Its world view also prompts the members to view an increase in the features which support that complexity, such as an agricultural surplus, as good. Together, a more complex culture's world view and the ongoing provision of resources can keep it functional and stable, to a point.

The process that results in a decrease in a culture's structural complexity (decrease in complexity) is called resimplification. It is indicated by reversals in a culture's complexity-indicating characteristics. Some of those characteristics leave a direct material record, such as a decrease in the number and size of urban centres. From this evidence, the other characteristics of resimplification, such as a decline in population size, can be inferred (Tainter 1988).

Most of the world's current population belongs to an industrial or industrializing (industrial) culture, or a more complex agricultural culture. Thus, the future of these cultural types will have a disproportionate influence on our global future. Our efforts to choose a realistic view of our possible future will be helped by looking at their pasts. Although industrial cultures are less than 200 years old, their younger histories merge seamlessly with the history of their complex ancestral agricultural cultures (Goldstone 2002). Thus, the long historical record of agricultural culture's will help us most with our choice.

History indicates that the more complex agricultural cultures all experienced resimplification. How did this occur? Is this the path that industrial cultures will follow too, or are they uniquely special? These questions will be answered in this and the next chapter.

In this chapter, three categories of features that contribute to historical resimplifications are defined and then illustrated by examples. These categories and examples are then used to broaden our appreciation of historical resimplifications and how they occurred. Our greater appreciation helps us incorporate the resimplification process into the descriptive model of the human-environment system discussed in chapter 11. This model is then used to provide a general account of cultural resimplification. The model and its account will be applied to our current circumstance in the next chapter.

A note of caution is in order. There are constraints on using history in this manner. For example, our knowledge of the past is sufficiently incomplete that we can't describe it fully or in detail. We can only reconstruct those parts for which we have sufficient evidence and understanding. Important difficult-to-fill gaps in our knowledge include how the average person saw the world around them and how they imagined their relationship to it. After all, they too made decisions that affected their culture's future. Written historical records could provide us with the easiest access to these details. Unfortunately, the authors of most records were members of the culture's elite. They were more likely to have the inclination and skills to make records, or the resources to direct others to do the job for them. The elite recorded what was of interest to them, which may not be of the most importance to most of the members of their culture's long-term future. On their own, written records can easily bias our interpretation of the past (Trigger 1989).

## Cultural Histories

A culture's history reveals many of the features that contributed to its resimplification. They can be divided into three broad generic categories that represent contributions from its local ecosystem (illustrated by Easter Island), from climate change (illustrated by the Greenland Norse) and from its world view (illustrated by the Xhosa). These categories overlap.

But the resimplification of a culture is a complex process that involves contributions from all three categories and complex interactions among them. An overview of those interactions and how they, collectively, contribute to a culture's resimplification is illustrated by the more complex Mayan culture. The example of Babylonia illustrates the same result, but focuses on the aspects of resimplification that are more difficult to appreciate: the dynamics of resimplification, and the range in time and space over which the contributions operate and interact. The Babylonian example is also relevant to Western culture because there are many links between the two cultures.

**Easter Island.** Easter Island (Rapa Nui) has a relatively simple history. It provides a compelling example of how the state of a culture's supporting ecosystem contributes to its resimplification.

Easter Island is an extremely remote Pacific island. Its nearest inhabited neighbour, Pitcairn Island, lies 2,075 km to the west. The island's 160 km$^2$ area is the top of a 750,000-year-old volcano that last erupted some 100,000 years ago. The region's subtropical climate provides the island with 1,250 mm of annual rainfall. Prior to the arrival of humans, the climate supported a subtropical to tropical broad leaf, hardwood forest ecosystem that included palm trees. The island originally supported nesting colonies of some 25 species of seabirds, which suggests that significant food, such as fish, was available from the surrounding ocean.

The first human settlers arrived from the west in ocean-going sailing canoes. They arrived no earlier than the first few centuries AD and no later than AD 900 (Anderson 2002). They brought with them their Polynesian cultural characteristics, which included a hierarchical, clan-based structure that emphasized interclan competition. The settlers also carried with them at least some of the staple Polynesian agricultural foods: taros, yams, sweet potatoes, coconuts, breadfruit, bananas, pigs, dogs, chickens and rats.

Shortly after their arrival, about half the seabird species on the island were no longer present or had become extinct. Some birds and their eggs were likely eaten by the settlers to supplement their diet while establishing their crops. Eggs were also likely eaten by the introduced rats.

By practising slash-and-burn agriculture the first settlers removed the forests that covered the more fertile soils near the coast and established the fields on which they grew their first crops. By the time 1100 rolled around, deforestation of the island was well underway. Shortly after 1100, the human population began to rise rapidly, accompanied by the spread of farming and habitation into the interior. At roughly the same time, the carving, moving and erecting of the larger stone statues, for which the island is justly famous, began.

Prior to the 1400s, the community appears to have been largely free of fighting, which is surprising for a Polynesian culture. However, after 1400, weapons started to appear. During the 1400s, agricultural production was still increasing, although by now deforestation was mostly complete. The decline in fishing yields had certainly started (Anderson 2002; Flenley and Bahn 2002).

By no later than the 1600s, the deforestation process had been completed to a degree that reforestation was impossible. The trees that could grow large were no longer present. The palms were extinct, apparently with help from the rats who ate the palm seeds. By this time, the nearshore marine resources were overexploited. The ability to fish further offshore was inhibited by the reduction in the availability of wood suitable for making boats.

Deforestation had other consequences: significant soil erosion, a reduction in available soil water, the loss of soil fertility and the significant reduction in their primary fuel supply. These factors probably contributed to the abandonment of the interior fields. By no later than the 1600s, the population had reached a maximum of at least 6,000 to 8,000 people (population density of 37 to 49 people/km$^2$). Coincidently, the building of statues had stopped.

Rather than stabilizing, the population soon began a rapid decline. The loss of soil fertility and, perhaps, a reduction in soil moisture meant that growing crops was a struggle, even with their quite sophisticated farming methods. Their inability to make large boats meant that whatever ocean resources existed could no longer be exploited to supplement their diet. The lack of boats also meant that leaving was not an option. The population decline was likely the result of malnourishment and starvation. The stressed culture started to disintegrate (Flenley and Bahn 2002).

The first Europeans to visit the now treeless island arrived in 1722. Shortly after that date, there were an estimated 2,000 (11 people/km$^2$) apparently undernourished people on the island. Observations at this time confirm that there was insufficient wood to build more than small leaky canoes with which to fish offshore.

Their cultural history fits into the above environment-focused framework. Soon after settlement, the island was divided into territories. Each was occupied by a clan with a hierarchical structure and a hereditary leader. A clan was partly distinguished by the specific goods and foods it produced. These goods were shared across the island through a highly organized cultural food and wealth redistribution system. The clans did compete with one another, at least through statue erection.

One apparent outcome from the large population of 1600s, and the likely associated social tensions, was that the island's clans reorganized themselves into two groups. During this transition, the basic nature of the island's hierarchical and clan structure remained intact, but the leaders were no longer hereditary. The new elite were chosen by a leadership competition between members of each group's warrior class.

In the process of reorganization, the clan exchange network disintegrated, and they no longer competed with one another through statue erection. Instead, they engaged in warrior activities in the form of interclan raids (which needed weapons) and territorial disputes.

At the same time, the clan-based religions were replaced with a single deity. The old statues were toppled, broken and defaced.

In essence, after their arrival on the island, the settlers had spent considerable energy attempting to re-establish and maintain their cultural world view's way of life and beliefs. Their endeavour ended in the late 1600s with a population crash and a major cultural transition. The outcome can be described as a cultural resimplification. By the time of a European visit in 1770, that process was complete (Flenley and Bahn 2002).

Many specific factors contributed to this outcome. However, underlying them all was the reality that once the island's ecosystems could no longer support its large human population and their culture's structure, the culture was forced to resimplify. The Easter Islanders' earlier disruption and destruction of their supporting ecosystem played a significant role in their later resimplification.

### Greenland Norse.
The history of the European Norse in Greenland illustrates a contribution to resimplification from climate change. In AD 985, during the period of the medieval climate warming, the Iceland Norse sailed from Iceland to found two farming settlements on Greenland's west coast. Near the southern tip of Greenland, they established the larger eastern settlement. Some 450 km by air to the northwest, they established the smaller western settlement. The climate and vegetation in both areas is subarctic, and the ocean close to the coast froze each winter.

The settled lands were located at the heads of fjords in the glacial valleys from which the nearby Greenland ice sheet had retreated at the end of the last ice age. As a result, the soil cover was thin, with only a few areas, mostly confined to the valley floors, suitable for farming. These areas supported indigenous grasses, Arctic shrubs and small patches of dwarf birch, willow and alder. The small trees were the only local wood available for fuel and construction.

The Norse who settled on Greenland, like the Easter Islanders in the Pacific, arrived with a farming culture. They thus brought with them their domestic animals, which included pigs, cattle, horses, sheep and goats (in that order of importance), and their hunting dogs. Their culture's structure was a rigid, militaristic hierarchy. They believed in a strong leader. They also placed high value on prestige, which they devoted considerable effort competing for. They were also recent converts to Christianity.

The Norse re-established their European way of life on Greenland, even though they were living at the northern limit of the biome that could support it. This is illustrated by the fact that they had settled in the only two places in Greenland with sufficient suitable pasturage to support their animals. The resulting marginal nature of their existence is indicated by their inability to grow cereals, their near elimination of the pigs and the reduced importance of cattle in their diet. Considering the significance of these two animals to their culture, these were major decisions.

The Norse survived on milk products and meat from their domestic animals, except the horses. They also hunted seals, which made up a significant portion of their diet, and caribou. Despite the presence of fish in the sea, the evidence suggests that the settlers rarely ate them. Because the opportunity to grow and store food was limited, their survival depended heavily on their success at completing the yearly round of seasonal activities that reflected their lives as subsistence farmers who also hunted.

The seasonal round began in the spring when they hunted for seals that could be caught using nets and clubs. In early summer, they would travel further north by boat to hunt walruses and narwhals, primarily for their tusks. These prestige goods would be exchanged for the iron and religious artefacts brought by the Icelandic and European Norse on their occasional summer visits to Greenland.

During the short summer, the Norse would cultivate and then prepare hay, which was crucial for the survival of their domestic animals over the long winter. This need indicates why the presence of adequate pasture had determined the locations of their settlements. Summer was also the time to milk goats, sheep and cows in order to produce cheese and skyr (similar to yoghurt).

In the fall, they would put up the hay for their barned animals, collect seabirds from the local colonies and hunt caribou, perhaps with their dogs. During the winter, they would live on their stored foods. Apparently, they neither hunted nor fished in the winter.

The constraints on their food resources were severe. Nevertheless, the Norse culture and their herding and hunting lifestyle was sufficiently flexible, robust and resilient for them to survive. However, this was only true as long as bad weather did not disrupt their hay making too severely or, when bad years did occur, they were followed by a few good years so that their herds could rebuild. Successive bad harvest years would make the settlements highly vulnerable to the variations in their alternative food sources of seals and caribou. Just how vulnerable is exemplified by the marginal energy content of their normal diets.

Around 1200, the population of the Norse Greenland settlements reached a maximum of about 2,000 to 4,000 people (800 to 1,000 in the western settlement). After this date, the population declined. By around 1340 to 1350, the western settlement was no longer inhabited. The eastern settlement hung on until about 1450, before it too became vacant.

Documents written in the eastern settlement suggest that, toward the end of the western settlement's existence, the Skraeling (Inuit in this case) were a problem. However, the archaeological record tells a different story. Although the exact sequence of events that occurred during the years when the western settlement was being abandoned are not known, the following sets the tone.

It is certain that, in the last years, the residents resorted to eating foods such as ptarmigan, hare and even their dogs. They were apparently unable to bridge the hungry time of late winter to early spring. The state of the farmhouses in their last year indicate that they did not migrate because portable treasures, such as a wooden cross, were left in their shuttered houses.

Many factors contributed to this state of affairs, but the base on which they all rested was long-term environmental change. A period of global cooling started at the end of the medieval warm period, shortly after the first settlers arrived. By the 1300s, the cooling had become the hemisphere-wide, if not global, cooling period called the Little Ice Age. This long-term cooling period is recorded in an ice core drilled through Greenland's ice cap at a site some 1,000 km to the north of the western settlement. Evidence of prolonged cooling is also provided in the west on Baffin Island, where, over this time period, tundra vegetation was buried by advancing ice.

When this proxy climate data from the field and the historical records from Iceland are combined, they indicate that, after the early 1300s and into the 1400s, there were many more multi-year periods of colder than normal weather, including the coldest years for centuries. The suspected short-term reason for this marked change was unusually frequent and powerful volcanic eruptions between 1275 and 1300, and between 1430 and 1455. The volcanic eruptions combined with the resulting sea-ice expansion to form a self-sustaining climate cooling feedback (Miller et al. 2012).

Connected, both directly and indirectly, to the cooling climate are the other factors that contributed to the settlement's failure. A longer, colder winter extended the time that the Norse had to survive on stored food. The corresponding shorter summer limited their ability to make hay. The delayed melting of the sea ice would have affected their ability to travel by boat to either hunt sea mammals or return to Iceland. Their location on the southern part of Greenland meant that, as the climate cooled, the Norse, like plants living on the top of a mountain, could not follow their climate niche southwards by migrating on land.

The Norse's ability to compensate for lower hay production by expanding their farms as the climate cooled had already been compromised as early as the 1100s. By then, virtually all available pasture was in use. They also operated their farms at maximum intensity, which led to over-utilization of the soil. Specifically, they allowed their domestic animals, especially goats, to overgraze the pasture.

At the same time, the Norse removed the woody plants for firewood. The deforestation of the few tree patches resulted in a fuel shortage. They compensated for it by burning manure and turf. But that degraded the soils and deprived it of fertilizer, thus exacerbating their food production problems.

Perhaps the Norse did not notice these changes to their supporting ecosystem because they occurred too slowly. Perhaps they did see the changes, but did not appreciate their significance. Or if they did, they may have thought of them as the normal price to pay for maintaining their lifestyle. In any event, the result was a steady loss of critical pasture through soil degradation and, ultimately, erosion.

The cooling climate made the voyage from Iceland and Europe more hazardous. That reality, along with changing tastes in Europe resulted in no ships being sent from Europe after 1369. The loss of trade likely contributed to the disappearance of at least the western settlement.

The strength with which the Greenland Norse held onto their European Norse heritage also made a contribution. Consider the presence, in the eastern settlement, of a stone church with a stained glass window imported from Europe. Consider too the records indicating that a heretic was burned to death in the settlement during the first decade of the 1400s.

The structure of the Greenland Norse's social hierarchy was a steep-sided pyramid, like that of the European Norse, as illustrated by the concentration of power and wealth in the local bishop, and by the archaeological record from the western settlement. It indicates that a large disparity in wealth existed between the few more prosperous farms and the abject poverty of the rest of the community's farms. This circumstance matches the more complete record of the Iceland Norse's social structure for that period. The mere fact that this disparity persisted among the Greenlanders despite them living in such a small, highly isolated community indicates the power of their cultural belief in a hierarchy. Their strong faith in their culture's world view may have given the Greenland Norse the fortitude to live under these harsh conditions, but it likely prevented the leaders and the members from making decisions that were more appropriate for their circumstances.

Even though the Inuit were present in the area of the western settlement and knew how to live well under the colder conditions, the Norse did not adopt Inuit practices. For example, the Norse did not adopt harpoon hunting, which would have allowed them to hunt ring seals from the sea ice during the winter. They thus lost the opportunity to access a food buffer during the hungry time, which would have helped them to survive. The presence of Norse artefacts in Inuit archaeological sites but almost no Inuit artefacts in Norse sites suggests a reason why. Their reticence to adopt Inuit practices was likely cultural.

Overall, despite their tenacity and hardiness, the Greenland Norse could not maintain their preferred lifestyle under the conditions being imposed on them by a cooling climate. However, it is also clear that the other two categories (supporting ecosystems and world view) also contributed significantly to the demise of the Greenland Norse. For example, like the Easter Islanders, the Norse overused their supporting ecosystem. The world view of the Norse also guided their day-to-day decisions to place too high a value on ideology and social standing while neglecting the pragmatic and environmental realities of their circumstance. They were thus closed to the ideas of the Inuit and were unable to avail themselves of alternate hard-time food buffers (Barlow et al. 1997; McGovern 2000; Vésteinsson et al. 2002). The role of a culture's world view in resimplification is more clearly illustrated by the Xhosa's history.

**Xhosa.** A part of Xhosa history illustrates how a culture's world view can contribute to its resimplification. The Xhosa of the 1800s lived on the grass-covered, rounded hills bordering the eastern coast of South Africa. This savannah biome, with its scattered trees, was supported by warm summers with limited rainfall and cool winters. At that time, the Xhosa were mostly living as Iron Age pastoralists and farmers within a rural, hierarchical, clan-based culture. Just as maize was central to Mayan culture, cattle were central to Xhosa culture. Cattle were food, money and symbols of prestige. They embodied Xhosa identity.

In the early 1800s, the Xhosa came in contact with the expanding British Empire. By 1853, the Xhosa were struggling to recover from the aftermath of losing their eighth border war with the British. Physically, they had lost land, resources and able males, so their ability to wage war was reduced. The latest war also added to the ongoing mental stresses from previous wars and the ongoing demands of the British. Dealing with these demands was a difficult and stressful task because the British had a habit of making promises but being inconsistent in holding to them. The wars and adjusting to their new circumstance steadily weakened the Xhosa's cultural framework.

On a more personal level, individual Xhosa were stuck with reconciling their indigenous world view with that of the European settlers arriving in the land. Some elements in the Xhosa religious views could easily be reconciled with Christianity's resurrection, apocalyptic and millennial teachings. For example, the idea of renewal was inherent in the Xhosa's cyclical view of time and in their belief that the spirits of their ancestors had the power to re-enact creation.

Other elements were less easily reconciled. For example, the Xhosa did not see the afterlife as separate from the present, as is the Christian belief. Although the ancestors may be in another world, they were always present and interested in today's goings on. To the Xhosa, disease, failure in war and disasters were linked to the displeasure of the ancestors and the practice of sorcery and witchcraft. Thus, when life was going particularly badly, they attempted to restore order and justice by seeking advice about how to placate the ancestors. Placating them included finding and removing witches. Help in completing these tasks was sought from those special people who had connections to the ancestors and the future.

In 1854, when bovine pneumonia, a fatal cattle disease, spread into Xhosa country, the presence of dying cattle prompted both a practical and a religious reaction. However, neither the killing or moving of cattle nor the routine spiritual cures, such as witch killings, were successful at stopping the spread of the disease. To make matters worse, some parts of Xhosaland were experiencing a loss of corn from both insects and insufficient rain. The crop failures likely heightened the desire for explanations and a more powerful response to their woes of cattle loss and dealing with the British.

When the former British governor of the area, George Cathcart, was killed in the Crimea in 1854, a number of Xhosa prophecies appeared. They said, in essence, that the Xhosa would be saved because the Russians would come and defeat the British. Many watched the ocean for their arrival. They were disappointed.

In the early 1850s, the Xhosa were certainly experiencing a cultural crisis. The degree to which the resilience of their culture was impaired was illustrated in 1855–1856 by the chiefs acquiescing to the British demand that magistrates be assigned to each chief. They were needed, the British said, to "assist" the chiefs in administering justice.

During this time, a young Xhosa girl named Nongqawuse was living in the household of a popular traditional seer, who was also well versed in Christian traditions. In early 1856, Nongqawuse was walking near a place on the Gxarha River that the Xhosa believed was endowed with special spiritual properties. While there, she experienced a visitation from the ancestors. They, she told her seer benefactor, were preparing to rise from the dead. A world of contentment and abundance was at hand.

However, the ancestors would not arise until the people had first killed their cattle, cut down their corn and stopped cultivating crops, because they had all become contaminated and were impure. The ancestors had also told her that the Xhosa need not worry about the practical consequences of their actions because, when the ancestors arose, they would also bring with them new cattle and corn in abundance. In fact, the people must prepare new grain bins and cattle pens for this abundance.

Although Nongqawuse's prophecy was not the first to report this message, hers garnered the greatest spontaneous grassroots acceptance. Belief spread most rapidly and widely in those areas hardest hit by the bovine pneumonia and by the loss of crops through drought and insects. However acceptance was not universal. There were many reasons dividing those who believed in Nongqawuse's prophecy from those who did not. Nonetheless, in 1856 the believers, led by the Xhosa king, Sarhili, began killing their cattle and stopped cultivating their fields.

When seen from the perspective of the culturally stressed Xhosa, their belief in her prophecy and their resulting actions were reasonable. Within this context, the prophecy provided them with a credible cultural explanation for their circumstance and culturally acceptable actions that could potentially restore their way of life. How they came to sustain their belief in this response is more difficult to appreciate.

The incentive to kill cattle and destroy crops certainly had its ups and downs. There were downs when the dates set for the ancestors and new cattle to appear passed uneventfully. And there were downs resulting from changes in the details of the conditions to be met before the ancestors returned. But none were sufficient to derail the widespread belief in the prophecy.

The momentum for cattle killing was certainly helped by King Sarhili's belief in it. The British too provided "ups" for the prophecy by their actions. For example, they attempted an ocean-based boat landing on the coast at the mouth of the Gxarha River, near where Nongqawuse received her messages from the ancestors. The attempt failed, which boosted the Xhosa's faith in the ancestors' power. The British drive to expand their empire also helped the prophecy because their relentless pressure further stressed the Xhosa culture. No wonder the hope that the Russians would arrive still persisted into 1856.

After most of the cattle were dead and the time for planting new seed had passed, the prophecy was sustained by the belief that the unbelievers were holding things up by not killing their cattle. By that time, the divide between believers and unbelievers had become a split in the foundation of the culture. Unbelievers became an intimidated minority whose ranks steadily shrank. Their cattle were stolen and killed, and their crops were destroyed by the believers. The unbelievers even requested help from the British, whose lack of response in effect gave support to the believers.

Finally, external reality intruded into the picture in the form of starvation. Yet, even as they died, many Xhosa still remained euphoric about their future resurrection. Those who chose life scattered in their search for food. The British used that opportunity to secure labour for their expanding colony.

The scale of the death and displacement is illustrated by the estimated change in the Xhosa population within that portion of Xhosaland known as British Kaffraria. The initial population of 105,000 dropped to 38,000 by December 1857, then to 26,000 by the end of 1858. At least 15,000 had died.

By the time the event was over, the culture was greatly weakened. The Xhosa were fragmented, disillusioned, dispersed and the power of the chiefs was greatly diminished. Many of them were dead or in jail (some on Robben Island where, over a 100 years later, Nelson Mandela, a Xhosa, would spend 27 years). Any hope of dealing with the British on a near-equal footing was gone. The Xhosa culture did not disappear, but the members did have to reconstruct both it and themselves (Mostert 1992; Peires 1989).

The Xhosa example illustrates features common to all cultures. For example, core aspects of a cultural world view that under normal circumstances serve the culture well (at least in the short term) can become a liability under some conditions or in the long term. This is especially the case if the world view poorly represents the external reality.

It also illustrates that cultures can and do become stressed. And, even when stressed, cultural inertia will still prompt many of the members to address their problems by looking for solutions that fall within the core of their culture's world view. Under these circumstances, the solution they choose may provide the needed mental comfort, but it may not be the most appropriate. Their decisions may even amplify the problem or push the culture too far outside the external reality boundaries within which they must live. The members' faith in their culture's world view could result in them being profoundly surprised by what, to them, is an unexpected outcome. This adds to their stress and further undermines the culture. A culture's world view can contribute to its own resimplification. The world views of the other examples of resimplification presented in this chapter should be considered in this light.

**Mayan.** The Maya's story, like that of Easter Island, has had a significant impact on Western European impressions of cultural resimplification and predictions about our future. They both act as a kind of cultural resimplification reference. When thinking of the Maya, we usually focus on their culture's collapse (AD 800–1000, the Terminal Classic Period) that ended their Classic period (AD 250–1000, during the European Middle Ages). However, when that event is viewed as part of their much longer history as a relatively complex agricultural culture, then a messier and more convoluted process of Mayan resimplification appears.

*Background.* The Maya occupied a 150,000 km$^2$ area of Central America that covered the southern part of the low-lying, sub-equatorial Yucatán Peninsula and the adjacent highlands to its south. The cooler highlands formed a relatively small part of the Mayan land base.

The majority was the hotter and more humid lowlands, whose seasonal 1,500–2,000 mm rainfall falls over eight months. Seasonality is so significant that, even though the area lies near the equator, there is no true rainforest, only a wet-dry, seasonal tropical forest.

The lowland vegetation cover grows on a thin organic layer typical of tropical soils. Under these conditions, nutrients are not stored in the soil but are quickly recycled into the standing vegetation. There they remain until the vegetation dies and falls to the ground to be quickly recycled.

The lowland's bedrock is dominated by limestone that, unlike most rocks, dissolves in acidic water. As a result, the mineral soils between the organic layer and bedrock are often unusually thin and slow forming. One exception is the valley bottoms, where thick soils have accumulated through erosion from hillsides.

The dissolution of limestone also controls the availability of water in the area. It creates a landscape pitted with caves (karst) into which surface water drains. As a result, surface rivers are rare, with most water flowing underground. Some karst hollows do form wetlands that can be seasonal or permanent. The amount of water available for plants to use during the seasonal and somewhat unpredictable dry months is further limited by the lack of mineral soils to hold near-surface soil water. Forest growth is severely constrained by this lack of surface and soil water (Webster 2002; Scarborough 2007; Gunn and Folan 2000). From a human perspective, both the soil and water characteristics of the area severely limit the area's agricultural potential.

The area of the southern Yucatán Peninsula that was occupied by the Maya during the Classic period first saw human activity as early as 7000 BC. Settled agriculture only started around 1400 BC to 1200 BC when maize and beans were grown. After about 650 BC, a significant increase in cultural complexity is suggested by the appearance of monument-type buildings. By 400 BC, numerals were being used.

Around AD 1, most of the basic elements of the Mayan culture during the later Classic period were in place. They are seen in the southern lowland area as complex aggregations of buildings, such as El Mirador, perhaps the largest settlement complex the Maya ever built. El Mirador thrived from 200 BC to AD 150. However, around AD 150 to AD 200, El Mirador, along with the neighbouring complex of Nakube and perhaps the greater region, were abandoned. Drought likely played a role. Resimplification was not new to the Maya of the Classic period.

Between AD 250 and AD 700 the Mayan population in the southern Yucatán Peninsula grew rapidly, as did the number of typical Mayan urban power centres. Among them were Tikal, Copán and Dos Pilas. At these centres, the remaining distinctive elements of the Classic period appeared, including the royal dynasties and the expanded role of monumental buildings.

The Classic period power centres were not well-defined cities. For example, they had no enclosing walls. They were more like royal courts: small concentrations of monuments and elite housing. Most settlements occupied less than 1 $km^2$ (100 ha), although a few reached between 4 $km^2$ (e.g., Tikal) and 9 $km^2$.

The ruler (usually a king) and the few elite, who lived among the monuments of the power centre, presided over the remaining 80% to 90% of the population, who were rural farmers. They lived on their farms, which were concentrated around the power centre in an area that could be walked across in a couple of days, at most. This small area, its people and the power centre formed an independent city state or kingdom.

At the height of the Classic period (Late Classic, AD 600–800), there were 40 to 50 power centres dotting the southern Yucatán Peninsula. The ruler of each centre had full control over a small farmed area around it but limited control over their extended surroundings and the other kingdoms it hosted. What influence a ruler had over their wider surroundings was based less on their control of territory and more on their role in the fine balancing of the power relationships between the various power centres. This role existed because Mayan power and politics was based on ancestry. The rulers of the centres were dependent on one another's mutual ability to reinforce their respective power and political relationships through the dynastic and noble lineages that linked them together. The kingdoms also relied on one another to act as limited trading partners with whom they could exchange their specialty resources.

Each kingdom was culturally unique in detail. For example, some erected many monuments with inscriptions, while others did not. Each had their own patron gods. Yet, there were many common elements, such as the presence of a monarch, and the same basic forms of farming, language, religious beliefs, pyramid temples and rituals. Thus, although the power centres did not collectively form a unified state, they were sufficiently alike and linked to collectively be called Mayan.

Like all cultures, the Maya needed to deal with uncertainty, maintain order, and explain both their world and their place in it. The Mayan city states all used the same basic elements to do so: that had a common cultural world view. For example, spirits infused all things. In the Mayan world view, their collective ills or misfortune resulted from personal behaviours and moral failings that displeased the creator gods and the ancestors.

The rulers, being closer to the great creator gods and the ancestors than the masses, bore special responsibilities. (The ruler may even have been thought of as a god.) They were obliged to conduct themselves in the manner necessary to keep the cosmos stable and in balance. This would ensure, for example, the timely arrival of the rains, the ongoing productivity of the fields and bountiful crops, especially maize. One way the rulers fulfilled their obligations and exerted their influence was to complete their ritual duties. These included sacrifices, which were needed to nourish the gods and ancestors as well as repay the debts mortals owed to them (Webster 2002).

A ruler was also responsible for building the required monuments to the ancestors and gods. A monarch was obliged to attack their state's enemies because they personified the supernatural forces that lay behind bad events. Victory in war helped neutralize or avoid such events (Freidel and Shaw 2000). Collectively, observing rituals, building monuments and making war had a hidden benefit: it unified the city state under the monarch.

Another important part of the Mayan world view was their concept of time. Life experiences were explained, at least in general, by nested, reoccurring cycles of time to which were attached events, including death and regeneration. This is in contrast to Western culture's more linear view of time.

The cyclicity of time allowed the Mayan prophet-priests to use their sophisticated calendars and celestial signs to fulfill the critical task of predicting the future. This ability helped the earthly world prepare for the future. It also gave the monarch, with the help of the people, an opportunity to influence or alter the future through the appropriate rituals and sacrifices.

But, at its heart, the cyclical Mayan calendar was based on agricultural needs, such as the need to predict the start of the rainy season. After all, the material survival of the Maya depended on the success of their staple food crop: maize. Unsurprisingly, maize also played a central role in their world view. For example, the square maize field was likened to a properly ordered community. Growing maize represented the essence of being a good person. In some sense, maize was considered to be, and was treated like, a god.

The Maya's annual success at producing maize was ecologically constrained. In any particular year, the maize yield depended on an adequate amount of rain falling at the right time on fertile soil. The nature of the soil also meant that to maintain its fertility over the long term, a long fallow was needed.

If all went well environmentally, then the state's land holding was capable of supporting a dense population for low effort. For example, if the population density was low (less than 40 people/km$^2$), then about 48 days of swidden (slash-and-burn) farming could provide enough food for a year. However, the denser the population, the greater the chance that a crop failure from factors such as drought, pests and soil depletion would result in food shortages. Just as importantly, greater density meant that there was less of a wild or even a cultivated buffer against bad times (Webster 2002).

The Classic Maya's ability to circumvent these basic constraints by intensifying agricultural production was constrained by their lack of draught animals and by the region's environmental features. For example, there were relatively few wet karst hollows that could provide the water needed to irrigate crops during the dry season. The Maya did practise water management, but their efforts seem to have been mainly directed at overcoming the water shortages experienced by the power centres. To this end, they built localized water systems within and immediately adjacent to the centres they were meant to serve (Scarborough 1996).

In addition, it seems that the efforts the Maya devoted to protecting their soil resources from erosion though forest maintenance and terracing were limited and clearly insufficient (Scarborough 2007; Dunning and Beach 1994). As a result, the loss of soils and the resulting impacts on the local hydrology likely played a significant role in altering Mayan history. For example, their ability to use local perennial (year-round) karst wetlands as sources of water would have been largely lost by about AD 250 because many became filled with silt washed in from the deforested and eroding hillsides (Dunning et al. 2002).

By the end of the Classic period, the regions around each power centre were intensely farmed and densely populated. By one account, within a 25 km radius around Tikal, there were an estimated 425,000 people at an average density of 216 people/km$^2$. This estimate may be high. A regional estimate suggested the maximum density was 145 people/km$^2$(Kennett et al. 2012). In either case, the density is extraordinarily high compared to 40 people/km$^2$ that could comfortably be fed by their supporting ecosystem. The high population density could have easily produced the levels of erosion and deforestation estimated from excavations. This increase in population and the related resource issues were likely accompanied by an increase in power intrigues, including warfare, between the various power centres (Webster 2002).

### The Resimplification Ending the Classic Period.

Starting around AD 750 in the western (inland) part of the southern lowlands, the elite aspects of the hierarchical cultural structure in one power centre after another started to disintegrate. This is indicated by, for example, the cessation of both monument building and the erecting of objects with dates inscribed on them.

Each centre seems to have resimplified in a unique manner whose details depended on a combination of factors. These included the local geographic and climatic conditions, the details of the centre's political structure, the nature of its cultural ideal and the relationships with its neighbours. Consider that the resimplification of some kingdoms seems to have been dominated more by warfare, for example Piedras Negras. Others, for example Tikal, apparently succumbed to more environmental factors, such as overpopulation, deforestation and soil erosion, as well as internal disruptions. The monarchs, nobles and commoners disappeared rapidly in some centres, for example Piedras Negras and Yaxchilan. In others, they disappeared quite slowly. For example, in Copán, the nobles lasted almost 200 years after the demise of the kingship, while the commoners lived in the area for much longer.

The process of resimplification also varied in its complexity and timing. For instance, Dos Pilas underwent a complicated and protracted resimplification involving population decline, forest clearance and regional rivalries, which included war. In contrast, at the same time Lamanai, roughly 200 km to the northeast, seems to have been little affected.

Part of the response of the people to unfolding events may have been migration to the northern lowlands because, starting around AD 700, a sudden influx of people into that area established centres such as Uxmal. Interestingly, despite the fact that the Mayan process of resimplification was, in a historical sense, comparatively rapid, the average citizen would, in many cases, only have noticed small changes in their lives.

They would have been more aware of the short-term variation in what was an inherently uneven process.

After AD 850 and generally within 150 to 250 years of their peak complexity, all of the southern centres were abandoned or in decay, and their populations had shrunk or were in decline. By 1000, few southern centres had significant populations. Certainly, there were no elites occupying their traditional roles at these centres. In the north, the newer power centres such as Uxmal and later Chichén Itzá appeared, blossomed and then repeated the resimplifications experienced in the south. For example, Chichén Itzá resimplified by about 1100. By this time, stone monuments were no longer being carved there.

The Maya and their culture did not disappear completely. The Spaniards, who explored the Yucatán Peninsula in the 1500s, found a much-reduced Mayan population. In northern Yucatán Peninsula, a small Mayan population living at a density of perhaps <20 people/km$^2$ were continuing many of the ancient Mayan cultural practices. These included living in power centres, building pyramids, warring and using prophecy tied to a cyclical calendar. But some things had changed. For example, the monarch was no longer divine or even semi-divine, and the power centres were smaller and less splendid.

During their exploration, the Spanish crossed the southern Yucatán Peninsula near the Mayan power centres of the Classic period. However, they were unaware of them because they had been abandoned and were being reclaimed by the jungle. The last Mayan kingdom, the Itza living in Nojpeten in northern Guatemala, near Tikal, was visited by Cortez in 1525. Only a few thousand people were living in an area that, a few hundred years before around AD 750, had supported a population of about five million. Nojpeten survived Cortez and the 1535 drought but was destroyed by the Spanish in 1697 (Webster 2002).

### Causes.

When looked at in detail the Mayan resimplification that ended the Classic period (Classic period resimplification) appears to have been a profoundly local phenomenon. Each centre experienced resimplification at a different time and in a different manner. However, when looked at on a regional scale and over a 200- to 250-year period, with attention to the contributions from their supporting ecosystems, climate change and their world view, then a different picture emerges. The Mayan resimplification was a complex process with many common and interconnected elements between the local experiences, such as population decline and reductions in the elite (Webster 2002; Scarborough 2007; Rice et al. 2004).

In overview, all three contribution categories played a significant role in the Maya's Classic period resimplification. The Maya's supporting ecosystem was involved because it constrained the cultivation of maize, the culturally endorsed primary food source. The Maya managed to feed their growing population by increasing agricultural production through practices such as widespread deforestation, reducing the fallow, and farming marginal soils. However, these practices led to exhaustion of the soil's limited fertility; soil erosion, including the silting of lakes; loss of the buffer provided by wild foods; a significant change to the hydrology of the area; and more opportunity for runaway fires.

Climate change contribute to the Mayan resimplification because growing maize depended on the seasonal rainfall. In wet years, the bumper harvests could support a growing population. However, too long a wet season makes slash-and-burn agriculture difficult. In contrast, drought reduced crop yields. When the population was small, both dry and wet years could be survived by switching to alternative wild foods or the remnants of cultivated foods.

Indeed, a high-resolution climate record shows that the timing of multidecadal droughts corresponds well with the abandonment of El Mirador around AD 150 to AD 200, and Chichén Itzá around 1100. Drought likely played a central role in these resimplifications. In between, periods of above-average rainfall correspond with the rise of the Classic period. Similarly, drying and short-term droughts around AD 750 likely contributed to the resimplification that ended the Classic period (Kennett et al. 2012).

The end of the Classic period found the Maya struggling to grow sufficient maize to feed themselves due to the likely combination of increased population size and density, decreased soil fertility, and dry years. The few remaining wild ecosystem remnants were likely unable to provide sufficient, alternative wild foods. This lack of food contributed to resimplification in a number of ways.

Trying to feed more mouths than the land can support leads to malnutrition. If the population is both unhealthy and living at a higher density, then diseases, such as yaws and tuberculosis, can spread more easily. The resulting underfed and/or ill population was likely less able to do the necessary, and increasing, amount of farm work needed to feed the population, thus amplifying their predicament.

A large, malnourished population is also less likely to be content with their lot. They could become disinterested, lose faith in the prevailing rulers or search for scapegoats. Throwing the bums out or migrating become tempting responses.

The Mayan culture's world view also contributed to its resimplification. Consider the Mayan belief that their collective ills or misfortune was caused by personal actions and moral failings that displeased the gods and ancestors. To avoid a bad future, each member, and especially their rulers, had to behave properly, complete the required rituals and, as needed, apply appropriate ritual measures to keep the gods and ancestors happy. The consequences of failure included the possibility of an apocalyptic outcome (Webster 2002; Rice et al. 2004).

This is the world view framework within which the Maya tried to culturally address external reality issues such as drought, siltation, lack of food, government malfunction and the impacts of overpopulation. When seen within this context, an increase in the Maya's cultural practices and rituals at a time of crisis is a logical solution, to a point (Freidel and Shaw 2000; Scarborough 2007). But, for some of the issues they faced, these efforts only widened their reality paradox thus amplifying their problems.

The repeated failure of the leaders and the people to resolve problems in a culturally acceptable manner could contribute to a general loss of faith in the leaders and their culture's world view. Rising dissatisfaction and discontent can contribute to cultural dysfunction. Like the response to malnutrition, simple, direct actions, such as migration or finding scapegoats, become realistic solutions.

The Mayan world view also helped create (precondition the appearance of) the circumstances that they had to later deal with. Consider its lack of guidance about overpopulation. In contrast, its guidance supported the power of the elite. It allowed them to allocate too many resources and too much personal energy to elite intrigues, both around court and between monarchs. This was most strongly expressed through endemic warfare.

War became a common activity in the Middle Preclassic period (1000 to 400 BC) and gained intensity into the AD 600s to 700s, the Late Classic period. By that time, it had become, in one researcher's words, "an almost pathological condition" in Mayan culture. It had become an ideologically, and practically essential, part of kingship. Ironically, in the short term, war helped to unify a kingdom and bind the region into a whole. But, in the long term, it diverted their energy and attention away from population and ecosystem problems and co-operative, long-term solutions (Webster 2002, 224).

One would imagine that, after their experiences, the survivors of the Classic resimplification would change their world view in order to avoid future resimplifications. However, the inertia to changing their world view was clearly considerable. The Classic period resimplification appears to have been a repeat of the earlier El Mirador resimplification, only on a grander scale. Consider too that the survivors of the Classic period resimplification who migrated north and built Chichén Itzá had rejected the idea of their monarchs being divine. But they did not alter their culture in ways that dealt more realistically with their external reality. Chichén Itzá also resimplified.

Elements of Classic period culture are still found in Mayan culture today. Are they still preconditioned to repeat the resimplification cycle of their ancestors? What about other cultures?

In overview, the resimplification at the end of the Mayan Classic period was the product of interacting changes in the complex human-environment system within which they lived. In particular, the interactions between the changes to their supporting ecosystem, the local climate and the nature of the Maya's world view created their circumstance and preconditioned their culture to resimplify. For example, the interactions reduced resource availability and prompted inappropriate decisions. The Maya's thus helped to create this situation and were victims of outside forces.

The lead up to resimplification took centuries. In contrast their culture's resimplification was rapid. The speed of the Classic period's resimplification suggests that interactions within the local human-environment system amplified the process. The Maya experienced, in quick succession, the failure of their supporting ecosystem and likely the climate supporting it, cultural stress, regional depopulation, and cultural reorganization. They experienced a rapid, unstoppable resimplification (Demarest et al. 2004; Webster 2002; Scarborough 2007; Rice et al. 2004).

**Babylonia.** There is a paucity of detailed information about Babylonia's 3,500-year history (Thompson 2004; van de Mieroop 2004). Despite this shortcoming, what is known provides critical insight into the process of resimplification. It uniquely reveals three particularly important features: a culture's world view routinely constrains its members' future-affecting decisions to within a narrow range; the level of a culture's complexity strongly influences the nature of its resimplification; and increasing cultural complexity doesn't preclude, and may even facilitate, a culture's resimplification. It also reveals that some of the trends and interactions leading to resimplification develop over the very long term, while some shorter-term events repeatedly appear in slightly different forms. Overall, Babylonia's history reinforces the idea that, although resimplifications are unique in detail, in overview they all, regardless of cultural complexity, involve similar types of interactions among the features of the culture's world view, its supporting ecosystem and the general environment.

There are a few reasons why Babylonia's history illustrates these resimplification features particularly well. Both Babylonia's long history as an identifiable cultural unit and its reliance, throughout its existence, on food crops produced by irrigation on a featureless plain provide an enduring background to its many resimplifications. This consistency allows the significance of cultural and environmental changes, such as increasing cultural complexity and climate changes, to be more easily seen and appreciated (Gunn and Folan 2000).

For Westerners, it also helps that we have links to Babylonian culture. For example, their contributions to our world view include defining parts of our zodiac (Gurshtein 1997); the writing of spoken sounds that led to the Western alphabet (Fischer 2001); religious thought, including contributions to the Christian Bible; and multiples of six for measuring time and angles. By late in their history, the Babylonians had formed a complex agricultural culture with which we can readily identify. No wonder that, since the 1800s, Babylonian archaeology has contributed significantly to Western culture's visions about the "natural" order to cultural "progress" and the nature of its "collapse."

These conclusions and observations about resimplification are presented in advance of Babylonia's story in order to facilitate the reading of its history. It is a long story whose faster reading more easily reveals the roles of long-term trends, interactions and the increasing cultural complexity. A slower reading reveals the short-term events, their diversity and connections to ourselves. This includes continuing the description of cultural complexity's rise (started with the story of Abu Hureyra in chapter 3) to a complexity level that Westerners can more easily relate to. The reading effort is worthwhile because, after all, understanding the process of increasing complexity is a prerequisite to understanding its resimplification. (Note that the framework for Babylonia's story given here relies heavily on the work of van de Mieroop, 2004.)

*Background.* Between the northeast corner of the Mediterranean Sea and Kuwait, on the Persian Gulf, lies a narrow, flat topographic low (figure 21). This lowland plain is bordered to the south by the Arabian Desert and to the north by mountains. The further up the slope from the lowlands toward the mountains, the higher the rainfall, which falls in winter. The biomes similarly change from semi-desert on the plain, to grassland with scrub and a few trees (steppe) in the mountains.

On the slopes and foothills of the mountains, the average rainfall can exceed 200 mm/year. Where it does, the rainfall-watered (dryland) farming can theoretically grow crops such as winter wheat. In reality, the annual variation in rainfall means that for long-term dryland agriculture to be successful, the average rainfall has to be greater than about 400 mm/year. The boundary of the area with adequate rainfall to grow wheat follows the trend of the mountains to form an arc called the Fertile Crescent. To the immediate south of the crescent, the dryer grasslands can support the sheep and goat flocks of nomadic to semi-nomadic pastoralists.

The boundary between the areas where dryland farming and pastoral agriculture can be practised moves up or down the mountain slopes as the climate becomes dryer or wetter, respectively. Severe, abrupt or long-lasting climate changes, such as a drought, can be clearly seen in the geological record. These types of events also appear in the archaeological record because they affect food production, to which people must respond.

Further to the south, on the desert to semi-desert lowland plain, rainfall is too low to sustain all but minimal grazing. However, this limitation is offset by the presence of the Tigris and Euphrates Rivers. They rise in the wetter mountains to the northwest, near the Mediterranean Sea, before flowing parallel to one another, southeastwards, along almost the entire length of the lowland plain before entering the Persian Gulf near Kuwait.

The general area between these two rivers is called Mesopotamia. At Baghdad, it can be divided into two ancient cultural regions. To the northwest lies ancient Assyria. To the southwest lies Babylonia. The Babylonian portion is a narrow (maximum width about 200 km), almost timberless, featureless alluvial plain

formed from the sediment deposited by the Tigris and Euphrates Rivers. Babylonia is a combination of ancient Akkad, nearest to Baghdad, and ancient Sumer, further to the southeast nearer to the Persian Gulf (figure 21) (van de Mieroop 2004; Redman 1978).

The Babylonian part of the plain is so flat that, even though Baghdad lies some 500 km from the sea, it is still only 30 m above sea level. It would be uninhabitable except that the perennial (year round) Tigris and Euphrates Rivers enable the fertile soils immediately adjacent to them to be irrigated, turning them into some of the most productive farmland in the region. The rivers also offer a transport advantage because they are navigable.

However, there are limits to the irrigation potential. Around Baghdad, both rivers are cut into the largely uninhabitable desert plateau that receives a total of about 150 mm of winter rainfall. Agriculture is limited to the narrow banks of these river valleys. From Baghdad to the Persian Gulf coast, the enormous volumes of fertile alluvium (silt and clay) deposited by both rivers has merged their deltas. As a result, the river channels immediately to the southeast of Baghdad are less deeply cut into the plain. In this area, it is possible, with sufficient intervention, to irrigate square fields over large areas.

The ancient city of Babylon was built further downstream from Baghdad at a site where the two rivers start to split into a braided network of meandering and moving river channels. From here to the sea, the channels can, at times, be elevated above the surrounding land by natural dikes or levees formed through the rapid deposition of silt. Irrigation is relatively easy to practise by breaching the levee and using canals to guide the water to elongated fields lined with furrows. However, these land characteristics also mean that the low-lying areas between the river braids and their levees can flood at the end of spring to form seasonal lakes and swamps, which add to the few that are more permanent.

The ancient city of Ur is located in the southeastern most part of Babylonia. Immediately downstream from it, the rivers form a marshy region that does not support agriculture. But it does provide other resources, such as fish and reeds for houses (van de Mieroop 2004; Redman 1978).

Farming began in southeastern Babylonia after 6000 BC. Winter irrigation was used to cultivate the early forms of domesticated cereal crops such as winter wheat and barley. Over time, the ancient Babylonians adopted other grain crops, such as millet. During the very hot, dry, windy and dusty summers, grain crops weren't grown. Instead, the Babylonians engaged in intensive, small-scale farming of other crops (Jacobsen 1982). These crops came to include fruits, such as dates, and vegetables, such as beans, chickpeas, leeks and onions.

There are constraints on using irrigation to grow crops in hot, dry environments. Doing so promotes the accumulation of salts, found in the irrigation water, into both the soils and the groundwater through a process called salinization. Its development is further promoted when the area is windy and underlain by impermeable clay-rich soils. Salinized areas can be recognized by the common occurrence of salt-tolerant vegetation. Well-developed salinization gives the desert the appearance of being covered with a dusting of snow. Crop plants struggle to survive and grow in salinized areas.

The area of Iraq that was Babylonia is such a hot, windy clay-rich region. Today, as then, the potential for salinization in this region increases rapidly from the northwest to the southeast, facilitated by a decrease in the permeability of the soil and the natural water table coming closer to the surface (van der Sluis and Hulsbos 1977). In the southeastern part of the region (Sumer in Babylonian times), under undisturbed (indigenous) conditions, there is a dynamic balance between evaporation, which concentrates salts in soils, and the slow flushing of the salts by rain and floods into the deeper groundwater, about 2 m below the surface. Over the long term, the water table is naturally kept at this depth through a combination of the indigenous, salt-tolerant, deep-rooted plants using and transpiring the groundwater, and the slow drying out of desert soils. However, the impermeable nature of the alluvium means that the groundwater does not move very far laterally, so it becomes highly saline.

Irrigation farming in Babylonia (especially in Sumer) disrupted this balance in a number of ways. When the original vegetation was used as fuel and fodder or removed to make irrigated fields, then the area's natural ability to lower the water table was decreased or lost. During irrigation, the amount of water needed to grow the preferred crops can bring the water table to within 0.5 m of the surface. At this shallow depth, the saline water is susceptible to being brought right up to the surface by either additional irrigation water or by capillary action. In both cases, evaporation will cause the salts to precipitate in the surface soil used to grow crops, salinizing it (van der Sluis and Hulsbos 1977; Jacobsen 1982). Over time, as salinization

increases, crop yields decline until they can no longer be grown.

Salinization played a significant role in Babylonia's history. The Babylonians did try to deal with it by leaving their fields fallow every other year. Some deep-rooted, salt-tolerant weeds had a chance to grow, reducing the water table and restoring nitrogen to the soils. In Sumer, the best of their farming practices, such as minimum irrigation and respecting the fallow, could have slowed the rate of salinization but probably not stopped it (van de Mieroop 2004; van der Sluis and Hulsbos 1977; Jacobsen 1982).

Agricultural productivity was constrained by other factors. These included pests, especially locusts, periods of frost and snow, and drought. Little is known about the changes in fertility of non-salinized Babylonian soils through time. These features form the background to Babylonia's story.

*Early Days.* Under these harsh climatic conditions, it is unsurprising that the Ubaid people only settled in Sumer around 6000 BC, roughly the time that Abu Hureyra was abandoned. To take full advantage of the natural resources in their new environment, the Ubaid had to modify their culture and their agricultural practices to accommodate its peculiarities (Redman 1992). For example, there is a roughly six-month delay between the winter rain falling in the mountains and that water arriving in Sumer's fields. Thus, when the water was most needed (during the fall sowing of winter grains in October–November), the rivers were at their lowest and irrigation was difficult. Conversely, when dry ground was needed for harvesting (during the spring harvest in April and May), the rivers were at their highest and the fields most subject to flooding. To overcome these difficulties required an irrigation system that exerted considerable control over the water, more control than the simpler irrigation systems in the foothills to the northwest had over that area's small, intermittent creeks.

Over millennia, the settlers developed irrigation and farming systems that took into account the variations in river flows, the movement of river channels, flooding and siltation. But, in order to build, repair and maintain their more complex irrigation works, the community members had no choice but to extensively co-operate and be managed. This was a prerequisite to success because work plans had to be made in advance and the labour needed to complete the work had to appear at the time they were required. In addition, for the irrigation system to function as intended, the members had to manage the water's distribution fairly and be able to predict the seasonal cycles.

These early Sumerians thus had to resolve both technical issues and make significant cultural changes. Consider that the building and managing of complex irrigation systems changes the relationship between people. Some people have to direct others when and how to complete particular activities. It also changes the relationship between the people and the land: they must control and modify it. For these relationships and activities to be acceptable their world view must be adjusted accordingly. The Ubaid thus began to display the characteristics of a more complex culture.

At the beginning of the Uruk period, around 4000 BC, the growing population of Sumer (southeastern Babylonia) lived in reed houses within permanent villages located near a river channel. There the land could be easily irrigated to produce an abundance of winter wheat and barley. In summer, vegetables and fruits were grown. Away from the rivers, on the land between the river channels, the residents hunted game, such as gazelle, or tended domestic herds of sheep, goats, cattle and donkeys.

The culture of the villages that lived in this environment could be described as territorial co-operatives. They were territorial because each village likely identified with specific lands, and they were co-operatives because they stored food centrally and had an organizational structure that could deal at least with the collection and distribution of the stored food. That organizational structure probably also dealt with collective irrigation issues, such as construction, maintenance and who could irrigate when (van de Mieroop 2004).

*Cities: Uruk.* With the passing of time and population growth, some of the villages became population centres. For example, around 3400 BC, one of the world's earliest cities, Uruk, was established in southeastern Babylonia (figure 21). It was also the largest aggregation of people at that time. It reached 100 ha (1 km$^2$) in size and, based on an estimated density of 200 people/ha, housed 20,000 people (Modelski 1997).

The city residents lived in tightly packed houses, clustered around one or more mud-brick temple buildings. The buildings, and the organization of the religious-government institution who used them, are thought to have arisen from the village's earlier co-operative structure that managed the communal food supply. In particular, the temple buildings likely had their origins in the buildings used by the village to store their communal food. Over time, the head priest is thought to have become the ruler of the urban centre, while specialists of lower ranks became the managers of temple affairs, including food storage and distribution.

The appearance of a city with its central temple and new management positions were accompanied by wider cultural changes. There was a general increase in the diversity of the culture's work functions and the skills needed to perform them. The 70%–90% of the population who worked as farmers required farming skills. Those who worked in the cities required different skills and more specialized tools. For example, around 3500 BC to 3100 BC, accounting using proto-cuneiform writing appeared (Nissen et al. 1993).

These changes to work affected how individuals saw themselves. Their distant ancestors' sense of identity and place had mostly been provided by kin relationships and landscape features. This was likely augmented by their newer, developing relationship to the city they lived in and to the work they performed. The changes to work also changed how they differentiated themselves from one another. The number of subgroups in the culture increased.

All of these adjustments facilitated the emergence of the more strongly defined social hierarchy and the greater structural complexity that would characterize Babylonian culture for millennia to come. This cultural transformation was reflected in architectural changes. Those higher in the hierarchy, such as the priests, lived in grander structures than those lower down.

The temple played a special role. The citizens had come to believe that both their personal and their city's welfare depended on the grace of the patron god to whom the temple was erected (Inanna in the case of Uruk). When the temple became a monument-like building, it literally embodied that relationship, thus strengthening the centralized power of those who built and controlled the temple.

By the middle 3000s BC, most of the basic aspects of the Babylonian culture were in place. These included a city, a leader with power over others, a monumental temple, specialization, social hierarchy and a congruent, supportive world view and identity. All of which was dependent on farmers producing agricultural products from their irrigated fields and the neighbouring pastoral lands.

Shortly after 3100 BC, Uruk was razed. The city's dominant influence on the surrounding area declined as the people dispersed to establish numerous rural villages in which local traditions re-emerged. The use of writing waned (van de Mieroop 2004; Hassan 1981). Although Uruk would rise again to become a prominent city, its period of dominance over Babylonia had ended.

The cause of Uruk's *resimplification* is unknown. Salinization was likely not an issue because, around 3500 BC, wheat and barley (which is more salt tolerant than wheat) were grown in equal proportions (Jacobsen and Adams 1958; Jacobsen 1982). There is some indirect evidence for a severe but short-lived (200-year) drought around 3200 BC to 3000 BC (Weiss and Bradley 2001; Cullen et al. 2000). This may have been the same event that is assumed to have driven the people of the Farafra oasis to migrate into the Nile valley (Hassan 1997). But drought alone does not fully explain the city's destruction. Neither does the slow, millennia-long climate cooling which began around this time. Maybe the citizens just became fed-up with the rulers or disillusioned with their culture's gods.

### City States: Ur, Nippur, Kish, Eridu and Uruk.
After 3100 BC, the cultural changes initiated in southeastern Babylonia during the rise of Uruk spread throughout Babylonia. The population continued to grow. By the high 2000s BC, there were about 35 urban centres, roughly evenly distributed across southeastern Babylonia. Each of these centres influenced or controlled, through a hierarchical structure, the irrigable areas within some 15 km around it. Thus the distance to and from those areas is similar to a hunter-gatherer's one-day maximum return journey from home of 30 km. (This is similar to the size of the Mayan city states.) Within this hinterland, nearly all the people were farmers.

Between about 2900 BC and 2350 BC, the urban proportion of the Babylonian population grew and stabilized at an all-time high (Thompson 2004). This is reflected in the size of the rebuilt Uruk that, by 2800 BC, had reached its maximum size of 400 ha (4 km$^2$). Its population is estimated at around 80,000 (Modelski 1997).

Some of the urban centres, including the cities of Ur, Nippur, Kish, Eridu and Uruk, exerted increasingly centralized control over their hinterland (figure 21). Each urban centre, its hinterland and the people living in it had become city states. It seems that the state was initially ruled by the priest in charge of the temple dedicated to the city's patron god. This situation was not to last.

Well-thought-out speculation suggests that the increase in population and the associated increase in land requirements led to the hinterlands of the various city states expanding till they met. Boundary disputes likely arose. A military leader was probably assigned to skirmish or feud, as needed, with rival city states over issues such as land boundaries, water use and irrigation works. Two parallel cultural organizations came into existence: one for religious matters, and one for military matters. Over time, the military position became permanent and the person holding it became the head of the city state.

Eventually the head person became a king (occasionally a queen), with a hierarchical organization below them. These rulers had the prerogative of establishing a dynasty. Under a monarch, the governing and the religious organizations remained separate. However, they were strongly linked together because the monarch required their rule to be endorsed by the priest, who acted on behalf of the city god.

Along with these organizational changes came complementary adjustments to the culture's technology and world view. The technology changes included the use of bronze, which first appears around 2900 BC. There were also alterations to writing, which by about 2400 BC, was being used to record speech and narratives as well as undertake accounting.

By this time, the Babylonian culture's world view was quite different from that of their hunter-gatherer ancestors. For example, the hunter-gatherer restraints on one person lording it over another had been replaced by a hierarchy of lording. Just how strong the members' acceptance of lording had become is illustrated by the royal funeral and burial practices at Ur between about 2600 BC and 2450 BC. (For reference, the great pyramid of Khufu [Cheops] in Egypt was being completed around 2500 BC.)

The divine ruler of the Ur city state, Queen Puabi (Pu-Abum), died at the age of 40. In keeping with the customs of the time, she was buried, likely within three days of her death, dressed in the cloths and adornments associated with the gods. A multi-day funeral celebration of music, wailing and feasting followed.

At some point in the proceedings, a number of the royal retainers were killed by a skull-puncturing blow to the back of the head. Their bodies were preserved and then dressed in the roles they played in life, such as soldiers or attendants. Some of the female attendants were dressed in cloths associated with the gods. Their adorned bodies, along with the objects they used to serve the dead royal, were arranged in a grave adjacent to her (Baadsgaard et al. 2011).

The retainers must have been aware of their fate, yet apparently submitted to their sacrifice willingly. They may have done so because they saw the opportunity as a great honour, not to be feared. After all, they had lived a good life in court and were now given the opportunity to serve their divine ruler in the afterlife (Wilford 2009). Perhaps their culture had already taken to heart the idea of learning to live with death, a lesson found in the *Epic of Gilgamesh*, whose written roots only go back to almost 2000 BC but whose spoken roots probably go back much further, to at least this period.

A mature city state was independent, had its own monarch, patron god, military force and unique characteristics. It built irrigation works, temples and palaces, managed trade, and kept records on clay tablets using the evolving cuneiform writing. It also competed with other city states, sometimes through war. These city states dominated Babylonia until the 1600s BC.

However, despite the political division into city states and their related ethnic differences, Babylonia was culturally quite homogeneous. The city states' cultural features had much in common. For example, the patron gods of all the city states belonged to the same pantheon; all the cities used the same basic script; they had a similar organizational hierarchy; all had a common way of utilizing the land; and their wealth still rested directly or indirectly on agricultural production from irrigated lands (van de Mieroop 2004). One immediately thinks of the Mayan cities and culture of the Classic period.

*Attempts at Unity: Akkad and Ur III.* During the period between 2334 BC and 2004 BC, the ruler of one city state occasionally managed to gain control over all of Babylonia by forming alliances and using military force. The first unifying attempt, called Akkad (ruled from the city of Kish and later Agade, near Kish), lasted from 2334 BC to 2192 BC before collapsing into confusion. Akkadian rule from Kish was replaced shortly thereafter by rulers from Ur in a unification known as the third dynasty of Ur (Ur III), which lasted from 2112 BC to 2004 BC.

Although both states pursued policies aimed at firmly centralizing political, administrative and ideological power in their respective city, neither could be called a territorial state. They are better described as one city state dominating the other city states who, although vassals, still retained their independent identity and organization. Thus, when the unity governments failed, the cities reverted to governing themselves as independent city states.

Although both of these unification attempts failed, they did result in significant cultural changes. Under Akkad, a stronger direct connection between the monarch and the gods (including the monarch being a god) was established. The political authority of the temple was reduced.

Physical changes under Akkad included expanding local irrigation works and securing resources from distant locations through various means, including the use of military forces. These resources included tax, tribute and access to trade routes. The places involved were diverse, including the Persian Gulf to the southeast and the dryland farming areas to the north and west, such as Tell Leilan in northwestern Mesopotamia (figure 21).

During the Ur III unification, there were similar significant undertakings in canal construction. Standardized weights measures and a writing system were in place by this time. The rulers managed to centralize the bureaucracy but only for a brief period (van de Mieroop 2004).

How did the Akkad and Ur III unification attempts come to end in cultural *resimplification*? The Babylonian world view held that the monarch ruled at the pleasure of the gods. The religious leaders emphasized that the calamities that befell the state, such as those that destroyed Akkad and Ur, were the result of impious deeds by the monarchs or their failure to fulfill obligations to their gods. If the gods were displeased, then they wreaked their revenge by, for example, sending invaders. For some of the calamities that befell them, the Babylonians believed that the reasons why were only known to the gods (Yoffee 1988b).

In contrast, researchers of today attribute the failure of Akkad to a combination of different factors. These include the strong opposition by the vassal city states to being ruled by a single central government; the stress on the Akkad government as it tried to expand its organizations into "foreign" lands hostile to the idea of unification; and the impact of invading or, more likely, in-migrating nomads (van de Mieroop 2004). There is also the suggestion that the Akkadian preoccupation with military expeditions outside Akkad meant that they neglected the difficulties associated with forming and running a centralized government.

Drought likely played a significant role. Evidence for drought is found at Tell Leilan (an Akkadian vassal), which is located in an area highly sensitive to changes in rainfall. Between 5000 BC and 2500 BC, Tell Leilan was a typical dryland farming and pastoral town.

However, from 2600 BC to 2200 BC, the town experienced a sixfold population growth, the building of a planned city, and membership in a local city state. These changes reflect a regional expansion trend of the time.

Starting around 2200 BC, the population of Tell Leilan, and an area within 15 km of it, rapidly declined by an estimated 20,000 people. It seems that they migrated out rather than died there. Tell Leilan and nearby settlements were abandoned and then immediately covered by the windblown products of desertification.

Within the limitations of the dating techniques, Tell Leilan's abandonment coincides with the onset of drought recorded by the area's desertification, a decline in the flow of the local rivers and an increase in wind. The drying may have lasted as long as 300 years (Weiss et al. 1993). Under these conditions, the production of food in this marginal farming area would have been drastically reduced, thus forcing the people to migrate (compare this with the Greenland Norse). The migrants could have been attracted to Babylonia because of the water in the Euphrates and Tigris Rivers.

Although not recorded in Babylonian written records, there is ample evidence for a severe, Eurasia-wide drought at this time. A temperature anomaly is recorded in the north Atlantic (deMenocal 2001). Ocean sediments from the Gulf of Oman reveal an increase in wind-borne, desert-derived carbonate particle that suggest a 300-year drying period around 2065 BC ± 150 years (Cullen et al. 2000). In central India, Lonar Lake sediments indicate one or more prolonged (multidecadal to multicentury) droughts between 2600 BC and 1900 BC (Prasad et al. 2014). Sediments from paleo-lake Kotla Dahar in the Indus River plain of northwest India indicate a drying period around 2100 BC (Dixit et al. 2014).

One or more severe droughts between 2200 BC and 2000 BC had significant impacts on other cultures. For example, in Egypt around 2150 BC, sand dunes encroached into the Nile valley at a time of abnormally low Nile floods. The Nile's reduced flow was likely caused by cooling and drying in the Nile headwaters. These events roughly coincide with Egypt's 40 years of famines, civil disorder and the loss of royal power to the provincial governors, a resimplification that started in 2180 BC (Hassan 1997).

It is unlikely that Akkad escaped a Eurasia-wide drought. It would have reduced the flow of the Tigris and Euphrates, thus affecting Akkad's ability to produce food through irrigation. A drought would also have reduced the pasture and crop wastage required to feed their domestic food animals.

Compounding Akkad's difficulties was the continued, millennia-long loss of food production to ongoing salinization in southeastern Babylonia (van der Sluis and Hulsbos 1977). For example, in the Girsu region, an increase in salinization is suggested by the change in the ratio of wheat to barley (figure 21). The proportion of wheat declined from 50% in 3500 BC to 16% by 2400 BC to 2% by 2100 BC. This declining trend continued until 1700 BC, after which wheat was no longer planted.

The change in grain yields follows this trend. For example, in 2400 BC near Girsu, barley yields were still high, at around 2,540 litres/ha. But, by 2100 BC, yields had dropped to 1,460 litres/ha. At this time, the Babylonians were recording the presence of salinized fields (Jacobsen and Adams 1958; Jacobsen 1982). Ironically, the expansion of irrigation works during both the Akkad and the Ur III unifications could have exacerbated the salinization process by making overwatering possible during a drought.

Interactions between salinization, irrigation changes, loss of pasture and perhaps an in-migration would have amplified their individual influences on food availability. The population was likely underfed. The economy would have been dragged into the mix because it was also highly dependent on the productivity of the land. For example, any lowering of agricultural production would have reduced the funds available to Akkad for other ventures, including irrigation maintenance and foreign military campaigns.

These issues alone would have sorely tested Akkad's inexperience at administering a more complex, unified Babylonia. When the drought arrived, it would have amplified any weaknesses in the state, making it more unstable. Adding to their woes were the vassal city states' striving for independence. Collectively these influences could have led to Akkad's resimplification.

After Akkad's resimplification in 2192 BC, times were still tough in at least some parts of Babylonia. This is suggested by Uruk's fate. By about 2100 BC, Uruk had shrunk to around 30 ha (0.3 km$^2$, 540 x 540 m) in size with a population of about 6,000 people. After this date, Uruk never became a town of any size (Modelski 1997).

Babylonia's Akkad resimplification was followed by the Ur III unification, which began in 2114 BC. It ended suddenly in 2004 BC for unknown reasons. Its resimplification can't be blamed on economic difficulties. Records from Ibbi-Sin's reign, nine years before Ur III resimplified, report that some 18 kg of gold and 75 kg of silver were requisitioned from the state coffers. The monarch was not broke (van de Mieroop 2004).

On the surface, Ur III's resimplification appears to have been caused by the city states' ongoing desire to reassert their independence and the lack of a leader powerful enough to keep them unified. On closer inspection, other possible contributing factors appear. Cuneiform tablets record a 15-fold rise in the price of grains, so high that these foods could no longer be fed to animals. The reason may have been a refusal of the provinces to pay their proportion of the harvest or tax. However, it is also possible that the very large domestic herds known for the period could have overgrazed the pasture (Jacobsen 1982). Perhaps a combination of these factors and ongoing salinization could have, like Akkad, reduced Ur III's ability to maintain or increase food production (van de Mieroop 2004).

Contradicting these explanations are Babylonian texts that blame the state's woes on invading nomads (Amorites) from the north. At least this is how the Ur III monarchs justified their building of the Amorite wall (2054–2030 BC) (figure 21) (Weiss et al. 1993). However, the Babylonians' overall positive relationship with those Amorites who had settled in Babylonia, and the general prejudice they had against semi-nomadic peoples in general, suggest that invaders or migrants, if they existed, were either scapegoats or not the sole cause of resimplification.

After Ur III's recorded resimplification in 2004 BC, economic activity remained high and urbanization remained dense. But the resimplification process continued, although more slowly. The cities re-established themselves as independent city states. But they also retained their Ur III culture. They saw themselves both as city states and as being closely linked together within a common territory under one ruler. For example, the city of Nippur (city god Enlil) was accepted by all city states as the religious capital, with Enlil being the head god. The ruler of the city state that held political control over Nippur was entitled to be acknowledged by the patron gods in the other city states as the monarch of both Akkad and Sumer (Babylonia).

This right to rule was supposed to circulate among the cities. But Babylon, Sippar, Uruk, Larsa and Isin squabbled over whose turn it was to produce the monarch of all Babylonia. They also occasionally fought each other to either become that ruler or remain independent from them. By the high 1800s BC, their squabbling had become open and, apparently, incessant warfare (as occurred with the Maya). Babylonia had fragmented into warring city states.

Whatever the real reasons, the sharp initial and subsequent drawn out resimplification of the Ur III unification had much in common with Akkad's resimplification. Of particular note was the ongoing difficulties Ur III's rulers experienced when governing the recently unified Babylonia, such as cities demanding independence and maintaining the irrigation works. The prohibitive costs of feeding grain to animals certainly indicates pressure on food supplies. Perhaps the state was weakened by the cities not paying their taxes, or by the limits to food production caused, at least in part, by salinization and overgrazing. A drought would have further limited the availability of food by reducing production and by enlarging the population through in-migration (Weiss et al. 1993).

***Attempts at Unity: Amorite.*** After the low 1900s BC, Babylonia was a collection of squabbling and then warring independent city states. In 1763 BC, Hammurabi of Babylon (patron god Marduk), conquered Larsa and successfully unified the city states, for the third time, now under an Amorite dynasty. Hammurabi integrated the cities tightly into a Babylon-based power structure, as illustrated by his codification of laws and their application across the land. This is one reason why Babylon remained the capital of Babylonia for the next 1,500 years. However, shortly after Hammurabi's death in 1750 BC, the unified Babylonia slowly started to resimplify, again. It continued to do so until the high 1500s BC.

The post-Hammurabi resimplification started in southeastern Babylonia. Cities that had previously been flourishing, such as Ur and Nippur, were rapidly abandoned. The southeastern population migrated northwestward. One can't help but think of the depopulation and migration following the Mayan culture's resimplification at the end of their Classic period (Webster 2002).

The disintegration of southeastern Babylonia and the migration shifted the cultural focus to northwestern Babylonia. Marduk, Babylon's city god, became the favoured god in Nippur's pantheon. The area around Babylon became Babylonia's heartland. Initially the northern cities flourished. For example, it was there that the *Epic of Gilgamesh* first appeared in writing and the first texts on mathematics appeared. But, around 1700 BC, indicators of economic decline appeared.

After 1750 BC, the Babylonian heartland held together for 155 years without, it seems, any major internal political discord. The state successfully fought off attacks from the Kassites as well as other opponents. Although the Babylonian state continued to function as a whole through this slow resimplification period, the mood of the written record reflects the regional-scale uncertainty and the violence of the times. Babylon increasingly existed in a void surrounded, as it was, by a sparsely inhabited region.

In 1595 BC, the Hittite king, Mursili, sacked Babylon. He soon withdrew, leaving the area leaderless. The ancient Babylonian political and social structures vanished. By this time, much of the population was rural.

The reasons for the Amorites' slow *resimplification* are not reliably known. There were a number of possible contributions. A well-documented factor whose roots go much further back in time might provide a reason. During Ur III's post-2004 BC resimplification, independent contractors rather than state "employees" were used to fulfill bureaucratic functions, such as the assigning and collecting of the farm-production quota. In the beginning, this method had the advantage of enabling the partly unified Babylonia to keep functioning during the instability associated with a series of rapidly changing rulers. However, during the Amorite unity, this method had a significant downside.

The contractors, who also happened to be Babylonia's elite, continued to set unrealistically high production targets for agricultural producers. Those targets, and an inevitable bad harvest, led to farmers taking out loans from the contractors to make up for the shortfall. Loans were accompanied by very high interest rates, for example 33% on grain and 20% on silver. Debts were often unrepayable, which led to chronic indebtedness and dissatisfaction among a significant proportion of the population.

A large, indebted population is not a tax base that will help keep a complex culture stable, while a dissatisfied people leads to a breakdown in order. To address these issues, the Babylonian monarchs, including Hammurabi, had to periodically annul the debts. But this process weakened the monarch (a proxy for the state) because they had to absorb the loss, while the contractors (the elite) became rich at the monarch's expense. More than a hundred years after Hammurabi's Amorite unification, Ammisaduqa, who ruled in the low

1600s BC, annulled debts. This indicates that the Babylonian rulers had been unable to permanently solve the problems of debt and powerful elites.

Reoccurring indebtedness suggests a significant cause for cultural destabilization. The account keepers of the time record significant numbers of people leaving their labouring jobs on the farms. Perhaps they were fleeing their debts. When widespread enough, this reaction could have contributed to mass migrations, rebellions and a general disintegration of the economic system, especially if the monarch was inherently weak (van de Mieroop 2004). This could have contributed to the slow resimplification of the Amorite unification.

Hammurabi's successful integration of the city states into a unified, centralized state controlled from Babylon likely also played a critical role in Babylonia's subsequent slow resimplification. The functioning of the unified state relied on resources being provided to Babylon by its state-controlled, regional-scale production and exchange system. If this system became dysfunctional, for example from a lack of food to distribute or ineffectual management, then the central authority in Babylon would have been weakened. However, Hammurabi's more complete integration of the provincial cities into a central government also meant that these cities could no longer resolve their differences with the ruler by simply reverting back to being independent city states. They were more likely to fail along with the central government.

The depopulation of southeastern Babylonia when its citizens migrated northwestward to the area around Babylon likely also played a part in the Amorite resimplification story. Their migration could have been a response to ongoing salinization. No wheat was grown there from 2100 BC to at least 1700 BC. The barley yield at Larsa was only 897 litres/ha (Jacobsen and Adams 1958; Jacobsen 1982). Migration could also have been driven by a shift in the river channels crossing the flat plain of southeastern Babylon. A large enough shift could leave a city and its lands dry, which would certainly have altered its liveability and ability to grow food. Perhaps a political rebellion and a harsh military response from the centre of power could have played a part. Whatever the reason for the migration, southeastern Babylonia never fully recovered.

In the short term, Babylon benefited from the shift in population and power to the north. But the migration may have been the forerunner of a longer-term decline in food availability. In northwestern Babylonia, salinization was not a problem, however food prices were still inflated. New methods of intensifying agricultural production appeared and new irrigation works were built. These features, together with salinization in the southeast, suggest that Babylonia, in general, did indeed suffer from a problem of food production, or too large a population for the northwest to alone feed (Jacobsen and Adams 1958; van de Mieroop 2004).

In summary, it seems that the resimplification of the Amorite unification was the product of a number of factors. In the southeast, ongoing salinization, perhaps helped by mismanagement by Babylon, may have led to a large proportion of a dissatisfied southeastern population abandoning their fields and cities to migrate into northwestern Babylonia. After the benefit of absorbing this migration waned, the inherent instability in the relationship between the elite, the cities, the monarch and the people on all levels, maybe made worse by food production issues, led to the slow breakdown of the Amorite unification. As the centre weakened, the inability of city states to revert to self-government in times of need left Babylonians with few options. Abandoning the cities and returning to rural living was a common solution.

Although the Amorite unification failed, its legacy did precondition the area for the appearance of a future territorial state. Babylonia was left with new canals and fortified cities. Most importantly it was left with the Amorite's unified administrative system that directed income to Babylon to pay for major projects: the idea of a central government was now a part of their culture's world view (van de Mieroop 2004). By the 1700s BC, Babylon became the centre of Babylonia and remained so for the next 1,200 years.

### Great Territorial State: Kassite.

The power void left by the 1595 BC Hittite invasion was eventually filled by the Kassites. Both the Amorites and the Kassites were semi-nomads. Since at least the 2100s BC, Amorites, and later Kassites, had lived on the pastures north and northeast of Babylonia for part of the year and in villages near Babylonian towns for the rest of the year. Some became permanent residents in Babylonian towns where, as was the case for many ethnic groups, they had become integrated but not absorbed into Babylonian culture.

From this position, the Kassites eventually started to exert political power. In 1475 BC, they ruled southeastern Babylonia. By the 1300s BC, they had unified Babylonia into a territorial state of equal status to the other territorial states now arising in the region. (*Territorial state* refers to a centralized administration and economy that unequivocally controls an extensive area with recognized boundaries.) These states would later become the region's great powers.

In the case of Babylonia, the Kassite administrative and economic system was centrally controlled to an unprecedented degree by the monarch, who resided in Babylon. The other major Babylonian cities still hosted political, religious and cultural life, but they no longer had political autonomy or an independent economic system. They and their hinterlands were now provinces. The monarch relied on their appointees to these provincial cities, who were thus the governors of the provinces, to supply the resources to run the territorial state.

Each provincial governor was the head of an agricultural organization that collected the crop harvests and animal products from their province. These goods were then redistributed to the provinces' dependants by the state institutions (e.g., the temple), with a portion being sent to the monarch in Babylon. Through this system, the small urban elite and the monarch became rich.

The monarch of Babylonia could afford to reconstruct the old cities as well as build a new capital at Dur-Kurigalzu. (Babylon, however, still retained its pre-eminent position.) Despite these administrative changes and much rebuilding, the level of urbanization still remained low. The minimum size of the Babylonian urban population at this time has been estimated at roughly equal to the urban population during the Uruk period of 3700 BC. The majority of the population lived rural lives (Thompson 2004).

During this territorial state period, the standardized version of the much older *Epic of Gilgamesh* and the famous Babylonian creation myth, in which Babylon was marked as the home for the gods on earth, were penned. Because Babylonia was culturally revered throughout the region, these and other Babylonian stories were found in all the major cities of the great powers, even in Egypt.

Regionally, from 1500 BC to 1100 BC, Babylonia was part of a near eastern world that had not been seen before. The large territorial states appeared: Kassite Babylonia, Hittite Anatolia (Hatti), Egypt, Assyria (including Mittani, which Assyria absorbed), Elam and Mycenae (Greece). These Bronze Age states rose and fell in unison, begging some coincident causes for their rise and demise that have relevance to Babylonia's history.

The rise of these great states has been portrayed as a chance coalescence of individual histories facilitated by their interaction. However, a more complete understanding emerges if one considers that both their rise and fall correlates with the growth and collapse of interconnected international-level diplomacy and trade. They formed a system that bound these states together.

In simplified form, contact between the expanding territorial states was inevitable. During these interactions, the monarchs of each great state treated one another as equals. However, at the same time, they competed with one another as rivals by constantly trying to expand or maintain their territories through warfare and intrigue.

To play this Bronze Age "Great Game," the respective monarchs relied heavily on diplomatic contact and trade. The diplomatic contact was needed to make treaties and facilitate the exchanging of status-maintaining luxury gifts. Trade was needed to provide each state with a continuous supply of those resources and highly valued products that each monarch felt was essential to demonstrate their power to their rivals, and to their own people. Luxury goods were also needed to maintain the extravagant lifestyle "expected" of the monarch and their supporting elites. This system of diplomacy and trade became a critical part of the quasi-stable symbiotic relationships among the great states. It helped to keep each of them in power and their administration functioning.

Although each of the great states was unique in detail, their basic characteristics had much in common. The states had arisen through military action, so their cultures were militaristic. The rulers all spent a considerable amount of wealth maintaining a large standing army to fight their rivals and outsiders, and to quell rebellions.

In order to boost their status and maintain power, each ruler spent vast sums building and rebuilding ever grander cities and monuments dedicated to themselves and the relevant gods. For the people of each state, these monuments embodied their culture's superior world view and the power of the builder, their ruler. Monument building was thus a critical part of the Great Game.

All of the states displayed an enormous discrepancy in access to wealth and power. A small minority of ultra-wealthy elite lived in the cities. A large majority of much poorer commoners lived, mostly, in the countryside.

The high cost of wars, maintaining an army, purchasing luxury goods, exchanging diplomatic gifts and building monuments was initially paid for by war booty. When that ran out, the great states relied on the efforts of their farmers and labourers. The way the wealth generation process functioned meant that all of the great states had to deal, to a greater or lesser degree, with the rising indebted servitude of their workers. However, the monarchs no longer felt there was any need to restore the social balance through debt reduction, as Babylonia had done between 2000 BC and 1600 BC, perhaps, in part, because many of the workers were conquered peoples. The people responded accordingly.

Because the indebted could expect no local respite for their condition, many fled their home state. The number of debt refugees was so significant that there were often clauses in the treaties between the great states stating that these fugitives were to be returned. After all, the loss of so many people weakened the state by reducing the crucial labour pool required to generate the needed wealth and supply the soldiers to form an army.

Between 1250 BC and 1100 BC, the Bronze Age great state system slowly disintegrated. Each of its member states underwent a decades-long resimplification (Cline 2014). Egypt and Babylonia maintained their complexity for the longest period, but they too eventually resimplified.

The end of Kassite Babylonia as a great state was signalled when Elam captured Babylon from the Kassites in 1155 BC. The Elamites did not occupy Babylon. Instead, the city of Isin provided the (puppet?) rulers to occupy the Babylonian throne. Under Elamite rule, Babylonia revived for a few decades but, by about 1100 BC, it went into a general decline (both northwest and southeast) that lasted until around 650 BC. Certainly from about 1200 BC until it formed an empire, Babylonia was no longer the cultural centre of the region and was subject to external influences of all kinds (van de Mieroop 2004).

Kassite Babylonia likely *resimplified* from being a territorial great state for reasons related to the near simultaneous resimplification of all the great states and the failure of their diplomacy and trade system. The historical records from the time suggest that the disintegration was caused by a widespread invasion. In the western great states, the perpetrators are commonly referred to as the "sea people" of the eastern Mediterranean. The archaeological evidence does not support this explanation.

Similarly, the Babylonians blamed their decline on the invasions by the semi-nomadic Arameans. These pastoralists from northern Syria migrated into the general area around the time of this resimplification. The Arameans did occasionally raid into Babylonia, perhaps as early as 1100 BC, but they did not occupy territory there until the 800s BC. If an earlier invasion did occur, it was unlikely to be the sole contribution to Babylonia's resimplification.

There is a more plausible explanation for the great states', and thus Babylonia's resimplification. The failures of each state were roughly synchronous because of their symbiotic relationship. Each state became mutually dependent on the other states to help keep the peace, make war within acceptable levels, demonstrate their power, and generate the wealth to keep their administrations functioning (features in common with the conditions before WWI).

After the booty from each great state's initial expansion had been spent, it became much harder to support themselves and their interconnected trading and diplomatic systems. In order to fulfill their obligations to this symbiotic relationship, each state had to have a large labour pool that could continually generate substantial wealth and staff large armies. But the limited size of each great state's often unwilling labour pool could not easily meet these obligations. This situation was made worse by the chronic indebtedness of the labour and their lack of respite from it, as discussed above. Because of these similarities and their mutual interdependence, one state's failure to generate wealth would likely soon be taken advantage of, or be experienced by, the other states. Any sign of weakness would also become an opportunity for the dissatisfied, displaced, deported or conquered groups within each great state, and for outsiders, to test a state's ruler.

The discovery, dispersal and increasing use of iron could also have contributed to the disruption. Unlike copper and tin, iron ore was plentiful. Iron metal is easier to produce than bronze, and a wider range of tools (including weapons) could be more easily manufactured from it. Iron also had a different, perhaps less mystical, quality than copper or bronze. Over time, as these features of iron were exploited and absorbed, the financial, power and perhaps even the religious or cultural belief aspects of the inherently unstable Bronze Age system could have been disrupted.

Kassite Babylonia was not immune. The long-term source of Babylonian wealth relied on farm production, which was managed by the provincial governors. If the producers could not meet their quota, then the temple would provide the farmers with advances and loans. This led to the same widespread, chronic indebtedness

seen in earlier Babylonian times, but now with no possibility of respite. Migrations and the dissatisfaction of the people was one result. This situation was likely exacerbated by the Kassite rulers' continuing the Isin dynasty's habit of granting large tracts of land to insiders in return for special services (van de Mieroop 2004).

Perhaps difficulties with labour were not the only factor preventing the Babylonians from producing the agricultural crops its rulers relied on to play the Bronze Age Great Game. There are no records discussing salinization conditions in southeastern Babylonia at this time, however its nature suggests that the yields in that area would still have been low. Central and northwestern Babylonia is less susceptible to salinization. The only references to it in these two areas are indirect: salinization is used in curses of the time. Perhaps salinization appeared in the centre and northwest between about 1300 BC and 900 BC (Jacobsen and Adams 1958; Jacobsen 1982).

There is fragmentary regionally dispersed evidence for a drought around this time. In Nubia, there is evidence of drought shortly after 1260 BC. Famines are thought to have occurred in Libya in 1214 BC (Hassan 1997).

A less well-documented environmental contribution that may have uniquely added to the Babylonian's difficulties was a shift in the channel of the Euphrates (Jacobsen and Adams 1958; van de Mieroop 2004). A city left high and dry by a shifted river channel would have difficulty to irrigating its surrounding fields, even in the northwest. Overall, it seems that Babylonia under the Kassites, along with the other great states, resimplified because their expensive way of life and the manner by which they were governed was inherently unstable.

The condition of resimplified Kassite Babylonia after 1100 BC reflected the region's new reality: chaos. From 1100 BC to 626 BC, Babylonia was what we today would call a failed state. Record keeping was rarely practised between 1055 BC and the 750s BC, indicating that Babylonia was economically weak. This is probably the result of the agricultural infrastructure, including the irrigation canals, disintegrating, and the lack of access to trade routes.

The long-term trend of declining urbanization that started around 2500 BC continued, with a sharp dip through 1100 BC. Around 1100 BC, the urban population had dwindled to roughly 25% of its size during the city-state era of 2500 BC. During that era, there were 13 cities that housed more than 10,000 people (50 ha). By 1100 BC, there were only two perhaps three: Babylon, Isin and Ur (Modelski 1997). Nippur may have become just a village around a crumbling ziggurat (Babylonian step pyramid). Most of the population now lived in villages. Some people reverted to a semi-nomadic lifestyle.

This state of affairs is unsurprising considering the political volatility. The remnants of the great Babylonian cities, particularly Babylon and Ur, were still the centre of a functioning Babylonian culture based on ancient traditions. But the political control the groups maintaining that culture could exercise did not extend much beyond the city walls. The Chaldeans, Arameans and remnants of the Kassites dominated the countryside, which was essentially independent from the cities. Aramaic was widely spoken. The various groups in the cities and countryside incessantly squabbled and warred among themselves, and with Assyria, over who controlled the Babylonian throne. Not that it mattered much who sat on it, as the defunct irrigation systems show.

Babylonia's regional significance was further reduced by other important regional developments that occurred during the unsettled, post-1100 BC times. One of them was the growing use of the linear alphabet with 22 characters, which began displacing cuneiform as the regional script. Another development was the displacement of bronze, the historical metal of choice, by the now more easily available iron. There were also the cultural changes that accompanied the regional-scale in-migrations of foreign people and its coincident internal population movements. However, in the short term, the most significant event for Babylonia was the rise of Assyria (van de Mieroop 2004).

*Assyria.* By 859 BC, Assyria, to the immediate northwest of Babylonia, had developed a highly militaristic culture that embraced the use of iron for weapons. By using them, Assyria re-conquered its previous great state territory and then expanded to dominate almost all of the Near East. From its capital at Nineveh, it managed an enormous empire that stretched from Egypt to Babylonia (figure 21), from which it had received a large, sudden influx of wealth.

Despite the Assyrians' new power, they did not completely take over Babylonia. Instead, it became a vassal state that the Assyrians controlled by influencing those in Babylon who claimed the throne. During the 700s and 600s BC, the Assyrians experienced difficulties controlling Babylonia, but their administrative method did not change.

Perhaps the Assyrians treated Babylonia in this manner to be pragmatic. They needed the state to be obedient, but they also recognized that they could not control Babylonia through standard military tactics. It was inhomogeneous, had a rebellious nature and the southeastern-most part of Babylonia was difficult to conquer or control because it was marshy. They also respected Babylonia's resiliency and long history, otherwise why would they sack Babylon in 689 BC and then promptly rebuild it?

The Assyrian empire reached its height around 640 BC under Assurbanipal. By 631 BC, at the end of his reign, the empire was crumbling in the east. Shortly after he died in 627 BC, the squabbles over the succession marked the start of a whirlwind of changes that was to be Assyria's resimplification.

In 627 BC, the ruler of Babylon, appointed by Assyria, mysteriously disappeared, and Babylonia was no longer under Assyria's control. By 626 BC, the Neo-Babylonian dynasty had arisen in Babylon. The year 616 BC saw a Neo-Babylonian king unite Babylonia under his rule and invade Assyria. An alliance between the Medes, a pastoral mountain people, and the Babylonians was formed. In 612 BC, the alliance took the Assyrian capital, Nineveh, and sacked it. By 610 BC, Babylonia controlled all Assyrian land.

Incredibly, 30 years after its height, the Assyrian empire was gone. Its urban characteristics had disappeared. All that was left of the cities was a remnant population living in small settlements atop massive mounds: the cities' remains (van de Mieroop 2004).

It is not exactly clear why Assyria *resimplified*. The most likely reasons lie within the empire itself. Assyrian culture was militaristic, with a highly hierarchical and centralized structure. Power and decision making rested almost exclusively with the leader. This meant that decision making was straightforward, but the management system could easily fail if the king was weak or overwhelmed with the decisions he had to make. Assyria also routinely experienced succession difficulties. The resulting squabbles inhibited the functioning of the bureaucracy. There was thus a lack of political and managerial resilience in the empire that made it inherently unstable.

That instability could have been exacerbated by Assyria's other basic characteristics. For example, in order for Assyria to both maintain a large army and pay for its perceived need to display its wealth, it had to receive a substantial, steady income. Any interruption in the flow of wealth would weaken the state (van de Mieroop 2004).

Once the booty from conquest was spent, Assyria had to rely on its empire's production capacity to supply the needed wealth. In this context, the use of deportees from conquered states to supply much of the labour and army recruits was a preconditioned weak link in its efforts to keep the system functioning. Both the unwilling, disaffected deportees and the peoples the Assyrians had conquered provided a pool of "enemies" that could quickly appear to complicate decision making and interrupt the supply of resources.

A possible environmental contribution to resimplification may have been a drought recorded in Northern Africa around 650 BC (Hassan 1996). It may have been severe enough to have affected the rest of the region. However, its contribution to Assyria's resimplification is still debated (Redman et al. 2007).

*Empire: Neo-Babylonia.* Despite this recent history, the Neo-Babylonians not only decided to occupy Assyria, but they proceeded to create their own empire. The Babylonians essentially stepped into a role left vacant by the resimplification of the Assyrian empire. They acquired the Assyrian empire, including what wealth was left, and then expanded it to the Egyptian border by force.

The Babylonians managed their empire by continuing Assyrian practices, such as deporting troublesome populations. For example, in 597 BC, Judah was turned into a province. A rebellion problem was solved by deporting a large part of the Judean population to Babylonia. Coincidently, the arrival of these deportees solved a labour shortage on the Babylonian farms.

With its new, suddenly acquired wealth, Neo-Babylonia went through a phase of extensive building. For example, in 600 BC, an 80 km long wall was built to the north of Babylon, ostensibly to keep out their former allies, the Medes, who were now a powerful, rival state (figure 21). Subsequently, Nebuchadnezzar II rebuilt Babylon. The city was 900 ha (9 km$^2$) in size, set within an 18 km long outer wall. Access to the city was through the spectacular Ishtar Gate that was faced with coloured, ceramic-tile symbolic images. Within the walls was a ziggurat so tall that it inspired the story of the Tower of Babel.

The Babylonians expanded their irrigation system to a degree that allowed larger tracts of land to produce more crops. The economy prospered through a mix of plunder, tribute and agricultural products. These goods were transported, sometimes by ships in the irrigation canals, to and from many areas outside Babylonia, including Egypt and Cyprus. A sophisticated trading system developed in which the traders seem to have been immune from either the changes in government or the changes in political boundaries. Regionally, Babylon was again seen as the epitome of wealth and majesty.

The Neo-Babylonian population was concentrated in the cities. Urbanization was higher than in any period since about 1700 BC. The people in the cities were of many ethnic backgrounds, in part because of the deportations from other states. Everyone paid taxes, including the monarch and indentured labour.

Despite the diversity of cultures, the new Babylonian rulers still maintained ancient Babylonian traditions and culture, as they understood them. Cultural inertia had ensured that, despite tremendous peripheral changes, the adjustments to Babylonian culture's core beliefs and practices over the previous few thousand years had occurred slowly. Thus, although Aramaic became the vernacular language, Akkadian was still the language of culture and administration. The cities were still centred on a temple, but city organization was no longer totally subservient to the monarch. The patron deity of Babylon was still Marduk.

Despite the sudden riches, or maybe because of them, the Babylonian empire existed for only 70 years. It came to an abrupt end in 539 BC when Cyrus, the Persian, marched into Babylon, thus conquering Babylonia (van de Mieroop 2004).

The short-term cause for Neo-Babylonia's *resimplification* may have been the unpopularity of the last ruler of the empire, Nabonidus. Traditionalists were dissatisfied with the king's lengthy stay outside of Babylon and his attempts at broad cultural changes. Their dissatisfaction helped the priesthood in their efforts to discredit Nabonidus.

The priests accused him of neglecting Marduk, the patron god. Disaffected priests tried to undermine Nabonidus by, for example, writing negative prophecies about the state's future. Ironically, the conqueror Cyrus later used these prophecies to convince the people that he was the legitimate heir to the Babylonian throne.

Perhaps these feelings and actions, plus the associated cultural inertia, were sufficient reasons for the empire to resimplify. Perhaps his inexperience at governing an empire and inability to keep the people "on side" was responsible. However, it seems more likely that other factors were involved.

Some of the generic problems that afflicted the Assyrian empire also affected the Babylonian empire's stability. For example, the empire's dependence on the conquered territories to maintain the flow of wealth. Understandably, the rebellious territories and their unwilling labour pools were not inclined to co-operate with their new masters (van de Mieroop 2004). Like Assyria, Babylonia had become a target for conquest by outsiders.

*Persians, Greeks and More.* The invasion by Cyrus saw the end of Babylonia's days as an autonomous political entity. It would never again be an independent state. Even though the subsequent Persian (Achaemenid) rulers continued Neo-Babylonian practices, Babylonia was not ruled by a local for centuries. Babylon's importance also diminished because the Persians made Susa their capital.

In 331 BC, the Persians were, in turn, defeated by Alexander the Great from Greece. He intended to make Babylon his capital, but he died in 323 BC. His generals carved up his empire so that the Near East was ruled by the Seleucid dynasty. Within Babylonia, the practices of the Persian government were continued but now with a Greek face.

During this long period of being conquered, there was no opportunity for Babylonians to reaffirm their cultures world view and authority structure. Over time, the remains of the ancient Babylonian culture were dispersed and absorbed into the surrounding cultures. The process was likely helped by Babylonians slowly mixing into the much larger population, which, by this time, included the people from the many cultures brought in over the years to supply labour for agricultural production (van de Mieroop 2004). The last known use of the cuneiform script was AD 75, after which no Babylonian language, belief or economic system was practised (Yoffee 1988b).

There is an afterword to the Babylonian story: today's ongoing connection to that past through irrigation and salinization. The post-Babylonian rulers, from the Greeks on, continued to build irrigation systems. For example, the area where the Diyala River meets the Tigris, just south of Baghdad, eventually became criss-crossed with a complex of irrigation dams and canals (figure 21). The high maintenance costs and management demands of this very complex irrigation system strained each successive central government. By around 1100, a combination of

the government's inability to maintain the irrigation works and incipient salinization of the surrounding fields, probably caused by irrigation, led to the abandonment of the southern Diyala region (Jacobsen and Adams 1958; Jacobsen 1982). In 1900, the southern Diyala was still salinized (Jacobsen 1982).

Today, in what was central and southeastern Babylonia, salinization is still occurring and still impacting agricultural production. For example, in 1960, the areas where summer vegetable and legume crops could be grown were limited. Those fields planted in barley commonly yielded only 500 to 1,500 litres/ha (van der Sluis and Hulsbos 1977). This is substantially less than the 2,500 litres/ha expected in unsalinized areas. In 2009, approximately 25,000 ha/year of productive, irrigated farmland was being lost to salinization (IRIN 2009).

The argument could be made that modern Iraq is an oil-rich country and could overcome these food and water problems. However, consider that technical fixes for salinization and the associated need to fertilize the soil are energy intensive and expensive to apply (van der Sluis and Hulsbos 1977). Similarly, the costs of repairing and maintaining the complex irrigation systems are substantial and ongoing, especially as wars continue to occur (Schnepf 2003).

In addition, increased water withdrawals from dams built in the headwaters of the Tigris and Euphrates Rivers have reduced the water available to grow food in Iraq. The presence of these dams has also likely increased the salinity of the irrigation water. Desertification is also becoming a problem as the few remaining trees in the area are being cut down (IRIN 2009). Along with these impacts come raised political tensions. The basic environmental and human characteristics that constrained Babylonia's future continue to constrain Iraq's future.

In overview, Babylonia, with its long history, was made possible by its rich, fertile environmental endowment. The early inhabitants took advantage of this fertility by settling and overcoming the difficulties of practising irrigation agriculture. Their success resulted in new issues. In the process of addressing them, the early Babylonians started down the road to becoming a more complex agricultural culture. On that journey, they increased their agricultural responsibilities, established a social hierarchy, established cities, applied technology, increased the population and changed the people's relationship to the land.

An increase in the complexity of the Babylonians' culture did not immunize it against possible resimplification. Indeed, a reoccurring theme of

Babylonian history was increasing cultural complexity followed by resimplification, which indicates that the increase likely contributed to it. In particular, it seems that they were unable to fulfill the increased responsibilities that come with an increase in cultural complexity, as exemplified by the increase in its ongoing reaffirmation and authority maintenance costs.

All three categories of the contributions to resimplification are represent in Babylonia's history. The contribution from their culture's world view is represented by issues such as indentured labour, the elite being contract managers, and invaders as scapegoats. Their local ecosystem's contribution is illustrated by salinization. The contribution from climate change is exemplified by severe droughts.

During the lead-up to a resimplification event, some of these contributions developed slowly over hundreds of years (e.g., cultural complexity, salinization), while others were short lived (e.g., a change in a river's course, a leader succession). They collectively contributed to the Babylonians no longer being able to maintain their culture's level of complexity. Although each of Babylonia's resulting resimplifications was unique in detail, in overview they displayed the same basic (generic) characteristics.

The Babylonians could have avoided forced resimplification if they had lived within the constraints imposed on them by their environmental endowment and their human characteristics. To achieve this goal, the Babylonians would have had to have recognized these constraints, absorbed them into their culture's world view, as they did the idea of central government, and included them in their day-to-day decisions. Clearly this did not happen.

In summary, the examples illustrate that historical agricultural cultures could experience resimplification regardless of their level of structural complexity. Each resimplification was unique in detail. However, in overview, each be described by a suite of interacting features from three categories (supporting ecosystem, climate, culture's world view). These interactions included delays and feedback. A historical culture's resimplification is a dynamic, complex process.

## Appreciating Resimplification

Our intuitive view of these historical examples can allow us to feel that we have a sufficiently deep appreciation for a culture's resimplification to provide a reasonable explanation for it. But our intuitive appreciation tends to be simple and based on our biased internal-reality view of the world, a view that can be significantly divorced from the external-reality view. Our appreciation changes after discussing, in this section, the biases and constraints in our innate thinking about resimplification. It also changes once an external-reality order in the apparent chaos of the resimplification process is pointed out. We come to appreciate that resimplification is a dynamic, complex process by which a culture's elevated level of structural (organizational) complexity is reduced, despite the members' efforts to maintain it.

<u>Collapse or Change?</u> When Westerners are asked about the above cultural histories, we tend to focus on what we believe ended a culture's grandeur and its civilizing traditions. We focus on those events that resulted in its spectacular monuments being abandoned to disintegrate in weed-infested fields, its failure or collapse. Consider our imaginings of the final days of the Easter Island culture, the Mayan culture and the Babylonian empire. We take the same view of the Roman Empire, and our own culture, as portrayed in movies.

When asked to explain such a cultural collapse, Westerners intuitively do so using one or two causes. A comprehensive review identified 11 broad types of these causes. Here are eight examples of those types: (1) depletion of resources (e.g., Easter Island, loss of trees and soil); (2) an insurmountable catastrophe (e.g., Norse, cooler temperatures); (3) insufficient response to circumstances (e.g., Xhosa, bovine pneumonia); (4) intruders (e.g., Babylonians, the Amorites); (5) class conflict, societal contradiction, elite mismanagement or misbehaviour (e.g., Kassite Babylonia, an indebted poor and a super-rich elite); (6) economic factors (e.g., Kassite Babylonia, Assyria and Rome, when the booty from conquest ran out); (7) social dysfunction; and (8) mystical events (e.g., Xhosa, rise of the ancestors and discord over prophecy) (Tainter 1988).

But trying to explain a collapse using only one or two causes runs into significant difficulties. Amongst them are the following: the selection of any particular cause can usually be countered by strong arguments for an alternate cause; a chosen cause can be seen to play contradictory roles; and a chosen cause may have occurred at an earlier time, but did not result in collapse at that time. For example, the growth in the Babylonian and the Mayan population can be seen as either providing the wealth to stimulate innovation and enhance complexity, or as a drain on resources leading to later collapse. Similarly, the Babylonians blamed barbarian invaders such as the Amorites for their issues, but the Babylonians were capable of repelling them in previous times. If a particular Babylonian collapse was caused by invaders, then there must have been an additional reason for that outcome (Bronson 1988). For example, the Babylonian's could no longer maintain the army needed to defeat the invaders. Treating collapse as a simple, deterministic cause-and-effect process is satisfying, but unsuccessful: context and complex interactions are important.

Similarly, a Westerner's choice of the word *collapse* and their selection of the events to explain it are both strongly influenced by our culture's world view of how the world works and how our complex industrial culture should function. We tend to admire those historical cultures that managed to maintain order (keep chaos away) by manipulating the environment and managing their members in ways we can identify with. Think of spectacular buildings and a strong leader. Not only do we relate to those cultures and their monuments, but their successes reaffirm our faith in Western culture's world view and its manipulation of the land.

At the same time, the ruins of these historical cultures embody Westerners' fears: they symbolize to us the potential for our culture to collapse. The ruins warn us that if we stray from the narrow path of civilization, stability and security then we too can suffer a rapid, de-civilizing cultural collapse, even an apocalypse. They also remind us of our belief that we can avoid such an outcome by placing our faith in the guidance provided by our culture's world view about staying on the path.

In overview, if we wish to meaningfully evaluate the examples of historical cultures and apply the knowledge we gain to our own culture, then we can't ignore the pervasive, powerful influence that our own culture's world view has on our perceptions and decision making about the past, present and future. In order to reduce this bias, the above examples will not be treated as illustrations of the loss of civilization or a collapse. They will be treated as examples of a forced reduction in a culture's structural complexity: a forced resimplification. This outcome is the result of the members' efforts to form or maintain their culture in a state that is at odds with the constraints on their lives.

**The Distorted Past.** There are factors other than our culture's world view that can distort our view of the past. As mentioned in the introduction, the historical record we use to think about the past is usually significantly incomplete and insufficiently detailed to provide a complete picture of a culture's resimplification (Thompson 2004). In particular, the general lack of historical information at a time scale relevant to those experiencing the resimplification provides much room for research, debate and speculation about its exact causes (Grossman 2002; Trigger 1989). Not discussed so far is how our innate manner of perceiving the world biases our view of the past and thus our perception of a historical culture's resimplification.

Consider that we inescapably view a historical resimplification from the present: we view it retrospectively. This view provides us with the sweep of time covering a resimplification, which allows us to more easily identify the slower, long-term influences, interactions and events that contributed to it. However, our process of perceiving this sweep compresses the time between a resimplification's events: time is telescoped. This telescoping of history makes it easier for us to discern the long-term trends that contributed to resimplification, but it also creates the impression that past events occurred more rapidly than they did. Telescoping also helps to obscure the short-term contributions. Overall, a past resimplification can seem to us to have followed a more simple and rapid process than it actually did: it can appear to us to be a collapse.

Telescoping also allows us to imagine that a person living through a resimplification would have experienced very exciting times, and that we should expect the same. However, a look at a historical resimplification from the perspective of a human lifespan indicates that, in many cases, a participant likely lived a life that, to them, appeared more normal than we imagine. Despite any complaints or worries about their future, most of their life was likely lived without much awareness of the ongoing, larger, longer-term resimplification trends going on around them.

In contrast, they were aware of their current circumstance and its parade of short-term events, such as a bad/bumper harvest year and political/social intrigue. After all, these short-term, spectacular events are the usual source of talking points about our lives and the state of our culture. Some of these short-term historical events may have contributed to resimplification, while others were just the noise of day-to-day living. Unless the members used a broader context to examine these short-term events, knowledge of them would have only provided limited reliable insight into their culture's longer-term, complex resimplification process underway in the background (Yoffee 1988a; Millon 1988).

Consider being born in a Mayan city during the terminal phase of the Classic period when food was in short supply. You would see hungry people, but that would appear to you to be a routine part of daily life, just as it was for those living in mid-1700s England. During your, say, 50 years of life, notable short-term events, such as another battle with a neighbouring city or this year's failed/bonanza crop, might prompt thoughts of a reportedly better past and hopes or fears about an uncertain more-distant future. But mostly life was spent dealing with the routine of day-to-day affairs, such as acquiring food and completing the required religious rituals.

Today, our retrospective view of a resimplification is similarly affected. When we find a well-defined, short-term event of immediate human interest, such as a great flood or battle, in the recorded history of a resimplified culture, we tend to treat them as highly significant. We forget that we, like the authors of that history, are drawn to these newsworthy events. We also forget that written history favours the perspectives of the elite, because they could make the records we read. Thus our thinking about resimplification tends to attach greater significance to short-term, newsworthy events found in the written historical records that are themselves biased both toward these types of events and the concerns of the elite. Care has to be taken when assigning a cause and explaining a culture's resimplification.

Consider too that we tend to label the short-term events we think are particularly important as "triggers" or "tipping points," or even "causes." Telescoping helps us link these "trigger" events into what we think of as a plausible story of a historical culture's resimplification. Although these events do have symbolic value, provide context and may have contributed to resimplification, they are usually not, on their own, causes or indicators of a resimplification. They are better thought of as cues indicating the possible background presence of a dynamic, complex resimplification process taking place over a wider range of time and space. A "trigger" event may be a cause but it is more usefully referred to as a marker event. In this role, it can prompt us to explore the event's context to see what significant longer-term

changes the culture was experiencing as well as the thresholds it was approaching at that time.

For example, the capture of Babylon by Elam in 1155 BC can be thought of as an event that triggered the end of the Kassite Babylonian great state. However, it more realistically represents just one marker event in a slow, ongoing resimplification process. The possibility of Babylon's capture on that date was preconditioned (but not predetermined) by a mix of factors and decisions that existed well before 1155 BC. These earlier contributions included the long-term use of irrigation with its salinization potential and the medium-term consequences of a cultural world view that favoured chronic worker indebtedness. It also included the short- to long-term repercussions of being unable to recognize or manage these types of problems and their causes in a way that ensured cultural viability. From this perspective, the Kassites' loss in the 1155 BC battle is just a marker event (a reminder or a symbol) of the complex culture's quasi-stable or unstable state. The full consequences of these earlier contributions were only revealed by the outcome of a resimplification process that continued long after the battle was lost.

Our search for possible causes of resimplification can also be biased because we are preferentially drawn to outcomes that we can interpret as easily avoidable: obvious mistakes in decision making. Examples might be the Easter Islanders cutting down all of the trees, or the Norse failing to adopt Inuit practices. In favouring these events as causes, we neglect telescoping and forget another aspect of our retrospective view of resimplification: our hindsight-biased view of the past. The members of historical cultures made the decisions we label as mistakes in *their* present, using *their* world view, not ours. Their flawed decisions, which we treat as causes of their resimplification, are more than simple mistakes.

When those long ago citizens looked into the future and made their decisions, they would have had, just as we do today, incomplete knowledge about what was to come and how it would arrive. The members bridged that critical information gap by relying on guidance from their culture's world view and the specialists who assisted with decision making. After all, a culture's world view provides the members with a general understanding of how the world works, the nature of the future, and cultural memories about how issues were dealt with in the past. The specialists' guidance is treated seriously because they are the people who the members believe possess the knowledge and skill needed to recognize and interpret the predictive signs, the things to consider in decision making, and the rules of thumb to use when making choices. Think Mande. Their guidance to the members was, of course, influenced by their human foibles and their culture's world view.

When seen in this context and from their perspective, many of the decisions made by the members of past cultures were, culturally, quite reasonable, for example, the Xhosa deciding to kill all of their cattle. It is only our long-term, somewhat disinterested, retrospective view which allows us to recognize that their decisions were, in the external reality, inappropriate, perhaps even mistakes. However, for us to expect the Xhosa to have come to the same conclusion as we have today, we have to ignore our advantage of a retrospective view of their circumstance. We must also expect them to have successfully completed something we struggle to do: broaden our understanding of our circumstance, question our cultural world view, and become aware of our reality paradox and the biases it introduces into our thinking. We should be wary of thinking that only the obvious mistakes we identify in our culture's decisions will contribute to its possible resimplification.

By taking the factors that distort our view of the past into account, we can more realistically consider historical, as well as possible, future resimplifications.

### Enhancing Our Appreciation.
We more fully appreciate a historical culture's resimplification if we take into account the influences that our culture's world view, our human foibles and the data constraints have on our perception of the event and our thinking about it. We are further helped if we think about the culture's resimplification from the perspective of the three contributing categories (supportive ecosystem, general environment and the culture's world view). In doing so, we learn from the examples that resimplification is a more complex, but less chaotic, process than it seems. Consider some of the common features and a few of the resulting patterns, themes, interactions and trends that the examples reveal.

The time scale over which a particular feature contributed directly to a culture's resimplification varied from a few years, such as kingship succession difficulties in Assyria, to many hundreds of years, such as salinization in Babylonia. Thus, the role and significance you assign to a particular feature's contribution depends on the scale and perspective you choose to look at it and at the resimplification. A long-term view provides a different understanding of the resimplification and the contributions to it than a short-term view.

When the short- to long-term contributions from the three categories to the culture's resimplification are considered collectively, they are seen to be interconnected, forming a foundational framework or context within which the resimplification occurred. The various contributions also form a hierarchy based on their significance to the resimplification process. The hierarchy is flexible because the significance of a feature's contribution can change over time. Consider how the relative significance of the climate's contribution to the Akkad resimplification changed from low to high when the climate switched from stable to drying. Significance can also change with the scale used to look at a resimplification. In the short-term view, the primary contributions to Akkad's resimplification seem to be human related. However, when viewed in the long-term view, they seem to be climate related. In reality they are both important.

The significance of a contribution is also affected by the context. For example, a sudden drought is less significant when the previous year's crop yields had filled the granaries, but more significant when long-term salinization leaves them partly empty. A drought is less significant when the culture's supporting wild ecosystem could act as a wild-food buffer, but more significant when, over the long term, the ecosystems have been transformed into fields, pasture or infrastructure.

As noted earlier, the characteristics of a culture's environmental endowment don't predetermine the features of a culture's world view or whether it will resimplify (Trigger 1989). The endowment just helps to precondition the members to make some choices rather than others, and to limit their options if they want their culture to be viable rather than resimplify. Similarly, the context within which resimplification occurred and the hierarchy of the contributions to that event did not predetermine a culture's resimplification. They too just outlined the particular constraints on the members' choice of options, if they had wished to avoid a forced cultural resimplification and favour cultural viability.

We find it easier to appreciate the complex resimplification process when it is described using words that we can more easily relate to. The familiar suite of generic summary terms previously used to describe our experience of the complex processes of cultural and personal change are adopted here: *drift*, *imposed*, *preconditioned*, *inertia*, *directed* and *unexpected*.

For example, some aspects of the Xhosa resimplification were preconditioned by their belief in the ability of their ancestors to re-enact creation. But other aspects were unexpected, such as the appearance of bovine pneumonia. A particular feature's contribution can be described by more than one term, depending on the context. For example, the British contribution to the Xhosa's resimplification was both unexpected and imposed. A term can also be common to many resimplification descriptions. For example, common to the Babylonian, Norse, Mayan and Easter Island resimplifications was their drift into the piecemeal destruction of their soils and their substantial cultural inertia to making changes that would avoid resimplification.

The usefulness of these easy-to-relate-to terms, the context, the hierarchy of influences and the contributing categories are illustrated by the change in our appreciation of Easter Island's resimplification when they are used to describe it. The future of the original Easter Island settlers and their descendants was constrained and preconditioned by the island's environmental endowment and the core features of the settler culture's world view (Kirch 2007). (Climate change, such as a drought, does not seem to have played a significant role in their future.) The seeds of the islander's future resimplification thus arrived with them, long before that event. Whether these seeds would bear fruit or not depended on the settlers' decisions. Of particular importance was how well their choices would reflect the environmental and social constraints imposed on their lives if their culture was to be viable. This description captures the essence of the context within which the islander's unfolding future can be more fully understood.

To continue, their reality paradox and their culture's world view preconditioned the islanders to fail to recognize the direction in which their day-to-day farming and procreation decisions were causing their culture and ecosystem to drift. The more the state of their supporting ecosystem changed, the further it moved up the hierarchy of influences on the islanders' future, and the harder it became to restore it to a more human-supporting state. Helped by inertia, they failed to make the necessary cultural changes in time. Eventually, from their perspective, they unexpectedly lost the ability to direct their future. They could neither change quickly enough, nor sufficiently enough, to avoid resimplification.

In overview, a flexible hierarchy of short- to long-term features from the three contributing categories (supportive ecosystem, general environment and the culture's world view) forms the context within which resimplification takes place. The process is, simultaneously, the product of powerful, longer-term features interacting with short-term features, both of which operate over a large range in space and involve feedback. Resimplification is a dynamic, complex process to which the members' day-to-day decisions make a significant contribution. Our appreciation for this resimplification process is helped by using general terms to think about it.

In summary, once we each take into account our culture's and our own innate influences on how we perceive and think about a historical resimplification, our appreciation for it changes. We can more easily relate to resimplification at a personal as well as a broader level. We also realize that the diversity of information about resimplification describes a dynamic, complex resimplification process that is a part of our human-environment system.

## Resimplification Model

An enhanced appreciation of resimplification reveals that it can be incorporated into the model of the dynamic, complex human-environment system discussed in chapter 11. This is done by first reviewing the model and then discussing how the features of resimplification can be fitted into it. The model is applied to round out our general understanding and appreciation of resimplification.

### The Human-Environment System Model Reviewed.

A reminder of the human-environment system's model is provided before adding in the resimplification process (figure 15). The human-environment system is divided into three primary factors: individual, environment and culture. The relationships between them are represented by a resource loop and a belief loop, which overlap and interact.

The belief loop represents the belief-related relationships that bind the individual members of a cultural group and their culture into a whole, as represented by its world view. For example, the members' decision making is guided by their culture's world view. But because the members are the carriers of their culture, their decisions shape their culture's world view. This forms an individual-member–cultural-worldview feedback (member-culture feedback).

The resource loop represents the interactions with our environment that bind the individual members and their cultural group to their environmental endowment and its functioning. For example, the members live a lifestyle that reflects their culturally supported relationship with their culture's environmental endowment. The members acquire the resources needed to live that lifestyle from that endowment. As they go about securing and using those resources, they make decisions. Those decisions can, collectively, change their lifestyle or their relationship with their environment, which completes the member-environment feedback.

The feedback connections within and between the two loops tie the members and their environment irrevocably together into a dynamic, complex human-environment system. People, as individuals, members of a culture and participants in the environment, respond to the system's dynamics by making decisions that are guided by their culture's world view. Through the belief loop, their decisions can alter their culture's world view, and through the resource loop, alter their environment. The resulting cultural and environmental changes become part of the system's dynamics.

A member's (personal) decisions are the product of their human decision-making foibles, their internal reality (including their personal world view) and the influence of their culture's world view. Because their internal reality and their culture's world view contain biases, inconsistencies and disconnects (divergences) from the external reality, so do their views and decisions. Indeed, these divergences are a ubiquitous part of the member's lifestyle and their relationship with their cultural group and their environment. The influence and consequences of these divergences are thus included in the feedbacks linking the three factors together.

When a member becomes aware of these divergences, they resolve them through a process of rationalization (figure 15). That process is a key part of the member's everyday decision making. Applying it can be as easy as ignoring the divergences. Rationalization can also be accomplished by the member (consciously or unconsciously) adjusting their beliefs, actions and demands in a manner that resolves the divergences, at least to the member's satisfaction. This may or may not be a resolution from the perspective of the external reality.

This model helps us to understand how a culture can come to increase its cultural complexity. That journey starts when the members (consciously or unconsciously) make issue-resolving decisions that (directly or indirectly)

increase the culture's structural complexity. Such decisions would, for example, favour increasing resource extraction, increasing environmental manipulation and/or increasing the population's size. If the rationalization of these decisions results in personal and cultural adjustments that facilitate the members continuing to make similar structural-complexity-increasing decisions, then the culture's complexity can continue to increase, if it can be supported by their environmental endowment. This outcome is illustrated in an earlier chapter by the Abu Hureyran's transitioning from a hunter-gathering to a farming culture. In this chapter, it was illustrated by early Babylonian history.

Under some circumstances, rationalization can result in a self-sustaining feedback that favours the members routinely increasing their culture's organizational complexity. Consider a complexity-resource feedback in the resource loop. In this feedback, an increase in resource production is achieved in ways that require increasing the human involvement in the resource production process, such as increasing management and applying technology. But making these changes increases the culture's complexity, which itself requires an increase in the amount of resources needed to sustain it, hence the feedback. If a more complex culture's world view is adjusted so that it favours the members solving their issues in ways that either increase their supply of resources or increase their culture's complexity, then the complexity-resource feedback becomes self-sustaining. Arguably, this was the case for Maya in the Classic period and for post-unity Babylonia. More generally, a culture can establish a self-sustaining complexity-increasing–issue-resolving feedback.

**Including Resimplification in the Model.** The reverse of an increase in a culture's organizational complexity is its reduction: the culture's resimplification. The model of the human-environment system in chapter 11 can represent the process of resimplification because it can accommodate resimplification's three categories of contributions, and their associated interactions. Resimplification's world-view category is included in the model's cultural factor. The supporting-ecosystem category (environmental endowment) and the general environment category are both included in the model's environment factor.

There was no contributing category for individual members in the discussion about the resimplification examples because of the previously discussed data limitations. However, as suggested by the above examples and by some of the 11 types of possible causes of resimplification presented above, it is easy to argue that such a category is implicit. After all, individuals are the basic decision makers in the human-environment system, and they are the carriers of their culture's world view. Their choices will strongly influence the occurrence and nature of resimplification. Think of how the decisions by Hammurabi and the indebted workers affected Babylonia's resimplification history. The generic role of the individual member in the resimplification process is represented by the model's individual factor.

With these few changes, the characteristics of the model's individual, cultural and environmental factors, and the feedback-containing dynamics of the relationship between them, can provide a culture-dependent description for the resimplification process. This is illustrated by first noting that if a complex culture is to remain functional over the long term (viable), then it must remain within human and environmental constraints. If a complex culture exceeds those constraints for long enough, it must resimplify. This can happen if the members drive, or outside forces place, a complex culture beyond its human and environmental stability limits, and then keep it there long enough for it to resimplify. How this can happen is illustrated by the interconnected features of diminishing returns, complexity and individual decision making.

*Finiteness and Diminishing Returns.* The members of a culture must continuously supply the resources needed to maintain its functioning (the resource loop). For a complex culture, those resources are required to do more than feed and clothe the members. They are also required to pay for its ongoing (world view and structure) reaffirmation and the authority maintenance costs, which includes those for the bureaucracy. In addition, a more complex culture's members generally have greater personal resource demands/expectations than simple cultures. Thus the per capita amounts of resources needed to keep a more complex culture functioning are much greater than for a simple culture.

The combination of a more complex culture's larger population and its greater per capita resource demands also mean that a complex culture has a much larger absolute demand for resources than a simple culture. Supplying these resources can be thought of as the means by which the members pay to maintain their culture's greater level of complexity. A resource supply shortage would limit the members' ability to pay that cost. If shortages are sufficiently severe and long-lasting (a shortfall), then their culture faces resimplification.

There are a number of reasons why such a resource shortfall could arise, as discussed in chapters 12, 14 and 15. A non-renewable resource may have been used up, or it may be present but not extractable. A renewable resource shortfall can arise when the rate at which it is being extracted exceeds its rate of renewal (i.e., when it is being exploited unsustainably). A shortfall can also arise if the resource-renewing processes are disrupted. It can also occur if the geographic distribution of the renewable resource changes. Think of the impact that a change in the distribution of rainfall will have on the availability of water.

There are other more subtle reasons why a shortfall could arise. Some resources, such as rainfall, are provided without us expending effort (energy/financial cost), but most aren't. A shortfall arises when the effort needed to supply a resource becomes too onerous: the resource can't be supplied.

The change in the effort needed to supply resources follows a long-term pattern. Simplistically, early in a (non-renewable or renewable) resource's production history, the amount supplied for a unit of our effort to produce it is near the maximum. After that maximum, the amount provided per unit of our effort declines. For a non-renewable resource, the decline to zero is inexorable. For a renewable resource, the decline can stabilize at a high value if the resource is exploited at a sustainable rate. Otherwise, its supply too will continue to decline.

An example of this reality for non-renewable resources is provided by the change over time in the energy for discretionary purposes (i.e., for our personal use, not for energy production) made available to Americans from the sum of US oil and gas production. Since the 1910s, there has been a steady decline in the discretionary energy provided by a unit of the energy expended to produce the sum of those fuels (i.e., a steady decline in the fuels' combined EROI) (Guilford et al. 2011). An example for renewable resources is the global ocean-fish catch. The size of the annual catch increased from at least the 1950s, until it peaked in the 1990s and then declined to the present. Over this period, the fishing effort rapidly increased (Pauly and Zeller 2016; Watson et al. 2012). As a result, the catch size per unit of fishing effort peaked around 1970. The fish catch returns per unit of effort have been diminishing ever since. Ocean fish are being exploited beyond their rate of renewal, beyond their environmentally sustainable limit.

The practical limits to the absolute amount of a resource that can be made available (exploited), the need to maintain the environmental functions providing renewable resources and the diminishing resource returns per unit of effort expended to secure them are all inescapable. They collectively constrain the level of a culture's complexity that its environmental endowment and thus its members can maintain indefinitely. When demand routinely exceeds supply, it warns of possible resimplification.

The phenomenon of diminishing returns applies to more than the effort needed to supply natural resources. It applies to other requirements needed to sustain a culture's complexity. Economists refer to diminishing returns as diminishing marginal returns or diminishing marginal utility. The term describes how a specific economic venture faces a steady decline in the profits for each new dollar invested in it. This occurs because, for example, the easy pickings occur at the beginning of a venture, while the competition and fixed costs (e.g., equipment maintenance, repair and replacement) increase as the venture ages.

When a complex culture's critical assets, production and services are related to the cash cost of providing them, then they too display diminishing returns. For example, in the United States, a declining benefit per unit of cost has been recorded for health improvement (measured by life expectancy), acquiring scientific information, bureaucracy's productivity and agriculture's productivity (Tainter 1988). It likely applies to the tangible assets of infrastructure, such as roads, water systems and irrigation works: at some point they all need to be replaced at considerable cost.

*Cultural Complexity.* The more complex a culture becomes, the more effort is needed by its members to maintain its complexity. But there is a level of complexity beyond which it can no longer be sustained: the culture is no longer viable. The presence of this complexity limit is illustrated by our efforts to intensify agricultural production.

By intensifying agriculture, humans become responsible for more than the growing crop. They become responsible for all of the ancillary services, such as providing tools, machines and energy; maintaining irrigation infrastructure; and developing new farming methods, such as pest control. Thus, an increase in agricultural intensity is accompanied by a rapid increase in the complexity of our agricultural production system. However, the amount of time, effort and resources that

can be devoted to fulfilling these responsibilities is limited. There is thus a complexity-related limit to the level of an agricultural production system's complexity that we can sustain. This, in turn, sets limits to the sustainable level of a culture's organizational complexity. Think of the cost of the irrigation works in the Diyala region to the rulers of Iraq just prior to AD 1000.

The constraints imposed on a complex culture by its complexity arise from more than material causes. Consider that an indicator of an increase in cultural complexity is an increase in the number of its subgroups (e.g., work specialties and social organizations). In order for the culture to remain functional, these subgroups must communicate with one another. Efforts to satisfy these communication requirements by increasing both the communication system's efficiency and effectiveness are inherently constrained, as discussed in chapter 3. Think first of the physical resources needed to build and run a communication system. Of particular interest here is the inherent slow speed with which we humans can meaningfully process communications. It limits our ability to increase effective communication. As a culture becomes more complex, the constraints on its ability to increase or maintain its communication system's efficiency and, especially, its effectiveness limit the sustainable level of the culture's complexity.

Overall, the nature of a culture's functioning constrains the level of its complexity that the members can maintain over the long term. For a complex culture to avoid resimplification, the members must, over the long term, stay within most of the human and environmental constraints on its functioning (e.g., resource availability, human abilities and human mental needs) and simultaneously discharge their complexity-related responsibilities. The more complex the culture becomes, the more difficult it is for the members to do so. This situation is made worse by diminishing returns (Tainter 2000). The less successful the members are, the more likely resimplification becomes.

When looked at from a different perspective, the limits to a culture's level of organizational complexity exist because, over the long term, the benefits from increasing its complexity are offset by the increase in effort and responsibilities (costs) needed to maintain that complexity. There is a level of complexity beyond which the effort and cost of maintaining or increasing a culture's complexity is counterproductive. If the members establish a complexity-promoting feedback, such as a complexity-resource feedback, then they precondition their culture to exceed its sustainable level of complexity. If the members make the feedback self-sustaining, as described above, then their culture's world view guides them to substantially exceed their culture's sustainable level of complexity (a type of overshoot) and stay there until it resimplifies (commonly referred to as a collapse).

*Individual Decisions.* An environmental event, such as an extended drought, or the impacts of other cultures, such as an invasion, can push a culture beyond its viability limits and keep it there for long enough that it resimplifies. Our enthusiasm for treating these types of externally imposed, unexpected events as the primary causes of a resimplification allows us to forget that the members of a culture are its decision makers. They make the decisions that will become the culture's internally generated contributions to its future state, including how their culture will respond to an external or unexpected influence on that state.

Each member makes a contribution because their personal awareness of their culture's current state, along with their attention to both it and its context, influences their decisions about these aspects. The members' individual decisions become part of their collective decision about how to respond to their culture's circumstance and to the (internal and external) influences on it. Thus each member's decisions contribute to, for example, how complex their culture will strive to become, its resource demands, how those demands will be satisfied and the (material and social) relationships among its members. An individual member's culturally influenced decisions are thus a key determiner of whether their culture will resimplify, how it will resimplify and at what level of complexity this will take place.

It is easy to imagine that a member would intuitively decide on a level of complexity for their culture and how it functions that was viable, and then ensure these conditions were not exceeded. If they were, then the members would quickly make decisions to voluntarily decrease their culture's complexity, or alter its functioning to ensure viability. But neither is usually the case.

This reality is represented in the descriptive model of the human-environment system. Consider the decision-making dynamics in the resource loop that results from a combination of our internal reality, the guidance from our culture's world view, our experience of the environment and our rationalization process (figure 15). For example, the members' internal reality and their

culture's world view includes divergences, which can strongly influence their perceived relationship with their culture's environmental endowment and thus their decisions that affect it. It takes time and effort to recognize these divergences, and then to make and rationalize the decisions needed to resolve/accommodate them in ways that will ensure the members and their culture stay within viability's environmental constraints.

The scale of that effort becomes more apparent when we also take into account the links between the resource loop and the belief loop, such as the member-culture feedback, the complexity-resource feedback and rationalization. Consider that, in the belief loop, many aspects of a more complex culture's world view, such as its religious or ideological beliefs, are primed to guide its members to make day-to-day decisions that are pro cultural complexity. For example, they support the members' efforts to alter the environment in order to obtain the resources needed to support their complex culture. This preconditions them to create a complexity-resource feedback and favour population growth.

In addition, in the belief loop, the member-culture feedback helps to create a cultural world view whose guidance the members can accept. This world view helps them to find a sense of belonging and purpose, while reducing their fears about what is always an uncertain future. In doing so, it strengthens their faith in their culture's world view and its guidance. This feedback gives the group stability, but it can also help to create and strengthen the divergences from the external reality in both their personal and their culture's world view. As a result, the guidance from either world view can help to veil the members' awareness of (and attention to) the human and environmental constraints imposed on their culture if it is to be viable.

These features, collectively, ensure that it takes time and effort for the members of a complex culture to become aware of and pay attention to the indicators of their culture's possible resimplification and their contributions to it. Even when they do, their culturally influenced day-to-day decisions still tend to downplay the significance of the indicators, and the human and environmental constraints they must live within if their culture is to be viable. Instead, their decisions tend to favour keeping their culture on a business-as-usual track. For example, a preferred response to a resource shortfall is to maintain or increase the supply, rather than reduce the demand. These tendencies, and the members' inertia to change, favour them maintaining their culture on its current trend long after that choice is contradicted by evidence of, for example, diminishing returns, environmental changes, the limits to increasing complexity, social dysfunction and the reality of resource depletion. Thus, counterintuitively, despite the advantages of living in a more complex culture, its members are predisposed to collectively direct their culture to drift into a future forced resimplification.

In overview, as these three subsections show, it is easy to include in the descriptive model of the dynamic, complex human-environment system those features that can contribute to a complex culture's resimplification, such as its complexity, resource finiteness, diminishing returns, and individual decision making. This should be unsurprising. After all, resimplification and the contributions to it are part of the human-environment system's functioning, which is represented in the model by the interactions among its three factors (individual, culture and environment) and described by the belief and resource loops, and rationalization.

Indeed, the model of that functioning helps to explain the overall nature of resimplification. It is a dynamic complex process that takes place over a wide range of scales and can include random or chaotic features. It occurs when the interactions, including feedback, among the guidance from a culture's world view, the members' decision making as individuals and the supporting environment's functioning place the culture outside of its environmental or human constraints, and hold it there. Resimplification is thus an integral part of its members' everyday lives and their culture's functioning.

The nature of that functioning ensures that the more complex cultures are inherently unstable and preconditioned to resimplify. However, their degree of instability, their susceptibility to resimplification, and the details of their resimplification vary with the level of the culture's complexity and the state of their environmental endowment. But, although the members are not the masters of their future, they are not slaves to it either. The members' (individual and collective) decisions play a key role in determining those features and thus their future.

A complex culture can avoid resimplification by its members ensuring that it stays within viability's human and environmental constraints on its functioning. Instead, its members tend to deal with issues in ways that exceed those constraints, for example, by establishing a complexity-resource feedback. Then the inescapable features of the human-environment system, such as resource supply limits, diminishing returns, the environment's functioning and complexity itself, collectively ensure that a culture functioning outside of those constraints can't be supported indefinitely. Eventually the members will be unable to fulfill all of their multiplying social, cultural and environmental responsibilities needed to maintain that functioning, and the culture will be forced to resimplify.

**Application.** By applying the upgraded descriptive model of the human-environment system to our knowledge about the historical examples of resimplification given above, we gain a more complete overview appreciation for the process of resimplification, as illustrated here.

The members of a simple culture living in a sparsely populated but environmentally well-endowed region can relatively easily address major issues, such as a resource shortfall or an internal group conflict, by migrating, splitting or changing how they live. The members can also make conscious or unconscious choices that resolve their issues in a manner that favours the appearance of greater cultural complexity, such as changing how they manage themselves or interact with their endowment. This choice is facilitated when a culture's supporting ecosystem is sufficiently densely populated, degraded or changed that there are few incentives or opportunities to make other choices. For example, environmental pressures likely influenced Abu Hureyra's hunter-gatherers to make the decisions that led them to become farmers, thus increasing the complexity of their culture.

A simple culture can become more complex, to some degree, without affecting the members' ability to maintain that level of cultural complexity indefinitely. For example, they can become nomadic pastoralists or subsistence farmers. But this is true only as long as certain conditions are satisfied. The culture's environmental endowment must be capable of continuously supplying the needed resources, such as water. The culture's world view must guide the members to live within their environmental constraints, such as soil fertility renewal rates, as well as their human constraints, such as engendering in the members a sense of personal belonging to the group. The culture's world view must also guide the members to fulfill the additional environmental and human responsibilities that come with greater complexity, and do so in a manner that will keep the culture viable. The fact that Neolithic farmers in Europe were, by as early as 6000 BC, practising mixed farming and using their animals' manure to fertilize some of their crop fields indicates that humans do have the intrinsic ability to appreciate and fulfill at least some of the increases in responsibility that come with an increase in cultural complexity (Bogaard et al. 2013).

As a simple culture becomes complex, both the major issues the members face, and the choices the members can make to deal with them, change. Consider that complex cultures are settled, at least to some degree, and have altered their environmental endowment. They have gained new skills while losing old ones. Thus, a complex culture may find it difficult or impossible to return to being a simpler culture, such as hunter-gathering. This happened to the Abu Hureyrans.

Consider too that as a simpler culture becomes more complex, its world view changes. In particular, increased complexity introduces beliefs, values and explanations into a culture's world view that guide the members to support their culture's complexity. It also preconditions them to favour resolving issues in ways that increase resource extraction and further increase their culture's complexity. Examples are the later Abu Hureyran and early Babylonian cultures.

A complex culture that continues to addresses its issues in this manner can successfully increase its structural complexity. But that increase doesn't eliminate the human-environment system's constraints on the level of a culture's sustainable complexity, as illustrated by the reality of diminishing returns, ultimate resource limits, complexity limits, and our decision-making foibles. Neither does it remove the need to maintain social and environmental buffers to accommodate unexpected events. An increase in complexity just changes the expression of these constraints or delays their emergence as issues. Thus, despite appearances to the contrary, even a more complex culture must, like a simple culture, respect the essence of the human and environmental constraints on how its members can live and function as a group. If a complex culture fails to respect them, such as making the complexity-resource feedback self-sustaining, then it must eventually cross limits that, if exceeded for long enough, will force the culture to face resimplification.

How long is "for long enough"? The answer is determined by a mix of features that includes the degree to which the members have disrupted their environmental endowment, the speed and direction of its ongoing change, and how stressed is their culture's functioning. The mix also takes in the level of their

culture's complexity and the responsibilities the members have acquired as a result. Also included is the appearance of unexpected/unaccounted-for features of living on the earth, such as invaders and drought. "For long enough" also varies with the members' culturally guided response to their new reality, in particular, how strongly they remain committed to their culture's currently preferred level of complexity, ways of satisfying their needs and relationship with its environmental endowment.

These features will also determine the nature of a culture's forced resimplification. Consider that the resimplification of less complex cultures, such as city states, can occur in a relatively ordered manner. For example, the resimplification of early Uruk saw the urbanites simply leave to re-establish villages nearby. Similarly, the resimplification of Akkad and Ur III that ended their attempts to unifying Babylonia's cities, allowed them to "easily" reverted back to being independent city states.

However, this easier resimplification is only possible if the relevant elements of the culture's world view remain intact and the local environment can still support the resimplified culture's population and world view. If not, then the forced resimplification is more disordered. Consider that the Amorite's strong unification of Babylon's cities contributed to its resimplification being a long-drawn-out affair: the cities could not easily return to being independent city states. Consider too that the Maya's resimplification at the end of the Classic period involved depopulation, including migration, in part because their supporting ecosystem had been severely disrupted.

This serves as a reminder that it takes time for a culture to resimplify and that its speed varies (like the process for becoming more complex). The Assyrian empire's resimplification in less than two decades was, in contrast to the Amorite resimplification, hyper-fast. Babylonia's long history also shows that ups and downs in the resimplification process are the norm, to the point that determining the beginning and the end of a resimplification can be a matter of choice.

Regardless of the speed, the resimplification of the well-established, more complex cultures is invariably a more convoluted process than for less complex cultures. After all, not only is their culture's structure more complex, but they have a longer record of success at resolving past issues in a manner that increases their complexity. As a result, their members have more faith in the complexity-favouring attributes of their culture's world view, which can include a complexity-resource feedback. The members are likely less aware that, in becoming more complex, they acquire a greater load of significant environmental responsibilities, which they will find difficult to fulfill.

The members are also less concerned with alternative, simpler ways of living. If they do have such alternatives in mind, the members likely lack or have forgotten the skills, and the associated beliefs, needed to adopt them. The well-established, more complex cultures are also more likely to have altered their environmental endowment to a degree that it can no longer support some of their preferred alternatives, such as a return to a (perhaps mythical) time of greater plenty or hunter-gathering.

Collectively, these features endow the established, more complex culture with considerable inertia to change and limited easy choices for change. The members thus tend to continue resolving their issues using short-term solutions that satisfy their complexity-resource feedback and maintain business as usual. For example, they can increase their practice of abundance-providing rituals; their use of technology, such as adopting iron tools; and their application of alternative methods to secure resources, such as plundering, deportations to secure labour and setting production quotas. In the process, they neglect the more appropriate longer-term responses. These tendencies are illustrated by the Babylonians for the period after the Amorite resimplification (from the Kassites forward).

But, despite their efforts, the more complex cultures can't, as mentioned, indefinitely overcome the human and environmental constraints they face when trying to maintain business as usual. After all, their efforts can't remove the constraints. They can only circumvent them, which simply changes the constraint's expression, transforms them or reveals new constraints. For example, when a culture takes up agriculture, it exchanges its environmental endowment's constraints on wild-food availability for the constraints associated with the members being responsible for a farmed ecosystem. When a more complex culture switches from using one kind of resource for tools, such as bronze, to another, such as iron, it also exchanges one set of constraints on a resource's supplies and the responsibilities that come with using it for another set. Thus, although a more complex culture can delay resimplification, it can only do so until its members are required to deal with too many constraints at once: until it has too many

responsibility balls to keep in the air. Or until it inevitably faces a critical insurmountable constraint, such as a gap in the supply of resources or an oversupply of people.

When this situation arises, a more complex culture displays increased signs of stress. These include struggling to provide sufficient resources to meet its wants, to pay for the reaffirmation and authority maintenance costs, and to meet all of its complexity related human and environment costs and responsibilities. The members reflect this stress by more of them experiencing difficulty meeting their basic human needs. They start to more vocally express their dissatisfaction as the differences between their needs/expectations and their day-to-day reality start to widen. This may be accompanied by the more common occurrence, and heightened intensity, of internal conflicts involving the interpretation of their culture's core values, beliefs and myths, not just its peripheral features.

In these early stages, the members might still have the option of deciding to respond in ways that would avoid the worst of a forced resimplification. However, it seems more likely that the members, supported by their leaders, would still try to follow their culture's guidance about how to meet their obligations, satisfy their expectations and mollify the gods/rest of the elite. The members are thus more likely to engage in last-resort efforts to maintain business-as-usual, such as living off the culture's natural and/or social capital. But the culture is then feeding on itself. Think of the last days of the Assyrian empire.

Eventually, the disconnect between a complex culture's striving to follow its world-view-sanctioned lifestyle and the pressure to change exerted by the external reality becomes too great, for too long. The culture starts to display the visible signs of the extremely painful forced resimplification we dread. These features include population declines and chaotic mass migrations, repeated episodes of violently throwing the ruling elite out, and conquest by outsiders. The details are determined by the specifics of the culture's world view and the state of their environmental endowment. Examples of a more complex culture's forced resimplification are those of Kassite Babylonia and the Neo-Babylonian empire.

In overview, a complex culture's process of resimplification is an integral part of its members' everyday lives and how the culture functions. It takes effort to avoid.

In summary, resimplification occurs as part of the dynamic, complex human-environment system's functioning: it is the product of the interactions, including feedback, among the system's cultural, individual and environmental factors. And, as expected, it takes place over a wide range of time and space.

The descriptive model of the human-environment system helps us to appreciate resimplification. It reminds us that the perspective, especially the time scale, that you choose to look at a historical resimplification influences your perception of it. Your perception is also biased by your personal and your culture's world view. The presence of these choices and influences means that, to fully appreciate a historical resimplification, you have to take them into account. The same is true of your ponderings about the possibility of a culture's future resimplification, including that of your culture.

The model helps us to appreciate that a complex culture can remain functional over the long term (viable) only if its members live within viability's human and environmental constraints (Dunbar 2003a). This limits the level of a culture's complexity that both the members can maintain and their environment can support. And it limits the members' average lifestyle. A complex culture that attempts to function outside of these constraints for too long will experience resimplification.

All cultures experience difficulties when trying to stay within their viability limits; however, the more complex cultures find doing so especially difficult. After all, the world view of a more complex culture guides its members to prefer solutions to day-to-day issues that maintain or increase their culture's complexity and responsibilities. But, the more complex a culture becomes, the more likely its world view will veil the members' awareness of both the constraints on their culture being viable and the indications that they are failing to respect them. At the same time, it is less likely that the members will be either guided or able to fulfill, on the external reality's terms, the increasing load of complexity-related social and environmental responsibilities needed to keep their increasingly complex culture functional. In addition, the members exhibit an inherent inertia to making changes to the core aspects of either their lifestyle or their culture's world view. Collectively, these features make it more difficult for the members to resolve their culture's issues in a manner that favours viability. Thus the more complex cultures are inherently unstable and prone to resimplify (Wright 2004).

As a complex culture nears the limit to its viable level of complexity, its members experience features such as rapidly diminishing returns for their efforts, resource supply limits, difficulties fulfilling responsibilities and social dysfunction. The members are increasingly forced to face resource, environmental, individual and cultural constraints that they can't circumvent: they can no longer maintain their culture's level of complexity. The culture must either stabilize or resimplify. But factors, such as inertia, tend to drive a culture, especially a more complex one, to continue on its current path (business as usual) for as long as possible, that is, until it is forced to resimplify.

This is more likely to occur when the members can no longer keep all of their resource, social, environmental, individual and cultural responsibility balls in the air, rather than when a specific event occurs. A particularly notable event is just a marker on the journey to resimplification. The nature and speed of the resimplification depends on the state of the human-environment system and the members' culturally influenced decisions. Similarly, the fate of a culture after resimplification depends on a number of factors: its remaining environmental endowment, the members' new awareness, cultural resources, and the changes they are willing to make to their personal and their culture's world view.

This chapter's summary starts by noting that at the end of the previous chapter questions were raised about the value of the global-scale mathematical models of the human-environment system. Of particular concern was their prediction that a future, global-scale, population-resource feedback can lead to a population-resource caused overshoot and collapse (figure 20). A common rejoinder was that we will likely muddle through just fine, especially because of our technology. To help us evaluate these two possible futures, historical examples of resimplification were discussed in this chapter, and the conclusions were included in the, now adjusted, descriptive model of the human-environment system from chapter 11.

The mathematical models' prediction of a global-scale, multicultural population-resource overshoot and collapse is valid and useful, but the historical resimplifications reveal two important limitations. The historical examples illustrate the first. By aggregating cultures, the mathematical models fail to represent the significance of the earth's cultural diversity. Think economic models. The examples reveal that, in detail, each historical resimplification was unique. The timing of a culture's resimplification and the members' experience of it was influenced by the culture's level of complexity, the state of its environmental endowment and its members' culturally influenced response to events. These characteristics also determined the culture's resimplification path and its post-resimplification future.

Consider the role that the Xhosa culture's world view played in its resimplification. The idea that the ancestors would arise from the dead springs to mind. Think too of the splits and discords among the Xhosa: some embraced the prophecy and forced compliance with the conditions on those that didn't. These human contributions can lead to resimplification occurring through social discord long before resource shortfalls do. Alternatively, social discord can contribute to the development of resource shortfalls that favour resimplification.

The second limitation concerns the path/process of resimplification. Although a culture's population-resource–induced resimplification may be broadly described as overshoot and collapse, the historical examples illustrate that the actual process and outcome is much messier and more complex than the mathematical models' smooth curve suggests. In particular, the ability of a culture and its supporting environment to at least partly recuperate and adapt can allow a culture to survive well beyond its theoretical collapse point.

The presence of both ups and downs were illustrated by Babylonia's long history of increases and decreases in complexity. Consider too the short-lived gains in complexity (efflorescences) that occurred in Europe and China between 1100 and 1800. These efflorescences were short-lived, in part, because the elite and powerful made decisions that benefited themselves but helped to stall the changes driving an efflorescence, thus eventually reversing it (Goldstone 2002). Similarly, the presence of short-lived losses in cultural complexity (setbacks) from which a culture can recover is illustrated by the late 1900s break-up of the Russian empire.

As a result of these variations, the members of a particular culture will find it difficult to decide on the importance of a specific trend or event they experience. Their interpretation and response to such trends and events will be significantly influenced by their personal and their culture's world view of how the world works, the time frame they use to evaluate it and their decision-making foibles. Consider how those Christians who believe in the apocalypse and rapture tend to interpret some of today's events (Nayeri 2017).

Thus, although mathematical models remind us that, in a finite system, a resource- and/or population-driven global-scale overshoot and collapse is possible, they do not provide a realistic description of the nature and timing of such a real-world event. In particular, although mathematical models do include our mental and social contributions to such an event, the way they do so is far too simplistic. Overshoot and collapse is unlikely to be the simple, largely population- and resource-driven affair that the models suggest. It is more likely to be a long-drawn-out and messy social-environmental process that affects the more complex cultures differently than the less complex ones. The broadness of the term cultural resimplification expresses this more messy, dynamic, multi-faceted complexity of the process much better than overshoot and collapse, which implies a more simple, black and white, rapid, one-way event.

The adjusted descriptive model of the human-environment system compensates for these deficiencies. It includes the insight gained from the historical examples of resimplification, from the global-scale mathematical models, and from the discussions about both the resources humans want and the state of the earth's environment that provides them. It also includes the complex, dynamic manner in which we make decisions, and how we (personally and culturally) change. The model is thus a tool that can be used to provide a general description of, and explanation for, the system's possible futures. It can also be used to provide and evaluate predictions about those futures.

Using the descriptive model indicates that if the earth's more complex cultures continue on their current path, then many will likely go through a resimplification process that, in broad view, is common to all cultures but, in detail, will be distinctly their own. Industrial cultures that form a complexity-resource or equivalent feedback are the most likely to resimplify. However, the other complex cultures are not immune because today's cultures are linked together through globalization and a common environment. It is likely that some cultural resimplifications will occur as part of a multicultural event.

How should the members of complex cultures respond to the possibility of resimplification? We could deal with it piecemeal (i.e., by muddling through). This is, apparently, the path historical cultures took to become more complex, and then later followed to experience forced resimplification. Having faith in muddling through is an easy choice, but in the long run, it can't meet, on its own, the hope that goes with it.

We could continue to maintain our hope and faith that our superior technology along with technological innovation will ensure we avoid resimplification. A culture's level of complexity is a proxy for the degree to which it relies on technology. History indicates that a high or increasing level of a culture's complexity is not a guarantee that it will avoid resimplification. Thus relying on technology is unlikely, on its own, to guarantee that a more complex culture will avoid resimplification. What matters is how and when we decide to, or not to, develop and apply technology.

We could select a response from the mass of available advice that was mentioned in this book's introduction. Our efforts to sift through that advice, or decide on the application of technology, are helped by using the adjusted descriptive model of the human-environment system because it prompts us to consider the human factors, and to do so much more thoroughly than the resource-focused mathematical models do. The descriptive model reminds us that we, as individual members of a more complex culture, are its decision makers. It is our individual decisions that will, significantly influence the members' collective contributions to our culture's possible resimplification, our recognition of that possibility and our response to both the possibility and the event. Whatever our decisions, they will be the product of our decision-making foibles, as influenced both by our experience of living in our culture's environmental endowment and by our culture's world view. Thus it is not just the response we select from those available that will determine our complex culture's future state but the beliefs and decision making behind our choice.

The adjusted descriptive model of the human-environment system can help us to recognize the signs favouring a future resimplification, and then help us make and implement the decisions that will avoid it. But, in order to do so, the model can't be used in the abstract. The broad insights it provides to us about resimplification must be made relevant to our present environmental and resource realities, and our current day-to-day experiences. The next chapter personalizes the lessons from this chapter by focusing on the individual member and tying the past to the present.

# CHAPTER 17

# RESIMPLIFICATION IN ACTION

History has provided us with enough information to outline a descriptive model of how some cultures come to increase their level of cultural complexity. This increase is indicated by characteristics such as a hierarchical organization, large population, large territory, monumental buildings, large urban centres and diverse work specialties. Those of us who are members of a more complex culture find it relatively easy to relate to an increase in a culture's complexity because it is part of our culture's origin story.

A complex culture that drives itself or is pushed to a point where it can no longer function experiences a reduction in its level of complexity: it resimplifies. A member of today's complex cultures can accept that resimplification is a part of a past culture's history. We treat the event as a precautionary tale. However, if we are to use this knowledge to help us consider the possibility that our *own* complex culture may resimplify, then we need to relate to resimplification in action. We need to treat it, at a personal level, as a dynamic process that is relevant to us today. This means personalizing and applying the intellectual insights gained from historical resimplifications and the updated model of the human-environment system (provided in chapter 16) in a contemporary context.

This chapter brings the process and experience of resimplification alive for the members of current complex cultures and then discusses the possibility of their culture resimplifying. The first section personalizes resimplification by looking at it through the eyes of a contemporary person. It uses the model to consider an aspect missing from the description of historical resimplifications: the contribution made by its individual members. The section ends by providing rules of thumb, based on historical cultures, for contemplating resimplification. The next section addresses the question of whether the industrial cultures of today are sufficiently similar to historical complex farming cultures for the rules to be applied to current cultures. The third section discusses whether technology can ensure that complex industrial cultures will avoid resimplification. The following section describes how the members of an agricultural culture avoided resimplification by changing themselves and their culture so that they lived more within their human and environmental constraints. The chapter ends by discussing the likelihood of current complex industrial cultures facing resimplification and how the members might respond to this possibility.

## Personalizing Resimplification

The descriptive model of the human-environment system, discussed in chapter 11 and updated in chapter 16, allows us to more easily recognize the connection between the functioning of complex cultures and the common features of historical resimplifications. The object of this section is to personalize that understanding and make it relevant to our circumstance today: that is, relevant to a member of a contemporary complex culture. A rules-of-thumb recipe for resimplification is provided.

**The Ingredients.** This subsection personalizes resimplification by discussing its three ingredients (environment, culture and individual) and their interactions. The ingredients are also the three factors previously used to model the human-environment system in a way that includes the understanding of human nature and the environment's functioning we gained in earlier chapters. Thus, by using these ingredients in conjunction with the model, the personalizing of resimplification will meaningfully include the individual, which was missing from the description of historical resimplifications in chapter 16. In addition, because resimplification occurs within the system's functioning, the discussion will also personalize the process of resimplification.

Personalization is helped by discussing the three ingredients and their interactions from the perspective of a member of a contemporary culture wishing to evaluate the possibility that their own contemporary culture might resimplify. Assume that they are trying to relate to the process of resimplification, appreciate the role that the individual members played in it and discover the rules of thumb for resimplification.

*Environment.* The resources (including services), such as water and water purification, that a culture needs to support itself are ultimately provided by a culture's supporting environment: its environmental endowment. We could conclude that the most important features of the culture's environmental endowment are the abundance and quality of the resources it provides. For example, compare the widespread thick, fertile soils of the Babylonian delta with the rare thin, less fertile soils of southern Greenland. Although true in the short term, this analysis is too simplistic. Over the long term, the culture also relies on its endowment's functioning because it renews those resources as they are used, even as the endowment itself changes over time. The features of a culture's environmental endowment thus constrain the culture's population size and lifestyle, and burdens them with the responsibility of ensuring that they don't disrupt its functioning.

This means that the characteristics of a culture's endowment must be reflected in its world view. After all, its guidance to the members will, for example, determine whether their resource demands will disrupt their endowment's functioning and thus reduce its ability to support them. More specifically, if a culture is to remain functional over the long term (viable), then its world view must guide its members to acknowledge their dependence on their environmental endowment's functioning and accept the responsibility of living within its constraints.

Hunter-gatherer cultures have a close relationship with their endowment. They have generally demonstrated their ability to recognize and respond appropriately to their endowment's dynamics and respect its constraints on their lives over the long term (Kelly 1995). Historical complex cultures have also, in the short term, displayed their resilience and adaptability to some environmental changes, such as drought. However, over the longer term, both the response of the members to natural environmental changes and their impacts on their endowment have generally contributed to their culture's resimplification (deMenocal 2001).

Considering the wide range of human skills and knowledge available to the more complex cultures, this outcome may seem surprising. It is less so after considering the complex nature of the constraints the environment imposes on the culture it hosts, especially when they try to manage it. Think responsibilities. This is illustrated by expanding the story of Easter Island's resimplification (discussed in chapter 16) to include other very remote Pacific islands. The focus will be on a critical link between farmers and their environment: the soil.

Each of the very remote Pacific islands has its own distinctive characteristics. Some of these islands have been occupied continuously since they were first settled (e.g., Easter Island). Others were settled briefly, then abandoned. Because the settlers on all of the islands were of Polynesian cultural heritage, it can be inferred that their initial culturally supported relationship with their island's environment were equivalent. This makes it easier to discover the environment's contributions to the islands' different histories and the differences that emerged among the settler cultures over the long term (Anderson 2002).

On a regional scale, the remote Pacific islands have similar geological characteristics: they are all volcanic islands that have a similar average rock composition (basaltic). The primary geological difference among the islands is their time of formation: it ranges from a few tens of millions of years ago to less than a million years ago. Thus, despite the differences in rainfall among the islands, the geologically older islands are usually more deeply weathered than the younger islands. As a result, the biologically active soils on the older islands contain less available phosphorous and nitrogen, which means that the plant growth and ecosystem characteristics on them are more nutrient constrained than on younger islands. These features constrain the soil's productivity when farmed.

Adding to the age and age-related differences among the remote islands is the variation in their geographical characteristics, such as size, topography, climate zone (especially rainfall) and the presence/absence of a reef. These details too constrain the nature and functioning of an island's land and ocean ecosystems (including their resilience, robustness and adaptability). As a result, each island has its own distinctive environmental endowment. How the relationships among a particular island's features collectively resulted in its distinctive endowment, and how that endowment influenced its post-settlement history, is illustrated by continuing to focus on the soil.

The ability of an island's environmental endowment to support settlers and their descendants is strongly influenced by its soils, in particular their stability and fertility. The susceptibility of an island's soils to erosion is significantly determined by the island's characteristics, such as its topography and rainfall. For example, on small, steep islands, erosion takes place more easily and rapidly than on larger, flatter ones. The type and amount of rainfall an island receives determines the amount and speed of the runoff and the type of ecosystem binding the soil together, both of which affect erosion rates. Farming increases the soil's erosion rate by disturbing its erosion-preventing ecosystem cover and altering the runoff path. That disruption occurs once when the fields are created by the destruction of the indigenous ecosystem, and then routinely thereafter by farming the fields.

As mentioned above, older, more weathered islands tend to have less fertile soils than younger islands. Regardless of their initial fertility, the rate of fertility replacement is influenced by the island's other characteristics. Consider islands of the same age: the greater an island's rainfall, the faster the weathering of its rocks and the creation of soils, and the greater the relative rate of fertility renewal. If large sea-bird colonies existed on an island, then their nitrogen-rich guano would have helped to maintain soil fertility. However, once the early settlers had eaten the birds and, with the help of the exotic rats, had consumed their eggs, this possibility would have no longer existed. The fertility of an island's soils is also maintained by the soil ecosystem's nutrient recycling system. By farmers removing the above ground portion of that ecosystem, they break or severely disrupt this recycling loop. The farmer must then accept the responsibility for replacing that lost service, if the farmed soil's fertility is to be maintained.

The complex interaction among an island's geological, geographical and ecosystem details results in it having a unique environmental endowment that both facilitates and constrains its ability to support humans and thus their culture's future. Despite these differences among the islands, the initial Polynesian settlers, as mentioned, all treated their endowment in a similar manner. Thus, the histories of the remote islands are, in a general sense, quite similar.

When that history is seen from the environment's perspective, the arrival of the exotic human species on an island was followed by its indigenous ecosystems being disrupted or destroyed by a mix of deforestation, soil erosion, species extinction and the environmental impacts of the exotic species that the humans introduced. If the humans' efforts to establish and propagate their favoured food exotics on the island were successful, then the human population grew. To feed their increasing numbers, the remnants of the island's original ecosystems experienced ongoing disruption and destruction. An island's ecosystems rarely recovered from these impacts to anywhere near its original state.

When seen from an individual human's perspective, the first Polynesian's to arrive on an island were lucky to have survived their amazing arduous ocean voyage. The settlers initially subsisted on the easiest-to-secure resources (flightless birds and easy to catch marine animals) until their farms could supply most of their food. To create their farms, the settlers laboured industriously to transform the island's ecosystems into fields suitable to support their culture's traditional agricultural food sources, which they had brought with them. These foods included rats, pigs, cassava and banana trees. If successful, then the settler population rapidly expanded. This was usually followed by a population crash and a cultural resimplification (Anderson 2002; Kirch 2007). If the culture was to survive, the members had to adapt to their island's changing conditions. This led to the diversification of the islands' cultures.

Both perspectives of Polynesians settling on remote Pacific islands highlight that their culture's guidance did not encourage them to fully consider the particulars of their island's environmental endowment: how it functioned and the constraints that imposed on how they could live. An example of the outcome is the Easter Island story, as told in chapter 16. A similar story can be told for the Polynesian settlers on another remote island, Mangaia. Because the cultures were the same but the characteristics of the two islands were different, a comparison of the two resimplification stories makes it easier to see the importance of an island's environmental endowment in determining the details of their respective resimplifications. The focus, again, is on the soil.

Mangaia is, like Easter Island, the top of a remote volcano. Its rainfall varies between 1,400 and 2,100 mm/year which is a little more than Easter Island receives. In other respects, Mangaia is quite different. It is much older, at 16.6–18.9 million years (My), and much smaller, at only 52 km$^2$. The island is also close enough to the equator to have a tropical climate and tropical ecosystems.

After settlement, Mangaia went through the same basic history as Easter Island: species extinction, overexploitation of natural resources, deforestation, erosion, the introduction of exotics and a human population expansion and crash. But the details are different.

Being an older, more deeply weathered island, Mangaia's environmental endowment at the time of settlement included nutrient-poor soils, low ecosystem resiliency and limited alternative wild foods. The island's landform and age meant that once a hill slope was disturbed or deforested for agriculture, then the thin, nutrient-containing organic soil could easily erode. After erosion, the regeneration of both soil and forest was extremely slow.

Thus, post-erosion, the only soils suitable for agriculture were those found in the depressions where the eroded soil had collected. This situation is similar to the Argolid in Greece and the Mayan Yucatán Peninsula. These fertile pockets made up only 2% of Mangaia's land base, a situation much worse than on Easter Island.

As the population grew to its peak and crashed, the Mangaians were increasingly faced with the need to adapt their culture to manage their extremely limited food-producing resources. They chose, like the Easter Islanders, to replace the hereditary rulers with military ones, and to adjust their religion and their culture's world view to support the new rulers and rituals (figure 15). In this newly rationalized culture, citizens were ruled by force and terror, including human sacrifice and, probably, cannibalism. Their lifestyle included fighting for control of the fertile land and its irrigation works (Kirch 1997; Kirch 2007). Easter Island is usually held up by Westerners as the epitome of a bad outcome from ecosystem disruption and cultural missteps. But Mangaia provides us with an even more apocalyptic lesson.

The Mangaians' failure to recognize the seriousness of their island's soil-related environmental constraints on their lives led to them experiencing what appears to us to have been an unnecessary environmental-endowment-related cultural resimplification. This soil factor played an important but less critical role in the Easter Island resimplification. This illustrates the more general conclusion that the characteristics of an island's environmental endowment played a critical role in determining the future of the people who settled on it.

The importance of a culture's endowment to its resimplification is not restricted to the Pacific islands. Consider the role that the salinization of Babylonia's soil, the loss of soil from the land settled by the Greenland Norse, and the erosion and likely nutrient depletion of the Maya's soil played in their resimplification. The characteristics of a culture's environmental endowment made a significant contribution to most, if not all, historical resimplifications.

In overview, the history of remote Pacific islands illustrates that the environment's contribution to an island culture's resimplification is the nature of its endowment to the settler's culture: the complex interacting mix of the endowment's geological, geographical and ecosystem features. Its resource limits and functioning imposed constraints on the settlers' decisions and burdened them with responsibilities when they took up farming.

A more meaningful appreciation of a particular historical resimplification emerges when the interaction between the endowment and the settlers is considered. After all, by settling, the members of the culture became participants in their environmental endowment's functioning. They formed a close interactive relationship with it, one in which they were neither powerless pawns nor its masters (deMenocal 2001).

When seen in this context, an island culture resimplified because the cultural relationship that developed between the settlers and the island's environmental endowment did not match the external-reality requirements for their culture to be viable. Although their endowment provided cues to the settlers that they were exceeding its constraints and failing to meet their environmental responsibilities, as illustrated by the fate of their soils, their culture's world view failed to prompt them to respond appropriately.

The members' contribution to the environmental aspects of their culture's resimplification was their lack of appreciation for their endowment's functioning. They did not recognize its cues and accept their responsibilities. Instead, they chose to live in ways that tried to overcome rather than live within their endowment's constraints: they maintained their older cultural practices, developed in a different environment.

It is no surprise that, in the longer term, the islands' complex agricultural cultures sleepwalked into an environmentally facilitated resimplification (figure 15). These conclusions are relevant to all historical resimplifications and the future of present cultures.

*Culture.* A human group's culture smooths the interactions among its members, ensures that the group's interests are protected and mediates the members' interactions with their environment. The culture accomplishes this by its world view guiding its members' decisions. But that world view is created by the collective decisions of the individual members. This member-culture feedback ensures that a culture is human biased, contains divergences, is deeply rooted in history and is subject to considerable inertia to change, as discussed in chapters 3 and 6. It is within this context that a culture's role in contributing to resimplification is most easily appreciated.

Consider the guidance a culture's world view routinely provides to its members about their relationship with their environment. As discussed above, if the culture is to be viable, then that guidance must reliably reflect the constraints that their environmental endowment imposes on the culture's functioning and thus on the members' decisions. To achieve this aim, the relevant environmental characteristics must become embedded into the culture's world view.

The hunter-gatherer Mande culture illustrates this process. This is the continuation of a discussion started in chapter 8. The Mande heartland is centred on the Niger River, near the Senegal River, in West Africa. The rainfall the area receives results in east-west zones of similar rainfall amounts, each of which supports a particular biome. Overall the rainfall decreases from north to south. Currently, the annual rainfall ranges from 100 mm/year in the north, which supports the Sahel semi-desert biome, to 1,000 mm/year in the south, which supports the savannah biome.

However, over both the short and the long term, the amount of rainfall that a region receives changes significantly. In the short term (years to decades), droughts are frequent. In the longer term (centuries), as the climate changes, the rainfall and the biome zones it supports move: northward as it becomes wetter and southwards as it dries. The distances covered can be considerable, as illustrated by rock paintings now found far to the north of the Sahel in parts of today's bone-dry, vegetation-free Sahara Desert. They depict a time when animals, such as giraffes, and activities, such as hunting, were common (Brooks 1989).

To survive in this highly variable, hash environment, the Mande had to pay close attention to their endowment's state and react appropriately to its changes. They were successful because their experience of past environmental conditions and the decisions that enabled them to live under them became a part of their cultural memory: their oral myths and legends. The Mande likely helped themselves to both remember their history and link it to their current circumstance by tying a particular myth, legend, event or circumstance to a specific feature of the land, such as a distinctive hill or a particular animal. These features could then act as symbolic memory cues and thinking guides.

Through these methods, the Mande's knowledge about how to live in their environment became both embedded into their culture's world view and part of their world. This ensured that their knowledge could do more than help them to make appropriate decisions today. It could be reliably transmitted down the generations and be available to help guide their decisions of their descendants.

When the Mande were faced with an environment-related decision, a trusted and skilled member of the group could use the culture's memory to help interpret the current signs and provide the relevant guidance. The interpreters were also expected to demonstrate wisdom: the guidance they provided had to be relevant to the

other aspects of the people's current circumstance. A degree of flexibility was thus inherent in their decision making. The Mande culture's structure and world view provided them with appropriate, dynamic guidance that enabled them to live in their harsh, shifting environment for thousands of years. It is thought they did so from at least 5000 BC (when they were hunter-gatherers) until closer to the time when they became a warrior state (Togola 2000; McIntosh 2000).

In contrast, the Easter Islanders and the Mangaians did not solve their ongoing resource depletion problems by appealing to their ancestors (as we infer they saw them) or fighting one another for dwindling resources (Flenley and Bahn 2002). Neither could the Mayan city states resolve their practical ecosystem problems of erosion and soil depletion by ignoring them, or by concentrating on their cultural rituals, wars, appeals to their gods and their sophisticated predictive astrological calendars (Freidel and Shaw 2000; Scarborough 2007). As these and the other examples of resimplification illustrate, the guidance provided by a historical complex culture's world view tends to have poorly reflected the environmental constraints on the members' lives, if their culture was to be viable.

It seems odd that inappropriate cultural guidance about environmental issues should be so widespread and persistent among complex cultures. After all, as discussed, a culture's world view is not carved in stone, and its members are both the carriers of their culture's world view and their culture's decision makers. They could thus adjust their culture's world view, like hunter-gatherer cultures, so that its guidance would be more appropriate for their environmental circumstance.

This is easy to suggest, but it is much more difficult, and much slower, to make the needed changes, even under optimal conditions. Some reasons were given in chapters 6 and 11. The basics are summarized and added to here. When a culture's members exploit their environment more intensely (e.g., by taking up farming), their culture's structural complexity increases, and its world view is adjusted to better reflect its members' new cultural reality. This includes adjusting its guidance about the appropriate relationship to have with their environmental endowment.

Unlike the hunter-gatherer Mande, farmers are guided to find more of their sense of meaning in altering their environment rather than in their close connection to its inherent features, functioning and constraints. But events such as a drought routinely remind the members of their ties to the land. Thus, their culture's world view can never fully scrap the memory of their close connection to, and reliance on, their environment and its functioning.

The presence of these two contradictory views of the environment (transform and revere) in their culture's world view results in the members routinely being provided with conflicting guidance about how to behave toward it. For example, the Babylonian culture's world view guided its members to redirect/control rivers in order to irrigate their fields. But it also guided them to venerate these sacred, life-giving rivers and their water by worshipping "water gods," such as Abzur and Marduk. The early complex farming cultures never fully resolved the conflict between transforming and revering their environmental endowment.

As a culture becomes more complex, the conflict between the cultural imperatives to transform versus revere their supporting environment becomes even more pronounced. Consider that, at some point, a culture becomes sufficiently complex that it forms a complexity-resource feedback. At this time it requires considerable resources to maintain that complexity, but in order to make them available from its environmental endowment, the culture must increase its complexity (see the resources loop in 15). It also requires its world view to be adjusted so that its guidance more strongly favours transforming the land to maintain the supply of resources. This resolves the contradiction between the culture's urge to transform versus revere their environmental endowment in favour of the short-term expedient of transforming it. The reminders of the external reality's constraints are sidelined, until the they become so forceful they can't be ignored.

Although it was possible for the more complex of historical cultures to have made the needed cultural adjustments to correct inappropriate cultural guidance about the environment and their relationship to it, the best they could actually accomplish was to partly address it. Consider that the Mayan, the Babylonian, and even the less complex Mangaian and Easter Islander cultures all adjusted their world views to somewhat deal with their soil issues, but it was insufficient to protect their soils over the long term. For example, they did not encourage population reductions or radical changes to their farming practices. In essence, they were unconsciously practising farming with a bias towards their culture's world view, rather than towards the constraints of their environmental endowment and its functioning on their lives. It is as though they felt, believed or received guidance that the only realistic solutions to their immediate issues were to either maintain business as usual or directly increase resource production.

The presence of this conundrum and the biased resolution can be seen in today's Western culture. Consider that the later complex Christian cultures of Western Europe inherited an expression of it through the Bible's old testament. Westerners could claim, as we saw fit, that Genesis either gives us the right/freedom to use the land however we want, or entrusts/burdens us with the responsibility of acting as stewards of God's creation (Lang 2002). As discussed in chapter 11, the bias is in favour of exploitation/transformation.

The guidance a historical culture's world view provided to its members affected more than their awareness and response to the environmental constraints on their lives (i.e., the members' relationship with their environment). It also affected their awareness and response to the human constraints on their lives (i.e., how they lived as a social group). Thus inappropriate guidance could also have far-reaching social consequences. The more complex a culture's structure became, the more difficult it became for its world view to guide the members to live within these human constraints and for the members to follow that guidance.

Consider that its guidance had to continue to ensure fairness, maintain cohesion, balance sharing with competition and balance individual freedom with group responsibilities, as it does in hunter-gatherer cultures. But it now had to do so in the presence of a social power hierarchy and among many more somewhat isolated cultural subgroups. For example, in a complex culture, some individuals could lord it over others, and subgroups could hold different views about what they deserved from the greater group. This meant that a complex culture's world view had to provide guidance that effectively prevented the abuse of power by the wealthy and powerful. Think corruption and greed. It also had to ensure that all subgroups had a voice and equitable access to resources. Think fairness. Providing and following such guidance is a challenging task in complex cultures.

Assume for the moment that a complex culture could provide appropriate social guidance to address these issues. For this guidance to be useful, it must be followed by most of the members. This is harder for a more complex culture to accomplish than we imagine.

Hunter-gatherer cultures prompt compliance with their culture's guidance through mutual social monitoring and social censure. These methods can work because the bands are small enough for members to know and be connected to one another in some way (e.g., kinship or agreements). They also had the opportunity to have routine contact with one another (Kelly 1995). Because the more complex cultures have much larger populations and many more distinct, somewhat isolated sub-groups than hunter-gatherer cultures, the hunter-gatherer's time-tested methods of prompting/correcting behaviours are insufficient to ensure compliance in complex cultures.

Complex cultures therefore have to augment their compliance procedures. Voluntary compliance is encouraged by obtaining the members' ongoing acceptance and support for both their culture's authority structure and world view. This is achieved through their complex culture's expanded reaffirmation and authority maintenance process, as discussed in chapter 3. Included in that expanded process is something not found in hunter-gatherer cultures: a social contract. In it, the members support their culture's power structure (its hierarchy) in exchange for services. The expected services can take many forms. One of them is the augmented procedures for enforcing compliance with the key aspects of their culture's world view. For example, as Babylonia's complexity increased, so did the use of formal laws to represent the guidance from their culture's world view and the "police" to enforce those laws, as we understand these terms. Think of Hammurabi's stela (an isolated stone column) with his laws of the land inscribed on it.

But the effectiveness of laws and police enforcement is constrained. Consider the difficulty of making a law that not only addresses a specific concern and is consistent with other laws but also reflects the powerful human urge for fairness. This task is made more difficult by the self-interest of the law makers and the potential for a complex culture's many subgroups to hold differing views about the need for a law or the law. The problem of fairness is also present during enforcement. Consider that the enforcers can abuse their power, especially if they are underpaid, undertrained and undercommitted to the greater good.

The history of Babylonia indicates that the issue of fairness likely played a significant role in its resimplifications. Today the lack of perceived fairness lies at the root of much of the dissatisfaction the members of complex culture's express about laws and their enforcement. This applies regardless of whether the laws deal with environmental responsibilities or social behaviours.

Unsurprisingly, the members must exert considerable effort to ensure that their complex culture's world view, and thus its guidance, favours their culture being functional over the long term and the members complying with it. This effort is illustrated by the short-term resimplification and subsequent recovery of the newly united ancient Egypt. A period of exceptionally low Nile floods began in 2200 BC, which resulted in a decline in food production. By 2180 BC, the low flow had helped to initiate 40 years of instability and disorder during which there were revolving-door claimants to the throne, popular revolts, famine and civil war. Egypt experienced a lack of revenue, a lack of irrigation system upkeep, weak cultural bonds between the newly united Upper and Lower Egypt, and doubts about the efficacy of the monarchs and the bureaucracy.

The normal floods returned in 2134 BC. But before the previous level of cultural complexity could be re-established, Egypt's institutions and structure had to go through a renaissance. This process of rehabilitation and reorganization included appointing more capable administrators, putting in place administrative reforms, strengthening the concept of justice and building irrigation projects (Hassan 1997). Of course all of these changes needed to be paid for using resources, which brings us back to the complexity-resource feedback and the environmental constraints on this complex culture that were waiting in the wings, again. To remain viable a culture's world view must appropriately deal with all of these constraints and conditions, simultaneously.

In overview, a culture's world view guides its members to adopt its preferred relationships with their environment, lifestyles, and methods of dealing with social and environmental issues. But, the more complex a culture becomes, the more difficult it is for its world view to provide guidance that recognizes and coherently deals with the human and environmental constraints on the members' lives. Consider that the culture faces increasingly complex social, resource and environmental issues whose resolution requires reconciling conflicting objectives. Think of the conflicts between the goals of maintaining resource supplies and protecting the environment's functioning (revere versus transform the environment).

By default, a complex culture's world view tends to favour resolutions to issues that result in an increase in resource use. But supplying them can increase the culture's complexity. The more a complex culture buys into this complexity-resource feedback, the less likely that its world view's guidance will appropriately guide its members to take into account both the environmental and the human constraints on its functioning. A complex culture's world view can thus contribute to the culture's inherent instability and susceptibility to resimplification (Tainter 1988).

*Individual.* Personalizing the resimplification of complex cultures would be incomplete without acknowledging that cultures are made by and for people. The decisions we attribute to a culture that led to its resimplification are, ultimately, the cumulative product of the decisions made by its individual members, from monarchs to peasants. A member's decisions are the result of their decision-making processes, which are applied within an environmental, historical and cultural context (Kelly 1995). Unfortunately, very little is known about the details of a historical individual's decision making, such as how they reacted to the specific influences on their lives. Any diaries they wrote that detailed their mental lives are unavailable to us.

Fortunately, as discussed in the previous chapter, for some historical cultures the overall context within which its members made their decisions and the general manner in which their culture subsequently resimplified are known. We also have a general understanding of our innate decision-making process, its biases and the influences on it, as discussed in part 1. Included in that understanding is the creation of our internal reality with its divergences from the external reality: the source of our reality paradox and cognitive dissonance experiences. This knowledge was summarized into the model of the human-environment system, described in chapter 11, which then provides the overall framework in which our decision making and change takes place.

If this general understanding of human decision-making and change is taken into account within the context of a historical complex culture's resimplification process, then it is possible to gain a more complete appreciation for how individual members contributed to their culture's resimplification. This two-part discussion does so by focusing first on the influences common to all individual decision making and then on the influences that are more relevant to the elite decision makers of more complex cultures.

A complex historical culture's future was significantly influenced by the decisions of its individual members. We might think that only those decisions about special events, such as deciding to go to war, would be of interest, but this is not the case. Just as influential were their day-to-day decisions. After all, these routine decisions largely determined the culture's ongoing resource demands, population size and environmental impacts: the practical basics of their culture's functioning. Also of interest were the inherent biases and constraints in their decisions.

Consider the biases that arose from their constrained (unconscious and conscious) awareness of both themselves and the world around them, and the limited attention they paid to what they were aware of, as discussed in chapter 6. Each member, regardless of their standing in their culture's social hierarchy, would have been most aware of and paid closest attention to those aspects that directly and immediately affected their day-to-day lives: their routine, human-centred activities and concerns. These would have included food acquisition, children, personal relationships, work duties and religious duties.

The details of each member's awareness, attention and focus would also have been influenced by the guidance from their culture's world view about which aspects of their material and mental lives to consider important and which to filter out. For example, cows were an extremely important aspect of the Xhosa culture's world view, just as barley and wheat were for the Babylonians, and corn for the Maya. Consider too the Greenland Norse's Christian focus on salvation after death and the current focus of Western culture's world view on progress.

However, in the past, like today, the members' awareness of a changing world and the attention they would have paid to it depended on more than their personal interests and their culture's guidance. It would have been affected by how the changes around them occurred. Consider a changing environment. If its changes occurred rapidly, directly affected people, or were spectacular or intense, then individuals were more likely to have been aware of it and to have paid more attention to it. Examples of such events are a sudden drought, pest outbreak or crop failure. Think of the Egyptians during the low Nile floods of 2200 BC.

In contrast, most members would likely have been unaware of a long-term environmental change that occurred slowly (over decades or longer) or was masked by routine, short-term environmental variations. The members who did notice hints of long-term change would likely have only noted it in passing. Those hints would have quickly become part of the background to the individual's life, to which they would have paid little, if any, direct attention during their decision making. For example, it seems that the Maya had limited awareness of ongoing soil erosion, or if they were aware, they gave little attention to it. This is also the case for the Greenland Norse and their soil erosion, and the Easter Islanders with deforestation. The Greeks in the Argolid were aware of erosion but paid only intermittent or ineffectual attention to it. The same might be said of the Babylonians and salinization.

In retrospect, it is easy to see that the members of these cultures could have responded more appropriately to their culture's pre-resimplification circumstance. It is easy to say that they would have done so if they had broadened their awareness of their complex world and paid attention to more than their immediate needs, their cultural imperatives and the more in-your-face changes. Easier said than done.

Unfortunately, humans have a limited innate ability to recognize the presence of this complexity and appreciate its significance, let alone consider it in their decision making. This is illustrated today by the difficulty even experts experience when trying to find a fault within a system of relatively low complexity, such as a broken down car (Gardner and Stern 1996). Thinking about complexity tends to overwhelm us, so we prefer to avoid dealing with it directly.

We use a similar approach when faced with making a decision, especially an unpleasant or difficult one. We prefer to make the easier and more emotionally satisfying decisions. They are usually the ones that require the least effort, such as simply trusting our intuition, choosing the familiar, relying on a plausible story we (consciously or unconsciously) make up, delaying the decision, following our culture's guidance or largely ignoring the decision/issue. These preferences help to explain the appeal of strongly held, simplistic, popular or culturally supported views about issues, and why our decisions tend to maintain the status quo.

We can also seek advice from someone we feel is more knowledgeable or qualified than us, such as an elder or a specialist. In complex cultures, the reliance on others to provide guidance and make decisions is partly formalized into a decision-making structure: the power hierarchy. Those lower in the structural hierarchy are required to defer the making of certain decisions to those higher up. However, both the practice of seeking advice and deferral can also be used to, consciously or by default, delay or avoid the effort of making or implementing a decision.

It is easy to infer that at least some of these techniques occurred in complex historical cultures. For example, individual Babylonians and Maya placed great store in the power of their monarchs and religious leaders to make and implement decisions that would deal with a drought or a crop failure in a satisfactory manner and prevent them occurring in the future. In turn, the leaders relied on the members accepting their decisions, including their need to lay claim to some of the resources the members produced. Without this support, it would be difficult for the leaders to make and implement the expected preventative and curative decisions.

This ever-present interdependence between ruler and ruled helped to unite the members of the culture and strengthen their faith in their culture's world view and structure. But it also led to the members making decision that we, in our retrospective view, could label as the product of a kind of culture-level groupthink. Consider Mayan peasants and elite feeling at ease as they participate in a ceremony involving human sacrifice because they felt it was needed to repay a debt to the gods and ensure the fertility of their fields. Similarly, an Easter Islander could have been content cutting down one of the island's last trees and a Babylonian royal retainer could have accepted being killed when their master died.

However, from a member's perspective, their decisions look different. According to their (personal and cultural) view of the world, they were fully aware of, and paying attention to, all that really mattered. They each had a familiar, perfectly logical and culturally acceptable explanation for their feelings and actions, and a handy scapegoat to blame if things went wrong. They made their decisions accordingly. Are we really any different?

No wonder great efforts were made to implement what the members thought were the most appropriate solutions to their issue. Consider the food issue. They conducted the ceremonies and completed the actions needed to increase next year's food supply. Other possible responses, such as population reduction (think of the Maya and Easter Islanders) and changes in food source (think of the Greenland Norse) seemed to have received little attention. These alternative responses may have been sidestepped, in part, because of the cultural influences on their routine decision making discussed above.

Consider the option of limiting or reducing the size of a population as a way to avoid a future food shortage: a contingency or buffer. We could expect that the parents of a historical complex culture's family treated the birth of their children as a positive event. Similarly, if there was any guidance from their culture's world view about births, we could expect that it generally favoured an increase in the population because that, in some way, indicated cultural success, such as being favoured by the gods. Thus, if either the general members or the elite of a historical culture, such as the Mayan, Easter Island and Mangaian cultures, had an opinion about their culture's increasing population, they were more likely to have interpreted it as normal, if not positive (Dean 2000). Without sufficient attention to the many constraints on population size, such as food supply and its variability, the members of both these cultures were left to discover, through experience, that beyond some population size and density, a population increase becomes maladaptive. As the records and excavations from historical complex cultures indicate, they commonly experienced food shortages. Thus some of the members' decisions contributed to their culture's resimplification.

The above discussion also hints that just making a decision is insufficient to affect a culture's future. If a decision, even if it is to do nothing, is to have any effect, it must be implemented. Implementation is, like decision making, a process that is constrained by personal, cultural and environmental factors. These factors can be common to both processes. This is the time to remember that although there may be a single final decision maker in a complex culture, the process of making and implementing their culture-affecting decision may involve many of the culture's members.

Consider that, despite their cultural unity, the individual members of a complex culture can still come up with a variety of different, yet culturally acceptable, interpretations of an issue. This potential for disagreement ensures that there is always a possibility for personal- and cultural-level tensions to arise about both a decision and its implementation. Both the excessive absence (think groupthink) or presence of disagreements can contribute to a culture's resimplification. The presence of the conflict between those grassroots Xhosa demanding that the preconditions in Nongqawuse's prophecy of renewal be satisfied as quickly as possible and those who were more skeptical of the prophecy and its required actions helped to tear the Xhosa culture apart. It appears that the Norse of Greenland experienced resimplification in part because there was an absence of disagreement about not adopting Inuit ways of living.

Consider too that the making and implementing of decisions takes time. It continues to inescapably pass between the point when the members become aware of an issue and the point when they finish implementing

their decision to deal with it. This means that both an individual's and their culture's response to an issue always lags behind the forces driving that issue (Inglehart 1990).

As a result, the making and implementing of decisions in a timely manner is a necessary part of successfully addressing issues. Human nature and complexity ensures that there are constraints on what *timely* means. One the one hand, if there is too long a delay in responding to an issue, then its inertia can result in it forcing a culture to resimplify. Think of climate warming or a dissatisfied population.

On the other hand, if the process of making and implementing a decision is too fast, then the outcome may not be what was intended: it can contribute to resimplification. Consider that there is an innate limit to the degree and speed at which the members, and thus their culture, can complete the changes associated with both making and implementing some decisions, and still remain functional, as discussed in chapter 6. For example, if change is too fast, then people tend to become stressed. They can burn out and start to lose their sense of cohesion and belonging. Some people cope with this stress by adjusting their individual sense of normal (whatever it is) to include beliefs and behaviours that are marginally acceptable to the culture or lie outside the constraints to be met for appropriate personal or cultural functioning. Once enough members feel this way, then the aggregate of their responses can result in "unexpected" changes to their behaviour and corresponding changes to their culture that can precondition resimplification.

For example, a common human response to stressful times is a heightened interest in the mystical aspects of their culture's world view (Myers 2001; White 1968a). This contention is illustrated by the Xhosa of the mid-1800s, the European witch trials of the 1500s (Roper 2004), and the appearance of a belief in child witches among the Congolese during the first decade of the 2000s (Dowden 2006). Could a heightened interest in unfounded conspiracy theories, simplistic solutions to issues or faith in Armageddon represent such a response among Westerners to an increase in stress during current times? In general, if the making and implementing of decisions neglects the need for an adequate but not excessive amount of time between awareness of an issue and completing the needed changes, then the outcome can contribute to a culture's resimplification.

The discussion so far may have created the impression that all of the members of a complex historical culture were equally responsible for the decisions that directed or allowed their culture to drift, unaware, into resimplification. Examples supporting these views are found in the early history of remote Pacific islands, the Greenland Norse and the Maya, whose members all, apparently, continued to steadfastly focus on their cultural imperatives and maintain life as usual as their environmental endowment disintegrated around them (Culbert 1988; Webster 2002). There is much truth to this view, but it is incomplete.

Some members do consciously strive to gather information so that they can make informed decisions directed at avoiding adverse outcomes, such as their culture's resimplification. Most importantly, in complex cultures, the members are not all equally responsible for making and implementing such decisions. Part of the essence of the more complex cultures is that much of the right, and thus the burden, of making and implementing specific culture-scale decisions is laid at the feet of the monarch and their supporting elite.

The elite of a historical complex culture were subject, as are today's elite, to the same general decision-making influences and biases as the rest of the members. In this sense, the elite were no different than the public. However their decisions made a more significant contribution to their culture's resimplification because the elite had both more decision-making responsibilities and the power to discharge them. In making them, the elite were subject to specific influences the public did not usually have to face. Three types of these overlapping elite-specific influences (information, hierarchy and implementation of decisions) are discussed next. The focus is on the ruling elite (the elite decision makers.)

Paying attention to *information* is a critical part of decision making. The elite decision makers in historical complex cultures were, like those of today, usually remote from the events and issues they were dealing with. Thus, rather than using their personal experiences of events to make their rulings, they relied on reported information, their personal opinions and trusted advisors (McIntosh et al. 2000).

The information in those reports was secured by establishing a hierarchical information gathering and storage network distributed across the lands of interest. The more complex the culture, the more complex and formal this network had to be if it was to satisfy the elite decision makers' information requirements. For example, later Babylonian monarchs appointed foreign representatives and created formal archives.

But the information that the elite decision makers received from this network was biased and contained errors. For example, the people tasked with gathering and reporting the information had culturally influenced perceptions about what was, and wasn't, important and what objectives should be met. They were also lower in the hierarchy, so their data collecting and report compiling (writing when it was available) efforts were both consciously and unconsciously guided by their boss's interests (Bella 1987).

Adding to these perceptual and processing biases was another basic bias. The act of compiling a report requires selecting, discarding, aggregating and summarizing data. These processes can inadvertently create the impression that there is more order in the topic or relationships than there is, while masking the lack of information (Norretranders 1998). This affects how the data is interpreted in the report. The resulting presence of biases and errors in the reports the elite decision makers used when making their decisions could have contributed to their culture's resimplification.

A second type of influence on the elite decision makers was their participation in a *hierarchy*. Consider its influence on the monarch's identification of issues and their contemplation of actions. If there was a single all-powerful leader, then decision making about an issue was relatively easy. However, simplicity brought with it the disadvantage that the monarch may not have been sufficiently skilled, interested, or even competent to recognize or deal with that issue. The Assyrian Empire's highly centralized structure with a single all-powerful leader contributed significantly to its instability and subsequent resimplification (van de Mieroop 2004).

To solve this problem, a historical culture might have surrounded its monarch with specialist advisors: think of a military or religious leader, or a lower tier of secondary decision makers. This would have provided the monarch with a selection of considered opinions. It would also have given them the opportunity to blame their own poor decisions on the incompetence of the advisors that surrounded them. But to do so, the monarch would have had to have conveniently forgotten that humans often prefer to receive advice and reports that they find acceptable, rather than those containing honest or balanced opinions. The generals in the German army experienced this rejection when Hitler was deaf to their observations that the campaign against Russia was failing and needed to be changed. Hitler's decisions contributed to the Germans losing the Russian campaign (Sereny 1995).

Alternatively, the monarch could have appointed a team to formally look at the potential problem from many perspectives and provide a recommendation. But this solution brings its own types of biases and errors. Consider that, for a number of reasons, the appointed team often consists of people who "get along" with one another, emotionally support one another and owe allegiance to the team leader and/or the monarch. This opens the door to groupthink. Consider too that the pressure to "get things done" can override the need to allocate sufficient time for a team with diverse opinions to complete its deliberations, thus negating the benefits of forming it.

Inevitably, the elite decision makers higher in the hierarchy were, and still are today, unavoidably surrounded by voices trying to influence their decisions. Many of these voices (who included the elite) were attempting to improve their personal power or wealth under the guise of helping the elite decision maker do the best for the culture's members as a whole (Reader 1990). The close presence of these excessively self-interested voices also diverted the attention of the elite decision maker away from the less self-interested advisors and the more important issues they should pay closer attention to (Goldstone 2002). The result could be decisions that favour resimplification.

For example, during the period from about 2000 BC to 1600 BC, some of Babylonia's government bureaucrats were contract workers. Their job was to ensure that the agricultural producers under their contractual control delivered the required goods to the state. When the producers had difficulty fulfilling onerous contracts, the contractors helped them out by providing loans for which they charged the producers exorbitant interest rates. The contractors thus enriched themselves but routinely drove the producers into chronic indebtedness. This weakened the ruling king because it forced him to periodically declare debt amnesties at his expense (van de Mieroop 2004).

In the light of these and other hierarchy-related influences on the elite decision makers, it is unsurprising that their decisions were often biased and misguided (i.e., inappropriate from the perspective of keeping the culture functioning over the long term). The elite decisions likely tended to heavily favour the preferences of the monarch and their inner circle, the non-decision-making elite who had the decision makers ear, and the culture's world view. At the same time, the elite decision makers tended to ignore the voiceless and the longer-term environmental or social constraints on the culture's functioning.

A third type of influence on an elite's decision making arose from the need to *implement* their decisions. Their ruling was more easily implemented if it was largely acceptable to the members and sufficient funds were available to pay for it. Success was more likely if these requirements were considered during the initial decision making.

The elite's rulings were easiest for the members to accept if they could voluntarily come to the conclusion that it was both in their personal interest and met their sense of fairness. We intuitively appreciate these two constraints on popular acceptance. It is less easy for us to appreciate a third condition: the ruling should be in keeping with the sense of identity and purpose (meaning) the members derived from their culture's world view.

We generally fail to fully appreciate the importance of this last constraint because we tend to underestimate the influence that our culture's world view exerts over our own decision making, including our evaluation of an elite's ruling. To remedy this oversight, imagine you are an appointed servant to Queen Puabi who ruled Babylonia during the Ur III period. As hard as it is to imagine today, when she died, you likely voluntarily accepted that you would be killed and interred with her, as discussed in chapter 16. You could see your death as fair, even if regrettable, because it was part of the commitment required of you by your culture when you became a royal servant and received that position's benefits (now and in the afterlife). You probably also felt this way because you accepted the guidance from the Babylonian world view about the nature of the afterlife and that the monarch was divine.

In contrast, as a Babylonian, you could have had a different reaction to a decision from on high that you felt was contrary to an important aspect of your culture's world view. For example, the last king of the Neo-Babylonian empire, Nabonidus (555–539 BC) made a number of culture-altering decisions. They were so unpopular that, in retaliation, the priesthood elite and the traditionalists are thought to have undermined the king's authority, perhaps to such an extent that they may have contributed to the conquest of the Babylonian empire by Cyrus, a Persian.

If the elite decision makers acknowledged to themselves that the members' evaluation of their rulings was influenced by their culture's world view, then they could improve the acceptability of their rulings in advance. The decision makers could, for example, justify a ruling by claiming that it aligned with or maintained the essence of the culture's world view, or that it built on or improved that world view through "necessary adjustments" (Inglehart 1990; Goldstone 2002). These claims would be boosted if a revered person, such as a god or founding figure, was found to support the ruling.

An example of these practices in action is provided by Babylonia's Ur III period (2112–2004 BC), during which Mesopotamia's many independent city states were merged into a single unified Babylonian state. This short-lived achievement is partly attributable to rulings that allowed each city to retain its chief god. But, at the same time, they and their families where placed under the supreme leadership of Enlil, the city god of Nippur. This made Nippur the religious centre of a unified Babylonia but kept the other cities happy (van de Mieroop 2004). A more current example is the habit of Western leaders to justify a ruling as "creating jobs" or "maintaining growth" regardless of whether this is actually true. Yet, even with this kind of help, rulings that required significant personal or cultural changes would always be difficult to implement, especially if the changes had to take place in a short period. Think of austerity today.

As mentioned above, the successful implementation of a ruling elite's decision is more likely if, in the main, it is inherently acceptable to the members and adequate funds are available. The funding factor has a wider ranging and more complex influence on implementing a decision than one would expect. The significance of this factor becomes clearer after considering its links to a key aspect of keeping a complex culture's functioning.

To remain unified, stable and adaptable, the members of a more complex culture must routinely reaffirm their faith in its world view and hierarchical power structure. The essence of this process is illustrated by the features of Babylonia's reaffirmation process during the Ur III period. The legitimacy of a Babylonian supreme monarch was established at the time of their ceremonial ascent to the throne and their receipt of the blessing from the god Enlil. The culture's institutions, legends, hierarchy (decision-making structure) and world view were reaffirmed at the monarch's ascent ceremonies because they were symbolically represented there, and in Enlil's temple. But the monarch's crowning was only the start of the ongoing process of reaffirmation. It continued throughout the year with activities such as monument building, the holding of annual religious ceremonies, the

provision of services to the people and the distribution of favours to a few (Wilson 1988; van de Mieroop 2004).

The reaffirmation process both gave the ruling elite the authority (legitimacy) to make decisions and preconditioned the members to more readily help to implement those decisions. But, to receive these benefits, the elite had to find a way to pay for the substantial costs of the reaffirmation process. They also had to pay for all of Babylonia's other fixed costs. These included maintaining the bureaucracy (salaries and equipment), maintaining infrastructure (especially the repairs, management and maintenance of irrigation works), supporting the elite and the army (salaries and rewards), and compensating for inefficiencies, theft and debt annulment.

A ruling elite, like those of Babylonia, has the power to secure the needed funds by demanding an increase in the agricultural production per person (increased taxes), enlarging the labour force (population growth), increasing trade, or forming an army that would create an empire or engage in plundering and slaving. They can also secure the funds by reducing the reaffirmation and maintenance costs (reduce services). But their power to do so rests on the co-operation of the members, which depends, in part, on the reaffirmation and authority maintenance efforts: a feedback is involved (Tainter 1988). Thus, if the elite wish to remain in power and their culture to remain functional, then their fundraising (including cost reduction) efforts cannot push the culture outside its human or environmental constraints for too long.

The consequences of the elite's preoccupation with just maintaining the flow of wealth could contribute to their culture experiencing resimplification. It could have contributed to the Babylonians losing the wealth-creating fertility of their soil to salinization, and the Maya losing theirs to erosion. The elite of the Bronze Age great states appear to have alienated their wealth-creating workforces, while the Assyrian elite created a dysfunctional wealth-grabbing empire.

In overview, by looking at the historical resimplifications of the more complex cultures from the perspective of their individual members, we more fully appreciate the critical role that their decisions played in it. Those decisions were made by both the regular members and by the elite. Some contributions were the product of routine, day-to-day decisions (made by all members), others were the result of culture-scale decisions (mostly made by the elite). The contributions resulting from some decisions were direct, most were indirect.

Regardless of their standing in the hierarchy, the members' decisions were the product of an interacting combination of their human characteristics, their environmental experiences and their culture's world view. In particular, there decisions were the product of their decision-making foibles as influenced by the divergences from the external reality found in both their personal and their culture's world view. Overall, it was the limitations and biases in the individual members' decision making that contributed to their historical culture's resimplification.

In summary, the process of resimplification is complex and occurs as part of the dynamic, complex functioning of the human-environment system. The effort in this section to personalize resimplification focused on the three factors (environment, culture and individual) used to describe the system (figure 15). In doing so, not only are the contributions of the factors themselves personalized, but so is the process of resimplification, and the role of the individual in it. This makes it easier for a member of a contemporary culture to contemplate both the possible resimplification of their own complex culture and their contribution to it.

An individual's contributions to resimplification are the result of their decisions. It can easily be inferred from the historical examples that those decisions were often at odds with the external reality. The interacting reasons why include the complexity of the human-environment system, the divergences in a member's internal reality and the divergences in their culture's world view. Consider too the member's decision-making foibles and the influence of their culture's guidance on them. As a result, the member displayed insufficient awareness and attention, at an appropriate scale, to their culture's circumstance and its environmental endowment's functioning. For example, they often failed to consider long-term limits when making their usually short-term culture-supported decisions (Van de Leeuw et al. 2000). Overall, when making their decisions, the members of a resimplified culture paid insufficient attention to the human and environmental constraints within which they and their descendants had to live, if they wished their culture to be viable.

A member also tended not to recognize that their culturally influenced decisions resulted in them and their culture exceeding those constraints. This is understandable because, usually, a member's personal decisions did not result in them making easy-to-recognize, direct cause-and-effect contributions to resimplification. Their contributions were mostly indirect, distant from themselves in time and place, and diffuse. This, in conjunction with their decision-making foibles and divergences, meant that the member was largely unaware of their contributions to their culture's resimplification.

Similarly, resimplification takes more than a lifetime to complete. Thus, those of its many slow- to fast-occurring changes that a member noticed or experienced could easily be treated by them as just another part of life's vagaries, explainable by their culture's world view, rather than as an ongoing, interconnected series of contributions to their culture's resimplification, in which they were participating (Kaufman 1988). The exceptions were the marker events, which they tended to focus on. However, when devoid of their wider, slower-changing context, these markers could be interpreted as harbingers of either a better or worse future.

Overall, despite their possible fears of a bleak future based on marker events, a member was likely largely unaware of their historical culture's unfolding resimplification or their personal contribution to it, which is unsurprising since it mostly took place in the background of their day-to-day life. A member of a contemporary complex culture should take these features and this context into account when evaluating whether their own culture could resimplify and what their personal contribution to it could be.

### A Recipe for Resimplification.
This subsection condenses the individual's view of historical resimplifications in the above sub section and merges it with the basic information from chapter 16 to form generic rules of thumb for resimplification. They make it easier for us to apply our thinking about resimplification to present day complex cultures. The more useful rules of thumb are those that remind us of how our human characteristics, such as our divergences and decision-making foibles, can contribute to resimplification. The selected rules are ordered from the most general to the more specific. Collectively, the rules of thumb are a recipe for resimplification.

The resimplification of a complex culture is a dynamic, complex process that occurs as part of the human-environment system's functioning. It involves interactions among the many variables represented by the individual members, their culture's world view and their supporting environment. Resimplification thus displays the characteristics of a complex system. In particular, its interactions occur over a wide variety of time and space and include feedback. It also displays characteristics of order, randomness and chaos. We tend to focus on the short-term aspects of the resimplification process, but the long-term is just as important.

For a complex culture to remain viable, it must stay within viability's human constraints (e.g., to feel we belong to a group and fairness) and environmental constraints (e.g., resource availability and the environment's functioning). The more complex a culture, the more difficult it is for the culture to stay within these constraints. Complex cultures can successfully function over the long term in only a few ways, but they can fail in many. A culture's complexity can contribute to its resimplification.

A complex culture can be formed from many different mixes of lifestyles and population sizes, but only a few combinations can be supported by the culture's environmental endowment and satisfy our human needs over the long term. The environment's support is limited in the short term by its initial endowment of resources to the culture. Its support is limited over the long term by both the endowment's functioning, which renews the resources, and how the endowment changes over time. A culture's ability to remain functional is limited by its endowment's least productive years.

The members of historical complex cultures that resimplified weren't stupid. They were out of their depth. The members' personal internal reality and their culture's world view that guided their decision making contained unrecognized divergences from the external reality. This, along with their decision-making foibles and the presence of complexity, made it difficult for the members to acknowledge the human and environmental constraints within which they had to live, to recognize the potential for resimplification, and to respond appropriately.

A complex culture's world view tends to guide its members to find a sense of meaning in their lives through their efforts to transform the land. It also tends to veil their environmental awareness and create a complexity-resource feedback. It tends not to remind the members that along with their more environmentally intrusive ways come greater, unavoidable environmental responsibilities.

From one perspective, a complex culture and its members drift, unaware, into resimplification. After all,

the individual members' decision-making foibles, culturally supported divergences and the environment's complexity ensures that they are largely unaware of the environmental and human constraints on their lives or the consequences of their actions. From a different perspective, the individual members impose resimplification on their culture. After all, they are the decision makers. They are the source of the excessive demand for resources associated with the complexity-resource feedback. It is they who ignore the human and environmental constraints on their and their culture's functioning. And the cues from their surroundings reminding them of this. Both perspectives are true.

Complex cultures act as if an increase in their complexity and the production of more resources will permanently reduce the uncertainty in their future. This is not the case (Kaufman 1988). Nor will greater complexity reduce the human and environmental constraints on the culture's functioning. It just changes the expression of those constraints and the times when they might appear. That changes the culture's options for the future.

A complex culture is inherently unstable. It can function in only a few ways but fail in many. The complexity of a complex culture can contribute to its resimplification in a number of ways. The more complex a culture, the more responsibilities the members must discharge to keep their culture functional. The members' ability to fulfill these responsibilities is constrained. Beyond some point, the more complex a culture becomes, the more inherently unstable it becomes.

When facing resimplification, some cultures may be able to stabilize or reverse the trend to resimplification. However, the more complex a culture, the more likely it will strive to maintain its complexity. Although a more complex culture has a greater capacity to maintain business-as-usual, beyond some point it can't stabilize and can only delay resimplification. The more complex cultures are preconditioned to make choices or to be exposed to situations that lead it to eventually face forced resimplification.

The process of resimplification starts long before the characteristics we associate with it became apparent. The single events we focus on as causes or trigger events, such as drought or a loss in war, are usually only marker events. They appear to us to be causes since, of all the features of resimplification, we most easily relate to marker events. A mix of the contributions to resimplification, such as climate change, complexity, too large a population, migrants, invaders, resource shortages, only become causes of resimplification if the members fail to recognize and respond appropriately to them. However, the faster and larger the contributions the more difficult it is for the culture to respond appropriately and the more unstable it becomes. Think of rapid climate warming.

The overshoot and collapse model is a useful representation of the process by which a more complex culture resimplifies through an excessive demand for resources to support its growing population and complexity. However, an actual resimplification is more multi-faceted, messier and more drawn out. In particular, overshoot and collapse under-represents the role of human decision making.

To avoid resimplification, a complex culture's world view must guide all of the culture's members (monarchs, elite and "the public") to recognize, accept and make decisions that enable them to live within the environmental and human constraints on their culture's functioning. And it must do so over the long term. However, it is the members' job to ensure that their culture's world view provides them with that guidance: a Catch 22 situation.

## Are Industrial Cultures Exempt?

The above recipe for cultural resimplification is based on the resimplification of historical cultures, none of which were industrial cultures. There are significant differences between industrial and agrarian cultures. For example, Western industrial culture has access to large amounts of energy, advanced science, incredible technology, vast knowledge and great wealth. There is a strong temptation to think that today's Western industrial cultures are thus not just significantly but fundamentally different from the historical agricultural cultures that resimplified. It is easy to conclude that the recipe for resimplification doesn't apply to at least Western industrial culture. Westerners can believe that we will avoid resimplification by muddling through or following a different path into the future.

Is this true? This section provides a personalized answer by considering how similar the essence of historical complex cultures is to that of today's industrial cultures. It also considers the degree to which the structural and functional features of the more complex historical complex cultures are present in today's industrial cultures. Although the focus will be on Western industrial culture, the conclusions drawn are applicable to other industrial or industrializing cultures.

**Culture's Essence.** After reading a description of a historical complex culture and its resimplification, we have to admit how easy it is for us to identify with the essence of the culture and its members' relationship to it. There is the need for their monarch to be legitimate; the use of ceremonies to reaffirm their identity; and the sense of awe, pride and power they felt in the presence of their monumental structures. Consider too their pain upon losing a battle; their desire for a benevolent, wise and fair leader; and their quest to maintain a steady supply of resources. We can relate to these generic features even though we may be skeptical about the specifics of those long-ago cultures, such as treating their king as divine or conducting rituals that include human sacrifice.

But we must also admit how easily we are distracted from the sameness of the essence by the differences in the details. We are helped to place the differences in perspective by remembering that the essence of all cultures lies mostly in its members' unconscious. That essence includes the idea of an identifying symbol, a belief about where being human resides and how to behave, and an authority hierarchy. In contrast, the members are aware of, work with and respond to their culture's pragmatic and symbolic details. They help the members to collectively identify with and function in their culture. Thus Westerners respond to our own country's flag and leaders, not someone else's. Westerners think of a consciousness that resides in our head rather than a soul that resides in the heart, as the ancient Egyptian's did.

With this in mind, it is easier to accept that the essence of our Western culture is more like that of past complex cultures that resimplified than we intuitively imagine. This is starkly illustrated by the events unfolding today on a small, extremely remote Pacific island within the Juan Fernández archipelago. Robinson Crusoe Island is the 47.6 km$^2$ top of a 4–5 My old volcano. It has a temperate climate and receives 1,150 mm of rainfall/year, which supports a temperate forest ecosystem at higher elevations and grasslands closer to the ocean. The island's extreme remoteness helps to explain why these ecosystems have among the highest number of endemic (unique) species per unit area of any island.

Robinson Crusoe Island could have supported the Polynesians, but it is so remote that it is one of the few Pacific islands they did not discover and try to settle. It was only discovered by the Spanish in 1574. Initially, the island experienced limited, intermittent European settlement. For example, in 1704, Alexander Selkirk, one of the possible models Defoe used for his positivist 1719 tale about Robinson Crusoe, started his four years of solitude there. (This positivism is ironic considering the story that follows.)

The initial settlers consumed those native resources that they could most easily exploit, such as wood from the indigenous trees. For meat, they relied on feral domestic animals abandoned by earlier visitors. By the mid-1800s, the island had permanent inhabitants and the speed of change had increased.

Along with more permanent settlement came more exotic fauna, 232 species to date, which compete with the 209 native species. The exotic flora and fauna include guava (for food), raspberries (for fencing) and the familiar rats, cats, cattle, goats, pigs, dogs and rabbits. Consider too that, up until at least a few years ago, the lifestyle of the approximately 600 locals (about 12.6 people/km$^2$) on Robinson Crusoe Island was characterized by their focus on those plants, animals and behaviours that matched the traditions of their Western cultural heritage. This bias is illustrated by the preference of the island's long-established families for growing exotic plants in their decorative gardens.

Today, erosion has affected roughly 75% of the island's surface. About 15% is covered in exotic shrubland, while only some 7% of the island is covered with intact indigenous forest. The overexploitation of indigenous flora and fauna, the introduction of exotic species and deforestation have, collectively, led to more than 75% of the 125 known endemic species facing extinction.

Overall, the settlement of Robinson Crusoe Island by Europeans has resulted in impacts on its environment that are the equivalent of those deduced for the Polynesian settlement of other similarly remote Pacific islands, such as Mangaia and Easter Island. In essence, the Europeans on Robinson Crusoe Island have been re-enacting, with a European touch, the Polynesians' disruption and destruction of the ecosystems on the islands they settled.

The path the island will follow into the future depends significantly on the islanders' future relationship with their island's environment (their environmental endowment) which, to some degree, will be influenced by outsiders. So far, the locals have yet to generate an action plan to deal with their disruption of the local ecosystems. They currently view a proposed national park as an imposition on their way of life.

But the object here isn't to point a finger at those individuals living on the island. It is to point out that Western culture and its members are quite capable of falling into the same trap as the Polynesians on Easter Island and Mangaia. From the discovery of Robinson Crusoe Island to the present, the members of Western culture have been displaying a personal awareness and following cultural guidance toward the island's environmental endowment that is at odds with its functioning and constraints. These are the same attributes that contributed to the outcomes on Easter Island and Mangaia. The Robinson Crusoe Islanders are us.

A twist to this story ties the island's current and future state even more closely to the wider Western culture. The most respected thinking in Western industrial culture about our ongoing disruption of Robinson Crusoe Island's ecosystems is found among the off-island scientific researchers. Some of them are working on a plan to conserve the island's ecosystems. Among their suggestions to achieve that goal is to sell seedlings of the endangered endemic species to tourists. Incredibly (given our previous discussions on exotics), the visitors would be expected to take the plants back to their off-island homes and propagate them as exotic souvenirs. Perhaps there is a cotton exposition someplace willing to take a few as decorative plants to complement the water hyacinth scourge introduced into the United States in 1884 (Anderson et al. 2002; Vargas and Reif 2009; Greimler et al. 2002; Dupont 2010).

Western industrial cultures certainly have a much greater scientific (intellectual) understanding of how our environment functions and the constraints it imposes on our lives than past cultures. However, as the Robinson Crusoe Island example illustrates, that factual knowledge plays less of a role in our everyday environment-affecting decision making than its significance demands and we claim. The reasons for this are familiar by now: the inappropriate guidance from Western culture's world view, our divergences and our decision-making foibles. They help to veil our awareness of the human and environmental constraints on our lives and bias the attention we pay to them. These are the same characteristics that contributed to the resimplification of historical complex cultures.

In overview, despite the differences in cultural details, the essence of current industrial cultures is much more similar to historical complex cultures that resimplified than we imagine. Western culture can certainly emulate the essence of historical resimplifications, but the details, of course, would be uniquely our own. This is perhaps best illustrated by the ultra-fast implosion of post-communist Russian society. The details unique to this event included alcohol- and drug-fuelled demographic contraction, emigration and the rapid spread of AIDS (Twigg 2002). Details we would not expect in a historical culture.

## Structure and Functioning.

The members of Western industrial culture can easily appreciate, at a personal level, much of the previously discussed basic features of a historical complex culture's structure and functioning. For example, the relationships between the ruled and the rulers form a social and power hierarchy, as discussed in chapter 3. Resources are extracted from the culture's environmental endowment in order to pay for the fixed costs of keeping the members unified and their culture functioning, as discussed in chapter 16. Those costs include the reaffirmation process (e.g., ceremonies and largesse, the authority maintenance costs (the bureaucracy and elite) and the military.

We can readily personalize these features because they are familiar to us. Consider the Western reaffirmation process with its elections and post-election "coronation" ceremonies, its annual celebrations of religious or cultural events, and its provision of services and subsidies to the people. These processes are paid for through taxes that are collected from us by a bureaucracy. Most of us are also personally familiar with the presence of a (public or private) hierarchy, having been a worker lower in its ranks with a boss lording it over us. Once pointed out, we can even personalize the complexity-resource feedback that links a complex culture's complexity to its resource demands, as described above.

Then there are some aspects of a historical complex culture's functioning that we can accept but we don't readily personalize. We can accept that their culture's functioning is the members' day-to-day decisions. Thus the members contributed to both the issues their culture faced and how their culture chose to resolve them. Over time, their decisions contributed to an increase in their culture's complexity. We can even accept that each member was likely largely unaware of how their decisions contributed to and influenced their culture's functioning and its resimplification.

Despite their relevance, we tend to distance ourselves from these features of a historical culture's functioning rather than personalize them. We do so in a number of ways. We tend to treat the members of a historical culture as a cultural group, but we treat ourselves as individuals. Most Westerners don't see ourselves as our culture's decision makers. Despite the rhetoric about endless personal opportunity and influence, the majority of us tend to see ourselves as largely powerless participants whose decisions carry little weight beyond their impact on our personal lives and local organizations. When things go wrong, we focus on those we think are the decision makers, the subgroups of the political and moneyed elite.

Similarly, we can personally accept that the awareness of members of a historical culture, such as the Easter Islanders, the Norse and the Maya, was veiled and biased by their divergences and their culture's world view. We can agree that their culture resimplified, in part, because the members' decisions failed to fully consider the human and environmental constraints on their lives. But, our conclusion that we are different means that, ironically, we don't personalize this information. There is little incentive for us to conclude that our personal awareness could be veiled and our decision making could fail to appropriately consider the human and environmental constraints on our lives.

At the same time, our attention to the obvious differences between past and present cultures allows us to think that we, as a group, are fundamentally different from them: from the historical cultures. That opinion is strengthened by our hindsight view of a historical culture. It telescopes the past, which masks the role of the individual, while using as a reference our culture's current state and its world view, both of which bias our interpretation/perception of the past.

Our resulting biased view of historical cultures allow us to more easily judge many of the decisions made by their members to be misguided, unsophisticated or illogical. This is certainly not how we think of ourselves, our decisions or our culture. We can think, "Sure we are somewhat similar to them, but we'd avoid making those obvious mistakes." And if those mistakes did happen, then it would surely be someone else's fault.

Overall, Westerners may intellectually acknowledge that, in general, the structure and functioning of current and historical complex cultures are basically the same as historical cultures. However, at the same time, we can just as easily conclude that the functioning of our complex culture is different from theirs. The differences can appear to Westerners to be sufficiently large that, despite our failings, we can feel confident our culture will avoid resimplification because we, unlike them, will make more appropriate decisions. And, especially with our wealth and technology, we will correct those that turned out to be mistakes.

Westerners' perceived separation/difference from the past is shown to be a mirage by considering how our commitment to our culture's world view can, as it did for historical cultures, favour our resimplification. We are guided by our culture's world view to believe that a better future will only arrive if we continue to respect its imperatives, such as economic progress (i.e., growth) and material progress (i.e., innovation and technology). But, in doing so, Westerners created our version of an amplifying complexity-resource feedback, as discussed in chapter 11 and 16, which favours future resimplification. Thus, ironically, by unreservedly following the guidance from our culture's world view about how to avoid resimplification, we also nurture the seeds of its resimplification.

Westerners' commitment to our culture's world view is seen when we encounter an issue or barrier. We may consider all kinds of solutions, but the default or preferred solutions are invariably those that favour us satisfying our culture's imperatives. For example, the preferred solutions to fuel shortages focus on directly or indirectly increasing the immediate supply, not reducing the long-term absolute demand. The solutions to global warming we find acceptable won't slow economic progress (growth) or our drive toward our better future. Think alternate energy sources. The acceptable solutions to financial issues are those that maintain growth, even if they require consumers and governments to take on debt. Think credit and quantitative easing.

We are also personally predisposed to continue making these kinds of decisions by our innate preference to sidestep difficult decisions. Think cognitive dissonance. This translates into a preference for easy-to-implement decisions that don't stray too far from our culturally supported social norms: business as usual.

We are being carried into the future by our commitment to our culture's world view. We can accept that the members of a historical culture contributed to its resimplification by unquestioningly following the guidance from their culture's world view, as illustrated by the Greenland Norse and the Xhosa, as discussed in chapter 16. But we don't personalize this knowledge. It doesn't become a reminder for us to question our culture's world view because we believe that our world view is fundamentally different from theirs.

We are forced to conclude that, even at the level of our personal awareness and decision making, the functioning of Western industrial culture and historical complex cultures are fundamentally the same. And any differences between the two are unlikely to ensure, on their own, that Western culture will automatically avoid resimplification. Any lingering reluctance to accept and personalize this conclusion should be removed by the following example.

The more complex of the historical complex cultures lacked a sufficiently effective cultural mechanism to restrain the members of the elite from breaking the trust between them and the ruled. Consider the long list of the wealthy and powerful in historical cultures who neglected their (cultural and environmental) responsibilities and, for selfish or self-centred reasons, contributed to the misdirection, stagnation or resimplification of their cultures (Goldstone 2002). For example, during the time of the near east's Bronze Age great states, Kassite Babylonia's elite lived lavish lives while much of the population was heavily indebted to them, that is, until Babylonia resimplified.

Compare the Kassite elite to the current Western world's ruling elite. They too created the framework that allowed the elite to enrich themselves at the public's personal and collective expense. Think of the global financial crisis of 2008 and its ongoing aftermath: an increase in personal and public indebtedness. Think too of the Panama and Paradise Papers, and the widening gap between the rich (the 1%) and the poor.

In overview, Westerners may acknowledge that the basics of the structure and functioning of current and historical complex cultures are the same, yet we prefer not to personalize this knowledge. We can still think that our personal awareness, attention and decision making is somehow fundamentally different from theirs and thus so is the functioning of our culture. It is sufficiently different, we believe, that if we follow the guidance from our culture's world view, we will inevitably avoid making the missteps that led to historical complex cultures being forced to resimplify. This conviction is misguided. If we wish to avoid resimplification, it is up to us to make the decisions that reflect the human and environmental constraints on our culture being functional and viable.

In summary, the essence of current industrial cultures and the more complex of historical agricultural cultures is the same. Similarly, the structure and functioning of past and present complex cultures are fundamentally the same. The members of current complex cultures are just as likely as the members of historical complex cultures to not appreciate where their current culture-supported decisions and actions are taking their culture over the longer term. The present members are also quite capable of ignoring the indications that unreservedly following some of their cultural imperatives is nurturing the seeds of their culture's resimplification. Industrial cultures too face the possibility of a future resimplification, and it will have the same basic features as historical resimplifications.

Today, the members of complex industrial cultures puzzle over the logic behind the actions of the members of historical complex cultures that led to their culture's resimplification. At the same time, we marvel at some of their accomplishments. If (when?) our complex cultures do resimplify, then future generations will likely shake their heads at some of our beliefs and behaviours that led to our resimplification, even as they marvel at some of our accomplishments.

They will likely point to the same lessons we discovered from past resimplifications but didn't personalize and thus learn from. They will conclude that we too thought that our gifts of consciousness and intellect, in conjunction with the (religious and technical) power under the command of our complex culture's world view, would automatically prevent our culture from resimplifying. But these features can't prevent a culture from being blind to the external reality or from making decisions that concentrate more on maintaining resource production than on living within our environmental and human constraints. As history indicates, these oversights are facilitators of a culture's resimplification.

The rules-of-thumb recipe for the resimplification of complex agricultural cultures applies to today's complex industrial cultures. We face the possibility of a resimplification for the same basic reasons as past more complex agricultural cultures. For our industrial cultures to avoid that outcome, we must consciously make and implement the decisions needed to live within our human and environmental constraints.

## Technology's Cure

The above conclusion is all well and good, but it neglects that industrial and industrializing cultures (industrial cultures) have access to highly sophisticated technology and its products (technology). They allow the members of these cultures to extensively manipulate the material world in neat and useful ways that were undreamt of by historical cultures. Isn't this difference sufficient to ensure that complex industrial cultures can and will avoid resimplification? Westerners certainly believe so.

This section will discuss our application of technology and our hope for it from three overlapping perspectives: technology's ability to overcome barriers, our cultural relationship with technology, and our personal relationship with technology. Western culture, which is the focus, will represent all industrial cultures.

**Overcoming Barriers.** Westerners argue strongly that innovation and the technology it creates has enabled us to become an industrial culture. It has allowed us to overcome all manner of barriers, such as being unable to fly, which we believe prevent us from reaching our full personal and cultural potential. If a series of barriers are encountered, then we believe they can be overcome by successively applying technology. For example, the limit to us flying is resolved by inventing the airplane. The limit to the speed of propeller-driven flight is resolved by inventing jets. We also believe that applying technology can similarly repair the harmful, unfavourable or undesirable direct and indirect consequences (adverse effects) from overcoming barriers. Think of the adverse effects from using the internal combustion engine for transport. However, there is more to technology overcoming barriers than we generally acknowledge.

At its most basic, technology is limited in its ability to overcome barriers by its reliance on resources to function. If Western industrial culture is to avoid resimplification by applying technology, then we must maintain a steady supply of the natural resources needed to build and run it. In our finite world, the stock of all but sunlight is finite, in a practical sense (i.e., the amounts are finite unless used at a rate below their rate of renewal), as discussed in chapters 12, 14 and 15. The appearance of resource shortages and their implication for the application of technology were extensively discussed in these chapters, so it is not discussed here. This subsection will illustrate the other constraints on technology overcoming barriers.

Consider the outcome from successively applying technology to overcome barriers. In the 1950s, an energy shortage developed in the industrialized countries. The electricity supply barrier was overcome by building power plants that turned coal into electricity, but they also produced the combustion products of gases and particulates. While the energy was being distributed to the consumers, the combustion products were being discarded through a smokestack into the atmosphere. There they contributed to the killer city smogs of the 1950s. That adverse effect (a barrier to health improvement) was partly reduced by increasing the height of the power plants' smokestacks and installing particulate scrubbers.

In the following decades, acid rain appeared in the countryside distant from the cities. A significant component of that acid rain was the combustion products, especially sulphur dioxide, coming from the coal- and oil-burning power plants. These products could reach the distant countryside because the plants' higher smokestacks dispersed the pollutants further downwind from the cities. To overcome that adverse effect (a barrier to forest health and timber production), power plants were ordered to install sulphur-scrubber technology on their smokestacks and to use low-sulphur coal.

Since then, it has become apparent that coal-fired power plants are having another adverse effect on us and our environment. The $CO_2$ in their smokestacks is contributing to climate warming and ocean acidification, which are disrupting the functioning of ecosystems worldwide. The coal-fired power plants now face this barrier to using coal to generate electricity.

One suggestion to overcome the $CO_2$ disposal barrier faced by coal-fired power plants is to capture their $CO_2$ emissions at the smokestack and store them underground. But this takes energy, which will be supplied by burning more coal, and points to the barrier of leakages. Alternative suggestions focus on developing "clean" coal technology or using natural gas.

Other solutions are part of a wider effort to deal with climate warming. They propose to address the adverse effects of all types of $CO_2$ emissions by applying technology (geoengineering) that will either remove $CO_2$ from the atmosphere or mitigate its impacts. However, many of those addressing the climate warming problem have repeatedly said that these solutions should not be considered a substitute for absolute reductions in $CO_2$ emissions (The Royal Society 2009). Despite those statements, a parade of technological solutions still passes by. They vary from removing the excess $CO_2$ from the atmosphere by seeding the ocean with iron in order to speed up the fixing of carbon by phytoplankton, to deploying carbon-dioxide-catching artificial trees and storing the captured $CO_2$ underground in geological formations (Boyd et al. 2000). They also include neutralizing the heating effect of the $CO_2$ by reflecting sunlight away from the earth

using a 100,000 km long cloud of tiny mirrors in space (Angel 2006).

The initial culturally sanctioned application of coal-burning technology to increase electricity production successfully overcame the barrier of a low energy supply. But it also resulted in a significant adverse effect and revealed a new, unanticipated barrier. Over the years, each subsequent application of technology to resolve a previous barrier/adverse effect also revealed a new barrier/adverse effect. In addition, each subsequent application of technology to deal with them consumed a little more of the energy we extract from the coal, thus reducing the energy we intended to make available for other purposes. Each subsequent application also added to our ongoing technology maintenance and future replacement costs: it added to our responsibilities. The application of the proposed technological solutions to the $CO_2$ disposal barrier will likely continue this trend.

When our culturally sanctioned application of technology is viewed from a broader, longer-term perspective and as a part of the human-environment system, then it routinely accomplishes more than we intended and incurs higher costs than we anticipated. Applying technology appears less to overcome barriers and more to either transform them into different barriers or circumvent them to reveal new barriers. The broader view also reveals that the distinction between barrier and adverse effect becomes blurred: today's adverse effect can become tomorrow's barrier and vice versa.

This view is expressed in summary form as a rule of thumb. Over the longer-term, the successive application of technology to deal with barriers and adverse effects inevitably exposes us to more of them and increases the effort needed to deal with them. It increases our responsibilities. The phenomena of diminishing returns and increasing responsibilities found in the human-environment model of resimplification should spring to mind.

There is a common belief among Westerners that many of the limitations to overcoming barriers, mentioned above, can be resolved by applying technology to increase efficiency. The limitations to this solution are illustrated by the telephone industry's application of technology to overcome barriers. In the early days of the telephone industry, your calls were routed to their destination by humans. They were replaced by more efficient switching circuits that require electricity to function. During the 1940s, the number of telephone calls within the United States increased rapidly. At that time, the energy demanded by the rooms filled with electro-mechanical switching circuits was predicted to require, within a few decades, a significant proportion of the US electricity budget. The search for a technological cure for this emerging barrier to communication ended with the invention of a more energy efficient electronic switch, called the transistor (Ayres 2005).

The smaller size of the transistor allowed more energy efficient switches to be packed into a smaller space. Soon a host of other transistor-using electronic devices, from computers to smart phones, appeared on the market. Today, companies are faced with satisfying a rapidly growing demand for phone and internet services. Their internet-searching services alone have to supply energy to the thousands of transistor switches in each of their increasing number of servers and search-engine computers crammed into ever-larger server centres. In 2006, the energy demand from these data centres accounted for 1.5% of the United States' electricity demand. In 2014, this had risen to 1.8% (70 TWh). The internet service providers are engaged in an ongoing struggle to find a fix for this increasing energy demand. So far, the rate of growth has been checked by increasing the efficiency of energy use by the data centres, but that era of significant increases in efficiency is coming to an end (Shehabi et al. 2016; Sverdlik 2016; Johnson 2009). Increased efficiency didn't overcome the barrier to communication posed by the energy required to power switching circuits, it just changed the barrier's expression. In addition, the growing variety and number of devices that use the energy-saving transistor has actually contributed to an absolute increase in their demand for energy.

Westerners' faith in the ability of technology to overcome barriers holds for more than their immediate material manifestation. Consider the neoliberal economic belief that economic growth will solve the inequality in living standard between the industrialized (the richer) and non-industrialized (the poorer) nations by lifting the poor to the level of the rich. Because energy is a requirement for economic growth and a proxy for current definitions of living standard, the neoliberal belief can be tested by asking, can technology supply the world's population with the amount of energy needed for everyone to live the lifestyle of the world's top 20% of energy consumers? A question like this was first asked in 1998 and then answered using the energy consumption data for 1990 (Goodland and Daly 1998). This discussion revisits that question using the 1990 energy reference.

The global energy use in 1990 was about 101,800 terawatt hours (TWh), which translates into an average of 19.3 megawatt hour per person (MWh/person) for the 1990 world population of 5.28 billion (EIA 2008; US Census Bureau 2008). In 1990, the top 20% of users consumed about 63.3 MWh/person. The remaining 80% used 8.4 MWh/person.

By 2025, the earth's human population is predicted to reach 8 billion (US Census Bureau 2008). For technology to supply everyone with 63.3 MWh/person in 2025, the global energy budget for that year would have to be 506,642 TWh: an amount about five times the total energy consumption for 1990. For this demand to be satisfied, the world's energy supply would have to grow by roughly 12,000 TWh/year between 1990 and 2025 (an average annual rate of 11%–12% of the 1990 energy usage). Is this a realistic rate?

By 2005, our global energy use had reached 135,342 TWh (EIA 2008). The actual average annual increase in energy supply from 1990 to 2005 was 2,236.1 TWh/year, or about 2.2% of the 1990 usage. By 2010, energy use had grown to 147,898 TWh at an average annual increase of about 2.3% of the 1990 usage (IEA 2012a). Since the 1990s, technology has increasingly struggled to meet this 2.3% growth rate, as indicated by our increasing use of more technically difficult to extract, low EROI fuels, such as tar sands oil. It seems highly unlikely that, by 2025, the application of technology can provide the energy needed to raise the world's living standard to anywhere near the level of the world's top 20% of energy users in 1990.

Many still choose to believe that, in the future, technology will be able to overcome this energy supply barrier after which we can all live an energy-rich lifestyle. Think fusion power. If we assume that fusion power will come to pass, then there is one other aspect of technology's resource-supplying ability to consider. What will the distribution of that energy look like?

A fivefold increase in global energy use took place between 1900 and 1990. That increase didn't result in everyone living anywhere near a similar energy-use lifestyle, as indicated above by the difference in consumption by 1990. That split in energy use had not changed by much in 2001. At this time US citizens used, on average, 92.81 MWh/person, while Bangladeshi citizens used 1.74 MWh/person (IEA 2003). Neither had it changed much by 2010 when US citizens used, on average, 83.15 MWh/person, while Bangladeshi citizens used 2.44 MWh/person (IEA 2012a).

Applying technology is unlikely to provide a fivefold increase in energy from the 1990 level by 2025 needed to ensure equitable access to energy at the level of the world's high energy users in 1990. Even if it did provide that amount, technology hasn't and cannot ensure that access to that energy would be equitable. It is up to us, the members and their elite leaders, not technology, to ensure that access to resources is equitable.

In overview, the idea that technology can unconditionally overcome the barriers to progress neglects that they do not exist in isolation but are interconnected aspects of the complex human-environment system. As a result, our application of technology to overcome barriers is constrained, and the outcome is more wide-ranging than we imagine. These constraints and outcomes only become clear when we treat the barriers and technology's application as part of that system.

When technology is evaluated in that context, we see that its application can overcome barriers, but the benefits from doing so are accompanied by costs (cash and adverse effects). There is always a trade-off to overcoming a barrier. The costs can be direct or indirect and need only appear over the long term, which makes it difficult to evaluate the trade-off. In addition, by technology overcoming a barrier, it doesn't remove it or render it irrelevant. Instead, technology delays its further appearance, circumvents it by transforming its nature and expression, and reveals new constraints. "Overcoming" a barrier also increases our responsibilities. The serial application of technology to overcome successive barriers doesn't change these conclusions; it simply blurs the distinction between adverse effects, responsibilities and barriers.

The most basic limitation to applying technology to overcome barriers is the finite supply of the resources needed to build and run it. Even if technology can increase our efficiency, it can't, on its own, permanently solve either this resource limitation or the other constraints to overcoming barriers. Neither can technology ensure that any undesirable outcomes, such as a disruption of the environment's resource-providing functions, that result from applying it will be avoided, mitigated or restored. Nor can it, on its own, solve social issues, such as the inequitable distribution of resources. These are our choices to make.

**Cultural Guidance.** As the above subsection hinted at, our decision to apply technology to address an issue is based on more than practical considerations. It is also based on the guidance we receive from our culture's world view about the appropriate relationship to have with technology and the issues it addresses. This subsection discusses that guidance by focusing on Western culture.

Westerners are innately inclined to take a short-term, optimistic view of the future. We are also socialized to follow the guidance from Western culture's world view. This includes its guidance about our relationship with technology.

So what is that guidance? In essence, it prompts us to believe that the best way to secure a better material and mental future is to continue both applying technology and favouring technological innovation (i.e., maintaining technological progress). It also advises us that we can satisfactorily mitigate and restore any adverse effects from applying technology by applying additional technology. Our efforts to deal with unwanted and unexpected (undesired) outcomes in this way will, we are guided to believe, spur innovation, improve technology and thus contribute to technological progress and a better future. In a nutshell, we are guided by our culture to place much of our hope for the future in technology.

As a result of this guidance, technologies can acquire significant political and financial importance as well as symbolic power. The mention of finance is a reminder of the strong culturally supported bond between technology and economics, as illustrated by the virtuous cycle (figure 14). This mutualistic-like relationship between the two reinforces our culture's guidance that both technological progress and economic growth are keys to our future personal and cultural well-being. Westerners are thus guided to preferentially defer to the needs of technology and the economy rather than the environment's functioning and, in some cases, basic human needs.

One way we express this is through our preference to resolve issues and overcome barriers by applying technology. This raises the question of how Western culture assures its members that technology's application will be a success and not a failure, both the first and the next time we apply it. This discussion continues and widens the deliberations in chapters 5 and 9.

Designers conduct lab and field tests of their technology and its products (technology) to discover those critical attributes required for their inventions to function as desired. However, bearing in mind the complexity of the human-environment system in which technology is applied, the designers' testing is unlikely to discover all of the many other ways in which the technology can fail, especially if it is, itself, complex. Neither does such testing consider all of the ways in which the technology can interact with our complex human-environment system (Morgan 1993; Gardner and Stern 1996). Testing is thus unlikely to reveal all of a technology's potential unintended, unwanted or unexpected failures and adverse effects. Think of CFCs, plastics and coal-fired electricity plants.

This makes it difficult to complete a meaningful evaluation of the future outcomes from applying a technology. How do we decide when the designers have done "enough" to ensure that a particular application will not result in "significant" adverse effects on humans or our environment? As a culture, we answer this question in a practical manner: by setting test standards, establishing assessment protocols, developing operating rules, introducing certification standards, formulating liability laws, assigning responsibility and calculating insurance costs. Each industrial culture believes that by meeting their version of these conditions (i.e., by exercising due diligence), they will have done "enough." Westerners, in particular, believe that our process of culturally sanctioning technology and its application will ensure that the risks will be so low that we can relax.

But this belief neglects something critical. Our methods of judging technology and the outcome from its application are strongly influenced by our culture's guidance about our relationship to technology, the issue and our environment. Its guidance thus influences both our formal and our personal choice of "enough" and "significant."

This cultural influence is seen as biases in our choices of the technologies (including their products) we hold dear and develop. It is also expressed in the design of the formal protocols and the choice of the thresholds we use for making our judgements about them, as illustrated by the safety testing of chemicals discussed in chapter 9. Think of the difficulty we face applying the precautionary principle and the emphasis on isolated short-term testing for toxic outcomes. Consider too the lag between the release of new technologies into the market place and our much later development and deployment of effective methods to test for their possible adverse effects, as previously illustrated for nanotechnology in chapter 13. Also consider that the culturally approved environmental assessments and product-testing processes gather masses of information and handle it properly but tend not to ask or answer the hard questions (Regal 1996). Think of the assessments of fossil fuel extraction and transportation projects that pay limited attention to the adverse effects of the associated greenhouse gas emissions.

Understandably most Westerners are unaware of these details. After all, it is difficult for us to personally evaluate technology, understand the limits to its testing and fully appreciate the potential adverse effects of its application. Instead, Westerners tend to accept our culture's guidance about technology: its application, evaluation and regulation. The result is our culturally supported relationship with technology (Regal 1996; Gardner and Stern 1996).

Our faith is sufficiently strong that we are unpleasantly surprised, even shocked, when the application of a technology that has satisfied the standards of "enough" and "significant" results in unintended, unexpected and unwanted outcomes (White 1967; Gordon 1978; Gardner and Stern 1996). Examples of the more serious "post-enough" failures are the space shuttle *Challenger* explosion (space engineering); the Chernobyl and Fukushima Daiichi nuclear power plant meltdowns (nuclear engineering); the Love Canal poisoning (chemical engineering); the Summitville and Mount Polley mining disasters (mining engineering); the Deepwater Horizon oil blowout (civil engineering); and any one of the innumerable recent cyber-attacks that steal money, breach our privacy and manipulate our decisions (computer engineering).

It is reasonable to expect that Westerners would react by expressing greater skepticism about technology and its future application. This is generally not the case because, after our initial outrage over a "post-enough" failure of technology has subsided, our culture's guidance exerts a strong influence on our judgement of the failure. If designing, operating or testing errors are discovered, then we have someone/something to blame, so we tend to conclude that after a few tweaks we can continue to use the technology. If no such error is detected, then the undesired outcome tends to be rationalized as an accident, an act of God, an unforeseeable event, a product of complexity, or a statistically unusual event, covered by insurance that is unlikely to be repeated. Bad luck is considered a more acceptable explanation than making a connection to our world view. In all cases, the undesired outcome is thought of as a normal part of learning. After all, accidents, such as oil spills, do happen (Gordon 1978).

Rather than an undesirable event prompting us to more closely question why we need to apply the technology or the basis of our belief in it, we prefer to put our faith in our ability to learn from our mistakes and improve our technology. We prefer to imagine that the next time technology is applied, it will function as intended and result in few adverse effects. Thus, although our culturally guided response to an undesirable event certainly tweaks the definition of "enough" and "significant," it does so in a way that doesn't disrupt, too much, what we think of as the orderly march of technological and economic progress. This is illustrated by the ongoing unregulated application of some technologies (or their products) despite their known significant adverse effects on humans or our environment. Historical examples are the long delay between knowing that lead was hazardous to health and its removal from paint, pipes and gasoline. Today, the adverse effects from tar sands oil production, some pesticides, social media and plastic use still await an effective response.

These features are not restricted to Western industrial cultures. Consider the efforts to minimize human responsibilities for adverse effects from nuclear events by waiving and limiting a company's liability (Daily Times Monitor 2009; American Nuclear Society 2005). Ask yourself why the Japanese gave the job for ensuring the safety of nuclear plants to the government agency responsible for promoting nuclear power?

An illustration of how thoroughly the guidance from Western culture's world view permeates our decision making and contributes to technology's undesired outcomes is provided by an Australian example. Occasionally, the population of a particular exotic mouse species in Australia reaches plague sizes. This was a specific problem that needed a specific solution. One idea was to permanently reduce the mouse population by routinely dispersing a mouse contraceptive in the affected area. In 1997, Australian scientists started conducting experiments to see if the mousepox virus (that normally only makes mice mildly ill) could be genetically modified to act as a self-dispersing carrier of the contraceptive. Concerns were raised that if the modified virus was to infect non-target species, then it could initiate an ecosystem disaster. Assurances were given that the mousepox virus was host specific (Anderson 1997). Concern for the spread of the contraceptive gene(s) themselves was apparently not raised.

In 2001, the same researchers reported that they had further genetically modified the mousepox virus in an attempt to improve its contraceptive action. To their surprise, the newly modified virus had gained the unexpected ability to shut down parts of the mouse's immune system, thus turning the virus into a very efficient mouse killer. That outcome stirred considerable debate about whether the experiment and its results should even be published. There was a fear that the knowledge of how to alter the virus could be used by terrorists to make a biological weapon from the similar human-infecting smallpox virus (Nowak 2001; Jackson et al. 2001). The results were published.

Soon after publication, a researcher in the United States enhanced the lethal nature of the altered mousepox virus. He then similarly altered the cowpox virus, which normally generates mild illness in humans. This researcher had plans to conduct further tests on the viruses in a US military lab's biosecure facilities.

Ironically, his intent was to advance the understanding of how to manufacture what were now, essentially, bioweapons. The tests were needed, apparently, to help the researcher determine the value of the modified viruses to bioterrorists. Following strong criticism of his experiments and of his justification for his work, the American researcher noted, in his defence, that the virus was non-contagious. Critics pointed out that he was unable to provide assurances that this state of affairs would not change with future modifications to the virus (MacKenzie 2003; Steinbruner and Harris 2003).

The mousepox example illustrates more than our culturally supported drive to develop and apply technology. It illustrates that, despite their training, skill and familiarity with a technology, those developing and applying it are not the people to rely on to inform us about it or fully and impartially evaluate it. After all, they have a considerable emotional (and possibly a financial) investment in developing and applying it. They are also culturally preconditioned to strongly believe in the promises for the future that they have made on behalf of their technology. They are proponents who tend to be focused more on the potential benefits (cash or other) of their project and less on its possible distant, undesired outcomes, such as adverse effects on humans and the environment (Regal 1996).

This makes itself apparent when the proponents issue a news release. It is routine to be informed that some breakthrough technology (its products or application) will eventually remove some barrier to progress and change our world for the better. We should not be surprised when, sometime in the future, we hear or read a newscast describing how that technology had either caused or contributed to a significant undesired outcome. Often these later reports use words such as *unusual* or *unexpected*, and they assure the reader that technology is being deployed to restore the original conditions (Gardner and Stern 1996; Perrow 2007). To hear these post-incident reports, we just have to wait for the appropriate time period to pass, as the people living around Fukushima, Japan, recently found out. Think too of plastics and social media.

In overview, Western culture's world view has a pervasive and significant influence on its members' relationship with technology. We are guided to see technology, like the economy to which it is closely tied, as a key to successfully dealing with issues and keeping us progressing toward our personal and our culture's visions of a better future.

This cultural influence can be seen in our personal and formal evaluation of both technology itself and the outcome from applying it. We are guided to ensure that our evaluation processes don't disrupt technological progress and the application of technology too much. In particular, our culture's influence affects the standards we use to decide if we have done "enough" to ensure technology is safe/functional and the criteria used when determining if the adverse effects from applying it are "significant." For example, we are guided to only conduct tests in isolation, over a short period, rather than cumulatively, in context, over the long term. If there are any "post-enough" failures or adverse effects, we are guided to rationalize them in ways that minimize their significance, such as labelling them as "unexpected" or an "opportunity to learn." Our cultural relationship with technology thus allows us to contribute to an undesired outcome, and still be surprised when it happens and that we contributed to it.

The development and application of technology can change our experience of the world. But, despite our culture's guidance, it can't change or remove the fundamental human and environmental constraints within which we must live. Our culturally supported hope that applying technology will, on its own, ensure our culture will avoid resimplification is unfounded.

**Personal Relationship.** Technology and its products (technology) can change both us and the world around us. But technology is a tool. It can't decide that we must make those changes. It is we who decide whether to accept a technology and the conditions under which it will be applied. Neither can technology, on its own, remove that decision-making prerogative from us. The general nature of our culturally influenced decision making has been discussed in earlier chapters, in particular chapter 2. The focus here is on our culturally influenced personal relationship with technology and

how it affects our decision about developing and applying it, and our response to the outcome.

From prehistory through to the present, a feedback interaction has existed between technology, individuals and their culture: a culture-technology feedback (Rudgley 1999). The individual members of a culture accept a technology and its products based on the hope that it will solve a perceived problem or satisfy a perceived need. As individuals adjust their lifestyle to better integrate the technology and its products into their lives, they change their culture in ways that better accommodate the technology's presence. This personal integration and cultural change preconditions the members and their culture to more easily accept related technologies and products, thus completing the feedback. Think of the oil-powered internal combustion engine. Its origins lie in the natural gas engine invented in 1876 by Nicolas Otto to, in part, provide a fixed source of power to craft shops. It now powers automobiles and chainsaws (Ferguson 1979).

But the mere presence of a technology doesn't preordain its acceptance by individuals or their culture. Technology can be rejected if the members perceive it to have no meaningful use: its application is not in their best interest, contradicts their world view or does not live up to the hope and faith they initially attached to it. The importance of our perception of technology is illustrated by Anabaptists, who can take a highly skeptical view of it and subject it to rigorous social evaluation (Huskins 1998).

In contrast, in British culture, starting in the early 1700s, technology and its facilitators (science and the economy) slowly took on a more important role. The British, and then Europeans, came to believe that technology and the material goods it produced would help them to achieve their ideal Christian vision of social and personal well-being. By the mid-1800s, the concept of technological progress had become a core part of both Western culture's world view and its members' sense of meaning and purpose (Goldstone 2002; Wilson 1988).

The post-WWI downturn in Westerners' belief in technological progress was followed by a resurgence into the mid-1900s, as captured in a 1945 view of the then expanding field of plastics. We would, it was stated, live in "a world in which nations are more and more independent of localized naturalised resources." After using plastics, we would be able to say "how much brighter and cleaner a world than that which preceded this plastic age" (Yarsley in Thompson et al. 2009, 152).

In 1959, similar pronouncements were made about the bright future heralded by the manufacture of CFCs (Briscoe 1959, 133).

Although the members of today's Western culture are more skeptical of technology than perhaps the Victorians were, we still believe in technological progress, as discussed in chapter 3. And much of our hope for a better personal and cultural future relies on the ongoing application of technology. Our unexpressed personal faith in technology and the implicit promises we make on technology's behalf can be seen in the hopeful but poorly considered statements made by everyone from environmentalists to business people. In essence, they contend that applying a particular technology, or its products, will ensure that our lives will be more comfortable or our environment better protected. Consider the promise in the early days of the internet that it would result in more than a connected society. We would become a more engaged and better informed society. There was little mention of a downside that was worthy of our attention.

A more personal example is our relationship with the newer, smaller, cheaper and more powerful electronic communication devices of today. We are told they will enable us to communicate faster (save time) with less effort and energy. We correspondingly tend to justify our purchase of such a device by expressing a desire to be more efficient and better informed. But, in our hearts, we also feel that the new electronic technology will allow us to personally achieve loftier goals. We have faith and even expect that such a new electronic device will help us socialize, fit in, become wiser and be happier. We can also believe that it will ensure we live more meaningful lives by, for example, allowing us to more easily address pressing social and environmental issues (Noble 1998; Strong 1995). The downsides are only emerging now.

One reason why we can have this type of personal relationship with technology is that we are socialized to it from childhood. Think advertising. Can you imagine embracing the Anabaptists' view of technology?

We can believe in our hopes for technology for another reason. When we consciously think of or work with a product of technology (or a technology), we primarily treat it as an object. But, at the same time, our unconscious is busy linking the product to our emotions and memories. These old and new unconscious associations affect our evaluation of the product and our decisions about it: for example, whether we will feel comfortable or uncomfortable using it.

The associations between feeling and product are stored in our memory. Thus, a later encounter with the product can prompt us to recall them from memory. If we react to those associations, then the product has symbolic value for us. Technologies and their products, like many objects in our lives, have symbolic value, so they have the power to unconsciously influence our decisions. The degree and nature of this influence depends on the strength of their symbolic value to us and the context in which the product is viewed. (It also depends on our culture's guidance.)

We should therefore think of our relationship with technology as both practical and symbolic in nature. For example, the Ford Model T was bought with more than transport in mind. Many were motivated to buy it for recreation because it symbolized roaming free on the open road (Ferguson 1979).

Similarly, today we purchase a specific vehicle for more than its ability to provide practical service. For us to feel good, a vehicle must symbolize key aspects of our sense of self or personal world view. Skeptical? Contemplate the context, such as mud slinger or family setting, in which an ad presents a vehicle for your consideration, and your reaction to it. The same is true for the electronic communication devices mentioned above. Our evaluation of them is influenced by both their practical and symbolic value to us.

The symbolic power of technology (including its products) joins with its practical features to collectively influence our faith in the abstract concept of technology. As symbols, they help to sustain our faith that technology will satisfy our material and psychological needs. As objects they help reaffirm our faith that technological progress will indeed provide us with a better life in the future.

We may intellectually imagine that our personal relationship with technology is based solely on our pragmatic, conscious assessment of its ability to transform the natural world in useful ways. In reality, we base our view of technology, its application and outcomes, on a mix of our culturally and symbolically influenced emotional and practical evaluations of it, which we conduct both consciously and unconsciously. Perhaps this is most easily appreciated by widening our view of technology's symbolic power.

Consider the monumental structures provided by our technology, such as dams and skyscrapers. They perform the same symbolic function that a steep Mayan step pyramid, a glistening white Egyptian pyramid and the spectacular Ishtar Gate at the entrance to the city of

Babylon performed for our ancestors. They all make their culture's world view tangible to their respective members, as discussed in chapter 3 (Wilson 1988). Similarly, even the every-day products provided by technology can make tangible to Westerners specific aspects of our culture's world view, such as the way things should be, or who we are and our vision of the future.

Overall, although technology is just a tool, our relationship with it means that it influences our lives through more than what it does. Technology and its products can become a physical extension of ourselves or even a part of our sense of self. They can symbolize to us, personally, our hopes and desires, and a path to fulfilling them. They can represent to us aspects of our culture's world view, including its guidance about our relationship with technology.

Because our view of technology is deeply buried in our psyche, along with the rest of our culture's world view, we are largely unaware of its powerful unconscious influence on our evaluations and decisions about its application. This means that for an individual to question the future application of technology, they must, to some degree, engage in questioning a core part of their and their industrial culture's world view, and their faith in it. No wonder that, as individuals, we prefer not to consciously conduct a comprehensive assessment of the need to apply a technology, the possibility of adverse effects from applying it or the use of alternatives. Nor do we question the value received from its application. Instead, we feel more comfortable relying mostly on our optimistic outlook and our culture-guided intuition about a technology (or its products), which we then treat as a sufficiently rational and objective assessment of its value.

Similarly, our common-sense predictions about the outcome from us (personally or culturally) applying a technology are largely made by imagining the degree to which we think using it will satisfy our desires and our culturally influenced visions of a better future. Our predictions are thus made more by projecting our hopes onto the technology and less by a dispassionate evaluation of the advantages and disadvantages. We tend not to critically weigh the current and future benefits, costs and responsibilities that come with its use (Ferguson 1979).

We can remind ourselves of the ubiquitous presence of these constraints and biases in our personal assessments of technology and of its influence on our lives (our relationship with it) if we pay attention to our reaction to reports about technological developments. For example, we could ask why researchers would consider it quite rational to investigate the possibility of using green, fossil-fuel-free, geothermal power to help extract bottom-of-the-barrel, low EROI, high-$CO_2$-emitting bitumen from the Athabasca tar sands (Vanderklippe 2009). We can also take note of the difference between the initial claims about a technology, or the optimistic hopes we held for it, and the actual outcome. An example is the contrast between the impressive technology that allows us to drill for oil below the bottom of the ocean and its use by BP that resulted in the Deepwater Horizon oil well blowout.

A more personal example is contrasting the fleeting happiness we can derive from an expensive collection of technological gadgets. It can't fully displace the lingering unhappiness we feel from the debt mountain created by their purchase or the nagging feeling that there has to be more to life (Kasser 2002). At these times of personal cognitive dissonance, our faith in technology competes with the realization that much of our sense of social-psychological well-being has little to do with technology or the goods it provides. It depends more on personal and social relationships, and a sense of belonging (McIntosh et al. 2000). In Faucheux's words, in the big picture human "identity and relations are not summed up in aggregate demands for goods. They are codified in complex social institutions and symbols, including religions, conventions of justice, tribal identities" (Faucheux 2005, 259).

In overview, by and large, the individual members of industrial cultures remain unconvinced that technology has influence over us and that we are not fully in control of it. They still believe that the increasing application of technology and the use of its products are central to their sense of social-psychological well-being and a better future. Presumably, they also believe that the members of hunter-gatherer cultures who lived over the last tens of thousands of years must have been a pretty miserable and dysfunctional lot.

In summary, our application of technology doesn't remove the barriers we perceive are preventing us from reaching our goals. It circumvents them, transforms them, changes their expression and reveals new ones. Although doing so has benefits, it also has costs: it disrupts the environment and burdens us with responsibilities. If we are to fully evaluate technology, then those costs have to be meaningfully included in our evaluation.

If we don't learn these lessons, then applying more technology won't eliminate the possibility of our industrial culture resimplifying. It will just change when and how it will arrive. Technology can even become a liability. Indeed, applying it can contribute to resimplification and its severity.

This view of technology contrasts strongly with the guidance provided by Western industrial culture's world view. To its members, technology has come to symbolize Western culture's hopes for the future. We tend to believe that merely having and applying it will allow us to reach for our culture's vision of a better future and save us from the fate of resimplification.

It is certainly heartening to imagine that the Maya or Easter Islanders could have easily avoided the tragedy of their resimplification if only they were in possession of enough technology. If only they had owned farming aids, such as farming tools pulled by draft animals, water pumps and fertilizer, then they could have solved their food shortage problem and survived droughts. This may have been true in the immediate term, but what about the longer term? If they had possessed enough technology, would their population growth, soil erosion and cultural preference for warring have ceased? To hope that the application of technology could have, alone, accomplished that is beyond optimistic.

Westerners are faced with analogous questions when we contemplate our future. Consider the prospect of fusion technology delivering its promise of unlimited, apparently environmentally benign and cheap energy. When the energy arrives, will we maintain economic growth, resource demands, population growth, and environmental disruption and destruction? Without the appearance of viability-favouring answers to these questions before fusion energy arrives, our long-term hope that it will usher in a better future is just another unrealistic dream. Think of the human and environmental constraints to be satisfied if a culture is to be viable.

There is a more realistic way to think of our relationship with technology. It is the knife, fork and spoon of consumption. The real knife, fork and spoon allow us to be genteel when we eat. But it won't prevent us from eating poor food, becoming overweight or choking on an unseen bone.

Similarly, technology allows us to manipulate our environment in neat and useful ways. But, it can't prevent us from degrading or destroying our ecosystem, or our fellow humans. Technology is a tool. It is neither the boon for society and the hope for the future that we

currently believe nor the villain claimed by some (Faucheux 2005).

It is we, its creators, who view technology as a symbol and impose the promise of a better future on it. It is we who treat it as a solution. But it is our decisions about developing and applying technology that will largely determine its real-life influence on our future.

If technology is to help us avoid resimplification, we first have to question what we imagine is a satisfying life and a desirable culture. We have to ask ourselves whether we can afford to spend the amount of environmental functioning it requires us to pay for the technological progress and material goods we believe are required to live that satisfying life and secure that cultural future. It is our responsibility to put in place a personal and a cultural world view that will guide us away from using our technology to convert the truly valuable into a few coloured electronic beads and toward living a viable life and establishing a viable culture.

## The Tikopians' Decisions

The historical evidence suggests that the least complex of complex cultures can relatively easily either stabilize or resimplify. In contrast, the more complex of complex cultures are unlikely to stabilize and will experience considerable difficulty reducing their complexity. The historical record suggests that they will experience some level of forced resimplification.

There is an alternative. They could change the trends they are following toward a forced resimplification so that at least its worst effects will be reduced. What would the members of a complex culture have to do to exercise this option?

Broadly speaking, they would have to reduce their personal divergences so that they become more aware of their human-environment system's functioning, and its human and environmental constraints on their culture becoming viable. They could then strive to live more within those constraints. Part of that effort would be taking their decision-making foibles into account. Part would be adjusting their culture's world view so that it guided them toward viability. In addition, for these changes to be effective, the members must make and implement them sufficiently early to ensure that their efforts are not rendered ineffective by either the time it takes to make the needed changes or the inertia maintaining the human-environment system on its current track. It is possible to implement this alternative, as illustrated by a personalized history of the Tikopian culture.

**History.** Tikopia is a tiny island in the Pacific Ocean. Its 4.6 km² of land is the top of a less than 80,000-year-old volcano. The climate is tropical, with an estimated 4,000 mm/year of rainfall.

Politically, Tikopia is part of the Solomon Islands. Physically, it is nearest to the Santa Cruz Islands. Although Tikopia is remote it is not isolated to the degree of Easter Island. Tikopia, lies about 140 km (a two-day journey by canoe) from its nearest neighbour, Anuta (0.37 km² in size with a population of about 300). In the past, occasional, long trading voyages to other islands did occur.

The first Polynesian settlers reached Tikopia around 900 BC by ocean-voyaging canoe. They found that, despite its small size, the combination of its rainfall and young nutrient-rich soils provided them with a substantial environmental endowment of tropical wild-plant foods (e.g., coconut). Its coral reef also supplied them with marine resources.

The initial history of Tikopia was similar to that of the other remote Pacific islands, including Easter Island and Mangaia, discussed above. The Polynesians who settled Tikopia initially subsisted on the island's wild foods while they established the crop plants (e.g., yams, taros, bananas) and animals (e.g., pigs, dogs) that they had brought with them. Once established, the growing population was mostly fed by farming their imported food plants using a slash-and-burn method, and local fish. Their diet was supplemented by wild food plants and protein from their domestic animals, especially pigs. Although the population was being fed, deforestation, soil erosion and overexploitation of their island's ecosystems resulted in their endowment being disrupted and some species (e.g., large birds) being driven to extinction. Marine foods were also overexploited. Thus, over time, the availability of wild foods declined.

Tikopia's later history was significantly different from that of the other remote islands. The Tikopians managed to adjust their way of life so that they could better accommodate the environmental and human constraints on their lives. Their split from the more common historical trajectory of the other island cultures may have started as early as 200 BC. But the adjustments only became distinct in the archaeological record around AD 1000. By this time, trees had become a more common source of food, and their efforts at marine resource conservation were apparent in the archaeological record. By 1200, there was a marked decline in the use of fire to clear land.

After 1200, those initial indicators of change in their agricultural practices were followed by major changes. New plant species were introduced (indicating outside contact) and pit storage was adopted to store excess foods. These changes preconditioned the Tikopians to later make further changes to their farming practices. By the mid 1600s, the moving fields that are part of slash-and-burn agriculture had been replaced by permanent plots planted with trees which supplied food and fibre. When they began routinely planting ground-level food crops under this tree canopy, they had fully adopted arboriculture as the appropriate farming method for their island. This production system reduced erosion, protected the soil's fertility and provided better buffering during food shortages, such as those caused by cyclones or drought.

After 1200 but prior to the mid-1600s, other cultural changes were underway. In particular, there were changes to the restrictions (taboos) on certain foods, perhaps, in part, for marine conservation purposes. They were followed by the introduction of methods to control their human population. In essence, the Tikopians introduced a zero-population-growth policy (Firth 1967).

Up until 1600, the prime source of protein was fish, supplemented by pigs. Some time in the mid-1600s to early 1700s, the Tikopians took the unprecedented step of destroying all of the dogs and pigs on the island. Pigs had played a central role in Tikopian culture, so their elimination was a significant marker event in their changing culture. After this event, their only source of animal protein was marine foods. Overall, during a period of 600 to 800 years, Tikopian culture underwent transformative changes.

During the 1800s, contact with Europeans became common. By this time, Tikopia was a large, self-sustaining forest-garden dedicated to feeding and clothing their deliberately limited population. Food storage and an ecosystem buffer helped to tide the people over in lean months or during disasters. Certainly, the post-1800s contact with the greater world changed the culture in some significant ways, such as the arrival of exotic diseases and Christianity. But the essence of their food production system and the associated cultural world view, as recorded in the 1930s, remained basically intact, until at least 2006.

By making the decisions that changed their culture, Tikopians avoided the final forced resimplification stage seen on other remote islands. Tikopia's population rose but never crashed. Neither was the rule by hereditary chiefs scrapped in favour of warrior leaders. Although much of the indigenous forest was replaced by a farmed forest, Tikopians avoided a level of ecosystem disruption and destruction that would have forced a drastic reduction in the human population to a size that the altered island could still support.

The life of the Tikopians would seem to be idyllic, until one is reminded of their responsibilities. Aside from the burden of managing marine and land ecosystem conservation, there was the constant need to balance population with food production. Both are difficult tasks.

Theoretically, the population could have been kept in balance with food production by the patriarchal head of the family exercising his privilege to distribute land to his extended family. But this control could not have changed the fact that Tikopia was a small island with a limited amount of agricultural land. And the balancing would have dominantly relied on the head's decision making and their ability to deal with lobbying.

Instead, the dominant, but not exclusive, method of balancing food with population was through a variety of culturally directed population control methods. The Tikopians implemented this method even though they had no numerical description of their population size nor of their food production. They relied on their practical, experience-based appreciation for the balance point between the two, and their acceptance of the ongoing need to limit population to maintain that balance (Firth 1967; Firth 1965).

Research into the Tikopian culture indicated that their intuitive idea of an appropriate population size lies near the 1,278 people recorded in 1929 and the 1,115 recorded in 1976. This implies that the island could support a surprisingly high density of about 240 people/km$^2$ (compare this with the estimated regional Mayan population in AD 800 of 145 people/km$^2$). The two most culturally acceptable methods for keeping the population in this range were both directed at limiting births to no more than two per couple. One method was celibacy. This required that the junior men in families with more than two children remain permanently unmarried. But they were permitted to have sex as long as no children were produced. The other method was coitus interruptus, which was practised by both celibate men, to avoid pregnancy, and married couples, to space and limit children.

The other less commonly used methods of managing the population focused on reducing its size. These culturally acceptable methods included abortion (rare), infanticide, ocean voyaging (undertaken by young males, supposedly for adventure but considered to be suicidal) and war (apparently used on one or two occasions) (Kirch 1997). After the 1800s, some of the traditional population control methods were displaced by the arrival of exotic diseases and outside pressure. After the 1950s, most traditional methods were replaced by emigration to take up cash labour.

Through these practices, the Tikopians managed to maintain a population size that was roughly in balance with food production. This is an extraordinary feat, considering that, just to make and implement the population control decisions, the Tikopians, both personally and culturally, had to have recognized, accepted and personalized the need for that control. In addition, for the practices to be maintained, they had to embed their population control obligation into the culture's world view to the level of a social norm.

The formal part of this embedding can be found in the proclamation at Rarokoka. It was the verbal expression of the behaviours needed to maintain law and order. When the proclamation was reiterated during an annual cultural reaffirmation ceremony, it reconfirmed the validity of those behaviours. One of them was the practice of population control in order to ensure that there was adequate food to feed their children. The proclamation included specific guidance: practise coitus interruptus to avoid pregnancy, and ensure that a family did not exceed its ideal size of two adults and two children (Firth 1967; Firth 1965).

How did the Tikopians manage to introduce these changes while other island cultures did not? Before answering this question, it is useful to decide whether their cultural choice to control their population was indeed an appropriate decision. In particular, did it result in them living in a manner better matched to their environmental endowment?

The test came in 1952–1953. In 1952, a cyclone, whose appearance was foreseeable but whose timing was unpredictable, devastated the island. That event was followed by a drought. In 1953, another cyclone struck. These events destroyed much of the Tikopians' food production system.

The test was made doubly severe by the cultural changes introduced earlier by Christian missionaries. Their influence reduced the use of traditional population control methods, as illustrated by the proclamation at Rarokoka not being recited after the 1910s. In particular, the missionaries insisted that celibate (unmarried) males could not have sex. They had to marry to have sex, which therefore meant children. Alternative population control methods were not introduced. As a result, between 1929 and 1952, the population grew at an average rate of 1.6%/year to reach 1,752 people in 1952, some 500 to 600 above the ideal assumed by the Tikopians.

After the 1952 cyclone, the Tikopians made an accurate assessment of their circumstance. They reacted appropriately to the devastation even to the point of ensuring that marriages (and hence births) were delayed until food production had recovered. However, after enduring three or four months of the subsequent drought-induced food scarcity, some people were near the end of their ability to endure near starvation. Some children died. The social order was stressed and started to break down. An external influence, in the form of some intervention in decision making and off-island food was needed to help stabilize the situation.

The 1953 cyclone was less severe. Despite the fact that the arboriculture food production system was still recovering from the events of 1952, the 1953 cyclone did not have much impact on it. Fortunately, there was no subsequent drought. Three month after the 1953 cyclone, the island was again producing sufficient food to support the Tikopians, thus demonstrating its robustness and resiliency.

Despite all of these strains, the Tikopians and their culture did survive the famine largely intact (Spillius 1957). However, immediately after the 1953 event, the Tikopians arranged for some of their members to emigrate to other islands, which reduced Tikopia's population by 13%. By 1976, the population was 1,115. In 2006, the population was estimated at around 1,100 and the soil fertility was found to be intact.

In retrospect, the impacts of the 1952 cyclone and food shortage indicated that, in that year, a population of 1,752 put the Tikopians' food production system near its Malthusian carrying capacity with an insufficient buffer to deal with a food shortage. (Remember the rule of thumb: the size of a species population that will endure through time is limited by the host ecosystem's least productive years.) If the Tikopians hadn't received outside help in the year following the 1952 events, then to prevent mass starvation or social collapse, they would have had to have resorted to their extreme methods of population control, such as ocean voyaging or killing. It is easy to conclude that their emphasis on population control was appropriate, because it, along with arboriculture, pig eradication and marine conservation, enabled them to live satisfying lives within their environmental and human constraints.

**Decision Making.** Back to the question of how the Tikopians came to make and implement their culture-changing decisions to eradicate pigs and dogs, conserve fish, establish arboriculture and introduce population control, while other Polynesian cultures, such as those on Mangaia and Easter Island, did not. We could conclude that the Tikopians, as individuals, were substantially different from other Polynesians. However, the anthropological studies suggest not.

These studies report that the Tikopians are, like other Polynesian cultures, a culturally homogeneous group. On Tikopia, they are divided into clans, each with a hereditary chief. Each member has equivalent access to most resources.

The island itself is divided into geographical districts in which members from any clan can live. Competition on Tikopia is evident by the considerable rivalry and hostility between these districts, which is basically the same system found on other islands settled by Polynesians. There is every indication that in the past Tikopians were quite capable of displaying the same decision-making foibles and having the same reality paradox experiences as the other Polynesian who settle on small islands.

It also seems highly unlikely that the Tikopians were consciously following a well-thought-out plan to reach their balance. After all, the relevant changes seem to have taken place over at least 600 years. Perhaps they came to their solution by chance? This assumption ignores the fact that they had sufficient knowledge, skills and resources to choose between options. For example, they could have chosen to intensify agricultural production rather than kill all their pigs and implement population controls. Thus, although chance probably played a role in determining their future, it seems not to have been the only force. We are thus more likely to gain an understanding of how the Tikopians came to live within their environmental and human constraints if we can appreciate the influences on their decision and their reaction to those influences.

Perhaps Tikopia's remoteness made a difference? Although not as remote as Easter Island, their level of isolation would have limited the number of realistic choices that the Tikopians could have made. For example, although they did trade with neighbouring islands for obsidian, Tikopia is sufficiently remote that trade could not routinely supplement or dominate their supply of food. This isolation, along with cyclones or drought, could have forcefully made the Tikopians aware that they were dependent on the natural resources they could produce locally. That is a nice idea, but the Mangaians' and Easter Islanders' more extreme isolation and their in-your-face exposure to the limits of their environmental endowment were unable to prompt them to appropriately change their world view. Thus, although remoteness may have influenced Tikopian decision making, on its own it is an insufficient explanation for their choices.

Another influence on the Tikopians' decision making is the size of their island. Many Polynesian islands are small, but Tikopia is tiny. It is sufficiently small that the estimated long-term, sustainable population size of about 1,200 is around the maximum size of hunter-gatherer bands. The smallness of the island also means that an individual could easily visit everyone else in their district, perhaps even all 1,200 people on the island. These two characteristics, plus the fact that they had no central authority, allowed the clan-divided farmers on Tikopia to use hunter-gatherer-like decision making and compliance methods, such as social monitoring, to govern themselves.

For example, although the clans had leaders, all of the members participated in some overt way in the decision making and cultural reaffirmation processes. Each Tikopian was also a part of an enforcement process because that process relied heavily on social monitoring and social pressure. Their economy also encouraged a feeling of participation and belonging in the group because its functioning was dominated by the value of social relations. This value was expressed and reaffirmed through food sharing with reciprocity, which played an important role in daily interactions and a central role in social functions. Their culture's functioning was also flexible.

Collectively, these physical and cultural features would have made it easier for an individual member to perceive of themselves and all other Tikopians as members of a single abstract unit. As a result, an individual could also feel that they had a vested interest in the group's welfare. In essence, their culture engendered in each individual a sense of commitment to more than a clan. Each member was also committed to the Tikopians as a group. These features made it easier to understand how the culture could change. But it leaves unanswered the question of how they came to make the specific decisions they did.

Somehow, the Tikopians were prompted to go beyond simply blaming, for instance, a cyclone, the gods or their neighbours for a nasty event, such as a famine. Somehow, they came to recognize that there

were long-term, external reality connections between their population size, their island's limited environmental endowment and food scarcity. They also recognized that these connections imposed external-reality constraints on their day-to-day, internal-reality-based decisions.

More importantly, they personalized this information. This enabled the Tikopians to make decisions that reduced their personal and thus their cultural divergences. In doing so, they expanded and adjusted their culture's world view to include the long-term external-reality functioning of their environment, their actual relationship to their environment, and the constraints it imposed on their decisions. One expression of this was the proclamation at Rarokoka.

The Tikopians may have been able to draw these conclusions, make the needed decisions and implement them because their island gave them the opportunity to do so. Their young island's rich, robust and resilient environmental endowment provided them with both the opportunity to live differently and allowed them to repeatedly make mistakes, and attempt to fix them. Their environmental endowment thus gave them the time (at least 600 years) to recognize their circumstance for what it was and then complete the slow process of adjusting their culture's world view so that it guided them to live more within their environmental and human constraints. This advantage was less available or unavailable to the settlers on many of the other remote islands.

This broad conclusion neglects some critical human aspects of cultural change. It is ultimately individual people who make the decisions that collectively sustain the complex process of cultural change. In this context, the Tikopians' abrupt, post-1600 decision to kill all the pigs was clearly not the moment they all suddenly decided to change. That marker event was merely an act that symbolized the irreversible cementing into the culture of their new way of securing food. Getting rid of the pigs was the collective end product of a series of individual decisions made over successive generations during the previous 600 plus years. During that time, each individual's day-to-day decisions were the product of the complex interplay between many cultural and environmental influences, which varied with time, context and individual perceptions. The collective outcome of these decisions both maintained the Tikopian's cultural world view and, through preconditioning, drift and direction, among other ways, changed it in a direction that, over time, resulted in its current features.

We are left to explain how the nature and direction of Tikopia's cultural change could have been sustained over many generations if it was based on the members' day-to-day decisions. An answer can be found in the series of cultural connections that sustained the rise of the notion of progress and individualism in the Western European world view for generations after 1600, as discussed in chapter 3. It seems likely that there must have been an equivalent "thread" or feedback that constrained the Tikopians' decisions and thus their direction of cultural change for generations.

We don't know what that thread was. However, known events in Tikopian history provide a speculative illustration of how the normal push and pull of the influences on personal decisions could have combined to favour the members' increased awareness and the longer-term cultural drift toward their culture's new world view. These events include soil erosion, the loss of marine resources, an increase in usable land, erosion control, their cultural view of food and the arrival of immigrants.

Tikopia's volcanic crater was breached by the sea to form a saline lagoon. On the steep slopes of the crater's rim above the lagoon, slash-and-burn agriculture resulted in its rich soil eroding into and partly filling it. The resulting degradation of an important marine food resource could have been partly offset by farming the new land.

But this practical balancing of food production neglects that these types of changes to the land and to land use also have personal and cultural impacts. Consider that the presence of new land raises the issue of access to it. And the loss of the marine resources in the silted up lagoon could have affected the fish harvesting rights of a family or clan. On a crowded island, these issues would demand a resolution in the social arena. Once a satisfactory resolution was found the ongoing presence of the steep crater walls would have acted as continual reminders of why they made it and the need to respect it.

For example, the desire to avoid the social consequences of the crater rim erosion could have provided an incentive for the Tikopians to question slash-and-burn agriculture and consider erosion abatement practices. Both issues would have been addressed by the Tikopians deciding to increase their dependence on trees for food. But forest trees are fixed and take years to grow. Dealing with this constraint would have preconditioned them to change their crop plots from temporary to permanent, and from open field

to arboriculture. In the process, their land tenure system could have drifted into its more recent form (more fixed than shifting).

It is reasonable to think that both the material and cultural changes resulting from the members' decisions about how to deal with practical issues could have contributed to adjustments in their wider view of their relationship with their environment. For example, their attention to erosion, trees and slash-and-burn farming, and the connections among them, were all likely accompanied by a change in their view of their relationship with the land from a shorter to a longer term. Perhaps the process of resolving food production and social issues that arose, for example, from having to farm the steep, erosion prone crater rim in order to feed themselves, and from the fallout of silting up the lagoon, had other consequences. It might have prompted them to the broader conclusion that there were limits to increasing agricultural intensity and that food resources were finite.

Another aspect of the thread sustaining the direction of Tikopia's change was, and still is, the Tikopians' use of food for more than sustenance. Sharing food is used to maintain social relations, while food taboos express aspects of their culture's world view. A change in their food taboos thus represents an adjustment in their culture's world view. A change that also alters their diet and their relationship to food organisms. This preconditions further changes to the culture's world view, such as requiring them to adjust their food-based ceremonies. In this context, the taboo could have been used in the past as a consciously wielded tool for the management of ocean resources, especially sustaining their conservation.

An additional aspect of the thread was the arrival of new Polynesian immigrants part way through the long process of the Tikopians' cultural adjustments. At least one group of the newcomers brought new plants with them. Perhaps they, or other immigrants, introduced new concepts of land use, food production and storage. Or maybe they just brought hard-learned lessons from another island where the old Polynesian ways had resulted in the usual bad outcome. Maybe Tikopia experienced a practical reminder of this when, on one occasion, land and food shortages apparently sparked a brief war between the island's older clans and a recently established immigrant clan. The immigrants were the winners in this social unrest. Perhaps these events prompted discussions about how to live together.

In overview, it seems that the Tikopians' cultural transition toward living within their human and environmental constraints was sustained by a complex combination of specific human (cultural, individual) and environmental factors. When threaded together, the interactions among them formed and maintained a transition-sustaining feedback. Consider this possibility within the model of the human-environmental system. Start with their developing recognition and acceptance of the constraints within which they must live. This preconditioned them to make decisions which would adjust their culture's world view in ways that would favour them living within those constraints. The guidance from their adjusted culture's world view favoured the Tikopians making decisions that recognized and accepted the external reality constraints on their lives, thus completing the feedback. This feedback was sustained by its prominence in their cultural reaffirmation process, successful social monitoring and pressure, and, probably, the presence of strong environmental cues. By the time the Tikopians made the decision to directly control their population and kill their pigs, their cultural transition to living within their human and environmental constraints was essentially complete.

Although the irrevocable act of killing off their pigs stands out, it is more of a marker during a long transition than a primary cause. In Tikopian oral history, the pigs were said to have been killed because they damaged crops. But pigs are a central part of Polynesian culture, just as cattle are for the Xhosa and money is for us. The act of eradicating the pigs was therefore likely driven by additional less obvious reasons. Potential factors include the need to deal with other social issues, such as access to land and resources, and absorbing immigrants.

By the time of the pig killing, it seems that the Tikopians had embedded into their culture's world view most of the lifeway features needed to live within their environmental and human constraints. In this context, the killing was a highly symbolic act. In later years, it could remind them of why they had changed their core beliefs and practices. It is the equivalent of Westerners deciding to symbolically abandon fossil fuels to mark our final rejection of the notion of progress (Borrie et al. 1957; Firth 1967; Kirch and Yen 1982; Firth 1983; Firth 1959; Mertz et al. 2010; Kirch 1997; Spillius 1957; Firth 1965; Kirch 2007).

In summary, the many personal and cultural adjustments the Tikopians made over the centuries brought their culture and their population size into line with the limits to their environment's long-term ability to support them, while at the same time staying within the human constraints on a functioning culture. These changes were neither solely the product of fully conscious decisions, nor chance events out of their control. The Tikopians were neither pawns nor masters of their future. They were active participants in a long, difficult and complex process of changing their personal perception of the world, their sense of their place in it and their culture's world view. Their environmental endowment helped this process because its resilience and abundance gave them the opportunity to make mistakes.

We might never know the exact combination of factors that raised the individual Tikopian's awareness of their circumstance and influenced their decisions to favour viability. But we do know that each member reduced their divergences, and accepted their (environmental and cultural) responsibilities. They made significant changes to their culture's world view that resulted in it guiding them to live more within the human and environmental constraints on their lives. We know these changes worked because they avoided the forced cultural resimplification of other Polynesian cultures (Firth 1967; Firth 1965).

## What's Next?

We make predictions about the future to help us prepare for it. Such predictions were made at the end of chapter 15. They are revisited here. This section divides common Western predictions about the future of our complex industrial culture into a few generic types, provides thumbnail descriptions of each and evaluates them by applying the knowledge gained in this and previous chapters. Because the members of industrial and industrializing cultures make up the vast majority of the world's population, this discussion also addresses predictions about our possible average global future (i.e., when disregarding regional-scale cultural and environmental differences.)

There is an optimistic business-as-usual prediction that Westerners will muddle through to a satisfactory, even a somewhat utopian, better future. In this prediction, we will deal with social and environmental issues when they arrive, so we believe, by applying innovation and technology in ways that will maintain business as usual: we will adapt. On the surface, and based on the last 150 years, this prediction sounds realistic. It is certainly the most likely to be acted upon because it simply requires us to continue having faith in our culture's world view, humans and their ingenuity.

But, to favour this "muddle through" prediction implicitly requires us to accept an illusionary view of ourselves and our world view. We must discount or reject our knowledge (discussed in previous chapters) about the biases in our decision making and the presence of divergences from the external reality in both our internal reality and our culture's world view. We would also have to reject the lessons of history and our poor, ongoing record as environmental managers, discussed in chapters 10, 11 and 16. To act upon this prediction, we would have to feel sufficiently confident that the future won't label us "those idiots of the past."

A version of the optimistic business-as-usual prediction is that "we can go it alone." This prediction relies heavily on the arguments that we have the ability to look after ourselves and the environment doesn't care, in an emotional sense, about its future. Thus we are free to focus intently on applying innovation and technology to maintain business as usual, regardless of our impacts on our environment.

To favour this prediction requires that we feel comfortable abandoning any obligations to present generations of species who may fall by the wayside. It also means accepting our greatly increased responsibilities for repairing or providing the environmental functions we have disrupted or destroyed, in particular those that provide the resources and services we and our culture depend on to function. To accept this burden, we have to believe that we can appropriately discharge those responsibilities. Adopting this prediction would thus require us to deny the strong evidence to the contrary. We have had limited success at recognizing, accepting and fulfilling our current responsibilities, and we thus do indeed rely heavily on our environment's functioning, as discussed in chapters 11, 14 and 15.

There is also a pessimistic/fatalistic business-as-usual prediction. It recognizes that our human characteristics and the inherent instability of the more complex cultures favour resimplification (Eisenstadt 1988). In this prediction, resimplification is inevitable because, when faced with critical decisions, we will either invariably make choices that drive the culture beyond some critical limit or fail to react in a timely and appropriate manner to signs of destabilization, or both. A common explanation for this point of view is that our dominant biological and mental characteristics are those of our hunter-gatherer ancestors, which are ill-matched to run a complex culture (Seligman and Csikszentmihalyi 2000). The conclusion from this argument is that all we can do is feel comfortable just carrying on, business as usual, while expecting an undesirable outcome.

But there are significant practical and emotional downsides from favouring this prediction. To hold it, we have to severely discount or even reject any claims to advanced knowledge, a powerful gift of consciousness, modernity and technology. An individual has to personally accept a severe blow to their sense of self-worth and pride in belonging to a complex culture. We have to give up hope.

There is a contrasting optimistic prediction that we will change the current business-as-usual trends. It recognizes that resimplification looms but believes that we have the capacity to avoid it. It imagines us becoming actively engaged in implementing a mix of personal, cultural and technological changes that will allow us to adjust ourselves and our cultures' world views in ways that will maintain both the environment's functioning and our current living standard. Typically, attention is focused on a particular current social or environmental issue.

To favour this prediction means accepting an often idealistic/simplistic view of how technology, we (as individuals) and our cultures function and change. Many of the suggested solutions and the predicted outcomes from implementing them are unrealistically straightforward or treated in isolation as mentioned in chapters 15 and 18. For example, our energy and related environmental issues can be resolved by us proactively promoting the generation of energy from clean energy sources (e.g., alternative and fusion energy) and an increase in efficiency. If we use this clean/green energy to power our new, energy efficient lifestyle, then our lives will be improved and the environment rescued: resimplification will be avoided. This optimistic prediction, like the other optimistic prediction, has overtones of a utopian future.

When contemplating the future, Westerners also pay attention to global-scale models of our complex world's functioning that explore its possible futures and possible solutions. The conclusions from these models are quite general in nature and mostly focused on material and population changes, as discussed in chapter 15. Think of overshoot and collapse. In particular, they largely ignore the foibles of human decision making and the nuances of personal and cultural change. Acting on solutions based heavily on the models will likely change the current business-as-usual trends, but are unlikely to avoid resimplification unless their limitations are taken into account.

There are variations to the above predictions, such as a more or less religious or scientific focus, but they don't change the essence of the following conclusion. All of these predictions provide us with some insight into the possible future of complex cultures. However, as pointed out, they all suffer from significant limitations and aspects that are at odds with the external reality. They can all be described as at least incomplete.

Is there a more comprehensive outlook for the future and a realistic action-based prediction to deal with it? When the discussions of this chapter and the updated model of the human-environment system in chapters 11 and 16 are combined, one emerges. It is unrealistic to expect us to make reliable, detailed predictions about the future of Western industrial culture, or any other industrial or industrializing culture for that matter, as discussed in chapter 5. However, the current global trends in resource usage and our ongoing environmental impacts collectively indicate that, in the future, these cultures are likely to face significant environmental and associated resource and social issues. The efforts their members are expected to exert when dealing with these issues will prompt an environmental (and social) response. The results will affect the members' way of life, and they will probably react to those impacts in ways that will likely maintain the current resimplification-favouring trends. This completes a feedback loop that favours the resimplification of the inherently unstable more complex cultures.

The current state of the biosphere (the environment's state plus our circumstance), the rate of the trends it is following and the inertia helping to keep it on those paths all suggest that there is only a limited amount of time for the members of complex cultures to avoid resimplification. Think of the speed of climate warming. To do so, they would have to expend the effort needed to change themselves and their decisions in ways that better recognize the human and environmental constraints on our lives. Considering the nature (manner and constraints) of human and cultural change, the amount of time available to do so seems too short. It thus seems reasonable to predict that, over the longer term, it is likely that at least Western industrial culture will experience resimplification.

However, industrial and industrializing cultures do have a choice about the type of resimplification they will experience. They could decide to continue following current trends that indicate a forced resimplification. Or they could reverse the trends in a manner and to a degree that will at least reduce the worst effects of forced

resimplification: voluntary resimplification. This is facilitated by the members simultaneously focusing on two objectives, immediately.

One objective would be to accept that resimplification is highly likely and probably unavoidable. Once accepted we could then prepare ourselves and our culture for the post-resimplification future. The task would be to change our culture's world view and our lifestyle at a realistic pace in a manner that prepares, preconditions perhaps, our descendants to live within their future environmental and human constraints. We could start that process by personalizing the reality that the earth's environment can exist without us, but we cannot exist without it being in a state to support us.

The other objective would be to recognize that we can make the decisions today that will start the process of our descendants living more within our human and environmental constraints in the future. We have the capacity to avoid the worst of forced resimplification. The task would be to change our personal internal reality by reducing our divergences and, as a result of those changes, adjust our culture's world view to more clearly reflect the external reality (reduce its divergences). We could start this process by admitting that we are currently out of our intellectual depth. We can also switch our focus away from a concern about the details of the future toward addressing today's issues in ways that meet the conditions known to favour a viable culture. With that acceptance and change, we are more likely to appropriately apply our skills to make the needed personal and cultural changes to favour voluntary resimplification and facilitate our culture becoming viable in the future.

These two objectives complement one another. By reaching for them simultaneously, we are more likely to realistically evaluate and appropriately react (materially, intellectually and emotionally) to the possibility and consequences of forced resimplification. If we place these objectives under the banner of "reach for viability," they become a call for action. This call is certainly the most difficult of all the prediction-based calls for action to implement. But, even if we fail, we would at least retain some sense of pride in ourselves, our humanity and our culture. Our children will be happy that we at least tried, and we pointed them in a realistic direction.

To make a breakthrough in scientific understanding about a problem requires that a researcher be willing to change the way they think about it. Similarly, if we wish to proactively deal with resimplification in a realistic manner, we have to change the way in which we think about our world and ourselves, and how we relate to our environment and one another. Thus the process of finding a way to live within our environmental and human constraints is more like the search for happiness. It is more a work in progress, a state of becoming, a never-ending task or quest rather than the simple deterministic solution we prefer.

We are helped in this task by the symbol of Tikopia. In contrast to Easter Island and Mangaia, the Tikopian agrarian culture made the changes needed to avoid a forced resimplification. But, before we treat Tikopia as a celebrity culture that discovered a simple solution to resimplification, we should remember a back-to-earth observation. They devoted time and effort to make and maintain the needed changes. To be successful, so will we.

If you choose to believe that we should be able to make decisions today that would change ourselves and our culture in ways that will enable us to avoid at least the worst of resimplification, then there is more for you to read.

# CHAPTER 18

# REACHING FOR VIABILITY

An outline of the story so far sets the context for this chapter. The earth's environment, humans as individuals, and their cultures have been discussed both in isolation, as complex systems in their own right, and as interacting pieces that form the dynamic, complex human-environment system (figure 15). As participants in the human-environment system, we both contribute to and are subject to its functioning. We are thus neither the puppets of the world around us nor, even in a human-dominated world, the masters of our destiny.

Certainly, the functioning of the human-environment system enables us to complete our lifeway, but it also constrains the decisions individuals can make if we wish humans and our cultures to remain functional over the long term (viable) (Hall et al. 1986). For example, our environment provides us with the resources and services (resources) we want. But their availability is constrained by both their finiteness and their rate of renewal. The renewal of resources, in turn, relies on the functioning of the environment's recycling and renewal processes. We are thus constrained to live in a manner that neither uses resources faster than they can be recycled and renewed nor disrupts the environmental functioning that provides those recycling and renewal services.

At present, the members of the more complex cultures don't, on a global scale, fully respect the system's human and environmental constraints on how we can live. For example, from our perspective, we are simply exploiting our environment in order to support our growing global population and its increasing average per person consumer and disposer wants. However, from the environment's perspective, there are too many people extracting too many resources and disposing of too much waste in ways that are disrupting or destroying, at all scales, many aspects of its functioning. In essence, we are effectively reducing our environment's ability to support us over the long term, while burdening ourselves with environmental management responsibilities that we are ill-equipped to fulfill.

Our insufficient concern/respect for the human and environmental constraints on our lives is understandable if we remember the limitations inherent in our brain's functioning. They are seen as biases in our decision-making processes (foibles), such as our anchoring, hindsight, self-serving, over-optimistic and group biases, and the bias in our analysis of risk (Weinstein 1989). Consider too that our internal reality, which influences our decision making, imperfectly represents the external reality: it contains divergences from the external reality.

The formation of our internal reality is also strongly influenced by that characteristic of a human group that binds the members together: our culture's world view. As we grow up, our culture's world view provides each of us, through our parents and our social group, with much of the basic guidance we receive about how the world works, what we should believe, how to behave and the things we should pay attention to. It even provides us with basic reasons why. Through this connection, our culture's world view also influences our decision making.

But a culture's world view is a dynamic creation of its members that is passed from generation to generation. The members of each generation are their culture's decision makers, so there is ample opportunity for them to change their culture's current world view in ways that result in it imperfectly representing the external reality. These cultural divergences are then absorbed by the youngest members into their internal reality as they grow up. The divergences are caught up in the member-culture feedback loop.

Overall, our personal awareness of and our attention to the dynamic, complex human-environment system in which we live our lives tend to be limited, biased, inaccurate and incomplete. This includes our respect for the system's human and environmental constraints on our lives. Our decisions reflect this. However, our consciousness tells us something different.

We are justly proud of our consciousness. But we are intuitively unaware that it is only a small part of our brain's internal reality and that it is affected by the same biases and influences as our brain's unconscious functioning. This means that our casual conscious inspection of the world around us allows us to believe that our perception of it is largely complete and true. We can also believe that the guidance we receive from our culture's representation of our world (its world view) is mostly reliable. We can therefore easily think of ourselves as being largely aware, logical beings making mostly rational decisions about a circumstance we basically understand. This, of course, is only partly true. Thus, when we do become aware of the difference between our internal reality and the external reality (the divergences), we experience a reality paradox or cognitive dissonance.

We find it difficult to deal with these experiences in a way that also prompts our thinking to better reflect the external reality. Consider too our inertia to change. It provides us with the sense of stability and continuity we need to remain functional. But it also makes it difficult for us to change in ways that remove the divergences contributing to our reality paradox and cognitive dissonance.

The member-culture feedback loop ensures that our personal inertia to change is reflected by our cultural group. The members' collective inertia to changing our culture's world view helps to provide the group with the stability and continuity needed for a culture to remain functional. But it also ensures that removing a significant divergence from our culture's world view can be difficult, so its divergence-biased guidance endures. Thus our current lifestyle and relationship with the environment, which reflect our personal and cultural divergences, also endure, and so does our insufficient concern/respect for the human and environmental constraints on our lives.

Thus, in practical terms, before a person can easily undergo a specific change they must be preconditioned/primed, either by themselves or by others, to make it, and their circumstances/context must favour the change occurring. In addition, if the resulting change is to favour viability then the information used and how it is being used must also favour viability. And, if a persuader/leader is involved, they must not only inspire viability but follow their own advice promoting it. The equivalent applies to cultural change.

In overview, the above discussion is the context within which we, as individual members of a more complex culture, perceive our surroundings and make those decisions that will play a significant role in determining our personal and our culture's future. Indeed, history (and this book) suggests that if the members of today's more complex cultures continue to neglect the human and environmental constraints on our lives and follow current trends, then we will likely experience a forced decline in our culture's complexity: a forced cultural resimplification. The historical record shows this is an unpleasant affair. It is an undesirable choice for a culture's future.

The members of today's more complex cultures could choose to avoid a forced resimplification completely by making all of the changes needed for them and their culture to become, and then remain, functional over the long term (i.e., viable). To do so, they would have to acknowledge the human-environment system's human and environmental constraints on their lives and decide to live within them. Unfortunately, when the current state of the environment and the inertia maintaining its trends are considered in combination with our difficulty dealing with our personal and cultural divergences, this choice is unlikely to be successful: there is insufficient time.

Fortunately, the members could still avoid the worst of a forced resimplification. They could decide to start a voluntary resimplification. Current generations would then be obliged to make sufficient changes to put them and their culture on a post-resimplification path to viability. In light of the alternative, this is not a failure. By striving to live a viable lifestyle in a viable culture, they could both avoid the worst of a forced resimplification and give their descendants the opportunity to continue voluntarily working towards viability.

This chapter discusses how we could reach for that goal. Because it is we, as our culture's decision makers, who must do the reaching, the discussion will take the perspective of an individual member of a culture. Think of yourself. Western culture will be used to both illustrate the discussions and represent all complex industrial and industrializing cultures. To simplify the discussions, the focus will be on the environmental rather than the human conditions for viability.

The first section discusses our relationship to viability. The next section provides a broad outline of a path we could follow toward viability and, in order to journey down it, the types of decisions and changes we will have to make, and the challenges we will face doing so. The following three sections deal with the practical details of following a path to viability. One section discusses the nature of the environmental conditions for viability and how we can represent them as objectives. When aggregated into the five environmental objective unit (five-objective unit), they act as a target we can aim for to become viability-favouring environmental trustees. Another section discusses three methods we can use to help us satisfy viability's conditions as represented, for example, by the five-objective unit. The last of the three sections illustrates how the three methods and the rules of thumb can, together, help us deal with the challenges we will face as we journey along a path to viability.

During these discussions, we will become more aware of the human conditions for viability. The significance of these conditions is made clear in the next-to-last section by the extra responsibilities that the elite of a complex culture must fulfill if their culture is to satisfy most, if not all, of the conditions for viability. The final section summarizes the chapter and provides an outlook for the future.

A reminder that *short* (or *immediate*) *term* refers to a period up to 15 years from 2015; *medium term*, up to 60 years; and *long term*, beyond 60 years.

## Viability

We are participants in the dynamic, complex human-environment system. If we conduct our thinking and decision making within that context, then our decisions about our possible future will be more valid and useful (meaningful). The system and its dynamics were previously illustrated from the environment's perspective in the element-cycle diagrams of figure 7 and from a human-focused system's perspective by the model in figure 15. However, when considering viability, it is more useful to look at the system from the perspective of the individual members. After all, it is we, as individuals, who will have to do the thinking, the decision making and the changing needed for us to live, and our culture to function, within viability's constraints.

This section illustrates the human-environment system from the individual's or their culture's perspective in a new diagram: the viability flower (figure 22). It is used to clarify our personal and cultural relationship with viability and make us aware of its associated constraints. We can then more fully appreciate the decisions we need to make in order to favour viability, and the difficulties in making them.

The flower represents the human-environment system and its functioning by breaking its three primary factors (individual, culture and environment) into components or aspects. These are the parts of the system's functioning that are important to us as an individual, a social species with a culture, or a member of an ecosystem (aspects of living, if you like). The diagram displays an individual's view (perception, opinion, belief) of each aspect as a flower petal not a point because our view of a particular aspect can vary. A petal thus represents the full range of our views about that aspect.

At its most basic, our views/opinions of the aspects being displayed on the flower can be divided into those that are consistent with the external reality and those that aren't. Our consistent views lie within the centre of the flower, as outlined by the dashed ring. Views that are most likely to lie in the centre are those that treat the many aspects of living as being interconnected (including feedbacks) into a single dynamic, complex human-environment system, as illustrated in figure 15. So are our views that recognize the human and environmental constraints on our lives and our cultures. Sustainability too could lie in the centre, but it is excluded because it is routinely used, as discussed below, in ways that express a meaning at odds with viability. Sustainability is better represented by a petal.

The boundary separating our external-reality-compliant views (in the centre) from our non-compliant views (outside of it) is a dashed circular ring rather than a single solid circular line, because the boundary is a gradational/fuzzy transition zone rather than a sharp one (figure 22). Think of where on the flower our views based on our intuition and our somewhat realistic imaginings can fall. The transitional nature of the boundary is also indicated by the gradual separation of the petals as they cross the centre's boundary.

The flower's petals can extend beyond the centre's boundary because our view of an aspect is not proactively forced to correspond with the external reality. After all, the processes forming our views occur in our heads, where they are subject to our decision-making foibles and the influence of both our internal reality (with its divergences) and our culture's world view (with its divergences). In addition, we are free to exercise our imagination and thus include in our view whatever we please. We can also hold our views as strongly as we like, which, as expected, also means that they can display differing degrees of inertia to being changed.

The more our personal view of an aspect differs from the external reality, the further our view lies away from the centre and toward the tip of that petal. For example, imagining that you are immortal places your view at the tip of the death petal. In this way, the flower's petals represent our divergences from the external reality.

As mentioned, the petals merge into the centre, where they represent those of our views that most accurately reflect the human-environment system and its functioning (the external reality). In contrast, the petals become increasingly separated from one another the further they are from the centre. This trend represents another important feature of our views. The more they differ from the external reality, the harder it is for us to weld them together into a coherent, consistent external-reality-based whole.

Consider the view held by libertarians that a complex culture should have little or no government. This view is at odds with the observation that an increase in the size of a cultural group is invariably accompanied by an increase in its organizational complexity, as discussed in chapter 3. A government lies at the heart of that complexity. It helps to keep the culture stable and functional (let alone viable) by replacing/augmenting the methods used by simpler cultures that are unsuited/insufficient for larger groups. By ignoring this, a libertarian can't form a coherent external-reality-based view of a complex culture or the system.

Certainly, we are free to try and live by our internal-reality views of the human-environment system. But if those views lie outside of the external-reality boundary, then the results of our efforts might not be to our liking. Think of king Midas, or those who act as if they are immortal or need little sleep.

By representing the dynamic complexity of the human-environment system from an individual's perspective, the diagram highlights the difference between our internal reality and the external reality (our divergences) (figure 22). This is important because our divergences make a significant contribution to our incomplete, somewhat disjointed and often biased views of the system and our relationship with it. Their presence explains much of our personal struggle, as mentioned in the introduction, to recognize, let alone satisfy, viability's conditions. Correspondingly, the flower reminds us that if our decisions are to favour viability, then our view of the system and its aspects must reliably reflect the external reality and viability's conditions: they must be located in centre of the flower. That is the place we should aim for when adjusting our personal world view to favour viability.

When making those adjustments, we should not forget the role of culture. A group of individuals has a culture whose world view influences their decisions. But the members are their culture's decision makers and its memory, so they can change their culture's world view. Thus a culture's world view and its members are involved in a member-culture feedback that influences how the members and their culture's world view changes.

A culture's world view, and thus its guidance, can be represented by the viability flower in the same manner as it represents an individual member's view (figure 22). The aspects of a culture's world view can lie on the petals either inside or outside of the external-reality centre. For example, guidance from a culture that practising specific rituals can alone resolve a food supply problem is unlikely to deliver the goods (i.e., it lies near the tip of the food petal). In this way, the viability flower also represents the divergences in a culture's world view from the external reality.

If enough features of a culture's world view fall outside of the centre and the culture's members follow its guidance for long enough, then the culture faces failure. For a more complex culture, this means a forced resimplification. Because the members are their culture's decision makers, they can avoid this outcome. If sufficient of them make the personal changes needed to live a more viable personal life, then they, through the member-culture feedback, will adjust their culture's world view so that it lies nearer to the flower's centre: its guidance will favour viability. Forced resimplification will be less likely and voluntary resimplification becomes possible.

However, to make the needed personal or cultural adjustments, a member must first make a judgement about where on the flower their (personal and culture's) views are currently located, and the direction they are heading: toward or away from viability. There are a number of factors that ensure a meaningful judgement is difficult to make. Consider the complexity of the human-environment system. It requires, for example, that we consider context when judging our personal view of an aspect. In this light, reconsider the earlier example of thinking that we are immortal. Although that view ignores the basic constraint of death, it is harmless if it is an isolated idle or make-believe view of death (i.e., you sit just outside the core on the death petal). However, when such a thought is backed by your religious belief in invincibility, then surely your days are numbered (i.e., you sit near the death petal's tip).

Similarly, we must consider context when making our judgement of a culture's world view/guidance. Reconsider the example of culture's guidance that specific rituals can solve a food issue. Where on the food petal we place this view depends on the context in which we judge it. For example, if judged in the short term, with a focus on its therapeutic value, then it can be placed near or in the centre. However, if judged in the long term, in the context of a long drought, then the guidance is more likely to be placed outside the centre.

More generally, disregarding context leads to an incomplete or biased evaluation of where an aspect of our personal or culture's world view sits on the flower and its direction of movement. This can result in us disregarding reality and thus becoming burdened with responsibilities we can't fulfill, or critically disrupting the functioning of our environmental endowment or culture. Where on the food petal should Western culture's guidance about applying technology to resolve food issues be placed? Remember that judgements relevant to viability are made in the context of a complex system, its feedbacks and the long term.

Making a meaningful judgement about our views is also subject to the human-environment system's dynamics. It ensures, for example, that with time the context in which a view should be judged can change. Thus, although viability's conditions do not change and our views may not, where our view falls on the flower can change.

There is one other factor that affects our judgement of our (personal and cultural) views. Our judgements are made relative to the centre of the flower, which represents both the external reality and the conditions for viability. Thus the making of meaningful judgements about our views requires that we have reliable knowledge about the external reality, viability's conditions and the state of the world. This is a challenge because checking for reliability, gaining familiarity with viability's constraints and ensuring consistency in using that information takes time and effort, as seen in the rest of this book. These factors ensure that meaningful judgements about our personal or our culture's views become more meaningful if they are expressed with conditions.

By aggregating our judgement of each aspect's location, we can estimate our personal (and our culture's) overall closeness to viability. By incorporating into that judgement an estimation of the direction our views are moving in and the significance of each aspect to future change, we can predict how close we will be to viability in the future. This can be done intuitively. It can also be completed theoretically using formal models similar to those used for resource predictions, such as the footprint model. The difficulty with these models is reliably representing the non-material interactions between the various aspects and our decision making, as previously discussed.

Overall, it would seem that striving toward viability is a time-consuming but straightforward task, until we remember our decision-making foibles. They also bias and limit our judgements of where we sit on the flower. It is easy for us to incorrectly place our views on the flower and thus fail to make the changes that favour viability. Thus accommodating our decision-making foibles when making decisions is a critical part of favouring viability. This too takes time and effort.

In summary, the viability flower illustrates the human-environment system and its functioning from the perspective of an individual member of a culture (figure 22). Its petals represent the range in our individual views/opinions about the aspects of living in the system,

relative to the external reality. Additionally, because of the connection between us and our culture, it also represents the views making up our culture's world view.

On the inner side of the ring marking the boundary of the flower's centre area lie our views that match the external reality and thus favour viability. The further away from the centre area a view is located on a petal, the greater the degree to which the view differs from the external reality/viability: the greater its degree of divergence. The divergences in an individual's internal reality precondition them to experience cognitive dissonance or a reality paradox. Similarly, divergences in a complex culture's view help to precondition its members as a group to experience an equivalent experience or even resimplification.

If we, as a member of a complex culture, wish our culture to avoid the worst of a forced resimplification, then we must adopt a lifestyle and a personal world view that favours viability. The first step is to identify our critical divergences. We can do so by making judgements about where our personal views of an aspect are located on the flower relative to its centre (i.e., relative to the external reality/viability) and the direction they are changing. We must do the same for our culture's world view.

But, to make meaningful judgements about a view, we have to have an appreciation for viability's conditions, the external reality, the relevant context and the dynamics of the human-environment system. We also have to have looked at a view from many perspectives. Valid and useful judgements about a view's location and direction of movement on the flower are invariably conditional.

Once our critical (personal and cultural) divergences are identified, we can reduce them by adjusting our views of our self, ourselves and the world around us so that they more realistically represent the external reality. If enough individual members do so, then we can, collectively, reduce the divergences in our culture's world view. In doing so, we also embed the conditions for viability into both our personal and our culture's world view, where they can guide us toward viability over the long term.

However, our making of both the judgements and the changes needed to favour viability is more difficult than this summary suggests. This is illustrated by colouring the flower in a rainbow of colours. When we realize that its attention-grabbing beauty has distracted us from the figure's guidance about divergences, then we will be served a reminder. Favouring viability also requires us to address the nature of our decision making: its foibles and the influences on it. We must deal with a self-referencing problem.

The brightly coloured viability flower informs us that success at meeting viability's conditions requires us to personally adopt them as goals (i.e., as references for decision making). We are also obliged to consistently and coherently reduce our (personal and our culture's) divergences and accommodate our decision-making foibles, and do so in ways that favour viability. Success at becoming viable depends on it.

## Achieving Viability

If a culture's world view is to favour viability, then the members must each devote the time and effort needed to make the personal changes to their lifestyle and world view that will enable them to satisfy viability's conditions. They will then, collectively, change their culture's world view to favour viability. But their existing culture's world view influences their efforts to change themselves. Thus the member-culture feedback is involved.

This illustrates why our (personal and collective) striving toward viability is most likely to be successful if we treat it as a complex task that occurs within the human-environment system. This is the context within which this section provides an outline of both a path that the members (personally and collectively) can follow toward viability and the nature of their journey down it. It discusses, in general terms, the types of decisions and changes they must make and the challenges they will face doing so. Later sections will discuss the particulars. Western culture is used to illustrate the discussion. The use of the phrase *decisions and changes* to describe this task might seem odd or contain a redundancy, but it acts as a reminder of the feedbacks between the two and thus the complex nature of the task, as illustrated in figure 15.

**The Path.** The current path that the members of a culture prefer to follow into the future is outlined by their current culture's world view and sustained by its member-culture feedback loop. The trends discussed in previous chapters provide ample evidence that Western culture is travelling down a business-as-usual path that leads to us living well outside of at least viability's environmental constraints. We are following a path toward forced resimplification.

However, we could follow an alternative path: one that leads toward viability. In order to switch to that path, we would have to know what to change in our personal world view, in our culture's world view and in the member-culture feedback so that they favour viability. Think of this as needing to know what to change in the belief and resource loops of figure 15, and how to make them, so that the loops will favour viability.

We can gain that knowledge by considering one of Western culture's futuristic visions of how to solve at least our environmental issues: moving to another planet. In this vision, Westerners find and move to a planet with a complete, functional human-friendly environment and an ecosystem that has a vacant human-friendly niche waiting to be occupied. It is the perfect place for humans to live a viable life, a return to Eden, if you like.

But what happens after Westerners settle on Earth 2? Will we recreate the current state of Earth 1's environment there? Will we reach for the goals of the acclimatization societies' ecosystem improvers? The decorative gardeners? The GMO specialists? The believers in infinite economic growth? Will our new population and its resource demands explode, as happened on Earth 1 after European immigrants arrived in the Americas? Or will we live our day-to-day lives in a manner that keeps Earth 2's environment sufficiently intact so that it will be able to support us, and the rest of the world's population, over the long haul? Will we live in a viable manner?

If we are to ensure that we won't recreate, on Earth 2, the issues from which we are trying to escape by leaving Earth 1, then we would have to address the need to live in a viable manner before we move to settle there. That means adjusting our current (personal and cultural) relationship with Earth 1's environment so that it favours viability. In other words, we must adopt viability's conditions as goals and then reach for them.

As discussed, in overview this requires us to make decisions and changes that reduce our (personal and our cultural) divergences and accommodate our decision-making foibles, and we must do so in ways that favour viability. Some of these changes will be specific, such as changing our belief about the connection between goods and lasting happiness. Others will be broad, such as changing our relationship with complexity (including the scales over which it operates) and our views about our tools of economics, science and technology. These decisions and changes, and the challenges of making them, are substantial. Addressing them will take time and effort. Overall, a path that leads toward viability is one that involves us undertaking a personal and cultural transition. Is this realistic?

History provides us with many examples of cultural transitions. Hunter-gatherer cultures transitioned to farming cultures, as illustrated by the Abu Hureyrans in chapter 3. During the Middle Ages, the cultural world views of European kingdoms guided their members to think of their personal identity, membership and role in society as being largely fixed by the station they were born into, such as a lord or a serf. After the late Middle Ages (late 1300s), that aspect of European culture's world view started to change. With each passing century, the cultural guidance the members received about their society and their place in it changed just a little. That slow cultural transition resulted in Europeans today being guided by the world view of their nation states to see themselves as largely autonomous individuals, free to choose their identity, membership and role in society (Eisenstadt 2000).

Western culture could go through such an unplanned transition again, but this time toward viability. Indeed Westerners, with our culture's support, believe strongly that we will indeed muddle through to such a satisfactory future. But our vision of muddling through barely, if at all, includes the changes needed to actually become viable or what is involved in making them. Instead, we tend to treat our indicators of civilization, such as technology, economic growth and the notion of progress, as reliable measures of our culture's ability to become and remain viable. We believe that by following our culture's guidance to maintain these indicators, we will automatically secure success, as illustrated by the presence of our futuristic vision of dealing with our issues by going into space. In essence, Westerners believe that our culture is "too big to fail" and powerful enough to ensure a brighter future.

History suggests otherwise. It seems highly unlikely that Western culture will transition to viability by either muddling along or by chance, especially if we consider the time constraints on implementing the needed changes. If Western culture is to strive to become more viable, then it will have to choose a different path.

Against this background, the Tikopians' success at becoming a more viable culture, as discussed in chapter 17, stands out. Could Westerners emulate the Tikopians' successful transition to viability? To answer, we need to know the secret to the Tikopians' anomalous success.

The members apparently interpreted, with varying degrees of awareness and knowing, a number of their day-to-day experiences to mean that they were living outside of viability's constraints. And they responded to many of them in a manner that favoured viability. The Tikopians made enough significant viability-favouring adjustments to their personal world view (think lifestyle) that they came to live more viable lives. Over time, their personal adjustments became embedded into their culture's world view, changing it. Its guidance could then prompt the members, and their descendants, to continue making decisions that favoured viability: they had created a viability-favouring feedback loop. The Tikopians may not have recognized that their decisions had reduced their personal and their culture's divergences, adopted viability's conditions as goals and accommodated our decision-making foibles in ways that favoured viability. But they had. They had met the conditions for viability laid out above.

Westerners could emulate the Tikopians by choosing to follow an equivalent path to viability. However, by making that choice, we would be agreeing to do more than emulate the Tikopians' efforts. We would be accepting responsibility for completing all of the needed decisions and changes with full awareness. We would be agreeing to consciously guide ourselves and our much more complex culture through a transition toward viability.

To ensure that we wouldn't get misdirected off the path to viability, we would, like the Tikopians, have to form a viability-favouring feedback loop. But, again, we would have to do so consciously. Fortunately, rather than creating a new one, we could, in keeping with the nature of cultural change, adjust an existing feedback loop. In this case, we could adjust one that is hindering us from becoming more viable and thus taking us down a path to forced resimplification. We could convert our complexity-resource loop into a viability-favouring feedback loop.

Although a difficult task, Westerners could, in theory, meet all of these conditions. There is a path to viability for Westerners to follow: consciously emulate Tikopia's transition. Now we need to find out what it takes to journey down it.

**The Journey Down the Path.** If Western culture is to beat the historical odds and complete a Tikopian-style transition to viability, then it is up to us, its members, to consciously make the decisions and complete the (specific and broad) changes required to journey down that path. Think of this task as us devoting the time and effort needed to adjust the belief and resource loops in figure 15 to favour viability. An overview of what this means is provided by discussing our need to remember viability's conditions, make personal changes and change how we interact with our surroundings.

*Remembering Viability's Conditions.* If a culture is to become and remain functional over the long term, then its members must remember and satisfy viability's many conditions/constraints. Think of the Mande. These conditions have been collectively referred to in this book as either the human-environment system's (human and environmental) constraints on how we can live, or the human and environmental conditions favouring viable living. They are also implied when using the word *viable* as in "a viable culture." By aggregating viability's conditions/constraints in this way, we can easily remind ourselves that they exist. But, unfortunately, we can only satisfy them if we remember the actual conditions/constraints: the details. Fortunately, the aggregation method can be adapted to help us remember smaller selections of the actual conditions/constraints that are relevant to addressing a particular issue in a viability-favouring manner. First, though, a reminder of the specific conditions/constraints the above overarching aggregates stand for.

Somewhat idealized examples of viability's human and environmental conditions/constraints relevant to each of the human-environment system's three primary factors (individual, culture and environment) are provided. An individual should be raised in a social and physical environment in which they experience few adverse childhood experiences (ACE), eat a healthy diet and engage in sufficient exercise (Felitti 2009; Felitti et al. 1998). An individual's living conditions should also provide them with the feelings of belonging, being respected and being treated fairly. The interconnections between these types of human conditions is provided, in overview, by the Dunedin Longitudinal Study and related analysis (Dunedin Longitudinal Study 2018; Poulton et al. 2015).

At the same time, a culture's world view should guide its members to favour being trustworthy, fair and cooperative as the default. It should also provide guidance about when to engage in self-interest, competition and anger, and how to do so in ways that favour viability. The culture's world view should inform the members of their responsibilities as well as their rights and freedoms. It should also restrain the power of the powerful.

Both the members' upbringing and their culture's world view should guide them to respect the functioning of their culture's environmental endowment. After all, it provides them with the resources and services that support them today, and they expect it to provide those resources in the future. To meet this expectation, the environment's long-term ability to self-manage its functioning must remain intact: an ecosystem's resilience, robustness and ability to self-adapt must be maintained, as detailed in earlier chapters.

The nature of the external reality can be expressed as rules of thumb. They have been used in this book as aids to help us remember the nature of the environment's structure and functioning and thus viability's environmental conditions/constraints. They are also used to express how we could relate to the environment in ways that favour us meeting those conditions. The equivalent rules can be used to help us remember and meet viability's human conditions/constraints, if we want to live a satisfying life that also contributes to our culture meeting viability's conditions.

These are examples of the rules regarding children. The parents should limit the number of children they raise. They should choose a lifestyle that neither detracts from their child's/children's basic material and psychological needs nor guides/preconditions their children to demand more resources from their culture's environmental endowment than it can continuously supply to them over the long term. Parents should strive to raise their offspring to appropriately fill their roles as members of a human community. This includes, in particular, encouraging them to continue their parents' long-term striving to live viable lives. More specific conditions/constraints and rules of thumb can also be found in the scientific literature.

There are rules of thumb that deal with the complexity of the human-environment system's functioning. These examples focus on the conditions/constraints for viability that can be easily missed. Consider that for the human species or a culture to be viable, it must be maintained in a functional state over the long term. This implies that we can know what the long-term future looks like, in advance, in sufficient detail to accommodate it in a timely manner. But we can't. The nature of a complex system's functioning ensures that there is always significant uncertainty in a prediction about the future, especially over the long term.

Here are some rules about dealing with this reality that also link the human and environmental constraints on viability together. The members must form the ability to self-adapt to the unexpected/unpredictable in ways that also favour viability. And they are obliged to live in ways and make decisions that routinely maintain and exercise that ability. They are also required to satisfy a complementary rule: to avoid making choices that drive their culture or its supporting environment toward their functional (stability) limits. To be effective, these rules have to be embedded into their culture's world view. In practical terms, these rules require the members to maintain a flexibility and focus in their everyday living and decision making that favours viability. For example, they should routinely add a viability-favouring "safety" margin to their decisions and establish ecosystem buffers for trying times. Think Mande and Tikopia.

In overview, in order to routinely address issues in a way that favours viability, we need to remember its conditions/constraints, select the relevant ones and satisfy them. We are helped to do so by treating the conditions/constraints and our generic viability-favouring responses as rules of thumb or as aggregates of them. We certainly need to embed the conditions/constraints into our culture's world view. By routinely living and making decisions that satisfy viability's conditions in a comprehensive, consistent and coherent manner, we favour viability.

*Making Personal Changes.* A key to successfully meeting viability's conditions is to consciously make the changes to our self and ourselves that facilitate that outcome, as discussed above. It is easier for us to make the needed changes if we apply our understanding of the process by which we change, gained in chapter 6.

When a particular personal change is viewed within the context of the complex, dynamic human-environment system given in figure 15, then it is seen to be part of a dynamic, complex process. Both that process and the change can be described using a mix of the generic terms applicable to complex systems, such as *inertia, drift, preconditioned, imposed* and *directed.* Making the personal changes needed to favour viability seem straightforward. However, when a change is viewed from the individual's perspective, as illustrated in figure 22, a complementary picture emerges. Although we have the capacity to make the change, complexity ensures that making it is more difficult than it appears and that there are many more changes to make than expected.

Consider that we can have trouble recognizing the need to change. For example, it is easier for us to accept the concept of viability than it is to accept that both *our*

personal and *our* culture's world view may not actually be viable. Consider too that there are also limits to those aspects of ourselves that can be changed, how they can be changed and how quickly they can be changed. We can't expect to change our basic human characteristics or the essence of how we or our cultures function. We can't alter the processes by which our internal reality and our cultures are created. We are stuck with our decision-making foibles and the influences on our decision making from our internal reality, our culture's world view and our environment's functioning. In the same vein, we are stuck with our general inertia to making personal changes, especially to our core. For example, we hold the primary features of our personal world view tightly to our chest, so they only change slowly.

With all of this in mind, we realize that the process of personal change is far more convoluted and wide-ranging (complex) than we imagine. For example, we believe that we can change a specific aspect of ourselves independently from the rest of ourselves and our surroundings. In reality, many of our habits and beliefs are far more closely tied to one another, to our past and to our culture than we think (Inglehart 1990).

Overall, consciously making a significant personal change is less about reaching for a fixed, isolated goal, such as finding lost keys (you find them or you don't). It is more like the broad, ongoing act of becoming, such as striving for happiness. We would therefore be wise to accept that the process of making the personal changes needed for our decisions to favour viability is more dynamic and complex than we think. And it takes more time and effort than we expect.

By accepting this, it becomes easier for us to recognize that our conscious efforts to make a personal change tend to focus on those adjustments that are immediately apparent to us, are easy to implement and rapidly produce the results we prefer. Understandably, we favour making a small adjustment to weakly held beliefs and behaviours on the periphery of our identity. We also tend to favour adjustments that are culturally acceptable. We particularly favour those adjustments that allow us to convince ourselves that we are adequately addressing the issue bothering us (i.e., the changes enable us to feel comfortable with ourselves), whether they are meaningfully addressing it or not.

As a result, when a feeling of cognitive dissonance or new-to-us knowledge prompts us to change, our immediate response tends to be one that allows us to avoid the need to change, such as rationalizing the feeling/knowledge away (Hock 1999c; Grube et al.

1994). For instance, as our personal environmental awareness increases, we may realize that our personal $CO_2$ emissions and our demand for resources are contributing more to environmental disruption and destruction than we thought, which makes us feel uncomfortable. But instead of making the personal adjustments needed to reduce our contributions, we find it easier to rationalize them. We can, for example, either justify to ourselves that our personal contributions are inconsequential, that the reason for a particular contribution is too important for us to forego it or that others are more to blame. We can also defer or delay making conscious changes on the grounds that there is no time to devote to them. After all, our busy lives are already filled, even dominated, by immediate obligations to our jobs, our family, our friends, our community and to answering electronic communications.

These characteristics are one reason why making a *significant* personal change usually occurs slowly, often through a series of steps (think drift and preconditioning) with the first one usually being, as just mentioned, small and easy. Fortunately, we can use this process to our advantage when deliberately making a significant or more difficult change. We can divide the change into easier-to-achieve intermediate goals. By sequentially making the smaller, directed adjustments to reach each intermediate goal, we can, with time and persistence, complete the larger change.

We are also helped in our efforts to consciously make personal changes that favour viability by remembering our decision-making foibles and the influences on them: how we make our decisions to change. Here is a reminder from the discussion in chapters 1 and 2. Our decision making can, for convenience, be divided into an unconscious and a conscious part. The unconscious part, the majority, is automatic, fast and inescapable. It usually produces certain (firm) but poorly considered decisions. The minority part, our familiar conscious (cognitive) logical and rational decision making, is slower and largely discretionary. It produces more considered decisions. But the conscious part is built on the unconscious part. We are reminded of this complex, unitary nature of our actual decision making by the swirl of feelings we experience when trying to resolve the conundrum of urges and will.

Our (conscious and unconscious) decision making serves us well to a point. But we should remember that the majority of our decision-making processes display characteristics that are unlike what we imagine. For example, our brain makes almost no use of formal statistics, but it does use rules of thumb (heuristics) that rely on our emotions and memories. Our brain's processes are also influenced by our awareness and attention, which we inherently use in a highly focused manner.

Consider too that our brain makes a decision by applying its processing abilities to what we can sense, to our memories (facts and feelings) and to our references. All of them suffer from deficiencies in their representation of the external reality. For example, our memories can change. Overall, the way we make decisions means that our decisions tend to be biased and incomplete. They tend to maintain our (historically established) self and favour our current concerns/desires.

We can't change our decision-making foibles, so we can't eliminate the appearance of biases and the differences between our personal internal reality and the greater external reality (divergences) that give rise to our cognitive dissonance and reality paradox experiences. And, the more we rely on our decision making to provide us with reliable decisions to guide personal change, the greater the burden we place on our conscious decision making to deal with our decision-making foibles and divergences. This is a significant burden. After all, it is always more difficult to draw a conclusion if we try to recognize and accommodate the biases and limitations in the available information, our decision-making process and the influences on it. We tend to avoid this burden by simply accepting the decision that comes into consciousness and then rationalize away any differences that would require effort to resolve. We display this tendency in many ways, such as tending to be overconfident in our conclusions.

Fortunately, by gaining knowledge about our decision-making foibles, we ease the burden of consciously dealing with them and the resulting divergences. In addition, we do have free will, so even though our ability to exercise it is constrained by our human characteristics, we do have some control over how we gather, interpret and use information. We can, for example, redirect our thinking and decision making so that it takes our decision-making foibles and the influences on them into account. And we can check our decisions, as discussed in chapters 4 and 5.

We are therefore able to take the context in which we view information into account. We can become more aware of how we and the complex human-environment system functions. We can recognize and reduce our divergences and limit their influence on our decisions.

We can take the process of personal change into account. It is therefore possible for us to make and implement the decisions that will change us so that we live in a manner that favours viability. It just takes time and effort.

In overview, our efforts to make the personal changes needed to favour viability are significantly constrained by how we make decisions, by how we change and by our culture's guidance. Fortunately, we can each gain the knowledge, experience and resolve required to make the decisions and personal changes needed for us to adopt viability's conditions as goals. We can change in ways that reduce our personal divergences and accommodate our decision-making foibles in a viability-favouring manner. By the members of a culture striving to make these changes, they can collectively change their culture's world view so that it provides guidance that favours viability. Thus their efforts to make these personal changes can tend, over time, to become easier and self-sustaining. Westerners can indeed make the changes needed to go through a personal transition to live a more viable life. The details are provided in later sections.

*Interacting with Our Surroundings.* We will make the majority of the decisions and changes needed to favour viability during our everyday living as we plan and implement our interactions with our surroundings. In order to favour viability, we will have to make both broader (foundational) and specific personal changes to those interactions. The nature of the broader changes is illustrated here.

Consider that Westerners' interactions with our surroundings often have unsatisfactory outcomes, despite our laudable efforts to do the right thing. For example, working conscientiously toward lowering a country's infant mortality rate hastens an increase in its population size, unless we simultaneously work toward lowering its above-replacement fertility rate. An increasing population results in an increasing demand for food and water that will more quickly exceed the capacity of the culture's environmental endowment to support all of its citizens. Over the long term, this outcome will negatively affect both the infants in the growing population and the state of the environment expected to comfortably support all of the country's more numerous citizens into the future. This long-term outcome runs counter to the stated aim of lowering child mortality. Short-term actions that don't realistically consider the long-term outcomes are unlikely to satisfy viability's conditions.

An equivalent environmental example is our effort to maintain wild pollinator populations by actions that include the nurturing of exotic flowering plants and/or pollinators, or require our long-term management of either indigenous ecosystems or pollinator lifeways. These actions may succeed in the short term, but in the long term they are counterproductive. They result in further disruption of the indigenous ecosystems and/or burden us with (short- or long-term) responsibilities we are unable to fulfill or will neglect. Engaging in these actions simply shifts or worsens the pollinator problem.

One reason for these outcomes is the absence of effective and meaningful coordination, guidance or oversight to ensure that Westerners' efforts to address an issue will indeed resolve it in a manner that, over the long term, is consistent, coherent and favours viability. Another reason is the perceived need to simultaneously reach for both the intended goal and satisfy the imperatives of the Western world view, even if the imperative doesn't favour viability. For example, our intent to protect the functioning of our environment through impact assessments and safety testing are often limited, distracted, misdirected and even subverted by the imperative to maintain continuous economic growth. Thus, despite the best of intentions, our efforts to address issues often incompletely reflect or contradict even the most basic of viability's constraints. Alternatively, our efforts can work at cross purposes to one another or simply transform rather than resolve an issue.

Some aspects of our interactions with our surroundings that don't favour viability can be dealt with by us making specific decisions and changes, such as reducing a particular divergence. Others, such as the lack of viability-favouring awareness, coordination, guidance or oversight described above, require us to make broader changes to our personal and/or our culture's world view. For example, this means making changes to our concepts, values, beliefs and goals, in general, and our relationship with complexity and our culture's tools, in particular. Westerners can think of what the words *adapt* and *sustainable* mean to us. These broad personal and cultural changes to our interactions with our surroundings are challenging to make and occur slowly.

In summary, the members of a complex culture could avoid the worst of a likely forced resimplification by undertaking a voluntarily resimplification. They start this process by deliberately (consciously) striving to live in a manner that emulates the essence of the Tikopians' cultural transition to viability. This will put them on a path that their descendants are more likely to successfully follow to viability.

In order to choose and follow that path, the members must make decisions and changes that favour viability. In particular, they will have to adopt viability's conditions as goals (references). They will also have to reduce their personal divergences and accommodate their decision-making foibles in a manner that favours viability.

The members will face challenges on their journey. There are the difficulties remembering viability's conditions and satisfying them coherently and consistently. The members' efforts to favour viability will occur as part of their day-to-day interactions with their surroundings. This means that their decisions and changes must deal with both specific and broad aspects, and involve both the members as individuals and as the carriers of their culture's world view. For example, they must deal with the dynamic complexity of personal change and the complexity of the human-environment system in which all of the pieces relevant to viability are linked to one another by feedbacks.

We can address these challenges by gaining the knowledge and practising the three methods needed to make viability-favouring decisions and changes. For example, we can become familiar with how we make decisions and change, and we can more easily satisfy the conditions for viability by aggregating them into objectives that we can aim for. These methods and their application will be discussed in more detail below.

By making the personal decisions and changes needed for us to live in ways that favour viability, we help our self and ourselves find the path to viability. In the process, we each contribute to changing our culture's world view and reducing its divergences so that it too guides us to find and then stay on the path to viability. We thus contribute to a viability-favouring feedback loop. In this way, we will help ourselves, our culture and our descendant to, over the long term, complete the transition to viability.

There *is* a realistic possibility that the members of Western industrial culture can consciously emulate the essence of the Tikopians' transition to viability. Although a challenging journey, it is easier, cheaper and more likely to be successful than moving cold turkey to Earth 2. How we can actually make the needed changes is discussed in the next three sections.

## Environmental Objectives (The Five-Objective Unit)

If Westerners want our complex culture to journey down a path to viability, then we, like all of the more complex cultures, are obliged to satisfy viability's human and environmental conditions. To simplify the discussion, this section will focus on the environmental conditions. These were chosen for two reasons. The environment is sufficiently connected to us that we can relate personally to its conditions. But, because our relationship with it is sufficiently impersonal, we can, relatively speaking, more easily make the decision to satisfy those conditions.

We are helped to satisfy those conditions if they are represented by practical objectives that we can relate to, acknowledge and strive toward. Five related and interconnected practical environmental objectives and the challenges of reaching for them are discussed. Think of these objectives in aggregate (the five-objective unit) as representing the features of the environmental factor in figure 15 that constrain viability. The five-objective unit represents the environmental conditions for viability.

**Accept Our Environment's Supremacy.** We each depend on our environment to provide us with food and water. The functioning of a more complex culture, such as an industrial one, depends on an ongoing supply of these and other resources. The members of a complex culture strive to satisfy all of its resource wants but their efforts are routinely hindered by what they perceive to be environment-related barriers. The members are guided by their culture's world view to overcome these barriers by applying their culture's tools.

However, when these barriers are seen from the environment's perspective, they are indicators or reminders to the members that the functioning of the environment places constraints on their lives. For example, a resource shortage is a reminder that there is a limit to the combination of the culture's population size and its average per capita resource use (its average lifestyle) that the environment's functioning can support over the long term (Cohen 1995). Although the members might think that, by applying their tools, they remove or overcome such a barrier, the complexity of the human-environment system ensures that they only change the expression of the environment's constraints on their lives. For example, when our ancestors took up agriculture, many of the environmental constraints on the availability of wild-food supplies were replaced with the constraints of being personally responsible for the functioning of their food-producing agricultural ecosystems.

Similarly, the sequential application of tools to overcome successive environmental barriers merely changes the expression of its constraints and reveals new ones. This was illustrated in the previous chapter by the use of coal to generate electricity. In general, each application of a culture's tools to overcome a barrier tends to disrupt the environment's functioning and burden the members with more environmental management responsibilities, thus adding to their culture's complexity (Costanza et al. 1997). We are limited in our ability to discharge those responsibilities in ways that favour viability.

It is possible for the members of a complex culture to both limit their environmental impacts and reduce their environmental management responsibilities: they can accept the environment's supremacy. But, if this solution is to be lasting, then they must do more than accept this reality in theory. The members are obliged to change their (personal and their culture's) relationship with the environment and its functioning (Rees 2003; Gardner and Stern 1996). This means, for example, switching from seeing the environment's features as isolated barriers to seeing them as reminders of its ultimate supremacy, and then acting accordingly.

The members of the more complex cultures experience great difficulty making these types of changes because its brings them into conflict with their culture's world view, which, at its most basic, guides them to modify the land. This cultural difficulty is illustrated by Westerners use of the word *adapt*. Westerners usually use it to express our intention to deal with emerging issues, particularly environmental and economic ones, by changing our tool use in ways that will allow us to continue both living our current lifestyle and reaching toward our culture's current objectives. For example, "adapting to a water shortage" means finding ways to make sufficient water available to maintain our business-as-usual water use. This meaning of *adapt* favours shorter-term goals, supports our culture's complexity-resource feedback loop and preferentially treats our environment's constraints as barriers to be overcome.

Although Westerners may accept the supremacy of our environment and its resulting constraints on our lives in theory, we rarely do so in practice. If we did, then *adapt* would indicate our intent to adjust our lifestyle and limit our tool use to live more within viability's environmental constraints (Ames and Maschner 2000). For example, we could adapt to a water shortage by voluntarily adjusting our lifestyle so that our demand for water better matches the environmental constraints on our water supply. This meaning of *adapt* favours longer-term goals, champions viability and acknowledges our environment's supremacy. Next time the word *adapt* is used, ask which meaning the user intends. The same applies to the word *sustainable*.

If Westerners wanted to accept the environment's supremacy (in the practical sense), then we would have to change both our personal and our culture's world view so that their guidance reflected that relationship (Gardner and Stern 1996; Oskamp 2000). Consider that these world views would have to guide us to find less of our sense of satisfaction, belonging and purpose in imposing a human sense of order or control on our environmental endowment. Instead, we would be encouraged to find more of our desired sense of meaning in the environment's innate order and self-management abilities. For example, less of our sense of meaning would come from a limited number of tree species regularly laid out in a tree plantation and more from that patchy distribution of more species in the original forest. Methods to help us make these changes will be discussed in a later section.

History indicates that if we fail to meaningfully accept our environment's supremacy, it will eventually impose its will upon us. A more complex culture will experience that future as a forced resimplification (Goodland and Daly 1998; Gardner and Stern 1996; Rees 1995). The Tikopians showed that it is possible to accept the environment's superiority, adjust to it and still live meaningful, satisfying lives.

**Maintain Our Environment's Functioning.** The environment is a dynamic, complex system that functions by its many variables interacting with one another in a variety of ways, including feedback, over a wide range of time, space and strengths. Its functioning results in the environment existing in a particular quasi-stable state. That functioning and state provides the living conditions for the earth's species.

We are one of those species and one of the environment's variables. Our effort to complete our lifeway, like those of all species, is facilitated by the environment's functioning and is subject to its constraints. Our efforts also contribute to and change its functioning. But there are limits to the amount and speed of that change, if the environment's functioning is to provide the living conditions we and the other species require. We are therefore obliged to ensure that how we live maintains the environment's functioning (and thus its state) within those limits.

This reality is imperfectly represented in the world views of the more complex cultures, and thus in their members' personal world view. These views tend to provide guidance to the members about the environment and their relationship with it that is too simplistic, relative to our ability to disrupt its functioning, especially if our decisions are to ensure that its functioning is maintained over the long term. Of particular note is the standard guidance that favours the members' efforts to transform the environment, and the high degree of the resulting changes to it that are considered acceptable. As a result, it is easy for the members of the more complex cultures to make and implement decisions that significantly disrupt the environment's functioning. This is the essence of our ongoing, long-term environment-related contribution to our complex cultures' possible resimplification.

We would find it easier to make decisions that maintained the environment's functioning if we paid more attention to the environmental principles and rules of thumb that are found throughout this book, for example, in chapters 7 and 8. By considering them, we are more likely to look at our circumstance from a number of environmentally relevant perspectives, as well as from different scales of time, space and impact intensities. We are also more likely to apply our resulting expanded knowledge when making decisions and evaluating their possible environmental effects (Folke et al. 1994). We are thus less likely to make our environment-affecting decisions based largely on short-term, narrow political, economic or personal imperatives and beliefs. For example, our interactions with an ecosystem are more likely to include efforts to retain its soil volume and functioning; retain its water purity and flow; protect its integrity and ability to self-adapt; and reduce the speed of human-induced climate change (i.e., maintain its functioning), than maximize production from it. We are then more likely to live off the interest from our natural capital than the capital itself, as we are today (Rees and Wackernagel 1994). Overall, our decisions are more likely to maintain the environment's functioning, and thus its state, in a condition that will support us over the long term, which favours viability.

The challenges that complex cultures will face when reaching for the objective of "maintaining the environment's functioning" is illustrated by Westerners' culturally supported preferred interpretation of what that phrase means. When seen within the context of our culturally perceived rights of dominion and stewardship over the environment, and our right to improve it (as discussed in chapter 11), it is unsurprising that the phrase is treated as guidance to become hands-on managers of the environment's functioning. By adopting the Western meaning, we take on the responsibility of managing those of the environment's existing functions we depend on, whether they are disrupted or not. We also become responsible for supplying the resources (including services) previously provided by functions we have too severely disrupted. But we neither easily recognize most of our resulting environmental management responsibilities nor tend to discharge those we do recognize in a viability-favouring manner, as concluded in chapter 10.

Examples include the highly disrupted state of most ocean fisheries. We have barely accepted our responsibilities as ocean-fisheries managers, let alone fulfilled them. We even struggle to recognize and appropriately fulfill our environmental responsibilities to our human-created intense agricultural ecosystems. Examples are pollinator declines and changes to soil functioning, such as erosion and excess nitrogen. Think too of climate warming, exotics and decorative gardens.

As hinted at when discussing the environment's supremacy, if the members of the more complex cultures are to strive toward the intended meaning of "maintaining the environment's functioning", they face the challenge of reducing our disruption of its functioning. Success requires us to severely curtail our manipulation of the environment and our overwhelming reliance on that ability to secure a viable future. In practical terms that means reducing our faith in our hands-on environmental management skills and reflecting that change in our decisions and activities. For Westerners this challenge is encapsulated, symbolized perhaps, in the need to change the meaning we currently give to *adapt*.

### Champion Our Environment's Self-Management.
Something that favours us meeting viabiliy's environmental conditions must substitute for a reduction in our faith in our environmental management abilities.

Humans, but especially the members of Western industrial cultures, have a powerful urge to feed wildlife, particularly birds (Galbraith et al. 2015). Observations and experiments have established that when humans provide supplementary food to a wild vertebrate, it has a better quality of life. But what are the long-term and collateral effects of that feeding?

One of the studies to address this question looked at song sparrow pairs, some of which received supplementary food from humans. As expected, providing extra food resulted in a sparrow pair producing more eggs, an outcome we judge as positive. However, the pair's balancing of the many variables involved in laying and raising this larger clutch of eggs had an unexpected consequence. The pair's male offspring had a smaller song repertoire.

From the male offspring's perspective, this is a less than positive outcome. Female sparrows use the male's song repertoire to assess his reproductive quality (i.e., how desirable a mate is he). Those males with a smaller song repertoire experience lower breeding success. Thus, although human feeding helps song sparrow parents achieve an increase in the number of offspring, it hinders the future breeding success of the next and following generations of the pair's male descendants (Zanette et al. 2009). That is the cost of the pleasure we gain from feeding them.

It took a specialist's study to make us aware that a human activity as simple as putting out extra food (at least for sparrows) can disrupt our environment's functioning. It illustrates that our incomplete and biased personal view of our environment's self-management abilities allows us to, despite our best intentions, routinely make decisions and act in ways that disrupt rather than meet the objective of maintaining our environment's functioning. The birdfeeder is a potent symbol of that reality (Daszak et al. 2000).

Westerners can counter our disruptive tendencies by striving to become champions of the environment's self-management abilities. This requires us to adjust our broader view of our relationship with our environment. The nature of those adjustments is revealed by considering today's general talk of us "needing to stabilize" the planet before it has changed so much that it switches states or can no longer support us. Instead, we could be discussing how we can adjust our beliefs and behaviours to "empower the planet to stabilize itself" within the desired stability range.

For this to happen, we must turn our belief in our ability to manage the environment's functioning on its head. One way to address this task is to acknowledge that our ability to fully identify the degree to which our activities will disrupt or destroy our complex environment's functioning in the medium to long term is limited. Similarly, despite the claims of environmental

mitigation and restoration proponents to the contrary, our ability to retroactively mitigate, restore or replace those aspects of the environment that we do recognize we have disrupted or destroyed is also severely limited. This means, in practical terms, that we should be questioning many of our casual or routine disruptions of the environment's functioning. And we should no longer consider our formal reactive environmental management methods to be either a preferred, or even an adequate, response to our impacts on the environment.

By completing these types of changes we can become proactive champions of our environment's self-management abilities. We can become, if you like, hands-off managers who use our managerial skill to facilitate/protect the environment's ability to self-maintain its functioning and self-adapting to change. In this role, we are more likely to favour decisions that limit our initial disruption and destruction of the environment. We thus reduce our need to acquire new hands-on management responsibilities. We also prompt ourselves to reduce our current management responsibilities by transferring them back to a more skilled manager: a functional environment.

**Conserve What We Have.** Many aspects of our biosphere face imminent, continuing or even accelerating human disruption and destruction. Environmental conservation is a long-term savings plan that aims to secure for posterity what remains of the environment's undisrupted self-management abilities. For an ecosystem, this means conserving its inherent robustness, resilience and adaptability.

For the members of Western culture to undertake meaningful conservation, we must resolve a conundrum. Consider Westerners' efforts to conserve an ecosystem. In order for those efforts to be meaningful, we should reduce or stop activities that disrupt or destroy the ecosystem. Those undesirable activities include continuing or increasing environmentally inappropriate access, maximizing future resource yields and allowing exploitation rates to exceed renewal rates (Begon et al. 1996). However, if Westerners are to easily approve of a particular conservation effort, then it must also be consistent with the activities the members consider to be acceptable within the ecosystem. More broadly, our conservation efforts are obliged to fit with our culturally supported view of our relationship with the environment. In the case of Western culture, its general guidance about that relationship favours us undertaking activities that exploit or transform ecosystems in order to maintain resource supplies and progress, or even to recreate. There are thus many opportunities for Westerners' conservation efforts to founder on the rocks of "acceptable" versus "meaningful" conservation.

Western culture conducts its formal attempts to resolve its "acceptable" versus "meaningful" conservation conundrum in an orderly fashion. We evaluate the human benefits and the environmental impacts we expect to result from undertaking a specific environment-affecting activity and then reconcile the two. The preferred techniques for doing so are variants of cost-benefit analysis, such as an environmental impact assessment. These techniques have many built-in biases that favour humans.

Consider that the choice of the activity, the identification of its impacts, the estimates of the benefits and costs, and the process of reconciling them all tend to be strongly biased by our culturally influenced, human-focused perceptions and decisions (Regal 1996). An expression of this human-favouring bias is Westerners' efforts to assign an economic (cash) value to all aspects of our environment, from its species to its functioning, as discussed in chapter 12. Consider too that behind a cost-benefit analysis lies a culture's world view that, for Western culture, places a higher cultural value on economic progress (growth) than on the environment's functioning. It also displays a greater preference for mitigating adverse impacts compared to avoiding them. Thus our evaluation and reconciliation process tends to favour us satisfying our culturally supported desires rather than us trying to live within the environment's constraints. The outcome, as discussed in earlier chapters, is that we usually end up doing what we want, now, while the environment experiences significant long-term disruption and destruction, which we pay for later.

The greater challenge that industrial cultures in general experience when trying to evaluate and reconcile human benefits with related environmental impacts is epitomized by the term *sustainable development*. In the Brundtland report, sustainable development was put forward as a reference standard or a goal (United Nations 1987). Because the functioning of our environment is a foundation of our culture's functioning, it is reasonable to expect that the notion of sustainable development would be based on ensuring environmental protection and conservation. This is not the case.

The Brundtland report's definition and discussion of sustainable development allows, with a focus on selected details, for the term to be used to express and justify a higher preference for sustaining some level of human-chosen consumption or growth (Wackernagel and Rees 1996). As a result, sustainability has itself taken on meanings that are contrary to its viability-favouring roots. Think of the sustainable yield for fish in chapter 10. For example, when economic assumptions form the primary basis for claiming that a decision favours sustainable development, then the term is, environmentally, nonsensical. *Sustainable development* has even been used to imply the absurd notion that striving for continual economic growth is compatible with respecting environmental constraints and meeting the conditions for both environmental and social viability. Hence the term *sustainable growth*. Similarly, *sustainable mining* is meaningless because the process deals with non-renewable (finite) resources.

This situation has been strengthened by the United Nations' selection of 169 targets to help achieve sustainable development. Although many of the individual targets favour the environmental objectives discussed here (and viability), there are also targets that favour a contrary outcome. This contradiction reflects an interpretation of the Brundtland report that is biased by the imperatives embedded in a complex culture's world view, such as favouring economic growth (United Nations 2015). Part of meaningfully reaching for the conservation objective requires adjusting, discounting or abandoning those of the United Nations' sustainable development goals and associated indicators that are not compatible with conservation, such as economic growth/development in its current form. Inevitably, this means the proponents of sustainable development must resolve its built-in contradictions that ensure it cannot reach its own overall goal, or simultaneously satisfy viability's personal and cultural conditions. *Sustainable development* should be superseded by *viability,* and *sustainable development goals* should be displaced by *viability's conditions.*

Would the deficiencies in the cost-benefit methods be addressed in a manner that favours viability if it was turned into a responsibility-benefit analysis? After all, a responsibility is more than a cost to be paid. In this discussion about the environment, a responsibility is a usually ongoing, inescapable (material and mental) obligation or duty we acquire when the benefits we gain from exploiting our environment result in us disrupting or destroying its functioning.

Adopting a responsibility-benefit analysis could help us to resolve conservation's acceptable-meaningful conundrum by reconciling the two in a way that recognizes viability. It would redirect our attention away from costs, which we can treat as externalities, toward responsibilities, which we can't. In doing so, the responsibility-benefit analysis reminds us that our difficulty resolving the conundrum has less to do with problems of materially balancing costs with benefits and more to do with our perception of our environment and our (mental and material) relationship with it.

More specifically, a responsibility-benefit analysis reminds us of the limitations to our personal awareness of both the environment's functioning and our relationship to it. It prompts us to pay attention to the roots of this deficiency: our internal reality, with its divergences; our decision-making foibles; and our culture's world view, with its divergences. For example, we think that the opposite of environmental disruption and destruction is their absence. A more appropriate opposite is the presence of a greater appreciation for our environment and our dependence on it: a greater respect and understanding based on a personalized connection to and awareness of it. Perhaps it could be called a hands-off respect. Compare shooting an animal or picking a flower to photographing or just admiring them.

Overall, a responsibility-benefit analysis would, by pointing out our limited ability to fulfill our environmental management responsibilities, discourage us from unnecessarily taking on those responsibilities. For example, can a mining company realistically fulfill their responsibility to monitor and maintain the integrity of a tailings dam for 300 years? A responsibility-benefit analysis would instead encourage us to pay attention to the environment's features, such as its self-management abilities, as well as the non-material values and benefits people attach to aspects of their local environment. It would encourage us to conserve what we have. The associated reduction in our disruption of our environment favours the continued availability of its resources and services over the long term (Folke et al. 1994).

Our conservation efforts will help us reach for the other environmental objectives because those efforts will favour us championing environmental self-management, which will reduce our disruption of our environment. This will help us to maintain our environment's functioning. The more successful we are at routinely maintaining its functioning, the easier it is to ensure environmental self-management and engage in ecosystem conservation. This completes a self-sustaining feedback loop favouring us meeting viability's environmental conditions.

**Reduce Our Total Resource Demand.** Our environment provides the resources (including services) that allow our culture to function: it supports us and our culture's world view. If a culture is to be viable, then the environment must provide those resources over the long term. But, for at least the more complex cultures, the ability of their supporting environment to do so has been negatively affected by the amounts being demanded and the manner in which their members supply, use and discard them. Complex cultures do recognize this situation. They tend to believe that they can maintain the supply of all resources and protect the environment's functioning by increasing their hands-on environmental management responsibilities. However, this belief is not supported by the outcome from their past management efforts: the current state of our already human-dominated planet.

An alternative approach is to strive toward the four objectives discussed so far. This would protect the environment's resource-providing functions, but it might also mean that the environment can no longer provide the amount and types of resources the more complex cultures currently want. To address this issue, a fifth environmental objective is added: reduce our total demand for resources (including services). From the environment's perspective, it aims to reduce our resource demands to a level that is compatible with the ability of the environment's functioning to provide resources over the long term. From a culture's perspective, it means reducing the types and amounts of resources being demanded to a level that enables the culture to satisfy viability's environmental conditions.

The fifth objective thus favours both protecting the environment and reaching for cultural viability, thus providing a link between the two. The required substantial reduction in the amount of resources can be achieved by a combination of a declining global population and a declining global average per capita demand for resources. The proportion of each varies across the globe because some industrial cultures have relatively small populations, but relatively large per capita resource demands, while others have the reverse.

A decline in the size of the global population will occur if couples voluntarily reduce the number of children they raise. This will happen when couples are willing, ready and able to control their fertility, especially if their choice is supported or encouraged by their culture's world view (Cohen 1995). However, biological reality decrees that the majority of the decline in the size of a population resulting from such a birth reduction process will take many decades to appear. This biological inertia is so strong that, even if an involuntary, catastrophic decline in the size of the global population occurred today, and even if it was followed by voluntary fertility controls targeted at replacement levels, the full decline would likely only be felt in the medium to long term (Bradshaw and Brook 2014).

A decline in our average global per capita demand for resources will occur if we each voluntarily make a personal commitment to consume and discard less. Assume that the industrial cultures, whose members consume most per capita, will make a larger reduction. Is expecting such a reduction realistic?

Imagine yourself seeking a commitment from a Westerner to reduce their demand for resources. You first mention that the resources they use are provided by our environment's functioning but that our provision, use and discarding of those resources is disrupting its functioning. You then ask the Westerner to bear that in mind when you ask them the following two personal questions: what do they think they can do to help protect/maintain the functioning of our environment over the long-term (>60 years)?, and its complement, how much stuff (resource consumption) is enough for them to live a satisfying life? (Durning 1992)? Westerners' answers commonly display reality-paradox-rich rationalizations. We find it extraordinarily difficult to provide a comprehensive, coherent answer to either question or an answer that is consistent between them. This is the case, in part, because the subject is complex. It also occurs, in part, because the concept of environmental protection conflicts with the strong, consumer-supporting guidance provided by our culture's world view. It encourages us to consume and allows consumption to play a central role in the formation of our personal identity, the development of our sense of belonging to a social group and the functioning of our economy.

Thus, if Westerners are to undertake a voluntary, environmentally significant reduction in our personal demand for resources, then we have to adjust our personal world view. If our efforts are to endure, then we must also adjust our culture's world view so that it too guides us toward a material lifestyle more closely matched to environmental realities. Making these personal and cultural adjustments is a slow process that requires effort. The voluntary path to a reduction in the per capita consumption of resources is a long, slow one.

There are other ways in which a resource demand reduction might occur. Some Western economists claim

that the needed reduction will occur automatically and painlessly. They expect the decline to occur after our basic infrastructure needs are satisfied.

For at least Western industrial cultures, this not the case. For example, the United States is well past the point of building the basic infrastructure and satisfying its citizens' basic per capita material needs. Yet their long-term per capita consumption of copper, for example, continues to increase, with no end in sight (Gordon et al. 2006). Think of electric cars.

Others argue strongly that the needed reduction in the absolute demand for resources can be achieved by increasing the efficiency of their use. Because the efficiency increases will be achieved, so the argument goes, by applying market incentives and technology the reduction in absolute resource demand will be painless for the Western consumer. Proponents of efficiency and its variants, such as dematerialization, find proof for their views in the steady innovation-related decline in the resources used to provide a specific good or service (an increase in the material efficiency of resource use). Think of lighting. Others point to the decline over time in Western countries' resource demands, carbon emissions and energy use, per unit GDP (i.e., an increase in the economic efficiency of resource use) (Ausubel 1996).

There are many flaws in the claim that the saving from increasing efficiency will automatically reduce our absolute resource demand, as discussed in chapters 12 and 17. They include the use of GDP as an efficiency reference, relative measures without reference to absolutes and conclusions based on data used in isolation rather than in a complete system or at a global scale (lifetime or lifecycle analysis). Critically, in one way or another, they all fail to pay attention to a key aspect: what will we decide to do with the savings.

Certainly, an increase in efficiency can help us reduce our average per capita demand for resources, and it may do so quite quickly. But this will only occur if efficiency's savings are dedicated to that end: that is, if they are metaphorically "put back into the ground." Currently, this is not the case.

Today, some of efficiency's savings are used to satisfy the needs of a growing human population. Where the rest is going and why is illustrated by Western culture's version of the virtuous (consumer-producer) cycle. It promotes an increase in the per capita demand for goods and provision of services (figure 14),. In providing them, the rest of the savings from an increase in efficiency are used up. The overall outcome is illustrated by recalling how the electricity savings provided by the more electrically efficient transistor combined with the consumer-producer cycle to contribute to an increase in electricity demand. Now is the time to remember that behind Western culture's drive for efficiency lies its imperative of economic growth and its support for the consumer-producer cycle.

Thus, even though it is implausible that efficiency on its own will reduce our demand for resources, when it is contemplated from the perspective of a consumer in the context of the culturally supported virtuous cycle, then it is no wonder that Westerners' faith in efficiency is so difficult to shake. After all, the cycle is fulfilling its promise of providing us with more for less. We, at least unconsciously, see the cycle and efficiency as together satisfying our culturally supported desires and expectations for a wide array of goods and services. In the process, the cycle makes concrete to us the validity of our culture's belief in progress. Think innovation and growth. Think sustainable development.

If Westerners were to take a broader view of efficiency, then the promise that an increase in it will automatically reduce our absolute resource demand is seen to be one that humans, not efficiency, has made. We are misleading ourselves. For example, in order to reduce our family's energy use, we buy homes that are more energy efficient (e.g., better insulated) and we buy devices that are more energy efficient. But our home energy use is unlikely to decrease if we also decide, as is happening in North America, that we want bigger, more brightly lit homes filled with many more electronic devices, on standby. Similarly, we might hope, even believe, that by increasing the efficiency of food use, we will automatically reduce the amount grown and thus the resources used to farm, transport and process our food, while at the same time feeding the hungry. But, like energy saving, there is no guarantee that the savings from reducing the amount of wasted food will be directed at those in need or be put back into the ground, as discussed in chapter 14.

More generally, the efforts directed at efficiency (Dietz and Rosa 1997), dematerialization (Ausubel 1996), economical usage (Wackernagel and Rees 1996), reuse, recycling, substitution, waste reduction and other such schemes do not automatically or even necessarily lead to a total reduction in either our resource use or our related impacts on the environment. That only occurs if we make the decision to put the savings back into the ground and ensure that it happens.

In contrast, the historical record shows that, over the last 100 years, the trend in our total global demand for most resources has increased. One reason is an increase in the global population. Another is an increase in our per capita demand.

In overview the objective of reducing our demand for resources could certainly be reached by some combination of a decline in our per capita use of resources, a decline in population and an increase in the material efficiency of resource use. However, a closer inspection of these options reveals that for such a decline in resource use too happen, we would have to question ourselves: How many children is appropriate? How much stuff is enough for a good life? Where do our resource savings go?

If we are to answer all of these questions in ways that will favour an absolute reduction in resource use, then we will also have to resolve the contradictions, inconsistencies and incompleteness in our culturally supported thinking and decision making. One example is Westerners treating the environmental impacts of our economic activities as externalities while at the same time acknowledging our dependence on our environment's functioning. Another is our belief that following our culture's imperative of growth (which requires an absolute increase in resource use) will secure our future because it will help us satisfy viability's environmental conditions (which require an absolute reduction in resource use).

More generally, to voluntarily reach for the objective of reducing our resource demands, we must change ourselves so that our resource-related decisions are consistently and coherently directed at that goal. And we must maintain our personal resource-reducing efforts over the long term. Doing so is compatible with our striving toward the other environmental objectives and satisfying them in a comprehensive, coherent and consistent manner.

In summary, if a more complex culture is to follow a path toward viability, then its members must personally satisfy viability's conditions. This section discussed viability's environmental conditions by representing its different aspects as objectives we can strive towards: accept our environment's supremacy, maintain our environment's functioning, champion our environment's self-management, conserve what we have, and reduce our total resource demand. These objectives don't exist in isolation. They are related, even connected to one another and are part of the human-environment system. They are thus most effective in helping us to satisfy viability's environmental conditions if we treat them as a package:

a five-objective unit representing the environmental conditions for viability. In practical terms, this means that when we reach for one environmental objective, the *intent* of the other four must also be satisfied.

By striving toward the five-objective unit, we direct our decision making and efforts at changing our self, ourselves and our culture in ways that favour viability. In essence, we each attempt to become an environmental trustee: a person who acts as a custodian or guardian charged with protecting our culture's environmental endowment for both present and future generations. This is in contrast to being a promoter, manager or steward charged with ensuring that the environment will primarily satisfy our culture's present wants (as opposed to needs), as is the case today. As we become an environmental trustee, we would each find it easier to gain more of our sense of meaning and purpose from a less transformed environment. On a global scale, this could help humans transition from the Industrial Age to the Age of the Environmental Trustee.

For our striving toward the five-objective unit to be successful, we have to make decisions and change ourselves in ways that favour viability. In particular, we are obliged to change our lifestyle and our environmental views so that they better reflect the external reality. This requires us to change our conception of the environment and our relationship with it, which is a challenge (Freidel and Shaw 2000; McIntosh et al. 2000). The methods we can use to help us make the needed decisions and changes are discussed in the next section.

## Three Methods to Favour Viability

We are our culture's decision makers, so it is we who must make both the decisions, and the specific and broad changes needed to satisfy viability's human and environmental conditions, as illustrated in 22. In particular, we are required to adopt viability's conditions as goals, for example the five-objective unit. We must also reduce our personal and cultural divergences, and accommodate our decision-making foibles in ways that favour viability.

This section discusses three methods/tools that can help us with those and all of the other aspects of our striving toward viability. These methods are as follows: increase our knowing, engage in nuanced thinking and decision making, and tweak our personal world view. Think of these methods as specialized parts of the rationalization process linking the belief and resource loops together in figure 15.

**Increase Our Knowing.** We can more easily make the needed decisions and changes if the information available to us better reflects the external reality. The scientific methods of gathering and checking information, whether applied formally or informally, provide us with the most widely available, reliable (independently reproducible) knowledge we have about the external reality. But just possessing reliable knowledge won't help us. To be successful, we must use that knowledge to change ourselves, personally, in ways that will achieve our goal of viability: we must use it to increase our knowing, as introduced in chapter 6.

We start to increase our knowing by attaching considered personal meaning or relevance to the reliable knowledge we have gathered about the external reality: by personalizing what we know. This can be more difficult to do than we expect because the knowledge we discover often repels, overwhelms or disheartens us. Therefore, part of personalizing knowledge is to come to emotional terms with it. Once the reliable knowledge we have gathered has been personalized, we can assimilate it into our personal world view to become part of who we are, part of our internal reality. (Although mentioned separately, these two processes can occur together.)

Reliable knowledge that has been personalized and assimilated changes our internal reality so that it more closely matches the external reality and our relationship with it. If we apply that information, and the awareness that comes with it, to ourselves (what we believe and how we live), then we have increased our knowing. Simplistically, after feeling the heat of flames, we know that fire burns. Deciding not to touch it shows that this knowledge has indeed been applied to ourselves: our decisions about how to live now better represent the environment and us for what they actually are. We have thus increased our knowing. This process is a key aspect of gaining wisdom and making wise decisions.

Our efforts to increase our knowing can be divided, for convenience, into two parts: increasing our knowing about ourselves, our self-knowing; and increasing our knowing about our environment, our environmental knowing. The two are interconnected as the example of flames illustrates.

Increasing our self-knowing focuses on skeptically gathering, personalizing and assimilating reliable information about who we are and how we relate to the world. Because that information is personally relevant and it may contradict what we currently imagine is or should be true, we can feel uncomfortable dealing with it. For example, you may be able to intellectually accept that humans innately tend to be overconfident in their judgements. But this is not the same as you concluding that you, personally, display that unflattering characteristic, and then deciding to take that knowledge into account in your decision making. There is generally a certain amount of discomfort involved in judging your own beliefs, decisions and actions, such as your relationship with our environment, your lifestyle or your efforts to change them. Think of why your daily to-do (will-do) list still continues to be rarely completed.

No wonder that the process of increasing our self-knowing displays features in common with the process of personal change through cognitive dissonance. Consider that we tend to, at least initially, avoid acknowledging/personalizing undesirable information about ourselves by rationalizing it away. We tend to overcome this avoidance in stages: it takes time and effort. If successful, we can assimilate that knowledge. Once it is assimilated, we can devote the time and effort needed to reflectively think about ourselves in the light of our newly assimilated knowledge (Gelter 2003). We are then in a better position to face any other personal conflicts/contradictions that the information may have brought to light and to make any needed adjustments to our sense of self and ourselves.

But, is it realistic to expect each of us to acknowledge who we *actually* are and then take it into account in our decision making? Can we really come to accept that we are at the same time co-operative and selfish (deWaal 1999), emotional and rational, conscious and in control but also unaware and not in control (Myers 2002a)? Is it possible for us to retain our basic human optimism and still develop a more realistic view about the future? Can we accept the fact that we are highly selective in our awareness and attention? Are we able to admit that we can and do focus on some things while ignoring equally important but different things? Are we really capable of making the tough decisions and trade-offs needed to question and change our sacred cows (Tetlock 2003)?

The answer to all is yes, because we are not required to change our basic human characteristics. We are only required to exercise our capacity to become more personally aware of them and take their existence to heart in our decision making: to acknowledge and account (accommodate or compensate) for our characteristics. Our efforts to increase our self-knowing does just that. We consciously prod ourselves to pay the needed attention to the actual nature of our perceptions, personal world view, thinking and decision making,

which includes their biases and the influences on them. And we can prompt ourselves to account for these characteristics when we make our decisions about how to live (Arkes 1981). We can, for example, replace untrained intuition with understanding, think skeptically (which implies independently verifying decisions and information) and take into account the limits to our self-knowledge (Myers 2002a).

An increase in our self-knowing thus helps us make the decisions and changes needed for us to adopt viability's conditions as goals. It helps us take our decision-making foibles and our internal reality's divergences into account in a manner that favours viability. And it helps us become aware of and account for the external influences on our decision making about viability, in particular our culture's world view with its divergences (Myers 2002a; Arkes 1981).

For example, it helps us deal with the specific challenge of feeling overwhelmed by too much change or choice that can arise when we try to make a particular consumer decision in a way that favours viability (Schwartz 2000). Consider too that it helps us deal with the broader challenges associated with making comprehensive and coherent viability-favouring decisions about complex issues for which there are no clear-cut answers, only tough choices or trade-offs. An example is the personal challenge we face when we reach for the five-objective unit. Overall, increasing our self-knowing can help us become environmental trustees.

An increase in our environmental knowing can similarly help us to make more environmentally realistic decisions about how to live. Reliable information about our environment's characteristics is available but, for a number of reasons, it can be a challenge to personalize and assimilate it into our personal world view. One reason is the environment's complexity, as described in chapters 7 and 8.

Consider too that our feelings play a significant role in how we relate to the world around us. For example, we are preferentially drawn to those things, people and thoughts that provide us with a sense of personal security, certainty, order, fairness, companionship, purpose, belonging and control (Myers 1999). We also tend to prefer an idealized or anthropomorphized view of the environment and our relationship to it. When we come face to face with its actual characteristics, the environment can seem to us to be overwhelmingly complex, chaotic and purposeless. It is easy for us to feel that the environment is arbitrary, cold, unforgiving, filled with uncertainty and against us. These feelings are far more negative and unsettling than we would emotionally prefer, which makes it more difficult for us to personalize and assimilate environmental knowledge.

This state of affairs can arise even in the members of more complex cultures because most of us live in urban centres where we are disconnected from our supporting environment. Our disconnect, and the related unsettled feelings toward the environment, tend to be enhanced by our complex culture's world view. For example, Westerners are guided to believe that the world is largely deterministic and has a hierarchical structure. We believe that exercising control over nature is a key characteristic of civilization. Think of the cultural support for transforming our environment. No wonder that our initial reaction to an environment-related event that contradicts this view is to look inward for a culturally supported explanation and a cure that can restore more positive feelings (i.e., provide comfort) (Taylor 1989; Raymo 1998). For example, after a river has flooded our property, we can restore pleasant feelings, a sense of order and control by diking and ditching it.

We increase our environmental knowing when we find ways to personalize and assimilate reliable knowledge about the environment and then use it to reduce both our disconnect from our environment and our unpleasant/unrealistic feelings about it. We can facilitate this process by spending time more closely observing and interacting with our environment on its, not our, terms (Strong 1995; Finger 1994). For example, we can engage in recreation activities that bring us closer to our environment but don't require us to disrupt or manipulate it (i.e., refrain from "consuming the wilderness").

At home, you can pay closer attention to the plants and animals around you. For example, you know that dogs have a good sense of smell and hearing. But have you spent time closely watching a dog sniffing and listening as it wanders about, and then reflected on that activity from the dog's perspective? If you do, you will have a greater appreciation for the dominant role that smell and hearing play in a dog forming and experiencing its reality.

Our benign personal experiences of the environment on *its* terms enables us to more easily place ourselves in the roots of the earth's plants and the paws of its animals, so to speak. And we can do so in a manner that is simultaneously rational and compassionate (reasoned empathy). This makes it easier for us to relate, at a personal level, to our human-environment system's actual characteristics rather than what we think they are or should be (Maser 1988).

By increasing our personalized and assimilated reliable knowledge about our environment, we build the capacity to live in a more environmentally realistic way. For example, we can more easily ask ourselves if our personal or our culture's world view is guiding us to treat organisms as a commodity or part of an ecosystem community, and whether that guidance is appropriate for a particular context and over the long term. We are more likely to prefer that organisms be left to participate in their indigenous host ecosystem's functioning rather than become a decorative exotic under our management. When we make those decisions, we will confirm that we have increased our environmental knowing.

In overview, by personalizing and assimilating reliable knowledge about the characteristics of humans and the environment, and then using it to make decisions about how we should live, we increase our self-knowing and our environmental knowing. By increasing our knowing, it is easier for us to adopt a lifestyle that better matches the external reality when treated as a functional whole. We have also helped ourselves to acquire some of the attributes of wisdom. We are then more likely to accept viability's conditions as goals, and to make the decisions and changes needed to satisfy them.

**Nuanced Thinking and Decision Making.** We are also helped in our striving toward viability if our thinking and decision making better reflect the complexity of how we and our environment function. Although we can't accomplish this goal by changing the basic processes by which we think and make decisions, we can change how we direct them. We can think and make decisions in a more nuanced manner (nuanced thinking). What does that require us to do?

The introductory chapter opened with me puzzling over how I can hold multiple, often contradictory opinions about our neighbouring forest. I also noted that I can switch between my opinions with little recognition of the contradictions among them. I started by seeing our forest as simply a place to be exploited for wood, then suggested we nurture our forest for the food and water it supplies. I also thought we should respect it and its inhabitants as having an unquestioned right to exist.

A resolution to my puzzle, as the previous chapters have tried to show, starts to appear with the realization that our forest and I are participants in a single, complex human-environment system. This is the context within which I have to think about myself, our forest and the relationship between us. But this contrasts with my culturally supported innate human tendency to perceive and think about our forest by focusing on one aspect at a time. As a result, I also tend to treat each aspect as representing a largely independent aspect of our forest, as illustrated by the viability flower (figure 22).

To fully resolve my puzzle, I am going to have to compensate for my innate focused and decisive manner of thinking. I will have to learn how to treat each of my opinions about our forest as arising from a different perspective of one dynamic, complex forest, in which I am a participant. In particular, when making decisions about our forest from one perspective, I will have to fully consider the links to the other possible perspectives I could take of the forest. I will have to engage in a more nuanced manner of thinking.

Nuanced thinking is a broad (multi-scale), comprehensive (multiperspective), flexible (open-to-change) manner of thinking and decision making about a topic/issue. Although nuanced thinking uses the techniques of scientific thinking, such as being skeptical (including being self-skeptical), it is more than scientific thinking. It is directed at more than acquiring factual knowledge. It is a striving to gain understanding and make wise decisions by treating oneself as both an objective and an emotional observer of an event and, at some level, a participant in it.

In order to engage in nuanced thinking and decision making about a particular topic, I must consciously take into account complexity and context. I must remember and consider that I am part of a dynamic, complex human-environment system in which I am simultaneously an individual, a member of a culture and a member of my culture's environmental endowment. Because I participate in the system, I both contribute to its functioning and am constrained by it. Similarly, the topic being considered is a part of the human-environment system, so despite my desire to, I can't treat it in a simple black and white manner. It, and the context in which it resides, is also complex, dynamic and grey. And I am, in some way, connected to it.

My contemplation of the topic has to consider many context-dependent views of it. In doing so, I must remember that contradicting observations and opinions about the topic can arise from differences in perspective as well as from errors. I am also obliged to accommodate in my thinking my internal reality's divergences, my decision-making foibles, the influence of my culture's world view (including its divergences) and the environment's complex functioning. And, to favour viability, I must always take the conditions for viability into account: they are a primary reference.

It is easy to suggest that we could each practise nuanced thinking by holding all of the relevant information (and concerns) in our consciousness simultaneously and then mentally linking them together. For example, I could hold all of the information about my three perceptions of our forest in my consciousness at the same time and then coherently merge them. This is extremely difficult for humans to do, even with practise and a trained intuition. And, as the discussion about the viability flower illustrates, the presence of any misinformation complicates the process significantly (figure 22).

The more realistic alternative is to use those parts of your knowledge you currently accept as reliable to create a *provisional* view of the topic of interest from one perspective. Then, over time, you would systematically upgrade and widen that view, one piece of reliable information or deduction at a time. This step can be repeated for each of the perspectives you take of the topic. By routinely consciously switching your mental attention from one perspective's widening view of the topic to another perspective, the connections between them come to light. This allows you to check for coherence between the different perspectives, which prompts you to revise your accepted knowledge and further adjust your understanding of the topic. Your appreciation for the topic as a whole grows.

This nuanced manner of thinking helps me resolve my puzzle about our forest. I no longer fixate on one perspective of our forest or try to decide which one of my perspectives is more correct or most important. Instead, I ask myself what other information is available, and I look for the connections between my perspectives and how the forest functions. At some point, I'm prompted to consider the many influences and biases that affect my choice of perspectives, the information I accept as reliable, and how I use it. Think of my culture's world view. Through this process, I gain a deeper understanding for how the forest functions, my relationship with it and why I think about the forest in the way I do. It is then easier for me to resolve my puzzling about our forest because I can place a question in the context in which it can be given answers that are consistent and coherent with one another, regardless of the perspective I use. In the process, my original firm conclusions and opinions becomes conditional.

The process of expanding or converting one's more familiar manner of focused, decisive thinking into a more nuanced manner of thinking is a slow work in progress that requires personal effort and persistence. Thus we can be prodded, but not pushed, into adopting nuanced thinking. Fortunately, if we are given the relevant information, appropriate guidance and support, and enough time to ponder, then it is possible for individuals to change their manner of thinking about themselves and the world around them from decisive to more nuanced.

For example, try focusing now on your views of an issue that has your attention, say the proposal to reduce our total resource demand by increasing our efficiency of resource use. Then ask yourself the following two questions: Firstly, can I say that I have considered this issue from a sufficient variety of perspectives, and then thought about my resulting view in a sufficiently broad manner, to reliably represent the complexity of the issue to someone else? Secondly, can I say that I have recognized and taken into account the (personal and cultural) biases in my decisions about it? A check with the discussions about this topic above and in earlier chapters will help you evaluate your answers.

Our children are our future. To help them adopt a more nuanced manner of thinking, we can introduce them to it early in their lives. For example, we can expose them to our personal efforts to practise it while providing some supporting explanations. We can, for instance, point out the limits to focused, decisive, deterministic thinking about a topic, and then apply the more broad and flexible manner of nuanced thinking to it. As our children mature, we can guide them to know when, and when not to, apply each method. If successful, our children should be better able to more appropriately deal with the complexity of the external reality that they will live in during their lifetime (Tetlock 2006). They will thus, more easily than us, reduce their internal realty's divergences and accommodate their decision-making foibles while they reach toward the five-objective unit.

But success for them, like us, will only come if they also work at increasing their self- and environmental knowing. Fortunately, the two methods are complementary parts of gaining the wisdom needed to make appropriate decisions. For example, are we a species whose unique characteristics make us superior to all others? Or are we just another species whose implied ordinariness means that we are worthless? If we can both increase our self-knowing about our need to feel special and practise nuanced thinking to moderate our tendency to focus strongly (especially on ourselves), then we can arrive at and accept the following answer.

Each of us is a member of a human community where our role as an individual is special. But each of us is also a member of the human species, which is part of the human-environment system. In this system, we are each personally subject to the same basic rules as every other living thing. In that context, we are ordinary. We are thus both special and ordinary. Which view is appropriate depends on the context or perspective we choose.

**Tweak Our Personal World View.** Our efforts to practise more nuanced thinking and increase our knowing change our personal world view. But our personal world view influences those efforts, creating a feedback loop. Similarly, as our personal world view changes, we influence our culture's world view. But our culture's world view has a strong influence on our personal world view, which completes the member-culture feedback loop. These feedbacks stabilize both our personal and our culture's world view. But, they can't guarantee that our culture's state will be viable and they constrain our efforts to make the personal and cultural changes needed to favour viability (figure 15). Our personal efforts would be helped if we could directly tweak our personal world view.

Surprisingly, there are opportunities to do so. For example, a complex system displays random and chaotic as well as deterministic features. Yet a member of at least Western culture tends to discuss and make decisions about that system using a deterministic vocabulary. By adjusting our vocabulary so that the words we use more accurately reflect the system's actual features, we tweak our personal world view in a way that boosts our efforts to practise more nuanced thinking and increase our knowing.

Such direct efforts to tweak our personal world view have already been used in this book. For example, when describing human responses to a circumstance, *tend* was used instead of *is* or *do*. It acts as a reminder that individually we, being a complex system, do not display a single deterministic response to a circumstance but a range. Yet when looked at as a group we collectively display a tendency toward a preference: we each have an above average probability of displaying it. An example of this usage can be found by rereading the third sentence in the above paragraph.

Other tweaks have been suggested, for example, instead of referring to sustainable development (or sustainability) when discussing a long-term general goal for the future, refer to viability, and take care to make the distinction between the two quite clear. When talking about evaluating a proposed environment-affecting action, refer to the need for a responsibility-benefit analysis rather than a cost-benefit analysis. Don't use the word *progress* as a synonym for *growth* or *development* of any kind, especially not economic.

A more personal example is discussed here. Instead of evaluating our personal (or other people's) thoughts and decisions about a human-environment topic using *right* and *wrong*, we can use *appropriate* and *inappropriate*. Their broader meaning reminds us to consider a topic within the wider, more complex context in which it usually exists: the dynamic, complex human-environment system. Using these two words thus helps us to think and make decisions about the topic in a more nuanced, rather than a black and white (simple, deterministic), manner.

The use of *appropriate* and *inappropriate* will also help us to apply our moral judgement in a more nuanced manner. For example, the making and then marketing for our amusement of an aquarium fish that glows in the dark (the GloFish) is certainly environmentally inappropriate: the change radically disrupts a species' lifeway and increases our environmental responsibilities (Mazoyer 2004). It should also be morally unacceptable in cultures that see themselves, whether for secular or religious reasons, as environmental trustees.

However, there is a constraint on the benefits we gain from adopting *appropriate* and *inappropriate*. Our knowledge about a complex topic and our appreciation for it is always incomplete to some degree. There will thus be situations where there is sufficient uncertainty that we are unable, despite our desire, to meaningfully assign *appropriate* and *inappropriate* to our thinking and decisions about a complex topic. We have to add a conditional term to our judgements. We could use *appears*, *seems* or an equivalent qualifier, but here the discussion turns to a broader generic word: *incomplete*.

Although useful, it takes effort to use *incomplete*. Consider that we often find the notion or sense of incompleteness unsettling. Think of the term *seeking closure*. This need for a sense of completeness is not a weakness. It is an inescapable aspect of how we innately and culturally deal with the reality of living in what, to us, is a sometimes unsettling, unpredictable complex world.

Consider that our brain's unconscious processing automatically tries to fill in any blanks in our perceptions or thinking before it becomes a conscious experience. Similarly, when a conscious thought or decision feels unsettlingly incomplete to us, we intuitively call first on our unconscious thinking and our culturally influenced personal world view to help us deal with it. We usually find (more accurately, we are provided by our brain with) a thought that is appealing and decisively fills the gap, as discussed in chapter 1. By accepting it, we gain both a logical and an emotional sense of completeness. But, if the thought hasn't been checked for reliability, it might just be a satisfying but misleading belief that removes the option of using *incomplete*. We are thus faced with a dilemma. Using the term *incomplete* can help us to more reliably reflect our limited understanding of a topic, but it can also leave us emotionally hanging. In contrast, relying on a belief provides us with the emotions we desire, but that likely won't help us to reliably reflect our limited understanding of a topic.

Fortunately, this issue can be resolved by us making another adjustment to our personal world view. We can accept that our need to form and hold beliefs is part of being human and that the beliefs embedded in our personal world view can provide us with the explanations and comfort we need to deal with unsettling experiences. We can also accept that our beliefs do not necessarily or always represent the external reality, which leaves the door open for us to question and adjust them.

We thus gain the benefit from our beliefs but avoid being trapped in them by blind faith (Moshman 1998). We also provide ourselves with the option of deciding whether our belief about a topic, especially a complex one, is appropriate, inappropriate or only provisional within a particular context, which we can indicate by attaching conditionals to our belief. We can thus help ourselves to use *incomplete* to represent our understanding of a topic. By tweaking our personal world view, the dilemma that results from the conflict between the benefits of our beliefs and the benefits of more accurately reflecting our understanding of a topic is made more manageable.

An illustration of this resolution in action is provided by a culture's formal religions and less formalized spiritual traditions (religions). The members of a religion rely on its beliefs (and associated explanations) to provide them with comfort and guidance, especially when they feel lost, unsettled or distressed (Barrett 1978). These religious beliefs can become part of a culture's world view, where they serve the same purpose. For example, religions of all stripes accept/advise that the missing/incomprehensible (the *incomplete*) is knowable only to their supreme being/force.

Consider Western culture. Although it is common to refer to Western culture as secular, religious beliefs continue to provide comfort and guidance to many. However, the source and context of these needs has changed in significant ways over the last 250 years. For example, Westerners have learned from the discoveries made by people such as Galileo, Einstein, Darwin, Wallace and Lylle that we occupy a tiny part of a vast material universe of great age. We are also faced with the knowledge that we are just one of many species interacting with one another in a dynamic, complex environmental community (Raymo 1998).

Westerners can find it difficult and unsettling to comprehend, let alone personalize and assimilate, this information. Both our effort to do so and the result can negatively affect our sense of self and our thinking about how we should live. For example, we can easily feel irrelevant, overwhelmed or confused, especially when we are guided by our culture to see ourselves as the superior species and deterministic order as a mark of civilization. This is the current context within which the beliefs of Western religions must provide their guidance and comfort.

The formal (traditional) Western religions could more easily provide that service if they could accommodate scientific information into their world view. This could be achieved by religious leaders tweaking their personal religious beliefs about our relationship with the world around us in ways that acknowledge scientific information, but retain the core of their religious beliefs. Think of this as their effort to bring their beliefs closer to the centre of the viability flower (figure 22). If the leaders made these personal adjustments, then they would prompt a similar tweaking in their religion's world view. The formal religions would then be better able to help Westerners live in a manner that recognizes and accepts the external reality's characteristics, such as the functioning of our environment and its constraints on our lives, for what they are. At the same time, the formal religions would be better able to provide the comfort and guidance to those who seek it from them. An example would be seeing the hand of God behind the order in our vast universe and the big bang.

Theoretically, the leaders of formal Western religions should have little trouble making this personal adjustment. After all, most religions directly connect their god or gods to the creation of the earth and how it, the environment and us function. This is the same earth that science describes (Gardner and Stern 1996).

Despite this commonality, some suggest that the nature of formal Western religions is fundamentally incompatible with scientific knowledge and unable to make the adjustments needed to accommodate it. This is not the case. Historically, Western science and formal Western religions are closely connected. They have common recent roots, and they can complement one another in their roles as cultural guides. As John Paul II said, "Science can purify religion from error and superstition, and religion can purify science from idolatry and false absolutes" (John Paul II 1988). Religion provides us with moral guidance and a sense of the sacred; science, the power to learn about ourselves and the external reality. They both help us to satisfy our needs. They can both help us reach for viability.

Indeed, the leaders of some formal Western religions have already adjusted their world view so that its guidance about their members' relationship with our environment better reflects our current circumstance. For example, both a significant proportion of the evangelical Southern Baptist Church's leaders and Pope Benedict XVI of the Roman Catholic Church consider that Genesis did not give Christians licence to subdue the earth. The Catholic position was confirmed by Pope Francis (Francis 2015). Both churches find in the Bible clear instructions that Christians are trustees of God's creation who are obliged to protect its integrity and live within our supporting environment's constraints (Haag 2006; WeGetIt 2009; RNS and ABP 2008; Baptist Creation Care 2009; Benedict XVI 2007; Benedict XVI 2008).

Admittedly, some formal Western religions do display significant inertia to accommodating scientific knowledge. Partly in response, many post-1950s Westerners have left the formal religions and are instead engaging in a search for spiritual meaning through nature (Inglehart 1990). Consider the New Age's earth-centred spiritual views (Myss 1996). They are epitomized by those who call the earth Gaia and treat it with great respect, even reverence. In one view, Gaia is seen as an organism striving for consciousness.

The idea of Gaia may seem radically different from the efforts of the more formal Western religions or past spiritual assemblies to provide guidance and comfort. But it too retains the essence of the extremely old religious and occult traditions about our relationship with our environment, some of which go back to at least the ancient Greeks (Katz 2005). Regardless of its roots, the New Age search for meaning through nature may be more accepted by the public today because it usually doesn't completely reject science or religion. Instead, it replaces religiosity with spirituality while, at the same time, merging faith and reason.

In overview, our decision making is strongly influenced by our internal reality (including our personal world view). This means that, by making external-reality-compliant adjustments to our personal world view at the appropriate time, we can help ourselves make the decisions and changes needed for us to satisfy viability's conditions. Tweaking our personal world view is especially powerful when the changes we make are compatible with our efforts to engage in nuanced thinking and to increase our knowing.

To make the appropriate tweaks, we have to ask ourselves questions such as, what modifications would help us live a skeptical spiritual life (Raymo 1998)? Which alterations would help us exhibit humble scientific thinking (Ziman 1978)? And which amendments would help us resolve, in a way that favours viability, the complex culture's conundrum of whether to revere or transform their environment? Inevitably, we must ask ourselves, what adjustments would help us become an environmental trustee?

The example discussed here was to replace *right* and *wrong* with *appropriate* and *inappropriate*. Conditionals could be used and the word *incomplete* could be added to reflect the limits to our knowledge of a complex world. But to gain the full power of *incomplete*, we have to adjust our view of belief to accommodate it. We should accept the need to form and hold beliefs but, to avoid being trapped by blind faith, we should remember that our beliefs do not always represent the external reality.

In summary, three methods to help us reach for viability were discussed: increasing our knowing about ourselves and our environment; practising nuanced thinking and decision making about our circumstance; and tweaking our personal world view so that it better reflects the nature of the external reality and our level of knowledge about it. By applying the methods during our day-to-day living, we help ourselves to accept viability's conditions as goals, and to make the decisions and changes needed for us to satisfy them. Applying the methods does not preclude or replace political and social action, it just makes it easier to keep them, and the application of technology, on the path to viability.

It takes time and effort to learn and successfully apply the three methods. However, they are linked to one another so that the successful application of one

makes it easier to apply the others. The benefits of applying them are self-reinforcing.

## Dealing with Challenges

We face inescapable challenges when making the decisions and changes needed to favour viability. They arise from omnipresent features in the background of our lives and in our interactions with our surroundings. This section illustrates how we can meet these challenges by routinely applying the three methods and rules of thumb during our day-to-day living. The challenges discussed are complexity, the scale of things, culture's tools and being a consumer-disposer. Think of dealing with these challenges as part of our effort to adjust the belief and resource loops in figure 15 so that they favour viability (figure 22).

**Complexity.** In our striving toward viability, we will have to routinely deal with dynamic, complex systems. Chief among them is the human-environment system, as discussed in chapter 11. It, like all complex systems, consists of many variables interacting with one another in a number of ways, including feedback, over a wide range of time, space and strength. Its functioning, and thus its changing, displays the characteristics of determinism, randomness and chaos. Overall, the system is quasi-stable.

Our innate characteristics mean that we find it a challenge to take these features of complexity into account during decision-making. Consider our preference to act on our first impressions, favour deterministic logic, focus narrowly, rely on our untrained intuition, and find solutions by trial and error. Our disconnect from complexity is illustrated by the significant difference between our simple intuitive speculations about what caused an aircraft accident and what studies reveal are the interacting mix of the contributions to it. Aircraft accidents are sufficiently complex that pilots and aircraft maintenance personnel, for example, have to expend considerable effort to ensure that their day-to-day decisions will not contribute to one (Shappell and Wiegmann 2000; Pole 2000).

Dealing with complexity is easier if we can outline the system relevant to the topic at hand. Think fossil fuel supplies. We can acquire information and gain an understanding of the system's functioning and changing by breaking it into isolated parts and then studying both the parts and the connections between them (i.e., by the process of reductionism). The information and understanding we gain can be integrated into a model of the system and its functioning. However, not only is this information somewhat incomplete, but the integrating process is itself imperfect. For example, both the use of inflexible scenarios and the need to simplify the data in order to integrate it can create a misleading sense of the system's order. Overall, the nature of the information, our understanding of the system's functioning and the process of integrating it invariably produces a model that is a biased and incomplete summary description of the complex system and its functioning. Think of the discounted major flaws in many economic models, such as infinite resources.

Fortunately, by applying the three methods, we are more likely to take these limitations into account, both when making the model and when using it for decision making. For example, by engaging in nuanced thinking, we are more likely to prompt ourselves to look at the diversity of the information being used and do so from a variety of perspectives, such as different scales of time, space and strengths (Weingart et al. 1997). We are also more likely to pay attention to the biases and limitations, such as simplification, associated with focusing on a particular perspective. We are thus more likely to consider the limitations of the model trying to represent the system as a whole.

We facilitate our application of the three methods when dealing with complexity if we remember the principles and rules of thumb about the features and functioning of a generic complex system. Here is a reminder of those rules, along with examples from the human-environment system that illustrate how they can help us. In keeping with the five-objective unit discussed above, the focus is on the environment.

Change in a complex system can be described using a mix of generic terms such as *preconditioning*, *drift*, *directed*, *random/chaotic* and *external*. These descriptors remind us that change in a complex system always includes knock-on (indirect) and feedback effects. As a result, our impact on a complex system will always extend beyond what we directly affect, and we can find it difficult to determine what our full impact is.

From a different perspective, the presence of inertia and multiple effects in a complex system mean that a present action of ours may not immediately solve the current problem it is directed at (Demeny 1990). And, although that action will influence the future of the system, it may not, in the long run, do so either exactly as or only as we intended (Allen 2004). Think and plan accordingly.

In particular, a decision intended to resolve a specific (social, cultural or environmental) issue must be made within the context of its host complex system, otherwise it is less likely to result in either the intended outcome or it enduring (Tainter 1996). Consider the citizens' efforts to solve their country's existing food shortage by focusing on increasing agricultural intensity through irrigation. This could satisfy their immediate food need, but not necessarily over the long term. For example, if the country has a high population growth rate, then agricultural intensification will only be successful in the short term. After all, both the area of land that can be irrigated and the volume of the renewable water supply are finite. And the water is also used for domestic and industrial purposes as well as irrigation. A meaningful, long-term solution to the food issue would include more than irrigation, it would include the complementary decisions to limit the population to a size that could be comfortably supported by the available water and land (Confino 2004; Rice 2006).

Uncertainty is an inescapable product of complexity. Thus predicting the details of a complex system's future with a high degree of certainty is essentially impossible. In addition, the more disrupted the system, and the further into the future one predicts, the more uncertain those predictions should become. For example, any predictions about the details of our highly disrupted environment's long-term future state are highly uncertain. We are helped to take this uncertainty into account during the making of predictions by applying the three methods. After all, they help us to focus less on making/using highly accurate, detailed predictions and more on making/using broader, conditional (if-then) predictions that consider the general features of the complex system's structure and functioning. Think rules of thumb.

More generally, we find it difficult to realistically take this ever-present uncertainty into account in our decision making. Think of our preference for certainty and completeness. We thus routinely use inherently uncertain decisions/predictions to set certain goals. We also prefer to believe that a rare, human-caused adverse event won't happen on our watch. If such an event did occur, then despite the system's complexity, we have faith that we will be able to easily mitigate or restore any disruption to its functioning.

We can ease the general difficulty of dealing with uncertainty by using the method of tweaking our personal world view to include uncertainty as an ever-present characteristic of the human-environment system.

We are then more likely to accept that we can neither fully identify nor remove uncertainty from our predictions/decisions by detailed studies, but that we can accommodate it during our decision making. Think buffers. Consider too that practising nuanced thinking helps us to systematically consider interactions that could lead to longer-term consequences.

Our efforts to apply the three methods and the rules of thumb when dealing with a complex system can be augmented by decision-making "techniques" designed to accommodate complexity's characteristics. Think human-environment system. The techniques include routinely establishing no-regrets policies (e.g., no environmental regrets) and applying the precautionary principle (Alley et al. 2003). Consider the precautionary principle.

It, like the Hippocratic oath in medicine, says that our actions should, firstly, do no harm. In attempting to meet this condition, we are forced to deal with the system's complexity and the inescapable presence of uncertainty and the "unexpected." We are also reminded to take into account the bias and errors in our decisions that result from the nature of our decision making and the influences on it. The precautionary principle should therefore form the overarching guide when making poorly constrained decisions about complex systems, regardless of whether our decisions are science-based or not. It should primarily be thought of as a proactive technique, one to be used before rather than after the details of a risk have been clearly identified and defined (Lyons et al. 2000; CEC 2000).

The use of the precautionary principle is illustrated by the European Union's decisions about penta brominated diphenyl ether (penta-BDE), a chemical almost exclusively used as a flame retardant in polyurethane foams. In Europe, there was a steady increase in the levels of penta-BDE in human breast milk over a 25-year period. Details of the resulting health effects were unknown. However, the available knowledge indicated that penta-BDE had the potential to disrupt the activity of the thyroid hormones responsible for brain development in children.

The EU could have waited for the required tests to be conducted, which would likely have been inconclusive in the short term. Instead, in light of the potential for harm, the EU decided to take risk-reduction steps by controlling penta-BDE before the tests were conducted. They applied the precautionary principle (Lyons et al. 2000).

Applying the precautionary principle to the human-environment system is useful because it, in essence, recognizes that the system's functioning, not humans, ultimately determines the constraints/conditions we must meet to live a viable lifestyle in a viable culture (Marx 1996). Consider that a culture's world view can provide guidance to its members about acceptable lifestyles and a satisfactory population size. But, to favour cultural viability, that guidance must ultimately satisfy the resource constraints/conditions for viability set by the functioning of the culture's complex environmental endowment. Similarly, a culture's world view can provide whatever guidance it likes about the acceptable concentrations of a compound in its members' bodies or in its environmental endowment. In reality, the threshold concentrations for the compound above which it produces adverse effects are set by the compounds place and action in the complex functioning of the human-environment system.

In overview, to be successful at making viability-favouring decisions, our decision making must accommodate the complexity of the world we live in. We are helped to include it by applying the three methods and the rules of thumb about the generic features of complex systems. Of course we must also remember viability's conditions, such as the environmental conditions represented by the five-objective unit.

When a person uses the phrase "it depends," we treat it as a mark of a person's indecision, evasiveness or lack of knowledge. However, when seen in the context of complexity, the person may simply be attempting to accommodate complexity. We should take care before deciding. The three methods can help with that evaluation too.

## Scale of Things.
If our decisions and changes are to favour viability, then we will have to deal with the wide range in scales over which the dynamic complex human-environment system functions and changes. This range is much larger than we can easily relate to. For example, the range in time over which change in the system occurs is a fraction of a second to millions of years, the range in space is submicron to global, and in strength from extremely weak to extremely strong. In contrast, our innate appreciation of time ranges from roughly a quarter of a second to a few decades, and in space from a few centimetres to regional. We have a bias in our attention toward changes that are strong. This subsection indicates how the three methods (nuanced thinking, increasing our knowing and tweaking our personal world view), in conjunction with rules of thumb, can help us broaden our consideration of the

wide range of scales over which a change occurs and thus our evaluation of its significance.

Consider that it takes on average around 200 years/cm for mineral soil to form (0.5 mm/10 years), with a range of between 80 to 400 years/cm (Pimentel et al. 1995). This should change our view of soil erosion and its depletion. But fully appreciating its significance and factoring it into our decision making is a slow process that can be easily sidetracked. Fortunately, with the help of rules of thumb about scale, the three methods can keep us on track.

The rule of thumb for the minimum time scale to use when considering change is 70 years. It is chosen because it takes into account our grandchildren (i.e., three generations). It also pushes us to take into account the time it takes for us (collectively) to intellectually recognize and respond to our disruption of our environment. For example, in the 1960s, unusually acidic rain was being detected far from urban areas. By the early 1970s, its cause had been traced to sulphur in the emissions from fossil fuel combustion, thus firmly establishing cause and effect. However, it still took the United States until 1990 (some 30 years later) to enact the Clean Air Act amendment needed to start dealing with the sulphur emissions (Oreskes and Conway 2010). Similarly, human-induced climate warming was recognized in the 1960s and accepted scientifically in the 1980s, but it is yet to be fully accepted by the policy makers and the public in the second decade of the 2000s (some 50 years later).

A time scale of 70 years also prompts us to take into account the time the environment takes to recover from our impacts. A good example is the time between the enactment of the Montreal protocol to decrease our release of ozone-layer-reducing chemicals into the atmosphere and the ozone layer re-establishing its pre-disruption thickness. Based on a recent estimate of the re-establishment date, that time is about 70 years (Dameris et al. 2006). Consider too how long eroded mineral soils can take to regenerate. Think of the Argolid.

The rule of thumb for a minimum space scale to use is the width and area of North America. It is about as large a multicultural, multi-biome area that a person can relate to at a basic level with just a moderate amount of effort. It also represents a common distance over which a local contribution to change can result in distant environmental changes. For example, the organochlorine pesticides emitted from Europe and the carcinogenic dioxins generated by backyard garbage incineration of chlorine-containing plastic south of the Canadian border in the United States meet in the Arctic to affect its ecosystems and people.

The minimum time and space scales can be used to help us consider a particular contribution to change. Not to be forgotten is the variation in its strength. We are reminded of this by a rule of thumb that the role of a particular contribution to change in the system is partly determined by a combination of its scales of time, space and strength. For example, both a strong but short-lived (spectacular) contribution in a small area and a weak but long-term (persistent) contribution over a large area can play important (but different) roles in environmental and social changes.

The importance of this rule becomes apparent when we consider that economists, engineers, politicians and the public focus on the environment's local, stronger, shorter-term local contributions to change, such as a volcanic eruption, earthquake and pest outbreak (Parmesan and Yohe 2003). On the other hand, ecologists, geochemists and geologists pay just as much additional attention to widespread, weak-but-persistent contributions to change, such as changes to soil formation, climate and pollution distribution. In general, we tend to ignore these widespread, slower, longer-term contributions to change even though they are usually the base on which the strong, shorter-term contributions ride, as described in chapter 8 (figure 11). For example, it is easier for us to recognize that a shortage of food is causing present hunger than to acknowledge that a slow, persistent population increase or climate change has made hunger's presence possible.

Some would argue that this concern for scale is excessive. After all, we now have regulations that require environmental impact testing and assessments. We also possess the advanced technology to conduct those tests and, we believe, reverse any inadvertent adverse impacts. Surely by using these tools we will ensure that environmental disruption on any human-relevant scale will be negligible. The holders of this view neglect that the use of our tools to test for and reverse adverse effects is constrained on many fronts, from cost to complexity. In particular, they neglect the limits to the time, space and strengths over which tests and assessments are conducted, and the constraints on the decision making needed to appropriately apply the results.

Consider the concentration level below which our short-term, single-chemical safety tests determine that a compound is safe. When we apply these levels to a real-world site, little, if any, attention is paid to the possibility of adverse effects on organisms from a chemical whose concentration lies below that safe level. The lack of attention can be a serious oversight. For example, an organism experiencing long-term exposure to sub-toxic concentrations may still suffer from significant, adverse, chronic disruptions to its endocrine system. Consider too that the organisms at the site may be exposed to not just one chemical but to a number of other natural and human-made chemicals. The organism may suffer from significant, cumulative adverse effects that result from this mix and the interactions among them. Think of organisms in the Arctic.

In the same vein, the environmental impact assessments of development projects, from inventions to mines, commonly focus on the area and people likely to be immediately and directly affected. It is normal to assess a project in relative isolation from its wider surroundings. These assessments commonly use short-term (1- to 2-year) tests and baseline studies, and pay limited attention to possible slow-forming, distant or cumulative adverse effects. A concern for scale is warranted.

Dealing with the challenges of scale is easier when the rules of thumb about it are used in conjunction with the three methods. Consider how they can, collectively, help us to recognize and deal with the influence that our human characteristics and our culture's world view have on the attention we pay to scale. For example, Western culture's world view guides its members to use the notion of progress, and its expected arrival, as a reference when evaluating a proposal or imagining the future. This allows the proponents of a new project, discovery or invention to promote, imply, if not promise that it will likely "soon" usher in a better future: a pitch that our human focus on the immediate finds appealing.

Westerners' resulting tendency to enthusiastically buy into new projects in the hope of a soon-to-arrive better future makes it difficult for us to think about a new project using more environmentally relevant scales of time, space and strength. Instead, we accept the short-term, small-area environmental tests and assessments. For example, in the 1940s, CFCs were hailed as wonder chemicals because initial testing of their physical and chemical properties were interpreted to show that they were safe and could bring huge benefits. We now know the full reality. Today we can think of nanotechnology, GMOs and tar sands oil production in the same vein. Efforts to face down the general feeling of excitement that accompanied the original announcement about these new technologies on the grounds that a more environmentally and socially relevant scale of safety and environmental-impact testing was needed had limited impact.

An appreciation and consideration of the range over which both the drivers of change and change itself take place is a critical part of making viability-favouring decisions. We can broaden our appreciation of scale by applying the three methods, especially if we remember the relevant rules of thumb. Here is a final one. Think globally, act (or, as needed, don't act) locally.

**Culture's Tools.** The world views of industrial and industrializing cultures guide, their members to believe that they will avoid resimplification because of their superior power and tools. Western culture is no exception. Westerners believe that we will sequentially apply various combinations of our tools to endlessly adapt to a changing world in a manner that will maintain business as usual. In more practical terms, we will use our tools to transform our environment and the resources it provides so that we can continue to live our culturally supported lifestyles and strive toward our culture's objective of progress, particularly economic growth.

However, treating our tools as solutions only seems reasonable if history is looked at through Western culture's rose-tinted glasses. This subsection indicates how applying the three methods can help us adopt a more realistic view of the tools of science, economy and technology. By doing so, we can help change our culture's world view and guidance about its tools so that their use favours viability. Each of the three subsections builds on a synopsis of the discussion in chapters 12, 4 and 5, and 17, respectively.

*Economics.* Westerners treat the tool of economics as an important cultural guide. We believe that if we allow economics to guide science to make the discoveries that technology can apply (with the help of innovation, of course), then economic growth will result. It, in turn, will supply the wealth needed to use science and technology to solve our issues. However, that promise for the future is one that economists, not the economy, has made. It is a promise that Westerners can believe in only when we neglect the many substantial limitations to applying economics, as was discussed in chapter 12 (van den Bergh et al. 2000).

Here is a reminder of the degree to which neoliberal economics, the current dominant brand, ignores how the environment functions and poorly considers our human characteristics. Its proponents assume, for example, that efficiency and substitution are capable of overcoming resource limits indefinitely. This leads to the assumption that resources are essentially infinite, in a non-renewable sense. Neoliberal economists still treat consumers as rational and assume that price and income have a greater capacity to change human behaviour than they do. These types of unrealistic assumptions and deductions about the external reality permeate neoliberal economic models and guidance. No wonder that it promotes short-term, narrow, human-focused decision making that ignores the human and environmental constraints within which we must all live (Hernstein 1990; Colander et al. 2009).

The promise that economics will save Westerners in the long run is a forlorn hope. Unfortunately, Western culture's world view still treats neoliberal guidance as reliable. As a result, the decisions that arise from its guidance are subject to only minimal and ineffectual scrutiny, as illustrated by the global financial crisis of 2008 (Pasquale 2015; Luyendijk 2015). By applying the three methods when considering the tool of economics, we can recognize these limitations and ensure that we only follow its guidance when it satisfies viability's conditions.

*Science.* The tool of science reveals to us the nature of our external reality. Westerners believe that this understanding is important because it will help us adapt, in the Western sense, to a changing world by identifying both the constraints on us reaching for our cultural objectives and how we might circumvent them. In particular, science will allow us to develop the technology needed to overcome the constraints we face when transforming our environment in order to maintain business as usual. Think fusion energy.

But science and its discoveries are of far greater value to industrial cultures than acting as technology's assistant. Science provides us with the opportunity to gain greater insight into ourselves. In particular, it allows us to more easily appreciate our internal reality and how it differs from the external one. Science can help us remove our divergences and deal with our cognitive dissonance and reality paradox experiences in ways that favour viability. In effect, science gives us the opportunity to recognize and take into account the biases in our personal perceptions and decisions without us having to change our innately (unchangeable) biased decision-making processes. It can help us gain wisdom and make wise decisions.

However, science is limited in its ability to help us in this way. Yes, it does provide us with information. But that knowledge is inherently uncertain because it is always surrounded by the great unknown and by those aspects of the topic that remain assumed or unconsidered. Neither is science empowered, by at least Western culture, to tell us how to live (Ziman 1978).

Science thus has a limited ability to provide us with a fail-proof solution to our environmental, personal and cultural issues, or to faithfully keep us on the road to some divine destiny. Neither can it tell us whether our decisions about applying the knowledge it provides are appropriate or not. It can only indicate the likely consequences of our choices, if we ask.

We alone are responsible for gathering scientific knowledge, applying it and evaluating the consequences. It is up to us to decide if we want science to help us discover the external reality of a situation. Only we can decide to integrate that scientific information, its uncertainty and its implications into our lives and decisions. Only we can decide to use it to adjust our internal reality and our culture's world view to favour viability (Costanza 1994). Science can help us to gain wisdom, but it cannot provide it to us.

Gro Brundtland, who promoted sustainable development, was right to say that cultures should ensure science contributes to their political decisions. But she failed to mention that science is limited in its ability to be a cultural guide (Brundtland 1997). For example, the scientific method of reductionism is a good research guide but a flop as an attitude toward nature (Raymo 1998).

Overall, if science is to help a culture satisfy viability's conditions, then its members must be prepared to change their personal and their culture's world view to better match any unpleasant facts science might uncover. Those facts include our biases, divergences, the limits to science, culture's influence on the practice of science and our extensive impacts on our environment's functioning. The three methods are ideally suited to helping us accept and deal with those facts.

*Technology.* The tool of technology is science in action and the handmaiden of economics. The application of technology allows us to manipulate the world in previously undreamed of ways. This ability helps us to satisfy our immediate personal wants and maintain our complex culture's current functioning.

Westerners, like the members of all industrial cultures, are proud of our ability to apply technology, and we take credit for the benefits. We do acknowledge that we are responsible for providing the resources required to build and run technological devices, and the time and effort needed to develop and maintain them. However, we prefer to neglect that we are also responsible for its unwanted and collateral effects, such as waste disposal issues and environmental disruptions.

Westerners also have a strong tendency to believe that technology will inevitably solve our problems. For example, we believe it will ensure that we will satisfy viability's conditions. But our hopes for technology are ones that we have made on its behalf, not ones that are inherent in its functioning or application. We neglect that the benefits we can receive from applying technology are constrained by its nature, how its devices function and the complexity of the world. For example, applying it rarely overcomes the constraints on our lives, it just transforms them and reveals new ones.

In essence, we fail to see that we have a dual relationship with technology. We see ourselves as the master of technology, but at the same time we place our hope in technology as if *it* can determine our destiny. In reality, technology is just a tool, as discussed in chapter 17.

Ultimately, the value of a tool like technology doesn't lie in our ability to build it or in what it can do when applied. It lies in the appropriateness of our decision to build and apply it. It is up to us to decide when, how and how much technology should or, more importantly, shouldn't be applied and what level of our resulting costs/responsibilities are reasonable. These decisions will determine the degree to which we will be the masters over technology or its obedient servants, and whether we will avoid or accept our responsibility for the unwanted effects from applying it (Wright 2004). The three methods can help us make those decisions in a way that favours viability.

In overview, by Westerners applying our culture's tools, we accomplish many things. However, our hope that their (isolated or sequential) application alone will unconditionally solve environmental, resource and social issues, while satisfying our wants and maintain current trends (e.g., economic growth) indefinitely, is unrealistic. The application of our tools will only help us to secure a viable future if we routinely constrain and direct their use to that end. For that to happen, both our personal and our culture's world view must guide our decisions about applying tools to favour viability. Which means that we, the decision makers, are

responsible for adjusting those views and our decision making to favour viability.

For Westerners, like the members of other more complex cultures, making these changes is particularly difficult. For example, our culture's current world view guides us to find our sense of meaning, purpose and place through our transformation of the environment and its resources. We transform them by applying our culturally respected tools. By doing so, we provide ourselves with a sense of stability and belonging. Fortunately, we can adjust this feedback loop so that it favours viability by taking the time and effort to apply the three methods to that end.

## Consumer-Disposer. Can our personal efforts to practise the three methods (nuanced thinking, increasing our knowing and tweaking our personal world view) help us ensure that our personal consuming and disposing decisions satisfy viability's conditions. In particular, can they help us make the decisions and changes needed to reach for the five-objective unit representing viability's environmental constraints? The following discussion will allow us to draw a conclusion.

The consumption of resources and the disposal of waste (referred to here as consumption, as the one follows the other) is a normal part of how humans live and our cultures function. However, when our consumption exceeds the ability of the environment to provide those resources (including services), or when our consumption has significantly disrupted the environment's functioning that provides them, then consumption has become over-consumption. For complex cultures, this point is reached when the members' consumption burdens them with resource-related (environmental or social) responsibilities that they are unwilling or unable to recognize, accept or fulfill. They and their culture are then on the path to forced resimplification.

The members of Western culture are currently over-consuming, as illustrated by our increasing environmental management responsibilities and the results of our poor record at fulfilling them. We could reduce our consumption by striving to satisfy the environmental objective of reducing our culture's total resource demand. As discussed above, we could do so by a mix of decreasing our culture's population, decreasing its per capita demand for resources and increasing its efficiency of resource use. But, as noted, success ultimately depends on how we personally answer a question tied to each of these actions, respectively: How many children is appropriate? How much stuff is enough for a good life? and Where do I want the resource savings from efficiency to go? If our answers are to favour a reduction in our culture's demand for resources, then our decision making must deal with the influences guiding us to maintain or increase our personal level of consumption. This section considers whether applying the three methods during our day-to-day living can help us reduce the power of those influences. It focuses on the influences that arise from Western culture's world view, consumer marketing and ourselves, and the feedbacks that sustain them.

The guidance from Western culture's world view favours consumption. A key reason is the prominent place it gives to economic growth. One expression of that prominence is the consumerism-producerism aspects of the self-sustaining virtuous cycle, as discussed in chapter 10. In it, mental, material, financial and energy resources are used by producers to provide the goods and services needed to satisfy a Western consumer's demands. By purchasing these goods, the consumer releases the funds needed to pay for the cost of providing the resources, producing the goods and paying those working in the producer-consumer economy. This completes the consumer-producer feedback loop. Among those being paid are the marketers, who maintain/enhance the consumer's demand for the goods and services being provided (figure 14). Overall, the cycle is geared toward increasing the demand for resources over time.

Western culture's world view favours consumption for another reason. When a resource shortage arises, the guidance from Western culture's world view favours the members resolving it by increasing the supply. At some point, the effort to satisfy a resource demand from a finite resource requires an increase in our culture's organizational complexity. The more complex the culture's organization becomes, the more resources it must use to stay functional. By focusing on increasing the supply of resources as a solution to issues, Western culture maintains a complexity-resource feedback loop that favours an increase in its resource demand.

Western culture's world view provides further consumption-favouring guidance to its members. As Westerners grow up, we absorb from our parents and surroundings our culture's consumer habits and beliefs, and the reasons why we should maintain them. For example, we are guided to believe that consumerism is an acceptable indulgence and an economically critical activity. But that guidance provides more than

permission to freely consume and dispose of goods (i.e., resources). Shopping and consumption are culturally acceptable methods for us to personally discover and confirm our identity and status (Campbell 2004). This belief/permission is further reinforced by consumer marketing. Throughout our adult lives, the consumer-culture habits and beliefs that we absorbed as we grew up are reinforced by social pressure and consumer marketing. Of course, consumer marketing is itself a culturally acceptable activity.

It is therefore easy for a Westerner to believe that the primary role of the environment in our lives is to supply us with the resources that enable us to discover and be who we are by living a consumer lifestyle. Think jobs and goods. If a Western consumer wishes to reduce their consumption (their culture's demand for resources), then they must deal with all of these consumer-favouring influences. The challenges involved in doing so are made clear by considering consumer marketing's influence on our consumer decisions. The three methods can help us deal with them.

A reminder of marketing's single-minded promotion of consumption and how it influences our thoughts and decisions is provided by examples from consumer advertising. First consider advertising in Canada during WWII. Its messages commonly suggested that buying the target product would help the war effort. Immediately after the war, advertising changed its tune. Its messages now suggested that Canadians could justify their purchase of the target product by treating it as their reward for having fought and won the war. Or it implied that purchasing the product was a right that was somehow tied to the political freedoms for which they had fought the war (Broad 2005).

Next consider VW's success at boosting the sales of its VW beetle during the 1960s. The company used advertising to associate the characteristics of the "bug" with the counter-culture values of the day. Ever since then, advertising has been successful at selling products by associating them with anti-establishment rebellion and individualistic values (Slade 2006).

Today's equivalent of these two examples is consumer advertising that urges Westerners to purchase "ecofriendly" cleaners, cosmetics, biofuels and electronic devices because doing so is environmentally responsible. However, lurking behind advertising's green guidance lies the enduring message: maintaining or increasing the consumption of something is normal. All that has changed between the war and now is the clothes the messenger wears to promote consumption.

Beyond some point, consumerism can't become green any more than it can reduce consumption.

These examples also remind us that consumer marketing (including its advertising) has inserted itself into Western culture's member-culture feedback loop. Marketing thus plays a key role in the process by which adult Westerners create/support our culture's world view, especially its consumer-supporting aspects. Think economic growth. Marketing also helps our children absorb those cultural aspects into their personal world view as they grow up, as discussed in chapters 2 and 10. Think advertising to children. And once absorbed, our consumer-supporting beliefs display considerable inertia to being changed.

Thus, despite Westerners' personal claims to the contrary, consumer marketing has a powerful, wide-ranging influence on our lives that extends well beyond reinforcing our consumer habits and a particular consumer-disposer decision. Consumer marketing is part of Western culture's virtuous cycle and found in both its complexity-resource and member-culture feedback loops. Marketing's influence helps to veil Westerners' awareness of the human and environmental constraints on our consumption and waste disposal choices (i.e., on our lifestyle). And it helps to limit the attention we pay to the environmental and social consequences of ignoring those constraints. Overall, consumer marketing alters both our personal and our culture's world view and contributes to the divergences present in both. If our efforts to reduce our personal consumption and disposal are to be successful, then we must deal with marketing's ongoing widespread insidious influence on our decisions and our culture's world view. And we must do so in ways that favour viability.

But not all aspects of Westerners' consuming and over-consuming decisions can be blamed on consumer marketing or our culture's world view. After all, we are the decision makers. In keeping with the Western notion of individualism, Westerners tend to conceive of ourselves as autonomous, rational, informed consumers. We therefore tend to believe that our decision to buy a specific product or service and dispose of the waste is a personal, freely and consciously made, logical choice. This is one reason why we can believe that consumer marketing has no effect on us.

In contrast, as discussed in chapter 2, marketers' success at influencing us depends on our consumer choices being made using foible-prone decision making processes that include a mix of social, emotional and rational aspects. We largely meet this expectation. We too, contrary to our beliefs, are responsible for our consumer and over-consuming decisions.

The resulting overall influence on our consumer decision making is illustrated by the difference between what we think an increase in our personal purchasing power can give us and what it actually does. Westerners imagine that an increase in personal wealth, which is a proxy for consumption, will significantly increase our social-psychological well-being, such as our self-reported happiness. This is rarely the case. Increasing our wealth has only a limited effect on our feelings of happiness (figure 16). In contrast, the feelings we prefer, including happiness, can more reliably be produced, increased and sustained by satisfying our social-psychological needs, such as being genuinely respected.

However, an increase in our personal wealth does provide us with the psychological benefit of an increase in our positive judgement about our life's circumstance (our satisfaction with our situation). Being wealthier enables us to make the judgement that we can have what we want. But the increase in the satisfaction we receive from a set increase in wealth declines rapidly as we become richer, as discussed in chapter 12. In essence, the richer we are, the greater the increase in wealth we need to increase our sense of satisfaction by a fixed amount.

In addition, regardless of our absolute personal wealth, the increase in the social-psychological benefit (i.e., our satisfaction plus happiness) that we receive from using our wealth to more freely consume is often short-lived. This occurs because of other psychological effects, such as rising expectations and misidentifying (including being misdirected by marketing) the sources of our pleasant feelings. For example, on the merry-go-round of the shopper's cycle, a purchase provides us with an immediate increase in social-psychological benefit. But that feeling soon fades. If we falsely imagine that a product is a source of our happiness, then as that feeling fades, it is replaced by the urge to make a further purchase in order to rekindle a feeling of happiness. When all of these factors are combined, it is no wonder that the consumption associated with an increase in wealth can quickly become over-consumption (Diener et al. 2010; Myers 2000; Easterlin 2003; Kasser 2002).

Applying the three methods can help us to accommodate the biases in our consumer-disposer decision making and compensate for the influences on it. For example, a popular rule of thumb to deal with our garbage is the three *r*'s: reduce, reuse, recycle. Currently we interpret them to mean reduce our discarding through recycling and reuse. But, in the long term, garbage management that favours viability will require us to focus on reducing the garbage we create in the first place: by reducing our consumption (Rathje and Murphy 1993a). This means adjusting or switching our lifestyle so that we consume less per person. We are helped with this task when the three *r*'s of garbage are joined by the three *a*'s of consumption: avoid, alternates, appropriate.

You can *avoid* unnecessary consumption by asking yourself pointed questions such as, do I really need this? and, if I buy this object, where will it be in six months? You can seek *alternates* to consumption by asking yourself questions such as, what is driving my desire for this product in the first place? and are there other ways to receive the material or psychological benefit I want or expect from this product? You can decide on *appropriate* consumption by asking, what is the personal and environmental price of this convenience? and, will this product help me reach for the environmental objectives in the five-objective unit? or, will it promote more consumption? (Huskins 1998).

These questions are part of answering the three basic consumer-lifestyle questions asked at the beginning of this subsection: How many children is appropriate? How much stuff is enough for a good life? Where do the resource savings (including time) from increasing efficiency go? They could be rounded off by including, what is actually important for living?

Applying the three methods in our day-to-day lives prompts us to ask these questions and then helps us answer them in ways that favour viability. After all, the methods favour us discounting or disregarding the factors promoting consumption (such as consumer marketing, some aspects of our culture's world view and certain of our human characteristics). For example, they help us to personally recognize and accept (rather than being told) that consumer products (goods and services) have a limited ability to satisfy our psychological and social needs. They also help us to realize that there are less consumptive ways to order the goals we place on our personal priority list for living. In particular, we can more easily accept that our long-term social-psychological well-being is better addressed by wanting a little than having a lot.

Applying the three methods also help us to avoid consumer traps. One example is the "landfill goods" trap: the purchasing of goods that appear to us to be both a bargain and able to quickly satisfy our immediate material or psychological needs. These are the goods that just as quickly leave us unsatisfied. They too soon end up in "storage" or in a garbage dump, seldom used, irreparably broken and either difficult to recycle or unrecyclable. Applying the methods can also help us to avoid the traps of the "drive for the new" cycle (think cell/smart phones), impulse shopping, rising expectations and the shopper's cycle. They help us to realize that a happy life, or even a happier life, can accompany a lifestyle that is less consumptive than Westerners aspire to.

Admittedly, as the doors to a big box store swoosh open before you, or as you watch the parade of enticing ads for stuff that shout at you from a screen, it is easy to come to the conclusion that we are stuck in a consumer culture feedback loop from which we can't really escape. But take heart! If we can be converted to consumerism, we can create the feedback that will convert us away from it (Kasser 2002).

Consider how the three methods, with the help of the three *a*'s and the three *r*'s, can adjust the meaning of an "informed consumer." It currently means someone who can get the best bang for their consumer buck. There is no implied constraint on consumption. Instead, an informed consumer could come to mean someone who, firstly, recognizes what really makes them happy and, secondly, accepts the environmental constraints on how they can live. This would be the context in which they would strive to get the most appropriate, viability-favouring consumer bang for their buck.

In overview, by applying the three methods while reaching for the objective of reducing our personal resource demands, we can more easily recognize the biases in our consumer-disposer decision making and the (cultural and commercial) consumption-favouring influences on it. We can then more easily accommodate or compensate (account) for them in ways that favour our resource demands staying below a level that our environment can sustain over the long term. In essence, by applying the three methods we prompt ourselves to create an environmentally affordable personal budget for our consumer-disposer wants. And we are more likely to stick to it because our focus will be more on a lifestyle that favours our mental quality of life rather than on our material standard of living. This allows us to treat a reduction in consumption as part of living a happier life, rather than a deprivation. Through these changes we can redirect our consumer-disposer decisions away from unnecessary/excessive consumption. Consciously applying the three methods while striving toward the objective of reducing our personal resource demand can indeed help us to reduce our personal level of consumption, and disposal, while improving our quality of life.

More is accomplished because we also help to create a (individual-culture) feedback loop that supports and sustains our personal efforts over the long term. This increase the chances that our resource saving will be used by our culture to reduce its total (absolute) demand for resources. We can further increase our chances of being successful if we use the three methods to help us simultaneously reach for the other four objectives representing the environmental conditions for viability.

In summary, we can more easily deal with the challenges we face in our personal efforts to favour viability if we apply the three methods in combination with rules of thumb. This was illustrated by examples of these challenges: complexity, the scale of things, culture's tools and being a consumer-disposer. The methods and rules can help us remember and accept the conditions for viability. They can also help us to make the decisions and the (specific and broad) changes needed to satisfy them.

If we apply the methods and rules while reaching for the five-objective unit representing viability's environmental conditions, for example, then they will help us to become environmental trustees. As we make the changes to become trustees, we will be exposed to the human conditions for viability. If we decide to apply the methods and rules to address those conditions, then we can more easily satisfy them. The methods and rules can indeed help us deal with the challenges we personally face when trying to satisfy both viability's human (individual and cultural) and environmental conditions.

If enough of a culture's members are successful at changing themselves to favour viability, then the members can, collectively, adjust their culture's world view to favour viability. They will thus create a feedback loop that will guide them and their culture to follow a path toward viability and keep it on that path over the long term. But, if their culture is complex, then the members' success at making these (personal and cultural) viability-favouring decisions and changes requires something else. The culture's elite, in their role of leaders, must actively promote striving for viability and support the efforts of those members trying to do so.

## The Elites' Responsibilities

The presence of a hierarchy in the more complex cultures ensures that some of their members have greater individual, moral, social, political and/or financial influence over their culture's functioning than the average member. This elite group includes politicians, the captains of industry, the wealthy, celebrities and the upper echelons of cultural institutions such as churches, the army, hospitals, NGOs and universities. They are, in the broader sense, their culture's leaders. Along with the power and perks of being a leader come responsibilities.

If a more complex culture is to become viable, then all of its members (elite and public) are required to adopt viability's conditions as goals. They must also reduce their personal and their culture's divergences, and accommodate our decision-making foibles, and do both in ways that favour viability. The elite, in their role as leaders, are each obliged to make additional viability-favouring decisions and changes to their lives. The elite's extra responsibilities and how they can be discharged are discussed in this section, using Western industrial culture as an example.

But first consider a few of the ways in which the elite of Western culture exercise their leadership powers. Some elite claim, exercise and maintain political power through Western culture's ongoing authority reaffirmation and maintenance process. One part of that process occurs through wealth-influenced interactions between the elite and the public (as individuals and as subgroups) that take place through social and traditional media. Another part occurs when the political elite meet the public's expectations about good governance. A part of the process also occurs through ballots, in their various forms. In all of these parts, the public can exercise some influence over their political leaders' decisions.

Those elite with no direct political power (extra-political elite) exercise their leadership, in part, by lobbying those who do. They can also exercise it through their power over the institutions/corporations and social groups that they control, head or participate in. They can also use their wealth, for example, by making significant donations to causes they think will further their aims. The extra-political elite thus exercise some of their leadership power in public, where it is subject to some influence from the members of the public, and some in private, where it isn't.

The elite of all stripes are cultural leaders who are, by their position, burdened with the responsibility of exercising their extra power in the greater public's best interest. However, they need not always do so. Indeed, some exercise their power in a dominantly self-serving manner. Think of self-enrichment, status seeking and preferences for ideologies that support these goals. Regardless of whether they exercise their power to satisfy these self-serving ends in public or secret, acting in this way corrupts a culture's functioning (MacLean 2017). The connections between the way the elite exercise their power and discharge their additional responsibilities, and the likelihood of their culture becoming viable will become apparent as this discussion unfolds.

If Westerners want our culture to become viable, then we must ensure that, for example, the environment can support our culture over the long term. Currently, Westerners' view of both our relationship with the environment and how we can treat it ensures that this is not the case. In keeping with how all complex culture's function, Westerners tend to think that this problem could be solved by our leaders simply enacting and enforcing laws. In particular, we look for laws that would require all decisions to more accurately reflect the reality of our environment's functioning and its constraints on the lives of us all. These laws would, for example, restrain us from disrupting or destroying our culture's environmental commons, such as air and water resources and ecosystem functioning (Ophuls and Boyan 1992; Hardin 1968; Hardin 1998). Certainly, just as laws and their enforcement are key to keeping a complex culture's social aspects functioning, they are key to keeping its environmental endowment functioning and its decisions favouring viability. However, the basic nature of a Western legal system means that the degree to which laws, on their own, can achieve these goals is limited.

Consider that our legal system, like all legal systems, is based on, and depends on, the cultural world view we inherited from our ancestors (McEvoy 1986; Scheiber 1970). However, as discussed in the introduction, Western culture's world view and thus its guidance contain divergences from the external reality. This is the guidance that influences our culture's present law makers. Those divergence-influenced laws can therefore poorly represent the nature and importance of the environment's functioning and the constraints it imposes on our lives, if the culture is to be viable.

In addition, even if a Western government did propose an environment-protecting law that was based on a realistic assessment of the environment's functioning, the law can be rendered ineffective by other influences on the lawmaking process. Think self-interest and lobbying. Similarly, once passed into law, the enforcement of the weakened law can be undercut by a lack of will, corruption and lobbying. Passing and enforcing laws that favour viability is more difficult than we expect.

A wonderful example of how environment protection laws, such as environmental management and impact assessment laws, can be imperfectly constructed or rendered ineffective in this way is provided by the historical saga of the Californian fisheries (McEvoy 1986). A more contemporary example is the Newfoundland cod fishery collapse of 1992. Before the collapse of the Newfoundland fishery, the Canadian government prided itself on having established laws that it believed would establish a comprehensive, science-based environmental management plan.

After the cod collapse, the plan was labelled as a classic failure of science-based fisheries management. It was described as being based on reductionist, top-down forms of coercive politics that considered rational humans to be the masters and owners of nature. The laws were changed.

The replacement laws were based on a new ethos. It provided for management plans in which ecosystems were seen as complex, non-linear, hierarchical, eco-social systems. Plans would be based on the understanding that humans are part of ecosystems. Management would thus occur through the participatory politics of inclusion and consensus. This mouthful of changes sounds positive until one looks closer. One idea considered to be compatible with these apparently new goals was to replace wild-fish stocks, such as the cod, with fish from ocean-based fish farms (Bavington and Sajay 2003; Bavington 2005).

Although the cod fishery management plan had changed after the collapse, the underlying cultural vision, as reflected in the laws and thus the management objectives, had not. The Canadian priority was, and still is, to ensure that ocean ecosystems provide as much resources as possible, as soon as possible, only now both humans and nature are being managed (Rowe and Rose 2017). This example illustrates how, despite the best of intentions, the environment can remain under-represented in decision making by the law makers, and how Western culture's complexity-resource feedback loop can remain intact.

The above discussion indicates that without deep and broad viability-favouring support from both the elite and the public it is unlikely that, from the environment's perspective, meaningful environmental protection laws will be passed and, if they were, that they will be meaningfully enforced. If Western culture is to rely on laws to ensure that its members' decision making favours viability, then the elite, as leaders of the culture, are burdened with the extra responsibilities of ensuring that both the elite and the public support viability. The elites' extra responsibilities require them to, in broad terms, direct the members to favour viability (e.g., adopt its constraints as goals), support the efforts of those that are, and oppose those whose actions don't favour viability. This is the case for the elite of all the more complex cultures.

There are constraints on how the elite can discharge their extra responsibilities in a socially meaningful manner. In particular, it is insufficient to just claim or promise to fulfill them. The elite should demonstrate to the public that they are fulfilling their extra responsibilities through their deeds. A few examples of what this means are discussed here.

The elite could discharge their directing responsibilities by publicly recognizing that, in the long-term view, we have been largely unsuccessful at fulfilling our environmental management responsibilities. They could attempt to correct that deficiency by using their directing tools, such as policies, laws and spending, in ways that unambiguously favour viability. For example, the elite could establish "no environmental regrets" policies and encourage the use of the precautionary principle (Alley et al. 2003). Doing so is especially important when evaluating or implementing decisions that affect our commons (Dietz et al. 2003; Ostrom et al. 1999).

The elite could demonstrate in many ways that they are providing mental, managerial and material support to those members already striving toward viability. For example, the elite could support those striving toward the five-objective unit by publicly questioning terms such as *sustainable development* and *sustainable growth*, and the notion of progress. They could treat technology, science and economics as tools, not cures. The elite could make decisions and voice opinions that recognize the practical aspects of striving toward viability. Here are some examples of those aspects expressed as rules of thumb. The less we disrupt our environment, the fewer our environmental management responsibilities will be. There is a limit to the resources that can be extracted from an endowment without disrupting its long-term functioning. A focus on maximizing the production of finite natural resources from an endowment is incompatible with maintaining its long-term functioning (Hughes 1975; Bavington 2005).

The elite could demonstrate that they are effectively dealing with those opposed to striving toward viability, and do so in ways that favour viability. How depends on which of the many forms this opposition can take. Some are deliberate, others ill-considered or inadvertent. Some are covert others are overt.

Consider those elite whose objections are based on views that are clearly divergent from the external reality: views that lie outside the centre of the viability flower (figure 22). An example is the elite who use neoliberal economic views to counter the idea that restraining business (as represented by the market) is necessary to help meet the five-objective unit. One such unfounded argument is that an unrestrained market and continued economic growth are compatible with the environment's functioning and can protect it.

These types of neoliberal arguments are divergent because its economic theory and business practices (its world view) do not inherently consider either the environment's actual functioning or its constraints on our lives, especially over the long term, as discussed in chapter 12. Think of externalities. By believing in its world view, economics-related environmental disruption and destruction can be accepted as the normal cost of maintaining growth, promoting competition and satisfying self-interest. For example, the tragedy of the commons can be accepted as part of rational economic behaviour or justified as an inconsequential or unexpected outcome (Hall et al. 2001).

The elite can demonstrate that they are dealing with this type of opposition in a viability favouring manner by focusing on the false, misleading or incomplete statements. In the above example, economists and the business elite must be pressed to demonstrate that their economic-based objections and alternative solutions are indeed valid and useful in the external reality. Are they based on an adequate amount of reliable field data, a sufficient understanding of how the human-environment system functions and the long-term constraints that functioning imposes on how we can live?

In this regard, those of the elite who are involved in the media have a special burden to bear. The media are a complex culture's communication tools, so it plays a key role in reaching for viability. The media elite would have to demonstrate their commitment to viability by ensuring that their media institutions routinely and accurately represent viability's human and environmental conditions and the challenges of meeting them. The media moguls could, for example, advise their employees that presenting two points of view about an environmental issue is not always balanced reporting. This is especially true when an untested opinion or comment not backed by data is presented as the equivalent (i.e., just as reliable) as a science-based "consensus." The media elite could, at least, report the details of a source's credentials and funding, and any conflicts of interest. And they should distinguish between testable conclusion and opinion. The media elite are also burdened with identifying fake news and limiting its spread by their organizations (Boykoff and Boykoff 2004; Tetlock 2006; Oreskes and Conway 2010).

The elite, in general, also bear the extra responsibility of dealing with those (elite or public) whose opposition to viability includes subverting, corrupting and co-opting the efforts toward it for their own ends (Hall et al. 1986; Goldstone 2002). After all, when these self-interested people/actions are present, the rest of the culture's members find it much more mentally and physically challenging to work toward living within their culture's human and environmental constraints (i.e., toward satisfying viability's conditions). They can lose their incentive. Beyond the mundane practical reasons why this is the case lies another surprising reason. The foundation of the public's willingness to follow an elite leader and strive toward goals such as viability is a sense of trust and fairness. By not restraining the self-interested, the elite undermine the members' sense of fairness and thus their trust.

The critical role that fairness (and trust) plays in the striving for viability is illustrated by the current size of the income gap between the business elite and the average worker. One measure of this gap is the ratio of the average (options realized) compensation provided to the CEOs of the top 350 revenue-earning firms in the United States to that of an average US worker. This CEO-worker compensation ratio changed from 20:1 in 1965, to 59:1 in 1989, 376:1 in 2000, 230:1 in 2010, and 286:1 in 2015 (Mishel and Sabadish 2017). In 1980, near the beginning of the rise of the neoliberal idea of shareholder value, the ratio was 42:1, while prior to that it was rarely greater than 30:1 (Anderson et al. 2010).

The recent large ratios reflect excessive CEO compensation packages. Some packages are so large they can even negatively affect a corporation's shareholders. For example, the average combined compensation for the top five senior executives from a selection of US public companies rose from 5% of corporate net earnings for the period 1993–1995 to 10% for 2001–2003. This indicates that a substantial share of these companies' earnings was dedicated to executive pay rather than to shareholders or other stakeholders (Bebchuk and Grinstein 2005).

Despite the considerable hand wringing by academic economists over whether these CEO compensation packages are justified on economic grounds, the answer to the fairness test is clear to the public. The packages are grossly unfair. And this evaluation is not made any more palatable by another blatant inconsistency between the theory and practice of neoliberal economic ideology: the habit of granting CEOs performance pay even if the business does poorly. Consider what happened to CEO pay during the global financial crisis of 2008 (Business Week 2002; Francis 2009). Collectively, these actions and events severely test the public's perception of our business elites' sense of fairness, and thus our trust in them.

This lack of fairness affects more than the public's relationship with the culpable economic/business elite. It spreads out like a contagion to the other elite/leaders because they are seen to be either in cahoots or to have done nothing to stop the abuse. In this way, the lack of fairness broadly undermines the public's trust in their elite leaders' intentions and their support for them (Bella et al. 1988). This affects the functioning of a complex culture and how its world view will change. For example, it affects the degree to which the public will respect the law and accept the risks to which they are exposed by the elites' decisions (Slovic 1990).

In particular, it affects the members' willingness to try and meet viability's conditions by making personally directed sacrifices, such as reduce our consumption of goods and services. It also affects the public's willingness to work with their leaders to make viability-favouring changes to the culture because we lack trust in the elite's intentions or believe that the implementation of the changes will be unfair. Consider, from this perspective, the example of the leaders' efforts to reduce carbon emissions by putting a price on carbon. The public can easily conclude that this effort is either unfair to them or just another way to disguise how the wealthy, the powerful and their corporations will benefit at the public's expense (Chan 2010). Beyond some point, a decline in both the public's sense of being fairly treated and their trust in the elites' intentions become more than minor distractions from the public's willingness to help reach for viability. They contribute to a culture's possible resimplification. Think of taxes, laws and access to resources that favour the elite: the 1%. Think of tax havens. Think of the Panama and Paradise Papers.

In summary, if a complex culture is to strive to become viable, then a significant proportion of its members must each adopt a viability-friendly lifestyle and work to satisfy viability's conditions. They must strive to reduce their personal divergences and accommodate their decision-making foibles in ways that favour viability. There is an additional condition to fulfill. Complex cultures have a power hierarchy whose upper echelons are filled with the culture's leaders, its elite. Because of their power, the elite must do more to address viability's conditions than the public. In broad terms, they are responsible for directing the members to favour viability, support the efforts of those who are striving to do so and oppose those whose aren't.

In practical terms, the elite's extra responsibilities include strenuously condemning and restraining all who abuse their privileged positions, take advantage of our foibles, spread fake news or misleading information, and exercise excessive self-interest (selfishness) (Kindleberger and Aliber 2005; Ho 2009). The elite should also demonstrate to the public that they are both trying to satisfy viability's conditions and doing so in ways that champion justice and fairness. This is essential if the public are to trust the elite and work with them to satisfy viability's conditions.

The members who are best suited to being a cultural leader (an elite) that favours viability are likely to be more environmentally and socially aware, pragmatic and more realistic than ideological. They also need to be more wise than smart. (Intelligence alone is not a guarantee that a person will make viability-favouring decisions.) They should be willing to ask of a decision, is it fair? and, is it appropriate? within the context of helping the culture and its members to become more viable.

With the support of a complex culture's elite who are striving to meet their extra responsibilities, the public will be more willing to devote the time and effort needed to satisfy viability's human and environmental conditions. Collectively, the members can create a viability-favouring feedback loop that will guide them and thus their culture to stay on a path to viability. With the co-operation of the elite, it is possible for Western culture to consciously emulate the essence of the Tikopian's cultural transition to viability, if they want to. They can undertake a voluntary resimplification.

## The Outlook

Each of us is an individual, a member of a cultural group, the carrier of its world view and its decision maker. And we are each a participant in the earth's environment. We are one of the three primary factors (individual, culture, environment) that form the dynamic, complex human-environment system. Within this system, our decisions contribute significantly to our personal, our culture's and our environment's future.

"What were they thinking?" we rhetorically ask of the Easter Islanders, the Norse and the Xhosa when we consider their choices that committed them and their cultures to a bleak future. Whatever each member's decisions were, they were provided by the innate, powerful decision-making ability of their brain as it worked with their experiences of the world. Thus, at some level, each member was guilty of causing their culture's difficulties because they failed to sufficiently consider the human and environmental constraints on their lives. Yet, they were also innocent because their decision making was limited by their innate decision-making foibles, the influence of their culture's inherited world view on their thinking, and their environment's complexity.

The members of today's industrial and industrializing cultures can be asked the same question. After all, in today's globalized world, there is ample evidence that our decisions are, collectively, keeping us on the same well-worn path to forced resimplification that previous complex agricultural cultures have travelled. So the answer would be the same.

Fortunately, the members of today's more complex cultures have the opportunity to ensure that their culture will avoid at least the worst of a forced resimplification: they can decide to voluntarily resimplify. Think of it as downsizing. It would be misleading to claim that this is an easy task because we now know there are no simple technical, scientific, economic or religious fixes to our current circumstance. Instead, we will have to personally take on the responsibility of consciously guiding ourselves and thus our culture through the slow transition toward personal and cultural viability.

To be successful, we will have to voluntarily adopt viability's human and environmental conditions as objectives/references and direct our everyday decisions and changes to satisfying them. We will have to work at reducing the significant divergences in both our personal and our culture's world view and, when required, make the effort needed to accommodate our decision-making foibles. In practical terms, this means consciously directing our awareness, attention and tool use to those ends, and ensuring that our decisions and changes are made with a sense of justice, fairness, reasoned empathy and principled pragmatism.

Those characteristics are also needed because we will face challenges. They can originate with us, such as self-interest and the nature of how we change, and with our culture, such as the influences of our culture's world view on our decision making. And there are general challenges, like those that arise from complexity, such as the connections among our consumption (and disposal) of resources, our lifestyle, our culture's world view and the state of the environment. Consider too that we will face difficult trade-offs, compromises and ever present uncertainty. We will also have to decide which decisions we should treat as serious, and which not. Overall, the degree to which we, and thus our complex culture, will become viable depends on our personal and collective success at dealing with all of this, over the long term, in a consistent and coherent manner that favours viability.

Your rising feeling that it is impossible to satisfy viability's conditions is tempered by the realization that striving toward it is not the same as working toward a fixed goal, such as finding lost keys. It is more of a slow process of becoming, like striving toward happiness or a voyage of discovery. Success does not require you to finish the task, just work steadily toward it. Therefore, as you strive to satisfy viability's conditions, you should focus less on where you are, how far you have to go, or the media reports of the latest deadline for acting. Focus more on steadily adjusting your beliefs and decisions so that your lifestyle, decisions and interactions with the world around you are more likely to satisfy viability's conditions than not. At the same time, help others, especially the next generation, to do the same. And enjoy the journey.

This chapter has provided a framework and tools that can help you to coherently and consistently discharge your responsibility of consciously changing yourself and your culture in ways that favour viability. The framework is the model of the human-environment system, the various views of it and the appreciation you have gained for its structure and functioning (figures 7, 15 and 22). One set of tools is the rules of thumb dealing with, for example, a complex system's functioning, our decision making and viability's constraints. Another tool is the three methods to consciously direct personal change: increase our knowing, tweak our personal world view, and practise nuanced thinking and decision making. They

are especially useful because applying them helps you to gain wisdom. In addition, objectives were provided to help direct your tool use. One is the five-objective unit that represents viability's environmental conditions: accept our environment's supremacy, maintain its functioning, champion its self-management, conserve what remains and reduce our total demand for resources.

Together the framework and tools help you to make the required decisions and changes, including overcoming the challenges, in a way that favours viability. For example, they help you to reduce the divergences in your personal and your culture's world views, and accommodate your decision-making foibles as you reach for the five-objective unit. By using the tools with the five-objective unit in mind, you will help yourself become an environmental trustee. (This process can also be used to address viability's human constraints.)

Through these means, each member of a culture can come, at least to some degree, to live a viability-favouring life. Think lifestyle. We can each contribute something toward our culture's efforts to successfully complete a cultural transition to viability.

The framework also highlights the additional responsibilities that the political, business, wealthy, military, religious and other cultural leaders (the elite) of a more complex culture bear because of the extra power they wield. The tools can help them discharge these responsibilities. They include directing the public to strive toward viability, creating the conditions that help us to do so and supporting those already working toward it. They also include restraining those (public or elite) who would obstruct, mislead, subvert, corrupt or co-opt the efforts of those striving toward viability. Think of excessive self-interest, misleading statements and fake news. If the elite use the framework and tools to help them discharge both their personal and their additional responsibilities, then they are more likely to be seen as just and fair by the public. We are then more likely to trust them and thus be willing to work with them to address cultural issues in ways that favour viability.

As we, the elite and the public, each strive to meet viability's human and environmental conditions, we will change our self and ourselves, including our personal world view, to favour viability. If enough of us strive to do so, then we can initiate our culture's transition to viability. In particular, we will, collectively, adjust our culture's world view so that its guidance favours us striving to meet viability's conditions. We will thus form a viability-favouring feedback loop that will sustain, over the longer-term, our personal and collective efforts to live more viable lives.

So take heart, we can each look in the mirror and consciously decide to personally live in a more viable manner. There is indeed a lifestyle we can adopt that is both satisfying and favours viability. We can make the personal decisions and changes needed to adopt viability's conditions as goals. We can reduce our divergences and accommodate our decision-making foibles, and do so in ways that favour viability.

We can, collectively, motivate ourselves by imagining a realistic viable future that lies within both the bounds of the external reality and a meaningful timeframe. And we can tell ourselves timeless stories that are relevant to both the present and that future. It is realistic to take inspiration from the Tikopians and believe that we can consciously emulate their slow cultural transition to personal and cultural viability. We can recognize and respond to the abundant signals prompting us to change our relationship with our environment in a way that favours viability. The "isms" we currently use to make our decisions can be tempered with a dose of human-environment system *realism*.

Even though we, the current generations, may devote considerable effort to reaching the goal of viability, we may not succeed in our lifetime. But, by choosing voluntary resimplification, we will leave an inheritance that will enable our descendants to more easily complete the transition to viability. Who could fault us for that? Not even God.

# FIGURES

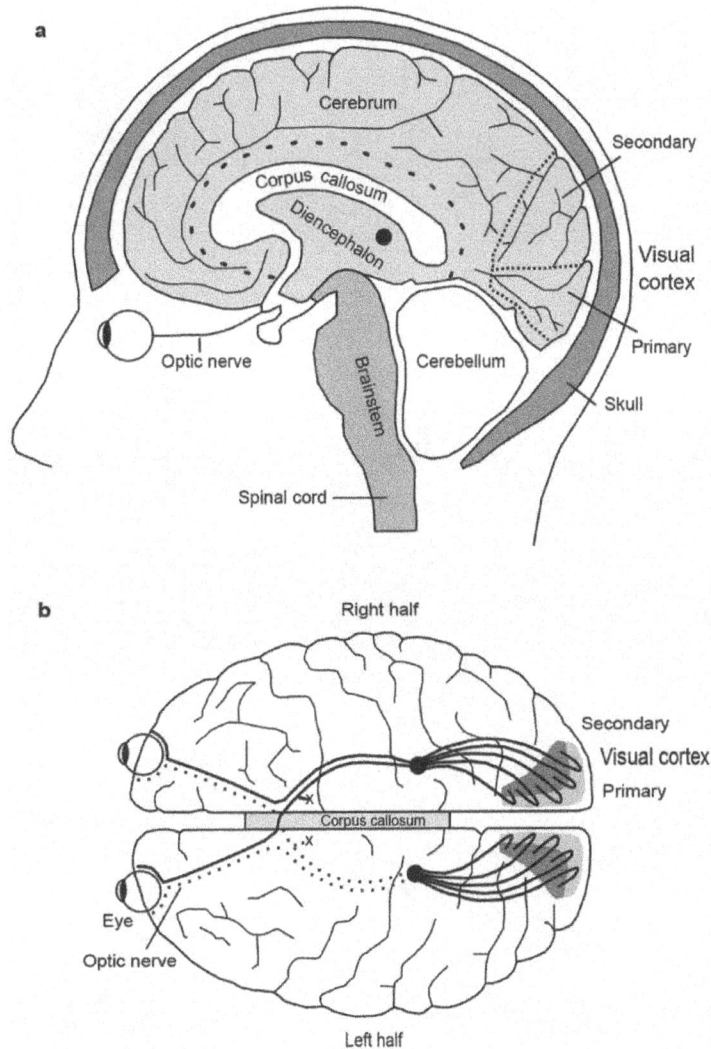

**Figure 1.** Three perspectives of the brain: structure, function and halves. *a* A vertical, front-to-back section shows the relationships between two of its structures: the forebrain (light grey) and the brain stem (medium grey). After Vander et al. (1998, fig. 9-8). The thick dashed line represents an equivalent non-human mammalian brain. *b* A top view showing the two halves of the brain and their primary join, the corpus callosum. After Gazzaniga (1998, p. 50). The function of sight is displayed because it occurs largely on a horizontal plane through the eyes. In *b*, the visual signals from the right and left halves of the eyes (solid and dotted lines respectively) cross to their respective half of the brain before reaching the geniculate nuclei (solid dot in *a* and *b*) at the base of the brain. From there, the signals travel to the primary (light grey) then secondary (dark grey) visual cortex at the back of the brain, after which we experience sight. (The visual cortex is outlined by a fine dotted line in *a*.) The X in *b* marks the site generating our circadian rhythms.

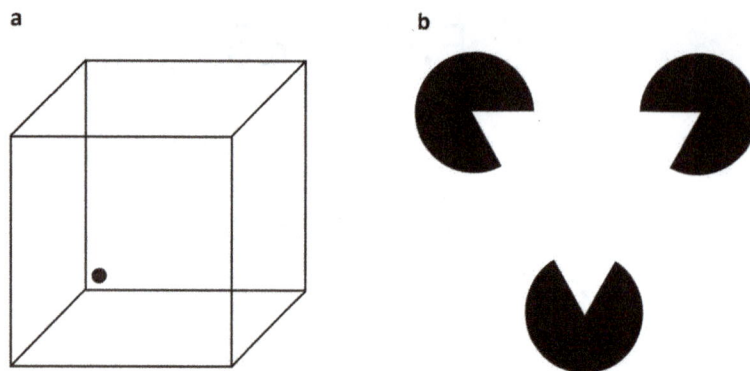

**Figure 2.** Optical illusions illustrate the limits to our visual perception. *a* When you look at the dot, it will appear to be at the bottom left of a cube, either on the outside face near to you, inside the cube and far from you, or switching from one to the other as your brain tries to decide which is the correct interpretation. *b* Can you see a triangle? Look closer. There is no solidly outlined triangle. The one you see is constructed by your brain from its interpretation of the cues it receives from the figure.

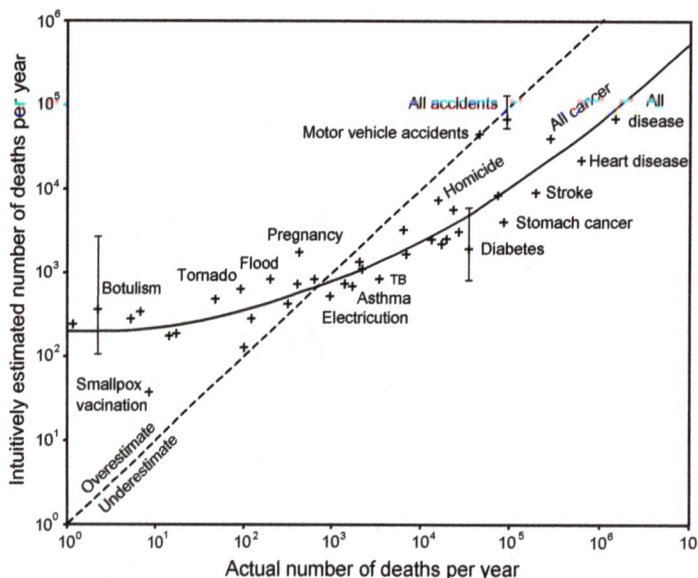

**Figure 3.** Our average estimate of the number of deaths per year from a cause compared to the actual number of deaths from that cause, for 41 causes. The uncertainty in the data is represented by the 25% and 75% bars drawn for three causes. The solid line fitted through the data reveals the systematic bias in our intuitive assessment of the frequency of death from these causes, which translated into a bias in our perception of our risk of dying from them. The figure illustrates a common feature of our intuitive methods of making sense of the world, such as using our recall of similar events to estimate their frequency (availability heuristic). After Slovic et al. (1982, fig. 2).

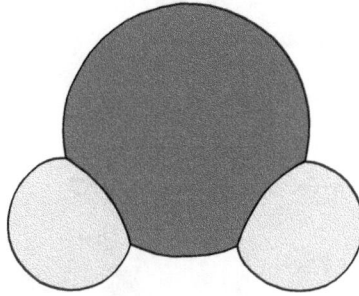

This is not a water molecule.

**Figure 4.** We tend to treat our diagrams and models of reality as though they are reality rather than being a specific, limited representation of reality. Hydrogen atoms are neither hard shelled nor grey. The figure illustrates that broadening our understanding of the world through science faces significant constraints because of the influence that the visual images of its results have on our understanding of them. After Coppola and Daniels (1999, fig. 3).

**Figure 5.** The self-organized, sorted patterned ground that arose from the freezing and thawing of unsorted (randomly ordered) glacial till at Kvadehuksletta, Spitsbergen. It illustrates that the presence of randomness in a system doesn't preclude it from displaying order at some scale, under some conditions. The full circle in the bottom right corner is approximately 2 m in diameter (Kessler and Werner 2003, fig. 1). Reprinted with permission from AAAS. Photo credit: Kessler, M. A., Murray, A. B. and Hallet, B.

**a**

ŠE System Š

System used to note capacity measures of grain, in particular barley.

$U_4$ System U

System used to note time and calendar units (twelve 30-day months to a year).

Sexagesimal System S

System used to count most discrete objects, for example, humans and animals, dairy and textile products, fish, wooden and stone implements, and containers.

**b**

Type of product: Barley. Therefore read using System Š not System S

Quantity of product: In System Š symbol N39 represents one barley measure. Total of 29,706 measures, approximately 135,000 litres at 4.8 lites per measure.

Accounting period: Therfore read using System U. Period of 37 months.

Name of the responsible official: Kushim.

Drawing of cuneiform tablet MSVO 3, 29

Function of the document (?): Final account (?) (inscribed over a partly erased sign.)

Use of Barley (?): Exchange (?).

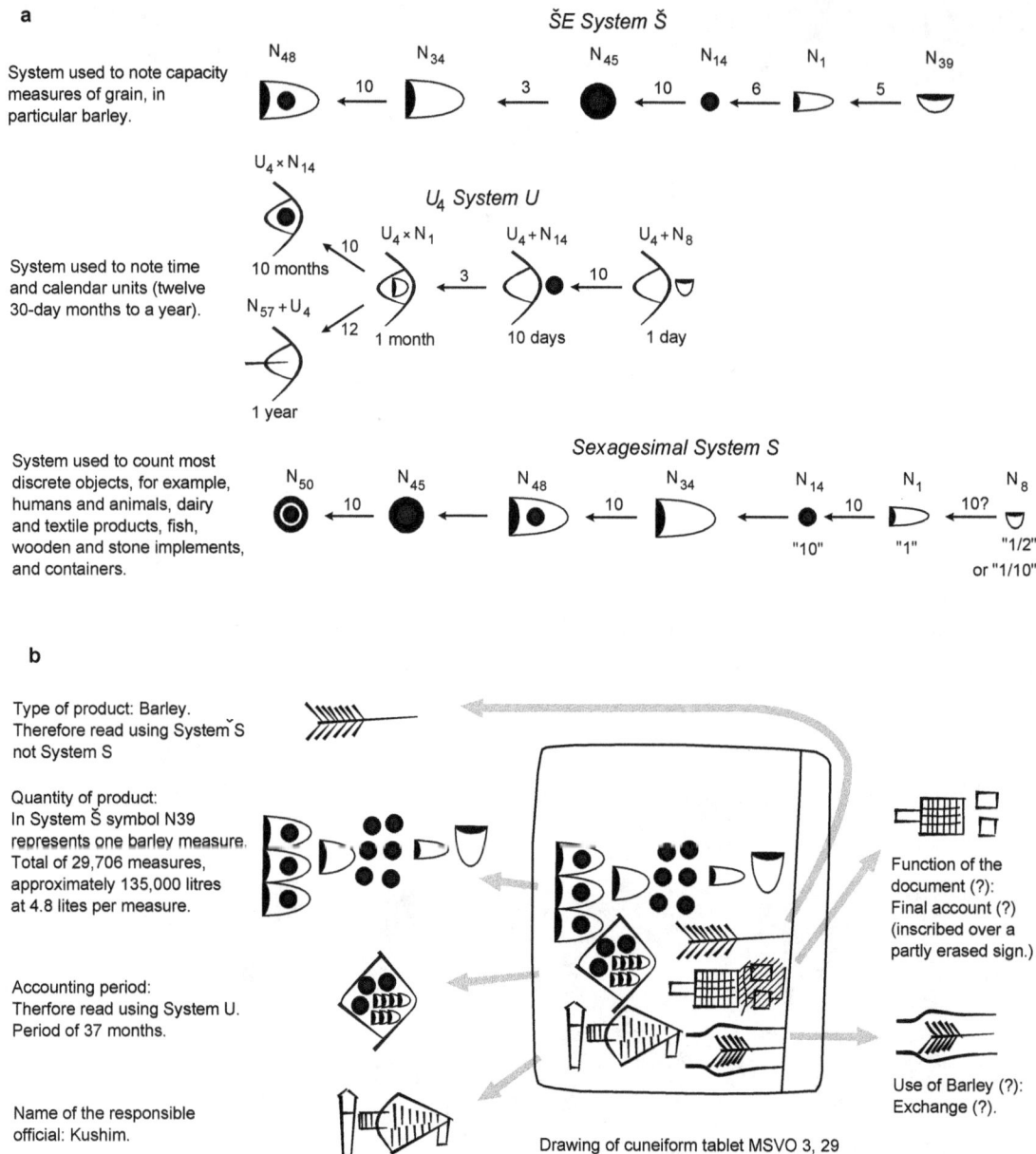

**Figure 6.** An illustration of the Babylonian accounting system and the details needed to use it. *a* Three of the more than 14 accounting systems from which a scribe would need to choose before recording an item. To make that choice, the scribe would have to know how items were used, measured and counted in real life. Note that the same count sign (e.g., the black dot N14) can be used in two systems but represent different values in each. Note too that in a system the spacing between the count signs varies, and that in each system the pattern of sign spacing is different. These features indicate that an abstract knowledge of numbers (a universal number theory) was not used to practise accounting. *b* An illustration of a Babylonian accounting record. The somewhat disorganized arrangement of the signs suggests that to read it, at least in the early stages of Babylonian accounting, the scribe needed to know more than the counting systems used to record items. They likely used the signs as cues to recall how the product was used, measured and counted in real life (Nissen et al. 1993, fig. 28). Copyright © 1995, Copyright 1995 History of Science Society, Inc.

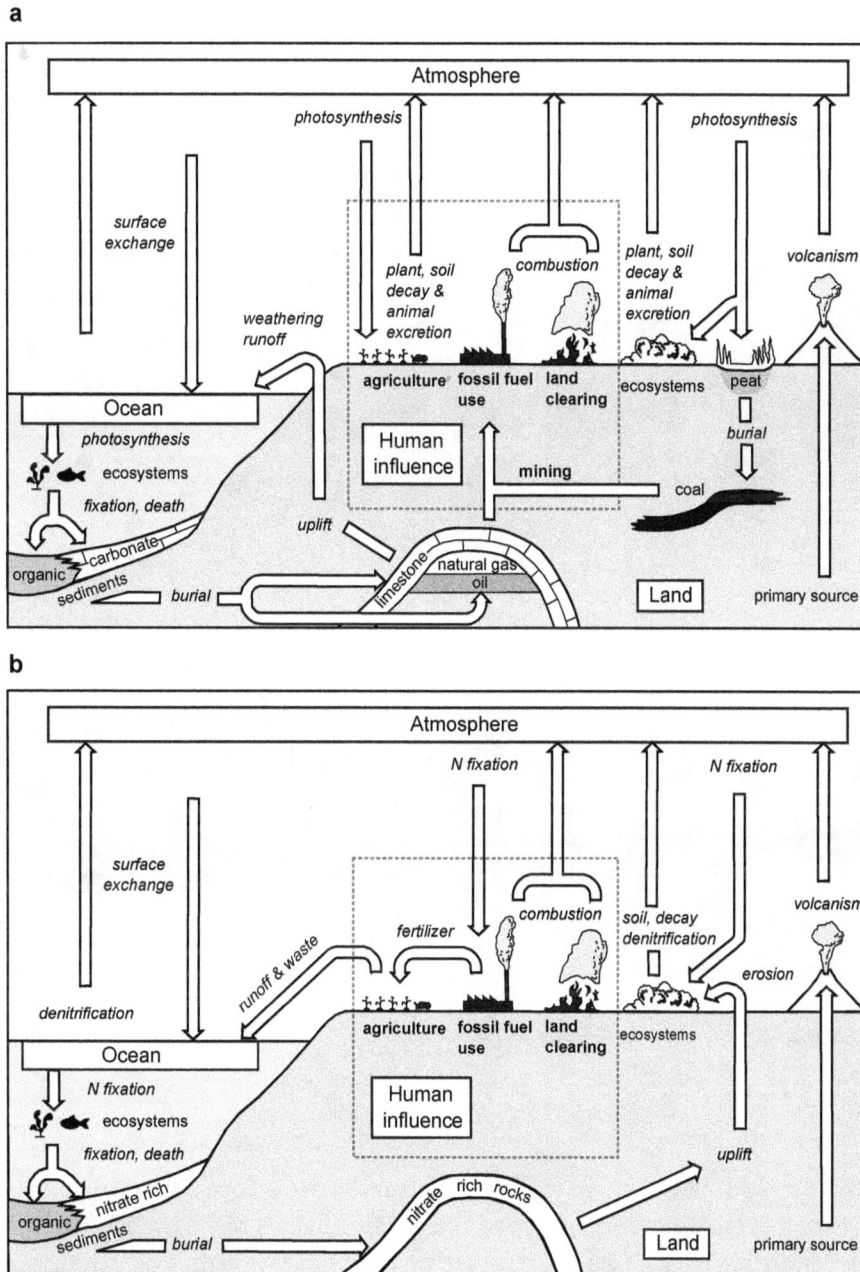

**Figure 7.** Simplified element cycles for carbon and nitrogen: a resource perspective of the human-environment system. The cycles are divided into atmosphere, ocean and land aspects. The dashed box outlines the human influence on the cycles. Italics represent processes and pathways. Non-italics outside the human influence box represent reservoirs; inside they represent activities. *a* Carbon cycle. There are many forms of carbon. Where the reservoir is oxygen rich (e.g., the atmosphere), the dominant form is carbon dioxide. Where oxygen is absent (e.g., peat swamps and underground), then hydrocarbons, such as methane and oil, can exist. After Strahler and Strahler (1997, fig. 20.12) *b* Nitrogen cycle. The most common form of nitrogen is N2 gas, which is found in the atmosphere. Other forms of nitrogen, such as nitrous oxide, ammonia, nitrate and carbon nitrogen compounds (e.g., protein), are relatively rare. After Strahler and Strahler (1997, fig. 20.14).

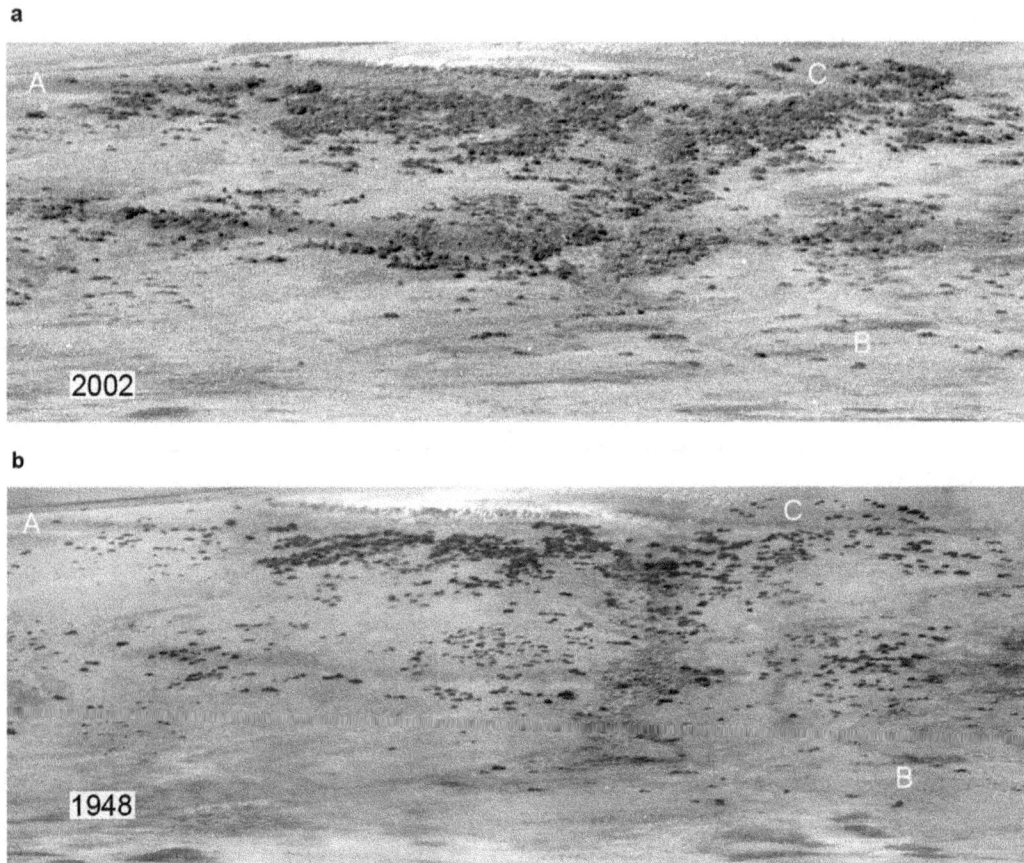

**Figure 8.** The shrubline at N 68°58'10.70", W 155°50'9.05" in the Colville River region of northern Alaska and its change over 54 years. It illustrates the patchiness of species distribution, the indistinct, gradational nature of a patch boundary, and the nature of patch change resulting from climate warming. The area changed through an increase in shrub size and density, and an increase in patch size and number. Both photographs after leaf-out. The letters *A*, *B* and *C* mark corresponding sites in each photograph (Tape 2010, fig. 15). *a* 2002 © Ken Tape. Reproduced with permission from the author. *b* 1948. Photo credit 1948 USGS/US Navy.

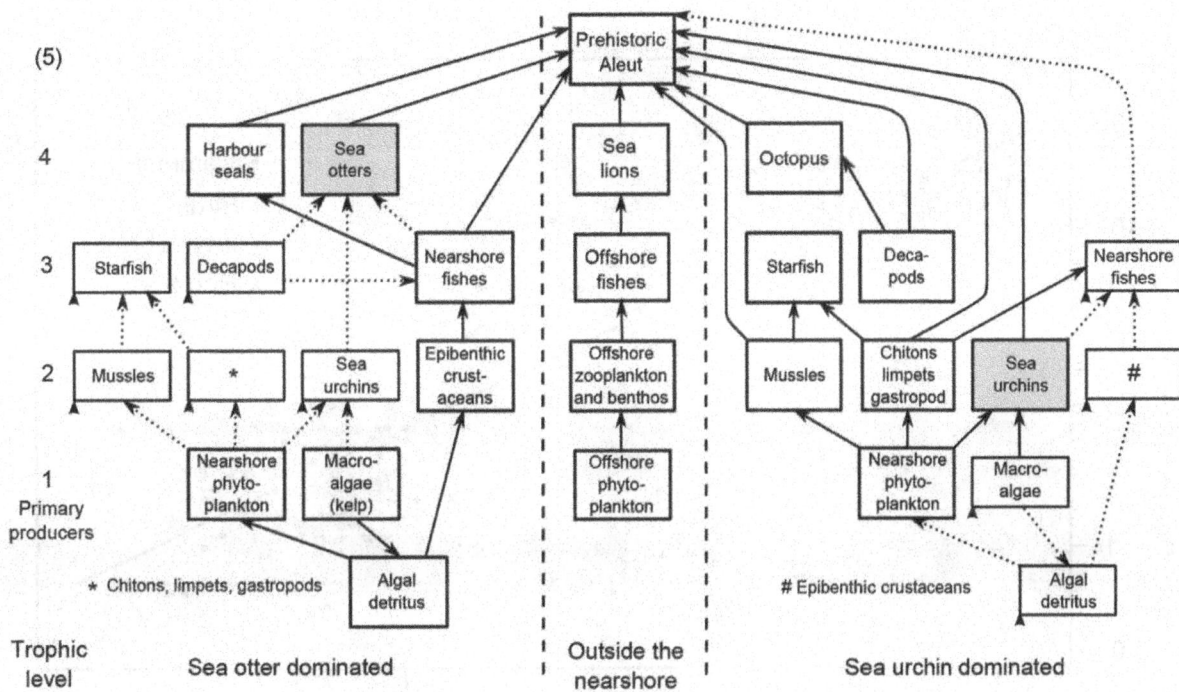

**Figure 9.** The nearshore ecosystem off the northwest Pacific coast of North America can exist in one of two states: sea otter dominated and sea urchin dominated. The trophic web for each state, as determined in the western Aleutian islands, is illustrated. The arrows indicate the direction of biomass or energy flow, and the boxes the relative mass of the noted species. The two states are distinguished by the differences in the relative importance of specific species and flows. A solid arrow represents a greater flow than a dashed arrow, and a large box represents a greater mass than a small box. The switch from the otter-dominated to the urchin-dominated state (grey species box) was triggered from the top down when otters were overharvested by the Aleut in prehistoric times, and by the fur traders in the 1700s and 1800s. They, and the orcas from the offshore ecosystem who are triggering the switch today, occupy trophic level 5. The term *benthic* refers to a deeper water ecosystem. *Epibenthic* refers to species almost in the deepwater ecosystem. A crustacean is an animal with a hard shell and legs divided in two parts. A decapod is a crustacean subgroup that includes shrimps, crabs and lobsters (Simenstad et al. 1978, fig. 1). Reproduced with permission from AAAS and the author.

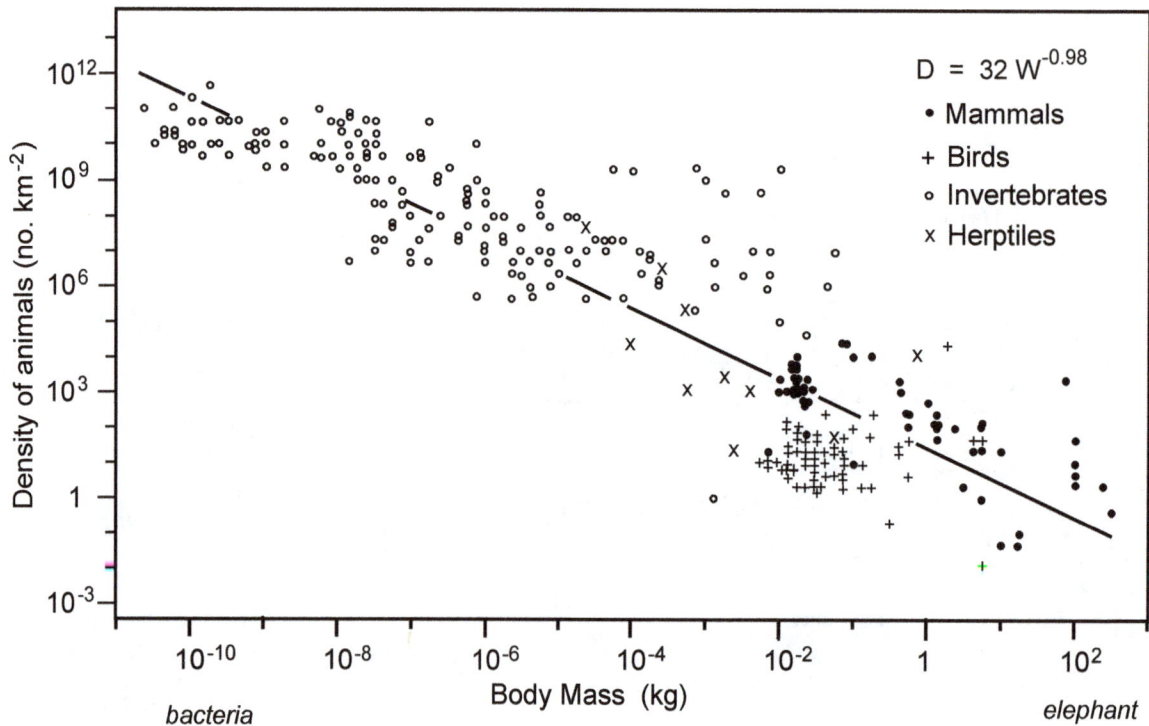

**Figure 10.** The allometric relationship between body mass and population density for mammals, birds, herptiles and invertebrates. A single curve can be fitted to all the data. The scattering represents data uncertainty as well as the (limited) flexibility available to a species to satisfy all of its allometic constraints. By applying the average body mass for hunter-gatherers of 59.5 kg, calculated from measurements, to the graph, their theoretical, allometrically constrained human population density was 0.58 people/km². Our current average population density is 49.6 people/km². After Peters (1983, fig. 10.3) (Note: Peters' formula changed to $D=32We^{0.98}$ using his p. 294.) Reproduced with permission from Cambridge University Press.

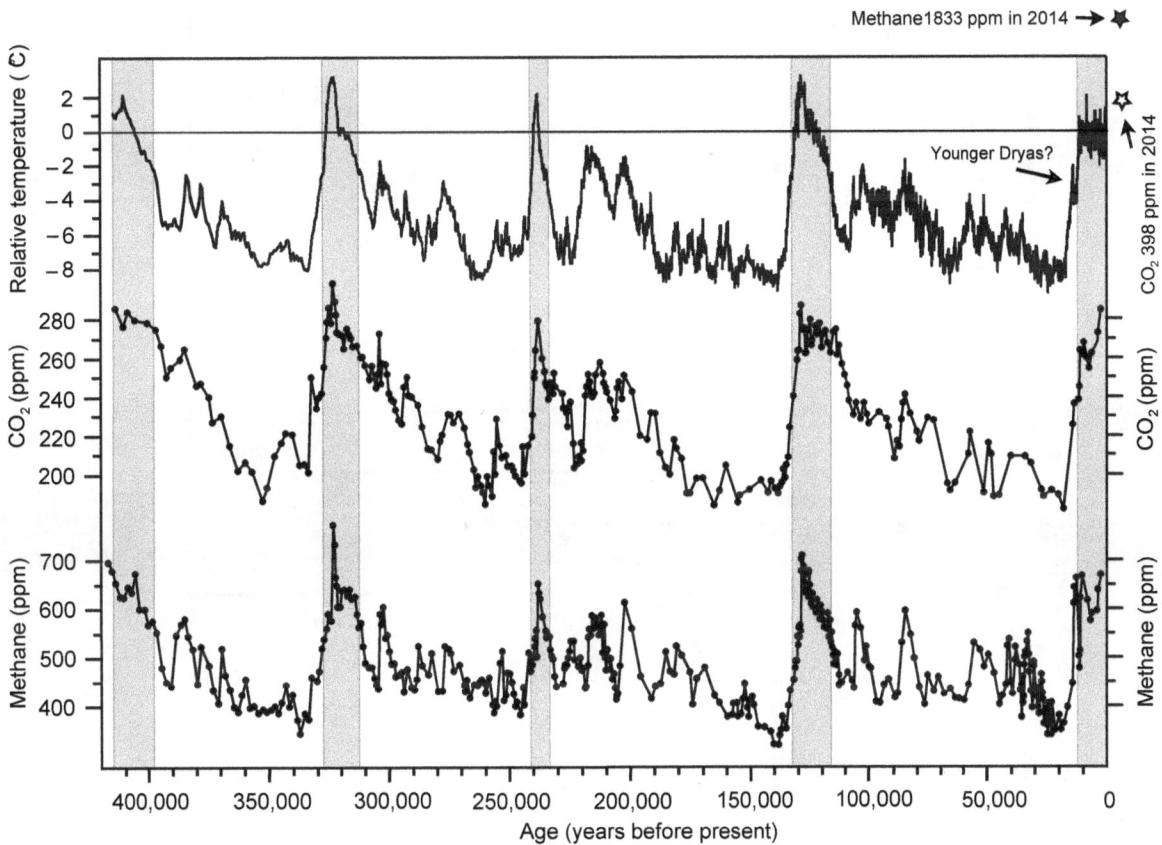

**Figure 11.** The variation in the atmosphere's absolute $CO_2$ and methane content, and the change in its temperature relative to the present, over the last 400,000 years, as recorded in an ice core recovered from the Antarctic ice sheet at Vostok. The grey zones are the interglacial warm periods, while the white are glacial periods. Note the asymmetric cyclical nature of the long-term temperature changes resulting from the rapid rise to and slower fall from the interglacial warmings. Riding on this curve are short, sharp abrupt temperature events. The 2014 values for $CO_2$ and methane far exceed the values for previous interglacials. Data from (WMO 2015). The average sampling interval for $CO_2$ and methane is around 1,000 years. The zero temperature reference is an average from the last quarter of the 1900s. After Petit et al. (1999, fig. 3).

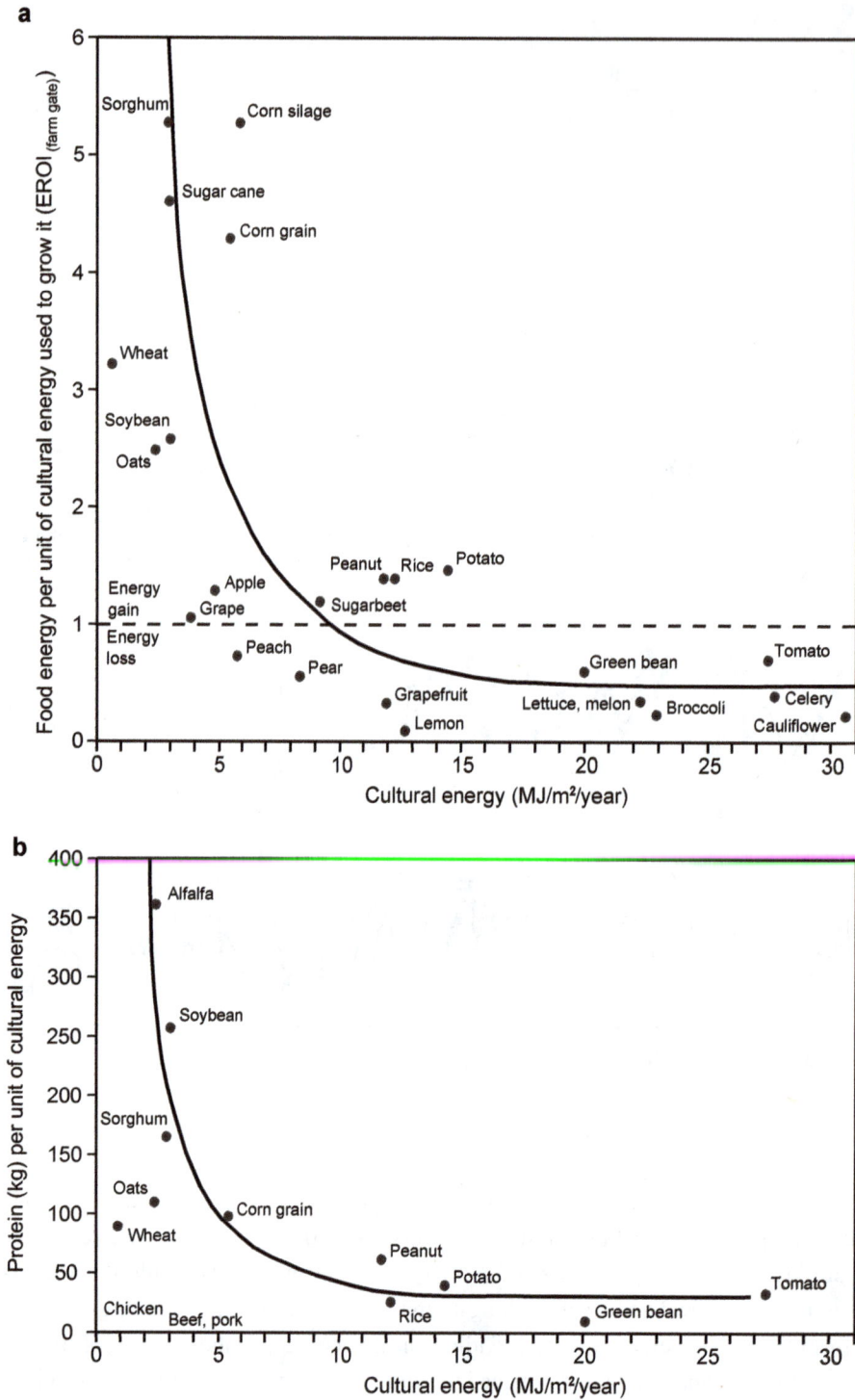

**Figure 12.** The relationship between the energy (*a*) or protein content (*b*) of foods produced in the United States during the early 1970s, and the cultural energy expended to provide them. *a* The *y*-axis energy ratio (EROI) shows that carbohydrate-rich foods can be produced at an energy gain (above 1 on the *y* axis), while vegetables are produced at a loss. *b* The *y*-axis ratio shows that many food plants produce protein with greater cultural energy efficiency than food animals. The location of soybeans and alfalfa indicates why they are used as an animal feed. Note: 3600 kJ = 1 kWh. After Heichel (1976, fig. 1 and 2).

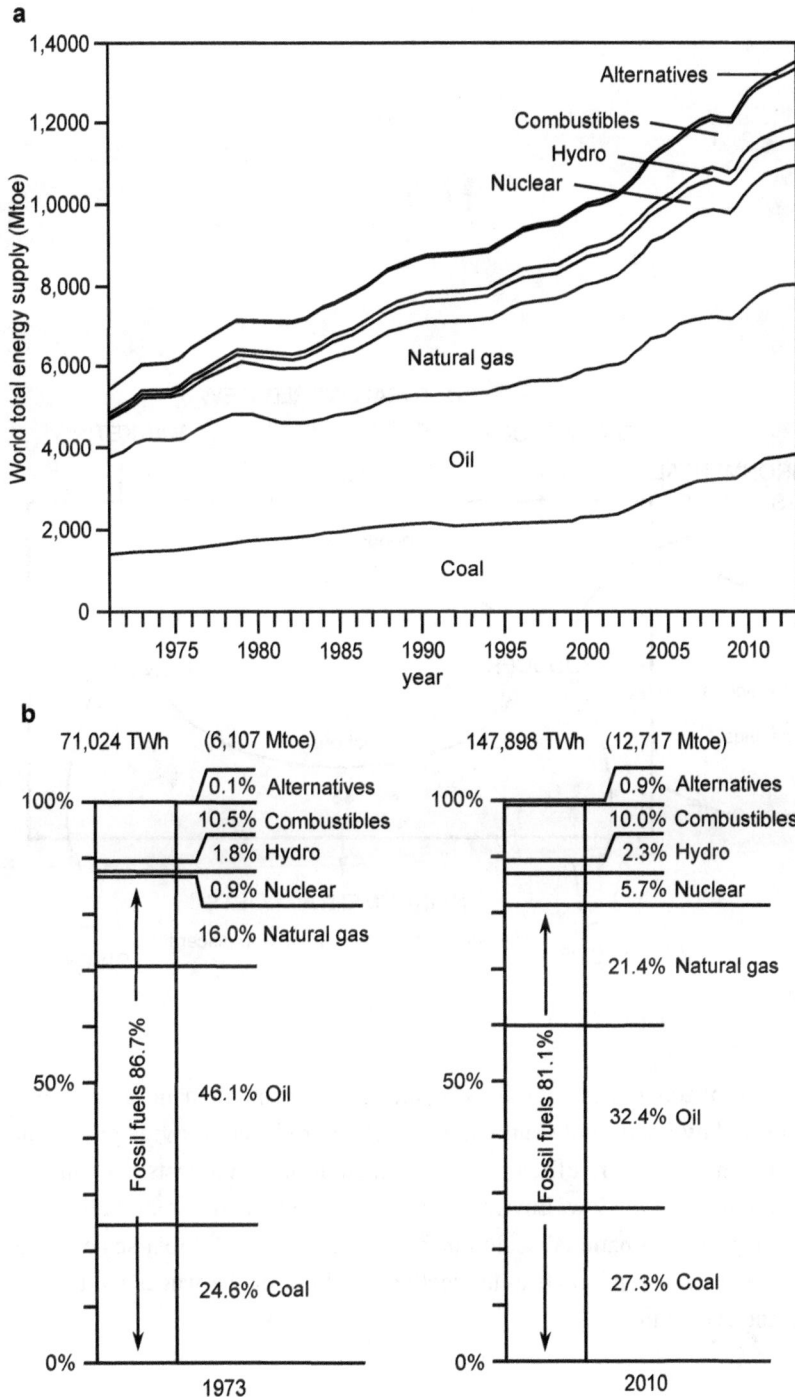

**Figure 13.** Global total primary energy supply from 1971 to 2013. The total supply closely matches the total energy demand. *a* The general increasing trend (roughly 3%) in the supply is needed to meet the increasing demand. Note the flattening of energy supply the year after the great financial crisis of 2008. After IEA (2015, p. 6). *b* Details for 1973 and 2010. After IEA (2012a, p. 6). Combustibles are wood (the majority), biofuels, waste, etc. Alternative energy sources are solar, wind, geothermal, tidal, etc. *Mtoe* is million tonnes of oil equivalent.

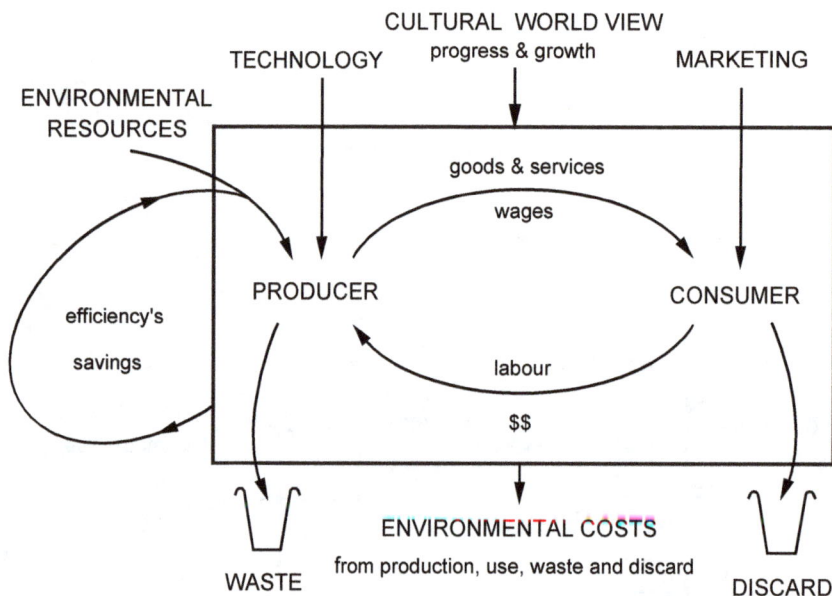

**Figure 14.** A broad view of economics' "virtuous" cycle of supply and demand. It is built around consumerism-producerism (consumers buy goods and supply labour, while producers provide goods and hire labour). Other factors involved in its functioning are efficiency savings, environmental costs, technology, and the influences of marketing and a culture's world view. Environmental costs are treated as externalities. Note that the virtuous cycle was originally applied at a national scale but is now applied at a global scale. Note, if replacing labour with machines and energy results in an absolute decline in labour or the wages paid, then there is less cash to spend on goods and the cycle fails.

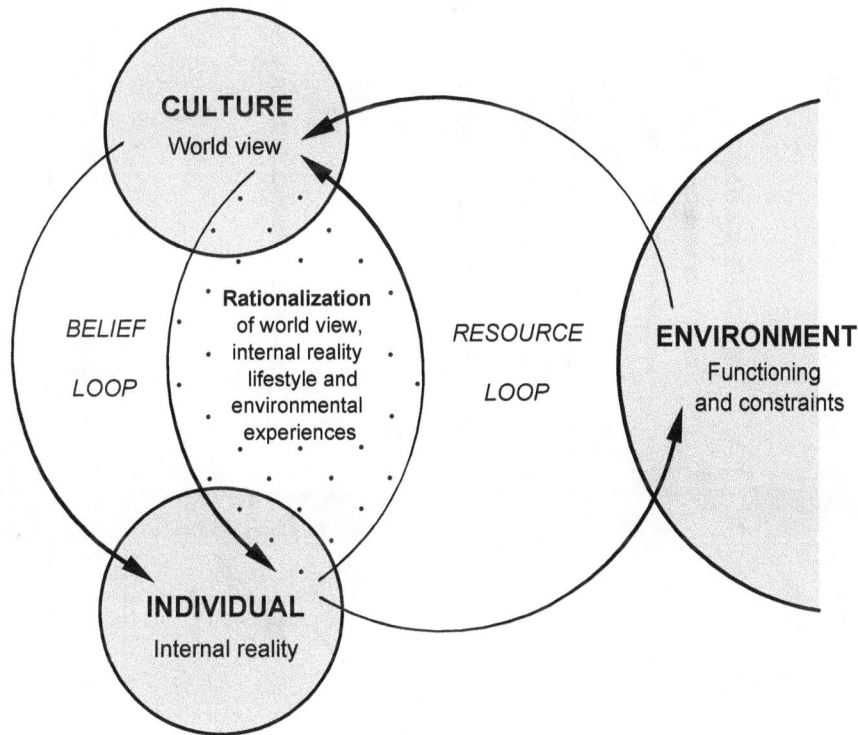

**Figure 15.** Our dominant influence on the biosphere's functioning transforms it into a dynamic complex human-environment system. This is a model of it from a system's perspective. The system is formed by the interactions (including feedbacks) among three factors: our cultures (represented by their world view), individuals (represented by our internal reality) and the environment (represented by its functioning and constraints). The system's functioning is expressed by the linked belief and resources loops. In the belief loop, our culture's world view guides our decision making, while our decisions induce the culture to change. In the resources loop, individuals, guided by their culture, use environmental resources to fulfill their culturally influenced wants and pay for their culture's reaffirmation and maintenance processes. In the process, they change the environment, which affects the supply of resources. The link between the two loops is our efforts to rationalize our thinking and decisions. In essence, this means rationalizing our internal reality with our culture's world view, our experience of the world and our wants.

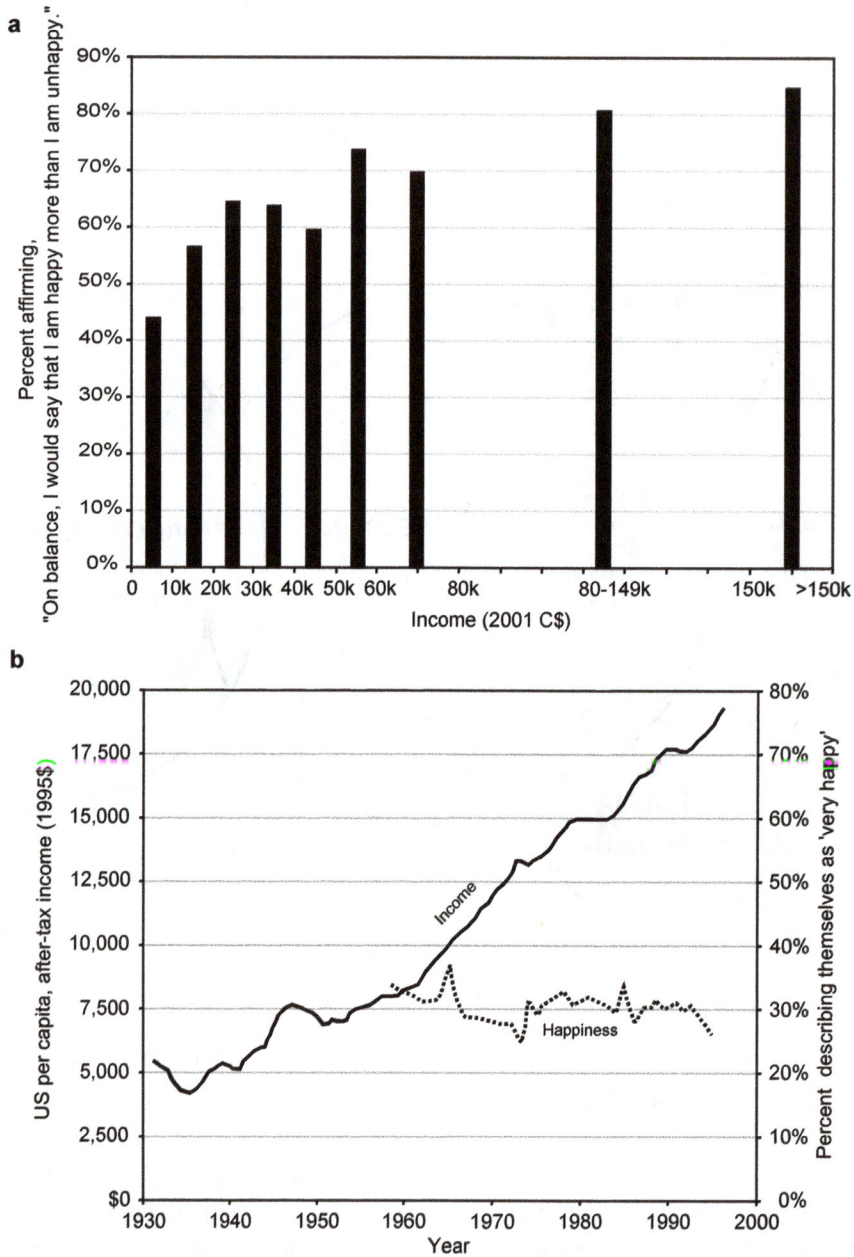

**Figure 16.** The relationship between our self-reported sense of happiness and financial indicators of personal wealth. *a* Self-reported happiness in Canada versus inflation-adjusted income for 2001. After Adams (2001, p. 172). *b* Self-reported happiness in the United States versus average inflation-adjusted personal income from 1959 to 1996. After Myers (1999, fig. 10.3).

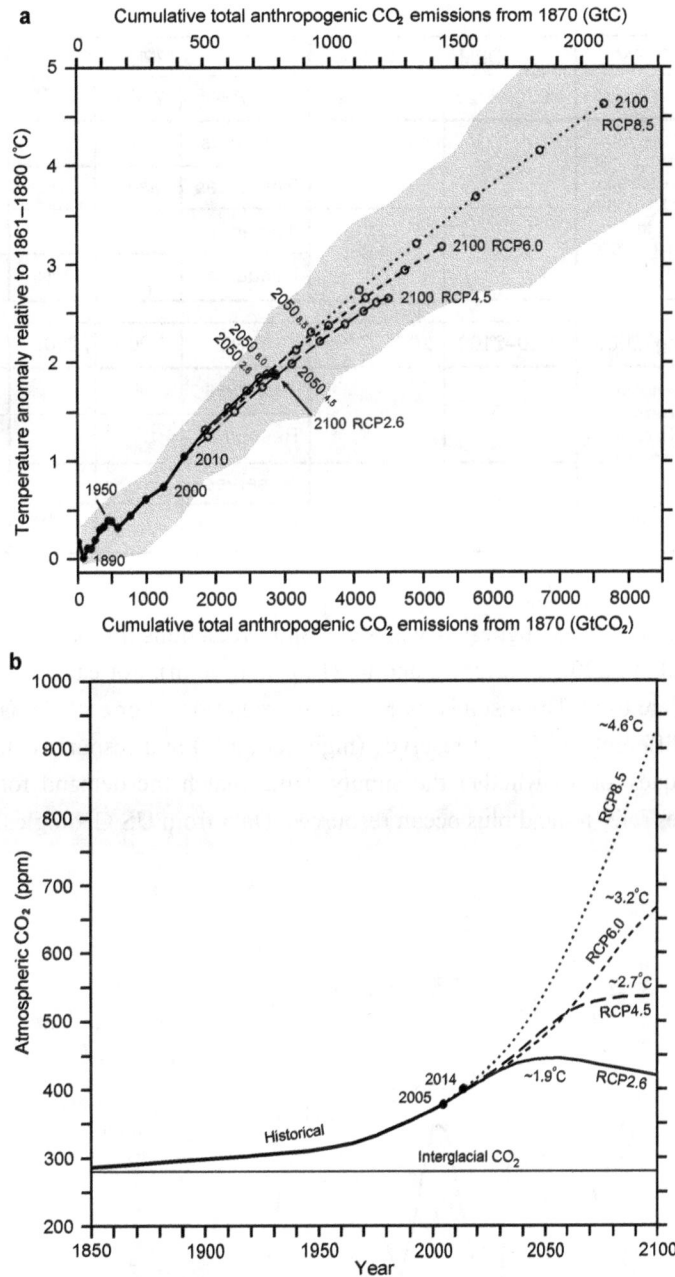

**Figure 17.** The contribution of our total past, present and possible future $CO_2$ emissions (only $CO_2$ sources) to the change in the global average surface temperature of the atmosphere (global warming) up to 2100. Our emission of other greenhouse gases, such as methane and $N_2O$, are excluded, so the estimated temperature change is a minimum. The historical record is shown to 2010 in *a*, and 2005 in *b*. After the historical record our possible future emissions and the resulting change in temperature up to 2100 are illustrated by four RCP scenarios. *a* The change in the temperature, relative to 1861–1880 (1870), expected to result from our total (historical plus RCP scenario) $CO_2$ emissions. The grey area represents the total uncertainty in the calculations. After IPCC (2013a, fig. SPM.10). *b* The change in the $CO_2$ content of the atmosphere over time for each RCP scenario. Only the two lowest RCP scenarios peak before 2100. After IPCC (2013b, fig. TS.19). The approximate temperatures are those expected at either the peak or 2100. The RCP2.6 curve is the source of the <2°C warming target. The measured value for 2014 and the interglacial value together indicate the ongoing deviation of today's trend from the geological past.

| TERM | DEMAND (Mt) | | SUPPLY (Mt) | | | BALANCE |
|---|---|---|---|---|---|---|
| Short (2005) | 2000–2020 | 2005–2020 | TYPE | 2000 | 2005 | |
| Production to increase from 13.2 Mt in 2000 to 24 Mt in 2020 | 372 | | Reserves | 340 | | almost |
| | 372 | | Resources | 1600 | | yes |
| | | 301 | Reserves | | 480 | yes |
| | | 301 | Resources | | 3000 | yes |

| TERM | DEMAND (Mt) | | SUPPLY (Mt) | | | BALANCE |
|---|---|---|---|---|---|---|
| Long (2006) | 2000–2100 | 2005–2100 | TYPE | 2000 | 2005 | |
| Production to increase to 170 kg/capita of in-service Cu in 2100 for 10 billion people | 2518 | | Reserves | 340 | | no |
| | 2518 | | Resources | 1600 (2300) | | no |
| | | 2437 | Reserves | | 480 | no |
| | | 2437 | Resources | | 3000 (3700) | yes |

**Figure 18.** Our copper future as suggested when demand predictions are balanced by supply estimates. The balancing is completed to 2020 (short term), and to 2100 (long term). For each term length, two different start dates, 2000 and 2005, are used. The result is four demand predictions. For each demand prediction there are two supply estimates of different reliability: reserves (high certainty) and resources (low certainty). The balance column answers the question of whether the supply could match the demand for eight circumstances. The numbers in parentheses refer to land plus ocean resources. Data from US Geological Survey (2002 –2007) and Gordon et al. (2006).

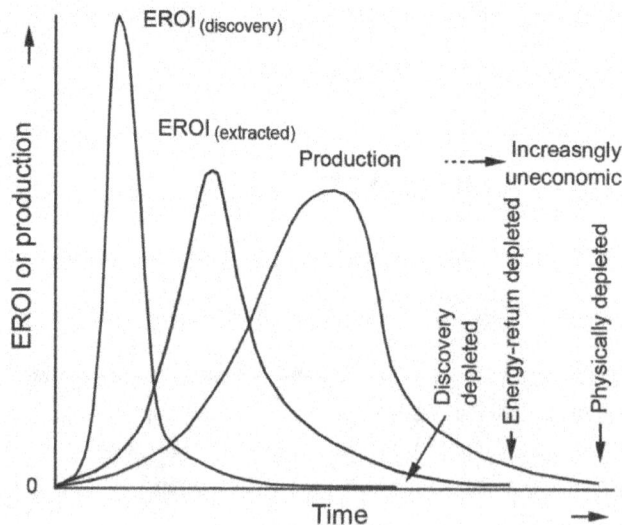

**Figure 19.** Idealized shapes and relative timings of the asymmetric curves for the regional- scale production of conventional oil and its energy return on (energy) invested (EROI) through time. Both of the EROI curves rise to their peak and then decline long before peak production is reached. The rise in production is slower than the fall, which includes a long post-peak production run out to physical depletion, at a low EROI$_{(extraction)}$. With a change in the shape of the production curve, this figure can be applied to conventional gas and both non-conventional oil and gas.

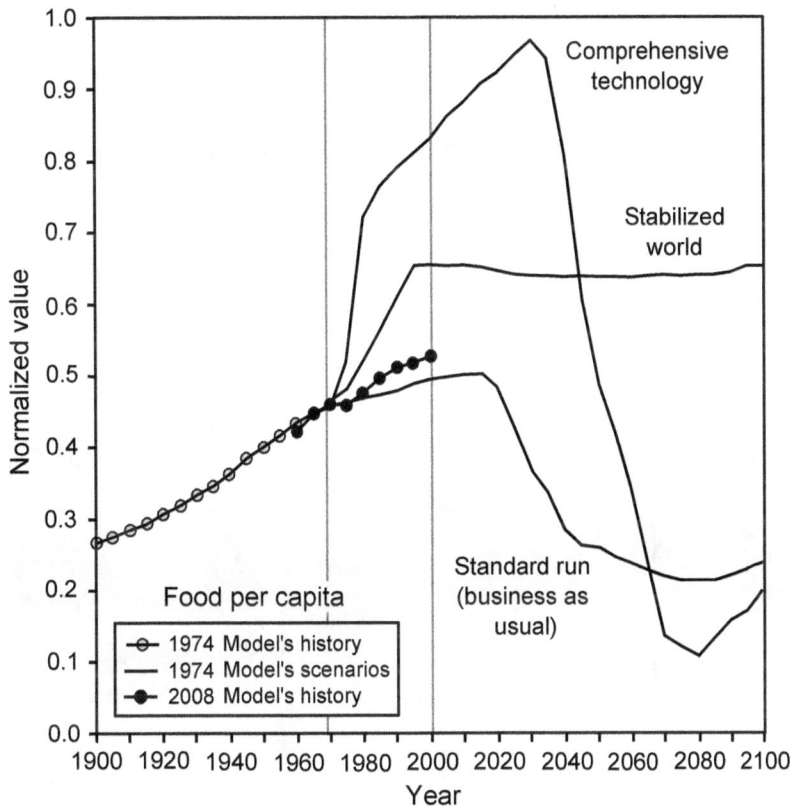

**Figure 20.** A 30-year retrospective assessment of the early 1970s Limits to Growth global model of the state of the earth focuses on food production per capita through time. The model's calculation of history up to ca. 1970 (grey dots) uses the historical data available in 1974. The post-1970 curves labelled *Comprehensive technology*, *Stabilized world* and *Standard run (business as usual)* are the model's 1974 output for three scenarios, representing our choices for the future after 1970. All but the stabilized-world scenario result in overshoot and collapse of food production. For the retrospective view, the model's calculation of history from 1960 to 2000 (black dots) uses the historical data available up to 2008. The result best fits the model's 1974 output for the business-as-usual scenario. The Limits to Growth model assumes finite resources and does not consider climate change. It is tempting to argue that the 2008 retrospective trend in food production lies above the 1974 projected trend because of the green revolution, but the model may be insufficiently sensitive to draw that conclusion. After Turner (2008, fig. 7). Reproduced with permission from CSIRO.

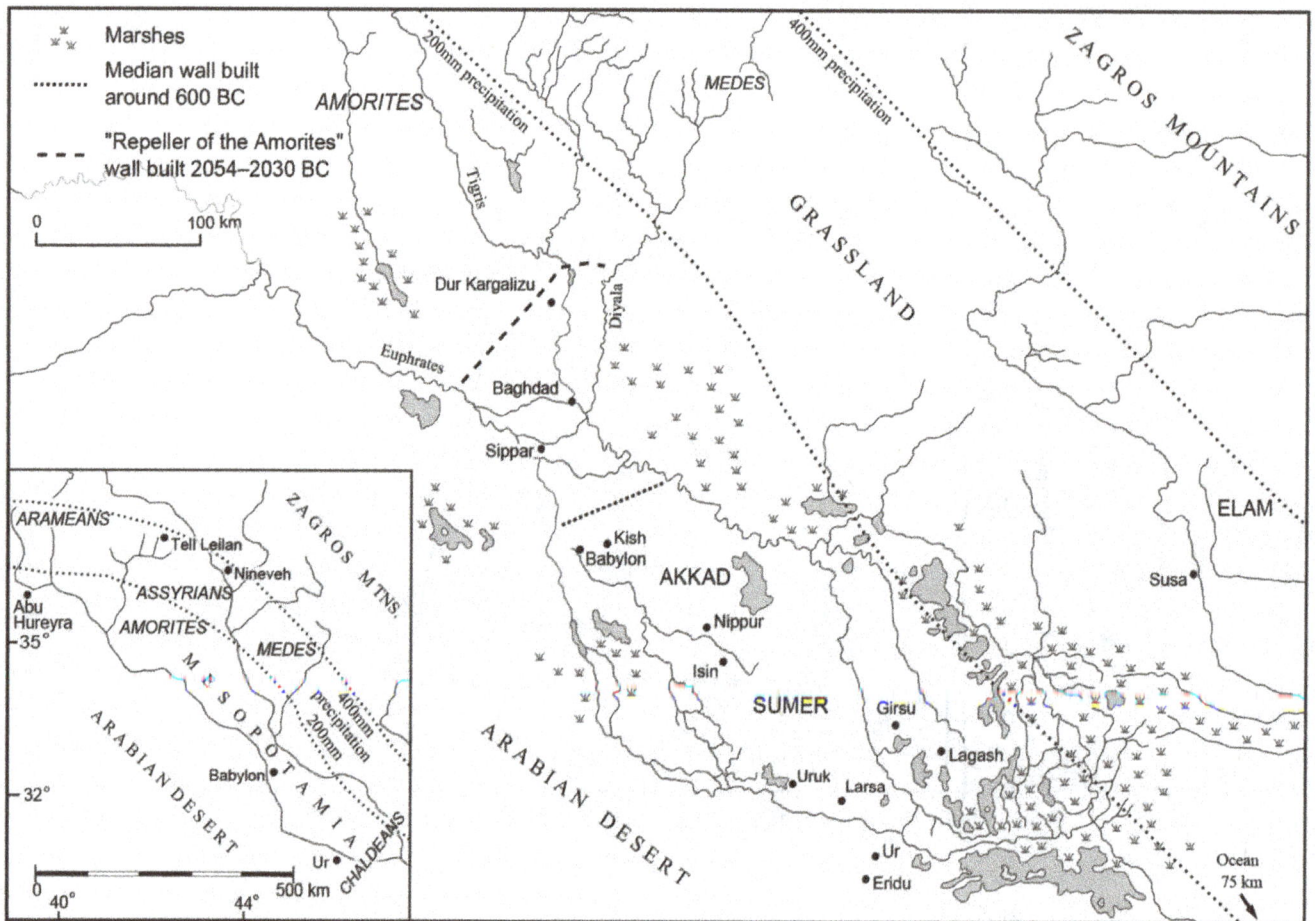

**Figure 21.** Peoples, states, place names and features of Babylonia and surroundings for the time period discussed in the text. Only a few of the river channels that flowed during the Babylonian period are shown. Some of them will likely have passed close to some of the noted cities. The precipitation contours represent present conditions. Wheat can be grown for precipitation >200 mm, but it can only be reliably grown if precipitation exceeds 400 mm. Irrigation is required for land receiving <200 mm of precipitation. After van de Mieroop (2004, fig. 13.2).

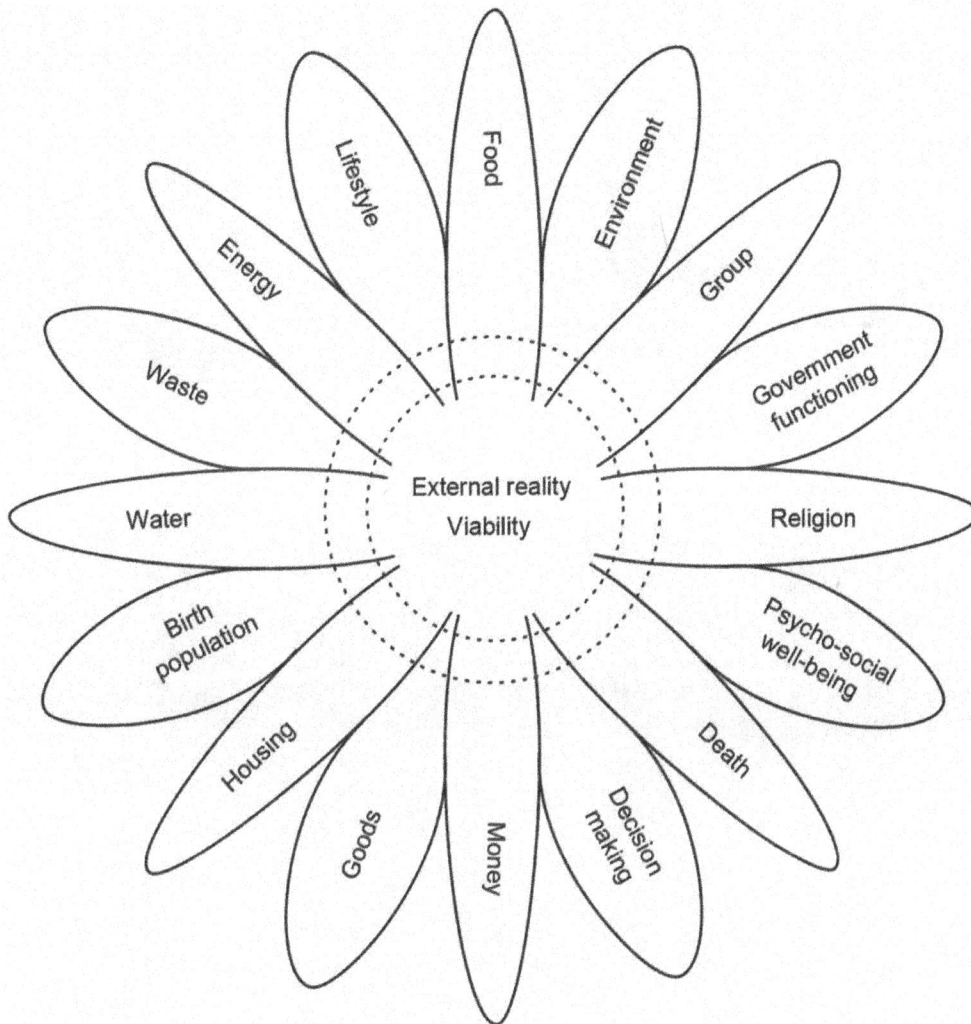

**Figure 22.** The viability flower: a human-centred perspective of the human-environment system. Each petal of the flower represents our personal or our culture's view of an aspect of the human-environment system that is important to us as an individual, as a social species with a culture, and as a member of an ecosystem. Collectively, the flower's petals (i.e., the flower) represent our personal or our culture's world view. The centre area of the flower, inside the dashed ring, represents the external-reality features of each aspect/petal. The external-reality view and our internal-reality view thus overlap in the centre area. The further outside the ring our view of an aspect falls, the greater the divergence between it and the external reality, and the harder it is for our personal, or our culture's, world view to be coherent and consistent. The boundary between the external reality and our internal one is gradational/fuzzy, not sharp, so it is marked by a ring not a line. The merging of the petals within the centre also represents the dynamic interconnections, like rationalization, and feedbacks that link these aspects together into the single human-environment system, as presented in figure 15.

# GLOSSARY

**adapt.** To change in order to accommodate a changing circumstance. The adapter achieves this through a mix of the two aspects of adapting: adjusting their surroundings or themselves. In Western culture, there is a strong preference to adapt by altering the world around us in order to maintain the status quo, rather than by adjusting our desires and expectations, our goals, to reflect the new circumstance. Thus Westerners' use of *adapt* invariably expresses our intention to adjust how we secure the environmental resources and services we need to maintain the essence of our current lifestyles and beliefs, rather than the alternative. *Mitigate* is often used with the same bias.

**adaptable.** The ability of a system, particularly an ecosystem, to adapt its functioning to accommodate change. See stability; functional.

**adjustments.** Small changes to a system, individual or culture, for example, to an individual's or a culture's world view. A series of adjustments results in larger changes.

**adsorb.** A chemical is adsorbed when it sticks to the outside of its host carrier. In contrast, when a chemical is absorbed, it enters into the carrier.

**adverse effects.** The harmful, unfavourable, undesirable direct and indirect impacts on humans or the environment. Adverse effects can arise from natural events, or from humans applying technology or following their beliefs as they live their lives. See *cumulative effects.*

**aerosols.** Airborne solid or liquid particles that are smaller than a micrometre.

**agricultural intensification.** The increase in the yield (the agricultural productivity) from existing farmland that is the result of changing farming practices and selecting higher yield plants and animals. More recently it has been achieved by genetically modifying plants and animals.

**albedo.** A measure of the amount of sunlight that the earth's surface, in part or in whole, reflects back into space like a mirror. Lighter-coloured surfaces, such as snow and ice, reflect more sunlight (high albedo) than darker surfaces, such as liquid water and most land plants (low albedo).

**allometric relationships.** The systematic relationships (correlations) within a group of species, that occur between pairs of their lifeway or shape characteristics. For example, among all species there is a correlation between their body mass and population density (figure 7). Although these correlations are not causes, they are predictive.

**antimicrobial drug resistance.** A disease-causing microbe is said to have developed resistance to the drug used to treat the disease when the dose needed to kill the microbe is also harmful to the patient.

**appropriate.** An expression of a judgement. We usually express our evaluation of our own or someone else's thinking or decisions using the terms *right* and *wrong.* But, in using these black and white terms, we treat the topic at hand as simple and deterministic when it is usually a part of the external reality's complex characteristics and functioning. In using them we also tend to neglect that our understanding of the topic is usually incomplete, biased and uncertain. Instead, by expressing our judgement using the broader terms *appropriate* and *inappropriate,* and favouring conditionals (e.g., *maybe, tends* and *if-,then*) rather than emphatics (e.g., *must, will*), we remind ourselves of this reality and the need to think in a more nuanced manner. We remind ourselves of a key feature of wisdom. We can also add *incomplete* to the terms we use. See *world view, personal; rationalization; wisdom.*

**assimilate.** To assimilate information is to accept it into our world view and make it part of our unconscious sense of self (our internal reality). Information needs to be assimilated if it is to have any long-term impact on our lives. The assimilation process is helped by repetition. It is also helped by the information being personalized and by the context in which it is presented. See *personalized knowledge*.

**bacteriophages.** Bacteria-attacking viruses.

**bioaccumulate.** The bioaccumulation of a chemical occurs when an organism or organic environment acts as a reservoir for it (e.g., PCBs accumulating in an organism's fat). A chemical bioaccumulates up the trophic web when a chemical-hosting organism lower in the food chain is consumed by an organism higher in the food chain that can itself act as a reservoir for the chemical.

**biome.** An ecosystem at a regional scale. See ecosystem.

**biosphere.** The thin layer at the earth's surface that supports life. The biosphere's boundaries lie a short distance above and below the earth's solid or liquid surface. See *environment*.

**bottom up.** Some aspects (features or processes) of a complex system can be treated as foundational, basic or lower order (at the bottom). Others can be treated as outcomes built on the foundation, or higher order (at the top). This hierarchical relationship can be used to describe, in directional terms, how a model of a complex system was built or how a trophic web regulates its ecosystem's functioning: from the bottom up or the top down. Feedback loops limit the value of this distinction by complicating the definition of the system's foundation.

**carbon intensity.** The carbon intensity of a fuel is the carbon emissions released when it is combusted, per unit of the energy released during combustion. Carbon intensity is usually expressed as $gCO_2$equivalents per megajoule.

**carrying capacity.** The maximum number of individuals of a species that its host community's resources can support. The minimum number of individuals needed for a population to reproduce plus raise functional adults is its minimum viable size. In between lies the possibility for a dynamically stable, viable population.

**city state.** A culture that has a centralized administration, but it only controls the immediate hinterland around the urban centre that houses the administration.

**climate.** The decadal (medium-term) averages of the weather variables at a location. See weather.

**climate warming.** An increase in the long-term average temperature of the atmosphere in a particular climate zone. Global climate warming (also often referred to as climate warming) refers to an increase in the average temperature of the earth's atmosphere as a whole.

**$CO_2$ equivalent.** The amount of $CO_2$ needed to trap the same amount of the earth's heat radiation as trapped by a given amount of another greenhouse gas. The total heat-trapping ability of the atmosphere is then represented by the sum of all of the $CO_2$ equivalents for all of the atmosphere's non-$CO_2$ greenhouse gases, plus its $CO_2$ content.

**cognitive dissonance.** The unpleasant feelings we experience when we become consciously aware of contradictions among our behaviours, beliefs, thoughts and experiences. They can stimulate us to consciously act in ways that will reduce those feelings. The degree to which our efforts will reduce the underlying reason for the dissonance can vary widely. See *rationalization*.

**collapse.** See overshoot; resimplification.

**collateral effects.** The peripheral effects that result from a behaviour (action) or belief. Collateral effects can be benign or harmful (adverse). They are usually unintended and unexpected. They are often indirect effects that are distant in time and space from the cause.

**commons.** Community assets to which everyone in a community has access. The term usually refers to environmental resources and services such as land, water, pasture and water purification.

**community.** An ecosystem at a scale relevant to an individual member. See *ecosystem*.

**complex system.** Complex systems, unlike simple systems, consist of many variables and the large number of interactions among them. These interactions, which exhibit a wide variety of linear to non-linear relationships, including feedback, take place over a wide range of time, space and intensities. The interactions also display discontinuities, thresholds and limits, and a flexible hierarchy of their significance to the system. Collectively, they link the variables together into a single complex system and form the overall mix of deterministic, random and chaotic characteristics that it displays: they determine how the system functions. They therefore also determine the degree of difficulty and uncertainty that we will experience when predicting the system's future.

**complexity-resource feedback.** A feedback in which an increase in a culture's organizational complexity requires it to provide more resources to pay for that complexity. But providing those resources requires an increase in the culture's organizational complexity.

**consciousness.** Our overall sense of awareness of ourselves and the world around us.

**constraints.** *Constraints* and *limits* are synonyms under most circumstances. In certain contexts, *constraint* can mean a conditional restraint, while a *limit* is absolute. An example is the constraint you experience when driving down a badly potholed road compared to the limit you experience when faced with a concrete barrier across the road.

**constructivism.** The scientific process of taking what is known about a complex problem and coherently tying it together. Computer modelling is a favoured tool. See *reductionism*.

**consumed water.** For a cultivated plant, it is sum of the water that it transpired while growing and the water that evaporated from the fields in which it grew (evapotranspired water). For a manufactured item, it is the sum of the water chemically used and evaporated during the production process. Consumed water is unavailable for further human use. Consumed water should not be confused with withdrawn water. See *withdrawn water*.

**conundrum.** A puzzling choice. See *dilemma*.

**conventional fossil fuels.** Oil and gas that can be extracted from their host deposits by simple drilling and pumping. They are easier and cheaper to extract than non-conventional fuels.

**correctness, prediction.** A prediction's correctness is how successful (accurate) it was in predicting the future or past, within the constraints attached to the prediction. See *reliability, prediction; usefulness, prediction*.

**cumulative effects.** The cumulative effect of human influences on the environment, an ecosystem, a species or a human community is the accumulation of their direct and indirect impacts on it, including the interactions between them, over both time and space. See *adverse effects*.

**depletion.** The point at which a fossil fuel deposit is depleted has several meanings. Physical depletion occurs when the fuel can no longer be physically extracted from the deposit. Energy depletion occurs when the energy gain from producing and combusting the fuel is the same or less than the energy cost of producing it. Economics usually determines when extraction stops, so that point can be referred to as economic depletion (figure 19). See *EROI; production curve*.

**destroyed ecosystem.** An ecosystem in which the population and lifeway of its species, and the interactions among them, (i.e., its functioning) have been disrupted to a degree that its resilience, robustness and adaptability are significantly reduced. The ecosystem is unlikely to recover to its pre-destroyed form and functioning. Its path into the future is highly uncertain. In the extreme, the ecosystem no longer exists. See *disrupted ecosystem*.

**dilemma.** A dilemma is a choice between two opposing options. See *conundrum*.

**diminishing returns.** In economics, it is the inescapable decline in the marginal utility of an activity. It is more broadly known and applied as the "law" of diminishing returns. For example, an ongoing set (given) increase in personal income is accompanied by a declining increase in an individual's judgement of their material standard of living. See *marginal utility*.

**dinoflagellate.** A single-celled (small) planktonic protist with rotating whips for propulsion. They often have cellulose shells. They can display the characteristics of both animals and plants. (Protists are a class of organism that have a cell with a nucleus but are not quite animal, plant or fungus.)

**disrupted ecosystem.** An ecosystem in which the population and lifeway of its species, and the interactions among them, (i.e., its functioning) have been affected to a degree that its resilience, robustness and adaptability have been significantly reduced. However, given enough time and suitable conditions, the ecosystem is likely to recover to its pre-disrupted form and functioning. Note the importance of the time scale used. There is a continuum between ecosystem disruption and destruction. See *destroyed ecosystem*.

**divergence.** The differences between our internal and the external reality. Divergences can reside in an individual, as part of our internal reality (including our personal world view), or in our culture, as part of its world view. Some divergences are trivial; others are life affecting. We may or may not be aware of their presence and influence on our perceptions and decisions. When we do become aware of them, we are prompted to resolve them, but our chosen method of doing so may not reduce their presence. See *reality paradox*; *cognitive dissonance*; *rationalization*.

**dominion.** Having dominion over our environment means to have the power or authority over it. The term has its roots in the Bible (e.g., Genesis 1:26–28).

**ecosystem.** As used here, an ecosystem is any sub-regional part of our environment that contains both the living (species) and the non-living features that are immediately relevant to treating that part of the environment as a functional system of life. See *community*; *biome*.

**efficiency.** The degree to which a goal is reached for a given level of effort or amount of resources. There are two common ways in which efficiencies are measured: economic efficiency, represented by GDP units per unit of natural resource; and material-use efficiency, represented by products produced per unit of natural resource. (They are often reported as their inverse: economic intensity [resource used/GDP] and resource use intensity [resource used/product produced].) An increase in efficiency is seen as an improvement, as positive, and as a reliable way to solve issues. However, whether this is true depends on the nature of our assessment. Was it conducted in the relevant context/complex system (e.g., resource, human-environment, social) at an appropriate time and space scale using a life-cycle analysis approach that considered how any savings would be used? In addition, the applicability and usefulness of economic efficiency is limited by its strong ties to the assumptions made by economics. See *life cycle*; *complex system*.

**elder.** A term of respect for an older person in a community that recognizes both their contribution to the community and the wisdom that comes with their years.

**embedded water.** See *virtual water*.

**endemic.** In ecology, it refers to a species that is consistently found in or restricted to a particular area.

**endocrine system.** Many of our body's functions are controlled by function-specific signalling agents (chemicals), called hormones, that operate at very low-concentrations. The production and distribution of hormones, and thus the control of the function, is managed by an individual's endocrine system.

**environment.** As used here, the word *environment* refers to the nature and or the functioning of some portion of the earth's system of life that includes the geophysical systems, such as climate, on which life depends. *Earth's environment* can be used as a synonym for *the biosphere*.

**environmental endowment.** The portion, and its nature, of the biosphere that supports a group of people. It is usually used when referring to a particular culture, but it can refer to a smaller or larger group, including the earth's human population. The nature (composition and functioning) of the group's endowment supports the members by providing them with the natural resources (including services), such as fresh water, energy sources and food, they need. But it constrains them (i.e., their lifestyle and belifs) by limiting the quantity and quality of those resources and their availability.

environmental manager. See *steward*.

**environmental services.** See *natural resources*.

**epigene.** The outer part of a gene. Changing the characteristics of the fluid in a cell, changes the epigene of the genes the cell hosts. Because the epigene mediates genetic processes, these changes can affect the expression of the genes.

**EROI.** The energy return on (energy) invested. The EROI is defined as the ratio of the number of energy units made available for each unit of energy consumed to make it available. It is a measure of the amount of energy we gain from expending energy to provide energy. For example, a fuel with an EROI of 5 means that 5 units of energy are supplied at a cost of 1 unit of energy. Thus, the equivalent of 20% of those 5 units is consumed to make the remaining 80% available for our discretionary use. If the EROI is $\leq 1$, then we gain no discretionary energy. The $EROI_{(discovery)}$ measures the EROI of a fuel at the time of its source's discovery; $EROI_{(extracted)}$, after extraction; $EROI_{(refinery)}$, after refining; and $EROI_{(on\ site)}$, at the site of use. The $EROI_{(electricity)}$ allows us to (meaningfully) compare the energy gained by generating electricity in different ways because it considers all of the energy costs.

**esoteric.** Esoteric knowledge and understanding is secret and meant to be available to a privileged (chosen) few. It implies that the holders of this knowledge have privilege and power, but they are also burdened with significant responsibilities. See *occult*.

**essence.** The basic or fundamental aspect of the topic. It can be thought of as its most important overall quality or its basic framework.

**eutrophication.** A substantial decline in a water body's dissolved oxygen content that is caused by a, usually, nutrient-promoted boom in the growth (such as algae) of aquatic plants. As the supply of nutrients increase, oxygen decreases and the ecosystem changes (including switching types) until, under hypoxic (ultra low oxygen) conditions, the body of water is almost free of life that uses dissolved oxygen.

**external reality.** The world and its functioning as it exists outside of our heads (i.e., regardless of our conscious awareness of it). When we die, the external reality continues. See *internal reality*.

**free market.** A market where the competition between the participants, which determines prices and wages, is unrestricted by (free of) government regulation and, some say, free of monopolies. The market's functioning is said to be self-regulating. See *neoliberal economics*.

**functional.** When the overall outcome of the interactions among a system's variables (its functioning) maintains it in a stable state for at least the short term, then it is said to be functional. For example, if, despite their diversity and differences, the overall behaviour of a group of people keeps them essentially unified and adaptable, and satisfying both the members' individual and their group's best interest over at least the short term, then the group and their culture can be said to be functional. If a complex system adapts to the influences on it in a manner that keeps it functional over the long term, then it is viable. If not, then it becomes non-functional and undergoes radical change or disappears. Thus, a functional system may or may not be viable. See *complex system, viable*; *stability*; *adaptable*.

**Gaia.** A term for the earth that attributes an incomprehensible (perhaps esoteric or occult) aspect to its self-regulating/self-adapting ability that may include it becoming conscious.

**GDP.** The gross domestic product (GDP) of a country is the annual sum of the country's cash flow from income, expenditures and production. It is the total value of the monetary transactions within the country's market. It, rather than net domestic product, is used by economists as a measure of a country's economic prosperity. Net domestic product is the income plus production minus

expenditures. An increase in a country's GDP is assumed to reflect an increase in both its economic and social prosperity. See *social welfare*.

**grace of God.** Among Christians, it is the love and favour God shows towards humans that make it possible for them to aspire to be more moral. See *providence of God*.

**greenhouse gas.** A gas that, as a constituent of the earth's atmosphere, has a heightened ability to trap the earth's radiation into space. This trapped radiation results in a warming of the atmosphere. Nitrogen and oxygen are poor greenhouse gases. Water vapour, carbon dioxide, methane, $N_2O$ and some CFCs and HFCs are very efficient greenhouse gases. The contribution that a particular greenhouse gas makes to the atmosphere's heat-trapping ability is determined by a combination of the radiation-trapping ability of the gas and its amount in the atmosphere. See *$CO_2$ equivalents*.

**heuristics.** A fancy word for rules of thumb. The affect heuristic describes how our decision making about a subject is influenced by where the feeling we associate with it is placed on a scale between positive (good) and negative (bad). The representative heuristic describes how our brain reasons about something by comparing it to a similar situation (the prototype) that it constructs from our memories. The availability heuristic combines the total number of instances of an event you can recall from memory with the ease (speed) with which you recalled them to decide on the frequency (the probability or likelihood) of the evaluated event's occurrence. This is the basis of our intuitive assessment of risk. See *rules of thumb*.

**home range, home territory.** The area over which an individual of a species normally moves to complete its yearly round of daily and seasonal activities.

**human-environment system.** Humans are just one of the many species making up the living part of the earth's environment. However, we have affected and are currently affecting all aspects of the environment's functioning to such a degree that we can think of our recent past, our present and our future as occurring within a human-environment system (figures 15 and 22).

**inflation adjusted.** A longer-lasting decline in the value (the purchase power) of money is called inflation. By mathematically manipulating economic measures, such as wages and the price of goods or services, the change in the value of money due to inflation can be neutralized or normalized: it can be adjusted for inflation.

**intelligence.** The ability to use knowledge. See *wisdom*.

**internal reality.** The world, including ourselves, and its functioning as it exists inside our heads. Our memories (including feelings), our innate references, and our sensings of both our body's state and our surroundings are used by our brain's innate decision-making processes to provide us with a culturally influenced reality: our internal reality. It resides mostly in our unconscious. We are only aware of those parts of our internal reality that we call our personal world view or that are our conscious experiences of ourselves and the world around us. Our personal world view is a broad reflection of our internal reality, while our conscious experiences are more a snapshot reflection of it. We express the presence of our internal reality both consciously, through considered words and actions, and unconsciously, through unconsidered words and actions. Although our internal reality dominates our conscious experience, it doesn't exclude the external reality: they can and do overlap. When they are different, we can experience a divergence or reality paradox. (figure 22). See *external reality*; *world view, personal*; *divergences*; *reality paradox*.

**Kew.** The site of the Royal Botanical Gardens in England.

**knowing.** Reliable knowledge that has been personalized and assimilated becomes part of who we are. If we apply that knowledge and the awareness that comes with it to change our view of ourselves, our view of our relationship to the world around us and thus our decisions, then we have increased our knowing. Simplistically, after feeling the heat of flames, we know that fire burns. Deciding not to touch it shows that this reliable knowledge has indeed been applied to ourselves. It also shows that our internal reality now more closely matches the external reality and our relationship with it. Increasing our knowing is a key aspect of gaining wisdom and making wise decisions: knowing when and how you should use what little you know. The degree of your knowing thus represents more than how much you know, whether you use knowledge routinely in your life, such as being able to count, or whether you have mastered a technique, such as the ability to build a space ship. It also reflects more than an increase in your awareness of yourself and your connection to the external reality.

**knowledge.** What you know; information, regardless of its usefulness or reliability, you are aware of. Knowledge exists in isolation, understanding doesn't. See *wisdom*.

**life cycle.** The biological cycle of birth, growth, reproduction and death undergone by all organisms. It is also applied to resources or products, in which case it refers to the cycle of their extraction/design, refining/construction, use, and discarding. The value of a life-cycle analysis of a resource or product depends on the boundaries of the system that are chosen and the accuracy of the data used. See *lifeway*; *EROI*; *efficiency*.

**lifeway.** The manner in which an individual of a species both biologically lives from birth to death and experiences their life. A lifeway is a life cycle to which the other aspects of living beyond simple biology have been added. It thus includes the details of nurturing and training its young, how it perceives the world around it and its interactions with the other members of its community and environment. See *life cycle*.

**limits.** See *constraints*.

**marginal utility (marginal benefit).** An economic term to describe the amount of benefit to be gained from a unit increase in the effort expended to gain the benefit: for example, the increase in cash benefit (the profit or return) from a unit increase in investment. The term has wider applicability.

**market.** Either an actual or imaginary place where wages and the prices of commodities are set by the participants buying and selling or negotiating them.

**median.** If the frequency distribution of a variable is asymmetric then the mode is the most frequent value of the variable (on the $x$ axis): the peak of the distribution curve. The mean is the average value of the variable, which can be weighted or unweighted. The median lies between the mean and the mode. It is the value of the variable that divides the distribution into two classes of equal ranges on the $x$ axis.

**member-culture feedback.** We absorb our culture's world view into our internal reality as we grow up, so we become its carrier. Understandably, when we make decisions as adults, we tend to follow the guidance from our culture's world view. But, as its carrier, we are also our culture's decision makers, which means that our collective decisions as adults adjust our culture's world view. But this is the cultural world view that our young absorb. It then guides their decisions as adults, which completes a member-culture feedback. This feedback, along with us being creatures of habit, stabilizes our personal and our culture's world view. However, it doesn't preclude change; it just results in inertia to us changing our personal and our culture's world view.

**microbes.** Bacteria and viruses.

**millennialist, millennialism.** The Christian belief that the successful advancement of humanity towards God will end with a thousand years (the millennium) of heaven on earth.

**mutualistic relationship.** A relationship between two species where their direct interaction with one another provides significant help to both of them in their effort to complete their respective lifeways: for example, the relationship between pollinator and pollinated. See *symbiotic relationship*.

**nanoparticles.** Usually refers to human-made particles in the nanometre size range.

**natural resources.** The materials and services provided to us by the functioning of the earth's environment.

**neoliberal economics.** The brand of neoclassical economics that favours a laissez-faire market (i.e., little or no government involvement in the economy). It is referred to as either neoliberal (the term used here), neoconservative, the Washington Consensus or libertarian economics. The neoliberal economists' view of how the market works is described as market fundamentalism. Adherents have a strong belief in free markets, free trade, globalization, monetary theories and market efficiency. They also strongly believe in economic growth and a focus on market-oriented, not government-controlled, solutions to economic and, most importantly, social and environmental issues. Effective solutions to issues are believed to incorporate deregulation, privatization, austerity, low taxes, motivation through financial incentives, a focus on economic growth, and only limited control of the economy through monetary policy. The neoliberal vision of the economy's workings currently dominates at least Western economic thinking and guidance.

**net primary productivity.** NPP is the amount of new plant biomass an ecosystem's plants produce each year.

**niche.** A species' niche is an abstract representation of its place and role in its host community. By implication, it also represents a view of the conditions needed for an individual of that species to complete its lifeway.

**normal, personal.** When all of our individual, longer-lasting references, views, visions, norms and beliefs are considered together, they can be treated as our usual starting point for decision making: our reference "normal." Associated with our normal are the shorter-term references created by our unconscious reasoning on the fly. Our normal is inherently biased and error prone because it contains divergences. See *divergences*.

**non-conventional fuels.** Oil and gas that can't be extracted from their host deposits by simple drilling and pumping. They require more difficult and expensive extraction methods, such as fracking their host or mining it.

**nuanced thinking.** A broad (multi-scale), comprehensive (multi-perspective), flexible (open-to-change) manner of thinking and making decisions about a subject. In order to engage in nuanced thinking, we must take a more context-dependent view of the subject, which requires us to accept that the world we live in is dynamic, complex and grey. Although nuanced thinking uses the techniques of scientific thinking, such as being skeptical (including being self-skeptical), it is more than scientific thinking. It is directed at more than acquiring factual knowledge. It is a striving to gain understanding and make wise decisions by treating oneself as both an objective and an emotional observer of an event and, at some level, a participant in it. Thus, to practise nuanced thinking also requires us to personalize and assimilate knowledge. And we must appreciate the influence that our human characteristics, such as our decision-making foibles, our culture's world view and divergences have on our thinking and decision making.

**occult.** Secret, hidden knowledge or practices, often of a mysterious (mystic) nature that are beyond human understanding (supernatural). It is used to single out specific practices, the occult arts, such as magic, Kabbalah, astrology and alchemy. See *esoteric*.

**overshoot.** Overshoot occurs when a sustained, growing demand for resources by a species significantly exceeds the amount of accessible resources and/or the ability to supply them. A point is reached where the lack of supply forces a drop in demand. Overshoot and collapse refers to a situation where the demand has so exceeded the supply that the result is not a resource shortage but a scarcity sufficient to significantly affect the species. It implies a population collapse.

**perception.** The conscious awareness of specific aspects (object or action) of the world around us. Perception is more than detecting that aspect, such as seeing it (registering its presence). Perception includes the feelings and meaning we attach to it: our interpretation. See *sensing*.

**permafrost.** Ground that is permanently frozen. Permafrost is found above the tree line in the polar climate zone and on mountains. A thin surface layer may thaw during the summer months.

**personalized knowledge.** Knowledge to which we have attached personal meaning or relevance. Personalized knowledge is more easily assimilated into our personal world view and more easily influences our decisions. This is in contrast to the knowledge that we view in an abstract, disconnected or intellectual manner. We are aware of it, but its influence on our decision making can be minimal.

**plankton.** Sub-millimetre, free-floating organisms found at any depth in a body of water. Phytoplankton are plants. They only live near the water's surface, to a depth that sunlight can penetrate.

**pollution.** The human-caused additions of gases, liquids and solids to the atmosphere, ocean and land, irrespective of their relative or absolute quantities, concentrations or impacts. Pollutants are divided into two types: additions of the environment's normal constituents (normal pollutants); and additions of human-made compounds that are not normally present in the environment (novel pollutants).

**preconditioning.** Preconditioning describes how events unevenly distributed in the past can, collectively, make a particular future change more probable, but not certain. These events can contribute to and facilitate specific future changes but not actually cause them. This is a basic feature of how complex systems change.

**production curve.** The amount of a metal or fuel produced from a deposit changes over time. A plot of the amount of production per year over time forms a production curve (figure 19). The general shape of the curve for oil and gas depends primarily on the properties of the fuel and the geology of the deposit. In contrast, the general shape for coal and metals depends on their distribution in a deposit and the geology of the deposit.

The details of a curve are determined by economic, political and technical factors.

**prosperity.** The economic prosperity of a country or an individual refers to their financial condition (access to or holdings of cash and cash equivalents), as measured by GDP, income or net worth. Economists consider that the measures of economic prosperity also reflect our social prosperity. In particular, a country's GDP is assumed to reliably reflect its population's quality of life, its social welfare. Similarly, an individual's net worth or income is assumed to reliably reflect an individual's level of social-psychological well-being.

**providence of God.** Among Christians, it is God's wisdom in management: his ability and wisdom to manage/interfere in earthly matters when needed. See *grace of God*.

**proxy.** In science, a proxy is a variable that can act as a substitute for a more difficult-to-measure or missing variable. Proxies substitute for a variable with different degrees of reliability.

**quality, energy.** In general, the quality of an energy source is its desirability and value to a culture. The economic quality of a fuel is its economic productivity, expressed in monetary terms.

**rationalization.** The overall process by which we personally (and collectively) use information to make sense of the world or an experience, and deal with an issue. Rationalization is accomplished by us (consciously or unconsciously) applying our constrained decision-making tools to what we know. That information includes facts and feelings, our personal world view, our culture's world view, our lifestyle and our past experiences. Rationalization includes dealing with the uncertainties, inconsistencies and/or disconnects in and among the information (its limitations) that can arise from missing and uncertain facts and the divergences between our internal and the external reality. When undertaking a rationalization, we are usually unaware of or selectively discount, excuse and ignore the limitations to what we know and the limitations of our decision-making tools. As a result, our decisions can appear (emphasis on *appear*) to us to be coherent and consistent. We can thus feel (emphasis on *feel*) that we have made meaningful sense of the world and an experience, or that we have appropriately dealt with an issue. But, from the perspective of the external reality, the outcome is usually just a cosmetic change to our internal reality and the expression of the issue, not a fundamental change to their essence. However, this is not always the case. If, during rationalization, we devote the required conscious effort to deal with or accommodate the limitations in our knowledge and decision making, then the outcome can be significant personal change and appropriate solutions to issues (figure 15). See *wisdom; nuanced thinking; knowing; world view, personal; appropriate.*

**RCP.** Representative concentration pathway. A scenario that represents our possible total CO$_2$ emissions (other greenhouse gases excluded) from all sources between 1870 and 2100, and the resulting change in the atmosphere's temperature (figure 17).

**reaffirmation and authority maintenance.** For a culture to remain unified, stable and adaptable (functional), its members must routinely reaffirm their faith in their culture's world view. Their reaffirmation process involves rituals and ceremonies in which the members participate. More complex cultures must also reaffirm their faith in their culture's hierarchical power (authority) structure. This reaffirmation becomes part of their world view reaffirmation process, which makes it more costly and complex. Complex cultures must also maintain the functioning of their authority structure, which is an unavoidable fixed cost. As a result, the reaffirmation and authority maintenance efforts of the more complex cultures are more costly and complex than those of the less complex and hunter-gatherer cultures. See *functional.*

**reality paradox.** A reality paradox occurs when you become aware of a baffling, even disturbing, contradiction or inconsistency between your perceptions, beliefs or expectations (your internal reality) and your actual circumstance (the external reality) that you can't in good faith dismiss. It is like a conundrum or dilemma with personal implications.

**reductionism.** Reductionism is a standard scientific practice of gaining knowledge by dividing a complex problem into its simpler constituent parts and studying each separately. See *constructivism.*

**reliability, prediction.** A prediction's reliability is a measure of the likelihood that it will be correct. Reliability is estimated by evaluating the methodology, uncertainty and conditional statements (the "ifs") that are a part of or relate to the prediction.

**reliable knowledge.** Reliable knowledge is independently reproducible knowledge. The scientific methods of gathering and checking information, whether applied formally or informally, provide us with the most widely available reliable knowledge we have about the external reality.

**religion.** A formalized way of thinking and living that relies on accepting the existence and superiority of a supernatural being, beings or force.

**reserve.** The reserve of a metal/fossil fuel in a deposit is the amount that is estimated, with a high degree of certainty, to be currently economic to mine/extract. The global reserve of a metal/fossil fuel is the sum of the reserves for all of its known deposits. See *resource.*

**resilience.** The ability of a system, particularly an ecosystem, to recover from a disturbance. See *stability; functional.*

**resimplification.** The reduction in the structural complexity of a more complex culture. It is the result of the members, usually involuntarily, adjusting to a circumstance that longer allows then to maintain their culture's current elevated level of structural complexity. Resimplification is a dynamic and complex process.)

**resource.** *Resource* has a number of meanings. *Natural resources* (or *resources*) is a general term for the natural materials and services the environment provides to us. However, when referring to an amount of a particular natural resource, then *resource* means the amount that will likely be produced plus the amount that has a low potential of being produced. It is thus an optimistic outlook for the resource's supply. In particular, the resource of a metal deposit is the combination of its reserves plus the amount of metal it is reasonable to expect would be mined in the future from the currently known sub and marginally economic parts of the deposit, and their geologically inferred extensions (i.e., the metal's resource includes amounts that likely won't be mined). The global resource of a metal refers to the sum of the resources from all its known deposits. The global resource for oil and gas can be reported in the same way as metals. But it is more meaningfully reported as the amount of oil and gas that is likely to be technologically feasible to extract/recover from known and undiscovered deposits. Whether the metal/fuel will be extracted/recovered depends on economic and other factors such as the EROI. See *reserve.*

**responsibilities.** The care and consideration we must exercise when making our decisions is the price we pay for receiving the benefits from living in a group (e.g., support and companionship) and for manipulating the environment (e.g., necessities, conveniences and comforts. Our human responsibilities are thus an inescapable companion to our human rights, hence the phrase "rights and responsibilities." Our responsibilities can be short and long term, and apply to humans or their environmental endowment. See *viability*.

**robustness.** The ability of a system, particularly an ecosystem, to withstand disturbance, disruption. See *stability*; *functional*.

**rules of thumb.** Simplifications and generalizations about the world and how it functions that are used in decision making: for example, a stereotype. They are useful aids to decision making, but only as long as they are applied as guides, not immutable, error-free recipes, and the results from applying them are not taken too literally. See *heuristics*.

**scarcity**. See *shortage*.

self-regulating market. See *free market*.

**self-sustaining.** A self-sustaining system or feedback is one whose functioning maintains its current state. Self-sustaining does not imply the equivalent of perpetual motion. A system or feedback is only self-sustaining for as long as the resources it must consume to maintain that state are available.

**sensings.** The signals our innate sensing mechanisms (sensors) detect. These sensings are used to inform us of an aspect of the world around us and the state of our body. See *perception*.

**shortage.** A lack of supply that is temporary or easily resolvable. A shortage becomes a scarcity when the shortage can't be easily filled: it becomes chronic. A shortage/scarcity can be thought of as either a resource supply issue or a demand issue. See *shortfall*.

**shortfall.** A resource shortfall is predicted to occur when the projected demand for a particular resource exceeds its projected supply. See *shortage*.

**social welfare.** A measure of the quality of life of a nation's citizens. It includes factors such as the quality of the country's natural environment, its level of crime, the availability of essential social services and the religious or spiritual aspects of life. It is not the same as standard of living, which is a measure of material necessities and comforts.

**social-psychological well-being.** Our mental as opposed to our physical sense of well-being. It can be divided into two largely independent components: our social-psychological feelings about ourselves, for example happiness, and our social-psychological judgement (evaluation) of our life circumstance, for example your satisfaction with your material standard of living.

**stable; stability.** When the functioning of a complex system, such as an ecosystem or a culture, keeps it within its functional limits, then it is stable. But a complex system is dynamic, so it exists in a dynamic, not static, stable state: a quasi-stable state. Its limits are also somewhat dynamic. The system can thus change and still remain stable, but only as long as the speed and degree of change does not push it beyond its functional limits. Within these limits, the system can even transition into a different quasi-stable state and remain functional. For example, the earth's climate system can transition from its glacial quasi-stable state to its interglacial quasi-stable state and back. However, if a complex system is pushed, or its functioning takes it outside of its dynamic functional limits for long enough, then it can become disrupted or be destroyed. See *functional; complex system; viable*.

**steward.** An environmental manager or steward is someone who acts to ensure that their culture's environmental endowment will continue to satisfy the members' present wants. See *trustee*.

**stratosphere.** The earth's upper or outer atmosphere that lies immediately above the troposphere. See *troposphere*.

**symbiotic relationship.** A symbiotic relationship exists between two species when the success of each of their efforts to remain functional depends on the other interacting with them (e.g., the relationship between the nitrogen-fixing *Rhizobium* bacteria and its host plant). See *mutualistic relationship*.

**symbol.** Objects or (human or animal) actions that have more meaning to us than can be inferred from their innate characteristics. They have become a summary representation of a human understanding. Symbols, which can have an objective and/or an emotional aspect and significance, help us to remember and recall that understanding. We can relate to symbols both consciously and unconsciously, and as individuals and as a group (e.g., a cultural symbol).

**taboos.** Traditional cultural restrictions or prohibitions that affect, for example, our relations with the environment, such as the harvesting and consumption of foods, and our interpersonal behaviours.

**territorial state.** A culture with a centralized administration and economy that controls an extensive land area and has many urban centres.

**top down.** See *bottom up.*

**trophic web.** An ecosystem's trophic web (food chain) represents all of the feeding interactions between its species: between the eaters and those being eaten. These feeding links and the resulting flow of food (energy) through the web allow it to be divided into trophic levels that have a hierarchical relationship to one another. At the foundation of the web are the plants that use sunlight to create plant matter. This is one of the ways in which the interactions between an ecosystem's species bind them into a whole.

**troposphere.** The earth's lower atmosphere. At the equator it extends to about 20 km above the earth's surface, while at the poles the distance is only 10 km. It is the place in the atmosphere where most weather is created. The jet stream occurs near the top of the troposphere.

**trustee.** A culture's environmental trustee is someone who acts as a custodian or guardian of their environmental endowment on behalf of present and future generations. See *steward.*

**ultimate supply.** The total global amount of past, present and future production of a metal/fossil fuel that will be supplied by applying existing and expected-to-be-developed extraction technology to its deposits until they are all depleted. An ultimate supply has meaning for conventional oil and natural gas. It can be applied to metals and coal but usually isn't because of the significant uncertainty in the data, especially the number and size of the undiscovered deposits.

**understanding.** As commonly used, *understand* means to grasp, comprehend, assume or relate to. There are a variety of ways in which one can understand. They range from disinterested to emotional, from reasoned to intuitive, from logic to empathy. It can occur within a narrow or broad context. The intended meaning should be apparent by its context. The most generally useful type of understanding occurs when we relate to information in a mix of ways within both its wider and narrow context. Relating to information in this manner provides for the greatest flexibility in appropriately checking and using it, especially when trying to apply it wisely. See *knowledge; knowing; wisdom; personalized knowledge, nuanced thinking.*

**usefulness, prediction.** The degree to which it will inform, rather than mislead, and improve our understanding. The degree to which a prediction will guide us towards more appropriate decisions. A prediction's usefulness is an extension of its validity. A prediction does not have to be reliable to be useful. See *reliable, prediction; validity, prediction.*

**validity, prediction.** A measure of a prediction's credibility. A valid prediction is based on information that is considered to be sufficiently relevant and reliable to support the prediction's reasoning and purpose. A valid prediction is never an unfounded speculation or guess. A prediction does not have to be reliable to be valid. See *reliable, prediction; useful, prediction.*

**viable.** A viable population, ecosystem or human culture is one that will continue to be functional over the long term. For that to happen, the system's functioning must respond to the influences on it in a manner that will keep it dynamically stable within its larger context. The details of the conditions that must be satisfied for it to be viable are wide-ranging. In overview, a viable culture is one whose world view guides the members to complete their lifeway in a manner that stays within viability's human (individual and group) and environmental constraints on a culture, and whose members promote and follow that guidance. A viable population is one whose dynamic size stays between its supporting environment's carrying capacity and its minimum functional size, while satisfying requirements such as genetic diversity, raising functional young and spatial distribution limits. A viable ecosystem is one whose functioning maintains its resilience, robustness and adaptability. *Sustainable* should not be confused with *viable. Sustainable* can refer to an aspect of a system continuing even though doing so will, in the longer term, render its host system unviable. See *disrupted ecosystem; destroyed ecosystem; resimplification; functional; stable; carrying capacity.*

**virtual water.** The sum of the water that is either consumed or used for waste-removal during an item's production. Also referred to as embedded water. See *consumed water.*

**virtuous cycle.** The product of consumerism-producerism. Traditionally, the virtuous cycle was described as a feedback in which the application of innovation and technology facilitated an increase in the production of goods and a lowering of the cost of production. These changes would result in cheaper goods and greater profit. Since a producer's income pays labour, steady or increasing wages and lower prices allows labour to demand more goods and services, which spurs producers to invest in innovation and technology. However, the useful-energy models of GDP indicate that the virtuous cycle is better described, at least until 2000, by profitable capital and labour combinations being replaced with more profitable capital and lower-priced energy combinations, a switch that is facilitated by the ability of innovation and technology to increase efficiency (figure 14). This, like the availability of cheaper labour, implies a different future for the cycle than traditionally expected. See *efficiency*.

**weather.** The short-term values of weather variables, such as sunlight, precipitation and temperature.

**wisdom.** Your knowledge is the information you are aware of. Intelligence is the ability to use that knowledge. Wisdom is the ability to know how and when to apply both in an appropriate manner. Wisdom is the skill to use knowledge as an aid to making assessments of many factors before deciding between, or delicately balancing, the accessible options. Wisdom includes the ability to know when not to use knowledge and when you don't have adequate knowledge. Wisdom is the ability to solve problems in a manner that takes into account the larger, longer-term, social, cultural, psychological and environmental context within which the problem is embedded. It can be thought of as the product of making the effort to be both rational and compassionate while looking at an issue from multiple perspectives. Wisdom favours appropriate decisions. See *appropriate*; *knowing*; *nuanced thinking*.

**withdrawn water.** Water extracted from rivers, lakes, dams and aquifers for human use. See *consumed water*.

**world view, cultural.** An aggregate of the characteristics, such as the structure, laws, beliefs and customs, to which a culture's members subscribe. It includes features such as the average member's views about how the world works; how to conduct themselves; the proper relationships to have with their environment and other members; the nature of the future; and the reasons why. A culture's world view thus guides its members how to live (figure 15). See *world view, personal*; *divergences*.

**world view, personal**. Our conscious expression of part of our internal reality. That expression reflects our highly simplified, imperfect, human-biased, internal-reality view of the external reality, which is influenced by our culture's world view and our personal experiences. A feedback links our personal to our culture's world view (figure 15). Our world view can come to better reflect the external reality if we adjust the terms we use to express it: the terms we use to think and talk about the world and our experiences. By doing so, we influence our decision making. See *member-culture feedback*; *world view, cultural*; *divergences*; *appropriate*.

# REFERENCES

Achard, F., Eva, H. D., Stibig, H., Mayaux, P., Gallego, J. et al. 2002. Determination of deforestation rates of the world's humid tropical forests. Science, vol. 297, pp. 999–1002.

Ackerman, F. 2007. The economics of atrazine. International Journal of Occupational and Environmental Health, vol. 13, pp. 441–449.

Adam, D. 2006. Scientists attack climate change denial. Guardian Weekly, 29 September, p. 14.

Adam, D. 2007. Scientists issue bleak forecast for warming world. Guardian unlimited, 6 April. http://environment.guardian.co.uk/print0.329770866-121568.00.html.

Adam, D. 2009a. Oil firm "funds climate deniers". Guardian Weekly, 10 July, p. 18.

Adam, D. 2009b. Coral reefs face a fading future. Guardian Weekly, 18 September, p. 31.

Adams, J. M. and Faure, H. (editors), QEN members. 1997. Review and atlas of palaeovegetation: Preliminary land ecosystem maps of the world since the last Glacial Maximum. Oak Ridge National Laboratory, TN, USA. http://www.esd.ornl.gov/ern/qen/adams1.html.

Adams, J., Maslin, M. A. and Thomas, E. 1999. Sudden climate transitions during the Quaternary. Progress in Physical Geography, vol. 23, no. 1, pp. 1–36.

Adams, M. 2001. Better happy than rich? Canadians, money and the meaning of life. Penguin Books, Toronto.

AESD. 2014. Oil sands operators' water use history. Alberta Environment and Sustainable Development, Oil Sands Information Portal. http://osip.alberta.ca/library/Dataset/Details/56.

AFSSA. 2008. Weakening, collapse and mortality of bee colonies. Agence Francaise de Sécurité Sanitaire des Aliments, November, pp. 1–155. http://www.afssa.fr/.

Akasofu, S. 2009. Two natural components of the recent climate change. http://people.iarc.uaf.edu/~sakasofu/pdf/two_natural_components_recent_climate_change.pdf.

Alerstam, T., Gudmundsson, G. A., Green, M. and Hendenstrom, A. 2001. Migration along orthdromic sun compass routes by arctic birds. Science, vol. 291, 12 January, pp. 300–303.

Al-Fattah, S. M. and Startzman, R. A. 2000. Forecasting world natural gas supply. Journal of Petroleum Technology, vol. 52, no. 5, pp. 62–72.

Allan, R. P. and Sodden, B. J. 2008. Atmospheric warming and the amplification of precipitation extremes. Science, vol. 321, pp. 1481–1484.

Alland, A. 1973. Evolution and human behaviour: An introduction to Darwinian anthropology. Anchor Books, New York.

Allanou, R., Hansen, B. G. and van der Bilt, Y. 1999. Public availability of data on EU high production volume chemicals. European Commission, Joint Research Centre, EUR 18996EN, pp. 1–24. http://ecb.jrc.it/

Allen, M. R. 2004. Blame game. Nature, vol. 432, pp. 551–552.

Allen, U. D. 2006. Public health implications for MRSA in Canada. Canadian Medical Association Journal, vol. 175, no. 2, pp. 181–162.

Alley, R. B., Marotzke, J., Nordhaus, W. D., Overpeck, J. T., Peteet, D. M. et al. 2003. Abrupt climate change. Science, vol. 299, pp. 2005–2010.

Allport, G. W., Bruner, J. S. and Jandorf, E. M. 1941. Personality under social catastrophe: Ninety life-histories of the Nazi revolution. Character and Personality, vol. 10, pp. 1–22.

Aluminium Institute. 2010. http:www.world-aluminium.org. Accessed 26 October, 2010.

American Nuclear Society. 2005. The Price-Anderson Act. American Nuclear Society, Background for position statement, pp. 1–4. www.ans.org.

American Psychological Association. 2006. Multitasking - Switching costs: Subtle "switching" costs cut efficiency, raise risk. American Psychological Association. www.psychologymatters.org/multitask0306.html. Accessed 20 March 2006.

Ames, K. M. and Maschner, H. D. G. 2000. Peoples of the northwest coast: Their archaeology and prehistory. Thomas and Hudson, London.

Anderegg, W. R., Prall, J. W., Harold, J. and Schneider, S. H. 2010. Expert credibility in climate change. Proceedings of the National Academy of Sciences, vol. 107, no. 27, pp. 12107–12109.

Anderson, A. 2002. Faunal collapse, landscape change and settlement in remote Oceania. World Archaeology, vol. 33, no. 3, pp. 375–390.

Anderson, A., Haberle, S., Rojas, G., Seelenfreund, A., Smith, I. et al. 2002. An archaeological exploration of Robinson Crusoe island Juan Fernandez archipelago, Chile. In Bedford, S., Sand, C. and Burley, D. (editors). Fifty years in the field. Essays in honour and celebration of Richard Shutler's archaeological career. New Zealand Archaeological Association, Monograph 25, pp. 239–249.

Anderson, C. 2000. Killed sea lions weren't at risk. Province, 23 April, pp. A31.

Anderson, D. 2001. Merchandize Advertising. Advertisement in Interior News, 17 January, p. A5.

Anderson, I. 1997. Alarm greets contraceptive virus. New Scientist, 26 April, p. 44.

Anderson, R. N. 1998. Oil production into the 21st century. Scientific American, March, vol. 278, no. 3, pp. 86–91.

Anderson, S., Collins, C., Pizzigati, S. and Shih, K. 2010. Executive excess 2010: CEO pay and the great recession. The 17th annual executive compensation survey. Institute for Policy Studies, Washington, pp. 1–31. http://www.ips-dc.org/reports/executive_excess_2010.

Angela, P. and Angela, A. 1993. The extraordinary story of human origins. Prometheus Books, Buffalo. Translator: Tonne, G. First published 1989.

Angel, R. 2006. Feasibility of cooling the Earth with a cloud of small spacecraft near the inner Lagrange point (L1). Proceedings of the National Academy of Sciences, vol. 103, no. 46, pp. 17184–17189.

Anton, S. C., Leonard, W. R. and Robertson, M. L. 2002. An ecomorphological model of the initial hominid dispersal from Africa. Journal of Human Evolution, vol. 43, pp. 773–785.

Appell, D. 2001. The new uncertainty principle. Scientific American, January, pp. 18–19.

Appell, D. 2006. Behind the hockey stick. Scientific American, March, pp. 34–35.

Ardis, L. 2001. Northern forest industry hurts most: NFPA. Interior News, 3 October, p. A7.

Arkes, H. R. 1981. Impediments to accurate clinical judgement and possible ways to minimize the impact. Journal of Consulting and Clinical Psychology, vol. 49, pp. 323–330.

Arkes, H. R. and Harkness, A. R. 1980. Effect of making a diagnosis on subsequent recognition of symptoms. Journal of Experimental Psychology: Human Learning and Memory, vol. 6, no. 5, pp. 568–575.

Arrhenius, S. A. 1896. On the influence of carbonic acid in the air upon the temperature of the ground. Philisophical Magazine and Journal of Science, Series 5, vol. 41, no. 251, pp. 237–276.

Asner, G. P., Knapp, D. E., Broadbent, E. N., Oliveira, P. J. C., Keller, M. et al. 2005. Selective logging in the Brazilian Amazon. Science, vol. 310, no. 5747, pp. 480–482.

The Associated Press. 2007. Climate change report warns of increased hunger, species extinction. 6 April, 2007. http://www.cbc.ca/world/story/2007/04/06/climate-un.html.

Ausubel, J. H. 1996. Can technology spare the earth? American Scientist, vol. 84, pp. 166–178.

AWIPMR. 2011. Arctic on the verge of record ozone loss. Alfred Wegener Institute for Polar and Marine Research. AWI press release, 14 March.

Ayres, R. U. 2005. Resources, scarcity, technology, and growth. In Simpson, R. D., Toman, M. A. and Ayres, R. U. (editors) 2005, pp. 33–53.

Ayres, R. U. and Warr, B. 2005. Accounting for growth: The role of physical work. Structural Change and Economic Dynamics, vol. 16, pp. 181–209.

Baadsgaard, A., Monge, J., Cox, S. and Zettler, R. L. 2011. Human sacrifice and intentional corpse preservation in the Royal Cemetery of Ur. Antiquity, vol. 85, no. 327, pp. 27–42.

Baia Mare Task Force. 2000. The cyanide spill at Baia Mare, Romania: Before, during and after. Summary of UNEP/OCHA assessment mission report March 2000. UNEP: The regional environmental centre for central and eastern Europe, June, pp. 1–8.

Baillie, J. E. M., Griffiths, J., Turney, S. T., Loh, J. and Collen, B. 2010. Evolution lost: Status and trends of the world's vertebrates. Zoological Society of London, United Kingdom, pp. 1–74.

Baillie, J. E. M., Hilton-Taylor, C. and Stuart, S. (editors). 2004. A global species assessment. 2004 IUCN red list of threatened species. IUCN—The world conservation union.

Baptist Creation Care. 2009. A Southern Baptist declaration on the environment and climate change. http://www.baptistcreationcare.org/node/1. Accessed 9 September, 2009.

Baraer, M., Mark, B. G., McKenzie, J. M., Condom, T., Bury, J. et al. 2012. Glacier recession and water resources in Peru's Cordillera Blanca. Journal of Glaciology, vol. 58, no. 207, pp. 134–150.

Baram, D. 2005. Drowning in money. Guardian Weekly, 25 March, p. 18.

Barash, P. D. 1986. The hare and the tortoise: Culture, biology and human nature. Viking, New York.

Barber, V. A., Juday, G. P. and Finney, B. P. 2000. Reduced growth of Alaskan white spruce in the twentieth century from temperature-induced drought stress. Nature, vol. 405, pp. 668–673.

Bark, J. P. 1995. Your skin: An owner's guide. Prentice Hall, New York.

Barlow, L. K., Sadler, J. P., Ogilvie, A. E. J., Bucjland, P. C., Amorosi, T. et al. 1997. Interdisciplinary investigations of the end of the Norse Western settlement in Greenland. Holocene, vol. 7, pp. 489–499.

Barnosky, A. D., Bell, C. J., Emslie, S. D., Goodwin, H. T., Mead, J. I. et al. 2004a. Exceptional record of mid-Pleistocene vertebrates helps differentiate climate from anthropogenic ecosystem perturbations. Proceedings of the National Academy of Sciences, vol. 101, no. 25, pp. 9297–9302.

Barnosky, A. D., Koch, P. L., Feranec, R. S., Wing, S. L. and Shabel, A. B. 2004b. Assessing the causes of late Pleistocene extinctions on the continents. Science, vol. 306, pp. 70–75.

Barnosky, A. D., Matzke, N., Tomiya, S., Wogan, G. O. U., Swartz, B. et al. 2011. Has the Earth's sixth mass extinction already arrived? Nature, vol. 471, pp. 51–57.

Bar-Oza, G., Zederb, M. and Holec, F. 2011. Role of mass-kill hunting strategies in the extirpation of Persian gazelle (Gazella subgutturosa) in the northern Levant. Proceedings of the National Academy of Sciences, vol. 108, no. 18, pp. 7345–7350.

Barr, D. B., Panuwet, P., Nguyen, J. V., Udunka, S. and Needham, L. L. 2007. Assessing exposure to atrazine and its metabolites using biomonitoring. Environmental Health Perspectives, vol. 115, pp. 1474–1478.

Barrett, J. H., Locker, A. M. and Roberts, C. M. 2004. The origins of intensive marine fishing in medieval Europe: The English evidence. Proceedings of the Royal Society of London, Series B, vol. 271, pp. 2417–2421.

Barrett, W. 1978. The illusion of technique: A search for meaning in a technological civilization. Anchor press/Doubleday, New York.

Barzun, J. 2000. From dawn to decadence: 1500 to the present. 500 Years of cultural life. Harper Colins, New York.

Basu, N., Cryderman, D. K, Miller, F. K., Johnston, S., Rogers, C. et al. 2013. Biomarkers of chemical exposure at Aamjiwnaang. McGill Environmental Health Sciences Lab Occasional Report 2013-1, pp. 1–13.

Bates, S. 2005. Faith leaders tackle the tsunami "problem". Guardian Weekly, 7 January, p. 12.

Baum, J. K., Myers, R. A., Kehler, D. G., Worm, B., Harley, S. J. et al. 2003. Collapse and conservation of shark populations in the northwest Atlantic. Science, vol. 299, pp. 389–392.

Bavington, D. 2005. Of fish and people: Managerial ecology in Newfoundland and Labrador cod fisheries. PhD dissertation, Wilfrid Laurier University. DAI 2006 66(11): 4133-A. DANR09915, pp. 1–293.

Bavington, D. and Sajay, S. 2003. How can we think beyond management when exploring the collapse of the northern cod fishery in Newfoundland and Labrador? CMS session, 3rd Critical Management Studies Conference, Lancaster, UK. 7–9 July, 2003.

BC Hydro. 2006. For generations. November–December.

BC MFR. 2008. Mountain pine beetle: Frequently asked questions. BC Ministry of Forest and Range. http://www.for.gov.bc.ca/hfp/mountain_pine_beetle/faq.htm#5. Last updated February 2008. Accessed September 2009.

Bearchell, S. J., Fraaije, B. A., Shaw, M. W. and Fitt, B. D. 2005. Wheat archive links long-term fungal pathogen population dynamics to air pollution. Proceedings of the National Academy of Sciences, vol. 102, no. 15, pp. 5438–5442.

Bebchuk, L. and Grinstein, Y. 2005. The growth of executive pay. Oxford Review of Economic Policy, vol. 21, no. 2, pp. 283–303.

Bechara, A., Damasio, H., Tranel, D. and Damasio, A. R. 1997. Deciding advantageously before knowing the advantageous strategy. Science, vol. 275, pp. 1293–1295.

Beckwith, H. 1997. Selling the Invisible: A field guide to modern marketing. Warner Books.

Begley, S. 2007. The truth about denial. Newsweek, 13 August. http://www.msnbc.msn.com/id/20122975/site/newsweek/

Begon, M., Harper, J. L. and Townsend, C. R. 1996. Ecology: Individuals, populations and communities. 3rd ed. Blackwell Science, Oxford.

Bejan, A. and Marden, J. H. 2006. Constructing animal locomotion from new thermodynamics theory. American Scientist, vol. 94, July–August, pp. 342–349.

Bella, D. A. 1987. Organizations and systematic distortion of information. Journal of Professional Issues in Engineering, vol. 113, no. 4, pp. 360–370.

Bella, D. A., Mosher, C. D. and Calvo, S. N. 1988. Technocracy and trust: Nuclear waste controversy. Journal of Professional Issues in Engineering, vol. 114, no. 1, pp. 27–39.

Belongia, E. A. and Schwartz, B. 1998. Strategies for promoting judicious use of antibiotics by doctors and patients. British Medical Journal, vol. 317, pp. 668–671.

Ben-Ami, Y., Koren, I., Rudich, Y., Artaxo, P., Martin, S. T. et al. 2010. Transport of Saharan dust from the Bod´el´e depression to the Amazon basin: A case study. Atmospheric Chemistry and Physics Discussions, vol. 10, pp. 4345–4372.

Benedict XVI. 2007. Papal Q-and-A session with priests. Part 1, 2 and 3. 24 July. http://zenit.org/article-20253?l=english, and http://zenit.org/article-20263?l=english.

Benedict XVI. 2008. Pope's Q and A session with clergy of Bressanone. Parts 1–5. 19 August, 2008. www.zenit.org/article-23405?l=english

Bentley, R. W. 2002. Global oil & gas depletion: An overview. Energy Policy, vol. 30, pp. 189–205.

Bentley, R., Booth, R., Burton, J., Coleman, M., Sellwood, B. et al. 2000. Perspectives on the future of oil. Energy Exploration and Exploitation, vol. 18, pp. 147–206.

Bentley, R., Miller, R., Wheeler, S. and Boyle, G. 2009. Review of evidence for global oil depletion. United Kingdom Energy Research Centre, Technical Report 7: Comparison of Global Supply Forecasts, REF UKERC/WP/TPA/2009/022, pp. 1–86.

Berger-Bächi, B. and McCallum, N. 2006. State of the knowledge of bacterial resistance. Injury, vol. 37, no. 2, Supplement 1, pp. S20–S25.

Bernard, P., Follorou, J. and Stroobants, J. 2006. The route to an Abidjan dump. Guardian Weekly, 20 October, p. 29.

Berryman, A. A. 2003. On principles, laws and theory in population ecology. Oikos, vol. 103, no. 3, pp. 695–701.

Bhattacharya, P., Lin, S., Turner, J. P. and Ke, P. C. 2010. Physical adsorption of charged plastic nanoparticles affects algal photosynthesis. Journal of Physical Chemistry, vol. 114, No, 39, pp. 16556–16561.

Bidleman, T. F., Patton, G. W., Walla, M. D., Hargrave, B. T., Vas, W. P. et al. 1989. Toxaphene and other organochlorines in Arctic ocean fauna: Evidence for atmospheric delivery. Arctic, vol. 42, no. 4, pp. 307–313.

Bintliff, J. 2002. Time, process and catastrophism in the study of Mediterranean alluvial history: A review. World Archaeology, vol. 33, no. 3, pp. 417–435.

BiPRO. 2010. Risk management evaluation Endosulfan. Long version - UNECE. Context. BiPRO, pp. 1–165.

BirdLife Global Seabird Programme. 2008. Albatross Task Force Annual Report 2007. Royal Society for the Protection of Birds, The Lodge, Sandy, Bedfordshire, UK, pp. 1–38.

BirdLife International. 2004. State of the world's birds, 2004: Indicators for our changing world. BirdLife International, Cambridge, UK, pp. 1–73.

BirdLife International. 2005. Diclofenac: The mystery solved. http://www.birdlife.net/action/science/species/asia_vulture_crisis/diclofenac.html.

Bjørnstad, O. N., Finkenstädt, B. F. and Grenfell, B. T. 2002. Dynamics of measles epidemics: Estimating scaling of transmission rates using a time series SIR. Ecological Monographs, vol. 72, no. 2, pp. 169–184.

Black, J. M. 1988. Preflight signaling in Swans: A mechanism for group cohesion and flock formation. Ethology, vol. 97, pp. 143–147.

Black, R. 2011. New warning on Arctic sea ice melt. BBC, 7 April, 2011. http://www.bbc.co.uk/news/science-environment-13002706.

Blaustein, A. R. and Johnson, P. T. 2003. Explaining frog deformities. Scientific American, February, pp. 60–65.

Blazer, V. S., Iwanowicz, L. R., Iwanowicz, D. D., Smith, D. R., Young, J. A. et al. 2007. Intersex (testicular oocytes) in smallmouth bass from the Potomac River and selected nearby drainages. Journal of Aquatic Animal Health, vol. 19, no. 4, pp. 242–253.

Blueweiss, L., Fox, H., Kudzma, V., Nakashima, D., Peters, R. et al. 1979. Relationships between body size and some life history parameters. Oecologia, vol. 37, pp. 257–272.

Bogaard, A., Fraser, R., Heaton, T. H. E., Wallace, M., Vaiglova, P. et al. 2013. Crop manuring and intensive land management by Europe's first farmers. Proceedings of the National Academy of Sciences, vol. 110, no. 31, pp. 12589–12594.

Borger, J. 2001. Mapmaking martyr. Guardian Weekly, 19 April, p. 20.

Bornehag, C., Lundgren, B., Weschler, C. J., Sigsgaard, T., Hagerhed-Engman, L. et al. 2005. Phthalates in indoor dust and their association with building characteristics. Environmental Health Perspectives, vol. 113, no. 10, pp. 1399–1404.

Bornehag, C., Sundell, J., Weschler, C. J., Sigsgaard, T., Lundgren, B. et al. 2004. The association between asthma and allergenic symptoms in children and phthalates in house dust: A nested case-control study. Environmental Health Perspectives, vol. 112, no. 4, pp. 1393–1397.

Borrie, W. D., Firth., R. and Spillius, J. 1957. The population of Tikopia, 1929 and 1952. Population studies, vol. 10, no. 3, pp. 229–252.

Boulter, M. 2002. Extinction: Evolution and the end of man. Columbia University Press, New York.

Boumans, R., Costanza, R., Farley, J., Wilson, M. A., Portela, R. et al. 2002. Modeling the dynamics of the integrated earth system and the value of ecosystems using the GUMBO model. Ecological Economics, vol. 41, no. 3, pp. 529–560.

Boyce, D. G., Lewis, M. R. and Worm, B. 2010. Global phytoplankton decline over the past century. Nature, vol. 466, pp. 591–596.

Boyd, P. W., Watson, A. J., Law, C. S., Abraham, E. R., Trull, T. et al. 2000. A mesoscale phytoplankton bloom in the polar Southern ocean stimulated by iron fertilization. Nature, vol. 407, pp. 695–702.

Boyd, R. and Richerson, P. 2000. Meme theory oversimplifies how culture changes. Scientific American, October, p. 67.

Boykoff, M. T. and Boykoff, J. M. 2004. Balance as bias: Global warming and the US prestige press. Global Environmental Change, vol. 14, pp. 125–136.

BP. 2014. BP Statistical Review of World Energy 2014, pp. 1–48. http://www.bp.com/statisticalreview.

Bradshaw, C. J. A. and Brook, B. A. 2014. Human population reduction is not a quick fix for environmental problems. Proceedings of the National Academy of Sciences, vol. 111, no. 46, pp. 16610–16615.

Brandt, A. R. 2011. Oil depletion and the energy efficiency of oil production: The case of California. Sustainability, vol. 3, no. 10, pp. 1833–1854.

Brandt, A. R., Englander, J. and Bharadwaj, S. 2013. The energy efficiency of oil sands extraction: Energy return ratios from 1970 to 2010. Energy, vol. 55, pp. 693–702.

Bright, C. 1998. Life out of bounds: Bioinvasion in a borderless world. Norton and company, New York.

Brin, S. and Page, L. 1998. The anatomy of a large-scale hypertextual web search engine. In: Seventh International World-Wide Web Conference (WWW 1998), April 14-18, 1998, Brisbane, Australia, pp. 1–20.

Briscoe, H. T. 1959. A short course in organic chemistry. The Riverside Press Cambridge, USA.

Bristow, D., Haynes, J-D., Sylvester, R., Frith, C. D. and Rees, G. 2005. Blinking suppresses the neural response to unchanging retinal stimulation. Current Biology, vol. 15, pp. 1296–1300.

Broad, G. 2005. Shopping for victory. Beaver, April–May, pp. 40–45.

Broecker, W. 1987. Unpleasant surprises in the greenhouse? Nature, vol. 32, July, pp. 123–126.

Bronson, B. 1988. The role of barbarians in the fall of states. In Yoffee, N. and Cowgill, W. (editors) 1988, pp. 196–218.

Brooks, G. E. 1989. Ecological perspectives on Mande population movements, commercial networks, and settlement patterns from the Atlantic wet phase (ca. 5500-2500 B.C.) to the present. History in Africa, vol. 16, pp. 23–40.

Brooks, T. M., Pimm, S. L. and Oyugi, J. O. 1999. Time lag between deforestation and bird extinction in tropical forest fragments. Conservation Biology, vol. 13, no. 5, pp. 1140–1150.

Brower, M. and Leon, W. 1999. The consumer's guide to effective environment choices. Three Rivers Press, New York.

Brown, A. W. A. and Pal, R. 1971. Insecticide resistance in arthropods. World Health Organization, Monograph series No. 38, 2nd ed., pp. 1–491.

Brown, D. 2005. Catch that slipped through the net. Guardian Weekly, 9 December, p. 13.

Brown, E. D. 2006. Antibiotic stops 'ping-pong' match. Nature, vol. 441, pp. 293–294.

Brown, J. H. and Maurer, B. A. 1989. Macroecology: The division of food and space among species on continents. Science, vol. 243, pp. 1145–1150.

Brown, P. 2000. Over fishing and global warming sink cod. Guardian Weekly, 30 July, p. 10.

Brown, P. 2004. Reform now or die, fishing industry warned. Guardian Weekly, 1 April, p. 9.

Brown, P. and Osborn, A. 2001. Ban imposed on North Sea cod fishing. Guardian Weekly, 1 February, p. 9.

Brown, W. A. 1998. The placebo effect. Scientific American, January, pp. 70–75.

Browne, J. 2006. A bigger picture of apes. Nature, vol. 439, p. 142.

Browne, M. A., Niven, S. J., Galloway, T. S., Rowland, S. J. and Thompson, R. C. 2013. Microplastic moves pollutants and additives to worms, reducing functions linked to health and biodiversity. Current Biology, vol. 23, no. 23, pp. 2388–2392.

Brundtland, G. H. 1997. The scientific underpinning of policy. Science, vol. 277, no. 5325, p. 457.

Bruner, J. 1992. Another look at New Look 1. American Psychologist, vol. 47, no. 6, pp. 780–783.

Bundy, A. 2005. Structure and functioning of the eastern Scotian Shelf ecosystem before and after the collapse of groundfish stocks in the early 1990s. Canadian Journal of Fisheries and Aquatic Sciences, vol. 62, no. 7, pp. 1453–1473.

Bureau of Economic Analysis. 2009. Tables. U S Government, Bureau of Economic Analysis. http://BEA.gov/national/nipaweb/selecttable.asp?selected=N.

Bureau of Economic Analysis. 2010. US domestic product. Gross output by industry. 30 September, 2010. http://bea.gov/industry/gpotables/gpo_action.cfm?anon=81259&table_id=26637&format_type=0

Bureau of Economic Analysis. 2014. Gross domestic product. US Government, Bureau of Economic Analysis, BEA current and historical data, Section D, Table 1.1.5, p. D-3. www.bea.gov/scb/pdf/2014/02%20February/D%20pages/D_Pages_cont.pdf.

Burke, E. J., Brown, S. J. and Christidis, N. 2006. Modeling the recent evolution of global drought and projections for the twenty-first century with the Hadley Centre climate model. Journal of Hydrometeorology, vol. 7, pp. 1113–1125.

Burke, J. 2013. Indian police chief's rape analogy causes outrage across country. Guardian,13 November, 2013. https://www.theguardian.com/world/2013/nov/13/prevent-rape-enjoy-it-india-police-chief.

Burkeman, O. 2002. Too darn hot as the water leaks away. Guardian Weekly, 25 April, p. 6.

Burkholder, J. M. 1999. The lurking perils of Pfiesteria. Scientific American, August, pp. 42–49.

Burnet, M. and White, D. O. 1972. Natural history of infectious disease. 4th ed. Cambridge University Press, Cambridge.

Business Week. 2002. CEOs: Why they're so unloved. BusinessWeek. 22 April, 2002, p. 1. http://www.businessweek.com/magazine/content/02_16/b3779125.htm

Buss, D. M. 2000. The evolution of happiness. American Psychologist, vol. 55, pp. 15–23.

Butler, C. C., Rollnick, S., Pill, R., Maggs-Rapport, F. and Stott, N. 1998. Understanding the culture of prescribing: Qualitative study of general practitioners' and patients' prescriptions of antibiotics for sore throats. British Medical Journal, vol. 317, pp. 637–642.

Caldwell, M. H. 2008. Molymania—the great molybdenum bull market. Creative Classics, Kelowna.

Callahan, G. N. 2009. Between XX and XY: Intersexuality and the myth of two sexes. Chicago Review Press, Chicago.

Callarotti, R. C. 2011. Energy return on energy invested (EROI) for the electrical heating of methane hydrate reservoirs. Sustainability, vol. 3, pp. 2105–2114.

Callaway, R. M., Brooker, R. W., Choler, P., Kikvides, Z., Lortie, C. J. et al. 2002. Positive interactions among alpine plants increase with stress. Nature, vol. 417, pp. 844–848.

Cameron, S. A. 2011. Patterns of widespread decline in North American bumble bee. Proceedings of the National Academy of Sciences, vol. 108, no. 2, pp. 662–667.

Campbell, C. 2004. "I shop therefore I know that I am.": The metaphysical foundations of modern consumerism. In Ekstrom, K. and Brembeck, H. (editors). Elusive Consumption. Berg, Oxford. Chapter 2, pp. 27–44.

Campbell, C. J. and Laherrere, J. H. 1998. The end of cheap oil. Scientific American, vol. 278, no. 3, pp. 78–83.

CAPP. 2011. Water use in Canada's oil sands. Canadian Association of Petroleum Producers, pp. 1–10. http://www.capp.ca/getdoc.aspx?DocId=161615&DT=NTV.

Carbon, C. and Bax, R. P. 1998. Regulating the use of antibiotics in the community. British Medical Journal, vol. 317, pp. 663–665.

Cardinale, B. J., Duffy, E. J., Gonzalez, A., Hooper, D. U., Perrings, C. et al. 2012. Biodiversity loss and its impact on humanity. Nature, vol. 486, pp. 59–67.

Carlson, S. 1985. A double blind test of astrology. Nature, vol. 318, pp. 419–425.

Carlson, S. 1999. Falling into chaos. Scientific American, November, pp. 120–121.

Carrell, S. 2012. Fishing skippers and factory fined nearly £1m for illegal catches: Police uncover "serious and organised" criminality in £63m scam to breach European fishing quotas. Guardian, 24 February. http://www.theguardian.com/environment/2012/feb/24/fishing-skippers-fined-illegal-catches.

Carrington, D. 2016. The quest for limitless energy. Guardian Weekly, 2 December, pp. 1 and 13.

Carson, R. 1962. Silent spring. Houghton Mifflin Co., New York.

Cauley, G. 2004. August 14, 2003 Blackout. North American Electric Reliability Council. Slide presentation, pp. 1–47.

CBC News. 2003. Flame retardants in Inuit breast milk. CBC News, 17 September, 2003.http://www.cbc.cz/news/story/2003/09/17/pollutants030917.html.

CDC. 2016. Estimated range of Aedes aegypti and Aedes albopictus in the United States, 2016. United States Centre for Disease Control. http://www.cdc.gov/zika/vector/range.html.

Ceballos, G., Ehrlich, P. R. and Dirzo, R. 2017. Biological annihilation via the ongoing sixth mass extinction signaled by vertebrate population losses and declines. Proceedings of the National Academy of Sciences, vol. 114, no. 30, E6089-E6096.

Ceballos, G., Ehrlich, P. R., Barnosky, A. D., García, A., Pringle, R. M. et al. 2015. Accelerated modern human-induced species losses: Entering the sixth mass extinction. Science Advances, vol. 1, no. 5, e1400253, pp. 1–5.

CEC. 2000. Communication from the commissions on the precautionary principle. The Commission of the European Communities, COM(2000) 1, 2 February, 2000, pp. 1–29. http://ec.europa.eu/dgs/health_consumer/library/pub/pub07_en.pdf.

Cello, J., Paul, A. V. and Wimmer, E. 2002. Chemical synthesis of poliovirus cDNA: Generation of infectious virus in the absence of natural template. Science, vol. 297, pp. 1016–1018.

Census of India. 2001. Census of India 2001, Census Reference Tables, A-Series-Primary Census Abstract. http://censusindia.gov.in/Census_Data_2001/Census_data_finder/A_Series/PCA.htm.

Census of India. 2011. Census of India 2011, Child sex ratio. http://www.census2011.co.in/sexratio.php.

Chadburn, S. E., Burke, E. J., Cox, P. M., Friedlingstein, P., Hugelius, G. et al. 2017. An observation-based constraint on permafrost loss as a function of global warming. Nature Climate Change, vol. 7, pp. 340–345.

Chaitin, G. 1988. Randomness in arithmetic. Scientific American, vol. 259, no. 1, pp. 80–85.

Chan, M. 2010. Ten ways to game the carbon market. Friends of the Earth, May, pp. 1–8. www.foe.org/sites/default/files/10waystogamethecarbonmarket.

Chapin, F. S., Zavaleta, E. S., Eviner, V. T., Naylor, R. L., Vitousek, P. M. et al. 2000. Consequences of changing biodiversity. Nature, vol. 405, pp. 234–242.

Chapman, J. L. and Reiss, M. J. 1992. Ecology principles and applications. Cambridge University Press, Cambridge.

Chapman, L. J. and Chapman, J. P. 1967. Genesis of popular but erroneous psychodiagnostic observations. Journal of Abnormal Psychology, vol. 72, pp. 193–204.

Charlton, A. 2012. Davos founder Klaus Schwab: Focus on jobs, morals. USA Today, 27 January. http://www.usatoday.com/money/world/story/2012-01-28/davos-founder-schwab-capitalism/52824354/1

Cheer, S. 1999. What do these women have in common? Marie Claire, October, pp. 98–105.

Chrisafis, A. 2007. Ophelia of the Seine. Guardian Weekly, 21 December, pp. 35–37.

Christensen, V., Piroddi, C., Coll, M., Steenbeek, J., Buszowski, J. et al. 2011. Fish biomass in the world ocean: A century of decline. The Fisheries Centre, Working paper series, Working paper #2011-06, pp. 1–19.

Cialdini, R. B. 2001. The science of persuasion. Scientific American, February, pp. 76–81.

Clark, E. 1988. The want makers. Lifting the lid off the world advertising industry: How they make you buy. Hodder and Stoughton, London.

Clark, L., Lawrence, A. J., Astley-Jones, F. and Gray, N. 2009. Gambling near-misses enhance motivation to gamble and recruit win-related brain circuitry. Neuron, vol. 61, no. 3, pp. 481–490.

Clarke, R. (editor). 1999. Overview GEO-2000: Global environment outlook 2000. United Nations Environment Program (UNEP). Earthscan, pp. 1–17.

Clarke, T. 2003. Pest resistance feared as farmers flout rules. Nature, vol. 424, p. 116.

Cleveland, C. J. 2005. Net energy from the extraction of oil and gas in the United States, 1954-1997. Energy, vol. 30, pp. 769–782.

Cleveland, C. J., Costanza, R., Hall, C. A. and Kaufmann, R. 1984. Energy and the U.S. Economy: A biophysical perspective. Science, vol. 225, pp. 890–897.

Cleveland, C. J. and Kaufmann, R. K. 1991. Forcasting ultimate oil recovery and its rate of production: Incorporating economic forces into the models of M. King Hubbert. Energy Journal, vol. 12, no. 2, pp. 316–345.

Cleveland, C. J. and O'Connor, P. A. 2011. Energy return on investment (EROI) of oil shale. Sustainability, vol. 3, no. 11, pp. 2307–2322.

Cline, E. H. 2014. 1177 B.C. The year civilization collapsed. Princeton University Press, Princeton.

Clout, M. 2001. Where protection is not enough: Active conservation in New Zealand. Trends in Ecology and Evolution, vol. 16, no. 8, pp. 415–416.

CNMAR. 1992. Environmental neurotoxicology. Committee on neurotoxicology and models for assessing risk. National Research Council, National Academy Press, Washington, DC.

Cohen, J. 2006. Extensively drug-resistant TB gets foothold in South Africa. Science, vol. 313, no. 5793, p. 1554.

Cohen, J. E. 1995. How many people can the earth support. W. W. Norton and Company.

Cohen, J., Chesnick, E. I. and Haran, D. 1982. Evaluation of compound probabilities in sequential choice. In Kahneman, D., Slovic, P. and Tversky, A. (editors) 1982, pp. 355–358.

Cohen, S. 2001. States of denial: Knowing about atrocities and suffering. Polity, Cambridge.

Cohen, S. N. and Shapiro, J. A. 1980. Transposable genetic elements. Scientific American, vol. 242, no. 2, pp. 40–49.

Colander, D., Föllmer, H., Haas, A., Goldberg, M., Juselius, K. et al. 2009. The financial crisis and the systemic failure of academic economics. Kiel Institute for the World Economy, Working paper 1489, pp. 1–17. http://www.irg-project.org/fileadmin/publications/PUBLIC/Events/Learning_from_Financial_Crisis.pdf.

Cole, M. and Cole, S. R. 2001. The development of children. 4th ed. Worth Publishers, New York.

Cole, M., Lindeque, P. K., Fileman, E., Clark, J., Lewis, C. et al. 2016. Microplastics alter the properties and sinking rates of zooplankton faecal pellets. Journal of Environmental Science and Technology, vol. 50, no. 6, pp. 3239–3246.

Collins, M. 2007. Ensembles and probabilities: A new era in the prediction of climate change. Philosophical Transactions of the Royal Society, Series A, vol. 365, pp. 1957–1970.

Coltman, D. W., O'Donoghue, P., Jorgenson, J. T., Hogg, J. T., Storbeck, C. et al. 2003. Undesirable evolutionary consequences of trophy hunting. Nature, vol. 426, pp. 655–658.

Colwell, R. R. 1996. Global climate and infectious disease: The cholera paradigm. Science, vol. 274, pp. 2025–2031.

Commoner, B., Woods Bartlett, P., Eisl, H. and Couchot, K. 2000. Long range air transport of dioxin from North American sources to ecologically vulnerable receptors in Nunavut, Arctic Canada. North American Commission for Environmental Cooperation, September 2000. Executive summary, pp. 1–8. http://www.cec.org.

Confino, J. 2004. Malawi villagers help sow the seeds of hope. Guardian Weekly, 10 December, p. 12.

Conradt, L. and Roper, T. J. 2003. Group decisions-making in animals. Nature, vol. 421, pp. 155–158.

Conway, G. and Toenniessen, G. 1999. Feeding the world in the twenty first century. Nature, vol. 402, Supplement, pp. C55–58.

Cook, A. J., Fox, A. J., Vaughan, D. G. and Ferrigno, J. G. 2005. Retreating glacier fronts on the Antarctic peninsula over the past half-century. Science, vol. 308, pp. 541–544.

Cook, D. R. 1987. A crisis for economic geologists and the future of the society. Economic Geology, vol. 82, pp. 792–804.

Cooke, R. 2003. Insider guide to the bunker mentality. Review of Until the final hour: Hitler's last secretary, by Junge, T. Guardian Weekly, 20 November, p. 17.

Coops, N. C. and Waring, R. H. 2011. A process-based approach to estimating lodgepole pine (Pinus contorta Dougl.) distribution in the Pacific Northwest under climate change. Climatic Change, vol. 105, no. 1-2, pp. 313–328.

Coppola, B. P. and Daniels, D. S. 1999. Mea Culpa: Formal education and the dis-integrated world. In Aerts, D., Gutwirth,

S., Smets, S. and Van Langehove, L. (editors). Science, technology and social change: The orange book of "Einstein meets Magritte". VUB University Press, Belgium, Kluwer Academic Publishers, Dortrecht. Volume 3, pp. 107–128.

Costanza, R. 1994. Three general policies to achieve sustainability. In Jansson, A., Hammer, M., Folke, C. and Costanza, R. (editors). Investing in natural capital: The ecological economics approach to sustainability. Island Press, Washington, DC. pp. 392–405.

Costanza, R., d'Arge, R., de Groot, R., Farber, S., Grasso, M. et al. 1997. The value of the world's ecosystem services and natural capital. Nature, vol. 387, pp. 253–260.

Costanza, R., Leemans, R., Boumans, R. M. J. and Gaddis, E. 2007. Integrated global models. In Costanza, R., Graumlich, L. and Steffen, W. (editors). Report of the 96th Dahlem workshop on an integrated history and future of people on earth. MIT Press, Cambridge, MA. pp. 417–443.

Costanza, R., Wainger, L., Folke, C. and Maler, K. 1993. Modeling complex ecological economic systems. Bioscience, vol. 43, no. 8, pp. 545–555.

Costerton, J. W. and Stewart, P. S. 2001. Battling biofilms. Scientific American, July, pp. 74–81.

COSTR. 2006. Executive summary and chapter 11: Large-scale multiproxy reconstruction techniques. Surface temperature reconstructions for the last 2,000 years. Committee on Surface Temperature Reconstructions for the Last 2,000 Years. National Research Council. NOAA, National Academies Press, pp. 1–160.

Courtenay, W. R. 1993. Biological pollution through fish introductions. In McKnight, B. N. (editor). Biological pollution: The control and impact of invasive exotic species. Indiana Academy of Science, Indianapolis. pp. 35–61.

Cox-Foster, D. and vanEngelsdorp, D. 2009. Saving the honeybee. Scientific American, April, pp. 40–47.

Coyle, D. 2014. GDP: A brief but affectionate history. Princeton University Press.

Craig, G. B. 1993. The diaspora of the Asian tiger mosquito. In McKnight, B. N. (editor). Biological Pollution: The control and impact of invasive exotic species, Indiana Academy of Science, Indianapolis. pp. 101–120.

Crouch, D. 2016. Robot mining trucks get rolling. Guardian Weekly, 3 June, p. 17.

Crowe, P. 2015. A Wall Street "Master of the Universe" nailed the millennial mindset with one simple story. http://www.businessinsider.com/wall-street-for-junior-bankers-2015-10. Accessed 10 July, 2017.

Crumley, C. L. 2000. From garden to globe: Linking time and space with meaning and memory. In McIntosh, R. J., Tainter, J. A. and McIntosh, S. K. (editors) 2000, pp. 193–208.

Csikszentmihalyi, M. 2000. If we are so rich why aren't we happy? American Psychologist, vol. 54, pp. 821–827.

CSPNA. 2007. The status of pollinators in North America. Committee on the Status of Pollinators in North America. National Academies Press, Washington, DC, pp. 1–307.

Culbert, T. P. 1988. The collapse of classic Mayan civilization. In Yoffee, N. and Cowgill, W. (editors) 1988, pp. 69–101.

Cullen, H. M., deMenocal, P. B., Hemming, S., Hemming, G., Brown, F. H. et al. 2000. Climate change and the collapse of the Akkadian empire: Evidence from the deep sea. Geology, vol. 28, pp. 379–382.

Cullon, D. L., Yunker, M. B., Alleyne, C., Dangerfield, N. J., O'Neill, S. et al. 2009. Persistent organic pollutants in Chinook salmon (oncorhychus tshawytscha): Implications for resident killer whales of British Columbia and adjacent waters. Environmental toxicology and chemistry, vol. 28, no. 1, pp. 148–161.

Cunningham, A. A., Garner, T. W. J., Aguilar-Sanchez, V., Banks, B., Foster, J. et al. 2005. Emergence of amphibian chytridiomycosis in Britain. Veterinary Record, vol. 157, pp. 386–387.

Currie, D. J. and Paquin, V. 1987. Large-scale biogeographical patterns of species richness in trees. Nature, vol. 329, pp. 326–327.

Curry, R., Dickson, B. and Yashayaev, I. 2003. A change in the freshwater balance of the Atlantic Ocean over the past four decades. Nature, vol. 426, pp. 826–829.

Daddi, B. 2009. U.S. advertising expenditure declined 143 percent in first half 2009. TNS Media Intelligence, news release, 19 September. http://www.tns-mi.com/news/09162009.htm.

Daily, G. C. and Ehrlich, P. R. 1992. Population, sustainability, and Earth's carrying capacity. BioScience, vol. 42, no. 10, pp. 761–771.

Daily Times Monitor. 2009. India may sign nuclear liability waiver. Daily Times Monitor, 3 April, p. 20. http://www.dailytimes.com.pk/default.asp?page=2009\04\03\story_3-4-2009_pg20_2.

Dainat, B., vanEngelsdorp, D. and Neumann, P. 2012. Colony collapse disorder in Europe. Environmental Microbiology Reports, vol. 4, no. 1, pp. 123–125.

Dale, M., Krumdieck, S and Bodger, P. 2011. Net energy yield from production of conventional oil. Energy Policy, vol. 39, pp. 7095–7102. http://dx.doi.org/10.1016/j.enpol.2011.08.021.

Damasio, R. A. 2000. The feeling of what happens: Body and emotion in the making of consciousness. Harcourt, Brace & Company, New York.

Dameris, M., Matthes, S., Deckert, R., Grewe, V. and Ponater, M. 2006. Solar cycle effects delays onset of ozone recovery. Geophysical Research Letters, vol. 33, L03806.

Damerow, P. 1998. Prehistory and cognitive development. In Langer, J. and Killen, M. (editors). Piaget, evolution and development. Mahwah, New Jersey, Erlbaum. Chapter 11, pp. 247–270.

Damon, W. 1999. Moral development of children. Scientific American, August, pp. 72–78.

Daniels, R., Boffey, C., Mogg, R., Bond, J. and Clarke, R. 2005. The potential for dispersal of herbicide tolerance genes from genetically-modified, herbicide-tolerant oilseed rape crops to wild relatives. DEFRA, report EPG 1/5/151, pp. 1–23.

Daniels, T. (editor). 1999. A doomsday reader: Prophets, predictors, and hucksters of salvation. New York University Press, New York. pp. 1–253.

Dansgaard, W., Johnsen, S. J., Clausen, H. B., Dahl-Jensen, D., Gundestrup, N. S. et al. 1993. Evidence for general instability of past climate from a 250 kyr ice-core record. Nature, vol. 364, pp. 218–220.

d'Arge, R. C. 1994. Sustenance and sustainability: How can we preserve and consume without major conflict? In Jansson, A., Hammer, M., Folke, C. and Costanza, R. (editors). Investing in natural capital: The ecological economics approach to sustainability. Island Press, Washington, DC. pp. 113–127.

Darimond, C., Carlson, S. M., Kinnison, M. T., Paquet, P. C., Reimchen, T. E. et al. 2009. Human predators outpace other agents of trait change in the wild. Proceedings of the National Academy of Sciences, vol. 106, no. 3, pp. 952–954.

Darwin, C. 1972. The voyage of the Beagle. Bantam Books, New York.

Dasgupta, S., Laplante, B., Meisner, C., Wheeler, D. and Yan, J. 2009. The impact of sea level rise on developing countries: A comparative analysis. Climatic Change, vol. 93, no. 3-4, pp. 379–388.

Daston, L. and Galison, P. 1992. The image of objectivity. Representations, vol. 40, pp. 81–128.

Daszak, P., Cunningham, A. A. and Hyatt, A. D. 2000. Emerging infectious diseases of wildlife—Threats to biodiversity and human health. Science, vol. 287, pp. 443–449.

Davis, M. B. and Shaw, R. G. 2001. Range shifts and adaptive responses to quaternary climate change. Science, vol. 292, no. 5517, pp. 673–679.

Davis, W. 1997. One River. Touchstone, New York.

Dawes, R. M. 1982. The robust beauty of improper linear models in decision making. In Kahneman, D., Slovic, P. and Tversky, A. (editors) 1982, pp. 392–407.

Dawes, R. M., Faust, D. and Meehl, P. E. 1989. Clinical versus actuarial judgement. Science, vol. 243, pp. 1668–1674.

Dawson, J. W. 2006. Godel and the limits of logic. +plus magazine, no. 39, pp. 1–11. http://plus.maths.org/issue39/features/dawson/. Adapted from Dawson, J. W. 1999. Scientific American, vol. 280, no. 6.

Day, R. H., Shaw, D. G. and Ignell, S. E. 1990. The quantitative distribution and characteristics of neuston plastic in the north Pacific Ocean, 1985–1988. Shomura, R. S. and Godfrey, M. L. (editors). Proceedings of the Second International Conference on Marine Debris, 2–7 April 1989, Honolulu, Hawaii, US Department of. Commerce, NOAA Tech. Memo. NMFS, NOAA-TM-NMFS-SWFC-54, pp. 247–266.

D'Costa, V. M., McGrann, K. M., Hughes, D. W. and Wright, G. D. 2006. Sampling the antibiotic resistome. Science, vol. 311, no. 5759, pp. 374–377.

De Beers. 1994. Consumer marketing. De Beers Annual report.

de Bresson, H. and Augereau, J. 2002. EU fishing reforms earn French wrath. Guardian Weekly, 6 June, p. 29.

de Leeuw, A. D. 1992. Environmental ethics and the British Columbia provincial fisheries program. A discussion document presented at the annual fisheries meeting, Yellow Point, Nanaimo, January, pp. 1–23.

de Vries, B. J. M. 2007. Scenarios: Guidence for an uncertain and complex world. In Costanza, R., Grqumlich, L. and Steffen, W. (editors). Sustainability or collapse?: An integrated history and future of people on earth. MIT Press, Cambridge, MA. pp. 389–397.

De Waal, F. B. M. 1995. Bonobo sex and society. Scientific American, vol. 272, pp. 82–88.

de Wit, M. and Stankiewicz, J. 2006. Changes in surface water supply across Africa with predicted climate change. Science, vol. 311, no. 5769, pp. 1917–1921.

Dean, J. S. 2000. Complexity theory and sociocultural change in the American southwest. In McIntosh, R. J., Tainter, J. A. and McIntosh, S. K. (editors) 2000, pp. 89–118. |

De'ath, G., Lough, J. M. and Fabricius, K. E. 2009. Declining coral calcification on the Great Barrier Reef. Science, vol. 323, pp. 116–119.

DEFRA. 2010. Food 2030. Britain, Department for Environment, Food and Rural Affairs, pp. 1–84.

Demarest, A. A., Rice, P. M. and Rice, D. S. 2004. The terminal classic in Maya lowlands: Assessing collapses, terminations and transformations. In Demarest, A. A., Rice, P. M. and Rice, D. S. (editors). The terminal classic in the Maya lowlands: Collapse transition and transformation. University Press of Colorado, Boulder. pp. 545–572.

deMenocal, P. B. 2001. Cultural responses to climate change during the late Holocene. Science, vol. 292, pp. 667–673.

Demeny, P. 1990. Tradeoffs between human numbers and material standards of living. In Davis, K. and Bernstein, M. S. (editors). Resources, environment, and population: Present knowledge, future options, Oxford University Press, New York. Supplement to Population and Development Review, V. 16, pp. 408–421.

deWaal, F. B. M. 1999. The end of nature versus nurture. Scientific American, December, pp. 2–7.

Diamond, J. 1997. Guns, germs, and steel: The fates of human societies. W. W. Norton, New York.

Diamond, J. 2005. Collapse: How societies choose to fail or succeed. Viking Press, New York

Dias, B. S. F., Raw, A. and Imperatri-Fonseca, V. L. 1999. Report on the recommendations of the workshop on the conservation and sustainable use of pollinators in agriculture with emphasis on bees. International Pollinators Initiative: The Sao Paulo declaration on pollinators. Brazilian Ministry of the Environment, Brazil, pp. 1–79.

Diaz, R. J. 2001. Overview of hypoxia around the world. Journal of Environmental Quality, vol. 30, no. 2, pp. 275–281.

Diaz, R. J. and Rosenberg, R. 2008. Spreading dead zones and consequences for marine ecosystems. Science, vol. 321, no. 5891, pp. 926–929.

Diener, E. 2000. Subjective well being: The science of happiness and a proposal for a national index. American Psychologist, vol. 55, pp. 34–43.

Diener, E., Ng, W., Harter, J. and Arora, R. 2010. Wealth and happiness across the world: Material prosperity predicts life evaluation, whereas psychosocial prosperity predicts positive feeling. Journal of Personality and Social Psychology, vol. 99, no. 1, pp. 52–61.

Dietz, T., Ostrom, E. and Stern, P. C. 2003. The struggle to govern the commons. Science, vol. 302, pp. 1907–1912.

Dietz, T. and Rosa, E. A. 1997. Effects of population and affluence on $CO_2$ emissions. Proceedings of the National Academy of Sciences, vol. 94, pp. 175–179.

Dixit, Y., Hodell, D. A. and Petrie, C. A. 2014. Abrupt weakening of the summer monsoon in northwest India 4100 yr ago. Geology, vol. 42, no. 4, pp. 339–343.

Dobson, A. P., Bradshaw, A. D. and Baker, A. J. M. 1997. Hopes for the future: Restoration ecology and conservation biology. Science, vol. 277, pp. 515–522.

Doll, R. and Hill, B. 1950. Smoking and carcinoma of the lung: Preliminary report. British Medical Journal, vol. 221, no. 2, pp. 739–748.

Doll, R. and Hill, B. 1954. The mortality of doctors in relation to their smoking habits: A preliminary report. British Medical Journal, vol. 228, no. 1, pp. 1529–1533.

Donaldson, J. 1996. The culture clash: A revolutionary new way to understand the relationship between humans and dogs. James and Kenneth, Berkley.

Doney, S. C. 2006. The danger of ocean acidification. Scientific American, March, pp. 58–65.

Doward, J. 2001. Hi-tech boom that turned to bust. Guardian Weekly, 13 September, p. 15.

Dowden, R. 2006. Small "witches" of Kinshasa. Guardian Weekly, 3 March, p. 30.

Doyle, R. 2005. The Lion's share. Scientific American, April, p. 30.

Dresselhaus, M. S. and Thomas, I. L. 2001. Alternate energy technologies. Nature, vol. 414, pp. 332–337.

Dubos, R. 1971. Man and his environment: Scope, impact, and nature. In Detwyler, T. R. (editor). Man's impact on environments. McGraw-Hill, New York. pp. 684–694.

Dukes, J. 2003. Burning buried sunshine: Human consumption of ancient solar energy. Climatic Change, vol. 61, pp. 31–44.

Dunbar, R. B. 2003b. Leads, lags and the tropics. Nature, vol. 421, pp. 121–122.

Dunbar, R. I. M. 1993. Coevolution of neocortical size, group size and language in humans. Behavioral and Brain Sciences, vol. 16, no. 4, pp. 681–735.

Dunbar, R. I. M. 2003a. The social brain: Mind, language, and society in evolutionary perspective. Annual Review of Anthropology, vol. 32, pp. 163–183.

Duncan, R. P., Boyer, A. G. and Blackburn, T. M. 2013. Magnitude and variation of prehistoric bird extinctions in the Pacific. Proceedings of the National Academy of Sciences, vol. 110, no. 16, pp. 6436–6441.

Dunedin Longitudinal Study. 2018. https://dunedinstudy.otago.ac.nz.

Dunning, N. P. and Beach, T. 1994. Soil erosion, slope management, and ancient terracing in the Maya lowlands. Latin American Antiquity, vol. 5, no. 1, pp. 51–69.

Dunning, N., Luzzadder-Beach, S., Beach, T., Jones, J. G., Scarborough, V. et al. 2002. Arising from the bajos: The evolution of a neotropical landscape and the rise of Maya civilization. Annals of the Association of American Geographers, vol. 92, no. 2, pp. 267–283.

Dupont, G. 2010. Rescue plan for Robinson Crusoe. Guardian Weekly, 5 February, p. 28.

Durham, M. 2004. A bitter taste of honey. Guardian Weekly, 6 August, p. 18.

Durning, A. T. 1992. How much is enough? The consumer society and the future of the earth. Norton and company, New York.

Dye, J. A., Venier, M., Zhu, L., Ward, C. R., Hites, R. A. et al. 2007. Elevated PBDE levels in pet cats: Sentinels for humans? Environmental Science and Technology, vol. 41, no. 18, pp. 6350–6356.

Dyke, A. S. and Prest, V. K. 1987. Late Wisconsinan and Holocene retreat of the Laurentide ice sheet. Geological Survey of Canada, Map No. 1702A. Scale 1:5000000.

Dzik, A. 1983. Snails, schistosomiasis, and irrigation in the tropics. Public Health, vol. 97, no. 4, pp. 214–217.

Easterlin, R. A. 2003. Explaining happiness. Proceedings of the National Academy of Sciences, vol. 100, no. 19, pp. 11176–11183.

Eaton, L., Flisher, A. J. and Aaro, L. E. 2003. Unsafe sexual behaviour in South African youth. Social Science and Medicine, vol. 56, pp. 149–165.

The Economist. 2016. From conflict to co-operation. The Economist, 16 February, pp. 1–6. http://www.economist.com/news/americas/21690100-big-miners-have-better-record-their-critics-claim-it-up-governments-balance.

Edwards, J. D. 1997. Crude oil and alternative energy production forecasts for the twenty first century: The end of the hydrocarbon era. American Association of Petroleum Geologists Bulletin, vol. 81, no. 8, pp. 1292–1305.

Edwards, W. 1982. Conservatism in human information processing. In Kahneman, D., Slovic, P. and Tversky, A. (editors) 1982, pp. 359–369.

Ehrlich, P. R. 1986. Extinction: What is happening and what needs to be done. In Elliott, D. K. (editor). Dynamics and extinction. John Wiley and Sons, NY. pp. 157–164.

EIA. 2000. Economic profile and trends: Chemical industry analysis briefs. US Energy Information Administration. http://www.eia.doe.gov/emeu/mecs/iab/chemicals/page1b.html.

EIA. 2008. World consumption of primary energy by energy types and selected country groups. US Energy Information Administration, International energy annual 2006, Table 18. Updated 2008. http:/www.eia.doe.gov/pub/international/iealf/table18.xls

EIA. 2011. International energy outlook 2011. Figure 28 data. US Energy Information Administration, report #:DOE/EIA-0484(2011), Washington. http://www.eia.gov/forecasts/ieo/liquid_fuels.cfm fig 28.

EIA. 2014a. US tight oil plays: Production and proved reserves, 2012-13. US Energy Information Administration, Annual Survey of Domestic Oil and Gas Reserves, 2012 and 2013, Table 2. http://www.eia.gov/naturalgas/crudeoilreserves/.

EIA. 2014b. Principal shale gas plays: Natural gas production and proved reserves, 2012-13. US Energy Information Administration, Annual Survey of Domestic Oil and Gas Reserves, 2012 and 2013, Table 4. http://www.eia.gov/naturalgas/crudeoilreserves/.

Eisenstadt, S. 1988. Beyond collapse. In Yoffee, N. and Cowgill, W. (editors) 1988, pp. 236–243.

Eisenstadt, S. N. 2000. Multiple modernities. Daedalus, vol. 129, no. 1, pp. 1–30.

Eisenstein, C. 2018. Climate: A new story. North Atlantic Press, Berkley.

Elliott, L. 2001. Pawns in a game that lacks any strategy. Guardian Weekly, 1 February, p. 14.

Elliott, L. 2005. The poor reap the whirlwind. Guardian Weekly, 9 September, p. 26.

Elliott, S. 2000. Two forecasters say the growth in industry revenue will slow moderately in the year to come. New York Times, 5 December, p. C12.

Ellis, K. 1973. Prediction and Prophecy. Wayland, London.

Ellison, P. T. 2001. On fertile ground: A natural history of human reproduction. Harvard University Press, Harvard.

emarketer. 2014. Total US ad spending to see largest increase since 2004: Mobile advertising leads growth; will surpass radio, magazines and newspapers this year. 2 July , 2014, https://www.emarketer.com/Article/Total-US-Ad-Spending-See-Largest-Increase-Since-2004/1010982.

emarketer. 2016. Worldwide ad spending growth revised downward: Annual gains in worldwide ad spending will hover around 6% throughout the forecast period. 21 April, 2016. https://www.emarketer.com/Article/Worldwide-Ad-Spending-Growth-Revised-Downward/1013858.

Emery, N. J. and Clayton, N. S. 2001. Effects of experience and social context on prospective caching strategies by scrub jays. Nature, vol. 414, pp. 443–446.

Enghoff, M. B. and Svensmark, H. 2008. The role of atmospheric ions in aerosol nucleation: A review. Atmospheric Chemistry and Physics, vol. 8, No, 16, pp. 7477–7508.

Englander, J. G., Bharadwaj, S. and Brandt, A. R. 2013. Historical trends in greenhouse gas emissions of the Alberta oil sands (1970-2010). Environmental Research Letters, vol. 8, 044036, pp. 1–7.

Environment Agency. 2006. Underground, under threat: The state of groundwater in England and Wales. British Government. November, pp. 1–24.

EPA. 2007. Atrazine: Chemical Summary. US Environmental Protection Agency, pp. 1–12. http://www.epa.gov/teach/chem_summ/Atrazine_summary.pdf, last revised 24 April, 2007. Accessed 14 July, 2014.

EPA. 2010. Asian carp and the Great Lakes. US Environmental Protection Agency. http://www.epa.gov/glnpo/invasive/asiancarp/.

EPA. 2011. Sulfur dioxide. US Environmental Protection Agency. Air trends: Sulfur dioxide. http/www.epa.gov/air/airtrends/sulfur.html.

EPIC community members (Wolff, E. leader). 2004. Eight glacial cycles from an Antarctic ice core. Nature, vol. 429, pp. 623–628.

ERCB. 2010. Alberta's energy reserves 2009 and supply/demand outlook 2009–2019. Alberta Energy Resources Conservation Board, Report ST98-2010, pp. 1–232.

ERCB. 2012. Alberta's energy reserves 2011 and supply/demand outlook 2012–2021. Alberta Energy Resources Conservation Board, Report ST98-2012, pp. 1–290.

Erdelyi, M. H. 1992. Psychodynamics and the unconscious. American Psychologist, vol. 47, no. 6, pp. 784–787.

Eriksen, M., Lebreton, L. C. M., Carson, H. S., Thiel, M., Moore, C. J. et al. 2014. Plastic pollution in the world's oceans: More than 5 trillion plastic pieces weighing over 250,000 tons afloat at sea. PLOS ONE, 10 December, pp. 1–15. doi:10.1371/journal.pone.0111913.

ESMFSTAG. 2006. Summary of the 2005 review of the financial security at the Equity Silver mine. Equity Silver Mine Financial Security Technical Advisory Group. Filed with Ministry of Energy Mines and Petroleum Resources, British Columbia, Canada.

Estes, J. A. and Palmisano, J. F. 1974. Sea otters: Their role in structuring near shore communities. Science, vol. 185, pp. 1058–1060.

Estes, J. A., Tinker, M. T., Doroff, A. M. and Burn, D. 2005. Continuing sea otter population declines in the Aleutian archipelago. Marine Mammal Science, vol. 21, no. 1, pp. 169–172.

Estes, J. A., Tinker, M. T., Williams, T. M. and Doak, D. F. 1998. Killer whale predation on sea otters linking oceanic and nearshore ecosystems. Science, vol. 282, pp. 473–476.

Evans, A. M. 1993. Ore geology and industrial minerals: An introduction. 3rd ed. Blackwell Scientific Publications, London, Chapter 1.

Evans, J. A. 2007. Resolving methane fluxes. New Phytologist, vol. 175, no. 1, pp. 1–4

Evans, R. 2010. Trafigura fined for dumping toxic waste. Guardian Weekly, 30 July, p. 4.

Evershed, R. M., Payne, S., Sherratt, A. G., Copley, M. S., Urem-Kotsu, D. et al. 2008. Earliest date for milk use in the Near East and southeastern Europe linked to cattle herding. Nature, vol. 455, no. 7212, pp. 528–531.

Evidence for Democracy. 2015. https://evidencefordemocracy.ca/en/about. Accessed 23 August 2015.

Fahrenthold, D. A. 2009. When either/or just won't do. Guardian Weekly, 30 October, pp. 32–33.

Fairchild, M. D. 2005. Human colour vision. Chapter 1, Color Appearance Models, 2nd ed. John Wiley & Sons, Ltd.

FAO. 2010. The state of world fisheries and aquaculture 2010. UN Food and Agriculture Organization, United Nations, Rome, pp. 1–218.

FAO. 2011. The state of the world's land and water resources for food and agriculture: Managing systems at risk. UN Food and Agriculture Organization, summary report, pp. 1–47.

FAO, WFP and IFAD. 2012. The state of food insecurity in the world 2012: Economic growth is necessary but not sufficient to accelerate reduction of hunger and malnutrition. UN Food and Agriculture Organization, Rome, pp. 1–65.

Farman, J. C., Gardiner, B. G. and Shanklin, J. D. 1985. Large loss of total ozone in Antarctica reveal season interaction. Nature, vol. 315, pp. 207–210.

Farmer, J. D., Patelli, P. and Zovko, I. J. 2005. The predictive power of zero intelligence in financial markets. Proceedings of the National Academy of Sciences, vol. 102, no. 6, pp. 2254–2259.

Farmer, P. and Kim, J. Y. 1998. Community based approaches to the control of multidrug resistant tuberculosis: Introducing "DOTS - plus". British Medical Journal, vol. 317, pp. 671–675.

Fast Company. 1999. How much is enough? Fast Company, no. 26, July–August, pp. 108–110.

Faucheux, S. 2005. The marvels and perils of modernity: A comment. In Simpson, R. D., Toman, M. A. and Ayres, R. U., (editors) 2005, pp. 250–260.

Faure, G. 1998a. Principles and applications of geochemistry. Chapter 22, Geochemical Cycles. Prentice Hall, Upper Saddle River.

Faure, G. 1998b. Principles and applications of geochemistry. Chapter 23, Chemistry of the atmosphere. Prentice Hall, Upper Saddle River.

Federle, M. and Bassler, B. 2003. Interspecies communication in bacteria. Journal of Clinical Investigation, vol. 112, no. 9, pp. 1291–1299.

Feely, R. A., Sabine, C. L., Lee, K., Berelson, W., Kleypas, J. et al. 2004. Impact of anthropogenic CO2 on the CaCO3 system in the oceans. Science, vol. 305, July, pp. 362–366.

Felitti, V. J. 2009. Adverse childhood experiences and adult health. Academic Pediatrics, vol. 9, pp. 131–132.

Felitti, V. J., Anda, R. F., Nordenberg, D., Williamson, D. F., Spitz, A. M. et al. 1998. Relationship of childhood abuse and household dysfunction to many of the leading causes of death in adults. The adverse childhood experiences (ACE) study. American Journal of Preventative Medicine, vol. 14, no. 4, pp. 245–258. http://www.cdc.gov/violenceprevention/acestudy/journal.html

Ferguson, E. S. 1979. Technology and its impact on society. Proceedings of the Tekniska museet symposia: Technology and its impact on society. Symposium No. 1, pp. 273–280.

Ferrer-i-Carbonell, A. and van den Bergh, C. J. M. 2004. A micro-econometric analysis of determinants of unsustainable consumption in the Netherlands. Ecological Economics, vol. 27, pp. 367–389.

Festa-Bianchet, M., Pelletier, F., Jorgenson, J. T., Feder, C. and Hubbs, A. 2014. Decrease in horn size and increase in age of trophy sheep in Alberta over 37 years. Journal of Wildlife Management, vol. 78, no. 1, pp. 133–141.

Festinger, L., Riecken, H. W. and Schachter, S. 1956. When Prophecy fails. University of Minnesota Press, Minneapolis.

Fickling, D. 2003. Losing Nemo: Vanuatu reef fear as film boosts fish trade. Guardian Weekly, 27 November, p. 4.

Finer, J. 2003. Overfishing makes for tough decisions. Guardian Weekly, 27 November, p. 33.

Finger, M. 1994. From knowledge to action? Exploring the relationships between environmental experiences, learning, and behaviour. Journal of Social Issues, vol. 50, no. 3, pp. 141–160.

Finke, D. L. and Denno, R. F. 2004. Predator diversity dampens trophic cascades. Nature, vol. 429, pp. 407–410.

Firestone, R. B., West, A., Kennett, J. P., Becker, L., Bunch, T. E. et al. 2007. Evidence for an extraterrestrial impact 12,900 years ago that contributed to the megafaunal extinctions and the Younger Dryas cooling. Proceeding of the National Academy of Science, vol. 104, no. 41, pp. 16016–16021.

Firth, R. 1959. Social change in Tikopia. George Allen and Unwin, London.

Firth, R. 1965. Primitive Polynesian economy. 2nd ed. Routledge and Kegan, London.

Firth, R. 1967. The work of the gods in Tokopia. 2nd ed. Humanities Press, New York.

Firth, R. 1983. We, the Tikopia. 2nd ed. Stanford University Press, Stanford.

Fischer, S. R. 2001. A history of writing. Reaktion books, London.

Fischetti, M. 2001. Drowning New Orleans. Scientific American, October, pp. 76–85.

Fischhoff, B. 1982. For those condemned to study the past: Heuristics and biases in hindsight. In Kahneman, D., Slovic, P. and Tversky, A. (editors) 1982, pp. 335–351.

Fischhoff, B., Slovic, P. and Lichtenstein, S. 1977. Knowing with certainty: The appropriateness of extreme confidence. Journal of Experimental Psychology: Human perception and performance, vol. 3, no. 4, pp. 552–564.

Fish, E. W., Shahrokh, D., Bagot, R., Caldji, C., Bredy, T. et al. 2004. Epigenetic programming of stress responses through variations in maternal care. Annals of the New York Academy of Science, vol. 1036, pp. 167–180.

Fleming, A. 1945. Nobel lecture on penicillin. 11 December, 1945. http://nobelprize.org/medicine/laureates/1945/fleming-lecture.pdf2.

Flenley, J. R. and Bahn, B. 2002. The enigmas of Easter Island: Island on the edge. Oxford University Press, Oxford.

Flückiger, J. 2008. Did You Say "Fast"? Science, vol. 321, no. 5889, pp. 650–651.

Fogel, R. 1994. The relevance of Malthus for the study of mortality today: Long run influences on health, mortality, labour force participation and population growth. In Lindahl-Kiessling, K. and Landberg, H. (editors). Population, economic development and the environment. Chapter 9, pp. 231–284.

Folke, C., Hammer, M., Costanza, R. and Jansson, A. 1994. Investing in natural capital—why, what and how? In Jansson, A., Hammer, M., Folke, C. and Costanza, R. (editors). Investing in natural capital: The ecological economics approach to sustainability. Island Press, Washington, DC. pp. 1–20.

Ford, J. 1983. How random is a coin toss? Physics today, vol. 36, no. 4, pp. 40–47.

Ford, J. 1989. What is chaos, that we should be mindful of it? In Davis, P. (editor). The new physics. Cambridge University Press, Cambridge. pp. 348–360.

Forero, J. 2009. Argentina's inflation "faked". Guardian Weekly, 28 August, p. 18.

Forrester, J. W. 1989. The beginning of system dynamics. International Meeting of the System Dynamics Society, Stuttgart, Germany, banquet talk, 13 July, 1989. Society archives D - 4165-1, pp. 1–16.

Francis, D. 1992. The imaginary Indian. Arsenal Pulp Press, Vancouver.

Francis, D. R. 2009. Should CEO pay restrictions spread to all corporations? Christian Science Monitor, 9 March. http://www.csmonitor.com/Money/2009/0309/p14s01-wmgn.html

Francis. 2015. Encyclical letter Laudato Si' of the holy father Francis on care for our common home. The Holy See, pp. 1–82. http://w2.vatican.va/content/dam/francesco/pdf/encyclicals/documents/papa-francesco_20150524_enciclica-laudato-si_en.pdf.

Franco, M., Orduñez, P., Caballero, B., Granados, J. A. T., Lazo, M. et al. 2007. Impact of energy intake, physical activity, and population-wide weight loss on cardiovascular disease and diabetes mortality in Cuba, 1980-2005. American Journal of Epidemiology, vol. 166, no. 12, pp. 1374–1380.

Frank, K. T., Petrie, B., Choi, J. S. and Leggett, W. C. 2005. Trophic cascades in a formerly cod-dominated ecosystem. Science, vol. 308, pp. 1621–1623.

Frank, M. C., Everett, D. L., Fedorenko, E. and Gibson, E. 2008. Number as a cognitive technology: Evidence from Pirahã language and cognition. Cognition, vol. 108, no. 3, pp. 819–824.

Frank, R. A., Roy, J. W., Bickerton, G., Rowland, S. J., Headley, J. V. et al. 2014. Profiling oil sands mixtures from industrial developments and natural groundwaters for source identification. Environmental Science and Technology, vol. 48, no. 5, pp. 2660–2670.

Fredrickson, O. A. (with East, B.). 2000. The silence of the north. Fitzhenry and Whiteside, Markham.

Freeport-McMoRan Copper and Gold Inc. 2007. Annual report ending 2006. Report to the US Securities and Exchange Commission, form 10-K, pp. 1–44 plus 3 appendices.

Freeth, T., Bitsakis, Y., Moussas, X., Seiradakis, J. H., Tselikas, A. et al. 2006. Decoding the ancient Greek astronomical calculator known as the Antikythera Mechanism. Nature, vol. 444, pp. 587–591.

Freidel, D. and Shaw, J. 2000. The lowland and Maya civilization: Historical consciousness and environment. In McIntosh, R. J., Tainter, J. A. and McIntosh, S. K. (editors) 2000, pp. 271–300.

Freise, J. 2011. The EROI of conventional Canadian natural gas production. Sustainability, vol. 3, no. 11, pp. 2080–2104.

Friedlingstein, P. and Solomon, S. 2005. Contributions of past and present human generations to committed warming caused by carbon dioxide. Proceedings of the National Academy of Sciences, vol. 102, no. 31, pp. 10832–10836.

Friedman, B. 2005. The moral consequences of economic growth. Knopf; New York.

Friedrich, M. J. 2002. Epidemic of obesity expands its spread to developing countries. Journal of the American Medical Association, vol. 287, pp. 1382–1386.

Frisbie, S. H., Ortega, R., Maynard, D. M. and Sarkar, B. 2002. The concentrations of arsenic and other toxic elements in Bangladesh's drinking water. Environmental Health Perspectives, vol. 110, no. 11, pp. 1147–1153.

Froese, R. 2011. Fishery reform slips through the net. Nature, vol. 475, p. 7.

Froese, R., Zeller, D., Kleisner, K. and Pauly, D. 2012. What catch data can tell us about the status of global fisheries. Marine Biology, vol. 159, no. 6, pp. 1283–1292.

Furtula, M., Danon, G., Bajic', V. and Lukacev, D. 2017. Energy consumption and equivalent emission of $CO_2$ at wood pellets production in Serbia. Thermal Science, pp. 1–12.

Gage, J. 1999. Colour and Meaning: Art, science and symbolism. University of California Press, Berkley.

Galbraith, J. A., Begga, J. R., Joneb, D. N. and Stanley, M. C. 2015. Supplementary feeding restructures urban bird communities. Proceedings of the National Academy of Sciences, vol. 112, no. 20, E2648-E2657.

Gallai, N., Salles, J., Settele, J. and Vaissière, B. 2009. Economic valuation of the vulnerability of world agriculture confronted with pollinator decline. Ecological Economics, vol. 68, pp. 810–821.

Ganopolski, A., Winkelmann, R. and Schellnhuber, H. J. 2016. Critical insolation-CO2 relation for diagnosing past and future glacial inception. Nature, vol. 529, pp. 200–203.

Gardner, D. 2011. Future babble: Why expert predictions are next to worthless and you can do better. McClelland and Stewart, Toronto.

Gardner, G. T. and Stern, P. C. 1996. Environmental problems and human behaviour. Allyn and Bacon, Needham Heights.

Garfield, S. 2005. Blowing smoke rings. Guardian Weekly, 13 May, pp. 26–27.

Garfield, S. 2016. Testaments to the illusion of control. Guardian Weekly, 25 November, pp. 26–30.

Garre, W. J. 1976. The psychotic animal: A psychiatrist's study of human delusion. Human Sciences Press, New York.

Gassmann, A. J., Petzold-Maxwell, J. L., Clifton, E. H., Dunbar, M. W., Hoffmann, A. M. et al. 2014. Field-evolved resistance by western corn rootworm to multiple Bacillus thuringiensis toxins in transgenic maize. Proceedings of the National Academy of Sciences, vol. 111, no. 14, pp. 5141–5146.

Gaston, K. J. 2000. Global patterns in biodiversity. Nature, vol. 405, pp. 220–227.

Gaze, W. H., Abdouslam, N., Hawkey, P. M. and Wellington, E. M. H. 2005. Incidence of class 1 integrons in a quaternary ammonium compound-polluted environment. Antimicrobial Agents and Chemotherapy, vol. 49, pp. 1802–1807.

Gazzaniga, M. 1998. The split brain revisited. Scientific American, July, pp. 50–55.

Gedney, N., Cox, P. M., Betts, R. A., Boucher, O., Huntingford, C. et al. 2006. Detection of a direct carbon dioxide effect in continental river runoff records. Nature, vol. 439, p. 793.

Gee, H. 2002. Clever thieves of a feather. Guardian Weekly, 27 December, p. 10.

Gelter, H. 2003. Why is reflective thinking uncommon? Reflective Practice, vol. 4, no. 3, pp. 337–344.

Gerhardt, S. 2004. Why love matters. Brunner-Routledge, Hove and New York.

Gerland, P., Raftery, A. E., Ševčíková, H., Li, N., Gu, D. et al. 2014. World population stabilization. Science, vol. 346, no. 626, pp. 234–237.

Germano, J. D. 1999. Ecology, statistics, and the art of misdiagnosis: The need for a paradigm shift. Environmental Reviews, vol. 7, pp. 167–190.

Geske, G. D., O'Neill, J. C., Miller, D. M., Wezeman, R. J., Mattmann, M. E. et al. 2008. Comparative analyses of N-Acylated homoserine lactones reveal unique structural features that dictate their ability to activate or inhibit quorum sensing. Chembiochem, vol. 9, no. 3, pp. 389–400.

Ghafur, A. K. 2010. An obituary—On the death of antibiotics! Journal of the Association of Physicians of India, vol. 58, pp. 143–144.

Ghebreyesus, T. A., Haile, M., Witten, K. H., Getachew, A., Yohannes, A. M. et al. 1999. Incidence of malaria among children living near dams in northern Ethiopia: Community based incidence survey. British Medical Journal, vol. 319, no. 7211, pp. 663–666.

Gibbons, M. 1999. Science's new social contract with society. Nature, vol. 402, Supplement, pp. C81–C84.

Gibbs, W. 2001a. Side splitting. Scientific American, January, pp. 24–26.

Gibbs, W. W. 2001b. All in the mind. Scientific American, October, p. 16.

Gilbert, D., Pinel, E. C., Wilson, T. D., Blumberg, S. J. and Wheatley, T. F. 1998. Immune neglect: A source of durability bias in affective forecasting. Journal of Personality and Social Psychology, vol. 75, pp. 617–638.

Gill, J. L., Williams, J. W., Jackson, S. T., Lininger, K. B. and Robinson, G. S. 2009. Pleistocene megafaunal collapse, novel plant communities, and enhanced fire regimes in North America. Science, vol. 326, pp. 1100–1103.

Gingerich, O. 1973. Copernicus and Tycho. Scientific American, vol. 229, no. 6, pp. 86–101.

Glaister, D. 2007. American beauty in danger. Guardian Weekly, 6 April, p. 21.

Gleick, P. H. 2001. Making every drop count. Scientific American, February, pp. 40–45.

Glennon, R. 2002. Water follies: Groundwater pumping and the fate of America's fresh waters. Island Press, Washington, DC.

Goldenberg, S. 2013a. US "dark money" funds sceptics. Guardian Weekly, 22 February, pp. 1, 4–5.

Goldenberg, S. 2013b. GM blow to US wheat exports. Guardian Weekly, 7 June, p. 12.

Goldstein, M., Rosenberg, M. and Cheng, L. 2012. Increased oceanic microplastic debris enhances oviposition in an endemic pelagic insect. Biology letters, vol. 8, no. 5, pp. 817–820.

Goldstone, J. 2002. Efflorescences and economic growth in world history: Rethinking the 'rise of the West' and the industrial revolution. Journal of World History, vol. 13, no. 2, pp. 323–389.

Goodland, R. and Daly, H. 1998. Imperatives for environmental sustainability: Decrease over consumption and stabilize population. In Polunin, N. (editor). Population and global security. Cambridge University Press, Cambridge. pp. 117–132.

Goodman, P. S. 2003. Unsafe recycling of computers grows in China. Guardian Weekly, 3 April, p. 33.

Goodspeed, P. 2008. Food crisis being felt around the world. National Post, 2 April. http://www.nationalpost.com/news/story.html?id=412984

Google. 2014. Split second search. http://static.googleusercontent.com/media/www.google.com/en//insidesearch/howsearchworks/assets/searchInfographic.pdf.

Goossens, H. and Sprenger, M. J. 1998. Community acquired infections and bacterial resistance. British Medical Journal, vol. 317, pp. 654–657.

Gopnik, A., Meltzoff, A. and Kuhl, P. 2001. The scientist in the crib: What early childhood learning tells us about the mind. Perennial, New York.

Gordon, J. E. 1978. Structures: Or why things don't fall down. Penguin books, London.

Gordon, R. B., Bertram, M. and Graedel, T. E. 2006. Metal stocks and sustainability. Proceedings of the National Academy of Sciences, vol. 103, no. 5, pp. 1209–1214.

Grabherr, G., Gottfried, M. and Pauli, H. 1994. Climate effects on mountain plants. Nature, vol. 369, p. 448.

Graedel, T. E. and Cao, J. 2010. Metal spectra as indicators of development. Proceedings of the National Academy of Sciences, vol. 107, no. 49, pp. 20905–20910.

Grant, P. 2003. Hydrogen lifts off - with a heavy load. Nature, vol. 424, No, 6945, pp. 129–130.

Gray, R. and Atkinson, Q. D. 2003. Language-tree divergence times support the Anatolian theory of Indo-European origin. Nature, vol. 426, pp. 435–439.

Graystock, P., Yates, K., Darvill, B., Goulson, D. and Hughes, W. O. H. 2013. Emerging dangers: Deadly effects of an emergent parasite in a new pollinator host. Journal of Invertebrate Pathology, vol. 114, pp. 114–119.

Grebmeier, J. M., Overland, J. E., Moore, S. E., Farley, E. V., Carnack, E. C. et al. 2006. A major ecosystem shift in the northern Bering sea. Science, vol. 311, pp. 1461–1464.

Green, R. E., Taggart, M. A., Das, D., Pain, D. J., Kumar, C. S. et al. 2006. Collapse of Asian vulture populations: Risk of mortality from residues of the veterinary drug diclofenac in carcasses of treated cattle. Journal of Applied Ecology, vol. 43, pp. 949–956.

Greenpeace. 1997. Du Pont: A case study in the 3D corporate strategy. Greenpeace position paper. 9th meeting of the Montreal protocol. http://archive.greenpeace.org/ozone/greenfreeze/moral97/6dupont.html.

Greenwald, A. G. 1992. New look 3: Unconscious cognition reclaimed. American Psychologist, vol. 47, no. 6, pp. 766–779.

Greimler, J., Stuessy, T. F., Swenson, U., Baeza, C. M. and Matthei, O. 2002. Plant invasions on an oceanic archipelago. Biological Invasion, vol. 4, no. 1-2, pp. 73–85.

Griffin, D. W., Garrison, V. H., Herman, J. R and Shinn, E. A. 2001. African desert dust in the Caribbean atmosphere: Microbiology and public health. Aerobiologia, vol. 17, no. 3, pp. 203–213.

Grossman, D. 2002. Parched turf battle. Scientific American, December, pp. 32–33.

Grube, J. W., Mayton, D. M. and Ball-Rokreach, S. J. 1994. Inducing change in values, attitudes, and behaviours: Belief system theory and the method of value self-confrontation. Journal of Social Issues, vol. 50, no. 4, pp. 153–173.

The Guardian Weekly. 2001. Precautions fail to halt the spread of GM elements in food chain. Guardian Weekly, 2 August, p. 25.

The Guardian Weekly. 2003. In Brief. Guardian Weekly, 18 December, p. 9.

Guilford, M. C., Hall, C. A. S., O'Connor. P. and Cleveland, C. J. 2011. A new long term assessment of energy return on investment (EROI) for U.S. oil and gas discovery and production. Sustainability, vol. 3, no. 10, pp. 1866–1887.

Gunn, J. D. and Folan, W. J. 2000. Three rivers: Subregional variations in earth system impacts in the Southwestern Maya lowlands (Candelaria, Usumacinta, and Champoton watersheds). In McIntosh, R. J., Tainter, J. A. and McIntosh, S. K. (editors) 2000, pp. 223–270.

Guralnik, D. B. (editor). 1982. Websters new world dictionary. 2nd college ed. Simon and Schuster.

Gurshtein, A. A. 1997. The origin of the constellations. American Scientist, vol. 85, pp. 264–273.

Guthrie, R. D. 2003. Rapid body size decline in Alaskan Pleistocene horses before extinction. Nature, vol. 426, pp. 169–171.

Guthrie, R. D. 2006. New carbon dates link climatic change with human colonization and Pleistocene extinctions. Nature, vol. 441, pp. 207–209.

Haag, A. 2006. Church joins crusade over climate change. Nature, vol. 440, 9 march, pp. 136–137.

Haberl, H., Erb, K. H., Krausmann, F., Gaube, V., Bondeau, A. et al. 2007. Quantifying and mapping the human appropriation of net primary productivity in earth's terrestrial ecosystems. Proceedings of the National Academy of Sciences, vol. 104, no. 31, pp. 12942–12947.

Haigh, J. D., Winning, A. R., Toumi, R. and Harder, J. W. 2010. An influence of solar spectral variations on radiative forcing of climate. Nature, vol. 467, pp. 696–699.

Hall, C. A. S. 2011. Synthesis to special issue on new studies in EROI (Energy Return on Investment), Sustainability, vol. 3, no. 10, pp. 2496–2499.

Hall, C. A. S., Balogh, S. and Murphy, D. J. R. 2009. What is the minimum EROI that a sustainable society must have? Energies, vol. 2, pp. 25–47.

Hall, C. A. S., Cleveland, C. J. and Kaufmann, R. 1986. Energy and resource quality: The ecology of the economic process. John Wiley and Sons, New York.

Hall, C. A. S., Dale, B. E. and Pimentel, D. 2011. Seeking to understand the reasons for different energy return on investment (EROI) estimates for biofuels. Sustainability, vol. 3, no. 12, pp. 2413–2432.

Hall, C., Lindenberger, D., Kummel, R., Kroeger, T. and Eichhorn, W. 2001. The need to reintegrate the natural sciences with economics. BioScience, vol. 51, no. 8, pp. 663–673.

Hall, E. T. 1981. Beyond Culture. Anchor books, New York.

Hall, M. 2016. Banning the "super" greenhouse gas. Deutsche Welle. 15 October, 2016. http://www.dw.com/en/banning-the-super-greenhouse-gas/a-36044849.

Hall, R. E. and Lieberman, M. 2008. Macroeconomics: Principles and applications. 4th ed. Thompson South-Western, Mason.

Halpern, B. S., Walbridge, S., Selkoe, K. A., Kappel, C. V., Micheli, F. et al. 2008. A global map of human impact on marine ecosystems. Science, vol. 319, pp. 948–952.

Hamann, A. and Wang, T. 2006. Potential effects of climate change on ecosystem and tree species distribution in British Columbia. Ecology, vol. 87, no. 11, pp. 2773–2786.

Hamilton, G. 2009. Protected areas out of bounds to loggers, Bell says. Vancouver Sun, 22 September. http://www.vancouversun.com/business/Protected+areas+bounds+loggers+Bell+says/2021570/story.html.

Hansen, J., Sato, M., Hearty, P., Ruedy, R., Kelley, M. et al. 2016. Ice melt, sea level rise and superstorms: evidence from paleoclimate data, climate modeling, and modern observations that 2°C global warming could be dangerous. Atmospheric Chemistry and Physics, vol. 16, no. 6, pp. 3761–3812.

Hansen, J., Sato, M., Kharecha, P., Russell, G., Lea, D. W. et al. 2007. Climate and trace gases, Philosophical Transactions of the Royal Society of London, Series A, vol. 365, pp. 1925–1954.

Hansen, M. J. 2010. The Asian carp threat to the Great Lakes. Great Lakes Fishery Commission. 9 February, 2010, pp. 1–8. http://www.glfc.org/fishmgmt/Hansen_testimony_aisancarp.pdf.

Haq, B. U. 1999. Natural gas deposits: Methane in the deep blue sea. Science, vol. 285, no. 5427, pp. 543–544.

Harari, Y. 2015. Sapiens: A brief history of humankind. McClelland & Stewart.

Hardin, G. 1968. The tragedy of the commons. Science, vol. 162, no. 3859, pp. 1243–1248.

Hardin, G. 1993. Second thoughts on "The tragedy of the commons". In Daly, H. and Townsend, K. (editors). Valuing the earth. MIT Press, Cambridge. Chapter 7, pp. 145–151.

Hardin, G. 1998. Extensions of "The tragedy of the commons". Science, vol. 280, no. 5364, pp. 682–683.

Hare, R. D. 1993. Without Conscience: The disturbing world of the psychopaths among us. The Guilford Press, New York.

Harl, K. W. 2009. Early medieval and Byzantine civilization: Constantine to crusades. http://www.tulane.edu/~august/H303/handouts/Population.htm.

Harpham, G. 2006. Science and the theft of humanity. American Scientist, vol. 94, July–August, pp. 296–298.

Harris, A. M. 2009. Bayer blamed at trial for crops "contaminated" by modified rice share business. Bloomberg, 4 November, 2009. http://www.bloomberg.com/apps/news?pid=newsarchive&sid=aT1kD1GOt0N0.

Harsch, M. A., Hulme, P. E., McGlone, M. S. and Duncan, R. P. 2009. Are treelines advancing? A global meta-analysis of treeline response to climate warming. Ecology Letters vol. 12, pp. 1040–1049.

Hart, C. W. 2008. J. Robert Oppenheimer: A faith development portrait. Journal of Religious Health, vol. 47, pp. 118–128.

Hart, M. 2001. Diamond: A journey to the heart of an obsession. Viking, Canada.

Hartung, T. and Rovida, C. 2009. Chemical regulators have overreached. Nature, vol. 462, pp. 1080–1081.

Harvey, P. H. and Zammuto, R. M. 1985. Patterns of mortality and age at first reproduction in natural populations of mammals. Nature, vol. 315, pp. 319–320.

Haseltine, E. 2000. Your stedicam. Discover, June, p. 108.

Hassan, F. A. 1980. The growth and regulation of human population in prehistoric times. In Cohen, M. N., Malpass, R. S. and Klein, H. G. (editors). Biosocial mechanisms of population regulation. Yale University Press, New Haven. Chapter 14, pp. 305–319.

Hassan, F. A. 1981. Demographic archaeology. Academic Press, New York.

Hassan, F. A. 1996. Abrupt Holocene climatic events in Africa. In Pwiti, G. and Soper, R. (editors). Aspects of African archaeology. University of Zimbabwe Publications, Harare. pp. 83–89.

Hassan, F. A. 1997. Nile floods and political disorder in early Egypt. In Dalfes, H. N., Kukula, G. and Weiss, H. (editors). Third Millenium BC abrupt climate change and the old world collapse. NATO ASI series 1, V. 49. Springer, Berlin. pp. 1–23.

Hassan, F. A. 2000. Environmental perception and human responses in history and prehistory. In McIntosh, R. J., Tainter, J. A. and McIntosh, S. K. (editors) 2000, pp. 121–140.

Hassan, R., Scholes, R. and Ash, N. (editors). 2005. Ecosystems and human well being: Current state and trends. Millenium ecosystem assessment, V. 1. Island Press, Washington. pp. 165–207.

Hawken, P. 1993. The ecology of commerce: A declaration of sustainability. Harper Business, New York.

Hawkey, P. M. 1998. The origins and molecular basis of antibiotic resistance. British Medical Journal, vol. 317, pp. 657–660.

Hawking, S. 2009.Steven Hawking. www.hawking.org.uk/disable/dindex.html. Accessed November 2009.

Hayes, T. B. 2004. There is no denying this: Diffusing the confusion about atrazine. BioScience, vol. 54, pp. 1138–1148.

Hayes, T. B., Collins, A., Lee, M., Mendoza, M., Noriega, N. et al. 2002a. Hermaphrodite, demasculinized frogs after exposure to the herbicide atrazine at low ecologically relevant doses. Proceedings of the National Academy of Sciences, vol. 99, no. 8, pp. 5476–5480.

Hayes, T. B., Khourya, V., Narayana, A., Nazir, M., Park, A. et al. 2010. Atrazine induces complete feminization and chemical castration in male African clawed frogs (Xenopuslaevis), Proceedings of the National Academy of Sciences, published online 1 March. doi:10.1073/pnas.0909519107

Hayes, T., Haston, K., Tsui, M., Hoang, A., Haeffele, C. et al. 2002b. Feminization of male frogs in the wild. Nature, vol. 419, pp. 895–896.

Heeney, J. L., Dalgleish, A. G. and Weiss, R. A. 2006. Origins of HIV and the evolution of resistance to AIDS. Science, vol. 313, pp. 462–466.

Heichel, G. H. 1976. Agricultural production and energy resources. American Scientist, vol. 64, pp. 64–72.

Heilbron, J. L. and Bynum, W. F. 1999. Plus ca change. Nature, vol. 402, Supplement, pp. C86–C88.

Heimann, M. 2010. How stable is the methane cycle? Science, vol. 327, no. 5970, pp. 1211–1212.

Heinberg, R. and Fridley, D. 2010. The end of cheap coal. Nature, vol. 468, pp. 367–369.

Henderson, H. 2005. The "Nobel prize" that isn't. Le Monde Diplomatique, February, p. 13.

Hendricks, R. C. and Bushnell, D. M. 2009. Synthetic and biomass alternative fueling in aviation. Mechanical Engineering Magazine. http://ntrs.nasa.gov/archive/nasa/casi.ntrs.nasa.gov/20090017985 _2009014996.pdf.

Hendricks, R. C., Daggett, D. L. and Bushnell, D. M. 2009. Synthetic and biomass alternative fueling for aviation. NASA AMES Green aviation workshop 25–26 April, slide presentation. http://event.arc.nasa.gov/Green-Aviation/home/pdf/NASA.AMES.Green.Aviation.Workshop.April2009.3.pdf.

Herlocker, C. E., Allison, S. T., Foubert, J. D. and Beggan, J. K. 1997. Intended and unintended overconsumption of physical, spatial, and temporal resources. Journal of Personality and Social Psychology, vol. 73, no. 5, pp. 992–1004.

Hernstein, R. J. 1990. Rational choice theory: Necessary but not sufficient. American Psychologist, vol. 45, pp. 356–367.

Herrington, A. 2008. The cure within: A history of mind-body medicine. W. W. Norton and Co., London.

Hertwig, R., Barron, G., Weber, E. U. and Erev, I. 2004. Decisions from experience and the effect of rare events. Psychological Science, vol. 15, pp. 534–539.

Hickie, B., Ross, P. S., Macdonald, R. W. and Ford, J. K. B. 2007. Killer whales (Orcinus orca) face protracted health risks associated with lifetime exposure to PCBs. Environmental Science and Technology, vol. 41, no. 18, pp. 6613–6619.

Hijiya, J. A. 2000. The "Gita" of J. Robert Oppenheimer. Proceedings of the American Philosophical society, vol. 141, no. 2, pp. 123–167.

Hill, M. 2002. Dollars under the waves. Above and Beyond, May–June, pp. 59–60.

Hillman, G., Hedges, R., Moore, A., Colledge, S. and Pettitt, P. 2001. New evidence of Lateglacial cereal cultivation at Abu Hureyra on the Euphrates. Holocene, vol. 11, no. 4, pp. 383–393.

Hixson, L. W. 2015. Internal TEPCO document reveals executives knew beefing up tsunami defenses was indispensable. http://enformable.com/2015/06/internal-tepco-document-reveals-executives-knew-beefing-up-tsunami-defenses-was-indispensable/. Accessed 19 July, 2015.

Ho, K. 2009. Liquidated. An enthographic study of Wall street. Duke University Press, Durham.

Hock, R. R. 1999a. One brain or two?: Gazzaniga, M. S. 1967. The split brain in man. Forty studies that changed psychology: Explorations into historical psychology. 3rd ed. Prentice Hall, New Jersey.

Hock, R. R. 1999b. Thanks for the memories!: Loftus, E. F. 1975. Leading questions and the eyewitness report. Forty studies that changed psychology: Explorations into historical psychology. 3rd ed. Prentice Hall, New Jersey.

Hock, R. R. 1999c. Thoughts out of tune: Festinger, L. and Carlsmith, J. M. 1959. Cognitive consequences of forced compliance. Forty studies that changed psychology: Explorations into historical psychology. 3rd ed. Prentice Hall, New Jersey.

Hock, R. R. 1999d. Who's crazy here anyway: Rosenhan, D. L. 1973. On being sane in an insane place. Forty studies that changed psychology: Explorations into historical psychology. 3rd ed. Prentice Hall, New Jersey.

Hodges, C. A. 1995. Mineral resources, environmental issues, and land use. Science, vol. 268, pp. 1305–1312.

Hoekstra, A. Y. and Chapagain, A. K. 2006. Water footprints of nations: Water use by people as a function of their consumption habits. Water Resource Management, doi:10.1007/s11269-006-9039-x.

Hoekstra, A. Y., Chapagain, A. K., Aldaya, M. M. and Mekonnen, M. M. 2011. The water footprint assessment manual: Setting the global standard. Earthscan, London, pp. 1–203.

Hoekstra, A. Y. and Mekonnen, M. M. 2012. The water footprint of humanity. Proceedings of the National Academy of Sciences, vol. 109, no. 9, pp. 3232–3237.

Hoffecker, J. F., Elias, S. A. and O'Rourke, D. H. 2014. Out of Beringia? Science, vol. 343, pp. 979–980.

Holton, G. 1996. Science and Progress Revisited. In Marx, L. and Mazlish, B. (editors) 1996, pp. 9–26.

Homer-Dixon, T. 1995. The ingenuity gap: Can the poor countries adapt to resource scarcity? Population and Development Review, vol. 21, no. 3, pp. 587–612.

Hong, S., Candelone, J-P., Patterson, C. C. and Bouton, C. F. 1994. Greenland ice evidence of hemispheric lead pollution two millennia ago by Greek and Roman civilizations. Science, vol. 265, pp. 1841–1843.

Hong, S., Candelone, J-P., Patterson, C. C. and Bouton, C. F. 1996. History of ancient copper smelting pollution during Roman and medieval times recorded in Greenland ice. Science, vol. 272, pp. 246–249.

Höök, M. and Aleklett, K. 2009. A review on coal to liquid fuels and its coal consumption. International Journal of Energy Research, published online at http://dx.doi.org/10.1002/er.1596.

Höök, M., Zittel, W., Schindler, J. and Aleklett, K. 2010. Global coal production outlooks based on a logistic model. Fuel, vol. 89, no. 11, pp. 3546–3558.

Howarth, R. W., Santoro, R. and Ingraffea, A. 2011. Methane and the greenhouse gas footprint of natural gas from shale. Climate Change, vol. 106, pp. 679–690.

Hsu, C. 1988. The role of the literati and of regionalism in the fall of the Han dynasty. In Yoffee, N. and Cowgill, W. (editors) 1988, pp. 176–195.

Huang, J., Pray, C. and Rozelle, S. 2006. Enhancing the crops to feed the poor. Nature, vol. 418, pp. 678–684.

Hubbert, M. K. 1973. Survey of world energy resources. Canadian Mining and Metallurgy Bulletin, vol. 66, no. 735, pp. 37–53.

Hughen, K. A., Eglinton, T. I., Xu, L. and Makou, M. 2004. Abrupt tropical vegetation response to rapid climate changes. Science, vol. 304, pp. 1955–1959.

Hughes, D. 2009. The energy issue: A more urgent problem than climate change? In Homer-Dixon, T. (editor). Carbon shift: How the twin crises of oil depletion and climate change will define the future. Random House, Canada. pp. 58–95, 218–219.

Hughes, D. 2014a. Drilling deeper: A reality check on U.S. government forecasts for a lasting tight oil and shale gas boom. Post Carbon Institute, pp. 1–308. www.postcarbon.org/ wp-content/ uploads/ 2014/ 10/ Drilling-Deeper_FULL.pdf.

Hughes, D. 2014b. BC LNG: A reality check. Global Sustainability Research Inc., pp. 1–19. http://www.watershedsentinel.ca/files/files/Hughes-BC-LNG-Jan2014.pdf accessed June 2015.

Hughes, D. J. 1975. Ecology in ancient civilizations. University of New Mexico Press, Albuquerque.

Hughes, J. D. 2013. Drill, baby, drill: Can unconventional fuels usher in a new era of energy abundance? Post Carbon Institute, Santa Rosa, California, USA, 2nd ed., pp. 1–178.

Humborg, C., Ittekkot, V., Cociasu, A. and Von Bodungen, B. 1997. Effect of Danube river dam on Black Sea biogeochemistry and ecosystem structure. Nature, vol. 386, pp. 385–388.

Hunt, G. and Gray, R. D. 2003. Diversification and cumulative evolution in New Caledonian crow tool manufacture. Proceeding of the Royal Society of London, Series B, vol. 270, no. 1517, pp. 867–874.

Hunter, L. M. 2000. The environmental implications of population dynamics. Rand, Santa Monica.

Huovinen, P. and Cars, O. 1998. Control of antimicrobial resistance: Time for action. British Medical Journal, vol. 317, p. 613.

Huskins, J. 1998. Mix and Match. In Savage, S. (editor). The Plain Reader. The Ballantine Publishing Group, New York.

Hutchings, J. A. 2000. Collapse and recovery of marine fishes. Nature, vol. 406, pp. 882–885.

IARC. 1999. Atrazine. International Agency for Research on Cancer, Monograph, vol. 73, monograph no. 73-8, pp. 59–113. http://monographs.iarc.fr/ENG/Monographs/vol73/mono73-8.pdf.

ICOLD. 2001. Tailings dams.Risk of dangerous occurrences: Lessons learnt from practical experiences. Bulletin 121, Published by United Nations Environment Programme (UNEP) Division of Technology, Industry and Economics (DTIE) and International Commission on Large Dams (ICOLD), Paris, pp. 1–144.

IEA. 2001. World energy outlook: Assessing today's supplies to fuel tomorrow's growth. International Energy Agency, Paris, pp. 1–421.

IEA. 2003. Key world energy statistics. International Energy Agency, Paris, pp. 1–78.

IEA. 2005 to 2013. Key world energy statistics for these years. International Energy Agency, Paris, pp. 1–80.

IEA. 2006. Key world energy statistics. International Energy Agency, Paris, pp. 1–80.

IEA. 2007. Key world energy statistics. International Energy Agency, Paris, pp. 1–82.

IEA. 2008. World energy outlook 2008: Executive summary. International Energy Agency, Paris, pp. 1–15.

IEA. 2009. Key world energy statistics. International Energy Agency, Paris, pp. 1- 80.

IEA. 2010a. World energy outlook 2010: Executive summary. International Energy Agency, Paris, pp. 1–14.

IEA. 2010b. World energy outlook 2010: Key graphs. International Energy Agency, Paris, pp. 1–10.

IEA. 2012a. Key world energy statistics. International Energy Agency, Paris, pp. 1–80.

IEA. 2012b. World energy outlook 2012. International Energy Agency, France, pp. 1–690.

IEA. 2015. Key world energy statistics. International Energy Agency, Paris, pp. 1–80.

IEA. 2017. Key world energy statistics. International Energy Agency, Paris, pp. 1–97.

Imhoff, M. L., Bounoua, L., Ricketts, T., Laucks, C., Harriss, R. et al. 2004. Global patterns in human consumption of net primary productivity. Nature, vol. 429, pp. 870–873.

Inglehart, R. 1990. Culture shift in advanced industrial society. Princeton University Press, Princeton.

Inglehart, R., Basanez, M. and Moreno, A. 1998. Human values and beliefs: A cross-cultural sourcebook. Political, religious, sexual and economic norms in 43 societies: Findings from the 1990-1993 world values survey. The University of Michigan Press, Ann Arbor.

IPCC. 1995. Climate change 1995. Intergovernmental Panel on Climate Change. 2nd assessment report, pp. 1–73.

IPCC. 2001. Report of the Working Group I of the Intergovernmental Panel on Climate Change. 3rd assessment report. Summary for policy makers, pp. 1–20.

IPCC. 2006. Principles governing IPCC work. Intergovernmental Panel on Climate Change, pp. 1–2. http://www.ipcc.ch/pdf/ipcc-principles/ipcc-principles.pdf.

IPCC. 2007a. Climate change 2007: The physical science basis. Summary for policy makers. Contribution of the Working Group I to the 4th assessment report of the Intergovernmental Panel on Climate Change, pp. 1–21.

IPCC. 2007b. Climate change 2007: Impacts, adaptation and vulnerability. Summary for policy makers. Working Group II contribution to the Intergovernmental Panel on Climate Change 4th assessment report, pp. 1–23.

IPCC. 2007c. Climate change 2007: Mitigation of climate change. Summary for policy makers. Working Group III contribution to the Intergovernmental Panel on Climate Change 4th assessment report. Cambridge, UK, pp. 1–35.

IPCC. 2007d. Summary for policymakers of the synthesis report of the 4th assessment report of the Intergovernmental Panel on Climate Change, Cambridge, UK, pp. 1–23.

IPCC. 2013a. Climate change 2013: The physical science basis. Summary for policy makers. Working Group I: Contribution to the Intergovernmental Panel on Climate Change Fifth assessment, pp. 1–36.

IPCC. 2013b. Compatible fossil fuel emissions simulated by the CMIP5 models for the four RCP scenarios. Climate Change 2013: Technical Summary, Working Group I. http://www.climatechange2013.org/images/report/WG1AR5_TS_FINAL.

IPCC. 2014a. Climate change 2014: Impacts, adaptation, and vulnerability. Summary for policy makers. Working Group II: Contribution to the Intergovernmental Panel on Climate Change 5th assessment, pp. 1–44.

IPCC. 2014b. Climate change 2014: Mitigation of climate change. Summary for policy makers. Working Group III: Contribution to the Intergovernmental Panel on Climate Change 5th assessment, pp. 1–33.

IRIN. 2008. Zimbabwe: Cholera feeds off a perfect storm. http://www.irinnews.org/PrintReport.aspx?ReportId=81699

IRIN. 2009. Iraq death knell for agriculture? IRIN, news release, 28 April, 2009. http://www.irinnews.org/report.sapx?reported=84142

Ivanhoe, L. F. 1995. Future world oil supplies: There is a finite limit. World Oil, October, pp. 77–88.

Jaccard, M. 2009. Peak oil and market feedbacks: Chicken little versus Dr. Pangeloss. In Homer-Dixon, T. (editor). Carbon Shift: How the twin crises of oil depletion and climate change will define the future. Random House, Canada. pp. 96–131, 220- 221.

Jackson, J. A. and Jackson, B. J. 2004. Ecological relationships between fungi and woodpecker cavity sites. Condor, vol. 106, no. 1, pp. 37–49.

Jackson, R. B., Canadell, J. G., Le Quéré, C., Andrew, R. M., Korsbakken, J. I. et al. 2015. Reaching peak emissions. Nature Climate Change, vol. 6, pp. 7–10.

Jackson, R. J., Ramsay, A. J., Christensen, C. D., Beaton, S., Hall, D. F. et al. 2001. Expression of mouse interleukin-4 by a recombinant ectromelia virus suppresses cytolytic lymphocyte responses and overcomes genetic resistance to mousepox. Journal of Virology, vol. 75, pp. 1205–1210.

Jacobsen, T. 1982. Salinity and irrigation agriculture in antiquity. Diyala basin archaeology projects: Report on essential results, 1957-58. Bibliotheca Mesopotamica, vol. 14, pp. 1–107.

Jacobsen, T and Adams, R. M. 1958. Salt and silt in ancient Mesopotamian agriculture. Science, vol. 128, no. 3334, pp. 1251–1258.

Jacobson, M. Z. 2009. Review of solutions to global warming, air pollution, and energy security. Energy and Environmental Science, vol. 2, pp. 148–173.

Jacobson, M. Z. and Delucchi, M. A. 2009. A path to sustainable energy by 2030. Scientific American, October, pp. 58–65.

Jacobson, M. Z., Delucchi, M. A., Ingraffea, A. R., Howarth, R. W., Bazouin, G. et al. 2014. A roadmap for repowering California for all purposes with wind, water, and sunlight. Energy, vol. 73, pp. 875–889.

Jacoby, L. L., Lindsay, D. S. and Toth, J. P. 1992. Unconscious influences revealed: Attention, awareness, and control. American Psychologist, vol. 47, pp. 802–808.

Jambeck, J. R., Geyer, R., Wilcox, C., Siegler, T. R., Perryman, M. et al. 2015. Plastic waste inputs from land into the ocean. Science, vol. 347, no. 6223, pp. 768–771.

James, T. Y., Litvintseva, A. P., Vilgalys, R., Morgan, J. A. T., Taylor, J. W. et al. 2009. Rapid global expansion of the fungal disease Chytridiomycosis into declining and healthy amphibian populations. PLoS Pathogens, vol. 5, no. 5, E10000458, pp. 1–12.

Jansen, B. 2003. Dangerous substances handle with care. A European perspective. EN magazine, vol. 6, pp. 4–6. http://osha.europa.eu/en/publications/magazine/6/

Jaward, F. M., Farrer, N. J., Harner, T., Sweetman, A. J. and Jones, K. C. 2004. Passive air sampling of PCBs, PBDEs, and organochlorine pesticides across Europe. Environmental Science and Technology, vol. 38, no. 1, pp. 34–41.

Jeroen, C. J. M., Van den Bergh, J. C. M. and Verbruggen, H. 1999. Spatial sustainability, trade and indicators: An evaluation of the 'ecological footprint'. Ecological Economics, vol. 29, no. 1, pp. 61–72

Jeschke, J. M. and Strayer, D. L. 2005. Invasive success of vertebrates in Europe and North America. Proceedings of the National Academy of Sciences, vol. 102, no. 20, pp. 7198–7202.

Jiang, Y., Dixon, T. H. and Wdowinski, S. 2010. Accelerating uplift in the North Atlantic region as an indicator of ice loss. Nature Geoscience, vol. 3, pp. 404–407.

John Paul II. 1988. Letter of his holiness John Paul II to reverend George V. Coyne, S.J., director of the Vatican observatory. http://www.vatican.va/holy_father/john_paul_ii/letters/1988/documents/hf_jp-ii_let_19880601_padre-coyne_en.html.

Johnsen, S. J., Clausen, H. B., Dansgaard, W., Fuhrer, K., Gundestrup, N. et al. 1992. Irregular glacial interstadials recorded in a new Greenland ice core. Nature, vol. 359, pp. 311–313.

Johnson, B. 2009. Energy-hungry internet is threatened by its own success. Guardian Weekly, 8 May, p. 3.

Johnson, N. K. and Cicero, C. 2004. New mitochondrial DNA data affirm the importance of Pleistocene speciation in North American birds. Evolution, vol. 58, no. 5, pp. 1122–1130.

Johnston, A. M. 1998. Use of antimicrobial drugs in veterinary practice. British Medical Journal, vol. 317, pp. 665–667.

Jones, L., Fredricksen, L. and Wates, T. 2002. Environmental indicators. Critical Issues Bulletin. 5th ed. Fraser Institute, Vancouver, pp. 1–136.

Jones, S. 1999. Darwin's Ghost: The origin of species updated. Doubleday Canada.

Jones, T. W. 2004. Using contemporary archaeology and applied anthropology to understand food loss in the American food system. Report. Bureau of Applied Research in Anthropology, University of Arizona, pp. 1–6. http://www.ce.cmu.edu/~gdrg/readings/2006/12/19/Jones_Using ContemporaryArchaeologyAndAppliedAnthropologyToUnderst andFoodLossInAmericanFoodSystem.pdf.

Joppa, L., Roberts, D. and Pimm, S. 2010. How many species of flowering plants are there? Proceedings of the Royal Society, Series B, published online 7 July, doi:10.1098/rspb.2010.1004.

Jowit, J. 2010. Scientists prune list of world's plants. Guardian Weekly, 1 October, pp. 30–31.

Jump, A. S. and Penuelas, J. 2005. Running to stand still: Adaptation and the response of plants to rapid climate change. Ecology Letters, vol. 8, pp. 1101–1020.

Junge, T. 2004. Until the final hour: Hitler's last secretary. Phoenix, London.

Jurewicz, J. and Hanke, W. 2011. Exposure to phthalates: Reproductive outcome and children health. A review of epidemiological studies. International Journal of Occupational Medicine and Environmental Health, vol. 24, no. 2, pp. 115–141.

Kahan, D. M., Peters, E., Wittlin, M., Slovic, P., Ouellette, L. L. et al. 2012. The polarizing impact of science literacy and numeracy on perceived climate change risks. Nature Climate Change, vol. 2, pp. 732–735.

Kahneman, D. 2003. A perspective on judgment and choice: Mapping bounded rationality. American Psychologist, vol. 58, no. 9, pp. 697–720.

Kahneman, D. and Tversky, A. 1982a. The simulation heuristic. In Kahneman, D., Slovic, P. and Tversky, A. (editors) 1982, pp. 201–208.

Kahneman, D. and Tversky, A. 1982b. Variants of uncertainty. In Kahneman, D., Slovic, P. and Tversky, A. (editors) 1982, pp. 509–520.

Kahneman, D. and Tversky, A. 1982c. On the psychology of prediction. In Kahneman, D., Slovic, P. and Tversky, A. (editors) 1982, pp. 4–68.

Kahneman, D. and Tversky, A. 1996. On the reality of cognitive illusions. Psychological Review, vol. 103, no. 3, pp. 582–591.

Kahneman, D., Slovic, P. and Tversky, A. (editors). 1982. Judgement under uncertainty: Heuristics and biases. Cambridge University Press, Cambridge.

Kaiser, J. 1998. Sea Otter declines blamed on hungry killers. Science, vol. 282, pp. 390–391.

Kaiser, J. 2004. Wounding earth's fragile skin. Science, vol. 304, pp. 1616–1618.

Kapoor, M. 2013. A crisis of culture: Valuing ethics and knowledge in financial services. The Economist Intelligence Unit, December 2013, pp. 1–30. http://www.economistinsights.com/sites/default/files/LON%20-%20SM%20-%20CFA%20WEB.pdf.

Karoly, D. J., Braganza, K., Stott, P. A., Arblaster, J. M., Meehl, G. A. et al. 2003. Detection of a human influence on north American climate. Science, vol. 302, pp. 1200–1203.

Kaser, G., Cogley, J. G., Dyurgerov, M. B., Meier, M. F. and Ohmura, A. 2006. Mass balance of glaciers and ice caps: Consensus estimates for 1961–2004. Geophysical Research Letters, vol. 33, no. 19, L19501.

Kasser, T. 2002. The high price of materialism. MIT Press, Cambridge.

Kast, B. 2001. Decisions, decisions. Nature, vol. 411, pp. 126–128.

Katz, D. S. 2005. The occult tradition: From the Renaissance to the present day. Jonathan Cape, London.

Kaufman, F. 2010. The food bubble. Harper's Magazine, July, pp. 27–34.

Kaufman, H. 1988. The collapse of ancient states and civilizations as an organizational problem. In Yoffee, N. and Cowgill, W. (editors) 1988, pp. 219–235.

Kaufman, L. 1992. Catastrophic change in a species rich freshwater ecosystem: Lessons from lake Victoria. Bioscience, vol. 42, pp. 846–858.

Kaufman, M. A. 2007. Climate change: "consensus?" Northern Miner, vol. 93, no. 33, 8 October.

Kaufmann, R. K. and Cleveland, C. J. 2001. Oil production in the lower 48 states: Economic, geological, and institutional determinants. Energy Journal, vol. 22, no. 1, pp. 27–49.

Keating, J. 2000. "Slaughter" of sea lions investigated. Province, 25 April, p. A3.

Keele, B. F., Van Heuverswyn, F., Li, Y., Bailes, E., Takehisa, J. et al. 2006. Chimpanzee reservoirs of pandemic and nonpandemic HIV-1. Science, vol. 313, pp. 523–526.

Keith, D. 2009. Dangerous abundance. In Homer-Dixon, T. (editor). Carbon Shift: How the twin crises of oil depletion and climate change will define the future. Random House, Canada. pp. 26–57, 216–218.

Kelly, J. 2005. The great mortality: An intimate history of the black death. Harper Collins.

Kelly, R. E., Cohen, L. J., Semple, R. J., Bialer, P., Lau, A. et al. 2006. Relationship between drug company funding and outcomes of clinical psychiatric research. Psychological Medicine, vol. 36, no. 11, pp. 1647–1656.

Kelly, R. L. 1995. The foraging spectrum: Diversity in hunter-gatherer lifeways. Smithsonian Institution Press, Washington, DC.

Kelly, T., Yang, W., Chen, C.-S., Reynolds, K. and He, J. 2008. Global burden of obesity in 2005 and projections to 2030. International Journal of Obesity, vol. 32, pp. 1431–1437.

Kendrick, T. D. 1956. The Lisbon earthquake. Methuen and Co., London.

Kennedy, T. A., Naeem, S., Howe, K. M., Knops, J. M., Tilman, D. et al. 2002. Biodiversity as a barrier to ecological invasion. Nature, vol. 417, pp. 636–638.

Kennett, D. J., Breitenbach, S. F. M., Aquino, V. V., Asmerom, Y., Awe, J. et al. 2012. Development and disintegration of Maya political systems in response to climate change. Science, vol. 338, no. 6108, pp. 788–791.

Kennett, D. J., Kennett, J. P., West, A., Mercer, C., Hee, S. S. Q. et al. 2009. Nanodiamonds in the younger Dryas boundary sediment layer. Science, vol. 323, p. 94.

Kerr, R. A. and Service, R. F. 2005. What can replace cheap oil - and when? Science, vol. 309, p. 101.

Kerry, L., Farris, K. L., Huss, M. J. and Zack, S. 2004. The role of foraging woodpeckers in the decomposition of ponderosa pine snags. Condor, vol. 106, no. 1, pp. 50–59.

Kessler, M. A. and Werner, B. T. 2003. Self-organization of sorted patterned ground. Science, vol. 299, pp. 380–383.

Keyfitz, N. 1991. Population and development within the ecosphere: One view of the literature. Population Index, vol. 57, no. 1, pp. 5–22.

Khush, G. K. 1999. Green revolution: Preparing for the 21st century. Genome, vol. 42, pp. 646–655.

Kidd, K. A., Blanchfield, P. J., Mills, K. H., Palace, V. P., Evans, R. E. et al. 2007. Collapse of a fish population after exposure to a synthetic estrogen. Proceedings of the National Academy of Sciences, vol. 104, no. 21, pp. 8897–8901.

Kidd, K. A., Paterson, M. J., Rennie, M. D., Podemski, C. L., Findlay, D. L. et al. 2014. Direct and indirect responses of a freshwater food web to a potent synthetic oestrogen. Philosophical Transactions of the Royal Society B, vol. 369, no. 1656, Paper 2013057, pp. 1–12.

Kidd, K. A., Schindler, D. W., Muir, D. C. G., Lockhart, W. L. and Hesslein, R. H. 1995. High concentrations of Toxaphene in fishes from a sub-Arctic lake. Science, vol. 269, pp. 240–242.

Kihlstrom, J. F. 1987. The cognitive unconscious. Science, vol. 237, pp. 1445–1452.

Kihlstrom, J. F. 1992. The psychological unconscious: Found lost regained. American Psychologist, vol. 47, no. 6, pp. 788–790.

Kim, D., Sexton, J. O. and Townshend, J. R. 2015. Accelerated deforestation in the humid tropics from the 1990s to the 2000s. Geophysical Research Letters, vol. 42, no. 9, pp. 3495–3501.

Kindleberger, C. P. and Aliber, R. Z. 2005. Manias, panics, and crashes: A history of financial crises. 5th ed. John Wiley and Son, Hoboken.

King, A. D., van Oldenborgh, G. J., Karoly, D. J., Lewis, S. C. and Cullen, H. 2015. Attribution of the record high Central England temperature of 2014 to anthropogenic influences. Environmental Research Letters, vol. 10, Article 054002, pp. 1–7.

Kinnard, C., Zdanowicz, C.M., Fisher, D.A., Isaksson, E., de Vernal, A. et al. 2011. Reconstructed changes in Arctic sea ice over the past 1,450 years. Nature, vol. 479, no. 7374, pp. 509–512.

Kirch, P. V. 1997. Microcosmic histories: Island perspectives on "global" change. American anthropologist, vol. 99, no. 1, pp. 30–42.

Kirch, P. V. 2007. Three islands and an archipelago: Reciprocal interactions between humans and island ecosystems in Polynesia. Earth and Environmental Science Transactions of the Royal Society of Edinburg, vol. 98, pp. 85–99.

Kirch, P. V. and Yen, D. E. 1982. Tikopia: The prehistory and ecology of a Polynesian outlier. Bernice P. Bishop Museum, Bulletin 238. Bernice P. Bishop Museum, Honolulu, Hawaii, pp. 1–396.

Kjeldsen, K. K., Korsgaard, N. J., Bjørk, A. A., Khan, S. A., Box, J. E. et al. 2015. Spatial and temporal distribution of mass loss from the Greenland Ice Sheet since AD 1900. Nature, vol. 528, pp. 396–400.

Klein, N. 2014. This changes everything: Capitalism vs. climate. Alfred A. Knopf: Toronto.

Klein, S. A. 2002. Libert's temporal anomalies: A reassessment of the data. Consciousness and Cognition, vol. 11, pp. 198- 214.

Kleisner, K., Zeller, D., Froese, R. and Pauly, D. 2013. Using global catch data for inferences on the world's marine fisheries. Fish and Fisheries, vol. 14, no. 3, pp. 293–311.

Kluser, S., Giuliani, G., De Bono, A. and Peduzzi, P. 2004. Caulerpa taxifolia, a growing menace for temperate marine environment. UNEP. Environment Alert Bulletin, January, pp. 1–4.
http://www.grid.unep.ch/product/publication/download/ew_caulerpa.en.pdf.

Knox, S. S., Jackson, T., Javins, B., Frisbee, S. J., Shankar, A. et al. 2011. Implications of early menopause in women exposed to perfluorocarbons. Journal of Clinical Endocrinology and Metabolism, vol. 96, no. 6, pp. 1747–1753.

Kolb, F. C. and Beaun, J. 1995. Blindsight in normal observers. Nature, vol. 377, pp. 336–339.

Koren, I., Kaufman, Y. J., Washington, R., Todd, M. C., Rudich, Y. et al. 2010. The Bodélé depression: A single spot in the Sahara that provides most of the mineral dust to the Amazon forest. Environmental Research Letters, vol. 1 , no. 1, 014005, pp. 1–5.

Kraft, J. C., Kayan, I. and Erol, O. 1980. Geomorphic reconstructions in the environs of ancient Troy. Science, vol. 209, pp. 776–782.

Kraft, J. C., Rapp, G. R., Kayan, I. and Luce, J. V. 2003. Harbour areas at ancient Troy: Sedimentology and geomorphology complement Homer's Iliad. Geology, vol. 31, pp. 163–166.

Krautkraemer, J. A. 2005. Economics of scarcity: The state of the debate. In Simpson, R. D., Toman, M. A. and Ayres, R. U. (editors) 2005, pp. 54–77.

Krebs, C. J., Boutin, S. and Boonstra, R. (editors). 2001. Ecosystem dynamics of the boreal forest. Oxford University Press, Oxford. pp. 1–511.

Kremen, C., Williams, N. M. and Thorp, R. W. 2002. Crop pollination from native bees at risk from agricultural intensification. Proceedings of the National Academy of Sciences, vol. 99, No, 26, pp. 16812–16816.

Krkosek, M., Ford, J. S., Morton, A., Lele, S., Myers, R. A. et al. 2007. Declining wild salmon populations in relation to parasites from farm salmon. Science, vol. 318, pp. 1772–1775.

Krummel, E. M., MacDonald, R. M., Kimpe, L. E., Gregory-Eaves, I., Demers, M. J. et al. 2003. Delivery of pollutants by spawning salmon. Nature, vol. 425, pp. 255–256.

Kubiszewski, I., Cleveland, C. and Endres, P. 2010. Meta-analysis of net energy return for wind power systems. Renewable Energy, vol. 35, pp. 218–225.

Kuhn, T. 1996. The structure of scientific revolutions. University of Chicago Press, Chicago. 3rd ed.

Kuipers, J., Maest, A., MacHardy, K. and Lawson, G. 2005. Comparison of predicted and actual water quality at hardrock mines: The reliability of predictions in environmental impact statements. Kuipers & Associates and Buka Environmental, pp. 1–196.

Kumar, K., Gupta, S. C., Baidoo, S. K., Chander, Y. and Rosen, C. J. 2005. Antibiotic uptake by plants from soil fertilized with animal manure. Journal of Environmental Quality, vol. 34, pp. 2082–2085.

Kumarasamy, K. K., Toleman, M. A., Walsh, T. R., Bagaria, J., Butt, F. et al. 2010. Emergence of a new antibiotic resistance mechanism in India, Pakistan, and the UK: a molecular, biological, and epidemiological study. Lancet Infectious Diseases, vol. 10, no. 9, pp. 597–602.

Kummel, R., Henn, J. and Lindenberger, D. 2002. Capital, labour, energy and creativity: Modeling innovation diffusion. Structural Change and Economic Dynamics, vol. 13, pp. 415–433.

Kunst-Wilson, W. R. and Zojonc, R. R. 1980. Affective discrimination of stimuli that cannot be recognised. Science, vol. 207, no. 4430, pp. 557–558.

Kurz, W. A., Dymond, C. C., Stinson, G., Rampley, G. J., Neilson, E. T. et al. 2008. Mountain pine beetle and forest carbon feedback to climate change. Nature, vol. 452, pp. 987–990.

Laherrere, J. 2003. Future of oil supplies. http://www.oilcrisis.com/laherrere/zurich.pdf.

Lamb, J. B., Willis, B. L., Fiorenza, E. A., Couch, C. S., Howard, R. et al. 2018. Plastic waste associated with disease on coral reefs. Science, vol. 359, No 6374, pp. 460-462.

Lambeck, K., Rouby, H., Purcell, A., Sun, Y. and Sambridge, M. 2014. Sea level and global ice volumes from the Last Glacial Maximum to the Holocene. Proceedings of the National Academy of Sciences, vol. 111, no. 4, pp. 5296–15303.

Land, E. H. 1959. Experiments in colour vision. Scientific American, vol. 200, no. 5, May, pp. 84–99.

Lang, B. 2002. The Hebrew god: Portrait of an ancient deity. Yale University Press, London.

Langer, E. J. 1982. The illusion of control. In Kahneman, D., Slovic, P. and Tversky, A. (editors) 1982, pp. 231–238.

Larsen, M. and Kaur, R. 2013. Signs of change?: Sex ratio imbalance and shifting social practices in Northern India. Economic & Political Weekly, vol. XLVIII, no. 35, pp. 45–52.

Lawrence, M. G. 2011. Asia under a high-level brown clouds. Nature Geoscience, vol. 4, no. 6, pp. 352–353.

Lazuly, P. 2003. Telling google what to think. Le Monde Diplomatique, November, pp. 12–13. http://mondediplo.com/2003/11/15google.

Lee, K. V., Steinhauer, N., Rennich, K., Wilson, M. E., Tarpy, D. R. et al. 2015. A national survey of managed honey bee 2013-2014 annual colony losses in the USA. Apidologie, vol. 46, No, 3, pp. 292–305.

Lee, R. B. 1980. Lactation, ovulation, infanticide, and women's work: A study of hunter-gatherer population regulation. In Cohen, M. N., Malpass, R. S. and Klein, H. G. (editors). Biosocial mechanisms of population regulation. Yale University Press, New Haven. pp. 321–348.

Legrand, C. 2009. Chile's troubled waters. Guardian Weekly, 11 September, p. 45.

Leigh, D. 2009a. Pollution cover-up by oil company fails. Guardian Weekly, 25 September, pp. 8–9.

Leigh, D. 2009b. Hoped to make a fortune, instead caused a disaster. Guardian Weekly, 25 September, pp. 8–9.

Leigh, D. and Hirsch, A. 2009. New evidence on Ivory Coast "killer waste". Guardian Weekly, 22 May, p. 11.

Lenton, T. M., Held, H., Kriegler, E., Hall, J. W., Lucht, W. et al. 2008. Tipping elements in the earth's climate system. Proceedings of the National Academy of Sciences, vol. 105, no. 6, pp. 1786–1793.

Lenzen, M. 2008. Life cycle energy and greenhouse gas emissions of nuclear energy: A review. Energy Conservation and Management, vol. 49, pp. 2178–2199.

Leonard, W. R. 2002. Food for thought: Dietary change was a driving force in human evolution. Scientific American, December, pp. 106–115.

Leonard, W. R. and Robertson, M. L. 1994. Evolutionary perspectives on human nutrition: The influence of brain and body size on diet and metabolism. American Journal of Human Biology, vol. 6, no. 1, pp. 77–88.

Leonard, W. R. and Robertson, M. L. 1997. Rethinking the energetics of bipedality. Current Anthropology, vol. 38, no. 2, pp. 304–309.

Leontief, W. 1982. Academic economics. Science, vol. 217, no. 4555, pp. 104 and 107.

Leslie, J. 2000. Running dry: What happens when the world runs out of fresh water? Harper's Magazine. July, pp. 37–52.

Lever, C. 1992. They dined on eland: The story of the Acclimatization societies. Quiller Press, London.

Levin, S. A. 2005. Self-organization and the emergence of complexity in ecological eystems. BioScience, vol. 55, no. 12, pp. 1075–1079.

Levine, H., Jørgensen, N., Martino-Andrade, A., Mendiola, J., Weksler-Derri, D. et al. 2017. Temporal trends in sperm count: a systematic review and meta-regression analysis. Human Reproduction Update, pp. 1–14. doi:10.1093/humupd/dmx022.

Leviticus, S., Antonov, J. and Boyer, T. 2005. Warming of the world ocean, 1955–2003. Geophysical Research Letters, vol. 32, L02604, doi:10.1029/2004GL021592.

Levy, S. B. 1998. Antimicrobial resistance: Bacteria on the defence. British Medical Journal, vol. 317, pp. 612–613.

Lewicki, P., Hill, T. and Czyzewska, M. 1992. Nonconscious acquisition of information. American Psychologist, vol. 47, pp. 796–801.

Lewis, P. 2017. "Everyone is distracted. All of the time". Guardian Weekly, 27 October, pp. 26–29.

Lexchin, J. R. 2005. Implications of pharmaceutical industry funding on clinical research. Annals of Pharmacotherapy, vol. 39, no. 1, pp. 194–197.

Lia, S., Leitheada, A., Moussaa, S. G., Liggioa, J., Morana, M. D. et al. 2017. Differences between measured and reported volatile organic compound emissions from oil sands facilities in Alberta, Canada. Proceedings of the National Academy of Sciences, vol. 114, no. 19, pp. E3756-E3765.

Libert, B. 1999. How does conscious experience arise? The neural time factor. Brain Research Bulletin, vol. 50, no. 5-6, pp. 339–340.

Lightman, A. and Gingerich, O. 1992. When do anomalies begin? Science, vol. 255, pp. 690–695.

Lilienfeld, S. O., Wood, J. M. and Garb, H. N. 2001. What's wrong with this picture? Scientific American, May, pp. 80–87.

Liljedahl, A. K., Boike, J., Daanen, R. P., Fedorov, A. N., Frost, G. V. et al. 2016. Pan-Arctic ice-wedge degradation in warming permafrost and its influence on tundra hydrology. Nature Geoscience, vol. 9, pp. 312–318.

Ling, L. L., Schneider, T., Peoples, A. J., Spoering, A. L., Engels, I. et al. 2015. A new antibiotic kills pathogens without detectable resistance. Nature, vol. 517, pp. 455–459.

Lips, K. R., Brem, F., Brenes, R., Reeve, J. D., Alford, R. A. et al. 2006. Emerging infectious disease and the loss of biodiversity in a Neotropical amphibian community. Proceedings of the National Academy of Sciences, vol. 103, no. 109, pp. 3165–3170.

Lister, R., Pelizzola, M., Dowen, R. H., Hawkins, R. D., Hon, G. et al. 2009. Human DNA methylomes at base resolution show widespread epigenomic differences. Nature, vol. 462, pp. 315–322.

Liu, J. and Diamond, J. 2005. China's environment in a globalizing world. Nature, vol. 435, pp. 1179–1186.

Liu, J., Daily, G. C., Ehrlich, P. R. and Luck, G. W. 2003. Effects of household dynamics on resource consumption and biodiversity. Nature, vol. 421, pp. 530–533.

Livermore, D. M., Macgowan, P. A. and Wale, M. C. J. 1998. Surveillance of antimicrobial resistance. British Medical Journal, vol. 317, pp. 614–615.

Livi-Bacci, M. 1997. A concise history of world population. Blackwell, Oxford.

Li, Y. F. and Macdonald, R. W. 2005. Sources and pathways of selected organochlorine pesticides to the Arctic and the effect of pathway divergence on HCH trends in biota: A review. Science of the Total Environment, vol. 342, pp. 87–106.

Loftus, E. F. 1997. Creating false memories. Scientific American, September, pp. 30–35.

Loftus, E. F. and Klinger, M. R. 1992. Is the unconscious smart or dumb? American Psychologist, vol. 47, no. 6, pp. 761–765.

Logothetis, N. K. 1999. Vision: A window on consciousness. Scientific American, November, pp. 14–21.

Long, S. P., Ainsworth, E. A., Leakey, A. D., Nösberger, J. and Ort, D. R. 2006. Food for thought: Lower-than-expected crop yield stimulation with rising CO2 concentrations. Science, vol. 312, no. 5782, pp. 1918–1921.

Lönnstedt, O. M. and Eklöv, P. 2016. Environmentally relevant concentrations of microplastic particles influence larval fish ecology. Science, vol. 352, no. 6290, pp. 1213–1216.

Lotto, R. P. and Purves, D. 2004. Perceiving colour. Review of Progress in Coloration and Related Topics. vol. 34, no. 1, pp. 12–25.

Lucas, R. 1972. Expectations and the neutrality of money. Journal of Economic Theory, vol. 4, no. 2, pp. 103–124.

Ludwig, D., Hilborn, R. and Walters, C. 1993. Uncertainty, resource exploitation, and conservation: Lessons from history. Science, vol. 260, pp. 17 and 36.

Lutz, W., Sanderson, W. and Scherbov, S. 2001. The end of world population growth. Nature, vol. 412, pp. 543–545.

Luyendijk, J. 2015. It's business as usual in our banking system. Guardian Weekly, 23 October, pp. 26–29.

Lynas, M. 2011. The God species: How humans can really save the planet. Forth Estate, London.

Lyons, G., Ahrens, A. and Salter-Green, E. 2000. An environmentalist's vision of operationalizing the precautionary principle in the management of chemicals. International Journal of Occupational and Environmental Health, vol. 6, no. 4, pp. 289–95.

Macalister, T. 2009. Crucial data "distorted" as global oil runs dry. Guardian Weekly, 13 November, pp. 1–2.

Macedo, I. C., Seabra, J. E. A. and Silva, J. E. A. R. 2008. Greenhouse gases emissions in the production and use of ethanol from sugarcane in Brazil: The 2005/2006 averages and a prediction for 2020. Biomass and Bioenergy, vol. 32, no. 7, pp. 582–595.

Mackenzie, C. A., Lockridge, A. and Keith, M. 2005. Declining sex ratio in a first nation community. Environmental Health Perspectives, vol. 113, no. 10, pp. 1295–1298.

MacKenzie, D. 2003. US develops lethal new viruses. New Scientist, 1 November, p. 6.

MacLean, N. 2017. Democracy in chains: The deep history of the radical right's stealth plan for America. Viking: New York.

Maharaj, N. and Dorren, G. 1995. The game of the rose: The third world in the global flower trade. International Books, Utrecht.

Maharaj, N. and Hohn, D. 2001. Fleurs du Mall. Harper's Magazine, February, pp. 66–67.

Mallon, M. 2002. Review of Sacred Hunt, by Pelly, D. F. Above and Beyond, May–June, p. 60.

Manire, C. A. and Gruber, S. H. 1990. Many sharks may be headed towards extinction. Conservation biology, vol. 4, pp. 10–11.

Mankiw, G. 2009. News flash:Economists agree. Greg Mankiw's Blog. http://gregmankiw.blogspot.ca/2009_02_01_archive.html.

Mann, M. E., Zhang, Z., Hughes, M. K., Bradley, R. S., Miller, S. K. et al. 2008. Proxy-based reconstructions of hemispheric and global surface temperature variations over the past two millennia. Proceedings of the National Academy of Science. vol. 105, no. 36, pp. 13252–13257.

Manning, R. 2004. The oil we eat. Harper's Magazine, February, pp. 37–45.

Mannion, A. M. 1999. Domestication and the origins on agriculture: An appraisal. Progress in Physical Geography, vol. 23, no. 1, pp. 37–56.

Marchant, J. 2006. In search of lost time. Nature, vol. 444, pp. 534–538.

Marcott, S. A., Shakun, J. D., Clark, P. U. and Mix, A. C. 2013. A reconstruction of regional and global temperature for the past 11,300 years. Science, vol. 339, no. 6124, pp. 1198–1201.

Marino, B. D. V. and Odum, H. T. 1999. Biosphere 2: Introduction and research progress. Ecological Engineering, vol. 13, pp. 3–14.

Marshall, L. G., Webb, S. D., Sepkowski, J. J. and Raup, D. M. 1982. Mammalian evolution and the great American interchange. Science, vol. 215, pp. 1351–1357.

Martineau, P. 1957. Motivation in advertising: Motives that make people buy. McGraw-Hill Book Company, New York.

Martinez-Austria, P. and Hofwegen, P. (editors). 2006. Synthesis of the 4th World Water Forum. World Water Council, Mexico, pp. 1–137.

Marx, L. 1996. The Domination of nature and the redefinition of progress. In Marx, L. and Mazlish, B. (editors) 1996, pp. 201–218.

Marx, L. and Mazlish, B. (editors). 1996. Progress: Fact or illusion? The University of Michigan Press, Ann Arbor.

Maser, C. 1988. The redesigned forest. R and E Miles, San Pedro.

Mattila, H. R., Rios, D., Walker-Sperling, V. E., Roeselers, G. and Newton, I. L. G. 2012. Characterization of the active microbiotas associated with honey bees reveals healthier and broader communities when colonies are genetically diverse. PLoS ONE, vol. 7, no. 3, e32962, pp. 1–11.

Maugeri, L. 2004. Oil: Never cry wolf—why the petroleum age is far from over. Science, vol. 304, p. 1114.

May, R. M. 1975. Energy costs of food gathering. Nature, vol. 225, p. 669.

Mayewski, P. A. and White, F. 2002. The ice chronicles. University Press of New England, Hanover.

Mazlish, B. 1996. Progress: A historical and critical perspective. In Marx, L. and Mazlish, B. (editors) 1996, pp. 27–44.

Mazlish, B. and Marx, L. 1996. Introduction. In Marx, L. and Mazlish, B. (editors) 1996, pp. 1–7.

Mazoyer, F. 2000. The science behind shopping. Le Monde Diplomatique, December, p. 12.

Mazoyer, F. 2004. Frankenfish and the future. Le Monde Diplomatique, January, p. 15.

McAllester, M. 2003. Uday's final days filled with anger, sense of loss. Edmonton Journal, 26 July, p. A3.

McCabe, P. 1998. Energy resources—Cornucopia or empty barrel? American Association of Petroleum Geologists Bulletin, vol. 82, no. 11, pp. 2110–2134.

McCallum, I. 2005. Ecological intelligence: Rediscovering ourselves in nature. Africa Geographic, Cape Town.

McCurry, J. 2015. Fukushima operator "knew of need to protect against tsunami but did not act". Guardian, 19 June. http://www.theguardian.com/world/2015/jun/18/fukushima-power-plant-operator-knew-need-protect-against-tsunami-japan-disaster. Accessed 19 July 2015.

McEvoy, A. E. 1986. The fisherman's problem: Ecology and law in the California fisheries, 1850–1980. Cambridge University Press, Cambridge.

McGhee, R. 2001. Ancient people of the arctic. UBC Press, Vancouver.

McGovern, T. H. 2000. The demise of Norse Greenland. In Fitzburg, W. W. and Ward, E. I. (editors). Vikings: The north Atlantic saga. Smithsonian Instituition Press, Washington. pp. 427–439.

McIntosh, R. J. 2000. Social memory in Mande. In McIntosh, R. J., Tainter, J. A. and McIntosh, S. K. (editors) 2000, pp. 141–180.

McIntosh, R. J., Tainter, J. A. and McIntosh, S. K. 2000. Climate, history, and human action. In McIntosh, R. J., Tainter, J. A. and McIntosh, S. K. (editors) 2000, pp. 1–44.

McIntosh, R. J., Tainter, J. A. and McIntosh, S. K. (editors). 2000. The way the wind blows: Climate change, history and human action. Columbia University Press, New York.

McKee, B. 2002. Solutions for the 21st century: Zero emissions technologies for fossil fuels. Technology status report. International Energy Agency, pp. 1–48.

McKellar, Q. A.1998. Antimicrobial Resistance: A veterinary perspective. British Medical Journal, vol. 317, pp. 610–611.

McNeill, J. R. and Winiwarter, V. 2004. Breaking the sod: Humankind, history, and the soil. Science, vol. 304, pp. 1627–1629.

McRae, D. 2016. Semenya return set to revive gender row. Guardian Weekly, 5 August, pp. 46–47.

Mead, M. N. 2005. Arsenic: In search of an antidote to a global poison. Environmental Health Perspectives, vol. 113, no. 6, June, pp. A379–A386.

Meadows, D. H., Meadows, D. L., Randers, J. and Behrens, W. W. 1972. The limits to growth. Universe Books, New York.

Meadows, D. L. 2007. Evaluating past forecasts: Reflections on one critique of the limits to growth. Sustainability or collapse? In Costanza, R., Graumlich, L. and Steffen, W. (editors). Report of the 96th Dahlem workshop on an integrated history and future of people on earth. MIT Press, Cambridge, MA. pp. 399–415.

Meinshausen, M., Meinshausen, N., Hare, W., Raper, S. C. B., Frieler, K. et al. 2009. Greenhouse-gas emission targets for limiting global warming to 2°C. Nature, vol. 458, pp. 1158–1163.

Meinzer, F. C. 2003. Functional convergence in plant responses to the environment. Oecologia, vol. 134, pp. 1–11.

Mekonnen, M. M. and Hoekstra, A. Y. 2010. The green, blue and grey water footprint of farm animals and derived animal products. UNESCO-IHE, Delft, The Netherlands, Value of Water Research Report Series, no. 48, pp. 1–50.

Mekonnen, M. M. and Hoekstra, A. Y. 2011. The green, blue and grey water footprint of crops and derived crop products. Hydrology and Earth System, vol. 15, pp. 1577–1600.

Menzie, W. D., Singer, D. A. and DeYoung, J. H. 2005. Mineral resources and consumption in the twenty-first century. In Simpson, R. D., Toman, M. A. and Ayres, R. U. (editors) 2005, pp. 33–53.

Merikle, P. M. 1992. Perception without awareness: Critical issues. American Psychologist, vol. 47, no. 6, pp. 792–795.

Mertz, O., Bruun, T. B., Fog, B., Rasmussen, K. and Agergaard, J. 2010. Sustainable land use in Tikopia: Food production and consumption in an isolated agricultural system. Singapore Journal of Tropical Geography, vol. 31, pp. 10–26.

Micklin, P. P. 1988. Desiccation of the Aral Sea: A water management disaster in the Soviet Union. Science, vol. 241, no. 4870, pp. 1170–1176.

Midgley, T. and Henne, A. L. 1930. Organic fluorides as refrigerants. Industrial and Engineering Chemistry, vol. 22, no. 5, p. 542.

Mill, J. 1848. Influence of the progress of society on production and distribution. Book IV, Principles of political economy

Miller, G. A. 1956a. Information and memory. Scientific American, vol. 195, no. 2, pp. 42–46.

Miller, G. A. 1956b. The magical number seven, plus or minus two: Some limits on our capacity for processing information. Psychological Review, vol. 63, no. 2, pp. 81–97.

Miller, G. H., Geirsdóttir, Á., Zhong, Y., Larsen, D. J., Otto-Bliesner, B. L. et al. 2012. Abrupt onset of the Little Ice Age triggered by volcanism and sustained by sea-ice/ocean feedbacks. Geophysical Research Letters, vol. 39, no. 2, L02708, pp. 1–5.

Miller, K. G., Kominz, M. A., Browning, J. V., Wright, J. D., Mountain, G. S. et al. 2005a. The Phanerozoic record of global sea-level change. Science , vol. 310, no. 5752, pp. 1293–1298.

Miller, L. G., Perdreau-Remington, F., Rieg, G., Mehdi, S., Perlroth, J. et al. 2005b. Necrotizing Fasciitis caused by community-associated methicillin-resistant Staphylococcus aureus in Los Angeles. New England Journal of Medicine, vol. 352, no. 14, pp. 1445- 53.

Miller, W. R. and C'de Baca, J. 2001. Quantum change: When epiphanies and sudden insights transform ordinary lives. The Guilford Press, New York.

Millon, R. 1988. The last years of Teotihuacan dominance. In Yoffee, N. and Cowgill, W. (editors) 1988, pp. 102–164.

Milonakis, D. and Fine, B. 2009. From political economy to economics: Method, the social and historical in the evolution of economic theory. Routledge, London.

Milton, K. and May, M. L. 1976. Body weight, diet and home range area in primates. Nature, vol. 259, pp. 459–462.

Mintz, S. W. 1989. Food and culture: An anthropological view. In Hirschoff, P. M. and Kotler, N. G. (editors). Completing the food chain: Strategies for combating hunger and malnutrition. Smithsonian Institution Press, Washington, DC. pp. 114–121.

Mishel, L. and Sabadish, N. 2017. CEO pay remains high relative to the pay of typical workers and high-wage earners. Economic Policy Institute, report, 20 July, pp. 1–24. http://www.epi.org/files/pdf/130354.pdf.

Mitchell, C. E. and Power, A. G. 2003. Release of invasive plants from fungal and viral pathogens. Nature, vol. 421, pp. 625–627.

Mitchell, L. A., Lauer, F. T., Burchiel, S. W. and McDonald, J. D. 2009. Mechanisms for how inhaled multiwalled carbon nanotubes suppress systemic immune function in mice. Nature Nanotechnology, vol. 4, pp. 451–456.

Mitchell, T. D. and Hulme, M. 1999. Predicting regional climate change: Living with uncertainty. Progress in Physical Geography, vol. 23, no.1, pp. 57–78.

Mittermeier, R. A., Wallis, J., Rylands, A. B., Ganzhorn, J. U., Oates, J. F. et al. (editors). 2009. Primates in peril: The World's 25 Most Endangered Primates 2008-2010. IUCN/SSC Primate Specialist Group (PSG), International Primatological Society (IPS), and Conservation International (CI), Arlington, VA., pp. 1–84.

Modelski, G. 1997. Cities of the ancient world: An inventory (3500 to 1200 BC). Paper presented at World System Historical Data Group, ISA Convention, Toronto, March 1997. Revised. http://faculty.washington.edu/modelski. See: World systems history.

Molden, D. (editor). 2007. Summary for decisionmakers. In Molden, D. (editor). Water for food, water for life: A comprehensive assessment of water management in agriculture. Earthscan, London. International Water Management Institute, Colombo. pp. 1–40.

Molleson, T. 1994. The eloquent bones of Abu Hureyra. Scientific American, vol. 271, August, pp. 70–75.

Molleson, T., Jones, K. and Jones, S. 1993. Dietary change and the effects of food preparation on microwear patterns in the late Neolithic of Abu Hureyra, northern Syria. Journal of Human Evolution, vol. 24, no. 6, pp. 455–468.

Monbiot, G. 2008. When will the oil run out? Guardian Weekly, 15 December. http://www.guardian.co.uk/business/2008/dec/15/oil-peak-energy-iea.

Montgomery, C. 2006. Nurturing doubt about climate change is big business. Globe and Mail, 12 August, pp. F4–F5.

Moore, A. M. T., Hillman, G. C. and Legge, A. J. 1975. The excavation of tell Abu Hureyra in Syria: A preliminary report. Proceedings of the Prehistoric Society, vol. 41, pp. 50–77.

Moore, A. M. T., Hillman, G. C. and Legge, A. J. 2000. Village on the Euphrates: From foraging to farming at Abu Hureyra. Oxford University Press, Oxford.

Moore, C. 2005. China brides are a diamond miner's best friend. Guardian Weekly, 2 September, p. 17.

Moore, C. J. 2008. Synthetic polymers in the marine environment: A rapidly increasing, long-term threat. Environmental Research, vol. 108, no. 2, pp. 131–139.

Moore, C. J., Moore, S. L., Leecaster, M. K. and Weisberg, S. B. 2001. A comparison of plastic and plankton in the north pacific gyre. Marine Pollution Bulletin, vol. 42, no. 12, pp. 1297–1300.

Morgan, M. G. 1993. Risk analysis and management. Scientific American, July, vol. 269, pp. 32–41.

Morgan, T. 1995. Theory versus empiricism in academic economics. Challenge, vol. 38, no. 6, pp. 46–51.

Morgenstern, N. 2001. Geotechnics and mine waste management - update. Seminar on safe tailing dam constructions. Gallivare, Sweden. 20–21 September. Technical papers, pp. 54–67.

Morin, H. and Ricard, P. 2009. Chemicals tests "will kill 54m lab animals". Guardian Weekly, 11 September, p. 5.

Morse, D. R., Lawton, J. H., Dodson, M. H. and Williamson, M. H. 1985. Fractal dimensions of vegetation and the distribution of arthropod body lengths. Nature, vol. 314, pp. 731–733.

Moshman, D. 1998. Cognitive development beyond childhood. In Kuhn, D. and Siegler, R. S. (editors). Handbook of child psychology, 5th ed. Vol. 2. Cognition, perception and language. Wiley, New York: pp. 947–978.

Mostert, N. 1992. Frontiers: The epic of South Africa's creation and the tragedy of the Xhosa people. Jonathan Cape, London.

Moy, D., Howard, W. R., Bray, S. G. and Trull, T. W. 2009. Reduced calcification in modern Southern ocean planktonic foraminifera. Nature Geoscience, vol. 2, pp. 276–280.

Mudd, G. M., Weng, Z. and Jowitt, S. M. 2013. A detailed assessment of global Cu trends and endowments. Economic Geology, vol. 108, no. 5, pp. 1163–1183.

Mudelsee, M. 2001. The phase relations among atmospheric $CO_2$ content, temperature and global ice volume over the past 420 ka. Quaternary Science Reviews, vol. 20, pp. 583–589.

Murphy, D. J. and Hall, C. A. S. 2010. Year in review-EROI or energy return on (energy)invested. Annals of the New York Academy of Sciences, vol. 1185, Issue: Ecological Economics Reviews, pp. 102–118.

Murphy, D. J., Hall, C. A.S., Dale, M. and Cleveland, C. 2011. Order from Chaos: A Preliminary Protocol for Determining the EROI of Fuels. Sustainability, vol. 3, no. 10, pp. 1888–1907.

Myers, D. G. 1999. Social psychology. 6th ed. McGraw-Hill College.

Myers, D. G. 2000. The funds, friends and faith of happy people. American Psychologist, vol. 55, no. 1, pp. 56–67.

Myers, D. G. 2001. The American paradox: Spiritual hunger in an age of plenty. Yale Nota Bene, New Haven.

Myers, D. G. 2002a. Intuition: Its powers and perils. Yale University Press, New Haven.

Myers, K. 1986. Understains: The sense and seduction of advertising. Comedia Series No 31, pp. 1–157.

Myers, N. 2002b. A convincing call for conservation. Review of The future of life, by Wilson, E. O. Science, vol. 295, pp. 447–448.

Myers, R. A. and Worm, B. 2003. Rapid worldwide depletion of predatory fish communities. Nature, vol. 423, pp. 280–283.

Myers-Smith, I. 2007. Shrub line advance in alpine tundra of the Kluane region: Mechanisms of expansion and ecosystems impacts. Arctic, vol. 60, no. 4, pp. 447–451.

Myers-Smith, I. H., Elmendorf, S. C., Beck, P. S. A., Wilmking, M., Hallinger, M. et al. 2015. Climate sensitivity of shrub expansion across the tundra biome. Nature Climate Change, 6 July, pp. 1–5. doi:10.1038/nclimate2697.

Myss, C. 1996. Anatomy of the spirit. Three Rivers Press, New York.

National Park Service. 2010. Glaciers/glacial features. Glacier National Park. http://www.nps.gov/glac/naturescience/glaciers.htm.

Nature. 2009. Of mice and men. Nature Nanotechnology, vol. 4, p. 395.

Navarro, E., Baun, A., Behra, R., Hartmann, N. B., Filser, J. et al. 2008. Environmental behavior and ecotoxity of engineered nanoparticles to algae, plants, and fungi. Ecotoxicology, vol. 17, no. 5, pp. 372–386.

Nayeri, D. 2017. The rapture: Yearning for the apocalypse. Guardian Weekly, 8 August, pp. 26–29.

NCD-RisC. 2016. Trends in adult body-mass index in 200 countries from 1975 to 2014: a pooled analysis of 1698 population-based measurement studies with 19·2 million participants. Lancet, vol. 387, no. 10026, pp. 1377-1396.

NCHS. 2008. Prevalence of overweight, obesity and extreme obesity among adults: United States, trends 1976–80 through 2005–2006. US Government, National Centre for Health Statistics, Health E-Stat. http://www.cdc.gov/nchs/data/hestat/overweight/overweight_adult.htm.

Neale, A., Miller, S. and Michaelsen, D. 2003. Overview of the acid rock drainage and overburden management program at PT Freeport Indonesia operations in Papua Province, Indonesia. Proceedings of the 6th International Conference on Acid Rock Drainage, pp. 107–110.

Nelsen, A. 2015. Carbon capture moves closer. Guardian Weekly, 17 April, p. 17.

Nelson, M., Burgess, T. L., Alling, A., Alvarez-Roma, N., Dempter, W. F. et al. 1993. Using a closed ecological system to study Earth's biosphere: Initial results from Biosphere 2. Bioscience, vol. 43, no. 4, pp. 225–236.

Nesse, R. M. and Williams, G. C. 1998. Evolution and the origins of disease. Scientific American, November, pp. 86–93.

Neumann, R. B., Ashfaque, K. N., Badruzzaman, A. B. M., Ali, M. A., Shoemaker, J. K. et al. 2009. Anthropogenic influences on groundwater arsenic concentrations in Bangladesh. Nature geosciences, vol. 3, pp. 46–52.

Newman, J. D. and Harris, J. C. 2009. The scientific contribution of Paul D. MacLean (1913–2007). Journal of Nervous and Mental disease, vol. 197, no. 1, pp. 3–5.

Newton, K., Côté, I. M., Pilling, G. M., Jennings, S. and Dulvy, N. K. 2005. Current and future sustainability of island coral reef fisheries. Current Biology, vol. 17, pp. 655–658.

Nicholls, R. J., Hoozemans, F. M. and Marchland, M. 1999. Increasing food risk and wetland losses due to global sea-level rise: Regional and global analysis. Global Environmental Change, vol. 9, pp. S69–S87.

Nilsson, C. and Berggren, K. 2000. Alterations of riparian ecosystems caused by river regulation. BioScience, vol. 50, No, 9, pp. 783–792.

Nilsson, C., Reidy, C. A., Dynesius, M. and Revenga, C. 2005. Fragmentation and flow regulation of the world's large river systems. Science, vol. 308, pp. 405–408.

Nisbett, R. E., Krantz, D. H., Jepson, C. and Fong, G. T. 1982. Improving inductive inference. In Kahneman, D., Slovic, P. and Tversky, A. (editors) 1982, pp. 445–459.

Nissen, H. J., Damerow, P. and Englund, R. K. 1993. Archaic bookkeeping: Early writing and techniques of economic administration in the ancient near east. Larsen, P. Translator. University of Chicago Press, Chicago.

Nobel, A. 1895. Full text of Alfred Nobel's will. http://www.nobelprize.org/alfred_nobel/will/will-full.html.

Noble, D. 1998. The religion of technology: The divinity of man and the spirit of invention. A. A. Knopf, New York.

Nobel Foundation. 2014. The Sveriges Riksbank Prize in Economic Sciences in memory of Alfred Nobel. The official website of the Nobel prize. http://www.nobelprize.org/nobel_prizes/economic-sciences/. Accessed 14 July, 2014.

Nohl, J. 1971. The black death: A chronicle of the plague compiled from contemporary sources. Abridged edition of 1926 original. Clarke, C. H. translator. Unwin, London.

Norgate, T.E. and Jahanshahi, S. 2010. Low grade ores-smelt, leach or concentrate? Minerals Engineering, vol. 23, no. 2, pp. 65–73.

Normile, M. A. and Price, J. 2004. The United States and the European Union—Statistical overview. Economic Research Service, USDA, US- EU Food and Agriculture Comparisons, WRS-04-04, pp. 1–13.

Norretranders, T. 1998. The user Illusion: Cutting consciousness down to size. Viking, New York.

Norwegian Ministry of Finance. 2006. Two companies—Wal-Mart and Freeport—are being excluded from the Norwegian Government Pension Fund – Global's investment universe. Norwegian Ministry of Finance, news release, 6 June, 2006. http://www.regjeringen.no/en/dep/fin/press-center/Press-releases/2006/Two-companies---Wal-Mart-and-Freeport---.html?id=104396&epslanguage=EN-GB.

Nowak, R. 2001. Disaster in the making. An engineered mouse virus leaves us one step away from the ultimate bioweapon. New Scientist, 13 January, p. 44.

NSIDC. 2005. Sea ice decline intensifies: Summer Arctic sea ice falls below average for forth year, winter ice sees sharp decline, spring melt starts earlier. National Snow and Ice Data Centre, press release, 28 September, 2005. http://nsidc.org/news/press/20050928_trendscontinue.html

NSIDC. 2010. Average monthly Arctic sea ice extent September 1979–2010. National Snow and Ice Data Centre. http://nsidc.org/images/arcticseaicenews/20100927_Figure3.png.

NSIDC. 2012. Arctic sea ice shatters previous low records; Antarctic sea ice edges to record high. National Snow and Ice Data Centre, press release, 2 October, 2012, pp. 1–2. http://nsidc.org/news/press/20121002_MinimumPR.html.

Nunn, P. 2007. Transmission of XDR TB in South Africa: Discussion of the global implications. Urgent issues in the developing world. 14th Conference on Retroviruses and Opportunistic Infections. 25 February, 2007. www.who.int/tb/features_archive/croi_feb07.pdf.

OBAC. 2008. Future forest products and fibre use strategy. Ominica Beetle Action Coalition, pp. 1–51. http://www.ominecacoalition.ca/Strategies/ForestAndFibre/pdf/OBAC-Forest-and-Fibre-Strategy.pdf.

OBAC. 2009. Alternate energy strategy brochure. Ominica Beetle Action Coalition. http://www.ominecacoalition.ca/Strategies/MineralsAndMining/pdf/OBAC-Mineral-Strategy-brochure.pdf.

Obbard, R. W., Sadri, S., Wong, Y. Q., Khitun, A. A., Baker, I. et al. 2014. Global warming releases microplastic legacy frozen in Arctic Sea ice. Earth's Future, vol. 2, pp. 1–6. doi:10.1002/2014EF000240.

O'Dor, R. K. 2003. The unknown ocean: Baseline report of the census of marine life program. Consortium for Oceanographic Research and Education. Washington, DC, October, pp. 1–28. http://www.coml.org/comlfiles/press/Baseline_Report_101603.pdf.

Ogodo, O. and Vidal, J. 2007. Draining the life out of Kenya lake. Guardian Weekly, 6 April, p. 5.

O'Hara, R. B. 2005. The anarchist's guide to ecological theory. Or, we don't need no stinkin' laws. Oikos, vol. 110, no. 2, pp. 390–393.

O'Loughlin, T. 2009. Mynah faces Australian cull. Guardian Weekly, 22 May, p. 7.

O'Niell, J. (chair). 2016. Tackling drug-resistant infections globally: Final report and recommendations. Review on antimicrobial resistance. UK Government. May, pp. 1–84.

Ophuls, W. and Boyan, A. S. 1992. Ecology and the politics of scarcity revisited: The unravelling of the American dream. Freeman, New York.

Oren, R., Ellsworth, D. S., Johnsen, K. H., Phillips, N., Ewers, B. E. et al. 2001. Soil fertility limits carbon sequestration by forest ecosystems in a CO2-enriched atmosphere. Nature, vol. 411, pp. 469–472.

Oreskes, N. and Conway, E. M. 2010. Merchants of doubt: How a handful of scientists obscured the truth on issues from tobacco smoke to global warming. Bloomsbury, New York.

Osborn, A. 2000. Drastic cuts in cod fishing. Guardian Weekly, 7 December, p. 10.

Osborn, A. 2002. Fishing faces doomsday scenario. Guardian Weekly, 31 October, p. 6.

Osborn, A. 2003. North Sea fish deal clinched. Guardian Weekly, 25 December, p. 5.

Oskamp, S. 2000. A sustainable future for humanity? How can psychology help? American Psychologist, vol. 55, pp. 496–508.

Ostrom, E., Burger, J., Field, C. B., Norgaard, R. B. and Policansky, D. 1999. Revisiting the commons: Local lessons, global challenges. Science, vol. 284, pp. 278–282.

Osvath, M. 2009. Spontaneous planning for future stone throwing by a male chimpanzee. Current Biology, vol. 19, no. 5, pp. R190–R191.

Oswald, A. J. 1997. Happiness and economic performance. Economic Journal, vol. 107 no. 445, pp. 1815–1831.

Otto-Bliesner, B. L., Marshall, S. J., Overpeck, J. T., Miller, G. H., Hu, A. et al. 2006. Simulating Arctic climate warmth and icefield retreat in the last interglaciation. Science, vol. 311, no. 5768, pp. 1751–1753.

Overpeck, J. T., Otto-Bliesner, B. L., Miller, G. H., Muhs, D. R., Alley, R. B. et al. 2006. Paleoclimatic evidence for future ice-sheet instability and rapid sea-level rise. Science, vol. 311, no. 5768, pp. 1747–1750.

Paddack, M. J., Reynolds, J. D., Aguilar, C., Appeldoorn, R S., Beets, J. et al. 2009. Recent region-wide declines in Caribbean reef fish abundance. Current Biology, vol. 19, no. 7, pp. 590–595.

Pandolfi, J. M., Bradbury, R. H., Sala, E., Hughes, T. P., Bjorndal, K. A. et al. 2003. Global trajectories of the long-term decline of coral reef ecosystems. Science, vol. 301, pp. 955–958.

Parajulee, A. and Wania, F. 2014. Evaluating officially reported polycyclic aromatic hydrocarbon emissions in the Athabasca oil sands region with a multimedia fate model. Proceedings of the National Academy of Sciences, doi:10.1073/pnas.1319780111.

Parmesan, C. and Yohe, G. 2003. A global coherent fingerprint of climate change impacts across natural systems. Nature, vol. 421, pp. 37–42.

Pasquale, F. 2015. The black box society. Harvard University Press, Cambridge.

Patterson, M. 2003. Water management and molybdenum treatment at the closed Noranda Inc.-Brenda Mines site, Peachland. Reclamation at closed mines where molybdenum is an issue. Proceedings of the 27th Annual British Columbia Mine Reclamation Symposium. The British Columbia Technical and Research Committee on Reclamation, pp. 1–12.

Patterson, S. 2010. The Quants: How a breed of math whizzes conquered Wall street and nearly destroyed it. Crown Business, New York.

Paull, D., Banks, G. and Ballard, C. 2006. Monitoring the environmental impact of mining in remote locations through remotely sensed data. Geocarto International, vol. 21, pp. 33–42.

Pauly, D. and Watson, R. 2003. Counting the last fish. Scientific American, July, pp. 42–47.

Pauly, D. and Zeller, D. 2016. Catch reconstructions reveal that global marine fisheries catches are higher than reported and declining. Nature communications, vol. 7, Article 10244, pp. 1–9.

Pauly, D., Belhabib, D., Blomeyer, R., Cheung, W. W., Cisneros-Montemayor, A. M. et al. 2013. China's distant-water fisheries in the 21st century. Fish and Fisheries, doi:10.1111/faf.12032, pp. 1–154.

PDAC. 2010. E3 program. Prospectors and Developers Association of Canada. http://www.pdac.ca/e3plus. Accessed 2010.

Pearce, F. 2010. Research red in tooth and claw. Guardian Weekly, 12 February, p. 50.

Pearson, H. 2002. Humanity: It's all in the mind. Nature news service, 14 April, 2001.

Pearson, H. 2006. Antibiotic faces uncertain future. Nature, vol. 441, pp. 260–261.

Pecl, G. T., Araújo, M. B., Bell, J. D., Blanchard, J., Bonebrake, T. C. et al. 2017. Biodiversity redistribution under climate change: Impacts on ecosystems and human well-being. Science, vol. 355, no. 6332, eaai9214. doi:10.1126/science.aai9214.

Peires, J. B. 1989. The dead will arise: Nongqawuse and the great Xhosa cattle killing movement of 1856–7. Raven Press, Johannesburg.

Pepto-Bismol. 2009. http://www.pepto-bismol.com/pepto-original-liquid.php.

Perlez, J. and Bonner, R. 2005. Below a mountain of wealth, a river of waste. New York Times, 27 December. http://www.nytimes.com/2005/12/27/international/asia/27gold.html?_r=1&pagewanted=print

Perloff, L. S. 1987. Social comparison and illusions of vulnerability. In Snyder, C. R. and Ford, C. R. (editors). Coping with negative life events: Clinical and social psychological perspectives. Plenum, New York. Chapter 9, pp. 217–242.

Perrow, C. B. 2007. The next catastrophe: Reducing our vulnerabilities to natural industrial, and terrorist disasters. Princeton University Press, Princeton.

Perry, A. L., Low, P. J. and Reynolds, J. D. 2005. Climate change and distribution shifts in marine fishes. Science, vol. 308, pp. 1912–1915.

Perry, C. T., Murphy, G. N., Kench, P. S., Smithers, S. G., Edinger, E. N. et al. 2013. Caribbean-wide decline in carbonate production threatens coral reef growth. Nature Communications, vol. 4, Article 1402, pp. 1–7.

Persinger, M. A., Bureau, Y. R. J., Peredery, O. P. and Richards, P. M. 1994. The senses presence as right hemispheric intrusions into the left hemispheric awareness of self: An illustrative case study. Perceptual and Motor Skills, vol. 78, pp. 999–1009.

Peters, H. 1983. The ecological implications of body size. Cambridge University Press, London.

Peterson, C. H., Rice, S. D., Short, J. W., Esler, D., Bodkin, J. L. et al. 2003. Long-Term Ecosystem Response to the Exxon Valdez Oil Spill. Science, vol. 302, no. 5653, pp. 2082–2086.

Petit, J. R., Jouzel, J., Raynaud, D., Barkov, N. I., Barnola, J. M. et al. 1999. Climate and atmospheric history of the past 420,000 years from the Vostok ice core, Antarctica. Nature, vol. 399, pp. 429–436.

Pezzey, C. V. and Toman, M. A. 2005. Sustainability and its economic interpretations. In Simpson, R. D., Toman, M. A. and Ayres, R. U. (editors) 2005, pp. 121–141.

Pham, C. K., Ramirez-Llodra, E., Alt, C. H. S., Amaro, T. and Bergmann, M. 2014. Marine litter distribution and density in European seas, from the shelves to deep basins. PLoS ONE, vol. 9, no. 4, e95839, pp. 1–13.

Phillips, K. 2008. Numbers racket: Why the economy is worse than we know. Harper's Magazine, May, pp. 43–47.

Pica, P., Lemer, C., Izard, V. and Dehaene, S. 2004. Exact and approximate arithmetic in an Amazonian indigene group. Science, vol. 306, pp. 499–503.

Pielou, E. C. 1991. After the ice age: The return of life to glaciated North America. University of Chicago Press, Chicago.

Pillay, D. and Zambon, M. 1998. Antiviral drug resistance. British Medical Journal, vol. 317, pp. 660–662.

Pimentel, D. and Patzek, T. 2005. Ethanol production using corn, switchgrass, and wood: Biodiesel production using soybean and sunflower. Natural Resources Research, vol. 14, no. 1, pp. 65–76.

Pimentel, D., Acquay, H., Biltonen, M., Rice, P., Silva, M. et al. 1992. Environmental and economic costs of pesticide use. Bioscience, vol. 42, no. 10, pp. 750–760.

Pimentel, D., Harvey, C., Resosudarmo, P., Sinclair, K., Kurz, K. et al. 1995. Environmental and economic costs of soil erosion and conservation benefits. Science, vol. 267, pp. 1117–1123.

Pimm, S. L. 1977. The value of everything. Nature, vol. 387, pp. 221–222.

Pimm, S. L. and Askins, R. A. 1995. Forest losses predict bird extinctions in eastern North America. Proceedings of the National Academy of Sciences, vol. 92, pp. 9343–9347.

Pimm, S. L., Russell, G. J., Gittleman, J. L. and Brooks, T. M. 1995. The future of biodiversity. Science, vol. 269, pp. 347–350.

Pineyro-Nelson, A., van Heerwaarden, J., Perales, H. R., Serratos-Hernander, J. A., Rangel, A. et al. 2008. Transgenes in Mexican maize: Molecular evidence and methodological considerations for GMO detection in landrace populations. Molecular Ecology, vol. 18, pp. 750–761.

Pitelka, L. F. and Plant Migration Workshop Group. 1997. Plant migration and climate change. American Scientist, vol. 85, pp. 464–473.

Plastics Europe. 2013. Plastics—the facts 2013: An analysis of European latest plastics production, demand and waste data. Plastics Europe, pp. 1–40. http://www.plasticseurope.org/Document/plastics-the-facts-2013.aspx.

Pleasance, E. D., Stephens, P. J., O'Meara, S., McBride, D. J., Meynert, A. et al. 2010. A small-cell lung cancer genome with complex signatures of tobacco exposure. Nature vol. 463, pp. 184–190.

Plotkin, H. 2000. People do more than imitate. Scientific American, October, p. 72.

Polanyi, M. 1968. Life's irreducible structure. Science, vol. 160, pp. 1308–1312.

Pole, K. 2000. To err is human: Addressing maintenance issues from a human factors perspective. Wings Magazine, Issue 4, Maintenance and overhaul supplement, pp. 15–17.

Popkin, B. M. 2002. An overview on the nutrition transition and its health implications: The Bellagio meeting. Public Health Nutrition, vol. 5, no.1A, pp. 93–103.

Popper, S. W., Lempert, R. J. and Bankes, S. C. 2005. Shaping the future, Scientific American, April, pp. 66–71.

Postel, S. 2001. Growing more food with less water. Scientific American, February, pp. 46–51.

Postel, S. L., Daily, G. C. and Ehrlich, P. R. 1996. Human appropriation of renewable fresh water. Science, vol. 271, pp. 785–788.

Potts, S. G., Biesmeijer, J. C., Kremen, C., Neumann, P., Schweiger, O. et al. 2010. Global pollinator declines: trends, impacts and drivers. Trends in Ecology and Evolution. vol. 25, no. 6, pp. 345–353.

Poulton, R., Moffitt, T. E. and Silva, P. A. 2015. The Dunedin Multidisciplinary Health and Development Study: overview of the first 40 years, with an eye to the future. Social Psychiatry and Psychiatric Epidemiology. vol. 50, no. 5, pp. 679–693.

Pradhan, A., Shrestha, D. S., McAloon, A., Yee, W., Haas, M. et al. 2011. Energy life-cycle assessment of soybean biodiesel revisited. Transactions of the American Society of Agricultural and Biological Engineers, vol. 54, no. 3, pp. 1031–1039.

Prasad, S., Anoop, A., Riedel, N., Sarkar, S., Menzel, P. et al. 2014. Prolonged monsoon droughts and links to Indo-Pacific warm pool: A Holocene record from Lonar Lake, central India. Earth and Planetary Science Letters, vol. 391, pp. 171–182.

Prichard, H. D., Arthern, R. J., Vaughan, D. and Edwards, L. A. 2009. Extensive dynamic thinning on the margins of the Greenland and Antarctic ice sheets. Nature, published online 23 September, doi:10.1038/nature08471.

Priede, I. G., Froese, R., Bailey, D. M., Bergstad, O. A., Collins, M. A. et al. 2006. The absence of sharks from abyssal regions of the world's oceans. Proceedings of the Royal Society, Series B, vol. 273, no. 1592, pp. 1435–1441.

Proven Selections. 2002. Proven Selections' sales tag for Bacopa (sutera) 'lilac king', labelled LXT0312 and CU0054D.

Quammen, D. 1998. Planet of weeds. Harper's Magazine, October, pp. 57–69.

Rader, R., Bartomeus, I., Garibaldi, L. A., Garratt, M. P. D., Howlett, B. G. et al. 2016. Non-bee insects are important contributors to global crop pollination. Proceedings of the National Academy of Sciences, vol. 113, no. 1, pp. 146–151.

Radetzki, M. 2010. Peak oil and other threatening peaks-chimeras without substance. Energy Policy. vol. 38, pp. 6566–6569.

Rahmstorf, S. 2000. Shifting seas in the greenhouse? Nature, vol. 399, pp. 523–524.

Ramanathan, V. and Ramana, M. V. 2003. Atmospheric brown clouds: Long range transport and climate impacts. EM, December, pp. 28–33. http://www-abc-asia.ucsd.edu/EM_Paper_20031_final.pdf.

Ramankutty, N., Evan, A., Monfreda, C. and Foley, J. A. 2008. Farming the planet: 1. Geographic distribution of global agricultural lands in the year 2000, Global Biogeochemical Cycles, vol. 22, GB1003.

Ramankutty, N. and Foley, J. 1999. Estimating historical changes in global land cover: Croplands from 1700 to 1992. Global Biogeochemical Cycles, vol. 13, pp. 997–1028.

Ramonet, I. 2001. Manufacturing desire. Le Monde Diplomatique, May, pp. 14–15.

Rathje, W. and Murphy, C. 1993a. Rubbish: The archaeology of garbage. Harper Perennial, New York..

Rathje, W. and Murphy, C. 1993b. Rubbish: The archaeology of garbage. In the new preface to the 2001 printing. Harper Perennial, New York.

Ratnieks, F. L. W. and Carreck, N. L. 2010. Clarity on Honey Bee Collapse? Science, vol. 327, no. 5962, pp. 152–153.

Raugei, M., Fullana-i-Palmer, P. and Fthenakis, V. 2012. The energy return on energy investment (EROI) of photovoltaics: Methodology and comparisons with fossil fuel life cycles. Energy Policy, vol. 45, pp. 576–582.

Raupach, M. R., Marland, G., Ciais, P., Le Quere, C., Canandell, J. G. et al. 2007. Global and regional drivers of accelerating CO2 emissions. Proceedings of the National Academy of Sciences, vol. 104, no. 24, pp. 10288–10293.

Ray, N. and Adams, J. M. 2001. A GIS based vegetation map of the world at the last glacial maximum (25,000–15,000 BP). Internet Archaeology11. http://intarch.ac.uk/journal/issue11/rayadams_toc.html.

Raymo, C. 1998. Skeptics and true believers: The exhilarating connection between science and religion. Doubleday Canada, Toronto.

Read, A. F. and Harvey, P. H. 1989. Life history differences among the eutherian radiations. Journal of Zoology, vol. 219, pp. 329–353.

Reader, J. 1990. Man on earth. Penguin books, London.

Redman, C. 1992. The impact on food production: Short term strategies and long term consequences. In Jackobsen, J. E. and Firor, F. (editor). Human impacts on the environment: Ancient roots and current challenges. Chapter 2, pp. 35–49.

Redman, C. 1999. Human impacts on ancient environments. University of Arizona Press, Tucson.

Redman, C. L. 1978. The Rise of Civilization: From early farmers to urban society in the ancient Near East. W. H. Freeman, San Francisco.

Redman, C. L., Hassan, F. A., Hole, F., Morais, J., Riedel, F. et al. 2007. Group report: Millennial perscectives on the dynamic interaction of climate, people and resources. In Costanza, R., Graumlich, L. and Steffen, W. (editors). Report of the 96th Dahlem workshop on an integrated history and future of people on earth. MIT Press, Cambridge, MA. Chapter 9, pp. 121–148.

Rees, D. A. 1993. Time for scientists to pay their dues. Nature, vol. 363, pp. 203–204.

Rees, W. E. 1995. Achieving sustainability: Reform or transformation? Journal of Planning Literature, vol. 9, no. 4, pp. 343–361.

Rees, W. E. 2003. Ecological footprints: A blot on the land. Nature, vol. 421, no. 6926, p. 898.

Rees, W. E. and Wackernagel, M. 1994. Ecological footprints and appropriated carrying capacity: Measuring the natural capital requirements of the human economy. In Jansson, A., Hammer, M., Folke, C. and Costanza, R. (editors). Investing in natural capital: The ecological economics approach to sustainability. Island Press, Washington, DC. pp. 362–388.

Regal, P. J. 1996. Metaphysics in genetic engineering: Cryptic philosophy and ideology in the "science " of risk analysis. In Van Dommelen, A. (editor). Coping with deliberate release: The limits to risk assessment. International Centre for Human and Public Affairs, Tillburg/Buenos Aires. pp. 15–32.

Regan, H. M., Lupia, R., Drinnan, A. N. and Burgman, M. A. 2001. The currency and tempo of extinction. American Naturalist, vol. 157, no. 1, pp. 1–10.

Regush, N. 1995. Brain storms and angels. Equinox, July–August, pp. 63–73.

Reichmann, J. R., Watrud, L. S., Lee, E. H., Burdick, C. A., Bollman, M. A. et al. 2006. Establishment of transgenic herbicide-resistant creeping bentgrass (Agrostis stolonifera L.) in nonagronomic habitats. Molecular Ecology, vol. 15, no. 13, pp. 4243–4255.

Reiss, B. 2001. The coming storm: Extreme weather and our terrifying future. Hyperion, New York.

Remer, L. A. 2006. Dust, fertilization and sources. Environmental Research Letters, vol. 1, no. 1, 011001, pp. 1–2.

Rescher, N. 1998. Predicting the future: An introduction to the theory of forecasting. State University of New York Press, Albany.

Reuters. 2008. Fish enlisted to fight West Nile spread by U.S. housing collapse. Globe and Mail, 13 June, p. B6.

Reynolds, D. B. 1999. The mineral economy: How prices and costs can falsely signal decreasing scarcity. Ecological Economics, vol. 31, pp. 155–166.

Rice, P. M., Demarest, A. A. and Rice, D. S. 2004. The terminal classic and the "classic Maya collapse" in perspective. In Demarest, A. A., Rice, P. M. and Rice (editors). The terminal classic in the Maya lowlands: Collapse transition and transformation. University Press of Colorado, Boulder. pp. 1–11.

Rice, X. 2006. High birthrate threatens to trap Africa in cycle of poverty. Guardian Weekly, 1 September, p. 1.

Rind, D. 2002. The Sun's role in climate variations. Science, vol. 296, pp. 673–677.

Rios, L. M., Moore, C. and Jones, P. R. 2007. Persistent organic pollutants carried by synthetic polymers in the ocean environment. Marine Pollution Bulletin, vol. 54, pp. 1230–1237.

Ripple, W. J., Wolf, C., Galetti, M., Newsome, T. M., Alamgir, M. et al. 2017. World scientists warning to humanity: A second notice. Bioscience, vol. 67, No, 12, pp. 1026–1028.

RNS and ABP. 2008. Southern Baptist group shifts position on climate. Christian Century, vol. 157, no. 7, p. 18.

Roach, M. 2008. Almost Human. National Geographic, April, pp. 124–145.

Robertson's Taxidermy. 2004. Robertson's Taxidermy Ltd. Sales brochure.

Rock, I. 1974. The perception of distorted figures. Scientific American, January, pp. 78–85.

Rockström, J., Steffen, W., Noone, K., Persson, Å., Chapin, F. S. et al. 2009. A safe operating space for humanity. Nature, vol. 461, pp. 472-475.

Rodrik, D. 2015. Economics rules: The rights and wrongs of the dismal science. Norton, New York.

Rogers, A.D. and Laffoley, D. d'A. 2011. International earth system expert workshop on ocean stresses and impacts. Summary report. International Programme on the State of the Ocean, Oxford, pp. 1–21. http://www.stateoftheocean.org/pdfs/1906_IPSO-LONG.pdf.

Rohrer, D. 2002. Misconceptions about incline speed for nonlinear slopes. Journal of Experimental Psychology: Human Perception and Performance, vol. 28, no. 4, pp. 963–973.

Root, T., Price, J. T., Hall, K. R., Schneider, S. H., Rosenzweig, C. et al. 2003. Fingerprint of global warming on wild animals and plants. Nature, vol. 4241, pp. 57–60.

Roper, L. 2004. Witch crazes: Terror and fantasy in Baroque Germany. Yale University Press, New Haven.

Rose, D. J. 1974. Energy policy in the U. S. Scientific American, vol. 230, no. 1, January, pp. 20–29.

Rosenberg, D. M., Berkes, F., Bodaly, R. A., Hecky, R. E., Kelly, C. A. et al. 1997. Large-scale impacts of hydroelectric development. Environmental Reviews, vol. 5, pp. 27–54.

Rosenkranz, P., Aumeier, P. and Ziegelmann, B. 2010. Biology and control of Varroa destructor. Journal of Invertebrate Pathology, vol. 103, pp. S96–S119.

Ross, L. and Anderson, C. A. 1982. Shortcomings in the attribution process: On the origin and maintenance of erroneous social assessments. In Kahneman, D., Slovic, P. and Tversky, A. (editors) 1982, pp. 129–152.

Ross, M. and Sicoly, F. 1979. Egocentric biases in availability and attribution. Journal of Personality and Social Psychology, vol. 37, pp. 322–337.

Rousmaniere, P. and Raj, N. 2007. Ship breaking in the developing world: Problems and prospects. International Journal of Occupational and Environmental Health, vol. 13, no. 4, pp. 359–368.

Rowe, S. and Rose, G. A. 2017. Don't derail cod's comeback in Canada. Nature, vol. 545, p. 412.

The Royal Society. 2009. Geoengineering the climate: Science, governance and uncertainty. Royal Society, report 10/09 RS1636, pp. 1–82.

Rudgley, R. 1999. Lost civilizations of the stone age. Arrow, London.

Runnels, C. N. 1995. Environmental degradation in ancient Greece. Scientific American, vol. 273, no. 3, pp. 72–75.

Ruppert, J. L. W., Travers, M. J., Smith, L. L., Fortin, M-J. and Meekan, M. G. 2013. Caught in the middle: Combined impacts of shark removal and coral loss on the fish communities of coral reefs. PLoS ONE, vol. 8, no. 9, e74648, pp. 1–9.

Rushe, D. 2011. US crackdown on insider deals. Guardian Weekly, 20 May, p. 18.

Ruskin, G. 1999. Why they whine: How corporations prey on our children. Mothering, November–December, pp. 41–50.

Russell, G. J., Diamond, J. M., Pimm, S. L. and Reed, T. M. 1995. A century or turnover: Community dynamics at three timescales. Journal of Animal Ecology, vol. 64, pp. 628–641.

Rutledge, D. 2011. Estimating long-term world coal production with logit and probit transforms. International Journal of Coal Geology, vol. 85, pp. 23–33.

Sacks, O. 1986. The man who mistook his wife for a hat. Picador, Sydney.

Sacks, O. 1996. The island of the colour-blind. Picador, Sydney.

Sanderman, J., Heng, T. and Fiske, G. J. 2017. Soil carbon debt of 12,000 years of human land use. Proceedings of the National Academy of Sciences, vol. 114, no. 36, pp. 9575–9580.

Sass, J. B. and Colangelo, A. 2006. European Union bans atrazine, while the United States negotiates continued use. International Journal of Occupational and Environmental Health, vol. 12, no. 3, pp. 260–267.

Scarborough, V. L. 1996. Reservoirs and watersheds in the central Maya lowlands. In Fedick, S. L. (editor). The managed mosaic: Ancient Maya agriculture and resource use. University of Utah Press, Salt Lake City. pp. 304–314.

Scarborough, V. L. 2007. The rise and fall of the ancient Maya: A case study in political ecology. In Costanza, R., Graumlich, L. and Steffen, W. (editors). Report of the 96th Dahlem workshop on an integrated history and future of people on earth. MIT Press, Cambridge, MA. pp. 51–59.

Scavia, D. and Donnelly, K. A. 2007. Reassessing hypoxia forecasts for the Gulf of Mexico. Environmental Science and Technology, vol. 41, no. 23, pp. 8111–8117.

Scheffer, M., Carpenter, S., Foley, J. A., Folke, C. and Walker, B. 2001. Catastrophic shifts in ecosystems. Nature, vol. 413, pp. 591–596.

Scheiber, N. 1970. At the borderland of law and economic history: The contribution of William Hurst. American Historical Review, vol. 75, no. 3, pp. 744–756.

Schindler, D. W. and Donahue, W. F. 2006. An impending water crisis in Canada's western prairie provinces. Proceedings of the National Academy of Sciences, vol. 103, no. 19, pp. 7210–7216.

Schmidt, C., Krauth, T. and Wagner, S. 2017. Export of Plastic Debris by Rivers into the Sea. Environmental Science & Technology, vol. 51, no. 21, pp. 12246–12253.

Schmitz, D. C., Schardt, J. D., Leslie, A. J., Dray, F. A., Osborne, J. A. et al. 1993. The ecological impact and management history of three invasive alien aquatic plant species in Florida. In McKnight, B. N. (editor). Biological pollution: The control and impact of invasive exotic species. Indiana Academy of Science, Indianapolis. pp. 173–194.

Schmookler, A. B. 1993. The illusion of choice: How the market economy shapes our destiny. State University of New York Press, Albany.

Schneider, W. and Bjorkland, D. F. 1998. Memory. In Kuhn, D and Siegler, R. S. (editors). Handbook of child psychology. 5th ed. Vol. 2, Cognition, perception and language. Wiley, New York. pp. 467–522.

Schnepf, R. 2003. Iraq's agriculture: Background and status. CRS report for Congress, order code RS21516, 13 May, 2003, pp. 1–6.

Schodde, R. 2010. The key drivers behind resource growth: An analysis of the copper industry over the last 100 years. MEMS Conference Mineral and Metal Markets over the Long Term. Joint Program with the SME annual meeting in Phoenix, 3 March, 2010.

Schott, G., Pachl, H., Limbach, U., Gundert-Remy, U., Lieb, K. et al. 2010. The financing of drug trials by pharmaceutical companies and its consequences: Part 2: A Qualitative, systematic review of the literature on possible influences on authorship, access to trial data, and trial registration and publication. Deutsches Ärzteblatt International, vol. 107, no. 17, pp. 295–301.

Schutter, O. 2009. Large-scale land acquisitions and leases: A set of core principles and measures to address the human rights challenge. UN Special Rapporteur on the right to food. 11 June, 2009, pp. 1–15.

Schuur, E. A. G., McGuire, A. D., Schädel, C., Grosse, G., Harden, J. W. et al. 2015. Climate change and the permafrost carbon feedback. Nature, vol. 520, pp. 171–179.

Schwartz, B. 2000. The tyranny of freedom. American Psychologist, vol. 55, no. 1, pp. 79–88.

Schwartz, J. D., Pallin, M. J., Michener, R. H., Mbabazi, D. and Kaufman, L. 2006. Effects of Nile perch, Lates Niloticus, on functional and specific fish diversity in Uganda's Lake Kyoga system. African Journal of Ecology, vol. 44, no. 2, pp. 145–156.

The Scotsman. 2005. Two factories raided after inquiry into illegal fish landings. The Scotsman, 28 September, http://www.scotsman.com/news/scotland/top-stories/two-factories-raided-after-inquiry-into-illegal-fish-landings-1-1098071.

Scott, J. P. 1997. Genetic analysis of social behaviour. In Siegel, N. I., Weisfeld, G. E. and Weisfeld, C. C. (editors). Uniting psychology and biology: Integrative perspectives on human development. American Psychological Association, Washington. pp. 131–144.

Seager, R. 2006. The source of Europe's mild climate. American Scientist, vol. 94, July–August, pp. 334–341.

Searle, J. 1995. The mystery of consciousness. Part 1: Review of The astonishing hypothesis: The scientific search for the soul, by Crick, F.; Shadows of the mind: A search for the missing science of consciousness, by Penrose, R.; The remembered present: A biological theory of consciousness, by Edelman, G. M.. Part 2: Review of Bright air, brilliant fire: On the matter of the mind, by Edelman, G. M.; Consciousness explained, by Dennett, D. C.; The strange, familiar, and forgotten: An astronomy of consciousness, by Rosenfield, I. New York Times, The New York Review of Books, part 1, 2 November, pp. 60–66; part 2, 16 November, pp. 54–61.

Seipp, C. 2001. Kids: The new captive market. Child Magazine, September, pp. 91–92 and 150–152.

Seligman, M. E. P. and Csikszentmihalyi, M. 2000. Positive psychology: An introduction. American Psychologist, vol. 55, no. 1, pp. 5–14.

Sell, B., Murphy, D. and Hall, C. A. S. 2011. Energy return on energy invested for tight gas wells in the Appalachian Basin, United States of America. Sustainability, vol. 3, no. 10, pp. 1986–2008.

Semeena, V. S. and Lammel, G. 2005. The significance of the grasshopper-effect on the atmospheric distribution of persistent organic substances. Geophysical Research Letters, vol. 32, L07804, doi:10.1029/2004GL022229,2005.

Sereny, G. 1995. Albert Speer: His battle with truth. Knopf, New York.

Shakhova, N., Semiletov, I., Salyuk, A., Yusupov, V., Kosmach, D. et al. 2010. Extensive methane venting to the atmosphere from sediments of the East Siberian shelf. Science, vol. 327, no. 5970, pp. 1246–1250.

Shapin, S. 1994. A social history of truth: Civility and science in seventeenth century England. University of Chicago Press, Chicago.

Shappell, S. S. and Wiegmann, D. A. 2000. The human factors analysis and classification system-HFACS. Department of Transport, Office of Aviation Medicine, DOT/FAA/AM-00/7,

February, 2000, pp. 1–19.
https://www.nifc.gov/fireInfo/fireInfo_documents/humanfactors
_classAnly.pdf

Shaw, G. 1995. The arctic haze phenomenon. Bulletin of the American Meteorological Society, vol. 76, no. 12, pp. 2403–2413.

Shehabi, A., Smith, S. J., Horner, N., Azevedo, I., Brown, R. et al. 2016. United States Data Center energy usage report. Lawrence Berkeley National Laboratory, Berkeley, California. LBNL-1005775, pp. 1–66.

Sheldon, K. M. and Mc Gregor, H. 2000. Extrinsic value orientation and the tragedy of the commons. Journal of Personality, vol. 68, pp. 383–411.

Shepherd, A., Ivins, E. R., Gerou, A., Barletta, V. R., Bentley, M. J. et al. 2012. A reconciled estimate of ice-sheet mass balance. Science, vol. 338, no. 611, pp. 1183–1189.

Sherwood, D. 2008. Mine disaster at Miramar: A story foretold. Tico Times, 18 January, pp. 1, 6, and 8. http://www.ticotimes.net/ and http://www.aida-americas.org/templates/aida/uploads/docs/Mine_disaster.pdf.

Shipman, P. 2002. Hunting the first hominid. American Scientist, vol. 90, January, pp. 25–27.

Siegel, J. 2005. Clues to the functions of mammalian sleep. Nature, vol. 437, pp. 1264–1271.

Siegenthaler, U., Stocker, T. F., Monnin, E., Lüthi, D., Schwander, J. et al. 2005. Stable carbon cycle-climate relationship during the late pleistocene. Science, vol. 310, no. 5752, pp. 1313–1317.

Silverstein, K. 1998. The radioactive boy scout. Harper's Magazine, November, pp. 59–72.

Simenstad, C. A., Estes, J. A. and Kenyon, K. W. 1978. Aleuts, sea otters and alternate stable-state communities. Science, vol. 200, no. 4340, pp. 403–411.

Simon, H. 1990. A mechanism for social selection and successful altruism. Science vol. 250, pp. 1665–1668.

Simpson, C. 1999. Colour: In the mind's eye. CBC Radio, Ideas, ID 9943, 13–14 September, 1999, pp. 1–27.

Simpson, R. D., Toman, M. A. and Ayres, R. U. 2005. Introduction : The "new scarcity". In Simpson, R. D., Toman, M. A. and Ayres, R. U. (editors) 2005, pp. 1–32.

Simpson, R. D., Toman, M. A. and Ayres, R. U. (editors). 2005. Scarcity and growth revisited: Natural resources and the environment in the new millennium. RFF Press, Washington, DC.

Simpson, S. 2001a. Fishy business. Scientific American, July, pp. 82–89.

Simpson, S. 2001b. Shrinking the dead zone. Scientific American, July, pp. 19–21.

Sims, G. T. 2007. Shell settles with Europe on overstated oil reserves. New York Times, 12 April. http://www.nytimes.com/2007/04/12/.

Sinclair, A. R. E. and Krebs, C. J. 2001. Trophic interactions, community, organization and Kluane ecosystems. In Krebs, C. J., Boutin, S. and Boonstra, R. (editors). Ecosystem dynamics of the boreal forest. Oxford University Press, Oxford. Chapter 3, pp. 25–48.

Singer, P. 2001. Writings on an ethical life. Harper Collins, New York.

Singh, M. 2008. Prime Minister Manmohan Singh on India's gender imbalance. Population and Development Review, vol. 34, no. 2, pp. 387–389.

Singh, R. P., Hodson, D. P., Huerta-Espino, J., Jin, Y., Bhavani, S. et al. 2011. The emergence of Ug99 races of the stem rust fungus is a threat to world wheat production. Annual Review of Phytopathology, vol. 49, no. 1, pp. 465–481.

SinoskiK. 2012. Cache Creek hopes Metro trash keeps on coming. Vancouver Sun, 2 October, 2012. https://www.pressreader.com/canada/vancouver-sun/20121002/281840050885936

Skinner, B. J. 1976. A second iron age ahead? American scientist, vol. 64, pp. 258–269.

Slade, G. 2006. Made to break: Technology and obsolescence in America. Harvard University Press, Cambridge.

Slovic, P. 1990. The legitimacy of public perceptions of risk. Journal of Pesticide Reform, spring, pp. 13–15.

Slovic, P. 1993. Perceived risk, trust, and democracy. Risk Analysis, vol. 13, No 6, pp. 675–682.

Slovic, P., Finucane, M., Peters, E. and MacGregor, D. G. 2002. The affect heuristic. In Gilovich, T., Griffin, D. and Kahneman, D. (editors). Intuitive judgment: Heuristics and biases. Cambridge University Press, New York. pp. 397–420.

Slovic, P. and Fischhoff, B. 1977. On the psychology of experimental surprises. Journal of Experimental Psychology: Human perception and performance, vol. 3, pp. 544–551.

Slovic, P., Fischhoff, B. and Lichtenstein, S. 1982. Facts versus fears: Understanding perceived risk. In Kahneman, D., Slovic, P. and Tversky, A. (editors) 1982, pp. 463–489.

Smith, A. 1776. The wealth of nations. Book IV. Chap 5. Paragraph 82, and Chapter 9. Paragraph 28.

Smith, C. 2008. Invasive plants of North Carolina. North Carolina Department of Transport, pp. 1–191.

Smith, D. M. 1998. Recent increase in the length of the melt season of perennial Arctic sea ice. Geophysical research letters, vol. 25, no. 5, pp. 655–658.

Smith, D. M., Cusak, S., Colman, A. W., Folland, C. K., Harris, G. R. et al. 2007. Improved surface temperature prediction for the coming decade from a global climate model. Science, vol. 317, pp. 796–799.

Smith, D., Schlaepfer, P., Major, K., Dyble, M., Page, A. E. et al. 2017. Cooperation and the evolution of hunter-gatherer storytelling. Nature Communications, vol. 8, Article number: 1853, pp. 1–9.

Smith, G. 2012. Why I am leaving Goldman Sachs. New York Times, 14 March. http://www.nytimes.com.

Smith, G. S. 2005. Human color vision and the unsaturated blue color of the daytime sky. American Journal of Physics, vol. 73, no. 7, pp. 590–597.

Smith, K. A. 2011. Edward Jenner and the small pox vaccine. Frontiers In Immunology, vol. 2, Article 21, pp. 1–6.

Smol, J. P., Wolfe, A. P., Birks, H. J. B., Douglas, M. S. V., Jones, V. J. et al. 2005. Climate-driven regime shifts in the biological communities of arctic lakes. Proceedings of the National Academy of Sciences, vol. 102, no. 12, pp. 4397–4402.

Smulders, S. 2005. Endogenous technological change, natural resources, and growth. Simpson, R. D., Toman, M. A. and

Ayres, R. U. (editors). Scarcity and growth revisited: Natural resources and the environment in the new millennium. RFF Press, Washington, DC.

Snow, A. 2009. Unwanted transgenes re-discovered in Oaxacan maize. Molecular Ecology, vol. 18, pp. 569–571.

Soares, A., Guieysse, B., Jefferson, B., Cartmell, E. and Lester, J. N. 2008. Nonylphenol in the environment: A critical review on occurrence, fate, toxicity and treatment in wastewaters. Environment International, vol. 34, no. 7, pp. 1033–1049.

Socolow, R. 1999. Nitrogen management and the future of food: Lessons from the management of energy and carbon. Proceedings of the National Academy of Sciences, vol. 96, pp. 6001–6008.

Solanki, S. K., Usoskin, I. G., Kromer, B., Schilssler, M and Beer, J. 2004. Unusual activity of the sun during recent decades compared to the previous 11,000 years. Nature, vol. 431, pp. 1084–1987.

Solomon, L. 2008. The Deniers. Richard Vigilante Books.

Sorrell, S., Speirs, J., Bentley, R., Brandt, B. and Miller, R. 2009. Global oil depletion: An assessment of the evidence for a near-term peak in global oil production. UK Energy Research Centre, pp. 1–45. http://www.ukerc.ac.uk/support/tiki-download_file.php?fileId=283.

Souder, W. 2006. It's not easy being green. Harper's magazine, August, pp. 59–66.

Spahni, R., Chappellaz, J., Stocker, T. F., Loulergue, L., Hausammann, G. et al. 2005. Atmospheric methane and nitrous oxide of the late Pleistocene from Antarctic ice cores. Science, vol. 310, no. 5752, pp. 1317–1321.

Sparling, B. 2001. Ozone depletion, history and politics. NASA. http://www.nas.nasa.gov/About/Education/Ozone/history.html.

Spatari, S., Bertram, M., Gordon, R. B., Henderson, K. and Graedel, T. E. 2005. Twentieth century copper stocks and flows in North America: A dynamic analysis. Ecological Economics, vol. 54, pp. 37–51.

Spillius, J. 1957. Natural disaster and political crises in Polynesian society: An exploration of operational research. Human Relations, vol. 10, no. 1, pp. 3–27.

Stahl, W. R. 1965. Organ weights in primates and in other mammals. Science, vol. 150, pp. 1039–1042.

Stainforth, D. A., Allen, M. R., Tredger, E. R. and Smith, L. A. 2007. Confidence, uncertainty and decision-support relevance in climate predictions. Philosophical Transactions of the Royal Society, Series A, vol. 365, pp. 2145–2161.

Stanford, J. D., Rohling, E. J., Hunter, S. E., Roberts, A. P., Rasmussen, S. O. et al. 2006. Timing of meltwater pulse 1a and climate responses to meltwater injections. Paleoceanography, vol. 21, PA4103, 9 pages. doi:10.1029/2006PA001340.

Stanwell-Fletcher, T. C. 1978. Driftwood Valley. Ballantine Books, New York.

Statoil. 2012. Advertisement promoting Athabasca tar sads oil extraction. Calgary airport and http://www.neversatisfied.statoil.com/article/902. October.

Staudenmaier, J. M. 1996. Denying the holy dark: The enlightenment ideal and the European mystical tradition. In Marx, L. and Mazlish, B. (editor) 1996, pp. 175–200.

Steadman, D. W. 1995. Prehistoric extinctions of Pacific island birds: Biodiversity meets zooarchaeology. Science, vol. 267, pp. 1123–1131.

Stein, B. A. and Flack, S. R. (editors). 1996. America's least wanted: Alien species invasions of US ecosystems. The Nature Conservancy, Arlington, Virginia. pp. 1–36.

Stein, R. 2003. Alert as bird flu infects boy. Guardian Weekly, 18 December, p. 27.

Steinacher, M., Joos, F., Frolicher, T. L., Plattner, G.-K. and Doney, S. C. 2009. Imminent ocean acidification in the Arctic projected with the NCAR global coupled carbon cycle-climate model. Biogeosciences, vol. 6, pp. 515–533.

Steinbruner, J. D. and Harris, E. D. 2003. When science breeds nightmares. International Herald Tribune, 3 December, p. 8.

Steinhauer, N., Rennich, K., Caron, D. M., Delaplane, K., Range, J. et al. 2016. Colony loss 20152016: Preliminary results. Beeinformed, May 2016. https://beeinformed.org/results/colony-loss-2015-2016-preliminary-results/.

Sterman, J. D, Siege, L. and Rooney-Varga, J. N. 2018. Does replacing coal with wood lower CO2 emissions? Dynamic lifecycle analysis of wood bioenergy. Environmental Research Letters, vol. 13, 15007, pp. 1–10.

Stern, P. C. 1993. A second environmental science: Human-environmental interactions. Science, vol. 260, pp. 1897–1899.

Stevens, G. C. and Fox, J. F. 1991. The causes of treeline. Annual Review of Ecological Systems, vol. 22, pp. 177–191.

Stokstad, E. 2004. Defrosting the carbon freezer of the north. Science, vol. 304, pp. 1618–1620.

Stokstad, E. 2007a. Deadly wheat fungus threatens world's breadbaskets. Science, vol. 315, no. 5820, pp. 1786–1787.

Stokstad, E. 2007b. The case of the empty hives. Science, vol. 316, pp. 970–972.

Stolarski, R. S. 2001. History of the study of atmospheric ozone. Ozone: Science and Engineering, vol. 23, no. 6, pp. 421–428.

Stone, R. 1999. Coming to grips with the Aral sea's grim legacy. Science, vol. 284, no. 5411, pp. 30–33.

Strahan, D. 2008. The great coal hole. New Scientist Magazine, no. 2639, 17 January.

Strahler, A. and Strahler, A. 1997. Physical Geography: Science and systems of the human environment. John Wiley and Sons, New York.

Strong, D. 1995. Crazy mountain: Learning from wilderness to weigh technology. State University of New York Press, Albany.

Struelens, M. J. 1998. The epidemiology of antimicrobial resistance in hospital acquired infections: Problems and possible solutions. British Medical Journal, vol. 317, pp. 652–654.

Stuart, S. N., Chanson, J. S., Cox, N. A., Young, B. E., Rodrigues, A. S. et al. 2004. Status and trends of amphibian declines and extinctions worldwide. Science, vol. 306, no. 5702, pp. 1783–1786.

Sturm, M., Racine, C. and Tape, K. 2001. Increasing shrub abundance in the Arctic. Nature, vol. 411, pp. 546–547.

Sullivan, B. J., Reid, T. A. and Bugoni, L. 2006. Seabird mortality on factory trawlers in the Falkland Islands and beyond. Biological Conservation, vol. 131, pp. 495–504.

Sverdlik, Y. 2016. Here's how much energy all US data centers consume. Data Centre Knowledge, 27 June, 2016. http://www.datacenterknowledge.com/archives/2016/06/27/heres-how-much-energy-all-us-data-centers-consume/.

Sveriges Riksbank. 2014. The Riksbank's Prize in Economic Sciences. http://www.riksbank.se/en/The-Riksbank/Economics-prize/. Accessed 14 July, 2014.

Swan, S. H. 2008. Environmental phthalate exposure in relation to reproductive outcomes and other health endpoints in humans. Environmental Research, vol. 108, pp. 177–184.

Swann, W. B. 1984. Quest for accuracy in person perception. A matter of pragmatics. Psychological Review, vol. 91, no. 4, pp. 457–477.

Swihart, R. K., Slade, N. A. and Bergstrom, B. J. 1988. Relating body size to the rate of home range use in mammals. Ecology, vol. 69, no. 2, pp. 393–399.

Syvitski, J., Vörösmarty, C., Kettner, A. and Green, P. 2005. Impact of humans on the flux of terrestrial sediment to the global coastal ocean. Science, vol. 308, pp. 376–380.

Taagepera, R. 1979. Size and duration of empires: Growth-decline curves, 600 B.C. to 600 A.D. Social Science History, vol. 3, no. 3-4, pp. 115–138.

Taibbi, M. 2011. Why isn't Wall Street in jail. Rollingstone, 3 March. www.rollingstone.com/politics/news/news/why-isnt-wall-street-in-jail-2011216.

Tainter, J. A. 1988. The collapse of complex societies. Cambridge University Press, Cambridge.

Tainter, J. A. 1996. Complexity, problem solving, and sustainable societies. In Costanza, R., Segura, O. and Martinez-Alier, J. (editors). Getting down to earth: Practical applications of ecological economics. Island Press, Washington, DC. pp. 61–76.

Tainter, J. A. 2000. Global change, history and sustainability. In McIntosh, R. J., Tainter, J. A. and McIntosh, S. K. (editors) 2000, pp. 331–356.

Tainter, J. A., Allen, T. F., Little, A. and Hoekstra, T. W. 2003. Resource transitions and energy gain: Contexts of organization. Conservation Ecology, vol. 7, no. 3, article 4. Online at http:///www.consecol.org/vol7/iss3/art4.

Taitz, L. 2002. Risky business. Sunday Times (South Africa), Lifestyles, 3 February, p. 14.

Talaga, T. 2010. Xstrata holds firm on decision to cut Timmins jobs. The Star, 16 April. http://www.thestar.com/business/article/796759--xstrata-holds-firm-on-decision-to-cut-timmins-jobs.

Tapparo, A., Marton, D., Giorio, C., Zanella, A., Soldà, L. et al. 2012. Assessment of the environmental exposure of honeybees to particulate matter containing neonicotinoid insecticides coming from corn coated seeds. Environmental Science and Technology, vol. 46, pp. 2592–2599.

Tattersall, I. 1998. Becoming human: Evolution and human uniqueness. Harcourt, Brace and Company, San Diego.

Taubes, G. 1998. Evolving a conscious machine. Discover, June, p. 73.

Taylor, K. C., Lamorey, G. W., Doyle, G. A., Alley, R. B., Grootes, P. M. et al. 1993. The "flickering switch" of late Pleistocene climate change. Nature, vol. 361, pp. 432–436.

Taylor, K. C., Mayewski, P. A., Alley, R. B., Brook, E. J., Gow, A. J. et al. 1997. The Holocene/Younger Dryas transition recorded at Summit, Greenland. Science, vol. 278, pp. 825–827.

Taylor, S. E. 1982. The availability bias in social perception and interaction. In Kahneman, D., Slovic, P. and Tversky, A. (editors) 1982, pp. 190–200.

Taylor, S. E. 1989. Positive illusions: Creative self deception and the healthy mind. Basic Books, New York.

Teller, J. T. and Leverington, D. W. 2004. Glacial lake Agassiz: A 5000 year history of change and its relationship to the O18 record in Greenland. Geological Society of America bulletin, vol. 116, no. 5-6, pp. 729–742.

Teodoru, C., Dimopoulos, A. and Wehrli, B. 2006a. Biogenic silica accumulation in the sediments of Iron Gate I Reservoir on the Danube River. Aquatic Sciences, vol. 68, pp. 469–481.

Teodoru, C., McGinnes, D. F., Wuest, A. and Wehrli, B. 2006b. Nutrient retention in the Danube's Iron Gate Reservoir. EOS, vol. 87, no. 38, pp. 385–400.

Tetlock, P. E. 2003. Thinking the unthinkable: Sacred values and taboo cognitions. Trends in Cognitive Sciences, vol. 7, no. 7, pp. 320–324.

Tetlock, P. E. 2006. Expert political judgement: How good is it? How can we know? Princeton University Press, Princeton.

Teuten, E. L., Saquing, J. M., Knappe, D. R. U., Barlaz, M. A., Jonsson, S. et al. 2009. Transport and release of chemicals from plastics to the environment and to wildlife. Philosophical Transactions of the Royal Society, Series B, vol. 364, pp. 2027–2045.

Thenkabail, P. S., Biradar, C. M., Turral, H., Noojipady, P., Li, Y. J. et al. 2006. An irrigated area map of the world (1999) derived from remote sensing. International Water Management Institute, Research report 105, pp. 1–67. http://www.imi.cgiar.org/publications/IWMI_Research_Reports/pdf/pub105.pdf.

Thomas, C. D., Cameron, A., Green, R. E., Bakkenes, M., Beaumont, L. J. et al. 2004. Extinction risk from climate change. Nature, vol. 427, pp. 145–148.

Thomas, D. H. 1991. Archaeaology: Down to earth. Harcourt Brace Janovich College Publishers, Fort Worth. Chapter 1: What is Americhanist archaeology? Chapter 2: The basics of anthropological archaeology.

Thomas, L. 1978. Hubris in science? Science, vol. 200, pp. 1459–1462.

Thompson, L., Mosley-Thompson, E., Davis, M. E., Henderson, K. A., Brecher, H. H. et al. 2002. Kilimanjaro ice core records: Evidence of Holocene climatic change in tropical Africa. Science, vol. 298, pp. 589–593.

Thompson, R. C., Moore, C. J., Vom Saal, F. S. and Swan, S. H. 2009. Plastics, the environment and human health: Current consensus and future trends. Philosophical Transactions of the Royal Society, Series B, vol. 364, pp. 2153–2166.

Thompson, R. C., Olsen, Y., Mitchell, R. P., Davis, A., Rowland, S. J. et al. 2004. Lost at sea: Where is all the plastic? Science, vol. 304, p. 838.

Thompson, W. R. 2004. Complexity, diminishing marginal returns and serial Mesopotamian fragmentation. Journal of World-System Research, vol. X, no. 3, Fall, pp. 613–652.

Tilling, R. L., Ridout, A., Shepherd, A. and Wingham, D. J. 2015. Increased Arctic sea ice volume after anomalously low melting in 2013. Nature Geoscience, vol. 8, pp. 643–646.

Tilman, D. 1999. The ecological consequences of changes in biodiversity: A search for general principles. Ecology, vol. 80, pp. 1455–1474.

Tilman, D., Cassman, K. G., Matson, P. A., Naylor, R. N. and Polasky, S. 2002. Agricultural sustainability and intensive production practices. Nature, vol. 418, pp. 671–676.

Tilman, D. and Polasky, S. 2005. Ecosystem goods and services and their limits: The roles of biological diversity and management practices. In Simpson, R. D., Toman, M. A. and Ayres, R. U. (editors) 2005, pp. 78–97.

Tivy, J. 1999. Biogeography: A study of plants in the ecosphere. 3rd ed. Longman, Harlow. p. 117.

Togola, T. 2000. Memories, abstractions, and conceptualization of ecological crisis in the Mande world. In McIntosh, R. J., Tainter, J. A. and McIntosh, S. K. (editors) 2000, pp. 181–192.

Travis, J. 2015. Making the cut: CRISPR genome-editing technology shows its power. Science, vol. 350, no. 6267, pp. 1456–1457.

Trigger, B. G. 1989. A history of archaeological thought. Cambridge University Press, Cambridge.

Trimble, S. W. and Crosson, P. 2000. U.S. soil erosion rates—Myth and reality. Science, vol. 289, pp. 248–250.

Tunved, P., Hansson, H-C., Kerminen, V-M., Ström, J., Dal Maso, M. et al. 2006. High natural aerosol loading over Boreal forests. Science, vol. 312, pp. 261–263.

Turner, G. 2008. A comparison of the limits to growth with thirty years of reality. Commonwealth Scientific and Industrial Research Organization (Australia). Socio-economics and the environment in discussion, CSIRO working paper series 2008-09, pp. 1–49.

Turnidge, J. 1998. What can be done about resistance to antibiotics. British Medical Journal, vol. 317, pp. 645–647.

Turunen, C. and Turunen, J. 2003. Development history and carbon accumulation of a slope bog in oceanic British Columbia, Canada. Holocene, vol. 13, no. 2, pp. 225–238.

Tversky, A. and Kahneman, D. 1981. The framing of decisions and the psychology of choice. Science, vol. 211, pp. 1453–1458.

Tversky, A. and Kahneman, D. 1982a. Judgements of and by representativeness. In Kahneman, D., Slovic, P. and Tversky, A. (editors) 1982, pp. 84–98.

Tversky, A. and Kahneman, D. 1982b. Causal schemas in judgements under uncertainty. In Kahneman, D., Slovic, P. and Tversky, A. (editors) 1982, pp. 117–128.

Tversky, A. and Kahneman, D. 1982c. Judgement under uncertainty: Heuristics and biases. In Kahneman, D., Slovic, P. and Tversky, A. (editors) 1982, pp. 3–20.

Twigg, J. 2002. What has happened to Russian society? In Kuchins, A. (editor). Russia after the fall. Carnegie Endowment for International Peace, Washington, DC. Chapter 8, pp. 147–162.

Twiggs, L. 2001. Smart woman, dumb sex. Femina, December, pp. 92–96.

Twitchell, J. B. 2000. Twenty ads that changed the world. Crown, New York. De Beers: A good campaign is forever, pp. 88–101; The Marlboro man: The perfect campaign, pp. 126–135.

Udvardy, M. D. 1977. The Audubon Society field guide to North American birds. Knopf, New York.

UN Water. 2007. Coping with water scarcity: Challenge of the twenty-first century. UN Water, Food and Agriculture Organization, pp. 1–29.

UNEP. 1999. Global environment outlook 2000. Earthscan, London, pp. 1–398.

UNEP. 2002. International effort to unearth the secrets of the soils. United Nations Environment Program, news release, 28 November, 2002. http://www.unep.org/Documents.Multilingual/Default.Print.asp?DocumentID=270&ArticleID=3180.

UNEP. 2007. Water. United Nations Environment Programme. State of the world report, Section B: State and trends of the environment: 1987–2007, Chapter 4, pp. 115–156.

United Nations. 1987. Towards sustainable development: Our common future: Report of the World Commission on Environment and Development. Published as Annex C to the General Assembly document A/42/427. Chapter 2, pp. 1–21. http://www.un-documents.net/our-common-future.pdf.

United Nations. 2003. Water for people, water for life. United Nations world water development report. Executive summary, WWDR_ex_summary_en, pp. 1–36.

United Nations. 2015. Transforming our world: The 2030 agenda for sustainable development. United Nations General Assembly, 21 October, 2015, A/RES/70/1 7th session, agenda items 15 and 116, pp. 1–35.

United Nations. 2016. Framework convention on climate change. Status of Ratification. http://unfccc.int/paris_agreement/items/9444.php.

US Census Bureau. 2008. Total midyear population for the world: 1950–2050. US Census Bureau, International database. http://www.census.gov/ipc/www/idb/worldpop.html.

US Census Bureau. 2016. Total midyear population for the world: 1950–2050. US Census Bureau, International database. http://www.census.gov/population/international/data/idb/worldpoptotal.php.

US Geological Survey. 1996–2008. Mineral commodity summaries for these years. US Geological Survey. http://minerals.usgs.gov/minerals/pubs/commodity/copper.

US Geological Survey. 2001. Mineral commodity summaries 2001. US Geological Survey, pp. 54–55. http://minerals.usgs.gov/minerals/pubs/commodity/copper/240302.pdf.

US Geological Survey. 2006. Mineral commodity summaries 2006. US Geological Survey, pp. 52–53. http://minerals.usgs.gov/minerals/pubs/commodity/copper/240302.pdf.

US Geological Survey. 2012. Copper statistics. In Historical statistics for mineral and material commodities in the United States, Kelly, T.D. and Matos, G.R. (compilers), US Geological Survey Data Series 140. http://pubs.usgs.gov/ds/2005/140/.

US Geological Survey. 2013. Mineral commodity summaries 2013. US Geological Survey, pp. 1–198.

http://minerals.usgs.gov/minerals/pubs/commodity/copper/mcs-2013-coppe.pdf.

USCPSOTF. 2004. Final report on the August 14 2003 blackout in the United States and Canada: Causes and recommendations. U.S.-Canada Power System Outage Task Force. April, 2014. pp. 1–228. https://reports.energy.gov/BlackoutFinal-Web.pdf.

USGS/USNavy. 1948. Col photo collection. U.S. Geological Survey. 11 August.

Van Andel, H. T., Runnels, C. T. and Pope, K. O. 1986. Five thousand years of land use and abuse in southern Argolid. Hesperia, vol. 55, pp. 103–128.

van Dam, J. A., Aziz, H. A., Sierra, M. A. A., Hilgen, F. J., van den Hoek Ostende, L. W. et al. 2006. Long-period astronomical forcing of mammal turnover. Nature, vol. 443, pp. 687–691.

Van de Leeuw, S. E. and The Archaeomedes research team. 2000. Land degradation as a socionatural process. In McIntosh, R. J., Tainter, J. A. and McIntosh, S. K. (editors) 2000, pp. 357–383.

van de Mieroop, M. 2004. A history of the ancient near east ca. 3000–323 BC. 2nd ed. Blackwell, Malden.

van den Bergh, C. J. M., Ferrer-i-Carbonell, A. and Munda, G. 2000. Alternative models of individual behaviour and implications for environmental policy. Ecological Economics, vol. 32, no. 1, pp. 43–61.

van der Sluijs, J. P., Amaral-Rogers, V., Belzunces, L. P., Bijleveld van Lexmond, M. F. I. J., Bonmatin, J-M. et al. 2014. Conclusions of the Worldwide Integrated Assessment on the risks of neonicotinoids and fipronil to biodiversity and ecosystem functioning. Environmental Science and Pollution Research, vol. 21, pp. 1–7.

van der Sluis, P. M. and Hulsbos, W. C. 1977. General introduction. In Dielman. P. J. (editor). Reclamation of salt affected soils in Iraq: Soil hydrology and agricultural studies. International Institute for Land Reclamation and Improvement (ILRI), The Netherlands. Chapter 1, pp. 1–26.

Van Helden, P. D., Victor, T. and Warren, R. M. 2006. The "Source" of Drug-Resistant TB Outbreaks. Science, vol. 314, no. 5798, p. 419.

van Mantgem, P. J., Stephenson, N. L., Byrne, J. C., Daniels, L. D., Franklin, J. F. et al. 2009. Widespread increase of tree mortality rates in the western United States. Science, vol. 323, no. 5913, pp. 521–524.

Van Meter, K. J., Basu, N. B., Veenstra, J. J. and Burras, C. L. 2016. The nitrogen legacy: emerging evidence of nitrogen accumulation in anthropogenic landscapes. Environmental Research Letters, vol. 11, Contribution 035014.

van Schaik, C. 2006. Why are some animals smart? Scientific American, April, pp. 64–71.

Vandenberg , L. N., Colborn, T., Hayes, T. B., Heindel, J. J., Jacobs, D. D. et al. 2012. Hormones and endocrine-disrupting chemicals: Low-dose effects and nonmonotonic dose responses. Endocrine Reviews, vol. 33, no. 3, pp. 378–455.

Vander, A., Sherman, J. and Luciano, D. 1998. Human Physiology: The mechanism of body function. McGraw-Hill, Boston. 7th ed. Chapter 8, Section D.

Vanderklippe, N. 2009. Researchers seek "preheater" for oil sands. Globe and Mail, 30 September, p. B09.

Vargas, R. and Reif, A. 2009. The structure, regeneration and dynamics of the original forest of Robinson Crusoe's island (Juan Fernández Archipelago, Chile): Guidelines for its restoration. XIII World Forestry Congress. WFC 2009, 18–23 October, 2009, pp. 1–13. http://www.waldbau.uni-freiburg.de/mitarbeiter/mitarbeiter_sammlung/vargas.

Velders, G. J. M., Fahey, D. W., Daniel, J. S., McFarland, M. and Andersen, S. O. 2009. The large contribution of projected HFC emissions to future climate forcing. Proceedings of the National Academy of Sciences, published online June, doi_10.1073_pnas.0902817106.

Vésteinsson, O., McGovern, T. H. and Keller, C. 2002. Enduring impacts: Social and environmental aspects of Viking age settlements in Iceland and Greenland. Archaeologia Islandica, vol. 2, pp. 98–136.

Villa, S., Vighi, M., Maggi, V., Finizio, A. and Bolzacchini, E. 2003. Histotrical trends of organochlorine pesticides in an alpine glacier. Journal of Atmospheric Chemistry, vol. 46, no. 3, pp. 295–311.

Visser, M. E. and Holleman, L. J. M. 2001. Warmer springs disrupt the synchrony of oak and winter moth phenology. Proceeding of the Royal Society B. vol. 268, no. 1464, pp. 289–294.

Vitousek, P. 1994. Beyond global warming: Ecology and global change. Ecology, vol. 75, no. 7, pp. 1861–1876.

Vitousek, P. M., D'antonio, C. M., Loopa, L. L. and Westbrooks, R. 1996. Biological invasions as global environmental change. American Scientist, vol. 84, no. 5, pp. 468–478.

Vitousek, P. M., Ehrlich, P. R., Ehrlich, A. H. and Matson, P. A. 1986. Human appropriation of the products of photosynthesis. Bioscience, vol. 36, pp. 368–373.

Vitousek, P. M., Mooney, H. A., Lubchenco, J. and Melillo, J. M. 1997. Human domination of earth's ecosystems. Science, vol. 277, pp. 494–499.

Vogel, G., Cohen, J. and Enserink, J. 2016. Zika virus: Your questions answered. Science, 29 January, 2016. http://www.sciencemag.org/news/2016/01/zika-virus-your-questions-answered.

Von Storch, H. and Stehr, N. 2002. Towards a history of ideas on anthropogenic climate change. In Behare, K. E. and Berger, W. H. (editors). Climate development and history of the north Atlantic realm. Proceedings of Hanse conference on climate and history. Springer Verlag, Berlin. pp. 17–24.

Vonk, J. E., Sánchez-García, L., van Dongen, B. E., Alling, V., Kosmach, D. et al. 2012. Activation of old carbon by erosion of coastal and subsea permafrost in Arctic Siberia. Nature, vol. 489, pp. 137–140.

Vörösmarty, C. J. and Sahagian, D. 2000. Anthropogenic disturbance of the terrestrial water cycle. BioScience, vol. 50, pp. 753–765.

Vörösmarty, C. J., Green, P., Salisbury, J. and Lammers, R. B. 2000. Global water resources: Vulnerability from climate change and population growth. Science, vol. 289, pp. 284–288.

Wackernagel, M. and Rees, W. 1996. Our ecological footprint. New Society Publishers, Gabriola.

Wackernagel, M., Schulz, N. B., Deumling, D., Linares, A. C., Jenkins, M. et al. 2002. Tracking the ecological overshoot of the human economy. Proceedings of the National Academy of Sciences, vol. 99, no. 14, pp. 9266–9271.

Wada, Y., van Beek, L. P. H., van Kempen, C. M., Reckman, J. W. T. M., Vasak, S. et al. 2010. A worldwide view of groundwater depletion. Geophysical Research Letters, vol. 37, L20402, pp. 1–5. doi:10.1029/2010GL044571

Wakefield, A. J., Murch, S. H., Anthony, A., Linnell, J., Casson, D. M. et al. 1998. RETRACTED: Ileal-lymphoid-nodular hyperplasia, non-specific colitis, and pervasive developmental disorder in children. Lancet, vol. 351, no. 9103, pp. 637–641. doi.org/10.1016/S0140-6736(97)11096-0.

Wald, M. L. 2007. Is ethanol the long haul? Scientific American, January, pp. 42–49.

Walker, D. 2000. Executive director's report. British Columbia Wildlife Federation. Outdoor Edge, V. 10, no. 4, Novemberl–December.

Walsh, B. 2005. Too much of a good thing. Time. 25 July. www.time.com/time/magazine/article/0,9171,1086195,00.html.

Walsh, T. R., Weeks, J., Livermore, D. M. and Toleman, M. A. 2011. Dissemination of NDM-1 positive bacteria in the New Delhi environment and its implications for human health: an environmental point prevalence study. Lancet Infectious Disease, vol. 11, no. 5, pp. 355–362.

Walter, K. M., Zimov, S. A., Chanton, J. P., Verbyla, D. and Chapin, F. S. 2006. Methane bubbling from Siberian thaw lakes as a positive feedback to climate warming. Nature, vol. 443, pp. 71–75.

Walther, G., Post, E., Convey, P., Menzel, A., Parmesan, C. et al. 2002. Ecological responses to recent climate change. Nature, vol. 416, pp. 389–395.

Walton, A. 2009. Provincial-level projection of the current mountain pine beetle outbreak: Update of the infestation projection based on the 2008 provincial aerial overview of forest health and revisions to the "model" (BCMPB.v6). Ministry of Forests and Range, British Columbia, 26 May, 2009, pp. 1–15. http://www.for.gov.bc.ca/hre/bcmpb/Year6.htm.

Wang, Y. and Beydoun, M. A. 2007. The obesity epidemic in the United States—Gender, age, socio-economic, racial/ethnic, and geographic characteristics: A systematic review and meta-regression analysis. Epidemiologic Reviews, vol. 29, no. 1, pp. 6–28.

Ward, A. J. W., Duff, A. J., Horsfall, J. S. and Currie, S. 2007. Scents and scents-ability: Pollution disrupts chemical social recognition and shoaling in fish. Proceedings of the Royal Society of London, Series B, vol. 274, pp. 101–105.

Wardle, D. A., Bardgett, R. D., Klironomos, J. N., Stala, H., van der Putten, W. H. et al. 2004. Ecological linkages between aboveground and belowground biota. Science, vol. 304, pp. 1629–1633.

Watrud, L. S., Lee, E. H., Fairbrother, A., Burdick, C., Reichman, J. R. et al. 2004. Evidence for landscape-level, pollen-mediated gene flow from genetically modified creeping bentgrass with CP4 EPSPS as a marker. Proceedings of the National Academy of Sciences, vol. 101, no. 40, pp. 14533–14338.

Watson, R., Cheung, W. W. L., Anticamara, J., Sumaila, R. U., Zeller, D. et al. 2012. Global marine yield halved as fishing intensity redoubles. Fish and Fisheries, doi:10.1111/j.1467-2979.2012.00483.x, pp. 1–11.

Watson, R. and Pauly, D. 2001. Systematic distortions in world fisheries catch trends. Nature, vol. 414, pp. 534–536.

Wax, E. 2006. A new way of death. Guardian Weekly, 14 April, p. 30.

Weart, S. 2003. The discovery of rapid climate change. Physics today, vol. 56, no. 8, pp. 30–36.

Weber, J., Halsall, C. J., Muir, D., Teixeira, C., Small, J. et al. 2010. Endosulfan, a global pesticide: a review of its fate in the environment and occurrence in the Arctic. Science of the Total Environment, vol. 408, no. 15, pp. 2966–2984.

Webster, D. 2002. The fall of the ancient Maya. Thames and Hudson, London.

WEC. 2013. World energy resources: 2013 Survey. World Energy Council, London, pp. 1–469.

WeGetIt. 2009. Declaration. http://www.we-get-it.org/declaration/. Accessed 1 September, 2009.

Weingart, P., Mitchell, S. D., Richerson, P. J. and Maasen, S. (editors). 1997. Human by nature: Between biology and the social sciences. Lawrence Erlbaum Associates, Mahwah. pp. 1–494.

Weinstein, N. D. 1989. Optimistic biases about personal risk. Science, vol. 246, pp. 1232–1233.

Weiss, H. and Bradley, R. S. 2001. What drives societal collapse? Science, vol. 291, pp. 609–610.

Weiss, H., Courty, M-A., Wetterstrom, W., Guichard, F., Senior, L. et al. 1993. The genesis and collapse of third millennium North Mesopotamian civilization. Science, vol. 261, pp. 995–1004.

Weiss, R. 2004. "Data quality" law is nemesis of regulation. 15 August, 2004, pp. 1–8. http://www.msnbc.msn.com/id/5716978/.

Welch, T. J., Fricke, W. F., McDermott, P. F., White, D. G., Rosso, M. et al. 2007. Multiple antimicrobial resistance in plague: An emerging public health risk. PLoS ONE, vol. 2, no. 3, pp. 1–6: e309. doi:10.1371/journal.pone.0000309.

Welker, M. 2014. Enacting the Corporation: An American mining firm in post-authoritarian Indonesia. Berkeley: University of California Press.

Wellcome Library. 2009. John Snow and the Broad Street pump. http://www.makingthemodernworld.org.uk/learning_modules/geography/05.TU.01/?section=2.

Westhoff, V. 1983. Man's attitude towards vegetation. In Holzner, W., Werger, M. J. A. and Ikusima. I. (editors). Man's impact on vegetation. Chapter 1: Man's attitude towards vegetation. Dr. W. Junk Publishers, The Hague. pp. 1–20.

White, J. A. and Plous, S. 1995. Self-enhancement and social responsibility: On caring more, but doing less, than others. Journal of Applied Social Psychology, vol. 25, no. 15, pp. 1297–1318.

White, L. 1967. The historical roots of our ecologic crisis. Science, vol. 155, pp. 1203–1207.

White, L. 1968a. Machina ex deo: Essays in the dynamism of western culture. MIT Press. Cambridge. Chapter 11, pp. 169–179.

White, L. 1968b. Machina ex deo: Essays in the dynamism of western culture. MIT Press. Cambridge. Chapter 6, pp. 95–105.

White, L. 1968c. Machina ex deo: Essays in the dynamism of western culture. MIT Press. Cambridge. Chapter 2, pp. 11–31.

White, R. 1996. The Nature of Progress: Progress and the Environment. In Marx, L. and Mazlish, B. (editors) 1996, pp. 121–140.

Whitehorn, P. R., Stephanie O'Connor, S., Wackers, F. L. and Goulson, D. 2012. Neonicotinoid pesticide reduces bumble bee colony growth and queen production. Science, vol. 336, no. 6079, pp. 351–352.

Whiten, A. and Boesch, C. 2001. The cultures of chimpanzees. Scientific American, January, pp. 60–67.

Whitfield, S. M., Bell, K. E., Philippi, T., Sasa, M., Bolanos, F. et al. 2007. Amphibian and reptile declines over 35 years at La Selva, Costa Rica. Proceedings of the National Academy of Sciences, vol. 104, no. 20, pp. 8352–8356.

Whitlock, C. and Bartlein, P. J. 1997. Vegetation and climate change in the northwest America during the past 125 ky. Nature, vol. 338, pp. 57–61.

WHO. 2014. Antimicrobial resistance: Global report on surveillance. World Health Organization, Geneva, pp. 1–233.

WHO. 2015. Plague—Madagascar. World Health Organization, Disease Outbreak News, 6 September, 2015, p. 1. http://www.who.int/csr/don/06-september-2015-plague/en/.

WHO/FAO. 2002. Diet, nutrition and the prevention of chronic diseases. Report of a joint WHO/FAO expert consultation. World Health Organization, Technical report series, No. 916, pp. 1–149.

Wikipedia. 2009. Sparta. http://en.wikipedia.org/wiki/Sparta.

Wilcove, D. S., Rothstein, D., Dubow, J., Phillips, A. and Losos, E. 1998. Quantifying threats to imperilled species in the United States. BioScience, vol. 48, no. 8, pp. 607–615.

Wilcox, C., Van Sebille, E. and Hardesty, B. D. 2015. Threat of plastic pollution to seabirds is global, pervasive, and increasing. Proceedings of the National Academy of Sciences vol. 112, no. 38, pp. 11899–11904.

Wilford, T. 2009. At Ur, ritual deaths that were anything but serene. New York Times, 27 October. http://www.nytimes.com/2009/10/27/science/27ur.htm.

Willerslev, E., Davison, J., Moora, M., Zobel, M., Coissac, E. et al. 2014. Fifty thousand years of Arctic vegetation and megafaunal diet. Nature, vol. 506, pp. 47–51.

Williams, C. 1998. Global environment and human intelligence. Globe, no. 42, April. http://www.nerc.ac.uk/ukgeroff/globe42.htm.

Williams, E. 2004. Energy intensity of computer manufacturing: Hybrid analysis combining process and economic input-output methods. Environmental Science and Technology, vol. 38, no. 22, pp. 6166–6174.

Williams, E. D., Ayres, R. U. and Heller, M. 2002. The 1.7 Kilogram microchip: Energy and material use in the production of semiconductor devices. Environmental Science and Technology, vol. 36, no. 24, pp. 5504–5510.

Williams, J. H., DeBenedictis, A., Ghanadan, R., Mahone, A., Moore, J. et al. 2012. The technology path to deep greenhouse gas emissions cuts by 2050: The pivotal role of electricity. Science, vol. 335, pp. 53–59.

Williams, R. J. and Ryan, M. J. 1998. Surveillance of antimicrobial resistance—an international perspective. British Medical Journal, vol. 317, p. 651.

Wilson, E. O. 1999a. Consilience: The unity of knowledge. Vintage Books, New York.

Wilson, E. O. 1999b. The diversity of life. W. W. Norton and company, New York.

Wilson, H. R., Blake, R. and Lee, S. 2001. Dynamics of travelling waves in visual perception. Nature, vol. 412, pp. 907–910.

Wilson, J. W., Ott, C. M., zu Bentrup, K. H., Ramamurthy, R., Quick, L. et al. 2007. Space flight alters bacterial gene expression and virulence and reveals a role for global regulator Hfq. Proceedings of the National Academy of Sciences, vol. 104, no. 41, pp. 16299–16304.

Wilson, P. J. 1988. The domestication of the human species. Yale University Press, New Haven.

Wise, R. 1998. The development of new antimicrobial agents. British Medical Journal, vol. 317, pp. 643–644.

Wise, R., Hart, T., Cars, O., Streulens, M., Helmuth, R. et al. 1998. Antimicrobial resistance: Is a major threat to public health. British Medical Journal, vol. 317, pp. 609–610.

WISE. 2012. Chronology of major tailings dam failures. http://www.wise-uranium.org/mdaf.html.

Wiseman, R. 2004. The Luck Factor. Arrow, London.

WMO. 2012. Greenhouse gas concentrations reach new record: WMO Bulletin highlights pivotal role of carbon sinks. United Nations, World Meteorological Organization, press pelease No. 965, 20 November, 2012, pp. 1–2.

WMO. 2014. Assessment for decision-makers: Scientific assessment of ozone depletion. World Meteorological Organization, Global Ozone Research and Monitoring Project, Report no. 56, Geneva, Switzerland, pp. 1–110.

WMO. 2015. The state of greenhouse gases in the atmosphere based on global observations through 2014. World Meteorological Organization, Greenhouse Gas Bulletin, no. 11, pp. 1–4. http://public.wmo.int/pmb_ged/ghg-bulletin_11_en.pdf..

Wolf, A., Cheney, I., Ellis, C. and Miller, J. K. 2006. King Corn. Ducurama films, 90 minutes. www.kingcorn.net.

Wolfe, D. W. 2001. Tales from the underground: A natural history of subterranean life. Perseus Publishing, Cambridge.

Wong, C. 2008. Environmental impacts of mountain pine beetle in the Southern Interior. British Columbia Ministry of Environment, Environmental Stewardship, 3 December, 2008, pp. 1–49. http://sibacs.com/.

Woodall, L. C., Sanchez-Vidal, A., Canals, M., Paterson, G. L. J., Coppock, R. et al. 2014. The deep sea is a major sink for microplastic debris. Royal Society Open Science, vol. 1, Article no. 140317, pp. 1–6.

Woods, A., Coates, K. D., Hamann, A. 2005. Is an unprecedented Dothistroma needle blight epidemic related to climate change? BioScience, vol. 55, no. 9, pp. 761–769.

World Bank. 2002. Overconsumption in a globalizing world: Is it a disease? Can it be cured? 27 February, 2002. http://www.worldbank.org/wbi/B-SPAN/sub-overconsumption.htm.

World Water Forums. 2002. Right to water: Frequently answered questions. http:/worldwaterforum5.org/index.php?Id=1748&L=0%22%25.

Worldwatch Institute. 2010. Vital signs 2010: The trends that are shaping out future. W. W. Norton, New York.

Worldwatch Institute. 2012. Vital signs 2012: The trends that are shaping out future. Island Press, Washington, DC.

Worm, B., Barbier, E. B., Beaumont, N., Duffy, J. E., Folke, C. et al. 2006. Impacts of biodiversity loss on ocean ecosystem services. Science, vol. 314, pp. 787–790.

Wright, A., Bai, G., Barrera, L., Boulahbal, F., Martín-Casabona, N. et al. 2006. Emergence of Mycobacterium tuberculosis with extensive resistance to second-line drugs—worldwide, 2000–2004. Morbidity and Mortality Weekly Report, vol. 55, no. 11, pp. 301–305.

Wright, J. S., Winter, W. L., Zeigler, S. K. and O'dea, P. N. 1984. Advertising. McGraw-Hill Ryerson Ltd., Toronto.

Wright, R. 2004. A short history of progress. House of Anansi Press, Toronto.

Wright, T. F., Toft, C. A., Enkerlin-Hoeflich, E., Gonzalez-Elizondo, J., Albornoz, M. et al. 2001. Nest poaching in Neotropical parrots. Conservation Biology, vol. 15, no. 3, pp. 710–720.

Yaden, D. B., Iwry, J., Slack, K. J., Eiechstaedt, J. C., Zhao, Y. et al. 2016. The overview effect: Awe and self-transcendent experience in space flight. Psychology of Consciousness: Theory, Research, and Practice, vol. 3, no. 1, pp. 1–11.

Yaritani, H. and Matsushima, J. 2014. Analysis of the energy balance of shale gas development. Energies, vol. 7, pp. 2207–2227.

Yip, M. 2001. Essay about Caulerpa taxifolia. Presented at Colloquial Meeting of Marine Biology 1, Salzburg, 6 May 1999. Revised April 2001. http://www.sbg.ac.at/ipk/avstudio/pierofun/ct/caulerpa.htm.

Yoffee, N. 1988a. Orienting collapse. In Yoffee, N. and Cowgill, W. (editors). 1988, pp. 1–19.

Yoffee, N. 1988b. The collapse of ancient Mesopotamian states and civilization. In Yoffee, N. and Cowgill, W. (editors) 1988, pp. 44–68.

Yoffee, N. and Cowgill, W. (editors). 1988. The collapse of ancient states and civilizations. University of Arizona Press, Tucson.

Youth Markets Alert. 1998. Directing the pitch: Do smart marketers to children target kids or their parents? Youth Markets Alert, 1 July. www.accessmylibrary.com/archive/435361-youth-markets-alert/july-1998.html.

Zanette, L., Clinchy, M. and Sung, H. 2009. Food-supplementing parents reduces their sons' song repertoire size. Proceedings of the Royal Society, Series B, vol. 276, pp. 2855–2860.

Zapiola, M. L., Campbell, C. K., Butler, M. D. and Mallory-Smith, C. A. 2008. Escape and establishment of transgenic glyphosate resistant creeping bentgrass Agrostis Stolonifera in Oregon, USA: A 4-year study. Journal of Applied Ecology, vol. 45, no. 2, pp. 486–494.

Zayed, A. and Packer, L. 2005. Complementary sex determination substantially increases extinction proneness of haplodiploid populations. Proceedings of the National Academy of Sciences, vol. 102, no. 30, pp. 10742–10746.

Zeidler, A. 2009. Influence and authenticity of l'Inconnue de la Seine, pp. 1–10. http://www.williamgaddis.org/recognitions/inconnue/index.shtml. Accessed 2009.

Zhang, Q., Jiang, X., Tong, D., Davis, S. J., Zhao, H. et al. 2017. Transboundary health impacts of transported global air pollution and international trade. Nature, vol. 543, pp. 705–709.

Zhang, X., Zwiers, F. W., Hegerl, G. C., Lambert, F. H., Gillett, N. P. et al. 2007. Detection of human influence on twentieth-century precipitation trends. Nature, vol. 448, no. 7152, pp. 461–465.

Ziman, J. 1978. Reliable knowledge: An exploration of the grounds for belief in science. Cambridge University Press, Cambridge.

# INDEX

www.ingramcontent.com/pod-product-compliance
Lightning Source LLC
Chambersburg PA
CBHW080351030426
42334CB00024B/2840